REPRINTS OF US-FLORAS

EDITED BY J. CRAMER

VOLUME 7

FLORA OF NEW MEXICO

BY

E. O. WOOTON

AND

PAUL C. STANDLEY

3301 LEHRE
VERLAG VON J. CRAMER
1972

WHELDON & WESLEY, LTD · STECHERT-HAFNER SERVICE AGENCY, INC.
CODICOTE, HERTS. NEW YORK, N. Y.

Printed in Germany
by Strauss & Cramer GmbH. 6901 Leutershausen
ISBN 3 7682 0745 5

FLORA OF NEW MEXICO.

By E. O. Wooton and Paul C. Standley.

INTRODUCTION.

This flora of New Mexico is a list of all the species of phanerogams and vascular cryptogams at present known to occur within the State, with keys to the families, genera, and species. Although we have examined all the herbarium material easily accessible and have endeavored to verify all published data, we know that the list is far from complete. Even in the most carefully explored areas of the eastern United States, species which have been overlooked are still coming to light and more careful study of more copious material is increasing the number of recognized species. Much more are additional species to be looked for within the 122,000 square miles embraced in the area of New Mexico, many large portions of which have never been visited by any botanist, while even the most familiar regions have not been thoroughly examined. Thus it is certain that as collectors extend their fields of exploration our present list of 2,975 species will be increased to far above 3,000. It is along the borders of the State that the greater number of additions will be found, especially in the southeastern and southwestern corners and in the high mountains along the Colorado line, but isolated mountain ranges in the interior probably hide endemic species still unknown.

Various short accounts of New Mexico and Arizona plants were published by the earlier botanists of the United States.[1] These, however, are too incomplete and disconnected to be of much use for identification purposes. Two or three more general works are available for use in New Mexico, but none is complete for any part of the State. The Botany of Western Texas, by Dr. J. M. Coulter,[2] contains descriptions of a majority of the plants of southeastern New Mexico, but the volume is not provided with keys to the species and the nomen-

[1] See, Paul C. Standley. A bibliography of New Mexican botany. Contr. U. S. Nat. Herb. **13**: 229–246. 1910.

[2] Contr. U. S. Nat. Herb. **2**. 1891–94.

clature is now antiquated. Dr. P. A. Rydberg's Flora of Colorado[1] is very satisfactory for use in the extreme northern part of the State. Even here, however, many plants will be found which have not been reported from Colorado and hence are not contained in that work, many of our Southwestern species seeming to reach the northern limit of their range just below the Colorado line. The new edition of Coulter's Rocky Mountain Flora, as revised by Prof. Aven Nelson, can be used in a limited way in northern New Mexico, but it will be found to describe only a fraction of our plants.

The material upon which this flora is based is chiefly that in the United States National Herbarium, in the herbarium of the New Mexico College of Agriculture and Mechanic Arts at Mesilla Park, and in the private herbarium of E. O. Wooton, lately acquired by the National Herbarium. In the National Herbarium are found sets of nearly all the larger New Mexican collections, both early and recent, such as those of Fendler, Bigelow, Wright, the first Mexican Boundary Survey, Heller, Wooton, Earle, Metcalfe, and Standley. These include duplicate types of most species that have been described from the State. Of particular value are the large collections made by Dr. E. A. Mearns in connection with the Mexican Boundary Survey of 1892 and 1893, and by members of the Biological Survey of the United States Department of Agriculture in connection with their studies of the fauna of New Mexico. There are also several smaller collections in the same herbarium of which no duplicates exist.

The herbarium of the Agricultural College contains probably the largest assemblage of New Mexican plants that has hitherto been gathered. Here are found not only sets of the more recent generally distributed collections, but several thousand plants collected by the present writers of which few duplicates were obtained. Local collectors in different parts of the State have forwarded collections from time to time, some of which are of great interest.

The Wooton herbarium contains duplicates of many of Mr. Wooton's collections deposited in the herbarium of the Agricultural College, besides many specimens not to be found elsewhere. It also includes sets of the plants collected by Dr. C. L. Herrick and Miss A. I. Mulford.

The New Mexican ranges given for the listed species are based upon the specimens in these herbaria. We have also examined New Mexican material of certain groups in the herbarium of the New York Botanical Garden, besides collections lent by Prof. T. D. A. Cockerell, now of Boulder, Colorado, and Miss Charlotte C. Ellis, formerly of Placitas, New Mexico.

The work of preparing the manuscript of the flora was carried on chiefly at the National Herbarium during the years 1910, 1911, and

[1] Colo. Agr. Exp. Sta. Bull. 100. 1906.

SMITHSONIAN INSTITUTION
UNITED STATES NATIONAL MUSEUM

CONTRIBUTIONS

FROM THE

UNITED STATES NATIONAL HERBARIUM

VOLUME 19

FLORA OF NEW MEXICO

By E. O. WOOTON and PAUL C. STANDLEY

WASHINGTON
GOVERNMENT PRINTING OFFICE
1915

WILLIAM MADISON RANDALL LIBRARY UNC AT WILMINGTON

BULLETIN OF THE UNITED STATES NATIONAL MUSEUM.

Issued June 24, 1915.

4

QK176
.W73
1972

PREFACE.

The present volume of the Contributions is devoted to a flora of New Mexico, by Mr. E. O. Wooton, of the United States Department of Agriculture, and Mr. Paul C. Standley, assistant curator, United States National Herbarium. Mr. Wooton was connected with the New Mexico College of Agriculture for twenty years, during which time he made extensive botanical collections in nearly all the counties of the State. Mr. Standley spent three years in botanical work at the same institution and has since revisited the State for the purpose of further studying its flora. This volume, therefore, is based very largely upon the collections made by the two authors, although all other available collections from New Mexico have been studied.

Only the flowering plants and vascular cryptogams of New Mexico are contained in the present work. Keys are given for the determination of the species as well as of the larger groups, so that the volume may be used as a field manual. At the same time the citations will enable those who have access to libraries to consult readily the original descriptions of the species.

The number of species treated is approximately 3,000. Notwithstanding the large amount of field work already accomplished, many remote districts in New Mexico are still imperfectly known botanically, so that eventually this number will doubtless be increased by several hundred species. The treatise in its present form, however, will be found to contain most of the plants growing spontaneously in those parts of the State thus far settled or frequently visited.

This is the fourth volume of the Contributions to be devoted to a State flora, the others being the Botany of Western Texas (volume 2), the Plant Life of Alabama (volume 6), and the Flora of Washington (volume 11).

FREDERICK V. COVILLE,
Curator of the United States National Herbarium.

159636

1912, although some preliminary work had been done previously at the New Mexico Agricultural College. Descriptions of most of the new species discovered in the course of the work have been published in a recent part of the Contributions from the United States National Herbarium.[1] Accounts of the Cactaceae and of the grasses and grass-like plants have appeared as bulletins of the New Mexico Agricultural Experiment Station.[2]

It is our intention to publish in the near future, in the Contributions from the United States National Herbarium, an account of the phytogeography of the State. This will include a discussion of the life zones and of the factors which influence them. There will also be a history of botanical exploration in New Mexico, and a discussion of other matters of botanical interest.

Under each species in the present volume we have cited the place of publication, to facilitate reference to the original description. No attempt has been made to give complete synonymy, the intention being rather to enter only names having some more or less direct bearing upon New Mexican botany. In citing data regarding habitat and zonal distribution, only conditions inside the State have been considered. In other States some of the plants often occur in habitats different from those we have indicated, although in all probability zonal distribution is practically constant for the same plant in whatever region it may grow.[3] The generic diagnoses have been drawn with only the New Mexican species in mind.

In the preparation of the flora we have received the assistance of many persons, of whose aid we wish to express our appreciation. We are especially indebted to the following for help in various ways: Dr. E. L. Greene, Dr. N. L. Britton, Dr. P. A. Rydberg, Dr. B. L. Robinson, Prof. M. L. Fernald, Dr. J. H. Barnhart, Dr. Ezra Brainerd, Mr. George V. Nash, Dr. J. K. Small, Mr. K. K. Mackenzie, Prof. T. D. A. Cockerell, Mr. Vernon Bailey, Mr. E. A. Goldman, and Mr. C. R. Ball, as well as several of our botanical associates in Washington. Many residents of New Mexico have assisted by collecting specimens and furnishing data concerning the distribution and uses of plants. Our sincerest thanks are extended to numerous citizens of the State who have always afforded all the assistance in their power to collecting expeditions, which would have been impossible or unfruitful without their labors so freely expended in our behalf.

---

[1] Contr. U. S. Nat. Herb. **16:** 109–196. 1913.

[2] Cacti in New Mexico. By E. O. Wooton. Bull. **78.** 1911. The grasses and grass-like plants of New Mexico. By E. O. Wooton and Paul C. Standley. Bull. **81.** 1912.

[3] For an account of life zones in New Mexico see, Bailey, Vernon. Life zones and crop zones of New Mexico. North American Fauna (U. S. Dept. Agr. Bur. Biol. Surv.) **35.** 1913.

SYSTEMATIC TREATMENT OF THE VASCULAR PLANTS.

SYNOPSIS OF THE LARGER GROUPS, WITH KEYS.

Subkingdom PTERIDOPHYTA.

Plants without flowers or seeds, producing spores, each of which, on germination, develops into a flat or irregular prothallium. The prothallia bear the reproductive organs (antheridia and archegonia). As a result of the fertilization of an egg in the archegonium by a sperm produced in the antheridia a fern or an allied plant is developed.

KEY TO THE ORDERS.

Leaves broad, entire or dissected; ferns or fernlike plants.
 Spores of 1 kind, borne in sporangia; plants not aquatic...1. **FILICALES** (p. 18).
 Spores of 2 kinds, borne in sporocarps; aquatics......2. **SALVINIALES** (p. 27).
Leaves narrow, scalelike or awllike; mosslike or rushlike plants.
 Sporangia in a terminal cone; stems hollow..........3. **EQUISETALES** (p. 28).
 Sporangia in the axils of small or leaflike bracts; stems solid.
 4. **LYCOPODIALES** (p. 29).

Subkingdom SPERMATOPHYTA.

Plants with flowers which produce seeds. Microspores (pollen grains) borne in the microsporangia (anther sacs) develop each into a tubular prothallium; a macrospore (embryo sac) develops a minute prothallium and, together with the macrosporangium (ovule) in which it is contained, ripens into a seed.

KEY TO THE CLASSES.

Ovules and seeds borne on the face of a bract or scale; stigmas wanting.
 1. **GYMNOSPERMAE** (p. 30).
Ovules and seeds borne in a closed cavity; stigmas present.
 2. **ANGIOSPERMAE** (p. 39).

Class 1. GYMNOSPERMAE.

KEY TO THE ORDERS.

Staminate and pistillate flowers both in aments; perianth none; trees or shrubs with needle-like or scalelike leaves..........................5. **PINALES** (p. 30).
Staminate flowers in aments; pistillate flowers single or in pairs; perianth present; shrubs with jointed stems, the leaves reduced to sheathing scales.
 6. **GNETALES** (p. 38).

Class 2. ANGIOSPERMAE.

KEY TO THE SUBCLASSES.

Cotyledon 1; stems endogenous; leaves parallel-veined.
 1. **MONOCOTYLEDONES** (p. 39).
Cotyledons normally 2; stems exogenous; leaves not parallel-veined, or rarely apparently so....................................2. **DICOTYLEDONES** (p. 154).

CONTENTS

	Page.
Introduction	9
Systematic treatment of the vascular plants	12
Synopsis of the larger groups, with keys	12
Annotated catalogue	18
Summary of larger groups, with numbers of genera and species	754
Geographic index	755
List of new genera, species, and hybrids, and new names	772
Index	775

Subclass 1. MONOCOTYLEDONES.

KEY TO THE ORDERS.

Perianth when present rudimentary or degenerate, often composed of bristles or mere scales, not corolla-like, sometimes wanting.
  Flowers in the axils of dry or chaffy, usually imbricated, bracts (scales or glumes). 10. **POALES** (p. 42).
  Flowers not in the axils of dry or chaffy bracts.
    Perianth of bristles or chaffy scales................7. **PANDANALES** (p. 39).
    Perianth fleshy or herbaceous, or wanting.
      Fruit baccate; endosperm present; plants 1 cm. broad or less, consisting merely of a flat thallus with 1 or more roots, floating. 11. **ARALES** (p. 124).
      Fruit drupaceous; endosperm wanting; aquatics with well-developed stems....................................8. **NAIADALES** (p. 39).
Perianth of 2 distinct series, the inner usually corolla-like.
  Gynœcium of distinct carpels............................9. **ALISMALES** (p. 41).
  Gynœcium of united carpels.
    Endosperm mealy...............................12. **XYRIDALES** (p. 125).
    Endosperm fleshy, horny, or cartilaginous.
      Ovary and fruit superior.........................13. **LILIALES** (p. 127).
      Ovary and fruit wholly or half inferior.
        Endosperm present; flowers regular.14. **AMARYLLIDALES** (p. 145).
        Endosperm wanting; flowers irregular..15. **ORCHIDALES** (p. 148).

Subclass 2. DICOTYLEDONES.

KEY TO THE ORDERS.

Corolla wanting.
  Calyx wanting.
    Herbs.
      Flowers monœcious or diœcious........30. **EUPHORBIALES** (p. 392).
      Flowers mainly perfect.
        Flowers spicate; styles wanting..........16. **PIPERALES** (p. 154).
        Flowers axillary; styles present. (Callitrichaceae) 30. **EUPHORBIALES** (p. 392).
    Trees or shrubs.
      Fruit many-seeded; seeds each with a tuft of hairs. 17. **SALICALES** (p. 154).
      Fruit 1-seeded; seeds without tufts of hairs ......42. **OLEALES** (p. 495).
  Calyx present, at least in the staminate or in the perfect flowers.
    Flowers, at least the staminate, in aments or ament-like spikes; fruit a nut or achene; trees or shrubs.
      Leaves simple; ovule pendulous and anatropous.19. **FAGALES** (p. 163).
      Leaves pinnate; ovule erect and orthotropous. 18. **JUGLANDALES** (p. 161).
    Flowers, at least the staminate, not in aments; fruit various; herbs, trees, or shrubs.
      Ovary inferior.
        Flowers, at least the staminate, in involucrate heads. (Ambrosiaceae) 50. **ASTERALES** (p. 618).

Flowers not in involucrate heads.
Fruit either a berry or a drupe or nutlike.
Stamens as many as the perianth segments and alternate with them or else fewer.
(Tetragoniaceae) **24. CHENOPODIALES** (p. 198).
Stamens as many as the perianth segments and opposite them or else twice as many.
(Families of) **37. MYRTALES** (p. 459).
Fruit a capsule.
Sepals as many as the cells of the ovary or half as many.
37. MYRTALES (p. 459).
Sepals (4 or 5) at least twice as many as the cells of the ovary.......(Saxifragaceae) **27. ROSALES** (p. 291).
Ovary superior.
Gynœcium of 1 carpel or several distinct carpels; stigma and styles of each solitary.
Carpels several.
Stamens inserted below the ovary.
(Families of) **25. RANALES** (p. 243).
Stamens inserted on the edge of a cup-shaped hypanthium.
(Families of) **27. ROSALES** (p. 291).
Carpels solitary.
Style lateral and oblique.
(Phytolaccaceae) **24. CHENOPODIALES** (p. 198).
Style axile, erect.
Ovary neither inclosed nor seated in a hypanthium or a calyx tube.
Flowers solitary in the axils of the leaves; aquatics.
(Ceratophyllaceae) **25. RANALES** (p. 243).
Flowers not solitary in the axils of the leaves; terrestrial plants. (Urticaceae) **20. URTICALES** (p. 174).
Ovary inclosed in or seated in a hypanthium or a calyx tube.
Stamens borne under the gynœcium.
(Allioniaceae) **24. CHENOPODIALES** (p. 198).
Stamens borne on the hypanthium or adnate to the calyx tube.......**36. THYMELAEALES** (p. 458).
Gynœcium of 2 or several united carpels; stigmas or styles 2 to several.
Ovary, by abortion, 1-celled and 1-ovuled.
Leaves with sheathing stipules.
23. POLYGONALES (p. 181).
Leaves exstipulate, or the stipules, if present, not sheathing.
Trees or shrubs....(Ulmaceae) **20. URTICALES** (p. 174).
Herbs or vines.
Stipules herbaceous; inflorescence spicate or racemose; leaf blades palmately veined.
(Cannabinaceae) **20. URTICALES** (p. 174).
Stipules scarious or hyaline or none; inflorescence cymose; leaf blades pinnately veined.
(Families of) **24. CHENOPODIALES** (p. 198).
Ovary several-celled, or with several placentæ, several-ovuled.
Stamens perigynous or epigynous, inserted on the margin of a hypanthium or a disk.
Fruit a samara...(Aceraceae) **31. SAPINDALES** (p. 405).

Fruit not a samara.
Fruit drupelike or berry-like; trees or shrubs.
32. **RHAMNALES** (p. 412).
Fruit a capsule; herbs.
22. **ARISTOLOCHIALES** (p. 181).
Stamens hypogynous, inserted under the gynœcium in the perfect flowers, not on a disk in the pistillate flowers.
Flowers monœcious or diœcious.
(Euphorbiaceae) 30. **EUPHORBIALES** (p. 392).
Flowers perfect.
Stamens tetradynamous.
(Brassicaceae) 26. **PAPAVERALES** (p. 260).
Stamens not tetradynamous.
24. **CHENOPODIALES** (p. 198).
Corolla present.
Petals more or less united.
Ovary inferior.
Stamens with their filaments free from the corolla.
Stamens 10; anther sacs opening by terminal pores or chinks.
(Vacciniaceae) 39. **ERICALES** (p. 486).
Stamens 5 or fewer: anther sacs opening by longitudinal slits.
48. **CAMPANULALES** (p. 612).
Stamens adnate to the corolla.
Ovary with 1 fertile cavity.
Flowers in involucrate heads.................50. **ASTERALES** (p. 618).
Flowers not in involucrate heads.........49. **VALERIANALES** (p. 617).
Ovary with 2 to many fertile cavities.
Plants tendril-bearing.
(Cucurbitaceae) 48. **CAMPANULALES** (p. 612).
Plants not tendril-bearing.
Ovules mostly on basal placentæ; plants parasitic.
21. **SANTALALES** (p. 177).
Ovules not on basal placentæ; plants not parasitic.
47. **RUBIALES** (p. 603).
Ovary superior.
Stamens free from the corolla.
Gynœcium of a single carpel..........(Families of) 27. **ROSALES** (p. 291).
Gynœcium of several united carpels.
Filaments distinct................(Families of) 39. **ERICALES** (p. 486).
Filaments united.
Stamens diadelphous......(Fumariaceae) 26. **PAPAVERALES** (p. 260).
Stamens monadelphous.
Anther sacs opening by slits.
(Oxalidaceae) 28. **GERANIALES** (p. 379).
Anther sacs opening by pores.
Calyx and corolla very irregular.....29. **POLYGALALES** (p. 390).
Calyx and corolla regular...(Families of) 39. **ERICALES** (p. 486).
Stamens partially adnate to the corolla.
Stamens as many as the lobes of the corolla and opposite them, or twice as many or more.
Ovary 1-celled.
Placentæ central or basal.................40. **PRIMULALES** (p. 490).
Placentæ parietal.......(Fouquieriaceae) 34. **HYPERICALES** (p. 427).

Ovary several-celled.
Upper portion of ovaries distinct. (Crassulaceae) **27. ROSALES** (p. 291).
Upper portion of ovaries united................**41. EBENALES** (p. 495).
Stamens as many as the lobes of the corolla and alternate with them or fewer.
Corolla scarious, veinless; fruit a pyxis..**46. PLANTAGINALES** (p. 602).
Corolla not scarious, veiny; fruit not a pyxis.
Carpels distinct, except sometimes at the apex.
Style terminal......................**44. ASCLEPIADALES** (p. 503).
Style basal........(Dichondraceae) **45. POLEMONIALES** (p. 513).
Carpels united.
Ovary 1-celled, with central placentæ...**43. GENTIANALES** (p. 497).
Ovary 2 or 3-celled or falsely 4-celled, or if 1-celled with parietal placentæ.......................**45. POLEMONIALES** (p. 513).
Petals distinct, at least at the base.
Carpels solitary, or several and distinct, or united only at the base.
Stamens on the margin of a hypanthium (this very small in some Saxifragaceae). **27. ROSALES** (p. 291).
Stamens at the base of the receptacle, hypogynous.
Flowers in monœcious heads........(Platanaceae) **27. ROSALES** (p. 291).
Flowers not in monœcious heads.
Plants with firm stems and leaves, not succulent..**25. RANALES** (p. 243).
Plants with succulent stems and leaves. (Crassulaceae) **27. ROSALES** (p. 291).
Carpels several and united.
Ovary inferior.
Stamens numerous.
Hypanthium produced beyond the ovary. (Families of) **37. MYRTALES** (p. 459).
Hypanthium not produced beyond the ovary.
Ovary partly inferior........(Hydrangeaceae) **27. ROSALES** (p. 291).
Ovary wholly inferior...................**35. OPUNTIALES** (p. 431).
Stamens not more than twice as many as the petals.
Styles wanting. (Stigmas sessile; aquatics.) (Gunneraceae) **37. MYRTALES** (p. 459).
Styles present.
Styles distinct.
Ovules solitary in each cell; fruit a drupe or of 2 to 5 more or less united achenes..........................**38. UMBELLALES** (p. 474).
Ovules several in each cell; fruit a capsule or a fleshy, many-seeded, berry.
Fruit, if dehiscent, valvate..(Families of) **27. ROSALES** (p. 291).
Fruit circumscissile. (Portulacaceae) **24. CHENOPODIALES** (p. 198).
Styles united or single.
Plants with tendrils; fruit a pepo. (Cucurbitaceae) **48. CAMPANULALES** (p. 612).
Plants without tendrils; fruit not a pepo.
Ovary exceeding the hypanthium, the top free. (Hydrangeaceae) **27. ROSALES** (p. 291).
Ovary inclosed in or surpassed by the hypanthium or adnate to it.
Ovules solitary in each cell.........**38. UMBELLALES** (p. 474).
Ovules several in each cell.
Ovary with parietal placentæ. (Loasaceae) **35. OPUNTIALES** (p. 431).
Ovary with central or basal placentæ. (Families of) **37. MYRTALES** (p. 459).

Ovary superior.
Stamens inserted on the margin of a disk or hypanthium (perigynous or hypogynous).
Stamens as many as the petals and opposite them.
Styles and upper part of the ovaries distinct; ovules and seeds many. (Saxifragaceae) 27. **ROSALES** (p. 291).
Styles united; ovules and seeds solitary or 2..32. **RHAMNALES** (p. 412).
Stamens as many as the petals and alternate with them or more.
Styles distinct................(Saxifragaceae) 27. **ROSALES** (p. 291).
Styles united.
Hypanthium cup-shaped or campanulate; disk obsolete or inconspicuous...........................37. **MYRTALES** (p. 459).
Hypanthium flat or obsolete; disk fleshy.
Plants with secreting glands in the bark.
(Rutaceae) 28. **GERANIALES** (p. 379).
Plants without secreting glands in the bark.
31. **SAPINDALES** (p. 405).
Stamens inserted at the base of the ovary or receptacle.
Stamens numerous.
Sepals valvate; filaments united.............33. **MALVALES** (p. 416).
Sepals imbricated; filaments various......26. **PAPAVERALES** (p. 260).
Stamens few, not over twice as many as the petals.
Stamens as many as the petals and opposite them.
Anther sacs opening by hinged valves.
(Berberidaceae) 25. **RANALES** (p. 243).
Anther sacs opening by slits.
Flowers monœcious..................30. **EUPHORBIALES** (p. 392).
Flowers perfect.
Ovules and seeds several or many; embryo coiled.
(Portulacaceae) 24. **CHENOPODIALES** (p. 198).
Ovules and seeds solitary; embryo straight.
(Plumbaginaceae) 40. **PRIMULALES** (p. 490).
Stamens as many as the petals and alternate with them or more, sometimes twice as many.
Stamens 6; petals 4; sepals 2 or 4.
(Families of) 26. **PAPAVERALES** (p. 260).
Stamens, petals, and sepals of the same number, or stamens more than the sepals or petals, then usually twice as many.
Ovary 1-celled.
Ovules and seeds on basal or central placentæ.
(Families of) 24. **CHENOPODIALES** (p. 198).
Ovules and seeds on parietal placentæ.
Stamens with united filaments (no staminodia).
33. **MALVALES** (p. 416).
Stamens with distinct filaments.
Staminodia present.(Parnassiaceae) 27. **ROSALES** (p. 291).
Staminodia wanting.
(Families of) 34. **HYPERICALES** (p. 427).
Ovary several-celled.
Stamens adnate to the gynœcium.
(Asclepiadaceae) 44. **ASCLEPIADALES** (p. 503).
Stamens not adnate to the gynœcium.
Filaments wholly or partly united.
Anthers opening by long slits.
(Families of) 28. **GERANIALES** (p. 379).

Anthers opening by pores.....29. **POLYGALALES** (p. 390).
Filaments distinct.
Anthers opening by pores.
(Families of) 39. **ERICALES** (p. 486).
Anthers opening by slits.
Stigmas and styles distinct and cleft, or foliaceous, or united by pairs..................30. **EUPHORBIALES** (p. 392).
Stigmas or styles all distinct or all united, neither cleft nor foliaceous.
Stamens 2.........................42. **OLEALES** (p. 495).
Stamens more than 2.
Ovules 2 or more in each carpel.
34. **HYPERICALES** (p. 427).
Ovules solitary in each carpel.
(Families of) 28. **GERANIALES** (p. 379).

ANNOTATED CATALOGUE.

Subkingdom PTERIDOPHYTA.

Order 1. FILICALES.

1. POLYPODIACEAE. Fern Family.

The only family of the order in New Mexico.

Notwithstanding the dryness of the climate, New Mexico has a considerable number of true ferns. With one exception they grow in the mountains. Most of the species occur in crevices or under overhanging rocks in the drier and warmer mountain ranges. A few of the more delicate ones live only in moist, cool forests in rich soil. A few others occur on high mountain peaks.

KEY TO THE GENERA.

Mature sori round or little elongated, appearing as separate small dots on the back of the frond.
Fronds once pinnate or pinnatifid, having few large pinnæ.
Sori furnished with an indusium; leaf margins spinulose ....10. PHANEROPHLEBIA (p. 26).
Sori naked; leaf margins not spinulose........13. POLYPODIUM (p. 27).
Fronds mostly twice pinnate or pinnatifid, having many small pinnules 1 cm. long or less.
Indusium superior, cordate or reniform, fixed at the sinus .... 9. DRYOPTERIS (p. 25).
Indusium inferior or lateral.
Indusium inferior, breaking at maturity into stellate lobes....................12. WOODSIA (p. 26).
Indusium lateral, thrown back at maturity as a delicate hood..................11. FILIX (p. 26).
Mature sori elongated, oblong to linear, mostly confluent.
Sori naked.
Sori scattered on the back of the frond, following the course of the veins, branching.... 1. BOMMERIA (p. 19).
Sori marginal, near the ends of the veins, sometimes covered at first by the reflexed edges of the pinnæ.................... 2. NOTHOLAENA (p. 19).

Sori with indusia.
Sori dorsal, not marginal.
Sori straight; fronds once pinnate; stipes dark-colored........................ 7. ASPLENIUM (p. 24).
Sori more or less curved; fronds twice pinnate; stipes stramineous....... 8. ATHYRIUM (p. 25).
Sori marginal, covered by reflexed edges of the pinnæ.
Reflexed margin discontinuous, appearing as separate large indusia........... 3. ADIANTUM (p. 21).
Reflexed margin continuous around each pinna.
An inner indusium present, making the covering of the sori double; fronds large, 40 to 100 cm. long.......................... 4. PTERIDIUM (p. 21).
Inner indusium wanting, the covering of the sori single; fronds in ours never over 30 cm. long.
Pinnules minute, beadlike, hairy (except in *C. wrightii*, which resembles the next genus), not coriaceous..... 5. CHEILANTHES (p. 21).
Pinnules larger, 3 mm. long or more, glabrous (except in *P. aspera*), coriaceous...... 6. PELLAEA (p. 23).

1. **BOMMERIA** Fourn.

Rootstocks creeping; fronds 5-angled, pinnate, hispid above, tomentose beneath; sori oblong or linear, following the course of the veinlets, exindusiate.

1. **Bommeria hispida** (Mett.) Underw. Bull. Torrey Club **29:** 633. 1902.
Gymnogramme hispida Mett.; Kuhn, Linnaea **36:** 72. 1869.
Gymnopteris hispida Underw. Native Ferns ed. 6. 84. 1900.
TYPE LOCALITY: Western Texas.
RANGE: Texas to New Mexico and Arizona.
NEW MEXICO: Bear Mountains; Organ Mountains; 5 miles east of San Lorenzo; Mimbres River; Silver City; Florida Mountains. Dry hills, in the Upper Sonoran Zone.

2. **NOTHOLAENA** R. Br. CLOAK FERN.

Sori marginal, at first round or oblong, soon confluent into a narrow naked band; veins free; fronds various.

Our species are of somewhat varied aspect, three of them (Eunotholaena) of distinct form, one resembling a Bommeria, and two others such that they might pass for Pellaeas. Some of them are very common in the dry rocky foothills, while two of the species are rare in our range.

KEY TO THE SPECIES.

Fronds covered more or less abundantly with scales or hairs, not farinose, once pinnate.
Fronds densely woolly beneath, the wool at first white, becoming ferruginous................................. 1. *N. bonariensis*.
Fronds scaly on both sides, the scales at first white, changing to darker.

Plants small, 10 to 15 cm. high; pinnæ rotund, entire or 2 or 3-toothed .... 2a. *N. sinuata integerrima.*

Plants larger, 20 to 30 cm. high; pinnæ oblong, sinuate, several-toothed .... 2. *N. sinuata.*

Fronds farinose beneath, neither hairy nor scaly.

Lower surface of fronds bright yellow; fronds pentagonal in outline, barely bipinnate .... 3. *N. hookeri.*

Lower surface of fronds white; fronds deltoid-ovate in outline, tripinnate or quadripinnate.

Rachises nearly straight; pinnules opposite, mostly simple, the terminal ones rarely lobed .... 4. *N. dealbata.*

Rachises and all their branches flexuous; pinnules alternate, the ultimate ones frequently 3-lobed .... 5. *N. fendleri.*

1. **Notholaena bonariensis** (Willd.) C. Chr. Ind. Fil. 6. 1905.

Acrostichum bonariense Willd. Sp. Pl. **5**: 114. 1810.

Cincinalis ferruginea Desv. Ges. Naturf. Freund. Berlin Mag. **5**: 311. 1811.

Notholaena ferruginea Hook. Journ. de Bot. **1**: 92. 1813.

Type locality: "Bonaria" (Argentina).

Range: Arizona and western Texas to Central and South America.

New Mexico: Organ and Dona Ana mountains. Dry hills, among rocks, in the Upper Sonoran Zone.

2. **Notholaena sinuata** (Swartz) Kaulf. Enum. Fil. 135. 1824.

Acrostichum sinuatum Swartz, Syn. Fil. 14. 1806.

Type locality: Peru.

Range: Arizona and western Texas to Mexico and South America.

New Mexico: Black Range; San Luis, Big Hatchet, Carrizalillo, and Bear mountains; Animas Valley; Tortugas Mountain; Florida, Organ, and Guadalupe mountains. Dry hills, in the Upper Sonoran Zone.

Reported from Las Lagunitas near Las Vegas by T. S. Brandegee.

2a. **Notholaena sinuata integerrima** Hook. Sp. Fil. **5**: 108. 1864.

Type locality: Mexico.

Range: Arizona and western Texas to Mexico.

New Mexico: Black Range; Big Hatchet Mountains; Organ Mountains; Tortugas Mountain; White Mountains; Queen; Lakewood. Dry hills, in the Lower and Upper Sonoran zones.

3. **Notholaena hookeri** D. C. Eaton in Wheeler, Rep. U. S. Surv. 100th Merid. **6**: 308. *pl. 30.* 1879.

Type locality: Western Texas.

Range: Arizona and western Texas to Mexico.

New Mexico: Socorro Mountain; Burro Mountains; Kingston; San Luis Mountains; Tres Hermanas; Florida Mountains; Dona Ana Mountains; Organ Mountains. Dry hills, in the Upper Sonoran Zone.

Also reported from Las Lagunitas, near Las Vegas, by T. S. Brandegee.

4. **Notholaena dealbata** (Pursh) Kunze, Amer. Journ. Sci. II. **6**: 82. 1848.

Cheilanthes dealbata Pursh, Fl. Amer. Sept. 671. 1814.

Notholaena nivea dealbata Davenp. Cat. Davenp. Herb. Suppl. 44. 1883.

Type locality: "On rocks on the banks of the Missouri."

Range: Nebraska and Missouri to New Mexico and Arizona and southward.

New Mexico: Burro Mountains; Big Hatchet Mountains; Lookout Mines; Tortugas Mountain; V Pasture. On limestone cliffs, dry hills, in the Lower and Upper Sonoran zones.

Reported from the following localities: Las Lagunitas near Las Vegas, *T. S. Brandegee;* San Domingo, *Bigelow;* Sandia Mountains, *Ferris.*

5. **Notholaena fendleri** Kunze, Farrnkr. **2**: 87. *pl. 136.* 1851.

TYPE LOCALITY: "In New Mexico." Type collected by Fendler.

RANGE: Wyoming to New Mexico and Arizona, and in northern Mexico.

NEW MEXICO: Santa Dona; Socorro; Cross L Ranch. Dry hills, in the Upper Sonoran Zone.

The type is Fendler's 1017a, collected in 1847 near Santa Fe. Although named from New Mexico, the species is very rare in the State, ranging mainly farther north.

3. ADIANTUM L. MAIDEN-HAIR FERN.

Sori marginal, short, covered by a flaplike reflexed portion of the edge of the pinnule, on the free but approximate tips of forking veins; fronds bipinnate; stipes slender, black, wiry; pinnules mostly obovate-cuneate, with a few incised teeth.

1. **Adiantum capillus-veneris** L. Sp. Pl. 1096. 1753. VENUS-HAIR FERN.

Adiantum modestum Underw. Bull. Torrey Club **28**: 46. 1901.

TYPE LOCALITY: "Habitat in Europa australi."

RANGE: Virginia and Florida, westward across the continent except in the extreme northwest.

NEW MEXICO: Eight miles northwest of Reserve; East Fork of the Gila; San Andreas Mountains; Kingston; South Spring River. Damp cliffs, in the Upper Sonoran and Transition zones.

The type of *Adiantum modestum* is Earle's 261 from South Spring River. If differs slightly from our other specimens in having broader, more rounded segments. When one examines a large series of specimens of *A. capillus-veneris* it is seen that it is a variable species and that *A. modestum* is hardly more than a local variation.

4. PTERIDIUM Scop. BRACKEN.

This is a coarse fern of almost world-wide distribution that occurs in the mountains of this State in parklike openings and beside small streams where the soil is rich and water plentiful. It is ordinarily not over 60 cm. high, but sometimes reaches a height of 2 meters. We have only one representative of the genus, the western or pubescent form.

1. **Pteridium aquilinum pubescens** Underw. Native Ferns ed. 6. 91. 1900.

TYPE LOCALITY: "Utah, California, and northward."

RANGE: Western North America from New Mexico to British Columbia.

NEW MEXICO: Tunitcha Mountains; Chama; Santa Fe and Las Vegas mountains; Sandia Mountains; Mogollon Mountains; White Mountains. Open slopes, in the Transition and Canadian zones.

5. CHEILANTHES Swartz. LIP FERN.

Sori terminal or nearly so on all the veins, at first very small and rounded, later confluent; indusium consisting of the reflexed margins of the pinnules, in ours (except one species) continuous all around the pinnule.

With the exception of *C. wrightii* our species belong to that division of the genus having very minute, beadlike segments with the whole margin reflexed. They grow in crevices of rocks and on ledges of cliffs in the mountains, generally between elevations of 1,450 and 2,100 meters.

KEY TO THE SPECIES.

Pinnules smooth; indusia not continuous........................ 1. *C. wrightii.*
Pinnules more or less pubescent or scaly; indusia continuous about the pinnules.
  Fronds tomentose, not scaly.
    Stipes densely tufted, at first woolly, becoming glabrate; fronds small, 10 cm. long or less.................. 2. *C. feei.*
    Stipes tufted, not so numerous, covered with brown tomentum and a few narrow scales; fronds larger, 20 to 40 cm. long.................................. 3. *C. eatoni.*
  Fronds scaly beneath.
    Fronds not at all tomentose, glabrous and bright green or with a few scales above........................... 4. *C. fendleri.*
    Fronds both tomentose and scaly beneath.
      Stipes tufted from a short thick rootstock; fronds tomentose to glabrate above, densely matted-woolly and scaly beneath..................... 5. *C. myriophylla.*
      Stipes scattered on a long slender rootstock; fronds white-tomentose above, very chaffy beneath, with cinnamon-brown scales.................. 6. *C. lindheimeri.*

1. **Cheilanthes wrightii** Hook. Sp. Fil. **2**: 87. *pl. 110. A.* 1858.

TYPE LOCALITY: Western Texas.

RANGE: Western Texas to southern Arizona and adjacent Mexico.

NEW MEXICO: Telegraph Mountains; Bear Mountains; Condes Camp. Upper Sonoran Zone.

2. **Cheilanthes feei** Moore, Ind. Fil. 38. 1857.

Myriopteris gracilis Fée, Gen. Fil. 150. 1850–2.

Cheilanthes gracilis Mett. Abh. Senckenb. Ges. Frankfurt **3**: 80. 1859–61, not Kaulf. 1824.

Cheilanthes lanuginosa Nutt.; Hook. Sp. Fil. **2**: 99. 1858, as synonym.

TYPE LOCALITY: "Habitat ad rupes circa Hillsboro, in America Septentr."

RANGE: Illinois and Minnesota to British Columbia, south to Arizona, Texas, and Mexico.

NEW MEXICO: On cliffs, throughout the State, at lower altitudes. Lower and Upper Sonoran zones.

This is probably the commonest fern in the State, occurring most frequently in crevices in the perpendicular faces of limestone cliffs, especially under projecting ledges. Its stipes are always short, and the fronds mostly lie flat against the rocks. It is not restricted to limestone, but is found much less frequently on other rocks.

3. **Cheilanthes eatoni** Baker in Hook. & Baker, Syn. Fil. 140. 1868.

Cheilanthes tomentosa eatoni Davenp. Cat. Davenp. Herb. Suppl. 49. 1883.

TYPE LOCALITY: Western Texas.

RANGE: Western Texas to Arizona and southward.

NEW MEXICO: Sierra Grande; San Mateo Peak; Sandia Mountains; Socorro; Mogollon Mountains; Black Range; Burro Mountains; San Luis Mountains; Dona Ana and Organ mountains; White and Capitan mountains; Tucumcari Mountain; Queen. In the drier mountains and foothills, Upper Sonoran and Transition zones.

4. **Cheilanthes fendleri** Hook. Sp. Fil. **2**: 103. *pl. 107. B.* 1858.

TYPE LOCALITY: New Mexico. Type collected by Fendler (no. 1015).

RANGE: Western Texas to Colorado, westward to California.

NEW MEXICO: Common in all the mountain ranges. Among rocks, in the Upper Sonoran and Transition zones.

5. **Cheilanthes myriophylla** Desv. Ges. Naturf. Freund. Berlin Mag. **5**: 328. 1811.
Cheilanthes villosa Davenp. Cat. Davenp. Herb. Suppl. 45. 1883.
TYPE LOCALITY: South America.
RANGE: Texas to Arizona and southward.
NEW MEXICO: Big Hatchet Mountains; North Percha Creek; Bishops Cap; Hanover Mountain; Sacramento Mountains. Upper Sonoran Zone.

6. **Cheilanthes lindheimeri** Hook. Sp. Fil. **2**: 101. *pl. 107. A.* 1858.
TYPE LOCALITY: Western Texas.
RANGE: Western Texas to southern Arizona and southward.
NEW MEXICO: Burro Mountains; Telegraph Mountains; Tres Hermanas; Florida Mountains; Dona Ana and Organ mountains. Among rocks on the lower slopes of the mountains, in the Upper Sonoran Zone.

6. PELLAEA Link. CLIFF BRAKE.

Sori intramarginal, terminal on the veins as dots, or decurrent, at length confluent, forming a marginal band; indusium formed by the reflexed margin of the pinnules, commonly broad and membranous.

Our species all belong to the division having coriaceous bluish green pinnules with inconspicuous veins, most of them having dark brown or glossy black stipes. They occur in crevices and under rocks in the drier mountains at altitudes below 2,000 meters.

KEY TO THE SPECIES.

Indusium narrow, concealed by the maturing sporangia; stipes pinkish-stramineous; rootstocks slender, widely creeping.. 1. *P. intermedia.*
Indusium broad, conspicuous; stipes dark brown to black; rootstocks short and thick, 2 to 3 cm. long.
 Fronds and stipes rough-hairy throughout.................... 2. *P. scabra.*
 Fronds and stipes glabrous.
  Pinnules obtuse or barely acute.
   Fronds pinnate above, bipinnate below; pinnules lanceolate or ovate-lanceolate, 5 to 20 mm. long.................................... 3. *P. atropurpurea.*
   Fronds quadripinnate below, simpler above; pinnules oval to cordate-ovate, 5 mm. long or less, very numerous.............................. 4. *P. pulchella.*
  Pinnules distinctly, although shortly, mucronate.
   Fronds narrowly oblong, bipinnate; pinnæ trifoliolate.................................... 5. *P. ternifolia.*
   Fronds broadly lanceolate to deltoid, bipinnate; pinnules numerous on each rachilla, the terminal one usually largest...................... 6. *P. mucronata.*

1. **Pellaea intermedia** Mett.; Kuhn, Linnaea **36**: 84. 1869.
TYPE LOCALITY: Mexico.
RANGE: Texas to Arizona.
NEW MEXICO: Black Range; Burro Mountains; Florida Mountains; Tortugas Mountain; Organ and San Andreas mountains. Dry hills, in the Upper Sonoran Zone.

2. **Pellaea scabra** C. Chr. Ind. Fil. 172. 1905.
Cheilanthes aspera Hook. Sp. Fil. **2**: 111. *pl. 108. A.* 1858, not Kaulf. 1831.
Pellaea aspera Baker in Hook. & Baker, Syn. Fil. 148. 1868.
TYPE LOCALITY: Western Texas.
RANGE: Western Texas to Arizona.
NEW MEXICO: Collected by the Mexican Boundary Survey (no. 1581) near the Copper Mines. Dry hills.

3. Pellaea atropurpurea (L.) Link, Fil. Hort. Berol. 59. 1841.

Pteris atropurpurea L. Sp. Pl. 1076. 1753.

TYPE LOCALITY: "Habitat in Virginia."

RANGE: Ontario and British Columbia to Georgia, Texas, Arizona, and California.

NEW MEXICO: Black Range; San Luis Mountains; Florida Mountains; Mangas Springs; Organ Mountains; highest point of the Llano Estacado; Queen. Thickets in the lower parts of the mountains, in the Upper Sonoran Zone.

4. Pellaea pulchella (Mart. & Gal.) Fée, Gen. Fil. 129. 1850–52.

Allosorus pulchellus Mart. & Gal. Nouv. Mém. Acad. Sci. Brux. **15:** 47. *pl. 10. f. 1.* 1842.

TYPE LOCALITY: "Dans la Cordillère au sud de Sola," Mexico.

RANGE: Western Texas to southeastern New Mexico, and adjacent Mexico.

NEW MEXICO: Queen (*Wooton*). Crevices of limestone rocks, dry hills, in the Upper Sonoran Zone.

5. Pellaea ternifolia (Cav.) Link, Fil. Hort. Berol. 59. 1841.

Pteris ternifolia Cav. Descr. Pl. 266. 1802.

TYPE LOCALITY: Andes of Peru.

RANGE: Western Texas to southern New Mexico and southward.

NEW MEXICO: Organ Mountain (*Wooton*). Upper Sonoran Zone.

This species is rare in New Mexico. We are doubtful of the determination of the Organ Mountain plant, since it is the only specimen collected at this station, although ferns have been collected there frequently and search has been made for the species. Our specimen is possibly a form of *P. mucronata*.

Doctor Underwood has reported a specimen from Socorro, collected in 1895 by Plank, and Mr. M. E. Jones reports having obtained it at Silver City in 1903. The species is not uncommon in Chihuahua.

6. Pellaea mucronata D. C. Eaton in Torr. U. S. & Mex. Bound. Bot. 233. 1859.

Allosorus mucronatus D. C. Eaton, Amer. Journ. Sci. II. **22:** 138. 1856.

Pellaea wrightiana Hook. Sp. Fil. **2:** 142. *pl. 115. B.* 1858.

TYPE LOCALITY: "Clefts of rocks in the hills near the bay of San Francisco, California."

RANGE: Kansas and Texas to Arizona and California and southward.

NEW MEXICO: Sandia Mountains; Socorro; Burro Mountains; Santa Rita; Florida Mountains; Dona Ana and Organ mountains. In the drier mountains and foothills, Upper Sonoran Zone.

This has usually been referred to as *P. wrightiana*. Wright's 2130 from Santa Rita is the type of *P. wrightiana*. It is one of the commonest species of the southern part of the State.

7. ASPLENIUM L. SPLEENWORT.

Sori oblong or linear, oblique, separate; indusia straight or very rarely curved, opening toward the midrib when single, toward each other when paired; veins all free.

KEY TO THE SPECIES.

Pinnæ 2 to 5, linear-cuneate; rachis green .... 1. *A. septentrionale.*

Pinnæ numerous, 10 to 30 pairs, oblong to oval; rachis brown or black.

  Plants tall, 10 to 25 cm. high; stipes black; pinnæ oblong .. 2. *A. resiliens.*

  Plants smaller, 15 cm. high or less; stipes purplish brown; pinnæ oval .... 3. *A. trichomanes.*

1. **Asplenium septentrionale** (L.) Hoffm. Deutschl. Fl. **2:** 12. 1795.
Acrostichum septentrionale L. Sp. Pl. 1068. 1753.
Belvisia septentrionalis Mirb. Hist. Nat. Pl. **4:** 65. 1803.
TYPE LOCALITY: "Habitat in Europae fissuris rupium."
RANGE: Black Hills of South Dakota to New Mexico and Arizona; also in Europe.
NEW MEXICO: Sierra Grande; highest point of the Llano Estacado; Cross L Ranch; Santa Rita; Ben Moore. Upper Sonoran Zone.
This grows in the crevices of rocks or beneath overhanging ledges. It is small and almost grasslike, so that it is easily overlooked.

2. **Asplenium resiliens** Kunze, Linnaea **18:** 331. 1844.
Asplenium parvulum Mart. & Gal. Nouv. Mém. Acad. Sci. Brux. **15:** 60. *pl. 15. f. 3.* 1842, not Hook. 1840.
TYPE LOCALITY: Mexico.
RANGE: Virginia and Florida to Kansas, Texas, and Arizona.
NEW MEXICO: Organ Mountains; Santa Rita; Florida Mountains. Upper Sonoran Zone.

3. **Asplenium trichomanes** L. Sp. Pl. 1080. 1753.
TYPE LOCALITY: "Habitat in Europae fissuris rupium."
RANGE: British America to Alabama, Texas, and Arizona.
NEW MEXICO: Las Vegas Mountains; Mogollon Mountains; Santa Rita; Organ Mountains. Damp slopes, Upper Sonoran to Transition Zone.

8. ATHYRIUM Roth.

Rootstocks stout; fronds large, oblong-ovate, twice pinnate; sori usually curved, oblong; indusium straight or curved, opening along the side nearest the midrib.

1. **Athyrium filix-foemina** (L.) Roth, Tent. Fl. Germ. **3:** 65. 1800. LADY FERN.
Polypodium filix-foemina L. Sp. Pl. 1090. 1753.
Asplenium filix-foemina Bernh. Neu. Journ. Bot. Schrad. **1**^{2}: 26. 1806.
TYPE LOCALITY: "Habitat in Europae frigidioris subhumidis."
RANGE: Throughout most of temperate North America; in New Mexico only in the mountains.
NEW MEXICO: Mogollon Mountains; Winsor Creek; Brazos Canyon. Transition Zone.
The lady fern is not common anywhere in the State, but has been found by a few collectors in cool, shaded canyons beside running streams.

9. DRYOPTERIS Adans.

Rootstocks stout and thick; fronds broadly oblong-lanceolate, bipinnatifid or bipinnate, 20 to 60 cm. long; sori dorsal, rounded, the indusium orbicular-reniform.

1. **Dryopteris filix-mas** (L.) Schott, Gen. Fil. 1834. MALE FERN.
Polypodium filix-mas L. Sp. Pl. 1090. 1753.
Aspidium filix-mas Swartz, Journ. Bot. Schrad. **1800**^{2}: 38. 1801.
TYPE LOCALITY: "Habitat in Europae sylvis."
RANGE: British America to Michigan, New Mexico, and California.
NEW MEXICO: Sierra Grande; Rito de las Frijoles; Las Vegas Mountains; Organ Mountains; Ruidoso Creek. Transition Zone.
The specimens here listed are doubtfully referred to this species, but they represent one of the forms which pass under the name. Further study may result in a change of name for the southwestern form. It is nowhere common in our range, but always seems well adjusted to its habitat wherever it occurs.

10. **PHANEROPHLEBIA** Presl.

Rootstock short and creeping; fronds pinnate, the pinnæ 10 to 16, usually auriculate at the base, serrate or incised; sori round, borne on the back of forking veins; indusium peltate, opening all around the margin.

1. **Phanerophlebia auriculata** Underw. Bull. Torrey Club **26**: 212. *pl. 359. f. 3, 4.* 1899.

Aspidium juglandifolium of authors, in part, not Kunze.

Type locality: "Cool damp cliffs, Mapula Mountains, Chihuahua."

Range: Mountains of southern Arizona and New Mexico and western Texas.

New Mexico: Organ Mountains. Transition Zone.

11. **FILIX** Adans.

Fronds oblong-lanceolate, 10 to 30 cm. long, 2 to 3-pinnatifid, thin, bright green; sori roundish, each borne on the back of a vein; indusium membranous, hoodlike, attached by a broad base on its inner side.

1. **Filix fragilis** (L.) Underw. Native Ferns ed. 6. 119. 1900. Brittle fern.

Polypodium fragile L. Sp. Pl. 1091. 1753.

Cystopteris fragilis Bernh. Journ. Bot. Schrad. **1**: 26. 1806.

Type locality: "Habitat in collibus Europae frigidioris."

Range: Throughout temperate North America, and in temperate regions around the world.

New Mexico: Common in all the mountains from the Black Range and White Mountains northward. Transition Zone.

12. **WOODSIA** R. Br.

Sori orbicular, borne on the back of simply forked, free veins; indusium inferior, thin, in ours conspicuous, breaking at the top and splitting into several laciniate lobes.

Ferns with much the aspect of the fragile fern, but the fronds stiffer and the divisions shorter, the indusial characters, also, different.

KEY TO THE SPECIES.

Fronds lanceolate; pinnæ short, triangular-lanceolate, not glandular .... 1. *W. mexicana.*

Fronds broader than lanceolate; pinnæ longer, the subdivisions broader, glandular-hairy .... 2. *W. plummerae.*

1. **Woodsia mexicana** Fée, Mém. Foug. **7**: 66. 1854.

Type locality: "Habitat in Republica Mexicana, prope San Angel."

Range: Western Texas to Arizona, south into Mexico.

New Mexico: Tunitcha Mountains; Chama; Winsors Ranch; Rio Pueblo; Sierra Grande; Magdalena Mountains; Mogollon Mountains; Organ Mountains; Gilmores Ranch. Transition Zone.

2. **Woodsia plummerae** Lemmon, Bot. Gaz. **7**: 6. 1882.

Woodsia obtusa glandulosa D. C. Eaton & Faxon, Bull. Torrey Club **9**: 50. 1882.

Type locality: "On the north side of a high peak of the Chirricahua Mountains," Arizona.

Range: New Mexico and Arizona.

New Mexico: Burro Mountains (*Rusby*).

13. POLYPODIUM L. Polypody.

Rootstocks elongated; fronds 5 to 20 cm. long, once pinnatifid into linear-oblong, obtuse or acute segments; sori rounded, exindusiate, borne at the ends of the veins midway between the margin and midrib.

1. **Polypodium hesperium** Maxon, Proc. Biol. Soc. Washington **13**: 200. 1900.
Type locality: Coyote Canyon, Lake Chelan, Washington.
Range: British Columbia to Arizona and New Mexico.
New Mexico: Sandia Mountains (*Miss C. C. Ellis*). Damp woods.
Miss Ellis reports that this is found in crevices and under rocks near Balsam, in Lagunita, and on ridges between the latter place and Las Huertas Canyon. The species should occur in some of the ranges in the western part of the State.

Order 2. SALVINIALES.

KEY TO THE FAMILIES.

Creeping plants with 4-parted petioled leaves of medium size.................................. 2. **MARSILEACEAE** (p. 27).
Minute floating plants with closely imbricated, lobed fronds.. 3. **SALVINIACEAE** (p. 27).

2. MARSILEACEAE.

1. MARSILEA L.

Herbaceous perennials growing in muddy places, with slender creeping stems and 4-foliolate long-petioled leaves; sporocarps borne at the base, in ours almost sessile, hard, reniform, 2-valved, several-celled, containing both kinds of spores.

A single species so far obtained in New Mexico, but others will probably be found growing about pools in the mountains.

A specimen in the U. S. National Herbarium obtained by one of the collectors of the Mexican Boundary Survey is determined as *M. uncinata* A. Br. The label shows nothing as to place or time of collection. The published report states that Doctor Bigelow obtained this species in New Mexico, without further locality. The specimen referred to is very small but is probably correctly determined.

1. **Marsilea vestita** Hook. & Grev. Icon. Fil. **2**: *pl. 159*. 1831.
Type locality: "Ad flumen Columbiam, ora occidentali Americae Septentrionalis."
Range: Arkansas and Texas to California, north to Washington and British Columbia.
New Mexico: Queen (*Wooton*). In mud.
The single station at which this plant was found was near the top of the Upper Sonoran Zone, but the same species was collected by Wright near San Elizario, Texas, which is Lower Sonoran, while the range given by most authors suggests the Transition.

3. SALVINIACEAE.

1. AZOLLA Lam.

Small floating plants with a more or less elongated and sometimes branching axis bearing leaves; sporocarps soft, thin-walled, two or more on a stalk, 1-celled; megasporangia containing 1 megaspore, the microsporangia bearing numerous microspores.

1. **Azolla caroliniana** Willd. Sp. Pl. **5**: 541. 1810.
Type locality: "Habitat in aquis Carolinae."
Range: New York to Florida, west to California and Oregon.
New Mexico: Animas Creek (*Metcalfe* 1110). Floating in still water.

Order 3. EQUISETALES.

4. EQUISETACEAE. Horsetail Family.

1. EQUISETUM L. Horsetail.

Plant body rushlike, with jointed, mostly hollow stems; leaves reduced to a whorl of scales forming a sheath at the nodes; sporangia forming a terminal cone composed of peltate scales bearing several sporangia; spores all alike, supplied with coiled elaters attached at the middle and coiled spirally about the spore; prothallia terrestrial, green, usually diœcious.

The family includes the plants which go under the name of "scouring rushes" or "horsetails," which, while very numerous in past ages of the world, are now reduced to a single genus.

KEY TO THE SPECIES.

Annual; plant of two forms, one spore-bearing the other vegetative; vegetative form much branched, with slender 4-angled branches; spore-bearing form not branched, brown......... 1. *E. arvense.*
Perennial, not dimorphous, if branched at all the branches similar to the main stems.
- Stems nearly smooth, the tubercles inconspicuous; sheaths spreading upward; teeth deciduous, leaving a ring of triangular black tips.................................. 2. *E. laevigatum.*
- Stems rough, the tubercles conspicuous; sheaths usually little or not at all spreading upward; teeth mostly adherent to the bases.
  - Stems generally less than 70 cm. high, frequently branched from the base.................................. 3. *E. hiemale.*
  - Stems generally taller, 1 meter high or more, very hard and rough, usually little or not at all branched......... 4. *E. robustum.*

1. Equisetum arvense L. Sp. Pl. 1061. 1753.

Type locality: "Habitat in Europæ agris, pratis."

Range: British America to Virginia, New Mexico, and California.

New Mexico: Taos; Rio Pueblo; Mogollon Mountains. Mountains, in the Transition Zone.

This is the common horsetail of the mountains, growing in very wet soil beside running water. It is usually associated with grasses, rushes, and sedges which cover the swampy meadows at elevations of 1,800 meters and more. Such meadows or marshy places usually go under the name of "ciénaga" (frequently corrupted to "siniky") or the diminutive "cienaguilla." The horsetail may be readily recognized in the vegetative state by its cluster of 4-angled jointed stems about 2 mm. in diameter, of a bright green color, that bear no proper leaves. The spore-bearing stalks are brown, 6 to 8 mm. in diameter, 10 to 20 cm. high, and bear their cones singly at the top. They appear early in the spring, shed their spores, and soon die.

2. Equisetum laevigatum A. Br. Amer. Journ. Sci. **46**: 87. 1844.

Smooth scouring rush.

Type locality: "On poor clayey soil, with Andropogon and other coarse grasses, at the banks of the river below St. Louis."

Range: New Jersey and Louisiana to British Columbia, California, and Texas.

New Mexico: Shiprock; Chama; Taos; Santa Fe and Las Vegas mountains; Ramah; Albuquerque; Mogollon Mountains; Mesilla Valley; Ruidoso Creek. In wet ground, in the Transition Zone, or lower, along streams.

This is the chief scouring rush of the mountains, its smooth, hollow, jointed stems being common along most of the mountain streams and in the cienagas. There is but one kind of stem produced; branching above the base is rare except when the plant is injured. Sometimes, though not frequently, it is somewhat branched from the base. The rather delicate texture and the somewhat spreading, smooth-topped, long sheaths tipped by a row of triangular black dots are characteristic.

3. Equisetum hiemale L. Sp. Pl. 1062. 1753. SCOURING RUSH.

TYPE LOCALITY: "Habitat in Europæ sylvis, asperis, uliginosis."

RANGE: North America north of Mexico.

NEW MEXICO: Reserve; Gilmores Ranch; near Las Vegas, on the Gallinas River; Rio Grande near Mesilla.

This is a common rush along the streams and ditches. The form here referred to is that spoken of as *E. hiemale intermedium* by Mr. A. A. Eaton.

4. Equisetum robustum A. Br. Amer. Journ. Sci. **46:** 88. 1844.

TYPE LOCALITY: "Islands of the Mississippi River in Louisiana."

RANGE: New Jersey and Louisiana, westward across the continent.

NEW MEXICO: Tunitcha Mountains; Cedar Hill; Mesilla; Mogollon Mountains. Wet ground, in the Lower Sonoran Zone.

The large scouring rush occurs not uncommonly along the rivers and irrigating ditches at the lower levels of the State. It may not be sufficiently distinct from *E. hiemale*.

Order 4. LYCOPODIALES.

5. SELAGINELLACEAE. Selaginella Family.

Mosslike terrestrial plants, usually only a few centimeters high; stems slender, branching, erect or trailing; leaves small and scalelike, arranged in 4 to many rows; sporangia 1-celled, globose, of two kinds, viz., megasporangia bearing 4 megaspores and microsporangia bearing many microspores, borne at the bases of the sporophylls, these differing little from foliage leaves.

1. SELAGINELLA Beauv.

KEY TO THE SPECIES.

Plants erect, tufted, with roots only on the lower part; leaves with long terminal bristles and numerous marginal hairs on each side; plants grayish green .... 1. *S. rupincola.*

Plants more or less prostrate, forming mats, mostly rooting along the stems; leaves various; plants grayish or bright green.

  Stems very short, 6 cm. long or less; strobiles erect, 4-angled, usually longer than the vegetative branches .... 2. *S. densa.*

  Stems longer, 10 cm. or more; strobiles various.

    Megaspores irregularly wrinkled; strobiles erect; leaves and short stems frequently much crowded .... 3. *S. wrightii.*

    Megaspores not wrinkled; strobiles hardly distinguishable from the vegetative parts.

      Stems very slender, wiry, terete; leaves small, appressed .... 4. *S. mutica.*

      Stems weaker; leaves lax, dark green .... 5. *S. underwoodii.*

Selaginella lepidophylla, the "resurrection plant," should be found in the Guadalupe Mountains near the southern boundary, or in the limestone mountains of the southwest corner.

There is a single specimen of a species closely allied to *S. arenicola* Underw. in the National Herbarium, the label of which states that it was collected at Las Vegas by Plank. There is some uncertainty as to whether the specimen is correctly labeled; for this reason it is not listed here. Collectors should look for such a species in that region and farther east and south.

1. **Selaginella rupincola** Underw. Bull. Torrey Club **25**: 129. 1898.

TYPE LOCALITY: "On perpendicular rocks, Organ Mountains." New Mexico. Type collected by Wooton (no. 124).

RANGE: Mountains of New Mexico and Arizona.

NEW MEXICO: San Luis Mountains; Dog Spring; Organ Mountains. On rocks and ledges, in the Upper Sonoran Zone.

2. **Selaginella densa** Rydb. Mem. N. Y. Bot. Gard. **1**: 7. 1900.

TYPE LOCALITY: "Little Rocky Mountains," Montana.

RANGE: Montana and western Nebraska to New Mexico.

NEW MEXICO: Winsors Ranch; Hillsboro Peak. In the Transition Zone or higher.

3. **Selaginella wrightii** Hieron. Hedwigia **39**: 298. 1900.

TYPE LOCALITY: Western Texas. Type, Wright's no. 828.

RANGE: Western Texas and New Mexico to Mexico.

NEW MEXICO: Lakewood; Las Vegas.

4. **Selaginella mutica** D. C. Eaton; Underw. Bull. Torrey Club **25**: 128. 1898.

TYPE LOCALITY: "New Mexico."

RANGE: Colorado to Arizona and New Mexico.

NEW MEXICO: Pecos; Canada Alamosa; Organ Mountains. Damp cliffs in the mountains, in the Upper Sonoran Zone.

5. **Selaginella underwoodii** Hieron. in Engl. & Prantl, Pflanzenfam. 1^4: 715. 1901.

Selaginella rupestris fendleri Underw. Bull. Torrey Club **25**: 127. 1898.

Selaginella fendleri Hieron. Hedwigia **39**: 303. 1900, not Baker, 1883.

TYPE LOCALITY: New Mexico. Type collected by Fendler (no. 1024).

RANGE: Colorado, New Mexico, and southward.

NEW MEXICO: Winsors Ranch; Folsom; Ramah; Mogollon Mountains; Black Range; Bear Mountain; Organ Mountains; White Mountains. On rocky walls and ledges in the mountains, in the Upper Sonoran, Transition, and Canadian zcnes.

Subkingdom SPERMATOPHYTA. Seed-bearing plants.

Class 1. GYMNOSPERMAE.

Order 5. PINALES.

KEY TO THE FAMILIES.

Leaves needle-like; carpellary scales with bracts, never peltate; ovules inverted; cones dry...... 6. **PINACEAE** (p. 30).

Leaves scalelike or awllike; carpellary scales without bracts, fleshy or peltate; ovules erect; cones berrylike.................................... 7. **JUNIPERACEAE** (p. 35).

6. PINACEAE. Pine Family.

Large evergreen trees with needle-shaped leaves; infertile flowers in short catkins; fertile flowers in scaly aments, becoming cones, with 2 or more ovules at the base of each scale; fertile scales numerous, spirally imbricated.

KEY TO THE GENERA.

Leaves fascicled, inclosed by sheaths at the base, at least when young; cones maturing the second year...... 1. PINUS (p. 31).

Leaves solitary, not sheathed; cones maturing the first year.

Branches rough with the persistent leaf bases; leaves quadrangular, falling off when dried; cone scales thin and persistent; cones pendulous... 2. PICEA (p. 33).
Branches smooth; leaves flat, persistent in dried specimens; cone scales and cones various.
Cones erect, the scales deciduous; bracts of the cones not exserted; leaves sessile, leaving circular scars.............................. 3. ABIES (p. 34).
Cones pendulous, the scales persistent; bracts of the cone scales conspicuously exserted, 3-parted; leaves petioled, leaving oval scars.................................... 4. PSEUDOTSUGA (p. 35).

1. PINUS L. PINE.

Large or small trees with needle-shaped leaves in fascicles of 2 or more, surrounded by a persistent or deciduous sheath at the base.

KEY TO THE SPECIES.

Leaves in fascicles of 2, short and curved, 3 to 4 cm. long; cones small, 4 to 5 cm. long; seeds not winged.................. 1. *P. edulis.*
Leaves in fascicles of 3 to 5; leaves, cones, and scales various.
Leaves in fascicles of 3 (rarely 4).
Leaves 4 cm. long or less.................................. 2. *P. cembroides.*
Leaves 6 cm. long or more.
Sheaths persistent and conspicuous; leaves 10 to 25 cm. long; cones 7 to 15 cm. long............ 3. *P. brachyptera.*
Sheaths deciduous; leaves 6 to 9 cm. long; cones 3 to 5 cm. long...................................... 4. *P. chihuahuana.*
Leaves in fascicles of 5.
Cones 10 to 18 cm. long, the scales with unarmed appendages; seeds with only rudimentary wings; leaves slender, 4 to 8 cm. long.
Leaves entire... 5. *P. flexilis.*
Leaves serrulate...................................... 6. *P. strobiformis.*
Cones 5 to 7 cm. long, the scales with armed appendages; seeds with conspicuous wings; leaves various.
Leaves short and stout, 2 to 4 cm. long, curved, crowded; cone scales with long weak spines; cones 6 to 7 cm. long........................ 7. *P. aristata.*
Leaves longer, 6 to 10 cm., straight, not crowded; cone scales with short and rigid spines; cones 5 to 6 cm. long.............................. 8. *P. arizonica.*

1. **Pinus edulis** Engelm. in Wisliz. Mem. North. Mex. 88. 1848. PINYON.
Caryopitys edulis Small, Fl. Southeast. U. S. 29. 1903.
Pinus cembroides edulis Voss, Mitt. Deutsch. Dendr. Ges. **16**: 95. 1907.

TYPE LOCALITY: "Not rare from the Cimarron to Santa Fe, and probably throughout New Mexico." Type collected by Wislizenus in 1847.

RANGE: Colorado and Utah to western Texas and northern Mexico.

NEW MEXICO: Common on low hills and high plains everywhere west of the Pecos, and in the mountains of the northeastern corner of the State. Upper Sonoran Zone.

A small, rather scraggy tree, 10 to 12 meters high or less, with rough, dark-colored bark, dark green leaves, and small, ovoid cones with the scales widely spreading when mature. The tree occurs in the drier foothills, associated with junipers and several evergreen oaks, at elevations of 1,500 to 2,150 meters, almost throughout the State. The wood is soft and decays rapidly, so that it is poor for firewood or fence posts and is but little used. Large quantities of the seeds are gathered every year to be eaten. They are very palatable, having a sweet flavor, especially after having been roasted. The tree is one of the most characteristic plants of the Upper Sonoran Zone, not occurring outside that division.

2. **Pinus cembroides** Zucc. Abh. Akad. Wiss. München **1**: 392. 1832.

TYPE LOCALITY: "Crescit in montibus altioribus imperii mexicana V. C. ad ecclesiam S. Crucis prope Sultepec."

RANGE: Mountains of southwestern New Mexico, southeastern Arizona, and southward.

NEW MEXICO: San Luis Mountains (*Goldman* 1408). Upper Sonoran Zone.

3. **Pinus brachyptera** Engelm. in Wisliz. Mem. North. Mex. 89. 1848.

YELLOW PINE.

Pinus engelmanni Torr. U. S. Rep. Expl. Miss. Pacif. **4**: 141. 1856.

Pinus ponderosa scopulorum Engelm. in S. Wats. Bot. Calif. **2**: 126. 1880.

Pinus scopulorum Lemmon, Gard. & For. **1897**: 183. 1897.

TYPE LOCALITY: "Mountains of New Mexico." Type collected by Wislizenus in 1847.

RANGE: Throughout the Rocky Mountains, from the northern boundary of the United States to northern Mexico.

NEW MEXICO: Common in all the mountain ranges of the State, which reach an altitude of 2,100 meters or more. Transition Zone.

This is the most common tree of New Mexico and Arizona, and constitutes perhaps two-thirds of the timber of the former State. It is certainly first in importance from the standpoint of quantity and quality of lumber. It occurs only in the mountains at elevations of from 1,800 to 2,850 meters, being associated with the pinyon in the lower edge of this belt, and with *Pinus flexilis* and Pseudotsuga near its upper limit, rarely forming pure forests. The older trees are frequently over 35 meters high and the trunks from 80 to 100 cm. in diameter. The bark loses its outer layers and becomes cut into irregular quadrangular segments, which are smooth and of light reddish or yellowish brown color. Younger trees, with trunks 45 cm. or less in diameter, have darker colored bark and are generally known to the lumbermen as a different tree—their "jack pine." Lumber made from the larger trees is usually spoken of as "Arizona" pine in distinction from "Texas" pine, and is regarded as the most valuable soft wood of the region.

The inner bark of this and other conifers was chewed or eaten by the Indians in earlier times when other food was wanting. To-day some of the tribes remove the bark from the trunks to secure an exudation of resin which they use in coating their wicker water bottles. Upon the Mescalero Apache Reservation one sees many trees killed by this girdling.

4. **Pinus chihuahuana** Engelm. in Wisliz. Mem. North. Mex. 103. 1848.

Pinus leiophylla chihuahuana Shaw, Publ. Arn. Arb. **1**: 14. 1909.

TYPE LOCALITY: Mountains of Chihuahua.

RANGE: Mountains of southern New Mexico and Arizona and southward.

NEW MEXICO: Animas and San Luis mountains. Transition Zone.

5. **Pinus flexilis** James in Long, Exped. **2**: 34. 1823. WHITE PINE.

Apinus flexilis Rydb. Bull. Torrey Club **32**: 598. 1905.

TYPE LOCALITY: "Arid plains subjacent to the Rocky Mountains, and * * * * up their sides to the region of perpetual frost."

RANGE: Northern Mexico to southern Alberta.

NEW MEXICO: Santa Fe and Las Vegas mountains; Sandia Mountains; Mogollon Mountains; Black Range; White and Sacramento mountains; Capitan Mountains. High mountains, chiefly in the Canadian Zone.

A medium-sized tree, 15 to 25 meters high, found only in the higher parts of the mountains, usually associated with the firs and spruces, at elevations of from 2,400 to 3,000 meters. It is not very abundant, although it is valued next to the yellow pine for its timber. The cone is large and pendent. The seeds are large for the genus and can be eaten like those of the pinyon, but they have thicker and harder shells.

6. **Pinus strobiformis** Engelm. in Wisliz. Mem. North. Mex. 102. 1848.

MEXICAN WHITE PINE.

Pinus ayacahuite brachyptera Shaw, Publ. Arn. Arb. **1**: 11. 1909, not *P. brachyptera* Engelm. 1848.

TYPE LOCALITY: Cosihuiriachi, Chihuahua.

RANGE: Northern Mexico to southern Arizona and New Mexico.

NEW MEXICO: Franeys Peak; San Luis Mountains. Mountains, in the Transition Zone.

A tree very similar to the preceding, nowhere abundant. It occurs within our area only in the southwestern corner of the State. Reports of its occurrence elsewhere in New Mexico doubtless refer to *Pinus flexilis*.

7. **Pinus aristata** Engelm. Amer. Journ. Sci. II. **34**: 331. 1862. FOXTAIL PINE.

TYPE LOCALITY: "Pikes Peak and high mountains of the Snowy Range," Colorado.

RANGE: Higher mountains of Colorado and northern New Mexico to Nevada and California.

NEW MEXICO: Pecos Baldy; Grass Mountain; Costilla Pass; Baldy. Canadian and Hudsonian zones.

A dark green, scrubby tree, 10 to 12 meters high or less, with short leaves curved toward the ends of the branches. It occurs only in the higher mountains at altitudes of 3,000 meters or more, and nowhere in abundance. On the higher peaks at or above timber line the plants are low and stunted, often spreading over the ground, forming what the Germans call "Krumholz." This is the result of the high velocity of the wind at these altitudes.

8. **Pinus arizonica** Engelm. U. S. Rep. Expl. Miss. Pacif. **6**: 261. 1878.

ARIZONA YELLOW PINE.

Pinus ponderosa arizonica Shaw, Publ. Arn. Arb. **1**: 24. 1909.

TYPE LOCALITY: "On the Santa Rita Mountains," Arizona.

RANGE: Mountains of northern Mexico and southern Arizona and New Mexico.

NEW MEXICO: Summit of Animas Peak (*Goldman* 1360). Transition Zone.

2. PICEA Link. SPRUCE.

Conical trees with short stiff sharp-pointed solitary leaves standing out in all directions from the stems; cones pendulous, their scales rather thin, persistent, the bracts shorter than the scales.

KEY TO THE SPECIES.

Young branches and leaf bases pubescent; cones short, 3 to 5 cm. long; leaves dull green, not glaucous........................ 1. *P. engelmanni.*

Young branches and leaf bases glabrous; cones longer, 5 to 9 cm. long; leaves on the older parts usually dark green, the young ones glaucous and light-colored................................ 2. *P. parryana.*

1. Picea engelmanni Parry; Engelm. Trans. Acad. St. Louis **2**: 212. 1863.
ENGELMANN SPRUCE.

Abies engelmanni Parry, loc. cit.

TYPE LOCALITY: "Higher parts of the Rocky Mountains, from New Mexico to the headwaters of the Columbia and Missouri rivers.

RANGE: British Columbia to New Mexico and Arizona.

NEW MEXICO: Sandia Mountains; West Fork of the Gila; Bonito. Higher mountains, Canadian and Hudsonian zones.

A conical tree 20 to 25 meters high or less, with smooth, thin, flaky bark, dark green foliage, and pendulous cones borne mostly on the uppermost branches. It occurs only in the higher mountains at 2,700 to 3,300 meters where there is permanent moisture, frequently forming dense pure forests. It is also found on the faces of cliffs and on the tops of mountains up to timber line, where it is generally straggling and dwarfed. When growing alone it is usually perfectly conical, bearing nearly horizontal branches almost to the ground. The cones are small and purplish until maturity, when they become dry and brown.

2. Picea parryana Parry, Gard. Chron. **11**: 334. 1879. COLORADO BLUE SPRUCE.

Abies parryana Engelm.; Parry, loc. cit.

TYPE LOCALITY: Not stated.

RANGE: Higher mountains of New Mexico and Arizona, northward to Wyoming.

NEW MEXICO: Chama; Winsors Ranch; Sandia Mountains; James Canyon; White Mountain Peak. Canadian and Hudsonian zones.

Very similar to the preceding, but the young leaves covered with a bloom which gives rise to the name of "blue spruce," and the bark thick and deeply furrowed. The range is similar to that of the Engelmann spruce, although usually at slightly lower levels, and the value of the timber is about the same. The lumber is in both cases rather poor, being weak and spongy, and full of knots. It is used to some extent for making boxes. The Colorado blue spruce is a much better tree for decorative purposes because of its color and also because it is a more rapid grower. It does well at Santa Fe and could, no doubt, be used in other places of similar elevation if properly cared for.

3. ABIES Link. FIR.

Large trees with spreading or ascending branches; leaves flat, blunt, short, so arranged as to make the branches appear flat; cones erect, cylindrical, borne near the top of the tree, their scales thin and deciduous.

KEY TO THE SPECIES.

Bark thin, smooth, corky .... 1. *A. arizonica.*
Bark thick, rough, not corky.
  Resin ducts of the leaves within the soft tissue, remote from the epidermis .... 2. *A. lasiocarpa.*
  Resin ducts near the epidermis, on the lower side of the leaf.... 3. *A. concolor.*

1. Abies arizonica Merriam, Proc. Biol. Soc. Washington **10**: 116. 1896.
CORK-BARK FIR.

TYPE LOCALITY: "West slope of San Francisco Mountain, Arizona."

RANGE: Higher mountains of Arizona and New Mexico.

NEW MEXICO: Twining; Sandia Mountains; Baldy; Baldy Peak, Mogollon Mountains. Hudsonian Zone.

A small conical tree growing in cooler situations in dense mixed forests, usually associated with spruce and aspen. It is easily recognized by its thin, smooth, white, corky bark, which persists after the tree has decayed.

2. **Abies lasiocarpa** (Hook.) Nutt. N. Amer. Sylv. **3:** 138. 1849.
Pinus lasiocarpa Hook. Fl. Bor. Amer. **2:** 163. 1842.
Type locality: "Interior of N. W. America."
Range: British America to Arizona and northwestern New Mexico.
New Mexico: Tunitcha Mountains; Brazos Canyon; Pecos Baldy. Mountains, in the Canadian Zone.

3. **Abies concolor** Lindl. Journ. Hort. Soc. Lond. **5:** 210. 1850. Balsam fir.
Pinus concolor Engelm.; Gord. & Glend. Pinet. 155. 1858.
Type locality: "Mountains of New Mexico."
Range: Oregon and California to Colorado and New Mexico.
New Mexico: Chama; Winsors Ranch; Trinchera Pass; Sandia Mountains; Mogollon Mountains; White and Sacramento mountains; Capitan Mountains. Mountains, in the Canadian and Hudsonian zones.

4. PSEUDOTSUGA Carr. Douglas spruce.

Large tree; leaves solitary, short-petiolate, flat, obtuse: cones ovate-oblong, pendulous, the bracts 3-parted, longer than the scales.

1. **Pseudotsuga mucronata** (Raf.) Sudw. Contr. U. S. Nat. Herb. **3:** 266. 1895.
Abies mucronata Raf. Atl. Journ. 120. 1832.
Abies douglasii Lindl. Penn. Cycl. **1:** 32. 1833.
Pseudotsuga douglasii Carr. Trait. Conif. ed. 2. 256. 1867.
Pseudotsuga taxifolia Britton, Trans. N. Y. Acad. **8:** 74. 1889.
Type locality: Mouth of Columbia River, Oregon.
Range: Alaska to Arizona and western Texas.
New Mexico: Common in all the higher mountains from the Las Vegas, Sacramento, and Organ ranges westward. Mountains, chiefly in the Canadian Zone.

This is the largest tree of the New Mexican mountains, in favorable situations over 60 meters high, with a trunk 2 meters or more in diameter. The bark is rough and dark-colored; the short (25 mm. or less) and obtuse leaves are arranged like those of the balsam fir. It may be most easily recognized by the cones, which are relatively small, composed of persistent thin scales, with the 3-parted bracts protruding a centimeter or more from beneath each scale. The tree occurs in mixed forests with yellow pine and the true spruces, at elevations ranging from 2,250 to 3,150 meters, sometimes reaching timber line. In the northern part of the State it often forms extensive pure stands in which there is little or no other vegetation. The lumber is of good quality. In cultivation the Douglas spruce makes an excellent decorative tree.

7. JUNIPERACEAE. Juniper Family.

Low trees or shrubs with much imbricated, short, scalelike or awllike leaves; cones composed of fleshy or peltate scales, without bracts; fruit berry-like, dehiscent or indehiscent.

KEY TO THE GENERA.

Cones dry, woody, dehiscent................................ 1. Cupressus (p. 35).
Cones fleshy, indehiscent.................................... 2. Juniperus (p. 36).

1. CUPRESSUS L. Cypress.

Small tree with short imbricated leaves; cones dry, woody, dehiscent, 6 to 8 mm. in diameter, composed of 6 to 8 peltate scales; seeds small, narrowly winged.

1. **Cupressus arizonica** Greene, Bull. Torrey Club **9:** 64. 1882. Arizona cypress.
Cupressus benthami arizonica Masters, Journ. Linn. Soc. Bot. **31:** 340. 1896.

TYPE LOCALITY: "On the mountains back of Clifton, in the extreme eastern part of Arizona."

RANGE: Mountains of southern Arizona and northern Mexico, coming into the southwestern corner of New Mexico.

NEW MEXICO: San Luis Mountains (*Mearns* 437, 560, 2244).

2. JUNIPERUS L. JUNIPER. CEDAR.

Large or small shrubs with awl-shaped or scalelike leaves; cones indehiscent, fleshy or fibrous; seeds 1 to 4, ovoid.

KEY TO THE SPECIES.

Leaves on mature branches not scalelike, 6 to 12 mm. long, smooth and shining above, glaucous beneath; a low shrub less that a meter high, often spreading .... 1. *J. sibirica.*

Leaves on mature branches scalelike, less than 5 mm. long, of the same color on both surfaces; large shrubs or small trees several meters high, never spreading.

  Seeds 3 or 4; branchlets smooth; leaves with a conspicuous resinous exudate; bark of the trunk broken into irregular quadrangular plates .... 2. *J. pachyphloea.*

  Seeds 1 or 2; branchlets mostly scaly; leaves not with a resinous exudate; bark shreddy or stringy.

    Fruit large, about 15 mm. in diameter .... 3. *J. megalocarpa.*

    Fruit small, 10 mm. in diameter or less.

      Branchlets slender, drooping; fruit 2-seeded; leaves 3-ranked .... 4. *J. scopulorum.*

      Branchlets rigid, erect; fruit mostly 1-seeded; leaves 2-ranked.

        Fruit large, 7 to 10 mm. long, oblong, brown and fibrous at maturity; leaves short and obtuse .... 5. *J. utahensis.*

        Fruit small, 5 to 7 mm. long, little if at all longer than thick, bluish, fleshy; leaves acute, long .... 6. *J. monosperma.*

1. Juniperus sibirica Burgsd. Anleit. Holz. no. 272. 1787. JUNIPER.

Juniperus communis sibirica Rydb. Contr. U. S. Nat. Herb. **3**: 533. 1896.

TYPE LOCALITY: Siberia.

RANGE: New Mexico to Alaska and Labrador.

NEW MEXICO: Chama; Santa Fe and Las Vegas mountains; Taos Mountains; Sandia Mountains. Deep woods, in the Canadian and Hudsonian zones.

2. Juniperus pachyphloea Torr. U. S. Rep. Expl. Miss. Pacif. **4**: 142. 1857. ALLIGATOR JUNIPER.

TYPE LOCALITY: Zuni Mountains, New Mexico.

RANGE: Arizona and western Texas to northern Mexico.

NEW MEXICO: Common from the Zuni Mountains, Black Range, Capitan Mountains, and Guadalupe Mountains southward and westward across the State. Low hills, in the Upper Sonoran Zone.

A round-topped tree 10 meters high or less, with a short, thick trunk, covered with thick, checkered bark which gives it its name of "alligator-bark juniper." On the cliffs at higher elevations it often attains a great age, developing a short and very thick trunk. The fruit is rather large for the genus, 8 to 10 mm. in diameter, ripening the second year. This is the common juniper in the southern part of the State in the foothills. The wood is used for fuel and to some extent for fence posts, although that of other species is preferred.

3. **Juniperus megalocarpa** Sudworth, For. & Irr. **13:** 307. 1907.

Sabina megalocarpa Cockerell, Muhlenbergia **3:** 143. 1908.

TYPE LOCALITY: "Midway between Alma and Frisco, about 3 miles above the 'Widow Kelley's' ranch on the San Francisco River," New Mexico.

RANGE: Known only from type locality.

A tree 9 to 15 meters high, the trunk 60 to 120 cm. in diameter; leaves in 3's, yellowish green.

This tree seems to have been first discovered in this same locality by Mr. Vernon Bailey of the Biological Survey, U. S. Department of Agriculture, who made some excellent photographs of it which we have seen.

4. **Juniperus scopulorum** Sarg. Gard. & For. **10:** 420. 1897.

ROCKY MOUNTAIN JUNIPER.

Sabina scopulorum Rydb. Bull. Torrey Club **32:** 598. 1905.

TYPE LOCALITY: Not definitely stated.

RANGE: British Columbia and Alberta to Arizona and western Texas.

NEW MEXICO: Coolidge; Rivera; Santa Fe; Pecos; Cebolla; Las Vegas; Stinking Lake; Mogollon Mountains; White Mountains. Open hills, in the Upper Sonoran Zone, often extending into the lower part of the Transition.

A beautiful though small tree, with dark green foliage and slender branches drooping at the ends. The fruit is small, blue, and succulent. The Rocky Mountain juniper occurs only in the higher mountains, associated sometimes with the common cedar (*Juniperus monosperma*), more often with pines. When growing alone it takes on a fine conical form with branches quite to the ground, rendering it an ideal tree for lawns.

5. **Juniperus utahensis** (Engelm.) Lemmon, Calif. Board For. Rep. **3:** 183. 1890.

UTAH JUNIPER.

Juniperus californica utahensis Engelm. Trans. Acad. St. Louis **3:** 588. 1877.

Sabina utahensis Rydb. Bull. Torrey Club **32:** 598. 1905.

TYPE LOCALITY: "Southern parts of Utah and into Arizona and Nevada."

RANGE: Wyoming to New Mexico, west to southeastern California.

NEW MEXICO: Tunitcha Mountains; Aztec; Carrizo Mountains; Frisco; Dona Ana Mountains. Open hills, in the Upper Sonoran Zone.

A stiff, upright, much branched tree, coming into New Mexico from the northwest. It differs from the next chiefly in the larger size and different color of its fruit. It is probably much more common than the citations would indicate.

6. **Juniperus monosperma** (Engelm.) Sarg. Silv. N. Amer. **10:** 89. 1889.

ONE-SEEDED JUNIPER.

Juniperus occidentalis monosperma Engelm. Trans. Acad. St. Louis **3:** 590. 1877.

Sabina monosperma Rydb. Bull. Torrey Club **32:** 598. 1905.

TYPE LOCALITY: "From Pikes Peak region of Colorado through west Texas and New Mexico to Arizona and California."

RANGE: Colorado to Nevada, south into Mexico.

NEW MEXICO: Common on foothills and high plains throughout the State. Upper Sonoran Zone.

This is the common juniper of the State. It is a low, much branched, frequently very scraggy tree, 4 to 8 meters high. Under favorable conditions it assumes an almost perfectly conical shape. The bark is gray and shreddy or stringy, the leaves of a rather yellowish green, and the fruit small and succulent. The wood does not decay readily and is much used for fence posts. It will no doubt prove of value as a decorative tree for lawns at elevations of from 1,800 to 2,250 meters.

We are unable to separate from this *Juniperus pinchotii* Sudworth.[1] Some of the material from the eastern side of the State should belong to that species. So far as

---

[1] For. & Irr. **13:** 307. 1907. The type came from Paloduro Canyon in the Panhandle of Texas.

we have been able to judge from the description and from the type material, the only difference suggested between the two is that the stumps left after *J. pinchoti* has been cut produce sprouts while those of *J. monosperma* do not, scarcely a substantial specific difference. As a matter of fact the stumps left after trees of the common cedar have been cut down often send up sprouts, just as they are said to do in this lately published species.

What is probably a form of *J. monosperma*, or possibly a distinct species, was described by Lemmon[1] as *Juniperus occidentalis gymnocarpa*. It is said to have the solitary seed partly exposed at the apex, hence the name. Mr. Lemmon states that this form is "abundant on the Sandia Mountains, near Albuquerque," New Mexico. No specimens have been seen by the writers. The same form has been collected near Fort Huachuca, Arizona, by Gen. T. E. Wilcox.

Order 6. GNETALES.

7a. EPHEDRACEAE. Joint-fir Family.

1. EPHEDRA L.

Shrubs 2 meters high or less, with slender terete striate stems; leaves reduced to small scarious bracts disposed in whorls at the nodes; flowers diœcious; fruit consisting of 1 or more seeds inclosed in few or many, chaffy, brownish or greenish scales.

Our species occur in the drier and lower parts of the State, on the sandy mesas, along arroyos, and on the rocky low slopes of the mountains, associated with mesquite, creosote bush, cactus, desert willow, and the like. A tea made by boiling the branches in water is used by the Mexicans and Indians as a remedy for venereal diseases and kidney affections. A chemical analysis shows a relatively high percentage of tannin in the stems. The shrubs are variously known as "popotillo," "cañatillo," "Mormon tea," and "Brigham Young weed," as also by several other names.

KEY TO THE SPECIES.

Leaf scales in 2's; cone scales few.
- Scales of the fruit acutish; fruit sharply angled; branches very numerous, erect, bright green .... 1. *E. viridis.*
- Scales rounded-obtuse; fruit scarcely angled; branches few, somewhat spreading, yellowish .... 2. *E. antisyphilitica.*

Leaf scales in 3's; cone scales numerous.
- Leaf scales 5 mm. long or less, merely acute, not acerose; fruit scabrous, less than 10 mm. long .... 3. *E. torreyana.*
- Leaf scales 8 to 10 mm. long, acerose; fruit smooth, 10 to 13 mm. long .... 4. *E. trifurca.*

1. Ephedra viridis Coville, Contr. U. S. Nat. Herb. **4**: 220. 1893.

TYPE LOCALITY: Near Crystal Spring, Coso Mountains, Inyo County, California.

RANGE: Southeastern California to Utah and western New Mexico.

NEW MEXICO: Western San Juan County; common. Mesas and low hills, in the Upper Sonoran Zone.

2. Ephedra antisyphilitica Meyer, Monogr. Ephedra 101. 1846.

TYPE LOCALITY: "Hab. in Mexici provincia orientali Coahuila, prope Laredo ad Rio del Norte."

RANGE: Colorado and Texas to Mexico.

NEW MEXICO: Bishops Cap; Tortugas Mountain. Mesas and dry hills, in the Lower and Upper Sonoran zones.

---

[1] Handbook of West-American cone-bearers 80. 1895.

3. Ephedra torreyana S. Wats. Proc. Amer. Acad. **14:** 299. 1899.

Type locality: "New Mexico to S. Utah."

Range: Colorado to California and Mexico.

New Mexico: Carrizo Mountains; Farmington; Santa Fe; Albuquerque; Nara Visa; Organ Mountains; San Andreas Mountains; Dona Ana Mountains; White Sands; Roswell. Plains and low hills, in the Lower and Upper Sonoran zones.

4. Ephedra trifurca Torr. in Emory, Mil. Reconn. 152. 1848.

Type locality: "From the region between the Del Norte and the Gila, and the hills bordering the latter river to the desert west of the Colorado."

Range: Colorado and Utah to northern Mexico.

New Mexico: Mangas Springs; Gila; San Antonio; Carrizalillo Mountains; Deming; Las Cruces; Organ Mountains. Plains and low hills, in the Lower and Upper Sonoran zones.

Class 2. ANGIOSPERMAE.

Subclass 1. MONOCOTYLEDONES.

Order 7. PANDANALES.

8. TYPHACEAE. Cattail Family.

1. TYPHA L. Cattail.

Tall marsh plant with creeping rootstocks and glabrous erect terete stems; leaves narrow, flat, striate; flowers monœcious, densely crowded in terminal spikes, the pistillate flowers below and the staminate above; ovary 1, stipitate, 1 or 2-celled.

1. Typha latifolia L. Sp. Pl. 971. 1753.

Type locality: "Habitat in paludibus Europae."

Range: Throughout most of North America; also in the Old World.

New Mexico: Farmington; Shiprock; Pecos; Mangas Springs; Fort Bayard; along the Rio Grande from Albuquerque to El Paso. In swamps and marshes, in the Lower and Upper Sonoran zones.

The Mexicans use the stems for a thatch upon which to lay mud roofs.

Order 8. NAIADALES.

KEY TO THE FAMILIES.

Gynœcium of distinct carpels; stigmas disklike or cuplike .... 9. **POTAMOGETONACEAE** (p. 39).
Gynœcium of united carpels; stigmas slender.. 10. **NAIADACEAE** (p. 41).

9. POTAMOGETONACEAE. Pondweed Family.

Aquatic herbs with jointed leafly stems; leaves sheathing at the base or stipulate; flowers perfect or unisexual, the perianth of 4 or 6 distinct valvate segments, or tubular, or none; stamens 1, 2, 4, or 6; ovaries 1 to 6, distinct, 1-celled, usually 1-ovuled; fruit indehiscent.

KEY TO THE GENERA.

Flowers perfect, spicate; stamens 4........................ 1. Potamogeton (p. 40).
Flowers monœcious, axillary; stamen 1.................. 2. Zanichellia (p. 40).

1. POTAMOGETON L. Pondweed.

Leaves all or only partly submerged, alternate, the blades broad or narrow; stipules more or less united and sheathing; flowers spicate; sepals and stamens 4; ovaries 4.

It is probable that we have more species in the State than are listed here. The material is difficult of collection and is usually neglected by collectors.

KEY TO THE SPECIES.

Leaves of two kinds, floating and submerged.
- Submerged leaves with blades; floating leaves elliptic.......... 1. *P. americanus.*
- Submerged leaves without blades; floating leaves oval........ 2. *P. natans.*

Leaves all submerged, narrow, sessile.
- Stipules free; spike continuous; fruits few.................... 3. *P. foliosus.*
- Stipules adnate to the petioles; spikes interrupted; fruits numerous.. 4. *P. interior.*

1. Potamogeton americanus Schlecht. & Cham. Linnæa **2**: 226. *pl. 6. f. 26.* 1827.

Potamogeton lonchites Tuck. Amer. Journ. Sci. II. **6**: 226. 1848.

Type locality: North America.

Range: In ponds and slow streams throughout North America except in the extreme northern part.

New Mexico: Collected by Fendler (no. 837), probably about Santa Fe.

2. Potamogeton natans L. Sp. Pl. 126. 1753.

Type locality: European.

Range: In still water throughout most of North America except the extreme north; also in Europe and Asia.

New Mexico: Tunitcha Mountains (*Standley* 7557).

3. Potamogeton foliosus Raf. Med. Repos. N. Y. n. ser. **5**: 354. 1808.

Potamogeton gramineum L. err. det. Michx. Fl. Bor. Amer. **1**: 102. 1803.

Potamogeton pauciflorus Pursh, Fl. Amer. Sept. 121. 1814.

Type locality: "Hab. in rivis affluente mari inundatis Carolinae inferioris."

Range: In streams and ponds nearly throughout North America.

New Mexico: Tularosa Creek near Aragon; Canada Creek at Ojo Caliente; Berendo Creek.

4. Potamogeton interior Rydb. Colo. Agr. Exp. Sta. Bull. **100**: 13. 1906.

Potamogeton marinus occidentalis Robbins; S. Wats. in King, Geol. Expl. 40th Par. **5**: 339. 1871.

Potamogeton filiformis occidentalis A. Benn. Ann. Cons. Jard. Genève **9**: 102. 1905.

Type locality: Colorado.

Range: Ontario and Northwest Territory to Utah and New Mexico.

New Mexico: Farmington; Cedar Hill; Carlsbad; Roswell.

The following species are represented by specimens the localities for which are uncertain but are probably in New Mexico or adjacent Texas:

Potamogeton pectinatus L. *Wright* 1894.
Potamogeton pusillus L. *Wright* 1896.

2. ZANICHELLIA L. Horned pondweed.

Leaves linear, mostly opposite, with sheathing stipules; flowers monœcious, sessile, axillary, the staminate ones consisting of a single stamen; ovaries 2 to 5, forming oblique oblong indehiscent nutlets in fruit.

1. **Zanichellia palustris** L. Sp. Pl. 969. 1753.

TYPE LOCALITY: "Habitat in Europae, Virginiae fossis, fluviis."

RANGE: In streams and ponds throughout North America except the extreme north; also in Eurasia.

NEW MEXICO: Tunitcha Mountains; Manguitas Spring; Salt Lake; Cienaga Ranch; Fort Tularosa; Roswell.

10. NAIADACEAE. Naias Family.

1. NAIAS L. NAIAS.

Slender branched aquatic, entirely submerged, with fibrous roots, numerous opposite or fasciculate leaves, and monœcious or diœcious, sessile or pedicellate, axillary, inconspicuous flowers; mature carpel solitary, sessile, ellipsoid, with a crustaceous pericarp.

1. **Naias guadalupensis** (Spreng.) Morong, Mem. Torrey Club 3[2]: 60. 1893.

Caulinia guadalupensis Spreng. Syst. Veg. **1**: 20. 1825.

TYPE LOCALITY: "Insula Guadalupa."

RANGE: Floating in water, Nebraska and Oregon to Florida and Tropical America.

NEW MEXICO: Lake La Jara (*Standley* 8274).

Order 9. ALISMALES.

KEY TO THE FAMILIES.

Petals and sepals similar; anthers long and narrow; carpels coherent .... 11. **JUNCAGINACEAE** (p. 41).
Petals and sepals unlike, the former white; anthers short and thick; carpels not coherent .... 12. **ALISMACEAE** (p. 42).

11. JUNCAGINACEAE. Arrow grass Family.

1. TRIGLOCHIN L. ARROW GRASS.

Perennial herbs with fleshy grasslike leaves clustered at the base of the scapelike stem; flowers small, spicate, with 3 ovate sepals and 3 similar petals; stamens 3 or 6; ovaries 3 or 6, united, the capsule splitting at maturity into 3 or 6 carpels.

KEY TO THE SPECIES.

Carpels 6; plants tall, 60 to 80 cm. high, stout .... 1. *T. maritimum*.
Carpels 3; plants low, 35 cm. high or less, slender .... 2. *T. palustre*.

1. **Triglochin maritimum** L. Sp. Pl. 339. 1753.

TYPE LOCALITY: "Habitat in Europae maritimis."

RANGE: Throughout the United States and in Mexico; also in Europe and Asia.

NEW MEXICO: Fitzgerald Cienaga; Mescalero Agency; Tularosa. Marshes, in the Transition Zone.

2. **Triglochin palustre** L. Sp. Pl. 338. 1753.

TYPE LOCALITY: "Habitat in Europae inundatis uliginosis."

RANGE: Widely distributed in North America; also in South America, Europe, and Asia.

NEW MEXICO: Grass Mountain; Rio Pueblo; Silver Spring Canyon. Wet ground, in the Transition and Canadian zones.

12. ALISMACEAE. Water-plantain Family.

Marsh herbs with fibrous roots, scapose stems, spongy petioles, and oval or sagittate leaf blades; leaves all radical; flowers perfect, monœcious, or diœcious; perianth of 3 herbaceous persistent sepals and as many white deciduous petals; stamens 6 or more; ovaries numerous, becoming 1-seeded achenes.

KEY TO THE GENERA.

Leaf blades ovate or oblong; all flowers perfect; carpels not winged; inflorescence paniculate............ 1. ALISMA (p. 42).
Leaf blades sagittate; all or part of the flowers unisexual; carpels winged; inflorescence raceme-like.
Lower flowers of the inflorescence pistillate; pedicels slender; leaves longer than broad............ 2. SAGITTARIA (p. 42).
Lower flowers of the inflorescence perfect; pedicels stout; leaves broader than long.............. 3. LOPHOTOCARPUS (p. 42).

1. ALISMA L. WATER-PLANTAIN.

Perennial with long-petioled leaves, ovate or oblong, acute blades, and 1 or 2 scapes terminating in a loose pyramidal panicle; flowers small; carpels numerous, in a simple circle on a flattened receptacle.

1. **Alisma plantago-aquatica** L. Sp. Pl. 342. 1753.
TYPE LOCALITY: "Habitat in Europae aquosis & ad ripas fluviorum, lacuum."
RANGE: Nearly throughout North America; also in Europe and Asia.
NEW MEXICO: Near Horace (*Wooton*). Wet ground.

2. SAGITTARIA L. ARROW HEAD.

Stoloniferous perennial herbs with long-petioled sheathing leaves with sagittate blades; stems simple, bearing a few whorls of flowers, the staminate flowers above, the pistillate below; ovaries many, on a globular receptacle, becoming flat membranous winged achenes.

1. **Sagittaria arifolia** Nutt.; J. G. Smith, Rep. Mo. Bot. Gard. **6**: 32. 1895.
TYPE LOCALITY: Oregon.
RANGE: British America southward through the western United States.
NEW MEXICO: San Juan Valley; Taos; Santa Fe; Belen; Reserve. Wet ground chiefly in the Upper Sonoran Zone.

3. LOPHOTOCARPUS Durand.

A perennial herb similar to the preceding, but the lower flowers of the inflorescence perfect instead of pistillate; leaves broadly sagittate.

1. **Lophotocarpus calycinus** (Engelm.) J. G. Smith, Mem. Torrey Club **5**: 25. 1894.
Sagittaria calycina Engelm. in Torr. U. S. & Mex. Bound. Bot. 212. 1859.
Sagittaria calycina maxima Engelm. loc. cit.
Sagittaria calycina media Engelm. loc. cit.
TYPE LOCALITY: "On the Red River, Louisiana."
RANGE: South Dakota and Delaware to Alabama and New Mexico.
NEW MEXICO: Mesilla (*Wooton* 74). Wet ground, in the Lower Sonoran Zone.

Order 10. POALES.

KEY TO THE FAMILIES.

Leaves 2-ranked; margins of sheaths not united; stems mostly hollow.................................. **13. POACEAE** (p. 43).
Leaves 3-ranked; margins of sheaths united; stems solid. **14. CYPERACEAE** (p. 110).

13. POACEAE. Grass Family.[1]

Fibrous-rooted annual or perennial herbs, often with rootstocks, with jointed, usually hollow, cylindrical stems and 2-ranked leaves, their blades parallel-veined, mostly long and narrow, their bases forming an open or rarely a closed sheath around the stem; inflorescence an open or spikelike panicle, a raceme, or a spike; flowers usually perfect, small, without a distinct perianth, arranged in spikelets, these consisting of an articulate axis (rachilla) and 3 to many 2-ranked bracts, the lower 2 (glumes) being empty, the succeeding 1 or more (lemmas) each containing in its axil a single flower subtended by a palea; stamens usually 3; pistil 1, with a 1-celled, 1-ovuled ovary, 2 styles, and plumose stigmas; fruit a caryopsis with a starchy endosperm and a small embryo.

KEY TO THE TRIBES.

Spikelets dorsally compressed, falling from the pedicels entire, 1-flowered, or sometimes with a rudimentary flower below the perfect one.
  Lemma and palea hyaline, much more delicate in texture than the glumes.
    Spikelets unisexual, the pistillate borne in the lower, the staminate in the upper part of the same spike........................**I. MAYDEAE.**
    Spikelets in pairs, one sessile, the other pedicellate, the former perfect, the latter perfect or with a staminate flower, often reduced to 1 or 2 scales.................................**II. ANDROPOGONEAE.**
  Lemmas, at least those of the perfect flowers, similar in texture to the glumes or thicker and firmer, never hyaline.
    Lemma and palea membranous; spikelets in groups of 3, these falling together from the continuous axis..........................**III. ZOYSIEAE.**
    Lemma and palea chartaceous to coriaceous, different in color and texture from the glumes; spikelets various..................**IV. PANICEAE.**
Spikelets laterally compressed, at least at maturity, the glumes usually persistent on the pedicel or rachis after the fall of the florets, 1 to many-flowered, the rudimentary flower, if any, usually uppermost.
  Spikelets in 2 rows, sessile or nearly so.
    Spikelets on one side of the continuous axis, forming one-sided spikes, these digitate or paniculate........................**VIII. CHLORIDEAE.**
    Spikelets alternate on opposite sides of a channeled, sometimes articulate, axis; spikes solitary.................................**X. HORDEAE.**
  Spikelets borne in an open or spikelike panicle or raceme, usually upon distinct pedicels.
    Spikelets with 1 perfect flower.
      No rudimentary or staminate floret below the perfect one. **V. PHALARIDEAE.**
      A pair of rudimentary or staminate florets below the perfect one. **VI. AGROSTIDEAE.**
    Spikelets 2 to many-flowered.
      Lemmas usually shorter than the glumes; awn dorsal or from between the teeth of the bidentate apex, usually bent.....**VII. AVENEAE.**
      Lemmas usually longer than the glumes; awn terminal (rarely dorsal in Bromus) and straight, or none.............**IX. FESTUCEAE.**

---

[1] For a more extended account of New Mexican grasses from an economic standpoint, see Wooton and Standley, N. Mex. Agr. Exp. Sta. Bull. **81**. 1912.

KEY TO THE GENERA.

Tribe I. MAYDEAE.

A single genus .... 1. TRIPSACUM (p. 49).

Tribe II. ANDROPOGONEAE.

- Spikelets all pedicellate, the longer pediceled one perfect, long-awned, the short-pediceled one staminate, awnless .... 2. TRACHYPOGON (p. 49).
- Some of the spikelets sessile, these perfect, the pedicellate spikelets staminate, sterile, or reduced to a pedicel.
  - Spikelets all awnless .... 3. ELYONURUS (p. 49).
  - Perfect spikelets awned.
    - Lower spikelets unlike the others .... 9. HETEROPOGON (p. 52).
    - Sessile spikelets all alike.
      - Racemes single; rachis joints with a cup-shaped appendage at the tip .... 4. SCHIZACHYRIUM (p. 49).
      - Racemes 2 or more; rachis joints not appendaged.
        - Rachis joints and pedicels sulcate, translucent .... 5. AMPHILOPHIS (p. 50).
        - Rachis neither sulcate nor translucent.
          - Some of the racemes sessile .... 6. ANDROPOGON (p. 51).
          - All the racemes pedunculate.
            - Pedicellate spikelets reduced to a pedicel .... 8. SORGHASTRUM (p. 52).
            - Pedicellate spikelets staminate .... 7. HOLCUS (p. 51).

Tribe III. ZOYSIEAE.

- Second glumes covered with hooked spines .... 10. NAZIA (p. 52).
- Second glumes not spiny .... 11. HILARIA (p. 53).

Tribe IV. PANICEAE.

- Spikelets involucrate.
  - Involucre a spiny bur, falling with the spikelets .... 20. CENCHRUS (p. 61).
  - Involucre of numerous bristles, persistent on the axis after the fall of the spikelets .... 19. CHAETOCHLOA (p. 60).
- Spikelets not involucrate.
  - Glumes, at least the second, awned or cuspidate .... 18. ECHINOCHLOA (p. 59).
  - Glumes not awned.
    - Spikelets lanceolate; fruit cartilaginous, not rigid, the white hyaline margins of the lemma not inrolled.
      - Spikelets densely covered with long silky hairs .... 12. VALOTA (p. 54).
      - Spikelets glabrous or nearly so.
        - Inflorescence of slender racemes, digitately arranged .... 13. SYNTHERISMA (p. 54).
        - Inflorescence a capillary panicle .... 14. LEPTOLOMA (p. 54).

Spikelets oval or obovate; fruit chartaceous, rigid, the lemma margins inrolled, not hyaline.
First glume present; spikelets panicled, rarely in racemes....................17. PANICUM (p. 56).
First glume obsolete; spikelets in racemes.
Spikelet with a swollen ringlike callus below; fruit awn-tipped...15. ERIOCHLOA (p. 54).
Spikelets without a callus; fruit not awned..........................16. PASPALUM (p. 55).

Tribe V. PHALARIDEAE.

Panicles dense and contracted; sterile lemmas minute.21. PHALARIS (p. 61).
Panicles loose and open; sterile lemmas inclosing staminate flowers..............................22. SAVASTANA (p. 61).

Tribe VI. AGROSTIDEAE.

Lemmas indurated at maturity, closely inclosing the grain.
Spikelets in pairs, one perfect, the other staminate or sterile (in spikelike panicles)............23. LYCURUS (p. 62).
Spikelets not in pairs, all perfect.
Lemma 3-awned (2 of the awns sometimes very small)..................................24. ARISTIDA (p. 62).
Lemmas 1-awned or awnless.
Awns twisted and bent ....25. STIPA (p. 65).
Awns not twisted, or wanting.
Lemmas narrow; awns, when present, persistent......................26. MUHLENBERGIA (p. 68).
Lemmas broad; awns deciduous.....27. ORYZOPSIS (p. 72).
Lemmas usually thin at maturity, at least more delicate than the glumes; grain loosely inclosed.
Glumes conspicuously compressed-keeled; panicle dense and spikelike, cylindrical.
Rachilla jointed above the glumes, these persistent; lemma awnless.................28. PHLEUM (p. 73).
Rachilla jointed below the glumes, these falling with the spikelets; lemma awned..29. ALOPECURUS (p. 73).
Glumes not compressed-keeled; panicles mostly open and spreading, rarely crowded and spikelike.
Panicle spikelike, elongated, 30 cm. long or more....................................30. EPICAMPES (p. 74).
Panicle not elongated.
Spikelets articulated below the glumes, falling entire.
Glumes awned; panicles dense......33. POLYPOGON (p. 77).
Glumes awnless; panicles open......34. CINNA (p. 78).
Spikelets articulated above the glumes, these persistent after the fall of the florets.

Lemmas pilose on the nerves.......31. BLEPHARONEURON (p. 74).
Lemmas not pilose on the nerves.
Lemmas 1-nerved; pericarp separating from the seed....32. SPOROBOLUS (p. 75).
Lemmas 3 to 5-nerved; pericarp adherent to the seed.
Rachilla prolonged behind the palea; lemma with a short awn on the back.............36. CALAMAGROSTIS (p. 79).
Rachilla not prolonged behind the palea; lemmas awnless.
Glumes longer than the floret............37. CALAMOVILFA (p. 80).
Glumes shorter than the floret.............35. AGROSTIS (p. 78).

Tribe VII. AVENEAE.

Awns attached between the teeth of the lemma, flattened...38. DANTHONIA (p. 80).
Awns dorsal, not flattened.
Grain adherent to the palea; spikelets mostly more than 10 mm. long.....................39. AVENA (p. 81).
Grain free; spikelets less than 10 mm. long.
Lemmas erose or shortly 2-lobed at the apex; panicles open..........................40. DESCHAMPSIA (p. 81).
Lemmas deeply 2-toothed at the apex, the teeth awn-pointed; panicles dense and congested..............................41. TRISETUM (p. 82).

Tribe VIII. CHLORIDEAE.

Spikelets unisexual, dissimilar; flowers monœcious or diœcious......................................42. BULBILIS (p. 82).
Spikelets all alike.
Spikelets with 2 to 4 perfect flowers.
Spikelets small, numerous, approximate; glumes thin..............................43. LEPTOCHLOA (p. 83).
Spikelets large, few, distant; glumes firm and thick..............................44. ACAMPTOCLADOS (p. 84).
Spikelets with 1, rarely 2, perfect flowers.
Rachilla jointed below the boat-shaped inflated glumes, the whole spikelet falling at maturity.........................45. BECKMANNIA (p. 84).
Rachilla jointed above the glumes, these persistent, not boat-shaped.
No sterile lemmas present above the perfect floret.
Plants with long stolons; spikelets numerous, crowded; spikes 2 to 6, digitate..................46. CAPRIOLA (p. 84).

Plants without stolons; spikelets few, scattered; spikes scattered along the central axis.........47. SCHEDONARDUS (p. 85).
One to several sterile lemmas above the perfect florets.
Spikes scattered along the central axis.........................48. BOUTELOUA (p. 85).
Spikes digitate, or crowded near the end of the stem.
Lemmas with a single awn or awnless...................49. CHLORIS (p. 87).
Lemmas 3-awned................50. TRICHLORIS (p. 88).

Tribe IX. FESTUCEAE.

Lemmas with numerous (9 or more) awnlike divisions or awned lobes................................51. PAPPOPHORUM (p. 88).
Lemmas with few lobes or entire.
Lemmas, at least those of the pistillate spikelets, 3-lobed and 3-awned......................52. SCLEROPOGON (p. 89).
Lemmas entire or at most 2-lobed.
Hairs on the rachilla or lemma very long, exceeding the lemma in length.
Rachilla hairy; lemma naked...........53. PHRAGMITES (p. 89).
Rachilla naked; lemma hairy...........54. ARUNDO (p. 89).
Hairs, if any, on the rachilla and lemma shorter than the latter.
Stigmas barbellate, on long styles; spikelets in 3's in the axils of the spinescent leaves; plants spreading, woolly when young................55. MUNROA (p. 90).
Stigmas plumose, sessile or on short styles; spikelets and plants various.
Lemmas 1 to 3-nerved.
Lateral nerves of the lemmas hairy.
Lemmas deeply 2-lobed....56. DASYOCHLOA (p. 90).
Lemmas entire or but slightly 2-lobed.
Inflorescence a short crowded raceme; leaf blades with cartilaginous margins; plants low and tufted.......57. ERIONEURON (p. 90).
Inflorescence a rather large panicle; leaf blades without cartilaginous margins; plants tall..58. TRIDENS (p. 91).
Lateral nerves of the lemmas glabrous.
Second glume very unlike the first, broadened upward...............59. SPHENOPHOLIS (p. 92).

Second glume similar to the first, not broadened upward.
Panicles narrow, dense and spikelike, the branches erect............60. KOELERIA (p. 92).
Panicles open, the branches spreading..............61. ERAGROSTIS (p. 93).
Lemmas 5 to many-nerved.
Spikelets with 2 or more of the upper glumes empty, broad and infolding each other..........62. MELICA (p. 95).
Spikelets with the upper glumes flower-bearing or narrow and abortive.
Stigmas plainly arising from below the apex of the ovary...63. BROMUS (p. 95).
Stigmas placed at or near the apex of the ovary.
Spikelets in 1-sided fascicles arranged in a glomerate or interrupted panicle......64. DACTYLIS (p. 97).
Spikelets in panicles of racemes.
Glumes more or less laterally compressed and keeled.
Flowers diœcious; lemmas coriaceous...........65. DISTICHLIS (p. 98).
Flowers monœcious, most of the flowers perfect; lemmas thin, scarious-margined.........66. POA (p. 98).
Glumes rounded on the back, at least below the middle.
Lemmas acute, pointed or awned at the apex...67. FESTUCA (p. 101).
Lemmas obtuse or acutish, usually toothed.
Lemmas distinctly 5 to 7-nerved; styles present...........68. PANICULARIA (p. 103).
Lemmas obscurely 5-nerved; styles none..............69. PUCCINELLIA (p. 104).

Tribe X. HORDEAE.

Spikelets usually single at the nodes of the rachis.
Glumes with their sides turned toward the rachis..70. AGROPYRON (p. 104).
Glumes with their backs turned toward the rachis.71. LOLIUM (p. 106).

Spikelets 2 to 6 at each joint of the rachis or if solitary the glumes arranged obliquely to the rachis.
Spikelets 1-flowered or with a rudimentary second flower....................................72. HORDEUM (p. 106).
Spikelets 2 to many-flowered.
Rachis of the spikes jointed, readily breaking into joints...........................73. SITANION (p. 107).
Rachis of the spikes continuous, not breaking into joints...........................74. ELYMUS (p. 108).

1. TRIPSACUM L.

Tall stout perennial with creeping rootstocks, broad flat leaves, and terminal digitate inflorescence, the spikes separating into joints at maturity; spikelets unisexual, the staminate in pairs at the joints of the rachis above, the pistillate solitary, embedded in each joint of the rachis below in the same inflorescence; glumes of the staminate spikelet subcoriaceous, those of the pistillate spikelet finally cartilaginous, the lemmas and paleas hyaline.

1. **Tripsacum lanceolatum** Rupr.; Fourn. Mex. Pl. **2:** 68. 1886.
TYPE LOCALITY: Aguas Calientes, Mexico.
RANGE: Southwestern New Mexico to southern Mexico.
NEW MEXICO: Guadalupe Canyon (*E. C. Merton* 2035).

2. TRACHYPOGON Nees.

Rather tall perennials with narrow leaves and usually solitary, long-exserted racemes; spikelets 1-flowered, in pairs at the nodes of the imperfectly jointed rachis, one nearly sessile, awnless, sterile, the other pedicellate, fertile, long-awned; glumes rigid, the outer large and inclosing the other; lemmas produced into long twisted geniculate awns.

1. **Trachypogon montufari** (H. B. K.) Nees, Agrost. Bras. 342. 1829.
Andropogon montufari H. B. K. Nov. Gen. & Sp. **1:** 184. 1816.
TYPE LOCALITY: "In aridis, apricis regni Quitensis prope Conocoto, Pintae et Villam Chilloensem Montufari."
RANGE: New Mexico and Arizona to Mexico and South America.
NEW MEXICO: Near White Water (*Mearns* 353). Dry hills.

3. ELYONURUS Humb. & Bonpl.

Low or tall annuals or perennials with rather rigid leaves and solitary terminal racemes; spikelets 1-flowered, awnless; first glume rigid or subcoriaceous, 2-toothed at the apex, the margins inflexed, more or less ciliate, with balsam-bearing lines between the lateral keels, the second a little shorter than the first, acute; lemma delicate and hyaline; palea minute or none; stamens 3; styles distinct.

1. **Elyonurus barbiculmis** Hack. in DC. Monogr. Phan. **6:** 339. 1889.
TYPE LOCALITY: Western Texas.
RANGE: Southern New Mexico and Arizona to western Texas and adjacent Mexico.
NEW MEXICO: Dog Spring (*Mearns* 2376). Dry hills.

4. SCHIZACHYRIUM Nees. SAGE GRASS.

Tall perennials, tufted or from rootstocks, with flat or involute leaves, and spikelike solitary racemes terminating the stem or its branches; spikelets in pairs at each node of the jointed and often hairy rachis, one sessile and fertile, the other pedicellate and sterile; glumes indurated, sometimes pubescent; lemma entire or 2-toothed at the apex, bearing a straight, contorted, or spiral awn; palea small, hyaline; stamens 1 to 3; styles distinct.

KEY TO THE SPECIES.

Hairs of the pedicels very few and short or none................ 1. *S. cirratum.*
Hairs of the pedicels long and silky, abundant.
  Peduncles long and slender, much exserted................ 2. *S. neomexicanum.*
  Peduncles short, stout, little if at all exserted................ 3. *S. scoparium.*

1. **Schizachyrium cirratum** (Hack.) Woot. & Standl. N. Mex. Agr. Exp. Sta. Bull. **81:** 30. 1912.

Andropogon cirratus Hack. Flora **1885:** 119. 1885.

TYPE LOCALITY: Western Texas.

RANGE: Mountains of southern New Mexico and Arizona and adjacent Mexico.

NEW MEXICO: Star Peak; near Silver City; Mangas Springs; Dog Spring. Upper Sonoran Zone.

2. **Schizachyrium neomexicanum** Nash; Woot. & Standl. N. Mex. Agr. Exp. Sta. Bull. **81:** 29. 1912.

Andropogon neomexicanus Nash, Bull. Torrey Club **25:** 83. 1898.

TYPE LOCALITY: White Sands, New Mexico. Type collected by Wooton.

RANGE: New Mexico.

NEW MEXICO: Crawfords Ranch; Organ Mountains; mountains west of Grants Station; White Sands; Buchanan. Plains and low hills, in the Lower Sonoran Zone.

3. **Schizachyrium scoparium** (Michx.) Nash in Small, Fl. Southeast. U. S. 59. 1903.

Andropogon scoparius Michx. Fl. Bor. Amer. **1:** 57. 1803.

TYPE LOCALITY: "Habitat in aridis sylvarum Carolinae."

RANGE: British America to Texas, Florida, and Mexico.

NEW MEXICO: Coolidge; San Lorenzo; Pecos; Clayton; Trout Spring; Taos; Raton Mountains; Sandia Mountains; Mogollon Mountains; White Mountains; Pecos Valley. Plains and low hills, in the Upper Sonoran and Transition zones.

5. AMPHILOPHIS Nash.

Tufted perennials with mostly flat leaves and showy, often silvery, white panicles, the axis short, making the panicle appear fanlike, or elongated; racemes usually numerous, the internodes with thickened margins, the median portion thin and translucent; pedicels ciliate with usually long hairs; first glume 2-keeled, the second 1-keeled; lemma hyaline, very narrow, stipelike, gradually merging into an awn; stamens 3; styles distinct.

KEY TO THE SPECIES.

Hairs on the rachis and pedicels shorter than the spikelets ...... 1. *A. wrightii.*
Hairs on the rachis and pedicels longer than the spikelets.
  Awns 10 mm. long or less; panicles usually small.......... 2. *A. saccharoides.*
  Awns more than 10 mm. long; panicles large................ 3. *A. barbinodis.*

1. **Amphilophis wrightii** (Hack.) Nash in Britton, Man. 71. 1901.

Andropogon wrightii Hack. Flora **1885:** 139. 1885.

TYPE LOCALITY: "New Mexico." Type collected by Wright (no. 2104).

RANGE: Southern New Mexico and Arizona and northern Mexico.

NEW MEXICO: Hillsboro (*Metcalfe* 1371). Dry hills, in the Upper Sonoran Zone.

2. **Amphilophis saccharoides** (Swartz) Nash; Woot. & Standl. N. Mex. Agr. Exp. Sta. Bull. **81:** 30. 1912.

Andropogon saccharoides Swartz, Prodr. Veg. Ind. Occ. 26. 1788.

TYPE LOCALITY: Jamaica.

RANGE: Western Texas and southern Arizona to Mexico and the West Indies.

NEW MEXICO: Mesilla Valley; Belen; Eagle Creek; Guadalupe Mountains; Lakewood; Carlsbad. Mesas and valleys, in the Lower Sonoran Zone.

3. **Amphilophis barbinodis** (Lag.) Nash in Small, Fl. Southeast. U. S. 65. 1903.
Andropogon barbinodis Lag. Gen. & Sp. Nov. 3. 1816.
TYPE LOCALITY: "H [abitat] in N [ova] H [ispania]."
RANGE: Southern New Mexico and Arizona and northern Mexico.
NEW MEXICO: Las Vegas Canyon; Black Range; Silver City; Burro Mountains; Mesilla Valley; Organ Mountains; Nara Visa; Buchanan; Knowles; Carlsbad. Dry hills and mesas, in the Lower and Upper Sonoran zones.

6. ANDROPOGON L. TALL SAGE GRASS.

Tall perennials, tufted or from elongated rootstocks, with flat or involute leaves and with spikelike racemes disposed in pairs or sometimes in 3's or more, terminating the stem or its branches; spikelets sometimes with a ring of short hairs at the base, in pairs at each node of the jointed and often hairy rachis, one sessile and fertile, the other pedicellate and sterile; glumes indurated, often pubescent; lemma entire or 2-toothed at the apex, awned or sometimes awnless; palea small, hyaline; stamens 1 to 3; styles distinct.

KEY TO THE SPECIES.

Second lemma of the sessile spikelet awnless, or with a short straight awn.......... 1. *A. hallii.*
Second lemma of the sessile spikelet with a long geniculate awn, more or less twisted at the base.
  Glumes of the sessile spikelet hispidulous all over; hairs of the rachis internodes 2 mm. long or less.......... 2. *A. furcatus.*
  Glumes of the sessile spikelet glabrous or nearly so except on the nerves; hairs of the rachis internodes 3 to 4 mm. long.......... 3. *A. chrysocomus.*

1. **Andropogon hallii** Hack. Sitzungsb. Akad. Wiss. Math. Naturw. (Wien) **89:** 127. 1884.
TYPE LOCALITY: Colorado.
RANGE: Montana and Nebraska to Kansas and Mexico.
NEW MEXICO: Near Portales; Buchanan; northeast of Clayton; mountains west of Las Vegas; Nara Visa; Arroyo Ranch. Plains, in the Upper Sonoran Zone.

2. **Andropogon furcatus** Muhl.; Willd. Sp. Pl. **4:** 919. 1806.
Andropogon provincialis furcatus Hack. in DC. Monogr. Phan. **5:** 442. 1889.
TYPE LOCALITY: "Habitat in America boreali."
RANGE: British America to Florida and New Mexico.
NEW MEXICO: Tunitcha Mountains; Dulce. Dry hills and plains, in the Upper Sonoran Zone.

3. **Andropogon chrysocomus** Nash in Britton, Man. 70. 1901.
TYPE LOCALITY: Stevens County, Kansas.[1]
RANGE: Kansas and Texas to New Mexico.
NEW MEXICO: Carrizo Mountains; White and Sacramento mountains. Plains and dry hills, in the Upper Sonoran and Transition zones.

7. HOLCUS L.

Tall perennial with numerous long rootstocks, broad flat leaves, and large terminal panicles; spikelets in pairs or 3's at the ends of the branches, one sessile and perfect, the others pedicellate and staminate, dorsally compressed, pubescent or glabrous; glumes indurated; lemma hyaline, awned or awnless; stamens 3; styles distinct.

[1] N. Amer. Fl. **17:** 120. 1912.

1. **Holcus halepensis** L. Sp. Pl. 1047. 1753. JOHNSON GRASS.

Andropogon halepensis Brot. Fl. Lusit **1:** 89. 1804.

Sorghum halepense Pers. Syn. Pl. **1:** 101. 1805.

TYPE LOCALITY: "Habitat in Syria, Mauritania."

RANGE: Native of the Old World, widely introduced into North America, frequent as a weed in cultivated fields.

NEW MEXICO: Nara Visa; Mangas Springs; Hillsboro; Gila; Deming; Mesilla Valley; Pecos Valley.

This is common in several parts of New Mexico, especially in the irrigated river valleys. So far it has not been introduced into the valley of the San Juan, but it is well established in those of the Rio Grande and Pecos. In some parts of the State it has been cultivated for hay. Unfortunately it is a very troublesome weed, and in the Rio Grande Valley has become a dangerous pest in alfalfa fields, taking possession of them and crowding out the less aggressive alfalfa.

8. SORGHASTRUM Nash. INDIAN GRASS.

Stout perennials with racemes arranged in open panicles; spikelets sessile at each joint of the slender rachis of the peduncled racemes, these reduced to 2 or 3 joints; sterile spikelets reduced to hairy pedicels; glumes indurated; sterile lemma hyaline, the fertile lemma reduced to hyaline appendages to the stout awn; palea obsolete.

1. **Sorghastrum nutans** (L.) Nash in Small, Fl. Southeast. U. S. 66. 1903.

Andropogon nutans L. Sp. Pl. 1045. 1753.

Andropogon avenaceus Michx. Fl. Bor. Amer. **1:** 58. 1803.

Sorghum nutans A. Gray, Man. 617. 1848.

TYPE LOCALITY: "Habitat in Virginia, Jamaica."

RANGE: British America to Arizona and Florida.

NEW MEXICO: Tesuque; Las Vegas; Clayton; East View; Kingston; Rio Mimbres. Plains, in the Upper Sonoran Zone.

9. HETEROPOGON Pers.

Coarse perennial with narrow leaves, compressed sheaths, and terminal solitary dense racemes; spikelets 1-flowered, in pairs at the rachis nodes, one sessile and fertile, the other pedicellate and staminate or sterile; glumes firm, convolute, awnless; lemma small, hyaline, awned; palea small and hyaline, or wanting; stamens 3; styles distinct.

1. **Heteropogon contortus** (L.) Roem. & Schult. Syst. Veg. **2:** 836. 1817.

Andropogon contortus L. Sp. Pl. 1045. 1753.

TYPE LOCALITY: "Habitat in India."

RANGE: Arizona and New Mexico to Tropical America; in tropical lands nearly around the world.

NEW MEXICO: Hillsboro; Tortugas Mountain; Organ Mountains. Dry hills, in the Lower and Upper Sonoran zones.

10. NAZIA Adans.

Diffusely branched annual with flat leaves and terminal spikelike inflorescence; spikelets in groups of 3 to several at each joint of the main axis, the uppermost in each fascicle sterile, 1-flowered; first glume minute or wanting; second glume rigid, exceeding the lemma, its back covered with hooked spines; lemma and palea hyaline; stamens 3; styles short and distinct; grain oblong, free.

1. **Nazia aliena** (Spreng.) Scribn. U. S. Dept. Agr. Div. Agrost. Bull. **17:** 28. 1899.

Lappago aliena Spreng. Neu. Entd. **3:** 15. 1822.

Tragus alienus Schult. Mant. **2:** 205. 1824.

TYPE LOCALITY: "Hab. in Brasilia."

RANGE: Western Texas and southern Arizona to Mexico, and throughout tropical America.

NEW MEXICO: Mogollon Mountains; Mangas Springs; Lake Valley; Socorro; Deming; Burro Mountains; Organ Mountains; Carrizozo. Dry sandy soil, in the Lower Sonoran Zone.

11. HILARIA H. B. K.

Cespitose or decumbent perennials, often stoloniferous, with flat or involute leaves and terminal solitary spikes; spikelets sessile, in groups of 3 at each joint of the flexuous continuous rachis, the groups falling off entire, the 2 outer or anterior spikelets staminate and 2 or 3-flowered, the posterior or inner one pistillate or hermaphrodite and 1-flowered; glumes firm, unequal, many-nerved, more or less connate below, entire at the apex or divided, usually unequally 2-lobed with 1 to several intermediate awns or awnlike divisions; lemmas narrow; stamens 3; styles united below; grain ovoid or oblong, free.

KEY TO THE SPECIES.

Base of glumes with black or purplish glands...................... 1. *H. cenchroides.*
Glumes not glandular.
  Glumes cuneate, awnless, the nerves divergent .... 2. *H. mutica.*
  Glumes linear or oblong, awned, the nerves approximate...... 3. *H. jamesii.*

1. **Hilaria cenchroides** H. B. K. Nov. Gen. & Sp. **1**: 117. *pl. 37*. 1816.

TEXAS CURLY MESQUITE GRASS.

TYPE LOCALITY: "Crescit in planitie montana regni Mexicani, inter Zelaya et Guanaxuato, locis subfrigidis, alt. 980 hexap."

RANGE: Western Texas and southwestern New Mexico and southward.

NEW MEXICO: Mangas Springs; Cook Spring. Dry hills, in the Upper Sonoran Zone.

2. **Hilaria mutica** (Buckl.) Benth. Journ. Linn. Soc. Bot. **19**: 62. 1881.

TOBOSA GRASS.

Pleuraphis mutica Buckl. Proc. Acad. Phila. **1862**: 95. 1863.

TYPE LOCALITY: "Northern Texas."

RANGE: Western Texas to southern Arizona and adjacent Mexico.

NEW MEXICO: Common on the plains and low hills from the Black Range and White Mountains southward; also collected by Bigelow at Laguna Colorado. Lower and Upper Sonoran zones.

Tobosa grass is one of the most important range grasses on the plains and mesas of southern New Mexico, being usually associated with black grama. Stock do not eat it after it has dried, because of its hard and somewhat woody stems, but they thrive upon it in late summer after the rains. It grows most frequently in flats that are sometimes flooded, being able to resist flooding for considerable periods. It is also very resistant to trampling.

3. **Hilaria jamesii** (Torr.) Benth. Journ. Linn. Soc. Bot. **19**: 62. 1881.

GALLETA GRASS.

Pleuraphis jamesii Torr. Ann. Lyc. N. Y. **1**: 148. *pl. 10*. 1824.

TYPE LOCALITY: "On the high plains of the Trap Formation at the sources of the Canadian River," Colorado or New Mexico. Type collected by James.

RANGE: Wyoming and Nevada to Texas.

NEW MEXICO: Abundant on the plains from the Mogollon Mountains, Engle, and the White Mountains northward and eastward. Plains, in the Upper Sonoran Zone.

Galleta grass occupies the same position in northern New Mexico as tobosa in the southern part. It is by far the most abundant and characteristic plant on the plains in the northwestern corner of the State, often forming practically pure stands for many miles. It is probably the second most valuable range grass of New Mexico, being an excellent forage plant, very persistent, and not easily killed by overstocking.

12. **VALOTA** Adans.

Tufted perennials with flat leaves and narrow or contracted, densely hairy panicles; spikelets numerous, articulated below the glumes, 1-flowered; glumes membranous, densely silky-pilose or long-ciliate on the margins, often acuminate, sometimes with a short bristle at the apex; lemma chartaceous, glabrous and shining, finally indurated; stamens 3; styles distinct.

1. **Valota saccharata** (Buckl.) Chase, Proc. Biol. Soc. Washington **19:** 188. 1906.

Panicum lachnanthum Torr. U. S. Rep. Expl. Miss. Pacif. 7[3]: 21. 1856, not Hochst. 1855.

Panicum saccharatum Buckl. Prel. Rep. Geol. Agr. Surv. Tex. App. 2. 1866.

Trichachne saccharatum Nash in Small, Fl. Southeast. U. S. 83. 1903.

TYPE LOCALITY: "Middle Texas."

RANGE: Colorado and Texas to Arizona and Mexico.

NEW MEXICO: Albuquerque; Mangas Springs; Black Range; Dog Spring; Dona Ana and Organ Mountains; Causey. Dry hills and plains, in the Lower and Upper Sonoran zones.

13. **SYNTHERISMA** Walt.

Annuals with branched culms, thin flat leaves, and subdigitate inflorescence; spikelets 1-flowered, lanceolate-elliptic, sessile or short-pediceled, solitary or in 2's and 3's in 2 rows on one side of a continuous, narrow or winged rachis, forming slender racemes, these aggregated toward the top of the culm; glumes 1 to 3-nerved, the first sometimes obsolete; sterile lemma 5-nerved, the fertile indurated, papillose-striate, with a hyaline margin.

1. **Syntherisma sanguinale** (L.) Dulac, Fl. Haut. Pyr. 77. 1867. CRABGRASS.

Panicum sanguinale L. Sp. Pl. 57. 1753.

Digitaria sanguinalis Scop. Fl. Carn. ed. 2. **1:** 52. 1772.

TYPE LOCALITY: "Habitat in America, Europa australi."

RANGE: Cultivated and waste grounds in nearly all parts of the United States, introduced from Europe.

NEW MEXICO: Galisteo; Animas Creek; Deming; Mesilla Valley; Guadalupe Mountains.

14. **LEPTOLOMA** Chase.

Tufted perennials with flat leaves and diffuse terminal panicles, these breaking away at maturity and becoming "tumbleweeds;" spikelets 1-flowered, fusiform, solitary on long capillary pedicels; first glume obsolete or minute, the second 3-nerved, nearly as long as the 5 to 7-nerved sterile lemma; fertile lemma indurated, papillose, with a hyaline margin, this not inrolled; grain free.

1. **Leptoloma cognatum** (Schult.) Chase, Proc. Biol. Soc. Washington **19:** 92. 1906. FALL WITCH GRASS.

Panicum cognatum Schult. Mant. **2:** 235. 1824.

Panicum autumnale Bosc; Spreng. Syst. Veg. **1:** 320. 1825.

TYPE LOCALITY: "In Carolina."

RANGE: New Hampshire and Florida to Minnesota, New Mexico, and Mexico.

NEW MEXICO: Organ Mountains; Knowles; Buchanan; Tortugas Mountain; Roswell. Dry soil, in the Lower and Upper Sonoran zones.

15. **ERIOCHLOA** H. B. K.

Annuals or perennials with usually flat leaves and terminal panicles composed of numerous somewhat one-sided racemes; spikelets 1-flowered, hermaphrodite; rachilla jointed below the glumes and expanded into a distinct ringlike callus; glumes 2, the first reduced to a ring at the articulation, the second equaling the lemma, membranaceous, more or less acuminate; lemma slightly indurated, mucronate or shortly awn-pointed; palea shorter than the lemma; stamens 3; styles distinct; grain included within the hardened lemma, free.

1. **Eriochloa polystachya** H. B. K. Nov. Gen. & Sp. **1**: 95. *pl. 31*. 1816.

TYPE LOCALITY: Near Guayaquil, Ecuador.

RANGE: Florida and Arizona to Mexico and Tropical America.

NEW MEXICO: Belen; Mesilla Valley; Pena Blanca; White Mountains. Moist ground, in the Lower and Upper Sonoran zones.

16. PASPALUM L.

Perennials, often stoloniferous, with flat leaves; spikelets 1-flowered, plano-convex, nearly sessile in 2 or 4 rows along one side of a continuous, narrow or dilated rachis, forming simple racemes, these either solitary or 2 or more, digitate or paniculate; rachilla jointed below the glumes; glumes awnless, obtuse, membranaceous, the first usually wanting (often present in *P. distichum*, shorter than the second); grain oblong, inclosed within the indurated lemma and palea.

KEY TO THE SPECIES.

Stems creeping, rooting at the nodes; inflorescence of 2 terminal spikes.......... 1. *P. distichum*.
Stems not creeping, mostly erect, not rooting at the nodes; panicle of more than 2 scattered spikes.
  Spikelets on short pedicels; plants almost glabrous, the slender hairs mostly confined to the leaf margins.......... 2. *P. ciliatifolium*.
  Spikelets sessile or nearly so; plants more or less villous with stiff hairs all over the leaves.......... 3. *P. bushii*.

1. **Paspalum distichum** L. Amoen. Acad. **5**: 391. 1759. JOINT GRASS.

TYPE LOCALITY: Jamaica.

RANGE: California and North Carolina to Florida; also in South America, East Indies, and Australia.

NEW MEXICO: Socorro; Kingston; Mangas Springs; Rincon; Cienaga Ranch; Apache Teju; Mesilla Valley; Organ Mountains. River valleys, especially in clay soil, in the Lower and Upper Sonoran zones.

A common weed in irrigated fields, difficult to exterminate because of its long, creeping stems.

2. **Paspalum ciliatifolium** Michx. Fl. Bor. Amer. **1**: 44. 1803.

Paspalum setaceum ciliatifolium Vasey, Contr. U. S. Nat. Herb. **3**: 17. 1892.

TYPE LOCALITY: "Habitat in Carolina."

RANGE: New Mexico and Texas to New Jersey and Florida; also in Mexico and South America.

NEW MEXICO: Arroyo Ranch, near Roswell (*Griffiths* 5734). Dry soil, in the Upper Sonoran Zone.

3. **Paspalum bushii** Nash in Britton, Man. 74. 1901.

TYPE LOCALITY: Missouri.

RANGE: Missouri to Kansas and eastern New Mexico.

NEW MEXICO: Northeast of Clayton; sands south of Melrose; Nara Visa. Plains, in the Upper Sonoran Zone.

17. PANICUM L. Panic grass.

Annuals or perennials; spikelets 1-flowered, or rarely with a staminate flower below the terminal perfect one, paniculate; glumes very unequal, the first often minute, the second subequal to the sterile lemma; fertile lemma and palea chartaceous-indurated, the nerves obsolete, the margins of the lemma inrolled; grain free within the rigid lemma and palea.

KEY TO THE SPECIES.

- Spikelets arranged in pairs in 1-sided racemes; plants spreading by long stolons .... 1. *P. obtusum.*
- Spikelets panicled; plants without stolons.
  - Leaf blades of two sorts, those of the stems broad and short.
    - Spikelets less than 3 mm. long; leaf blades thin; sheaths glabrous or sparsely hispid .... 2. *P. helleri.*
    - Spikelets more than 3 mm. long; leaf blades firm; at least some of the sheaths hispid .... 3. *P. scribnerianum.*
  - Leaf blades all alike.
    - Annuals.
      - Inflorescence of several more or less secund, spike-like racemes.
        - Spikelets strongly reticulate-veined, glabrous .... 4. *P. fasciculatum reticulatum.*
        - Spikelets not reticulate-veined, pubescent and copiously papillose-hirsute .... 5. *P. arizonicum.*
      - Inflorescence a more or less diffuse panicle.
        - First glume very short, not over one-fourth the length of the second; sheaths glabrous .... 6. *P. dichotomiflorum.*
        - First glume longer, half as long as the second or more; sheaths pubescent.
          - Panicles somewhat drooping .... 7. *P. miliaceum.*
          - Panicles erect.
            - Panicles large, more than half the length of the entire plant .... 8. *P. barbipulvinatum.*
            - Panicles small, not over one-third the length of the plants.
              - First glume more than three-fourths the length of the second; spikelets 4 mm. long .... 9. *P. pampinosum.*
              - First glume half to two-thirds the length of the second; spikelets not over 3.3 mm. long .... 10. *P. hirticaule.*
    - Perennials.
      - Stems neither bulbous nor rhizomatous.
        - Sterile palea enlarged and indurated at maturity; glumes acute .... 11. *P. hians.*
        - Sterile palea not enlarged; glumes acuminate .... 12. *P. hallii.*

Stems bulbous at the base or rhizomatous.
Glumes acuminate.
Spikelets 3 to 5 mm. long; first glume acuminate to cuspidate..........13. *P. virgatum.*
Spikelets 6 to 8 mm. long; first glume acute..........................14. *P. havardii.*
Glumes obtuse or merely acute, never acuminate.
Culms from a rootstock, not bulbous....15. *P. plenum.*
Culms from enlarged bulbous bases.
Leaf blades over 5 mm. wide; culms usually over 1 meter high....16. *P. bulbosum.*
Leaf blades less than 5 mm. wide; culms usually less than 1 meter high......................16a. *P. bulbosum sciaphilum.*

1. Panicum obtusum H. B. K. Nov. Gen. & Sp. **1:** 98. 1816.

VINE MESQUITE GRASS.

Brachiaria obtusa Nash in Britton, Man. 77. 1901.

TYPE LOCALITY: "Crescit in planitie montana regni Mexicani prope Guanaxuato et Burras, in humidis, alt. 1,080 hexap."

RANGE: Colorado and Missouri to Texas, Arizona, and Mexico.

NEW MEXICO: Common from Gallup, Santa Fe, Las Vegas, and Clayton southward across the State. Plains and river valleys, in the Lower and Upper Sonoran zones.

2. Panicum helleri Nash, Bull. Torrey Club **26:** 572. 1899.

Panicum pernervosum Nash, Bull. Torrey Club **26:** 576. 1899.

TYPE LOCALITY: Kerrville, Kerr County, Texas.

RANGE: Missouri and Louisiana to Texas and New Mexico.

NEW MEXICO: West Fork of the Gila; Sierra Grande. Plains and low hills, in the Upper Sonoran Zone.

3. Panicum scribnerianum Nash, Bull. Torrey Club **22:** 421. 1895.

TYPE LOCALITY: Pennsylvania.

RANGE: Washington and Maine to California, New Mexico, Texas, and Maryland.

NEW MEXICO: Las Vegas (*Cockerell*). Dry fields, in the Upper Sonoran Zone.

4. Panicum fasciculatum reticulatum (Torr.) Beal, Grasses N. Amer. **2:** 117. 1896.

Panicum reticulatum Torr. in Marcy, Expl. Red Riv. 299. 1852.

TYPE LOCALITY: "Main fork of Red River," Texas.

RANGE: Texas and Arizona to Mexico.

NEW MEXICO: Socorro (*Plank* 38). Dry fields.

5. Panicum arizonicum Scribn. & Merr. U. S. Dept. Agr. Div. Agrost. Circ. **32:** 2. 1901.

TYPE LOCALITY: On mesas near Camp Lowell, Santa Cruz Valley, Arizona.

RANGE: New Mexico and southern California to northern Mexico.

NEW MEXICO: Mogollon Creek; Mangas Springs; Burro Mountains; Hillsboro; mesa west of Organ Mountains. Dry hills and sandy mesas, in the Lower and Upper Sonoran zones.

6. Panicum dichotomiflorum Michx. Fl. Bor. Amer. **1:** 48. 1803.

TYPE LOCALITY: "Hab. in occidentalibus montium Alleghanis."

RANGE: Maine and Nebraska to Florida and Texas, and in California; also in Mexico, the West Indies, and South America.

NEW MEXICO: Las Cruces (*Plank* 29). Lower and Upper Sonoran zones.

7. **Panicum miliaceum** L. Sp. Pl. 58. 1753. HOG MILLET.

Milium panicum Mill. Gard. Dict. no. 1. 1759.

Milium esculentum Moench, Meth. Pl. 203. 1794.

TYPE LOCALITY: "Habitat in India."

RANGE: Native of the Old World, introduced rather sparingly in the United States; often cultivated.

NEW MEXICO: Flora Vista; Gilmores Ranch.

8. **Panicum barbipulvinatum** Nash, Mem. N. Y. Bot. Gard. **1**: 21. 1900.

TYPE LOCALITY: "Yellowstone Park; Lower Geyser Basin."

RANGE: British Columbia and Wisconsin to California and Texas.

NEW MEXICO: Common throughout the State except along the Pecos Valley and eastward. Sandy fields, in the Lower and Upper Sonoran zones.

9. **Panicum pampinosum** Hitchc. & Chase, Contr. U. S. Nat. Herb. **15**: 66. 1910.

TYPE LOCALITY: "On range reserve, altitude 2,600 feet, Wilmot, Arizona."

RANGE: Southern New Mexico and Arizona.

NEW MEXICO: Organ Mountains; Grant County. Dry hills, in the Upper Sonoran Zone.

10. **Panicum hirticaule** Presl, Rel. Haenk. **1**: 308. 1830.

TYPE LOCALITY: Acapulco, Mexico.

RANGE: New Mexico and southern California to Mexico.

NEW MEXICO: Southwest corner of the State, north to Mangas Springs, east to the Organ Mountains. Dry hills and plains, in the Lower and Upper Sonoran zones.

11. **Panicum hians** Ell. Bot. S. C. & Ga. **1**: 118. 1816.

TYPE LOCALITY: South Carolina or Georgia.

RANGE: South Carolina and Florida to Texas and New Mexico.

NEW MEXICO: Las Cruces (*Plank* 6). Damp ground, in the Lower Sonoran Zone.

12. **Panicum hallii** Vasey, Bull. Torrey Club **11**: 64. 1884.

TYPE LOCALITY: Dry hills, Austin, Texas.

RANGE: Texas to Arizona, south into Mexico.

NEW MEXICO: Cross L Ranch; Mangas Springs; Deming; Las Cruces; Organ Mountains; Buchanan; Roswell; Carlsbad; Queen. Dry hills and mesas, in the Lower and Upper Sonoran zones.

13. **Panicum virgatum** L. Sp. Pl. 59. 1753. SWITCH GRASS.

Panicum giganteum Scheele, Linnæa **22**: 340. 1849.

TYPE LOCALITY: "Habitat in Virginia."

RANGE: Wyoming and Maine to Florida and Arizona, south into Mexico and the West Indies.

NEW MEXICO: Farmington; Pecos; Tesuque; Grant County; Organ Mountains; Ruidoso Creek; Roswell. Plains and low hills, in the Upper Sonoran Zone.

14. **Panicum havardii** Vasey, Bull. Torrey Club **14**: 95. 1887.

TYPE LOCALITY: Guadalupe Mountains, Texas.

RANGE: Western Texas to New Mexico and Mexico.

NEW MEXICO: Las Vegas; Roswell. Plains and low hills, in the Upper Sonoran Zone.

15. **Panicum plenum** Hitchc. & Chase, Contr. U. S. Nat. Herb. **15**: 80. 1910.

TYPE LOCALITY: Mangas Springs, New Mexico. Type collected by Metcalfe (no. 739).

RANGE: Texas to southern Arizona and northern Mexico.

NEW MEXICO: Mangas Springs; Organ Mountains. Dry hills and canyons, in the Upper Sonoran Zone.

16. Panicum bulbosum H. B. K. Nov. Gen. & Sp. **1:** 99. 1815.

Panicum maximum bulbosum Vasey in Wheeler, Rep. U. S. Surv. 100th Merid. **6:** 295. 1878.

TYPE LOCALITY: "Crescit in Novae Hispaniae scopulosis et frigidis juxta Santa Rosa, Los Joares et Guanaxuato, inter 1,070 et 1,360 hexap."

RANGE: Arizona and New Mexico to Mexico.

NEW MEXICO: Las Vegas; Carpenter Creek; Animas Valley; Burro Mountains; Copper Mines; Organ Mountains; West Fork of the Gila; Ruidoso Creek. Canyons and open slopes, in the Upper Sonoran Zone.

16a. Panicum bulbosum sciaphilum (Rupr.) Hitchc. & Chase, Contr. U. S. Nat. Herb. **15:** 83. 1910.

Panicum sciaphilum Rupr.; Fourn. Mex. Pl. **2:** 19. 1881.

Panicum bulbosum minor Vasey, U. S. Dept. Agr. Div. Bot. Bull. **8:** 38. 1889.

TYPE LOCALITY: Sierra de Yavesia, Mexico.

RANGE: Arizona and New Mexico to Mexico.

NEW MEXICO: Mangas Springs; Las Vegas; Organ Mountains; Gray. Canyons and low hills, in the Upper Sonoran Zone.

18. ECHINOCHLOA Beauv.

Coarse annuals with compressed sheaths, long flat leaves, and terminal panicles of stout racemes; spikelets 1-flowered, with sometimes a staminate flower below the perfect terminal one, nearly sessile in one-sided racemes; glumes unequal, spiny-hispid, mucronate; sterile lemma similar, awned from the apex, inclosing a hyaline palea; fertile lemma and palea chartaceous, acuminate; margins of the glume inrolled except at the summit.

KEY TO THE SPECIES.

Spikes simple.. 1. *E. colonum.*
Spikes compound.
  Awns about 25 mm. long.. 2. *E. crus-galli.*
  Awns 2 mm. long or less.. 3. *E. zelayensis.*

1. Echinochloa colonum (L.) Link, Hort. Berol. **2:** 209. 1833. JUNGLE RICE.

Panicum colonum L. Syst. Nat. ed. 10. 870. 1759.

TYPE LOCALITY: Jamaica.

RANGE: Wet ground and cultivated fields, Kansas and Virginia, southward throughout tropical America.

NEW MEXICO: Deming; Mesilla Valley; Organ Mountains; Gavilan Canyon. Wet ground and cultivated fields, in the Lower and Upper Sonoran zones.

The subspecies *zonalis* is a form with transverse purplish bands upon the leaves. It is common with the typical form, and is especially abundant among those plants that have grown in dry soil. Both are frequent as weeds in cultivated ground.

2. Echinochloa crus-galli (L.) Beauv. Ess. Agrost. 53. 1812. BARNYARD GRASS.

Panicum crusgalli L. Sp. Pl. 56. 1753.

TYPE LOCALITY: "Habitat in Europae et Virginiae cultis."

RANGE: In waste ground in the warmer parts of North America, and nearly around the world.

NEW MEXICO: Cedar Hill; Mangas Springs.

3. Echinochloa zelayensis (H. B. K.) Schult. Mant. **2:** 269. 1824.

Oplismenus zelayensis H. B. K. Nov. Gen. & Sp. **1:** 108. 1816.

TYPE LOCALITY: Near Zelaya, Querétaro, Mexico.

RANGE: Southwestern United States to South America.

NEW MEXICO: Common in waste and cultivated land throughout the State.

19. CHAETOCHLOA Scribn.

Annuals or perennials with flat leaves and bristly spikelike panicles; spikelets hermaphrodite, 1-flowered, or sometimes with a staminate flower below the hermaphrodite terminal one, surrounded by few or many persistent awnlike branches springing from the pedicels below the articulation of the spikelets; glumes awnless; stamens 3; styles distinct; grain included within the hardened lemma and palea, free.

KEY TO THE SPECIES.

Panicles dense and thick.
- Spikelets arranged singly in racemes; 5 to 16 bristles at the base of each spikelet.......... 1. *C. glauca.*
- Spikelets clustered but not in racemes; 1 to 3 bristles at base of each spikelet.......... 2. *C. viridis.*

Panicles slender, interrupted.
- Leaf blades more than 5 mm. wide; spikes with distinct, somewhat spreading branches below.......... 3. *C. grisebachii ampla.*
- Leaf blades less than 5 mm. wide; spikes not branched below, merely interrupted.......... 4. *C. composita.*

1. Chaetochloa glauca (L.) Scribn. U. S. Dept. Agr. Div. Agrost. Bull. **4:** 39. 1897.
PIGEON GRASS.

Panicum glaucum L. Sp. Pl. 56. 1753.
Setaria glauca Beauv. Ess. Agrost. 51. 1812.
TYPE LOCALITY: "Habitat in India."
RANGE: A native of the Old World, widely introduced into North America, in New Mexico still rare.
NEW MEXICO: Kingston; Mesilla Valley; Roswell.

2. Chaetochloa viridis (L.) Scribn. U. S. Dept. Agr. Div. Agrost. Bull. **4:** 39. 1897.
GREEN FOXTAIL.

Panicum viride L. Sp. Pl. ed. 2. 83. 1762.
Setaria viridis Beauv. Ess. Agrost. 51. 1812.
TYPE LOCALITY: "Habitat in Europa australi."
RANGE: Native of Europe, widely introduced into North America; in New Mexico a common weed in cultivated fields.
NEW MEXICO: Common in waste and cultivated ground in nearly every part of the State.

3. Chaetochloa grisebachii ampla Scribn. & Merr. U. S. Dept. Agr. Div. Agrost. Bull. **21:** 36. 1900.

TYPE LOCALITY: Organ Mountains, New Mexico. Type collected by G. R. Vasey.
RANGE: New Mexico and Arizona to northern Mexico.
NEW MEXICO: Mangas Springs; Hillsboro; Rio Frisco; Burro Mountains; Organ Mountains; Tortugas Mountain. Damp canyons, in the Upper Sonoran Zone.

4. Chaetochloa composita (H. B. K.) Scribn. U. S. Dept. Agr. Div. Agrost. Bull. **4:** 39. 1897.

Setaria composita H. B. K. Nov. Gen. & Sp. **1:** 111. 1816.
TYPE LOCALITY: "Crescit regione calidissima prope Cumana et Bordones, in Nova Andalusia: item in ripa fluminis Magdalenae prope Guarumo et in sylvis Orinocensibus juxta Esmeralda."
RANGE: Colorado and Arizona to Texas and Mexico; also in South America.
NEW MEXICO: Cross L Ranch; Albuquerque; Chiz; Animas Creek; Mangas Springs; Lake Valley; Aden; Rincon; Burro Mountains; Mesilla Valley; Guadalupe Mountains; Pecos Valley. River valleys and low hills, in the Lower and Upper Sonoran zones.

20. CENCHRUS L. Sand bur.

Annuals or perennials with spreading or erect culms and few or many more or less crowded "burs" in terminal spikes; spikelets 1-flowered, hermaphrodite, 1 to 4 together with an ovoid or globular involucre of rigid, more or less connate bristles forming spiny burs or false capsules, these sessile or nearly so in simple terminal spikes or racemes, falling with the spikelets; glumes awnless; grain free within the slightly hardened lemma and palea.

1. **Cenchrus carolinianus** Walt. Fl. Carol. 79. 1788.

Type locality: Carolina.

Range: Maine and Minnesota to Florida and New Mexico, and southward.

New Mexico: Waste and cultivated ground at lower elevations throughout the State; common. In sandy soil, in the Lower and Upper Sonoran zones.

Sand burs are the most pernicious weeds of the State. They are often abundant in cultivated ground, where, unless steps are taken to exterminate them, they spread rapidly. In alfalfa fields they often become so numerous as to render the hay valueless. The spines of the burs are extremely sharp and will pierce the uppers of shoes. After they have entered one's flesh they usually break off when an attempt is made to extract them.

21. PHALARIS L. Canary grass.

Annuals or perennials with flat leaves and densely flowered spikelike or capitate inflorescence; spikelets 1-flowered, strongly flattened laterally; rachilla jointed above the glumes; glumes awnless, equal, boat-shaped, usually winged on the keel; lemmas hard and shining in fruit, closely investing the grain and palea.

KEY TO THE SPECIES.

Glumes not winged; inflorescence a narrow panicle.............. 1. *P. arundinacea.*
Glumes winged; inflorescence spikelike......................... 2. *P. caroliniana.*

1. **Phalaris arundinacea** L. Sp. Pl. 55. 1753. Reed Canary grass.

Type locality: "Habitat in Europae subhumidis ad ripas lacuum."

Range: British America to Nevada, New Mexico, and New Jersey; also in Europe and Asia.

New Mexico: Chama (*Standley* 6806). Wet ground, in the Transition Zone.

2. **Phalaris caroliniana** Walt. Fl. Carol. 74. 1788. Southern Canary grass.

Type locality: South Carolina.

Range: California and South Carolina to New Mexico and Florida.

New Mexico: Burro Mountains; Agricultural College. Moist ground.

22. SAVASTANA Schrank.

Fragrant perennials with flat leaves and usually rather small pyramidal terminal panicles; spikelets 3-flowered, the terminal floret hermaphrodite, the others staminate; rachilla jointed above the glumes; glumes nearly equal, about the length of the spikelet, acute, smooth; lemmas about equaling the glumes, awnless or short-awned; stamens in the male florets 3, in the hermaphrodite floret 2; styles distinct, rather long; grain free.

1. **Savastana odorata** (L.) Scribn. Mem. Torrey Club **5**: 34. 1894. Vanilla grass.

Holcus odoratus L. Sp. Pl. 1048. 1753.

Hierochloe borealis Roem. & Schult. Syst. Veg. **2**: 513. 1817.

Type locality: "Habitat in Europae frigidioris pascuis humentibus."

Range: British America to New Mexico, Wisconsin, and New Jersey.

NEW MEXICO: Trout Spring; Pecos Baldy. Meadows, in the Hudsonian and Arctic-Alpine zones.

23. LYCURUS H. B. K.

Cespitose erect perennial with narrow or convolute leaves and densely flowered cylindrical spikelike terminal panicles; spikelets 1-flowered, usually in pairs; glumes nerved, the nerves often produced into awns; lemma 3-nerved, awned, broader and longer than the glumes; palea 2-nerved, 2-keeled; stamens 3; styles short, distinct; grain included within the glumes, free.

1. Lycurus phleoides H. B. K. Nov. Gen. & Sp. **1**: 141. *pl. 45.* 1816.

TEXAS TIMOTHY.

TYPE LOCALITY: "Crescit in temperatis Mexici, inter Guanazuato et Temescatio et in radicibus aridissimi montis La Buffa, alt. 1030 hexap."

RANGE: Western Texas and southern Arizona to Mexico.

NEW MEXICO: Abundant from the Mogollon Mountains and Santa Fe and Las Vegas mountains southward and eastward across the State. Dry hills, in the Lower and Upper Sonoran zones.

Texas timothy is abundant on the dry hills of the southern part of the State. It is less common in the north. It grows in bunches and is a rather important range grass in some sections.

24. ARISTIDA L. NEEDLE GRASS.

Tufted annuals or perennials with narrow leaves; spikelets 1-flowered, on long or short slender pedicels, in more or less expanded terminal panicles; rachilla articulated above the glumes and produced into a hard obconical hairy callus below the lemma but not extending beyond it; glumes more or less unequal, acute or bristle-pointed, slightly keeled; lemma somewhat firmer in texture, closely rolled around the floret and the usually short palea, terminating in a usually trifid awn; grain slender, tightly inclosed in the hardened lemma but free from it.

KEY TO THE SPECIES.

Annual .... 1. *A. bromoides.*
Perennials.
    Plants widely divaricate-branched, the branches of the panicle rigid and straight.
        Awns 3, all of about the same length .... 2. *A. divaricata.*
        Awns apparently 1, the lateral ones short or obsolete ...... 3. *A. schiediana.*
    Plants with erect or at most rather weakly spreading stems.
        Glumes nearly equal.
            Plants stout and strict, 30 cm. high or more; pedicels short, straight; glumes conspicuously awned .... 4. *A. arizonica.*
            Plants slender, 20 cm. high or less, rather spreading; pedicels slender, sinuous; glumes acuminate, not awned .... 5. *A. havardii.*
        Glumes very unequal, the first usually about half as long as the second.
            Mature lemma not tapering upward, the neck of about the same diameter as the base; second glume considerably longer than the lemma, the latter smooth; awns 6 to 8 cm. long .... 6. *A. longiseta.*
            Mature lemmas tapering upward into a slender neck; second glume barely surpassing the lemma, usually shorter, the lemmas usually scabrous; awns usually much less than 6 cm. long.

Spikelets small, 10 mm. long or less, the awns never more than twice as long as the lemma.
Panicle strict, congested, never spreading... 7. *A. vaseyi.*
Panicle spreading, at least in age.
Panicles always spreading, the pedicels very weak and sinuous; awns merely spreading.................. 8. *A. micrantha.*
Panicles at first congested, finally spreading, the pedicels rigid, straight, ascending; awns strongly divergent............................... 9. *A. pansa.*
Spikelets large, 15 mm. long or more, the awns usually more than twice as long as the lemma.
Panicles simple or nearly so.
Panicles strict, the spikelets numerous and congested, relatively small; pedicels short...................... 7. *A. vaseyi.*
Panicles lax, the spikelets few, scattered, about 15 mm. long; pedicels elongated...............................10. *A. fendleriana.*
Panicles compound.
Culms stout; panicles rigidly erect; pedicels straight.......................11. *A. wrightii.*
Culms weaker and more slender; panicles laxly spreading; pedicels very slender, often curved..............12. *A. purpurea.*

1. **Aristida bromoides** H. B. K. Nov. Gen. & Sp. **1**: 122. 1816.

SIX-WEEKS NEEDLE GRASS.

TYPE LOCALITY: "In montibus regni Quitensis, juxta Tambo de Guamote et Llanos de Tiocaxas, alt. 1,600 hexap."

RANGE: Texas and Arizona to Mexico; also in South America.

NEW MEXICO: Santa Fe; Cross L Ranch; Cerrillos; Chama River; Algodones; Socorro; Mangas Springs; Black Range; Deming; Dona Ana and Organ mountains; White Sands; White Mountains; Guadalupe Mountains; Knowles; Roswell. Dry plains and hills, in the Lower and Upper Sonoran zones.

2. **Aristida divaricata** Humb. & Bonpl.; Willd. Enum. Pl. 99. 1809.

Aristida humboldtiana Trin. & Rupr. Mém. Acad. St. Pétersb. VI. Sci. Nat. **5**[1]: 118. 1842.

TYPE LOCALITY: "Habitat in Mexico."

RANGE: Arizona and western Texas to Mexico.

NEW MEXICO: Cross L Ranch; Texline; Gallinas Mountains; Black Range; Deming; Dona Ana and Organ mountains; Eagle Creek; Lake Arthur. Dry plains and hills, in the Lower and Upper Sonoran zones.

3. **Aristida schiediana** Trin. & Rupr. Mém. Acad. St. Pétersb. VI. Sci. Nat. **5**[1]: 120. 1842.

Aristida divergens Vasey, Contr. U. S. Nat. Herb. **3**: 48. 1892.

TYPE LOCALITY: "Mexico: prope Jalapam."

RANGE: Arizona and western Texas to Mexico.

NEW MEXICO: Socorro; Bear Mountain; Copper Mines; Organ Mountains; plains south of Roswell. Upper Sonoran Zone.

4. **Aristida arizonica** Vasey, Bull. Torrey Club **13:** 27. 1886.

Type locality: "Arizona."

Range: Arizona and New Mexico to western Texas.

New Mexico: Clayton; Santa Fe; Las Vegas; near Jewett Gap; Grant County; Buchanan; Leachs; Knowles; White Mountains. Plains and low hills, in the Upper Sonoran and Transition zones.

5. **Aristida havardii** Vasey, Bull. Torrey Club **13:** 27. 1886.

Type locality: "Western Texas."

Range: Arizona and western Texas to Mexico.

New Mexico: Gallinas Mountains; Albuquerque; Fort Bayard; Grant; Bonito Crossing; Gilmores Ranch; Gray; Carlsbad; Arroyo Ranch. Hills and plains, in the Upper Sonoran and Transition zones.

6. **Aristida longiseta** Steud. Syn. Pl. Glum. **1:** 420. 1854. Wiregrass.

Type locality: New Mexico, probably near or at Santa Fe. Type collected by Fendler (no. 978).

New Mexico: Common throughout the State.

7. **Aristida vaseyi** Woot. & Standl. N. Mex. Agr. Exp. Sta. Bull. **81:** 55. 1912; Contr. U. S. Nat. Herb. **16:** 113. 1913.

Aristida reverchoni augusta Vasey, Contr. U. S. Nat. Herb. **3:** 46. 1892.

Type locality: Comanche Peak, Texas.

Range: Western Texas and southern New Mexico.

New Mexico: Socorro; mountains west of San Antonio; Mangas Springs; Tortugas Mountain; Pena Blanca; Alamogordo. Plains and low hills, in the Lower Sonoran Zone.

8. **Aristida micrantha** (Vasey) Nash in Small, Fl. Southeast. U. S. 117. 1903.

Aristida purpurea micrantha Vasey, Contr. U. S. Nat. Herb. **3:** 47. 1892.

Type locality: Western Texas.

Range: Western Texas and southern New Mexico.

New Mexico: Carlsbad (*Smith*). Dry plains, in the Upper Sonoran Zone.

9. **Aristida pansa** Woot. & Standl. Contr. U. S. Nat. Herb. **16:** 112. 1913.

Type locality: Tortugas Mountain near Las Cruces, New Mexico. Type collected by Wooton, October 6, 1904.

Range: Southern New Mexico to Mexico.

New Mexico: Tortugas Mountain. Dry hills, in the Lower Sonoran Zone.

10. **Aristida fendleriana** Steud. Syn. Pl. Glum. **1:** 420. 1855.

Aristida longiseta fendleriana Merr. U. S. Dept. Agr. Div. Agrost. Circ. **24:** 5. 1901.

Type locality: Santa Fe, New Mexico. Type collected by Fendler (no. 973).

Range: Colorado and Texas to Arizona.

New Mexico: Tunitcha Mountains; Farmington; Carrizo Mountains; Santa Fe; northeast of Clayton; 25 miles south of Gallup; Rito Quemado; 35 miles south of Torrance; near Suwanee; Copper Mines; Mimbres and Cooks Spring. Dry hills and plains, in the Upper Sonoran Zone.

11. **Aristida wrightii** Nash in Small, Fl. Southeast. U. S. 116. 1903.

Type locality: Dallas, Texas.

Range: Texas and New Mexico.

New Mexico: Atarque de Garcia; Deming; Organ Mountains; Dona Ana Mountains; Buchanan; Redlands. Dry hills and mesas, in the Lower and Upper Sonoran zones.

12. **Aristida purpurea** Nutt. Trans. Amer. Phil. Soc. n. ser. **5:** 145. 1837. Purple needle grass.

Type locality: "On grassy plains of the Red River, in arid situations."

Range: Arizona and western Texas to Mexico.

New Mexico: Twenty-five miles south of Gallup; Clayton; Cross L Ranch; Albuquerque; Suwanee; Socorro; Texline; Mogollon Mountains; Tortugas Mountain; Buchanan; Knowles; Arroyo Ranch; east of Carlsbad; Mesilla. Dry plains and hills, in the Lower and Upper Sonoran zones.

The subspecies *laxiflora* is a form with very slender curved pedicels and is probably the same as *A. subuniflora* Nash,[1] the type of which we have not seen. It occurs wherever the type grows, and may be recognized by the fewer spikelets on very slender curved pedicels and the usually longer awns.

25. STIPA L. Porcupine grass.

Tufted perennials with mostly narrow or involute leaves, and terminal, usually open panicles; spikelets 1-flowered; rachilla articulated above the glumes and produced below the lemma into a strong bearded obconical sharp-pointed callus; glumes thin, membranaceous, subequal, acute or bristle-pointed; lemma narrow, subcoriaceous, closely investing the floret, terminating in a twisted and geniculate simple awn articulated with the apex; grain terete, closely enveloped by the indurated lemma.

KEY TO THE SPECIES.

Glumes 2 cm. long or more.
- Awns plumose .... 1. *S. neomexicana.*
- Awns not plumose.
  - Base of panicle usually included in the sheaths; lemmas 8 to 12 mm. long .... 2. *S. comata.*
  - Base of panicle exserted; lemmas more than 12 mm. long.
    - Lemmas 12 to 15 mm. long .... 3. *S. tweedyi.*
    - Lemmas 20 to 25 mm. long .... 4. *S. spartea.*

Glumes 15 mm. long or less.
- Panicles loose and open.
  - Lemmas 4 mm. long or less.
    - Awns 60 to 80 mm. long, curled above the joint; lemmas about 3 mm. long .... 5. *S. tenuissima.*
    - Awns 15 mm. long or less, not curled; lemmas about 4 mm. long .... 6. *S. fimbriata.*
  - Lemmas 6 to 9 mm. long.
    - Glumes broad; awns about 25 mm. long .... 7. *S. pringlei.*
    - Glumes narrow; awns 40 to 50 mm. long .... 8. *S. eminens.*
- Panicles narrow, dense, spikelike.
  - Glumes scarious, prominently nerved.
    - Awns long-hairy below .... 9. *S. speciosa.*
    - Awns not long-hairy.
      - Lemmas 5 mm. long or less, long-hairy near the apex .... 10. *S. lettermanii.*
      - Lemmas more than 5 mm. long, equally hairy throughout .... 11. *S. viridula.*
  - Glumes firm, thick, not prominently nerved.
    - Lemmas 4 to 5 mm. long .... 12. *S. minor.*
    - Lemmas 8 to 10 mm. long.
      - Panicles slender; stems low and slender; leaves narrow .... 13. *S. scribneri.*
      - Panicles stout and dense; stems tall and stout; leaves broad .... 14. *S. vaseyi.*

---

[1] Nash in Small, Fl. Southeast. U. S. 116. 1903. The type was collected in New Mexico in 1881, by G. R. Vasey.

1. **Stipa neomexicana** (Thurb.) Scribn. U. S. Dept. Agr. Div. Agrost. Bull. **17:** 132. 1899. NEW MEXICAN PORCUPINE GRASS.

Stipa pennata neomexicana Thurb.; Vasey, U. S. Dept. Agr. Div. Bot. Bull. **12**[2]: no. 81. 1891.

TYPE LOCALITY: New Mexico.

RANGE: Colorado and Texas to California.

NEW MEXICO: North of Ramah; mountains north of Santa Rita; Albuquerque; 10 miles north of Santa Fe; Las Vegas Hot Springs; Silver City; Rio Mimbres; east of Alamogordo; Arroyo Ranch. Dry hills, in the Upper Sonoran Zone.

2. **Stipa comata** Trin. & Rupr. Mém. Acad. St. Pétersb. VI. Sci. Nat. **5**[1]: 75. 1842.

TYPE LOCALITY: "Carlton House Fort ad fl. Saskatchawan."

RANGE: Alaska and Alberta to California and New Mexico.

NEW MEXICO: Tunitcha Mountains; San Lorenzo; Chama; Horse Spring; Agua Azul; Santa Fe; Torrance; Clayton; Pecos; Raton Mountains; Nara Visa; Jewett Gap; Little Creek. Plains and low hills, in the Upper Sonoran Zone.

Both this and the preceding are valuable range grasses, especially because they grow in the spring when other forage is scarce. Neither, however, reproduces well, but either is soon killed by overstocking and replaced by the needle grasses.

3. **Stipa tweedyi** Scribn. U. S. Dept. Agr. Div. Agrost. Bull. **11:** 47. 1898.

Stipa comata intermedia Scribn. Bot. Gaz. **11:** 171. 1886. not *S. intermedia* Trin. 1842.

TYPE LOCALITY: Junction Butte, Yellowstone Park.

RANGE: Washington and Alberta to Arizona and New Mexico.

NEW MEXICO: Tunitcha Mountains (*Standley* 7676). Open slopes, in the Upper Sonoran Zone.

4. **Stipa spartea** Trin. Mém. Acad. St. Pétersb. VI. Math. Phys. Nat. **1:** 82. 1830.

TYPE LOCALITY: North America.

RANGE: British America to Illinois and New Mexico.

NEW MEXICO: Sierra Grande (*Standley* 6223). Plains and prairies, in the Upper Sonoran Zone.

5. **Stipa tenuissima** Trin. Mém. Acad. St. Pétersb. VI. Sci. Nat. **2**[1]: 36. 1836.

TYPE LOCALITY: Chile.

RANGE: New Mexico and Arizona to Mexico and South America.

NEW MEXICO: Socorro (*Plank* 44).

6. **Stipa fimbriata** H. B. K. Nov. Gen. & Sp. **1:** 126. 1816. PINYON GRASS.

Oryzopsis fimbriata Hemsl. Biol. Centr. Amer. Bot. **3:** 538. 1885.

TYPE LOCALITY: "Crescit in alta planitie Mexicana inter Burras et Guanaxuato; item in scopulosis prope Mina de Villapando, inter 1050 et 1330 hexap."

RANGE: Arizona to western Texas and southward.

NEW MEXICO: Bear Mountains, Burro Mountains, Organ Mountains, and Guadalupe Mountains and southward across the State. Dry hills, in the Upper Sonoran and Transition zones.

7. **Stipa pringlei** Scribn. Contr. U. S. Nat. Herb. **3:** 54. 1892.

TYPE LOCALITY: Mexico or Arizona.

RANGE: Southern New Mexico and Arizona to northern Mexico.

NEW MEXICO: West Fork of the Gila (*Metcalfe* 557).

8. **Stipa eminens** Cav. Icon. Pl. **5:** 42. *pl. 467. f. 1.* 1799.

TYPE LOCALITY: Near Chalma, Mexico.

RANGE: New Mexico to southern Mexico.

NEW MEXICO: Kingston; Mangas Springs; Big Hatchet Mountains; Organ and Dona Ana mountains. Upper Sonoran Zone.

9. **Stipa speciosa** Trin. & Rupr. Mém. Acad. St. Pétersb. VI. Sci. Nat. **5**[1]: 45. 1842.
TYPE LOCALITY: Chile.
RANGE: California to New Mexico and Mexico; also in South America.
NEW MEXICO: Carrizo Mountains (*Standley* 7515.) Dry hills, in the Upper Sonoran Zone.

10. **Stipa lettermanii** Vasey, Bull. Torrey Club **13**: 53. 1886.
Stipa viridula lettermanii Vasey, Contr. U. S. Nat. Herb. **3**: 50. 1892.
TYPE LOCALITY: Idaho.
RANGE: Idaho and Wyoming to Utah and New Mexico.
NEW MEXICO: Chama; Santa Fe. Hills and meadows, in the Upper Sonoran and Transition zones.

11. **Stipa viridula** Trin. & Rupr. Mém. Acad. St. Pétersb. VI. Sci. Nat. **2**[1]: 39. 1836.
TYPE LOCALITY: North America.
RANGE: British America to Kansas, Utah, and New Mexico.
NEW MEXICO: Trout Spring; Taos; Santa Fe Canyon; mountains near Las Vegas; Raton Mountains; Cross L Ranch; El Rito Creek. Transition Zone.

12. **Stipa minor** (Vasey) Scribn. U. S. Dept. Agr. Div. Agrost. Bull. **11**: 46. 1898.
Stipa viridula minor Vasey, Contr. U. S. Nat. Herb. **3**: 50. 1892.
TYPE LOCALITY: Kelso Mountain, Colorado.
RANGE: Montana to Utah and New Mexico.
NEW MEXICO: Chama; Winsors Ranch; mouth of Indian Creek. Mountains, in the Transition Zone.

13. **Stipa scribneri** Vasey, Bull. Torrey Club **11**: 125. 1884.
TYPE LOCALITY: Dry hillsides at Santa Fe, New Mexico. Type collected by George Vasey in 1884.
RANGE: Colorado and New Mexico.
NEW MEXICO: Santa Fe; Pecos; Bear Mountain; near Ruidoso. Mountains and low hills, in the Upper Sonoran and Transition zones.

14. **Stipa vaseyi** Scribn. U. S. Dept. Agr. Div. Agrost. Bull. **11**: 46. 1898.
SLEEPY GRASS.
Stipa viridula robusta Vasey, Contr. U. S. Nat. Herb. **3**: 50. 1892, not *S. robusta* Nutt. 1842.
TYPE LOCALITY: "Texas and Mexico."
RANGE: Idaho to Mexico and Texas.
NEW MEXICO: Dulce; Santa Fe; Raton Mountains; Glorieta Mountains; Ramah; Winsor Creek; Las Vegas; Gila Hot Springs; White Mountains. Mountains, in the Upper Sonoran and Transition zones.

A very abundant grass in meadows at middle elevations. In the Sacramento-White Mountain region this is known as "sleepy grass," and is said to have a narcotic effect upon animals that eat it, especially horses. While neither of the writers has had an opportunity of personally corroborating this statement, it is vouched for by many reliable residents of the region. The narcotic effects of the plant, which are said to appear soon after it has been eaten, are indicated by drowsiness or sleep on the part of the affected animals, which continues often for 48 hours and sometimes results in death. Animals which have been reared in these mountains never eat sleepy grass, but those which are brought into the region will consume it because of its luxuriance and freshness, unless restrained. Strangely enough, the dried grass does not have a narcotic effect. The grass is abundant in other regions of the State, but nowhere besides in this one range is it reputed to have narcotic properties. Where it is not eaten by stock it grows vigorously and spreads rapidly, especially when other plants are killed by overstocking.

26. MUHLENBERGIA Schreb.

Perennials or rarely annuals, with small spikelets; culms simple or much branched; leaves long or short, flat or involute; panicles narrow and spikelike or open and widely spreading; spikelets 1-flowered; rachilla jointed above the glumes, forming a very short and usually hairy callus below the lemma but not extending beyond it; glumes membranaceous or hyaline, 1 to 3-nerved or nerveless, usually unequal and shorter than the lemma, acute or mucronate; lemma narrow, smooth, or more or less pilose below, 3 or 5-nerved, awned from the acute apex or from between the teeth of the bidentate apex; awn straight or flexuous; palea thin, 2-nerved; stamens 3; grain closely enveloped by the lemma.

KEY TO THE SPECIES.

Plants tall, 1 meter high or more, stout; panicle 25 to 35 cm. long. 1. *M. emersleyi.*
Plants lower, less than 1 meter high, mostly less than 60 cm., more slender; panicles shorter, less than 25 cm. long.
  Panicles open and spreading.
    Plants diffusely branched throughout, weakly ascending or decumbent.......... 2 *M. porteri.*
    Plants erect, branched only at the base.
      Secondary branches of the panicle clustered; leaves stiff and spiny-pointed.......... 3. *M. pungens.*
      Secondary branches of the panicle single; leaves neither stiff nor spiny-pointed.
        Basal leaves 5 cm. long or less, strongly recurved.......... 4. *M. gracillima.*
        Basal leaves more than 5 cm. long, not recurved.
          Awns short, 4 mm. long; leaf blades 5 to 10 cm. long; panicles green.......... 5. *M. arenicola.*
          Awns long, 10 to 15 mm.; leaf blades about 20 cm. long; panicles dark purple.......... 6. *M. rigida.*
  Panicles narrow and spikelike.
    Annual, 5 to 15 cm. high.......... 7. *M. schaffneri.*
    Perennials, mostly more than 15 cm. high.
      Glumes subulate; plants with leafy branches and long scaly rhizomes.
        Glumes about as long as the lemma, not awned. 8. *M. mexicana.*
        Glumes longer than the lemma, awned.
          Lemmas only slightly villous.......... 9. *M. racemosa.*
          Lemmas covered with long white hairs.... 10. *M. comata.*
      Glumes lanceolate to ovate; plants various.
        Lemmas awnless or with very short awns.
          Glumes under half as long as the lemmas.. 11. *M. squarrosa.*
          Glumes more than half as long as the lemmas.
            Glumes acute, not awned.
              Panicles on long peduncles.......... 15. *M. thurberi.*
              Panicles partly included in the sheaths.......... 16. *M. repens.*
            Glumes awned.
              Lemmas hairy below.......... 12. *M. lemmoni.*

Lemmas scabrous or glabrous.
Panicles dense, obtuse, 5 to 10 mm. wide .... 13. *M. wrightii.*
Panicles rather lax, tapering at the apex, less than 5 mm. wide .... 14. *M. cuspidata.*
Lemmas with conspicuous long awns.
Leaf sheaths very broad at the base and papery, loose, not closely investing the stems.
Second glume 3-toothed; lemma pubescent at the top .... 17. *M. trifida.*
Both glumes acute or acuminate; lemmas pubescent only below ...... 18. *M. virescens.*
Leaf sheaths not broad and papery, closely investing the stems.
Spikelets on long slender pedicels.
Second glume entire at the apex . 19. *M. affinis.*
Second glume sharply 3 to 5-toothed .... 20. *M. subalpina.*
Spikelets on short stout pedicels, or sessile.
Awns about 5 mm. long; stems stout; internodes long ..... 21. *M. acuminata.*
Awns about 20 mm. long; stems slender, wiry; internodes short .... 22. *M. monticola.*

1. **Muhlenbergia emersleyi** Vasey, Contr. U. S. Nat. Herb. **3:** 66. 1892.

Muhlenbergia vaseyana Scribn. Rep. Mo. Bot. Gard. **10:** 52. 1899.

TYPE LOCALITY: "Rocky Cañon, Arizona."

RANGE: Western Texas to southern Arizona and adjacent Mexico.

NEW MEXICO: Silver City; Mogollon Mountains; Santa Rita; Animas Valley; Organ Mountains; Dona Ana Mountains. Dry hills, in the Upper Sonoran Zone.

2. **Muhlenbergia porteri** Scribn.; Beal, Grasses N. Amer. **2:** 259. 1898.

MESQUITE GRASS.

TYPE LOCALITY: Texas.

RANGE: Colorado and western Texas to California and Mexico.

NEW MEXICO: Mangas Springs; Albuquerque; Organ Mountains; Tortugas Mountain; Mesilla Valley; Dona Ana Mountains; Jarilla; Arroyo Ranch. Hills and mesas, in the Lower and Upper Sonoran zones.

Mesquite grass receives its name from the fact that, in the southern part of the State, where it is very common, it is nearly always found growing in the shade of mesquite bushes, its slender, lax stems often clambering over them. Cattle are very fond of it and will force their way into the mesquite to reach the grass.

3. **Muhlenbergia pungens** Thurb. Proc. Acad. Phila. **1863:** 78. 1864.

PURPLE HAIR GRASS.

TYPE LOCALITY: Colorado.

RANGE: Utah and Nebraska to Arizona and Texas.

NEW MEXICO: Carrizo Mountains; Zuni Reservation; Chama River; Los Pilares; on the San Juan; White Sands. Sandhills and on plains, in the Lower and Upper Sonoran zones.

4. **Muhlenbergia gracillima** Torr. U. S. Rep. Expl. Miss. Pacif. **4:** 155. 1856.

RING GRASS.

TYPE LOCALITY: "Llano Estacado, and near the Antelope hills of the Canadian River," Texas or New Mexico. Type collected by Bigelow.

RANGE: Colorado and Kansas to Texas and New Mexico.

NEW MEXICO: Sierra Grande; Nara Visa; El Rito Draw; Las Vegas; Albuquerque; Pecos; Roy; Clayton; Socorro; Santa Fe; Llano Estacado; Mangas Springs; Buchanan; Deming; Dona Ana; Guadalupe Mountains; Fort Stanton; Gray. Plains and low hills, in the Upper Sonoran Zone.

This is very common on the plains of the northern part of the State. Its habit of growth is peculiar: the plants are low and form dense circular mats; after a time the center of the mat dies and a sort of "fairy ring" is left.

5. **Muhlenbergia arenicola** Buckl. Proc. Acad. Phila. **1862:** 91. 1863.

TYPE LOCALITY: "Arid places in Western Texas."

RANGE: Western Texas and southern New Mexico.

NEW MEXICO: Deming; Redlands; Hope; Lake Arthur; Rio San Jose; near Suwanee; Gila Hot Springs; Arroyo Ranch; Jornada del Muerto. Dry hills, in the Lower and Upper Sonoran zones.

6. **Muhlenbergia rigida** (H. B. K.) Kunth, Rév. Gram. **1:** 63. 1829.

Podosaemum rigidum H. B. K. Nov. Gen. & Sp. **1:** 129. 1816.

TYPE LOCALITY: Near Guanajuato, Mexico.

RANGE: New Mexico and Arizona to Mexico.

NEW MEXICO: Five miles east of San Lorenzo on Mimbres River (*Metcalfe* 1447).

7. **Muhlenbergia schaffneri** Fourn. Mex. Pl. **2:** 85. 1881.

TYPE LOCALITY: "Prope Tacubaya," Mexico.

RANGE: New Mexico and Arizona to Mexico.

NEW MEXICO: Trujillo Creek; Mogollon Creek; Organ Mountains. Dry slopes, in the Upper Sonoran Zone.

8. **Muhlenbergia mexicana** (L.) Trin. Gram. Unifl. 189. 1824.

Agrostis mexicana L. Mant. Pl. **1:** 31. 1767.

TYPE LOCALITY: "Habitat in America calidiore."

RANGE: British America to Tennessee and New Mexico; also in Mexico.

NEW MEXICO: West Fork of the Gila; Winsors Ranch. Damp ground, in the Transition Zone.

9. **Muhlenbergia racemosa** (Michx.) B. S. P. Prel. Cat. N. Y. 67. 1888.

Agrostis racemosa Michx. Fl. Bor. Amer. **1:** 53. 1803.

Polypogon glomeratus Willd. Enum. Pl. 87. 1809.

Muhlenbergia glomerata Trin. Gram. Unifl. 191. 1824.

TYPE LOCALITY: "Habitat in ripis sabulosis inundatis fluminis Mississippi."

RANGE: British America to New Mexico, Missouri, and New Jersey.

NEW MEXICO: Cedar Hill; Dulce; Las Vegas; Pecos; Raton Mountains; Sandia Mountains; Mangas Springs; Animas Creek; Mesilla Valley; Little Creek; Gilmores Ranch. Damp woods, Lower Sonoran to Transition Zone.

10. **Muhlenbergia comata** (Thurb.) Benth. Journ. Linn. Soc. Bot. **19:** 83. 1881.

Vaseya comata Thurb. Proc. Acad. Phila. **1863:** 79. 1863.

TYPE LOCALITY: "Plains of Nebraska."

RANGE: California to Colorado and New Mexico.

NEW MEXICO: Winsors Ranch (*Standley* 4359). Open slopes, in the Transition Zone.

11. Muhlenbergia squarrosa (Trin.) Rydb. Bull. Torrey Club **36:** 531. 1909.

Vilfa squarrosa Trin. Mém. Acad. St. Pétersb. VI. Sci. Nat. **3**[1]**:** 100. 1840.

TYPE LOCALITY: North America.

RANGE: British America to California and Mexico.

NEW MEXICO: Tunitcha Mountains; Chama; Ensenada; Sandia Mountains; Rio Pueblo; Pecos; Grants Station. Transition Zone.

12. Muhlenbergia lemmoni Scribn. Contr. U. S. Nat. Herb. **1:** 56. 1890.

TYPE LOCALITY: Ballinger, Runnels County, Texas.

RANGE: Western Texas to Arizona and Mexico.

NEW MEXICO: Organ Mountains (*Wooton*). Hillsides.

13. Muhlenbergia wrightii Vasey; Coulter, Man. Rocky Mount. 409. 1885.

TYPE LOCALITY: "New Mexico."

RANGE: Colorado to Mexico.

NEW MEXICO: Baldy; Johnsons Mesa; Trout Springs; El Rito Creek; Las Vegas; near Datil; near La Jara; Kingston; Winter Folly. Damp mountain slopes, in the Transition Zone.

14. Muhlenbergia cuspidata (Torr.) Rydb. Bull. Torrey Club **32:** 599. 1905.

Vilfa cuspidata Torr.; Hook. Fl. Bor. Amer. **2:** 238. 1839.

TYPE LOCALITY: "Banks of the Saskatchawan, near the Rocky Mountains."

RANGE: British America to New Mexico and Missouri.

NEW MEXICO: Pecos; Cross L Ranch; Kingston. Open slopes, in the Upper Sonoran Zone.

15. Muhlenbergia thurberi (Scribn.) Rydb. Bull. Torrey Club **32:** 601. 1905.

Sporobolus thurberi Scribn. U. S. Dept. Agr. Div. Agrost. Bull. **11:** 48. 1898.

TYPE LOCALITY: Plaza Larga, New Mexico. Type collected by Bigelow.

RANGE: Colorado to New Mexico.

NEW MEXICO: Plaza Larga; Carrizo Mountains; Eagle Creek. Dry hills, in the Upper Sonoran Zone.

16. Muhlenbergia repens (Presl) Hitchc. in Jepson, Fl. Calif. **1:** 111. 1912.

APAREJO GRASS.

Sporobolus repens Presl, Rel. Haenk. **1:** 241. 1830.

Vilfa utilis Torr. U. S. Rep. Expl. Miss. Pacif. **5:** 365. 1857.

Sporobolus utilis Scribn. U. S. Dept. Agr. Div. Agrost. Bull. **17:** 171. 1899.

Muhlenbergia utilis Rydb.; Woot. & Standl. N. Mex. Agr. Exp. Sta. Bull. **81:** 74. 1912.

TYPE LOCALITY: Mexicó.

RANGE: Western Texas to southern California.

NEW MEXICO: Kingston; Mangas Springs; Salinas; Tularosa; Fort Bayard; Thornton; Raton Mountains; Deming; Mesilla Valley. Plains and low hills, in the Lower and Upper Sonoran zones.

Aparejo grass receives its name from its use by the Mexicans in packing their "aparejos" or pads which are a substitute for pack saddles. It is a rather aggressive weed in the lower Rio Grande Valley, crowding out grasses and other plants.

17. Muhlenbergia trifida Hack. Repert. Nov. Sp. Fedde **8:** 518. 1910.

TYPE LOCALITY: Michoacán, Mexico.

RANGE: Western Texas and Colorado to California and Mexico.

NEW MEXICO: Santa Fe and Las Vegas mountains; Johnsons Mesa; Raton Mountains; Tunitcha Mountains; Chama; Grants Station; Mogollon Mountains; Black Range; Burro Mountains; San Luis Mountains; Organ Mountains; White Mountains. Open hills and in canyons, in the Upper Sonoran and Transition zones.

18. Muhlenbergia virescens (H. B. K.) Kunth, Rév. Gram. **1:** 64. 1829.

Podosaemum virescens H. B. K. Nov. Gen. & Sp. **1:** 132. 1816.

TYPE LOCALITY: "Crescit locis asperis, excelsis regni Mexicani prope Santa Rosa de la Sierra et Puerto de Varientos, alt. 1350 hexap."

RANGE: Arizona and New Mexico to Mexico.

NEW MEXICO: Northwestern New Mexico (*Palmer*); Ben More (*Bigelow*). Mountains, in the Transition Zone.

19. Muhlenbergia affinis Trin. Mém. Acad. St. Pétersb. VI. **6**[2]: 301. 1845.

Muhlenbergia metcalfi Jones, Contr. West. Bot. **14:** 12. 1912.

TYPE LOCALITY: "Toluco," Mexico.

RANGE: Southern New Mexico and Arizona and southward.

NEW MEXICO: Santa Rita Mountain; Fort Bayard; Filmore Canyon; Mangas Springs; near Silver City. Dry hills, in the Upper Sonoran Zone.

The type of *M. metcalfi* (*metcalfei?*) is Metcalfe's 1485, from Santa Rita Mountain.

20. Muhlenbergia subalpina Vasey, Descr. Cat. Grasses U. S. 40. 1885.

Muhlenbergia gracilis breviaristata Vasey in Wheeler, Rep. U. S. Surv. 100th Merid. **6:** 284. 1874.

TYPE LOCALITY: Twin Lakes, Colorado.

RANGE: Wyoming to New Mexico.

NEW MEXICO: Northern New Mexico (*George Vasey*). Mountains, in the Transition Zone.

21. Muhlenbergia acuminata Vasey, Bot. Gaz. **11:** 337. 1886.

TYPE LOCALITY: "New Mexico." Type collected by Wright (no. 1993).

RANGE: Western Texas to Arizona.

NEW MEXICO: Kingston; Mangas Springs; Filmore Canyon. Dry hills, in the Upper Sonoran Zone.

22. Muhlenbergia monticola Buckl. Proc. Acad. Phila. **1862:** 91. 1863.

Muhlenbergia neomexicana Vasey, Bot. Gaz. **11:** 337. 1886.

Muhlenbergia pringlei Scribn. Trans. N. Y. Acad. **14:** 25. 1894.

TYPE LOCALITY: "Northwestern Texas."

RANGE: Western Texas to Arizona and Mexico.

NEW MEXICO: Santa Fe and Las Vegas mountains; Albuquerque; Watrous; Grant; Mangas Springs; Kingston; Reserve; Dog Spring; Socorro; Organ Mountains; Dona Ana Mountains; White Mountains. Dry slopes, in the Upper Sonoran Zone.

The type of *M. neomexicana* was collected in New Mexico.

27. ORYZOPSIS Michx.

Slender perennials with flat or involute leaves and loosely flowered, spreading or narrow panicles; spikelets 1-flowered, hermaphrodite; rachilla jointed above the glumes and not produced behind the palea, usually extended below the lemma into a short obtuse callus; glumes nearly equal, obtuse or acuminate; lemma a little shorter than the glumes, rather broad, cartilaginous, terminated by a slender deciduous awn; grain free.

KEY TO THE SPECIES.

Lemmas covered with long silky hairs, these much exceeding them in length .... 1. *O. hymenioides.*

Lemmas glabrous or short-pubescent.

- Leaves slender, involute; spikelets small, 2.5 to 4 mm. long .. 2. *O. micrantha.*
- Leaves broad and flat; spikelets large, 6 to 8 mm. long ...... 3. *O. asperifolia.*

1. **Oryzopsis hymenioides** (Roem. & Schult.) Ricker, Contr. U. S. Nat. Herb. **11**: 109. 1906. SAND BUNCHGRASS.

Stipa hymenioides Roem. & Schult. Syst. Veg. **2**: 339. 1817.

Eriocoma cuspidata Nutt. Gen. Pl. **1**: 40. 1818.

Oryzopsis cuspidata Benth.; Vasey, U. S. Dept. Agr. Spec. Rep. **63**: 23. 1883.

TYPE LOCALITY: "Ad litora fluvii Missouri."

RANGE: Washington and Alberta to Nebraska and Mexico.

NEW MEXICO: Carrizo and Tunitcha mountains; Farmington; Tierra Amarilla; Santa Fe; Canjilon; Ramah; Zuni; Pecos; Torrance; Albuquerque; Reserve; Mesilla; White Sands. Dry hills, in the Lower and Upper Sonoran zones.

The seeds of this grass were formerly gathered and used for food by the Zuni. The ground seeds were eaten alone, or mixed with corn meal and made into dumplings.

2. **Oryzopsis micrantha** (Trin. & Rupr.) Thurb. Proc. Acad. Phila. **1863**: 78. 1863.

Urachne micrantha Trin. & Rupr. Mém. Acad. St. Pétersb. VI. Sci. Nat. **5**[1]: 16. 1842.

TYPE LOCALITY: North America.

RANGE: Montana and Saskatchewan to Nebraska, Arizona, and Mexico.

NEW MEXICO: Raton; Sierra Grande; Tierra Amarilla; Manguitas Spring; Canjilon; Coolidge; Glorieta; Raton Mountains; Santa Fe; Pecos. Dry hills and plains, in the Upper Sonoran Zone.

3. **Oryzopsis asperifolia** Michx. Fl. Bor. Amer. **1**: 51. 1803. MOUNTAIN RICE.

Urachne asperifolia Trin. Gram. Unifl. **1**: 174. 1824.

TYPE LOCALITY: "Hab. a sinu Hudsonis ad Quebec, per tractus montium."

RANGE: British America to New Mexico and Pennsylvania.

NEW MEXICO: Winsor Creek (*Standley* 4206). Deep woods, in the Canadian and Hudsonian zones.

28. PHLEUM L. TIMOTHY.

Perennials with simple erect culms, flat leaves, and dense, terminal, cylindrical or oblong, spikelike panicles; spikelets 1-flowered; rachilla jointed above the glumes, not prolonged beyond the floret; glumes 2, compressed-carinate, equal, usually ciliate on the keels, abruptly mucronate or shortly awn-pointed; lemma shorter than the glumes, thin, truncate, awnless, rather loosely inclosing the grain; stamens 3; styles distinct.

KEY TO THE SPECIES.

Spikes elongate-cylindric; awns less than half as long as the glumes... 1. *P. pratense.*

Spikes short, ovoid or oblong; awns about half as long as the glumes.. 2. *P. alpinum.*

1. **Phleum pratense** L. Sp. Pl. 59. 1753. TIMOTHY.

TYPE LOCALITY: "Habitat in Europae versuris & pratis."

RANGE: Fields and meadows nearly throughout North America, introduced from Europe and often cultivated for hay; also in Europe and Asia.

NEW MEXICO: Chama; Raton; Cedar Hill; Fort Bayard; Santa Fe; Ruidoso Creek; Gilmores Ranch.

2. **Phleum alpinum** L. Sp. Pl. 59. 1753. MOUNTAIN TIMOTHY.

TYPE LOCALITY: "Habitat in Alpibus."

RANGE: Alaska and British America to California, Arizona, and New Hampshire; also in Europe and South America.

NEW MEXICO: Tunitcha Mountains; Chama; Santa Fe and Las Vegas mountains. Meadows, Canadian to Arctic-Alpine Zone.

29. ALOPECURUS L. MARSH FOXTAIL.

Annuals or perennials with erect or ascending culms, flat leaves, and densely flowered cylindrical spikelike terminal panicles; spikelets 1-flowered, strongly flattened; rachilla jointed below the glumes; glumes equal, awnless, more or less ciliate, especially along the keel, usually connate at the base; lemma obtuse, hyaline, usually awned on the back, the margins connate near the base, forming a short tube; palea none; stamens 3; styles usually distinct.

KEY TO THE SPECIES.

Spikelets about 3 mm. long; lemma shorter than the glumes....... 1. *A. aristulatus.*
Spikelets 6 to 7 mm. long; lemma longer than the glumes......... 2. *A. agrestis.*

1. **Alopecurus aristulatus** Michx. Fl. Bor. Amer. **1:** 43. 1803.

Alopecurus geniculatus aristulatus Torr. Fl. North & Mid. U. S. **1:** 97. 1824.

TYPE LOCALITY: "Hab. in paludosis Canadae."

RANGE: British America to California, New Mexico, and Florida.

NEW MEXICO: Tunitcha Mountains; Cedar Hill; Chama; Ramah; Gallo Spring; Taos; Santa Fe and Las Vegas mountains; Mogollon Mountains; Rio Mimbres; White and Sacramento mountains. Wet soil, often about the edge of water, in the Transition Zone.

2. **Alopecurus agrestis** L. Sp. Pl. ed. 2. 89. 1762. SLENDER FOXTAIL.

TYPE LOCALITY: "Habitat in Europa australi."

RANGE: Native of Europe and Asia; introduced in many places in the United States.

NEW MEXICO: Agricultural College (*Cockerell*).

30. EPICAMPES Presl.

Tall perennials with very long spikelike many-flowered panicles; spikelets small, 1-flowered; glumes membranaceous, slightly unequal, convex on the back, carinate, often finely 3-nerved; lemmas 3-nerved, obtuse or emarginate, a little shorter than or about equaling the glumes, tipped with a slender, usually short awn; stamens 3; styles short, distinct; grain included within the lemmas, free.

KEY TO THE SPECIES.

Inflorescence spikelike .... 1. *E. rigens.*
Inflorescence paniculate .... 2. *E. stricta.*

1. **Epicampes rigens** Benth. Journ. Linn. Soc. Bot. **19:** 88. 1881.

TYPE LOCALITY: California.

RANGE: Western Texas to Arizona and southward.

NEW MEXICO: Berendo Creek; Mogollon Creek; Mangas Springs; Deming. Dry hills, in the Upper Sonoran Zone.

2. **Epicampes stricta** Presl, Rel. Haenk. **1:** 235. *pl. 39.* 1830.

TYPE LOCALITY: Mexico.

RANGE: Western Texas to southern Arizona and southward.

NEW MEXICO: West Fork of the Gila; Crawfords Ranch; Silver City; near White Water; Socorro; Organ Mountains. Dry hills and canyons, in the Upper Sonoran Zone.

31. BLEPHARONEURON Nash.

Tufted perennial with flat leaves and loosely flowered open panicles; spikelets 1-flowered; glumes 1-nerved, glabrous, the second about as long as the lemma, this 3-nerved, the nerves densely pilose for nearly their whole length, the midnerve often excurrent at the apex; palea as long as the lemma, 2-nerved, densely pilose between the nerves; stamens 3; styles 2, distinct.

1. **Blepharoneuron tricholepis** (Torr.) Nash, Bull. Torrey Club **25:** 88. 1898.

Vilfa tricholepis Torr. U. S. Rep. Expl. Miss. Pacif. **4:** 155. 1856.

Sporobolus tricholepis Coulter, Man. Rocky Mount. 411. 1885.

TYPE LOCALITY: Sandia Mountains, New Mexico. Type collected by Bigelow.

RANGE: Utah and Colorado to Arizona and Mexico.

New Mexico: Tunitcha Mountains; Coolidge; Horsethief Canyon; Albuquerque; Johnsons Mesa; Rio Pueblo; Trout Spring; Raton Mountains; Chama; Sandia Mountains; Fort Bayard; Mangas Springs; Socorro. Open slopes, in the Transition and Canadian zones.

32. SPOROBOLUS R. Br. Dropseed.

Annuals or perennials with small spikelets; spikelets 1-flowered, pedicellate, in narrow or broad panicles; glumes rounded or slightly keeled, awnless, obscurely nerved or nerveless, usually unequal; lemma equaling or exceeding the glumes, not awned; palea as long as the lemma or longer; stamens 3; styles short, distinct.

KEY TO THE SPECIES.

Panicles narrow, spikelike.
- Plants tall, robust, 1 meter high, erect........................ 1. *S. giganteus.*
- Plants low, slender, 60 cm. high or less, spreading or ascending........................ 2. *S. strictus.*

Panicles branched and spreading.
- Annual, 20 cm. high or less........................ 3. *S. confusus.*
- Perennials, usually more than 20 cm. high.
  - Plants with long scaly rootstocks; glumes about equal.
    - Panicles 8 cm. long or less; stems rigid although slender........................ 4. *S. auriculatus.*
    - Panicles 13 to 20 cm. long; stems weak, often elongated........................ 5. *S. asperifolius.*
  - Plants without long scaly rootstocks; glumes very unequal.
    - Sheaths naked or sparingly ciliate in the throat.
      - Plants 30 cm. high or less; spikelets long-pedicellate; sheaths villous........................ 6. *S. texanus.*
      - Plants more than 50 cm. high; spikelets short-pedicellate; sheaths not villous.
        - Plants less than 90 cm. high; panicles open, with comparatively few spikelets; glumes nerved........................ 7. *S. airoides.*
        - Plants 100 to 180 cm. high; panicles rather narrow, with very numerous spikelets; glumes not nerved........................ 8. *S. wrightii.*
    - Sheaths with a conspicuous tuft of hairs in the throat.
      - Sheaths pubescent; leaf blades divergent; panicles 8 cm. long or less; plants not more than 30 cm. high, slender........................ 9. *S. nealleyi.*
      - Sheaths almost or quite glabrous; leaf blades not divergent; panicles 15 to 30 cm. long or more; plants mostly 60 cm. high or more, stout.
        - Panicles exserted, spreading, sometimes somewhat nodding, the lower branches about as long as the upper ones......12. *S. flexuosus.*
        - Panicles mostly included in the sheaths, rarely if at all spreading, the lower branches longer than the upper.
          - Lemmas about equaling the glumes, acute to obtuse, less than 2 mm. long........................10. *S. cryptandrus.*
          - Lemmas much longer than glumes, long-acuminate, 5 to 6 mm. long......11. *S. asper.*

1. **Sporobolus giganteus** Nash, Bull. Torrey Club **25:** 88. 1898.

SANDHILL DROPSEED.

Sporobolus cryptandrus giganteus Jones, Contr. West. Bot. **14:** 11. 1912.

TYPE LOCALITY: On the White Sands, New Mexico. Type collected by Wooton (no. 394).

RANGE: Southern New Mexico.

NEW MEXICO: Gila Hot Springs; Socorro; Sabinal; Mesilla Valley; White Sands; south of Carrizozo; Arroyo Ranch. Sandhills, in the Lower Sonoran Zone.

2. **Sporobolus strictus** (Scribn.) Merr. U. S. Dept. Agr. Div. Agrost. Circ. **32:** 6. 1901.

Sporobolus cryptandrus strictus Scribn. Bull. Torrey Club **9:** 103. 1882.

TYPE LOCALITY: Banks of the Rillita, near Camp Lowell, Arizona.

RANGE: Colorado to Arizona and Mexico.

NEW MEXICO: Throughout the State except in the southeastern corner. Dry hills and plains, in the Lower and Upper Sonoran zones.

3. **Sporobolus confusus** (Fourn.) Vasey, Bull. Torrey Club **15:** 293. 1888.

Vilfa confusa Fourn. Mex. Pl. **2:** 101. 1881.

TYPE LOCALITY: "In devexis arenosis montis ignivomi Jorullo," Mexico.

RANGE: Washington to Texas and Mexico.

NEW MEXICO: Ensenada; Santa Fe and Las Vegas mountains; Mogollon Mountains; Mangas Springs; Black Range; San Luis Mountains; Animas Valley; Tortugas Mountain; Organ Mountains; White Mountains. Dry hills and canyons, in the Lower and Upper Sonoran zones.

4. **Sporobolus auriculatus** Vasey, Contr. U. S. Nat. Herb. **3:** 64. 1892.

TYPE LOCALITY: Texas.

RANGE: Western Texas and southern New Mexico.

NEW MEXICO: Albuquerque; Cross L Ranch; Farmington; Chama; Deming; Socorro; Carrizozo; White Sands; Chosa Springs; Lake Arthur; Hope; Roswell; Dona Ana Mountains. Plains, in the Lower and Upper Sonoran zones.

5. **Sporobolus asperifolius** (Nees & Mey.) Thurber in S. Wats. Bot. Calif. **2:** 269. 1880.

Vilfa asperifolia Nees & Mey. Mém. Acad. St. Pétersb. VI. Sci. Nat. 4^1: 95. 1840.

TYPE LOCALITY: "Chile; Rio Mayno; Copiapo."

RANGE: British America to California, New Mexico, and Missouri; also in South America.

NEW MEXICO: Albuquerque; Carrizo Mountains; Cedar Hill; Dulce; Pecos; Alamillo; Kingston; Mangas Springs; Mesilla Valley; Roswell; White Sands; Dona Ana Mountains. Valleys and plains, often in alkaline soil, in the Lower and Upper Sonoran zones.

6. **Sporobolus texanus** Vasey, Contr. U. S. Nat. Herb. **1:** 57. 1890.

TYPE LOCALITY: Screw Bean, Presidio County, Texas.

RANGE: Western Texas and southern New Mexico.

NEW MEXICO: Carlsbad; along the Pecos near Roswell. Dry plains.

7. **Sporobolus airoides** Torr. U. S. Rep. Expl. Miss. Pacif. 7^3: 21. 1856.

BUNCHGRASS.

Agrostis airoides Torr. Ann. Lyc. N. Y. **1:** 151. 1824.

TYPE LOCALITY: "On the branches of the Arkansas, near the Rocky Mountains," Colorado.

RANGE: Washington and Nebraska to California and New Mexico.

NEW MEXICO: Carrizo Mountains; Aztec; McCarthys Station; Santa Fe; Ojo Caliente; Algodones; Belen; Zuni; Socorro; Cliff; Mangas Springs; Mesilla Valley; White Sands. Open plains and dry slopes, in the Lower and Upper Sonoran zones.

One of the important range grasses of the State, on the plains. It is able to endure considerable amounts of alkali.

8. **Sporobolus wrightii** Munro; Scribn. Bull. Torrey Club **9**: 103. 1882. SACATON.

TYPE LOCALITY: Near Pantano, Arizona.

RANGE: Western Texas to southern Colorado and northern Mexico.

NEW MEXICO: Socorro; Fort Bayard; Dog Spring; Deming; Mangas Springs; Gila Hot Springs; Las Vegas; Buchanan; Carrizozo; Carlsbad. Dry hills and plains, in the Upper Sonoran Zone.

An important range grass. Both this and the preceding are often cut for hay.

9. **Sporobolus nealleyi** Vasey, Bull. Torrey Club **15**: 49. 1888, name only; Contr. U. S. Nat. Herb. **1**: 57. 1890. NEALLEY'S DROPSEED.

TYPE LOCALITY: Brazos Santiago, Texas.

RANGE: Western Texas to New Mexico.

NEW MEXICO: Pecos; near Suwanee; Las Cruces; White Sands; Round Mountain; plains 35 miles south of Torrance; Roswell. Dry plains and hills, in the Lower and Upper Sonoran zones.

10. **Sporobolus cryptandrus** (Torr.) A. Gray, Man. 576. 1848.

Agrostis cryptandrus Torr. Ann. Lyc. N. Y. **1**: 151. 1824.

TYPE LOCALITY: "On the Canadian River," Colorado?

RANGE: Washington and Maine to Arizona and Texas.

NEW MEXICO: Common throughout the State. Plains and dry slopes, in the Lower and Upper Sonoran zones.

11. **Sporobolus asper** (Michx.) Kunth, Enum. Pl. **1**: 210. 1833.

Agrostis asper Michx. Fl. Bor. Amer. **1**: 52. 1803.

TYPE LOCALITY: "Habitat in collibus rupibusque regionis Illinoensis."

RANGE: Minnesota and Nebraska to New England, south to Texas and Florida.

NEW MEXICO: Pecos (*Standley* 5313). Open slopes, in the Upper Sonoran Zone.

12. **Sporobolus flexuosus** (Thurb.) Rydb. Bull. Torrey Club **32**: 601. 1905.

Sporobolus cryptandrus flexuosus Thurb. Contr. U. S. Nat. Herb. **3**: 62. 1892.

TYPE LOCALITY: "Dry western plains, Colorado, New Mexico, Arizona to Texas."

RANGE: Nevada to Texas and Mexico.

NEW MEXICO: Carrizo Mountains; Farmington; Socorro; Albuquerque; Deming; Tortugas Mountain; White Sands; Mesilla Valley; Organ Mountains. Sandy soil, in the Lower and Upper Sonoran zones.

33. POLYPOGON Desf. BEARDGRASS.

Mostly annuals with decumbent or erect stems, flat leaves, and densely flowered terminal panicles; spikelets 1-flowered, hermaphrodite; glumes nearly equal, usually broader above, entire or 2-lobed, awned; lemma much smaller than the glumes, thin, hyaline, entire, emarginate, or bifid, awned, the awn slender, straight, or geniculate and twisted below; stamens 1 to 3; styles short, distinct; stigmas plumose; grain free.

KEY TO THE SPECIES.

Glumes notched at the apex; awns very long, concealing the spikelets.. 1. *P. monspeliensis*.

Glumes attenuate at the apex; awns short, not concealing the spikelets.. 2. *P. littoralis*.

1. **Polypogon monspeliensis** (L.) Desf. Fl. Atlant. **1**: 67. 1800.

Alopecurus monspeliensis L. Sp. Pl. 61. 1753.

TYPE LOCALITY: "Habitat Monspelii."

RANGE: British America to Mexico; also in Europe.

NEW MEXICO: Shiprock; Farmington; Sandia Mountains; Albuquerque; Socorro; Kingston; Mangas Springs; Mesilla Valley; Fort Bayard; Alamogordo. In wet ground, in the Upper Sonoran and Transition zones.

2. **Polypogon littoralis** (With.) J. E. Smith, Comp. Fl. Brit. ed. 2. 13. 1816.

Agrostis littoralis With. Bot. Arr. Veg. Brit. ed. 3. **2**: 129. 1796.

TYPE LOCALITY: "Wells, on the Norfolk coast," England.

RANGE: British America to California and New Mexico and the Gulf Coast; also in the Old World.

NEW MEXICO: Berendo Creek; Albuquerque; Alamogordo; Carrizo Mountains. Wet ground.

34. CINNA L. WOOD REED GRASS.

Tall perennials with numerous flat leaves and with many-flowered nodding panicles; spikelets 1-flowered; rachilla jointed below the glumes as well as above them, produced below the floret into a short smooth stipe and usually extending behind the palea as a slender naked bristle; lemmas similar to the glumes, 3-nerved, obtuse, usually with a very short subterminal awn; palea apparently 1-nerved, keeled; stamen 1; styles short, distinct; grain free.

1. **Cinna latifolia** (Trevir.) Griseb. in Ledeb. Fl. Ross. **4**: 435. 1853.

Agrostis latifolia Trevir.; Göpp. Beschr. Bot. Gart. Breslau 82. 1830.

Cinna pendula Trin. Mém. Acad. St. Pétersb. VI. Sci. Nat. **4**^{1}: 280. 1841.

TYPE LOCALITY: Not ascertained.

RANGE: Alaska and British America to Oregon, New Mexico, and North Carolina.

NEW MEXICO: Sandia Mountains (*Wooton*). Transition Zone.

35. AGROSTIS L.

Annuals or usually perennials with small spikelets in open panicles; spikelets 1-flowered; rachilla jointed above the glumes, not produced beyond the floret; glumes equal or nearly so, acute, longer than the floret; lemma rather broad, less firm than the glumes, usually obtuse, awnless or with a slender dorsal awn; palea hyaline, much shorter than the lemma or wanting; stamens 3; grain inclosed in the lemma but free.

KEY TO THE SPECIES.

Panicles dense or very narrow.
  Culms decumbent at the base, with long creeping stolons rooting at the nodes; panicles short and thick .... 1. *A. stolonifera*.
  Culms erect, without stolons; panicles long and narrow .... 2. *A. exarata*.
Panicles loose and spreading.
  Branches of the panicle 7 to 10 cm. long .... 3. *A. hiemalis*.
  Branches of the panicle 4 to 6 cm. long.
    Palea minute and inconspicuous .... 4. *A. idahoensis*.
    Palea half as long as the lemma.
      Lemma usually not awned; branches of the panicle ascending .... 5. *A. alba*.
      Lemma with an awn of about the same length; branches of the panicle spreading or reflexed .... 6. *A. rosei*.

1. **Agrostis stolonifera** L. Sp. Pl. 62. 1753. WATER BENTGRASS.

Agrostis verticillata Vill. Prosp. Pl. Dauph. 16. 1779.

TYPE LOCALITY: "Habitat in Europa."

RANGE: California and Texas to Mexico; also in South America, Europe, and Asia.

NEW MEXICO: Carrizo Mountains; Farmington; Sandia Mountains; Santa Fe; Las

Vegas Hot Springs; Burro Mountains; Socorro; Fort Bayard; Berendo Creek; Rincon; Cloverdale; Mesilla Valley; Organ Mountains; Malones Ranch; Roswell. Wet ground and borders of streams, in the Upper Sonoran and Transition zones.

2. **Agrostis exarata** Trin. Gram. Unifl. 207. 1824.

TYPE LOCALITY: "Unalaschka."

RANGE: Alaska and British America to Mexico; also in Siberia.

NEW MEXICO: Tunitcha Mountains; Ramah; Winsor Creek; Pecos; Rio Pueblo; Las Vegas; Cross L Ranch; Fort Bayard; Rio Mimbres; Chiz; Lower Plaza; Deming; Santa Fe; Burro Mountains; Organ Mountains; Gilmores Ranch. Wet ground, in the Transition Zone.

3. **Agrostis hiemalis** (Walt.) B. S. P. Prel. Cat. N. Y. 68. 1888. HAIR GRASS.

Cornucopiae hiemalis Walt. Fl. Carol. 73. 1788.

Agrostis scabra Willd. Sp. Pl. **1:**370. 1799.

TYPE LOCALITY: Carolina.

RANGE: Throughout most of North America.

NEW MEXICO: Tunitcha Mountains; Santa Fe and Las Vegas mountains; Sandia Mountains; Grants Station; Inscription Rock; Mogollon Mountains; Mimbres River; White and Sacramento mountains. Meadows and woods, in the Transition and Canadian zones.

4. **Agrostis idahoensis** Nash, Bull. Torrey Club **24:** 42. 1897.

Agrostis tenuis Vasey, Bull. Torrey Club **10:** 21. 1883, not Sibth. 1794.

TYPE LOCALITY: Forest, Nez Perces County, Idaho.

RANGE: Washington and Montana to California and New Mexico.

NEW MEXICO: El Rito Creek (*Wooton* 2989). Damp woods, in the Transition Zone.

5. **Agrostis alba** L. Sp. Pl. 63. 1753. REDTOP.

TYPE LOCALITY: "Habitat in Europae nemoribus."

RANGE: British America, southward to Mexico.

NEW MEXICO: El Rito Creek; Santa Fe; Albuquerque; Zuni Reservation; Indian Creek; Farmington; Cedar Hill; Fort Bayard; Animas Creek; White Mountains. Wet meadows, in the Upper Sonoran and Transition zones.

6. **Agrostis rosei** Scribn. & Merr. U. S. Dept. Agr. Div. Agrost. Bull. **24:** 21. 1901.

TYPE LOCALITY: Sierra Madre, Zacatecas, Mexico.

RANGE: Southwestern New Mexico to central Mexico.

NEW MEXICO: Cloverdale (*Mearns* 462).

36. CALAMAGROSTIS Adans. REED BENTGRASS.

Tall perennials with small spikelets in many-flowered terminal panicles; spikelets 1-flowered; rachilla produced above the floret into a short, usually hairy pedicel or bristle; glumes nearly equal, awnless, usually exceeding the lemma; lemma surrounded at the base by numerous hairs, these sometimes equaling or exceeding it in length, awned on the back usually from below the middle; palea more than half the length of the lemma, faintly 2-nerved; stamens 3; styles distinct; grain inclosed by the lemma and palea and more or less adherent.

KEY TO THE SPECIES.

Panicles open, the lower branches spreading or drooping; spikelets greenish .... 1. *C. canadensis.*

Panicles dense, the branches erect or ascending; spikelets strongly tinged with purple .... 2. *C. hyperborea americana.*

1. **Calamagrostis canadensis** (Michx.) Beauv. Ess. Agrost. 15. 1812.
Arundo canadensis Michx. Fl. Bor. Amer. **1**: 73. 1803.
TYPE LOCALITY: Canada.
RANGE: British America to Oregon, New Mexico, Ohio, and New Jersey.
NEW MEXICO: Ponchuelo Creek; Winsor Creek. Wet ground, in the Transition and Canadian zones.

2. **Calamagrostis hyperborea americana** Vasey; Kearney, U. S. Dept. Agr. Div. Agrost. Bull. **11**: 41. 1898.
Deyeuxia neglecta americana Vasey; Macoun, Cat. Can. Pl. **4**: 206. 1888.
TYPE LOCALITY: Donald, Columbia Valley, British Columbia.
RANGE: British America to Oregon, New Mexico, and Vermont.
NEW MEXICO: Tunitcha Mountains; Harveys Upper Ranch. Damp woods, in the Canadian Zone.

37. CALAMOVILFA Scribn. SAND GRASS.

Rather tall rigid perennials with loosely spreading panicles; spikelets 1-flowered; rachilla jointed above the glumes but not prolonged beyond the floret, the callus densely bearded; glumes laterally compressed, keeled, chartaceous, awnless, unequal, acute; lemma 1-nerved, acute; stamens 3; styles distinct.

KEY TO THE SPECIES.

Panicle branches erect; spikelets about 6 mm. long.................. 1. *C. longifolia.*
Panicle branches spreading; spikelets 8 mm. long.................... 2. *C. gigantea.*

1. **Calamovilfa longifolia** (Hook.) Hack. True Grasses 113. 1890.
Calamagrostis longifolia Hook. Fl. Bor. Amer. **2**: 241. 1840.
TYPE LOCALITY: "Saskatchawan."
RANGE: British America to New Mexico and Indiana.
NEW MEXICO: Near Texline (*Griffiths* 5650). Plains, in the Upper Sonoran Zone.

2. **Calamovilfa gigantea** (Nutt.) Scribn. & Merr. U. S. Dept. Agr. Div. Agrost. Circ. **35**: 2. 1901.
Calamagrostis gigantea Nutt. Trans. Amer. Phil. Soc. n. ser. **5**: 143. 1837.
TYPE LOCALITY: "On the sandy banks of Great Salt river of the Arkansas."
RANGE: Sandy soil, Kansas to Arizona.
NEW MEXICO: A single specimen seen, without definite locality.

38. DANTHONIA DC. WILD OAT GRASS.

Low cespitose perennials with simple, spreading or narrow panicles; spikelets several-flowered, the uppermost flower imperfect or rudimentary; rachilla jointed above the glumes; glumes 2, much exceeding the lemmas, these rounded on the back, 2-toothed or bifid, awned between the teeth, the awn formed by an extension of the 3 middle nerves of the lemma.

KEY TO THE SPECIES.

Lemmas pubescent only on the margin and at the base............. 1. *D. intermedia.*
Lemmas pubescent on the back as well as on the margin.
    Glumes 15 to 20 mm. long.. 2. *D. parryi.*
    Glumes 10 mm. long or less...................................... 3. *D. spicata.*

1. **Danthonia intermedia** Vasey, Bull. Torrey Club **10**: 52. 1883.
TYPE LOCALITY: "California, Rocky Mountains, Plains of British America to Mount Albert, Lower Canada."
RANGE: British America to California and New Mexico.

New Mexico: Horsethief Canyon (*Standley* 4880). Meadows, in the Canadian Zone.

2. Danthonia parryi Scribn. Bot. Gaz. **21**: 133. 1896.
Type locality: Colorado.
Range: Colorado and New Mexico.
New Mexico: Grass Mountain (*Standley* 4371). Meadows, in the Canadian Zone.

3. Danthonia spicata (L.) Beauv.; Roem. & Schult. Syst. Veg. **2**: 690. 1817.
Avena spicata L. Sp. Pl. 80. 1753.
Type locality: "Habitat in Pennsylvania."
Range: British America to New Mexico, Louisiana, and North Carolina.
New Mexico: Harveys Upper Ranch; West Fork of the Gila. Damp woods, in the Canadian Zone.

39. AVENA L. Oats.

Annuals or perennials with rather large spikelets variously paniculate; spikelets 2 to 6-flowered; rachilla jointed above the glumes, bearded below the lemmas; glumes 2, unequal, membranaceous, longer than the lemmas, these rounded on the back, 5 to 9-nerved, often bidentate at the apex, with a long dorsal twisted awn; grain pubescent, at least at the apex, frequently adherent to the lemma or palea.

KEY TO THE SPECIES.

Glumes shorter than the lemmas; panicles lax, somewhat nodding; lemmas hairy at the base.. 1. *A. striata.*
Glumes longer than the lemmas; panicles open; lemmas often hairy up to the base of the awn.. 2. *A. fatua.*

1. Avena striata Michx. Fl. Bor. Amer. **1**: 73. 1803.
Type locality: "Hab. in sinu Hudsonis per tractus montium ad Canadam."
Range: British America to New Mexico and Pennsylvania.
New Mexico: Ponchuelo Creek (*Standley* 4185). Meadows, in the Canadian Zone.

2. Avena fatua L. Sp. Pl. 80. 1753. Wild oats.
Type locality: "Habitat in Europae agris inter segetes."
Range: Native of Europe and Asia, widely introduced into the United States; especially common in grain fields.
New Mexico: Shiprock; Carrizo Mountains; Dulce; Cedar Hill; Cleveland; Taos; Pecos; Mora.

40. DESCHAMPSIA Beauv.

Annuals or perennials with flat or convolute leaves and rather small shining spikelets in terminal or lateral, narrow or loose panicles; spikelets mostly 2-flowered; rachilla hairy, jointed above the glumes and prolonged beyond the upper floret as a hairy bristle; glumes 2, thin and scarious, acute or obtuse, nearly equal; lemmas subhyaline, 4-nerved, truncate and more or less regularly 2 to 4-toothed, awned on the back, the awn slender, twisted below; palea narrow, 2-nerved; grain oblong, free.

KEY TO THE SPECIES.

Plants low, 20 to 40 cm. high; glumes 4 mm. long; awns much longer than the lemmas.. 1. *D. alpicola.*
Plants tall, 60 to 100 cm.; glumes 3 to 3.5 mm. long; awns little if at all longer than the lemmas.. 2. *D. caespitosa.*

1. Deschampsia alpicola Rydb. Bull. Torrey Club **32**: 601. 1905.
Type locality: Mountain meadows, Pikes Peak, Colorado.
Range: Wyoming and Utah to northern New Mexico.

New Mexico: Truchas Peak; El Rito Creek; Las Vegas Range; near the head of the Nambe. Meadows, in the Arctic-Alpine Zone.

2. **Deschampsia cespitosa** (L.) Beauv. Ess. Agrost. 91, 160. 1812.

Aira cespitosa L. Sp. Pl. 64. 1753.

Type locality: "Habitat in Europae pratis cultis & fertilibus."

Range: Arctic America to California, Arizona, Illinois, and New Jersey; also in Europe and Asia.

New Mexico: North of Ramah; Chama; Tunitcha Mountains; Harveys Upper Ranch; Rio Pueblo; Spirit Lake; Silver Spring Canyon. Meadows, Transition to Hudsonian Zone.

41. TRISETUM Pers. False oats.

Cespitose perennials or rarely annuals, with flat leaves and dense, spikelike or narrow, loose panicles; spikelets 2-(rarely 3 to 5-)flowered; rachilla hairy or naked, jointed above the glumes and between the florets, produced beyond the upper flower as a usually hairy bristle; glumes 2, awnless, carinate, unequal, usually longer than the lemmas; lemmas subhyaline, carinate, cleft or 2-toothed at the apex, the teeth sometimes produced into slender awns, awned between or a little below the teeth; awns twisted and usually geniculate; palea narrow, 2-toothed; grain smooth, free.

KEY TO THE SPECIES.

Panicles slender, interrupted; plants slender.................... 1. *T. interruptum.*
Panicles dense and crowded, not interrupted; plants stout.
  Leaf blades and sheaths long-hairy; upper part of the stem densely pubescent.................................... 3. *T. spicatum.*
  Leaf blades and sheaths glabrous or the lowest sheath short-pubescent with reflexed hairs; stems glabrous or slightly scabrous in the inflorescence................ 2. *T. montanum.*

1. **Trisetum interruptum** Buckl. Proc. Acad. Phila. **1862**: 100. 1863.

Type locality: Middle Texas.

Range: Western Texas and southern New Mexico.

New Mexico: Bishops Cap (*Wooton*). Upper Sonoran Zone.

2. **Trisetum montanum** Vasey, Bull. Torrey Club **13**: 118. 1886.

Type locality: Not stated.

Range: Wyoming to northern New Mexico.

New Mexico: Winsors Ranch; Cowles; Rio Pueblo; mountains near Las Vegas; Eagle Creek. Meadows, in the Transition and Canadian zones.

3. **Trisetum spicatum** (L.) Richt. Pl. Eur. **1**: 59. 1890.

Aira spicata L. Sp. Pl. 63. 1753.

Aira subspicata L. Syst. Veg. ed. 10. 873. 1759.

Trisetum subspicatum Beauv. Ess. Agrost. 88. 1812.

Type locality: "Habitat in Lapponiae Alpibus."

Range: Arctic America to California, New Mexico, and New Hampshire; also in Europe.

New Mexico: Pecos Baldy; Truchas Peak; Jemez Mountains. Meadows, in the Arctic-Alpine Zone.

42. BULBILIS Raf. Buffalo grass.

Creeping or stoloniferous perennial with narrow flat leaves and unlike staminate and pistillate flowers borne on the same or different plants; staminate spikelets 2 or 3-flowered, sessile in 2 rows along the short one-sided spikes, the glumes obtuse, unequal, the lemmas larger, 3-nerved, the palea 2-nerved; stamens 3; pistillate spikelets 1-flowered, in nearly capitate one-sided spikes scarcely exserted from the broad sheaths of the upper leaves, the glumes 2, or the first sometimes wanting, 3-toothed at the apex, the lemma narrow, hyaline, entire or bifid at the apex; styles distinct; grain free.

1. **Bulbilis dactyloides** (Nutt.) Raf.; Kuntze, Rev. Gen. Pl. 763. 1891.
Sesleria dactyloides Nutt. Gen. Pl. **1**: 65. 1818.
Buchloe dactyloides Engelm. Trans. Acad. St. Louis **1**: 432. 1859.
TYPE LOCALITY: "On the open grassy plains of the Missouri."
RANGE: North Dakota and Minnesota to Arkansas and Mexico.
NEW MEXICO: Sierra Grande; Nara Visa; Raton; Santa Fe; Coolidge; Pecos; Logan; Buchanan; Mesilla Park. Plains, in the Upper Sonoran Zone.

43. LEPTOCHLOA Beauv.

Mostly tall annuals with flat leaves and elongated simple panicles composed of numerous more or less spreading, slender spikes scattered along the main axis; spikelets 2 to several-flowered, sessile in 2 rows along one side of the slender and often numerous branches; rachilla jointed above the glumes, these 2-keeled, awnless or very short-awned; lemmas carinate, 3-nerved, acute, awnless or very short-awned or 2 or 3-toothed, mucronate or short-awned between the teeth; palea 2-keeled.

KEY TO THE SPECIES.

Spikelets 2.5 mm. long or less, broad, 2 to 4-flowered.
Sheaths pilose.. 1. *L. filiformis.*
Sheaths not pilose.................................... 2. *L. nealleyi.*
Spikelets 3 mm. long or more, narrow, 5 to 12-flowered.
Lemmas pubescent at the base; annual.................. 3. *L. fascicularis.*
Lemmas glabrous; perennial............................ 4. *L. dubia.*

1. **Leptochloa filiformis** (Lam.) Beauv. Ess. Agrost. 71. 1812.
Festuca filiformis Lam. Tabl. Encycl. **1**: 191. 1791.
Eleusine mucronata Michx. Fl. Bor. Amer. **1**: 65. 1803.
Leptochloa mucronata Kunth, Rév. Gram. **1**: 91. 1835.
TYPE LOCALITY: "Ex Amer. Merid."
RANGE: Virginia and Florida to California and Mexico; also in the West Indies and southern Asia.
NEW MEXICO: Hillsboro; Mesilla Valley. Sandy fields, in the Lower Sonoran Zone.

2. **Leptochloa nealleyi** Vasey, Bull. Torrey Club **12**: 7. 1885.
TYPE LOCALITY: Texas.
RANGE: Western Texas and southeastern New Mexico.
NEW MEXICO: Carlsbad (*Tracy* 8191). Plains.

3. **Leptochloa fascicularis** (Lam.) A. Gray, Man. 588. 1848.
Festuca fascicularis Lam. Tabl. Encycl. **1**: 189. 1791.
Festuca procumbens Muhl. Descr. Gram. 160. 1817.
Uralepis composita Buckl. Proc. Acad. Phila. **1862**: 94. 1863.
Diplachne procumbens Nash in Britton, Man. 128. 1901.
TYPE LOCALITY: "Ex Amer. Merid."
RANGE: Maryland and Florida to South Dakota and Mexico.
NEW MEXICO: Salt Lake; Socorro; Mesilla Valley; Roswell; Carlsbad. Sandy fields, in the Lower Sonoran Zone.
The type of *Uralepis composita* was collected in New Mexico by Woodhouse.

4. **Leptochloa dubia** (H. B. K.) Nees, Syll. Pl. Ratisb. **1**: 4. 1824. SPRANGLE.
Chloris dubia H. B. K. Nov. Gen. & Sp. **1**: 169. 1815.
Diplachne dubia Scribn. Bull. Torrey Club **10**: 30. 1883.

TYPE LOCALITY: "Crescit in apricis subhumidis prope rupem porphyreticam el Penon, in convalle Mexicana, alt. 1168 hexap."

RANGE: Arizona and western Texas to Florida and Mexico.

NEW MEXICO: Mangas Springs; near White Water; Dog Spring; near Silver City; Organ Mountains; Tortugas Mountain; Carlsbad. Dry hills, in the Lower and Upper Sonoran zones.

44. ACAMPTOCLADOS Nash.

Tufted perennial with stiff stems, involute leaves, and a panicle of scattered, distant, widely spreading, rigid branches; spikelets scattered, singly disposed in 2 rows, sessile, 4 to 6-flowered; glumes subequal, acuminate, the first 1-nerved, the second usually 3-nerved; lemmas 3-nerved, acute, indurated in fruit; palea compressed, the 2 nerves ciliolate, gibbous at the base; stamens 3; styles distinct.

1. **Acamptoclados sessilispicus** (Buckl.) Nash in Small, Fl. Southeast. U. S. 140. 1903.

Eragrostis sessilispica Buckl. Proc. Acad. Phila. **1862**: 97. 1863.

Diplachne rigida Vasey, U. S. Dept. Agr. Div. Bot. Bull. **12**: *pl. 41*. 1891.

TYPE LOCALITY: Near Austin, Texas.

RANGE: Kansas and Texas to eastern New Mexico.

NEW MEXICO: Sands south of Melrose; Nara Visa. Plains, in the Upper Sonoran Zone.

45. BECKMANNIA Host. SLOUGH GRASS.

Tall erect plant with flat leaves and terminal elongated inflorescence; spikelets 1 or 2-flowered, compressed, imbricated in 2 rows along one side of the rachis of the short spikes; glumes narrow, boat-shaped, obtuse or abruptly pointed, nearly equal; lemmas narrow, acute or mucronate; palea hyaline, 2-keeled; stamens 3; styles short, distinct; grain oblong, inclosed within the rigid fruiting lemma and palea, free.

1. **Beckmannia erucaeformis** (L.) Host, Icon. Gram. Austr. **3**: 5. 1805.

Phalaris erucaeformis L. Sp. Pl. 55. 1753.

TYPE LOCALITY: "Habitat in Siberia, Russia, Europa australi."

RANGE: British America to California, New Mexico, and Iowa; also in Europe and Asia.

NEW MEXICO: Farmington; Chama; Grants Station; Zuni. In marshes and along streams, in the Transition Zone.

46. CAPRIOLA Adans.

Low creeping perennial with short flat leaves and slender spikes digitate at the apex of the erect branches; spikelets 1-flowered, awnless, sessile in 2 rows along one side of a slender continuous axis, forming unilateral spikes; glumes narrow, keeled, usually acute; lemma broader, usually slightly longer than the glumes, obtuse, pilose on the keel and margins; palea about the length of the lemma, 2-keeled; stamens 3; styles distinct; grain free.

1. **Capriola dactylon** (L.) Kuntze, Rev. Gen. Pl. **2**: 764. 1891. BERMUDA GRASS.

Panicum dactylon L. Sp. Pl. 58. 1753.

Cynodon dactylon Pers. Syn. Pl. **1**: 85. 1805.

TYPE LOCALITY: "Habitat in Europa australi."

RANGE: Native of the Old World, widely introduced in southern North America, often cultivated as a lawn grass.

NEW MEXICO: Mesilla Valley.

Bermuda grass is often employed in New Mexico for lawns. It is especially valuable for this purpose in the southern part of the State, since it is resistant to heat and drought. Sometimes it becomes a troublesome weed in cultivated fields.

47. SCHEDONARDUS Steud. TEXAN CRABGRASS.

Low, diffusely branched perennial with short narrow leaves and slender paniculate spikes; spikelets 1-flowered, hermaphrodite, sessile, scattered along one side of the slender rachis of the widely spreading spikes; rachilla jointed above the glumes, these narrow, slightly unequal, membranaceous; lemmas longer than the glumes, membranaceous, becoming somewhat rigid, acuminate or minutely mucronate; stamens 3; styles distinct; grain inclosed within the rigid lemmas and palea but free.

1. **Schedonardus paniculatus** (Nutt.) Trel. Rep. Ark. Geol. Surv. **1888**^{4}: 236. 1891.
Lepturus paniculatus Nutt. Gen. Pl. **1**: 81. 1818.
Schedonardus texanus Steud. Syn. Pl. Glum. **1**: 146. 1855.
TYPE LOCALITY: "On dry saline plains, near Fort Mandan, on the Missouri."
RANGE: Manitoba and Saskatchewan to New Mexico and Texas.
NEW MEXICO: From the Mogollon and White Mountains northward and eastward throughout the State. Dry hills and plains, in the Upper Sonoran Zone.

48. BOUTELOUA Lag. GRAMA GRASS.

Low annuals or perennials, with narrow, flat or convolute leaves and few or many unilateral spikelets nearly sessile along a common rachis; spikelets 1 or 2-flowered, numerous, crowded and closely sessile in 2 rows along one side of a continuous flattened rachis, this usually projecting beyond the spikelets; rachilla articulated above the glumes, the continuation beyond the hermaphrodite floret usually bearing a few rudimentary glumes and 3 awns; glumes unequal, the lower smaller, keeled; lemma broader, 3-nerved, 3 to 5-toothed or cleft; palea 2-nerved and 2-toothed; grain free.

KEY TO THE SPECIES.

Spikes numerous, 5 to 60; spikelets few, usually less than 12.
  Spikes 30 to 60, each with 4 to 10 spikelets .... 1. *B. curtipendula.*
  Spikes 5 to 11, each with 3 to 6 spikelets .... 2. *B. radicosa.*
Spikes few, 1 to 6; spikelets numerous, 25 or more.
  Annuals.
    Spikes solitary; plants low, tufted .... 3. *B. procumbens.*
    Spikes more than one; plants various.
      Spikelets closely appressed to the rachis, forming a cylindrical spike .... 4. *B. aristidoides.*
      Spikelets crowded on one side of the rachis, making it one-sided.
        Plants 30 cm. high or more, the stems erect .... 5. *B. parryi.*
        Plants 10 to 15 cm. high, the stems spreading .. 6. *B. barbata.*
  Perennials.
    Spikes loose, more or less cylindric; lower part of stems densely woolly .... 7. *B. eriopoda.*
    Spikes with more numerous crowded spikelets, one-sided; stems not woolly.
      Glumes smooth or slightly roughened .... 8. *B. breviseta.*
      Glumes stiff-hairy.
        Spikes 3 to 5, short and broad; rachis extended much beyond the spike .... 9. *B. hirsuta.*
        Spikes 1 to 3, mostly 2, long and narrow; rachis but slightly extended .... 10. *B. gracilis.*

1. **Bouteloua curtipendula** (Michx.) Torr. in Emory, Mil. Reconn. 154. 1848.
TALL GRAMA.
Chloris curtipendula Michx. Fl. Bor. Amer. **1**: 59. 1803.
Bouteloua racemosa Lag. Var. Cienc. **2**^{4}: 141. 1805.

Atheropogon curtipendulus Fourn. Mex. Pl. **2:** 138. 1881.

TYPE LOCALITY: "Hab. in aridis regionis Illinoensis ad *Wabast* et in rupibus ad *prairie du rocher.*"

RANGE: British America to New Jersey, California, and Mexico.

NEW MEXICO: Common throughout the State. Plains and hillsides, in the Upper Sonoran and Transition zones. An important range grass in some parts of the State.

2. **Bouteloua radicosa** (Fourn.) Griffiths, Contr. U. S. Nat. Herb. **14:** 411. 1912.

Dinebra bromoides H. B. K. Nov. Gen. & Sp. **1:** 172. *pl. 51.* 1816, not *Bouteloua bromoides* Lag. 1816.

Atheropogon radicosus Fourn. Mex. Pl. **2:** 140. 1881.

TYPE LOCALITY: Mexico.

RANGE: California and New Mexico to Mexico.

NEW MEXICO: Mangas Springs; Burro Mountains; Mogollon Creek. Dry hills, in the Upper Sonoran Zone.

3. **Bouteloua procumbens** (Durand) Griffiths, Contr. U. S. Nat. Herb. **14:** 364. 1912. SIX-WEEKS GRAMA.

Chloris procumbens Durand, Chlor. Sp. 1808.

Bouteloua prostrata Lag. Gen. & Sp. Nov. 5. 1816.

Bouteloua pusilla Vasey, Bull. Torrey Club **11:** 6. 1884.

TYPE LOCALITY: Not ascertained.

RANGE: Colorado and Arizona to Mexico.

NEW MEXICO: Cedar Hill; Tierra Amarilla; Chama; Santa Fe; El Rito Creek; Ensenada; Las Vegas; Grants; Pecos; Roy; Kingston; West Fork of the Gila; White Mountains. Sandy soil, in the Upper Sonoran Zone.

The type of *B. pusilla* was collected at Kingman by G. R. Vasey, in 1881.

4. **Bouteloua aristidoides** (H. B. K.) Griseb. Fl. Brit. W. Ind. 537. 1864. SIX-WEEKS GRAMA.

Dinebra aristidoides H. B. K. Nov. Gen. & Sp. **1:** 171. 1816.

Triathera aristidoides Nash in Small, Fl. Southeast. U. S. 137. 1903.

TYPE LOCALITY: "Crescit in asperis frigidis convallis Tolucensis, alt. 1320 hexap."

RANGE: California and western Texas to Mexico and South America.

NEW MEXICO: Bear Mountain; Deming; Socorro; Dog Spring; Organ Mountains; Las Cruces. Dry plains and hills, in the Lower Sonoran Zone.

5. **Bouteloua parryi** (Fourn.) Griffiths, Contr. U. S. Nat. Herb. **14:** 381. 1912. SANDHILL GRAMA.

Chondrosium parryi Fourn. Mex. Pl. **2:** 150. 1881.

Bouteloua polystachya vestita S. Wats. Proc. Amer. Acad. **18:** 177. 1883.

Bouteloua vestita Scribn. Contr. U. S. Nat. Herb. **2:** 531. 1894.

TYPE LOCALITY: Near San Luis Potosí, Mexico.

RANGE: Western Texas to southern Arizona and Mexico.

NEW MEXICO: Mesilla Valley; mesa west of Organ Mountains; Jarilla Junction. Mesas, in the Lower Sonoran Zone.

6. **Bouteloua barbata** Lag. Var. Cienc. **2**[4]: 141. 1805. SIX-WEEKS GRAMA.

Chondrosium polystachyum Benth. Bot. Voy. Sulph. 56. 1844.

Bouteloua polystachya Torr. U. S. Rep. Expl. Miss. Pacif. **5**[2]: 366. 1857.

TYPE LOCALITY: Described from cultivated plants.

RANGE: California and Utah to Mexico.

NEW MEXICO: Carrizo Mountains; San Juan Valley; Chama River; Albuquerque; Socorro; Mangas Springs; Santa Rita; Deming; Black Range; Dog Spring; Mesilla Valley; Organ Mountains; White Sands; Pecos Valley. Sandy fields and mesas, in the Lower and Upper Sonoran zones.

7. **Bouteloua eriopoda** Torr. U. S. Rep. Expl. Miss. Pacif. **4**: 155. 1856.
BLACK GRAMA.

Chondrosium eriopodum Torr. in Emory, Mil. Reconn. 154. 1848.

TYPE LOCALITY: Along the Rio Grande, New Mexico. Type collected by Emory.

RANGE: Arizona and western Texas to Mexico.

NEW MEXICO: Common throughout the State except in the extreme northeast. Dry hills and plains, in the Lower and Upper Sonoran zones.

This is one of the most valuable range grasses in the southern part of New Mexico.

8. **Bouteloua breviseta** Vasey, Contr. U. S. Nat. Herb. **1**: 58. 1890.

TYPE LOCALITY: Screw Bean, Presidio County, Texas.

RANGE: Western Texas and southern New Mexico.

NEW MEXICO: White Sands; Lakewood; Carlsbad. Dry plains, in the Lower Sonoran Zone.

9. **Bouteloua hirsuta** Lag. Var. Cienc. **2**[4]: 141. 1805. HAIRY GRAMA.

Chondrosium hirtum H. B. K. Nov. Gen. & Sp. **1**: 176. *pl. 59.* 1816.

Chondrosium foeneum Torr. in Emory, Mil. Reconn. 154. *pl. 12.* 1848.

TYPE LOCALITY: Described from cultivated plants.

RANGE: Colorado and Nebraska to Mexico and Florida.

NEW MEXICO: Pecos; Clayton; Nara Visa; Silver City; Socorro; Torrance; Organ Mountains; Dona Ana Mountains; Leachs; Buchanan. Dry hills, in the Lower and Upper Sonoran zones.

The type of *Chondrosium foeneum* was collected by Emory along the Rio Grande.

10. **Bouteloua gracilis** (H. B. K.) Lag.; Steud. Nom. Bot. ed. 2. **1**: 219. 1840.
BLUE GRAMA.

Chondrosium gracile H. B. K. Nov. Gen. & Sp. **1**: 176. *pl. 58.* 1816.

Atheropogon oligostachyum Nutt. Gen. Pl. **1**: 178. 1818.

Bouteloua oligostachya Torr.; A. Gray, Man. ed. 2. 553. 1853.

TYPE LOCALITY: "Crescit in crepidinibus et devexis montis porphyritici La Buffa de Guanaxuato Mexicanorum, alt. 1270 hexap."

RANGE: British America to California, Missouri, and Mexico.

NEW MEXICO: Common throughout the State except at lower levels. Meadows and hillsides, in the Upper Sonoran and Transition zones.

Blue grama is undoubtedly the most valuable forage plant of New Mexico. It occurs generally on the higher plains and lower mountains at altitudes of from 1,800 to 2,400 meters, often forming nearly pure stands. When not molested it becomes knee-high, and a large field in such a condition is a beautiful sight because of the myriads of purple spikes. The grass is resistant to overgrazing and is able to spread rapidly when not too closely grazed.

This has generally been known as *Bouteloua oligostachya*, a name which, unfortunately, lacks priority.

49. CHLORIS Swartz.

Mostly perennials with flat leaves and rather showy inflorescence of 2 to many digitate spikes; spikelets 1-flowered, awned, sessile in 2 rows along one side of a continuous rachis, forming unilateral spikes; rachilla jointed above the glumes and produced beyond the palea, bearing 1 or more empty rudimentary awned glumes; glumes unequal, lanceolate, acute, somewhat keeled; lemma narrow or broad, 1 to 3-nerved, acute to truncate or emarginate or 2-lobed, often ciliate on the back or margins, the middle nerve usually prolonged into an awn; grain free.

KEY TO THE SPECIES.

Spikes slender, naked or interrupted at the base; panicle of more than a single verticel of spikes.............................. 1. *C. verticillata.*

Spikes stout, spikelet-bearing to the base; panicle of a single terminal verticel of spikes.

Lemma conspicuously hairy, long-villous on the nerves and margins.. 2. *C. elegans.*

Lemma not conspicuously hairy, the pubescence very short or none.

Lemma 3-nerved, obovate-cuneate, the apex rounded..... 3. *C. brevispica.*

Lemma 7-nerved, broadly triangular, very shortly awned.. 4. *C. cucullata.*

1. **Chloris verticillata** Nutt. Trans. Amer. Phil. Soc. n. ser. **5**: 150. 1837.

TYPE LOCALITY: "On the sandy banks of the Arkansas, near Fort Smith."

RANGE: Kansas and Texas to New Mexico.

NEW MEXICO: Pecos; Clayton; Redlands; Socorro; Nambe Valley. Dry plains and hills, in the Upper Sonoran Zone.

2. **Chloris elegans** H. B. K. Nov. Gen. & Sp. **1**: 165. 1816.

TYPE LOCALITY: "Inter Mexico et Queretaro."

RANGE: Texas and Arizona to Mexico.

NEW MEXICO: Common from Silver City, Socorro, and Roswell southward. Plains and river valleys, in the Lower and Upper Sonoran zones.

Often a troublesome weed in alfalfa fields.

3. **Chloris brevispica** Nash, Bull. Torrey Club **25**: 438. 1898.

TYPE LOCALITY: Nueces County, Texas.

RANGE: Western Texas to southeastern New Mexico.

NEW MEXICO: Roswell (*Griffiths* 5710, 5746). Sandy soil.

4. **Chloris cucullata** Bisch. Ann. Sci. Nat. III. Bot. **19**: 357. 1853.

CROWFOOT GRAMA.

TYPE LOCALITY: "Hab. Mexico boreali. Semina in provincia *Tamaulipas* prope *Matamoros* lecta absque nomine misit Dr. Engelmann, 1849."

RANGE: Western Texas and southern New Mexico to Mexico.

NEW MEXICO: Roswell; Carlsbad. Plains.

50. TRICHLORIS Fourn.

Erect perennials with flat leaves and with many slender spikes digitate or closely approximate at the apex of the culm, forming dense oblong panicles; spikelets 1 to 3-flowered, hermaphrodite, sessile in 2 series along the continuous rachis of the unilateral spikes; rachilla articulated above the glumes and prolonged above the hermaphrodite flowers, terminating in an awned rudimentary floret; glumes unequal, 1-nerved, membranaceous, the second short-awned; lemmas membranaceous, 3-nerved, 3-awned, the awns erect, subequal or the lateral ones much shorter; palea hyaline, 2-keeled; stamens 3; styles distinct; grain narrow, oblong, free.

1. **Trichloris fasciculata** Fourn.; Benth. Journ. Linn. Soc. Bot. **19**: 102. 1881.

TYPE LOCALITY: Not stated.

RANGE: Western Texas to New Mexico and Mexico.

NEW MEXICO: Mesilla Valley. Sandy mesas, in the Lower Sonoran Zone.

51. PAPPOPHORUM Schreb.

Cespitose perennial with narrow, usually convolute leaves and spikelike panicles; spikelets 1 or 2-flowered; rachilla jointed above the glumes, these persistent, membranaceous, acute, carinate, nerveless, or with 1 to 3 nerves on each side; lemmas broad at the base, subcoriaceous, obscurely many-nerved, unequally divided into 9 to 23 awnlike lobes; palea rather broad, 2-keeled near the margins.

1. **Pappophorum wrightii** S. Wats. Proc. Amer. Acad. **18:** 178. 1883.

TYPE LOCALITY: Western Texas or southern New Mexico.

RANGE: Western Texas and southern Arizona and southward.

NEW MEXICO: Bear Mountain; Kingston; Cerrillos; Socorro; Dog Spring; Las Cruces; Organ Mountains; Carrizozo. Dry hills, in the Lower and Upper Sonoran zones.

52. SCLEROPOGON Phil. FALSE NEEDLE GRASS.

Perennial, cespitose, often stoloniferous grasses with nearly simple panicles; spikelets unisexual, the two kinds unlike, 2 to many-flowered; staminate spikelets many-flowered, the glumes narrow, acute, 3-nerved, awnless, unequal, the lemmas sometimes minutely 3-toothed at the apex, the palea narrow and rigid; stamens 3; pistillate spikelets 1 to many-flowered, the glumes persistent, very unequal, the lemmas rigid, narrow, the 3 nerves produced into very long slender divergent twisted awns; styles distinct, elongated; grain free, narrow, elongated.

1. **Scleropogon brevifolius** Phil. Anal. Univ. Chile **34:** 205. 1870.

TYPE LOCALITY: Chile.

RANGE: Arizona and western Texas to Mexico and South America.

NEW MEXICO: Carrizo Mountains; Hillsboro; Albuquerque; Socorro; Tucumcari; Dona Ana Mountains; Organ Mountains; Deming; Carrizozo; White Mountains; Pecos Valley. Dry hills and plains, in the Lower and Upper Sonoran zones.

Very common on the mesas of southern New Mexico and of considerable importance as a range grass.

53. PHRAGMITES Trin. CARRIZO.

Tall reedlike perennial with stout leafy culms and large terminal panicles; spikelets loosely 3 to 7-flowered; rachilla jointed above the glumes and between the florets, clothed with long silky hairs; lowest floret staminate or neutral, the others fertile; glumes unequal, lanceolate, acute, shorter than the florets; lemmas glabrous, very narrow, acuminate; grain free.

1. **Phragmites phragmites** (L.) Karst. Deutsch. Fl. 379. 1880–3.

Arundo phragmites L. Sp. Pl. 81. 1753.

Phragmites communis Trin. Fund. Agrost. 134. 1820.

TYPE LOCALITY: "Habitat in Europae lacubus fluviis."

RANGE: Nearly throughout the United States and in Mexico; also in Europe and Asia.

NEW MEXICO: Farmington; Cimarron; Canada Alamosa; Copper Mines; Mimbres River; Mesilla Valley; Round Mountain; Roswell. In wet ground, especially in river valleys, in the Lower and Upper Sonoran zones.

54. ARUNDO L.

Tall (2 to 3 meters or more) perennial with flat leaves and ample terminal panicles; spikelets 2 to many-flowered; rachilla jointed above the glumes and between the florets, smooth; florets crowded, fertile, or the upper or lower staminate; glumes 2, narrow, subequal, 3-nerved, smooth, acute or acuminate, about the length of the spikelet; lemmas membranaceous, 3-nerved, 2-toothed at the apex, mucronate between the teeth, long-pilose on the back; palea hyaline, 2-keeled; grain smooth, free.

1. **Arundo donax** L. Sp. Pl. 81. 1753. GIANT REED.

TYPE LOCALITY: "Habitat in Hispania, Galloprovincia."

RANGE: Western Texas and southern New Mexico to Mexico, probably naturalized; also in the Old World.

New Mexico: Mesilla Valley. Wet ground.

This grass, the largest of all those found in New Mexico, is frequent along ditches in the Rio Grande Valley, where it has probably been introduced.

55. MUNROA Torr.

Low, diffusely much branched annual with short sharp-pointed leaves clustered at the ends of the branches; spikelets 2 to 4-flowered, 3 to 5 together and nearly sessile in the axis of the floral leaves; rachilla jointed above the glumes; glumes lanceolate, acute, hyaline, 1-nerved; lemmas longer, 3-nerved, entire, retuse, or 3-cleft, the midnerve or all the nerves excurrent as short mucronate points; palea hyaline, 2-keeled; stamens 3; styles distinct, elongated; grain inclosed within the lemma, free.

1. **Munroa squarrosa** (Nutt.) Torr. U. S. Rep. Expl. Miss. Pacif. **4:** 158. 1856.

Crypsis squarrosa Nutt. Gen. Pl. **1:** 49. 1818.

Type locality: "On arid plains near the 'Grand Detour' of the Missouri, almost exclusively covering thousands of acres."

Range: Alberta and South Dakota to Arizona and Texas.

New Mexico: Common throughout the State. Dry plains and low hills, in the Lower and Upper Sonoran zones.

56. DASYOCHLOA Willd.

Low, densely tufted, often creeping perennial, with very narrow, somewhat rigid leaves and crowded spikelets in clusters of 3 to 6, equaled or exceeded by the upper leaves; spikelets several-flowered, sessile; glumes unequal, keeled; lemmas thin, densely hairy below, deeply bilobate, awned from between the rounded lobes; stamens 3.

1. **Dasyochloa pulchella** (H. B. K.) Willd.; Steud. Nom. Bot. ed. 2. **1:** 484. 1840.

Triodia pulchella H. B. K. Nov. Gen. & Sp. **1:** 155. *pl. 47.* 1816.

Type locality: "In subfrigidis, siccis, apricis regni Mexicani inter Guanaxuato, Mina de Belgrado et Cubilente, alt. 1050 hexap."

Range: Western Texas to Arizona, south into Mexico.

New Mexico: Shiprock; Carrizo Mountains; Albuquerque; Mangas Springs; Socorro; Tortugas Mountain; Mesilla Valley; Orogrande; Roswell. Sandy mesas, in the Lower and Upper Sonoran zones.

57. ERIONEURON Nash.

Tufted perennials with thick linear leaves having white margins, and dense, contracted, almost capitate panicles; spikelets several to many-flowered; glumes narrow, acuminate; lemmas broad, 3-nerved, pubescent on the nerves below and sometimes on the body of the lemma at the base, the apex acuminate, entire or slightly 2-toothed, the awn terminal or arising between the minute teeth; stamens 3; style short, distinct.

1. **Erioneuron pilosum** (Buckl.) Nash in Small, Fl. Southeast. U. S. 144. 1903.

Uralepis pilosa Buckl. Proc. Acad. Phila. **1862:** 94. 1863.

Sieglingia pilosa Nash in Britt. & Brown, Illustr. Fl. **3:** 504. 1898.

Type locality: "Middle Texas."

Range: Kansas and Colorado to New Mexico and Texas.

New Mexico: Farmington; Pecos; Knowles; Torrance; Buchanan; Las Vegas Hot Springs; Cross L Ranch; Mangas Springs; Dayton; Gray; Guadalupe Mountains; Roswell. Dry hills and plains, in the Upper Sonoran Zone.

58. TRIDENS Roem. & Schult.

Perennials with flat or involute leaves and open or contracted, sometimes spikelike inflorescence; spikelets 3 to many-flowered, the flowers perfect or the upper staminate; glumes keeled, obtuse to acuminate, usually shorter than the lemma; lemmas 3-nerved, the midnerve or all the nerves excurrent, pilose, the apex entire or shortly 2-toothed; palea compressed, 2-keeled; stamens 3; styles short, distinct.

KEY TO THE SPECIES.

Lemmas not pilose on the back; glumes considerably surpassing the lower florets; spikelets subcompressed, oblong, in a rather dense spikelike panicle .... 1. *T. albescens.*
Lemmas pilose on the back, at least at the base; glumes barely as long as the lowest florets, or shorter; spikelets various.
Spikelets terete; sterile lemma not ciliate, neither lobed nor awned .... 2. *T. muticus.*
Spikelets compressed; sterile lemma ciliate, deeply bilobate, with an intermediate awn.
Lemmas only slightly cleft at the apex, the lobes narrow, acute; spikelets 10 mm. long .... 3. *T. avenaceus.*
Lemmas cleft half their length, the lobes obtuse; spikelets 6 to 8 mm. long .... 4. *T. nealleyi.*

1. **Tridens albescens** (Vasey) Woot. & Standl. N. Mex. Agr. Exp. Sta. Bull. **81:** 129. 1912.

Triodia albescens Vasey, U. S. Dept. Agr. Div. Agrost. Bull. **12**[2]**:** 33. 1891.
Sieglingia albescens Kuntze; L. H. Dewey, Contr. U. S. Nat. Herb. **2:** 538. 1894.
Rhombolytrum albescens Nash in Britton, Man. 129. 1901.
TYPE LOCALITY: "Texas and New Mexico."
RANGE: Kansas to Texas and eastern New Mexico.
NEW MEXICO: Mesilla Valley; Carlsbad; Roswell. Dry plains, in the Lower and Upper Sonoran zones.

2. **Tridens muticus** (Torr.) Nash in Small, Fl. Southeast. U. S. 143. 1903.

Tricuspis mutica Torr. U. S. Rep. Expl. Miss. Pacif. **4:** 156. 1856.
Triodia mutica Benth.; S. Wats. Proc. Amer. Acad. **18:** 180. 1883.
TYPE LOCALITY: Laguna Colorado, New Mexico.
RANGE: Western Texas and eastern New Mexico.
NEW MEXICO: Socorro; Laguna Colorado; Cross L Ranch; Tortugas Mountain; Roswell. Dry hills, in the Lower and Upper Sonoran zones.

3. **Tridens avenaceus** (H. B. K.) Hitchc. Contr. U. S. Nat. Herb. **17:** 357. 1913.

Triodia avenacea H. B. K. Nov. Gen. & Sp. **1:** 156. *pl. 48.* 1816.
Triodia grandiflora Vasey, Contr. U. S. Nat. Herb. **1:** 59. 1890.
Sieglingia avenacea grandiflora L. H. Dewey, Contr. U. S. Nat. Herb. **2:** 538. 1894.
Tridens grandiflorus Woot. & Standl. N. Mex. Agr. Exp. Sta. Bull. **81:** 129. 1912.
TYPE LOCALITY: "In convalle Mexicana inter montem Chapultepec et Penol de los Banos."
RANGE: Western Texas to Arizona and southward.
NEW MEXICO: Kingston (*Metcalfe* 1334).

4. **Tridens nealleyi** (Vasey) Woot. & Standl. N. Mex. Agr. Exp. Sta. Bull. **81:** 129. 1912.

Triodia nealleyi Vasey, Bull. Torrey Club **15:** 49. 1888.
Sieglingia nealleyi L. H. Dewey, Contr. U. S. Nat. Herb. **2:** 538. 1894.
TYPE LOCALITY: Western Texas.

RANGE: Western Texas and southern New Mexico.

NEW MEXICO: Tortugas Mountain (*Wooton* 2018). Dry fields, in the Lower Sonoran zone.

59. SPHENOPHOLIS Scribn.

Rather slender tufted perennials with flat leaves and narrow terminal many-flowered panicles; spikelets 2 or 3-flowered; rachilla jointed above the glumes and between the florets and produced above the upper floret into a slender naked bristle; glumes slightly shorter than the florets, the first very narrow or linear and 1-nerved, the second broadly obovate, 3-nerved, with rather broad scarious margins; lemmas obtuse, usually awnless; palea narrow, 2-nerved; styles distinct, short; grain narrow, free.

KEY TO THE SPECIES.

Second glume not much, if at all, wider than the lemmas, obtuse or acute .... 1. *S. pallens.*
Second glume much wider than the lemmas, rounded or truncate and somewhat cucullate at the apex.
  Intermediate nerves of the second glume faint; leaves narrow; panicle very narrow, dense and spikelike .... 2. *S. obtusata.*
  Intermediate nerves of the second glume almost as prominent as the principal ones; leaves wide; panicle longer and broader, loose .... 3. *S. robusta.*

1. **Sphenopholis pallens** (Spreng.) Scribn. Rhodora **8:** 145. 1906.
Aira pallens Spreng. Mant. Fl. Hal. 36. 1807.
Koeleria pennsylvanica DC. Cat. Hort. Monsp. 117. 1813.
Eatonia pennsylvanica A. Gray, Man. ed. 2. 558. 1856.
TYPE LOCALITY: Not stated.
RANGE: British America to New Mexico, Texas, and Georgia.
NEW MEXICO: Mouth of Mora River; Albuquerque; Santa Fe Canyon 9 miles east of Santa Fe. Wet ground, in the Transition Zone.

2. **Sphenopholis obtusata** (Michx.) Scribn. Rhodora **8:** 144. 1906.
Aira obtusata Michx. Fl. Bor. Amer. **1:** 62. 1803.
Eatonia obtusata A. Gray, Man. ed. 2. 558. 1856.
TYPE LOCALITY: "Habitat in aridis a Carolina ad Floridam."
RANGE: British America to Oregon, Arizona, and Florida.
NEW MEXICO: Farmington; Carrizo Mountains; Albuquerque; Zuni; Socorro; Kingston; Organ Mountains. In wet ground, especially along ditch banks, in the Upper Sonoran Zone.

3. **Sphenopholis robusta** (Vasey) Heller, Muhlenbergia **6:** 12. 1910.
Eatonia obtusata robusta Vasey, Contr. U. S. Nat. Herb. **3:** 190. 1895.
Eatonia robusta Rydb. Bull. Torrey Club **32:** 602. 1905.
TYPE LOCALITY: Mullen, Nebraska.
RANGE: Washington and Nebraska to Arizona.
NEW MEXICO: Rio Mimbres; Mesilla. Damp meadows, in the Lower and Upper Sonoran zones.

60. KOELERIA Pers. JUNE GRASS.

Cespitose perennials with usually flat narrow leaves and densely flowered spikelike panicles; spikelets 2 to 4-flowered, compressed; rachilla jointed above the glumes, these unequal, keeled, somewhat shorter than the lemmas; lemmas membranaceous, faintly 3 to 5-nerved, obtuse, acute, or mucronate-pointed; palea hyaline, acute, 2-nerved, about as long as the lemma; stamens 3; styles very short.

1. **Koeleria cristata** (L.) Pers. Syn. Pl. **1**: 97. 1805.

Aira cristata L. Sp. Pl. 63. 1753.

TYPE LOCALITY: "Habitat in Angliae, Helvetiae siccioribus."

RANGE: British America to Arizona, Kansas, and Pennsylvania; also in Europe.

NEW MEXICO: Common in all the mountain ranges. Open slopes and in woods, in the Transition Zone.

61. ERAGROSTIS Beauv.

Annuals or perennials with simple or branched culms; spikelets 2 to many-flowered, the uppermost floret imperfect; rachilla jointed but sometimes not disarticulating until after the fall of the lemmas; glumes more or less unequal, usually shorter than the lemma; lemmas glabrous, obtuse or acute, awnless, 3-nerved, the lateral nerves often faint; paleas shorter than the lemmas, often persistent, 2-nerved.

KEY TO THE SPECIES.

Annuals.
- Spikelets broad, more than 2 mm. wide .... 1. *E. megastachya.*
- Spikelets narrow, 1.5 mm. wide or less.
  - Plants low, spreading, 30 cm. high or less; leaves narrow; spikelets many-flowered; plants of cultivated fields and river valleys .... 2. *E. pilosa.*
  - Plants tall, 30 to 100 cm.; leaves broad; spikelets few-flowered; plants usually found in the mountains.
    - Panicles spreading, often nearly 30 cm. long; spikelets 5 to 8 mm. long .... 4. *E. mexicana.*
    - Panicles contracted, 10 cm. long or less; spikelets 3 to 6 mm. long .... 3. *E. limbata.*

Perennials.
- Plants with rigid scaly rootstocks; leaves pungent-pointed .... 5. *E. obtusiflora.*
- Plants tufted, without rootstocks; leaves not pungent-pointed.
  - Spikelets crowded, on very short branches .... 6. *E. secundiflora.*
  - Spikelets not crowded, on long slender branches.
    - Panicles narrow and elongated, the branches long and flexuous, erect or nearly so; lateral nerves of the lemmas evident .... 7. *E. trichodes.*
    - Panicles rather open, the branches spreading or ascending, rather rigid; lateral nerves of the lemmas evident or obscure.
      - Lateral nerves of the lemmas faint; plant of the mountains .... 8. *E. lugens.*
      - Lateral nerves of the lemmas very prominent; on the plains of the eastern part of the State .... 9. *E. pectinacea.*

1. **Eragrostis megastachya** (Koel.) Link, Hort. Berol. **1**: 187. 1827.

STINK GRASS.

Briza eragrostis L. Sp. Pl. 70. 1753, not *Poa eragrostis* L. op. cit. 68 (=*Eragrostis eragrostis*).

Poa megastachya Koel. Descr. Gram. 181. 1802.

Eragrostis major Host, Icon. Gram. Austr. **4**: 14. *pl. 24.* 1809.

Eragrostis poaeoides megastachya A. Gray, Man. ed. 5. 631. 1867.

TYPE LOCALITY: European.

RANGE: Nearly throughout the United States; naturalized from Europe.

NEW MEXICO: Cedar Hill; Santa Fe; Pecos; Las Vegas Hot Springs; Mangas Springs; Dog Spring; Berendo Creek; West Fork of the Gila; Organ Mountains; Tularosa; Roswell; Mesilla Valley; Carlsbad; Texico. Waste ground.

2. Eragrostis pilosa (L.) Beauv. Ess. Agrost. 162. 1812.

Poa pilosa L. Sp. Pl. 68. 1753.

TYPE LOCALITY: "Habitat in Italia."

RANGE: Nearly throughout the United States; also in the Old World.

NEW MEXICO: Farmington; Carrizo Mountains; Santa Fe; Las Vegas; Albuquerque; Fort Bayard; near White Water; Mogollon Mountains; Mesilla Valley; Roswell; Gilmores Ranch; Tularosa; Texline. Waste places and in meadows, in the Lower and Upper Sonoran zones.

3. Eragrostis limbata Fourn. Mex. Pl. **2**: 116. 1886.

Eragrostis neomexicana Vasey, Contr. U. S. Nat. Herb. **2**: 542. 1894.

TYPE LOCALITY: Mexico.

RANGE: Western Texas to California, and southward.

NEW MEXICO: Organ Mountains. Dry hills, in the Upper Sonoran Zone.

4. Eragrostis mexicana (Lag.) Link, Hort. Berol. **1**: 190. 1827.

Poa mexicana Lag. Gen. & Sp. Nov. 3. 1816.

TYPE LOCALITY: "Hab. in Imperio Mexicana."

RANGE: Western Texas to southern California and southward.

NEW MEXICO: Gallinas Mountains; Raton; Las Vegas; Torrance; Albuquerque; Deming; Dog Spring; Mogollon Mountains; Animas Creek; Organ Mountains; Mesilla Valley; Leachs; Queen; Roswell; White Mountains. Dry hills and canyons, in the Upper Sonoran Zone.

5. Eragrostis obtusiflora (Fourn.) Scribn. U. S. Dept. Agr. Div. Agrost. Bull. **8**: 10. 1897. MEXICAN SALTGRASS.

Brizopyrum obtusiflorum Fourn. Mex. Pl. **2**: 120. 1881.

TYPE LOCALITY: Mexico.

RANGE: Arizona and New Mexico to Mexico.

NEW MEXICO: Las Playas (*Wooton*). Alkaline flats, in the Lower Sonoran Zone.

6. Eragrostis secundiflora Presl, Rel. Haenk. **1**: 276. 1830.

Poa interrupta Nutt. Trans. Amer. Phil. Soc. n. ser. **5**: 146. 1837, not Lam. 1791.

Poa oxylepis Torr. in Marcy, Expl. Red Riv. 301. *pl. 19*. 1854.

TYPE LOCALITY: Mexico.

RANGE: Texas and New Mexico to southern Mexico.

NEW MEXICO: Nara Visa; Melrose; Dora; Clayton; Arroyo Ranch; Texline. Plains, in the Upper Sonoran Zone.

7. Eragrostis trichodes (Nutt.) Nash, Bull. Torrey Club **22**: 465. 1895.

Poa trichodes Nutt. Trans. Amer. Phil. Soc. n. ser. **5**: 146. 1837.

TYPE LOCALITY: "In bushy prairies and open alluvial lands, Arkansas."

RANGE: Nebraska and Ohio to New Mexico and Tennessee.

NEW MEXICO: Gray; Queen. Dry soil, in the Upper Sonoran Zone.

8. Eragrostis lugens Nees, Agrost. Bras. **2**: 505. 1829.

TYPE LOCALITY: "Habitat ad Monte-Video et in confinibus Paraguayanis."

RANGE: Texas and Arizona to Mexico and South America.

NEW MEXICO: Kingston; Mangas Springs; near White Water; San Luis Mountains; Dona Ana Mountains; Organ Mountains; Round Mountain. Dry hills, in the Upper Sonoran Zone.

9. Eragrostis pectinacea (Michx.) Steud. Syn. Pl. Glum. **1**: 272. 1855.

Poa pectinacea Michx. Fl. Bor. Amer. **1**: 69. 1803.

TYPE LOCALITY: "Hab. in arvis Illinoensibus."

RANGE: Illinois and Massachusetts to New Mexico, Texas, and Florida.

NEW MEXICO: Near Causey (*Wooton*). Plains and dry fields, in the Upper Sonoran Zone.

62. **MELICA** L. Melic grass.

Perennials with usually soft flat leaves and with rather large spikelets in lax or dense, usually narrow panicles, or sometimes in simple racemes; spikelets 2 to several-flowered, terete or slightly flattened; rachilla jointed above the glumes and between the fertile florets, usually bearing 2 or 3 empty glumes at the apex; glumes unequal, membranaceous, awnless, 3 to 5-nerved; lemmas larger, rounded on the back, 7 to 13-nerved, scarious-margined, awnless or short-awned; palea broad, 2-keeled; stamens 3; styles distinct; grain free.

1. **Melica porteri** Scribn. Proc. Acad. Phila. **1885:** 44. *pl. 1. f. 17, 18.* 1885.
Melica mutica parviflora Porter in Port. & Coult. Syn. Fl. Colo. 149. 1874.
Melica parviflora Scribn. Mem. Torrey Club **5:** 50. 1894.
Type locality: Glen Eyrie, Colorado.
Range: Colorado and Kansas to Arizona and Texas.
New Mexico: Chama; Santa Fe; Sandia Mountains; Las Vegas; Winsors Ranch; Mogollon Mountains; Silver City; Organ Mountains; White and Sacramento mountains. Damp woods, in the Transition and Canadian zones.

63. **BROMUS** L. Brome grass.

Annuals or perennials with flat leaves and rather large, erect or pendulous spikelets; spikelets few to many-flowered, slightly or strongly flattened laterally, paniculate or rarely racemose; rachilla jointed above the glumes and between the florets; glumes unequal, acute or the second short-awned, 1 to 5-nerved, shorter than the lemmas; lemmas keeled or rounded on the back, 5 to 9-nerved, usually 2-toothed at the apex and awned from the back just below the point or from between the teeth, or sometimes awnless; palea a little shorter than the lemma, 2-keeled; stamens usually 3; stigmas sessile, plumose; grain sulcate, adherent to the palea.

KEY TO THE SPECIES.

Lemmas compressed-carinate at the base.
  Lemmas appressed-villous.
    Sheaths more or less villous .... 1a. *B. marginatus latior.*
    Sheaths glabrous or nearly so .... 1b. *B. marginatus seminudus.*
  Lemmas smooth or scabrous.
    Leaves and sheaths conspicuously pubescent .... 2. *B. unioloides.*
    Leaves glabrous, the sheaths sometimes slightly pubescent.
      Awns inconspicuous; leaves narrow .... 3. *B. polyanthus.*
      Awns conspicuous; leaves wide .... 3a. *B. polyanthus paniculatus.*
Lemmas not compressed-carinate but rounded, at least at the base.
  Lemmas glabrous or scabrous.
    Sheaths pubescent.
      Panicles dense, contracted; plants low, 40 cm. high or less .... 4. *B. hordeaceus glabrescens.*
      Panicles loose, more or less spreading; plants more than 50 cm. high .... 5. *B. racemosus.*

Sheaths glabrous.
Spikelets laterally compressed, ovate-lanceolate... 6. *B. secalinus.*
Spikelets terete, narrowly oblong.................. 7. *B. inermis.*
Lemmas more or less pubescent.
Pubescence unevenly distributed on the margins and dorsal surface of the lemmas..................... 8. *B. richardsoni.*
Pubescence about equally distributed on margins and dorsal surface of the lemmas.
Sheaths densely villous.......................... 9. *B. lanatipes.*
Sheaths glabrous or nearly so.
Glumes pubescent; tall coarse plant.........10. *B. porteri.*
Glumes glabrous; weak leafy plant...........11. *B. frondosus.*

1a. **Bromus marginatus latior** Shear, U. S. Dept. Agr. Div. Agrost. Bull. **23:** 55. 1900.

TYPE LOCALITY: Walla Walla, Washington.

RANGE: Washington and Wyoming to Arizona and New Mexico.

NEW MEXICO: North of Ramah; Santa Fe; East Fork of the Gila; Mangas Springs. Canyons, in the Upper Sonoran and Transition zones.

1b. **Bromus marginatus seminudus** Shear, U. S. Dept. Agr. Div. Agrost. Bull. **23:** 55. 1900.

TYPE LOCALITY: On open mountain side 5 miles above Wallowa Lake, Oregon.

RANGE: Washington and Montana to California and New Mexico.

NEW MEXICO: Baldy; Sandia Mountains; Water Canyon; James Canyon; White Mountains. Meadows, in the Transition Zone.

2. **Bromus unioloides** (Willd.) H. B. K. Nov. Gen. & Sp. **1:** 151. 1816.

Festuca unioloides Willd. Hort. Berol. **1:** 3. *pl. 3.* 1816.

TYPE LOCALITY: "Habitat in Carolina."

RANGE: South Carolina and Florida to Texas, also in Mexico and South America; introduced in other parts of the United States.

NEW MEXICO: Mangas Springs; Agricultural College.

3. **Bromus polyanthus** Shear, U. S. Dept. Agr. Div. Agrost. Bull. **23:** 56. *f. 34.* 1900.

Bromus multiflorus Scribn. U. S. Dept. Agr. Div. Agrost. Bull. **13:** 46. 1898, not Weig. 1772.

TYPE LOCALITY: Battle Lake, Sierra Madre Mountains, Wyoming.

RANGE: Oregon and Montana to Utah and New Mexico.

NEW MEXICO: Harveys Upper Ranch; Santa Fe; Johnsons Mesa; Rio Pueblo; Barranca; Chama; Tunitcha Mountains; Las Vegas; Silver City; Patterson; James Canyon; Organ Mountains. Shaded canyons, in the Transition Zone.

3a. **Bromus polyanthus paniculatus** Shear, U. S. Dept. Agr. Div. Agrost. Bull. **23:** 56. *f. 35.* 1900.

TYPE LOCALITY: West Mancos Canyon, Colorado.

RANGE: Utah and Colorado to Arizona and New Mexico.

NEW MEXICO: Pecos; Glorieta; Baldy; Inscription Rock; Chiz; Mogollon Mountains; Gilmores Ranch; James Canyon; Gray. Damp woods and thickets, in the Transition Zone.

4. **Bromus hordeaceus glabrescens** (Coss.) Shear, U. S. Dept. Agr. Div. Agrost. Bull. **23:** 20. 1900.

Bromus mollis glabrescens Coss. Fl. Env. Paris 654. 1845.

TYPE LOCALITY: Paris, France.

RANGE: Native of Europe, widely introduced in the United States.

NEW MEXICO: Willow Creek (*Wooton*).

5. **Bromus racemosus** L. Sp. Pl. ed. 2. 114. 1762.

TYPE LOCALITY: "Habitat in Anglia."

RANGE: Native of Europe, adventive in many places in the United States.

NEW MEXICO: Albuquerque; Mesilla Valley.

6. **Bromus secalinus** L. Sp. Pl. 76. 1753. CHEAT.

TYPE LOCALITY: "Habitat in Europae agris secalinis arenosis."

RANGE: Native of Europe, a common weed in many parts of North America, especially in grain fields.

NEW MEXICO: Mangas Springs.

7. **Bromus inermis** Leyss. Fl. Hal. 16. 1761. HUNGARIAN BROME GRASS.

TYPE LOCALITY: "Habitat in pratis succulentis fertilissimis *im Fürstengarten in den Pulverweiden* frequens."

RANGE: Native of Europe, locally established in the United States.

NEW MEXICO: Farmington; Mesilla Valley.

8. **Bromus richardsoni** Link, Hort. Berol. **2:** 281. 1833.

Bromus ciliatus scariosus Scribn. U. S. Dept. Agr. Div. Agrost. Bull. **13:** 46. 1898.

TYPE LOCALITY: Described from plants grown from seed sent from western North America.

RANGE: British America to Arizona and New Mexico.

NEW MEXICO: Sandia Mountains; Rio Pueblo; Trout Spring; Beulah; Tunitcha Mountains; Hillsboro Peak; Organ Mountains; Cloudcroft; White Mountains. Thickets in the mountains, in the Transition and Canadian zones.

9. **Bromus lanatipes** (Shear) Rydb. Colo. Agr. Exp. Sta. Bull. **100:** 52. 1906.

Bromus porteri lanatipes Shear, U. S. Dept. Agr. Div. Agrost. Bull. **23:** 37. 1900.

TYPE LOCALITY: Idaho Springs, Colorado.

RANGE: Colorado and New Mexico to California.

NEW MEXICO: Carrizo Mountains; Sandia Mountains; Glorieta; Johnsons Mesa; Santa Fe; Water Canyon; East Fork of the Gila; Organ Mountains; Gray; White Mountains. Damp thickets, in the Transition Zone.

10. **Bromus porteri** (Coulter) Nash, Bull. Torrey Club **22:** 512. 1895.

Bromus kalmii porteri Coulter, Man. Rocky Mount. 425. 1885.

TYPE LOCALITY: Twin Lakes, Colorado.

RANGE: Montana and South Dakota to Arizona and New Mexico.

NEW MEXICO: Dulce; Chama; Albuquerque; Glorieta; Raton Mountains; Pescado Spring; Ramah; Johnsons Mesa; Mogollon Creek; Fort Bayard; Organ Mountains; Tularosa Creek; Gilmores Ranch. Damp thickets, in the Transition Zone.

11. **Bromus frondosus** (Shear) Woot. & Standl. N. Mex. Agr. Exp. Sta. Bull. **81:** 144. 1912.

Bromus porteri frondosus Shear, U. S. Dept. Agr. Div. Agrost. Bull. **23:** 37. *f. 20.* 1900.

TYPE LOCALITY: Mangas Springs, New Mexico. Type collected by J. G. Smith.

RANGE: New Mexico and Arizona.

NEW MEXICO: Raton; Ponchuelo Creek; Santa Fe Canyon; Mangas Springs; Mogollon Creek; Organ Mountains; San Luis Mountains. Damp canyons, in the Upper Sonoran and Transition zones.

64. DACTYLIS L. ORCHARD GRASS.

Perennial with flat leaves and narrow glomerate panicles; spikelets 3 to 5-flowered, nearly sessile in dense fascicles; rachilla jointed above the glumes and between the florets; glumes unequal, 1 to 3-nerved, sharply keeled, acute; lemmas 5-nerved, shortly awn-pointed, strongly compressed and keeled, ciliate on the keel; palea a little shorter than the lemma, 2-keeled; stamens 3; styles distinct; grain narrow, oblong, free.

1. Dactylis glomerata L. Sp. Pl. 71. 1753.

TYPE LOCALITY: "Habitat in Europae cultis ruderatis."

RANGE: Widely distributed in North America, introduced from Europe and often cultivated.

NEW MEXICO: Shiprock; Cedar Hill; Chama; Winsors Ranch; Mesilla Valley.

65. DISTICHLIS Raf. SALTGRASS.

Rigid erect stoloniferous perennial with dense panicles of rather few compressed spikelets; spikelets 8 to 16-flowered, diœcious; glumes carinate, acute, shorter than the lemmas; lemmas broader, 3 to many-nerved, acute, rigid; palea 2-keeled, equaling or shorter than the lemmas; stamens in the staminate flower 3; styles thickened at the base, rather long, distinct; grain closely enveloped in the thickened and coriaceous base of the palea.

1. Distichlis spicata (L.) Greene, Bull. Calif. Acad. **2**: 415. 1887.

Uniola spicata L. Sp. Pl. 71. 1753.

TYPE LOCALITY: "Habitat in Americae borealis maritimis."

RANGE: Throughout the United States and in Mexico.

NEW MEXICO: Farmington; Dulce; El Rito; Algodones; Las Palomas; Nambe Valley; near Cliff; Rincon; White Sands; Mesilla Valley. Saline soil, in the Lower and Upper Sonoran zones.

66. POA L. BLUEGRASS.

Annuals or perennials with usually flat leaves and with a paniculate inflorescence; spikelets 2 to 6-flowered, the uppermost floret rudimentary; rachilla jointed above the glumes, these herbaceous, lanceolate or ovate, 1 to 3-nerved, keeled, persistent; lemmas herbaceous or membranaceous, lanceolate or ovate, acute or obtuse, awnless, 5-nerved, carinate, falling with the 2-keeled palea and a joint of the rachilla, the dorsal or marginal nerves usually soft-hairy and often with a tuft of cobwebby hairs at the base; stamens 3; styles distinct.

KEY TO THE SPECIES.

Annuals.
  Plants low, 10 to 20 cm. high; branches of the panicle spreading .... 1. *P. annua.*
  Plants taller, 15 to 50 cm. high; branches of the panicle erect .... 2. *P. bigelovii.*
Perennials.
  Cobweb at the base of the flowers wanting; spikelets acute at the base; lemmas 5 mm. long or more.
    Spikelets only slightly compressed; lemmas rounded at the apex.
      Lemmas merely slightly scabrous; plants bright green .... 11. *P. laevigata.*
      Lemmas strigose below, scabrous above; plants yellowish green .... 12. *P. lucida.*
    Spikelets strongly compressed; lemmas acute.
      Ligules long, 5 to 7 mm., acute or acuminate .. 13. *P. longiligula.*
      Ligules short, less than 5 mm. long, rounded or truncate at the apex.
        Panicles very narrow and contracted, long-peduncled .... 14. *P. longipedunculata.*

Panicles open, broad, on long or short peduncles.
Panicles very short, 5 cm. or less; plants less than 30 cm. high; leaves smooth beneath, scabrous above .... 15. *P. brevipaniculata.*
Panicles longer, 8 to 15 cm.; plants 30 to 60 cm. high; leaves variously pubescent.
Glumes nearly equal, 3-nerved; leaves smooth beneath, scabrous above .... 16. *P. arida.*
Glumes unequal, the first 1-nerved, the second 3-nerved; leaves scabrous beneath, hispid-puberulent above .... 17. *P. fendleriana.*
Cobweb present at the base of the flowers, though sometimes scanty; lemmas acute (except in *P. compressa*) and usually strongly keeled; plants with horizontal rootstocks; spikelets and lemmas various.
Intermediate nerves of the lemmas faint or obsolete.
Stems compressed; panicles narrow, open .... 8. *P. compressa.*
Stems not compressed; panicles various.
Branches of the panicles reflexed .... 9. *P. aperta.*
Branches of the panicles not reflexed .... 10. *P. interior.*
Intermediate nerves of the lemmas conspicuous.
Panicles with numerous many-flowered spikelets, the branches in fruit ascending (the lower ones in 3's and 4's); lemmas acutish .... 3. *P. pratensis.*
Panicles usually with few-flowered spikelets, the branches reflexed or spreading in fruit; lemmas very acute.
Spikelets few, usually purplish; branches of the panicle few, solitary or in pairs.
Intermediate nerves of the lemmas long-hairy .... 4. *P. arctica.*
Intermediate nerves of the lemmas glabrous, the hairs on the principal nerves copious and spreading .... 5. *P. reflexa.*
Spikelets numerous, green; branches of the panicle numerous, the lower ones often in 3's and 4's.
Lemmas only slightly pubescent on the keel below .... 6. *P. occidentalis.*
Lemmas copiously white-pubescent on the back below, villous on the marginal nerves and keel .... 7. *P. tracyi.*

1. Poa annua L. Sp. Pl. 68. 1753. LOW SPEARGRASS.

TYPE LOCALITY: "Habitat in Europa ad vias."

RANGE: Nearly throughout the United States; also in Europe and Asia.

NEW MEXICO: Santa Fe; Ponchuelo Creek. Moist soil, in the Upper Sonoran and Transition zones.

2. **Poa bigelovii** Vasey & Scribn. Contr. U. S. Nat. Herb. **1:** 270. 1893.

TYPE LOCALITY: "New Mexico", probably near Santa Fe. Type collected by Fendler (no. 931).

RANGE: Colorado and Texas to California.

NEW MEXICO: Winsors Ranch; Watrous; Santa Fe; Glorieta; Taos; Organ Mountains; White Mountains. Meadows, in the Upper Sonoran and Transition zones.

3. **Poa pratensis** L. Sp. Pl. 67. 1753. KENTUCKY BLUEGRASS.

TYPE LOCALITY: "Habitat in Europae pratis fertilissimis."

RANGE: Nearly throughout North America; also in Europe and Asia.

NEW MEXICO: Tunitcha Mountains; Chama; Santa Fe Canyon; Truchas Peak; Rio Pueblo; Reserve; Cloudcroft; Raton; Albuquerque; White Mountains. Meadows and woods, in the Transition Zone.

4. **Poa arctica** R. Br. Suppl. App. Parry's Voy. 288. 1824.

TYPE LOCALITY: Melville Island.

RANGE: New Mexico and Colorado to Washington, Alaska, and Labrador.

NEW MEXICO: Truchas Peak (*Standley* 4835). Meadows, in the Arctic-Alpine Zone.

5. **Poa reflexa** Vasey & Scribn. Contr. U. S. Nat. Herb. **1:** 276. 1893.

Poa leptocoma reflexa Jones, Contr. West. Bot. **14:** 15. 1912.

TYPE LOCALITY: Kelso Mountain, near Torrey Peak, Colorado.

RANGE: Oregon and Montana to New Mexico.

NEW MEXICO: A single collection seen (*Fendler* 920), probably from near Santa Fe. Wet meadows, in the Transition Zone.

6. **Poa occidentalis** Vasey, Contr. U. S. Nat. Herb. **1:** 274. 1893.

Poa platyphylla Nash & Rydb. Colo. Agr. Exp. Sta. Bull. **100:** 44. 1906.

TYPE LOCALITY: Las Vegas, New Mexico. Type collected by G. R. Vasey.

RANGE: Colorado and New Mexico.

NEW MEXICO: Sierra Grande; Baldy; Winsors Ranch; Sandia Mountains; Las Vegas; Santa Fe Canyon; White and Sacramento mountains. Damp meadows and thickets, in the Transition and Canadian zones.

7. **Poa tracyi** Vasey, Contr. U. S. Nat. Herb. **1:** 276. 1893.

TYPE LOCALITY: On mountain sides at Raton, New Mexico. Type collected by Tracy in 1887.

RANGE: Northern New Mexico and southern Colorado.

NEW MEXICO: Raton. Hillsides, in the Transition Zone.

8. **Poa compressa** L. Sp. Pl. 69. 1753. ENGLISH BLUEGRASS.

TYPE LOCALITY: "Habitat in Europae et Americae septentrionalis siccis, muris, tectis."

RANGE: Native of Europe and Asia, widely naturalized in the United States.

NEW MEXICO: Harveys Upper Ranch; Raton.

9. **Poa aperta** Scribn. & Merr. U. S. Dept. Agr. Div. Agrost. Circ. **35:** 4. 1901.

TYPE LOCALITY: Telluride, Colorado.

RANGE: Mountains of Colorado and New Mexico.

NEW MEXICO: Top of Sandia Mountains (*Wooton*).

10. **Poa interior** Rydb. Bull. Torrey Club **32:** 604. 1905.

TYPE LOCALITY: Headwaters of Clear Creek and Crazy Woman River, Wyoming.

RANGE: Western British America to northern New Mexico.

NEW MEXICO: Jemez Mountains; top of Sandia Mountains. Damp meadows, in the Canadian Zone.

11. **Poa laevigata** Scribn. U. S. Dept. Agr. Div. Agrost. Bull. **5**: 31. 1897.
Poa laevis Vasey, Contr. U. S. Nat. Herb. **1**: 273. 1893, not Borb. 1877.
TYPE LOCALITY: Montana.
RANGE: Washington and Montana to New Mexico.
NEW MEXICO: Bell; Fitzgerald Cienaga. Meadows.

12. **Poa lucida** Vasey, Contr. U. S. Nat. Herb. **1**: 274. 1893.
TYPE LOCALITY: Mountain sides near Georgetown, Colorado.
RANGE: Wyoming and South Dakota to New Mexico.
NEW MEXICO: Fitzgerald Çienaga (*Wooton*).

13. **Poa longiligula** Scribn. & Williams, U. S. Dept. Agr. Div. Agrost. Circ. **9: 3.** 1899.
TYPE LOCALITY: Silver Reef, Utah.
RANGE: Oregon and South Dakota to California and New Mexico.
NEW MEXICO: Aztec (*Baker* 87, 204). Low hills, in the Upper Sonoran Zone.

14. **Poa longipedunculata** Scribn. U. S. Dept. Agr. Div. Agrost. Bull. **11**: 54. *pl. 11.* 1898.
TYPE LOCALITY: Summit of Sheep Mountain, Laramie, Albany County, Wyoming.
RANGE: Wyoming to New Mexico.
NEW MEXICO: Raton; Barranca; Santa Fe Canyon. Hillsides, in the Upper Sonoran and Transition zones.

15. **Poa brevipaniculata** Scribn. & Williams, U. S. Dept. Agr. Div. Agrost. Circ. **9**: 2. 1899.
TYPE LOCALITY: Table Peak, Colorado.
RANGE: Colorado and New Mexico.
NEW MEXICO: Las Vegas Hot Springs; Gallinas River. Plains, in the Upper Sonoran Zone.

16. **Poa arida** Vasey, Contr. U. S. Nat. Herb. **1**: 270. 1893.
Poa fendleriana arida Jones, Contr. West. Bot. **14**: 14. 1912.
TYPE LOCALITY: Socorro, New Mexico. Type collected by G. R. Vasey in 1881.
RANGE: Utah and Kansas to New Mexico.
NEW MEXICO: Socorro; Las Vegas Hot Springs; Roswell. Plains and low hills, in the Upper Sonoran Zone.

17. **Poa fendleriana** (Steud.) Vasey, Illustr. N. Amer. Grasses **2**: 74. 1893.
MUTTON GRASS.
Eragrostis fendleriana Steud. Syn. Pl. Glum. **1**: 278. 1855.
Uralepis poaeoides Buckl. Proc. Acad. Phila. **1862**: 94. 1863.
TYPE LOCALITY: New Mexico, probably near Santa Fe. Type collected by Fendler (no. 932).
RANGE: Colorado and New Mexico to California.
NEW MEXICO: Tierra Amarilla; Chama; Las Vegas Hot Springs; Santa Fe; Pecos Baldy; Bear Canyon; San Antonio; Mangas Springs; Burro Mountains; Magdalena Mountains; San Luis Mountains; Organ Mountains; Little Creek. Meadows and canyons, in the Upper Sonoran and Transition zones.
The type of *Uralepis poaeoides* is Fendler's 932, which is also the type of *Eragrostis fendleriana*.

67. FESTUCA L. Fescue.

Annuals or perennials of various habit; spikelets several-flowered, pedicellate in narrow and dense or loose and spreading panicles; rachilla jointed above the glumes and between the florets; glumes more or less unequal, narrow, acute; lemmas rounded on the back, at least below, acute or tapering into a straight awn, faintly 3 to 5-nerved, not webbed at the base; stamens 3; styles very short, distinct; grain elongated, furrowed, frequently adnate to the palea.

KEY TO THE SPECIES.

Annuals or biennials.
  Spikelets loosely 1 to 5, rarely 6-flowered .... 1. *F. pacifica.*
  Spikelets densely 8 to 13-flowered .... 2. *F. octoflora.*
Perennials.
  Glumes thin, the second 1-nerved, or 3-nerved only at the base; ligules long and acuminate .... 3. *F. thurberi.*
  Glumes firm, the second 3 to 5-nerved; ligules various.
    Plants low, less than 30 cm. high .... 4. *F. brachyphylla.*
    Plants tall, 60 cm. high or more.
      Leaf blades very narrow, filiform, involute, grayish green .... 5. *F. arizonica.*
      Leaf blades wide, flat, bright green.
        Spikelets narrowly oblong, 3 to 5-flowered .... 6. *F. sororia.*
        Spikelets ovate or oblong, 6 to 11-flowered .... 7. *F. elatior.*

1. **Festuca pacifica** Piper, Contr. U. S. Nat. Herb. **10:** 12. 1906.

TYPE LOCALITY: Pullman, Washington.

RANGE: British Columbia to California and New Mexico.

NEW MEXICO: Ledges of Gallinas River, near Las Vegas (*Cockerell*). Upper Sonoran Zone.

2. **Festuca octoflora** Walt. Fl. Carol. 81. 1788.

Festuca tenella Willd. Sp. Pl. **1:** 419. 1797.

TYPE LOCALITY: Carolina.

RANGE: Throughout most of temperate North America.

NEW MEXICO: Santa Fe; Farmington; Sierra Grande; Las Vegas; Mogollon Mountains; Socorro; Organ Mountains; Roswell. Upper Sonoran Zone.

3. **Festuca thurberi** Vasey in Wheeler, Rep. U. S. Surv. 100th Merid. **6:** 292. *pl. 29.* 1879.

Festuca scabrella vaseyana Hack.; Beal, Grasses N. Amer. **2:** 603. 1896.

TYPE LOCALITY: Twin Lakes, Colorado.

RANGE: Wyoming to northern New Mexico.

NEW MEXICO: Beattys Cabin; Sandia Mountains; Baldy. Meadows, in the Canadian Zone.

4. **Festuca brachyphylla** Schult. Mant. **3:** 646. 1827.

Festuca brevifolia R. Br. Suppl. App. Parry's Voy. 188. 1824, not Muhl. 1817.

Festuca ovina brevifolia S. Wats. in King, Geol. Expl. 40th Par. **5:** 389. 1871.

TYPE LOCALITY: Melville Island.

RANGE: British America to New Mexico and Vermont.

NEW MEXICO: Las Vegas Range; Winsors Ranch; Truchas Peak; Baldy; Albuquerque. Meadows, in the Hudsonian and Arctic-Alpine zones.

5. **Festuca arizonica** Vasey, Contr. U. S. Nat. Herb. **1:** 277. 1893.

ARIZONA FESCUE.

TYPE LOCALITY: Near Flagstaff, Arizona.

RANGE: Utah and Colorado to Arizona and New Mexico.

NEW MEXICO: Sierra Grande; Raton Mountains; Chama; Grants Station; Winsors Ranch; Trout Spring; Johnsons Mesa; Rio Pueblo; San Lorenzo; West Fork of the Gila; White Mountains. Open slopes, in the Canadian Zone.

6. **Festuca sororia** Piper, Contr. U. S. Nat. Herb. **16:** 197. 1913.
TYPE LOCALITY: Rincon Mountains, Arizona.
RANGE: Colorado to New Mexico and Arizona. Mountains, in the Transition Zone.
NEW MEXICO: Tunitcha Mountains (*Standley* 7714); Hillsboro Peak (*Metcalfe* 1236).

7. **Festuca elatior** L. Sp. Pl. 75. 1753. MEADOW FESCUE.
Festuca pratensis Huds. Fl. Angl. 37. 1762.
TYPE LOCALITY: "Habitat in Europae pratis fertilissimis."
RANGE: Native of Europe, widely introduced into the United States, sometimes cultivated for hay.
NEW MEXICO: Harveys Upper Ranch; Winsors Ranch; Mesilla Valley.

68. PANICULARIA Fabr. MANNA GRASS.

Tall aquatic perennials with flat leaves and usually diffuse terminal panicles; spikelets few to many-flowered, terete or slightly flattened, in narrow or spreading panicles; rachilla jointed between the florets, usually smooth; glumes unequal, shorter than the lemmas, obtuse or acute, 1 to 3-nerved; lemmas smooth or scabrous, rounded on the back, herbaceous except at the scarious and usually blunt apex, 5 to 9-nerved, the nerves usually prominent; palea a little shorter than the lemma, 2-nerved; stamens 2 or 3; styles short, distinct; grain oblong, smooth, free, or when dry slightly adherent to the palea.

KEY TO THE SPECIES.

Spikelets linear, 12 mm. long or more .... 1. *P. borealis.*
Spikelets ovate or oblong, 6 mm. long or less.
  Spikelets 3 mm. long or less; branches of the panicle drooping.... 2. *P. nervata.*
  Spikelets 4 to 6 mm. long; branches of the panicle ascending or spreading .... 3. *P. grandis.*

1. **Panicularia borealis** Nash, Bull. Torrey Club **24:** 348. 1897.
Glyceria borealis Piper, Fl. Palouse 27. 1901.
TYPE LOCALITY: Van Buren, Maine.
RANGE: Alaska and Maine to California, New Mexico, and New York.
NEW MEXICO: Tunitcha Mountains (*Standley* 7540). In swamps, in the Canadian Zone.

2. **Panicularia nervata** (Willd.) Kuntze, Rev. Gen. Pl. **2:** 783. 1891.
Poa nervata Willd. Sp. Pl. **1**: 389. 1797.
Glyceria nervata Trin. Mém. Acad. St. Pétersb. VI. Math. Phys. Nat. **1:** 365. 1830.
TYPE LOCALITY: "Habitat in America boreali."
RANGE: British America to California, Mexico, and Florida.
NEW MEXICO: Chama; Tunitcha Mountains; Winsors Ranch; Rio Pueblo; Santa Fe Canyon; Mogollon Creek; Wheelers Ranch. In wet ground, in the Transition Zone.

3. **Panicularia grandis** (S. Wats.) Nash in Britt. & Brown, Illustr. Fl. ed. 2. **1:** 265. 1913.
Poa aquatica americana Torr. Fl. North. & Mid. U. S. **1:** 108. 1824.
Glyceria grandis S. Wats. in A. Gray, Man. ed. 6. 667. 1890.
Panicularia americana MacM. Met. Minn. Val. 81. 1892.
TYPE LOCALITY: Northeastern United States.
RANGE: British America to Nevada, New Mexico, and Tennessee.
NEW MEXICO: Pecos; Taos; Las Vegas; Albuquerque; Middle Fork of the Gila. Wet ground, in the Transition Zone.

69. PUCCINELLIA Parl. MEADOW GRASS.

Cespitose perennial with flat leaves and narrow terminal panicles; spikelets as in Panicularia but with usually smaller lemmas, the nerves less distinct or obscure.

1. Puccinellia airoides (Nutt.) Wats. & Coult. in A. Gray, Man. ed. 6. 668. 1890.

Poa airoides Nutt. Gen. Pl. 1: 68. 1818.

Panicularia distans airoides Scribn. Mem. Torrey Club 5: 54. 1894.

TYPE LOCALITY: "In depressed situations around the Mandan village, on the Missouri."

RANGE: British America to Nevada and New Mexico.

NEW MEXICO: Shiprock; Farmington; Arroyo Pecos near Las Vegas. Wet ground, in the Upper Sonoran Zone.

70. AGROPYRON Gaertn. WHEAT GRASS.

Tall perennials with erect simple culms and often bearded terminal spikes; spikelets 3 to many-flowered, closely sessile and single at each notch of the axis; rachilla articulated above the glumes under each lemma; glumes 2, narrower and usually shorter than the lemmas, acute or awned; lemmas rounded on the back or slightly keeled above, 5 to 7-nerved, acute or awned from the apex, rarely obtuse; palea 2-keeled, bristly-ciliate on the keels; grain pubescent at the apex, usually adherent to the palea.

KEY TO THE SPECIES.

Rachis of the spike breaking up at maturity, the joints falling with the spikelets.. 1. *A. scribneri.*
Rachis of the spike continuous, not breaking up at maturity.
  Awns of the lemmas conspicuous.
    Awns erect.
      Awns short, not exceeding the body of the lemma; spikes usually tinged with purple.......... 2. *A. violaceum.*
      Awns long, usually exceeding the body of the lemma; spikes not purplish tinged.
        Stems stout; spikes 7 to 10 mm. wide, usually unilateral; spikelets (excluding the awns) 12 to 15 mm. long.................... 3. *A. richardsoni.*
        Stems slender; spikes about 5 mm. wide, seldom unilateral; spikelets (excluding the awns) about 10 mm. long.......... 4. *A.* [illegible]
    Awns divergent.
      Spikelets subterete, more or less crowded.......... 5. *A. bakeri.*
      Spikelets flattened, distant.
        Leaves scabrous above........................ 6. *A. arizonicum.*
        Leaves glabrous............................ 7. *A. spicatum.*
  Awns of the lemmas not conspicuous.
    Plants bright green, not glaucous.
      Plants cespitose; spikelets small, few-flowered..... 8. *A. tenerum.*
      Plants stoloniferous; spikelets large, many-flowered. 9. *A. pseudorepens.*
    Plants conspicuously glaucous.
      Sheaths pubescent..............................10. *A. palmeri.*
      Sheaths glabrous.
        Lemmas scabrous or glabrous..................11. *A. smithii.*
        Lemmas pubescent...........................12. *A. molle.*

1. **Agropyron scribneri** Vasey, Bull. Torrey Club **10:** 128. 1893.
Elymus scribneri Jones, Contr. West. Bot. **14:** 20. 1912.
TYPE LOCALITY: Montana.
RANGE: Montana to northern Arizona and New Mexico.
NEW MEXICO: Top of Las Vegas Range (*Cockerell*). Meadows in the mountains, in the Arctic-Alpine Zone.

2. **Agropyron violaceum** (Hornem.) Lange, Consp. Fl. Groenland. **3:** 155. 1880.
Triticum violaceum Hornem. Fl. Dan. *pl. 2044*. 1832.
TYPE LOCALITY: Southern Greenland.
RANGE: British America to Pennsylvania and New Mexico.
NEW MEXICO: Rio Pueblo; Sandia Mountains; White Mountain Peak.

3. **Agropyron richardsoni** (Trin.) Schrad.; Shear, U. S. Dept. Agr. Div. Agrost. Bull. **4:** 29. 1897.
Triticum richardsoni Trin. Linnaea **12:** 467. 1838.
TYPE LOCALITY: "America borealis arctica?."
RANGE: British Columbia and New England to California and northern New Mexico.
NEW MEXICO: Beattys Cabin (*Standley* 4856). Meadows, in the Canadian Zone.

4. **Agropyron caninum** (L.) Roem. & Schult. Syst. Veg. **2:** 756. 1817.
Triticum caninum L. Sp. Pl. 86. 1753.
TYPE LOCALITY: "Habitat in Europae sepibus."
RANGE: Idaho and Nova Scotia to New Mexico and North Carolina; also in Europe.
NEW MEXICO: Truchas Peak (*Standley* 4831). Meadows, in the Hudsonian Zone.

5. **Agropyron bakeri** E. Nels. Bot. Gaz. **38:** 378. 1904.
TYPE LOCALITY: Near Pagosa Peak, southern Colorado.
RANGE: Mountains of southern Colorado and northern New Mexico.
NEW MEXICO: Baldy; Sandia Mountains; White Mountain Peak. Hudsonian and Arctic-Alpine zones.

6. **Agropyron arizonicum** Scribn. & Smith, U. S. Dept. Agr. Div. Agrost. Bull. **4:** 27. 1897.
TYPE LOCALITY: Rincon Mountains, Arizona.
RANGE: Mountains of Arizona and New Mexico to northern Mexico.
NEW MEXICO: Laguna; Jemez Mountains; Bear Mountain; Mogollon Mountains; Organ Mountains; San Luis Mountains; Alamogordo. Transition Zone.

7. **Agropyron spicatum** (Pursh) Scribn. & Smith, U. S. Dept. Agr. Div. Agrost. Bull. **4:** 33. 1897.
Festuca spicata Pursh, Fl. Amer. Sept. 83. 1814.
Agropyron divergens Nees in Steud. Syn. Pl. Glum. **1:** 347. 1854.
TYPE LOCALITY: "Camp Chopunnish," opposite Kamiah, Idaho.
RANGE: Washington and Montana to Colorado and New Mexico.
NEW MEXICO: Bear Mountain; Socorro. Upper Sonoran and Transition zones.

8. **Agropyron tenerum** Vasey, Bot. Gaz. **10:** 258. 1885. SLENDER WHEAT GRASS.
TYPE LOCALITY: Fort Garland, Colorado.
RANGE: British America to New Mexico and New Hampshire.
NEW MEXICO: Dulce; Raton; Las Vegas; Torrance; Johnsons Mesa; Glorieta; Chama; Cubero; Albuquerque; North Percha Creek; Mangas Springs; White Mountains. Mountains, in the Upper Sonoran and Transition zones.
This is a rather valuable range grass in the mountains and is often cut for hay.

9. **Agropyron pseudorepens** Scribn. & Smith, U. S. Dept. Agr. Div. Agrost. Bull. **4:** 34. 1897.
Agropyron tenerum pseudorepens Jones, Contr. West. Bot. **14:** 19. 1912.

TYPE LOCALITY: Texas.

RANGE: British Columbia and Montana to Arizona and Texas.

NEW MEXICO: Tunitcha Mountains; Farmington; Carrizo Mountains; Pecos; Sandia Mountains; Santa Fe and Las Vegas mountains; Raton Mountains; Albuquerque; Fort Bayard; Socorro; Mogollon Creek; Roy; White and Sacramento mountains. Moist hillsides, in the Upper Sonoran and Transition zones.

10. **Agropyron palmeri** (Scribn. & Smith) Rydb.; Woot. & Standl. N. Mex. Agr. Exp. Sta. Bull. **81**: 153. 1912.

Agropyron spicatum palmeri Scribn. & Smith, U. S. Dept. Agr. Div. Agrost. Bull. **4**: 33. 1897.

Agropyron smithii palmeri Jones, Contr. West. Bot. **14**: 18. 1912.

TYPE LOCALITY: Arizona.

RANGE: Arizona and New Mexico.

NEW MEXICO: Wheelers Ranch; Coolidge; Santa Fe; Agua Azul.

11. **Agropyron smithii** Rydb. Mem. N. Y. Bot. Gard. **1**: 64. 1900.

COLORADO BLUESTEM.

Agropyron glaucum occidentale Scribn. Trans. Kans. Acad. **9**: 119. 1885.

Agropyron occidentale Scribn. U. S. Dept. Agr. Div. Agrost. Circ. **27**: 9. 1900.

TYPE LOCALITY: Kansas.

RANGE: Washington and Wisconsin to Arizona and Texas.

NEW MEXICO: Chama; Shiprock; Dulce; Carrizo Mountains; Sierra Grande; Tunitcha Mountains; Pecos; Cowles; Taos; Johnsons Mesa; Raton; Coolidge; Puertecito; Gallo Spring; Mangas Springs; Reserve; Socorro; Nogal; White Mountains. Plains and meadows, in the Upper Sonoran and Transition zones.

Colorado bluestem is a valuable range grass, occurring in great abundance on the foothills and the higher plains.

12. **Agropyron molle** (Scribn. & Smith) Rydb. Mem. N. Y. Bot. Gard. **1**: 65. 1900.

Agropyron spicatum molle Scribn. & Smith, U. S. Dept. Agr. Div. Agrost. Bull. **4**: 33. 1897.

Agropyron smithii molle Jones, Contr. West. Bot. **14**: 18. 1912.

TYPE LOCALITY: "Saskatchewan to Colorado and New Mexico, and westward to Idaho and Washington."

RANGE: Washington and Saskatchewan to northern New Mexico.

NEW MEXICO: Raton; Chama; Farmington. Plains and low hills, in the Upper Sonoran and Transition zones.

71. LOLIUM L. RYE GRASS.

Annuals or perennials with simple erect culms, flat leaves, and simple terminal spikes; spikelets several-flowered, solitary, sessile in alternate notches of the continuous rachis, one edge of each spikelet placed against the rachis; rachilla jointed between the florets; glumes 1 (2 in the terminal spikelet), shorter than or exceeding the florets; lemmas rounded on the back, 5 to 7-nerved, obtuse, acute, or awned; palea 2-keeled; stamens 3; styles very short, distinct; grain smooth, adherent to the palea.

1. **Lolium perenne** L. Sp. Pl. 83. 1753.

TYPE LOCALITY: "Habitat in Europa ad agrorum versuras solo fertili."

RANGE: Native of Europe, introduced in many parts of North America.

NEW MEXICO: Chama; Santa Fe; Agricultural College.

72. HORDEUM L.

Annuals or perennials with terminal cylindrical spikes of awned spikelets; spikelets 1-flowered, 2 or 3 together at each joint of the rachis, sessile or on very short pedicels; rachilla articulated above the glumes and continued behind the palea of the central spikelet into a naked bristle; glumes 2, narrowly lanceolate, subulate, or setaceous, rigid, persistent; lemmas lanceolate, rounded on the back, obscurely 5-nerved above, usually awned; palea shorter than the lemma, 2-keeled; stamens 3; styles very short, distinct; grain sulcate, adherent to the palea.

KEY TO THE SPECIES.

Plants glaucous throughout.. 1. *H. murinum.*
Plants not glaucous.
  Glumes 3 to 6 cm. long.. 2. *H. jubatum.*
  Glumes 1 to 2 cm. long.
    Awns spreading; spikes yellowish........................ 3. *H. caespitosum.*
    Awns erect; spikes reddish or brownish green............ 4. *H. nodosum.*

1. Hordeum murinum L. Sp. Pl. 85. 1753. WALL BARLEY.
TYPE LOCALITY: "Habitat in Europae locis ruderatis."
RANGE: Native of Europe, widely naturalized in the United States.
NEW MEXICO: Mangas Springs; Mesilla Valley.

2. Hordeum jubatum L. Sp. Pl. 85. 1753. SQUIRREL-TAIL GRASS.
TYPE LOCALITY: "Habitat in Canada."
RANGE: Alaska and British America to California, New Mexico, and Missouri.
NEW MEXICO: Farmington; Carrizo Mountains; Tunitcha Mountains; Chama; Raton; Sierra Grande; Magdalena Mountains; Pecos; Torrance; Rio Pueblo; Mora; Pescado Spring; Santa Fe; Kingston; White Mountains. Plains and meadows, in the Upper Sonoran and Transition zones.
Often a troublesome weed in cultivated ground.

3. Hordeum caespitosum Scribn. in Pammel, Proc. Davenport Acad. **7**: 245. 1899.
TYPE LOCALITY: Edgemont, South Dakota.
RANGE: Wyoming and South Dakota to northern New Mexico.
NEW MEXICO: Farmington (*Standley* 6904). Dry hills and plains, in the Upper Sonoran Zone.

4. Hordeum nodosum L. Sp. Pl. ed. 2. 126. 1762. MEADOW BARLEY.
TYPE LOCALITY: "Habitat in Italia, Anglia."
RANGE: Temperate North America, Asia, and Europe.
NEW MEXICO: Tunitcha Mountains; Chama; Ramah; Grants Station; El Rito Creek; Rio Pueblo. Wet ground, in the Transition Zone.

73. SITANION Raf.

Cespitose perennials with mostly flat leaves and terminal bearded spikes; spikelets usually 2, sometimes 3 or 1, at each joint of the rachis, 2 to several-flowered; glumes many-parted from near the base or merely bifid, or subulate and entire, awned; lemmas terminating in a single long awn, or trifid and 3-awned; palea as long as the lemma, entire, bidentate, or 2-awned.

KEY TO THE SPECIES.

Glumes bifid from about the middle, the lobes abruptly divergent.
  Sheaths long-villous.. 1. *S. molle.*
  Sheaths not villous.
    Glumes 3 to 4 cm. long.................................. 2. *S. caespitosum.*
    Glumes 2 to 3 cm. long.................................. 3. *S. rigidum.*
Glumes entire, subulate-setaceous.
  Culm leaves long and flexuous.................................. 4. *S. longifolium.*
  Culm leaves short, rigid, spreading.
    Lemmas 10 mm. long, glaucous............................ 5. *S. brevifolium.*
    Lemmas 7 mm. long, soft-pubescent....................... 6. *S. pubiflorum.*

1. **Sitanion molle** J. G. Smith, U. S. Dept. Agr. Div. Agrost. Bull. **18:** 17. 1899.

Type locality: East side Buffalo Pass, Larimer County, Colorado.

Range: Colorado and New Mexico.

New Mexico: Craters; north of Ramah; Box S Spring; Chama. Open slopes, in the Upper Sonoran and Transition zones.

2. **Sitanion caespitosum** J. G. Smith, U. S. Dept. Agr. Div. Agrost. Bull. **18:** 16. 1899.

Type locality: Near Cliff, New Mexico. Type collected by J. G. Smith in 1897.

Range: Southwestern New Mexico.

New Mexico: Near Cliff; Mangas Canyon.

3. **Sitanion rigidum** J. G. Smith, U. S. Dept. Agr. Div. Agrost. Bull. **18:** 13. 1899.

Type locality: Cascade Mountains, Washington.

Range: Washington and California to Wyoming and New Mexico.

New Mexico: Summit of Organ Peak (*Standley*).

4. **Sitanion longifolium** J. G. Smith, U. S. Dept. Agr. Div. Agrost. Bull. **18:** 18. 1899.

Type locality: Near Silverton, Colorado.

Range: Wyoming and Kansas to Texas and Nevada.

New Mexico: Abundant from the Mogollon Mountains and Organ Mountains northward to Las Vegas and westward across the State. Plains and rocky hills, in the Upper Sonoran and Transition zones.

5. **Sitanion brevifolium** J. G. Smith, U. S. Dept. Agr. Div. Agrost. Bull. **18:** 17. 1899.

Elymus brevifolius Jones, Contr. West. Bot. **14:** 20. 1912.

Type locality: Tucson, Arizona.

Range: Washington to Arizona and New Mexico.

New Mexico: Rio Pueblo; Sandia Mountains; Tierra Amarilla; Santa Fe Canyon; Duran; Chama; Grants; Gallo Spring; Mangas Springs; Middle Fork of the Gila; Organ Mountains; Gilmores Ranch. Plains and hillsides, in the Upper Sonoran and Transition zones.

6. **Sitanion pubiflorum** J. G. Smith, U. S. Dept. Agr. Div. Agrost. Bull. **18:** 19. 1899.

Type locality: Tucson, Arizona.

Range: Colorado to New Mexico and Arizona.

New Mexico: Carrizo Mountains; Tunitcha Mountains; Cedar Hill; Sierra Grande; Raton; Las Vegas Hot Springs; Santa Fe; San Augustine Plains; Animas Creek; Reserve; Roswell. Upper Sonoran Zone.

74. ELYMUS L. Wild rye.

Tall erect perennials with flat leaves and closely flowered terminal spikes; spikelets 2 to 6-flowered, the uppermost imperfect, sessile, mostly in pairs, at the alternate notches of the continuous or jointed rachis, forming terminal spikes; rachilla jointed above the glumes and between the florets; glumes 2, nearly equal, rigid, narrow, 1 or 3-nerved, acute or awn-pointed, persistent; lemmas shorter than the glumes, rounded on the back, obscurely 5-nerved, obtuse, acute, or awned from the apex; palea a little shorter than the lemma, 2-keeled; stamens 3; styles short; grain adherent to the lemmas and paleas.

KEY TO THE SPECIES.

Lemmas not awned or with very short awns.
- Glumes aristiform or narrowly subulate; spikelets usually 2 at each joint.......... 1. *E. triticoides*.
- Glumes lanceolate-subulate; spikelets usually single........ 2. *E. simplex*.

Lemmas long-awned.
  Spikes narrow; spikelets erect.
    Leaves 7 to 15 mm. wide, spreading; glumes lanceolate, acuminate to short-awned .... 3. *E. glaucus.*
    Leaves less than 5 mm. wide, mostly erect; glumes narrowly linear-lanceolate, long-awned .... 4. *E. macounii.*
  Spikes broad; spikelets spreading.
    Lemmas glabrous .... 7. *E. brachystachys.*
    Lemmas pubescent.
      Lemmas hirsute or villous .... 5. *E. canadensis.*
      Lemmas strigose-hispidulous or scabrous .... 6. *E. robustus.*

1. **Elymus triticoides** Buckl. Proc. Acad. Phila. **1862:** 99. 1863.

TYPE LOCALITY: "Rocky Mountains."

RANGE: Washington and California to Arizona and New Mexico.

NEW MEXICO: White Mountain Peak; Mesilla Valley. Lower Sonoran to the Transition Zone.

2. **Elymus simplex** Scribn. & Williams, U. S. Dept. Agr. Div. Agrost. Bull. **11:** 57. 1898.

TYPE LOCALITY: On banks of Green River, Wyoming.

RANGE: Oregon and Wyoming to northern Arizona and New Mexico.

NEW MEXICO: Carrizo Mountains (*Standley* 7466). Dry plains and hills, in the Upper Sonoran Zone.

3. **Elymus glaucus** Buckl. Proc. Acad. Phila. **1862:** 99. 1863.

Elymus americanus Vasey & Scribn.; Macoun, Cat. Can. Pl. **2:** 245. 1888.

Elymus sibiricus americanus Wats. & Coult. in A. Gray, Man. ed. 6. 673. 1890.

TYPE LOCALITY: "Columbia River."

RANGE: Alaska and California to Texas and the Great Lakes.

NEW MEXICO: Chama; Tunitcha Mountains; Winsors Ranch; Johnsons Mesa. Mountains, in the Transition Zone.

4. **Elymus macounii** Vasey, Bull. Torrey Club **13:** 119. 1886.

TYPE LOCALITY: "Great Plains of British America."

RANGE: Manitoba and Saskatchewan to Nebraska and New Mexico.

NEW MEXICO: Albuquerque (*Tracy*). Upper Sonoran Zone.

5. **Elymus canadensis** L. Sp. Pl. 83. 1753.

TYPE LOCALITY: "Habitat in Canada."

RANGE: British America to New Mexico and Texas.

NEW MEXICO: Farmington; Raton Mountains; Pecos; Santa Fe; Las Vegas; Pescado Spring; Kingston; Sabinal; Mesilla Valley; White Mountains. Damp ground, Lower Sonoran to Transition Zone.

6. **Elymus robustus** Scribn. & Smith, U. S. Dept. Agr. Div. Agrost. Bull. **4:** 37. 1897.

TYPE LOCALITY: Illinois.

RANGE: Montana and Illinois to New Mexico.

NEW MEXICO: Mangas Springs.

7. **Elymus brachystachys** Scribn. & Ball, U. S. Dept. Agr. Div. Agrost. Bull. **24:** 47. *f. 21.* 1901.

TYPE LOCALITY: Oklahoma.

RANGE: South Dakota and Michigan to Mexico.

NEW MEXICO: Black Range. Moist ground.

14. CYPERACEAE. Sedge Family.

Grasslike or rushlike herbs; stems usually solid; roots fibrous; leaves narrow, with closed sheaths, the whole leaf sometimes reduced to a sheath; flowers perfect or unisexual, arranged in spikelets, one in the axil of each scale, the spikelets solitary or clustered, 1 to many-flowered; perianth of bristles or wanting; stamens 1 to 3; styles 2 or 3; fruit a lenticular or trigonous achene.

KEY TO THE GENERA.

Flowers all unisexual, usually in separate spikes.
Achenes inclosed in a perigynium; glumes 1-flowered. 8. CAREX (p. 116).
Achenes not inclosed in a perigynium; glumes 2-flowered.................................. 9. KOBRESIA (p. 124).
Flowers all, or at least part of them, perfect; spikelets similar.
Scales of the spikelets 2-ranked; spikelets more or less flattened... 1. CYPERUS (p. 110).
Scales of the spikelets imbricated spirally in several ranks; spikelets not flattened.
Perianth bristles much elongated, woolly......... 7. ERIOPHORUM (p. 116).
Perianth bristles short or wanting.
Spikelets 1 to 4-flowered; plants large, about 1 meter high, leafy.................... 2. CLADIUM (p. 112).
Spikelets several to many-flowered; plants mostly low.
Base of the style persistent, enlarged.
Leaves reduced to sheaths; spikelets solitary....................... 3. ELEOCHARIS (p. 112).
Leaves not reduced; spikelets several, mostly paniculate............ 4. STENOPHYLLUS (p. 114).
Base of the style deciduous, enlarged or narrow.
Perianth consisting of bristles........ 5. SCIRPUS (p. 114).
Perianth of a single hyaline scale.... 6. HEMICARPHA (p. 116).

1. CYPERUS L.

Tufted or simple-stemmed annuals or perennials, 50 cm. high or less, with basal leaves and triangular stems, the flowers in headlike clusters or unequally branched umbels subtended by leaflike bracts; spikelets flattened or cylindric; glumes deciduous, or if persistent the spikelets falling entire, 2-ranked; flowers perfect; perianth none; stamens 1 to 3; achene without a tubercle.

KEY TO THE SPECIES.

Annual; plants small, 5 to 15 cm. high; tips of the bracts subulate, conspicuously reflexed; inflorescence capitate............. 1. *C. inflexus.*
Perennials (rarely annual); plants taller, more than 15 cm. high; tips of the bracts mostly erect, rarely spreading; inflorescence various.
Spikelets ovate to ovate-oblong (4 to 5 mm. long, crowded at the ends of the rather long, subequal rays of the umbel). 2. *C. cyrtolepis.*
Spikelets linear to narrowly oblong.
Spikelets narrowly oblong; scales of the flowers not overlapping, especially in fruit (very strongly nerved).
Inflorescence crowded, subcapitate.................. 3. *C. fendlerianus.*

Inflorescence a compound umbel with unequal rays.
Glumes as broad as long, 2 mm. long or less, mucronulate, usually green.................. 4. *C. rusbyi.*
Glumes twice as long as broad, 3 mm. long, acuminate into a spreading awn, yellowish brown.................................... 5. *C. schweinitzii.*
Spikelets linear; scales of the flowers overlapping from one-half to two-thirds their length.
Spikelets deciduous as a whole when mature.
Spikelets with few, usually 2 or 3, flowers...... 6. *C. uniflorus.*
Spikelets with 6 to 9 flowers...................... 7. *C. speciosus.*
Scales of the spikelets falling from the rachilla.
Rachilla narrowly winged, the wings adnate; plants stout, stoloniferous; spikelets loosely clustered............................. 8. *C. esculentus.*
Wings of the rachilla not adnate, forming scales anterior to the flower; plants and spikelets various.
Spikelets densely crowded; flowers numerous, about 20 to the spikelet; scales of the flower not bordered with red; plants stout.................................. 9. *C. erythrorhizos.*
Spikelets fewer, loosely clustered; flowers 12 to the spikelet or less; scales red-margined; plants slender..............10. *C. sphacelatus.*

1. **Cyperus inflexus** Muhl. Descr. Gram. 16. 1817.
TYPE LOCALITY: Pennsylvania.
RANGE: British America south to Mexico.
NEW MEXICO: Cedar Hill; Shiprock; West Fork of the Gila; Santa Rita; San Luis Mountains; Organ Mountains. Moist ground, in the Upper Sonoran Zone.

2. **Cyperus cyrtolepis** Torr. & Hook. Ann. Lyc. N. Y. **3:** 436. 1836.
TYPE LOCALITY: "Texas."
RANGE: Oklahoma to Texas and Arizona.
NEW MEXICO: A single specimen, without locality, seen.

3. **Cyperus fendlerianus** Boeckel. Linnaea **35:** 520. 1868.
TYPE LOCALITY: Near Santa Fe, New Mexico. Type collected by Fendler (no. 865).
RANGE: Arizona and western Texas to Mexico.
NEW MEXICO: Tunitcha Mountains; Santa Fe and Las Vegas mountains; Wagon Mound; Magdalena; Mangas Springs; Black Range; San Luis Mountains; Organ Mountains; White and Sacramento mountains. Open slopes, in the Upper Sonoran and Transition zones.

4. **Cyperus rusbyi** Britton, Bull. Torrey Club **11:** 29. 1884.
TYPE LOCALITY: Near Silver City, New Mexico. Type collected by Rusby in 1880.
RANGE: Mountains of New Mexico and western Texas.
NEW MEXICO: West Fork of the Gila; Silver City; Animas Valley; San Luis Mountains; Organ Mountains; Arroyo Ranch; Gray; Elida; Queen. Upper Sonoran and Transition zones.

5. **Cyperus schweinitzii** Torr. Ann. Lyc. N. Y. **3:** 276. 1836.
TYPE LOCALITY: Dry sand on the shore of Lake Ontario, near Greece, Monroe County, New York.

RANGE: British America to New Mexico and Kansas.

NEW MEXICO: Clayton; Elida; Arroyo Ranch; Nara Visa. Sandy soil, in the Upper Sonoran Zone.

6. **Cyperus uniflorus** Torr. & Hook. Ann. Lyc. N. Y. **3:** 431. 1836.

TYPE LOCALITY: Texas.

RANGE: Western Texas and southern New Mexico.

NEW MEXICO: Organ Mountains; sands south of Melrose. Dry soil, in the Upper Sonoran Zone.

7. **Cyperus speciosus** Vahl, Enum. Pl. **2:** 364. 1806.

Cyperus michauxianus Schult. Mant. **2:** 123. 1824.

TYPE LOCALITY: "Habitat in Virginia."

RANGE: Throughout most of the United States.

NEW MEXICO: Mangas Springs; Roswell. Wet ground.

7a. **Cyperus speciosus squarrosus** Britton, Bull. Torrey Club **13:** 214. 1886.

TYPE LOCALITY: "New Mexico."

This differs from the species in having the scales spreading to recurved, and reddish. Reported from New Mexico by Dr. N. L. Britton. Based on Fendler's 870, which probably came from near Santa Fe.

8. **Cyperus esculentus** L. Sp. Pl. 45. 1753. NUT GRASS.

Cyperus phymatodes Muhl. Descr. Gram. 23. 1817.

TYPE LOCALITY: "Habitat Monspelii, inque Italia, Oriente."

RANGE: British America and southward throughout the United States and tropical America; also in the Old World.

NEW MEXICO: Las Vegas; Albuquerque; Hillsboro; San Luis Mountains; Mesilla Valley; Belen; White Mountains; Gray; Roswell. Wet ground, in the Lower and Upper Sonoran zones.

9. **Cyperus erythrorhizos** Muhl. Descr. Gram. 20. 1817.

TYPE LOCALITY: Pennsylvania.

RANGE: Throughout most of the United States.

NEW MEXICO: Mesilla Valley. Wet ground, in the Lower Sonoran Zone.

10. **Cyperus sphacelatus** Rottb. Descr. Pl. 26. 1786.

TYPE LOCALITY: Surinam.

RANGE: In the southern and southwestern United States and in tropical America.

NEW MEXICO: Organ Mountains (*Wooton* 620). Upper Sonoran Zone.

2. CLADIUM R. Br. SAW GRASS.

Coarse leafy perennial with cylindric stems about a meter high; spikelets small, in large, much branched, terminal panicles; glumes overlapping, the lower empty, the middle with unisexual flowers, the uppermost with perfect flowers; perianth none; stamens 2 or 3; styles not persistent; achenes ovoid to globose, smooth or longitudinally ridged.

1. **Cladium jamaicense** Crantz, Inst. Herb. **1:** 362. 1766.

Cladium effusum Torr. Ann. Lyc. N. Y. **3:** 374. 1836.

TYPE LOCALITY: Jamaica.

RANGE: Virginia and Florida to Texas and New Mexico; also in the West Indies.

NEW MEXICO: Roswell. In shallow water.

3. ELEOCHARIS R. Br. SPIKE RUSH.

Annual or perennial scapose herbs, 15 to 30 cm. high or more, the leaves reduced to basal sheaths, the solitary terminal spikes without subtending bracts; stems cylindric, flattened, or angular, erect; spikelets small; perianth of 1 to 12 bristles; stamens 2 or 3; base of style swollen, persistent as a tubercle on the lenticular or 3-angled achene.

KEY TO THE SPECIES.

Style branches 2.
  Annuals; bristles shorter than the achenes; spikes oblong-cylindric; tubercle broad and low.................... 1. *E. engelmanni.*
  Perennial by rootstocks; bristles longer than achenes; spikes and tubercles various.
    Plants stout; tubercles conic-triangular................ 2. *E. palustris.*
    Plants slender; tubercles almost cylindrical............ 3. *E. glaucescens.*
Styles branches 3.
  Plants very small, 3 to 10 cm. high; fruit obovoid-oblong, with numerous longitudinal ridges and finer transverse ones. 6. *E. acicularis.*
  Plants larger, 20 cm. high or more; fruit various.
    Tubercles constricted at the base, clearly distinct from the achene; plants slender, with slender rootstocks.. 4. *E. montana.*
    Tubercles apparently confluent with the achene, cylindric; plants stouter, not stoloniferous............. 5. *E. rostellata.*

1. **Eleocharis engelmanni** Steud. Syn. Pl. Glum. **2:** 79. 1855.
TYPE LOCALITY: St. Louis, Missouri.
RANGE: New England to California.
NEW MEXICO: West Fork of the Gila (*Metcalfe* 589). In wet soil.

2. **Eleocharis palustris** (L.) Roem. & Schult. Syst. Veg. **2:** 151. 1817.
Scirpus palustris L. Sp. Pl. 47. 1753.
TYPE LOCALITY: European.
RANGE: Throughout North America except in the extreme northern part.
NEW MEXICO: Chama; Farmington; Jewett; Mule Creek; Mesilla Valley. In wet soil, in the Lower and Upper Sonoran zones.

3. **Eleocharis glaucescens** (Willd.) Schult. Mant. **2:** 89. 1824.
Scirpus glaucescens Willd. Enum. Pl. 76. 1809.
Eleocharis palustris glaucescens A. Gray, Man. 558. 1848.
TYPE LOCALITY: "Habitat in America boreali."
RANGE: Throughout North America except in the extreme north.
NEW MEXICO: Santa Fe Creek; Pecos; Las Vegas. Wet soil, in the Upper Sonoran Zone.

4. **Eleocharis montana** (H. B. K.) Roem. & Schult. Syst. Veg. **2:** 153. 1817.
Scirpus montanus H. B. K. Nov. Gen. & Sp. **1:** 226. 1816.
TYPE LOCALITY: "In monte Quindiu," Colombia.
RANGE: Colorado to California, southward to South America.
NEW MEXICO: Zuni Reservation; Las Vegas; Bear Canyon; Rio Pueblo; Wheelers Ranch; Berendo Creek; Rincon; Apache Teju; Mesilla Valley; Silver Spring Canyon; Mangas Springs. Wet soil, in the Lower and Upper Sonoran and the Transition zones.

5. **Eleocharis rostellata** Torr. Fl. N. Y. **2:** 347. 1843.
Scirpus rostellatus Torr. Ann. Lyc. N. Y. **3:** 318. 1836.
TYPE LOCALITY: Penn Yan, New York.
RANGE: Throughout North America except in the extreme northern part.
NEW MEXICO: Grant County; plains north of the White Sands. In wet soil, in the Lower Sonoran Zone.

6. Eleocharis acicularis (L.) Roem. & Schult. Syst. Veg. **2**: 154. 1817.

Scirpus acicularis L. Sp. Pl. 48. 1753.

TYPE LOCALITY: European.

RANGE: Throughout North America except in the extreme northern part; also in the Old World.

NEW MEXICO: Tunitcha Mountains; West Fork of the Gila; Cloverdale. Wet soil.

Eleocharis capitata R. Br. and *E. atropurpurea* (Retz.) Kunth may come into New Mexico, as they occur very close to our borders.

4. STENOPHYLLUS Raf.

Small grasslike annuals, 15 cm. high or mostly less, with basal leaves and umbellate or capitate flower clusters of small spikelets subtended by 1 to several bracts; flowers perfect; glumes overlapping; perianth none; stamens 2 or 3; style swollen at the base and persistent; achenes 3-angled or lenticular.

KEY TO THE SPECIES.

Achenes longitudinally ribbed and transversely roughened; plants 10 to 15 cm. high; spikelets solitary or umbellate on the same plant.. 1. *S. capillaris.*

Achenes rugose; plants 8 cm. high or less; spikelets solitary at the summit of the culm and also at the bases of the leaves........ 2. *S. funckii.*

1. Stenophyllus capillaris (L.) Britton, Bull. Torrey Club **21**: 30. 1894.

Scirpus capillaris L. Sp. Pl. 49. 1753.

Fimbristylis capillaris A. Gray, Man. 530. 1848.

TYPE LOCALITY: "Habitat in Virginia, Aethiopia, Zeylona."

RANGE: Throughout North America except in the extreme northern part; also in the Old World.

NEW MEXICO: Mogollon Mountains; Santa Rita; San Luis Mountains; Organ Mountains. Wet ground, in the Upper Sonoran Zone.

2. Stenophyllus funckii (Steud.) Britton, Bull. Torrey Club **21**: 30. 1894.

Isolepis funckii Steud. Syn. Pl. Glum. **2**: 91. 1855.

Scirpus heterocarpus S. Wats. Proc. Amer. Acad. **18**: 171. 1883.

TYPE LOCALITY: Venezuela.

RANGE: Arizona and New Mexico, southward through tropical America to Bolivia.

NEW MEXICO: West Fork of the Gila (*Metcalfe* 661). Wet ground.

5. SCIRPUS L. BULRUSH.

Annuals or perennials, sometimes small and grasslike, sometimes tall (1 meter or more), with reduced basal leaves or sheaths; spikelets cylindric or somewhat flattened, spirally imbricated, in terminal clusters, single, capitate, or umbellate, subtended by 1 to several bracts; flowers perfect; perianth of 1 to 6 bristles (rarely none); stamens 2 or 3; style not swollen at the base; achenes triangular or lenticular.

KEY TO THE SPECIES.

Involucral bracts 1 or 2 or none.

- Spikelets solitary, terminal; involucral bracts none............ 1. *S. pauciflorus.*
- Spikelets several, seemingly lateral; involucral bracts 1 or 2.
  - Culms terete; involucral bracts 2.......................... 5. *S. occidentalis.*
  - Culms triangular; involucral bract 1, seeming to be a prolongation of the culm.
    - Involucral bract short, 3 cm. long or less, barely exceeding the spikelets, these generally 4 to 6, crowded; leaves about one-fifth the length of the culm.................................. 2. *S. olneyi.*

Involucral bract 4 to 10 cm. long, much exceeding the spikelets, these few, frequently only 1; leaves half as long as the culm or more .... 3. *S. americanus.*

Involucral bracts of several flat leaves much exceeding the compound umbellate inflorescence.

Culms triangular; spikelets large, 10 to 20 mm. long, light yellowish brown; inflorescence a simple umbel or in young plants capitate .... 4. *S. brittonianus.*

Culms terete; spikelets small, 2 to 7 mm. long, greenish; inflorescence a once or twice compound umbellate cluster with numerous unequal rays.

Style branches 2; achenes rounded on the back; inflorescence twice compound; spikelets not capitate...... 6. *S. microcarpus.*

Style branches 3; achenes angled on the back; inflorescence generally once compound; spikelets densely capitate at the ends of the rays.................... 7. *S. atrovirens.*

1. **Scirpus pauciflorus** Lightf. Fl. Scot. 1078. 1777.

Eleocharis pauciflorus Link, Hort. Berol. 1: 284. 1827.

TYPE LOCALITY: Highlands of Scotland.

RANGE: British America to New York, New Mexico, and California; also in Europe.

NEW MEXICO: A single specimen without locality seen.

It is probable that the plant is not uncommon in the mountains of New Mexico but has been overlooked by collectors.

2. **Scirpus olneyi** A. Gray, Bost. Journ. Nat. Hist. 5: 238. 1845.

TYPE LOCALITY: In a salt marsh on the Seekonk River, Rhode Island.

RANGE: Across the United States.

NEW MEXICO: Salt Lake; Santa Rita; Dog Spring; Round Mountain. Wet alkaline soil, in the Upper Sonoran Zone.

3. **Scirpus americanus** Pers. Syn. Pl. 1: 68. 1805.

Scirpus pungens Vahl, Enum. Pl. 2: 255. 1808.

TYPE LOCALITY: "Hab. in Carolina inferiore."

RANGE: Throughout North America; also in South America.

NEW MEXICO: Farmington; Carrizo Mountains; Taos; San Juan; Wheelers Ranch; Berendo Creek; Rincon; Mesilla Valley. In swamps, in the Lower and Upper Sonoran zones.

4. **Scirpus brittonianus** Piper, Contr. U. S. Nat. Herb. 11: 157. 1906.

Scirpus campestris Britton in Britt. & Brown, Illustr. Fl. 1: 267. 1896, not Roth, 1800.

Scirpus robustus campestris Fernald, Rhodora 2: 241. 1900.

TYPE LOCALITY: "On wet prairies and plains, Manitoba and Minnesota to Nebraska, Kansas, and Mexico, west to Nevada."

RANGE: As under type locality.

NEW MEXICO: Carrizo Mountains; Farmington; Salt Lake; Mesilla Valley; near Carrizozo; Roswell. Wet ground, in the Lower and Upper Sonoran zones.

5. **Scirpus occidentalis** (S. Wats.) Chase, Rhodora 6: 68. 1904.

Scirpus lacustris occidentalis S. Wats. Bot. Calif. 2: 218. 1880.

TYPE LOCALITY: San Diego County, California.

RANGE: British Columbia and California to New England.

NEW MEXICO: Shiprock; Farmington; Gallo Spring; Mangas Springs; Berendo Creek; Mesilla Valley; Roswell; Carrizozo. Wet ground, in the Lower and Upper Sonoran zones.

6. Scirpus microcarpus Presl, Rel. Haenk. **1:** 195. 1828.

Scirpus lenticularis Torr. Ann. Lyc. N. Y. **3:** 328. 1836.

Scirpus sylvaticus digynus Boeckel. Linnaea **36:** 727. 1870.

TYPE LOCALITY: Nootka Sound, Vancouver Island.

RANGE: British America and New England to California, Utah, and New Mexico.

NEW MEXICO: Chama; West Fork of the Gila; Mimbres River. Wet ground, in the Transition Zone.

7. Scirpus atrovirens Muhl. Descr. Gram. 43. 1817.

TYPE LOCALITY: Pennsylvania.

RANGE: Northeastern Atlantic States west to Alberta, south in the Rocky Mountains to New Mexico.

NEW MEXICO: Pecos (*Standley* 5104). Upper Sonoran Zone.

6. HEMICARPHA Nees.

Low tufted grasslike annual, 10 cm. high or less, with erect or spreading, slender leaves and small, terminal, headlike or solitary spikelets with 1 to 3 leaflike bracts surrounding and much exceeding them; glumes spirally imbricated, deciduous; perianth wanting; stamen 1; achene obovoid-oblong, little compressed, brown.

1. Hemicarpha micrantha (Vahl) Britton, Bull. Torrey Club **21:** 34. 1894.

Scirpus micranthus Vahl, Enum. Pl. **2:** 254. 1806.

Hemicarpha subsquarrosa Nees in Mart. Fl. Bras. **2**[1]: 61. 1842.

TYPE LOCALITY: Given doubtfully as South America.

RANGE: Nearly throughout North America and in South America.

NEW MEXICO: Albuquerque (*Bigelow*). Wet ground.

7. ERIOPHORUM L. COTTON GRASS.

Perennial from a rootstock, the culms erect; spikelets in a terminal umbel subtended by an involucre of one or more leaves; flowers perfect; perianth of numerous white bristles, these soft and cotton-like, much exserted; style 3-cleft; achenes obovoid, 3-angled, light brown.

1. Eriophorum angustifolium Roth, Tent. Fl. Germ. **1:** 24. 1788.

TYPE LOCALITY: Germany.

RANGE: Alaska and Newfoundland to Maine, Illinois, and northern New Mexico.

NEW MEXICO: Costilla Valley (*Wooton*). Bogs.

8. CAREX L. SEDGE.

Perennial grasslike plants with 3-ranked leaves and mostly 3-angled culms; flowers unisexual, monœcious or diœcious; perianth wanting; stamens 3; pistillate flowers a single pistil with 2 or 3 stigmas, in a saclike perigynium, this completely inclosing the achene; achenes 3-angled or lenticular.

A very large genus of which the following listed species probably represent only a part of those indigenous to New Mexico. Collectors rarely take the trouble to examine the plants unless their attention is particularly called to them. There are no doubt several species common in the high mountains of the northern part of the State which have not been collected.

The writers are under special obligations to Mr. K. K. Mackenzie for assistance in the preparation of an account of this genus. Mr. Mackenzie identified most of our material and prepared the key to the species.

KEY TO THE SPECIES.

Achenes lenticular; stigmas 2; terminal spike partly pistillate or if staminate the lateral spikes short or heads diœcious. (VIGNEA.)
- Spikes mostly staminate at the base.
  - Perigynia not wing-margined. (STELLULATAE.)
    - Perigynia with very short beaks, widely spreading at maturity........ 13. *C. interior.*
    - Perigynia with long beaks, appressed........ 14. *C. bolanderi.*
  - Perigynia wing-margined. (OVALES.)
    - Beak of the perigynium flattened and margined to the tip.
      - Scales strongly tinged with reddish brown........ 21. *C. wootoni.*
      - Scales little if at all tinged with reddish brown.
        - Perigynia thin, lanceolate (at least two and one-half times as long as wide)........ 19. *C. scoparia.*
        - Perigynia thick, ovate........ 20. *C. festucacea.*
    - Beak of the perigynium slender, nearly terete and scarcely margined at the apex.
      - Several of the bracts conspicuously exceeding the head........ 18. *C. tenuirostris.*
      - Bracts inconspicuous.
        - Perigynia 2.5 to 3.5 mm. long; culms smooth beneath the head........ 15. *C. subfusca.*
        - Perigynia 4.5 to 6 mm. long; culms rough beneath the head.
          - Perigynia about 4.5 mm. long, ovate........ 16. *C. festiva.*
          - Perigynia 4.5 to 6 mm. long, lanceolate........ 17. *C. ebenea.*
- Spikes staminate at apex or some spikes wholly staminate.
  - Perigynia little compressed, whitish-puncticulate. (TENELLAE.)........ 12. *C. disperma.*
  - Perigynia strongly compressed, not whitish-puncticulate.
    - Culms one to few together, the rootstocks long and creeping.
      - Perigynia wing-margined, the beak bidentate. (ARENARIAE.)........ 5. *C. siccata.*
      - Perigynia not wing-margined, the beak obliquely cut, or bidentate in age. (DIVISAE.)
        - Culms smooth above; rootstocks slender........ 1. *C. douglasii.*
        - Culms rough beneath the head; rootstocks stout.
          - Perigynia chestnut, scarcely sharp-edged, the beak about one-fifth the length of the body at maturity........ 4. *C. simulata.*
          - Perigynia brownish or blackish, sharp-edged, tapering into a beak about half the length of the body.
            - Spikes with one to several perigynia, the heads appearing a mass of straw-colored scales; staminate flowers conspicuous; perigynium body at maturity about 2 mm. wide........ 2. *C. latebrosa.*

Spikes with about 10 perigynia, concealed by inconspicuous scales, these tinged with green or brown; staminate flowers inconspicuous; perigynium body at maturity less than 1.5 mm. long........................ 3. *C. camporum.*

Culms cespitose, the rootstocks at most short-creeping.

Spikes numerous, in a more or less compound head. (MULTIFLORAE.)....................11. *C. agrostoides.*

Spikes less than 10, in a simple head.

Perigynia strongly nerved, the beak exceeding the body. (STENORHYNCHAE.)...10. *C. stipata.*

Perigynia weakly nerved, the beak not exceeding the body. (MUHLENBERGIANAE.)

Scales conspicuously tinged with reddish brown............................ 6. *C. occidentalis.*

Scales at most faintly tinged with reddish brown.

Perigynia obliquely cut or shallowly bidentate, weakly serrulate........ 9. *C. rusbyi.*

Perigynia deeply bidentate, strongly serrulate.

Spikes with few perigynia; sheaths tight, inconspicuously septate-nodulose...................... 7. *C. neomexicana.*

Spikes with several to many perigynia: sheaths soon loose, easily breaking, conspicuously septate-nodulose...................... 8. *C. gravida.*

Achenes triangular or lenticular; if lenticular the lower lateral spikes elongated and terminal spikes staminate. (EUCAREX.)

Spike solitary.

Perigynia coriaceous, glabrous; rootstocks long-creeping. (NITIDAE.)....................................22. *C. obtusata.*

Perigynia not coriaceous, puberulent; culms densely cespitose. (FILIFOLIAE.)..........................23. *C. filifolia.*

Spikes more than one.

Perigynia puberulent, triangular or suborbicular in cross section, long-stipitate, 2-ribbed. (MONTANAE.)

Basal spikes absent..................................24. *C. heliophila.*

Basal spikes numerous.

Perigynium body suborbicular; staminate spike 2.5 mm. wide; bract shorter than the culm; blades 1.5 to 2.5 mm. wide................25. *C. geophila.*

Perigynium body oval; staminate spike 1.5 mm. wide; bract normally exceeding the culm; blades 0.75 to 1.5 mm. wide..............26. *C. pityophila.*

Perigynia differing from the above section in one or more particulars.

Pistillate spikes drooping on slender penduncles; perigynia strongly beaked, not bidentate. (CAPILLARES.).................................27. *C. capillaris.*

Pistillate spikes erect or, if drooping, the perigynia differing from above.
Achenes normally lenticular and the stigmas 2.
Perigynia golden yellow at maturity; spikes few-flowered; plants low, slender. (BICOLORES.)..............................28. *C. aurea.*
Perigynia not golden yellow at maturity; spikes many-flowered; plants tall or stout. (RIGIDAE.)
Perigynium beak strongly bidentate; perigynia ribbed..........................33. *C. nebraskensis.*
Perigynium beak, if present, not bidentate; perigynia various.
Lowest bract not exceeding the inflorescence..............................29. *C. scopulorum.*
Lowest bract exceeding the inflorescence.
Perigynia pale green, finely many-nerved..........................30. *C. kelloggii.*
Perigynia greenish straw-colored, few-nerved or nerveless.
Leaf blades flat and canaliculate, the edges serrulate above only; perigynia at most obscurely nerved..31. *C. variabilis.*
Leaf blades, at least the lower, plicate, the margins revolute, the edges serrulate throughout; perigynia few-nerved...............32. *C. emoryi.*
Achenes triangular; stigmas 3.
Perigynia beakless or very shortly beaked.
Terminal 2 or 3 spikes staminate. (TRACHYCHLAENAE.) ....37. *C. ultra.*
Terminal spikes pistillate above, staminate below. (ATRATAE.)
Perigynia 2.5 mm. long or less, little compressed, the margins not appearing winglike..................34. *C. halleri.*
Perigynia more than 2.5 mm. long, strongly compressed, the margins winglike.
Spikes all closely sessile, contiguous, forming a dense lobed head.....35. *C. nova.*
Lateral spikes peduncled, distant, usually nodding..................36. *C. bella.*
Perigynia strongly beaked, the beak deeply bidentate
Perigynia or sheaths pubescent. (HIRTAE.)
Perigynia pubescent, the teeth short; sheaths glabrous..............38. *C. lanuginosa.*
Perigynia glabrous, the teeth long; sheaths pubescent............39. *C. atherodes.*

Perigynia and sheaths not pubescent.
Perigynia closely ribbed; pistillate spikes nodding. (PSEUDO-CYPEREAE.) .... 40. *C. hystricina.*
Perigynia coarsely ribbed; pistillate spikes erect. (PHYSOCARPAE.). 41. *C. rostrata.*

1. **Carex douglasii** Boott in Hook. Fl. Bor. Amer. **2**: 213. 1840.
Carex fendleriana Boeckel. Linnaea **39**: 135. 1875.
TYPE LOCALITY: Northwest coast of North America.
RANGE: Wyoming to Utah and New Mexico.
NEW MEXICO: Chama; Santa Fe; Ponchuelo Creek; Nambe Valley. Open meadows, in the Transition Zone.

2. **Carex latebrosa** Mackenz. Bull. Torrey Club **34**: 603. 1908.
Carex gayana hyalina Bailey, Proc. Amer. Acad. **22**: 135. 1886.
TYPE LOCALITY: Sonora, Mexico.
RANGE: Nevada, New Mexico, and northern Mexico.
NEW MEXICO: Cienaga Ranch; Berendo Creek. Upper Sonoran Zone.

3. **Carex camporum** Mackenz. Bull. Torrey Club **37**: 244. 1910.
TYPE LOCALITY: "Columbia River."
RANGE: British Columbia to California and New Mexico.
NEW MEXICO: Chama (*Standley* 6753).

4. **Carex simulata** Mackenz. Bull. Torrey Club **34**: 604. 1908.
TYPE LOCALITY: Chug Creek, Albany County, Wyoming.
RANGE: Washington to Montana, Colorado, and New Mexico.
NEW MEXICO: Cienaga Ranch (*Wooton*).
Mr. Mackenzie also refers here Fendler's 881, collected somewhere about Santa Fe.

5. **Carex siccata** Dewey, Amer. Journ. Sci. **10**: 278. 1826.
TYPE LOCALITY: Westfield, Massachusetts.
RANGE: British America to California, Arizona, and New Mexico.
NEW MEXICO: Pecos Baldy; Winsor Creek. Arctic-Alpine Zone.

6. **Carex occidentalis** Bailey, Mem. Torrey Club **1**: 5. 1889.
Carex muricata americana Bailey, Proc. Amer. Acad. **22**: 140. 1886.
TYPE LOCALITY: Santa Rita Mountains, Arizona.
RANGE: Montana to Colorado and New Mexico.
NEW MEXICO: Santa Fe; Winsors Ranch; Magdalena Mountains; Mangas Springs; Manguitas Spring; Eagle Peak; Sierra Grande; Tierra Amarilla. Damp ground, chiefly in the Transition Zone.

7. **Carex neomexicana** Mackenz. Bull. Torrey Club **34**: 153. 1907.
TYPE LOCALITY: Santa Rita del Cobre on the Rio Mimbres, New Mexico.
RANGE: New Mexico and Arizona.
NEW MEXICO: Organ Mountains; Santa Rita.

8. **Carex gravida** Bailey, Mem. Torrey Club **1**: 5. 1889.
TYPE LOCALITY: "Northern Illinois * * * to northwestern Iowa."
RANGE: Ohio and Illinois to South Dakota, Oklahoma, and northeastern New Mexico.
NEW MEXICO: Sierra Grande (*Standley* 6069). Upper Sonoran Zone.

9. **Carex rusbyi** Mackenz. Smiths. Misc. Coll. **65**[7]: 2. 1915.
TYPE LOCALITY: Yavapai county, Arizona.
RANGE: New Mexico and Arizona.
NEW MEXICO: Organ Mountains (*Wooton*).

10. Carex stipata Muhl.; Willd. Sp. Pl. **4:** 233. 1805.

TYPE LOCALITY: Pennsylvania.

RANGE: British America, south through the United States.

NEW MEXICO: Middle Fork of the Gila (*Wooton*).

11. Carex agrostoides Mackenz. Bull. Torrey Club **34:** 607. 1908.

TYPE LOCALITY: Luna, northwest of the Mogollon Mountains, Socorro County, New Mexico. Type collected by Wooton, July 28, 1900.

RANGE: Mountains of western New Mexico and of Arizona.

NEW MEXICO: Luna; Guadalupe Canyon; Mangas Springs. Transition Zone.

12. Carex disperma Dewey, Amer. Journ. Sci. **8:** 266. 1824.

TYPE LOCALITY: Massachusetts.

RANGE: British America to Pennsylvania, New Mexico, and California.

NEW MEXICO: Santa Fe and Las Vegas mountains. Transition Zone.

13. Carex interior Bailey, Bull. Torrey Club **20:** 426. 1893.

TYPE LOCALITY: "Bogs and swamps in the interior country from Maine to Minnesota and Kansas."

RANGE: Maine to Minnesota, Florida, and New Mexico.

NEW MEXICO: Winsors Ranch (*Standley* 4254). Transition Zone.

14. Carex bolanderi Olney, Proc. Amer. Acad. **7:** 393. 1868.

Carex deweyana bolanderi Boott in S. Wats. Bot. Calif. **2:** 236. 1880.

TYPE LOCALITY: "California, Yosemite Valley, and Mariposa Bigtree grove."

RANGE: British Columbia to California, east to Idaho and western New Mexico.

NEW MEXICO: Mogollon Creek (*Metcalfe* 286). Damp ground.

15. Carex subfusca Boott in S. Wats. Bot. Calif. **2:** 234. 1880.

Carex macloviana subfusca Kükenth. in Engl. Pflanzenreich **38:** 197. 1909.

TYPE LOCALITY: "Lake Tahoe (*Kellogg*), and near Virginia City, Nevada, *Bloomer*."

RANGE: Washington to California, east to western New Mexico.

NEW MEXICO: Bear Mountains (*Rusby* 423). Wet ground.

16. Carex festiva Dewey, Amer. Journ. Sci. **29:** 246. *pl. w. f. 71.* 1836.

TYPE LOCALITY: "At Bear Lake and on the Rocky Mountains."

RANGE: British America to Mexico.

NEW MEXICO: Winsor Creek; Mogollon Creek; Ruidoso; Truchas Peak; Chama; Tierra Amarilla. Wet ground, in the Transition Zone.

17. Carex ebenea Rydb. Bull. Torrey Club **28:** 266. 1901.

TYPE LOCALITY: Pikes Peak, Colorado.

RANGE: British Columbia and Alberta to Utah and New Mexico.

NEW MEXICO: Truchas Peak; Pecos Baldy. Meadows, in the Arctic-Alpine Zone.

18. Carex tenuirostris Olney, Amer. Nat. **8:** 214. 1874.

TYPE LOCALITY: Western Wyoming.

RANGE: Rocky Mountains, south to northwestern New Mexico.

NEW MEXICO: Tunitcha Mountains (*Standley* 7626).

19. Carex scoparia Schkuhr; Willd. Sp. Pl. **4:** 230. 1805.

TYPE LOCALITY: "Habitat in America boreali."

RANGE: British America to Washington, New Mexico, and Florida.

NEW MEXICO: West Fork of the Gila (*Metcalfe* 577.) Transition Zone.

20. Carex festucacea brevior (Dewey) Fernald, Proc. Amer. Acad. **37:** 477. *pl. 3. f. 49–51.* 1902.

Carex straminea brevior Dewey, Amer. Journ. Sci. **11:** 158. 1826.

TYPE LOCALITY: "In Missouri."

RANGE: British Columbia to New England, south to Arkansas and northern New Mexico.

NEW MEXICO: Chama; Sierra Grande. Transition Zone.

21. Carex wootoni Mackenz. Smiths. Misc. Coll. **65**[7]: 1. 1915.

TYPE LOCALITY: San Francisco Mountains, New Mexico.

RANGE: New Mexico and Arizona.

NEW MEXICO: Chama; Sawyers Peak; Winter Folly.

22. Carex obtusata Liljebl. Bih. Svensk. Vet. Akad. Handl. **14:** 69. *pl. 4.* 1793.

TYPE LOCALITY: "Habitat Oelandiae, prope Kjoping, in locis apricis, arenosis, rarius."

RANGE: British America to Colorado and New Mexico.

NEW MEXICO: Winsor Creek (*Standley*). Transition Zone.

23. Carex filifolia Nutt. Gen. Pl. **2:** 204. 1818.

TYPE LOCALITY: "Dry plains and gravelly hills of the Missouri."

RANGE: British America to Nebraska, New Mexico, and California.

NEW MEXICO: Santa Fe; Las Vegas. Dry plains, in the Upper Sonoran Zone.

24. Carex heliophila Mackenz. Torreya **13:** 15. 1913.

TYPE LOCALITY: Open prairie, near Lee's Summit, Jackson County, Missouri.

RANGE: New Mexico and Wyoming to Iowa and Illinois.

NEW MEXICO: Chama; Tierra Amarilla; Nutritas Creek; Glorieta; Raton. Open plains and hillsides, in the Upper Sonoran and Transition zones.

25. Carex geophila Mackenz. Bull. Torrey Club **40:** 546. 1913.

TYPE LOCALITY: Tierra Amarilla, Rio Arriba County, New Mexico. Type collected by W. W. Eggleston (no. 6584).

RANGE: Mountains of western New Mexico.

NEW MEXICO: Tierra Amarilla; Chama; Mogollon Mountains.

26. Carex pityophila Mackenz. Bull. Torrey Club **40:** 545. 1913.

TYPE LOCALITY: Southeast of Tierra Amarilla, Rio Arriba County, New Mexico. Type collected by W. W. Eggleston (no. 6605).

RANGE: Northern New Mexico.

NEW MEXICO: Near Tierra Amarilla. Upper Sonoran Zone.

27. Carex capillaris elongata Olney; Fernald, Proc. Amer. Acad. **37:** 509. 1902.

TYPE LOCALITY: Twin Lakes, Colorado.

RANGE: British America to New Mexico and New York.

NEW MEXICO: Truchas Peak; Grass Mountain. Canadian and Hudsonian zones.

28. Carex aurea Nutt. Gen. Pl. **2:** 205. 1818.

TYPE LOCALITY: On the shores of Lake Michigan.

RANGE: British America to Pennsylvania, New Mexico, and Wyoming.

NEW MEXICO: Chama: Grass Mountain; Winsors Ranch. Meadows, in the Transition and Canadian zones.

29. Carex scopulorum Holm, Amer. Journ. Sci. **14:** 421. *f. 1–6.* 1902.

TYPE LOCALITY: Clear Creek Canyon, Colorado.

RANGE: Wet ground, Washington and Montana to Colorado.

So far not collected in New Mexico, but doubtless to be found on the high mountains in the northern part of the State.

30. Carex kelloggii Boott in S. Wats. Bot. Calif. **2:** 240. 1880.

TYPE LOCALITY: In the Sierra Nevada at Alta, California.

RANGE: Alaska to California, east to Utah and New Mexico.

NEW MEXICO: Chama (*Standley* 6844).

31. Carex variabilis Bailey, Mem. Torrey Club **1:** 18. 1889.
Type locality: Colorado.
Range: Montana and Idaho to New Mexico.
New Mexico: Ponchuelo Creek; Spirit Lake; Pecos Baldy; Truchas Peak. Transition to Arctic-Alpine Zone.

32. Carex emoryi Dewey; Torr. U. S. & Mex. Bound. Bot. 230. 1859.
Type locality: "On the upper Rio Grande." Type collected by Bigelow.
Range: Southern New Mexico and western Texas.
New Mexico: Mesilla Valley (*Metcalfe*). Along ditches, in the Lower Sonoran Zone.

33. Carex nebraskensis Dewey, Amer. Journ. Sci. II. **18:** 102. 1854.
Carex nebraskensis praevia Bailey, Bot. Gaz. **21:** 3. 1896.
Carex jamesii Torr. Ann. Lyc. N. Y. **3:** 398. 1836, not Schwein. 1824.
Type locality: Nebraska.
Range: Washington and Nebraska to New Mexico.
New Mexico: Taos.

34. Carex halleri Gunn. Fl. Norveg. no. 849. 1766–72.
Type locality: Mountains of Norway.
Range: British America to Oregon, Minnesota, and New Mexico; also in Europe and Asia.
New Mexico: Ponchuelo Creek (*Standley* 4183). Damp meadows, Canadian to Arctic-Alpine Zone.

35. Carex nova Bailey, Lond. Journ. Bot. **26:** 322. 1888.
Type locality: "Mountains of Wyoming and Colorado and southward."
Range: Wyoming and New Mexico to California.
New Mexico: Pecos Baldy; Truchas Peak; Spirit Lake. Meadows and damp woods, in the Hudsonian and Arctic-Alpine zones.

36. Carex bella Bailey, Bot. Gaz. **17:** 152. 1892.
Type locality: "Mountains, Colorado, Utah, and Arizona.
Range: Wyoming to Utah, New Mexico, and Arizona.
New Mexico: Near the head of the Nambe (*Standley* 4434). Canadian Zone.

37. Carex ultra Bailey, Proc. Amer. Acad. **22:** 83. 1886.
Carex spissa ultra Kükenth. in Engl. Pflanzenreich **38:** 422. 1909.
Type locality: Southern Arizona.
Range: Southern Arizona and New Mexico.
New Mexico: Head of Guadalupe Canyon near Cloverdale (*Mearns* 377, 487).

38. Carex lanuginosa Michx. Fl. Bor. Amer. **2:** 175. 1803.
Carex filiformis latifolia Boeckel. Linnaea **41:** 309. 1876.
Carex lasiocarpa lanuginosa Kükenth. in Engl. Pflanzenreich **38:** 748. 1909.
Type locality: "Ad lacus Mistassins," Canada.
Range: British America to Pennsylvania, New Mexico, and California.
New Mexico: Chama; Ponchuelo Creek; Winsors Ranch; Las Vegas; Jewett; Magdalena Mountains; Grant County; Gilmores Ranch. Wet ground, in the Transition Zone.

39. Carex atherodes Spreng. Syst. Veg. **3:** 828. 1826.
Type locality: Arctic America.
Range: British America to New York, Nebraska, and New Mexico.
New Mexico: Near Fort Defiance (*Palmer*).

40. Carex hystricina Muhl.; Willd. Sp. Pl. **4:** 282. 1805.

TYPE LOCALITY: "Habitat in humidis Pennsylvaniae."

RANGE: British America to Georgia, Nebraska, and New Mexico.

NEW MEXICO: North Percha Creek (*Metcalfe* 1122).

41. Carex rostrata Stokes in With. Bot. Arr. Veg. Brit. ed. 2. **2:** 1059. 1787.

TYPE LOCALITY: "Bogs of Isla, and on Bentelkerny and Breadalbane," England.

RANGE: British America to California, New Mexico, and New York; also in Europe.

NEW MEXICO: Truchas Peak; Silver Spring Canyon; Chama. Bogs, Transition to Arctic-Alpine Zone.

9. KOBRESIA WILLD.

Slender tufted perennial with erect culms and solitary spikes; spikelets 1 or 2-flowered, spicate; perigynium none; stigmas 3; achenes obtusely 3-angled, sessile.

1. Kobresia bellardi (All.) Degland in Loisel. Fl. Gall. **2:** 626. 1807.

Carex bellardi All. Fl. Pedem. **2:** 264. *pl. 92. f. 2.* 1785.

Kobresia scirpina Willd. Sp. Pl. **4:** 205. 1805.

Elyna spicata Schrad. Fl. Germ. **1:** 155. 1806.

Elyna bellardi Koch, Linnaea **21:** 616. 1848.

TYPE LOCALITY: European.

RANGE: Arctic America, south in the Rocky Mountains to northern New Mexico; also in Europe and Asia.

NEW MEXICO: Costilla Valley (*Wooton*). Wet ground.

Order 11. ARALES.

15. LEMNACEAE. Duckweed Family.

Small unattached aquatics, the plant body a thallus, rooting from beneath.

KEY TO THE GENERA.

Roots several from a prominently nerved thallus............ 1. SPIRODELA (p. 124).
Root solitary from a faintly nerved thallus................. 2. LEMNA (p. 142).

1. SPIRODELA SCHLEID.

Thallus disk-shaped, 7 to 12-nerved, 2 to 10 mm. long, bearing a single cluster of 4 to 16 elongated roots; spathe saclike; ovary 2-ovuled; fruit unknown.

1. Spirodela polyrhiza (L.) Schleid. Linnaea **13:** 392. 1839.

Lemna polyrhiza L. Sp. Pl. 970. 1753.

TYPE LOCALITY: "Habitat in Europae paludibus fossis."

RANGE: In still water throughout most of North America and in the Old World.

Reported by Mr. J. R. Watson from "ponds about the fair grounds, Albuquerque." [1]

2. LEMNA L. DUCKWEED.

Thallus disk-shaped, usually provided with a central nerve, with or without lateral ones, each with a single root; ovary with 1 to 6 ovules; fruit ovoid, more or less ribbed.

KEY TO THE SPECIES.

Fronds long-stipitate, narrowly oblong, 6 to 10 mm. long, mostly submerged, often forming large masses........................... 1. *L. trisulca.*
Fronds not stipitate, broadly elliptic to obovate, 3 mm. long or less, floating.
  Fronds obovate, 3 mm. long; fruit more or less lenticular........ 2. *L. minor.*
  Fronds elliptic or oblong, 2 mm. long or less; fruit elongated...... 3. *L. minima.*

[1] Bull. Univ. N. Mex. **49:** 94. 1908.

1. **Lemna trisulca** L. Sp. Pl. 970. 1753.
TYPE LOCALITY: "Habitat in Europe sub aquis pigris puris."
RANGE: Throughout North America, Asia, and Europe.
NEW MEXICO: Mountains west of San Antonio; Mimbres. Floating in water.

2. **Lemna minor** L. Sp. Pl. 970. 1753.
TYPE LOCALITY: "Habitat in Europae aquis quietis."
RANGE: Nearly cosmopolitan.
NEW MEXICO: Santo Domingo; Sycamore Creek; mountains northeast of Santa Rita; Mule Creek; Nutritas Creek. Floating in water.

3. **Lemna minima** Phil. Linnaea **33**: 239. 1864, name only; Hegelm. Lemn. 138. 1868.
TYPE LOCALITY: Chile.
RANGE: Southwestern United States to South America.
NEW MEXICO: West Fork of the Gila (*Metcalfe* 407). Floating in water.

Order 12. XYRIDALES.

KEY TO THE FAMILIES.

Calyx and corolla free, of very different members; stamens free .... 16. **COMMELINACEAE** (p. 125).
Calyx and corolla partly united, of similar members; stamens partly adnate to the perianth. 17. **PONTEDERIACEAE** (p. 126).

16. COMMELINACEAE. Spiderwort Family.

Herbs with simple or branched stems and fibrous or fleshy roots; leaves sheathing at the base, the uppermost often dissimilar and forming a spathe about the flowers; flowers blue or purple; sepals 3, persistent; petals 3; stamens 6, hypogynous; capsule 2 or 3-celled.

KEY TO THE GENERA.

Perfect stamens 3 or 2; bracts spathelike; petals dissimilar; filaments naked .... 1. COMMELINA (p. 125).
Perfect stamens 6 or 5; bracts like the foliage leaves; petals similar; filaments hairy .... 2. TRADESCANTIA (p. 126).

1. COMMELINA L. DAYFLOWER.

Perennial herbs with tuberous roots in clusters, sheathing petioles, and linear, more or less succulent leaves; flowers blue, open for only a few hours in the morning.

KEY TO THE SPECIES.

Floral bracts abruptly long-acuminate, 3 to 6 cm. long, glabrous or puberulent; stems frequently simple, never much branched; petals all blue .... 1. *C. dianthifolia.*
Floral bracts short, 2 cm. long or less, usually with long divergent hairs on the sides; stems much branched; one petal white.... 2. *C. crispa.*

1. **Commelina dianthifolia** Delile in Red. Liliac. 7: *pl. 390.* 1801.
Commelina linearis Benth. Pl. Hartw. 27. 1839.
Commelina linearis longispatha Torr. U. S. & Mex. Bound. Bot. 224. 1859.
TYPE LOCALITY: Described from cultivated plants.
RANGE: New Mexico and Arizona to Mexico.
NEW MEXICO: Common in all the higher mountains from the Las Vegas Mountains to the Capitan Mountains and westward across the State. Open slopes, in the Transition Zone.

The type of *C. linearis longispatha* came from the Copper Mines.

2. Commelina crispa Wooton, Bull. Torrey Club **25:** 451. 1898.

Commelina virginica of many authors, not L.

TYPE LOCALITY: At the base of the Organ Mountains, New Mexico. Type collected by Wooton (no. 545).

RANGE: Colorado and Missouri to New Mexico.

NEW MEXICO: Santa Rita; Dog Spring; Tortugas Mountain; Organ Mountains; Jarilla Junction; Knowles; Nara Visa; Carrizozo; Orogrande. Dry plains and hills, in the Lower Sonoran Zone.

2. TRADESCANTIA L. SPIDERWORT.

Perennial herbs with simple or branched erect stems, narrow elongated leaves, and showy flowers in terminal umbel-like cymes subtended by leaflike bracts; sepals 3, distinct, herbaceous; petals 3, sessile, blue.

KEY TO THE SPECIES.

Plants small, 25 cm. high or less; sheaths strongly pubescent, especially along the margins; roots tuberous, the tubers attached along a creeping rootstock........................ 1. *T. pinetorum.*

Plants larger, 30 to 60 cm. high; sheaths glabrous; roots somewhat fleshy, fascicled.

Plants sparingly branched, green; sepals and pedicels densely glandular-pubescent............................ 2. *T. occidentalis.*

Plants much branched, glaucous; sepals and pedicels sparingly glandular.. 3. *T. scopulorum.*

1. Tradescantia pinetorum Greene, Erythea **1:** 247. 1893.

Tradescantia tuberosa Greene, Bot. Gaz. **6:** 185. 1881, not Roxb. 1798.

TYPE LOCALITY: Pinos Altos Mountains, New Mexico. Type collected by E. L. Greene.

RANGE: New Mexico and Arizona.

NEW MEXICO: Mogollon Mountains; West Fork of the Gila; San Luis Mountains; Tularosa Creek. Upper Sonoran and Transition zones.

2. Tradescantia occidentalis Britton, Mem. N. Y. Bot. Gard. **1:** 87. 1900.

Tradescantia virginica occidentalis Britton in Britt. & Brown, Illustr. Fl. **1:** 377. 1896.

TYPE LOCALITY: "Wisconsin to Missouri, Texas, and New Mexico."

RANGE: As above.

NEW MEXICO: Hills near Clayton (*Bartlett* 235, 252). Upper Sonoran Zone.

3. Tradescantia scopulorum Rose, Contr. U. S. Nat. Herb. **5:** 205. 1899.

TYPE LOCALITY: Santa Catalina Mountains, Arizona.

RANGE: Western Nebraska and Montana to Arizona and western Texas.

NEW MEXICO: Gallup; Zuni Reservation; Organ Mountains; East Canyon. Dry hills, in the Upper Sonoran Zone.

17. PONTEDERIACEAE. Pickerel-weed Family.

1. HETERANTHERA Willd. MUD PLANTAIN.

Succulent herb with branched stems, numerous oval or ovate long-petioled leaves, and small white or blue flowers in 1-flowered spathes; lobes of the perianth linear; stamens 3, equal; fruit an ovoid many-seeded capsule.

1. Heteranthera limosa (Swartz) Willd. Ges. Naturf. Freund. Berlin Mag. **3:** 439. 1801.

Pontederia limosa Swartz, Prodr. Veg. Ind. Occ. 57. 1788.

TYPE LOCALITY: "Jam. Hispaniola."

RANGE: Virginia to New Mexico, southward through Tropical America.

NEW MEXICO: Santa Rita; Middle Fork of the Gila. In mud or shallow water.

Order 13. LILIALES.

KEY TO THE FAMILIES.

Styles wanting.................................... **18. CALOCHORTACEAE** (p. 127).
Styles present.
  Styles distinct .... **19. MELANTHACEAE** (p. 128).
  Styles united.
    Capsules septicidal; petals and sepals very unlike.
      18. CALOCHORTACEAE (p. 127).
    Capsules loculicidal; sepals and petals nearly alike.
      Sepals and petals chaffy...................... **20. JUNCACEAE** (p. 130).
      Sepals and petals not chaffy.
        Shrubby plants with caudices, or trees.
          21. DRACAENACEAE (p. 135).
        Herbs with bulbs, corms, or rootstocks.
          Plants with elongated horizontal rootstocks.
            22. CONVALLARIACEAE (p. 138).
          Plants with bulbs or corms or short erect rootstocks.
            Flowers in umbels, at first included in and later subtended by a scarious involucre....**23. ALLIACEAE** (p. 140).
            Flowers solitary or racemose (in Leucocrinum by the shortening of the axis apparently umbellate), without involucres.
              Plants from bulbs or corms....**24. LILIACEAE** (p. 143).
              Plants from elongated tuberous roots.
                25. ASPHODELACEAE (p. 144).

18. CALOCHORTACEAE. Mariposa lily Family.

1. CALOCHORTUS Pursh. Mariposa lily.

Low bulbous plants with narrow grasslike leaves; flowers large, showy, pale yellow, lilac, or bright yellow, borne on slender glabrous scapes.

Several species are not rare in cultivation.

KEY TO THE SPECIES.

Flowers bright yellow; plants low, 8 to 20 cm.; gland at the base of the petals longer than broad; anthers obtuse................. 1. *C. aureus.*
Flowers pale yellow to lilac; plants taller, 20 to 40 cm.; glands and anthers various.
  Anthers obtuse; glands orbicular or nearly so; petals 35 mm. long or less.. 2. *C. nuttallii.*
  Anthers very acute; glands much broader than long; petals 35 to 40 mm. long.. 3. *C. gunnisonii.*

1. Calochortus aureus S. Wats. Amer. Nat. **7**: 303. 1873.

Type locality: "On sand-cliffs, Southern Utah."

Range: Southern Utah to Arizona and northwestern New Mexico.

New Mexico: Fort Wingate; Gallup; Farmington. Open hills, in the Upper Sonoran Zone.

One of the lowest species of the genus, seldom exceeding a height of 20 cm.; leaves slender, very long for the size of the plant, often recurved; probably the handsomest of our species.

2. Calochortus nuttallii Torr. & Gray, U. S. Rep. Expl. Miss. Pacif. **2**: 124. 1855.

TYPE LOCALITY: "Summit of Noble's Pass, Sierra Nevada."

RANGE: Montana to New Mexico, west to California.

NEW MEXICO: Mangas Springs; Silver City; Tunitcha Mountains. Transition and Canadian zones.

A taller plant than the preceding and with less handsome but more numerous lilac flowers.

3. Calochortus gunnisonii S. Wats. in King, Geol. Expl. 40th Par. **5**: 346. 1871.

Calochortus gunnisonii perpulcher Cockerell, Bot. Gaz. **29**: 281. 1900.

TYPE LOCALITY: Rocky Mountains of Colorado.

RANGE: Montana and Wyoming to New Mexico and Arizona.

NEW MEXICO: Chama; Santa Fe and Las Vegas mountains; Chusca Mountains; Ramah; Mogollon Mountains. Meadows in the mountains, in the Transition and Canadian zones.

Similar to the last in general appearance but with different anthers and glands. The plants from the Santa Fe and Las Vegas mountains have larger and yellower flowers than those from other parts of New Mexico. To this form Prof. T. D. A. Cockerell gave the subspecific name *perpulcher*. His type was collected at Harveys Ranch.

19. MELANTHACEAE. Bunch-flower Family.

Perennial caulescent or scapose herbs, with elongated or bulblike rootstocks; leaves alternate, often all basal; flowers polygamous or diœcious, regular, in terminal spikes, racemes, or panicles, or solitary; perianth usually inconspicuous; sepals and petals distinct or nearly so; filaments often adnate to the base of the sepals and petals; ovary superior or slightly inferior; styles distinct.

KEY TO THE GENERA.

Stems tall, 1 meter high or more, from rootstocks; leaves large, oval........ 1. VERATRUM (p. 128).
Stems low, less than 50 cm. high, from elongated bulbs; leaves linear.
  Perianth segments without glands, narrowly linear, 2 to 3 mm. long, greenish........ 2. SCHOENOCAULON (p. 129).
  Perianth segments gland-bearing, not linear, 5 mm. long or more, white.
    Ovary partly inferior; glands obcordate; flowers 1 cm. long or more, not crowded..... 3. ANTICLEA (p. 129).
    Ovary superior; glands obovate or semiorbicular; flowers less than 1 cm. long, crowded........ 4. TOXICOSCORDION (p. 130).

1. VERATRUM L. SKUNK CABBAGE.

Tall coarse perennial herb, 1.5 meters high or less, from thick rootstocks; leaves broad, sessile, strongly veined; flowers rather large, in a broad terminal panicle; perianth of 6 similar distinct elliptic-oblong segments; capsules 3-beaked, the persistent styles divergent.

1. Veratrum speciosum Rydb. Bull. Torrey Club **27**: 531. 1900.

TYPE LOCALITY: Bridger Mountains, Montana.

RANGE: Montana and Washington to California and New Mexico.

NEW MEXICO: Tunitcha Mountains; Winsor Creek; Pecos Baldy; Willow Creek; White Mountain Peak. Wet meadows in the mountains, Transition to Hudsonian Zone.

A common and very characteristic plant of the mountains of the northern part of the State, often thickly covering large areas of open marshy land. It is sometimes eaten by sheep, with fatal results. The common name given in the books is "false hellebore," but in New Mexico it is always known as "skunk cabbage," although it is very unlike the plant which bears that name in the eastern United States.

2. SCHOENOCAULON A. Gray.

Low plant with a slender scape from a black fibrous-coated elongated bulb; leaves all radical, pale green, long, grasslike; flowers perfect, pale green, almost sessile in a spikelike raceme; capsules about 12 mm. long, with 4 to 6-seeded cells.

1. **Schoenocaulon drummondii** A. Gray; Hook. & Arn. Bot. Beechey Voy. 388. 1841.

Schoenocaulon texanum Scheele, Linnaea **25**: 262. 1852.

TYPE LOCALITY: Southwestern Texas.

RANGE: Western Texas and southern New Mexico to Mexico.

NEW MEXICO: Ten miles west of Roswell (*Wooton*). Dry hills and plains.

3. ANTICLEA Kunth.

Glabrous herbs from tunicated bulbs, the stems scapose, or bearing a few leaves; flowers of medium size, ochroleucous, greenish; perianth segments similar, each bearing an obcordate gland near the base; inflorescence open, loose, few-flowered.

KEY TO THE SPECIES.

Inflorescence paniculate, widely branched, glaucous; pedicels slender, divergent, 2 or more times the length of the subtending bracts; petals about 5 mm. long .... 1. *A. porrifolia.*
Inflorescence racemose, sometimes with a few short branches below, green; pedicels stout, erect or ascending, of about the same length as the subtending bracts; petals 5 to 8 mm. long.
  Perianth segments 7 to 8 mm. long, 7 to 13-nerved .... 2. *A. elegans.*
  Perianth segments 5 to 6 mm. long, 3 to 7-nerved .... 3. *A. coloradensis.*

1. **Anticlea porrifolia** (Greene) Rydb. Bull. Torrey Club **30**: 273. 1903.

Zygadenus porrifolius Greene, Bull. Torrey Club **8**: 123. 1881.

TYPE LOCALITY: "Mogollon Mountains, near the summits," New Mexico. Type collected by Greene in 1881.

RANGE: Southwestern New Mexico to Chihuahua.

NEW MEXICO: Mogollon Mountains; Lookout Mines. Mountains, in the Canadian Zone.

2. **Anticlea elegans** (Pursh) Rydb. Bull. Torrey Club **30**: 273. 1903.

Zygadenus elegans Pursh, Fl. Amer. Sept. 241. 1814.

Zygadenus dilatatus Greene, Pl. Baker. **1**: 51. 1901.

TYPE LOCALITY: "On the waters of the Cokahlaishkit river, near the Rocky mountains."

RANGE: Alaska and Saskatchewan to Nevada and New Mexico.

NEW MEXICO: Chama; Santa Fe and Las Vegas mountains; Baldy; White Mountains. Damp woods, in the Canadian and Hudsonian zones.

3. **Anticlea coloradensis** Rydb. Bull. Torrey Club **30**: 273. 1903.

Zygadenus coloradensis Rydb. Bull. Torrey Club **27**: 534. 1900.

TYPE LOCALITY: Idaho Springs, Colorado.

RANGE: Utah and Colorado to northwestern New Mexico.

NEW MEXICO: Tunitcha Mountains (*Standley* 7554). Meadows in the mountains, in the Transition Zone.

4. TOXICOSCORDION Rydb. DEATH CAMASS.

Plants much as in the preceding genus, but the flowers smaller and much more numerous, the perianth segments with obovate or semiorbicular glands, and the ovary wholly superior instead of partly inferior.

KEY TO THE SPECIES.

Plants stout; leaves 10 to 15 mm. wide; inflorescence paniculate, widely branching.. 1. *T. paniculatum.*
Plants slender; leaves about 5 mm. wide (strongly falcate); inflorescence racemose, little or not at all branched........... 2. *T. falcatum.*

1. **Toxicoscordion paniculatum** (S. Wats.) Rydb. Bull. Torrey Club **30:** 272. 1903.

Zygadenus paniculatus S. Wats. in King, Geol. Expl. 40th Par. **5:** 343. 1871.

TYPE LOCALITY: "Oregon and Washington Territory."

RANGE: Montana and Washington to California and northwestern New Mexico.

NEW MEXICO: Carrizo Mountains (*Matthews*). Open hills, in the Upper Sonoran Zone.

2. **Toxicoscordion falcatum** Rydb. Bull. Torrey Club **30:** 272. 1903.

Zygadenus falcatus Rydb. Bull. Torrey Club **27:** 536. 1900.

TYPE LOCALITY: Fort Collins, Colorado.

RANGE: Colorado and northwestern New Mexico.

NEW MEXICO: Aztec (*Baker* 260). Open hills, in the Upper Sonoran Zone.

20. JUNCACEAE. Rush Family.

Grasslike plants, annuals or perennials, tufted or from rootstocks, with terete solid stems; leaves various, the sheaths open or closed, the margins sometimes produced into auriculate ligule-like organs, the blades flat or terete or wanting; inflorescence of terminal heads, spikes, or panicles, usually bracted; flowers regular, mostly complete, 3-merous; sepals and petals 3 each, more or less glumelike; stamens 3 or 6; ovary superior, 1 or 3-celled, forming a 1-celled or 3-celled capsule with 3 to many seeds.

KEY TO THE GENERA.

Leaf sheaths open; capsules 1 or 3-celled; seeds many........ 1. JUNCUS (p. 130).
Leaf sheaths closed; capsules 1-celled; seeds 3................. 2. JUNCOIDES (p. 134).

1. JUNCUS L. RUSH.

Chiefly perennial herbs of wet soil, with pithy or hollow, usually simple stems; leaf sheaths open; flowers cymose or glomerate, small, greenish or brownish; capsule 3-celled or rarely 1-celled, the seeds numerous, often appendaged.

KEY TO THE SPECIES.

Lower bract of the inflorescence terete, erect, appearing as an elongation of the stem; inflorescence apparently lateral.
- Flowers few, 1 to 5, one of them subsessile, the others pediceled... 1. *J. drummondii.*
- Flowers several, in a more or less compound panicle.
  - Plants slender; bracts extending considerably beyond the inflorescence; basal sheaths without blades... 2. *J. balticus.*
  - Plants stout; bracts short, extending little if at all beyond the inflorescence; uppermost basal sheath bearing a scapiform blade............................. 3. *J. mexicanus.*

Lower bracts not appearing as a continuation of the stems, or if so channeled on the upper surface; inflorescence terminal.
Leaves septate, sometimes equitant, the septa sometimes poorly developed and hard to see in dried material.
Leaves terete, not equitant.
Capsules narrowly lanceolate; inflorescence with short branches; flowers echinate-spreading, or the lowest of the head reflexed.
Heads 7 to 8 mm. in diameter; leaf blades erect; petals usually longer than the sepals.... 4. *J. nodosus.*
Heads more than 10 mm. in diameter; leaf blades usually spreading; petals shorter than the sepals.................................. 5. *J. torreyi.*
Capsules oblong; inflorescence with elongated branches (in *J. mertensianus* a single head); flowers erect or ascending.
Heads several; leaves terete; seeds not caudate.. 6. *J. badius.*
Heads solitary or rarely 2 or 3; leaves somewhat flattened; seeds mostly caudate.......... 7. *J. mertensianus.*
Leaves equitant, laterally flattened, with one edge toward the stem.
Flower clusters numerous, small, 5 to 12-flowered, generally light-colored....................... 8. *J. brunnescens.*
Flower clusters few, larger, 15 to 25-flowered, usually dark-colored.
Perianth segments green-margined; ligules usually not auricled..................... 9. *J. parous.*
Perianth segments fuscous or dark brown; ligules produced into small auricles......10. *J. saximontanus.*
Leaves neither septate nor equitant.
Leaves hollow; flowers few, in small heads.
Stems leafy only at base; perianth about 4 mm. long; lower bracts of inflorescence membranous....................................11. *J. triglumis.*
Stems leafy throughout; perianth segments 5 to 6 mm. long; lower bracts foliaceous.............12. *J. castaneus.*
Leaves not hollow; flowers numerous.
Flowers not bracteolate, in true heads on branches of the inflorescence; leaves broad and grass-like...13. *J. longistylis.*
Flowers bracteolate, inserted singly on the branches of the inflorescence; leaves narrowly linear, flat, or subterete and channeled.
Annual; stems branched.......................14. *J. bufonius.*
Perennials; stems simple.
Auricles cartilaginous, yellowish brown; bracts usually elongated, much exceeding the inflorescence...........15. *J. dudleyi.*
Auricles scarious or membranaceous; bracts usually much shorter, hardly exceeding the inflorescence.

Auricles conspicuously produced beyond the point of insertion.
Stems stout; leaves short and broad; perianth 4 mm. long, scarious at base; cymes open....................16. *J. brachyphyllus.*
Stems slender; leaves long and narrow; perianth 3.5 to 4 mm. long, scarious to the apex; cymes dense ....17. *J. confusus.*
Auricles scarcely produced beyond the point of insertion.
Perianth segments about equaling the capsule, 3 to 4 mm. long......................18. *J. interior.*
Perianth segments mostly exceeding the capsule, 4 to 5 mm. long......................19. *J. arizonicus.*

1. **Juncus drummondii** E. Mey. in Ledeb. Fl. Ross. **4:** 235. 1853.
Juncus arcticus Willd. err. det. Hook. Fl. Bor. Amer. **2:** 189. 1838.
TYPE LOCALITY: "At a great elevation on the Rocky Mountains."
RANGE: British America to California and northern New Mexico; also in Europe and Asia.
NEW MEXICO: Spirit Lake; Truchas Peak. Bogs, in the Hudsonian and Arctic-Alpine zones.

2. **Juncus balticus** Willd. Ges. Naturf. Freund. Berlin Mag. **3:** 298. 1809.
TYPE LOCALITY: "An den sandigen Meeresufern bei Warnemünde," Germany.
RANGE: Alaska and British America to California, New Mexico, and New York; also in Europe and Asia.
NEW MEXICO: Cedar Hill; Dulce; Chama; Farmington; Taos; Baldy; Winsors Ranch; Jewett; Gallo Spring; Lone Mountain; Berendo Creek; Mesilla Valley; Silver Spring Canyon. Wet ground, in the Upper Sonoran and Transition zones.

3. **Juncus mexicanus** Willd.; Roem. & Schult. Syst. Veg. **7:** 178. 1829.
Juncus balticus mexicanus Parish, Muhlenbergia **6:** 119. 1910.
TYPE LOCALITY: "In Mexico."
RANGE: Texas and Arizona and southward.
NEW MEXICO: Cloverdale; White Sands; Carrizozo; Malones Ranch. Wet ground, in the Lower Sonoran Zone.

4. **Juncus nodosus** L. Sp. Pl. ed. 2. 466. 1762.
TYPE LOCALITY: "Habitat in America septentrionali."
RANGE: British America to Nevada and Virginia.
NEW MEXICO: Farmington; near Pecos; Castle Rock. Wet ground, in the Upper Sonoran Zone.

5. **Juncus torreyi** Coville, Bull. Torrey Club **22:** 303. 1895.
Juncus nodosus megacephalus Torr. Fl. N. Y. **2:** 326. 1843.
Juncus megacephalus Wood, Bot. & Flor. ed. 2. 724. 1861, not Curtis, 1835.
TYPE LOCALITY: "On the shores of Lake Ontario."
RANGE: British Columbia and New York to California and Texas.
NEW MEXICO: Shiprock; Farmington; Dulce; Las Vegas; Zuni Reservation; Albuquerque; Pecos; Berendo Creek; Silver City; Mesilla Valley; Carrizozo; Mescalero Agency. Wet ground, Lower Sonoran to Transition Zone.

6. Juncus badius Suksd. Deutsch. Bot. Monatschr. **19:** 92. 1901.

Juncus truncatus Rydb. Bull. Torrey Club **31:** 399. 1904.

TYPE LOCALITY: "Im Falkenthal im westl. Teil von Klickitat County," Washington.

RANGE: Washington and Wyoming to northern New Mexico.

NEW MEXICO: Tunitcha Mountains (*Standley* 7565). Wet ground, in the Transition Zone.

7. Juncus mertensianus Bong. Mém. Acad. St. Pétersb. VI. Math. Phys. Nat. **2:** 167. 1832.

TYPE LOCALITY: Sitka, Alaska.

RANGE: Alaska to California and New Mexico.

NEW MEXICO: Spirit Lake (*Standley* 4397). Bogs, in the Hudsonian Zone.

8. Juncus brunnescens Rydb. Bull. Torrey Club **31:** 400. 1904.

Juncus xiphioides montanus Engelm. Trans. Acad. St. Louis **2:** 481. 1868, in part.

TYPE LOCALITY: Pagosa Springs, Colorado.

RANGE: Nevada and Colorado to Arizona and New Mexico.

NEW MEXICO: Tunitcha Mountains; Carrizo Mountains; mountains west of Grants Station; Pecos; Bear Mountains; Mogollon Mountains; Rio Mimbres; Cloverdale; Guadalupe Canyon; Ruidoso Creek. Wet meadows, in the Transition Zone.

9. Juncus parous Rydb. Bull. Torrey Club **31:** 401. 1904.

TYPE LOCALITY: Fort Garland, Colorado.

RANGE: Colorado and New Mexico.

NEW MEXICO: Dulce; mountains west of Grants Station; Rio Pueblo; Middle Fork of the Gila; Silver Spring Canyon; Organ Mountains; White Mountains; Carrizozo. Wet meadows, in the Transition Zone.

10. Juncus saximontanus A. Nels. Bull. Torrey Club **29:** 401. 1902.

Juncus xiphioides montanus Engelm. Trans. Acad. St. Louis **2:** 481. 1868, in part.

TYPE LOCALITY: Colorado.

RANGE: British America to New Mexico.

NEW MEXICO: Middle Fork of the Gila; Chama; White Mountain Peak. Wet meadows in the mountains, in the Transition Zone.

11. Juncus triglumis L. Sp. Pl. 328. 1753.

TYPE LOCALITY: "Habitat frequens in Alpibus Lapponicis, Tauro Rastadiensi."

RANGE: British America to New Mexico and New York; also in Europe and Asia.

NEW MEXICO: Truchas Peak (*Standley* 4764). Wet meadows, in the Hudsonian and Arctic-Alpine zones.

12. Juncus castaneus J. E. Smith, Fl. Brit. **1:** 383. 1800.

TYPE LOCALITY: "In paludosis alpinis Scotiae, solo micaceo. On Ben Lawer."

RANGE: British America to Colorado and New Mexico; also in the Old World.

NEW MEXICO: Truchas Peak (*Standley* 4771, 4770). Meadows, in the Arctic-Alpine Zone.

13. Juncus longistylis Torr. U. S. & Mex. Bound. Bot. 223. 1859.

TYPE LOCALITY: Near the Copper Mines, New Mexico. Type collected by Bigelow.

RANGE: British America to California and New Mexico.

NEW MEXICO: Dulce; Tunitcha Mountains; Carrizo Mountains; Chama; Winsors Ranch; Santa Fe; Rio Pueblo; Jewett Spring; Copper Mines; White Mountains. Wet ground, in the Transition and Canadian zones.

14. Juncus bufonius L. Sp. Pl. 328. 1753.

TYPE LOCALITY: European.

RANGE: Nearly cosmopolitan.

NEW MEXICO: Chama; north of Ramah; Santa Fe Creek. Wet ground, in the Upper Sonoran and Transition zones.

15. Juncus dudleyi Wiegand, Bull. Torrey Club **27:** 524. 1900.

TYPE LOCALITY: Truxton, New York.

RANGE: Washington and Maine to Mexico.

NEW MEXICO: Winsors Ranch; Santa Fe; Las Vegas; Pecos; Ramah; Jewett Spring; Bear Mountains. Meadows, in the Transition Zone.

16. Juncus brachyphyllus Wiegand, Bull. Torrey Club **27:** 519. 1900.

TYPE LOCALITY: "Upper Platte," Colorado. This was originally cited as "Arkansas," which is altogether wrong.

RANGE: Mountains, Idaho to Colorado and New Mexico.

We have seen no material of this, but it is probable that one of the cotypes, although cited as coming from "Arkansas," really came from within our limits.[1]

17. Juncus confusus Coville, Proc. Biol. Soc. Washington **10:** 127. 1896.

TYPE LOCALITY: In an irrigated meadow, North Park, Colorado.

RANGE: Montana and Wyoming to Colorado and northern New Mexico.

NEW MEXICO: Tunitcha Mountains (*Standley* 7547). Wet ground, in the Transition Zone.

18. Juncus interior Wiegand, Bull. Torrey Club **27:** 516. 1900.

TYPE LOCALITY: Richmond, Illinois.

RANGE: Wyoming and Illinois to New Mexico and Missouri.

NEW MEXICO: Johnsons Mesa; Mogollon Creek; McKinneys Park; Kingston; Gilmores Ranch. Meadows, in the Upper Sonoran and Transition zones.

19. Juncus arizonicus Wiegand, Bull. Torrey Club **27:** 517. 1900.

TYPE LOCALITY: Copper Mines, New Mexico. Type collected by Thurber.

RANGE: Colorado and Arizona to Texas.

NEW MEXICO: Chama; Sierra Grande; Taos; Bear Mountains; Lorenzo Spring; Organ Mountains; San Luis Mountains; White Mountains. Meadows, in the Upper Sonoran and Transition zones.

2. JUNCOIDES Adans. WOOD RUSH.

Slender perennial grasslike herbs, often hairy, with flat leaves, the leaf sheaths closed; flowers small, spicate, glomerate, or umbellate; capsule 1-celled, 3-seeded.

KEY TO THE SPECIES.

Flowers on slender pedicels in corymbiform panicles............ 1. *J. parviflorum.*
Flowers in crowded spikelike clusters.
  Spikelets peduncled, forming a corymb........................ 2. *J. intermedium.*
  Spikelets subsessile, forming a compound spike.............. 3. *J. spicatum.*

1. Juncoides parviflorum (Ehrh.) Coville, Contr. U. S. Nat. Herb. **4:** 209. 1893.

Juncus parviflorus Ehrh. Beitr. Naturk. **6:** 139. 1791.

Luzula parviflora Desv. Journ. Bot. Schrad. **1:** 144. 1808.

TYPE LOCALITY: European.

RANGE: Alaska and British America to California, New Mexico, and New York.

NEW MEXICO: Ponchuelo Creek; Pecos Baldy. Meadows, Transition to Arctic-Alpine Zone.

2. Juncoides intermedium (Thuill.) Rydb. Bull. Torrey Club **32:** 610. 1905.

Juncus intermedius Thuill. Fl. Env. Paris ed. 2. 178. 1799.

Juncus multiflorus Ehrh.; Hoffm. Deutschl. Fl. ed. 2. **1:** 169. 1800, not Retz. 1795.

---

[1] See, Bartlett, H. H. Rhodora **11:** 156. 1909.

TYPE LOCALITY: Near Paris, France.

RANGE: British America to California and New Mexico.

NEW MEXICO: Winsors Ranch (*Standley* 4167). Damp woods, in the Canadian Zone.

3. Juncoides spicatum (L.) Kuntze, Rev. Gen. Pl. **2:** 725. 1891.

Juncus spicatus L. Sp. Pl. 330. 1753.

Luzula spicata DC. & Lam. Fl. Franç. **3:** 161. 1805.

TYPE LOCALITY: "Habitat in Lapponiae Alpibus."

RANGE: Temperate North America; also in Europe and Asia.

NEW MEXICO: Pecos Baldy; Truchas Peak; Baldy. Meadows, in the Arctic-Alpine Zone.

21. DRACAENACEAE. Yucca Family.

Shrubby plants or trees with woody caudices copiously furnished with narrow rigid leaves; flowers in racemes or panicles terminating scapes or scapelike stems; perianth greenish or white, the sepals and petals similar; gynœcium of 3 united carpels; ovary superior, 1 to 3-celled; styles united, sometimes very short or obsolete during anthesis; ovules 2 to several or many in each cell; fruit a loculicidal capsule, or berry-like and indehiscent.

KEY TO THE GENERA.

Flowers perfect.. 1. YUCCA (p. 135).
Flowers diœcious or polygamo-diœcious.
  Flowers polygamo-diœcious, in open panicles; ovary 3-celled; stamens included...................... 2. NOLINA (p. 137).
  Flowers diœcious, in dense panicles; ovary 1-celled; stamens exserted............................... 3. DASYLIRION (p. 138).

1. YUCCA L. YUCCA.

Thick-stemmed (in several species the stems short and mostly subterranean) perennials with narrow, mostly rigid, sharp-pointed leaves and large panicles or racemes of white campanulate flowers; fruit a 3-celled capsule, this dry or sometimes baccate and fleshy.

KEY TO THE SPECIES.

Leaves 10 mm. wide or less.
  Stems conspicuous in old plants, reaching a height of 3 to 4 meters, naked below, clothed with a tuft of leaves above; inflorescence a much branched panicle....... 1. *Y. elata.*
  Stems short, mostly subterranean, covered with leaves to the base; inflorescence racemose, sometimes with a few branches.
    Flowers large, 6 cm. long or more; style oblong, white.. 2. *Y. baileyi.*
    Flowers small, 4 cm. long or less; style swollen at the base, greenish.
      Leaves narrow, 6 mm. wide or less, very thick, sparsely filiferous............................ 3. *Y. glauca.*
      Leaves broader, 8 to 10 mm. wide, thin, abundantly filiferous....................................... 4. *Y. neomexicana.*
Leaves broader, 15 to 50 mm. wide.
  Fruit dehiscent; plants acaulescent......................... 5. *Y. harrimaniae.*
  Fruit indehiscent; plants caulescent or acaulescent.
    Stems short, 20 cm. high or less, leafy to the base; perianth segments narrowly lanceolate, 5 to 8 cm. long; fruit large, 12 to 15 cm. long, very pulpy.... 6. *Y. baccata.*

Stems taller, 1.5 to 5 meters high; perianth segments elliptic, 2 to 4 cm. long; fruit smaller, 10 cm. long or less, only slightly pulpy.
Leaves rigid, rough, yellowish green; filaments coarse and grayish.......................... 7. *Y. macrocarpa.*
Leaves flexible, smooth, bluish green, glaucous; filaments, when present, fine, usually brownish.................................. 8. *Y. schottii.*

1. **Yucca elata** Engelm. Bot. Gaz. **7**: 17. 1882. PALMILLA.

Yucca angustifolia radiosa Engelm. in King, Geol. Expl. 40th Par. **5**: 496. 1871.

Yucca angustifolia elata Engelm. Trans. Acad. St. Louis **3**: 50. 1873.

Yucca radiosa Trel. Rep. Mo. Bot. Gard. **3**: 163. 1892.

TYPE LOCALITY: "Extending from West Texas to Utah, Arizona and Northern Mexico."

RANGE: Southern Arizona to western Texas, southward into Mexico.

NEW MEXICO: Fort Bayard; Mimbres River; Dog Spring; Cambray; Hachita; Deming; mesa west of Organ Mountains; White Sands; Alamogordo; Mescalero Agency; Mesquite Lake. Mesas, in the Lower Sonoran Zone.

This is the common narrow-leaved Yucca of southern New Mexico, known as "palmilla," or "soapweed." The roots, termed "amole," are often used as a substitute for soap. The plant has considerable decorative value, but because of its large roots is difficult to transplant. It is one of the most abundant and characteristic plants of the Lower Sonoran Zone.

2. **Yucca baileyi** Woot. & Standl. Contr. U. S. Nat. Herb. **16**: 114. 1913.

TYPE LOCALITY: Dry slope in pine woods in the Tunitcha Mountains, New Mexico. Type collected by Standley (no. 7638).

RANGE: Northwestern New Mexico and northeastern Arizona.

NEW MEXICO: Tunitcha Mountains; Carrizo Mountains; Chusca Mountains. Dry hills and low mountains, in the Transition Zone, extending down into the Upper Sonoran.

3. **Yucca glauca** Nutt. Fraser's Cat. no. 89. 1813. SOAPWEED.

Yucca angustifolia Pursh, Fl. Amer. Sept. 227. 1814.

TYPE LOCALITY: "Collected 1,600 miles up the Missouri, about lat. 49°."

RANGE: South Dakota and Wyoming to Missouri and New Mexico.

NEW MEXICO: Raton; Farmington; Sierra Grande; Rosa; Albuquerque; Fairview; San Augustine Plains; Horse Camp; Pecos. Plains and low hills, chiefly in the Upper Sonoran Zone.

This is the common Yucca of the northern and eastern parts of New Mexico, where it is often very abundant. The leaves have been used in the manufacture of stable brooms. The fruits of this, as well as of some of the other dry-fruited species, were cooked and eaten by some of the Indians.

4. **Yucca neomexicana** Woot. & Standl. Contr. U. S. Nat. Herb. **16**: 115. 1913.

TYPE LOCALITY: On a volcanic hill about half a mile north of Des Moines, Union County, New Mexico. Type collected by Standley (no. 6208).

RANGE: Known only from type locality, in the Upper Sonoran Zone.

5. **Yucca harrimaniae** Trel. Rep. Mo. Bot. Gard. **13**: 59. *pl. 28, 29, 83. f. 10.* 1902.

TYPE LOCALITY: Helper, Utah.

RANGE: Southern Utah and Colorado to northeastern Arizona and northwestern New Mexico.

NEW MEXICO: Carrizo Mountains (*Standley* 7314). Dry hills, in the Upper Sonoran Zone.

6. **Yucca baccata** Torr. U. S. & Mex. Bound. Bot. 221. 1859. DATIL.

TYPE LOCALITY: High table lands between the Rio Grande and the Gila, New Mexico.

RANGE: New Mexico to Colorado and Nevada.

NEW MEXICO: Farmington; Raton; Carrizo Mountains; Tunitcha Mountains; Hurrah Creek; Santa Fe Canyon; Crawfords Ranch; Socorro; Fairview; Rincon; Carrizalillo Mountains; Florida Mountains; Organ Mountains; Burro Mountains; Bear Mountains. Dry hills and high plains, in the Upper Sonoran Zone.

The species is the low, stiff-leaved Yucca of the rocky ridges and mesas at the bases of the mountains. It is the largest flowered of our species. The fruit, too, is characteristic, somewhat resembling the eastern pawpaw in general appearance. The Indians of New Mexico slice the ripe fruit and dry it in the sun for use in winter. When fresh, it has a peculiar sweet taste and is quite palatable.

7. **Yucca macrocarpa** (Torr.) Engelm. Bot. Gaz. **6**: 224. 1881. PALMA.

Yucca baccata macrocarpa Torr. U. S. & Mex. Bound. Bot. 221. 1859.

TYPE LOCALITY: On the plains of western Texas near the Limpio.

RANGE: Western Texas to southern Arizona and southward.

NEW MEXICO: Silver City; Fort Bayard; Las Cruces; Tortugas Mountain. Mesas and plains, in the Lower Sonoran Zone.

This the common broad-leaved Yucca or "dagger" of the mesas of the southern part of the State. It is used not a little for decorative purposes in this region and is very effective. It is easily transplanted and under cultivation becomes 5 to 6 meters high. The leaves are used extensively by the various Indians, notably the Apaches, in their basketry. By using different parts of the leaves, different colors are secured for forming designs, the outer part of the leaf being greenish yellow and the inner white.

8. **Yucca schottii** Engelm. Trans. Acad. St. Louis **3**: 46. 1873.

TYPE LOCALITY: Upper Santa Cruz River in southern Arizona.

RANGE: Southwestern New Mexico, southern Arizona, and northern Mexico.

NEW MEXICO: Indian Canyon, Animas Mountains; San Luis Mountains. Lower Sonoran Zone.

A little-known arborescent species with smooth, glaucous leaves and pubescent inflorescence. It is known with us only in the extreme southwest corner of the State.

2. NOLINA Michx. BEARGRASS.

Coarse-leaved perennials, the leaves linear, serrulate; inflorescence of a stout, nearly naked stem, paniculately branched above; flowers polygamo-diœcious, small, with whitish oblong-lanceolate segments; stamens included; fruit indehiscent, thin-walled, with subglobose seeds.

KEY TO THE SPECIES.

Leaves 6 mm. wide or less, the edges smooth........................ 1. *N. greenei.*
Leaves 6 to 12 mm. wide, scabrous on the edges..................... 2. *N. microcarpa.*

1. **Nolina greenei** S. Wats.; Trel. Proc. Amer. Phil. Soc. **50**: 418. 1911.

TYPE LOCALITY: Between the Purgatory and Apeshipa rivers, north of Trinidad, Colorado.

RANGE: Southeastern Colorado to New Mexico.

NEW MEXICO: San Miguel County; White Mountains. Dry hills, in the Upper Sonoran Zone.

2. **Nolina microcarpa** S. Wats. Proc. Amer. Acad. **14**: 247. 1879.

TYPE LOCALITY: "S. Arizona (Rock Cañon; Rothrock, n. 278)."

RANGE: Southwestern New Mexico to southern Arizona and southward.

NEW MEXICO: Mimbres River; Big Hatchet Mountains; Silver City; San Luis Mountains; Dog Mountains; Burro Mountains; Mogollon Creek; Lake Valley; Magdalena Mountains. Dry plains and low hills, in the Lower and Upper Sonoran zones.

A specimen collected by Bailey at San Rafael probably belongs here, although the margins of the leaves have much fewer teeth.

The leaves of this plant were used by the Indians in former times in weaving baskets and mats. They also furnish a fairly good quality of fiber, which may some day be utilized in making cordage.

3. DASYLIRION Zucc. SOTOL.

Diœcious perennials with thick short stems, numerous strap-shaped spiny-margined leaves, and very numerous small white flowers borne in tall narrow panicles.

The bases of the leaves form a round head about the thick stems, when the ends have been cut or burned off, and these are used for feeding stock. These heads are roasted by the native people and used for food and for the manufacture of a drink called "sotol" which contains from 40 to 50 per cent of alcohol. It has been found practicable to manufacture commercial alcohol from the plant.

KEY TO THE SPECIES.

Prickles of the leaves mostly recurved; leaves green............ 1. *D. leiophyllum.*
Prickles of the leaves directed forward; leaves somewhat glaucous.. 2. *D. wheeleri.*

1. **Dasylirion leiophyllum** Engelm.; Trel. Proc. Amer. Phil. Soc. **50**: 433. 1911.

TYPE LOCALITY: Presidio, Texas.

RANGE: Western Texas to southern New Mexico and southward.

NEW MEXICO: Central; Florida Mountains; Big Hatchet Mountains. Dry hills, in the Lower Sonoran Zone.

2. **Dasylirion wheeleri** S. Wats. in Wheeler, Rep. U. S. Surv. 100th Merid. **6**: 272. 1879.

TYPE LOCALITY: Southern Arizona.

RANGE: Western Texas to southern Arizona.

NEW MEXICO: San Mateo Mountains; Kingston; Mangas Springs; Big Hatchet Mountains; San Luis Mountains; Rincon; mesa near Las Cruces; Organ Mountains; White Mountains. Dry hills, in the Lower and Upper Sonoran zones.

22. CONVALLARIACEAE. Lily-of-the-valley Family.

Perennial herbs arising from rootstocks, never with bulbs or corms; leaves alternate (in ours cauline), sometimes reduced to scales; flowers perfect, in terminal racemes or panicles or axillary in small clusters; perianth segments distinct or more or less united at the base; pistil 3-parted; fruit a fleshy berry.

KEY TO THE GENERA.

Leaves reduced to scales; branches numerous, filiform..... 1. ASPARAGUS (p. 139).
Leaves not reduced; stems sparingly branched or simple, not filiform.
  Perianth segments united into a tube................ 2. SALOMONIA (p. 139).
  Perianth segments distinct.
    Flowers in terminal racemes or panicles.......... 3. VAGNERA (p. 139).
    Flowers terminal or opposite the leaves, solitary or in few-flowered clusters.
      Flowers in terminal, few-flowered clusters.... 5. DISPORUM (p. 140).
      Flowers solitary opposite the leaves.......... 4. STREPTOPUS (p. 139).

1. ASPARAGUS L. Asparagus.

Tall perennial with much branched stems from thick matted rootstocks; branchlets capillary, often referred to as leaves, the true leaves reduced to small scales; flowers small, greenish yellow, axillary, on jointed pedicels.

1. **Asparagus officinalis** L. Sp. Pl. 313. 1753.
Type locality: "Habitat in Europae arenosis."
New Mexico: Farmington; Santa Fe; Mesilla Valley.
The cultivated asparagus thrives in New Mexico and is a not uncommon escape in the valleys.

2. SALOMONIA Heist. Solomon's seal.

Perennial herbs with simple erect stems from creeping rootstocks; leaves sessile or clasping; flowers axillary, nodding, greenish, on jointed pedicels; ovary 3-celled, with 2 to 6 ovules in each cell; berry black or blue.

1. **Salomonia cobrensis** Woot. & Standl. Contr. U. S. Nat. Herb. **16:** 113. 1913.
Type locality: Copper Mines, New Mexico.
Range: Mountains of southwestern New Mexico.
New Mexico: Copper Mines; near Kingston.

3. VAGNERA Adans. False Solomon's seal.

Low plants with running rootstocks, leafy stems, alternate, sessile, lanceolate or elliptic leaves, small, inconspicuous, paniculate or racemose flowers, and reddish fruit.

KEY TO THE SPECIES.

Flowers paniculate; leaves elliptic to oval.......................... 1. *V. amplexicaulis.*
Flowers in a simple raceme; leaves lanceolate.................. 2. *V. stellata.*

1. **Vagnera amplexicaulis** (Nutt.) Greene, Bot. San Fran. Bay 316. 1894.
Smilacina amplexicaulis Nutt. Journ. Acad. Phila. **7:** 58. 1834.
Smilacina racemosa amplexicaulis S. Wats. in King, Geol. Expl. 40th Par. **5:** 345. 1871.
Type locality: "In the valleys of the Rocky Mountains about the sources of the Columbia River."
Range: British Columbia and Montana to California and New Mexico.
New Mexico: Santa Fe and Las Vegas mountains; Sandia Mountains; Tunitcha Mountains; Chama; Mogollon Mountains; Black Range; Organ Mountains; White and Sacramento mountains. Damp woods, in the Transition and Canadian zones.
We have specimens from the Mogollon Mountains in which the leaves are abundantly variegated with white.

2. **Vagnera stellata** (L.) Morong, Mem. Torrey Club **5:** 114. 1894.
Convallaria stellata L. Sp. Pl. 316. 1753.
Smilacina stellata Desf. Ann. Mus. Paris **9:** 52. 1807.
Type locality: Canada.
Range: British America to Pennsylvania, New Mexico, and California.
New Mexico: Santa Fe and Las Vegas mountains; Magdalena Mountains; Chama; Mogollon Mountains; Black Range; White Mountains; Sierra Grande. Damp woods, in the Transition and Canadian zones.
Some of our specimens may represent *Vagnera liliacea* Greene, but we are unable to separate the two species by any constant character.

4. STREPTOPUS Michx. Twisted-stalk.

Perennial from a creeping rootstock, with branched stems and small axillary flowers; perianth segments acute; fruit a red globose many-seeded berry.

1. **Streptopus amplexifolius** (L.) DC. & Lam. Fl. Franç. **3:** 174. 1805.
Uvularia amplexifolia L. Sp. Pl. 304. 1753.
TYPE LOCALITY: "In Bohemiae, Saxoniae, Delphinatus montibus."
RANGE: British America to Arizona and Pennsylvania.
NEW MEXICO: Winsor Creek (*Standley* 4200). Damp woods, in the Canadian Zone.

5. DISPORUM Salisb.

Low herb with creeping rootstocks, erect branched stems, and sessile ovate thin leaves; flowers small, solitary on slender terminal peduncles; perianth narrowly campanulate; fruit a 3 to 6-seeded red berry.

1. **Disporum trachycarpum** (S. Wats.) Benth. & Hook. Gen. Pl. **3:** 832. 1883.
Prosartes trachycarpa S. Wats. in King, Geol. Expl. 40th Par. **5:** 344. 1871.
TYPE LOCALITY: Colorado.
RANGE: British America to South Dakota and New Mexico.
NEW MEXICO: Carrizo Mountains; Tunitcha Mountains; Chama. Damp woods, in the Transition and Canadian zones.

23. ALLIACEAE. Onion Family.

Perennial scapose herbs with scaly or reticulate-coated bulbs; leaves few, narrow, basal; flowers in terminal umbels, at first inclosed in and finally subtended by a scarious involucre; perianth segments all alike, petaloid, mostly conspicuous, persistent, becoming scarious in fruit; stamens 6; fruit a dry 3-celled capsule.

KEY TO THE GENERA.

Perianth segments nearly free (in ours pinkish, fading lighter); capsule deeply lobed, sometimes crested; plants strong-scented .... 1. ALLIUM (p. 140).
Perianth segments united for one-third their length or more; capsule not lobed nor crested; plants not strong-scented.
  Perianth campanulate or funnelform, about 1 cm. long, bluish purple .... 2. DIPTEROSTEMON (p. 143).
  Perianth salverform, 3 cm. long, the limb white with pronounced greenish midribs .... 3. MILLA (p. 143).

1. ALLIUM L. ONION.

Strong-scented herbs with narrow leaves and 1 to several scapes from a coated bulb; flowers in umbels, sometimes replaced by bulblets; perianth of 6 petaloid, nearly free segments; fruit a deeply lobed 3-celled capsule.

KEY TO THE SPECIES.

Outer bulb coats strongly reticulated, the veins separating into a mat of fibers.
  Scapes bulblet-bearing .... 7. *A. sabulicola.*
  Scapes not bulblet-bearing.
    Capsules crested.
      Plants tall, 25 to 40 cm.; perianth segments 5 mm. long, bright rose pink; pedicels slender ..... 8. *A. geyeri.*
      Plants lower, 10 to 15 cm., stouter; perianth segments 7 or 8 mm. long, pale pink with prominent midveins; pedicels stout .... 9. *A. deserticola.*

Capsules not crested.
Plants slender; pedicels 10 to 12 mm. long; perianth segments 4 to 5 mm. long, pale..............10. *A. helleri.*
Plants stout; pedicels 13 to 16 mm. long; perianth segments 6 to 7 mm. long, bright pink.......11. *A. nuttallii.*
Outer bulb coats scaly, not reticulate, the veins never separating into fibers.
Bulbs without rootstocks; umbels erect; perianth segments acute or acuminate; stamens not exserted.
Capsule and ovary not crested........................ 4. *A. scaposum.*
Capsule and ovary crested.
Perianth segments oblong-lanceolate; plants low, 10 to 12 cm.; bulb coats dark chestnut brown. 5. *A. bigelovii.*
Perianth segments ovate-lanceolate; plants taller, 18 to 30 cm.; bulb coats lighter colored...... 6. *A. palmeri.*
Bulbs arising from rootstock.
Umbels erect; perianth segments acute; stamens not exserted; rootstocks long and slender; bulbs usually solitary................................ 3. *A. rhizomatum.*
Umbels cernuous; perianth segments obtuse; stamens exserted; rootstocks short and thick; bulbs usually clustered.
Leaves not carinate, 3 to 6 mm. wide in dried specimens; flowers numerous................ 1. *A. recurvatum.*
Leaves carinate, 2 mm. wide or less; flowers few.... 2. *A. neomexicanum.*

1. Allium recurvatum Rydb. Mem. N. Y. Bot. Gard. **1:** 94. 1900.

TYPE LOCALITY: Indian Creek, Montana.

RANGE: South Dakota and British Columbia to New Mexico.

NEW MEXICO: Chama; Santa Fe and Las Vegas mountains; Kingston; White and Sacramento mountains. Open meadows, in the Transition and Canadian zones.

2. Allium neomexicanum Rydb. Bull. Torrey Club **26:** 541. 1899.

TYPE LOCALITY: Organ Mountains, New Mexico. Type collected by Wooton, October 14, 1894.

RANGE: Colorado to New Mexico and Arizona.

NEW MEXICO: Tunitcha Mountains; Abiquiu Peak; Laguna Blanca; mountains west of Grant; West Fork of the Gila; San Luis Mountains; Organ Mountains; Las Huertas Canyon. Open slopes, in the Transition Zone.

3. Allium rhizomatum Woot. & Standl. Contr. U. S. Nat. Herb. **16:** 114. 1913.

TYPE LOCALITY: Gila Hot Springs, New Mexico. Type collected by Wooton, August 20, 1900.

RANGE: Known only from the type locality, in the Transition Zone.

4. Allium scaposum Benth. Pl. Hartw. 26. 1840.

TYPE LOCALITY: "Secus rivulos, Aguas Calientes," Mexico.

RANGE: Western Texas to southern Arizona and southward.

NEW MEXICO: Sixteen Spring Canyon (*Wooton*). Transition Zone.

Doctor Watson included this species with those having reticulate bulb coats, but all the specimens we have seen (ten or a dozen sheets), including some to which he refers, have scaly bulb coats, the inner ones very thin and white or hyaline, the outer somewhat thicker, yet light-colored. His illustration in the Botany of King's Survey, plate 38, was no doubt made with a compound microscope, since the markings are not visible under a hand lens. The illustration of the flower is excellent.

5. Allium bigelovii S. Wats. in King, Geol. Expl. 40th Par. **5:** 487. *pl. 38. f. 8, 9.* 1871.

Type locality: Cooks Spring, New Mexico.

Range: New Mexico and Arizona.

We have seen no further specimens of this from New Mexico.

6. Allium palmeri S. Wats. in King, Geol. Expl. 40th Par. **5:** 487. *pl. 37. f. 10, 11.* 1871.

Type locality: Northwestern New Mexico. Type collected by Palmer.

Range: Southern Utah to northern New Mexico and Arizona.

New Mexico: Known only from the northwest corner of the State. Upper Sonoran Zone.

7. Allium sabulicola Osterhout, Bull. Torrey Club **27:** 539. 1900.

Allium arenicola Osterhout, Bull. Torrey Club **27:** 506. 1900, not Small, 1900.

Type locality: In sandy soil on the bank of the Chama River at Chama, New Mexico. Type collected by Osterhout.

Range: New Mexico.

New Mexico: Spirit Lake; West Fork of the Gila; Fitzgerald Cienaga. Wet places in the mountains, from the Transition to the Hudsonian Zone.

Our plants all agree in having several whitish ovoid bulblets, ovate acuminate sepals, and reticulated bulbs, but they are in every case much larger plants than the original description indicates. They are certainly not *A. rubrum* Osterhout and we do not believe that Nelson [1] is right in reducing them to *A. nuttallii.*

8. Allium geyeri S. Wats. Proc. Amer. Acad. **14:** 227. 1879.

Allium reticulatum var. *β* S. Wats. in King, Geol. Expl. 40th Par. **5:** 486. 1871.

Allium dictyotum Greene, Pl. Baker. **1:** 52. 1901.

Type locality: Stony banks of the Kooskooskie River, Idaho.

Range: New Mexico to British Columbia.

New Mexico: Sierra Grande; Tierra Amarilla; Sandia Mountains; mountains west of Grants Station; White and Sacramento mountains. Transition and Canadian zones.

9. Allium deserticola (Jones) Woot. & Standl. Contr. U. S. Nat. Herb. **16:** 114. 1913.

Allium reticulatum deserticola Jones, Contr. West. Bot. **10:** 30. 1902.

Type locality: "On the adobe plains of eastern Utah, south of the Uintas and western Colorado and southward to Texas."

Range: As above; probably also in northern Mexico.

New Mexico: Aztec; Carrizalillo Mountains; Organ Mountains. Upper Sonoran Zone.

This is the largest flowered wild onion we have in the State. The perianth segments are pale pinkish to white, with a darker midrib, fading to a dry papery envelope in fruit. The plant occurs with us in the foothills of the more arid mountains.

10. Allium helleri Small, Fl. Southeast. U. S. 264. 1903.

Type locality: Southern Texas.

Range: Nebraska and Colorado to Texas and Arizona.

New Mexico: Las Vegas; Winsors Ranch; Bear Mountain; mountains east of Gila River; Copper Mines; Burro Mountains. Plains and low hills, in the Transition Zone.

11. Allium nuttallii S. Wats. Proc. Amer. Acad. **14:** 227. 1879.

Allium mutabile var. *β* S. Wats. in King, Geol. Expl. 40th Par. **5:** 487. 1871.

Type locality: "Kansas, Texas and New Mexico."

Range: Kansas and Colorado to Texas and Arizona.

---

[1] In Coulter, New Man. Rocky Mount. 114. 1909.

New Mexico: Glorieta; West Fork of the Gila; near Fort Defiance; Nara Visa. Low hills, in the Upper Sonoran Zone.

2. **DIPTEROSTEMON** Rydb.

Flowers few, umbellate on unequal rays, bluish purple, broadly funnelform, with a short tube; stamens 6, the inner wing-appendaged.

1. **Dipterostemon pauciflorus** (Torr.) Rydb. Bull. Torrey Club **39:** 111. 1912.
Brodiaea capitata pauciflora Torr. U. S. & Mex. Bound. Bot. 218. 1859.
Dichelostemma pauciflorum Standley, Contr. U. S. Nat. Herb. **13:** 179. 1910.
Type locality: Near the Copper Mines, New Mexico. Type collected by Bigelow.
Range: Southwestern New Mexico and southern Arizona.
New Mexico: Mangas Springs; Santa Rita. Upper Sonoran Zone.

3. **MILLA** Cav.

Flowers white, the perianth segments with greenish midribs, salverform with a narrowly turbinate tube, usually 2 to each scape; stamens nearly sessile, the anthers fixed by the base; capsules oblong-obovate, sessile.

1. **Milla biflora** Cav. Icon. Pl. **2:** 76. *pl. 196.* 1794.
Type locality: "Habitat in Imperio Mexicano."
Range: Southern New Mexico and Arizona and southward.
New Mexico: Animas Valley (*Mearns* 2513). Low hills.
Both this and the preceding plant are well worthy of cultivation and would probably thrive in southern New Mexico.

24. LILIACEAE. Lily Family.

Perennial, mainly caulescent herbs, with bulbs or corms; leaves alternate or whorled, sometimes basal or apparently basal; flowers solitary or in terminal racemes, corymbs, or panicles; perianth conspicuous and showy; sepals and petals similar, sometimes partly united; gynœcium of 3 united carpels; ovary superior, 3-celled; styles united; fruit a loculicidal capsule, globular or elongated.

KEY TO THE GENERA.

Perianth segments united into a long tube; flowers white.. 1. Leucocrinum (p. 143).
Perianth segments distinct or nearly so; flowers white or colored.
  Bulbs tunicated; flowers white...................... 4. Lloydia (p. 144).
  Bulbs scaly; flowers not white.
    Flowers large, 6 or 7 cm. long; perianth segments clawed................................ 2. Lilium (p. 144).
    Flowers small, 2 cm. long or less; perianth segments not clawed........................ 3. Fritillaria (p. 144).

1. **LEUCOCRINUM** Nutt. White mountain lily.

Plants acaulescent, with numerous leaves from a short rootstock and a cluster of fleshy roots; flowers few to many from the crown, white, 3 to 5 cm. long.

1. **Leucocrinum montanum** Nutt.; A. Gray, Ann. Lyc. N. Y. **4:** 110. 1848.
Type locality: "In planitiebus altis fluminis Platte."
Range: Oregon and South Dakota to Nevada and northern New Mexico.
New Mexico: Dulce; Raton. Open slopes.

2. LILIUM L. Lily.

Stems tall, with all but the uppermost leaves scattered; leaves linear-lanceolate; perianth campanulate, showy, reddish orange spotted with purple inside; capsules subcylindric, attenuate at the base.

Our species is one of our handsomest native plants. It occurs only occasionally in moist places in the higher mountains. It is well worthy of cultivation and would doubtless do well in gardens at elevations of 2,000 meters or more.

1. **Lilium umbellatum** Pursh, Fl. Amer. Sept. 229. 1814.
Lilium montanum A. Nels. Bull. Torrey Club **26:** 6. 1899.
Lilium philadelphicum montanum Cockerell, Univ. Mo. Stud. Sci. **2**[2]: 92. 1911.
Type locality: "On the banks of the Missouri."
Range: Ohio to Alberta, south to Arkansas and New Mexico.
New Mexico: Chama; Santa Fe and Las Vegas mountains; Fresnal. Open woods, in the Transition Zone.
We are unable to separate *Lilium montanum* from this, since the characters of the narrowness of the leaves and the number of flowers do not hold for New Mexican material.

3. FRITILLARIA L. Fritillaria.

Slender plant 20 to 40 cm. high, with leafy stems, each bearing 1 to 6 flowers; bulbs of numerous thick scales; perianth campanulate, of 6 equal, dull purple segments; styles united to the middle.

1. **Fritillaria atropurpurea** Nutt. Journ. Acad. Phila. **7:** 54. 1834.
Type locality: "On the borders of the Flat-Head river."
Range: Oregon and North Dakota to California and New Mexico.
New Mexico: Carrizo Mountains (*Matthews*).

4. LLOYDIA Salisb.

Low plants, 5 to 15 cm. high, with leafy 1-flowered stems; bulbs upon an oblique rhizome, covered by the persistent bases of the leaves; perianth segments spreading, white with purple veins.

1. **Lloydia serotina** (L.) Sweet, Hort. Brit. ed. 2. 527. 1830.
Anthericum serotinum L. Sp. Pl. ed. 2. 444. 1762.
Lloydia alpina Salisb. Trans. Hort. Soc. Lond. **1:** 328. 1812.
Type locality: "In alpibus Angliae, Helvetiae, Taureri rastadiensis, Wallaesiae."
Range: Arctic regions southward to Washington and New Mexico; also in the Old World.
New Mexico: Hermits Peak; Pecos Baldy; top of Las Vegas Range. Meadows, in the Arctic-Alpine Zone.

25. ASPHODELACEAE. Asphodel Family.

1. ANTHERICUM L.

A low plant with naked stems (sometimes with 1 or 2 small leaves) from a thick cylindric fleshy-fibrous root; leaves linear, grasslike; flowers yellow, on jointed pedicels; capsules oblong, with several flattened seeds in each cell.

1. **Anthericum torreyi** Baker, Journ. Linn. Soc. Bot. **15:** 318. 1876.
Echeandia terniflora angustifolia Torr. U. S. & Mex. Bound. Bot. 219. 1859.
Hesperanthes torreyi S. Wats. Proc. Amer. Acad. **14:** 241. 1879.

TYPE LOCALITY: Copper Mines, New Mexico.

RANGE: New Mexico and Arizona, southward into Mexico.

NEW MEXICO: San Ignacio; Hop Canyon; Las Vegas Mountains; Mogollon Mountains; Burro Mountains; Black Range; White and Sacramento mountains. Mountains, in the Transition Zone.

Order 14. AMARYLLIDALES.

KEY TO THE FAMILIES.

Stamens 6; leaves not 2-ranked .... 26. **AMARYLLIDACEAE** (p. 145).
Stamens 3; leaves 2-ranked .... 27. **IRIDACEAE** (p. 147).

26. AMARYLLIDACEAE. Amaryllis Family.

Perennials with bulbs or corms or sometimes with fibrous roots; leaves basal; flowers regular or irregular, solitary or corymbose; andrœcium of 6 stamens inserted on an epigynous disk or at the throat of the tube opposite the sepals and petals; ovary inferior, 3-celled; styles united; fruit a 3-celled capsule or berry.

KEY TO THE GENERA.

Leaves spiny-toothed and spine-tipped; plants with elongated caudices .... 1. AGAVE (p. 145).
Leaves not spiny-toothed; plants with bulbs............... 2. ATAMOSCO (p. 147).

1. AGAVE L. CENTURY PLANT.

Long-lived perennials with a cluster of numerous thick fleshy basal leaves and a tall flower stalk, this either nearly spicate or paniculate and with numerous thick divergent branches; perianth persistent, tubular-funnelform, parted into numerous narrow, nearly equal divisions; anthers linear, versatile; fruit an oblong coriaceous 3-celled capsule containing numerous flat black seeds.

Agave americana is an introduced species very common in cultivation in the southern part of the State. It is the common "maguey" of the Mexicans, who use the sap taken from the developing flower stalk for making "pulque," "mescal," and "tequila." It is not cultivated far north of Las Cruces, and even here the leaves are sometimes frosted in the winter and rarely the whole plant killed.

KEY TO THE SPECIES.

Leaves not spiny-margined, filiferous, 1 cm. broad or less, tapering upward.. 1. *A. schottii.*
Leaves bearing hooked spines along the margins, not filiferous, 4 cm. wide or more, generally broadest about the middle.
  Leaves few, 10 to 15, 20 to 30 cm. long, yellowish green; panicle with very short branches, spikelike in appearance.. 2. *A. lechuguilla.*
  Leaves more numerous, 30 or more, 20 to 100 cm. long, deep green or bluish green, glaucous; panicles with spreading longer branches.
    Stamens inserted near the middle of the corolla tube; leaves deep green, 5 to 12 cm. wide, generally 40 to 60 cm. long, sometimes much longer........... 3. *A. palmeri.*

Stamens inserted at the base of the corolla segments; leaves bluish green, glaucous, closely imbricated, broader, 8 to 14 cm. wide, usually 20 to 45 cm. long.

Leaves broad, 10 to 14 cm. wide, 30 cm. long or more; panicles large and widely spreading; flowers 8 to 9 cm. long.............................. 4. *A. parryi.*

Leaves of same relative proportions but smaller, 15 to 20 cm. long and 5 to 8 cm. broad; panicles with few branches; flowers mostly about 6 cm. long.................................... 5. *A. neomexicana.*

1. **Agave schottii** Engelm. Trans. Acad. St. Louis **3:** 305. 1875.

Agave geminiflora sonorae Torr. U. S. & Mex. Bound. Bot. 214. 1859.

TYPE LOCALITY: Sierra del Pajarito, southern Arizona.

RANGE: Southern Arizona, southwestern New Mexico, and adjacent Mexico.

NEW MEXICO: Guadalupe Canyon (*Mearns* 575).

2. **Agave lechuguilla** Torr. U. S. & Mex. Bound. Bot. 213. 1859. LECHUGUILLA.

TYPE LOCALITY: "Mountains near El Paso, and along the Rio Grande downward."

RANGE: Low hills and dry plains, western Mexico and southern New Mexico and southward, in the Lower Sonoran Zone.

Miss Mulford reported this from the Organ Mountains as having been collected May 18, 1851. Neither of the writers has seen the plant in this range, but it may occur at the southern end, where little collecting has been done. A single plant from the north end of the Franklin Mountains, just on the boundary between New Mexico and Texas, is growing in the garden at the Agricultural College. It is said to occur along the southern border farther east as well.

The species is of economic importance as a fiber plant in northern Mexico, where it is used extensively in making cordage. The short caudex is used as a substitute for soap, one form of the "amole" found on the market.

3. **Agave palmeri** Engelm. Trans. Acad. St. Louis **3:** 319. 1875.

TYPE LOCALITY: Mountains of southern Arizona.

RANGE: Southeastern Arizona and southwestern New Mexico and southward.

NEW MEXICO: Florida Mountains; Cloverdale; San Luis Mountains; Animas Mountains; La Luz Canyon. Dry hills, in the Upper Sonoran Zone.

Miss Mulford reports finding a plant of this species a few miles from Fort Bayard, and that must be about its northern limit.

4. **Agave parryi** Engelm. Trans. Acad. St. Louis **3:** 311. 1875.

Agave americana latifolia Torr. U. S. & Mex. Bound. Bot. 213. 1859.

TYPE LOCALITY: Near the Copper Mines, New Mexico.

RANGE: Southern Arizona and New Mexico and southward.

NEW MEXICO: Fierro; Big Hatchet Mountains; Lake Valley; Burro Mountains; Florida Mountains; Bear Mountains; 5 miles north of Reserve; Mogollon Creek. Low hills, in the Lower and Upper Sonoran zones.

This is the common "mescal" of western New Mexico. It has considerable decorative value and, while never as large as *A. americana*, reaches sufficient size to warrant its use in large urns and in other positions in formal gardening.

This, like the other larger plants of the genus, was used by the Indians in making mescal. The thick leaves were cooked in large pits made in the ground and lined with stones, which were first fired, then filled with the plant. It is from their preparation of this article of food that the Mescalero Apaches receive their name.

5. **Agave neomexicana** Woot. & Standl. Contr. U. S. Nat. Herb. **16:** 115. *pl. 48.* 1913.

TYPE LOCALITY: Organ Mountains, New Mexico. Type collected by Standley (no. 541).

RANGE: Mountains of southern New Mexico.

NEW MEXICO: Tortugas Mountain; Organ and San Andreas mountains.

2. ATAMOSCO Adans. ATAMASCO LILY.

Low plant with large tunicated bulbs, slender grasslike leaves, and rather large (3 or 4 cm. in diameter) yellow flowers borne singly upon a stout fleshy scape; capsules large and deeply 3-lobed.

1. **Atamosco longifolia** (Hemsl.) Cockerell, Canad. Ent. **1901:** 283. 1901.

Zephyranthes longifolia Hemsl. Diag. Pl. Mex. 55. 1880.

TYPE LOCALITY: New Mexico. Type collected by Wright (no. 1904).

RANGE: Western Texas to southern Arizona, south into Mexico.

NEW MEXICO: Mesa near Las Cruces; Lordsburg; Animas Valley. Dry hills and mesas, in the Lower Sonoran Zone.

27. IRIDACEAE. Iris Family.

Perennial, mostly caulescent herbs with bulblike or elongated rootstocks; leaves equitant, 2-ranked; flowers regular or irregular, solitary or in clusters from spathelike bracts; perianth usually showy; sepals and petals often very unlike, distinct, or united below; stamens 3, adnate to the perianth opposite the sepals; gynœcium of 3 united carpels; ovary inferior; styles distinct; fruit a loculicidally 3-valved capsule.

KEY TO THE GENERA.

Flowers yellow.. 1. OREOLIRION (p. 147).
Flowers blue or white.
    Styles alternate with the stamens; leaves narrow, less than 5 mm. wide.............................. 2. SISYRINCHIUM (p. 147).
    Styles opposite or arching over the stamens; leaves broad, 10 mm. wide or more.................. 3. IRIS (p. 148).

1. OREOLIRION Bicknell.

An erect perennial, 25 to 50 cm. high, with flat, grasslike, conspicuously nerved leaves; roots clustered, somewhat fleshy; flowers large, 30 mm. in diameter, yellow; capsules oblong, 12 to 14 mm. high.

In general appearance this plant is much like the species of Sisyrinchium, but the yellow flowers enable one to distinguish it readily.

1. **Oreolirion arizonicum** (Rothr.) Bicknell.

Sisyrinchium arizonicum Rothr. Bot. Gaz. **2:** 125. 1877.

TYPE LOCALITY: Willow Spring, Arizona.

RANGE: Southern Arizona and New Mexico.

NEW MEXICO: Mogollon Mountains; Black Range.

2. SISYRINCHIUM L. BLUE-EYED GRASS.

Slender perennial grasslike plants with numerous erect leaves, winged stems, and small blue flowers, occurring in the higher mountains in moist meadows and along streams.

KEY TO THE SPECIES.

Outer bracts of the inflorescence about twice as long as the inner..... 1. *S. campestre.*
Outer bracts of about the same length as the inner.
  Perianth 7 to 10 mm. long; plants somewhat glaucous, the stems clustered; bracts broad, 10 to 20 mm. long; stems flexuous, often ascending.. 2. *S. demissum.*
  Perianth 10 to 14 mm. long; plants more slender, bright green, the stems mostly solitary, erect, straight; bracts 16 to 32 mm. long.. 3. *S. occidentale.*

1. **Sisyrinchium campestre** Bicknell, Bull. Torrey Club **26**: 341. 1899.

TYPE LOCALITY: "Wisconsin to North Dakota, south to Louisiana, Oklahoma and the mountains of New Mexico."

RANGE: As under type locality.

NEW MEXICO: Chama; Santa Fe and Las Vegas mountains; Sierra Grande; Sacramento Mountains. Transition Zone.

2. **Sisyrinchium demissum** Greene, Pittonia **2**: 69. 1890.

TYPE LOCALITY: "In moist meadows at the base of Bill Williams Mountain Arizona, and also near Flagstaff."

RANGE: Arizona to western Kansas.

NEW MEXICO: Las Vegas; mountains west of Grants Station; Zuni; Barranca; Mogollon Mountains; Black Range; Chavez; Socorro; White Mountains. Meadows, in the Transition Zone.

3. **Sisyrinchium occidentale** Bicknell, Bull. Torrey Club **26**: 447. 1899.

TYPE LOCALITY: "Idaho and Nevada to Colorado and North Dakota."

RANGE: As under type locality.

NEW MEXICO: Near Pecos; Iron Creek, Mogollon Mountains; north of El Vado. Upper Sonoran Zone.

3. **IRIS** L. BLUE FLAG.

Plants 30 to 70 cm. high, with long, flat, somewhat glaucous leaves arising from a thickened rootstock; flowers large, very showy, sweet-scented, pale blue.

1. **Iris missouriensis** Nutt. Journ. Acad. Phila. **7**: 58. 1834.

TYPE LOCALITY: "Towards the sources of the Missouri."

RANGE: British America south to California, Arizona, and New Mexico.

NEW MEXICO: Santa Fe and Las Vegas mountains; Sandia Mountains; Tunitcha Mountains; Chama; Sierra Grande; Manguitas Spring; Black Range; White and Sacramento mountains. Meadows, in the Transition and Canadian zones.

Order 15. ORCHIDALES.

28. ORCHIDACEAE. Orchis Family.

Herbaceous plants, perennial by bulbs or thickened roots, sometimes parasitic; leaves entire, from mere sheathing bracts to broadly ovate; flowers sometimes conspicuous, in ours usually small, of bizarre forms especially adapted to insect pollination; corolla of two similar lateral petals and a third (the lip or labellum) very different one, this frequently spurred or saccate; stamens gynandrous, with usually only one anther; pollen in small coherent masses (pollinia); ovary inferior; fruit a capsule.

KEY TO THE GENERA.

Anthers 2; lip a large inflated sack........................ 1. CYPRIPEDIUM (p. 149).
Anthers only one; lip various in different genera.
Flowers solitary, scapose........................... 2. CYTHEREA (p. 150).
Flowers several, racemose or spicate.
Plants without green leaves; stems glandular-pubescent............................... 3. CORALLORHIZA (p. 150).
Plants with green leaves; stems not glandular (except in Peramium), usually glabrous.
Leaves rosulate; stems very short.......... 4. PERAMIUM (p. 150).
Leaves not rosulate, scattered along the stems; stems 10 cm. long or more.
Leaves 1 or 2.
Leaves 1, elliptic to oval; racemes many-flowered; flowers maroon or green............... 5. ACHROANTHES (p. 151).
Leaves 2, opposite, reniform; racemes laxly few-flowered; flowers greenish............ 6. OPHRYS (p. 152).
Leaves several.
Inflorescence loosely racemose; flowers few, large, 25 to 35 mm. long; capsules reflexed. 7. EPIPACTIS (p. 152).
Inflorescence spicate, strict; flowers numerous, small, 10 to 18 mm. long; capsules erect.
Spikes twisted; spur wanting..10. IBIDIUM (p. 154).
Spikes not twisted; spur present.
Lip bifid; bracts of inflorescence very conspicuous........ 8. COELOGLOSSUM (p. 152).
Lip entire; bracts of inflorescence usually not very conspicuous.................. 9. LIMNORCHIS (p. 152).

1. CYPRIPEDIUM L. LADY'S-SLIPPER.

Broad-leaved plants arising from thickened fascicled roots; flowers usually solitary, on long peduncles, showy, bright yellow, with purple spots on the saccate lower petal.

1. **Cypripedium veganum** Cockerell & Barker, Proc. Biol. Soc. Washington **14:** 178. 1901.

TYPE LOCALITY: Sapello Canyon, Las Vegas Range, New Mexico.

RANGE: Mountains of northern New Mexico and southern Colorado.

NEW MEXICO: Santa Fe and Las Vegas mountains; Mogollon Mountains; Cloudcroft. Damp woods, in the Canadian and Hudsonian zones.

The dried specimens from New Mexico agree with Colorado material collected by Baker and by Coulter and referred to *C. pubescens* Willd., but these are slightly different from *C. pubescens* material from the Eastern States. It is likely that the plant of the Rocky Mountains is *C. veganum*.

The specimen in the National Museum deposited by Professor Cockerell disagrees with his description in two particulars: neither leaves nor stems are glabrous, but both are sparsely and coarsely pubescent though not glandular; and the leaves are entirely too broad to be called less than elliptic-lanceolate. The specimen shows only three upper leaves. Specimens collected both by Standley and by Snow show the lower leaves as elliptic.

The writers have heard fairly reliable reports of the occurrence of another species of Cypripedium in the mountains east of Santa Fe. The plant has been observed in Santa Fe Canyon and in the mountains east of the Pecos. It is said to have a white lip splotched with purple. We have been unable to procure material of it, and it is possible that the plant belongs to some other genus.

2. **CYTHEREA** Salisb. CALYPSO.

A low herb, 10 to 15 cm. high, with a single showy rose-colored nodding flower at the end of a slender bracted stem; bracts narrowly oblong, clasping, acuminate; single radical leaf broadly elliptic, with numerous veins.

1. **Cytherea bulbosa** (L.) House, Bull. Torrey Club **32:** 382. 1905.
Cypripedium bulbosum L. Sp. Pl. 951. 1753.
Calypso borealis Salisb. Parad. Lond. *pl. 89*. 1806.
Cytherea borealis Salisb. Trans. Hort. Soc. Lond. **1:** 301. 1812.
Calypso bulbosa Oakes, Cat. Vt. Pl. 28. 1842.
TYPE LOCALITY: "In Lapponia, Russia, Sibiria."
RANGE: Alaska and British America, south to Maine, Michigan, and New Mexico; also in Europe and Asia.
NEW MEXICO: Hermits Peak; Winsor Creek; Sandia Mountains. Deep woods.

3. **CORALLORHIZA** R. Br. CORAL ROOT.

Stems stout, simple, erect, from a cluster of coral-like rootstocks (whence the generic name); leaves represented only by membranous sheaths; whole plant without green coloring matter; flowers purplish, the white lip usually spotted with purple.

Our species are found only in moist, shaded, usually cool woods, where they are very striking because of the absence of green coloring.

KEY TO THE SPECIES.

Spur present at the summit of the ovary; lip 3-lobed................ 1. *C. multiflora.*
Spur absent; lip entire.. 2. *C. vreelandii.*

1. **Corallorhiza multiflora** Nutt. Journ. Acad. Phila. **3:** 138. *pl. 7*. 1823.
Corallorhiza grabhami Cockerell, Torreya **3:** 140. 1903.
TYPE LOCALITY: "From New England to Carolina."
RANGE: Alaska and British America to Florida and California.
NEW MEXICO: Winsor Creek; Harveys Upper Ranch; Tunitcha Mountains; Chama; East Canyon. In woods, in the Canadian Zone.

2. **Corallorhiza vreelandii** Rydb. Bull. Torrey Club **28:** 271. 1901.
TYPE LOCALITY: Veta Mountains, Colorado.
RANGE: Colorado and northern New Mexico.
NEW MEXICO: Horsethief Canyon; Sandia Mountains.
Reported from the vicinity of Pecos by Professor Cockerell, the specimens identified by Rydberg.
Metcalfe's 1513 from the Black Range seems to represent another and possibly undescribed species. Our material is entirely insufficient for diagnosis.

4. **PERAMIUM** Salisb. RATTLESNAKE PLANTAIN.

Low plants, 10 to 25 cm. high, with basal rosettes of somewhat fleshy, often variegated, ovate or oblong-ovate leaves; flowers on a stout scape, this glandular-viscid, twisted; flowers whitish, small; roots somewhat fleshy.

KEY TO THE SPECIES.

Lip of the corolla evidently saccate, the margins recurved; plants low, 10 to 14 cm. high; leaves 20 to 25 mm. long............. 1. *P. ophioides.*
Lip scarcely saccate, the margins incurved; plants taller, 15 to 30 cm. high; leaves 40 to 60 mm. long......................... 2. *P. decipiens.*

1. Peramium ophioides (Fernald) Rydb. in Britton, Man. 302. 1901.

Goodyera ophioides Fernald, Rhodora **1**: 6. 1899.

TYPE LOCALITY: Not definitely stated.

RANGE: British America to New Mexico, South Dakota, and North Carolina.

NEW MEXICO: Winsor Creek; Upper Pecos. Damp woods, in the Canadian Zone.

2. Peramium decipiens (Hook.) Piper, Contr. U. S. Nat. Herb. **11**: 208. 1906.

Spiranthes decipiens Hook. Fl. Bor. Amer. **2**: 203. 1839.

Goodyera menziesii Lindl. Gen. Sp. Orchid. 492. 1840.

Peramium menziesii Morong, Mem. Torrey Club **5**: 124. 1894.

TYPE LOCALITY: Lake Huron.

RANGE: British America to New York, New Mexico, and California.

NEW MEXICO: Winsor Creek; Harveys Upper Ranch; Sandia Mountains; Tunitcha Mountains. Damp woods, in the Canadian Zone.

The two species grow together, and one is likely to be overlooked because of its similarity in general appearance to the other.

5. ACHROANTHES Raf. ADDER'S MOUTH.

Low herbs from solid bulbs, with 1 or 2 leaves and 1 to several scales at the base of the stem; flowers small, green or purplish, in a terminal raceme or spike; sepals spreading, separate; petals filiform or linear, spreading; lip cordate or auriculate at the base.

KEY TO THE SPECIES.

Flowers greenish, in a very dense spike; divisions of the perianth oblong or oblong-lanceolate to ovate......................... 1. *A. montana.*
Flowers purplish, in a loosely flowered spike; divisions of the perianth linear or linear-lanceolate............................. 2. *A. porphyrea.*

1. Achroanthes montana (Rothr.) Greene, Pittonia **2**: 183. 1891.

Microstylis montana Rothr. in Wheeler, Rep. U. S. Surv. 100th Merid. **6**: 264. 1878.

TYPE LOCALITY: Mount Graham, Arizona, at an elevation of 2,800 meters.

RANGE: In the mountains of Arizona and New Mexico.

NEW MEXICO: Gallinas Planting Station (*Bartlett* 324). Deep woods, in the Canadian Zone.

This is a considerable extension of range for the species. Heretofore it has been known in the United States only from Arizona. Specimens in the National Herbarium are from Mount Graham and the Rincon and Huachuca Mountains of that State.

2. Achroanthes porphyrea (Ridley) Woot. & Standl. Contr. U. S. Nat. Herb. **16**: 116. 1913.

Microstylis purpurea S. Wats. Proc. Amer. Acad. **18**: 195. 1883, not Lindl. 1840.

Microstylis porphyrea Ridley, Journ. Linn. Soc. Bot. **24**: 320. 1888.

Achroanthes purpurea Greene, Pittonia **2**: 184. 1891.

TYPE LOCALITY: In Tanners Canyon, Huachuca Mountains, southern Arizona.

RANGE: Mountains of southern Arizona and New Mexico.

NEW MEXICO: Cloudcroft (*Wooton*).

6. OPHRYS L. TWAYBLADE.

Stems slender and delicate, 10 to 20 cm. high, from fibrous creeping roots; flowers small, greenish, in few-flowered racemes; leaves 2, opposite, reniform, thin, near the top of the stem.

1. **Ophrys nephrophylla** Rydb. Bull. Torrey Club **32:** 610. 1905.
Listera nephrophylla Rydb. Mem. N. Y. Bot. Gard. **1:** 108. 1900.
TYPE LOCALITY: Spanish Basin, Montana.
RANGE: Alaska and Oregon to Montana and New Mexico.
NEW MEXICO: Horsethief Canyon; Upper Pecos. Damp woods, in the Canadian and Hudsonian zones.

7. EPIPACTIS R. Br. HELLEBORINE.

A rather tall coarse-leaved plant from a creeping rootstock; inflorescence racemose; flowers few, pediceled, conspicuously bracteate; capsule reflexed at maturity.

1. **Epipactis gigantea** Dougl.; Hook. Fl. Bor. Amer. **2:** 202. *pl. 202.* 1839.
TYPE LOCALITY: "N. W. America. On the subalpine regions of the Blue and Rocky mountains."
RANGE: Washington and California to Texas.
NEW MEXICO: Mimbres; Grand Canyon of the Gila; Mangas Springs. Damp woods, in the Transition Zone.

8. COELOGLOSSUM Hartman. BRACTED ORCHIS.

Stems erect, rather stout, succulent, from a bifid fusiform tuber; leaves oblong-elliptic to lanceolate, the lower obtuse, the upper acute; inflorescence a few-flowered spike with conspicuous lanceolate spreading bracts.

1. **Coeloglossum bracteatum** (Willd.) Parl. Fl. Ital. **3:** 409. 1858.
Orchis bracteata Willd. Sp. Pl. **4:** 34. 1805.
Habenaria bracteata R. Br. in Ait. Hort. Kew. ed. 2. **5:** 192. 1813.
TYPE LOCALITY: "Habitat in Pennsylvania."
RANGE: British America south to North Carolina and New Mexico.
NEW MEXICO: Hillsboro Peak; Upper Pecos River; Winsors Ranch. Cold woods.

9. LIMNORCHIS Rydb. BOG ORCHIS.

Erect herbaceous perennials, with succulent greenish stems arising from elongated rootlike tubers and bearing slender, more or less crowded spikes of inconspicuous greenish or white flowers.

The plants occur in cool, moist situations in shaded thickets in rich soil. They have usually been referred to the genus Habenaria and are so treated in the latest revision of the genus.[1] We prefer the treatment of Doctor Rydberg,[2] which is followed here so far as it relates to New Mexican species.

KEY TO THE SPECIES.

Leaves short, 3 to 7 cm. long, the lowest usually largest......... 1. *L. brevifolia.*
Leaves much longer, 8 to 20 cm. long, the lowest shorter than those along the middle of the stem.
  Flowers white or nearly so; spur and lip various.
    Lip linear, not at all dilated at the base, 8 mm. long; spur over 10 mm. long; spike long, lax, slender... 4. *L. sparsiflora.*

---

[1] Ames, Oakes. Studies in the family Orchidaceae, fasc. 4.

[2] Rydberg, P. A. The American species of Limnorchis and Piperia north of Mexico. Bull. Torrey Club **28:** 605. 1901.

Lip lanceolate, dilated at the base; spur various.
Spur about equaling the short lip.................. 6. *L. borealis.*
Spur nearly two-thirds longer than the rather long lip.. 2. *L. thurberi.*
Flowers greenish or purplish; spur shorter than or about equaling the lip, this lanceolate, 4 to 5 mm. long.
Petals purplish; spur one-half to two-thirds as long as the lip, conspicuously saccate, slightly curved... 5. *L. purpurascens.*
Petals greenish; spur almost equaling the lip, clavate, curved.. 3. *L. viridiflora.*

Mr. Ames refers a specimen collected on the Pecos River, August 6, 1898 (*G. E. Coghill* 147), to *L. dilatata.* Doctor Rydberg cites a specimen of *L. ensifolia* from Silver City, collected in 1880 by E. L. Greene. We have seen neither of these specimens. *L. dilatata* is a northeastern species, ranging only as far west as Nebraska, excluding the specimen mentioned. *L. ensifolia* is closely related to *L. sparsifolia,* and is reported from the same region; it differs in having a shorter and denser spike, the upper sepals larger, and the bracts shorter and broader; its leaves are noticeably different in shape in material we have seen.

1. **Limnorchis brevifolia** (Greene) Rydb. Bull. Torrey Club **28:** 631. 1901.
Habenaria brevifolia Greene, Bot. Gaz. **6:** 218. 1881.
TYPE LOCALITY: "In dry ground under *Pinus ponderosa,* Pinos Altos Mts.," New Mexico. Type collected by Greene (no. 369).
RANGE: Mountains of southern New Mexico, southward into Mexico.
NEW MEXICO: Pinos Altos Mountains; White Mountains; Iron Creek; Sacramento Mountains. Canadian and Transition zones.

2. **Limnorchis thurberi** (A. Gray) Rydb. Bull. Torrey Club **28:** 624. 1901.
Habenaria thurberi A. Gray, Proc. Amer. Acad. **7:** 389. 1868.
TYPE LOCALITY: Arizona.
RANGE: California and Arizona to the mountains of western New Mexico.
NEW MEXICO: Mogollon Creek (*Metcalfe* 282).

3. **Limnorchis viridiflora** (Cham.) Rydb. Bull. Torrey Club **28:** 616. 1901.
Habenaria borealis viridiflora Cham. Linnaea **3:** 28. 616. 1828.
TYPE LOCALITY: "Unalascha."
RANGE: Alaska, southward to the mountains of Colorado and New Mexico, eastward to Nebraska and South Dakota.
NEW MEXICO: Santa Fe and Las Vegas mountains. Canadian Zone.

4. **Limnorchis sparsiflora** (S. Wats.) Rydb. Bull. Torrey Club **28:** 631. 1901.
Habenaria sparsiflora S. Wats. Proc. Amer. Acad. **12:** 276. 1877.
TYPE LOCALITY: "Common in the Sierra Nevada and mountains of Northern California."
RANGE: Oregon and California to the mountains of western New Mexico.
NEW MEXICO: Mogollon Mountains (*Rusby* 399, in part).
Mr. Ames refers to this species a specimen from "spring at Twin Sisters near Silver City." This may be the *L. ensifolia* Rydb., reported from New Mexico.

5. **Limnorchis purpurascens** Rydb. Bull. Torrey Club **28:** 269. 1901.
TYPE LOCALITY: Iron Mountain, Colorado.
RANGE: Mountains of Colorado and New Mexico.
NEW MEXICO: Santa Fe and Las Vegas mountains.

6. **Limnorchis borealis** (Cham.) Rydb. Bull. Torrey Club **28:** 621. 1901.
Habenaria borealis Cham. Linnaea **3:** 28. 1828.
TYPE LOCALITY: "Unalaschka."
RANGE: Alaska to Colorado and northern New Mexico.
NEW MEXICO: Chama (*Standley* 6643). Bogs, in the Transition and Canadian zones.

10. IBIDIUM Salisb.

Stems erect, from tuberous roots, bearing few leaves near the base; flowers small, white, spurless, spicate, the spikes twisted; sepals and petals all more or less connivent into a hood.

1. **Ibidium strictum** (Rydb.) House, Bull. Torrey Club **32:** 381. 1905.
Gyrostachys stricta Rydb. Mem. N. Y. Bot. Gard. **1:** 107. 1900.
TYPE LOCALITY: Indian Creek, Montana.
RANGE: Alaska and Newfoundland to Pennsylvania, California, and northern New Mexico.
NEW MEXICO: Costilla Valley (*Wooton*). Bogs.

Subclass DICOTYLEDONES.

Order 16. PIPERALES.

29. SAURURACEAE. Lizard's-tail Family.

1. ANEMOPSIS Hook. YERBA MANSA.

Perennial herb with long stolons; leaves subcoriaceous, elliptic-oblong or oblong, pellucid-punctate, petioled, mostly basal; flowers very small, crowded on a simple involucrate conic spadix; involucral bracts petal-like, white; ovary solitary, immersed in the rachis; seeds oblong, puncticulate.

1. **Anemopsis californica** Hook. & Arn. Bot. Beechey Voy. 390. *pl. 92.* 1841.
Houttuynia californica Benth. & Hook.; S. Wats. Bot. Calif. **2:** 483. 1880.
TYPE LOCALITY: California.
RANGE: California to Utah and New Mexico, south into Mexico.
NEW MEXICO: Albuquerque; Mogollon Mountains; Mangas Springs; Berendo Creek; Belen; Rincon; Dog Spring; Mesilla Valley; above Tularosa. Wet alkaline meadows, chiefly in the Lower Sonoran Zone.

The plants form large and conspicuous patches in wet places, especially in alkaline soil. The form found in New Mexico, Arizona, and Chihuahua differs from the typical Californian plant in being smaller and nearly or quite glabrous, and in having the involucral bracts shorter than the spadix.

Order 17. SALICALES.

30. SALICACEAE. Willow Family.

Trees or shrubs with simple alternate deciduous leaves; flowers diœcious, in catkins; bracts of the aments scalelike; perianth none; stamens 1 to several; ovary 1-celled; stigmas 2; fruit a small capsule; seeds very numerous, small, comose.

KEY TO THE GENERA.

Bracts incised; disk cup-shaped; stamens numerous; winter buds with several scales.................................... 1. POPULUS (p. 155).
Bracts entire; disk represented by one or two small glands; stamens few, generally less than 5; winter buds with a single scale.................................... 2. SALIX (p. 156).

1. POPULUS L. Cottonwood.

Trees with rough light-colored bark and scaly resinous buds; leaves usually long-petioled, somewhat coriaceous, with prominent veins; flowers in pendulous aments, appearing before the leaves; seeds with a conspicuous white coma (the "cotton").

This is the genus containing the common cottonwoods of the State and the less well known aspen or "quaking asp" of the higher mountains. They are all rather short-lived trees and grow in stations where the soil is at least moderately wet, preferring the broad river valleys, where one species (*P. wislizeni*) is almost the only tree, or locations besides mountain streams or springs. The aspen is a characteristic plant of the Canadian Zone. Three of the species here mentioned are used more or less extensively and effectively as shade trees, and might well be used a great deal more. The wood of all species is light and spongy and not valuable for posts or firewood, although frequently used for these purposes for lack of something better.

The silver-leaf poplar (*Populus alba*), the Lombardy poplar (*P. italica*), and the Carolina poplar (*P. deltoides*) are cultivated in many localities in the State, and prove very satisfactory, though short-lived, shade trees. Doctor Britton states that *P. mexicana* S. Wats. occurs in New Mexico, but we have seen no material like the Mexican plant.

KEY TO THE SPECIES.

Petioles flattened laterally; leaves broad, deltoid to rotund.
- Leaves broadly ovate to rotund, abruptly short-acuminate, 3 to 5 cm. long and broad, paler beneath; small tree of the high mountains........ 1. *P. aurea.*
- Leaves broadly deltoid, acuminate, 5 to 8 cm. long and 6 to 10 cm. broad, of the same color on both surfaces; large tree of the lower valleys........ 2. *P. wislizeni.*

Petioles terete, or channeled on the upper surface; leaves narrower, ovate to narrowly lanceolate.
- Leaves ovate to ovate-lanceolate, 6 to 10 cm. long, 3 to 5 cm. broad, rather coarsely crenate, both surfaces of the same color........ 3. *P. acuminata.*
- Leaves broadly to narrowly lanceolate, 7 to 15 cm. long and 2 to 4 cm. wide, finely serrate with blunt teeth, much paler beneath........ 4. *P. angustifolia*

1. Populus aurea Tidestrom, Amer. Mid. Nat. **2**: 35. 1911. Quaking aspen.

Populus tremuloides aurea Daniels, Univ. Mo. Stud. Sci. **2**[2]: 98. 1911.

Type locality: Vicinity of Mount Carbon, Colorado.

Range: New Mexico and Arizona to British America.

New Mexico: Common in all the higher mountain ranges. Canadian Zone.

The aspen is a slender, white-barked tree found along streams and on cool slopes of the mountains, or in shaded canyons, associated with firs and spruces, occasionally forming pure forests covering small areas. It is the first tree to take possession of burned areas, completely covering them before the conifers establish themselves. In pure stands the trees are usually very close together, and, the trees being short-lived, such forests soon become a dense tangle of fallen timber. The foliage is thin and scanty, and notwithstanding the number of trees their shade is never dense.

2. Populus wislizeni (S. Wats.) Sarg. Man. Trees N. Amer. 165. 1905.

Valley cottonwood.

Populus fremontii wislizeni S. Wats. Amer. Journ. Sci. III. **15**: 3. 1878.

Type locality: "From S. California to the Rio Grande."

Range: Colorado to western Texas and northern Mexico.

New Mexico: San Juan Valley; Santa Fe; Zuni; Mogollon Mountains; Burro Mountains; Black Range; Deming; Socorro; Mesilla Valley; above Tularosa. Along the larger streams, in the Lower and Upper Sonoran zones.

The common "valley cottonwood," as it is called by those who are acquainted with the "mountain cottonwood," is perhaps the best known tree of the State. It is doubtless the most common shade tree of New Mexico, being used almost everywhere. It is very common along the broad flood plains of the Rio Grande and the San Juan, where it forms "bosques" of considerable extent. Besides its use as firewood and for fence posts, straight trunks are used by the Mexicans for the "vigas" or rafters of their houses. When stripped of their bark, the trees do not decay rapidly.

3. Populus acuminata Rydb. Bull. Torrey Club **20**: 50. 1893.

Type locality: Carter Canyon, Scotts Bluff County, Nebraska.

Range: Montana to New Mexico.

New Mexico: Carrizo Mountains; Farmington; near Alma; mountains west of San Antonio; Cliff; Fort Bayard; Dog Spring; Kingston; Deming. Canyons and river valleys, in the Upper Sonoran Zone.

This species grows in situations similar to those in which *P. wislizeni* is found, besides extending farther up into the mountains. At Deming and Silver City it is used as a shade tree along with the valley cottonwood and is probably equally valuable for that purpose.

4. Populus angustifolia James in Long, Exped. **1**: 497. 1823.

Mountain cottonwood.

Type locality: Rocky Mountains.

Range: British America to Nebraska, Utah, and New Mexico.

New Mexico: Common in all the mountain ranges from the Black Range, Organ Mountains, and White Mountains northward. Canyons and along streams, in the Transition and Upper Sonoran zones.

The mountain cottonwood or "narrow-leaved cottonwood" grows naturally in the mountains along streams, sometimes attaining a great size. It is also common along the valley of the San Juan in San Juan County. It is a rapid grower like its congeners, and is worthy of much more extensive use than has so far been accorded it.

2. SALIX L. Willow.

Shrubs or small trees, from a few centimeters to several meters high; leaves from narrowly linear to short-elliptic or obovate; flowers in aments, appearing before or with the leaves; perianth a single scale; stamens few; pistil single, with a gland at the base of the ovary; the stigmas short; fruit a capsule, containing numerous very small hairy seeds.

The plants of this genus occur at all levels in the State, but are always found where the ground water is abundant and near the surface during the growing season.

KEY TO THE SPECIES.

Leaves usually only about 3 times as long as broad, elliptic-oval to obovate, never narrowly lanceolate or linear; capsules hairy (except in *S. monticola*).
 Styles obsolete, or less than 0.5 mm. long.
  Alpine plants less than 10 cm. high.................... 1. *S. saximontana*.
  Small trees or shrubs, much more than 10 cm. high, at middle elevations in the mountains.
   Aments slender, lax; scales pale; stigmas very short; leaves elliptic-lanceolate, acute............ 2. *S. bebbiana*.

Aments stout, dense; scales dark; stigmas long, slender; leaves obovate, obtuse or abruptly acute.................................. 3. *S. scouleriana.*
Styles elongated, 1 mm. long or more.
Leaves glabrous on both surfaces, bright green; aments closely sessile.................................. 4. *S. chlorophylla.*
Leaves pubescent, sometimes sparingly so; aments on leafy stems.
Capsules glabrous.................................. 7. *S. monticola.*
Capsules tomentose.
Leaves glabrate on the upper surface, densely sericeous beneath......................20. *S. subcaerulea.*
Leaves pubescent on both surfaces, not densely sericeous beneath.
Plants 40 to 150 cm. high................ 6. *S. glaucops.*
Plants less than 10 cm. high.............. 5. *S. petrophila.*
Leaves several times as long as broad, linear to elongate-lanceolate or oblong-lanceolate; capsules glabrous (or weakly villous in nos. 10, 11, 12).
Scales not pale yellow, mostly brownish and persistent.
Leaves broadly lanceolate; young branches not glaucous; capsules distinctly pedicellate.................. 8. *S. cordata watsoni.*
Leaves narrowly oblong-lanceolate; young branches very glaucous; capsules subsessile............... 9. *S. irrorata.*
Scales pale yellow, deciduous.
Stamens 2, hairy below; leaves more or less canescent, linear, remotely denticulate, or sometimes entire; capsules more or less hairy.
Capsules 3 to 4 mm. long; leaves 1 to 3 cm. long, finely pubescent............................12. *S. taxifolia.*
Capsules 5 to 7 mm. long, glabrate; leaves 5 to 10 cm. long.
Leaves bright green and glabrate, at least above, denticulate; capsules 7 mm. long, on long pedicels.........................10. *S. fluviatilis.*
Leaves canescent, entire or sometimes denticulate; capsules smaller, 5 mm. long, on a short pedicel or sessile.
Capsules sessile...........................11. *S. exigua.*
Capsules stipitate.........................18. *S. argophylla.*
Stamens 3 or more, hairy below; leaves bright green above, lanceolate, finely serrulate; capsules pubescent or glabrous.
Petioles and leaf blades glandular.
Leaves long-acuminate, only slightly paler beneath, thin.........................13. *S. fendleriana.*
Leaves short-acuminate, glaucous beneath, somewhat coriaceous....................14. *S. lasiandra.*
Petioles and leaf blades not glandular.
Leaves paler beneath.
Capsules short-stipitate; stamens usually 3..................................17. *S. bonplandiana.*
Capsules long-stipitate; stamens 5 to 9....19. *S. amygdaloides.*

Leaves of the same color on both surfaces.
Leaves long-lanceolate; a compact, spreading tree..........................16. *S. nigra.*
Leaves shorter; a straggling tree or shrub..15. *S. wrightii.*

1. **Salix saximontana** Rydb. Bull. N. Y. Bot. Gard. **1**: 261. 1899.

TYPE LOCALITY: Grays Peak, Colorado.

RANGE: New Mexico to Montana and westward.

NEW MEXICO: Taos Mountains; Truchas Peak. Alpine summits, in the Arctic-Alpine Zone.

A densely cespitose plant, less than 10 cm. high, with small, elliptic, obtuse or acute, entire leaves, these green above, glaucous beneath, on slender petioles. In New Mexico known only from the tops of the highest peaks in the northern part of the State.

2. **Salix bebbiana** Sarg. Gard. & For. **8**: 463. 1895.

Salix perrostrata Rydb. Bull. N. Y. Bot. Gard. **2**: 163. 1901.

TYPE LOCALITY: British America.

RANGE: From New Mexico northward and eastward.

NEW MEXICO: Santa Fe and Las Vegas mountains; Raton; Catskill; El Rito Creek; Chama; Zuni Mountains; Mogollon Mountains; White Mountains. Along streams in the mountains at middle elevations, in the Transition Zone.

3. **Salix scouleriana** Barratt; Hook. Fl. Bor. Amer. **2**: 145. 1839.

Salix flavescens Nutt. N. Amer. Sylv. **1**: 65. 1842.

Salix nuttallii Sarg. Gard. & For. **8**: 463. 1895.

TYPE LOCALITY: "North West America, on the Columbia. *Dr. Scouler.* Fort Vancouver. *Tolmie.*"

RANGE: New Mexico to Alberta and westward.

NEW MEXICO: Tunitcha Mountains; Sierra Grande; Beulah; Zuni Mountains; Eagle Peak; Black Range; Magdalena Mountains; Cloudcroft. Along mountain streams, in the Transition and Canadian zones.

Similar to the next in appearance, but the leaves of a different shape and the characters of the inflorescence conspicuously different. It comes into our range from the west, while *S. bebbiana* comes into New Mexico from the east.

4. **Salix chlorophylla** Anderss. Vet. Akad. Handl. Stockholm **6**: 138. 1867.

TYPE LOCALITY: Western Canada.

RANGE: New Mexico, Utah, and California and northward.

NEW MEXICO: Pecos Baldy; top of Las Vegas Range; Taos Mountains. In the higher mountains, in the Arctic-Alpine Zone.

A low branching shrub, only a few, often not more than two decimeters high; leaves thin, elliptic-ovate, 2 to 3 cm. long, usually entire and acute, paler and slightly glaucous beneath.

5. **Salix petrophila** Rydb. Bull. N. Y. Bot. Gard. **1**: 268. 1899.

Salix arctica petraea Anderss. in DC. Prodr. **16**: 287. 1864.

TYPE LOCALITY: "In summis Rocky Mountains."

RANGE: New Mexico, Colorado, Utah, and northward.

NEW MEXICO: Truchas Peak; Taos Mountains. Among rocks on alpine summits, in the Arctic-Alpine Zone.

A low, creeping plant, 10 cm. high or less, with glabrous, yellowish or brown stems and elliptic, green leaves 3 cm. long or less, found only on very high peaks in the northern part of the State.

6. **Salix glaucops** Anderss. in DC. Prodr. **16**: 281. 1868.

Salix seemanii Rydb. Bull. N. Y. Bot. Gard. **2**: 164. 1901.

Salix wyomingensis Rydb. Bull. Torrey Club **28**: 271. 1901.

TYPE LOCALITY: Rocky Mountains.

RANGE: New Mexico to Canada, and westward.

NEW MEXICO: Taos Mountains; Upper Pecos River. Boggy places in the high mountains, in the Hudsonian Zone.

A low shrub, 40 to 150 cm. high, known in New Mexico only from the tops of high mountains in the northern part on the State.

7. **Salix monticola** Bebb in Coulter, Man. Rocky Mount. 336. 1885.

Salix padophylla Rydb. Bull. Torrey Club **28:** 499. 1901.

TYPE LOCALITY: Golden, Colorado.

RANGE: New Mexico to Alberta.

NEW MEXICO: Santa Fe and Las Vegas mountains. At high levels in the mountains, in the Transition and Canadian zones.

A shrub 3 to 6 meters high, with reddish brown, glabrous stems and elliptic-oblong to broadly oblanceolate leaves 5 to 7 cm. long. It comes into our region from the north and is now known only from the mountains between Santa Fe and Las Vegas.

8. **Salix cordata watsoni** Bebb in S. Wats. Bot. Calif. **2:** 86. 1880.

Salix flava Rydb. Bull. Torrey Club **28:** 273. 1901.

Salix watsoni Rydb. Bull. Torrey Club **33:** 157. 1906.

TYPE LOCALITY: Near Carson City, Nevada.

RANGE: Rocky Mountains, westward to the Pacific coast.

NEW MEXICO: Pecos; Atarque de Garcia; White Mountains. Along streams, in the Transition Zone.

9. **Salix irrorata** Anderss. Öfv. Svensk. Vet. Akad. Förh. **15:** 117. 1858.

TYPE LOCALITY: "Hab. in Mexico nova." Type collected by Fendler, probably about Santa Fe (no. 812).

RANGE: Western Texas to southwestern Arizona, northward to Colorado.

NEW MEXICO: Chama; Santa Fe and Las Vegas mountains; Zuni; Sandia Mountains; San Mateo Mountains; Magdalena Mountains; Mogollon Mountains; Fort Bayard; Dog Spring; Organ Mountains; White Mountains. Transition Zone.

This is one of two very common shrubby willows found beside mountain streams at middle elevations almost throughout the State. Rarely it takes the form of a low tree, but it is usually a shrub 3 to 4 meters high or less. The catkins appear before the leaves, from rather large buds borne on glaucous stems, and are closely followed by short, narrowly elliptic-oblong leaves, 2 or 3 cm. long. The summer foliage consists of numerous, thin, narrowly oblong-lanceolate, abruptly acute, entire or serrate leaves 15 cm. long or less, dark green above and very glaucous beneath. The stems may retain their glaucous coat, but are usually brown.

10. **Salix fluviatilis** Nutt. N. Amer. Sylv. **1:** 73. 1842.

Salix interior Rowlee, Bull. Torrey Club **27:** 273. 1900.

Salix linearifolia Rydb. in Britton, Man. 316. 1901.

TYPE LOCALITY: "The immediate border of the Oregon a little below its confluence with the Wahlamet."

RANGE: Idaho to New Mexico, eastward across the continent; not common in the Rocky Mountain region.

NEW MEXICO: Fort Bayard (*Blumer* 124). Upper Sonoran Zone.

Very rare in our range, most of the specimens that have been referred here belonging, probably, to *S. exigua*.

11. **Salix exigua** Nutt. N. Amer. Sylv. **1:** 75. 1842. SANDBAR WILLOW.

Salix stenophylla Rydb. Bull. Torrey Club **28:** 271. 1901.

TYPE LOCALITY: "Territory of Oregon."

RANGE: Rocky Mountain Region and westward,

NEW MEXICO: Farmington; Tierra Amarilla; Tunitcha Mountains; Santa Fe; Pecos; Magdalena; Gila; Kingston; near Carlisle; White Mountains; Organ Mountains; Mesilla Valley. Common along streams, from the Lower Sonoran to the Transition Zone.

Salix thurberi Rowlee is a form with longer leaves that are noticeably dentate. In our opinion it is not essentially different from *S. exigua* as defined above.

This is the common shrubby willow which grows in sandy soil, on ditch banks, and to some extent beside streams, in the mountains at the lower levels throughout the State. It is exceedingly variable in the size of the leaves and the degree of pubescence of the whole plant. The leaves are always narrowly linear and acute. The Indians and Mexicans use the stripped branches in basketry.

12. Salix taxifolia H. B. K. Nov. Gen. & Sp. **2**: 22. 1817.

TYPE LOCALITY: "Colitur in hortis Mexicani, Queretari, Zelayae, alt. 900–1200 hex."

RANGE: Arizona and New Mexico to Mexico.

NEW MEXICO: Deer Creek (*Goldman* 1441).

Closely related to *S. exigua*, but with shorter leaves and very pubescent capsules. It is known to us only from a single specimen from the extreme southwestern corner of the State.

13. Salix fendleriana Anderss. Öfv. Svensk. Vet. Akad. Förh. **15**: 115. 1858.

TYPE LOCALITY: "Hab. in Mexico nova." Type collected by Fendler, probably about Santa Fe (no. 816).

RANGE: Northern New Mexico, northward and westward to the Pacific Coast.

NEW MEXICO: Nutritas Creek below Tierra Amarilla (*Eggleston* 6634, 6637). Transition Zone.

The species seems to be much more abundant farther north.

14. Salix lasiandra Benth. Pl. Hartw. 335. 1849. WESTERN BLACK WILLOW.

TYPE LOCALITY: "Ad flumen Sacramento," California.

RANGE: Western and northern New Mexico to California, and northward.

NEW MEXICO: Chama; Santa Fe Canyon; north of Ramah; Mogollon Mountains. Transition Zone.

This is one of the two shrubby willows found along the borders of mountain streams at middle elevations. The stems are brown and shining; the leaves are somewhat coriaceous, resembling a cottonwood leaf, almost perfectly lanceolate, acuminate, finely serrate, shining dark green above and pale or somewhat glaucous beneath, 8 to 12 cm. long. It is sometimes associated with *S. irrorata*, but our material shows it only from the western and northern parts of the State.

15. Salix wrightii Anderss. Öfv. Svensk. Vet. Akad. Förh. **15**: 115. 1858.

TYPE LOCALITY: "Hab. in Nova Mexico." The type is Wright's 1877 and did not come from New Mexico but from the banks of the Rio Grande, in Texas or Chihuahua, or from Lake Santa Maria, Chihuahua.

RANGE: From New Mexico southward into western Texas and Chihuahua, westward to Arizona and Sonora.

NEW MEXICO: Cross L Ranch; Albuquerque; Socorro; Kingston; Mangas Springs; Mesilla Valley; Roswell. Lower and Upper Sonoran zones.

The species is the common narrow-leaved willow tree of New Mexico. It occurs mostly along water courses (or near acequias or flooded bottoms) at the lower levels. The mature leaves are light green, of about the same color on both surfaces, narrowly lanceolate, with numerous small teeth. Small, narrowly oblong-lanceolate, acute leaves appear with the flowers. The trunk is sometimes 20 cm. in diameter and is covered with rough gray bark. The branches are slender, greenish yellow, and not reflexed, even on old trees.

16. **Salix nigra** Marsh. Arb. Amer. 139. 1785. BLACK WILLOW.

TYPE LOCALITY: North America.

RANGE: California to Colorado and New Mexico, and eastward.

NEW MEXICO: Gila; Dog Spring; Emory Spring; Grant County; White Sands. Lower and Upper Sonoran zones.

It is very probable that further study of what is here considered *S. nigra* in New Mexico will show that it belongs to *S. wrightii*, which is the common plant of the type. The leaves of our specimens are firmer than those of the eastern form and usually broader.

17. **Salix bonplandiana** H. B. K. Nov. Gen. & Sp. **2:** 24. *pl. 101, 102*. 1817.

TYPE LOCALITY: "In Regno Mexicano, locis opacatis prope Moran, Cabrera, Omitlan et Pachuca, alt. 1270–1350 hexap."

RANGE: Arizona and New Mexico to Mexico.

NEW MEXICO: San Luis Mountains (*Mearns* 2189, 2434).

18. **Salix argophylla** Nutt. N. Amer. Sylv. **1:** 71. 1842.

TYPE LOCALITY: "On the Boise River, toward its junction with the Shoshonee," Idaho.

RANGE: Washington and Oregon to Idaho and New Mexico.

NEW MEXICO: Chama; Shiprock; Reserve. Along streams, in the Upper Sonoran and Transition zones.

With us a shrub 2 to 3 meters high.

19. **Salix amygdaloides** Anderss. Proc. Amer. Acad. **4:** 53. 1858.

TYPE LOCALITY: Fort Pierre, South Dakota.

RANGE: British Columbia and Quebec, southward to New York, Texas, and Oregon.

NEW MEXICO: Shiprock; Farmington. Along streams, in the Upper Sonoran Zone.

This becomes a tree 10 meters high.

20. **Salix subcaerulea** Piper, Bull. Torrey Club **27:** 400. 1900.

Salix covillei Eastwood, Zoe **5:** 80. 1900.

Salix pachnophora Rydb. Bull. Torrey Club **31:** 402. 1904.

TYPE LOCALITY: Powder River Mountains, in wet meadows near the head of Eagle Creek, Oregon.

RANGE: Oregon and Montana to California and New Mexico.

NEW MEXICO: Along Willow Creek, Rio Arriba County (*Standley* 6702). Along streams and in wet meadows, in the Transition and Canadian zones.

A tree 5 meters high or less.

Order 18. JUGLANDALES.

31. JUGLANDACEAE. Walnut Family.

A small family of large or small trees and large shrubs, of considerable economic importance on account of the value of their wood for various purposes and because of their edible seeds generally called "nuts"; leaves pinnately compound, the leaflets mostly large; flowers monœcious, the sterile flowers in catkins, the fertile solitary or few together in short spikes.

The family contains the well known black walnut and the English walnut of commerce, the hickory nut, and the pecan. Only a single genus occurs native in our range, but the pecan is cultivated in a few localities.

52576°—15——11

1. JUGLANS L. WALNUT.

Strong-scented trees or shrubs; buds few-scaled or naked; flowers in simple pendulous catkins from the branches of the previous year; calyx 3 to 6-cleft; stamens 12 to 40, with short filaments; styles 2, short; fruit drupaceous, the exocarp fibrous-fleshy, indehiscent, the endocarp and irregularly roughened "nut" with an edible embryo.

There are at least two species of Juglans in New Mexico and possibly a third. The two do not occur together so far as any records show or as we know. The small bushy species occurs only in the southeastern part of the State in the Lower Sonoran Zone, while the tree grows in the mountains, mostly in the Transition or just at the top of the Upper Sonoran Zone. Mr. O. B. Metcalfe once collected some nuts from a tree in the mountains of the western part of the State which were much larger than those of either of the species mentioned here. These were not accompanied by any other material, so that we are unable to characterize the species.

KEY TO THE SPECIES.

Large shrub 4 to 6 meters high, branching from the ground, rarely if ever with a noticeable trunk.................................. 1. *J. rupestris.*

Tree 8 meters high or more, with a trunk often 3 meters high and 30 to 50 cm. in diameter.. 2. *J. major.*

1. Juglans rupestris Engelm. in Sitgreaves, Rep. Zuñi & Colo. 171. *f. 15.* 1854.

TYPE LOCALITY: "New Mexico."

RANGE: Western Texas and southeastern New Mexico.

NEW MEXICO: Guadalupe Mountains; east slope of the Sacramento Mountains; Pecos Valley. Along streams, in the Lower Sonoran Zone.

This walnut is a large branching shrub, in New Mexico usually less than 5 meters high, with several stems from one root, branched down to the ground. It is not uncommon along the Pecos and its tributaries in the southeastern part of the State and may go some distance back up the tributaries toward the mountains. It is common on the Pecos and the Devils River in western Texas, where it reaches a larger size but still retains the shrubby habit. Its leaflets (6 to 12 pairs) are lanceolate, acuminate, subfalcate, with oblique bases, entire or with relatively few teeth which are rendered more inconspicuous because the margin is revolute. It has been confused with the other New Mexican species by many writers and collectors, although Doctor Torrey had a good conception of the two when he described them. He evidently believed them to be distinct and assigned one of them (*J. major*) to subspecific rank, as he says, "for the present," probably because of the statements of some of the collectors who had not seen both. The species is easily recognized by its smaller size, shrubby habit, smaller and more numerous leaflets, and very small, thick-walled nut only 10 to 15 mm. in diameter.

2. Juglans major (Torr.) Heller, Muhlenbergia 1: 50. 1900.

Juglans rupestris major Torr. in Sitgreaves, Rep. Zuñi & Colo. 171. *pl. 16.* 1854.

TYPE LOCALITY: "Western New Mexico." This is certainly Arizona.

RANGE: New Mexico and Arizona and southward.

NEW MEXICO: Magadalena Mountains; Burro Mountains; Mangas Springs; Fort Bayard; Black Range; Dog Spring; Animas Mountains; White Mountains. Mountains, in the Transition Zone.

In his original description of this species Doctor Torrey says that "Dr. Woodhouse found the plant in western New Mexico and Dr. Bigelow collected it at the Copper Mines." "Western New Mexico" at that time included Arizona. The plant common in the region of the Copper Mines is a good-sized tree, not infrequently with a trunk 3 to 4 meters to the first branch and 40 to 50 centimeters in diameter, with a large rounded top 12 to 15 meters high. Its leaflets are broadly lanceolate, 6 to 9 cm. long, not revolute-margined but serrate. The nut is 20 to 25 mm. in diameter.

This species has been confused with a Californian one (*J. californica* S. Wats.) from which it is said, by those who know both, to be distinct. The Californian plant is found in the Sacramento Valley.

Our native walnuts, this species in particular, are often known by the native name of "nogal."

Order 19. FAGALES.

KEY TO THE FAMILIES.

Staminate and pistillate flowers in aments; fruit never with a bur or cup.............................. 32. **BETULACEAE** (p. 163).

Staminate flowers in aments, the pistillate often solitary; fruit with a bur or cup.................... 33. **FAGACEAE** (p. 164).

32. BETULACEAE. Birch Family.

Monœcious or rarely diœcious trees or shrubs with alternate simple leaves and deciduous stipules; sterile flowers in catkins; fertile flowers clustered, spicate, or in scaly catkins; fruit a 1-celled and 1-seeded nut with or without a foliaceous involucre.

KEY TO THE GENERA.

Ovary inclosed in a bladdery bag.............................. 1. OSTRYA (p. 163).

Ovary not inclosed in a bladdery bag.

Stamens 2; bracts of the mature pistillate aments membranous, usually 3-lobed, deciduous with the nut....... 2. BETULA (p. 163).

Stamens usually 4; bracts of the mature pistillate aments thickened and woody, erose or toothed, persistent.... 3. ALNUS (p. 164).

1. OSTRYA Scop. HOP HORNBEAM.

A small tree; sterile flowers consisting of several stamens in the axil of each bract; fertile flowers a pair to each deciduous bract, inclosed in a bractlet, this in fruit becoming a bladdery bag, the involucres forming a kind of strobile resembling that of the hop.

1. **Ostrya baileyi** Rose, Contr. U. S. Nat. Herb. **8:** 293. 1905.

TYPE LOCALITY: Guadalupe Mountains, Texas.

RANGE: Known only from the type locality.

The type was collected only two miles from the New Mexico line, and the species, with but little doubt, occurs at the north end of the range in New Mexico.

2. BETULA L. BIRCH.

Small tree or large shrub with slender stems; sterile flowers 3 to each shield-shaped scale of the catkin; fertile flowers 2 or 3 to each 3-lobed bract, the bracts thin, deciduous; fertile catkins ovoid to cylindric.

1. **Betula fontinalis** Sarg. Bot. Gaz. **31:** 239. 1901.

Betula microphylla fontinalis Jones, Contr. West. Bot. **12:** 77. 1908.

TYPE LOCALITY: "On the Sweetwater, one of the branches of the Platte."

RANGE: British America to Colorado and New Mexico.

NEW MEXICO: San Juan Valley; Tunitcha Mountains; Paquate. Along streams, in the Upper Sonoran and Transition zones.

3. ALNUS Hill. ALDER.

Shrubs or small trees with thin toothed leaves; sterile catkins with 4 or 5 bractlets and 3 flowers upon each scale; fertile catkins ovoid or ellipsoid, the scales each subtending 2 flowers and a group of 4 small scales, the latter becoming woody in fruit, wedge-obovate.

KEY TO THE SPECIES.

Leaves rounded to truncate at the base, somewhat lobed, ovate to broadly oblong; stamens 4.................................. 1. *A. tenuifolia.*
Leaves usually cuneate or at least narrowed at the base, seldom lobed, the younger ones lanceolate, the older elliptic or oblong; stamens 1 to 3, usually 2.......................... 2. *A. oblongifolia.*

1. **Alnus tenuifolia** Nutt. N. Amer. Sylv. **1**: 32. 1842.

TYPE LOCALITY: "On the borders of small streams within the Range of the Rocky Mountains, and afterwards in the valleys of the Blue Mountains of Oregon."

RANGE: British America to California and New Mexico.

NEW MEXICO: Tunitcha Mountains; Cedar Hill; Chama; Santa Fe and Las Vegas mountains. Along streams, in the Transition Zone.

The powdered bark of the alder, together with ashes of *Juniperus monosperma* and a decoction of *Cercocarpus montanus*, were used by the Navahos in preparing a red dye for wool.

2. **Alnus oblongifolia** Torr. U. S. & Mex. Bound. Bot. 204. 1859.

Alnus acuminata H. B. K. err. det. many authors.

TYPE LOCALITY: Banks of the Mimbres and near Santa Barbara, New Mexico. Type collected by Wright (no. 1864).

RANGE: Southern Arizona and New Mexico.

NEW MEXICO: Magdalena Mountains; Mogollon Mountains; Black Range. Along streams, in the Transition Zone.

33. FAGACEAE. Beech Family.

1. QUERCUS L. OAK.

Low shrubs or large trees with rough bark on the older stems and hard tough wood; leaves chlorophyll green and deciduous, or bluish or grayish green and persistent almost or quite until the appearance of the leaves of the following season, of various shapes, size, and texture, generally short-petioled, mostly more or less stellate-pubescent at some time; flowers monœcious, the staminate usually in slender pendulous aments, the pistillate solitary or in few-flowered spikelike aments, appearing with the leaves; fruit (acorn) a nut varying in shape and size with the species, the cup being also of varying size and shape.

The treatment here given follows that of Doctor Rydberg,[1] and much of the work was done in consultation with him, while examining a rather extended series of New Mexican specimens. The species listed cover the material at New York and Washington and that in the herbarium of the New Mexico Agricultural College. With the use of this material is combined the field experience of Doctor Rydberg and the authors, extending over a number of years of careful study of the genus.

The attitude here assumed is that forms represented by numerous individuals that are easily distinguishable in the field and herbarium are worthy of separate names. Whether one calls them species or subspecies matters little; we prefer the former and the forms are so treated here.

[1] The Oaks of the Continental Divide. Bull. N. Y. Bot. Gard. **2**: 187. 1901.

There are several well-marked groups of closely-related species, the most conspicuous of which is that clustered about *Q. gambelii*, including *Q. utahensis*, *Q. submollis*, *Q. gunnisonii*, *Q. vreelandii*, *Q. novomexicana*, and *Q. leptophylla*, all of which have green deciduous leaves of much the same texture and outline with varying degrees of pubescence. Another group is that consisting of shrubs of small or large size (never forming trees) of the higher mountains, having more or less persistent blue green leaves—*Q. undulata*, *Q. fendleri*, *Q. rydbergiana*. Yet another well-marked group contains the low trees of the southern part of the State, occurring among the rocks and canyons of the drier and hotter mountains. These are *Q. grisea*, *Q. arizonica*, and *Q. reticulata*, the last being a large tree in the mountains of New Mexico. A single chestnut oak, known from two stations, is *Q. muhlenbergii*, a most unexpected find.

The affiliations of the other species are not so easily seen, each species standing more or less by itself in New Mexico.

Acorns of the different oaks were formerly used by the Indians as food. They were boiled or roasted or sometimes dried and ground into flour.

KEY TO THE SPECIES.

Acorns sericeous-tomentose inside, maturing the second year.... 1. *Q. hypoleuca.*

Acorns not sericeous-tomentose inside, maturing the first year.

Leaves bluish, grayish, or yellowish green (never bright chlorophyll green), more or less coriaceous, mostly persisting until the appearance of new leaves, hence the plant leafy all the time.

Leaves not persisting; medium-sized shrub............ 2. *Q. fendleri.*

Old leaves persisting till after the appearance of the young ones; shrubs or trees.

Mature plants shrubs, never trees.

Plant about 1 meter high, with very small acorns and leaves........................ 4. *Q. rydbergiana.*

Plants more than a meter high, the leaves and acorns large.

Leaves fulvous beneath; cup turbinate.... 6. *Q. turbinella.*

Leaves not fulvous beneath; cup hemispheric.

Leaves only moderately coriaceous, neither spinulose-toothed nor crisped........................ 3. *Q. undulata.*

Leaves strongly coriaceous, much crisped and spinulose-toothed... 5. *Q. pungens.*

Mature plants trees (shrubby forms immature, usually not fruiting).

Scales of the cup thin, only slightly corky-thickened on the back; mature leaves yellowish green.

Leaves of the same color on both surfaces.. 7. *Q. emoryi.*

Leaves fulvous beneath, especially when young............................ 8. *Q. wilcoxii.*

Scales of the cup corky-thickened on the back; leaves fulvous beneath, glabrate above.

Acorns large; mature leaves all more or less conspicuously toothed (resembling those of *Q. fendleri*).................. 9. *Q. confusa.*

Acorns of medium size; only the younger leaves conspicuously toothed.
Leaves large, obovate, strongly reticulate; teeth small and numerous....13. *Q. reticulata.*
Leaves of medium size, oblong, only slightly reticulate, entire or with few coarse teeth.
Leaves and twigs of the year glabrous in age; leaves rarely with any teeth..........................10. *Q. oblongifolia.*
Leaves permanently and densely stellate-pubescent beneath, as also the twigs of the year; leaves various.
Cup shallow; acorn acute.........12. *Q. arizonica.*
Cup deep, covering one-third the acorn; acorn truncate or obtuse.11. *Q. grisea.*
Leaves chlorophyll green, not coriaceous, deciduous in the fall, hence the plants leafless in the winter.
Leaves coarsely serrate-toothed with numerous teeth from base to apex, not truly lobed................14. *Q. muhlenbergii.*
Leaves more or less sinuately lobed.
Low shrubs, never forming trees; leaves small, 7 cm. long or less.
Lobes few and shallow, appearing as a few large teeth; some of the leaves obovate in outline................................15. *Q. media.*
Lobes deep and more numerous; leaves oblong in outline.
Acorns very large, 25 mm. long; a plant of the southeastern sandhills...........16. *Q. havardii.*
Acorns small, 10 mm. long or less, racemose; plant of the mountains of the northern part of the State.......................17. *Q. venustula.*
Taller shrubs or trees with large, deeply lobed leaves mostly 10 cm. long or more.
Mature leaves soft-pubescent and almost velvety beneath.
Scales of the cup thin, little thickened on the back; leaves distinctly obovate in outline..........................18. *Q. submollis.*
Scales of the cup thickened on the back; leaves mostly oblong, only slightly broadened upward...................19. *Q. utahensis.*
Mature leaves not velvety beneath, usually glabrate, sometimes slightly pubescent, especially on the veins.
Cup saucer-shaped, covering less than one-fourth of the acorn...................20. *Q. vreelandii.*
Cup hemispheric, covering one-third to half the acorn.
Acorns ovoid, acute; cup covering about half the acorn................24. *Q. gambelii.*

Acorns barrel-shaped, obtuse; cup various.
Mature leaves thin, large, obovate, cuneate, dark green above; acorn very short, frequently more than half in the cup.................21. *Q. leptophylla.*
Mature leaves firm, deeply lobed; acorns longer, about one-third in the cup.
Leaves oblong, lobed half way to the midrib, dull-colored; lobes usually simple.................22. *Q. gunnisonii.*
Leaves obovate, lobed more than half way to the midrib, dark green above; lobes frequently again lobed...................23. *Q. novomexicana.*

1. **Quercus hypoleuca** Engelm. Trans. Acad. St. Louis **3:** 384. 1876.

WHITE-LEAF OAK.

TYPE LOCALITY: Arizona.

RANGE: Southwestern New Mexico, southeastern Arizona, and adjacent Mexico.

NEW MEXICO: Common from the Black Range and the Mogollon Mountains south to the Mexican border. Low dry mountains, in the Upper Sonoran Zone.

One of the two easily recognizable species of the State, occurring only in the mountains of the southwestern part. It becomes a tree 10 meters high or occasionally higher, but is frequently found as a small bush forming clumps. The leaves are characteristic, being very thick and leathery, oblong-lanceolate, entire or with a few coarse teeth near the apex, yellowish green and glabrous above, densely white-woolly beneath. The tree is well worth cultivation for decorative purposes.

2. **Quercus fendleri** Liebm. Overs. Dansk. Vid. Selsk. Forh. **1854:** 170. 1854.

FENDLER OAK.

Quercus undulata A. DC. in DC. Prodr. **16²:** 23. 1864, in part.

Quercus undulata pedunculata A. DC. in DC. Prodr. **16²:** 23. 1864.

Quercus undulata Sarg. Silv. N. Amer. **8:** 75. 1895, in part.

TYPE LOCALITY: New Mexico, probably near Santa Fe. Type collected by Fendler (no. 805).

RANGE: Southern Colorado, northern New Mexico, and Arizona, and in the Panhandle region of Texas.

NEW MEXICO: Santa Fe and Las Vegas mountains; Raton; El Rito Creek; Ramah; Sandia Mountains; East View; Gallinas Mountains; White and Sacramento mountains; Buchanan; Duran. Drier mountains, in the Transition Zone.

This is very near *Quercus undulata*, with which it is usually geographically associated, being separated from that species merely by size of the parts and the persistence of the leaves. It is practically impossible to distinguish ordinary herbarium specimens showing leaves and fruit. Doctor Rydberg's key puts them in two different subsections on the ground of persistence of leaves, thus throwing *Q. undulata* next *Q. pungens*, which has a very different zonal distribution in New Mexico.

3. **Quercus undulata** Torr. Ann. Lyc. N. Y. **2:** 248. 1828.

Quercus undulata jamesii Engelm. Trans. Acad. St. Louis **3:** 382. 1876.

TYPE LOCALITY: "Sources of the Canadian and the Rocky Mountains," Colorado or New Mexico.

RANGE: Northern New Mexico and Arizona and southern Colorado, and western Texas.

NEW MEXICO: Glorieta; 25 miles south of Gallup; Pajarito Park; East View; Gallinas Mountains; Buchanan; Duran; Guadalupe Mountains; Sierra Grande; Organ Mountains. Drier mountains, in the Transition Zone, extending down into the Upper Sonoran.

What is here accepted as *Quercus undulata* is a low, straggling shrub 1 to 3 meters high, with small oblong leaves 3 to 5 cm. long, their margins sinuate-dentate, the teeth few and distinctly cuspidate but not spinulose. The leaves are firm but not coriaceous, and Doctor Rydberg believes them to be blue green, although from the type specimen and the description it is impossible to determine this. However, this is the common type of plant having the other characteristics ascribed to the species that is to be found in eastern Colorado and northeastern New Mexico, the region from which the type came.

The plant here accepted is one of the two common shrubs having blue green leaves in the mountains of the northern part of the State. It also occurs as a low shrub high up on the peaks of the dry, rocky mountains of the southern part, a thousand feet or more above the common live oaks of that region. The acorns are rather small, 10 to 15 mm. long, in a thickened, hemispherical cup.

4. **Quercus rydbergiana** Cockerell, Torreya **3**: 7. 1903.

Quercus undulata rydbergiana Cockerell, Torreya **3**: 86. 1903.

Type locality: Las Vegas Hot Springs, New Mexico. Type collected by Cockerell.

Range: Mountains of the north central part of New Mexico.

New Mexico: Las Vegas Mountains; Cebolla Springs. Transition Zone.

A small bush, 1 meter high or less, with small (2 to 4 cm. long), oblong, bluish green leaves with a few coarse sinuate lobelike teeth. The acorns are very small, less than 1 cm. long, in a shallow cup whose scales are very small, numerous, and somewhat thickened on the back.

This certainly is a relative of what is here regarded as *Q. undulata*, and Professor Cockerell may be right in reducing it to a subspecies of that, but it is more easily separable from *Q. undulata* than is *Q. fendleri* and is more distinct than the various species or subspecies, as one chooses to consider them, that are grouped around *Q. gambelii*. There is little doubt that the various Rocky Mountain species hybridize readily, as seems to be the case with the eastern members of the genus.

5. **Quercus pungens** Liebm. Overs. Dansk. Vid. Selsk. Forh. **1854**: 171. 1854.

Quercus undulata wrightii Engelm. Trans. Acad. St. Louis **3**: 382. 1876, in part.

Quercus undulata pungens Engelm. Trans. Acad. St. Louis **3**: 392. 1876.

Type locality: "Texas & Nov. Mexico.—California."

Range: Western Texas, New Mexico, southeastern Arizona, and adjacent Mexico.

New Mexico: Sandia Mountains; Mangas Springs; Silver City; Black Range; Big Hatchet Mountains; Dona Ana Mountains; mountains west of San Antonio; Carrizalillo Mountains; Organ Mountains; Queen; Socorro Mountain. Dry, rocky mountains, in the Upper Sonoran Zone.

A scrubby bush, 2 to 3 meters high, with small, coriaceous, spiny-toothed leaves on rather slender branches. Doctor Rydberg may be right in his belief that it is most closely related to *Q. undulata*, but it seems to be allied with *Q. toumeyi* and *Q. turbinella*. It is possible that the specimens referred in this treatment to *Q. turbinella* more properly belong to this species.

6. **Quercus turbinella** Greene, W. Amer. Oaks **1**: 37. 1889.

Type locality: Mountains of Lower California.

Range: Lower California to southwestern New Mexico and adjacent Mexico.

New Mexico: Bear Mountain; Socorro; Magdalena Mountains; Cook Spring. Dry hills, in the Upper Sonoran Zone.

A shrub (or low tree ?) with small (1 to 3 cm. long), oblong, elliptic, or oval leaves, bluish green above, fulvous beneath, sinuate-dentate wth spiny teeth. The acorn is elongated, acute, with a turbinate cup whose scales are only slightly thickened.

It is possible, not to say probable, that further study in the field will show that true *Q. turbinella*, which was named from the Californian peninsula, does not come into New Mexico at all.

7. Quercus emoryi Torr. in Emory, Mil. Reconn. 152. 1848. BLACK OAK.

Quercus hastata Liebm. Overs. Dansk. Vid. Selsk. Förh. **1854:** 171. 1854.

TYPE LOCALITY: "Common on the elevated country between the Del Norte and the Gila," New Mexico. The type specimen is from Pigeon Creek (Las Palomas), and was collected by Emory.

RANGE: Mountains of southwestern New Mexico, southeastern Arizona, extreme western Texas, and adjacent Mexico.

NEW MEXICO: Kingston; Bear Mountains; Animas Mountains; San Luis Mountains; Fort Bayard. Upper Sonoran Zone, occasionally extending down into the Lower Sonoran.

This is the common black oak of the southwestern part of the State and is easily recognizable. It deserves its name, since the bark is black and thick. The leaves are pale yellowish green, of about the same color on both surfaces, more or less yellowish brown pubescent on the main nerves, oblong, flat, not crispate, coarsely sinuate-dentate with spinulose teeth. The acorns are small and acute, with a shallow cup having pale yellowish brown scales not thickened on the back. They are produced early in the season and are much appreciated by the animals of the region. The species shows a tendency to hybridize.

Quercus emoryi × pungens.

A specimen from the Rio Frisco, near Alma, collected in 1906 by Vernon Bailey (no. 1058), has the acorn cup of *Q. pungens*, the acorn elongated and acute as in *Q. emoryi*, while the leaves are intermediate between those of the two species.

8. Quercus wilcoxii Rydb. Bull. N. Y. Bot. Gard. **2:** 227. 1901.

TYPE LOCALITY: Fort Huachuca, Arizona.

RANGE: Mountains of southern Arizona, southwestern New Mexico, and adjacent Mexico.

NEW MEXICO: San Luis Mountains; Animas Peak; Bear Mountain; Bullards Peak. Upper Sonoran Zone.

Mature plants of this species are medium-sized trees, though the young plants often are low and shrubby and form a moderately thick growth on the mountain sides. It is probable that some of the material here referred to *Q. pungens* is from young plants of *Q. wilcoxii*. The latter species reaches only the extreme southwestern border of the State. Mature leaves on fruiting trees are mostly elliptic and abruptly acute, very coriaceous, and with involute margins. Leaves on sterile shoots are crisped and have several coarse, triangular, strongly spiny teeth. All the leaves are distinctly yellow to tawny beneath when young, but the pubescence disappears, leaving them whitish or pale. The leaves are a yellowish or grayish green when growing.

The species includes the material from southeastern Arizona and the adjacent country which has passed as *Q. chrysolepis*. It is readily recognized by the acorns, the Californian species having an acorn easily three times as large as that of *Q. wilcoxii*, with a very much thickened cup.

9. Quercus confusa Woot. & Standl. Contr. U. S. Nat. Herb. **16:** 116. 1913.

TYPE LOCALITY: On Ruidoso Creek, 5 miles east of Ruidoso Post Office, New Mexico. Type collected by Wooton, August 5, 1901.

RANGE: White Mountains of New Mexico, in the Upper Sonoran Zone.

A moderately large tree, 5 to 7 meters high, with oblong, sinuate-dentate leaves almost velvety beneath with yellowish stellate hairs; acorns 20 to 23 mm. long, barrel-shaped, obtuse, about 3 times as long as the cup.

This species is most nearly related to *Q. fendleri*, from which it differs in being a tree, having still larger leaves (persistent?) of the same general type, and in having a larger acorn. It occurs at a lower level than is common for *Q. fendleri*, being at home in the Upper Sonoran instead of the Transition Zone, although the latter sometimes comes into the Upper Sonoran.

10. Quercus oblongifolia Torr. in Sitgreaves, Rep. Zuñi & Colo. 173. 1853.

TYPE LOCALITY: "Western New Mexico." Arizona was a part of New Mexico at this time and, as the expedition started from what is now extreme western New Mexico, this locality must have been in western Arizona.

RANGE: Western and southern Arizona, southeastern California, southwestern New Mexico, and adjacent Sonora.

NEW MEXICO: Dog Spring; Guadalupe Canyon. Mountains, in the Upper Sonoran Zone.

This species has frequently been confused with *Q. undulata*, *Q. grisea*, and *Q. arizonica*. The characters used in the key will separate these species at once. This is not at all closely related to the first-named species, but very near the other two. When mature it is a low, spreading tree of the live-oak type with oblong leaves which are wholly glabrous, as are the young twigs.

11. Quercus grisea Liebm. Overs. Dansk. Vid. Selsk. Forh. **1854:** 171. 1854.

Quercus undulata grisea Engelm. Trans. Acad. St. Louis **3:** 393. 1877.

TYPE LOCALITY: "Texas. Nov. Mexico pr. el Paso." The type is Wright's 665 from western Texas.

RANGE: Western Texas, New Mexico, southeastern Arizona, and adjacent Mexico.

NEW MEXICO: Sandia Mountains; Santa Clara Canyon; Magdalena Mountains; Bear Mountain; Florida Mountains; near Hermosa; Organ Mountains; Guadalupe Mountains; White Mountains; Llano Estacado; San Luis Mountains; Kingston; Burro Mountains. Drier, rocky foothills of the mountains, in the Upper Sonoran Zone.

New Mexico seems to be the region in which *Q. grisea* and *Q. arizonica* meet, the former coming in from Texas and the latter from Arizona. They are closely related species, possibly too closely for convenient separation, but there are slight differences in the general form of the trees, hard to describe but moderately easy to see, and the acorns are noticeably different.

Generally speaking, *Q. grisea* is a low scrubby tree (young ones which do not yet bear forming much of the scrub oak of the lower slopes of the mountains in the southern part of the State), small groups of which growing in open canyons or on slopes frequently give the impression of an old apple orchard. *Q. arizonica* is usually a larger tree, though never with a very tall trunk. It is commonly much branched from near the base and wide spreading.

Quercus grisea is variously confused by different authors with *Q. undulata*, a low shrub of the mountains of northern New Mexico and southern Colorado, and with *Q. oblongifolia*, a tree from farther west.

Quercus grisea × emoryi.

A large, round-topped tree with dark gray trunk and limbs, and slender young twigs with dense, yellowish, stellate pubescence; young leaves yellowish green, becoming gray-green and glabrous above; most of the leaves oblong, entire, some with a few coarse, spinulose teeth, their texture subcoriaceous, thinner than in either of the species; young fruit with the cup of *Q. grisea*.

Collected on the Rio Frisco near Lone Pine, in 1904, by E. O. Wooton (no. 3115). This may prove to be a new species, rather than a hybrid.

12. Quercus arizonica Sarg. Gard. & For. **8:** 92. 1895. ARIZONA OAK.

TYPE LOCALITY: Southern Arizona.

RANGE: Southern New Mexico and Arizona and adjacent Mexico.

NEW MEXICO: Santa Clara Canyon; Mogollon Mountains; Bear Mountain; Black Range; Burro Mountains; Big Hatchet Mountains; San Luis Mountains; Lordsburg; Animas Mountains; Organ Mountains; Oscuro Mountains; Capitan Mountains; White Mountains. Lower parts of drier mountains, in the Upper Sonoran Zone.

This and *Q. grisea* are the common live oak *trees* of the drier and lower mountains of the southern part of the State. They are commonly found among the rocks and open canyons of the mountains, associated with junipers and pinyon. The leaves are exceedingly variable in form, from oblong-elliptic, flat, and entire to sinuate-dentate with the large teeth more or less spine-tipped and decidedly crisped. The texture is always subcoriaceous, and the living leaves are bluish green (never chlorophyll green) above and glabrate, not shining, duller and stellate-pubescent beneath, with prominent veins. The leaf approximates typical *Q. pungens* on the one side and *Q. reticulata* on the other. The leaves of *Q. arizonica* and *Q. grisea* are hardly distinguishable, although those of the latter are usually smaller and less sinuate-dentate. The acorns are noticeably different: In *Q. arizonica* the cup is shallow and covers only the lower fifth of the rather slender acute acorn, while the acorn of *Q. grisea* is barrel-shaped, shorter, and almost truncate, the cup covering fully one-third of the acorn, and the scales being much more noticeably corky-thickened.

Quercus arizonica × grisea.

At Van Pattens Camp in the Organ Mountains there is a single tree, growing with others of *Q. grisea* and *Q. arizonica*, which it is impossible to distinguish from the latter by vegetative characters, but the acorn of which is very peculiar. It is of the general barrel shape of *Q. grisea* and truncate, but is as long as the largest *Q. arizonica*, and the cup is deeper than in either of the species and twice as much thickened. It was impossible to find more than the one tree with this kind of fruit in the region, although the two species are common there.

13. Quercus reticulata Humb. & Bonpl. Pl. Aequin. **2:** 40. 1809.

TYPE LOCALITY: "Habitat in montibus aridis Novae Hispaniae [Mexico], inter Guanajuato et Santa Rosa."

RANGE: Mountains of extreme southwestern New Mexico, southeastern Arizona, and northern Mexico.

NEW MEXICO: Florida Mountains; Animas Mountains; San Luis Mountains; Mogollon Mountains. Lower and Upper Sonoran zones.

This species is a large tree in Mexico, but in our range is a straggling bush only a few meters high. It is somewhat closely related to *Q. arizonica*, but typical leaves are considerably larger, obovate, merely repand-dentate with rather small teeth, and strongly reticulate. They are dull green above and paler beneath, and the veins below are covered with yellowish brown pubescence.

14. Quercus muhlenbergii Engelm. Trans. Acad. St. Louis **3:** 391. 1877.

CHESTNUT OAK.

Quercus prinus acuminata Michx. Hist. Chênes Amer. no. 5. *pl. 8.* 1801.

Quercus acuminata Sarg. Gard. & For. **8:** 93. 1895.

TYPE LOCALITY: Pennsylvania.

RANGE: Vermont to Minnesota, Florida, and eastern New Mexico.

NEW MEXICO: East base of Capitan Mountains (*Bailey* 141). Upper Sonoran or lower part of the Transition Zone.

Mr. Vernon Bailey says this plant is common on Coyote Creek near Guadalupita, but specimens collected there are not at hand. He also collected it in the Guadalupe Mountains of Texas near the New Mexico border in 1901. These three stations extend the range of the species some hundreds of miles westward and add another type of oak to those of our State. The material corresponds very well with specimens from Kansas and Arkansas, but the leaves are somewhat smaller and relatively longer petioled than in plants from farther east. The acorns are about typical as to shape, but slightly smaller than those of the eastern tree.

15. Quercus media Woot. & Standl. Contr. U. S. Nat. Herb. **16:** 116. 1913

TYPE LOCALITY: Glorieta, New Mexico. Type collected by Wooton, August 24, 1910.

RANGE: Northeastern New Mexico.

NEW MEXICO: Glorieta; Oak Canyon, Folsom. Transition Zone.

A low shrub, 1 to 3 meters high, with oblong to obovate, sinuate-dentate leaves; acorns small, 10 to 13 mm. long, acute.

Assuming that *Q. undulata* is a species with bluish green, persistent leaves, this species resembles it in nearly all particulars except that its leaves are bright chlorophyll green and probably deciduous. This would make it intermediate between the two groups of the region—the blue green leaved species, which it resembles in habit and shape of leaf, and the green-leaved species, which it resembles in color and texture of leaves and time of shedding them. It might be a hybrid, but the plant was very common about Glorieta, forming numerous clumps of bushes a rod or so in diameter, and Mr. Howell's plant from Folsom is almost a perfect match from a similar region farther east.

16. **Quercus havardii** Rydb. Bull. N. Y. Bot. Gard. **2:** 213. 1901.

SHINNERY OAK.

TYPE LOCALITY: Sand hills on the Staked Plains, Texas. Type collected by Havard (no. 51).

RANGE: Western Texas and eastern New Mexico.

NEW MEXICO: Roswell; sand hills 35 miles east of Carlsbad. Sand hills, in the Lower Sonoran Zone.

This is a low shrub, rarely over 70 cm. high, forming a tolerably thick growth on the sand hills of southeastern New Mexico east of the Pecos Valley. The plant is generally spoken of as "shinnery" or "shin-oak" and the sands it covers go by the name of "shinnery sands." Its leaves are bright green and deciduous, glabrous above, somewhat paler and velutinous beneath, oblong in form, 6 cm. long or less, coarsely sinuate-lobed or dentate with a few teeth. The acorn is the largest borne by any species in New Mexico, the cup hemispheric, the upper scales slightly or not at all thickened, acuminate, the lower ones somewhat thickened; the acorn 20 mm. long and three-fourths as broad, obtuse.

17. **Quercus venustula** Greene, W. Amer. Oaks **2:** 69. 1890.

TYPE LOCALITY: "Mountains of southern Colorado and northern New Mexico; plentiful near Trinidad, and also on higher mountains farther southward."

RANGE: Southern Colorado and northern New Mexico.

NEW MEXICO: Cross L Ranch; Santa Fe and Las Vegas mountains; White Mountains. Mountains, in the Transition Zone.

A low shrub, with deciduous leaves; these small (3 to 4 cm. long), oblong, with a few rounded coarse teeth or lobes, green above, paler and pubescent beneath. The acorns are small, rather numerous, and racemose, the cup hemispheric, 6 or 7 mm. in diameter, covering almost half the acute acorn. Rare in the mountains of the northern part of the State.

18. **Quercus submollis** Rydb. Bull. N. Y. Bot. Gard. **2:** 202. 1901.

TYPE LOCALITY: Arizona.

RANGE: Mountains of eastern Arizona and western New Mexico.

NEW MEXICO: Carrizo Mountains; south of Gallup; north of Datil; Ramah.

A low tree or shrub with deciduous green leaves, which are elliptic-obovate in outline, with rounded, mostly two-toothed lobes and rather deep sinuses, the lower surfaces velvety-pubescent. The young twigs are reddish brown; the acorn cup is depressed-hemispheric, with scales scarcely thickened, and the acorn is barrel-shaped and obtuse.

19. **Quercus utahensis** (A. DC.) Rydb. Bull. N. Y. Bot. Gard. **2:** 202. 1901.

Quercus stellata utahensis A. DC. in DC. Prodr. 16^2: 22. 1864.

TYPE LOCALITY: "Inter Salt Lake et Sierra Nevada."

RANGE: Mountains of Utah, Colorado, Arizona, and New Mexico.

NEW MEXICO: Common in all the mountain ranges. Middle elevations in the Transition Zone.

This is the common white oak throughout the State, perhaps dividing that distinction in the southern part with *Q. novomexicana.* It is frequently a shrub growing in clumps and, in the mountains of the southern part of the State, equally often a solitary tree. It often has a trunk 30 to 40 cm. in diameter, with a large spheroidal top, reaching a height of 10 meters or more. The leaves are bright green above, paler and almost velvety beneath, deciduous in late autumn, turning yellow before falling. In outline they are broadly oblong-obovate, deeply lobed, the lobes rounded at the apex. The acorn is of medium size, 15 to 20 mm. long, obtuse, in a thickened, hemispheric cup.

20. **Quercus vreelandii** Rydb. Bull. N. Y. Bot. Gard. **2**: 204. 1901.

TYPE LOCALITY: Mesa near La Veta, Colorado.

RANGE: Mountains of southern Colorado and northern New Mexico.

NEW MEXICO: Chama (*Baker* 280). In the Transition Zone.

The specimens upon which the species is included are doubtfully referred here by Doctor Rydberg himself.

21. **Quercus leptophylla** Rydb. Bull. N. Y. Bot. Gard. **2**: 205. 1901.

TYPE LOCALITY: Tributaries of Turkey Creek, Colorado.

RANGE: Mountains of southern Colorado and northern New Mexico.

NEW MEXICO: Johnsons Mesa; Trinchera Pass; Folsom; Chama; White Mountains. Transition Zone.

A rough, scraggy tree with dark bark, very crooked limbs, small top, large obovate dark green leaves, and a short acorn about half covered by the cup. It grows commonly in clusters on the sides of canyons in the mountains, or on the high mesas of the northern and eastern parts of the State. The acorns are rarely abundantly produced.

22. **Quercus gunnisonii** (Torr.) Rydb. Bull. N. Y. Bot. Gard. **2**: 206. *pl. 26. f. 3.* 1901.

Quercus alba gunnisonii Torr. U. S. Rep. Expl. Miss. Pacif. **2**[1]: 130. 1855.

Quercus undulata gunnisonii Engelm. Trans. Acad. St. Louis **3**: 382. 1876.

TYPE LOCALITY: "Coochetopa Pass, Sierra San Juan," Colorado.

RANGE: Mountains of southern Colorado and Utah and northern New Mexico and Arizona.

NEW MEXICO: Tunitcha Mountains; Cedar Hill; Dulce; Raton Mountains; Folsom; Pecos; East View; Luna Valley; Gallinas Mountains. Upper Sonoran Zone.

One of the common shrubby oaks of the northern part of the State, resembling *Q. gambelii,* coming into our range from Colorado and Utah. The leaves are oblong in outline, sinuately lobed, green, deciduous, not velvety beneath. The acorn is obtuse, barrel-shaped, in a much thickened and rather deep cup. In general appearance it approaches most closely *Q. utahensis,* but is to be distinguished from that species by the absence of a velvety pubescence on the back of the mature leaves.

23. **Quercus novomexicana** (A. DC.) Rydb. Bull. N. Y. Bot. Gard. **2**: 208. 1901.

Quercus douglasii novomexicana A. DC. in DC. Prodr. **16**[2]: 24. 1864.

Quercus nitescens Rydb. Bull. N. Y. Bot. Gard. **2**: 207. 1901.

TYPE LOCALITY: Santa Fe, New Mexico. Type collected by Fendler (no. 809).

RANGE: Mountains of New Mexico, Colorado, and Utah.

NEW MEXICO: Carrizo Mountains; Santa Fe and Las Vegas mountains; Sandia Mountains; Ramah; East View; Mogollon Mountains; Black Range; Organ Mountains; White and Sacramento mountains. Transition Zone.

A shrub in the mountains of the northern part of the State, but frequently becoming a good-sized tree in the southern part. The mature leaves are about the largest among the deciduous-leaved species, being frequently more than 10 cm. long and half as broad. They are of a rich dark green and glabrous on the upper surface and much paler beneath. In outline the leaves are broadly elliptic-obovate, deeply pinnate-lobed, the rounded, open sinuses reaching three-fourths the way to the midrib. The lobes are broadly oblong to triangular, rounded toward the acute or obtuse apex, and many of them bilobate. The acorns are large, with hemispheric cups having moderately thickened scales.

A queer form which seems to be most nearly related to this species is shown in Standley's 4755 from Winsors Ranch, where it is common on rich hillsides. The leaves are divided to within 2 or 3 mm. of the midrib, the segments reduced in number and size; in a few cases the leaf is reduced to a long-oblanceolate form less than 1 cm. wide at the obtuse tip, without lobes of any kind.

24. Quercus gambelii Nutt. Journ. Acad. Phila. II. **1**: 179. 1848. GAMBEL OAK.

Quercus douglasii gambelii A. DC. in DC. Prodr. **16²**: 23. 1864, in part.

Quercus undulata gambelii Engelm. Trans. Acad. St. Louis **3**: 382. 1876.

TYPE LOCALITY: Banks of the Rio Grande, New Mexico, west of Santa Fe. Type collected by Gambel.

RANGE: Mountains of northern New Mexico and southern Colorado.

NEW MEXICO: Chama; Tierra Amarilla; Canjilon; Santa Fe and Las Vegas mountains; Johnsons Mesa; Sandia Mountains; Zuni Mountains; East View. At middle elevations, in the Upper Sonoran and Transition zones.

The original description of this species applies to a shrubby, deciduous, green-leaved oak with an acute acorn, from the region of Santa Fe. Such an oak is to be found in that region, although it is by no means the common form. Most of the deciduous, green-leaved oaks of New Mexico and Colorado have obtuse or truncate acorns, and are to be found listed here under other names. If one were to consider all the white oaks of the State as belonging to a single species, it should be called *Q. gambelii*; but there are numerous easily recognizable variants of that type and our judgment as to their proper recognition is expressed in this treatment.

Order 20. URTICALES.

KEY TO THE FAMILIES.

Fruit a samara or drupe, sometimes nutlike.........34. **ULMACEAE** (p. 174).
Fruit an achene.
  Flowers on the outside or inside of a receptacle; fruits forming syncarps; sepals accrescent, enveloping the achenes......35. **MORACEAE** (p. 175).
  Flowers not on a receptacle; fruits not forming syncarps; sepals neither thick and juicy nor enveloping the achenes.
    Style or stigma 1; ovule erect; filaments inflexed in the bud.................36. **URTICACEAE** (p. 176).
    Styles or stigmas 2; ovule pendulous; filaments erect in bud..................37. **CANNABINACEAE** (p. 177).

34. ULMACEAE. Elm Family.

1. CELTIS L. HACKBERRY.

A small tree or large shrub; leaves ovate or ovate-lanceolate, abruptly acuminate, reticulate, cordate and very unequal at the base; flowers greenish, axillary, the fertile solitary or in pairs, appearing with leaves; calyx 5 or 6-parted, persistent; stamens 5 or 6; ovary 1-celled, with a single ovule.

1. **Celtis reticulata** Torr. Ann. Lyc. N. Y. **2:** 247. 1824.

Type locality: "Base of the Rocky Mountains," Colorado or New Mexico.

Range: Colorado to Arizona and Texas and southward.

New Mexico: Sierra Grande; Santa Rita; Burro Mountains; Black Range; Florida Mountains; Guadalupe Canyon; Organ Mountains; Guadalupe Mountains; White Mountains. Dry hills and canyons, in the Upper Sonoran Zone.

The berries of this tree are edible and were often eaten by the Indians.

35. MORACEAE. Mulberry Family.

1. MORUS L. Mulberry.

Ours a small scraggy tree with alternate ovate small (3 to 6 cm. long) serrate leaves, these usually 3 to 5-lobed, acute; flowers diœcious, small and inconspicuous, green; fruit technically a "multiple fruit," consisting of a cylindrical or oblong cluster of separate 1-seeded berries, the whole appearing to be a single fruit.

Morus alba L. is extensively cultivated in New Mexico as a shade tree and for its fruit. The trees are of two kinds, staminate and pistillate, the former being much more desirable as shade trees. This species occasionally occurs as an escape.

1. **Morus microphylla** Buckl. Proc. Acad. Phila. **1862:** 8. 1863.

Morus vernonii Greene, Leaflets **2:** 115. 1910.
Morus vitifolia Greene, op. cit. 116.
Morus goldmanii Greene, op. cit. 117.
Morus betulifolia Greene, op. cit. 117.
Morus canina Greene, op. cit. 118.
Morus albida Greene, op. cit. 118.
Morus crataegifolia Greene, op. cit. 119.
Morus radulina Greene, op. cit. 119.
Morus confinis Greene, op. cit. 119.

Type locality: "Western Texas."

Range: Texas to Arizona, southward into Mexico.

New Mexico: Mangas Springs; Burro Mountains; Black Range; Dog Spring; Little Florida Mountains; Dona Ana Mountains; Organ Mountains; Ruidoso; Queen. Dry hills and canyons, in the Upper Sonoran Zone.

We are unable to distinguish from typical *Morus microphylla* the numerous plants to which Doctor Greene's names were given. There is some slight variation in the outline of the leaves, but, as shown by the extraordinary variation in those of individuals of *Morus rubra*, the species of this genus can not be separated by leaf form alone. This inconstancy is strongly realized when mature leaves and those from young sprouts of the same tree are compared. There seems to be no variation in pubescence.

The type of *Morus vernonii* came from the Chisos Mountains of western Texas; that of *M. vitifolia* from the Dona Ana Mountains (*Wooton & Standley* in 1906); that of *M. goldmanii* from the Little Florida Mountains (*Goldman* in 1908); that of *M. betulifolia* from the Organ Mountains (*Standley* in 1906); that of *M. canina* from Dog Spring (*Mearns* in 1892); that of *M. albida* from Berendo (misspelled Berend in Doctor Greene's citation of the locality) Creek (*Metcalfe* in 1904); that of *M. crataegifolia* from the Blue River, southeastern Arizona; that of *M. radulina* from Beaver Creek, Arizona; and that of *M. confinis* from Santa Rita Mountains, southern Arizona.

The species is a stunted, irregular tree 5 meters high or less, with small, scabrous leaves. The red fruit is palatable when ripe, having a pleasant acid flavor. The trees occur chiefly in arroyos of the foothills and on the drier slopes of the mountains.

36. URTICACEAE. Nettle Family.

Usually coarse; monœcious, diœcious, or polygamous herbs, often armed with stinging hairs; leaves simple, opposite or alternate; flowers inconspicuous, greenish, in axillary simple or compound cymes.

KEY TO THE GENERA.

Plants armed with stinging hairs; leaves opposite; inflorescence not involucrate .... 1. URTICA (p. 176).
Plants without stinging hairs; leaves opposite or alternate; inflorescence various.
  Flower clusters not involucrate; leaves opposite .... 2. BOEHMERIA (p. 176).
  Flower clusters surrounded by an involucre; leaves alternate .... 3. PARIETARIA (p. 177).

1. URTICA L. NETTLE.

Coarse annual or perennial herbs armed with stinging hairs; leaves opposite, toothed; flowers in axillary cymes, these often panicled; achenes flattened.

Our species are inconspicuous plants found chiefly in moist, shaded places in the mountains.

KEY TO THE SPECIES.

Teeth of the leaves ovate, strongly directed forward; stems armed with rather few stinging hairs, otherwise glabrous; leaves lanceolate .... 1. *U. gracilis.*
Teeth of the leaves broadly triangular, salient, not strongly directed forward; stems armed with numerous stinging hairs, strigose; leaves commonly ovate .... 2. *U. gracilenta.*

1. **Urtica gracilis** Ait. Hort. Kew. **3**: 341. 1789.

TYPE LOCALITY: "Native of Hudson's Bay."

RANGE: British America to Arizona, Texas, and Louisiana.

NEW MEXICO: Carrizo Mountains; Chama; Santa Fe and Las Vegas mountains; Clayton; Sandia Mountains; Mogollon Mountains; White Mountains. Damp woods and canyons, in the Transition Zone.

2. **Urtica gracilenta** Greene, Bull. Torrey Club **8**: 122. 1881.

TYPE LOCALITY: Mimbres Mountains, New Mexico. Type collected by E. L. Greene.

RANGE: New Mexico and Arizona.

NEW MEXICO: Mogollon Mountains; Organ Mountains; White Mountains. Damp thickets, in the Upper Sonoran and Transition zones.

2. BOEHMERIA Jacq. FALSE NETTLE.

A coarse stout unarmed perennial herb, 30 to 80 cm. high, with opposite, petioled, pubescent, coarsely serrate, lanceolate leaves; flowers in axillary spikes; stems finely pubescent.

1. **Boehmeria scabra** (Porter) Small, Fl. Southeast. U. S. 358. 1903.

Boehmeria cylindrica scabra Porter, Bull. Torrey Club **16**: 21. 1889.

TYPE LOCALITY: "Crawford and Lancaster counties," Pennsylvania.

RANGE: New York to Michigan, Florida, and New Mexico.

NEW MEXICO: Roswell (*Earle* 265).

3. PARIETARIA L. Pellitory.

Low annuals with alternate thin petioled entire leaves; inflorescence axillary, surrounded by an involucre of 2 to 6 more or less united bracts; achenes nearly terete.

KEY TO THE SPECIES.

Involucre 2 to 3 times as long as the flowers; stems simple or sparingly branched.................................... 1. *P. pennsylvanica.*
Involucre about equaling the flowers or very slightly surpassing them; stems much branched at the base, stouter........ 2. *P. obtusa.*

1. **Parietaria pennsylvanica** Muhl.; Willd. Sp. Pl. **4:** 955. 1806.

Type locality: Pennsylvania.

Range: British America to Florida and Mexico.

New Mexico: North Percha Creek; Sierra Grande. Damp ground, in the Upper Sonoran Zone.

2. **Parietaria obtusa** Rydb. in Small, Fl. Southeast. U. S. 359. 1903.

Type locality: Southern Utah.

Range: Colorado and Utah to Texas and Arizona.

New Mexico: Gila Hot Springs; mountains west of San Antonio; Organ Mountains. Damp ground, chiefly in the Upper Sonoran Zone.

37. CANNABINACEAE. Hemp Family.

1. HUMULUS L. Hop.

A climbing perennial herb; flowers diœcious, the staminate in loose axillary panicles, the pistillate in short axillary spikes; bracts foliaceous, imbricated; leaves palmately 3 to 5-lobed; fruiting calyx and other parts of the plant covered with yellow resinous dots.

1. **Humulus lupulus neomexicanus** Nels. & Cockerell, Proc. Biol. Soc. Washington **16:** 45. 1903.

Type locality: Beulah, New Mexico.

Range: Wyoming to New Mexico.

New Mexico: Mountains west of Grant; Chama; Santa Fe and Las Vegas mountains; Sandia Mountains; Mogollon Mountains; Black Range; White Mountains. Thickets, in the Transition Zone.

Order 21. SANTALALES.

KEY TO THE FAMILIES.

Leaves opposite; fruit a berry; tree parasites......**38. LORANTHACEAE** (p. 177).
Leaves alternate; fruit a drupe or nut; root parasites.**39. SANTALACEAE** (p. 181).

38. LORANTHACEAE. Mistletoe Family.

Evergreen plants parasitic on shrubs and trees, yellowish or brownish green; branches dichotomous, the joints swollen; leaves opposite, thick, entire, often reduced to connate scales; flowers small and inconspicuous, greenish, diœcious; sepals 2 to 5; stamens of the same number and inserted on the sepals; ovary inferior, 1-celled; fruit a berry with a viscid endocarp.

KEY TO THE GENERA.

Berry compressed, fleshy, opaque; anthers 1-celled; leaves all reduced to connate scales............ 1. Razoumofskya (p. 177).
Berry globose, pulpy, semitransparent; anthers 2-celled; leaves usually large and foliaceous............... 2. Phoradendron (p. 179).

1. RAZOUMOFSKYA Hoffm.

Small stout branched plants, with little or no greenish tinge, brownish or yellowish, parasitic on the Pinaceae; branches 4-angled, glabrous; leaves scalelike; fruit a compressed berry containing a single seed, maturing the second year.

In general appearance the plants resemble *Phoradendron juniperinum*, but they are usually much smaller. They are of little value for decorations because of their lack of foliaceous leaves, but the berries of some species are handsome.

KEY TO THE SPECIES.

Stems stout, 2 to 5 mm. in diameter; plants large, 6 to 20 cm. high. On *Pinus brachyptera* .... 1. *R. cryptopoda.*

Stems slender, 1 to 2 mm. in diameter; plants usually much smaller, 6 cm. high or less.

- Plants greenish brown; accessory branches of fruiting specimens mostly leaf-bearing. On *Pinus edulis* .... 2. *R. divaricata.*
- Plants greenish yellow; accessory branches of fruiting specimens flower-bearing.
  - Fruit nearly truncate at the apex, obovoid, not manifestly stipitate. On *Pinus flexilis* .... 3. *R. cyanocarpa.*
  - Fruit rounded or acutish at the apex, ellipsoid, evidently stipitate.
    - Plants very slender, small, 3 cm. high or less. On Pseudotsuga .... 4. *R. douglasii.*
    - Plants stouter, larger, 4 to 6 cm. high. On *Picea engelmanni* .... 5. *R. microcarpa.*

1. Razoumofskya cryptopoda (Engelm.) Coville, Contr. U. S. Nat. Herb. **4:** 192. 1893.

Arceuthobium robustum Engelm. Mem. Amer. Acad. n. ser. **4:** 59. 1849, nomen nudum.

Arceuthobium cryptopodum Engelm. Bost. Journ. Nat. Hist. **6:** 214. 1850.

Razoumofskya robusta Kuntze, Rev. Gen. Pl. **2:** 587. 1891.

Type locality: "Santa Fe, only on *Pinus brachyptera.*" Type, Fendler's no. 283.

Range: Colorado to New Mexico and Arizona.

New Mexico: Common on the yellow pine (*Pinus brachyptera*), wherever this is found.

2. Razoumofskya divaricata (Engelm.) Coville, Contr. U. S. Nat. Herb. **4:** 192. 1893.

Arceuthobium divaricatum Engelm. in Wheeler, Rep. U. S. Surv. 100th Merid. **6:** 254. 1878.

Type locality: "On Nut-pines (*P. edulis* and *monophylla*) from southern Colorado through New Mexico to Arizona."

Range: Colorado to Arizona and New Mexico.

New Mexico: Tunitcha Mountains; Cedar Hill; Santa Fe; Mogollon Creek; Telegraph Mountains; Fort Bayard; Burro Mountains. On *Pinus edulis.*

3. Razoumofskya cyanocarpa A. Nels.; Rydb. Colo. Agr. Exp. Sta. Bull. **100:** 101. 1906.

Arceuthobium cyanocarpum A. Nels. in Coulter, New Man. Rocky Mount. 146. 1909.

Type locality: "Parasitic on *Pinus flexilis*, from Wyo. to Colo."

Range: Wyoming to New Mexico.

New Mexico: Hillsboro Peak; East Canyon. On *Pinus flexilis.*

4. **Razoumofskya douglasii** (Engelm.) Kuntze, Rev. Gen. Pl. **2**: 587. 1891.

Arceuthobium douglasii Engelm. in Wheeler, Rep. U. S. Surv. 100th Merid. **6**: 253. 1878.

Type locality: "On *Pseudotsuga douglasii* from New Mexico (on Santa Fe River, Rothrock, No. 69, 1874) to Utah, Parry, Siler, and Northern Arizona, Camp Apache, G. K. Gilbert (109), 1873."

Range: Idaho to Arizona and New Mexico.

New Mexico: Santa Fe Canyon; Tunitcha Mountains. On *Pseudotsuga mucronata.*

5. **Razoumofskya microcarpa** (Engelm.) Woot. & Standl.

Arceuthobium douglasii microcarpum Engelm. in Wheeler, Rep. U. S. Surv. 100th Merid. **6**: 253. 1878.

Type locality: "Parasitic on *Picea Engelmanni,* found by Mr. Gilbert in 1873 (100 and 102) in the Sierra Blanca, Arizona."

Range: Arizona and New Mexico.

New Mexico: West Fork of the Gila (*Metcalfe* 493). On Picea.

The specific name is misleading, for the fruit seems to be no smaller than in related species.

2. PHORADENDRON Nutt. Mistletoe.

Plants parasitic on the branches of trees and shrubs; stems brittle, woody, jointed, much branched; leaves entire, thick and firm (or reduced to scales), persistent; flowers small and inconspicuous, monœcious, in jointed axillary spikes; calyx usually 3-parted; fruit a semitransparent berry crowned with the persistent sepals.

The various species of Phoradendron found in New Mexico are among our commonest and most conspicuous plants, being found upon one or more kinds of trees in almost every locality. They are true parasites, growing from seeds deposited, usually by birds, on the branches of trees. The mistletoe does great injury to trees, especially the valley cottonwood, where planted for shade. Some species have been observed on cultivated fruit trees. The plants when covered in the winter with their handsome white berries are extensively used as Christmas greens. In some parts of the State large quantities of mistletoe are gathered and sold each year, the larger amount being shipped east.

KEY TO THE SPECIES.

Leaves reduced to scalelike ovate bracts 2 mm. long or less. (On Juniperus.) .... 1. *P. juniperinum.*

Leaves foliaceous, 6 to 60 mm. long.

  Leaves elliptic, 6 to 12 mm. long, 4 mm. wide or less; spikes of the inflorescence very short, few-flowered. (On Juniperus.) .... 2. *P. bolleanum.*

  Leaves broader and larger, more than 12 mm. long, 10 mm. wide or more; spikes long and many-flowered.

    Leaves abruptly contracted into a short stout petiole, usually orbicular or orbicular-oblong, densely pubescent. (On oaks and other hard-wood trees.) 3. *P. orbiculatum.*

    Leaves attenuate to the longer petiole, obovate or oblanceolate, not densely pubescent, usually glabrous.

      Leaf blades conspicuously yellowish green, oblong or obovate. (Usually on *Populus wislizeni.*). 4. *P. macrophyllum.*

      Leaf blades only slightly yellowish, oblanceolate.. 5. *P. flavescens.*

1. Phoradendron juniperinum Engelm. Mem. Amer. Acad. n. ser. **4:** 58. 1849.

JUNIPER MISTLETOE.

TYPE LOCALITY: "Parasitic on the two kinds of shrub cedar (Juniperus) which grow on the hills and elevated plains about Santa Fe, and on no other tree." Type collected by Fendler (no. 281).

RANGE: Oregon and California to Colorado and Texas, southward into Mexico.

NEW MEXICO: Coolidge; Santa Fe; Canjilon; Magdalena; Mogollon Mountains; Burro Mountains; Fort Bayard; White Mountains; Queen; Albuquerque; mountains west of San Antonio; Cedar Hill; Tunitcha Mountains; Carrizo Mountains; Tierra Amarilla. On species of Juniperus.

The berries are handsome, but the plant is not suitable for decorative purposes because of the lack of leaves.

2. Phoradendron bolleanum (Seem.) Eichl. in Mart. Fl. Bras. **5**^{2}**:** 134. 1868.

Viscum bolleanum Seem. Bot. Voy. Herald 295. *pl. 63.* 1856.

Phoradendron pauciflorum Torr. U. S. Rep. Expl. Miss. Pacif. **4:** 134. 1856.

TYPE LOCALITY: "Sierra Madre," Mexico.

RANGE: California to southern New Mexico, south into Mexico.

NEW MEXICO: Carrizalillo Mountains; San Luis Mountains; near Dog Spring. On species of Juniperus.

3. Phoradendron orbiculatum Engelm. Mem. Amer. Acad. n. ser. **4:** 59. 1849.

Phoradendron flavescens orbiculatum Engelm. Bost. Journ. Nat. Hist. **6:** 212. 1850.

TYPE LOCALITY: "On different species of Quercus; on *Q. nigra*, sterile hills of Arkansas (Engelm.); on several oaks, San Felipe, Texas (*Lindheimer*)."

RANGE: New Jersey to Missouri, Texas, and Arizona.

NEW MEXICO: Fort Bayard; Berendo Creek; Dona Ana and Organ mountains; San Luis Mountains; Dog Spring; Magdalena; Guadalupe Mountains. Usually on oaks.

This species, so far as we have seen it, is found upon nothing but oaks. It differs from the following species chiefly in the different form of its leaves, brighter green color, smaller size, and more abundant pubescence. It occurs, as do its hosts, only in the mountains and foothills, never coming down into the valleys. Two of our specimens, one from the Organ Mountains and one from Berendo Creek, have narrower, lanceolate or elliptic, acutish leaves. Possibly they represent a different species.

4. Phoradendron macrophyllum (Engelm.) Cockerell, Amer. Nat. **34:** 293. 1900.

Phoradendron flavescens macrophyllum Engelm. in Wheeler, Rep. U. S. Surv. 100th Merid. **6:** 252. 1878.

TYPE LOCALITY: "They grow on soft woods (Ash, Willow, Poplar, Sycamore, and Sapindus) on the Gila and Benita Rivers, and extend into Southern California."

RANGE: Western Texas to Arizona and California.

NEW MEXICO: Mangas Springs; Silver City; Rincon; Mesilla Valley. On various trees and shrubs.

The species is very abundant in the southern and southwestern parts of the State, especially in the Rio Grande Valley, growing chiefly on the valley cottonwood (*Populus wislizeni*), although it has been found on the tornillo and on cultivated plums. So heavily loaded are the cottonwood trees at times that they appear to have as dense foliage in winter as in summer. Of course the trees soon succumb to such exhaustive attacks by the pest. The Spanish name is "muérdago."

This is the species preferred for commercial purposes. It is superior to *P. orbiculatum* because of its larger size, more abundant, larger berries, and generally handsomer appearance.

5. Phoradendron flavescens (Pursh) Nutt.; A. Gray, Man. ed. 2. 383. 1856.
Viscum flavescens Pursh, Fl. Amer. Sept. 114. 1814.
TYPE LOCALITY: Not stated.
RANGE: New Jersey to Missouri, south to Florida and New Mexico.
NEW MEXICO: Mesilla Valley, on cultivated ash trees (*Standley* 6377).

39. SANTALACEAE. Sandalwood Family.

1. COMANDRA Nutt. BASTARD TOADFLAX.

A low glaucous herbaceous perennial; leaves alternate, sessile, lanceolate to linear, entire; flowers greenish white, in terminal and axillary clusters; perianth campanulate, the limb 3 to 5-lobed, persistent; fruit spherical, 1-seeded.

1. Comandra pallida A. DC. in DC. Prodr. **14:** 636. 1857.
Comandra pallida angustifolia Torr. U. S. & Mex. Bound. Bot. 185. 1859.
TYPE LOCALITY: "Prope Clearwater," Idaho.
RANGE: British America to California and Texas.
NEW MEXICO: Barranca; Magdalena Mountains; Burro Mountains; Kingston; Mesilla Valley; Organ Mountains; San Augustine Plains; Tunitcha Mountains; Chama; Raton; Nara Visa. Parasitic on the roots of various plants, Lower Sonoran to the Transition Zone.

Order 22. ARISTOLOCHIALES.

40. ARISTOLOCHIACEAE. Birthwort Family.

1. ARISTOLOCHIA L.

A prostrate perennial with slender tomentulose stems; leaves alternate, narrowly hastate, long-attenuate, with mostly divergent auricles; flowers solitary, axillary, small; calyx tube broadly arcuate; stamens 5, the sessile anthers adnate to the short 5-lobed style; pod 5-celled.

Some species of the genus have large and showy flowers and are cultivated as decorative plants. Our species is a small and inconspicuous herb of the southwestern deserts.

1. Aristolochia watsoni Woot. & Standl. Contr. U. S. Nat. Herb. **16:** 117. 1913.
Aristolochia brevipes acuminata S. Wats. Proc. Amer. Acad. **18:** 148. 1883, not *A. acuminata* Lam. 1783.
TYPE LOCALITY: Mexico or Arizona.
RANGE: Southwestern New Mexico and southern Arizona to northern Mexico.
NEW MEXICO: Guadalupe Canyon (*Mearns* 697). Dry hills.

Order 23. POLYGONALES.

41. POLYGONACEAE. Buckwheat Family.

Herbaceous or suffruticose annuals or perennials with alternate, or sometimes opposite or verticillate leaves, the stipules forming a sheath or wanting; inflorescence cymose, capitate, racemose, spicate, or panicled; flowers small, mostly perfect; perianth of 2 to 6 segments, the inner ones sometimes petaloid; stamens 2 to 9; pistil solitary; fruit a lenticular or angled achene.

KEY TO THE GENERA.

Flowers subtended by an involucre; stamens 9; sheaths wanting.................................... 1. ERIOGONUM (p. 182).
Flowers not involucrate; stamens 4 to 8; sheaths present.
  Stigmas tufted.
    Sepals 4; stigmas 2; achenes lenticular, winged; leaves orbicular-reniform.................. 2. OXYRIA (p. 190).
    Sepals 6; stigmas 3; achenes 3-angled, not winged; leaves not orbicular-reniform, elongated or hastate.................................... 3. RUMEX (p. 191).
  Stigmas capitate.
    Leaf blades jointed at the base; filaments, at least the inner, dilated.
      Herbs; flowers fascicled; sepals not winged... 5. POLYGONUM (p. 193).
      Shrub; flowers solitary; inner sepals winged.. 4. GONOPYRUM (p. 193).
    Leaf blades not jointed at the base; sheaths not 2-lobed; filaments slender.
      Sheaths cylindric, truncate.................. 6. PERSICARIA (p. 195).
      Sheaths oblique, more or less open on the side facing the leaf.
        Sepals, at least the outer ones, keeled or winged; stems twining............ 8. BILDERDYKIA (p. 197).
        Sepals neither keeled nor winged; stems erect.
          Perennials with fleshy rootstocks, mostly basal leaves, and simple stems.................. 7. BISTORTA (p. 197).
          Annual with fibrous roots, cauline leaves, and branched stems.. 9. FAGOPYRUM (p. 197).

1. ERIOGONUM Michx.

Low annuals or perennials, herbaceous, or somewhat woody at the base; leaves entire, basal or scattered along the stem; inflorescence various; flowers involucrate; involucre 4 to 8-toothed or lobed, mostly many-flowered; calyx with 6 divisions (valves), colored, corolla-like; stamens 9; achenes triangular, sometimes winged.

The genus is one of the largest in our State. Representatives are found almost everywhere, especially at lower elevations. Strangely enough, although among our commonest plants, no common name seems to have been given them, probably because the plants are usually small and inconspicuous. Some species, however, are rather handsome.

KEY TO THE SPECIES.

Fruit winged, at least above; large coarse perennials.
  Wings extending the entire length of the fruit; perianth glabrous.
    Involucres strigose.................................... 1. *E. alatum.*
    Involucres glabrous.................................... 2. *E. triste.*
  Fruit winged only above the middle; perianth pubescent.
    Leaves loosely and coarsely tomentose beneath, commonly acutish; involucres 3 to 5 mm. long........ 3. *E. hieracifolium.*
    Leaves very densely and finely tomentose beneath, rounded-obtuse; involucres 2 to 3 mm. long...... 4. *E. pannosum.*

Fruit merely angled, never winged; perennials or annuals.
  Annuals.
    Involucres pubescent.
      Basal leaves linear or linear-oblanceolate; involucres on glabrous peduncles 15 to 50 mm. long..... 5. *E. pharnaceoides.*
      Basal leaves broadly oblong, ovate, or orbicular; peduncles less than 15 mm. long, more or less pubescent.
        Valves cordate at the base, orbicular or nearly so; plants not densely white-tomentose.
          Perianth yellow, tinged with red; stems 10 to 20 cm. high, much branched at the base; lobes of the involucre triangular to lanceolate, acute.............. 6. *E. abertianum.*
          Perianth white, tinged with pink; stems tall, 30 to 50 cm., sparingly branched, usually simple at the base; lobes of the involucre oblong or spatulate, mostly obtuse....................... 7. *E. pinetorum.*
        Valves tapering to the base, spatulate or obovate; plants densely white-tomentose.
          Stems simple below, ending in a flat-topped cyme; perianths 3 mm. long.. 8 *E. annuum.*
          Stems much branched throughout, the inflorescence of many slender secund racemes; perianth 1.5 mm. long or less.
            Plants erect, 30 to 50 cm. high; flowers bright rose pink................. 9. *E. polycladon.*
            Plants widely and densely diffuse, about 10 cm. high; flowers paler and smaller.....................10. *E. densum.*
    Involucres glabrous.
      Bracts of the inflorescence leaflike.
        Stems and leaves glabrous; leaves tapering to the flat margined petioles; peduncles filiform.................................11. *E. salsuginosum.*
        Stems and leaves more or less pubescent; leaves abruptly contracted into the terete petioles; peduncles stout................12. *E. divaricatum.*
      Bracts of the inflorescence scalelike.
        Peduncles abruptly reflexed in age; leaves extending up the stems 5 to 10 cm.; flowers white...........................13. *E. cernuum.*
        Peduncles erect or ascending, never reflexed; leaves basal; flowers white or colored.
          Perianth yellow; pedicels filiform.
            Leaves not tomentose.................14. *E. trichopodum.*
            Leaves tomentose beneath.............15. *E. wetherillii.*
          Perianth white or pimk; pedicels filiform or stout.
            Pedicels filiform; plants tall, erect, 20 to 50 cm. high..................16. *E. subreniforme.*
            Pedicels stout; plants low, spreading, less than 15 cm. high............17. *E. rotundifolium.*

Perennials.
Perianth narrowed into a long stipelike base, pubescent.
Perianth whitish; styles hairy at least to the middle....18. *E. jamesii.*
Perianth bright yellow; styles hairy only at the base....19. *E. bakeri.*
Perianth not narrowed into a stipelike base, pubescent or glabrous.
Ovaries and fruit pubescent.
Stems leafy; perianth 6 mm. long....................20. *E. longifolium.*
Stems not leafy, scapelike; perianth 3 mm. long or less.
Inflorescence an open cyme 10 to 15 cm. high; involucres 2 mm. high........................21. *E. leucophyllum.*
Inflorescence a congested cyme less than 3 cm. high; involucres 3 to 5 mm. high..............22. *E. lachnogynum.*
Ovaries and fruit glabrous or nearly so.
Involucres in a headlike cluster......................23. *E. ovalifolium.*
Involucres in open cymes.
Cymes one-sided and spikelike.
Leaves basal, abruptly narrowed at the base or subcordate; perianth 5 mm. long............24. *E. racemosum.*
Leaves scattered along the woody branches, tapering at the base; perianth 3 mm. long....25. *E. wrightii.*
Cymes dichotomous or trichotomous.
Perianth deep dark red; stems and involucres glabrous; plants tall and stout...............26. *E. atrorubens.*
Perianth white, pink, or yellow, never dark red; stems and involucres glabrous or pubescent; plants various.
Perianth densely sericeous.
Leaves elliptic, acute, tomentose............27. *E. havardii.*
Leaves orbicular, obtuse, not tomentose.....28. *E. inflatum.*
Perianth glabrous.
Perianth bright yellow; leaves glabrous (all basal, orbicular or broadly ovate).........29. *E. gypsophilum.*
Perianth white or pink; leaves tomentose, at least beneath.
Leaves all basal, orbicular or obovate, the stems densely cespitose; leaves permanently and densely white-tomentose on the upper surface..................30. *E. tenellum.*
At least the lower part of the stem leafy; stems usually not densely cespitose; leaves glabrate on the upper surface.
Flowering branches leafy for only a few centimeters at the base, the peduncles relatively long.
Stems and involucres tomentulose throughout; lobes of the involucre acute; leaves narrowly oblong, flat..............................31. *E. ainsliei.*
Branches of the inflorescence and involucres glabrous; lobes of the involucre broadly obtuse; leaves various.

Involucres in the forks of the inflorescence sessile. (Leaves revolute) .... 32. *E. nudicaule.*
Involucres in the forks of the branches distinctly peduncled, at least the lower ones.
Leaves linear, revolute .... 33. *E. tristichum.*
Leaves linear-oblong, flat .... 34. *E. lonchophyllum.*
Flowering branches leafy up to the inflorescence, the peduncles relatively short.
Leaf blades broad, 1 cm. wide or more, oblong to oval, obtuse.
Involucres 4 to 5 mm. long .... 35. *E. fendlerianum.*
Involucres 2 to 2.5 mm. long.
Branches of the inflorescence strongly divaricate .... 36. *E. divergens.*
Branches of the inflorescence ascending .... 37. *E. corymbosum.*
Leaf blades narrow, 5 mm. wide or less, linear or spatulate, acute.
Leaf blades oblanceolate, flat; inflorescence densely branched, 20 cm. high or less, tomentulose .... 38. *E. effusum.*
Leaf blades linear, revolute; inflorescence sparingly branched, 5 cm. high or less, glabrous or tomentulose.
Inflorescence glabrous; leaves 16 to 35 mm. long .... 39. *E. leptophyllum.*
Inflorescence tomentulose; leaves 15 mm. long or less .... 40. *E. simpsonii.*

1. **Eriogonum alatum** Torr. in Sitgreaves, Rep. Zuñi & Colo. 168. *pl. 8.* 1854.
TYPE LOCALITY: On the Zuni River, New Mexico.
RANGE: Nebraska and Wyoming to Texas and Arizona.
NEW MEXICO: Common from the Black Range to the White Mountains and northward to the Colorado line. Dry slopes, in the Upper Sonoran and Transition zones.

2. **Eriogonum triste** S. Wats. Proc. Amer. Acad. **10**: 347. 1875.
TYPE LOCALITY: Kane County, southern Utah.
RANGE: Utah and Colorado to Arizona and New Mexico.
NEW MEXICO: Farmington (*Standley* 7123); Upper Canadian (*Bigelow*). Dry hills, in the Upper Sonoran Zone.

3. **Eriogonum hieracifolium** Benth. in DC. Prodr. **14**: 6. 1853.
TYPE LOCALITY: Western Texas.
RANGE: Southern New Mexico and western Texas.
NEW MEXICO: Mountains southeast of Patterson; Horse Spring; Gila Hot Springs; White Mountains; Fort Stanton; Queen. Dry hills, in the Upper Sonoran Zone.

4. **Eriogonum pannosum** Woot. & Standl. Contr. U. S. Nat. Herb. **16**: 118. 1913.
TYPE LOCALITY: Organ Mountains, New Mexico. Type collected by G. R. Vasey, August, 1881.

Range: Organ Mountains of New Mexico.

New Mexico: Organ Mountains. Dry hillsides, in the Upper Sonoran Zone.

5. **Eriogonum pharnaceoides** Torr. in Sitgreaves, Rep. Zuñi & Colo. 167. *pl. 11.* 1854.

Type locality: Arizona.

Range: New Mexico and Arizona.

New Mexico: Fort Tularosa; Fort Bayard; Mogollon Mountains. Dry hills, in the Upper Sonoran Zone.

6. **Eriogonum abertianum** Torr. in Emory, Mil. Reconn. 151. 1848.

Eriogonum cyclosepalum Greene, Muhlenbergia **6**: 1. 1910.

Type locality: "On the upper waters of the Arkansas."

Range: New Mexico and western Texas to Arizona and Chihuahua.

New Mexico: Hillsboro; Big Hatchet Mountains; San Luis Mountains; Lake Arthur; Dona Ana Mountains; San Marcial; Lake Valley; Socorro; mesa west of the Organ Mountains. Dry sandy mesas and hillsides, in the Lower Sonoran Zone.

Apparently the plant has not been collected recently anywhere near the type locality. It is altogether possible that there is an error in the citation of the locality of Abert's specimen. Torrey further says that the species is "Very common in the region between the Del Norte and the Gila." The plant is one of the first to bloom in the spring in southern New Mexico. The type locality of *E. cyclosepalum* is New Mexican.

7. **Eriogonum pinetorum** Greene, Muhlenbergia **6**: 3. 1909.

Eriogonum abertianum neomexicanum Gandog. Bull. Soc. Bot. Belg. **42**: 185. 1906.

Type locality: Black Range, Sierra County, New Mexico. Type collected by Metcalfe (no. 1327).

Range: Western Texas to Arizona.

New Mexico: Mangas Springs; Mogollon Mountains; Fort Bayard; Kingston; Lordsburg; Dog Spring; Mesilla Valley; Organ Mountains. Dry mesas and hillsides, in the Lower and Upper Sonoran zones.

The type of *E. abertianum neomexicanum* is Wooton's no. 427 from the Organ Mountains.

8. **Eriogonum annuum** Nutt. Trans. Amer. Phil. Soc. n. ser. **5**: 164. 1837.

Type locality: "On the banks of the Great Salt River of Arkansas, and near the confluence of the Kiawesha and Red Rivers."

Range: Montana and South Dakota to Texas and northern Mexico.

New Mexico: Inscription Rock; Clayton; Buchanan; Fort Cummings; Mesilla Valley, plains west of Roswell; Causey; Nara Visa. Dry plains and hillsides, in the Lower and Upper Sonoran zones.

9. **Eriogonum polycladon** Benth. in DC. Prodr. **14**: 16. 1856.

Eriogonum polycladon crispum Gandog. Bull. Soc. Bot. Belg. **42**: 196. 1906.

Type locality: Western Texas.

Range: Arizona and western Texas.

New Mexico: Albuquerque; Fort Tularosa; Mangas Springs; Fort Bayard; Gila Hot Springs; Organ Mountains; Dog Spring. Dry hills and canyons, in the Upper Sonoran Zone.

The type of *E. polycladon crispum* is Wooton's 460 from the Organ Mountains.

10. **Eriogonum densum** Greene, Pittonia **3**: 17. 1896.

Type locality: Mountains of New Mexico, near Santa Rita del Cobre. Type collected by Greene, September 21, 1880.

Range: Mountains of southwestern New Mexico.

New Mexico: Near Santa Rita; Bear Mountain. In the Upper Sonoran Zone.

11. Eriogonum salsuginosum (Nutt.) Hook. Journ. Bot. Kew Misc. **5**: 264. 1853.
Stenogonum salsuginosum Nutt. Journ. Acad. Phila. II. **1**: 170. 1848.

TYPE LOCALITY: "Bare saline hills of the Colorado of the West, in the Rocky Mountains."

RANGE: Wyoming to Utah and New Mexico.

NEW MEXICO: Aztec; Carrizo Mountains; Farmington. Dry hills, in the Upper Sonoran Zone.

12. Eriogonum divaricatum Hook. Journ. Bot. Kew Misc. **5**: 265. 1853.

TYPE LOCALITY: "On saline clayey soils, within the high calcareous hills of the Upper Colorado."

RANGE: Wyoming to Arizona and New Mexico.

NEW MEXICO: Carrizo Mountains; near Zuni. Dry hills, in the Upper Sonoran Zone.

13. Eriogonum cernuum Nutt. Journ. Acad. Phila. II. **1**: 162. 1848.

TYPE LOCALITY: "On the plains of the Oregon and in the Rocky Mountains."

RANGE: Montana and Idaho to Arizona and New Mexico.

NEW MEXICO: Aztec; Farmington; Salt Lake Crater; Tierra Blanca; Zuni; Pajarito Park; Tunitcha Mountains; Carrizo Mountains; Dulce; Stinking Lake. Dry slopes, in the Upper Sonoran Zone.

14. Eriogonum trichopodum Torr. in Emory, Mil. Reconn. 151. 1848.

TYPE LOCALITY: "Eastern slope of the Cordilleras of California."

RANGE: Southern California to western Texas, south into Mexico.

NEW MEXICO: Mesa west of the Organ Mountains; Mesilla Valley. Dry hills and plains, in the Lower Sonoran Zone.

The species was published originally under the name of *trichopes*, but this was later corrected to the name used here. This is one of the commonest plants in the lower Rio Grande Valley and on the bordering mesas. The rosette of basal leaves frequently drys up and breaks away while the plant is still in flower.

15. Eriogonum wetherillii Eastw. Proc. Calif. Acad. II. **6**: 319. 1896.

TYPE LOCALITY: At the base of sandstone cliffs along the San Juan River, Utah.

RANGE: Southern Utah to northeastern Arizona and northwestern New Mexico.

NEW MEXICO: Near the Carrizo Mountains (*Standley* 7474). Dry, rocky hills, in the Upper Sonoran Zone.

16. Eriogonum subreniforme S. Wats. Proc. Amer. Acad. **12**: 260. 1877.

TYPE LOCALITY: "Arizona, S. Utah."

RANGE: Arizona and Utah to western New Mexico.

NEW MEXICO: Zuni (*Wooton*). Upper Sonoran Zone.

17. Eriogonum rotundifolium Benth. in DC. Prodr. **14**: 21. 1856.

TYPE LOCALITY: Western Texas.

RANGE: Western Texas and southern New Mexico to Chihuahua.

NEW MEXICO: Albuquerque; Dog Spring; Mesilla Valley; mesa west of Organ Mountains; plains south of the White Sands; Lake Valley; east of Hachita. Dry mesas, in the Lower Sonoran Zone.

18. Eriogonum jamesii Benth. in DC. Prodr. **14**: 7. 1856.
Eriogonum jamesii neomexicanum Gandog. Bull. Soc. Bot. Belg. **42**: 190. 1906.

TYPE LOCALITY: "In montibus Scopulosis ad fontes fl. Platte."

RANGE: Colorado to New Mexico and Arizona.

NEW MEXICO: Common throughout the State in the mountains and on high plains. Open slopes, chiefly in the Transition Zone.

The type of *E. jamesii neomexicanum* is Wooton's 385 from the White Mountains.

19. Eriogonum bakeri Greene, Pl. Baker. **3**: 15. 1901.

Eriogonum jamesii flavescens S. Wats. Proc. Amer. Acad. **12**: 255. 1877.

Eriogonum vegetius A. Nels. Bull. Torrey Club **31**: 239. 1904.

TYPE LOCALITY: Black Canyon, Colorado.

RANGE: Wyoming to New Mexico.

NEW MEXICO: White and Sacramento mountains. Meadows, in the Transition Zone.

Our New Mexican specimens are rather larger and more robust than the typical form and their flowers seem to be of a brighter yellow. Within our limits the species is known only from the White and Sacramento mountains, which are far removed from the usual range in Colorado and northern Arizona.

20. Eriogonum longifolium Nutt. Trans. Amer. Phil. Soc. n. ser. **5**: 164. 1837.

Eriogonum texanum Scheele, Linnaea **22**: 150. 1849.

TYPE LOCALITY: "On the ledges of the Cadron rocks, and in denudated prairies from Arkansas to Red River."

RANGE: Southern Missouri to Texas and New Mexico.

NEW MEXICO: Leachs (*Wooton*). Dry plains, in the Upper Sonoran Zone.

21. Eriogonum leucophyllum Woot. & Standl. Contr. U. S. Nat. Herb. **16**: 118. 1913.

TYPE LOCALITY: Lakewood, New Mexico. Type collected by Wooton, August 6, 1909.

RANGE: Known only from type locality.

22. Eriogonum lachnogynum Torr. in DC. Prodr. **14**: 8. 1856.

TYPE LOCALITY: New Mexico.

RANGE: Utah and Arizona to New Mexico and Kansas.

NEW MEXICO: Banks of the Cimarron; Llano Estacado; Buchanan; Perico. Plains, in the Upper Sonoran Zone.

The type was collected by Fendler (no. 765) in 1847, either about Santa Fe or farther east.

23. Eriogonum ovalifolium Nutt. Journ. Acad. Phila. **7**: 50. 1834.

TYPE LOCALITY: "Sources of the Missouri."

RANGE: Washington and Montana to California and New Mexico.

NEW MEXICO: Carrizo Mountains; Aztec; Farmington. Dry hills, in the Upper Sonoran Zone.

24. Eriogonum racemosum Nutt. Journ. Acad. Phila. II. **1**: 161. 1848.

Eriogonum orthocladon Torr. in Sitgreaves, Rep. Zuñi & Colo. 167. *pl. 9*. 1854.

TYPE LOCALITY: "Colorado of the West."

RANGE: Colorado and Utah to Arizona and Texas.

NEW MEXICO: Chama; north of Ramah; Inscription Rock; Santa Fe; Sandia Mountains; Glorieta; Carrizo Mountains; Tunitcha Mountains. Open slopes, in the Upper Sonoran and Transition zones.

The type of *E orthocladon* was collected in the Zuni Mountains.

25. Eriogonum wrightii Torr. in DC. Prodr. **14**: 15. 1856.

TYPE LOCALITY: Western Texas.

RANGE: Arizona, New Mexico, and western Texas, southward into Mexico.

NEW MEXICO: On the San Juan; Mangas Springs; Kingston; Animas Valley; White Mountains; Organ Mountains; Sandia Mountains. Dry hills and canyons, in the Upper Sonoran Zone.

26. Eriogonum atrorubens Engelm. in Wisliz. Mem. North. Mex. 108. 1848.

TYPE LOCALITY: Cosihuiriachi, Chihuahua.

RANGE: Chihuahua and southeastern New Mexico.

NEW MEXICO: San Luis Mountains (*Mearns* 568, 2123, 2463).

27. **Eriogonum havardii** S. Wats. Proc. Amer. Acad. **18:** 194. 1883.

TYPE LOCALITY: Chenate Mountains, western Texas.

RANGE: Southern New Mexico and western Texas.

NEW MEXICO: Plains 35 miles south of Torrance; west of Roswell; Queen; near Elk. Plains and low hills, in the Upper Sonoran Zone.

28. **Eriogonum inflatum** Torr. in Frém. Rep. Exped. Rocky Mount. 317. 1845.

TYPE LOCALITY: "On barren hills in the lower part of North California."

RANGE: Colorado and northwestern New Mexico to southern California.

NEW MEXICO: Carrizo Mountains (*Standley* 7346, 7486). Dry hills and plains, in the Upper Sonoran Zone.

The branches are commonly swollen and inflated at the nodes, but not invariably so.

29. **Eriogonum gypsophilum** Woot. & Standl. Contr. U. S. Nat. Herb. **16:** 118. *pl. 49.* 1913.

TYPE LOCALITY: On a hill southwest of Lakewood, New Mexico. Type collected by Wooton, August 6, 1909.

RANGE: Known only from type locality.

The plants grow in nearly pure gypsum.

30. **Eriogonum tenellum** Torr. Ann. Lyc. N. Y. **2:** 241. 1827.

TYPE LOCALITY: "Near the Rocky Mountains," Colorado or New Mexico.

RANGE: Colorado and Utah to western Texas.

NEW MEXICO: Clayton; Buchanan; 10 miles west of Roswell; Knowles. Dry plains, in the Upper Sonoran Zone.

31. **Eriogonum ainsliei** Standley, Contr. U. S. Nat. Herb. **16:** 117. 1913.

TYPE LOCALITY: Cimarron, New Mexico. Type collected by C. N. Ainslie, September 10, 1909.

RANGE: Northeastern New Mexico.

NEW MEXICO: Cimarron; Raton Mountains; Colfax. Plains and low hills, in the Upper Sonoran Zone.

32. **Eriogonum nudicaule** (Torr.) Small, Bull. Torrey Club **33:** 54. 1906.

Eriogonum effusum nudicaule Torr. U. S. Rep. Expl. Miss. Pacif. **4:** 132. 1856.

TYPE LOCALITY: Pine and cedar woods, near Galisteo, New Mexico. Type collected by Bigelow.

RANGE: Northern New Mexico.

NEW MEXICO: Near Galisteo; Placitas.

33. **Eriogonum tristichum** Small, Bull. Torrey Club **33:** 55. 1906.

TYPE LOCALITY: Rosa, New Mexico.

RANGE: Northern New Mexico and southern Colorado.

NEW MEXICO: Rosa; Cedar Hill; Dulce; Chupadero. Dry hills, in the Upper Sonoran Zone.

In the original publication of the species the State is given incorrectly as Colorado.

34. **Eriogonum lonchophyllum** Torr. & Gray, Proc. Amer. Acad. **8:** 173. 1870.

TYPE LOCALITY: On the Rio Blanco, New Mexico. Type collected by Newberry.

RANGE: Northern New Mexico and southern Colorado.

NEW MEXICO: Rio Blanco; near Tierra Amarilla.

Rydberg in the Flora of Colorado [1] describes the plant, in the key, as having the involucres in the forks of the branches sessile; in our specimen of the type collection they are conspicuously peduncled. To this plant the native people give the name of "cola de ratón."

---

[1] Colo. Agr. Exp. Sta. Bull. **100:** 103. 1906.

35. Eriogonum fendlerianum (Benth.) Small, Bull. Torrey Club **33:** 55. **1906.**

Eriogonum microthecum fendlerianum Benth. in DC. Prodr. **14:** 18. 1856.

TYPE LOCALITY: New Mexico.

The type is Fendler's 767, collected probably somewhere about Santa Fe.

36. Eriogonum divergens Small, Bull. Torrey Club **33:** 55. 1906.

Eriogonum corymbosum divaricatum Torr. & Gray, U. S. Rep. Expl. Miss. Pacif. 2^1: 129. 1855, not *E. divaricatum* Hook. 1853.

TYPE LOCALITY: "Near springs on Green River."

RANGE: Colorado and Utah to Arizona and New Mexico.

NEW MEXICO: Mesa La Vaca; Bad Lands. Dry hills and plains, in the Upper Sonoran Zone.

37. Eriogonum corymbosum Benth. in DC. Prodr. **14:** 17. 1856.

TYPE LOCALITY: "Prope Grand-River," Colorado.

RANGE: Colorado and northern New Mexico.

NEW MEXICO: On the San Juan River; Cerrillos; Logan; Farmington. Dry hills and plains, in the Upper Sonoran Zone.

38. Eriogonum effusum Nutt. Journ. Acad. Phila. II. **1:** 164. 1848.

TYPE LOCALITY: Rocky Mountains.

RANGE: Montana and Nebraska to Colorado and New Mexico.

NEW MEXICO: Near Belen; Farmington. Dry hills, in the Upper Sonoran Zone.

39. Eriogonum leptophyllum (Torr.) Woot. & Standl. Contr. U. S. Nat. Herb. **16:** 118. 1913.

Eriogonum effusum leptophyllum Torr. in Sitgreaves, Rep. Zuñi & Colo. 168. 1854.

TYPE LOCALITY: Rio Zuni, New Mexico.

RANGE: Northwestern New Mexico.

NEW MEXICO: Grants; Tunitcha Mountains; Carrizo Mountains. Dry hills, in the Upper Sonoran Zone.

40. Eriogonum simpsonii Benth. in DC. Prodr. **14:** 18. 1856.

TYPE LOCALITY: "In Sierra de Tunecha (Novi-Mexici)."

RANGE: Colorado to northern New Mexico and Arizona.

NEW MEXICO: Gallup; Fort Wingate; Atarque de Garcia; east of Ramah; near McIntosh; Pajarito Park; Carrizo Mountains. Dry hills and plains, in the Upper Sonoran Zone.

The "Sierra de Tunecha" is doubtless the Tunitcha Mountains.

2. OXYRIA Hill. MOUNTAIN SORREL.

A low alpine perennial with round-reniform, long-petioled, chiefly basal leaves; flowers small, greenish, in panicled racemes on a slender scape; sepals 4, unchanged in fruit, usually reddish; stamens 6; achene thin, lenticular, surrounded by a broad veined wing.

1. Oxyria digyna (L.) Hill, Hort. Kew. 158. 1768.

Rumex digynus L. Sp. Pl. 337. 1753.

TYPE LOCALITY: "In Alpibus Lapponicis, Helveticis, Wallicis."

RANGE: Alaska and Greenland, south to New England, New Mexico, and California; also in Europe and Asia.

NEW MEXICO: Brazos Canyon; Upper Pecos River; Wheeler Peak. Wet meadows, Canadian to Arctic-Alpine Zone.

3. RUMEX L. Dock.

Coarse perennial herbs with leafy stems (usually most of the leaves basal); stipules united to form more or less hyaline sheaths (ocreæ); flowers numerous, small, greenish, perfect, polygamous, or diœcious, in simple or compound racemes or paniculate; inner sepals (valves) becoming enlarged in fruit, persistent.

KEY TO THE SPECIES.

Plants low, 10 to 30 cm. high, slender, diœcious; leaves hastate .... 1. *R. acetosella.*
Plants taller, 30 to 60 cm. high, stout, polygamous or monœcious; leaves never hastate.
  Inner sepals (valves) much enlarged in fruit, 10 to 15 mm. long and broad, mostly reddish .... 2. *R. hymenosepalus.*
  Valves of medium size, less than 10 mm. long, greenish or brown.
    Valves without callosities on the back.
      Leaves large, 50 cm. long or less, ovate or oblong-ovate, cordate .... 3. *R. occidentalis.*
      Leaves small, 10 cm. long or less, narrowly elliptic-lanceolate, attenuate at the base .... 7. *R. ellipticus.*
    One or more of the valves bearing callosities on the back.
      Valves entire or nearly so.
        Leaves crispate .... 5. *R. crispus.*
        Leaves flat.
          Leaves elliptic-lanceolate or broader; valves ovate, usually only one bearing a callosity .... 6. *R. altissimus.*
          Leaves narrowly oblong-lanceolate; valves triangular-ovate, usually all three bearing callosities .... 8. *R. mexicanus.*
      Valves distinctly dentate.
        Teeth of the valves broadly triangular, the teeth 2 or 3 times as long as broad .... 4. *R. britannica.*
        Teeth of wings long and narrow, as long as broad or much longer.
          Perennial; leaves large, 30 cm. long or more, oblong to ovate, cordate .... 9. *R. obtusifolius.*
          Annual; leaves much smaller, 10 to 15 cm. long, narrowly oblong-lanceolate, truncate or rounded at the base ... 10. *R persicarioides.*

Rumex berlandieri should come into New Mexico; we have seen specimens from El Paso, Texas.

1. Rumex acetosella L. Sp. Pl. 338. 1753. SHEEP SORREL.

TYPE LOCALITY: "Habitat in Europae pascuis & arvis arenosis."

RANGE: A native of Europe, introduced into many parts of the United States as a field weed.

NEW MEXICO: Harveys Upper Ranch; Gallinas Planting Station; Sandia Mountains.

2. Rumex hymenosepalus Torr. U. S. & Mex. Bound. Bot. 177. 1859. CAÑAIGRE.

TYPE LOCALITY: "Sandy soils from El Paso to the Canyons of the Rio Grande."

RANGE: Western Texas to New Mexico and Arizona, and northern Mexico.

NEW MEXICO: Aztec; Farmington; Mangas Springs; Las Cruces; Gray; mountains west of San Antonio; Hillsboro. Sandy soil, in the Lower and Upper Sonora zones.

Cañaigre is common on the sandy mesas of the southern part of the State, where it is a rather conspicuous plant in the latter part of the winter and in early spring, being about the only green thing to be seen. It commences to grow in January or February, and is ready to bloom by March. It withstands the cold night temperatures well. The tuberous roots, resembling dahlia tubers or, somewhat less, sweet potatoes, contain a relatively high percentage of tannin. An industry was once established in a small way at Deming dependent upon the extraction of tannin from these roots for the preparation of a tanning fluid. The product proved very satisfactory and a European market was found which would take the output, but the supply of tubers was insufficient and people could not be induced to grow them. The plants grow very slowly. The tubers have long been used by the native people for the tanning of skins; they were also used by the Navahos in dying wool yellow.

3. **Rumex occidentalis** S. Wats. Proc. Amer. Acad. **12**: 253. 1876.

TYPE LOCALITY: "From Alaska to northern California, eastward to Saskatchewan and Labrador, and southward in the mountains to Colorado and New Mexico."

RANGE: As under type locality.

NEW MEXICO: Pecos Baldy; Middle Fork of the Gila; James Canyon; White Mountains; Costilla Valley; Brazos Canyon. Wet ground, especially along streams, in the Canadian and Transition zones.

A thick-stemmed, large-leaved dock, not uncommon along streams in the mountains. It is usually 60 to 90 cm. high, with a large compound panicle, and the fruits are often reddish when ripe.

4. **Rumex britannica** L. Sp. Pl. 334. 1753.

TYPE LOCALITY: "Habitat in Virginia."

RANGE: Northeastern United States and Canada, extending into Colorado, Utah, and New Mexico.

NEW MEXICO: Near Fort Bayard (*Blumer* 115). Wet ground.

5. **Rumex crispus** L. Sp. Pl. 335. 1753. YELLOW DOCK.

TYPE LOCALITY: "Habitat in Europae suculentis."

RANGE: An introduced weed, common throughout the United States, in cultivated fields and waste ground.

NEW MEXICO: Common nearly throughout the State.

The plant is often a troublesome weed in alfalfa fields and along ditch banks. The leaves are sometimes gathered and cooked as "greens."

6. **Rumex altissimus** Wood, Class-book 477. 1855. PALE DOCK.

TYPE LOCALITY: "Marshy prairies and borders of streams, Indiana."

RANGE: Massachusetts to Colorado, Maryland, Texas, and New Mexico.

NEW MEXICO: Mangas Springs. Damp ground.

7. **Rumex ellipticus** Greene, Pittonia **4**: 234. 1900.

TYPE LOCALITY: In fields and along river banks at Roswell, New Mexico. Type collected by Earle (no. 272).

RANGE: Known only from the Pecos Valley of New Mexico.

NEW MEXICO: Roswell; near Lake Arthur. Lower Sonoran Zone.

The species is doubtfully distinct from *R. altissimus*.

8. **Rumex mexicanus** Meisn. in DC. Prodr. **14**: 45. 1856.

TYPE LOCALITY: "In Mexico circa Leon."

RANGE: Throughout the Rocky Mountains and across the continent northward, extending south into Mexico.

NEW MEXICO: Common except along the eastern side of the State. In cultivated fields and along ditches, from the Lower Sonoran to the Transition Zone.

Along ditches and streams and in fields wherever crops are cultivated in the State, this is the common dock. It is very resistant to alkali and often occurs in alkali spots. It has been confused with the seacoast plant *R. salicifolius* Weinm.,[1] a species of restricted distribution on the Californian coast.

9. Rumex obtusifolius L. Sp. Pl. 335. 1753. BITTER DOCK.

TYPE LOCALITY: "Habitat in Germania, Helvetia, Gallia, Anglia."

RANGE: A native of Europe, introduced into many parts of America.

NEW MEXICO: Kingston (*Metcalfe* 1099).

A large, thick-stemmed dock, 60 to 90 cm. high, resembling *R. occidentalis* in general appearance, but with broader leaves. It is readily distinguished from that species by the structure of the fruiting calyx. In this the valves have 3 to 5 thin, spinelike teeth on each side and smooth callosities, while in *R. occidentalis* the callosities are absent and the margin of the valves is entire or remotely denticulate.

10. Rumex persicarioides L. Sp. Pl. 335. 1753. GOLDEN DOCK.

TYPE LOCALITY: "Habitat in Virginia."

RANGE: Throughout temperate North America.

NEW MEXICO: Mangas Springs; mountains southeast of Patterson; Cliff; Dulce Shiprock; Farmington. Along streams and ditches, in the Upper Sonoran Zone.

4. GONOPYRUM Fisch. & Mey.

Low shrub, 1 meter high or less, with stout erect stems; leaves linear, glaucous, fleshy, jointed to the ocreæ; flowers perfect, solitary, on jointed pedicels; sepals 5, white, the 3 inner developing wings at maturity, the 2 outer reflexed; achene elliptic, oblong, brown, pointed at both ends.

1. Gonopyrum americanum Fisch. & Mey. Mém. Acad. St. Pétersb. VI. Sci. Nat. **4**[1]: 144. 1845.

Polygonella ericoides Engelm. & Gray, Bost. Journ. Nat. Hist. **5**: 231. 1845.

TYPE LOCALITY: Texas.

RANGE: Georgia to Arkansas, eastern Texas, and central New Mexico.

NEW MEXICO: Tijeras Canyon (*C. R. Ellis*). Sandy soil, in the Upper Sonoran Zone.

This is a most remarkable extension of range for a plant not known heretofore from any station west of eastern Texas. It occurs, however, even in the southeast, only locally. Our specimens seem to agree very well with eastern material, but they are rather fragmentary; perhaps if they were more ample some difference might be discovered.

5. POLYGONUM L. KNOTWEED.

Annuals with slender stems branching near the base, prostrate or erect; leaves small, alternate, elliptic to linear-lanceolate, obtuse or acute, entire, the upper generally much reduced; ocreæ hyaline, at length lacerate, not fringed; flowers axillary in short few-flowered clusters, sometimes solitary; calyx of 5 or 6 greenish sepals with white or pink margins; stamens 8 or fewer, at least the inner with dilated filaments; achenes 3-angled, surrounded by the persistent calyx.

---

[1] See, Fernald, M. L. The representatives of *Rumex salicifolius* in eastern America. Rhodora **10**: 71. 1908.

KEY TO THE SPECIES.

Inflorescence aggregated at the ends of the branches; plants small, less than 10 cm. high .... 1. *P. watsoni.*

Inflorescence of small scattered axillary clusters; stems 20 cm. long or more.

Stems prostrate (except where the plants are much crowded, there weakly ascending).

Leaves thin, not prominently veined, bright green; ocreæ not conspicuous .... 2. *P. aviculare.*

Leaves thick, with rather prominent veins, pale or glaucous; ocreæ conspicuous .... 3. *P. buxiforme.*

Stems erect.

Achenes not deflexed.

Upper leaves little reduced; plants bright green .... 4. *P. erectum.*

Upper leaves much reduced, bractlike; plants bright or yellowish green.

Upper leaves or bracts not subulate, mostly elliptic or lanceolate; achenes dull .... 5. *P. ramosissimum.*

Upper leaves reduced to subulate bracts; achenes black and shining .... 6. *P. sawatchense.*

Achenes deflexed.

Upper bracts much reduced, subulate; lower leaves mostly narrowly lanceolate or linear .. 7. *P. douglasii.*

Upper bracts broader, oblong to lanceolate; lower leaves broader, elliptic to oblong-lanceolate .. 8. *P. montanum.*

1. **Polygonum watsoni** Small, Mem. Bot. Columb. Coll. **1**: 138. *pl. 56.* 1895.

TYPE LOCALITY: "Washington to Montana; south to California and Colorado."

RANGE: As under type locality.

NEW MEXICO: Tunitcha Mountains; Chama. Wet ground, in the Transition Zone.

2. **Polygonum aviculare** L. Sp. Pl. 362. 1753.

TYPE LOCALITY: "Habitat in Europae cultis ruderatis."

RANGE: Common throughout North America except in the extreme north; also in the Old World.

NEW MEXICO: Santa Fe and Las Vegas mountains; Dulce; White Mountains; Roswell. Upper Sonoran and Transition zones.

A common dooryard weed at middle levels in the mountains. The plant is rather variable in general aspect, being influenced by the conditions under which the individuals have grown.

3. **Polygonum buxiforme** Small, Bull. Torrey Club **33**: 56. 1906.

Polygonum littorale Link, err. det. Small, Mem. Bot. Columb. Coll. **1**: 102. 1895.

TYPE LOCALITY: Not stated.

RANGE: Throughout North America except in the extreme north.

NEW MEXICO: Farmington; Tunitcha Mountains; Las Vegas; Bernalillo; Maxwell City; Santa Fe; Kingston; Mangas Springs; Mesilla Valley; Sacramento and White mountains. Damp ground, Lower Sonoran to Transition Zone.

This species seems too near the preceding, and it may be that Doctor Robinson's treatment of it as a subspecies is correct. Usually this plant is less green, a little stouter, and the ocreæ are more conspicuous, but the differences are more of degree than kind.

4. **Polygonum erectum** L. Sp. Pl. 363. 1753.

TYPE LOCALITY: "Habitat in Philadelphia."

RANGE: Throughout the United States as far west as the Rockies.

NEW MEXICO: Chama; Sierra Grande. Waste ground.

5. **Polygonum ramosissimum** Michx. Fl. Bor. Amer. **1:** 237. 1803.

TYPE LOCALITY: "Hab. in regione Illinoensi."

RANGE: British America to California, eastward to the Atlantic Coast.

NEW MEXICO: Farmington; Dulce; Ojo Caliente; Santa Fe; Pecos; Frisco; Mangas Springs; Mesilla Valley; Tularosa Creek. Lower Sonoran to Transition Zone.

A common weed of roadsides and waste ground, often occurring in cultivated fields.

6. **Polygonum sawatchense** Small, Bull. Torrey Club **20:** 213. 1893.

TYPE LOCALITY: Sawatch Range, Colorado.

RANGE: Washington and North Dakota to New Mexico.

NEW MEXICO: Carrizo Mountains; Dulce; Stinking Lake; Sierra Grande; Grass Mountain; Gilmores Ranch. Meadows, in the Transition Zone.

7. **Polygonum douglasii** Greene, Bull. Calif. Acad. **1:** 125. 1885.

TYPE LOCALITY: "From the Saskatchewan to British Columbia, and southward everywhere in the mountains to the borders of Mexico."

RANGE: As under type locality.

NEW MEXICO: Tunitcha Mountains; Chama; Ramah; Cross L Ranch; Luna; Burro Mountains; West Fork of the Gila; Mimbres River; White Mountains. Transition Zone.

This is the western equivalent of *Polygonum tenue* of the Eastern States and is separated from that species by two characters which are not readily apparent. The ripe fruit is deflexed, but much of the ordinary herbarium material fails to show the character because the plants are too young. The other character is generally more certainly present though harder to make out. *P. tenue* has three parallel veins in the leaf, while *P. douglasii* has only a midvein. Our plant occurs in forests at middle altitudes in the mountains. It is closely related to the next species as well.

8. **Polygonum montanum** (Small) Greene, Pl. Baker. **3:** 13. 1901.

Polygonum tenue latifolium Engelm. Proc. Acad. Phila. **1863:** 75. 1864.

Polygonum douglasii latifolium Greene, Bull. Calif. Acad. **1:** 125. 1885.

Polygonum douglasii montanum Small, Mem. Bot. Columb. Coll. **1:** 118. 1895.

TYPE LOCALITY: Rocky Mountains of Colorado.

RANGE: New Mexico and Arizona to California and northward in the higher mountains.

NEW MEXICO: Upper Pecos River; Beattys Cabin. Canadian Zone.

6. PERSICARIA Adans. SMARTWEED.

Annual or perennial herbs; leaves alternate, the blades entire; ocreæ cylindric, membranous, naked or fringed with bristles; flowers in terminal or axillary spikelike racemes; calyx white, greenish, or rose-colored, persistent in fruit; sepals mostly 5; stamens 4 to 8, the filaments not dilated; achenes lenticular or trigonous, usually black.

KEY TO THE SPECIES.

Racemes terminal only, usually solitary.
  Leaves elliptic, obtuse or acute; spikes 12 to 24 mm. long; pedicels glabrous........ 1. *P. amphibia.*
  Leaves lanceolate to ovate, acuminate; spikes 30 mm. long or more; pedicels hispid, often glandular........ 2. *P. muhlenbergii.*
Racemes axillary as well as terminal, numerous.
  Sheaths without marginal bristles.
    Styles included........ 3. *P. lapathifolia.*
    Styles conspicuously exserted........ 4. *P. longistyla.*

Sheaths with marginal bristles.
  Racemes oblong or cylindric, densely flowered; perianth not punctate, usually pink or rose-colored........ 5. *P. persicaria.*
  Racemes slender, loosely flowered; perianth white or green, copiously punctate........................ 6. *P. punctata.*

Doctor Small reports *Persicaria persicarioides* (L.) Small from New Mexico. The species is much like *P. persicaria*, but is slightly larger throughout and perennial, while the latter is annual.

1. **Persicaria amphibia** (L.) S. F. Gray, Nat. Arr. Brit. Pl. **2**: 268. 1821.

Polygonum amphibium L. Sp. Pl. 361. 1753.

Polygonum ccccineum Muhl. Cat. Pl. 40. 1813.

TYPE LOCALITY: "Habitat in Europa."

RANGE: British America to California, New Mexico, and New Jersey; also in Europe and Asia.

NEW MEXICO: Tunitcha Mountains; Dulce Lake. Floating in water, in the Transition Zone.

2. **Persicaria muhlenbergii** (Meisn.) Small; Rydb. Colo. Agr. Exp. Sta. Bull. **100**: 111. 1906.

Polygonum amphibium muhlenbergii Meisn. in DC. Prodr. **14**: 116. 1856.

Polygonum muhlenbergii S. Wats. Proc. Amer. Acad. **14**: 245. 1879.

Persicaria rothrockii Greene, Leaflets **1**: 45. 1904.

TYPE LOCALITY: North America.

RANGE: Throughout North America from the Arctic regions to central Mexico.

NEW MEXICO: Mesilla Valley. Wet ground, in the Lower Sonoran Zone.

3. **Persicaria lapathifolia** (L.) S. F. Gray, Nat. Arr. Brit. Pl. **2**: 270. 1821.

Polygonum lapathifolium L. Sp. Pl. 360. 1753.

TYPE LOCALITY: "Habitat in Gallia."

RANGE: Throughout most of North America and in Europe; with us possibly introduced, but now widely distributed.

NEW MEXICO: Common throughout the State. Low, wet ground in the Lower and Upper Sonoran zones.

One of the commonest weeds in cultivated fields. It is most abundant in the Rio Grande Valley, especially where the ditch water overflows.

4. **Persicaria longistyla** Small, Fl. Southeast. U. S. 377. 1903.

Polygonum longistylum Small, Bull. Torrey Club **21**: 169. 1894.

TYPE LOCALITY: New Mexico.

RANGE: Missouri and Louisiana to New Mexico.

NEW MEXICO: Pitt Lake; north of Melrose. Wet ground.

The type is Fendler's 749, collected probably somewhere about Santa Fe.

5. **Persicaria persicaria** (L.) Small, Fl. Southeast. U. S. 378. 1903.

Polygonum persicaria L. Sp. Pl. 361. 1753.

TYPE LOCALITY: "Habitat in Europae cultis."

RANGE: British America to California and Florida, south into Mexico; also in the Old World.

NEW MEXICO: Farmington; Santa Fe; Indian Creek; Pajarito Park; Gallinas Canyon; Gila Hot Springs; Pecos; Sandia Mountains. Waste and wet ground.

6. **Persicaria punctata** (Ell.) Small, Fl. Southeast. U. S. 379. 1903.

Polygonum punctatum Ell. Bot. S. C. & Ga. **1**: 445. 1817.

TYPE LOCALITY: South Carolina and Georgia.

RANGE: British America to California, southward into South America and the West Indies.

NEW MEXICO: Mesilla Valley; Mimbres River; Cloverdale. Wet ground.

7. BISTORTA L. Bistort.

Herbaceous, perennial, alpine or subalpine plants, glabrous, bright green; stems simple; leaves radical and cauline, oblong to linear; sheaths never ciliate; inflorescence sometimes showy, of a single terminal spicate raceme; stamens 8 or 9, exserted; styles usually 3-parted, exserted.

KEY TO THE SPECIES.

Racemes not viviparous (i. e. not bulblet-bearing), oblong, 10 to 20 mm. thick .... 1. *B. bistortoides.*
Racemes viviparous below, linear, 5 to 8 mm. thick .... 2. *B. vivipara.*

1. **Bistorta bistortoides** (Pursh) Small, Bull. Torrey Club **33**: 57. 1906.
Polygonum bistortoides Pursh, Fl. Amer. Sept. 271. 1814.
Type locality: "In low grounds on the banks of the Missouri, called *Quamash-flats.*"
Range: Arctic America to California and New Mexico.
New Mexico: Santa Fe and Las Vegas mountains; Mogollon Mountains. Meadows in the mountains, Canadian to Arctic-Alpine Zone.

2. **Bistorta vivipara** (L.) S. F. Gray, Nat. Arr. Brit. Pl. **2**: 268. 1821.
Alpine bistort.
Polygonum viviparum L. Sp. Pl. 360. 1753.
Type locality: "Habitat in Europae subalpinis pascuis duris."
Range: British America to New Mexico and New England; also in Europe and Asia.
New Mexico: Santa Fe and Las Vegas mountains; Rio Pueblo; Baldy. Meadows, in the Arctic-Alpine Zone.

8. BILDERDYKIA Dum. Black bindweed.

A twining annual with ovate, hastate, acute to acuminate leaves, the upper ones narrower; stems rough-angled; inflorescence axillary, slender, interrupted, of compound racemes bearing reduced leaves.

1. **Bilderdykia convolvulus** (L.) Dum. Fl. Belg. 18. 1827.
Polygonum convolvulus L. Sp. Pl. 364. 1753.
Tiniaria convolvulus Webb & Moq. in Webb & Berth. Hist. Nat. Canar. **3**[2]: 221. 1836–50.
Type locality: "Habitat in Europae agris."
Range: Nearly throughout North America, except in the extreme north.
New Mexico: Tunitcha Mountains; Chama; Farmington; Winsors Ranch; Pecos; Mogollon Mountains; Kingston; Mesilla Valley; White Mountains; Maxwell City; Sandia Mountains. Waste ground and fields, chiefly in the Transition Zone.

9. FAGOPYRUM Gaertn. Buckwheat.

Annual with hastate leaves scattered along the stems; ocreæ fugacious; flowers in terminal or axillary racemes; calyx not keeled.

1. **Fagopyrum fagopyrum** (L.) Karst. Deutsch. Fl. 522. 1880–83.
Polygonum fagopyrum L. Sp. Pl. 364. 1753.
Fagopyrum esculentum Moench, Meth. Pl. 290. 1794.
Type locality: "Habitat in Asia."
New Mexico: Balsam Park, Sandia Mountains (*Ellis* 273).
Common in cultivation and frequently escaped in North America.

Order 24. CHENOPODIALES.

KEY TO THE FAMILIES.

Fruit a capsule, dehiscent by apical or longitudinal valves.
- Ovary several-celled; corolla wanting........47. **AIZOACEAE** (p. 228).
- Ovary 1-celled; corolla usually present.
  - Sepals 2..............................48. **PORTULACACEAE** (p. 229).
  - Sepals 4 or 5.
    - Sepals distinct; petals not clawed; ovary sessile......................49. **ALSINACEAE** (p. 234).
    - Sepals united; petals clawed; ovary more or less distinctly stipitate...50. **SILENACEAE** (p. 240).

Fruit a utricle, achene, or anthocarp, indehiscent, circumscissile, or bursting irregularly.
- Fruit an anthocarp, the achene surrounded by the tube of the corolla-like calyx ...45. **ALLIONIACEAE** (p. 216).
- Fruit not an anthocarp.
  - Fruit an achene or berry...............46. **PHYTOLACCACEAE** (p. 228).
  - Fruit a utricle.
    - Stipules present, scarious...........44. **CORRIGIOLACEAE** (p. 216).
    - Stipules wanting.
      - Bracts scarious.................43. **AMARANTHACEAE** (p. 209).
      - Bracts not scarious.............42. **CHENOPODIACEAE** (p. 198).

42. CHENOPODIACEAE. Goosefoot Family.

Annual or perennial herbs or shrubs; leaves usually simple, alternate, sometimes much reduced; flowers perfect or unisexual, small, apetalous, the sepals sometimes wanting, replaced in the pistillate flower by a pair of scales, these becoming variously modified in fruit; sepals 5 or fewer, the stamens as many and opposite them; pistil 1, with a single ovule; fruit an achene or utricle.

The family is a very important one in the arid regions, where representatives are numerous both as to species and individuals. They seem to be particularly adapted to bright sunlight and dry soil, and are tolerant of alkali. Several of the species are important forage plants, a few are eaten by man, and several are troublesome weeds.

KEY TO THE GENERA.

Embryo spirally coiled; leaves fleshy (except in no. 2), linear or awl-shaped.
- Shrubs, 1 to 3 meters high, with monœcious flowers; staminate flowers spicate, without a perianth; pistillate flowers solitary, axillary; fruiting calyx winged.. 1. SARCOBATUS (p. 199).
- Herbs, at most suffrutescent, the stems 150 cm. high or less; flowers perfect; fruiting calyx winged or naked.
  - Fruiting calyx winged; leaves spiny; plants becoming tumbleweeds...................... 2. SALSOLA (p. 199).
  - Fruiting calyx not winged; leaves fleshy; not tumbleweeds.............................. 3. DONDIA (p. 200).

Embryo annular; leaves mostly flat and broad (linear in no. 7, scalelike in no. 4).
- Stems and branches jointed (younger parts terete and and very succulent); leaves scalelike.......... 4. ALLENROLFEA (p. 201).

Stems not jointed; leaves never scalelike, mostly broad and flat.
Flowers monœcious or diœcious.
Pericarp and plant densely hairy........... 5. EUROTIA (p. 201).
Pericarp not hairy; plant more or less scurfy.. 6. ATRIPLEX (p. 201).
Flowers perfect.
Fruit dorsally flattened (narrowly winged), exserted from the calyx................ 7. CORISPERMUM (p. 205).
Fruit not dorsally flattened, inclosed in the calyx.
Fruiting calyx transversely winged.
Flowers paniculate; leaves broad, flat, toothed...................... 8. CYCLOLOMA (p. 206).
Flowers axillary; leaves terete.......12. KOCHIA (p. 209).
Fruiting calyx not winged.
Sepals and stamens each 1.......... 9. MONOLEPIS (p. 206).
Sepals 3 to 5; stamens 1 to 5.
Fruiting calyx fleshy, reddish; plants glabrous...........10. BLITUM (p. 206).
Fruiting calyx herbaceous, greenish; plants mostly mealy or scurfy................11. CHENOPODIUM (p. 206).

1. SARCOBATUS Nees. GREASEWOOD.

A divaricately branched shrub with linear fleshy leaves; staminate flowers naked, in aments; pistillate flowers with a saccate calyx adherent at the 2-lipped apex to the base of the stigmas; calyx laterally margined with an erect 2-lobed border, this finally becoming a broad membranous wing.

1. **Sarcobatus vermiculatus** (Hook.) Torr. in Emory, Mil. Reconn. 149. 1848.

Batis ? vermiculatus Hook. Fl. Bor. Amer. 2: 128. 1838.

Fremontia vermicularis Torr. in Frém. Rep. Exped. Rocky Mount. 95. *pl. 3*. 1845.

TYPE LOCALITY: "Common on the barren grounds of the Columbia, and particularly near salt marshes."

RANGE: Washington to Montana, Arizona, and New Mexico.

NEW MEXICO: Carrizo Mountains; San Juan Valley; Gallup; Zuni; Puertecito; Patterson. Alkaline soil, in the Upper Sonoran Zone.

This shrub grows to be 2 to 3 meters high, though the commoner form is lower, probably as a result of browsing. The leaves are bright green, terete, and succulent; the young branches are pale yellowish white and rigidly divaricate, the shorter branchlets thornlike. Also known as "chico" or "chico bush."

2. SALSOLA L.

Annual herb, densely branched, with rigid awl-shaped leaves; flowers perfect, with 2 bractlets; calyx 5-parted, the segments finally horizontally winged on the back; stamens usually 5; styles 2; flowers sessile, axillary.

1. **Salsola pestifer** A. Nels. in Coulter, New Man. Rocky Mount. 169. 1909.
RUSSIAN THISTLE.

Salsola tragus of American authors, not L.

TYPE LOCALITY: Not stated.

RANGE: Widely introduced as a weed in North America; a native of the Old World.

NEW MEXICO: Common at lower altitudes throughout the State.

One of the commonest introduced weeds on waste lands, along roadsides, and to some extent in fields on the open range. In some places it covers cultivated fields so closely as to appear like a sown crop. It was first noticed about fifteen years ago at Lamy by Prof. T. D. A. Cockerell, and was called to the attention of one of us at that time. It is now to be found in practically every locality in the State except in the higher mountains. The common name is misleading, since the plant resembles a thistle in no way except in being spiny. The plants when dry break off at the ground and are blown about by the wind as tumbleweeds.[1]

3. DONDIA Adans. QUELITE SALADO.

Succulent, more or less clammy herbs or suffrutescent plants with inconspicuous flowers and fruits; leaves terete, alternate; flowers sessile in the axils of leaflike bracts; calyx 5-parted, inclosing the fruit; stamens 5; seed vertical or horizontal.

KEY TO THE SPECIES.

Annuals; one or 2 of the sepals keeled; leaves broadest at the base.
  Plant depressed, spreading .... 1. *D. depressa.*
  Plant erect .... 2. *D. erecta.*
Perennials; none of the sepals keeled; leaves narrowed at the base.
  Stems and leaves pubescent; leaves broad, short, stout; plants usually woody at the base .... 3. *D. suffrutescens.*
  Stems and leaves glabrous; leaves narrow, long, slender; plants not woody at the base .... 4. *D. moquini.*

1. Dondia depressa (Pursh) Britton in Britt. & Brown, Illustr. Fl. **1:** 585. 1896.
Salsola depressa Pursh, Fl. Amer. Sept. 197. 1814.
Suaeda depressa S. Wats. in King, Geol. Expl. 40th Par. **5:** 294. 1871.
TYPE LOCALITY: "On the volcanic plains of the Missouri."
RANGE: Montana and Saskatchewan to Missouri and northwestern New Mexico.
NEW MEXICO: Farmington (*Standley* 6896). Alkaline soil, in the Upper Sonoran Zone.

2. Dondia erecta (S. Wats.) A. Nels. Bot. Gaz. **34:** 364. 1902.
Suaeda depressa erecta S. Wats. Proc. Amer. Acad. **9:** 90. 1874.
Suaeda erecta A. Nels. in Coulter, New Man. Rocky Mount. 169. 1909.
TYPE LOCALITY: Kern County, southern California.
RANGE: British America to California and New Mexico.
NEW MEXICO: Farmington; Shiprock; Albuquerque; south of Roswell. Alkaline soil, in the Upper Sonoran Zone.

3. Dondia suffrutescens (S. Wats.) Heller, Cat. N. Amer. Pl. 3. 1898.
Suaeda suffrutescens S. Wats. Proc. Amer. Acad. **9:** 88. 1874.
TYPE LOCALITY: "From Western Texas to Southern California and Northern Mexico, in saline plains."
RANGE: As under type locality.
NEW MEXICO: White Mountains; Mesilla Valley; Tularosa; White Sands. Lower and Upper Sonoran Zones.

This is frequently called "yerba de burro" by the Mexican laborers in the southern part of the State, but this probably results from a confusion of the plant with the true burro weed (*Allenrolfea occidentalis*), which is much less common in the region.

4. Dondia moquini (Torr.) A. Nels. Bot. Gaz. **34:** 363. 1902.
Chenopodina moquini Torr. U. S. Rep. Expl. Miss. Pacif. 7^3: 18. 1856.
Suaeda torreyana S. Wats, Proc. Amer. Acad. **9:** 88. 1874.

---

[1] See also, Wooton, E. O. The Russian Thistle. N. Mex. Agr. Exp. Sta. Bull. **15**. 1895.

Suaeda moquini A. Nels. in Coulter, New Man. Rocky Mount. 170. 1909.

TYPE LOCALITY: Mountain on the west shore of the Salt Lake, Utah.

RANGE: Wyoming and Colorado to California and New Mexico.

NEW MEXICO: Carrizo Mountains; Llano Estacado; Farmington; Alamogordo; White Sands; south of Roswell; Las Mitas. Alkaline soil, in the Lower and Upper Sonoran Zones.

4. ALLENROLFEA Kuntze. BURRO WEED.

Succulent erect much-branched perennial, somewhat woody at the base; leaves scalelike, broadly triangular; flowers in dense spikes, in threes in the axils of the spirally ranked bracts.

1. **Allenrolfea occidentalis** (S. Wats.) Kuntze, Rev. Gen. Pl. **2:** 546. 1891.

Halostachys occidentalis S. Wats. in King, Geol. Expl. 40th Par. **5:** 293. 1871.

Spirostachys occidentalis S. Wats. Proc. Amer. Acad. **9:** 125. 1874.

TYPE LOCALITY: "About Great Salt Lake and in alkaline valleys westward to the sinks of the Carson and Humboldt Rivers."

RANGE: Utah and Nevada to Arizona and western Texas.

NEW MEXICO: Socorro; Mesilla Valley; White Sands; above Tularosa; Roswell. Alkaline soil, in the Lower Sonoran Zone.

A very peculiar, almost leafless, shrubby halophyte. The young branches are terete, of a pronounced green (though sometimes glaucous), and very succulent. It grows to a height of about 150 cm. and is usually conspicuous for color alone among the gray or brown plants with which it is commonly associated. It is sparingly eaten by burros, hence the common name.

5. EUROTIA Adans. WINTER FAT.

A low, stellately tomentose shrub; leaves alternate, entire, linear to narrowly linear-lanceolate; flowers small, clustered, axillary and subspicate; calyx 4-parted; stamens 4.

1. **Eurotia lanata** (Pursh) Moq. Chenop. 81. 1840.

Diotis lanata Pursh, Fl. Amer. Sept. 602. 1814.

TYPE LOCALITY: "On the banks of the Missouri, in open prairies."

RANGE: New Mexico and Arizona, northward to Oregon and Manitoba.

NEW MEXICO: Common nearly throughout the State. Dry hills and plains, in the Upper Sonoran Zone.

The plant is highly prized by stockmen, particularly those who raise sheep, because it furnishes a good feed when other kinds are scarce—hence the common name.

6. ATRIPLEX L. SALT BUSH.

Monœcious or diœcious, mealy or scurfy annuals or perennials; staminate flowers bractless, variously clustered; pistillate flowers subtended by 2 persistent bracts. these becoming variously enlarged, thickened, and coalescent in fruit; leaves flat, alternate or opposite.

The genus contains several species which are of value as forage plants, most of them being eaten more or less. One (*A. expansa*) is a common tumbleweed in the cultivated lands in certain parts of the State. The plants occur mostly in open flats, preferring rather compact soils, and all of them will tolerate considerable alkali in the soil.

KEY TO THE SPECIES.

Annuals.

Leaves narrow, oblong to oblanceolate; plants mostly erect, the branches not widely spreading.

Plants low, 40 cm. high or less; fruiting bracts completely united, orbicular, flattened, the margins with numerous teeth; leaves small, 2 cm. long or less .... 1. *A. elegans.*

Plants tall, 70 to 100 cm. high; fruiting bracts united only at the base, triangular-ovate, few-toothed; leaves larger, 3 to 5 cm. long .... 2. *A. wrightii.*

Leaves ovate-lanceolate or triangular-ovate to subhastate; branches mostly spreading.

Bracts united only at the base .... 3. *A. hastata.*

Bracts united to about the middle.

Inflorescence tomentose as well as scurfy .... 4. *A. powellii.*

Inflorescence merely scurfy.

Leaves subcordate; scales loose and mealy .... 5. *A. saccaria.*

Leaves rounded or acute at the base; scales appressed, somewhat silvery.

Leaves coarsely repand-toothed nearly throughout .... 6. *A. rosea.*

Leaves toothed only near the base or entire (subhastate).

Leaves all except the lowest sessile .... 7. *A. expansa.*

Leaves all or nearly all conspicuously petioled.

Bracts naked or with a few short broad tubercles; some of the leaves truncate at the base. 8. *A. argentea.*

Bracts covered with long narrow tubercles; all leaves narrowed at the base .... 9. *A. caput-medusae.*

Perennials.

Stems prostrate, slender; flowers monœcious .... 10. *A. semibaccata.*

Stems erect, mostly stout; flowers mostly diœcious.

Fruiting bracts united, completely surrounding the seed, 10 mm. in diameter or more.

Bracts winged on the back and margin, forming a 4-winged pericarp, the margins shallowly if at all toothed .... 11. *A. canescens.*

Bracts becoming thick and spongy, with numerous rigid teeth .... 12. *A. acanthocarpa.*

Fruiting bracts united only at the base, 5 to 8 mm. in diameter.

Stems woody, widely branched, spinescent; leaves crowded.

Bracts rounded, entire; pistillate flowers stipitate; leaves rounded .... 13. *A. confertifolia.*

Bracts acutish, dentate; pistillate flowers sessile; leaves mostly elliptic .... 14. *A. collina.*

Stems mostly herbaceous, woody only at the base, not spinescent; leaves not crowded.

Fruiting bracts not crested (thick, not nerved); leaves 2 to 3 cm. long .... 15. *A. sabulosa.*

Fruiting bracts crested on the back; leaves small, 1 cm. long or less.

Bracts only slightly tuberculate; upper leaves very small; plants slender ... 16. *A. greggii.*

Bracts densely covered with long narrow appendages; all leaves large; plants stout .... 17. *A. cuneata.*

1. **Atriplex elegans** (Moq.) D. Dietr. Syn. Pl. **5**: 537. 1852.

Obione elegans Moq. in DC. Prodr. **13²**: 113. 1849.

TYPE LOCALITY: "In regni Mexicani Sonore alta."

RANGE: Western Texas to California, southward into Mexico.

NEW MEXICO: Mesilla Valley; Cienaga Ranch; near Duncan. Dry fields, in the Lower Sonoran Zone.

A common dooryard and wayside weed in the southern part of the State, readily recognized by the characteristic lenticular, many-toothed, small fruiting bracts, which are borne in great abundance.

2. **Atriplex wrightii** S. Wats. Proc. Amer. Acad. **9**: 113. 1874.

TYPE LOCALITY: "New Mexico and Arizona."

RANGE: Southern New Mexico, Arizona, California, and adjacent Mexico.

NEW MEXICO: Mule Creek; Gila; Mangas Springs. Sonoran zones.

A tall coarse annual weed of the southwestern part of the State, common in the Gila bottoms. The terminal paniculate spikes of staminate flowers, somewhat scanty fruit, and leaves glabrate above serve to separate it from our other allied New Mexican species.

3. **Atriplex hastata** L. Sp. Pl. 1053. 1753.

Atriplex carnosa A. Nels. Bot. Gaz. **34**: 361. 1902.

TYPE LOCALITY: European.

RANGE: Montana and Nebraska to Kansas and New Mexico and along the Atlantic coast; also in Europe and Asia.

NEW MEXICO: Farmington; Bloomfield; Dulce; Rio San Jose. Alkaline soil, in the Upper Sonoran Zone.

4. **Atriplex powellii** S. Wats. Proc. Amer. Acad. **9**: 114. 1874.

TYPE LOCALITY: Arizona.

RANGE: Utah and Colorado to New Mexico and Arizona.

NEW MEXICO: South of Gallup; Zuni; Aztec; Dulce. Plains, in the Upper Sonoran Zone.

A common plant of the open alkaline flats of the northwestern part of the State. The stems are rather slender and the inflorescence consists of crowded axillary glomerules subtended by reduced leaves at the ends of the stems. The leaves are from ovate to narrowly lanceolate, acute or short-acuminate, glabrate above, white-scurfy beneath, decurrent into a petiole almost as long as the blade, entire, with conspicuous veins.

5. **Atriplex saccaria** S. Wats. Proc. Amer. Acad. **9**: 112. 1874.

Atriplex cornuta Jones, Proc. Calif. Acad. II. **5**: 718. 1895.

Atriplex argentea cornuta Jones, Contr. West. Bot. **11**: 21. 1903.

TYPE LOCALITY: Desert plains of southern Wyoming or northern Utah.

RANGE: Wyoming, Colorado and Utah to northern Arizona and New Mexico.

NEW MEXICO: Carrizo Mountains (*Standley* 7359). Dry hills, in the Upper Sonoran Zone.

6. **Atriplex rosea** L. Sp. Pl. ed. 2. 1493. 1763.

Atriplex spatiosa A. Nels. Bot. Gaz. **34**: 360. 1902.

TYPE LOCALITY: European.

RANGE: Utah and Wyoming to New Mexico and Kansas.

NEW MEXICO: Chama; Farmington; Carrizo Mountains; Agricultural College. Dry hills and in river valleys, in the Lower and Upper Sonoran zones.

7. **Atriplex expansa** S. Wats. Proc. Amer. Acad. **9**: 116. 1874.

TYPE LOCALITY: "From New Mexico and Southern Colorado to Southern California."

RANGE: As under type locality.

NEW MEXICO: Mesilla Valley; Alamillo; Albuquerque. Lower Sonoran Zone.

A common tumbleweed in the southern part of the State. Without the fruit, which appears rather late in the season, it may be confused with certain species of Chenopodium. Herbarium specimens are easily confused with *A. argentea*, but that is a smaller, whiter plant, with prevailingly petioled leaves.

8. Atriplex argentea Nutt. Gen. Pl. **1:** 198. 1818.

TYPE LOCALITY: "On sterile and saline places near the Missouri."

RANGE: British Columbia and North Dakota to New Mexico.

NEW MEXICO: Carrizo Mountains; Farmington; Rio San Jose. Plains and valleys, in the Upper Sonoran Zone.

9. Atriplex caput-medusae Eastw. Proc. Calif. Acad. II. **6:** 316. 1896.

TYPE LOCALITY: Not stated, but probably in southeastern Utah along the San Juan River.

RANGE: Southern Utah and Colorado to northwestern New Mexico.

NEW MEXICO: San Juan Valley. Dry hills, in the Upper Sonoran Zone.

10. Atriplex semibaccata R. Br. Prodr. Fl. Nov. Holl. 406. 1810.

Atriplex flagellaris Woot. & Standl. Contr. U. S. Nat. Herb. **16:** 119. 1913.

TYPE LOCALITY: Tasmania.

RANGE: Australia and Tasmania; adventive from New Mexico to southern California.

NEW MEXICO: Mesilla Valley; Alamorgordo. Dry fields, in the Lower Sonoran Zone.

11. Atriplex canescens (Pursh) Nutt. Gen. Pl. **1:** 197. 1818. SHAD SCALE.

Calligonum canescens Pursh, Fl. Amer. Sept. 370. 1814.

Obione canescens angustifolia Torr. U. S. & Mex. Bound. Bot. 121. 1859.

Atriplex canescens angustifolia S. Wats. Proc. Amer. Acad. **9:** 120. 1874.

Atriplex angustior Cockerell, Proc. Davenport Acad. **9:** 7. 1902.

TYPE LOCALITY: "In the plains of the Missouri, near the Big-bend."

RANGE: North Dakota to Arizona and northern Mexico.

NEW MEXICO: Common nearly throughout the State. Dry plains, in the Lower and Upper Sonoran zones.

This is one of the commonest shrubs over the plains, in arroyos, and in the lower valleys of the Sonoran zones throughout the State. It is variously called "chamiso," "chamis," and "sagebrush," the last name being used mostly by newcomers who think that name applies to any grayish shrub. There are two forms common in the southern part of the State. One of them is a plant about 100 to 150 cm. high with short obovate or elliptic leaves, flowering generally in June and fruiting in late August and September. The other is a taller plant, frequently 2 meters high or more, with narrowly oblong-oblanceolate leaves, flowering and fruiting a month to 6 weeks earlier. The latter is the form described as *A. angustior* by Professor Cockerell. This seems also to be the form to which Dr. Rydberg applies the name *tetraptera*.[1] The same writer maintains *Atriplex occidentalis* as a distinct species, but the characters depended upon to separate it seem not to hold in New Mexican material.

The plant is of considerable value as a forage plant wherever it grows, being browsed extensively by cattle, sheep, and goats, particularly in winter and early spring when other forage is scarce. It tolerates large quantities of alkali, but also grows in soils where there is little or none, hence it is not always an indicator of alkali.

12. Atriplex acanthocarpa (Torr.) S. Wats. Proc. Amer. Acad. **9:** 117. 1874.

Obione acanthocarpa Torr. U. S. & Mex. Bound. Bot. 183. 1859.

TYPE LOCALITY: Plains between the Burro Mountains, New Mexico. **Type collected** by Bigelow.

---

[1] Bull. Torrey Club **39:** 311. 1912.

RANGE: Arizona to western Texas, southward into Mexico.

NEW MEXICO: Providencia Lake (*Wooton*). Dry plains, in the Lower Sonoran Zone.

A low shrub, 30 to 70 cm. high, with a peculiar burlike fruit and white-scurfy leaves and stems. It occurs in the alkaline flats of the extreme southern part of the State.

13. **Atriplex confertifolia** (Torr.) S. Wats. Proc. Amer. Acad. **9:** 119. 1874.

Obione confertifolia Torr. in Frém. Rep. Exped. Rocky Mount. 318. 1845.

Obione spinosa Moq. in DC. Prodr. **13²:** 108. 1849.

Atriplex spinosa D. Dietr. Syn. Pl. **5:** 536. 1852.

TYPE LOCALITY: Borders of the Great Salt Lake, Utah.

RANGE: Idaho and Wyoming to northern Mexico.

NEW MEXICO: Western McKinley and San Juan counties. Dry plains and lower hills, in the Upper Sonoran Zone.

A low shrub, seldom more than 50 cm. high, forming dense clumps. Many of the branches end in spinose points. The leaves and stems are dull whitish-scurfy. The leaf blades are broadly ovate to almost rotund and are thickly crowded on the branches.

14. **Atriplex collina** Woot. & Standl. Contr. U. S. Nat. Herb. **16:** 119. 1913.

TYPE LOCALITY: Dry hills near the north end of the Carrizo Mountains, northeast corner of Arizona. Type collected by Standley (no. 7481).

RANGE: Northwestern New Mexico, northeastern Arizona, and western Colorado.

NEW MEXICO: Carrizo Mountains. Dry plains, in the Upper Sonoran Zone.

15. **Atriplex sabulosa** Jones, Contr. West. Bot. **11:** 21. 1903.

TYPE LOCALITY: Winslow, Arizona.

RANGE: Southwestern Colorado to northern New Mexico and Arizona.

NEW MEXICO: Farmington; Tiznitzin; Gallup; near Horace; Carrizo Mountains; east of Deming. Dry plains and low hills, in the Upper Sonoran Zone.

A low suffrutescent plant with generally numerous stems 30 to 50 cm. high. The leaves are broadly obovate or oval, entire, short-petioled, 3 to 5 cm. long or less, and are, like the stems, thickly whitish-scurfy. The plant is browsed more or less by cattle, sheep, and goats. It is very common in alkaline spots in the northwestern part of the State.

16. **Atriplex greggii** S. Wats. Proc. Amer. Acad. **9:** 118. 1874.

TYPE LOCALITY: Cerros Bravos, Mexico.

RANGE: Western Texas and southern New Mexico and southward.

NEW MEXICO: Burro Mountains (*Mexican Boundary Survey* 1215a).

17. **Artiplex cuneata** A. Nels. Bot. Gaz. **34:** 357. 1902.

TYPE LOCALITY: Emery, Utah.

RANGE: Utah and Colorado to northern Arizona and northwestern New Mexico.

NEW MEXICO: Shiprock; Farmington. Dry hills, in the Upper Sonoran Zone.

A low shrub about 50 cm. high.

7. CORISPERMUM L. BUGSEED.

Widely spreading tumbleweeds with linear spinescent-tipped leaves, inconspicuous axillary flowers, and peculiar lenticular "buglike" fruit.

KEY TO THE SPECIES.

Spikes slender and lax; lower bracts narrower than the fruit....... 1. *C. nitidum.*
Spikes stout, thick, dense; all the bracts much wider than the fruit. 2. *C. marginale.*

1. **Corispermum nitidum** Kit.; Schult. Oesterr. Fl. ed. 2. **1:** 7. 1814.

Corispermum hyssopifolium microcarpum S. Wats. Proc. Amer. Acad. **9:** 123. 1874.

TYPE LOCALITY: Hungary.

RANGE: Illinois to North Dakota, New Mexico, and Texas.

NEW MEXICO: Farmington; Willard; Mesilla Valley; Carrizo Mountains. Dry fields, in the Lower and Upper Sonoran zones.

2. Corispermum marginale Rydb. Bull. Torrey Club **30:** 247. 1903.

TYPE LOCALITY: Albuquerque, New Mexico. Type collected by C. L. Herrick in 1894.

RANGE: Wyoming to New Mexico.

NEW MEXICO: Thornton; Grants; Albuquerque.

8. CYCLOLOMA Moq.

Erect annual with alternate, thin, oblanceolate or oblong, saliently toothed leaves; flowers very small, solitary, axillary, in open panicles; calyx 5-cleft, the lobes carinate, becoming closely appressed and developing a broad transverse membranous wing; pericarp pubescent; seed lenticular.

1. Cycloloma atriplicifolium (Spreng.) Coulter, Mem. Torrey Club **5:** 143. 1894.

Salsola atriplicifolia Spreng. Mant. Fl. Hal. **1:** 35. 1811.

Salsola platyphylla Michx. Fl. Bor. Amer. **1:** 174. 1803.

TYPE LOCALITY: North America.

RANGE: Ontario to Montana, Arkansas, and Arizona.

NEW MEXICO: Mesilla Valley; Nara Visa; Arch; Elida; Shiprock; Roswell. Sandy soil, in the Lower and Upper Sonoran zones.

9. MONOLEPIS Schrad.

A low spreading annual with petioled, lanceolate, hastately lobed leaves; flowers in axillary clusters; calyx of a single persistent fleshy unappendaged sepal; pericarp persistent upon the flattened seed.

1. Monolepis nuttalliana (Roem. & Schult.) Engelm. Trans. Amer. Phil. Soc. n. ser. **12:** 206. 1861.

Blitum chenopodioides Nutt. Gen. Pl. **1:** 4. 1818, not Lam. 1783.

Blitum nuttallianum Schult. Mant. **1:** 65. 1822.

Monolepis chenopodioides Moq. in DC. Prodr. **13**^{2}: 85. 1849.

TYPE LOCALITY: "On arid soils near the banks of the Missouri."

RANGE: Washington and Minnesota to California and Texas.

NEW MEXICO: El Rito Creek; Carrizo Mountains; Tunitcha Mountains; Chama; Tierra Amarilla. Open slopes, in the Upper Sonoran and Transition zones.

10. BLITUM L. BLITE.

A glabrous light green annual; leaves alternate, hastate, petioled; flowers small, crowded in axillary capitate clusters; calyx fleshy in fruit, becoming bright red at maturity.

1. Blitum capitatum L. Sp. Pl. 4. 1753.

Chenopodium capitatum S. Wats. Bot. Calif. **2:** 48. 1880.

TYPE LOCALITY: "Habitat in Europa: praesertim in comit. Tyrolensi."

RANGE: British America to New Jersey, New Mexico, and California; also in the Old World.

NEW MEXICO: Tunitcha Mountains; Santa Fe and Las Vegas mountains; Sandia Mountains; Mogollon Mountains; White and Sacramento mountains. Damp woods, in the Transition Zone.

11. CHENOPODIUM L. GOOSEFOOT. LAMB'S QUARTERS.

Annual herbs with alternate, often fleshy, petioled leaves; flowers small, green, sessile, in axillary, terminal, or panicled spikes; lobes of the perianth usually keeled or crested; stamens 5; pericarp membranous, closely investing the lenticular seed.

Seeds of the different species of Chenopodium were formerly collected by the Indians, ground or parched, and used in making cakes or porridge.

KEY TO THE SPECIES.

Leaves glandular, sweet-scented, pinnately lobed; embryo not forming a complete ring.
- Lobes of the leaves rounded or broadly oblong, more or less toothed........ 1. *C. botrys.*
- Lobes of the leaves lanceolate, entire........ 2. *C. cornutum.*

Leaves never glandular or sweet-scented, sinuately lobed to entire; embryo forming a complete ring.
- Leaves large, 7 to 20 cm. long, with large divergent acute lobes; seeds 2 mm. in diameter; leaves bright green.. 3. *C. hybridum.*
- Leaves smaller, less than 7 cm. long, with small, obtuse or acute, never divergent lobes, or entire; seeds 1.5 mm. or less in diameter; leaves pale or bright green.
  - Calyx lobes not carinate; inflorescence mostly axillary, shorter than the leaves; stems usually prostrate; leaves elliptic-lanceolate, coarsely toothed, conspicuously whitish beneath........ 4. *C. glaucum.*
  - Calyx lobes carinate; at least the upper panicles exceeding the leaves; stems never prostrate; leaves various.
    - Plants with a very strong unpleasant odor when crushed; leaves triangular-hastate, like the stems densely mealy........ 5. *C. watsoni.*
    - Plants with no strong nor disagreeable odor, odorless or nearly so; leaves various.
      - Leaves entire, linear to oblong.
        - Leaves linear to narrowly oblong, acute... 6. *C. leptophyllum.*
        - Leaves oblong or broadly so, obtuse....... 7. *C. oblongifolium.*
      - Leaves more or less toothed, hastate, ovate or triangular-ovate.
        - Plants densely mealy, appearing whitish.
          - Plants tall and stout, little branched below, erect; leaves conspicuously sinuate-toothed.......... 8. *C. album.*
          - Plants low, 30 cm. high or less, densely and diffusely branched, the branches spreading; leaves small, entire except for the hastate lobes at the base........... 9. *C. incanum.*
        - Plants only very slightly mealy, bright green.
          - Plants stout, erect; leaves thick, conspicuously sinuate-toothed.....10. *C. paganum.*
          - Plants lower, weak, often somewhat decumbent; leaves thin, entire except for the pair of hastate lobes at the base ....11. *C. fremontii.*

1. **Chenopodium botrys** L. Sp. Pl. 219. 1753. JERUSALEM OAK.

TYPE LOCALITY: "Habitat in Europae australis arenosis."

RANGE: A native of the Old World, introduced in many parts of North America.

NEW MEXICO: Upper Rio Tesuque; Cedar Hill.

2. **Chenopodium cornutum** (Torr.) Benth. & Hook. Gen. Pl. **3:** 51. 1880.

Teloxys cornuta Torr. U. S. Rep. Expl. Miss. Pacif. **4:** 129. 1856.

TYPE LOCALITY: Rocky places, Hurrah Creek, New Mexico. Type collected by Bigelow.

RANGE: Colorado, New Mexico, and Arizona to Mexico.

NEW MEXICO: Tunitcha Mountains; Santa Fe and Las Vegas mountains; Gallinas Mountains; Taos; Chloride; Mogollon Mountains; Santa Rita; San Luis Mountains; Organ Mountains; White and Capitan mountains. Open slopes, in the Upper Sonoran and Transition zones.

3. **Chenopodium hybridum** L. Sp. Pl. 219. 1753. MAPLE-LEAVED GOOSEFOOT.

TYPE LOCALITY: "Habitat in Europae cultis."

RANGE: British America southward; also in the Old World.

NEW MEXICO: Cross L Ranch; Sandia Mountains. Damp woods.

4. **Chenopodium glaucum** L. Sp. Pl. 220. 1753. OAK-LEAVED GOOSEFOOT.

TYPE LOCALITY: "Habitat in Europae fimeta."

RANGE: Widely introduced into North America as a weed, in many places apparently native.

NEW MEXICO: Farmington; Shiprock; Ojo Caliente; mountains southeast of Patterson; Mesilla Valley. Alkaline soil, in the Upper Sonoran Zone.

5. **Chenopodium watsoni** A. Nels. Bot. Gaz. **34:** 362. 1902.

Chenopodium olidum S. Wats. Proc. Amer. Acad. **9:** 95. 1874, not Curtis, 1787.

TYPE LOCALITY: "Colorado to Salt Lake Valley and southward into New Mexico and Arizona."

RANGE: As under type locality.

NEW MEXICO: Mule Creek; Mangas Springs; Mesilla Valley; Dulce. Lower and Upper Sonoran zones.

6. **Chenopodium leptophyllum** (Moq.) Nutt.; S. Wats. Proc. Amer. Acad. **9:** 94. 1874.

Chenopodium album leptophyllum Moq. in DC. Prodr. **13**[2]: 71. 1849.

TYPE LOCALITY: "In Nova California (Nuttall); Laplatte, Gordon."

RANGE: Washington and Saskatchewan to Missouri and Arizona.

NEW MEXICO: Carrizo and Tunitcha mountains; San Juan Valley; Zuni; Des Moines; Pecos; Patterson; Mangas Springs; near White Water; Mesilla Valley; Roswell. Dry hills and plains, in the Lower and Upper Sonoran zones.

7. **Chenopodium oblongifolium** (S. Wats.) Rydb. Bull. Torrey Club **33:** 137. 1906.

Chenopodium leptophyllum oblongifolium S. Wats. Proc. Amer. Acad. **9:** 95. 1874.

TYPE LOCALITY: "From Colorado to New Mexico."

RANGE: North Dakota and Wyoming to Missouri and Arizona.

NEW MEXICO: Wingfields Ranch; Mesilla Valley; No Agua. Lower Sonoran to Transition Zone.

This is too closely related to *C. leptophyllum*, differing chiefly in the broader, oblong, obtuse leaves. It is perhaps as distinct as most of the related species of Chenopodium.

8. **Chenopodium album** L. Sp. Pl. 219. 1753. LAMB'S QUARTERS.

TYPE LOCALITY: "Habitat in agris Europae."

RANGE: Widely introduced as a weed in North America.

NEW MEXICO: Cultivated and waste ground throughout the State. The young plants are gathered and cooked as greens. Among the native people they are known by the name of "quelite."

9. **Chenopodium incanum** (S. Wats.) Heller, Pl. World **1**: 23. 1897.

Chenopodium fremontii incanum S. Wats. Proc. Amer. Acad. **9**: 94. 1874.

TYPE LOCALITY: "Colorado and New Mexico."

RANGE: Colorado to New Mexico, Arizona, and western Texas.

NEW MEXICO: Farmington; Carrizo Mountains; Sierra Grande; Nara Visa; Santa Fe; Mule Creek; Silver City Draw; Mesilla Valley; White Mountains. Dry hills and plains, in the Lower and Upper Sonoran zones.

10. **Chenopodium paganum** Reichenb. Fl. Germ. 579. 1830.

Chenopodium viride of many authors, not L. 1753.

TYPE LOCALITY: Germany.

RANGE: Native of Europe, widely introduced into North America.

NEW MEXICO: Sandia Mountains; Harveys Upper Ranch; White and Sacramento mountains.

11. **Chenopodium fremontii** S. Wats. in King, Geol. Expl. 40th Par. **5**: 287. 1871.

TYPE LOCALITY: "Collected by Frémont on the North Platte."

RANGE: Montana and South Dakota to Arizona and northern Mexico.

NEW MEXICO: Farmington; Chama; Tunitcha Mountains; Carrizo Mountains; Glorieta; Santa Fe; West Fork of the Gila; Mineral Creek; Organ Mountains; Agricultural College; White and Sacramento mountains. Upper Sonoran and Transition zones.

12. KOCHIA Roth.

Low perennial, 20 cm. high or less, from a woody base; stems numerous, simple, erect; leaves terete, fleshy; flowers solitary or clustered in the axils; perianth tomentose, persistent, the lobes transversely winged; stamens 5, usually exserted; ovary tomentose; seed horizontal.

1. **Kochia americana** S. Wats. Proc. Amer. Acad. **9**: 93. 1874.

TYPE LOCALITY: "Foothills and valleys from northern Nevada to southern Wyoming and southward to Arizona and south Colorado."

RANGE: Wyoming and Colorado to California and northwestern New Mexico.

NEW MEXICO: Carrizo Mountains (*Standley* 7468). Dry hills, in the Upper Sonoran Zone.

Kochia scoparia Schrad., an annual species, has been cultivated at Albuquerque, and probably will be found escaped.

43. AMARANTHACEAE. Amaranth Family.

Herbaceous-stemmed, erect, diffuse, or prostrate annuals or perennials with alternate or opposite exstipulate leaves, and with apetalous flowers in crowded, axillary or terminal, bracted heads or simple or paniculately branched spikes; sepals scarious or herbaceous; stamens 5 or fewer (staminodia present in some), mostly hypogynous; pistil simple, the ovary mostly 1-seeded; fruit a utricle or pyxidium.

KEY TO THE GENERA.

Anthers 4-celled; leaves alternate; plants mostly glabrous, never conspicuously white-hairy.
- Perianth present in all flowers; bracts not much enlarged and not cordate in fruit............ 1. AMARANTHUS (p. 210).
- Perianth wanting in pistillate flowers; floral bracts much enlarged and broadly cordate in fruit.. 2. ACANTHOCHITON (p. 213).

Anthers 2-celled; leaves opposite (mainly basal in Gossypianthus); at least the younger parts of the plant white-hairy.
  Pubescence of stellate hairs.......................... 3. CLADOTHRIX (p. 213).
  Pubescence not stellate.
    Stems erect; fruiting calyx with spiny crests... 4. FROELICHIA (p. 214).
    Stems decumbent or prostrate (except in one species of Gomphrena); fruiting calyx not with spiny crests.
      Stamens perigynous.......................... 8. BRAYULINEA (p. 216).
      Stamens hypogynous.
        Stems erect or decumbent; heads of flowers pedunculate or terminal, large, 2 to 3 cm. in diameter.... 5. GOMPHRENA (p. 214).
        Stems prostrate; flower clusters sessile, axillary, small, 1 cm. in diameter or less.
          Staminodia wanting; cauline leaves elliptic-lanceolate, 15 mm. long or less, appearing as bracts subtending the flower clusters............. 6. GOSSYPIANTHUS (p. 215).
          Staminodia present (shorter than the filaments); cauline leaves obovate-spatulate, petiolate, 15 to 30 mm. long, not bractiform....... 7. ALTERNANTHERA (p. 215).

1. AMARANTHUS L. AMARANTH. PIGWEED.

Rather coarse annuals; flowers monœcious or diœcious, with 5 or fewer hyaline or greenish scalelike sepals, subtended by spine-tipped greenish or reddish bracts, borne in axillary clusters or in terminal paniculately branched spikes; petals wanting; stamens 5 or fewer; pistil solitary, with 2 or 3 styles; fruit a single-seeded indehiscent or circumscissile pyxis.

The leaves and seeds of the amaranths were formerly used for food by the Indians.

KEY TO THE SPECIES.

Flowers diœcious (in long terminal paniculate spikes).
  Bracts of the pistillate flowers much longer than the flowers, aristate-pungent.................................. 1. *A. palmeri.*
  Bracts of the pistillate flowers of the same length as the flowers, not pungent.. 2. *A. torreyi.*
Flowers monœcious, in short, clustered, terminal or axillary spikes.
  Sepals usually fewer than 5; plants low, usually spreading or ascending; inflorescence mostly of axillary clusters.
    Sepals 4 or 5; plants prostrate; seeds about 1.6 mm. in diameter..11. *A. blitoides.*
    Sepals 3; plants erect or spreading; seeds 0.8 mm. in diameter.
      Plants glabrous or glabrate; leaves plane............12. *A. graecizans.*
      Plants viscid-pubescent; leaves crispate............13. *A. pubescens.*

Sepals 5; plants tall, stout, erect; inflorescence mostly of dense terminal spikes.
- Sepals contracted near the base .... 3. *A. pringlei.*
- Sepals not contracted at the base.
  - Stamens uniformly 3.
    - Seed obovate .... 4. *A. obovatus.*
    - Seed orbicular.
      - Sepals obtuse, purplish, firm; spikes leafy, interrupted .... 5. *A. wrightii.*
      - Sepals acute, whitish, scarious; spikes naked, dense .... 6. *A. powellii.*
  - Stamens 5, or rarely fewer.
    - Plants densely viscid; bracts about 3 times as long as the sepals .... 7. *A. bracteosus.*
    - Plants not viscid; bracts twice as long as the sepals or shorter.
      - Sepals obtuse; spikes stout, erect .... 8. *A. retroflexus.*
      - Sepals acute; spikes slender, drooping.
        - Inflorescence green .... 9. *A. hybridus.*
        - Inflorescence red .... 10. *A. paniculatus.*

1. **Amaranthus palmeri** S. Wats. Proc. Amer. Acad. **12:** 274. 1877.

TYPE LOCALITY: Larkins Station, San Diego County, California.

RANGE: Western Texas to the Pacific coast, southward into Mexico.

NEW MEXICO: Common from the Mogollon Mountains and Socorro to the White Mountains and Pecos Valley and southward across the State. Lower and Upper Sonoran zones.

A tall, coarse native weed, usually 50 to 100 cm. high, occasionally reaching 250 cm., common in fields, on ditch banks, and along roadsides. The staminate plants are usually rather slender, and the terminal spike is frequently weak and drooping, sometimes considerably elongated. The pistillate plants are usually branched near the base and sometimes spread considerably, while the spikes are very dense and elongated, ultimately becoming very spiny from the fruiting bracts.

In some localities the plant is considered valuable as stock feed and has been cut and cured for hay when at the right stage of growth. It is said to cause bloating in cattle when eaten in too great abundance while the plants are young and succulent.

2. **Amaranthus torreyi** (A. Gray) Benth.; S. Wats. Bot. Calif. **2:** 42. 1880.

Amblogyne torreyi A. Gray, Proc. Amer. Acad. **5:** 167. 1861.

TYPE LOCALITY: Western Texas.

RANGE: Nebraska to Nevada, southward into New Mexico, western Texas, and Mexico.

NEW MEXICO: Arroyo Ranch (*Griffiths* 5702). Upper Sonoran Zone.

The species is probably more or less common along the eastern tier of counties of the State, but we know of only the single collection cited above.

3. **Amaranthus pringlei** S. Wats. Proc. Amer. Acad. **21:** 476. 1886.

TYPE LOCALITY: On rocky hills near Chihuahua, Mexico.

RANGE: Western Texas to Nevada, southward into Mexico.

NEW MEXICO: Mangas Springs; Berendo Creek; Mineral Creek; Organ Mountains; Dog Spring. Upper Sonoran Zone.

This is much rarer than the other species of the genus, occurring mostly in the foothills and lower mountains and not appearing as a weed.

4. **Amaranthus obovatus** S. Wats. Proc. Amer. Acad. **12:** 275. 1877.

TYPE LOCALITY: Copper Mines, New Mexico. Type collected by Wright (no. 1748, in part).

RANGE: Southwestern New Mexico to southern California.

NEW MEXICO: Santa Rita.

Easily recognized by the obovate seed.

5. **Amaranthus wrightii** S. Wats. Proc. Amer. Acad. **12**: 275. 1877.

TYPE LOCALITY: Copper Mines, New Mexico. Type collected by Wright (no. 1748, in part).

RANGE: Western Texas to southern Arizona.

NEW MEXICO: Santa Rita; Mineral Creek.

6. **Amaranthus powellii** S. Wats. Proc. Amer. Acad. **10**: 347. 1875.

TYPE LOCALITY: Arizona.

RANGE: Western Texas, Colorado, New Mexico, Arizona, and northern Mexico.

NEW MEXICO: Tunitcha Mountains; Winsors Ranch; Rio Alamosa; Chama; Pecos; Mule Creek; Trujillo Creek; Fort Bayard; Organ Mountains; San Luis Mountains; White and Sacramento mountains. Lower and Upper Sonoran zones.

A common weed in waste and cultivated ground.

7. **Amaranthus bracteosus** Uline & Bray, Bot. Gaz. **19**: 314. 1894.

Amaranthus viscidulus Greene, Pittonia **3**: 344. 1898.

TYPE LOCALITY: "New Mexico," probably about Santa Fe. Type collected by Fendler (no. 735).

RANGE: New Mexico.

NEW MEXICO: Pecos; White Mountains; Silver City Draw. Open hills, in the Upper Sonoran and Transition zones.

The type of *A. viscidulus* was collected in the White Mountains by Wooton (no. 300).

8. **Amaranthus retroflexus** L. Sp. Pl. 991. 1753.

TYPE LOCALITY: "Habitat in Pennsylvania."

RANGE: Widely scattered in fields and waste land in North America.

NEW MEXICO: Waste and cultivated ground nearly throughout the State.

9. **Amaranthus hybridus** L. Sp. Pl. 990. 1753.

TYPE LOCALITY: "Habitat in Virginia."

RANGE: Waste ground nearly throughout North America except in the extreme north.

NEW MEXICO: San Juan Valley.

10. **Amaranthus paniculatus** L. Sp. Pl. ed. 2. 1406. 1763.

Amaranthus hybridus paniculatus Uline & Bray, Mem. Torrey Club **5**: 145. 1894.

TYPE LOCALITY: "Habitat in America."

RANGE: Temperate and tropical North America.

NEW MEXICO: Shiprock; Zuni; Mesilla Valley. Chiefly a weed in cultivated or waste ground.

This is easily distinguished from all our other species by the reddish color of the leaves and inflorescence. With us it seems to have escaped from cultivation and it is becoming naturalized in various places.

11. **Amaranthus blitoides** S. Wats. Proc. Amer. Acad. **12**: 273. 1877.

TYPE LOCALITY: "Frequent in the valleys and plains of the interior, from Mexico to N. Nevada and Iowa, and becoming introduced in some of the Northern States eastward."

RANGE: New York to Montana, Louisiana, and California, introduced eastward.

NEW MEXICO: Common in dry fields and waste ground, in the Lower and Upper Sonoran zones.

A low spreading weed, forming thick circular mats on waste ground, also occurring in gardens and fields.

12. Amaranthus graecizans L. Sp. Pl. 990. 1753. TUMBLEWEED.

Amaranthus albus L. Syst. Nat. ed. 10. **2:** 1268. 1759.

TYPE LOCALITY: "Habitat in Virginia."

RANGE: A common weed in temperate and subtropical North America.

NEW MEXICO: Common throughout the State.

A widely spreading tumbleweed, common in the drier and warmer cultivated parts of the State. Young plants are leafy and rather succulent, but in age the stems become rigid, yellowish, and covered with the very numerous spiny fruiting bracts and later, scalelike leaves which are also spiny-tipped. The dead plants form part of the pile of tumbleweeds commonly seen along the fences, where they are associated with the Russian thistle and bugseed.

13. Amaranthus pubescens (Uline & Bray) Rydb. Bull. Torrey Club **39:** 313. 1912.

Amaranthus graecizans pubescens Uline & Bray, Bot. Gaz. **19:** 317. 1894.

TYPE LOCALITY: Silver City, New Mexico. Type collected by Greene in 1880 (no. 185).

RANGE: New Mexico and Arizona.

NEW MEXICO: South of Santa Fe; Mule Creek; Cliff; Fort Bayard; chalk hills near Parkers Well. Open hills, in the Upper Sonoran Zone.

This seems to be one of the most distinct forms of the genus and should certainly receive specific rank. It differs decidedly from *A. graecizans* in its lower habit, its pubescence, and the crisped leaves.

2. ACANTHOCHITON Torr.

An erect branching annual, with glabrous, green and white striped stems and alternate lanceolate aristate-tipped leaves; flowers diœcious or sometimes monœcious, the staminate with 5 sepals but bractless, the pistillate without sepals, subtended by a cordate clasping scale, this accrescent and spiny in fruit.

1. Acanthochiton wrightii Torr. in Sitgreaves, Rep. Zuñi & Colo. 170. 1853.

TYPE LOCALITY: Western Texas.

RANGE: Western Texas to Arizona.

NEW MEXICO: Shiprock; Chama River; San Marcial; Deming; Mesilla Valley; Organ Mountains; Jarilla Junction; Sabinal. Sandhills, in the Lower and Upper Sonoran zones.

The plant might easily be taken for an Amaranthus when in flower, but the fruiting plant is very strongly marked and easily recognized by the enlarged bracts. It is a common garden and roadside weed on sandy soils in the southern part of the State.

3. CLADOTHRIX Nutt.

Diffusely spreading or ascending herbaceous annuals or perennials, densely covered with white stellate pubescence; leaves petiolate, ovate to obovate; flowers very small, yellow, in small axillary clusters.

KEY TO THE SPECIES.

Perennial; stems erect or ascending; leaves truncate or rounded at the base .... 1. *C. suffruticosa.*

Annual; stems prostrate; leaves attenuate at the base............ 2. *C. lanuginosa.*

1. Cladothrix suffruticosa (Torr.) Benth. & Hook.; S. Wats. Bot. Calif. **2:** 43. 1880.

Alternanthera ? suffruticosa Torr. U. S. & Mex. Bound. Bot. 181. 1859.

TYPE LOCALITY: Mountains near Frontera and between the Pecos and the Limpio, western Texas.

RANGE: Southern New Mexico and western Texas.

NEW MEXICO: Tortugas Mountain; Bishops Cap. Dry hills, in the Lower Sonoran Zone.

2. Cladothrix lanuginosa Nutt.; Moq. in DC. Prodr. **13**[2]**:** 360. 1849.

Achyranthes lanuginosa Nutt. Trans. Amer. Phil. Soc. n. ser. **5:** 166. 1837.

TYPE LOCALITY: "On the sand beaches of Great Salt River, Arkansas."

RANGE: Kansas and Texas to Arizona and Mexico.

NEW MEXICO: Common nearly throughout the State. Dry plains and fields, chiefly in the Lower Sonoran Zone.

4. FROELICHIA Moench.

Erect branching white-woolly plants, 30 to 100 cm. high, with opposite obovate, oblanceolate, or lanceolate leaves; flowers in terminal spikes; bracts of the inflorescence yellowish or blackish, glabrous; fruiting calyx variously winged and toothed, covered with long cottony hairs.

KEY TO THE SPECIES.

Calyx tube with a lateral crest of distinct spines at maturity, the faces tuberculate; plants low, slender, 20 to 50 cm. high...... 1. *F. gracilis.*

Calyx tube with lateral crests of toothed wings at maturity, the faces each with a spine at the base; plants taller, stout, 60 to 120 cm. high.. 2. *F. campestris.*

1. Froelichia gracilis Moq. in DC. Prodr. **13**[2]**:** 420. 1849.

TYPE LOCALITY: "In Texas."

RANGE: Nebraska to Arizona and Texas.

NEW MEXICO: Las Vegas; Bear Mountain; Mangas Springs; Kingston; Organ Mountains; Nara Visa; Hanover Mountains. Dry hills, in the Lower and Upper Sonoran zones.

2. Froelichia campestris Small, Fl. Southeast. U. S. 397. 1903.

TYPE LOCALITY: Oklahoma.

RANGE: Minnesota to Colorado, Arizona, and Texas.

NEW MEXICO: Hurrah Creek, Hillsboro; Santa Rita; San Luis Mountains; Tortugas Mountain; Organ Mountains. Dry hills, in the Lower and Upper Sonoran zones.

5. GOMPHRENA L. GLOBE AMARANTH.

Erect or prostrate, annual or perennial herbs, hirsute or villous; leaves sessile or short-petioled, entire; flowers in large, often petioled heads, with white or pinkish bracts; flowers perfect, the calyx 5-parted or 5-cleft, often villous below.

KEY TO THE SPECIES.

Annual; heads subtended by leaves; plants tall, 20 to 50 cm........ 1. *G. nitida.*

Perennials; heads not subtended by leaves; plants low, cespitose, less than 10 cm. high.

Plants densely pubescent, gray or whitish; peduncles short, scarcely if at all exceeding the nearly sessile cauline leaves.. 2. *G. caespitosa.*

Plants sparingly pubescent, green; peduncles long, much exceeding the petiolate cauline leaves..................... 3. *G. viridis.*

The common globe amaranth or bachelor's button (*Gomphrena globosa* L.) is often planted for ornament in the State. It is a larger plant than any of our native species and has larger heads of crimson to white flowers.

Torrey reported[1] *Gomphrena tuberifera* Torr. from New Mexico, but this is probably incorrect, since the species ranges much farther south in Texas.

1. **Gomphrena nitida** Rothr. in Wheeler, Rep. U. S. Surv. 100th Merid. **6:** 233. 1879.

TYPE LOCALITY: Chiricahua Mountains, southern Arizona.

RANGE: Arizona and southwestern New Mexico.

NEW MEXICO: Kingston; San Luis Mountains; Organ Mountains; Gila Hot Springs. Dry hills, in the Lower and Upper Sonoran zones.

A strict annual with erect opposite branches, mostly sessile leaves, and flowers in a terminal head subtended by a pair of bractiform leaves. The scales are frequently tinged with red, adding to the similarity to the cultivated bachelor's button.

2. **Gomphrena caespitosa** Torr. U. S. & Mex. Bound. Bot. 181. 1859.

TYPE LOCALITY: "Gravelly plains near the Organ Mountains, New Mexico; also at the Copper Mines and near Mimbres." Type collected by Bigelow.

RANGE: Southwestern New Mexico to southern Arizona.

NEW MEXICO: Kingston; Dog Spring; Organ Mountains; Water Canyon. Dry hills, in the Lower and Upper Sonoran zones.

This little plant usually first appears as a rosette of oblanceolate to obovate, white-hairy leaves, with ovoid heads of small yellow flowers subtended by silvery white bracts. Later the slender and decumbent stems with smaller opposite leaves appear, and the plant spreads to 10 to 20 cm. in diameter. In the type locality it is rarely larger than this, probably because it is eaten by stock.

3. **Gomphrena viridis** Woot. & Standl. Contr. U. S. Nat. Herb. **16:** 120. 1913.

TYPE LOCALITY: Hanover Mountain, Grant County, New Mexico. Type collected by J. M. Holzinger.

RANGE: Southwestern New Mexico and adjacent Arizona and Mexico.

NEW MEXICO: Hanover Mountain; San Luis Mountains. Dry hills, in the Upper Sonoran Zone.

6. GOSSYPIANTHUS Hook.

Perennial herb with procumbent woolly stems; radical leaves elongate-spatulate, subcoriaceous; cauline leaves ovate, sessile, silky-woolly; flowers perfect, with 2 or 3 deciduous bracts; calyx of 5 strongly pilose sepals; stamens 3.

1. **Gossypianthus lanuginosus** (Poir.) Moq. in DC. Prodr. **13²:** 337. 1849.

Paronychia lanuginosa Poir. in Lam. Encycl. Suppl. **4:** 303. 1816.

TYPE LOCALITY: San Domingo.

RANGE: Oklahoma and New Mexico to tropical America.

NEW MEXICO: Cabra Springs; between Anton Chico and Las Vegas.

7. ALTERNANTHERA Forsk.

Herb with prostrate stems; leaves oval or obovate, narrowed into a petiole; flowers in dense heads; bracts conspicuous, white; sepals 5; stamens 5.

1. **Alternanthera repens** (L.) Kuntze, Rev. Gen. Pl. **2:** 540. 1891.

Achyranthes repens L. Sp. Pl. 205. 1753.

TYPE LOCALITY: "Habitat in Turcomannia."

RANGE: South Carolina to California and South America, in waste ground; also in the Old World.

NEW MEXICO: Florida Mountains (*Jones*).

[1] U. S. & Mex. Bound. Bot. 181. 1859.

8. BRAYULINEA Small.

A prostrate lanate leafy herb from a perennial root, forming thick mats; leaves opposite, ovate, entire; flowers minute, perfect, axillary; calyx campanulate, with a 5-lobed limb; stamens 5; fruit an indehiscent utricle.

1. **Brayulinea densa** (Humb. & Bonpl.) Small, Fl. Southeast. U. S. 394. 1903.
Illecebrum densum Humb. & Bonpl.; Roem. & Schult. Syst. Veg. **5**: 517. 1819.
Guilleminea densa Moq. in DC. Prodr. **13**^{2}: 338. 1849.
Type locality: "In America Merid."
Range: Texas and New Mexico to tropical America.
New Mexico: Water Canyon; Mangas Springs; San Luis Mountains; Organ Mountains; Queen; Kingston; Santa Rita. Dry hills, in the Upper Sonoran Zone.

44. CORRIGIOLACEAE. Whitlow-wort Family.

1. PARONYCHIA Adans. Whitlow-wort.

Low herbaceous perennials, lignescent at the base; leaves often acerose, with conspicuous scarious stipules; flowers solitary or clustered, mostly apetalous; sepals and stamens 5, the former aristate; fruit a utricle.

KEY TO THE SPECIES.

Flowers solitary; elliptic leaves scarcely exceeding the bracts; plants densely pulvinate .... 1. *P. pulvinata.*
Flowers clustered; linear leaves much longer than the bracts; plants not pulvinate, 10 cm. high or more .... 2. *P. jamesii.*

1. **Paronychia pulvinata** A. Gray, Proc. Acad. Phila. **1863**: 58. 1864.
Type locality: Rocky Mountains of Colorado.
Range: Wyoming to Utah and New Mexico.
New Mexico: Truchas Peak; Wheeler Peak. Open slopes, in the Arctic-Alpine Zone.

2. **Paronychia jamesii** Torr. & Gray, Fl. N. Amer. **1**: 170. 1838.
Type locality: Rocky Mountains.
Range: Nebraska and Colorado to Texas and New Mexico.
New Mexico: Bear Mountain; Organ Mountains; west of Roswell; Knowles; Berendo Creek; south of Torrance; Buchanan; Nara Visa. Dry soil, in the Upper Sonoran Zone.

45. ALLIONIACEAE. Four-o'clock Family.

Annual or perennial herbs with usually dichotomous stems, the joints often swollen; leaves opposite or alternate, usually entire, exstipulate, petiolate or sessile, the opposite ones often very unequal; flowers regular, perfect or sometimes unisexual, mostly subtended by bracts forming a calyx-like involucre; perianth of only a calyx, this usually colored and corolla-like; stamens 1 to many; anthers 2-celled, opening by longitudinal fissures; ovary 1-celled, superior but surrounded by the calyx tube, sessile or short-stalked; stigma usually capitate; ovule solitary, erect; fruit an anthocarp, indehiscent, either fleshy, leathery, or hard, either angled, ribbed, grooved, or winged.

KEY TO THE GENERA.

Involucre gamophyllous, calyx-like.
  Involucre 3-parted; fruit lenticular, with lateral wings usually armed with teeth............ 1. WEDELIELLA (p. 218).
  Involucre 5-parted; fruit oblong, turbinate, or almost spherical, never with toothed wings.
    Fruit with 5 prominent ribs; involucre enlarged and membranous in fruit, with 1 to 3 flowers; perianth usually campanulate............................ 2. ALLIONIA (p. 219).
    Fruit smooth or slightly 5-angled; involucre not membranous, usually not enlarged in fruit.
      Involucre 1-flowered, campanulate; tube of the perianth long and slender.... 3. MIRABILIS (p. 221).
      Involucres with 3 or several flowers; perianth various.
        Involucre rotate, 3-flowered; perianth small, short-funnelform, almost campanulate.................. 4. ALLIONIELLA (p. 222).
        Involucre campanulate or tubular, with usually more than 3 flowers; perianth large, funnelform, with a long and conspicuous tube........................ 5. QUAMOCLIDION (p. 222).
Involucre polyphyllous, consisting of 1 to many small bracts, these often deciduous.
  Stigmas linear; inner cotyledon abortive.
    Wings completely encircling the fruit, hyaline, membranous; flowers mostly tetramerous.................................... 6. TRIPTEROCALYX (p. 222).
    Wings not completely encircling the fruit, thick and coriaceous or wanting; flowers pentamerous............................ 7. ABRONIA (p. 223).
  Stigmas spherical or hemispherical; cotyledons more or less unlike but neither abortive.
    Flowers in heads, these surrounded by a regular many-bracted involucre; stamens long-exserted; fruit turbinate, 10-ribbed. 8. NYCTAGINIA (p. 224).
    Flowers not in heads, each flower with an involucre of 1 to 3 bracts; fruit various.
      Fruit conspicuously winged.............. 9. SELINOCARPUS (p. 224).
      Fruit not winged.
        Fruit asymmetrical, gibbous; inflorescence racemose................10. CYPHOMERIS (p. 225).
        Fruit symmetrical, never gibbous; inflorescence various.
          Perianth 10 cm. long, with a long and very slender tube; flowers axillary...........11. ACLEISANTHES (p. 225).
          Perianth smaller, 2 cm. long or less, the tube much shorter and less conspicuous or wanting.

Fruit 10-ribbed; leaves thick and leathery, mostly basal; perianth funnel-form.................12. ANULOCAULIS (p. 226)
Fruit with 5 or fewer angles; leaves thin, scattered along the stems; perianth campanulate.....13. BOERHAAVIA (p. 226).

1. WEDELIELLA Cockerell.

Annual or perennial herbs with prostrate stems; leaves opposite, petiolate; involucre of 3 oval sepal-like bracts united at the very base, solitary on axillary peduncles; perianths purplish red, rarely white, oblique, 3 in each involucre; fruit leathery, winged on each side, smooth on the inner side, with 2 parallel rows of stipitate glands on the outer side, these often concealed by the incurved, entire or toothed wings.

KEY TO THE SPECIES.

Annual; stems only slightly pubescent, or glabrous; wings of the fruit not incurved, with numerous slender teeth; leaves oblong, glaucous beneath, obtuse, somewhat crispate.......... 1. *W. glabra.*
Perennials; stems strongly viscid; wings of the fruit incurved; leaves usually ovate, mostly acutish, not glaucous beneath, plane.
  Wings of the fruit toothed.................................. 2. *W. incarnata.*
  Wings of the fruit entire.................................. 2a. *W. incarnata anodonta.*

1. Wedeliella glabra (Choisy) Cockerell, Torreya **9**: 167. 1909.
Allionia incarnata glabra Choisy in DC. Prodr. **13**^{2}: 435. 1849.
Wedelia glabra Standley, Contr. U. S. Nat. Herb. **12**: 332. 1909.
TYPE LOCALITY: "Circa Mexicum."
RANGE: Arizona to western Texas, south into Mexico.
NEW MEXICO: Carrizo Mountains; Farmington; Zuni; Pecos; Santa Fe; Albuquerque; Mangas Springs; Mesilla Valley; Gray; south of Roswell. Sandy plains, in the Lower and Upper Sonoran zones.

2. Wedeliella incarnata (L.) Cockerell, Torreya **9**: 167. 1909.
Allionia incarnata L. Syst. Nat. ed. 10. **2**: 890. 1759.
Wedelia incarnata Kuntze, Rev. Gen. Pl. **2**: 533. 1891.
TYPE LOCALITY: "Juxta Cumana urbem, in silvis arenosis," Venezuela.
RANGE: Colorado to Arizona, western Texas, and Mexico; also in the West Indies and South America.
NEW MEXICO: Burro Mountains; Rincon; Carrizalillo Mountains; Organ Mountains; Mesilla Valley; Alamogordo; Lake Arthur; south of Roswell. Dry hills, in the Lower and Upper Sonoran zones.
A plant from Highrolls is an albino form. Analogous forms are known in other Allioniaceae.

2a. Wedeliella incarnata anodonta (Standley) Cockerell, Torreya **9**: 167. 1909.
Wedelia incarnata anodonta Standley, Contr. U. S. Nat. Herb. **12**: 333. 1909.
TYPE LOCALITY: Plains of western New Mexico. Type collected by Rusby (no. 355).
RANGE: New Mexico and Arizona.
NEW MEXICO: Shiprock; Albuquerque; Valverde. Lower and Upper Sonoran zones.

2. ALLIONIA L.

Perennial herbs, glabrous or pubescent; leaves opposite, often thick and fleshy, green or glaucous, petiolate or sessile; involucre gamophyllous, 5-lobed, enlarged in fruit; flowers 1 to 5 in each involucre; perianth white to crimson, short-funnelform or campanulate, oblique; stamens 2 to 5; fruit clavate, 5-angled or 5-ribbed.

KEY TO THE SPECIES.

Fruit deeply 5-lobed; flowers deep red, often cleistogamous; perianth deeply lobed or somewhat bilabiate.
- Plants tall, sparingly branched; involucres 3-fruited; flowers seldom cleistogamous; leaves linear ...... 1. *A. coccinea.*
- Plants low, diffusely branched; involucres 1-fruited; flowers usually cleistogamous; leaves filiform ...... 2. *A. linearifolia filifolia.*

Fruit 5-angled, not lobed; flowers whitish to purplish red, never deep red; perianth shallowly lobed, never bilabiate.
- Leaves narrowly oblong to ovate.
  - Stems hirsute throughout ...... 3. *A. hirsuta.*
  - Stems not hirsute.
    - All leaves sessile or nearly so ...... 4. *A. lanceolata.*
    - All leaves except the uppermost petiolate.
      - Leaves and stems pubescent throughout ...... 5. *A. comata.*
      - Leaves glabrous; stems pubescent only above ...... 6. *A. melanotricha.*
- Leaves linear or nearly so.
  - Plants glabrous throughout or with a few appressed hairs on the involucres and pedicels; fruit glabrous.
    - Peduncles and involucres glabrous ...... 7. *A. glabra.*
    - Peduncles and involucres sparingly hairy ...... 7a. *A. glabra recedens.*
  - Plants copiously pubescent, at least on the branches of the inflorescence; fruit pubescent.
    - Inflorescence axillary or of few-flowered clusters at the ends of the branches.
      - Lobes of the involucre obtuse; plants slender, erect, little branched ...... 8. *A. pinetorum.*
      - Lobes of the involucre acute; plants stout, spreading, much branched ...... 9. *A. bodini.*
    - Inflorescence paniculate or corymbose, well developed.
      - Stems hirsute throughout ...... 10. *A. subhispida.*
      - Stems glabrous below.
        - Stems very stout, simple or sparingly branched, erect; leaves very glaucous and fleshy, sessile ...... 11. *A. linearis.*
        - Stems much branched, slender, not stiffly erect; leaves not glaucous or but slightly so.
          - Leaves sessile, thick, pale green; stems low, spreading or ascending, diffusely branched; branches of the inflorescence densely viscid-hairy; perianth pinkish ...... 12. *A. diffusa.*
          - Leaves distinctly petioled, bright green, thin; branches of inflorescence merely viscid-puberulent; perianth bright purplish red ...... 13. *A. divaricata.*

1. **Allionia coccinea** (Torr.) Standley, Contr. U. S. Nat. Herb. **12**: 339. 1909.
Oxybaphus coccineus Torr. U. S. & Mex. Bound. Bot. 169. 1859.
Mirabilis coccinea Benth. & Hook. Gen. Pl. **3**: 3. 1880.
TYPE LOCALITY: "Hillsides, copper mines, and on the Mimbres," New Mexico. Type collected by Wright (no. 1723).
RANGE: Southern New Mexico and Arizona to Sonora.
NEW MEXICO: Kingston; Mogollon Creek; Silver City; Burro Mountains; Eagle Peak; Hanover Mountain. Dry hills and rocky canyons, in the Upper Sonoran Zone.

2. **Allionia linearifolia filifolia** Standley, Contr. U. S. Nat. Herb. **16**: 120. 1913.
Allionia gracillima filifolia Standley, Contr. U. S. Nat. Herb. **12**: 340. 1909.
TYPE LOCALITY: Mangas Springs, New Mexico. Type collected by Wooton, August 17, 1902.
RANGE: Known only from type locality.

3. **Allionia hirsuta** Pursh, Fl. Amer. Sept. 728. 1814.
Calymenia pilosa Nutt. Gen. Pl. **1**: 26. 1818.
Oxybaphus hirsutus Sweet, Hort. Brit. **1**: 334. 1825.
TYPE LOCALITY: "In upper Louisiana."
RANGE: Wyoming and Minnesota to New Mexico and Oklahoma.
NEW MEXICO: Raton Mountains; Colfax. Upper Sonoran Zone.

4. **Allionia lanceolata** Rydb. Bull. Torrey Club **29**: 691. 1902.
TYPE LOCALITY: Estes Park, Larimer County, Colorado.
RANGE: Wyoming and Missouri to New Mexico and Oklahoma.
NEW MEXICO: Johnsons Mesa (*Wooton*). Transition Zone.

5. **Allionia comata** Small, Fl. Southeast. U. S. 407. 1903.
Oxybaphus nyctagineus pilosus A. Gray in Torr. U. S. & Mex. Bound. Bot. 174. 1859, not *Allionia pilosa* Nutt. 1818.
TYPE LOCALITY: Probably New Mexico.
RANGE: New Mexico and Arizona.
NEW MEXICO: Silver City; Magdalena; near the Mimbres; East Canyon. Dry hills, in the Upper Sonoran Zone.

6. **Allionia melanotricha** Standley, Contr. U. S. Nat. Herb. **12**: 351. 1909.
Oxybaphus melanotrichus Weatherby, Proc. Amer. Acad. **45**: 425. 1910.
TYPE LOCALITY: Barfoot Park, Chiricahua Mountains, Arizona.
RANGE: New Mexico and Arizona, south into Mexico.
NEW MEXICO: Santa Fe and Las Vegas mountains; Rio Pueblo; Sandia Mountains; Chama; Mogollon Mountains; Black Range; San Luis Mountains; Organ Mountains; White and Sacramento mountains. Meadows in the mountains, in the Transition Zone.

7. **Allionia glabra** (S. Wats.) Kuntze, Rev. Gen. Pl. **2**: 533. 1891.
Oxybaphus glaber S. Wats. Amer. Nat. **7**: 301. 1873.
TYPE LOCALITY: Kanab, Utah.
RANGE: Southern Utah to Arizona and New Mexico.
NEW MEXICO: Cedar Hill; Nara Visa. Dry hills, in the Upper Sonoran Zone.

7a. **Allionia glabra recedens** (Weatherby) Standley, Contr. U. S. Nat. Herb. **13**: 406. 1911.
Oxybaphus glaber recedens Weatherby, Proc. Amer. Acad. **45**: 425. 1910.
TYPE LOCALITY: Chihuahua, between Casas Grandes and Sabinal.
RANGE: Western Texas and southern New Mexico to Chihuahua.
NEW MEXICO: Albuquerque: Brockmans Ranch; Mesilla Valley; Roswell. Lower Sonoran Zone.

8. **Allionia pinetorum** Standley, Contr. U. S. Nat. Herb. **12:** 344. 1909.

TYPE LOCALITY: Gilmores Ranch, on Eagle Creek, White Mountains, New Mexico. Type collected by Wooton and Standley (no. 3896).

RANGE: Known only from type locality, in the Transition Zone.

9. **Allionia bodini** (Holzinger) Morong, Mem. Torrey Club **5:** 354. 1894.

Oxybaphus bodini Holzinger, Contr. U. S. Nat. Herb. **1:** 287. 1893.

TYPE LOCALITY: Pueblo, Colorado.

RANGE: Wyoming and South Dakota to Arizona and Texas.

NEW MEXICO: Carrizo Mountains; Sierra Grande; Organ Mountains. Plains and low hills, in the Upper Sonoran Zone.

10. **Allionia subhispida** (Heimerl) Standley, Contr. U. S. Nat. Herb. **16:** 120. 1913.

Mirabilis linearis subhispida Heimerl, Ann. Cons. Jard. Genève **5:** 186. 1901.

Allionia linearis subhispida Standley, Contr. U. S. Nat. Herb. **12:** 342. 1909.

TYPE LOCALITY: Capitan Mountains, New Mexico. Type collected by Earle (no. 383).

RANGE: New Mexico.

NEW MEXICO: South of San Rafael; Atarque de Garcia; Magdalena; Capitan Mountains; Gray. Low hills, Upper Sonoran Zone.

11. **Allionia linearis** Pursh, Fl. Amer. Sept. 728. 1814.

Calymenia angustifolia Nutt. Gen. Pl. **1:** 26. 1818.

Oxybaphus angustifolius Sweet, Hort. Brit. **1:** 334. 1826.

Allionia montanensis Osterhout, Muhlenbergia **1:** 39. 1906.

TYPE LOCALITY: "In Upper Louisiana."

RANGE: Wyoming and Illinois to Texas and Mexico.

NEW MEXICO: Dulce; Sierra Grande; Raton; Farmington; Pecos; Zuni Reservation; Mangas Springs; Rio Frisco; Socorro; Gila Hot Springs; Colfax; Taos; Organ Mountains; Dog Spring; Capitan Mountains; White Mountains. Dry plains and low hills, in the Upper Sonoran Zone.

12. **Allionia diffusa** Heller, Minn. Bot. Stud. **2:** 33. 1898.

TYPE LOCALITY: Dry gravelly hills, 10 miles west of Santa Fe, New Mexico. Type collected by Heller (no. 3740).

RANGE: Wyoming to Arizona and western Texas.

NEW MEXICO: Carrizo Mountains; Stinking Lake; west of Santa Fe; Johnsons Mesa; Albuquerque; Sierra Grande; Roy; Mangas Springs; Kingston; Organ Mountains; Eagle Creek; near Carrizo. Dry plains and low hills, in the Lower and Upper Sonoran zones.

13. **Allionia divaricata** Rydb. Bull. Torrey Club **29:** 681. 1902.

TYPE LOCALITY: Durango, Colorado.

RANGE: Utah and Colorado to Mexico.

NEW MEXICO: Tunitcha Mountains; Cedar Hill; Pecos; Santa Fe Canyon; Chusca Mountains; Sandia Mountains; Chama; West Fork of the Gila. Meadows in the mountains, in the Upper Sonoran and Transition zones.

3. MIRABILIS L. FOUR-O'CLOCK.

Perennial dichotomous-branching herb with opposite, entire, deltoid, sessile or petiolate, thin leaves; flowers solitary in a calyx-like involucre; perianth white, with a very long slender tube and broad limb; stamens 5, unequal; fruit leathery, 5-angled.

1. **Mirabilis wrightiana** A. Gray; Britt. & Kearn. Trans. N. Y. Acad. **14:** 28. 1894.

TYPE LOCALITY: Not stated.

RANGE: Rocky canyons, western Texas to southern Arizona and southward.

NEW MEXICO: Mogollon Mountains and Black Range, south to the Mexican border. Upper Sonoran and Transition zones.

4. ALLIONIELLA Rydb.

Low, much branched herb with weak, prostrate or ascending, viscid branches; leaves opposite, petiolate; involucres loosely paniculate, rotate and enlarged at maturity, 5-lobed; perianths 3 in each involucre, short-funnelform; stamens 3; fruit ellipsoid, smooth or obscurely tuberculate, glabrous.

KEY TO THE SPECIES.

Stems pubescent throughout.. 1. *A. oxybaphoides.*
Stems glabrate below, slightly puberulent above.............. 1a. *A. oxybaphoides glabrata.*

1. **Allioniella oxybaphoides** (A. Gray) Rydb. Bull. Torrey Club **29:** 687. 1902.
Quamoclidion oxybaphoides A. Gray, Amer. Journ. Sci. II. **15:** 320. 1853.
Mirabilis oxybaphoides A. Gray in Torr. U. S. & Mex. Bound. Bot. 173. 1859.
TYPE LOCALITY: East of El Paso, Texas.
RANGE: Utah and Colorado to Arizona and western Texas.
NEW MEXICO: Near Pecos; Sierra Grande; Raton; Fort Wingate; Santa Fe Creek; Bear Mountain; Kingston; Organ Mountains; Gray. Dry hills, Upper Sonoran Zone.

1a. **Allioniella oxybaphoides glabrata** (Heimerl) Standley, Contr. U. S. Nat. Herb. **12:** 357. 1909.
Mirabilis oxybaphoides glabrata Heimerl, Ann. Cons. Jard. Genève **5:** 180. 1901.
TYPE LOCALITY: Capitan Mountains, New Mexico. Type collected by Earle (no. 399).
RANGE: With the type.
NEW MEXICO: Capitan Mountains; Gallinas Mountains.

5. QUAMOCLIDION Choisy.

Low, diffusely branched, perennial herb with glabrous petiolate ovate leaves; involucre gamophyllous, calyx-like; flowers large, purplish red, several in each involucre; perianth with a thick, rather long tube and a wide spreading limb; stamens 5, exserted; fruit oblong, smooth, glabrous.

1. **Quamoclidion multiflorum** Torr.; A. Gray, Amer. Journ. Sci. II. **15:** 321. 1853.
Oxybaphus multiflorus Torr. Ann. Lyc. N. Y. **2:** 237. 1828.
Mirabilis multiflora A. Gray in Torr. U. S. & Mex. Bound. Bot. 173. 1859.
TYPE LOCALITY: "About the forks of the Platte," Colorado.
RANGE: Colorado to Arizona and western Texas.
NEW MEXICO: Common throughout the State except in the extreme southwest and southeast. Plains and low hills, in the Lower and Upper Sonoran zones.

6. TRIPTEROCALYX Hook.

Much branched annuals with fleshy lanceolate unequal petiolate leaves; involucral bracts 4 to 6, surrounding a head of numerous flowers; perianth showy, with a long slender tube and rather broad limb, white to bright pink or greenish white; fruit a hard spindle-shaped body, completely surrounded by 2 to 4 broad thin reticulate-veined wings.

KEY TO THE SPECIES.

Flowers 2 cm. long or less, greenish; peduncles shorter than the leaves.. 1. *T. micranthus.*
Flowers more than 2 cm. long, pink or white; peduncles longer than the leaves.
  Perianth pink; fruit 20 to 28 mm. long; plants stout, with erect stems; bracts narrowly ovate............................ 2. *T. cyclopterus.*
  Perianth white; fruit less than 20 mm. long; plants more slender, usually spreading; bracts narrowly lanceolate.. 3. *T. wootoni.*

1. **Tripterocalyx micranthus** (Torr.) Hook. Journ. Bot. Kew Misc. **5**: 261. 1853.
Abronia micrantha Torr. in Frém. Rep. Exped. Rocky Mount. 96. 1845.
TYPE LOCALITY: "Near the mouth of Sweet Water river."
RANGE: Montana and Nebraska to Nevada and New Mexico.
NEW MEXICO: Albuquerque; opposite San Juan. Plains and low hills, in the Upper Sonoran Zone.

2. **Tripterocalyx cyclopterus** (A. Gray) Standley, Contr. U. S. Nat. Herb. **12**: 329. 1909.
Abronia cycloptera A. Gray, Amer. Journ. Sci. II. **15**: 319. 1853.
Abronia carnea Greene, Pittonia **3**: 343. 1898.
TYPE LOCALITY: On the Rio Grande, New Mexico.
RANGE: Western Texas and southern New Mexico to Chihuahua.
NEW MEXICO: Near Albuquerque; Pecos River; Chavez; Rincon; Deming; Mesilla Valley. Sandy fields and mesas, in the Lower Sonoran Zone.
The type of *A. carnea* was collected near Las Cruces (*Wooton* 59).

3. **Tripterocalyx wootoni** Standley, Contr. U. S. Nat. Herb. **12**: 329. 1909.
TYPE LOCALITY: Near Ojo Caliente, Zuni Reservation, New Mexico. Type collected by Wooton, July 20, 1906.
RANGE: Northwestern New Mexico and northeastern Arizona.
NEW MEXICO: Western parts of Valencia, McKinley, and San Juan counties. Dry plains and foothills, in the Upper Sonoran Zone.

7. ABRONIA Juss. SAND VERBENA.

Annual or perennial herbs with erect or prostrate, glabrous or pubescent stems; leaves opposite, petiolate; flowers in a head surrounded by numerous or few distinct thin bracts; perianth white or red, with an elongated tube and a rather narrow 5-lobed limb; stamens 3 to 5, included; fruit leathery, with 3 to 5 wings or sometimes merely lobed and not winged.

KEY TO THE SPECIES.

Plants nearly acaulescent, with a short thick, caudex; leaves narrowly oblong or linear; perennial .... 1. *A. bigelovii.*
Plants with long stems with conspicuous internodes; leaves broader; perennials or annuals.
  Perennials; bracts ovate or oblong; stems erect or spreading, never prostrate; perianth white.
    Fruit biturbinate, tapering at both ends .... 2. *A. fragrans.*
    Fruit turbinate, not tapering above.
      Stems densely viscid-hirsute; bracts 10 to 15 mm. long .... 3. *A. fendleri.*
      Stems only puberulent or almost glabrous; bracts 8 mm. long or less .... 4. *A. ramosa.*
  Annuals; bracts lanceolate; stems prostrate; perianth purplish red.
    Leaves mostly ovate, rounded or broadly cuneate at the base; seeds lanceolate, 2 to 2.5 mm. long .... 5. *A. torreyi.*
    Leaves narrowly lanceolate, narrowed at the base; seeds ovate, 1.5 mm. long .... 6. *A. angustifolia.*

1. **Abronia bigelovii** Heimerl, Smiths. Misc. Coll. **53**: 197. 1908.
TYPE LOCALITY: Near Galisteo, New Mexico. Type collected by Bigelow.
RANGE: Known only from type locality, in the Upper Sonoran Zone.

2. **Abronia fragrans** Nutt.; Hook. Journ. Bot. Kew Misc. **5**: 261. 1853.

TYPE LOCALITY: "On loamy, sandy, firm banks, within the high drift-sand hills of the Lower Platte."

RANGE: Nebraska and Wyoming to New Mexico.

NEW MEXICO: Fort Wingate; Farmington; Cimarron; Lamy; Willard; Stanley; Clayton; Nara Visa; Roswell. Plains, in the Upper Sonoran Zone.

3. **Abronia fendleri** Standley, Contr. U. S. Nat. Herb. **12**: 324. *pl. 43*. 1909.

TYPE LOCALITY: Santa Fe, New Mexico. Type collected by Fendler (no. 739).

RANGE: New Mexico and northern Chihuahua.

NEW MEXICO: Farmington; Santa Fe; Coolidge; Chama River; Lamy; Socorro; Jornada del Muerto; Tortugas Mountain. Dry plains, in the Lower and Upper Sonoran zones.

The specimens from the southern Rio Grande Valley differ from the typical form in having a much denser, yellowish pubescence. They are erect, and stouter than the plant about Santa Fe, which is much branched and spreading.

4. **Abronia ramosa** Standley, Contr. U. S. Nat. Herb. **12**: 321. *pl. 39*. 1909.

TYPE LOCALITY: Holbrook, Arizona.

RANGE: Northeastern Arizona and northwestern New Mexico.

NEW MEXICO: Carrizo Mountains; Shiprock. Sandy plains, in the Upper Sonoran Zone.

5. **Abronia torreyi** Standley, Contr. U. S. Nat. Herb. **12**: 319. *pl. 38*. 1909.

TYPE LOCALITY: Mesilla, Dona Ana County, New Mexico. Type collected by Wooton (no. 11).

RANGE: Southern New Mexico to Arizona and Chihuahua.

NEW MEXICO: Emory's 55th Monument; Mesilla Valley. Sandy fields, in the Lower Sonoran Zone.

6. **Abronia angustifolia** Greene, Pittonia **3**: 344. 1898.

Abronia turbinata stenophylla Heimerl, Ann. Cons. Jard. Genève **5**: 190. 1901.

TYPE LOCALITY: White Sands, Dona Ana County, New Mexico. Type collected by Wooton (no. 157).

RANGE: Known only from the type locality. Lower Sonoran Zone.

The type collection of *A. turbinata stenophylla* is the same as that of *A. angustifolia*.

8. NYCTAGINIA Choisy.

Low perennial herb with a thick fleshy root; leaves opposite, the blades deltoid or triangular-hastate, petiolate, the margins entire or toothed; flowers capitate, numerous, surrounded by an involucre of thin narrow bracts; perianth long-funnelform, deep red; fruit leathery, turbinate, 10-costate.

1. **Nyctaginia cockerellae** A. Nels. Proc. Biol. Soc. Washington **16**: 29. 1903.

TYPE LOCALITY: Near Roswell, New Mexico. Type collected by Mrs. T. D. A. Cockerell.

RANGE: Western Texas and southeastern New Mexico.

NEW MEXICO: Lower Pecos Valley. Lower Sonoran Zone.

9. SELINOCARPUS A. Gray.

Low branched perennial herbs; leaves opposite, sessile or petiolate, thick, often fleshy; flowers solitary in the axils or clustered at the ends of the branches; bracts small, inconspicuous; perianth funnelform, variously colored, with a short or long tube and a narrow or broad limb; stamens 2 to 5, exserted; fruit with 3 to 5 prominent wings.

KEY TO THE SPECIES.

Perianth 10 mm. long or less, with scarcely any tube; leaves broadly ovate .... 1. *S. chenopodioides.*
Flowers 15 mm. long or more, with a conspicuous tube; leaves ovate or lanceolate.
Leaves lanceolate, thick and fleshy .... 2. *S. lanceolatus.*
Leaves ovate, not fleshy .... 3. *S. diffusus.*

1. **Selinocarpus chenopodioides** A. Gray, Amer. Journ. Sci. II. **15:** 262. 1853.

TYPE LOCALITY: "Valleys from Providence Creek to the Rio Grande," Texas.

RANGE: Western Texas to southern Arizona and southward.

NEW MEXICO: Socorro; Albuquerque; above Rincon; Lordsburg; Mesilla Valley; mesa west of Organ Mountains; plains south of the White Sands; Alamogordo. Dry mesas and hills, in the Lower Sonoran Zone.

2. **Selinocarpus lanceolatus** Wooton, Bull. Torrey Club **25:** 304. 1898.

TYPE LOCALITY: South of the White Sands, New Mexico. Type collected by Wooton (no. 389).

RANGE: New Mexico.

NEW MEXICO: El Rito; near Suwanee; White Sands. Strongly alkaline soil, plains, in the Lower Sonoran Zone.

3. **Selinocarpus diffusus** A. Gray, Amer. Journ. Sci. II. **15:** 262. 1853.

TYPE LOCALITY: "Rocky hills and valleys from the Pecos to the Limpio," Texas.

RANGE: Western Texas to New Mexico.

NEW MEXICO: Acoma; Socorro; south of Carrizozo; Delaware Creek; Turneys Ranch, Dona Ana County. Dry hills, in the Lower Sonoran Zone.

10. CYPHOMERIS Standley.

Slender erect perennial herb, woody at the base; leaves thick and fleshy, glaucous, ovate or triangular, entire, petiolate; flowers in bracted racemes; perianth red, funnelform, with a short narrow tube expanding gradually into a broad limb; fruit gibbous, glaucous, 10-ribbed.

1. **Cyphomeris gypsophiloides** (Mart. & Gal.) Standley, Contr. U. S. Nat. Herb. **13:** 428. 1911.

Lindenia gypsophiloides Mart. & Gal. Bull. Acad. Sci. Brux. **10**[1]**:** 358. 1843.

Boerhaavia gibbosa Pavon; Choisy in DC. Prodr. **13**[2]**:** 457. 1849.

Boerhaavia gypsophiloides Coulter, Contr. U. S. Nat. Herb. **2:** 354. 1894.

TYPE LOCALITY: "Dans les plaines á mimosées et á cactées de Tehuacan de las Granadas, á environ 5,000 pieds," Mexico.

RANGE: New Mexico and western Texas to Mexico.

NEW MEXICO: Organ Mountains; La Luz Canyon; Carlsbad. Rocky canyons, in the Upper Sonoran Zone.

11. ACLEISANTHES A. Gray.

Low perennial from a woody base; leaves opposite, thick and fleshy, petiolate, long-attenuate at the apex; flowers axillary or terminal, mostly solitary; involucre of 2 or 3 small bracts; perianth white, with a very long slender tube and a broad limb; stamens 2 to 5, unequal, often exserted; fruit narrowly ellipsoid, 5-angled or 5-ribbed.

1. **Acleisanthes longiflora** A. Gray, Amer. Journ. Sci. II. **15:** 261. 1853.

TYPE LOCALITY: Valley of the Limpio, Texas.

RANGE: Western Texas to southeastern New Mexico, southern California, and southward.

NEW MEXICO: Delaware Creek; mouth of Dark Canyon, Guadalupe Mountains. Dry hills, in the Lower Sonoran Zone.

52576°—15——15

12. **ANULOCAULIS** Standley.

Stout perennial herb; leaves mostly basal, nearly orbicular, lacerate-margined or dentate, thick and leathery; stems much branched, glabrous but with viscid rings about the internodes; flowers in small clusters, sessile or pedicellate; perianth funnelform, with a long tube; fruit biturbinate, 10-ribbed.

1. **Anulocaulis leiosolenus** (Torr.) Standley, Contr. U. S. Nat. Herb. **12:** 375. 1909.

Boerhaavia leiosolena Torr. U. S. & Mex. Bound. Bot. 172. 1859.

Type locality: "In gypseous soil, Great Cañon of the Rio Grande, 70 miles below El Paso," Texas.

Range: Western Texas and southeastern New Mexico.

New Mexico: Near Lakewood (*Wooton*). In gypsum soil, in the Lower Sonoran Zone.

13. **BOERHAAVIA** L.

Slender annual or perennial herbs, glabrous or pubescent, often with glandular rings about the internodes; leaves opposite, unequal, petiolate or sessile; flowers small, variously arranged, each usually subtended by 1 or 2 minute bracts; perianth campanulate, 5-lobed; stamens 1 to 5, exserted or included; fruit clavate to obpyramidal, 3 to 5-ribbed or angled, or sometimes with 3 to 5 low thick wings.

KEY TO THE SPECIES.

Perennials.
- Leaves linear or narrowly linear-lanceolate .... 1. *B. tenuifolia.*
- Leaves broadly ovate.
  - Flowers solitary, on long slender pedicels.
    - Fruit glabrous; flowers about 1 mm. broad .... 4. *B. organensis.*
    - Fruit viscid; flowers 3 to 5 mm. wide .... 5. *B. gracillima.*
  - Flowers short-pediceled, in umbels.
    - Stems hirsute below .... 2. *B. ixodes.*
    - Stems glabrous or puberulent below .... 3. *B. viscosa oligadena.*

Annuals.
- Flowers in slender spikes or racemes.
  - Ribs of the fruit 4; fruit truncate or very obtuse at the apex; bracts large, persistent; stems mostly erect, sparingly branched .... 6. *B. wrightii.*
  - Ribs of the fruit 5; fruit rounded at the apex; bracts small, deciduous; stems decumbent or ascending, much branched .... 7. *B. torreyana.*
- Flowers not spicate or racemose.
  - Fruit rounded at the apex, subtended by large reddish persistent bracts; plants glandular .... 8. *B. purpurascens.*
  - Fruit truncate at the apex; bracts minute, green, deciduous; plants glabrous or nearly so.
    - Leaves thick and fleshy, very glaucous, especially beneath, oblong; flowers in simple umbels, short-pediceled; stems decumbent or ascending .... 9. *B. intermedia.*
    - Leaves thin, not conspicuously glaucous, paler beneath, broadly ovate; flowers in compound umbels; stems erect ....10. *B. erecta thornberi.*

1. **Boerhaavia tenuifolia** A. Gray; Coulter, Contr. U. S. Nat. Herb. **2**: 355. 1894.
Type locality: Camp Charlotte, Ixion County, Texas.
Range: Western Texas and southeastern New Mexico.
New Mexico: Guadalupe Mountains. Alkaline plains, in the Lower Sonoran Zone.

2. **Boerhaavia ixodes** Standley, Contr. U. S. Nat. Herb. **13**: 423. 1911.
Type locality: Vicinity of Chihuahua, Mexico.
Range: New Mexico and southern California to Mexico.
New Mexico: Gila Valley (*Greene*). Lower Sonoran Zone.

3. **Boerhaavia viscosa oligadena** Heimerl, Ann. Cons. Jard. Genève **5**: 189. 1901.
Boerhaavia ramulosa Jones, Contr. West. Bot. **10**: 40. 1902.
Type locality: Organ Mountains, New Mexico. Type collected by Wooton (no. 421).
Range: Arizona to Florida, south into Mexico.
New Mexico: Mangas Springs; Black Range; Florida Mountains; Organ Mountains; Mesilla Valley. Dry plains and fields, in the Lower Sonoran Zone.

4. **Boerhaavia organensis** Standley, Contr. U. S. Nat. Herb. **12**: 385. 1909.
Type locality: Filmore Canyon, Organ Mountains, New Mexico. Type collected by Wooton, October 23, 1904.
Range: Known only from type locality, in the Upper Sonoran Zone.

5. **Boerhaavia gracillima** Heimerl, Bot. Jahrb. Engler **11**: 86. 1889.
Type locality: "In territorio Mexicano."
Range: Western Texas and southern New Mexico and southward.
New Mexico: Organ Mountains. Dry hills, in the Lower Sonoran Zone.

6. **Boerhaavia wrightii** A. Gray, Amer. Journ. Sci. II. **15**: 322. 1853.
Boerhaavia bracteosa S. Wats. Proc. Amer. Acad. **20**: 370. 1885.
Type locality: "Pebbly hills near El Paso."
Range: Western Texas to Nevada and southward.
New Mexico: Mesa west of Organ Mountains. Sandy mesas, in the Lower Sonoran Zone.

7. **Boerhaavia torreyana** (S. Wats.) Standley, Contr. U. S. Nat. Herb. **12**: 385. 1909.
Boerhaavia spicata torreyana S. Wats. Proc. Amer. Acad. **24**: 70. 1889.
Type locality: "Texas, New Mexico, and Arizona."
Range: Western Texas to southern Arizona.
New Mexico: Chama River; Albuquerque; Silver City; Florida Mountains; Deming; Tortugas Mountain; Mesilla Valley; south of Roswell. Sandy mesas, in the Lower Sonoran Zone.

8. **Boerhaavia purpurascens** A. Gray, Amer. Journ. Sci. II. **15**: 321. 1853.
Type locality: Stony hills near the copper mines of Santa Rita, New Mexico. Type collected by Wright (no. 1725).
Range: New Mexico and Arizona and southward.
New Mexico: Carlisle; Mogollon Mountains; Hanover Mountain. Mesas and low hills, in the Upper Sonoran Zone.

9. **Boerhaavia intermedia** Jones, Contr. West. Bot. **10**: 41. 1902.
Type locality: El Paso, Texas.
Range: Western Texas to southern California and southward.
New Mexico: Plains of the Rio Gila; Organ Mountains; Mesilla Valley. Plains and low hills, in the Lower Sonoran Zone.

10. Boerhaavia erecta thornberi (Jones) Standley, Contr. U. S. Nat. Herb. **12:** 381. 1909.

Boerhaavia thornberi Jones, Contr. West. Bot. **12:** 72. 1908.

TYPE LOCALITY: Tucson, Arizona.

RANGE: Arizona and New Mexico and southward.

NEW MEXICO: Mangas Springs; Guadalupe Canyon. Low hills, in the Upper Sonoran Zone.

From *B. erecta*, which ought to be found in New Mexico, this differs slightly in its more strict habit and not dotted leaves. The leaves of *B. erecta* are conspicuously dotted beneath.

46. PHYTOLACCACEAE. Pokeweed Family.

1. RIVINA L.

A low erect branching herb with ovate leaves; flowers small, white or rose-colored, in axillary and terminal racemes; calyx 4-parted; stamens 4 to 8; fruit a reddish berry containing a solitary seed.

1. Rivina portulaccoides Nutt. Trans. Amer. Phil. Soc. n. ser. **5:** 167. 1837.

TYPE LOCALITY: "On the alluvial lands of the Verdigris river, near its confluence with the Arkansas."

RANGE: New Mexico and Arizona to Texas and northern Mexico.

NEW MEXICO: Guadalupe Canyon (*Mearns* 683).

47. AIZOACEAE. Carpetweed Family.

Fleshy or succulent herbs with usually opposite leaves and no stipules; ovary and capsule 2 to several-celled; stamens and petals sometimes numerous; petals wanting in our genera.

KEY TO THE GENERA.

Sepals distinct; leaves scarcely fleshy; stamens 3 to 5; capsules 3-celled, many-seeded .... 1. MOLLUGO (p. 228).
Sepals united below; leaves fleshy; stamens 5 to 60; capsules 1 to 5-celled.
  Capsules 3 to 5-celled, many-seeded; leaves linear to oblong-lanceolate; stamens 5 to 60 .... 2. SESUVIUM (p. 229).
  Capsules 1-celled, with few seeds; leaves round-obovate; stamens 6 to 10 .... 3. TRIANTHEMA (p. 229).

1. MOLLUGO L. CARPETWEED.

Slender annuals with erect or prostrate much branched stems; leaves linear or narrowly oblanceolate, thin; sepals 5, white inside; stamens 5 and alternate with the sepals or 3 and alternate with the 3 cells of the ovary; stigmas 3.

KEY TO THE SPECIES.

Cauline leaves linear, glaucous; plants small, less than 10 cm. in diameter, erect .... 1. *M. cerviana.*
Cauline leaves narrowly oblanceolate, bright green; plants larger, 20 to 70 cm. in diameter, prostrate .... 2. *M. verticillata.*

1. Mollugo cerviana (L.) Seringe in DC. Prodr. **1:** 392. 1824.

Pharnaceum cervianum L. Sp. Pl. 272. 1753.

TYPE LOCALITY: "Habitat Rostockii, in Russia, Hispania."

RANGE: Texas to California and Mexico; also in the tropics of the Old and New World.

NEW MEXICO: Mesa west of the Organ Mountains; Filmore Canyon; Chama River. Dry sandy plains, in the Lower Sonoran Zone.

2. **Mollugo verticillata** L. Sp. Pl. 89. 1753. CARPETWEED.

TYPE LOCALITY: "Habitat in Africa, Virginia."

RANGE: Throughout most of North America; also in the Old World.

NEW MEXICO: Tierra Blanca; Mangas Springs; San Luis Mountains; Organ Mountains. Waste and cultivated ground.

2. SESUVIUM L. SEA PURSLANE.

Fleshy herb with prostrate stems; leaves linear to oblong-lanceolate, fleshy, glaucous; calyx 5-parted, purplish inside; stamens 5 to 60; styles 3 to 5.

1. **Sesuvium sessile** Pers. Syn. Pl. 2: 39. 1807.

TYPE LOCALITY: Not stated.

RANGE: Kansas and California, southward to tropical America.

NEW MEXICO: Rio Grande Valley, from Socorro southward. In alkaline soil, in the Lower Sonoran Zone.

3. TRIANTHEMA L.

Annual with ascending stems; leaves round-obovate, somewhat fleshy, bright green; calyx 5-parted, colored within; stamens 6 to 10; style 1.

1. **Trianthema portulacastrum** L. Sp. Pl. 223. 1753.

Trianthema monogyna L. Mant. Pl. 1: 69. 1767.

TYPE LOCALITY: "Habitat in Jamaica, Curassao."

RANGE: Florida to Texas and Lower California, and southward; also in the West Indies.

NEW MEXICO: Mesilla Valley; Alamogordo. In sandy soil, in the Lower Sonoran Zone.

An inconspicuous herb, found on dry sandy plains and sometimes in cultivated fields in the extreme southern part of the State. Often after the summer rains the plants come up so thickly as completely to cover the ground.

48. PORTULACACEAE. Purslane Family.

Annual or perennial herbs, sometimes woody at the base; leaves fleshy; flowers regular, asymmetrical, the sepals fewer than the petals; stamens opposite the petals when of the same number, often indefinite; sepals 2; petals 5 or sometimes none; stamens usually 5 to 20; styles 2 to 8; pod 1-celled.

KEY TO THE GENERA.

Ovary partly inferior; capsule circumscissile.............. 1. PORTULACA (p. 230).
Ovary superior, capsules various.
  Plants woody at the base or throughout; capsules with a 6-valved endocarp.............................. 2. TALINOPSIS (p. 230).
  Plants herbaceous, rarely woody near the base; capsules without endocarp.
    Sepals deciduous.............................. 3. TALINUM (p. 231).
    Sepals persistent.
      Capsule circumscissile near the base......... 6. OREOBROMA (p. 233).
      Capsule opening by 3 valves at the apex.
        Leaves mostly basal; roots fleshy; plants not stoloniferous.................... 4. CLAYTONIA (p. 233).
        Leaves scattered along the stems; roots slender, not fleshy; plants stoloniferous.............................. 5. CRUNOCALLIS (p. 233).

1. PORTULACA L. Purslane.

Fleshy annuals or perennials with terete or flattened leaves; calyx 2-cleft, the tube cohering with the ovary; petals 5, rarely 6, fugacious; stamens 7 to 20; style 3 to 8-parted; pod 1-celled, many-seeded, the upper part separating as a lid.

KEY TO THE SPECIES.

Leaves terete, hairy in the axils.
  Petals carmine to purple, 2 to 4 mm. long; seeds blackish... 1. *P. pilosa.*
  Petals yellow or copper-colored; seeds black or red.
    Petals 6 to 10 mm. long; seeds blackish.................. 2. *P. suffrutescens.*
    Petals about 2 mm. long; seeds red........................ 3. *P. parvula.*
Leaves flat, not hairy in the axils.
  Petals acute; capsule with a crownlike border near the mouth. 4. *P. lanceolata.*
  Petals notched; capsule without a crownlike border.
    Sepals acute; styles 5 to 7; seeds granulate.............. 5. *P. oleracea.*
    Sepals obtuse; styles 3 or 4; seeds sharply tuberculate.. 6. *P. retusa.*

1. **Portulaca pilosa** L. Sp. Pl. 445. 1753.
Type locality: "Habitat in America meridionali."
Range: Florida to Missouri and California and southward.
New Mexico: Tortugas Mountain. Sandy soil, in the Lower Sonoran Zone.

2. **Portulaca suffrutescens** Engelm. Bot. Gaz. **6**: 236. 1881.
Type locality: Western New Mexico at the Copper Mines. Type collected by Wright (no. 874).
Range: Western Texas to Arizona and southward.
New Mexico: Santa Rita; Telegraph Mountains; Mangas Springs; Kingston; Tortugas Mountain; Organ Mountains. Dry hills, in the Lower Sonoran Zone.

3. **Portulaca parvula** A. Gray, Proc. Amer. Acad. **22**: 274. 1887.
Type locality: "On the plains of W. Texas and New Mexico, and in Mexico."
Range: Western Texas to southern Arizona and southward.
New Mexico: Mesilla Valley (*Wooton*). Lower Sonoran Zone.

4. **Portulaca lanceolata** Engelm. Bost. Journ. Nat. Hist. **6**: 154. 1850.
Type locality: "Granite region of the Liano, in Western Texas."
Range: Texas to Arizona.
New Mexico: Berendo Creek; north of Deming. Lower Sonoran Zone.

5. **Portulaca oleracea** L. Sp. Pl. 445. 1753. Common Purslane.
Type locality: "Habitat in Europa australi, India, Insula Ascensionis, America."
Range: Common in cultivated ground throughout most of North America, extending into South America; in the Old World.
New Mexico: Cultivated and waste ground throughout the State.

6. **Portulaca retusa** Engelm. Bost. Journ. Nat. Hist. **6**: 154. 1850.
Type locality: "Granite region of the Liano in western Texas."
Range: Arkansas to Texas and Arizona.
New Mexico: Rio Alamosa; Gilmores Ranch; Tortugas Mountain. Sandy soil, Lower Sonoran to Transition Zone.

2. TALINOPSIS A. Gray.

A low shrub, 60 cm. high or less, with slender branches; leaves linear, nearly terete; flowers few, purple, in a terminal cyme; fruit about 24 mm. long, covered by the persistent calyx.

1. **Talinopsis frutescens** A. Gray, Pl. Wright. 1: 15. *pl. 3*. 1852.

TYPE LOCALITY: Mountain valleys, seventeen miles east of the Rio Grande, Texas.

RANGE: Western Texas and New Mexico to Mexico.

NEW MEXICO: Tortugas Mountain. Dry rocky hills, in the Lower Sonoran Zone.

3. TALINUM Adans.

Low glabrous perennials; leaves fleshy, terete to broad and flat, basal or cauline; stamens 5 to 30; style 3-lobed; capsules subglobose or oblong, with numerous shining seeds.

KEY TO THE SPECIES.

Flowers axillary.
- Leaves flat, not crowded; seeds conspicuously costate.
  - Petals yellow; leaves linear; stems suffrutescent at the base, stiff, almost erect; capsule globose, 4 to 5 mm. in diameter.......... 2. *T. angustissimum.*
  - Petals orange to orange scarlet; leaves widest at or near the middle; stems fleshy, ascending or spreading; capsules ovoid, 5 to 7 mm. long.......... 3. *T. aurantiacum.*
- Leaves terete or slightly flattened, crowded; seeds smooth.
  - Petals about 10 mm. long; leaves short, 12 mm. long or less, comparatively broad; sepals obtuse or nearly so; pedicels stout, much shorter than the sepals.......... 4. *T. brevifolium.*
  - Petals about 15 mm. long; leaves longer, 12 to 20 mm., slender; sepals acute; pedicels slender, equaling or longer than the sepals.......... 5. *T. pulchellum.*

Flowers in terminal cymes, these sometimes panicled.
- Leaves flat, obovate; flowers in panicled cymes.......... 1. *T. paniculatum.*
- Leaves terete; flowers in terminal cymes.
  - Inflorescence shorter than the leaves or at most not exceeding them; plants low, 5 to 8 cm. high; flowers yellow.......... 6. *T. humile.*
  - Inflorescence much exceeding the leaves; plants larger, 8 to 30 cm. high; flowers pink to purplish red.
    - Sepals orbicular; capsules globose or nearly so; pedicels slender, usually twice as long as the capsules or more; stamens 10.......... 7. *T. longipes.*
    - Sepals acute or acuminate, ovate to lanceolate; capsules oblong; pedicels stout, little if at all exceeding the capsules; stamens 5.
      - Inflorescence congested, most of the flowers nearly sessile; sepals purplish; leaves stout, not noticeably narrowed at the base.......... 8. *T. confertiflorum.*
      - Inflorescence loose, all the flowers conspicuously pediceled; sepals green; leaves slender, much narrowed at the base.......... 9. *T. parviflorum.*

Talinum calycinum Engelm. should be found in the northeastern part of the State; it is possible that its type locality is inside our borders. In habit it resembles *T. parviflorum*, but it has much larger flowers and 30 or more stamens.

1. **Talinum paniculatum** (Jacq.) Gaertn. Fruct. & Sem. **2:** 219. 1791.
Portulaca paniculata Jacq. Enum. Pl. Carib. 22. 1762.
Talinum patens Willd. Sp. Pl. **2:** 863. 1799.
TYPE LOCALITY: West Indies.
RANGE: Texas and Arizona to tropical America, and in the West Indies.
NEW MEXICO: Guadalupe Canyon; near White Water.

2. **Talinum angustissimum** (A. Gray) Woot. & Standl. Contr. U. S. Nat. Herb. **16:** 120. 1913.
Talinum aurantiacum angustissimum A. Gray, Pl. Wright. **1:** 14. 1852.
TYPE LOCALITY: Bottoms of Live Oak Creek, and on the San Felipe, Texas.
RANGE: Western Texas to Arizona.
NEW MEXICO: Fort Cummings; Florida Mountains; Dog Spring; Tortugas Mountain; Organ Mountains; south of Roswell; Dexter. Dry hills, in the Lower and Upper Sonoran zones.

This plant has long been confused with *T. aurantiacum*, a larger, stouter, more succulent plant with larger, differently colored flowers. It is difficult to distinguish the two in herbarium specimens, but no one can confuse them in the field. This species has also been confused with *T. lineare* H. B. K., a different plant found only in central and southern Mexico.

3. **Talinum aurantiacum** Engelm. Bost. Journ. Nat. Hist. **6:** 153. 1850.
TYPE LOCALITY: "On the Sabinas, and more abundantly on the Liano, rare about New Braunfels, Texas."
RANGE: Western Texas to Arizona.
NEW MEXICO: Santa Rita; Fort Cummings; Mogollon Mountains; Hillsboro; Florida Mountains; Tortugas Mountain; Organ Mountains; Carrizozo; Queen. Dry hills and plains, in the Lower and Upper Sonoran zones.

4. **Talinum brevifolium** Torr. in Sitgreaves, Rep. Zuñi & Colo. 156. 1854.
Talinum brachypodum S. Wats. Proc. Amer. Acad. **20:** 355. 1885.
TYPE LOCALITY: On the Little Colorado, Arizona.
RANGE: Northern Arizona and New Mexico.
NEW MEXICO: Fort Wingate; north of Ramah. Dry hills, in the Upper Sonoran Zone.

The type of *T. brachypodum* was collected near Laguna by Mr. and Mrs. J. G. Lemmon.

5. **Talinum pulchellum** Woot. & Standl. Contr. U. S. Nat. Herb. **16:** 121. 1913.
TYPE LOCALITY: Near Queen, New Mexico. Type collected by Wooton, August 2, 1909.
RANGE: New Mexico.
NEW MEXICO: Queen; near Pecos; Van Pattens. Dry hills, in the Upper Sonoran Zone.

6. **Talinum humile** Greene, Bot. Gaz. **6:** 183. 1881.
TYPE LOCALITY: On a rocky table-land near the southern base of the Pinos Altos Mountains, New Mexico. Type collected by Greene, August 11, 1880.
RANGE: Known only from the type locality.

7. **Talinum longipes** Woot. & Standl. Contr. U. S. Nat. Herb. **16:** 120. 1913.
TYPE LOCALITY: Tortugas Mountain, New Mexico. Type collected by Wooton, August 27, 1894.
RANGE: Southern New Mexico.
NEW MEXICO: Tortugas Mountain. Dry hills, in the Lower Sonoran Zone.

8. **Talinum confertiflorum** Greene, Bull. Torrey Club **8**: 121. 1881.

TYPE LOCALITY: Pinos Altos Mountains, New Mexico. Type collected by Greene.

RANGE: New Mexico.

NEW MEXICO: Sandia Mountains; Mogollon Mountains; Hanover Hills; Organ Mountains.

9. **Talinum parviflorum** Nutt.; Torr. & Gray, Fl. N. Amer. **1**: 197. 1838.

TYPE LOCALITY: "On rocks, Arkansas."

RANGE: Minnesota to Colorado, New Mexico, and Texas.

NEW MEXICO: Las Vegas; Pecos; Hop Canyon; Tortugas Mountain; Nara Visa; near Dora. Open hills, in the Lower and Upper Sonoran zones.

4. CLAYTONIA L. Spring beauty.

Succulent perennial herbs with fleshy roots and narrow basal or cauline leaves; flowers white to pink, in naked, loose, terminal, simple or paniculate racemes; sepals 2, persistent; style 3-cleft; capsules 3-valved, with 6 or fewer seeds.

KEY TO THE SPECIES.

Plants with rounded corms; basal leaves 1 or 2, not spatulate....... 1. *C. lanceolata.*
Plants with fleshy taproots; basal leaves numerous, spatulate....... 2. *C. megarrhiza.*

1. **Claytonia lanceolata** Pursh, Fl. Amer. Sept. 175. 1814.

TYPE LOCALITY: "On the Rocky Mountains."

RANGE: British Columbia and Wyoming to California and northern New Mexico.

NEW MEXICO: Pass southeast of Tierra Amarilla; Willow Creek. Moist ground, in the Transition Zone.

2. **Claytonia megarrhiza** (A. Gray) Parry; S. Wats. Bibl. Ind. 118. 1878.

Claytonia arctica megarrhiza A. Gray, Amer. Journ. Sci. II. **33**: 406. 1862.

TYPE LOCALITY: Rocky Mountains of Colorado.

RANGE: Washington and Alberta to Colorado and northern New Mexico.

NEW MEXICO: Top of Truchas Peak; Wheeler Peak. Arctic-Alpine Zone.

5. CRUNOCALLIS Rydb.

Slender stoloniferous perennial, the stems rooting at the nodes; leaves oblong or oblanceolate, fleshy; petals white, 3 or 4 times as long as the sepals; stamens 3 to 5; ovary 3-ovuled.

1. **Crunocallis chamissonis** (Ledeb.) Rydb. Bull. Torrey Club **33**: 139. 1906.

Claytonia chamissonis Ledeb.; Spreng. Syst. Veg. **1**: 790. 1826.

Montia chamissonis Greene, Fl. Franc. 180. 1891.

TYPE LOCALITY: Unalaska.

RANGE: Alaska, British Columbia, and Minnesota to California and New Mexico.

NEW MEXICO: Ponchuelo Creek; Chama. Wet ground, in the Transition Zone.

6. OREOBROMA Howell.

Low perennial with fleshy thick roots and numerous narrow basal leaves; stems mostly 1-flowered, bearing a pair of reduced bractlike leaves; sepals 2, entire, ovate, obtuse; petals large, pink or reddish; capsule circumscissile near the base.

1. **Oreobroma nevadensis** (S. Wats.) Howell, Erythea **1**: 33. 1893.

Calandrinia nevadensis S. Wats. Proc. Amer. Acad. **8**: 623. 1873.

Lewisia nevadensis Robinson in A. Gray, Syn. Fl. **1**^{1}: 268. 1895.

TYPE LOCALITY: "Subalpine region of Wahsatch and East Humboldt Mountains," Nevada.

RANGE: Washington and Oregon to California and northern New Mexico.

NEW MEXICO: Vicinity of Tierra Amarilla (*Eggleston* 6478, 6543). Open slopes, in the Transition Zone.

49. ALSINACEAE. Chickweed Family.

Low slender annuals or perennials, with opposite simple leaves and small white flowers; inflorescence cymose or the flowers solitary and axillary; sepals 4 or mostly 5, distinct, or slightly united at the base; petals not clawed, entire, emarginate, or deeply cleft and parted, often wanting; fruit a capsule, dehiscing by longitudinal valves.

KEY TO THE GENERA.

Stipules present though sometimes fugacious.
  Styles simple below .... 1. DRYMARIA (p. 234).
  Styles branched to the base .... 2. TISSA (p. 235).
Stipules wanting.
  Petals 2-cleft or 2-parted.
    Capsules short-ovate or oblong; styles usually 3... 3. ALSINE (p. 235).
    Capsules long-cylindric, often curved; styles usually 5 .... 4. CERASTIUM (p. 236).
  Petals entire or emarginate.
    Styles as many as the sepals and alternate with them .... 5. SAGINA (p. 238).
    Styles fewer than the sepals, or when of the same number opposite them.
      Seeds with a basal membrane (strophiole) at the hilum .... 6. MOEHRINGIA (p. 238).
      Seeds not strophiolate.
        Capsules with twice as many valves as styles .... 7. ARENARIA (p. 238).
        Capsules with the same number of valves as styles .... 8. ALSINOPSIS (p. 240).

1. DRYMARIA Willd.

Low slender erect or prostrate annuals or perennials with inconspicuous flowers; leaves with stipules, these sometimes fugacious; sepals and petals 5, divided; stamens usually 5; capsule 3-valved.

KEY TO THE SPECIES.

Cauline leaves broadly ovate.
  Plants erect, glandular, bright green; leaves abruptly acuminate .... 1. *D. fendleri.*
  Plants prostrate, glabrous and glaucous; leaves obtuse....... 2. *D. pachyphylla.*
Cauline leaves linear.
  Cauline leaves pseudoverticillate .... 3. *D. sperguloides.*
  Cauline leaves opposite.
    Plants tall, 10 to 15 cm., erect; sepals acute; pedicels usually several times as long as the calyx............ 4. *D. tenella.*
    Plants low, 4 cm. or less, depressed; sepals obtuse; pedicels as long as the calyx or shorter................. 5. *D. depressa.*

1. Drymaria fendleri S. Wats. Proc. Amer. Acad. **17**: 328. 1882.

TYPE LOCALITY: New Mexico. Type collected by Fendler (no. 60).

RANGE: New Mexico and Arizona, southward into Mexico.

NEW MEXICO: La Cuesta; Kingston; Mogollon Mountains; Organ Mountains; White Mountains; Gray. Moist canyons, Upper Sonoran and Transition zones.

2. **Drymaria pachyphylla** Woot. & Standl. Contr. U. S. Nat. Herb. **16**: 121. 1913.

TYPE LOCALITY: Dry plains south of the White Sands, New Mexico. Type collected by Wooton (no. 405).

RANGE: Southern New Mexico and western Texas.

NEW MEXICO: South of the White Sands; Parkers Well. Dry plains, in the Lower Sonoran Zone.

3. **Drymaria sperguloides** A. Gray, Mem. Amer. Acad. n. ser. **4**: 11. 1849.

TYPE LOCALITY: "Valley of Santa Fe Creek in the mountains, in a plain grazed by cattle and horses," New Mexico. Type collected by Fendler (no. 55).

RANGE: New Mexico and Arizona to western Texas.

NEW MEXICO: Santa Fe; Mogollon Mountains; Santa Rita; Sandia Mountains; Animas Valley; San Luis Mountains; White Mountains. Open slopes, in the Upper Sonoran and Transition zones.

4. **Drymaria tenella** A. Gray, Mem. Amer. Acad. n. ser. **4**: 12. 1849.

TYPE LOCALITY: "Shady places, in woodland in the mountain region, 8 miles west of Las Vegas," New Mexico. Type collected by Fendler (no. 56).

RANGE: New Mexico, southward into Mexico.

NEW MEXICO: Las Vegas; Mogollon Mountains; Sawyers Peak. Transition Zone.

5. **Drymaria depressa** Greene, Leaflets **1**: 153. 1905.

TYPE LOCALITY: Sawyers Peak, Black Range, New Mexico. Type collected by Metcalfe (no. 1430).

RANGE: Known only from the type locality.

2. TISSA Adans. Sand spurry.

Branched annual with fleshy linear scarious-stipulate leaves; styles and capsule valves each 3.

1. **Tissa rubra** (L.) Britton, Bull. Torrey Club **16**: 127. 1899.

Arenaria rubra L. Sp. Pl. 423. 1753.

TYPE LOCALITY: European.

RANGE: Adventive from Eurasia in many parts of the United States.

NEW MEXICO: Albuquerque (*Herrick*).

3. ALSINE L. STARWORT.

Slender low annuals or perennials; flowers solitary or cymose; sepals 4 or 5; petals white, 4 or 5, deeply cleft, sometimes wanting; stamens 8 or 10 or fewer; styles 3, rarely 4 or 5; capsule ovoid, 1-celled, opening by twice as many valves as there are styles.

KEY TO THE SPECIES.

Leaves ovate, conspicuously petioled.
- Leaves cordate or subcordate, all long-petioled .... 1. *A. cuspidata.*
- Leaves rounded or narrowed at the base, the uppermost sessile.. 2. *A. media.*

Leaves linear to narrowly lanceolate, sessile.
- Plants more or less viscid .... 3. *A. jamesiana.*
- Plants not at all viscid.
  - Petals minute or none; branches of the inflorescence reflexed .... 4. *A. baicalensis.*
  - Petals equaling or exceeding the sepals; branches of the inflorescence ascending.
    - Leaves broadest above the middle, narrowed at the base .... 5. *A. longifolia.*

Leaves broadest near the base.
Leaves narrowly linear-lanceolate, light green; flowers numerous........................ 6. *A. longipes.*
Leaves lanceolate, usually bluish green; flowers few, often solitary........................ 7. *A. laeta.*

1. **Alsine cuspidata** (Willd.) Woot. & Standl.
Stellaria cuspidata Willd.; Schlecht. Ges. Naturf. Freund. Berlin Mag. 7: 196. 1816.
TYPE LOCALITY: Near Quito, Peru.
RANGE: New Mexico and Texas to Mexico and South America.
NEW MEXICO: Organ Mountains. Moist canyons, in the Transition Zone.

2. **Alsine media** L. Sp. Pl. 272. 1753. CHICKWEED.
TYPE LOCALITY: "Habitat in Europae cultis."
RANGE: A native of Europe, widely introduced into North America.
NEW MEXICO: Santa Fe; Vermejo Park; Sandia Mountains. Frequently a weed in gardens.

3. **Alsine jamesiana** (Torr.) Heller, Cat. N. Amer. Pl. ed. 2. 4. 1900.
Stellaria jamesiana Torr. Ann. Lyc. N. Y. 2: 169. 1827.
TYPE LOCALITY: "Rocky Mountains," Colorado.
RANGE: Wyoming and California to New Mexico.
NEW MEXICO: Chama; Santa Fe and Las Vegas mountains; Sandia Mountains; Magdalena Mountains; Black Range; White and Sacramento mountains. Damp thickets, in the Transition and Canadian zones.

4. **Alsine baicalensis** Coville, Contr. U. S. Nat. Herb. 4: 70. 1893.
Stellaria umbellata Turcz. Fl. Baical. 1: 236. 1842–5, not *Alsine umbellata* Lam. 1778.
TYPE LOCALITY: "In alpe Nuchu-Daban," Baical Mountains, Siberia.
RANGE: Oregon and Montana to California and New Mexico; also in Asia.
NEW MEXICO: Above Winsors Ranch; Spirit Lake. Wet ground, in the Canadian and Hudsonian zones.

5. **Alsine longifolia** (Muhl.) Britton, Mem. Torrey Club 5: 150. 1894.
Stellaria longifolia Muhl.; Willd. Enum. Pl. 479. 1809.
TYPE LOCALITY: "Habitat in Pennsylvania."
RANGE: British America to Maryland and New Mexico.
NEW MEXICO: Above Winsors Ranch; Harveys Upper Ranch. Damp thickets, in the Canadian Zone.

6. **Alsine longipes** (Goldie) Coville, Contr. U. S. Nat. Herb. 4: 70. 1893.
Stellaria longipes Goldie, Edinburgh Phil. Journ. 6: 327. 1822.
TYPE LOCALITY: "Woods near Lake Ontario."
RANGE: British America to California, New Mexico, and New England.
NEW MEXICO: Chama (*Standley* 6757). Damp ground, in the Transition Zone.

7. **Alsine laeta** (Richards.) Rydb. Mem. N. Y. Bot. Gard. 1: 144. 1900.
Stellaria laeta Richards. Bot. App. Frankl. Journ. 738. 1823.
TYPE LOCALITY: Not stated.
RANGE: British America to Nevada and New Mexico.
NEW MEXICO: Sierra Blanca (*Turner* 52). Hudsonian and Arctic-Alpine zones.

4. CERASTIUM L. MOUSE-EAR CHICKWEED.

Low annuals or perennials with narrow leaves; flowers cymose, sometimes solitary; sepals usually 5; petals of the same number, 2-cleft, often wanting; stamens 10 or fewer; styles usually 5, opposite the sepals; fruit 1-celled, elongated, often curved, opening at the summit by twice as many teeth as there are styles.

KEY TO THE SPECIES.

Perennials; petals about twice as long as the sepals.
  Lower leaves oblong or oblanceolate, obtuse.......... 1. *C. beeringianum.*
  Lower leaves linear or linear-lanceolate, acute.
    Leaves all linear or linear-lanceolate................ 2. *C. scopulorum.*
    Leaves of the inflorescence ovate.................. 3. *C. oreophilum.*
Annuals; petals only slightly or not at all exceeding the sepals.
  Lower leaves sericeous................................ 4. *C. sericeum.*
  None of the leaves sericeous.
    Pedicels in fruit 1 to 3 times as long as the calyx, usually straight or nearly so................ 5. *C. brachypodum.*
    Pedicels in fruit 5 times as long as the calyx or more, strongly curved............................ 6. *C. longipedunculatum.*

1. Cerastium beeringianum Schlecht. & Cham. Linnaea **1:** 62. 1826.

TYPE LOCALITY: "Ad sinus Eschscholzii et bonae spei."

RANGE: British America to Arizona and New Mexico.

NEW MEXICO: Upper Pecos River (*Bartlett* 178). Damp woods.

2. Cerastium scopulorum Greene, Pittonia **4:** 298. 1901.

TYPE LOCALITY: La Plata Mountains, Colorado.

RANGE: Wyoming to New Mexico.

NEW MEXICO: Rio Pueblo; Winsors Ranch; Magdalena Mountains; Organ Mountains; White and Sacramento mountains. Damp woods, in the Transition Zone.

3. Cerastium oreophilum Greene, Pittonia **4:** 297. 1901.

TYPE LOCALITY: Foothills of the Rocky Mountains near Fort Collins, Colorado.

RANGE: Colorado and New Mexico to California.

NEW MEXICO: Northwestern corner of the State (*Palmer*).

4. Cerastium sericeum S. Wats. Proc. Amer. Acad. **20:** 354. 1885.

TYPE LOCALITY: Huachuca Mountains, Arizona.

RANGE: Southern New Mexico and Arizona.

NEW MEXICO: West Fork of the Gila (*Wooton*).

A most striking species, at once distinguished from all our others by the dense sericeous pubescence of the lower leaves.

5. Cerastium brachypodum (Engelm.) Robinson in Britton, Mem. Torrey Club **5:** 150. 1894.

Cerastium nutans brachypodum Engelm.; A. Gray, Man. ed. 5. 94. 1867.

TYPE LOCALITY: Illinois.

RANGE: South Dakota and Montana to Texas and Arizona, south into Mexico.

NEW MEXICO: Santa Fe and Las Vegas mountains; Mogollon Mountains; Burro Mountains; Hillsboro; White Mountains. Damp woods, in the Transition Zone.

6. Cerastium longipedunculatum Muhl. Cat. Pl. 46. 1813.

Cerastium nutans Raf. Préc. Somiolog. 36. 1814.

TYPE LOCALITY: Pennsylvania.

RANGE: British America to North Carolina and Arizona.

NEW MEXICO: Mogollon Mountains; Burro Mountains; Sawyers Peak; Organ Mountains; White Mountains. Damp woods, in the Transition Zone.

5. SAGINA L. PEARLWORT.

Low herbs, 4 cm. high or less; leaves awl-shaped; flowers small, terminating the stems or branches; sepals 4 or 5; petals 4 or 5, not divided, often wanting; capsule 4 or 5-valved.

1. **Sagina saginoides** (L.) Britton, Mem. Torrey Club **5**: 151. 1894.
Spergula saginoides L. Sp. Pl. 441. 1753.
Sagina linnaei Presl, Rel. Haenk. **2**: 14. 1835.
Alsinella saginoides Greene, Fl. Franc. 125. 1891.
TYPE LOCALITY: "Habitat in Gallia, Sibiria."
RANGE: Greenland and Alaska to New Mexico; also in the Old World.
NEW MEXICO: Winsors Ranch; Chama. Damp meadows, in the Transition Zone.

6. MOEHRINGIA L.

Weak stoloniferous perennials with few axillary or terminal, small, white flowers; ovary at first 3-celled; seeds few, smooth, appendaged at the hilum; leaves lanceolate to oval.

KEY TO THE SPECIES.

Petals longer than the obtuse sepals; leaves elliptic-oblong to oval, mostly obtuse; stems terete .... 1. *M. lateriflora.*
Petals shorter than the acute or acuminate sepals; leaves lanceolate, acute; stems angled .... 2. *M. macrophylla.*

1. **Moehringia lateriflora** (L.) Fenzl, Versuch Alsin. 18. 1833.
Arenaria lateriflora L. Sp. Pl. 423. 1753.
TYPE LOCALITY: "Habitat in Sibiria."
RANGE: British America to New Jersey, Utah, and New Mexico; also in Asia.
NEW MEXICO: Winsors Ranch; Tierra Amarilla. Damp woods, in the Transition Zone.

2. **Moehringia macrophylla** (Hook.) Torr. in Wilkes, U. S. Expl. Exped. **15**: 246. 1874.
Arenaria macrophylla Hook. Fl. Bor. Amer. **1**: 102. 1830.
TYPE LOCALITY: "North-West America, in shady woods."
RANGE: British America to Vermont, New Mexico, and California.
NEW MEXICO: Santa Fe Canyon (*Heller* 3690). Damp woods, in the Transition Zone.

7. ARENARIA L. SANDWORT.

Slender annuals or perennials with more or less diffusely branched stems; leaves flat to subulate; flowers in open or capitate cymes or solitary in the axils; sepals 5, often ribbed; petals 5, white, entire or rarely notched; stamens normally 10; styles 3; capsule globose to oblong, opening by twice as many valves as there are styles.

KEY TO THE SPECIES.

Leaves narrowly linear, more or less rigid or pungent.
- Plants glabrous; cymes open, many-flowered .... 1. *A. eastwoodiae.*
- Plants glandular, at least on the pedicels; cymes various.
  - Calyx and pedicels densely glandular; lower leaves 50 to 100 mm. long, erect or ascending .... 2. *A. fendleri.*
  - Clayx glabrous, the pedicels only slightly glandular; lower leaves 15 mm. long or less, mostly divergent. 3. *A. aculeata.*

Leaves neither narrowly linear nor pungent.
Pedicels divergent; sepals about 3.5 mm. long.
Plants low and spreading, the stems less than 10 cm. long; leaves ovate or ovate-lanceolate, less than 1 cm. long .... 4. *A. polycaulos.*
Plants tall, the stems 20 to 30 cm. long; leaves oblong or linear-oblong, over 1 cm. long .... 5. *A. confusa.*
Pedicels ascending; sepals about 5 mm. long.
Pedicels several times as long as the calyx; stems about 15 cm. long, much branched; leaves glabrous ...... 6. *A. mearnsii.*
Pedicels about equaling the calyx; stems low, usually less than 10 cm. long, depressed; leaves cinereous-puberulent .... 7. *A. saxosa.*

1. **Arenaria eastwoodiae** Rydb. Bull. Torrey Club **31**: 406. 1904.

TYPE LOCALITY: Grand Junction, Colorado.

RANGE: Southwestern Colorado to northeastern Arizona and northwestern New Mexico.

NEW MEXICO: Carrizo Mountains; Cedar Hill. Dry hills, in the Upper Sonoran Zone.

2. **Arenaria fendleri** A. Gray, Mem. Amer. Acad. n. ser. **4**: 13. 1849.

TYPE LOCALITY: Prairies 5 miles west of Las Vegas, New Mexico. Type collected by Fendler (no. 57).

RANGE: Wyoming to Arizona and New Mexico.

NEW MEXICO: Santa Fe and Las Vegas Mountains; Raton; Sierra Grande; Jemez Mountains; Johnsons Mesa; Mogollon Mountains; Hillsboro Peak; Organ Mountains; White Mountains. Meadows, in the Transition and Canadian zones.

3. **Arenaria aculeata** S. Wats. in King, Geol. Expl. 40th Par. **5**: 40. 1871.

Arenaria congesta aculeata Jones, Proc. Calif. Acad. II. **5**: 626. 1895.

TYPE LOCALITY: Fremonts Pass, East Humboldt Mountains, Nevada.

RANGE: Oregon to Nevada, Utah, and northwestern New Mexico.

NEW MEXICO: Western San Juan County. Sandy soil, in the Upper Sonoran Zone.

4. **Arenaria polycaulos** Rydb. Bull. Torrey Club **31**: 406. 1904.

TYPE LOCALITY: Silverton, Colorado.

RANGE: Colorado to New Mexico and Arizona.

NEW MEXICO: Harveys Upper Ranch (*Standley* 4724). Mountains, in the Canadian Zone.

5. **Arenaria confusa** Rydb. Bull. Torrey Club **28**: 275. 1901.

TYPE LOCALITY: White Mountains, New Mexico.

RANGE: Colorado to New Mexico and Arizona.

NEW MEXICO: Common in all the higher mountains. Damp woods, in the Transition and Canadian zones.

6. **Arenaria mearnsii** Woot. & Standl. Contr. U. S. Nat. Herb. **16**: 121. 1913.

TYPE LOCALITY: San Luis Mountains, New Mexico. Type collected by E. A. Mearns (no. 2216).

RANGE: Southwestern New Mexico, southeastern Arizona, and adjacent Mexico.

NEW MEXICO: San Luis Mountains.

7. **Arenaria saxosa** A. Gray, Pl. Wright. **2**: 18. 1853.

TYPE LOCALITY: Stony hills at the Copper Mines, New Mexico. Type collected by Wright (no. 865).

RANGE: Southern New Mexico and Arizona.

NEW MEXICO: Mogollon Mountains; Santa Rita.

8. ALSINOPSIS Small.

Densely tufted perennials, mostly less than 5 cm. high; leaves narrow, usually subulate; flowers solitary in the axils or in terminal cymes; sepals 5; petals 5, entire or emarginate; stamens usually 10; styles normally 3; capsules slightly longer than broad, opening by as many valves as there are styles.

KEY TO THE SPECIES.

Sepals obtuse .... 1. *A. obtusiloba.*
Sepals acute .... 2. *A. propinqua.*

1. **Alsinopsis obtusiloba** Rydb. Bull. Torrey Club **33:** 140. 1906.
Arenaria obtusa Torr. Ann. Lyc. N. Y. **2:** 170. 1827, not All. 1785.
TYPE LOCALITY: "On the higher parts of the Rocky Mountains."
RANGE: British America to Utah and New Mexico.
NEW MEXICO: Santa Fe and Las Vegas mountains; White Mountain Peak. Meadows, in the Arctic-Alpine Zone.

2. **Alsinopsis propinqua** (Richards.) Rydb. Bull. Torrey Club **33:** 140. 1906.
Arenaria propinqua Richards. Bot. App. Frankl. Journ. 17. 1823.
Arenaria verna aequicaulis A. Nels. Bull. Torrey Club **26:** 352. 1899.
Arenaria aequicaulis A. Nels. in Coulter, New Man. Rocky Mount. 185. 1909.
TYPE LOCALITY: Barren grounds from Point Lake to the Arctic Sea.
RANGE: British America to Utah and northern New Mexico.
NEW MEXICO: Top of Las Vegas Range above Sapello Creek (*Cockerell*). High mountains, in the Arctic-Alpine Zone.

50. SILENACEAE. Pink Family.

Herbaceous annuals or perennials with opposite, exstipulate, mostly sessile leaves; flowers in cymes, sometimes thyrsiform; calyx tubular with a short limb; petals clawed; fruit a capsule, dehiscent by longitudinal valves.

KEY TO THE GENERA.

Calyx strongly 5-angled or 5-ribbed, or both.
  Calyx cylindric, not angled; perennials .... 1. SAPONARIA (p. 240).
  Calyx ovoid, angled; annuals .... 2. VACCARIA (p. 241).
Calyx ribs usually 10, at least twice as many as the teeth.
  Styles 5, alternate with the foliaceous calyx teeth .. 5. AGROSTEMMA (p. 242).
  Styles 3 to 5, opposite the short, not foliaceous, calyx teeth.
    Styles mostly 3; capsules usually septate at the base .... 3. SILENE (p. 241).
    Styles 5; capsule 1-celled to the base .... 4. WAHLBERGELLA (p. 242).

1. SAPONARIA. L.

Perennial herb with stout, mostly simple, very leafy stems and large corymbose pink flowers; calyx 5-toothed, obscurely nerved; petals 5, long-clawed; ovary 1-celled or partially 2 to 4-celled; styles 2; capsule dehiscent by 4 short apical teeth.

1. **Saponaria officinalis** L. Sp. Pl. 408. 1753. BOUNCING BET.
TYPE LOCALITY: "Habitat in Europa media."
RANGE: A native of Europe, often established as a roadside weed in North America.
NEW MEXICO: Farmington (*Standley* 6873).
Well established along ditch banks in this locality.

2. VACCARIA Medic. Cowherb.

A glabrous annual; flowers in corymbed cymes; calyx ovoid, 5-angled; stamens 10; styles 2; petals pale red; leaves ovate-lanceolate, glabrous.

1. **Vaccaria vaccaria** (L.) Britton in Britt. & Brown, Illustr. Fl. **2**: 18. 1897.
Saponaria vaccaria L. Sp. Pl. 409. 1753.
Type locality: "Habitat inter segetes Galliae, Germaniae."
New Mexico: Ensenada; Pecos; Beulah; Cleveland.

3. SILENE L. Catchfly.

Annual or perennial herbs; flowers solitary or in cymes; calyx 5-toothed, 10 to many-nerved; stamens 10; styles 3, rarely 4; capsule 1-celled, opening by 3 or 6 teeth; petals usually with a scale at the base of the blade.

KEY TO THE SPECIES.

Annuals.
Stems not villous; calyx 4 to 6 mm. long; corolla pink .... 1. *S. antirrhina.*
Stems densely villous; calyx 8 to 10 mm. long; corolla white ... 7. *S. noctiflora.*
Perennials.
Plants nearly acaulescent, densely matted, less than 5 cm. high .. 2. *S. acaulis.*
Plants caulescent, not matted, 20 cm. high or more.
Petals bright scarlet .... 3. *S. laciniata.*
Petals white or purplish.
Inflorescence axillary; plants very slender .... 6. *S. menziesii.*
Inflorescence thyrsoid-paniculate or spicate; plants stout.
Leaves densely viscid, not reduced above; flowers almost sessile .... 4. *S. wrightii.*
Leaves not viscid or but slightly so, much reduced above; flowers conspicuously pediceled .... 5. *S. pringlei.*

1. **Silene antirrhina** L. Sp. Pl. 419. 1753. Sleepy catchfly.
Type locality: "Habitat in Virginia, Carolina."
Range: Throughout temperate North America.
New Mexico: Santa Fe; Las Vegas; Sandia Mountains; Hillsboro; Farmington; Sierra Grande; mountains west of San Antonio; Mogollon Mountains; Organ Mountains; White Mountains. Dry fields, in the Upper Sonoran and Transition zones.

2. **Silene acaulis** L. Sp. Pl. ed. 2. 603. 1762. Moss campion.
Type locality: "Habitat in alpibus Lapponicis, Austriacis, Helveticis, Pyrenaeis."
Range: Arctic America to New Hampshire and Arizona; also in the Old World.
New Mexico: Pecos Baldy; Wheeler Peak. Meadows, in the Arctic-Alpine Zone.

3. **Silene laciniata** Cav. Icon. Pl. **6**: 44. *pl. 564.* 1301.
Silene greggii A. Gray, Pl. Wright. **2**: 17. 1853.
Melandryum laciniatum greggii Rohrb. Monogr. Silen. 232. 1868.
Melandryum greggii Rohrb. Linnaea **36**: 256. 1869.
Type locality: "Habitat in Pachuca, Real del monte et Acapulco," Mexico.
Range: California to western Texas, southward into Mexico.
New Mexico: Tunitcha Mountains; mountains west of Grant; Hop Canyon; Mogollon Mountains; Hillsboro Peak; Florida Mountains; San Luis Mountains; Organ Mountains; White and Sacramento mountains; Capitan Mountains. Damp woods, in the Transition and Canadian zones.

52576°—15——16

This species is one of the handsomest plants of our mountains, showy because of its bright scarlet flowers. Most of our specimens might be placed under *greggii*, but there seems to be no reason for maintaining this as a subspecies. It differs from the typical form only in its wider leaves and there seems to be no constancy in this character.

4. **Silene wrightii** A. Gray, Pl. Wright. **2**: 17. 1853.

Melandryum wrightii Rohrb. Linnaea **36**: 253. 1869.

TYPE LOCALITY: Crevices of rocks, mountain sides near the Copper Mines, New Mexico. Type collected by Wright (no. 862).

RANGE: Southwestern New Mexico.

NEW MEXICO: Mogollon Creek; Santa Rita.

One of the rarest species of the genus, for apparently it has been collected but twice.

5. **Silene pringlei** S. Wats. Proc. Amer. Acad. **23**: 269. 1888.

Silene concolor Greene, Leaflets **1**: 153. 1905.

TYPE LOCALITY: Cool slopes at the base of the cliffs in the Sierra Madre, Chihuahua.

RANGE: New Mexico and Arizona, southward into Mexico.

NEW MEXICO: Higher mountains throughout the State. Damp woods, in the Transition and Canadian zones.

There seems to be no essential difference between this and *Silene scouleri*, a species of the northwest. Probably *S. pringlei* should be treated as a synonym of that species.

The type of *Silene concolor* is from Iron Creek (*Metcalfe* 1482).

6. **Silene menziesii** Hook. Fl. Bor. Amer. **1**: 90. *pl. 30.* 1830.

TYPE LOCALITY: "North-West coast of America."

RANGE: British America to California and northern New Mexico.

NEW MEXICO: Near Chama (*Standley* 6765). Moist woods, in the Canadian Zone.

7. **Silene noctiflora** L. Sp. Pl. 419. 1753. NIGHT-FLOWERING CATCHFLY.

TYPE LOCALITY: "Habitat in Suecia, Germania."

RANGE: Native of Europe, widely naturalized in North America.

NEW MEXICO: Balsam Park, Sandia Mountains (*Ellis* 364).

4. WAHLBERGELLA Fries.

Herbaceous perennials; styles 5, rarely 4, the capsule opening by as many or twice as many teeth; petals in ours wanting or inconspicuous.

1. **Wahlbergella drummondii** (Hook.) Rydb. Bull. Torrey Club **39**: 318. 1912.

Silene drummondii Hook. Fl. Bor. Amer. **1**: 89. 1830.

Lychnis drummondii S. Wats. in King, Geol. Expl. 40th Par. **5**: 37. 1872.

TYPE LOCALITY: "Plains of the Saskatchewan."

RANGE: British America to Arizona and New Mexico.

NEW MEXICO: Rio Pueblo; Ponchuelo Creek; Tunitcha Mountains; Chama. Damp meadows, in the Transition Zone.

5. AGROSTEMMA L.

Pubescent biennial with linear-lanceolate leaves; flowers large, showy, reddish purple; calyx lobes linear, foliaceous; stamens 10; styles 5, alternate with the calyx lobes.

1. **Agrostemma githago** L. Sp. Pl. 435. 1753. CORN COCKLE.

Lychnis githago Scop. Fl. Carn. ed. 2. **1**: 310. 1772.

TYPE LOCALITY: "Habitat inter Europae segetes."

RANGE: Native of Europe and Asia, widely introduced into grain fields and waste ground in North America.

NEW MEXICO: Balsam Park, Sandia Mountains (*Ellis* 382).

Order 25. RANALES.

KEY TO THE FAMILIES.

Stamens 6; anther sacs opening by hinged valves .... 53. **BERBERIDACEAE** (p. 258).
Stamens numerous; anther sacs opening by slits.
    Flowers monœcious, minute, sessile; anthers with hornlike appendages; plants submerged.................. 51. **CERATOPHYLLACEAE** (p. 243).
    Flowers usually perfect; anthers not with hornlike appendages; mostly land plants, rarely aquatics............. 52. **RANUNCULACEAE** (p. 243).

51. CERATOPHYLLACEAE. Hornwort Family.

1. CERATOPHYLLUM L. HORNWORT.

A submerged aquatic; leaves whorled, finely dissected; flowers minute, axillary, sessile, monœcious, without floral envelopes but surrounded by an 8 to 12-cleft involucre; fertile flowers consisting of a simple 1-celled ovary, the sterile with 10 to 20 stamens, the anthers large, sessile; fruit a smooth achene, with a long persistent beak.

1. **Ceratophyllum demersum** L. Sp. Pl. 992. 1753.
TYPE LOCALITY: "Habitat in Europae fossis majoribus sub aqua."
RANGE: Throughout temperate North America, and in Europe.
NEW MEXICO: Mangas Springs; San Augustine Ranch. In slow-flowing streams.

52. RANUNCULACEAE. Crowfoot Family.

Herbs or sometimes woody plants, with colorless, often bitter juice and polypetalous or apetalous flowers, the calyx often corolla-like; stamens numerous; pistils many or few, rarely single, distinct; flowers regular or irregular; sepals 3 to 15; fruit of dry pods or of achenes or berries.

KEY TO THE GENERA.

Carpels with several ovules; fruit a follicle or berry.
    Flowers irregular.
        Posterior sepal spurred....................... 1. DELPHINIUM (p. 244).
        Posterior sepal hooded, helmet-shaped or boat-shaped............................ 2. ACONITUM (p. 247).
    Flowers regular.
        Petals conspicuous, produced into a spur at the base; leaves ternately compound; flowers showy......................... 3. AQUILEGIA (p. 248).
        Petals inconspicuous or none, not spurred; leaves simple or ternate; flowers showy or small.
            Fruit of follicles; leaves simple; flowers solitary......................... 4. CALTHA (p. 249).
            Fruit a berry; leaves ternately compound; flowers racemose................. 5. ACTAEA (p. 249).
Carpels 1-ovuled; fruit an achene.
    Petals usually present.
        Sepals spurred; annuals; leaves all basal; receptacle in fruit elongate-cylindric.... 6. MYOSURUS (p. 249).

Sepals not spurred; perennials or biennials; leaves not all basal; receptacle merely conic or short-cylindric.
Achenes transversely wrinkled; petals white; plants floating; leaves dissected into capillary segments..... 7. BATRACHIUM (p. 250).
Achenes not transversely wrinkled; petals yellow, at least outside; plants not floating; leaves not dissected into capillary segments.
Achenes not ribbed.................. 8. RANUNCULUS (p. 251).
Achenes longitudinally ribbed...... 9. HALERPESTES (p. 253).
Petals wanting, but the sepals often petal-like.
Sepals imbricated in the bud; leaves alternate or only those subtending the inflorescence opposite.
Flowers numerous, not subtended by opposite or verticillate bracts.
Seed pendulous; leaves compound ..10. THALICTRUM (p. 254).
Seed erect; leaves simple...........16. TRAUTVETTERIA (p. 258).
Flowers solitary or few, subtended by opposite or verticillate bracts.
Styles short, not elongated in fruit..11. ANEMONE (p. 255).
Styles much elongated in fruit (plumose).......................12. PULSATILLA (p. 255).
Sepals valvate in the bud; leaves all opposite.
Flowers cymose-paniculate, diœcious or polygamo-diœcious. (Stamens and sepals spreading)..................13. CLEMATIS (p. 256).
Flowers solitary, perfect.
Stamens erect; sepals thick, more or less convergent; staminodia wanting......................14. VIORNA (p. 256).
Stamens spreading; sepals thin, spreading; staminodia often present........................15. ATRAGENE (p. 258).

1. DELPHINIUM L. LARKSPUR.

Perennial herbs with leafy stems; sepals 5, irregular, petal-like, the upper one prolonged into a spur; petals 4, irregular, the upper continued as a short spur, enveloped in the spur of the calyx; pistils 1 to 5, forming many-seeded pods in fruit; flowers in terminal racemes.

KEY TO THE SPECIES.

Plants low, 50 cm. high or mostly less.
Leaves not crowded at the base of the stem, the cauline like the basal ones; flowers dark blue.
Plants viscid; lower pedicels not elongated......... 2. *D. alpestre.*
Plants not viscid; lower pedicels elongated......... 1. *D. nelsonii.*
Leaves forming a cluster about the base of the stem, the cauline ones few or none; flowers blue to whitish.
Flowers whitish or very pale blue.................. 3. *D. camporum.*

Flowers blue.
Leaves with numerous narrow segments; inflorescence short; flowers small, crowded....... 4. *D. confertiflorum.*
Leaves with few broad segments; inflorescence scapose; flowers large and scattered....... 5. *D. scaposum.*
Plants tall, 1 meter high or more, with leafy stems.
Segments of the leaves more or less oblong to linear, abruptly acute, only the basal segments cuneate; pubescence of short, curled, closely appressed hairs, rarely if ever glandular.
Bracts of the inflorescence expanded and at least the lowest resembling foliar leaves................ 6. *D. amplibracteatum.*
Bracts of the inflorescence narrowly linear.
Plants robust, 2 meters high or more............ 9. *D. robustum.*
Plants of medium height, about 1 meter tall.
Ultimate segments of the leaves oblong..... 7. *D. scopulorum.*
Ultimate segments of the leaves narrowly linear............................... 8. *D. tenuisectum.*
Segments of the leaves narrowly diamond-shaped in outline, with acute or acuminate apex and cuneate base; pubescence of spreading hairs, more or less glandular, at least in the inflorescence (except in no. 10).
Flowers dull brownish or greenish purple.
Leaf segments 10 to 20 mm. broad...............12. *D. sapellonis.*
Leaf segments narrow, 7 mm. wide or less.......11. *D. sierrae-blancae.*
Flowers deep blue.
Sepals acuminate...............................13. *D. cockerellii.*
Sepals not acuminate, at least a part of them obtuse.
Leaf segments broad; leaves yellowish green.14. *D. macrophyllum.*
Leaf segments very narrow; leaves dark green above..........................10. *D. novomexicanum.*

1. **Delphinium nelsonii** Greene, Pittonia **3**: 92. 1896.

TYPE LOCALITY: Wyoming.

RANGE: Washington and Nebraska to Utah and northern New Mexico.

NEW MEXICO: Chama; Tierra Amarilla; Stinking Lake. Open slopes, in the Transition Zone.

2. **Delphinium alpestre** Rydb. Bull. Torrey Club **29**: 146. 1902.

TYPE LOCALITY: Mountains northwest of Como, Colorado.

RANGE: Colorado and northern New Mexico.

NEW MEXICO: Timber line above Baldy; Wheeler Peak. Meadows, in the Arctic-Alpine Zone.

3. **Delphinium camporum** Greene, Erythea **2**: 183. 1894.

Delphinium wootoni Rydb. Bull. Torrey Club **26**: 583. 1899.

TYPE LOCALITY: "A plant of dry sandy plains along the eastern base of the whole Rocky Mountain range, apparently from British America to Mexico."

RANGE: Colorado to Texas and Arizona.

NEW MEXICO: Nara Visa; Mangas Springs; flats near Nutt; Organ Mountains; Jornada del Muerto. Dry plains and hills, in the Upper Sonoran Zone.

The type of *D. wootoni* was collected in the Organ Mountains.

4. **Delphinium confertiflorum** Wooton, Bull. Torrey Club **37**: 33. 1910.

TYPE LOCALITY: Mountains 15 miles southeast of Patterson, New Mexico, near Culbertsons Ranch. Type collected by Wooton, August 16, 1900.

RANGE: Known only from the type locality, in the Transition Zone.

5. **Delphinium scaposum** Greene, Bot. Gaz. **6**: 156. 1881.

TYPE LOCALITY: Hill country between the Gila and San Francisco rivers, New Mexico. Type collected by Greene.

RANGE: Colorado to New Mexico and Arizona.

NEW MEXICO: Coolidge; Chusca Mountains; Defiance; between the Gila and Frisco rivers. Transition Zone.

6. **Delphinum amplibracteatum** Wooton, Bull. Torrey Club **37**: 35. 1910.

TYPE LOCALITY: N Bar Ranch, Mogollon Mountains, New Mexico. Type collected by Wooton.

RANGE: Known only from the type locality, in the Transition Zone.

7. **Delphinium scopulorum** A. Gray, Pl. Wright. **2**: 9. 1853.

Delphinium calophyllum Greene; Wooton, Bull. Torrey Club **37**: 36. 1910, as synonym.

TYPE LOCALITY: Mountain ravines near the Mimbres, New Mexico. Type collected by Wright (no. 842).

RANGE: New Mexico and Arizona.

NEW MEXICO: Wheelers Ranch; Agua Fria Spring; Kingston; Fort Bayard; Hanover Mountain. Transition Zone.

8. **Delphinium tenuisectum** Greene, Erythea **2**: 184. 1894.

TYPE LOCALITY: "Chihuahua, cool banks of ravines, plains at the base of the Sierra Madre."

RANGE: New Mexico, southward into Mexico, possibly also in Arizona.

NEW MEXICO: Mountains west of Grant; Santa Rita; Mogollon Mountains. Transition Zone.

9. **Delphinium robustum** Rydb. Bull. Torrey Club **28**: 276. 1901.

TYPE LOCALITY: Wahatoya Creek, below the Spanish Peaks, Colorado.

RANGE: Montana to New Mexico.

NEW MEXICO: Baldy; Johnsons Mesa; Chusca Valley; Sierra Grande; Raton. Meadows, in the Transition Zone.

10. **Delphinium novomexicanum** Wooton, Bull. Torrey Club **37**: 37. 1910.

TYPE LOCALITY: Near Cloudcroft, Otero County, New Mexico. Type collected by Wooton, July 31, 1899.

RANGE: White and Sacramento mountains of New Mexico, in the Transition Zone.

11. **Delphinium sierrae-blancae** Wooton, Bull. Torrey Club **37**: 38. 1910.

TYPE LOCALITY: White Mountain Peak, New Mexico. Type collected by Wooton, August 1, 1901.

RANGE: White Mountains of New Mexico, in the Hudsonian Zone.

12. **Delphinium sapellonis** Cockerell, Bot. Gaz. **34**: 453. 1902.

TYPE LOCALITY: Sapello Canyon, near Beulah, New Mexico. Type collected by Cockerell.

RANGE: Northern and central New Mexico.

NEW MEXICO: Santa Fe and Las Vegas mountains; Sandia Mountains. Along streams, in the Transition and Canadian zones.

13. Delphinium cockerellii A. Nels. Bot. Gaz. **42:** 51. 1906.

TYPE LOCATION: Baldy Mountains, Elizabethtown, New Mexico.

RANGE: Southern Colorado and northern New Mexico.

NEW MEXICO: Elizabethtown; Santa Fe and Las Vegas mountains. Hudsonian Zone.

14. Delphinium macrophyllum Wooton, Bull. Torrey Club **37:** 40. 1910.

TYPE LOCALITY: Hillsboro Peak, Black Range, Sierra County, New Mexico. Type collected by Metcalfe (no. 1311).

RANGE: Mountains of southwestern New Mexico.

NEW MEXICO: Mogollon Mountains; Black Range; San Mateo Mountains. Hudsonian Zone.

2. ACONITUM L. MONKSHOOD.

Tall perennials with palmately cleft or dissected leaves and showy paniculate or racemose flowers; sepals 5, irregular, the upper one helmet-shaped, hooded, larger than the others; upper petals 2, small, spur-shaped, long-clawed, concealed under the helmet; other petals 6 or less, very small or wanting; pistils 3 to 5; pods several-seeded.

KEY TO THE SPECIES.

Front line of the hood strongly concave; beak long, porrect, almost horizontal .... 1. *A. porrectum.*

Front line of the hood almost straight; beak directed downward.

Flower ochroleucous .... 4. *A. lutescens.*

Flowers blue.

Leaf segments all acute, narrow, much dissected; pedicels much shorter than the flowers .... 2. *A. mogollonicum.*

Leaf segments mostly obtuse, abruptly acuminate, very broad, little dissected; pedicels about equaling the flowers .... 3. *A. arizonicum.*

1. Aconitum porrectum Rydb. Bull. Torrey Club **29:** 150. 1902.

Aconitum robertianum Greene, Repert. Nov. Sp. Fedde **7:** 6. 1909.

TYPE LOCALITY: Coffee Pot Spring, Colorado.

RANGE: Damp woods, Wyoming to New Mexico.

NEW MEXICO: Chama; Santa Fe and Las Vegas mountains; Sandia Mountains; White Mountain Peak; Costilla Valley. Transition to Hudsonian Zone.

The type of *Aconitum robertianum* was collected on Pecos Baldy (*Bailey* 583).

2. Aconitum mogollonicum Greene, Repert. Nov. Sp. Fedde **7:** 5. 1909.

TYPE LOCALITY: West Fork of the Gila, Mogollon Mountains, New Mexico. Type collected by Metcalfe (no. 518).

RANGE: Southwestern New Mexico.

NEW MEXICO: West Fork of the Gila; Hillsboro Peak.

3. Aconitum arizonicum Greene, Repert. Nov. Sp. Fedde **7:** 5. 1909.

TYPE LOCALITY: Santa Rita Mountains, southern Arizona.

RANGE: Southern Arizona and New Mexico.

NEW MEXICO: Mountains near Holts Ranch (*Wooton*).

4. Aconitum lutescens A. Nels. Bot. Gaz. **42:** 51. 1906.

TYPE LOCALITY: Cummins, Wyoming.

RANGE: Wyoming to New Mexico.

NEW MEXICO: Tunitcha Mountains; Beulah. Damp woods, in the Transition Zone.

3. AQUILEGIA L. Columbine.

Perennial herbs with twice or thrice ternately compound leaves with lobed leaflets; flowers large and showy; sepals 5, regular, colored like the petals; petals 5, similar, each with a short spreading lip produced backward into a long hollow spur; pistils 5, with slender styles; pods erect, many-seeded, with long filiform tips.

KEY TO THE SPECIES.

Flowers nodding, red, with more or less yellow; spurs 2 cm. long or less.
  Sepals greenish; spurs slender, 2 to 3 times as long as the sepals.. 1. *A. elegantula.*
  Sepals red; spurs thick, only slightly if at all longer than the sepals, never twice as long.............................. 2. *A. formosa.*
Flowers erect, yellow or blue; spurs 3 to 7 cm. long.
  Flowers yellow; spurs 5 to 7 cm. long........................ 3. *A. chrysantha.*
  Flowers blue or bluish; spurs 3 to 4 cm. long.................. 4. *A. caerulea.*

1. **Aquilegia elegantula** Greene, Pittonia **4:** 14. 1899. Red columbine.

Type locality: "Southern Colorado, in Slide Rock Cañon, and on the flanks of Mt. Hesperus in spruce woods."

Range: Colorado and New Mexico.

New Mexico: Santa Fe and Las Vegas mountains; Sandia Mountains; White and Sacramento mountains. Transition and Canadian zones.

This and the following resemble the eastern *Aquilegia canadensis.* They occur only in the higher mountains, in damp woods, usually on the faces of cliffs. Both species are very handsome and might well be cultivated in gardens in the higher parts of the State.

2. **Aquilegia formosa** Fisch.; DC. Prodr. **1:** 50. 1824. Red columbine.

Type locality: Kamchatka.

Range: Alaska to California, Utah, and New Mexico; also in Siberia.

New Mexico: Mountains west of Grant; Sandia Mountains; Kingston; Bear Mountain; Luna. Damp woods, in the Transition Zone.

3. **Aquilegia chrysantha** A. Gray, Proc. Amer. Acad. **8:** 621. 1873. Yellow columbine.

Aquilegia leptocera flava A. Gray, Pl. Wright. **2:** 9. 1853.

Type locality: Organ Mountains, New Mexico. Type collected by Wright (no. 1306).

Range: New Mexico and Arizona.

New Mexico: Sandia Mountains; Mogollon Mountains; Organ Mountains; Fresnal. Shaded canyons, in the Transition Zone.

Although the plant is properly of the Transition Zone, it extends much farther down in moist, shaded places in the canyons. With its large, golden yellow flowers it is one of the most beautiful of our Southwestern plants. In the Organ Mountains it is abundant in several places, sometimes growing on the shaded faces of cliffs and sometimes about the edges of small pools.

4. **Aquilegia caerulea** James in Long, Exped. **2:** 15. 1825. Rocky Mountain columbine.

Type locality: On the divide between the Platte and the Arkansas, Colorado.

Range: Montana to Utah and New Mexico.

New Mexico: Tunitcha Mountains; Chama; Santa Fe and Las Vegas mountains; Sandia Mountains. Woods and open meadows, Transition to Hudsonian Zone.

This is the State flower of Colorado and no other State has one so beautiful. Few indeed are the flowers of the Rockies that can compare with this in beauty. The great blossoms, sometimes six inches in diameter, look like bits of fallen sky, and when the plants cover acres of meadow, as they sometimes do, no words can be found to do them justice. On the Upper Pecos the Rocky Mountain columbine is very abundant, especially in the parks near timber line. At Harveys Ranch in the Las Vegas Mountains they were seen in great quantity in a grain field. Forms with light-colored, almost white, flowers are not uncommon.

4. CALTHA L.

Glabrous perennial with oblong or oblong-ovate obtuse bright green leaves; sepals 5 to 9, petal-like, white inside, bluish without; pistils 5 to 10, with scarcely any styles; pods compressed, spreading.

1. **Caltha leptosepala** DC. Reg. Veg. Syst. **1**: 310. 1818. ELKSLIP.

Caltha leptosepala rotundifolia Huth, Helios **9**: 68. 1891.

Caltha rotundifolia Greene, Pittonia **4**: 80. 1899.

Caltha chionophila Greene, loc. cit.

TYPE LOCALITY: Prince William Sound, Alaska.

RANGE: British America, along the Rockies to northern New Mexico.

NEW MEXICO: Santa Fe and Las Vegas mountains. Wet meadows, in the Hudsonian and Arctic-Alpine zones.

The plant is very common on the higher peaks of the northern part of the State, growing in marshes about the small mountain lakes and in wet ground around snowbanks. The flowers are very handsome, the sepals being almost white inside and bluish outside, in color suggesting some of the water lilies.

5. ACTAEA L. COHOSH.

Perennial, with broad, twice or thrice ternately compound leaves, the leaflets cleft and toothed; flowers in a short terminal raceme rising above the leaves; sepals 4 or 5, caducous; petals 4 to 10, small, spatulate, clawed, whitish; stamens numerous, with slender, whitish or greenish filaments; pistil single, with a sessile depressed 2-lobed stigma.

1. **Actaea viridiflora** Greene, Pittonia **2**: 108. 1890.

TYPE LOCALITY: "In open rocky places among the pine and spruce woods of Mt. San Francisco, Arizona."

RANGE: Colorado to Arizona and New Mexico.

NEW MEXICO: Tunitcha Mountains; Chama; Dulce; Santa Fe and Las Vegas mountains; Sandia Mountains; Mogollon Mountains; Hillsboro Peak; White and Sacramento mountains. Transition to Hudsonian Zone.

It is possible that more than one species of the genus occurs in the State, but we are unable to find any means of separating them, at least in herbarium material. In a given locality forms with red and white berries occur, often growing together. The plants are found on moist slopes, usually in thick shade. When in fruit they are rather showy.

6. MYOSURUS L. MOUSETAIL.

Small annuals with linear entire basal leaves and simple 1-flowered scapes; sepals 5, spurred or appendaged at the base; petals 5, small, narrow, sometimes wanting, clawed; stamens 5 to 20; achenes very numerous, spicate on a filiform receptacle.

KEY TO THE SPECIES.

Mature carpels with a whitish, cellular or cartilaginous, cupulate border around the laterally compressed keel.
  Scapes elongated, 60 to 120 mm. long; heads of achenes 30 to 40 mm. long; beaks not exceeding the diameter of the back of the achene.. 1. *M. cupulatus.*
  Scapes short, less than 5 mm. long; heads 13 mm. long or less; beaks much exceeding the diameter of the back of the achenes.. 2. *M. egglestonii.*
Carpels not with a cupulate border.
  Achenes flat on the back, tipped with a very short appressed beak, or this wanting.................................. 3. *M. minimus.*
  Achenes strongly carinate, tipped with a long subulate ascending beak.. 4. *M. aristatus.*

1. **Myosurus cupulatus** S. Wats. Proc. Amer. Acad. **17**: 362. 1882.

TYPE LOCALITY: "Arizona, hills between the Gila and San Francisco rivers."

RANGE: Arizona and New Mexico.

NEW MEXICO: Bear Mountain; Hillsboro. Wet soil, in the Upper Sonoran Zone.

2. **Myosurus egglestonii** Woot. & Standl. Contr. U. S. Nat. Herb. **16**: 123. 1913.

TYPE LOCALITY: On a mesa on the road between Tierra Amarilla and Park View, Rio Arriba County, New Mexico. Type collected by W. W. Eggleston (no. 6472).

RANGE: Known only from type locality.

3. **Myosurus minimus** L. Sp. Pl. 284. 1753.

TYPE LOCALITY: "Habitat in Europae collibus apricis aridis."

RANGE: British America to California and Florida; also in Europe and Africa.

NEW MEXICO: Mimbres Valley; near Tierra Amarilla. Moist soil, in the Upper Sonoran Zone.

4. **Myosurus aristatus** Benth. Lond. Journ. Bot. **6**: 458. 1847.

TYPE LOCALITY: "Moist places in the Cordillera of Chili at Los Patos, Province of Coquimbo, elev. 11,200 feet above the level of the sea."

RANGE: Washington and Montana to California and New Mexico; also in South America.

NEW MEXICO: Tunitcha Mountains; Sandia Mountains. In mud, in the Transition Zone.

7. BATRACHIUM S. F. Gray. WATER CROWFOOT.

Aquatic or subaquatic perennials with dissected submersed leaves having many filiform segments, and occasionally with a few dilated emersed ones; peduncles solitary, opposite the leaves; petals white; achenes not margined, rugose.

KEY TO THE SPECIES.

Flowers small, the petals less than 5 mm. long, oblong-obovate; stamens 5 to 12.. 1. *B. drouetii.*
Flowers larger, the petals 5 to 7 mm. long, broadly obovate; stamens numerous.
  Segments of the leaves rather short, 10 to 15 mm. long, scarcely collapsing when withdrawn from the water.. 2. *B. trichophyllum.*
  Segments of the leaves longer, 15 to 30 mm. long, flaccid, collapsing when withdrawn from the water........... 3. *B. flaccidum.*

1. **Batrachium drouetii** (F. Schultz) Nyman, Bot. Not. 98. 1852.
Ranunculus drouetii F. Schultz, Arch. Fl. France et Allem. 10. 1842.
TYPE LOCALITY: European.
RANGE: Alaska to Vermont, Rhode Island, and Lower California, in the Rocky Mountains south to New Mexico.
NEW MEXICO: Tunitcha Mountains; Taos; Negrito Creek; Middle Fork of the Gila. In streams and ponds, in the Transition Zone.

2. **Batrachium trichophyllum** (Chaix) Bosch, Prodr. Fl. Bat. 5. 1850.
Ranunculus trichophyllus Chaix; Vill. Prosp. Pl. Dauph. **1:** 335. 1786.
Ranunculus aquatilis trichophyllus A. Gray, Man. ed. 5. 40. 1867.
Batrachium aquatile trichophyllum Cockerell, Univ. Mo. Stud. Sci. **2**[2]**:** 122. 1911.
TYPE LOCALITY: European.
RANGE: British America to North Carolina, California, and Mexico; also in Europe and Asia.
NEW MEXICO: Las Vegas; Vermejo Park. In water, in the Transition Zone.

3. **Batrachium flaccidum** (Pers.) Rupr. Fl. Cauc. 15. 1869.
Ranunculus flaccidus Pers. Ann. Bot. Usteri **14:** 39. 1795.
Ranunculus trichophyllus flaccidus A. Gray, Syn. Fl. **1**[1]**:** 21. 1895.
TYPE LOCALITY: Not stated.
RANGE: British America to North Carolina and Lower California.
NEW MEXICO: Canada Creek; Pecos River; Mimbres River; Dulce. In streams and ponds.

8. RANUNCULUS L. BUTTERCUP.

Annual or perennial herbs with alternate cauline leaves and numerous basal ones; flowers solitary or corymbose, yellow or white; sepals usually 5; petals commonly conspicuous; achenes capitate, numerous, usually smooth.

KEY TO THE SPECIES.

Sepals black-hairy .... 1. *R. macauleyi.*
Sepals not black-hairy.
  Basal leaves entire.
    Plants with long stolons rooting at the nodes; petals less than 5 mm. long .... 10. *R. reptans.*
    Plants not stoloniferous; petals more than 5 mm. long, often 10 mm. or more .... 11. *R. ellipticus.*
  Basal leaves not entire.
    None of the leaves cleft; plants very succulent, stoloniferous .... 2. *R. hydrocharoides.*
    At least part of the leaves cleft; plants less succulent, seldom stoloniferous.
      Stems floating or submerged .... 3. *R. purshii.*
      Stems neither floating nor submerged.
        Achenes glabrous.
          Petals short, 3 to 4 mm. long, about equaling the sepals; cauline leaves long-petioled .... 4. *R. eremogenes.*
          Petals longer, 8 to 10 mm., much longer than the sepals; cauline leaves nearly sessile.
            Petals oblong, 8 to 10 mm. long; basal leaves parted almost to the base .... 5. *R. nudatus.*

Petals broadly obovate, not more than 8 mm. long; basal leaves not cleft, oblong to deltoid-ovate.. 6. *R. subsagittatus.*

Achenes pubescent.

Part of the basal leaves undivided; stems and leaves with only a few soft and inconspicuous hairs.

Petals 5 to 6 mm. long; head of achenes oblong .... 7. *R. inamoenus.*

Petals 3 to 5 mm. long; head of achenes cylindric ....12. *R. micropetalus.*

All of the leaves cleft; stems and leaves densely hirsute.

Head of carpels oblong; petals not longer than the sepals; plants mostly erect.................. 8. *R. pennsylvanicus.*

Head of carpels globose; petals longer than the sepals; plants low and spreading .... 9. *R. macounii.*

1. Ranunculus macauleyi A. Gray, Proc. Amer. Acad. **15:** 45. 1880.

Type locality: Rocky Mountains in San Juan County, Colorado.

Range: Colorado and northern New Mexico.

New Mexico: Pecos Baldy; Truchas Peak. Meadows, in the Arctic-Alpine Zone.

A most handsome species with large bright yellow flowers set off by rich brown sepals. The plants grow up to the very edge of the snow banks.

2. Ranunculus hydrocharoides A. Gray, Mem. Amer. Acad. n. ser. **5:** 306. 1861.

Type locality: Wet marshes, Mabibi, Sonora.

Range: Southwestern New Mexico and southern Arizona.

New Mexico: Cloverdale; Middle Fork of the Gila. In swamps.

3. Ranunculus purshii Richards. Bot. App. Frankl. Journ. 741. 1823.

Type locality: "Wooded country from latitude 54° to 64° north."

Range: Alaska to Nova Scotia, Oregon, and New Mexico.

New Mexico: Near Fort Defiance (*Palmer*).

4. Ranunculus eremogenes Greene, Erythea **4:** 121. 1896.

Ranunculus sceleratus eremogenes Cockerell, Univ. Mo. Stud. Sci. **2**[2]: 124. 1911.

Type locality: "Of wet springy places and margins of pools in the West American desert regions, from along the eastern base of the Colorado Rocky Mountains, through the Great Basin, and to southeastern Oregon and northeastern British America."

Range: British America to California and New Mexico.

New Mexico: Gallo Spring; San Juan; Mangas Springs; Wheelers Ranch; Farmington. In marshes, in the Upper Sonoran Zone.

5. Ranunculus nudatus Greene, Leaflets **1:** 211. 1906.

Type locality: Burro Mountains, New Mexico. Type collected by Metcalfe (no 198).

Range: Southwestern New Mexico.

New Mexico: Burro Mountains; Santa Rita. In wet soil.

6. Ranunculus subsagittatus (A. Gray) Greene, Pittonia **2:** 59. 1890.

Ranunculus arizonicus subsagittatus A. Gray, Proc. Amer. Acad. **21:** 370. 1886.

Type locality: "North Arizona in De la Vergne Park of the San Francisco Mountains, in wet ground."

Range: Arizona and New Mexico.

NEW MEXICO: Chusca Mountains; Willow Creek; Middle Fork of the Gila. Wet soil, in the Transition Zone.

Although originally described as a subspecies of *R. arizonicus*, this may be distinguished at once by its broadly obovate petals and less dissected leaves.

7. **Ranunculus inamoenus** Greene, Pittonia 3: 91. 1896.

TYPE LOCALITY: "Common in the whole Rocky Mountain region, at middle elevations."

RANGE: Montana to Utah and New Mexico.

NEW MEXICO: Chama; Santa Fe and Las Vegas mountains; Ensenada; Rio Pueblo; Sandia Mountains; White and Sacramento mountains. Wet meadows, in the Transition Zone.

8. **Ranunculus pennsylvanicus** L. f. Suppl. Pl. 272. 1781.

TYPE LOCALITY: "Habitat in Pennsylvania."

RANGE: Nova Scotia to Washington, Georgia, and New Mexico.

NEW MEXICO: Middle Fork of the Gila. Damp woods, in the Transition Zone.

9. **Ranunculus macounii** Britton, Trans. N. Y. Acad. 12: 3. 1892.

TYPE LOCALITY: "Banks of rivers from Canada to near the mouth of the Mackenzie River lat. 65°; and from the shores of Hudson's Bay to the Pacific."

RANGE: British America to Iowa and New Mexico.

NEW MEXICO: Tunitcha Mountains; San Juan Valley; Rio Pueblo; Chama; Winsors Ranch; White and Sacramento mountains. Damp woods, in the Upper Sonoran and Transition zones.

10. **Ranunculus reptans** L. Sp. Pl. 549. 1753.

TYPE LOCALITY: "Habitat in Suecia, Russia, ad ripas lacuum."

RANGE: British America to New Jersey and New Mexico; also in the Old World.

NEW MEXICO: Tunitcha Mountains (*Standley* 7606). In mud, in the Transition Zone.

11. **Ranunculus ellipticus** Greene, Pittonia 2: 110. 1890.

TYPE LOCALITY: Not definitely stated.

RANGE: British Columbia and Montana to California and northern New Mexico.

NEW MEXICO: Vicinity of Chama and Tierra Amarilla. Wet ground, in the Transition Zone.

12. **Ranunculus micropetalus** (Greene) Rydb. Bull. Torrey Club 29: 158. 1902.

Ranunculus affinis micropetalus Greene, Pittonia 2: 110. 1890.

TYPE LOCALITY: San Francisco Mountain, Arizona.

RANGE: Colorado and Utah to Arizona and northern New Mexico.

NEW MEXICO: Chama (*Standley* 6620). Damp meadows, in the Transition Zone.

9. HALERPESTES Greene.

A low slender glabrous plant with long runners; scapes with 1 to 7 flowers; leaves clustered, rounded-ovate or reniform, crenate; petals 5 to 8, yellow; carpels thin-walled, striate, in a cylindric head.

1. **Halerpestes cymbalaria** (Pursh) Greene, Pittonia 4: 208. 1900.

Ranunculus cymbalaria Pursh, Fl. Amer. Sept. 392. 1814.

Oxygraphis cymbalaria Prantl in Engl. & Prantl, Pflanzenfam. 3^2: 63. 1891.

TYPE LOCALITY: Saline marshes near Onondaga, New York.

RANGE: Alaska to California, New Mexico, and Arizona, and eastward; also in South America and Asia.

NEW MEXICO: Common except along the eastern side of the State. Wet ground, in the Sonoran and Transition zones.

10. THALICTRUM L. Meadow rue.

Low or tall perennial herbs with alternate, twice or thrice ternately compound leaves, the petioles dilated at the base; inflorescence corymbose or paniculate, often polygamous or diœcious; sepals 4 or 5, petal-like, greenish, inconspicuous; petals none; achenes 4 to 15, grooved or ribbed or inflated.

KEY TO THE SPECIES.

Flowers perfect; plants small, 20 cm. high or less.............. 1. *T. cheilanthoides.*
Flowers diœcious or polygamous; plants large, 30 to 100 cm. high.
  Flowers polygamous; achenes not flattened, thick-walled; ribs thick, separated by acute grooves; cauline leaves sessile.. 2. *T. dasycarpum.*
  Flowers diœcious; achenes flattened, thin-walled; ribs thin, separated by wide rounded grooves; cauline leaves petiolate.
    Leaflets very large, 20 to 40 mm. long; achenes symmetrical or nearly so................................ 3. *T. occidentale.*
    Leaflets seldom more than 15 mm. long, usually less; achenes asymmetrical.
      Leaflets large; achenes almost twice as long as broad; pedicels and fruit glandular.................. 4. *T. fendleri.*
      Leaflets small; achenes but little longer than broad; pedicels and fruit usually glabrous.......... 5. *T. wrightii.*

1. Thalictrum cheilanthoides Greene, Leaflets **2**: 89. 1910.

Type locality: Pecos Baldy, New Mexico. Type collected by Standley (no. 4324).

Range: Known only from type locality, in the Arctic-Alpine Zone.

This is possibly not distinct from *T. alpinum* L.

2. Thalictrum dasycarpum Fisch. & Lall. in Fisch. & Mey. Ind. Sem. Hort. Petrop. **8**: 72. 1841.

Type locality: "Hab. in America boreali, ubi sub expeditione Frankliniana semina collecta sunt."

Range: British America to northern New Mexico.

New Mexico: Aztec (*Griffin*). Wet ground.

3. Thalictrum occidentale A. Gray, Proc. Amer. Acad. **8**: 372. 1872.

Type locality: Vancouver Island.

Range: British America to California and New Mexico.

New Mexico: Four miles south of Tijeras (*Herrick*). Mountains, in the Transition Zone.

4. Thalictrum fendleri Engelm. in A. Gray, Mem. Amer. Acad. n. ser. **4**: 5. 1849.

Type locality: Damp, shady places in the mountains around Santa Fe, New Mexico. Type collected by Fendler (no. 13).

Range: Wyoming to Arizona and New Mexico.

New Mexico: Carrizo Mountains; Tunitcha Mountains; Chama; Raton; Johnsons Mesa; Sierra Grande; Upper Pecos; Mogollon Mountains. Meadows and thickets in the mountains, Transition and Canadian zones.

5. Thalictrum wrightii A. Gray, Pl. Wright. **2**: 7. 1853.

Thalictrum fendleri wrightii Trel. in A. Gray, Syn. Fl. 1^1: 16. 1895.

Type locality: Mountain ravine at Santa Cruz, Sonora, Mexico.

Range: New Mexico and Arizona and southward.

New Mexico: North of Ramah; Pinos Altos Mountains; Luna; Wheelers Ranch; Hanover Mountain; Organ Mountains; White and Sacramento mountains. Mountains, in the Transition Zone.

11. ANEMONE L. Anemone.

Perennial herbs with mostly radical compound leaves; cauline leaves 2 or 3 together, forming an involucre remote from the flower; peduncles 1-flowered, solitary or umbellate; sepals few or many, petal-like; petals none; achenes pointed, flattened; styles short, not plumose; staminodia wanting; ovule pendulous.

KEY TO THE SPECIES.

Achenes glabrate in age.. 1. *A. canadensis.*
Achenes densely villous.
  Plants low, 20 cm. high or less; sepals 8 or more, colored.... 4. *A. sphenophylla.*
  Plants tall, 40 to 120 cm. high; sepals 5 or 6, white.
    Head of carpels globose or nearly so; leaf segments linear; styles filiform, usually deciduous; involucral leaves short-petioled.................. 2. *A. globosa.*
    Head of carpels cylindric; leaf segments cuneate or lanceolate; styles subulate, persistent; involucral leaves long-petioled.............................. 3. *A. cylindrica.*

1. **Anemone canadensis** L. Syst. Nat. ed. 12. **3:** App. 231. 1768.
Type locality: "Habitat in Pennsylvania."
Range: British America to Maryland and New Mexico.
New Mexico: Beulah; Gilmores Ranch. Woods, in the Transition Zone.

2. **Anemone globosa** Nutt.; Pritz. Linnaea **15:** 673. 1841.
Type locality: "In planitie fluminis Platte, et in vallibus montium rupestrium in lat. 42°."
Range: British America to California and New Mexico.
New Mexico: Santa Fe and Las Vegas mountains. Woods, in the Transition Zone.

3. **Anemone cylindrica** A. Gray, Ann. Lyc. N. Y. **3:** 221. 1836.
Type locality: Dry pine barrens, near Oneida Lake, New York.
Range: British America to New Jersey and Arizona.
New Mexico: Dulce; Santa Fe and Las Vegas mountains; Johnsons Mesa; Raton; Sierra Grande; White Mountains. Damp woods, in the Transition Zone.

4. **Anemone sphenophylla** Poepp. Fragm. Syn. Chile 27. 1833.
Type locality: "In Chile boreal. collibus graminos. ad Concon."
Range: Texas to California and southward.
New Mexico: Bishops Cap; Hillsboro; Florida Mountains; Tortugas Mountain. Dry hills, in the Lower Sonoran Zone.

12. PULSATILLA Adans. Pasque flower.

A low perennial, silky-villous throughout; flowers large and showy, purplish blue to white, usually with abortive stamens answering to petals; carpels numerous, capitate, with long styles, these in fruit becoming long feathery tails.

1. **Pulsatilla hirsutissima** (Pursh) Britton, Ann. N. Y. Acad. **6:** 217. 1891.
Clematis hirsutissima Pursh, Fl. Amer. Sept. 385. 1814.
Anemone nuttalliana DC. Reg. Veg. Syst. **1:** 193. 1818.
Type locality: "On the plains of Columbia river."
Range: British America to Washington, New Mexico, and Illinois.
New Mexico: Chama; Tierra Amarilla; Sandia Mountains; White Mountains. Meadows, in the Transition Zone.

13. CLEMATIS L. VIRGIN'S BOWER.

Perennial, more or less woody vines; flowers small, numerous, paniculate, diœcious or the pistillate with a few sterile stamens; sepals petal-like, white, thin, spreading; petals wanting.

KEY TO THE SPECIES.

Tails of the carpels 7 to 10 cm. long; panicle with few (usually less than 10) flowers; leaflets small, 30 mm. long or less........ 1. *C. drummondii.*
Tails of the carpels 4 cm. long or less; panicle many-flowered; leaflets 35 to 70 mm. long.
  Leaflets loosely pubescent on both surfaces, acute, never long-attenuate, the lobes coarsely crenate with obtuse teeth; achenes attenuate to the tails........................ 2. *C. neomexicana.*
  Leaflets usually entirely glabrous, mostly long-attenuate, the lobes incised with acute teeth or entire; achenes abruptly contracted into the tails........................ 3. *C. ligusticifolia.*

1. Clematis drummondii Torr. & Gray, Fl. N. Amer. **1:** 9. 1838.

TYPE LOCALITY: Texas.

RANGE: Texas to Arizona, south into Mexico.

NEW MEXICO: Tortugas Mountain; Organ Mountains; Otis. Dry hills and canyons, in the Lower Sonoran Zone.

The native name is "barba de chivo."

2. Clematis neomexicana Woot. & Standl. Contr. U. S. Nat. Herb. **16:** 122. 1913.

TYPE LOCALITY: San Luis Mountains, New Mexico. Type collected by E. A. Mearns (no. 2136).

RANGE: Southwestern New Mexico and adjacent Arizona and Mexico.

NEW MEXICO: San Luis Mountains; Organ Mountains. Dry canyons in the mountains, Upper Sonoran Zone.

3. Clematis ligusticifolia Nutt.; Torr. & Gray, Fl. N. Amer. **1:** 9. 1838.

TYPE LOCALITY: "Plains of the Rocky Mountains."

RANGE: British America to California and New Mexico.

NEW MEXICO: Common, except on the plains of the eastern part of the State. Lower and Upper Sonoran zones.

Clematis orientalis L. (*Clematis crux-flava* Cockerell [1]) is sometimes cultivated in New Mexico, and is reported to have escaped near Las Vegas. The nature of the plant is such, however, that it is not likely to become a permanent part of our flora.

14. VIORNA Reichenb. LEATHER FLOWER.

Herbaceous or woody perennials with erect or climbing stems; leaves pinnate; flowers large, solitary on long peduncles, usually nodding; sepals thick, erect, mostly dull purple; petals none; anthers linear.

KEY TO THE SPECIES.

Stems herbaceous, erect; plants mostly 1-flowered.
  Sepals conspicuously dilated at the apex; plants permanently villous.. 1. *V. eriophora.*
  Sepals not dilated; plants glabrate in age.
    Plants stout; leaf segments large, 20 to 40 mm. long........ 2. *V. scottii.*
    Plants slender; leaf segments small, 5 to 15 mm. long....... 3. *V. bakeri.*

[1] Science n. ser. **10:** 898. 1899.

Stems woody, at least at the base, climbing or reclining; flowers several or numerous.
  Leaflets strongly reticulate-veined .... 4. *V. filifera.*
  Leaflets not reticulate-veined.
    Division of the leaves linear .... 5. *V. arizonica.*
    Divisions of the leaves lanceolate or broader.
      Divisions of the leaves acute, lanceolate; tails of the achenes plumose .... 6. *V. bigelovii.*
      Divisions of the leaves obtuse, ovate or oblong; tails of the achenes glabrate, or pubescent near the base.. 7. *V. palmeri.*

1. **Viorna eriophora** Rydb. Colo. Agr. Exp. Sta. Bull. **100:** 141. 1906.
Clematis eriophora Rydb. Bull. Torrey Club **29:** 154. 1902.
TYPE LOCALITY: Vicinity of Horsetooth, Colorado.
RANGE: Wyoming and Utah to northern New Mexico.
NEW MEXICO: Nutritas Creek below Tierra Amarilla (*Eggleston* 6492). Hills, in the Transition Zone.

2. **Viorna scottii** (Porter) Rydb. Colo. Agr. Exp. Sta. Bull. **100:** 141. 1906.
Clematis scottii Porter in Port. & Coult. Syn. Fl. Colo. 1. 1874.
Clematis douglasii scottii Coulter, Man. Rocky Mount. 3. 1885.
TYPE LOCALITY: Soda Springs, 35 miles west of Canyon City, Colorado.
RANGE: South Dakota and Wyoming to northern Mexico.
NEW MEXICO: Chama; Raton; Las Vegas. Open hills, in the Transition Zone.

3. **Viorna bakeri** (Greene) Rydb. Colo. Agr. Exp. Sta. Bull. **100:** 141. 1906.
Clematis bakeri Greene, Pittonia **4:** 147. 1900.
TYPE LOCALITY: On hillsides among scrub oaks, at Los Pinos, southern Colorado.
RANGE: Mountains of southern Colorado and northern New Mexico.
NEW MEXICO: Fort Wingate (*R. V. Shufeldt*). Upper Sonoran and Transition zones.

4. **Viorna filifera** (Benth.) Woot. & Standl. Contr. U. S. Nat. Herb. **16:** 123. 1913.
Clematis filifera Benth. Pl. Hartw. 285. 1848.
TYPE LOCALITY: "Prope Leon," Mexico.
RANGE: New Mexico, southward into Mexico.
NEW MEXICO: Ruidoso Creek; Roswell; Guadalupe Mountains. Upper Sonoran and Transition zones.

5. **Viorna arizonica** Heller, Muhlenbergia **1:** 40. 1904.
Clematis arizonica Heller, Bull. Torrey Club **26:** 547. 1899.
TYPE LOCALITY: Rocky slopes of Walnut Canyon, near Flagstaff, Arizona.
RANGE: Northern Arizona and northwestern New Mexico.
NEW MEXICO: Ramah (*Wooton*).

6. **Viorna bigelovii** (Torr.) Heller, Muhlenbergia **6:** 96. 1910.
Clematis bigelovii Torr. U. S. Rep. Expl. Miss. Pacif. **4:** 61. 1856.
TYPE LOCALITY: Sandia Mountains, New Mexico. Type collected by Bigelow in October, 1853.
RANGE: Mountains of New Mexico.
NEW MEXICO: Tunitcha Mountains; Sandia Mountains; White Mountains. Transition Zone.

7. **Viorna palmeri** (Rose) Woot. & Standl. Contr. U. S. Nat. Herb. **16:** 123. 1913.
Clematis palmeri Rose, Contr. U. S. Nat. Herb. **1:** 118. 1891.
TYPE LOCALITY: Fort Apache, Arizona.
RANGE: Southern New Mexico and Arizona.
NEW MEXICO: Mimbres River (*Metcalfe* 1044).

52576°—15——17

15. ATRAGENE L. Virgin's bower.

Vines, woody at least below, with compound leaves; peduncles bearing single, rather large flowers; sepals thin, widely spreading; some of the outer filaments enlarged and petaloid.

1. **Atragene pseudoalpina** (Kuntze) Rydb. Bull. Torrey Club **29:** 157. 1902.
Clematis pseudoatragene pseudoalpina Kuntze, Verh. Bot. Ver. Brand. **26:** 160. 1884.
Clematis occidentalis albiflora Cockerell, Bot. Gaz. **29:** 281. 1900.
Type locality: Colorado.
Range: Colorado and New Mexico.
New Mexico: Tunitcha Mountains; Dulce; Tierra Amarilla; Raton; Zuni Mountains; Santa Fe and Las Vegas mountains; Sandia Mountains; Mogollon Mountains; Black Range; White and Sacramento mountains. Damp woods, in the Transition and Canadian zones.

16. TRAUTVETTERIA Fisch. & Mey.

Perennial herb with alternate, 2 or 3-ternately compound leaves; flowers in corymbs or panicles, often polygamous or diœcious; sepals 4 or 5, petal-like, whitish, soon deciduous; petals wanting; achenes numerous, in a head, compressed; ovule ascending.

1. **Trautvetteria grandis** Nutt.; Torr. & Gray, Fl. N. Amer. **1:** 37. 1839.
Trautvetteria media Greene, Leaflets **2:** 192. 1912.
Type locality: "Shady woods of the Oregon."
Range: British Columbia to Idaho, California, and New Mexico.
New Mexico: Santa Fe and Las Vegas mountains; Sandia Mountains; Mogollon Mountains. Transition and Canadian zones.

A rather showy plant with its panicles of white flowers and its large leaves, growing in bogs and along the edges of streams. The type of *T. media* is Metcalfe's 517 from the West Fork of the Gila. It differs in no perceptible way from typical *grandis*.

53. BERBERIDACEAE. Barberry Family.

Shrubs (one species low) with alternate, exstipulate, simple or compound, more or less spiny-toothed leaves; flowers yellow, in racemes, the pedicels mostly opposite; perianth segments distinct, free; anthers opening by uplifted valves; pistil simple; fruit a berry.

KEY TO THE GENERA.

Stems spiny; leaves simple.................................. 1. Berberis (p. 258).
Stems not spiny; leaves compound.......................... 2. Odostemon (p. 259).

1. BERBERIS L. Barberry.

A low shrub, 40 to 80 cm. high, with simple, elliptic to oblanceolate, fascicled leaves with weakly spiny-toothed margins; stems spiny; flowers crowded in short reflexed racemes; fruit a scarlet ellipsoidal berry about the size of a currant.

1. **Berberis fendleri** A. Gray, Mem. Amer. Acad. n. ser. **4:** 5. 1849.
Type locality: "Santa Fe Creek, at the foot of steep and rocky banks, near the water," New Mexico. Type collected by Fendler (no. 15).
Range: Colorado and northern New Mexico.
New Mexico: Dulce; Tierra Amarilla; Brazos; Sandia Mountains; Pajarito Park; Santa Fe and Las Vegas mountains. Open hillsides, in the Transition Zone.

A not uncommon shrub of open slopes in the mountains of the northern part of the State. It would be well worth cultivation as a hedge plant.

2. ODOSTEMON Raf.

Shrubs with compound leaves, the stems without spines; leaflets mostly coriaceous, persistent, sinuate-dentate with few or many spiny teeth; flowers in rather loose racemes, the parts, except the pistil, in 6's; fruit a few-seeded berry.

KEY TO THE SPECIES.

Low shrub, 10 to 30 cm. high, with but few leaves; leaflets with numerous small teeth.......... 1. *O. repens.*
Tall shrubs, often 150 cm. high or more, with numerous leaves; leaflets with few coarse teeth.
  Leaves trifoliolate; fruit red.......... 2. *O. trifoliolatus.*
  Leaves with 5 to 7 leaflets; fruit variously colored.
    Leaflets usually 7, oblong-ovate, bright green, mostly more than 3 cm. long.......... 5. *O. wilcoxii.*
    Leaflets 5, lanceolate, glaucous, 3 cm. long or mostly less.
      Fruit juicy, not inflated at maturity, blood red; terminal leaflet long-attenuate, comparatively narrow.......... 3. *O. haematocarpus.*
      Fruit dry and inflated at maturity, dark blue; terminal leaflet acute, broad.......... 4. *O. fremontii.*

1. Odostemon repens (Lindl.) Cockerell, Univ. Mo. Stud. Sci. 2[2]: 125. 1911.
OREGON GRAPE.

Berberis repens Lindl. in Edwards's Bot. Reg. **14:** *pl. 1176.* 1828.
Berberis nana Greene, Pittonia **3:** 98. 1896.

TYPE LOCALITY: "A native of the north-western part of North America."

RANGE: British Columbia and Wyoming to California and New Mexico.

NEW MEXICO: Tunitcha and Carrizo Mountains; Dulce; Chama; Ramah; Santa Fe and Las Vegas mountains; Zuni Mountains; Sandia Mountains; Black Range; Luna; Sacramento Mountains. Shaded hillsides, in the Transition and Canadian zones.

A decoction of the leaves and branches of this plant was used by the Navahos in treating rheumatism.

2. Odostemon trifoliolatus (Moric.) Heller, Muhlenbergia **7:** 139. 1912.

Berberis trifoliolata Moric. Pl. Amer. Rar. 113. *pl. 69.* 1841.

TYPE LOCALITY: "Hab. in Republica Mexicana, *inter Laredo et Bejar.*" This is now Texas. The type was collected by Berlandier.

RANGE: Western Texas to Arizona.

NEW MEXICO: Near Hermanas; Carrizalillo Mountains. Dry hills, in the Lower Sonoran Zone.

3. Odostemon haematocarpus (Wooton) Heller, Muhlenbergia **7:** 139. 1912.

Berberis haematocarpa Wooton, Bull. Torrey Club **25:** 304. 1898.

TYPE LOCALITY: Mescalero Agency in the White Mountains, New Mexico. Type collected by Wooton (no. 376).

RANGE: New Mexico and southern Arizona.

NEW MEXICO: Manzano and Sandia Mountains; Gallinas Mountains; Black Range; Carlisle; Carrizalillo Mountains; Organ and San Andreas mountains; White and Sacramento mountains; Guadalupe Mountains. Dry hillsides, in the Upper Sonoran Zone.

This is a fairly common shrub on the lower slopes of the mountains in the southern part of the State. The berries are bright blood red, pleasantly acid to the taste, and are used for making jellies. The shrub is evergreen and is well worth cultivation for decorative purposes.

4. Odostemon fremontii (Torr.) Rydb. Bull. Torrey Club **33**: 141. 1906.

Berberis fremontii Torr. U. S. & Mex. Bound. Bot. 30. 1859.

TYPE LOCALITY: On the tributaries of the Rio Virgen, southern Utah.

RANGE: Colorado and Utah to northern Arizona and New Mexico.

NEW MEXICO: Las Vegas (*B. E. Fernow*). Upper Sonoran Zone.

Judging from the leaflets alone this specimen belongs here. It is an incomplete one and it is impossible to be certain as to its identity. The locality from which it comes is within what may be expected to be the range of *O. fremontii* rather than *O. haematocarpus*. The species should be looked for along the western border of the State north of the middle, since it occurs in the mountains of adjacent Arizona.

5. Odostemon wilcoxii (Kearney) Heller, Muhlenbergia **7**: 139. 1912.

Berberis wilcoxii Kearney, Trans. N. Y. Acad. **14**: 29. 1894.

TYPE LOCALITY: Fort Huachuca, Arizona.

RANGE: Southwestern New Mexico, southeastern Arizona, and adjacent Mexico.

Although we have seen no New Mexican specimens of this it is said by Dr. E. A. Mearns[1] to grow in the San Luis Mountains.

Order 26. PAPAVERALES.

KEY TO THE FAMILIES.

Sepals 2 or 3; endosperm present.
  Flowers regular; stamens numerous............ 54. **PAPAVERACEAE** (p. 260).
  Flowers irregular; stamens 6 (diadelphous).... 55. **FUMARIACEAE** (p. 262).
Sepals 4, rarely more; endosperm wanting.
  Capsules 2-celled; stamens 6, tetradynamous, rarely 2 or 4........................ 56. **BRASSICACEAE** (p. 263).
  Capsules 1-celled; stamens various.
    Leaves compound; flowers showy; capsules with 2 parietal placentæ...... 57. **CAPPARIDACEAE** (p. 289).
    Leaves simple; flowers inconspicuous; capsules with 3 to 6 parietal placentæ.......................... 57a. **RESEDACEAE** (p. 291).

54. PAPAVERACEAE. Poppy Family.

Herbaceous annuals or perennials, with watery or thickened colored sap; leaves exstipulate, variously pinnatifid or dissected; flowers perfect, regular or irregular; sepals fugacious; petals 4, early deciduous; stamens distinct, hypogynous, with slender filaments; fruit a 1-celled capsule.

KEY TO THE GENERA.

Coarse spiny plants; leaves once or twice pinnatifid; flowers large, white.......................... 1. ARGEMONE (p. 261).
Slender glabrous plants; leaves usually more dissected; flowers yellow (except in *Papaver somniferum*, there usually red).
  Fruit a long slender striate pod splitting lengthwise.................................. 2. ESCHSCHOLZIA (p. 262).
  Fruit a globose capsule opening by pores at the top.................................. 3. PAPAVER (p. 262).

[1] Bull. U. S. Nat. Mus. **56**: 91. 1907.

1. ARGEMONE L. PRICKLY POPPY.

Coarse herbaceous biennials or perennials, 40 to 90 cm. high, frequently much branched, spiny throughout, with alternate, pinnatifid or bipinnatifid, more or less glaucous leaves; sap thickened; flowers large, 10 cm. in diameter or less; petals thin and delicate, falling readily, white; stamens numerous, forming a conspicuous yellow center; fruit a 4-valved capsule with numerous seeds.

KEY TO THE SPECIES.

Stems spiny and hispid .... 1. *A. hispida.*
Stems spiny but not hispid.
  Spines of the capsule herbaceous near the base and furnished
    with small spines or hairs .... 2. *A. squarrosa.*
  Spines of the capsule not herbaceous below, simple.
    Valves of the capsule sparsely spiny; horns of the sepals
      pyramidal or terete, smooth on the outer surface.... 3. *A. intermedia.*
    Valves of the capsule densely armed; horns of the sepals
      dilated, spiny on the outer surface.................. 4. *A. platyceras.*

1. **Argemone hispida** A. Gray, Mem. Amer. Acad. n. ser. **4**: 5. 1849.

Enomegra hispida A. Nels. Key Pl. Rocky Mount. 27. 1902.

TYPE LOCALITY: Low, sandy places around Santa Fe, New Mexico. Type collected by Fendler (no. 16).

RANGE: Wyoming to Utah and northern New Mexico.

NEW MEXICO: Santa Fe; Raton; Sierra Grande. Dry plains, in the Upper Sonoran Zone.

This is the common species farther north, coming into our range from that direction.

2. **Argemone squarrosa** Greene, Pittonia **4**: 68. 1901.

TYPE LOCALITY: Near Gray, Lincoln County, New Mexico. Type collected by Miss Josephine Skehan (no. 79).

RANGE: Central New Mexico.

NEW MEXICO: Gray; Fort Stanton. Upper Sonoran Zone.

3. **Argemone intermedia** Sweet, Hort. Brit. ed. 2. 585. 1830.

TYPE LOCALITY: Mexico.

RANGE: Nebraska and Kansas to Texas and Arizona, southward into Mexico.

NEW MEXICO: Organ Mountains; near Glorieta; headwaters of the Pecos. Dry plains and hills, in the Upper Sonoran Zone.

4. **Argemone platyceras** Link & Otto, Icon. Pl. Rar. **1**: 85. 1828.

Argemone pleiacantha Greene, Repert. Nov. Sp. Fedde **6**: 161. 1908.

Argemone platyceras pleiacantha Fedde in Engl. Pflanzenreich **40**: 285. 1909.

TYPE LOCALITY: "In Mexico, in Confre de Perote prope Hacienda de la Laguna."

RANGE: Wyoming and Nebraska to Arizona and Texas, southward into Mexico.

NEW MEXICO: Chiz; Silver City; Mangas Springs; Kingston; north base of Animas Peak; Cloverdale; Rio Fresnal; La Luz Canyon. On dry plains and hillsides, in the Upper Sonoran and sometimes the Transition zone.

This is the common species at middle levels throughout the State, occurring in wide arroyos, on flats along the foothills, and in open parks in the mountains. Within its altitudinal range it is frequently a conspicuous range weed, indicating an advanced stage of deterioration resulting from overstocking. It produces an abundance of seed which is eaten freely by doves. It usually goes under the name of "thistle" in New Mexico. The type of *A. pleiacantha* was collected near Kingston by Metcalfe.

2. ESCHSCHOLZIA Cham. CALIFORNIA POPPY.

Smooth, slender, more or less glaucous annuals with finely dissected leaves and colorless sap; flowers bright yellow to orange, 5 cm. or less in diameter; sepals coherent at the tip, caducous; stamens numerous; pod elongate-linear, 10-nerved.

1. Eschscholzia parvula (A. Gray) Cockerell, Bot. Gaz. **26**: 279. 1898.
Eschscholzia douglasii parvula A. Gray, Pl. Wright. **2**: 10. 1853.
Eschscholzia scapifera Fedde, Notizbl. Bot. Gart. Berlin **4**: 153. 1904.
TYPE LOCALITY: "Among rocks, on mountains near El Paso."
RANGE: Western Texas to Arizona, south into Mexico.
NEW MEXICO: Mangas Springs; Florida Mountains; Carrizalillo Mountains; near White Water; Organ Mountains; Tortugas Mountain. Dry hills and mesas, in the Upper Sonoran Zone.
A beautiful little short-lived annual, appearing among rocks along the mountain foothills. It is well worth cultivation in gardens throughout the State, where it would doubtless grow readily.

3. PAPAVER L. POPPY.

Annual or perennial herbs with a white juice; leaves various, pubescent or glabrous, glaucous or bright green; petals mostly 4; sepals 2; stigmas united in a flat crown resting upon the summit of the ovary; fruit ovoid to globose, opening by pores under the edge of the stigmas.

KEY TO THE SPECIES.

Low perennial, 15 cm. or less, hairy, not glaucous; leaves pinnately cleft; petals yellow .... 1. *P. coloradense.*
Tall annual, 50 to 80 cm., glaucous, not hairy; leaves not pinnately cleft; petals white to red, never yellow .... 2. *P. somniferum.*

1. Papaver coloradense Fedde, Repert. Nov. Sp. Fedde **7**: 256. 1909.
Papaver nudicaule coloradense Fedde, Repert. Nov. Sp. Fedde **7**: 256. 1909.
TYPE LOCALITY: Grays Peak, Colorado.
RANGE: High peaks of Colorado and New Mexico.
NEW MEXICO: Taos Mountains (*Bailey* 853). Arctic-Alpine Zone.
A rare plant of high peaks in the northern part of the State, coming into our range from Colorado. It may be recognized by its small, bright yellow flowers, borne singly on hairy, scapelike peduncles.

2. Papaver somniferum L. Sp. Pl. 508. 1753. COMMON POPPY.
TYPE LOCALITY: "Habitat in Europae australioris ruderatis."
NEW MEXICO: Mesilla Valley.
The opium poppy has escaped from cultivation at several places in the Mesilla Valley and persists from year to year.

55. FUMARIACEAE. Fumitory Family.

1. CAPNOIDES Adans.

Short-lived perennial or biennial herbs with watery juice, compound, usually finely dissected leaves, and racemose yellow or pink flowers; sepals 2, small; corolla irregular, one of the outer pair of petals spurred at the base; stamens 6, in 2 groups opposite the outer petals; capsule 2-valved, linear-oblong; seeds lenticular, shining, black.

KEY TO THE SPECIES.

Corolla purplish pink; plants tall, about 1 meter high.......... 4. *C. brandegei.*
Corolla bright yellow; plants low, usually less than 30 cm. high.
  Pods not torulose, slightly curved; pedicels erect or ascending; racemes many-flowered, dense; spur about equaling the body of the flower; stems stout, mostly erect.. 1. *C. montanum.*
  Pods torulose, incurved-ascending; pedicels reflexed; racemes few-flowered, lax; spur about half as long as the body; stems slender, weakly ascending or prostrate.
    Bracts narrowly lanceolate to oblong, 8 mm. long or usually less; petioles without pinnæ near the base.. 2. *C. aureum.*
    Bracts rather broadly oblanceolate, 12 to 25 mm. long; a pair of pinnæ present almost at the base of the petiole.. 3. *C. euchlamydeum.*

1. **Capnoides montanum** (Engelm.) Britton, Mem. Torrey Club **5**: 166. 1894.
Corydalis montana Engelm.; A. Gray, Mem. Amer. Acad. n. ser. **4**: 6. 1849.
Corydalis aurea occidentalis Engelm.; A. Gray, Man. ed. 5. 62. 1867.
TYPE LOCALITY: Rocks, Santa Fe Creek, New Mexico. Type collected by Fendler (no. 17).
RANGE: South Dakota to Utah, Arizona, and Texas, south into Mexico.
NEW MEXICO: Santa Fe Creek; Burro Mountains; Mangas Springs; Black Range; Mesilla Valley; Organ Mountains; near Gray; mountains west of San Antonio. Open slopes, in the Lower and Upper Sonoran zones.

2. **Capnoides aureum** (Willd.) Kuntze, Rev. Gen. Pl. **1**: 14. 1891.
Corydalis aurea Willd. Enum. Pl. 740. 1809.
TYPE LOCALITY: "Habitat in Canada."
RANGE: British America to Pennsylvania, Texas, and California.
NEW MEXICO: Chama; Tierra Amarilla; Santa Fe and Las Vegas mountains; Zuni; Magdalena Mountains; Sandia Mountains; Mogollon Mountains; Black Range; San Luis Mountains; Organ Mountains. Damp thickets, in the Transition Zone.

3. **Capnoides euchlamydeum** Woot. & Standl. Contr. U. S. Nat. Herb. **16**: 122. 1913.
TYPE LOCALITY: Cloudcroft, Sacramento Mountains, New Mexico. Type collected by Wooton, August 8, 1899.
RANGE: Damp woods in the Sacramento and White mountains of New Mexico, in the Transition Zone.

4. **Capnoides brandegei** (S. Wats.) Heller, Cat. N. Amer. Pl. 4. 1898.
Corydalis brandegei S. Wats. Bot. Calif. **2**: 430. 1880.
TYPE LOCALITY: "Mountains of southern Colorado and in the Wahsatch."
RANGE: Utah and Colorado to northern New Mexico.
NEW MEXICO: Chama (*Standley* 6706). Damp canyons, in the Canadian Zone.

56. BRASSICACEAE. Mustard Family.

Herbaceous annuals, biennials, or perennials, sometimes with woody base, with watery, acrid or pungent sap; leaves alternate; flowers in mostly terminal racemes, generally small; sepals 4, deciduous; petals 4, rarely wanting; stamens 6, tetradynamous, rarely 2 or 4; ovary 2-celled by a thin partition, rarely 1-celled; fruit a silique or silicle. (The fruit is necessary for the determination of genera and species.)

KEY TO THE GENERA.

Pods indehiscent, 1-celled, with perforated wing-margin .... 1. THYSANOCARPUS (p. 266).
Pods dehiscent (except in Dithyraea and Raphanus), 2-celled, not wing-margined.
  Pods stipitate, terete. (Anthers sometimes curved or twisted.)
    Pods long-stipitate; stipes slender .... 2. STANLEYA (p. 266).
    Pods short-stipitate; stipes stout.
      Sepals spreading in anthesis, soon deciduous .... 3. STANLEYELLA (p. 267).
      Sepals erect or ascending in anthesis. (Stigma entire or indistinctly lobed, the lobes expanded over the valves.)
        Stigma conical; outer sepals gibbous at the base .... 4. HESPERIDANTHUS (p. 267).
        Stigma truncate; sepals scarcely gibbous at the base.
          Septum of pod with a strong midrib .... 5. PLEUROPHRAGMA (p. 267).
          Septum of the pod without midrib .... 6. THELYPODIUM (p. 268).
  Pods not stipitate, of various shapes.
    Anthers sagittate at the base, spirally curved.
      Pods terete; flowers small, more or less irregular; calyx not urceolate .... 7. HETEROTHRIX (p. 268).
      Pods flattened parallel to the partition; flowers large, regular; calyx urceolate .... 8. EUKLISIA (p. 268).
    Anthers neither sagittate nor spirally curved.
      Pods more or less flattened contrary to the narrow septum.
        Pods didymous; plants densely stellate-pubescent.
          Seeds solitary in each cell, the halves of the pod falling with the contained seeds; pods strongly flattened .... 9. DITHYRAEA (p. 269).
          Seeds 1 or 2 in each cell, not retained by the valve; pods inflated .... 10. PHYSARIA (p. 270).
        Pods not didymous; plants not densely stellate (except in Nerisyrenia).
          Pods oblong-linear, little compressed; plants strongly stellate-pubescent .... 11. NERISYRENIA (p. 270).
          Pods orbicular, oval, or cuneate, strongly flattened; plants not stellate-pubescent.
            Cells 1-seeded; pods broadest at the base.

Pods ovate-cordate, acute at the apex, neither winged nor retuse........12. CARDARIA (p. 271).

Pods orbicular or ovate, retuse or notched at the apex, usually winged....13. LEPIDIUM (p. 271).

Cells 2-seeded; pods broadest at the summit, more or less truncate.

Pods more or less winged; plants glabrous; cotyledons accumbent.....14. THLASPI (p. 272).

Pods wingless; plants with branched hairs; cotyledons incumbent.....15. BURSA (p. 273).

Pods of various shapes, never compressed nor flattened contrary to the septum.

Pods flattened parallel to the septum.

Valves of the pods elastically dehiscent; seeds in one row...........17. CARDAMINE (p. 276).

Valves of the pods not elastically dehiscent; seeds in one or two rows.

Pods short-oblong (in some species almost oval), frequently spirally twisted...................18. DRABA (p. 276).

Pods elongate-linear, never twisted.19. ARABIS (p. 279).

Pods not flattened nor compressed in any direction (sometimes slightly compressed at the apex in Lesquerella).

Pods short, globose, ovoid, or short-cylindric.

Valves of the pod nerved; cotyledons incumbent...............30. CAMELINA (p. 288).

Valves not nerved; cotyledons accumbent.

Pubescence conspicuously stellate; seeds flattened.........16. LESQUERELLA (p. 274).

Pubescence not stellate; seeds terete.......................24. RADICULA (p. 283).

Pods longer, terete or quadrangular.

Pods conspicuously beaked.

Pods terete, moniliform (indehiscent)......................20. RAPHANUS (p. 280).

Pods quadrangular.

Beak of the pod flat, swordlike..21. ERUCA (p. 281).

Beak of the pod elongated, conic or 4-angled..........22. BRASSICA (p. 281).

Pods not beaked, or at most only tipped by the persistent style or stigma.

Pods quadrangular by the thickening of the midribs of the valves.
  Plants glabrous; cauline leaves clasping .... 29. CONRINGIA (p. 288).
  Plants pubescent; cauline leaves not clasping .... 23. CHEIRINIA (p. 281).
Pods terete or nearly so, the midrib wanting or but little thickened.
  Petals conspicuously lobed ..... 28. DRYOPETALON (p. 288).
  Petals not lobed.
    Leaves, at least some of them, twice pinnately dissected .... 27. SOPHIA (p. 286).
    Leaves not twice pinnate.
      Seeds in two rows .... 24. RADICULA (p. 283).
      Seeds in one row.
        Glabrous perennial with creeping rootstock... 25. SCHOENOCRAMBE (p. 285).
        Plants pubescent with simple or branched hairs (one species glabrous annual), without rootstocks... 26. SISYMBRIUM (p. 285).

Selenia dissecta Torr. probably grows in the southeast corner of the State. The type locality is in western Texas very near the New Mexican boundary.

1. THYSANOCARPUS Hook.

Slender annual with flattened disk-shaped wing-margined 1-celled pods, the wing perforated or toothed like a cogwheel.

1. Thysanocarpus amplectens Greene, Pittonia **3:** 87. 1896.

TYPE LOCALITY: Southwestern New Mexico. Type collected by Greene, April 16, 1880.

RANGE: Known only from the type locality.

2. STANLEYA Nutt.

Stout perennial herbs with entire or pinnatifid leaves; flowers rather large, in elongated racemes, crowded in bud; calyx narrow, spreading; petals long-clawed, yellow; anthers linear or curved, the filaments elongated; silique subterete, elongated, long-stipitate.

KEY TO THE SPECIES.

Blades of the petals linear-oblong to elliptic; flowers bright yellow.. 1. *S. arcuata.*
Blades of the petals rounded-oval; flowers ochroleucous............. 2. *S. albescens.*

1. Stanleya arcuata Rydb. Bull. Torrey Club **29:** 232. 1902.

TYPE LOCALITY: Unionville Valley, Nevada.

RANGE: Wyoming and California to northwestern New Mexico.

NEW MEXICO: Western San Juan County. Dry hills and plains, in the Upper Sonoran Zone.

2. **Stanleya albescens** Jones, Zoe **2**: 17. 1891.
Type locality: "On the Moencoppa," Arizona.
Range: Western Colorado and northwestern New Mexico to Arizona.
New Mexico: Northwestern corner of the State (*Palmer*). Upper Sonoran Zone.

3. STANLEYELLA Rydb.

A tall branched biennial; leaves thin, the lower lyrately pinnatifid, the upper entire; sepals thin, petaloid, white, oblong or linear, spreading or reflexed in anthesis; petals white, with spatulate blades tapering into short claws; pods slender, terete, with short stipes and styles; stigmas truncate or nearly so.

1. **Stanleyella wrightii** (A. Gray) Rydb. Bull. Torrey Club **34**: 435. 1907.
Thelypodium wrightii A. Gray, Pl. Wright. **1**: 7. 1852.
Type locality: Pass of the Limpio, western Texas.
Range: Colorado and Utah to New Mexico and Arizona.
New Mexico: Dulce; Hurrah Creek; Magdalena Mountains; Mogollon Mountains; Organ Mountains; White Mountains; Gray; Raton. Hillsides, in the Transition Zone.

4. HESPERIDANTHUS (Robinson) Rydb.

Erect slender glabrous perennial herb with glaucous foliage, the stems corymbosely branched above; basal leaves obovate, toothed, the cauline ones linear, entire; sepals firm, erect, the outer strongly saccate at the base, purple; petals purple, with obovate blades; stigma conic or ovate, neither truncate nor bilobate; pods terete, linear, short-stipitate.

1. **Hesperidanthus linearifolius** (A. Gray) Rydb. Bull. Torrey Club **34**: 434. 1907.
Streptanthus linearifolius A. Gray, Mem. Amer. Acad. n. ser. **4**: 7. 1849.
Thelypodium linearifolium S. Wats. in King, Geol. Expl. 40th Par. **5**: 25. 1871.
Type locality: Mountainous regions from Santa Fe to Las Vegas, New Mexico. Type collected by Fendler (no. 24).
Range: Colorado to northern Mexico.
New Mexico: Gallup; Santa Fe and Las Vegas mountains; Sierra Grande; Gallinas Mountains; Raton; Mogollon Mountains; Burro Mountains; Black Range; San Luis Mountains; Tortugas Mountain; Organ Mountains; White and Sacramento mountains; Guadalupe Mountains. Canyons and thickets, in the Transition Zone.

5. PLEUROPHRAGMA Rydb.

Glabrous biennials with paniculate inflorescence; leaves thick, entire, the basal oblanceolate, the cauline linear-lanceolate, sessile; sepals ascending, thin, somewhat petaloid; petals white, on slender claws; pods slender, terete, torulose, tapering to a short stipe below and to a slender style above; stigma entire; septum with a strong rib.

KEY TO THE SPECIES.

Stipes about 1 mm. long; inflorescence short...................... 1. *P. integrifolium.*
Stipes 2 to 3 mm. long; inflorescence elongated.................. 2. *P. gracilipes.*

1. **Pleurophragma integrifolium** (Nutt.) Rydb. Bull. Torrey Club **34**: 433. 1907.
Pachypodium integrifolium Nutt.; Torr. & Gray, Fl. N. Amer. **1**: 96. 1838.
Thelypodium integrifolium Endl.; Walp. Repert. Bot. **1**: 172. 1842.
Type locality: "Elevated plains of the Rocky Mountains, toward the Oregon, as far as Wallahwallah."
Range: Nebraska and Washington to California and New Mexico.
New Mexico: Farmington (*Wooton* 2783, *Standley* 7158). Damp ground, in the Upper Sonoran Zone.

2. Pleurophragma gracilipes (Robinson) Rydb. Bull. Torrey Club **34**: 433. 1907.
Thelypodium integrifolium gracilipes Robinson in A. Gray, Syn. Fl. **1**[1]: 176. 1895.
TYPE LOCALITY: Southwestern Colorado.
RANGE: Southwestern Colorado to northern Arizona and northwestern New Mexico.
NEW MEXICO: Carrizo Mountains; Cedar Hill. Wet ground, in the Upper Sonoran and Transition zones.

6. THELYPODIUM Endl.

Slender glabrous biennial about 40 cm. high, with triangular-lanceolate auriculate-clasping leaves and elongated racemes; petals white, purplish-tinged, 5 mm. long or less; pods slender, 4 to 6 cm. long, somewhat divergent, arcuate; septum without a midrib; style truncate.

1. Thelypodium vernale Woot. & Standl. Contr. U. S. Nat. Herb. **16**: 128. 1913.
TYPE LOCALITY: Low mountains west of San Antonio, Socorro County, New Mexico. Type collected by Wooton (no. 3847).
RANGE: Known only from the type locality, in the Upper Sonoran Zone.

7. HETEROTHRIX Rydb.

Slender biennials, pubescent below with stellate or branched hairs; basal leaves oblanceolate, toothed; cauline leaves lance-linear or linear, entire; racemes elongated, slender; calyx somewhat oblique, the lower sepals longer than the upper, all ascending; petals spatulate, indistinctly clawed; pods slender, terete, sessile; stigma minute, entire or slightly lobed.

KEY TO THE SPECIES.

Sepals 3 mm. long or less, mostly green; siliques 20 to 35 mm. long, erect or ascending.......... 1. *H. micrantha*.
Sepals 4 to 5 mm. long, purple; siliques 50 to 70 mm. long, widely spreading.......... 2. *H. longifolia*.

1. Heterothrix micrantha (A. Gray) Rydb. Bull. Torrey Club **34**: 435. 1907.
Streptanthus micranthus A. Gray, Mem. Amer. Acad. n. ser. **4**: 7. 1849.
Thelypodium micranthum S. Wats. Proc. Amer. Acad. **17**: 321. 1882.
TYPE LOCALITY: Margins of Santa Fe Creek, New Mexico. Type collected by Fendler (no. 23).
RANGE: Colorado to Arizona and western Texas, southward into Mexico.
NEW MEXICO: Santa Fe and Las Vegas mountains; Mogollon Mountains; Black Range; Organ Mountains. Damp slopes, in the Transition Zone.

2. Heterothrix longifolia (Benth.) Rydb. Bull. Torrey Club **34**: 435. 1907.
Streptanthus longifolius Benth. Pl. Hartw. 10. 1839.
Thelypodium longifolium S. Wats. in King, Geol. Expl. 40th Par. **5**: 25. 1871.
TYPE LOCALITY: Mexico.
RANGE: New Mexico and Arizona to Mexico.
NEW MEXICO: Santa Fe and Las Vegas mountains; Tunitcha Mountains; La Jara; Sandia Mountains; Mogollon Mountains; Black Range; White and Sacramento mountains; Capitan Mountains. Transition Zone.

8. EUKLISIA (Nutt.) Rydb.

Annual or biennial herbs with glabrous glaucous leaves, the basal more or less toothed, the cauline linear to cordate-clasping; flowers rather large, 1 cm. long or less, in elongated terminal racemes; calyx urceolate; petals long-clawed, the blades undulate-crisped; anthers sagittate; siliques 5 to 7 cm. long or less, flattened parallel to the partition; seeds winged.

KEY TO THE SPECIES.

Cauline leaves lance-linear, narrowed at the base; sepals about 4 mm. long.. 1. *E. longirostris.*
Cauline leaves oblong to cordate, clasping; calyx 8 mm. long or more.
  Basal leaves toothed, the cauline obtuse; sepals purplish...... 2. *E. crassifolia.*
  Basal leaves pinnatifid, the cauline acute; sepals yellow...... 3. *E. valida.*

1 **Euklisia longirostris** (S. Wats.) Rydb. Bull. Torrey Club **33:** 142. 1906.
Arabis longirostris S. Wats. in King, Geol. Expl. 40th Par. **5:** 17. *pl. 2.* 1871.
Streptanthus longirostris S. Wats. Proc. Amer. Acad. **25:** 127. 1890.
TYPE LOCALITY: "Growing in alkaline soil at the Steamboat Springs near Washoe City, about Humboldt Lake, Nevada, and on Stansbury Island in Salt Lake."
RANGE: Washington to Nevada and New Mexico.
NEW MEXICO: Aztec; Albuquerque. Upper Sonoran Zone.

2. **Euklisia crassifolia** (Greene) Rydb. Bull. Torrey Club **33:** 142. 1906.
Streptanthus crassifolius Greene, Pittonia **3:** 227. 1897.
TYPE LOCALITY: "Frequent in the mountain districts of eastern California southward, and in adjacent Nevada and Arizona."
RANGE: California and Utah to New Mexico and Arizona.
NEW MEXICO: Carrizo Mountains; Aztec. Dry hills, in the Upper Sonoran Zone.

3. **Euklisia valida** (Greene) Woot. & Standl. Contr. U. S. Nat. Herb. **16:** 125. 1913.
Disaccanthus validus Greene, Leaflets **1:** 225. 1906.
Disaccanthus mogollonicus Greene, loc. cit.
Disaccanthus luteus Greene, loc. cit.
TYPE LOCALITY: El Paso, Texas.
RANGE: Western Texas and southern New Mexico.
NEW MEXICO: Mountains west of San Antonio; Silver City; Upper Corner Monument; Tres Hermanas; Tortugas Mountain; Bishops Cap; Kingston. Dry hills, in the Lower Sonoran Zone.

In the type of *Disaccanthus luteus* the flowers are of a deeper yellow than in the plant of the Rio Grande region. Following his description of *D. mogollonicus* Doctor Greene says: "All white-flowered material from New Mexico from Las Cruces to the upper Gila belongs here." As a matter of fact none of these plants have white flowers, but both the calyx and corolla are a pale, clear yellow. The type of *Disaccanthus mogollonicus* was collected in the Mogollon Mountains (Greene in 1881); that of *D. luteus* came from Kingston (*Metcalfe* 1593).

9. DITHYRAEA Haw. SPECTACLE-POD.

Erect branching canescent annuals with entire or pinnatifid leaves; flowers rather large, white, 6 to 8 mm. long, in elongated terminal racemes; silicles laterally flattened, each cell nearly orbicular, containing a single seed, indehiscent but separating at maturity from the persistent septum.

KEY TO THE SPECIES.

Pods stellate-pubescent, not strongly veined; cauline leaves with sinuate teeth, at least near the base.......................... 1. *D. wislizeni.*
Pods glabrous, conspicuously reticulate-veined; cauline leaves entire.. 2. *D. griffithsii.*

1. **Dithyraea wislizeni** Engelm. in Wisliz. Mem. North. Mex. 96. 1848.
Biscutella wislizeni Brewer & Wats. Bot. Calif. **1:** 48. 1878.
TYPE LOCALITY: Sandy soil near Valverde and Fray Cristobal, New Mexico. Type collected by Wislizenus in 1846.

RANGE: Colorado and Utah to Texas and northern Mexico.

NEW MEXICO: San Juan Valley; Zuni; Mimbres; Sabinal; Florida Mountains; Noria; Mesilla Valley; Organ Mountains; Jarilla Mountains; Melrose; 35 miles west of Roswell. Dry hills and mesas, in the Lower and Upper Sonoran zones.

This is one of the first plants to bloom in the spring, continuing in flower until summer. It is abundant in many parts of the State, especially on sandhills and gravelly mesas.

2. **Dithyraea griffithsii** Woot. & Standl. Contr. U. S. Nat. Herb. **16**: 124. 1913.

TYPE LOCALITY: Arroyo Ranch near Roswell, New Mexico. Type collected by David Griffiths (no. 5687).

RANGE: New Mexico.

NEW MEXICO: Arroyo Ranch; Zuni. Plains, in the Lower and Upper Sonoran zones.

10. PHYSARIA A. Gray. DOUBLE BLADDER-POD.

A low stellate-pubescent silvery-canescent cespitose perennial with obovate-spatulate leaves; racemes few-flowered; petals linear, yellow; siliques membranous, the two cells inflated-globose and joined by the narrow septum.

1. **Physaria newberryi** A. Gray in Ives, Rep. Colo. Riv. 6. 1861.

TYPE LOCALITY: Near Tegua, Arizona (one of the Moqui villages).

RANGE: Nevada and Colorado to New Mexico and Arizona.

NEW MEXICO: Western San Juan and McKinley counties. Dry hills and plains, in the Upper Sonoran Zone.

11. NERISYRENIA Greene.

Canescent herbaceous perennials 20 to 40 cm. high, branching from the base, densely covered with soft stellate hairs; racemes somewhat elongated; flowers white, purplish-tinged, about 1 cm. long; siliques 1 to 2 cm. long, oblong, compressed contrary to the septum, appearing somewhat quadrangular, tipped by the persistent style.

KEY TO THE SPECIES.

Leaves linear, entire .... 1. *N. linearifolia.*
Leaves obovate or oblanceolate, toothed or pinnatifid .... 2. *N. camporum.*

1. **Nerisyrenia linearifolia** (S. Wats.) Greene, Pittonia **4**: 225. 1900.

Greggia linearifolia S. Wats. Proc. Amer. Acad. **18**: 191. 1883.

Parrasia linearifolia Greene, Erythea **3**: 75. 1895.

TYPE LOCALITY: Presidio on the Rio Grande, Texas.

RANGE: Western Texas and southern New Mexico.

NEW MEXICO: White Sands; plains 35 miles south of Torrance; Lakewood; Los Mitos. In gypsum soil, in the Lower Sonoran Zone.

2. **Nerisyrenia camporum** (A. Gray) Greene, Pittonia **4**: 225. 1900.

Greggia camporum A. Gray, Pl. Wright. **1**: 8. *pl. 1.* 1852.

Parrasia camporum Greene, Erythea **3**: 75. 1895.

TYPE LOCALITY: "High prairies and calcareous hills, at the head of the San Felipe," Texas.

RANGE: Western Texas to New Mexico, south into Mexico.

NEW MEXICO: Upper Corner Monument; Tortugas Mountain; plains south of White Sands; Tres Hermanas; White Mountains; Jarilla Mountains; Organ Mountains. Dry hills and plains, in the Lower Sonoran Zone.

12. CARDARIA Desv. HOARY CRESS.

1. **Cardaria draba** (L.) Desv. Journ. de Bot. **3**: 163. 1813.
Lepidium draba L. Sp. Pl. 645. 1753.
TYPE LOCALITY: "Habitat in Germania, praesertim Austria, Gallia, Italia."
NEW MEXICO: Mesilla Valley.

An introduced weed having the appearance of a Lepidium, but the pods ovate-cordate, acute at the apex. It probably occurs elsewhere in the State, since the seeds are often distributed with those of garden or field crops or with grass seeds. It has been well established in the Mesilla Valley in alfalfa fields for several years.

13. LEPIDIUM L. PEPPERGRASS.

Herbaceous annuals or short-lived perennials with pinnatifid or simple leaves; flowers of medium or small size (in one of our species apetalous), with white petals, in crowded racemes, these elongating in fruit; stamens often fewer than 6; siliques orbicular or oblong, strongly flattened contrary to the septum, dehiscent, sometimes wing-margined; seeds flattened, solitary in each valve.

KEY TO THE SPECIES.

Style conspicuous, equaling or exceeding the wing margins of the fruit.
  Stems pubescent throughout; all the cauline leaves pinnatifid. 1. *L. thurberi.*
  Stems glabrous, at least below; upper cauline leaves entire.
    Branches of the inflorescence pubescent; lower cauline leaves pinnatifid, the upper linear-oblong........ 2. *L. eastwoodiae.*
    Branches of the inflorescence glabrous; all the cauline leaves entire, linear............................ 3. *L. alyssoides.*
Style obsolete, at least much shorter than the wing margins of the fruit.
  Petals wanting.. 4. *L. apetalum.*
  Petals present.
    Stems and leaves glabrous except sometimes below. (Plants bright green.)............................ 5. *L. medium.*
    Stems and leaves pubescent.
      Fruit glabrous; stems puberulent, erect.............. 6. *L. hirsutum.*
      Fruit pubescent; stems villous, spreading............ 7. *L. lasiocarpum.*

1. **Lepidium thurberi** Wooton, Bull. Torrey Club **25**: 259. 1898.
TYPE LOCALITY: Lava, New Mexico. Type collected by Wooton (no. 672).
RANGE: Southwestern New Mexico to Arizona.
NEW MEXICO: Lava; Cienaga Ranch; Silver City; Dog Spring; near White Water; Hatchet Ranch; Reserve. Upper Sonoran Zone.

2. **Lepidium eastwoodiae** Wooton, Bull. Torrey Club **25**: 258. 1898.
TYPE LOCALITY: Mescalero Agency, White Mountains, New Mexico. Type collected by Wooton.
RANGE: New Mexico and southern Colorado.
NEW MEXICO: Farmington; Chama; Cerrillos; Sandia Mountains; Gallinas Mountains; Galisteo; Organ Mountains; Sacramento and White mountains. Mountains, in the Upper Sonoran and Transition zones.

A common plant in the mountains of the central and southern parts of the State. It is a short-lived perennial corymbosely branched above, with numerous racemes of bright white flowers. The foliage is of a rather pronounced green (rarely yellowish green), and the lower leaves are pinnately divided, the upper leaves simple, oblong-lanceolate, flat.

3. Lepidium alyssoides A. Gray, Mem. Amer. Acad. n. ser. **4:** 18. 1849.

TYPE LOCALITY: "Mountain valleys, from Santa Fe eastward to Rabbit's Ear Creek," New Mexico. Type collected by Fendler (no. 46).

RANGE: Colorado to Texas and Arizona.

NEW MEXICO: Common nearly everywhere from the Pecos River westward across the State. Plains and hills, in the Lower and Upper Sonoran zones.

Nearly related to the last, but as seen growing it is noticeably yellowish green, the flowers are not so bright a white, the leaves are narrower, and the basal ones are less divided.

4. Lepidium apetalum Willd. Sp. Pl. **3:** 439. 1800.

TYPE LOCALITY: "In Siberia."

RANGE: British Columbia and New England, southward throughout the United States; also in Asia; probably introduced in New Mexico.

NEW MEXICO: Aztec; Agricultural College; Santa Fe; Pajarito Park.

5. Lepidium medium Greene, Erythea **3:** 36. 1895.

Lepidium intermedium A. Gray, Pl. Wright. **2:** 15. 1853, not A. Rich. 1847.

TYPE LOCALITY: Ravines of the Organ Mountains, New Mexico. Type collected by Wright (no. 1320).

RANGE: Missouri and Texas to California.

NEW MEXICO: Rio Pueblo; Sandia Mountains; Mangas Springs; Mimbres River; Florida Mountains; Mesilla Valley; Organ Mountains. Plains and hills, in the Lower and Upper Sonoran zones, rarely extending into the Transition.

6. Lepidium hirsutum Rydb. Bull. Torrey Club **39:** 322. 1913.

Lepidium intermedium pubescens Greene, Bot. Gaz. **5:** 157. 1881, not *L. pubescens* Desv. 1814.

Lepidium medium pubescens Robinson in A. Gray, Syn. Fl. 1^1: 127. 1895.

TYPE LOCALITY: Mangas Springs, New Mexico. Type collected by E. L. Greene, in 1880.

RANGE: Southwestern New Mexico and adjacent Arizona.

NEW MEXICO: Farmington; Chama; Winsors Ranch; Bear Mountain; Middle Fork of the Gila; Cliff; Pecos. Dry hillsides.

7. Lepidium lasiocarpum Nutt.; Torr. & Gray, Fl. N. Amer. **1:** 115. 1838.

Lepidium wrightii A. Gray, Pl. Wright. **2:** 15. 1853.

TYPE LOCALITY: "Near St. Barbara, Upper California."

RANGE: Washington to California and Texas.

NEW MEXICO: Carrizo Mountains; Farmington; Upper Corner Monument; Mesilla Valley; Tortugas Mountain; Organ Mountains. Dry fields, in the Lower and Upper Sonoran zones.

A low, spreading, more or less hirsute annual or winter annual, with stout divergent stems, pinnatifid leaves, and inconspicuous flowers followed quickly by the numerous somewhat crowded silicles. It is one of the few true "spring flowers" of the southern part of the State. Often it is a weed in irrigated land.

14. THLASPI L. PENNY CRESS.

Low annuals or perennials with a rosette of oblanceolate-spatulate leaves with few teeth; cauline leaves clasping; flowers white, purple-tinged, rather large for the family, 5 to 7 mm. long; siliques obovate to almost obcordate; style slender, remaining on the persistent septum.

KEY TO THE SPECIES.

Pods emarginate, the sinus narrow .... 1. *T. coloradense.*
Pods truncate or nearly so, the sinus broad and open.
  Cauline leaves ovate, 18 to 25 mm. long; pedicels slender, 8 to 10 mm. long; pods 5 mm. wide .... 2. *T. fendleri.*
  Cauline leaves oblong, 15 mm. long or less; pedicels stout, 6 mm. long or shorter; pods not more than 2.5 mm. wide.
    Stems 20 to 30 cm. high, slender; cauline leaves broader than the basal ones; sepals green .... 3. *T. glaucum.*
    Stems 10 cm. high or less, stouter; cauline leaves not broader than the basal ones; sepals purplish ..... 4. *T. purpurascens.*

1. **Thlaspi coloradense** Rydb. Bull. Torrey Club **28**: 280. 1901.

TYPE LOCALITY: Bald Mountain near Pikes Peak, Colorado.

RANGE: Colorado and New Mexico.

NEW MEXICO: Chama; Las Vegas Hot Springs; Sierra Grande. Mountains, in the Transition Zone.

2. **Thlaspi fendleri** A. Gray, Pl. Wright. **2**: 14. 1853.

TYPE LOCALITY: Organ Mountains, New Mexico. Type collected by Wright (no. 1322).

RANGE: New Mexico.

NEW MEXICO: Cooney; Hillsboro Peak; Organ Mountains. Canyons in the mountains, in the Transition Zone.

Perhaps, judging from the specific name and from the citation of synonymy, Fendler's 44 should be taken as the type of this species. Doctor Gray's description, however, evidently applies to the very different plant of southern New Mexico, and Wright's specimens are the only ones cited. If Fendler's plant were taken as the type, the specimens here listed should have a new name.

3. **Thlaspi glaucum** A. Nels. Bull. Torrey Club **25**: 275. 1898.

Thlaspi alpestre glaucum A. Nels. Wyom. Agr. Exp. Sta. Bull. **28**: 84. 1896.

TYPE LOCALITY: La Plata Mines, Wyoming.

RANGE: Idaho to central New Mexico.

NEW MEXICO: Winsors Ranch; San Mateo Mountains; Kelly; Cloudcroft; Chama; Pecos Baldy. Meadows under pine trees, in the Transition and Canadian zones.

4. **Thlaspi purpurascens** Rydb. Bull. Torrey Club **28**: 281. 1901.

TYPE LOCALITY: Arizona.

RANGE: Colorado to Arizona and New Mexico.

NEW MEXICO: Hermits Peak; Stinking Lake. Mountains, in the Transition Zone.

15. **BURSA** Weber. SHEPHERD'S-PURSE.

Branched annual with a rosette of narrow pinnatifid leaves; flowers small, white; silique obcordate-triangular, flattened contrary to the partition.

1. **Bursa bursa-pastoris** (L.) Web. Prim. Fl. Hols. 41. 1780.

Thlaspi bursa-pastoris L. Sp. Pl. 647. 1753.

Capsella bursa-pastoris Medic. Pflanzengat. **1**: 85. 1792.

TYPE LOCALITY: "Habitat in Europae cultis ruderatis."

NEW MEXICO: Winsors Ranch; Kingston; Las Huertas Canyon; Gallinas Canyon; Pecos; Mesilla Valley; Gilmores Ranch; Chama.

A very common weed in the Eastern States, introduced from Europe. So far it is rather rare in New Mexico, but occurs occasionally in irrigated fields and in gardens. Doubtless it will become more common. The plant may be recognized by the triangular-cuneate silicle which gives rise to the popular as well as the Latin name.

52576°—15——18

16. LESQUERELLA S. Wats. Bladder-pod.

Low herbaceous annuals or perennials, stellate or lepidote-hoary; leaves entire or repand-toothed; flowers mostly yellow, purplish white in one species; pods globose to ovoid, inflated, with nerveless valves; seeds several, in two rows in each cell; septa and styles persistent.

KEY TO THE SPECIES.

Flowers purplish white .... 1. *L. purpurea.*
Flowers yellow.
  Plants tall, 30 cm. or usually more, erect; pedicels reflexed from the base; annual .... 2. *L. aurea.*
  Plants lower, usually less than 20 cm., ascending, erect, or prostrate; pedicels erect, or reflexed from about the middle, never from the base; perennials or rarely annuals.
    Capsules lepidote-stellate.
      Pods compressed at the top.
        Plants densely cespitose, 6 cm. high or less; cauline leaves crowded .... 6. *L. intermedia.*
        Plants not densely cespitose, 15 cm. high; cauline leaves remote .... 7. *L. valida.*
      Pods not compressed at the summit.
        Capsules pointed at the apex, 7 to 8 mm. long; cauline leaves narrowly oblanceolate, acutish .... 3. *L. montana.*
        Capsules rounded at the apex, 4.5 mm. long or less; cauline leaves various.
          Capsules 3 mm. long; cauline leaves broadly oblanceolate .... 4. *L. lata.*
          Capsules 4.5 mm. long; cauline leaves linear-oblong or linear-oblanceolate .... 5. *L. rectipes.*
    Capsules glabrous.
      Plants annual; pedicels reflexed above the middle in age .... 8. *L. gordoni.*
      Plants perennial (sometimes blossoming the first year); pedicels erect, stout.
        Flowers not exceeding the leaves; pedicels arising from a common point, or axillary. (Plants densely cespitose, 4 to 6 cm. high; basal leaves linear-oblanceolate) .... 9. *L. praecox.*
        Flowers much exceeding the leaves, in more or less elongated racemes.
          Plants low, 5 cm. high or less; basal leaves ovate to orbicular, as broad as long .... 10. *L. ovalifolia.*
          Plants taller, usually more than 10 cm.; basal leaves broadly oblanceolate or narrower, always much longer than broad.
            Plants white with very dense lepidote pubescence; leaves linear or linear-oblanceolate, thick .... 11. *L. fendleri.*
            Plants greener, the pubescence much less dense; leaves oblanceolate or spatulate, thin .... 12. *L. pinetorum.*

1. **Lesquerella purpurea** (A. Gray) S. Wats. Proc. Amer. Acad. **23:** 253. 1888.
Vesicaria purpurea A. Gray, Pl. Wright. **2:** 14. 1853.
TYPE LOCALITY: "Stony hills near El Paso." Type collected by Wright (no. 1320).
RANGE: Western Texas to Arizona, south into Mexico.
NEW MEXICO: Big Hatchet Mountains; Florida Mountains; Organ Mountains. Canyons, in the Upper Sonoran Zone.
One of the rather few early spring flowers occurring in the mountains of the southern part of the State.

2. **Lesquerella aurea** Wooton, Bull. Torrey Club **25:** 260. 1898.
TYPE LOCALITY: South Fork of Tularosa Creek, 3 miles east of the Mescalero Agency, White Mountains, New Mexico. Type collected by Wooton (no. 245).
RANGE: Mountains of southern New Mexico.
NEW MEXICO: Luna; White and Sacramento mountains. Transition Zone.
The largest of our species. Its flowers are a deep yellow, abundantly borne on erect or ascending stems, followed by the globose fruit on strongly recurved pedicels.

3. **Lesquerella montana** (A. Gray) S. Wats. Proc. Amer. Acad. **23:** 251. 1888.
Vesicaria montana A. Gray, Proc. Acad. Phila. **1863:** 58. 1863.
TYPE LOCALITY: "From the Middle Mountains," Colorado.
RANGE: Wyoming to New Mexico.
NEW MEXICO: Pecos River; Sierra Grande; Raton. Hillsides, in the Upper Sonoran Zone.

4. **Lesquerella lata** Woot. & Standl. Contr. U. S. Nat. Herb. **16:** 126. 1913.
TYPE LOCALITY: Lincoln National Forest, New Mexico. Type collected by Fred G. Plummer in 1903.
RANGE: White Mountains of New Mexico.

5. **Lesquerella rectipes** Woot. & Standl. Contr. U. S. Nat. Herb. **16:** 127. 1913.
TYPE LOCALITY: Northwestern New Mexico. Type collected by C. C. Marsh (no. 81).
RANGE: New Mexico.
NEW MEXICO: South of Atarque de Garcia; banks of the Rio Grande, 19 miles west of Santa Fe; near East View; Tunitcha Mountains. Dry plains and hills, in the Upper Sonoran Zone.

6. **Lesquerella intermedia** (S. Wats.) Heller, Pl. World **1:** 22. 1897.
Lesquerella alpina intermedia S. Wats. Proc. Amer. Acad. **23:** 251. 1888.
TYPE LOCALITY: Hills west of Santa Fe, New Mexico. Type collected by Fendler (no. 38).
RANGE: Northern New Mexico and southern Colorado.
NEW MEXICO: Santa Fe; Wheelers Ranch. Upper Sonoran Zone.

7. **Lesquerella valida** Greene, Pittonia **4:** 68. 1899.
TYPE LOCALITY: Gray, New Mexico. Type collected by Miss Josephine Skehan (no. 89).
RANGE: Known only from the type locality.

8. **Lesquerella gordoni** (A. Gray) S. Wats. Proc. Amer. Acad. **23:** 253. 1888.
Vesicaria gordoni A. Gray, Bost. Journ. Nat. Hist. **6:** 149. 1850.
TYPE LOCALITY: On the Canadian, in the Raton Mountains, New Mexico or Colorado.
RANGE: Western Texas to Arizona.
NEW MEXICO: Mangas Springs; Mesilla Valley; Gray; Kingston. Dry fields, in the Lower Sonoran Zone.
A common spring annual in the southern part of the State, where it occurs associated with Sophias and other spring crucifers.

9. Lesquerella praecox Woot. & Standl. Contr. U. S. Nat. Herb. **16:** 126. 1913.

TYPE LOCALITY: New Mexico. Type collected by Bigelow.

RANGE: Central New Mexico.

NEW MEXICO: Gallinas Mountains; Cabra Springs. Open hills, in the Upper Sonoran Zone.

From *L. fendleri*, its nearest relative, this plant is at once distinguished by its lower, densely cespitose habit and its few pedicels which are surpassed by the leaves. The general appearance of the two is very different.

10. Lesquerella ovalifolia Rydb. in Britt. & Brown, Illustr. Fl. **2:** 137. *f. 1749.* 1897.

TYPE LOCALITY: Kimball County, Nebraska.

RANGE: Nebraska to New Mexico.

NEW MEXICO: Nara Visa; Sandia Mountains; San Mateo Mountains. Plains and open hills, in the Upper Sonoran Zone.

11. Lesquerella fendleri (A. Gray) S. Wats. Proc. Amer. Acad. **23:** 254. 1888.

Vesicaria fendleri A. Gray, Mem. Amer. Acad. n. ser. **4:** 9. 1849.

Vesicaria stenophylla A. Gray, Bost. Journ. Nat. Hist. **6:** 149. 1850.

Lesquerella stenophylla Rydb. Colo. Agr. Exp. Sta. Bull. **100:** 155. 1906.

TYPE LOCALITY: On the smaller hills around Santa Fe, New Mexico. Type collected by Fendler (no. 40).

RANGE: Colorado and Arizona to western Texas.

NEW MEXICO: Farmington; Nara Visa; Socorro Mountain; Santa Fe; Carrizalillo Mountains; Florida Mountains; Organ Mountains; Roswell; Lake Arthur; Round Mountain; Queen; mountains west of San Antonio. Dry rocky hills, in the Lower and Upper Sonoran zones.

This is the common perennial silvery-white Lesquerella found in the drier soil of the foothills of the mountains and the higher rocky mesas. It usually blooms early in the spring, but if the season is especially dry it flowers after the rains begin.

12. Lesquerella pinetorum Woot. & Standl. Contr. U. S. Nat. Herb. **16:** 126. 1913.

TYPE LOCALITY: On a dry hillside under pine trees at Gilmores Ranch on Eagle Creek, White Mountains, New Mexico. Type collected by Wooton & Standley (no. 3460).

RANGE: White Mountains of New Mexico, in the Transition Zone.

17. CARDAMINE L.

A herbaceous glabrous perennial 30 to 60 cm. high; leaves simple, broadly ovate-cordate, sparingly repand-dentate; flowers white, 5 to 8 mm. long; siliques rather stout, ascending, on long pedicels; valves nerveless; seeds in one row.

1. Cardamine cordifolia A. Gray, Mem. Amer. Acad. n. ser. **4:** 8. 1849.

TYPE LOCALITY: Margin of Santa Fe Creek, in the mountains, New Mexico. Type collected by Fendler (no. 28).

RANGE: Wyoming to Arizona and New Mexico.

NEW MEXICO: Chama; Santa Fe and Las Vegas mountains; White Mountains. Along streams and in marshes, in the Transition and Canadian zones.

18. DRABA Dill. WHITLOW GRASS.

Low annuals, biennials, or perennials, with alternate entire or toothed leaves and yellow or white flowers in crowded racemes; sepals short, broad, obtuse; petals obovate or spatulate, entire or somewhat notched; styles short or obsolete; pubescence simple or branched; siliques flattened parallel to the septum, flat or twisted; seeds not winged.

KEY TO THE SPECIES.

Winter annuals; styles obsolete.
  Petals yellow; leaves extending well up on the stem........ 1. *D. montana.*
  Petals white or wanting; leaves clustered at the base of the stem.
    Petals wanting or very small............................ 2. *D. micrantha.*
    Petals conspicuous.
      Leaves all entire; pedicels clustered at the end of the stem.................................. 3. *D. coloradensis.*
      Leaves toothed; fruit in an elongated raceme....... 4. *D. cuneifolia.*
Perennials, or occasionally biennials, sometimes flowering the first year; style conspicuous, 1 mm. long or more.
  Basal leaves long-ciliate, the hairs simple or nearly so.
    Stems pubescent; cauline leaves pubescent on both surfaces....................................... 5. *D. streptocarpa.*
    Stems glabrous; cauline leaves glabrous on the faces, usually ciliate.................................. 6. *D. tonsa.*
  Basal leaves not long-ciliate; hairs branched.
    Petals white... 7. *D. cana.*
    Petals bright yellow.
      Stems equally leafy throughout, the cauline leaves larger than the basal ones. (Plants large, usually 20 cm. high or more).
        Stems solitary, simple below.................... 9. *D. helleriana.*
        Stems clustered, usually branched below.......10. *D. patens.*
      Stems with reduced leaves or almost naked; basal leaves much larger and more conspicuous than the cauline ones.
        Basal leaves oblanceolate or obovate, 15 to 25 mm. wide, toothed; stems almost naked..11. *D. mogollonica.*
        Basal leaves oblanceolate or narrower, less than 5 mm. wide, entire; stems with more numerous leaves.
          Roots slender; stems finely stellate-pubescent................................ 8. *D. neomexicana.*
          Roots thick and woody; stems long-pubescent or glabrous.
            Leaves glabrous or nearly so, acute or acutish; stems glabrous, slender.13. *D. gilgiana.*
            Leaves all pubescent and ciliate, obtuse; stems pubescent, stout...........12. *D. petrophila.*

1. **Draba montana** S. Wats. Proc. Amer. Acad. **14**: 289. 1879.

TYPE LOCALITY: South Park, Colorado.

RANGE: Colorado and New Mexico.

NEW MEXICO: Rio Pueblo (*Wooton*). Mountains, in the Canadian Zone.

2. **Draba micrantha** Nutt.; Torr. & Gray, Fl. N. Amer. **1**: 109. 1838.

Draba caroliniana micrantha A. Gray, Man. ed. 5. 72. 1867.

TYPE LOCALITY: Open plains and rocky places about St. Louis, Missouri, and in Arkansas.

RANGE: Washington and Illinois to New Mexico and Texas.

NEW MEXICO: Nutritas Creek below Tierra Amarilla (*Eggleston* 6495). Open slopes, in the Upper Sonoran and lower parts of the Transition Zone.

3. Draba coloradensis Rydb. Bull. Torrey Club **31:** 555. 1904.

Type locality: Fort Collins, Colorado.

Range: Colorado and northern New Mexico.

New Mexico: Gallinas River below Las Vegas; La Cueva; near Tierra Amarilla. Plains, in the Upper Sonoran Zone.

A species much resembling the next, but usually a smaller plant with larger flowers.

4. Draba cuneifolia Nutt.; Torr. & Gray, Fl. N. Amer. **1:** 108. 1838.

Type locality: Grassy places around St. Louis, Missouri.

Range: Illinois to Alabama, New Mexico, and California.

New Mexico: Magdalena Mountains; Mangas Springs; mountains west of San Antonio; Florida Mountains; Organ Mountains; Tortugas Mountain. Open hillsides, in the Upper Sonoran Zone.

Not uncommon in the drier mountains of the southern part of the State, appearing early in the spring if there has been rain or snow. If the moisture is scanty, it commences to bloom as soon as the first two leaves are formed, and the plants will be only 5 or 6 cm. high. With abundant water and good soil in the crevices of a rock, it sometimes is 15 to 20 cm. high and much branched at the base.

5. Draba streptocarpa A. Gray, Amer. Journ. Sci. II. **33:** 242. 1862.

Type locality: Rocky cliffs bordering the upper Clear Creek, Colorado.

Range: Colorado and New Mexico.

New Mexico: Las Vegas Range; Johnsons Mesa; Sierra Grande. High mountains, in the Arctic-Alpine Zone.

A high-mountain species with pale yellowish flowers and pubescent leaves, stems, and inflorescence, the sepals sparingly ciliate and the pods scabrous.

6. Draba tonsa Woot. & Standl. Contr. U. S. Nat. Herb. **16:** 125. 1913.

Type locality: Hermits Peak in the Las Vegas Mountains, New Mexico. Type collected by Snow.

Range: Las Vegas Mountains, New Mexico.

7. Draba cana Rydb. Bull. Torrey Club **29:** 241. 1902.

Type locality: Morley, Alberta.

Range: British America to Colorado and New Mexico.

New Mexico: Truchas Peak; Pecos Baldy. Meadows, in the Arctic-Alpine Zone.

8. Draba neomexicana Greene, Pittonia **4:** 18. 1899.

Draba aurea stylosa A. Gray, Amer. Journ. Sci. II. **33:** 242. 1862, in part.

Type locality: Mountains back of Santa Fe, New Mexico. Type collected by Fendler (no. 43).

Range: Mountains of northern New Mexico.

New Mexico: Jemez Mountains; Santa Fe and Las Vegas mountains. Chiefly in the Arctic-Alpine Zone.

9. Draba helleriana Greene, Pittonia **4:** 17. 1899.

Draba aurea stylosa A. Gray, Amer. Journ. Sci. II. **33:** 243. 1862, in part.

Draba stylosa Heller, Pl. World **1:** 23. 1897, not Turcz. 1854.

Draba pinetorum Greene, Pittonia **4:** 18. 1899.

Type locality: Canyon 4 miles east of Santa Fe, New Mexico. Type collected by Heller (no. 3669).

Range: Mountains of New Mexico.

New Mexico: Santa Fe and Las Vegas mountains; Rio Pueblo; Sandia Mountains; Mogollon Mountains; Black Range. Transition Zone.

The type of *D. pinetorum* came from the Pinos Altos Mountains (*Greene* in 1880).

10. Draba patens Heller, Bull. Torrey Club **26:** 624. 1899.

TYPE LOCALITY: White Mountains, New Mexico. Type collected by Wooton (no. 275).

RANGE: White and Sacramento mountains of New Mexico, in the Transition Zone.

11. Draba mogollonica Greene, Bot. Gaz. **6:** 157. 1881.

TYPE LOCALITY: Northward slopes of the Mogollon Mountains, New Mexico. Type collected by E. L. Greene, April 18, 1880.

RANGE: Mountains of southwestern New Mexico.

NEW MEXICO: Bear Mountain; Mangas Springs; Magdalena Mountains; Mogollon Mountains. Upper Sonoran Zone.

The plant is a perennial but blooms the first year. The pods are typically glabrous, but sometimes they are pubescent.

12. Draba petrophila Greene, Pittonia **4:** 17. 1899.

TYPE LOCALITY: Ledges of the Santa Rita Mountains, southern Arizona.

RANGE: Mountains of Arizona and New Mexico.

NEW MEXICO: San Luis Mountains (*Mearns* 2206).

13. Draba gilgiana Woot. & Standl. Contr. U. S. Nat. Herb. **16:** 124. 1913.

TYPE LOCALITY: Organ Peak in the Organ Mountains, New Mexico. Type collected by Wooton & Standley, September 23, 1906.

RANGE: Mountains of southern New Mexico.

NEW MEXICO: Organ Mountains; Tortugas Mountain. Transition Zone.

19. ARABIS L. Rock cress.

Annuals or perennials, glabrous or pubescent, with entire or toothed leaves more or less rosulate at the base, the cauline leaves usually smaller, sessile, sometimes clasping; flowers white, rose-colored, or purple; siliques long-linear, flattened parallel to the septum, the valves one-nerved; seeds winged or marginless.

KEY TO THE SPECIES.

Pods erect.
- Seeds broadly winged; none of the leaves coarsely hirsute, the basal ones sometimes with a few hairs; plants glaucous........ 1. *A. oxyphylla.*
- Seeds not winged; at least the lower leaves coarsely hirsute; plants bright green........ 2. *A. ovata.*

Pods reflexed or spreading.
- Leaves hispid on the margins, the faces usually stellate-pubescent........ 3. *A. fendleri.*
- Leaves not hispid on the margins, but the faces stellate-pubescent.
  - Pods pubescent; flowers 12 or 13 mm. long........ 4. *A. formosa.*
  - Pods glabrous; flowers 6 mm. long or less.
    - Sepals stellate-pubescent.
      - Leaves finely and densely stellate-pubescent... 5. *A. eremophila.*
      - Leaves coarsely and loosely stellate-pubescent.. 8. *A. consanguinea.*
    - Sepals glabrous.
      - Pods 45 to 60 mm. long, 1 mm. wide, green, curved upward........ 6. *A. angulata.*
      - Pods 35 mm. long, 1.5 mm. wide, purple, straight or curved downward........ 7. *A. porphyrea.*

1. **Arabis oxyphylla** Greene, Pittonia **4:** 196. 1900.
TYPE LOCALITY: Empire, Colorado.
RANGE: Wyoming to Utah and New Mexico.
NEW MEXICO: Tierra Amarilla; Chama; Baldy; Grass Mountain. Meadows, in the Transition Zone.

2. **Arabis ovata** (Pursh) Poir. in Lam. Encycl. Suppl. **5:** 557. 1817.
Turritis ovata Pursh, Fl. Amer. Sept. 438. 1814.
TYPE LOCALITY: "On rocks, Pennsylvania to Virginia."
RANGE: British America to Georgia and California.
NEW MEXICO: Tunitcha Mountains; Chama; Santa Fe and Las Vegas mountains; Raton; Sandia Mountains; Grosstedt Place; White and Sacramento mountains. Damp woods, in the Transition Zone.

3. **Arabis fendleri** (S. Wats.) Greene, Pittonia **3:** 156. 1897.
Arabis holboellii fendleri S. Wats. in A. Gray, Syn. Fl. 1^1: 164. 1895.
Arabis gracilenta Greene, Pittonia **4:** 194. 1900.
TYPE LOCALITY: New Mexico. Type collected by Fendler, probably near Santa Fe (no. 27).
RANGE: Colorado and New Mexico.
NEW MEXICO: Santa Fe; Winsors Ranch; Chama; Tunitcha Mountains. Open slopes.
The type of *A. gracilenta* was collected near Santa Fe by Heller.

4. **Arabis formosa** Greene, Pittonia **4:** 198. 1900.
TYPE LOCALITY: Hills about Aztec, New Mexico. Type collected by Baker (no. 345).
RANGE: Known only from type locality, in the Upper Sonoran Zone.

5. **Arabis eremophila** Greene, Pittonia **4:** 194. 1900.
TYPE LOCALITY: Peach Springs, northern Arizona.
RANGE: Northern Arizona and New Mexico.
NEW MEXICO: Aztec (*Baker* 343). Upper Sonoran Zone.

6. **Arabis angulata** Greene, Contr. U. S. Nat. Herb. **16:** 123. 1913.
TYPE LOCALITY: Mangas Springs, New Mexico. Type collected by Metcalfe (no. 12).
RANGE: Known only from type locality.

7. **Arabis porphyrea** Woot. & Standl. Contr. U. S. Nat. Herb. **16:** 123. 1913.
TYPE LOCALITY: Dry hills near the Cueva on the west side of the Organ Mountains, New Mexico. Type collected by Wooton & Standley, April 25, 1907.
RANGE: Southern New Mexico.
NEW MEXICO: Magdalena Mountains; Continental Divide west of Patterson; South Percha Creek; Organ Mountains. Dry hills, in the Upper Sonoran Zone.

8. **Arabis consanguinea** Greene, Pittonia **4:** 190. 1900.
TYPE LOCALITY: Los Pinos, southern Colorado.
RANGE: Mountains of Colorado and northern New Mexico.
NEW MEXICO: Carrizo Mountains; near Tierra Amarilla. Upper Sonoran and Transition zones.

20. RAPHANUS L. RADISH.

Tall annual, often with a thickened root, the leaves pinnatifid; flowers pale purple, pod thick, oblong, tapering upward, indehiscent.

1. **Raphanus sativus** L. Sp. Pl. 669. 1753.
TYPE LOCALITY: Not stated.
NEW MEXICO: James Canyon (*Wooton*).
The common cultivated radish is not rare as an escape in some of the Eastern States, and is well established in California.

21. ERUCA Mill.

Branched annual with pinnately lobed or pinnatifid leaves and yellowish flowers; siliques linear-oblong, long-beaked, the valves 3-nerved, concave; seeds in 2 rows in in each cell.

1. **Eruca eruca** (L.) Britton in Britt. & Brown, Illustr. Fl. ed. 2. **2**: 192. 1913.
Brassica eruca L. Sp. Pl. 667. 1753.
Eruca sativa Mill. Gard. Dict. ed. 8. no. 1. 1768.
TYPE LOCALITY: "Habitat in Helvetia."
RANGE: Native of Europe, introduced into many parts of North America.
NEW MEXICO: Mesilla Valley.

22. BRASSICA L. MUSTARD.

Coarse annuals, 60 to 100 cm. high or larger, with simple or pinnately lobed leaves 10 to 15 cm. long; flowers yellow, in terminal elongated racemes; siliques elongated, 4-angled, beaked; seeds in one row in the cell.

KEY TO THE SPECIES.

Plants glabrous; pedicels 10 to 20 cm. long........................... 1. *B. juncea.*
Plants hispid; pedicels about 5 mm. long............................. 2. *B. arvensis.*

1. **Brassica juncea** (L.) Coss. Bull. Soc. Bot. France **6**: 609. 1859. INDIAN MUSTARD.
Sinapis juncea L. Sp. Pl. 668. 1753.
TYPE LOCALITY: "Habitat in Asia."
NEW MEXICO: Fresnal; Gilmores Ranch; Las Cruces; Ponchuelo Creek; Santa Fe; Espanola; Las Vegas; Shiprock; Agricultural College.
Not uncommon in cultivated fields; widely introduced into North America from Europe.

2. **Brassica arvensis** (L.) B. S. P. Prel. Cat. N. Y. 5. 1888. CHARLOCK.
Sinapis arvensis L. Sp. Pl. 668. 1753.
Brassica sinapistrum Boiss. Voy. Bot. Esp. **2**: 39. 1839–45.
TYPE LOCALITY: "Habitat in agris Europae."
NEW MEXICO: Mesilla Valley.
Introduced into grain fields, gardens, and waste ground in many parts of the United States.

23. CHEIRINIA Link. WESTERN WALLFLOWER.

Coarse biennials or perennials with harsh pubescence of branched appressed hairs; leaves alternate, entire or toothed, simple; flowers large for the family, 6 to 20 mm. long, in long terminal racemes; sepals oblong, one pair saccate; petals long-clawed, yellow, brownish, or maroon; siliques subterete or more or less strongly 4-angled.

KEY TO THE SPECIES.

Flowers small, less than 1 cm. long.
  Plants tall and slender; basal leaves fugacious; plant of the high mountains.................................... 3. *C. inconspicua.*
  Plants low and stout, with persistent basal leaves; of the gravelly mesas of the southern part of the State......... 2. *C. desertorum.*
Flowers large, more than 1 cm. long.
  Petals orange, reddish brown, or purplish maroon.............. 6. *C. wheeleri.*
  Petals light golden yellow.
    Basal leaves, at least, silvery white; cauline leaves very narrow.. 1. *C. bakeri.*

Basal leaves, as well as the whole plant, grayish. (Plants taller).
Pods widely spreading; plants very stout............ 7. *C. aspera.*
Pods erect or ascending; plants slender.
Claws of the petals one-half longer than the sepals. 5. *C. elata.*
Claws of the petals little or not at all longer than the sepals................................ 4. *C. asperrima.*

1. Cheirinia bakeri (Greene) Rydb. Bull. Torrey Club **39:** 324. 1912.
Cheiranthus aridus Greene, Pittonia **4:** 198. 1900, not A. Nels. 1899.
Cheiranthus bakeri Greene, Pittonia **4:** 235. 1900.
Erysimum bakeri Rydb. Bull. Torrey Club **33:** 141. 1906.

TYPE LOCALITY: Dry hills among nut pines and cedars at Aztec, New Mexico. Type collected by Baker (no. 350).

RANGE: Southwestern Colorado to western New Mexico and adjacent Utah and Arizona.

NEW MEXICO: Aztec; Carrizo Mountains; Florida Mountains; mountains west of San Antonio. Dry hills, in the Upper Sonoran Zone.

A low (20 to 40 cm.) plant with sinuate-dentate leaves, occuring in the northwestern part of the State. The type of *Cheiranthus aridus* was collected at Aztec by Baker.

2. Cheirinia desertorum Woot. & Standl. Contr. U. S. Nat. Herb. **16:** 125. 1913.

TYPE LOCALITY: Near Hachita, New Mexico. Type collected by Wooton, June 16, 1906.

RANGE: Known only from type locality, in the Lower Sonoran Zone.

3. Cheirinia inconspicua (S. Wats.) Rydb. Bull. Torrey Club **39:** 323. 1912.
Erysimum parviflorum Nutt.; Torr. & Gray, Fl. N. Amer. **1:** 95. 1838, not Pers. 1807.
Erysimum asperum inconspicuum S. Wats. in King, Geol. Expl. 40th Par. **5:** 24. 1871.

TYPE LOCALITY: Diamond Valley, Nevada.

RANGE: New Mexico and Kansas and northward.

NEW MEXICO: Dulce; White Mountains; Cloudcroft. Upper Sonoran and Transition zones.

4. Cheirinia asperrima (Greene) Rydb. Bull. Torrey Club **39:** 324. 1912.
Cheiranthus asperrimus Greene, Pittonia **3:** 133. 1896.
Erysimum asperrimum Rydb. Bull. Torrey Club **33:** 141. 1906.

TYPE LOCALITY: None given, but plains of Wyoming indicated.

RANGE: South Dakota to Arizona and New Mexico.

NEW MEXICO: Santa Fe; Las Vegas; Raton; Farmington; Ramah; Nara Visa; Sandia Mountains; Magdalena Mountains; Mogollon Mountains; Black Range; Organ Mountains; White and Sacramento mountains; Capitan Mountains. Hills, in the Upper Sonoran and Transition zones.

This and the next are the common plants of the middle elevations in the mountains of the State. It is the plant which has ordinarily passed as *Erysimum asperum*, but it is much slenderer than that species, the leaves are mostly entire, and the pods are usually less stout and are erect or ascending, never strongly divaricate.

5. Cheirinia elata (Nutt.) Rydb. Bull. Torrey Club **39:** 323. 1912.
Erysimum elatum Nutt.; Torr. & Gray, Fl. N. Amer. **1:** 95. 1838.
Cheiranthus elatus Greene, Pittonia **3:** 135. 1896.

TYPE LOCALITY: "Grassy situations by the banks of the Wahlamet."

RANGE: North Dakota, Montana, and Washington, southward to the the mountains of Colorado and New Mexico, westward to California.

NEW MEXICO: Santa Fe and Las Vegas mountains; Sandia Mountains; Organ Mountains; Jarilla Mountains; White Mountains. On hills, in the Transition Zone.

The species strongly resembles the preceding, but may easily be recognized by the elongated claws of the petals. It sometimes branches somewhat profusely at the top, usually having but a single stem from a root.

6. **Cheirinia wheeleri** (Rothr.) Rydb. Bull. Torrey Club **39:** 324. 1912.

Erysimum wheeleri Rothr. in Wheeler, Rep. U. S. Surv. 100th Merid. **6:** 64. 1879.

Erysimum asperum alpestre Cockerell, Bull. Torrey Club **18:** 168. 1891.

Cheiranthus wheeleri Greene, Pittonia **3:** 135. 1896.

Erysimum alpestre Rydb. Bull. Torrey Club **28:** 277. 1901.

TYPE LOCALITY: Camp Grant, Arizona.

RANGE: In the higher mountains of Arizona, New Mexico, and Colorado.

NEW MEXICO: Santa Fe and Las Vegas mountains; Tunitcha Mountains; Sierra Grande; Mogollon Mountains; Hanover Mountains; Sandia Mountains; White and Sacramento mountains. Upper Transition to Arctic-Alpine Zone.

This is the slender species of our mountains in the coniferous timber. At the higher elevations the petals are nearly always reddish brown or maroon, drying purplish; lower down they are yellow or orange. The plant is 70 cm. high or less, with sometimes 2 or 3 erect stems from a root. The siliques are the longest borne by any of our species, frequently 12 to 13 cm. long, slender, nearly terete, somewhat divaricate, but at first erect.

7. **Cheirinia aspera** (Nutt.) Rydb. Bull. Torrey Club **39:** 323. 1912.

Cheiranthus asper Nutt. Gen. Pl. **2:** 69. 1818.

Erysimum asperum DC. Reg. Veg. Syst. **2:** 505. 1821.

TYPE LOCALITY: "On the plains of the Missouri, commencing near the confluence of White River."

RANGE: Saskatchewan to Arkansas and New Mexico.

NEW MEXICO: Sierra Grande; Castle Rock; Vermejo Park. Plains and low hills, in the Upper Sonoran Zone.

24. RADICULA Hill.

Glabrous or hispid annual or perennial herbs with yellow or white flowers; sepals flat, nearly equal at the base; petals short-clawed; siliques usually subterete, not compressed, short; style short and thick; valves one-nerved; seeds in two rows, not flattened.

KEY TO THE SPECIES.

Petals white; aquatic plant, immersed and rooting; leaves pinnate. 1. *R. nasturtium-aquaticum.*
Petals yellow; terrestrial or marsh plants; leaves pinnatifid.
  Perennials with rootstocks; leaf segments acute or acutish, the terminal one lanceolate.
    Segments of the leaves toothed; style 0.5 mm. long...... 2. *R. sylvestris.*
    Segments of the leaves entire; style 1.5 mm. long.
      Pods papillose.................................... 3. *R. calycina.*
      Pods not papillose................................ 4. *R. sinuata.*
  Annuals or biennials; leaf segments obtuse, the terminal broadly oblong to ovate.
    Pods spherical. (Plants low, diffuse, glabrous).......... 5. *R. sphaerocarpa.*
    Pods oblong or elongated, never spherical.
      Stems more or less hirsute............................ 6. *R. hispida.*
      Stems glabrous.
        Stems erect, sparingly branched; pods cylindric, 8 to 10 mm. long......................... 7. *R. terrestris.*
        Stems spreading or ascending, much branched; pods short-oblong or ovoid, 3 mm. long or less.................................. 8. *R. obtusa.*

1. **Radicula nasturtium-aquaticum** (L.) Britten & Rendle, Journ. Bot. Brit. & For. **14:** 99. 1907. WATER CRESS.

Sisymbrium nasturtium-aquaticum L. Sp. Pl. 657. 1753.

Nasturtium officinale R. Br. in Ait. f. Hort. Kew. ed. 2. **4:** 110. 1812.

Roripa nasturtium Rusby, Mem. Torrey Club **3:** 5. 1893.

TYPE LOCALITY: "Habitat in Europa & America septentrionali ad fontes."

RANGE: Throughout most of temperate North America; also in Europe.

NEW MEXICO: Farmington; Pecos; Santa Fe; Las Vegas; Mogollon Mountains; Berendo Creek; Gilmores Ranch; Roswell. In streams, in the Upper Sonoran and Transition zones.

2. **Radicula sylvestris** (L.) Druce, Ann. Scott. Nat. Hist. **1906:** 219. 1906. YELLOW CRESS.

Sisymbrium sylvestre L. Sp. Pl. 657. 1753.

Nasturtium sylvestre R. Br. in Ait. f. Hort. Kew. ed. 2. **4:** 110. 1812.

TYPE LOCALITY: "Habitat in Helvetiae, Germaniae, Galliae ruderatis."

RANGE: A native of Europe, introduced into many parts of North America.

NEW MEXICO: Mogollon Mountains.

3. **Radicula calycina** (Engelm.) Greene, Leaflets **1:** 113. 1905.

Nasturtium calycinum Engelm. in Warren, Prel. Rep. Nebr. Dak. 156. 1855–7.

Roripa calycina Rydb. Mem. N. Y. Bot. Gard. **1:** 175. 1900.

TYPE LOCALITY: "Sandy bottoms of Yellowstone River; Fort Sarpy to Fort Union."

RANGE: Washington and Oregon to Colorado and New Mexico.

NEW MEXICO: Shiprock; Zuni; Cross L Ranch; Moriarity; Socorro; Mesilla; Patterson; Ruidoso Creek. Wet ground, in the Upper Sonoran and Transition zones.

4. **Radicula sinuata** (Nutt.) Greene, Leaflets **1:** 113. 1905.

Nasturtium sinuatum Nutt.; Torr. & Gray, Fl. N. Amer. **1:** 73. 1838.

Roripa sinuata Hitchc. Spr. Fl. Manhattan 18. 1894.

TYPE LOCALITY: "Banks of the Oregon and its tributaries."

RANGE: British America to Arkansas and New Mexico.

NEW MEXICO: Coolidge; along the Rio Grande west of Santa Fe; Glorieta; N Bar Ranch; Chama; Sierra Grande. Wet ground, in the Upper Sonoran and Transition zones.

5. **Radicula sphaerocarpa** (A. Gray) Greene, Leaflets **1:** 113. 1905.

Nasturtium sphaerocarpum A. Gray, Mem. Amer. Acad. n. ser. **4:** 6. 1849.

Roripa sphaerocarpa Britton, Mem. Torrey Club **5:** 170. 1894.

TYPE LOCALITY: Low places along Santa Fe Creek, New Mexico. Type collected by Fendler (no. 21).

RANGE: Illinois to Wyoming, California, and Arizona.

NEW MEXICO: Tunitcha Mountains; Santa Fe and Las Vegas mountains. Wet ground, in the Transition Zone.

6. **Radicula hispida** (Desv.) Heller, Muhlenbergia **7:** 123. 1912.

Brachylobus hispidus Desv. Journ. de Bot. **3:** 183. 1814.

Nasturtium hispidum DC. Reg. Veg. Syst. **2:** 201. 1821.

Roripa hispida Britton, Mem. Torrey Club **5:** 169. 1894.

TYPE LOCALITY: "Habitat in Pennsylvania."

RANGE: British America to Arizona and Florida.

NEW MEXICO: Santa Fe; Pecos; Wheelers Ranch; Mimbres River; Mangas Springs. Wet ground, in the Upper Sonoran Zone.

7. **Radicula terrestris** (R. Br.) Woot. & Standl.

Sisymbrium amphibium palustre L. Sp. Pl. 657. 1753.

Nasturtium terrestre R. Br. in Ait. f. Hort. Kew. ed. 2. **4:** 110. 1812.

Nasturtium palustre DC. Reg. Veg. Syst. **2:** 191. 1821.

Roripa palustris Besser, Enum. Pl. 27. 1821.
TYPE LOCALITY: European.
RANGE: British America to California, New Mexico, and North Carolina; also in Europe and Asia.
NEW MEXICO: Alamo Viejo (*Mearns* 12).

8. **Radicula obtusa** (Nutt.) Greene, Leaflets **1:** 113. 1905.
Nasturtium obtusum Nutt.; Torr. & Gray, Fl. N. Amer. **1:** 74. 1838.
TYPE LOCALITY: "Banks of the Mississippi."
RANGE: British Columbia to California, Michigan, and Texas.
NEW MEXICO: San Juan Valley; Inscription Rock; Ramah; Mogollon Mountains; White and Sacramento mountains; Mesilla Valley. Wet ground.

25. SCHOENOCRAMBE Greene.

Perennial glabrous green herb with long horizontal rootstocks; stems simple or branched, slender; cauline leaves linear, mostly entire; flowers rather large, yellow; siliques 3 to 6 cm. long, slender, terete, suberect, on short spreading pedicels.

1. **Schoenocrambe linifolia** (Nutt.) Greene, Pittonia **3:** 127. 1896.
Nasturtium linifolium Nutt. Journ. Acad. Phila. **7:** 12. 1834.
Erysimum linifolium Jones, Proc. Calif. Acad. II. **5:** 622. 1895.
TYPE LOCALITY: "Head of Salmon River, in dry soils," Montana.
RANGE: Montana to British Columbia, Utah, and New Mexico.
NEW MEXICO: Carrizo Mountains (*Matthews*).

26. SISYMBRIUM L.

An anomalous remnant which needs further study and generic segregation. The species here assembled show little relationship to each other.

KEY TO THE SPECIES.

Pubescence of branched hairs.................................... 1. *S. diffusum.*
Pubescence of unbranched hairs, or none.
  Plants pubescent; flowers bright yellow; pedicels erect........ 3. *S. officinale leiocarpum.*
  Plants glabrous; flowers white or ochroleucous; pedicels spreading or ascending.
    Plants glaucous; leaves entire........................... 2. *S. vaseyi.*
    Plants green; leaves pinnatifid.......................... 4. *S. altissimum.*

1. **Sisymbrium diffusum** A. Gray, Pl. Wright. **1:** 8. 1852.
TYPE LOCALITY: "Pass of the Limpia, in crevices of rocks on the mountains," Texas. Type collected by Wright (no. 10).
RANGE: Western Texas to California, south into Mexico.
NEW MEXICO: Hop Canyon; Mangas Springs; Mogollon Mountains; Black Range; Florida Mountains; Organ Mountains. Dry hills and canyons, in the Upper Sonoran Zone.

2. **Sisymbrium vaseyi** S. Wats. in A. Gray, Syn. Fl. **1**[1]**:** 138. 1895.
Thelypodium vaseyi Coulter, Contr. U. S. Nat. Herb. **1:** 30. 1890, in part.
TYPE LOCALITY: Mountains west of Las Vegas, New Mexico. Type collected by G. R. Vasey in 1881.
RANGE: Mountains of New Mexico.
NEW MEXICO: Santa Fe and Las Vegas mountains; Sandia Mountains; Graham; White and Sacramento mountains. Transition Zone.
A handsome plant of the open parks at higher elevations in the mountains. Its masses of white flowers make it very attractive, especially when, as often happens, it is combined with brighter colored flowers.

3. Sisymbrium officinale leiocarpum DC. Prodr. **1**: 191. 1824. HEDGE MUSTARD.

TYPE LOCALITY: "In Carolina merid. et Teneriffa."

NEW MEXICO: Gilmores Ranch (*Wooton & Standley* 3642).

A common introduced weed in many parts of North America, still rare in New Mexico.

4. Sisymbrium altissimum L. Sp. Pl. 659. 1753.

TYPE LOCALITY: "Habitat in Italia, Gallia, Siberia."

NEW MEXICO: San Juan Valley.

A native of Europe, widely introduced into the United States, a noxious weed in the Northwest.

27. SOPHIA Adans. TANSY MUSTARD.

Annuals, more or less stellate-pubescent; leaves once, twice, or thrice pinnately parted into mostly small segments; flowers small, in terminal racemes; petals usually yellow, white in one species; racemes elongated in fruit; siliques from one-half to one and one-half times the length of the pedicels; seeds in apparently one row in some species, really from alternate funiculi from two lateral placentæ in each cell, mostly in two evident rows.

KEY TO THE SPECIES.

Petals white; leaves nearly all thrice pinnately parted .... 1. *S. ochroleuca.*
Petals yellow; leaves mostly once or twice pinnately parted.
  Plants appearing glabrous, really sparingly stellate-pubescent, green.
    Pedicels erect, like the pods, seemingly appressed to the rachis .... 2. *S. procera.*
    Pedicels divergent; pods erect or curved.
      Inflorescence glandular-pubescent; pods longer than the pedicels.................................. 3. *S. incisa.*
      Inflorescence merely sparingly stellate-pubescent, not glandular; pods shorter than the pedicels.. 4. *S. serrata.*
  Plants canescent, thickly and persistently stellate-pubescent, grayish green.
    Plants tall, 80 to 120 cm.; segments of the leaves large, some of them obtuse; sepals yellow.
      Inflorescence glandular-pubescent, not canescent... 5. *S. adenophora.*
      Inflorescence canescent with stellate hairs, like the rest of the plant.......................... 6. *S. obtusa.*
    Plants lower, 30 to 60 cm. high; leaf segments mostly very small; sepals purplish.
      Plants slender, sparingly branched; inflorescence glabrous.................................. 7. *S. glabra.*
      Plants stout, much branched; inflorescence glandular or stellate-pubescent.
        Plants divergently much branched from the base; inflorescence strongly glandular, not stellate-pubescent; petals equaling the sepals. 8. *S. halictorum.*
        Plants with more erect stems; inflorescence stellate-pubescent, sometimes sparingly glandular; petals longer than the sepals....... 9. *S. andrenarum.*

1. Sophia ochroleuca Wooton, Bull. Torrey Club **25**: 455. 1898.

TYPE LOCALITY: Mesilla Park, New Mexico. Type collected by J. D. Tinsley.

RANGE: Southern New Mexico, probably also in adjacent Arizona and Mexico.

NEW MEXICO: Mesilla Valley; Grant County; Gray. Dry fields, in the Lower Sonoran Zone.

This is a common spring weed in the cultivated fields of the valleys of southern New Mexico. It is easily recognized by its finely dissected leaves and dull white flowers.

2. **Sophia procera** Greene, Pittonia 4: 199. 1900.

TYPE LOCALITY: "Common in the open pine woods of the Colorado Rocky Mountains, at 8,000 or 9,000 feet altitude."

RANGE: Mountains of New Mexico, Colorado, and Utah, and northward.

NEW MEXICO: Sandia Mountains. Transition Zone.

A tall, green plant, 150 cm. high or more, with leaves not very finely dissected, resembling *S. incisa*, but with siliques and pedicels erect, thus bringing the pods close to the rachis. It comes into our range from the north and west, where it is apparently common.

3. **Sophia incisa** (Engelm.) Greene, Pittonia 3: 95. 1896.

Sisymbrium incisum Engelm. in A. Gray, Mem. Amer. Acad. n. ser. 4: 8. 1849.

TYPE LOCALITY: "Banks of streams in New Mexico; Santa Fe Creek and Mora River." Type collected by Fendler (nos. 29, 30, 31).

RANGE: Mountains of New Mexico, northward to Wyoming.

NEW MEXICO: Santa Fe and Las Vegas mountains; Tunitcha Mountains; Chama; Mogollon Mountains; White and Sacramento mountains. Transition Zone.

This and the next are the common species found on the timber-covered mountains, growing along streams, in the open parks, and in similar locations in the Transition Zone.

4. **Sophia serrata** Greene, Leaflets 1: 96. 1904.

TYPE LOCALITY: Black Range, New Mexico. Type collected by Metcalfe (no. 1069).

RANGE: New Mexico.

NEW MEXICO: El Rito Creek; Stinking Lake; Chama; Santa Fe and Las Vegas mountains; Sandia Mountains; Mogollon Mountains; Black Range; Carrizo Mountains; White Mountains. Mountains, in the Transition Zone.

5. **Sophia adenophora** Woot. & Standl. Contr. U. S. Nat. Herb. 16: 127. 1913.

TYPE LOCALITY: Head and Wilson Ranch south of Mule Creek, northwestern Grant County, New Mexico. Type collected by Wooton, July 13, 1900.

RANGE: Southern New Mexico.

NEW MEXICO: Mogollon Mountains; San Augustine Plains. Upper Sonoran and Transition zones.

6. **Sophia obtusa** Greene, Leaflets 1: 96. 1904.

TYPE LOCALITY: Black Range, New Mexico. Type collected by Metcalfe (no. 1074).

RANGE: Mountains of western New Mexico and adjacent Arizona.

NEW MEXICO: Gallup; Santa Fe; Silver City; Mogollon Mountains; Bear Mountain; Mimbres River; 8 miles west of Durfeys Well; Magdalena; Pescado Spring. At lower levels, in the Upper Sonoran Zone.

A tall canescent herb about 1 meter high, not uncommon in arroyos or along creeks in the western part of the State.

7. **Sophia glabra** Woot. & Standl. Contr. U. S. Nat. Herb. 16: 127. 1913.

TYPE LOCALITY: Organ Mountains, New Mexico. Type collected by Wooton & Standley, March 21, 1907.

RANGE: Organ Mountains of New Mexico, in the Upper Sonoran Zone.

8. **Sophia halictorum** Cockerell, Bull. Torrey Club **25:** 460. 1898.

TYPE LOCALITY: Mesilla Park, New Mexico. Type collected by Cockerell.

RANGE: New Mexico and northward.

NEW MEXICO: Santa Fe; Sierra Grande; Aztec; Las Vegas; Zuni; mountains west of San Antonio; Organ Mountains; Mesilla Valley. Sandy valleys and dry plains, in the Lower Sonoran Zone.

A low spreading canescent plant with purplish stems, inconspicuous flowers, finely divided leaves, even the uppermost pinnatifid, and copiously glandular-pubescent inflorescence.

9. **Sophia andrenarum** Cockerell, Bull. Torrey Club **28:** 48. 1901.

TYPE LOCALITY: Mesilla Park, New Mexico. Type collected by Cockerell.

RANGE: New Mexico and Arizona to Colorado and Utah.

NEW MEXICO: Mesilla Valley; Gray. Dry fields, in the Lower and Upper Sonoran zones.

9a. **Sophia andrenarum osmiarum** Cockerell, Bull. Torrey Club **28:** 48. 1901.

TYPE LOCALITY: Mesilla Park, New Mexico. Type collected by Cockerell.

RANGE: With the species.

NEW MEXICO: Organ Mountains; Magdalena; mountains west of San Antonio; Mesilla Valley; Alamo Viejo.

This is similar to the species, but the inflorescence is not glandular and is pubescent throughout with branched hairs.

28. DRYOPETALON A. Gray.

Annual, 30 to 60 cm. high, with runcinate clustered radical leaves, few and smaller cauline ones, and corymbosely branching stems hispid below, ending in crowded racemes of bright white flowers; petals 6 mm. long, the limb pinnately 5 to 7-lobed; siliques terete, long and slender, crowded.

1. **Dryopetalon runcinatum** A. Gray, Pl. Wright. **2:** 12. *pl. 11.* 1853.

TYPE LOCALITY: Mountains, near Lake Santa Maria, Chihuahua. Type collected by Wright (no. 1314).

RANGE: Southern Arizona and New Mexico and northern Mexico.

NEW MEXICO: Organ Mountains. Canyons, in the Upper Sonoran Zone.

The species is known to us in New Mexico only from the Organ Mountains, where it is a common spring plant growing among the rocks. It should occur in the mountains of the southwestern corner of the State, and probably does, but no collector has been there at the proper time of the year to find it.

29. CONRINGIA Link. HARE'S-EAR MUSTARD.

Tall glabrous annual with broad sessile entire clasping cauline leaves; flowers pale yellow; pods long, spreading, linear, 4-angled; seeds oblong, in 1 row in each cell.

1. **Conringia orientalis** (L.) Dum. Fl. Belg. 123. 1827.

Brassica orientalis L. Sp. Pl. 666. 1753.

Conringia perfoliata Link, Enum. Pl. **2:** 172. 1822.

TYPE LOCALITY: "Habitat in Oriente."

NEW MEXICO: Des Moines (*Standley* 6215).

A native of Europe, introduced into the United States. It seems to be well established in this one locality in New Mexico.

30. CAMELINA Crantz. FALSE FLAX.

Erect pubescent annual with entire or slightly toothed leaves, the cauline ones with clasping auriculate bases; flowers small, yellow, racemose; fruit obovoid, slightly flattened; seeds several or numerous in each cell, marginless.

1. **Camelina microcarpa** Andrzej.; DC. Syst. Veg. **2**: 517. 1821.

TYPE LOCALITY: European.

RANGE: Native of Europe, introduced in waste ground in many parts of the United States.

NEW MEXICO: Pecos National Forest (*Louis Rudolph*).

57. CAPPARIDACEAE. Caper Family.

Annuals, 1 meter high or less, with watery juice usually of an unpleasant odor; leaves palmately trifoliolate; flowers rather large, in terminal crowded racemes; sepals 4; petals 4, entire or emarginate; stamens 6 or more, not tetradynamous, mostly long-exserted; fruit 1-celled, 2-valved, of various forms, sometimes long-stipitate, the valves separating from the filiform placentæ.

KEY TO THE GENERA.

Pods large, 3 to 7 cm. long, terete.
- Stamens 12 to 24; petals dull white; plants mostly clammy; pods sessile or short-stipitate .... 1. POLANISIA (p. 289).
- Stamens 6; petals purplish or yellow; plants glabrous; pods long-stipitate .... 2. PERITOMA (p. 290).

Pods short, 1 cm. long or less, irregular (on long slender stipes; flowers yellow).
- Valves of the pods cymbiform or elongate-conic; pods several-seeded .... 3. CLEOMELLA (p. 290).
- Valves of the pods ellipsoid, indurate, reticulate; pods 2-seeded .... 4. WISLIZENIA (p. 290).

1. POLANISIA Raf. CLAMMY WEED.

Coarse branching clammy viscid-pubescent herbs, 40 to 70 cm. high, with trifoliolate leaves and terminal crowded racemes of dull whitish flowers; stamens long-exserted, purplish; leaflets elliptic-obovate, entire, obtuse; inflorescence with crowded unifoliolate leaflike bracts; fruit 10 cm. long or less, terete, with numerous large seeds.

KEY TO THE SPECIES.

Petals 12 mm. long or less, often purplish; filaments not exceeding 20 mm.; seeds rough .... 1. *P. trachysperma*.

Petals more than 15 mm. long, sulphur-yellow; filaments 35 to 50 mm. long; seeds smooth .... 2. *P. uniglandulosa*.

1. **Polanisia trachysperma** Torr. & Gray, Fl. N. Amer. **1**: 669. 1840.

TYPE LOCALITY: Texas.

RANGE: British America to Nevada, Texas, and Missouri.

NEW MEXICO: Farmington; Santa Fe; Zuni; Tucumcari; Sabinal; Albuquerque; Perico; Pajarito Park; mesa west of Organ Mountains. Dry hills and plains, in the Upper Sonoran Zone.

2. **Polanisia uniglandulosa** (Cav.) DC. Prodr. **1**: 242. 1824.

Cleome uniglandulosa Cav. Icon. Pl. **4**: 3. *pl. 386*. 1797.

TYPE LOCALITY: "Habitat in Nova-Hispania praesertim in Acapulco."

RANGE: New Mexico and western Texas, southward into Mexico.

NEW MEXICO: Mogollon Mountains; Burro Mountains; Mangas Springs; Black Range; Dog Spring; Organ Mountains; Three Rivers. Dry plains and hills, in the Lower and Upper Sonoran zones.

A common plant of the drier mountains, the arroyos, and the sandhills of the southern part of the State. It is never very abundant in any one spot, but is rather widely distributed.

52576°—15——19

2. PERITOMA DC.

Coarse glabrous branching annuals, 1 meter high or less; leaves alternate, trifoliolate; leaflets lanceolate to elliptic, acute or obtuse, entire; petals yellow, rose purple, or rarely white; stamens 6, long-exserted; pods stipitate, cylindric, 10 cm. long or less, with few or many large seeds.

KEY TO THE SPECIES.

Petals rose purple, rarely white; capsules 5 to 10 cm. long; seeds numerous.. 1. *P. serrulatum.*
Petals yellow; capsules 2 cm. long or less; seeds 6 or fewer....... 2. *P. breviflorum.*

1. Peritoma serrulatum (Pursh) DC. Prodr. **1**: 237. 1824.

ROCKY MOUNTAIN BEE PLANT.

Cleome serrulata Pursh, Fl. Amer. Sept. 441. 1814.

Cleome integrifolia Torr. & Gray, Fl. N. Amer. **1**: 122. 1838.

TYPE LOCALITY: "On the banks of the Missouri."

RANGE: Saskatchewan and Idaho to Arizona and Missouri.

NEW MEXICO: North of Gallup; Chama; Zuni; Santa Clara Canyon; Santa Fe; Las Vegas; Pecos; Folsom; Frisco; Gila; Mangas Springs; White Mountains. Hills and plains, in the Upper Sonoran and Transition zones.

A common range weed, occupying considerable areas of land that has been overstocked. The flowers supply large quantities of nectar, which fact gives the common name.

Peritoma sonorae (A. Gray) Rydb. has been reported from New Mexico, but we have seen no specimens. It may be distinguished from *P. serrulatum* by its small pods, less than 15 mm. long, and by having its sepals distinct instead of united at the base.

2. Peritoma breviflorum Woot. & Standl. Contr. U. S. Nat. Herb. **16**: 128. 1913.

TYPE LOCALITY: Dry, stony hills about Shiprock, New Mexico. Type collected by Standley (no. 7282).

RANGE: Northwestern New Mexico and northeastern Arizona, probably in adjacent Utah and Colorado.

NEW MEXICO: San Juan Valley. Dry hills and plains, in the Upper Sonoran Zone.

3. CLEOMELLA DC.

Erect annual, 30 to 60 cm. high, with trifoliolate leaves; leaflets oblong or spatulate obovate; petals yellow; stamens 6, exserted; pods long-stipitate, several-seeded, the valves obliquely conic.

1. Cleomella longipes Torr. Journ. Bot. Kew Misc. **2**: 255. 1850.

TYPE LOCALITY: "Valley near San Pablo, Chihuahua, and near San Francisco, San Luis Potosi, Mexico."

RANGE: Western Texas to Arizona and Sonora.

NEW MEXICO: Dog Spring (*Mearns* 2379).

A rather uncommon plant of the Southwest. The peculiar small, 2-valved pod with conical valves and the long stipes serve to distinguish the genus from our others.

4. WISLIZENIA Engelm.

Erect, much branched annual; leaves trifoliolate, the leaflets oblong to obovate; petals small, yellow; stamens 6, exserted; pods long-stipitate, small, 2-seeded, the valves ellipsoid, indurate, reticulate.

1. **Wislizenia refracta** Engelm. in Wisliz. Mem. North. Mex. 99. 1848.

TYPE LOCALITY: "On the upper crossing of the Rio Grande, near El Paso," Texas.

RANGE: Western Texas and southern New Mexico.

NEW MEXICO: Mesilla Valley. Dry fields, in the Lower Sonoran Zone.

Occurs rather abundantly in the Rio Grande Valley in the extreme southern part of the State. It is readily recognized by the very small, 2-seeded, 2-valved fruit, each valve of which is indurated and reticulated, and closely invests the seed.

57a. RESEDACEAE. Mignonette Family.

1. DIPETALIA Raf.

Low branching herb, somewhat succulent, with numerous linear entire leaves; flowers small and inconspicuous, in terminal spikes; sepals 4; petals 2, entire or lobed; stamens 3 to 8; pod 4-beaked, about 3 mm. in diameter, opening at the summit.

1. **Dipetalia subulata** (Webb & Berth.) Kuntze, Rev. Gen. Pl. 1: 39. 1891.

Resedella subulata Webb & Berth. Hist. Nat. Canar. 1: 107. 1836.

TYPE LOCALITY: "Circa Portum Caprarum in insula Fuerteventura, et in Lancerotta circa oppidum Arecife."

RANGE: Southern California to western Texas, southward into Mexico; also in Asia and Africa.

NEW MEXICO: Near La Luz; Range Reserve, Dona Ana County. Alkaline soil.

Order 27. ROSALES.

KEY TO THE FAMILIES.

Flowers irregular.
  Fruit indehiscent, armed with spines; leaf blades simple; stipules wanting......69. **KRAMERIACEAE** (p. 336).
  Fruit a legume or loment; leaf blades compound; stipules usually present.
    Upper petals inclosed by the lateral ones in the bud....................68. **CASSIACEAE** (p. 332).
    Upper petal inclosing the lateral ones in the bud....................70. **FABACEAE** (p. 336).
Flowers regular or nearly so.
  Endosperm wanting or scant.
    Flowers monœcious, in dense clusters..63. **PLATANACEAE** (p. 304).
    Flowers perfect or if not perfect not in capitate clusters.
      Carpels several or numerous or if solitary becoming an achene.
        Carpels distinct, free from the hypanthium; fruit of achenes, follicles, or drupelets................64. **ROSACEAE** (p. 305).
        Carpels united, inclosed by and adnate to the hypanthium; fruit a pome.....65. **MALACEAE** (p. 321).
      Carpels solitary, not becoming achenes.
        Ovary 2-ovuled; fruit a drupe; leaves simple...........66. **AMYGDALACEAE** (p. 324).
        Ovary several-ovuled; fruit a legume; leaves pinnate..67. **MIMOSACEAE** (p. 327).

Endosperm present, usually copious and fleshy; stipules usually wanting.
  Herbs.
    Carpels as many as the sepals; plants succulent............**58. CRASSULACEAE** (p. 292).
    Carpels fewer than the sepals; plants seldom or never succulent.
      Staminodia wanting; carpels 2 or rarely 3, distinct or only partly united......**59. SAXIFRAGACEAE** (p. 294).
      Staminodia present; carpels 3 or 4, wholly united into a 1-celled gynœcium....**60. PARNASSIACEAE** (p. 298).
  Shrubs or trees.
    Leaves opposite; fruit a leathery capsule, more or less adnate to the hypanthium..........**61. HYDRANGEACEAE** (p. 298).
    Leaves alternate; fruit various.
      Fruit of thin-walled follicles, free from the hypanthium. (Stipules present. Opulaster)........**64. ROSACEAE** (p. 305).
      Fruit a berry; hypanthium adnate to and prolonged beyond the ovary.......**62. GROSSULARIACEAE** (p. 301).

58. CRASSULACEAE. Orpine Family.

Succulent herbaceous annuals or perennials, 30 cm. high or less, with flowers in cymes or 1-sided racemes; flowers mostly perfect (except in one genus), symmetrical and regular; sepals and petals 5, the former somewhat united at the base; stamens 10; carpels 5; follicles 1-celled, dehiscent along the ventral suture.

KEY TO THE GENERA.

Inflorescence axillary.................................... 1. CLEMENTSIA (p. 292).
Inflorescence terminal, cymose.
  Flowers yellow or purplish, never white, polygamous; plants usually 20 to 30 cm. high................. 2. RHODIOLA (p. 293).
  Flowers white or tinged with pink, rarely yellow, perfect; plants usually less than 10 cm. high....... 3. SEDUM (p. 293).

1. CLEMENTSIA Rose.

Perennial, 10 to 30 cm. high, glabrous, with usually numerous stems; leaves flat, entire or toothed; flowers in axillary racemes or cymes, the petals rose or white, twice as long as the calyx; follicles erect, with spreading tips.

1. **Clementsia rhodantha** (A. Gray) Rose, Bull. N. Y. Bot. Gard. **3:** 3. 1903.
Sedum rhodanthum A. Gray, Amer. Journ. Sci. II. **33:** 405. 1862.
TYPE LOCALITY: Rocky Mountains of Colorado.
RANGE: Montana to Arizona and northern New Mexico.
NEW MEXICO: Pecos Baldy (*Bailey* 614). Meadows, in the Arctic-Alpine Zone.

2. RHODIOLA L.

Perennials with woody, somewhat branching, thickened rootstocks; leaves flat and comparatively thin; flowers diœcious or polygamous, 4 or 5-parted, in compact terminal cymes; carpels erect; styles very short or none.

KEY TO THE SPECIES.

Petals yellowish, obtuse .... 1. *R. neomexicana.*
Petals purplish, acuminate .... 2. *R. polygama.*

1. **Rhodiola neomexicana** Britton, Bull. N. Y. Bot. Gard. **3**: 38. 1903.
TYPE LOCALITY: White Mountain Peak, New Mexico. Type collected by Wooton.
RANGE: White Mountains of New Mexico, in the Hudsonian Zone.

2. **Rhodiola polygama** (Rydb.) Britt. & Rose, Bull. N. Y. Bot. Gard. **3**: 39. 1903.
Sedum polygamum Rydb. Bull. Torrey Club **28**: 283. 1901.
TYPE LOCALITY: West Spanish Peak, Colorado.
RANGE: Mountains of Colorado and New Mexico.
NEW MEXICO: Pecos Baldy; Baldy; Jemez Mountains. Transition and Arctic-Alpine zones.

3. SEDUM L. STONECROP.

Fleshy, mostly glabrous herbs, erect or decumbent, with alternate entire leaves and perfect flowers in terminal cymes; calyx 5-parted; petals 5, yellow, white, or pinkish; stamens 10; carpels 5, distinct; fruit consisting of 5 follicles.

KEY TO THE SPECIES.

Petals yellow .... 6. *S. stenopetalum*
Petals white or pink.
  Leaves terete or nearly so .... 1. *S. stelliforme.*
  Leaves oblong, lanceolate, or spatulate, flattened.
    Leaves lanceolate to ovate or oblong, broadest at or near the base .... 2. *S. cockerellii.*
    Leaves spatulate to obovate, narrowed at the base.
      Plants glabrous .... 3. *S. wrightii.*
      Plants pubescent or puberulent-granular, at least above.
        Leaves turgid, the basal ones smooth .... 4. *S. wootoni.*
        Leaves flattish, at least the basal ones papillose.. 5. *S. griffithsii.*

1. **Sedum stelliforme** S. Wats. Proc. Amer. Acad. **20**: 365. 1885.
TYPE LOCALITY: Huachuca Mountains, southern Arizona.
RANGE: New Mexico and Arizona to Chihuahua.
NEW MEXICO: Hillsboro Peak; San Francisco Mountains; Mogollon Mountains.

2. **Sedum cockerellii** Britton, Bull. N. Y. Bot. Gard. **3**: 41. 1903.
TYPE LOCALITY: Tuerto Mountain, east of Santa Fe, New Mexico. Type collected by Cockerell.
RANGE: Known only from type locality.

3. **Sedum wrightii** A. Gray, Pl. Wright. **1**: 76. 1852.
TYPE LOCALITY: "Hills near the San Pedro River, in crevices of rocks, and summit of mountains near El Paso," Texas.
RANGE: Western Texas and southern New Mexico to northern Mexico.
NEW MEXICO: Black Range; Eagle Creek, White Mountains.

4. Sedum wootoni Britton, Bull. N. Y. Bot. Gard. **3:** 44. 1903.

TYPE LOCALITY: Organ Mountains, New Mexico. Type collected by Wooton.

RANGE: New Mexico and Arizona.

NEW MEXICO: Santa Fe and Las Vegas mountains; Sandia Mountains; Hillsboro Peak; Organ Mountains; White Mountains. Rocky cliffs of the mountains, in the Transition Zone.

5. Sedum griffithsii Rose, N. Amer. Fl. **22:** 71. 1905.

TYPE LOCALITY: Santa Rita Mountains, Arizona.

RANGE: Southern Arizona and New Mexico.

NEW MEXICO: Bear Mountain; Tularosa River, Socorro County; Burro Mountains; San Luis Mountains; Organ Mountains. Transition Zone.

6. Sedum stenopetalum Pursh, Fl. Amer. Sept. 324. 1814.

Sedum lanceolatum Torr. Ann. Lyc. N. Y. **2:** 205. 1827.

Sedum stenopetalum rubrolineatum Cockerell, Bull. Torrey Club **18:** 169. 1891.

TYPE LOCALITY: "On rocks on the banks of Clarck's river and Kooskoosky."

RANGE: Alberta and Nebraska to California and northern New Mexico..

NEW MEXICO: Upper Pecos (*Maltby & Coghill* 183).

59. SAXIFRAGACEAE. Saxifrage Family.

Perennial herbs with more or less scapelike flower-bearing stems; leaves mostly basal in a rosette about a shortened and thickened or slender and elongated axis; leaves simple, entire, toothed, or lobed, the cauline, when present, of slightly different shape; flowers perfect, solitary or in simple or paniculately branched cymes; hypanthium usually well developed, of various shapes; flowers 5-parted, rarely 4-parted; stamens as many or twice as many as the sepals; gynœcium of 2 (rarely 3 or 4) carpels; ovary partially or wholly inferior; fruit of capsules or follicles.

KEY TO THE GENERA.

Placentæ parietal, sometimes nearly basal.
- Flower stalk lateral from a stout scaly rootstock; gynœcium 2-carpellary .... 1. HEUCHERA (p. 294).
- Flower stalk axial from a slender bulbiferous rootstock; gynœcium 3-carpellary .... 5. LITHOPHRAGMA (p. 297).

Placentæ axial.
- Hypanthium well developed, accrescent, at maturity longer than the sepals; leaves 5-lobed.. 2. SAXIFRAGA (p. 296).
- Hypanthium only slightly developed, unchanged at maturity; leaves not lobed.
  - Inflorescence scapiform, not leafy nor bracteate. 3. MICRANTHES (p. 296).
  - Inflorescence leafy and bracteate .... 4. LEPTASEA (p. 297).

1. HEUCHERA L. ALUM ROOT.

Cespitose perennials with mostly basal, broadly oval to rotund, cordate leaves arising from the thickened rhizomatous stems, these covered by the bases of the petioles; flowers in elongated, scapelike, narrow or spreading panicles, dull greenish white or rose-purplish; calyx tube turbinate or campanulate, the limb 5-parted; petals entire, small; stamens 5; styles 2; capsule 2-beaked, about half inclosed in the calyx tube.

KEY TO THE SPECIES.

Stamens much shorter than the sepals; flowers dull greenish white.
    Petioles not hirsute; hypanthium 2 to 3.5 mm. high....... 1. *H. parvifolia.*
    Petioles hirsute; hypanthium 3 to 5 mm. high.
        Petals spatulate, almost clawless, not exceeding the sepals.. 2. *H. novomexicana*
        Petals obovate-spatulate, distinctly clawed, exceeding the sepals....................................... 3. *H. wootoni.*
Stamens longer than the sepals; flowers more or less rose-colored.
    Hypanthium deeply turbinate, fully twice as long as broad and nearly twice as long as the sepals............... 4. *H. leptomeria.*
    Hypanthium campanulate, turbinate only at the base, not more than half longer than broad and scarcely surpassing the sepals.
        Hypanthium with the sepals 5 mm. long; plants tall, 15 to 20 cm. high............................... 5. *H. versicolor.*
        Hypanthium 4 mm. long or less; plants less than 15 cm. high.
            Hypanthium 4 mm. long; inflorescence secund, dense... 6. *H. pulchella.*
            Hypanthium 3 mm. long; inflorescence not secund, loose....................................... 7. *H. nana.*

1. **Heuchera parvifolia** Nutt.; Torr. & Gray, Fl. N. Amer. **1:** 581. 1840.

Heuchera flavescens Rydb. N. Amer. Fl. **22:** 114. 1905.

TYPE LOCALITY: "Blue Mountains of Oregon."

RANGE: Oregon and Alberta to Arizona and New Mexico.

NEW MEXICO: Chama; Stinking Lake; Raton; Santa Fe and Las Vegas mountains; Magdalena Mountains; Organ Mountains. Damp woods, in the Transition and Canadian zones.

This species and the next two usually occur in rich soil on cool shady hillsides under trees in the mountains of the State at middle or high elevations. They closely resemble each other in general appearance and are never very abundant or conspicuous.

The type of *H. flavescens* was collected in Santa Fe Canyon (*Heller* 3693).

2. **Heuchera novomexicana** Wheelock, Bull. Torrey Club **17:** 200. 1890.

TYPE LOCALITY: Santa Rita, New Mexico. Type collected by Wright (no. 1098).

RANGE: Mountains of New Mexico and Arizona.

NEW MEXICO: Santa Rita; Sawyers Peak; Rio Apache. Transition and Canadian zones.

3. **Heuchera wootoni** Rydb. N. Amer. Fl. **22:** 113. 1905.

TYPE LOCALITY: Gilmores Ranch, White Mountains, New Mexico. Type collected by Wooton (no. 283).

RANGE: Damp woods, White and Sacramento mountains of New Mexico, in the Transition Zone.

4. **Heuchera leptomeria** Greene, Leaflets **1:** 112. 1905.

TYPE LOCALITY: Organ Mountains, New Mexico. Type collected by Wooton, September 17, 1893.

RANGE: Moist slopes, Organ Mountains of New Mexico, in the Transition Zone.

This pretty plant grows in crevices of rocks on bold rocky cliffs where there is water from a seep or spring. Its rootstocks are thick and crowded, bearing numerous radiating leaves and delicate, pale rose-colored flowers.

5. Heuchera versicolor Greene, Leaflets **1**: 112. 1905.

TYPE LOCALITY: Damp shady bluffs in the Black Range, New Mexico. Type collected by Metcalfe (no. 1203).

RANGE: Mountains of southwestern New Mexico.

NEW MEXICO: Hillsboro Peak; Copper Mines; Mogollon Road. Transition Zone.

6. Heuchera pulchella Woot. & Standl. Contr. U. S. Nat. Herb. **16**: 130. 1913.

TYPE LOCALITY: Crevices of rocks on the summit of the Sandia Mountains, New Mexico. Type collected by Wooton, August 4, 1910.

RANGE: Mountains of central New Mexico.

NEW MEXICO: Sandia Mountains; headwaters of the Pecos.

7. Heuchera nana (A. Gray) Rydb. N. Amer. Fl. **22**: 111. 1905.

Heuchera rubescens nana A. Gray, Pl. Wright, **2**: 64. 1853.

TYPE LOCALITY: Santa Rita, New Mexico. Type collected by Wright.

RANGE: New Mexico and Arizona to northern Mexico.

NEW MEXICO: Frisco Canyon near Luna; Santa Rita. Mountains.

2. SAXIFRAGA L.

A small herbaceous perennial less than 20 cm. high, the stems arising from a rootstock; leaves alternate, 5 to 7-lobed; inflorescence of a single terminal white flower and a cluster of bulblets; follicles mostly undeveloped.

1. Saxifraga cernua L. Sp. Pl. 403. 1753.

TYPE LOCALITY: "Habitat in Alpibus Lapponicis frequens."

RANGE: Arctic regions south to Labrador, and in the Rocky Mountains to northern New Mexico; also in the Old World.

NEW MEXICO: Top of Truchas Peak (*Bailey* 649). Arctic-Alpine Zone.

3. MICRANTHES Haw.

Perennial acaulescent herbs with very short caudices and solitary or numerous scapes; leaves basal, long-petiolate or decurrent into a short and broad petiole; flowers small, white, on crowded or paniculately branched cymes; hypanthium flat, shorter than the calyx; follicles more or less divergent.

KEY TO THE SPECIES.

Leaf blades orbicular-reniform, on petioles several times as long; scapes 15 to 50 cm. high, widely branched above .... 1. *M. arguta*.

Leaf blades lanceolate to ovate or oblong, on short petioles; scapes 25 cm. high or less, the inflorescence congested.

Bracts, calyx, scape, and lower surface of the leaves dark purplish red; inflorescence open .... 2. *M. eriophora*.

Whole plant green; inflorescence compact .... 3. *M. rhomboidea*.

1. Micranthes arguta (D. Don) Small, N. Amer. Fl. **22**: 147. 1905.

Saxifraga arguta D. Don, Trans. Linn. Soc. Bot. **13**: 356. 1822.

TYPE LOCALITY: "Habitat ad oras occidentales Americae septentrionalis."

RANGE: British Columbia and Montana to California and New Mexico.

NEW MEXICO: Santa Fe and Las Vegas mountains; Mogollon Mountains. Wet slopes and along streams, Transition to Hudsonian Zone.

2. Micranthes eriophora (S. Wats.) Small, N. Amer. Fl. **22**: 142. 1905.

Saxifraga eriophora S. Wats. Proc. Amer. Acad. **17**: 372. 1882.

TYPE LOCALITY: Santa Catalina Mountains, Arizona.

RANGE: Mountains of southern Arizona and New Mexico.

NEW MEXICO: Organ Peak (*Wooton*). Transition Zone.

3. **Micranthes rhomboidea** (Greene) Small, N. Amer. Fl. **22**: 136. 1905.
Saxifraga rhomboidea Greene, Pittonia **3**: 343. 1898.
TYPE LOCALITY: Colorado Rocky Mountains.
RANGE: Montana to New Mexico.
NEW MEXICO: Wheeler Peak; Santa Fe and Las Vegas mountains. Moist ground, Transition to Arctic-Alpine Zone.

4. LEPTASEA Haw.

More or less matted herbaceous perennials, with well developed leafy caudices and leafy flower stalks a few centimeters high; leaves simple, alternate, thickish, often clustered at the base; flowers few or rather numerous; sepals 5; petals 5, usually clawed, white or yellow; stamens 10; ovary mostly superior; carpels united to above the middle; follicles erect with spreading tips.

KEY TO THE SPECIES.

Petals white, dotted with yellow and purple.................. 1. *L. austromontana.*
Petals yellow.
  Leaf blades spine-tipped................................ 2. *L. flagellaris.*
  Leaf blades not spine-tipped............................ 3. *L. chrysantha.*

1. **Leptasea austromontana** (Wiegand) Small, N. Amer. Fl. **22**: 153. 1905.
Saxifraga austromontana Wiegand, Bull. Torrey Club **27**: 389. 1900.
Saxifraga cognata E. Nelson, Bot. Gaz. **30**: 118. 1900.
TYPE LOCALITY: Not definitely stated.
RANGE: Alberta and British Columbia to Washington, Montana, and New Mexico.
NEW MEXICO: Santa Fe and Las Vegas mountains; Hillsboro Peak; White Mountains. Damp cliffs, in the Transition and Hudsonian zones.

2. **Leptasea flagellaris** (Willd.) Small, N. Amer. Fl. **22**: 154. 1905.
Saxifraga flagellaris Willd.; Sternb. Rev. Saxifr. 25. 1810.
TYPE LOCALITY: Caucasus.
RANGE: Arctic regions, south in the Rocky Mountains to Arizona and New Mexico.
NEW MEXICO: Baldy; Wheeler Peak; Pecos Baldy. Arctic-Alpine Zone.

3. **Leptasea chrysantha** (A. Gray) Small, N. Amer. Fl. **22**: 152. 1905.
Saxifraga chrysantha A. Gray, Proc. Amer. Acad. **12**: 83. 1877.
TYPE LOCALITY: "High alpine region of the Colorado Rocky Mountains, especially abundant on Torrey's and Gray's peaks."
RANGE: Colorado and northern New Mexico.
NEW MEXICO: Wheeler Peak; Truchas Peak; Pecos Baldy; Bartlett Ranch; Mountains, in the Arctic-Alpine Zone.

5. LITHOPHRAGMA Nutt.

Slender perennial with low simple stems and bulblet-bearing rootstocks; leaves mostly basal, ternately divided, the divisions again subdivided; hypanthium campanulate; petals white, clawed, digitately cleft, longer than the sepals; stamens 10, included; gynœcium 1-celled, 3-valved.

1. **Lithophragma australis** Rydb. N. Amer. Fl. **22**: 86. 1905.
TYPE LOCALITY: Rocky hillsides, Cedar Creek, Arizona.
RANGE: Wyoming to Arizona and New Mexico.
NEW MEXICO: Hills near Tierra Amarilla (*Eggleston* 6442, 6579). Open slopes, in the Transition Zone.

60. PARNASSIACEAE. Flower-of-Parnassus Family.

1. PARNASSIA L. FLOWER-OF-PARNASSUS.

Glabrous perennial herbs with short rootstocks and scapiform stems; leaves entire, mostly basal and petioled, the single cauline leaf sessile; flowers solitary, terminating the scape; sepals 5, green; petals 5, white, conspicuously veined; stamens 5, alternate with the petals and with the 5 clusters of gland-bearing staminodia; capsule 1-celled, with 3 or 4 valves; seeds numerous, winged.

KEY TO THE SPECIES.

Petals fimbriate on the sides near the base.......................... 1. *P. fimbriata.*
Petals entire.. 2. *P. parviflora.*

1. **Parnassia fimbriata** König, Ann. Bot. Kön. & Sims **1:** 391. 1805.
TYPE LOCALITY: "On the coast of northwest America."
RANGE: Alaska and Alberta to California and northern New Mexico.
NEW MEXICO: Truchas Peak (*Standley* 4768). Wet ground, in the Arctic-Alpine Zone.

2. **Parnassia parviflora** DC. Prodr. **1:** 320. 1824.
TYPE LOCALITY: North America.
RANGE: British America to South Dakota, Utah, and New Mexico.
NEW MEXICO: Rio Pueblo; headwaters of the Pecos; White Mountains. Wet ground, in the Transition Zone.

61. HYDRANGEACEAE. Hydrangea Family.

Low or tall widely branching shrubs with opposite branches; leaves opposite, exstipulate, simple, more or less persistent, entire or toothed; flowers perfect, with mostly conspicuous white or yellowish petals, solitary or cymose; calyx of 4 or 5 sepals surmounting the hypanthium; stamens numerous, the filaments slender or sometimes stout, then appendaged; ovary partly inferior; fruit a woody capsule.

KEY TO THE GENERA.

Flowers in cymes, very numerous; sepals and petals 5; stamens 10.
  Plants large, often 2 meters high; leaves large, toothed 1. EDWINIA (p. 298).
  Plants small, depressed; leaves small, entire....... 2. FENDLERELLA (p. 299).
Flowers solitary or in 2 or 3-flowered clusters; sepals and petals 4 or 5, usually 4; stamens 8 or numerous (15 to 60).
  Filaments appendaged; flowers uniformly 4-parted; stamens 8.................................. 3. FENDLERA (p. 299).
  Filaments not appendaged; flowers occasionally 5-parted; stamens 15 to 60.................... 4. PHILADELPHUS (p. 300).

1. EDWINIA Heller.

A rather large shrub, often 2 meters high, with opposite branches and brownish, partly deciduous bark; leaves deciduous, thin, ovate, petiolate, serrate, 10 cm. long or less, bright green above, pale or whitish-tomentulose beneath; flowers in crowded cymes, white, 5-parted.

1. **Edwinia americana** (Torr. & Gray) Heller, Bull. Torrey Club **24:** 477. 1897.
Jamesia americana Torr. & Gray, Fl. N. Amer. 1: 593. 1840.

TYPE LOCALITY: "Along the Platte or the Canadian River, near the Rocky Mountains."

RANGE: Wyoming to Utah and New Mexico.

NEW MEXICO: Jemez Mountains; Santa Fe and Las Vegas mountains; Manzano and Sandia mountains; Magdalena Mountains; San Mateo Mountains; Mogollon Mountains; White and Sacramento mountains. Mountains, in the Transition and Canadian zones.

This shrub is common along the mountain streams, often with its roots in running water. It also occurs high up on the mountain peaks. While well worth cultivation, it is doubtful whether it would endure the conditions of the ordinary garden.

2. FENDLERELLA Heller.

A low, much branched shrub 50 to 60 cm. high, with grayish young branches and small lanceolate leaves 1 cm. long or less; flowers small, about 5 mm. long, white, in cymose several-flowered clusters; hypanthium decidedly turbinate; capsule considerably exceeding the calyx.

1. **Fendlerella cymosa** Greene, Contr. U. S. Nat. Herb. **16:** 129. 1913.

TYPE LOCALITY: Huachuca Mountains, southern Arizona.

RANGE: Arizona and New Mexico to northern Mexico.

NEW MEXICO: San Luis Mountains; Organ Mountains; San Andreas Mountains. Dry slopes, in the Upper Sonoran Zone.

3. FENDLERA Engelm. & Gray.

Shrubs 2 meters high or less, much branched, with grayish bark, rather small leaves, and white or pink-tinged flowers borne in great profusion; leaves entire, mostly sessile; flowers 4-parted; filaments of the 8 stamens flattened and with 2 narrow appendages at the top extending beyond the anthers.

KEY TO THE SPECIES.

Leaves narrow, linear-elliptic or narrowly linear-lanceolate, usually falcate, glabrous or nearly so .... 1. *F. falcata.*
Leaves broader, elliptic to ovate-lanceolate, not falcate, copiously pubescent.
  Leaves more or less white-tomentose beneath .... 2. *F. tomentella.*
  Leaves green beneath, not tomentose .... 3. *F. rupicola.*

1. **Fendlera falcata** Thornber, Contr. U. S. Nat. Herb. **16:** 129. 1913.

TYPE LOCALITY: Tunitcha Mountains, New Mexico. Type collected by Standley (no. 7806).

RANGE: Colorado to Arizona and New Mexico.

NEW MEXICO: Tunitcha Mountains; Cedar Hill; Carrizo Mountains; Embudo; Cloudcroft. Dry hills, in the Upper Sonoran Zone.

2. **Fendlera tomentella** Thornber, Contr. U. S. Nat. Herb. **16:** 129. 1913.

TYPE LOCALITY: Canyon of the Blue River near Coopers Ranch, Graham County, Arizona.

RANGE: Colorado to New Mexico and Arizona.

NEW MEXICO: Hurrah Creek; Mangas Springs. Dry hills, in the Upper Sonoran Zone.

3. Fendlera rupicola A. Gray, Pl. Wright. **1:** 77. 1852.

TYPE LOCALITY: "On perpendicular rocks of the Guadalupe, above New Braunfels," Texas.

RANGE: Arizona to western Texas and southward.

NEW MEXICO: Magdalena Mountains; Sandia Mountains; Burro Mountains; Animas Mountains; Big Hatchet Mountains; Florida Mountains; Organ Mountains; Gray. Dry hills, in the Upper Sonoran Zone.

A beautiful shrub, occurring in the drier mountains of the State, among rocks. It has never been cultivated, so far as we can learn, but it is certainly as handsome as the commonly grown species of Philadelphus.

4. PHILADELPHUS L. MOCK ORANGE.

Freely branching shrubs 2.5 meters high or less, mostly with conspicuous white flowers; leaves small, 2 cm. long or less, elliptic-lanceolate to ovate; flowers on short pedicels, mostly solitary; sepals and petals 4, rarely 5; stamens numerous, 15 to 60; ovary about two-thirds inferior.

KEY TO THE SPECIES.

Petals acute, ochroleucous; stamens about 15.................. 1. *P. mearnsii.*
Petals rounded at the apex, white; stamens 25 to 60.
  Hypanthium externally glabrous to strigose.................. 2. *P. microphyllus.*
  Hypanthium densely pubescent, silvery white.
    Leaves hirsute beneath, the pubescence loose, the blades 20 to 35 mm. long.................. 3. *P. argyrocalyx.*
    Leaves silky-strigose beneath, the pubescence close and dense, the blades 10 to 15 mm. long.................. 4. *P. argenteus.*

1. Philadelphus mearnsii W. H. Evans; Rydb. N. Amer. Fl. **22:** 174. 1905.

TYPE LOCALITY: Near the Upper Corner Monument, Grant County, New Mexico. Type collected by E. A. Mearns (no. 36).

RANGE: Known only from the type locality.

2. Philadelphus microphyllus A. Gray, Mem. Amer. Acad. n. ser. **4:** 54. 1849.

TYPE LOCALITY: Santa Fe Creek, on sunny and steep sides of the mountains, between rocks, 11 miles above Santa Fe, New Mexico. Type collected by Fendler (no. 266).

RANGE: Southern Colorado to Arizona and New Mexico.

NEW MEXICO: Carrizo Mountains; Santa Fe Mountains; Sandia Mountains; mountains west of Grants Station; Magdalena Mountains; Hillsboro Peak; San Mateo Mountains. Open slopes, in the Transition Zone.

This species and the next two should do well in cultivation at levels of 1,500 meters or more, in open porous soils, if supplied with sufficient water.

3. Philadelphus argyrocalyx Wooton, Bull. Torrey Club **25:** 452. 1898.

Philadelphus ellipticus Rydb. N. Amer. Fl. **22:** 174. 1905.

TYPE LOCALITY: On Eagle Creek, in the White Mountains, New Mexico. Type collected by Wooton (no. 524).

RANGE: White and Sacramento mountains of New Mexico, in the Transition Zone.

There seems to be no essential difference between the types of *P. argyrocalyx* and *P. ellipticus*. The type of the latter bears on its label the legend "Mesilla Park." This certainly is wrong, for no Philadelphus is found nearer Mesilla Park than in the Organ Mountains 10 or 12 miles away. No such plant as is represented by the type of *P. ellipticus* has ever been found in the Organs by either of the writers. However, the type of Dr. Rydberg's species exactly matches a specimen collected by Wooton on Ruidoso Creek in the White Mountains, June 30, 1895. Comparing the two specimens in every respect, there is practically no doubt that the two are part of the same collection from near the type locality of *P. argyrocalyx.* Evidently some mistake was made in sending out the specimen which became the type of the new species, resulting in the mixing of labels.

4. **Philadelphus argenteus** Rydb. N. Amer. Fl. **22**: 171. 1905.
Type locality: Fort Huachuca, Arizona.
Range: Southeastern Arizona and southwestern New Mexico.
New Mexico: Burro Mountains; Santa Rita; Animas Peak.

62. GROSSULARIACEAE. Gooseberry Family.

Erect or spreading shrubs, often with bristly or spiny stems; leaves alternate, simple, petiolate, broadly ovate to rotund, usually palmately veined, more or less lobed and toothed; inflorescence terminal on short, lateral, sometimes leafless branches, racemose, or the raceme reduced to a single flower; flowers regular, perfect (rarely unisexual); hypanthium elongated, short, or obsolete; sepals, petals, and stamens 5, alternate; ovary 1-celled; fruit a berry.

KEY TO THE GENERA.

Stems without spines or bristles (except in *R. montigenum*); pedicels jointed beneath the ovary; fruit breaking from the pedicel.................................. 1. Ribes (p. 301).
Stems with nodal spines, with or without extranodal bristles; pedicels not jointed beneath the ovary; fruit not breaking from the pedicels............... 2. Grossularia (p. 303).

1. RIBES L. Currant.

Unarmed shrubs (*R. montigenum* spiny and bristly) with palmately veined, mostly lobed leaves; flowers in several-flowered racemes; pedicels jointed beneath the ovary; ovary not spiny, sometimes glandular; hypanthium tubular to campanulate, sometimes obsolete; fruit breaking from the pedicel.

KEY TO THE SPECIES.

Stems armed with spines; leaves pubescent or glandular-hairy; berries bright red.. 1. *R. montigenum.*
Stems unarmed; leaves pubescent or glabrous; berries red or black.
  Hypanthium obsolete, the sepals slightly united at the base... 2. *R. coloradense.*
  Hypanthium evident (very short in *R. wolfii*).
    Anthers with a conspicuous cup-shaped apical gland.
      Hypanthium 3 or 4 times as long as broad; fruit red.. 3. *R. inebrians.*
      Hypanthium less than twice as long as broad; fruit black.. 4. *R. mescalerium.*
    Anthers with at most a mere callus at the apex.
      Hypanthium smooth, 3 or more times as long as thick; leaves involute in vernation.................... 5. *R. aureum.*
      Hypanthium hairy, less than 3 times as long as broad; leaves plicate in vernation.
        Leaves with amber-colored glands on both surfaces; hypanthium and calyx together 10 mm. long.................................. 6. *R. americanum.*
        Leaves without glands on the upper surface; hypanthium and calyx together 5 mm. long or less.................................. 7. *R. wolfii.*

1. Ribes montigenum McClatchie, Erythea **5:** 38. 1897.

Ribes nubigenum McClatchie, Erythea **2:** 80. 1894, not Phil. 1857.

Ribes lentum Coville & Rose, Proc. Biol. Soc. Washington **15:** 28. 1902.

TYPE LOCALITY: "On summit of Mt. San Antonio, 10,000 ft. altitude, among dry exposed rocks," California.

RANGE: British Columbia and Montana to California, Arizona, and New Mexico.

NEW MEXICO: Wheeler Peak; Santa Fe and Las Vegas mountains; Jemez Mountains; Sandia Mountains; White Mountains. Exposed slopes and summits, chiefly in the Hudsonian Zone.

The fruit is smooth in the New Mexico plant, and edible.

2. Ribes coloradense Coville, Proc. Biol. Soc. Washington **14:** 3. 1901.

Ribes laxiflorum coloradense Jancz.; Vilm. & Bois, Frutic. Vilm. Cat. Prin. 137. 1904.

TYPE LOCALITY: Marshall Pass, Colorado.

RANGE: Mountains of Colorado and northern New Mexico.

NEW MEXICO: Taos Mountains. Hudsonian Zone.

A decumbent or prostrate shrub occurring on the higher mountain peaks of the northern part of the State. Herbarium specimens resemble those of *R. wolfii*, but may be distinguished from that species by the deeper lobing and sharper teeth of the leaves, and by the slightly smaller fruit which is not glaucous. Its habit would well distinguish the growing plant.

3. Ribes inebrians Lindl. in Edwards's Bot. Reg. **18:** *pl. 1471.* 1832.

Ribes pumilum Nutt.; Rydb. Colo. Agr. Exp. Sta. Bull. **100:** 177. 1906.

TYPE LOCALITY: Described from cultivated plants.

RANGE: California, Idaho, and South Dakota to Nebraska, New Mexico, and Arizona.

NEW MEXICO: Zuni; Tunitcha Mountains; Baldy; Taos; Johnsons Mesa; Santa Fe and Las Vegas mountains; Dulce; Raton; Sierra Grande; San Mateo Mountains; Magdalena Mountains; San Francisco Mountains; Mogollon Mountains. Woods and canyons, in the Transition Zone.

The common currant in our mountains at middle elevations. It is easily recognized by the small, crowded, almost orbicular leaves, and glandular-viscidulous stems. The berries are small, about 5 or 6 mm. in diameter, glandular, dull scarlet or yellowish red, and insipid.

4. Ribes mescalerium Coville, Proc. Biol. Soc. Washington **13:** 196. 1900.

TYPE LOCALITY: Fresnal, Otero County, New Mexico. Type collected by Wooton, July 21, 1899.

RANGE: Mountains of southern New Mexico and western Texas.

NEW MEXICO: Datil; White and Sacramento mountains. Transition Zone.

A black currant occurring in the mountains of the south-central part of the State among pines. It is usually a larger bush than the preceding with larger leaves, not quite so glandular-hairy, and rarely viscidulous, while the fruit is much larger, frequently 1 cm. in diameter, black instead of red, and even more insipid.

5. Ribes aureum Pursh, Fl. Amer. Sept. 164. 1814. GOLDEN CURRANT.

TYPE LOCALITY: "On the banks of the rivers Missouri and Columbia."

RANGE: British America and South Dakota to California and New Mexico.

NEW MEXICO: McCarthy Station; Trout Spring; Mangas Springs; Mimbres River; Nogal Canyon; Farmington. Transition Zone.

This is a handsome plant because of the abundance of its perfumed yellow blossoms. It is recognized by its long, tubular, yellow flowers and the glabrous leaves of characteristic outline and dissection.

6. Ribes americanum Mill. Gard. Dict. ed. 8. no. 4. 1768.

Ribes floridum L'Her. Stirp. Nov. 4. 1785.

TYPE LOCALITY: Pennsylvania.

RANGE: British America to Virginia, Nebraska, Colorado, and northern New Mexico.

NEW MEXICO: Trout Spring, Gallinas Canyon (*Cockerell*). Damp woods.

7. Ribes wolfii Rothr. Amer. Nat. **8:** 358. 1874.

Ribes mogollonicum Greene, Bull. Torrey Club **8:** 121. 1881.

TYPE LOCALITY: "Twin Lakes and Mosquito Pass, Colorado Territory."

RANGE: Colorado and Utah to New Mexico and Arizona.

NEW MEXICO: Wheeler Peak; Santa Fe and Las Vegas mountains; Jemez Mountains; Sandia Mountains; Magdalena Mountains; Mogollon Mountains; White Mountains. Damp woods, in the Transition and Hudsonian zones.

This is the more common black currant of the higher parts of the mountains. It is a shrub, sometimes 3 meters high, with rather large, 5-lobed leaves, the lobes broadly ovate, the sinuses not deeply cut, and the margin crenate-serrate, the teeth small; the fruit which is edible but insipid is about 1 cm. in diameter, generally with a bloom.

The type of *Ribes mogollonicum* was collected in the Mogollon Mountains by E. L. Greene, in 1881.

2. GROSSULARIA Mill. GOOSEBERRY.

Spreading shrubs with numerous stems armed at the nodes with simple or 3-forked spines; leaves broadly ovate to rotund, rather deeply 3 to 5-lobed, the lobes coarsely crenate; racemes few-flowered; pedicels not jointed beneath the ovary; ovary and fruit spiny, hairy, or smooth; hypanthium evident; fruit not separating from the pedicel.

KEY TO THE SPECIES.

Ovary densely bristly, the bristles developing into sharp stout spines in fruit.. 1. *G. pinetorum.*

Ovary smooth, not spiny in fruit.

- Styles glabrous; leaves small, 20 mm. in diameter or less, on petioles as long or shorter, crowded; young stems densely spiny, the spines usually stout, often 1 cm. long, divergent and curved; flowers copiously ciliate and somewhat glandular outside....................................... 2. *G. leptantha.*
- Styles hairy near the base; leaves larger, mostly more than 20 mm. in diameter, on rather slender petioles longer than the blades, not so numerous as in the preceding; young stems mostly smooth, the spines short, often deflexed, 6 mm. long or less; flowers almost glabrous outside...... 3. *G. inermis.*

1. Grossularia pinetorum (Greene) Coville & Britton, N. Amer. Fl. **22:** 217. 1908.

Ribes pinetorum Greene, Bot. Gaz. **6:** 157. 1881.

TYPE LOCALITY: "In woods of *Pinus ponderosa*, in the higher elevations of the Pinos Altos and Mogollon Mountains," New Mexico. Type collected by Greene.

RANGE: New Mexico and Arizona.

NEW MEXICO: Zuni Mountains; San Mateo Peak; Magdalena Mountains; Mogollon Mountains; Black Range; White and Sacramento mountains. Woods in the mountains, Transition Zone.

A large shrub, often 2 meters high or more, common in the pine-covered areas of the mountains of the southern half of the State. It may be recognized by the large and very spiny fruits, 1 cm. or more in diameter, these at first red but purplish at maturity, borne singly. They have a pleasant acid flavor when ripe, but are hard to eat on account of the spines, and partly ripe berries are very astringent. The flowers are orange-colored.

2. **Grossularia leptantha** (A. Gray) Coville & Britton, N. Amer. Fl. **22:** 219. 1908.
Ribes leptanthum A. Gray, Mem. Amer. Acad. n. ser. **4:** 53. 1849.
Ribes leptanthum veganum Cockerell, Proc. Biol. Soc. Washington **15:** 99. 1902.

TYPE LOCALITY: "Rocky banks of the Rio del Norte, and ravines near Santa Fe," New Mexico. Type collected by Fendler (no. 254).

RANGE: Colorado and Utah to New Mexico and Arizona.

NEW MEXICO: Taos; Santa Fe and Las Vegas mountains; Raton Mountains; Sierra Grande; El Rito; Sandia Mountains; Magdalena Mountains; Puertecito; Mogollon Mountains; Santa Rita; Organ Mountains. Canyons and in woods, in the Upper Sonoran and Transition zones.

A very spiny shrub, from 1 to 1.5 meters high, occurring in the mountains almost throughout the State, at middle elevations. It is distinguished by the very spiny young branches which are light brown, by the long, curved, nodular spines, and the small, crowded, more or less glandular leaves whose petioles are usually shorter than the blades. The fruit is edible.

The type of *Ribes leptanthum veganum* was collected by Cockerell along the Gallinas River below Las Vegas.

3. **Grossularia inermis** (Rydb.) Coville & Britton, N. Amer. Fl. **22:** 224. 1908.
Ribes inerme Rydb. Mem. N. Y. Bot. Gard. **1:** 202. 1900.
Ribes purpusi Koehne; Blankinship, Mont. Agr. Coll. Sci. Stud. **1:** 64. 1905.

TYPE LOCALITY: Slough Creek, Yellowstone National Park.

RANGE: British Columbia and Montana to California, Utah, and New Mexico.

NEW MEXICO: Santa Fe and Las Vegas mountains; Chama; Mogollon Mountains. Woods, in the Transition Zone.

This gooseberry occurs in the mountains of the northern part of the State. It may be recognized by its rather slender young branches with few, small nodular spines, and the slender-petioled leaves which are commonly very numerous. The berries are wine-colored, about 8 mm. in diameter, smooth, and of good flavor.

63. PLATANACEAE. Sycamore Family.

1. PLATANUS L. SYCAMORE.

A large tree, 10 to 15 meters high, the bark deciduous in thin brittle plates, brownish, the young bark white or pale greenish; leaves large, 15 to 25 cm. in diameter, deeply 5-lobed, the lobes triangular-lanceolate, acuminate, densely tomentose, especially when young; flowers monœcious, in racemes of 3 to 5 spherical heads along an elongated peduncle; fruiting heads 20 to 25 mm. in diameter, on pedicels half as long; achenes glabrous, about 6 mm. long, exceeding the basal hairs.

1. **Platanus wrightii** S. Wats. Proc. Amer. Acad. **10:** 349. 1875.

TYPE LOCALITY: "In southeastern Arizona near the San Pedro."

RANGE: Southern New Mexico and Arizona to Sonora.

NEW MEXICO: Bear Mountain; Mogollon Mountains; Black Range; Animas Peak; Guadalupe Canyon; Dog Spring. Upper Sonoran Zone.

The sycamore is usually found along the streams and in the mouths of rocky canyons near the bases of the mountains. It would probably grow in cultivation, but, so far as known to us, it has not been tried in the State.

64. ROSACEAE. Rose Family.

Herbs, shrubs, or trees with alternate stipulate leaves (stipules often fugacious) and perfect flowers; hypanthium saucer-shaped, spherical, turbinate, or tubular, often margined by a disk bearing the stamens; sepals and petals normally 5, rarely of a different number, the petals wanting in one genus; stamens numerous, sometimes reduced to 5; carpels 1 to many, dry or fleshy, dehiscent in a few genera; fruit of achenes, follicles, or druplets (in some genera the receptacle accrescent).

KEY TO THE GENERA.

Hypanthium constricted at the throat, fleshy or prickly, wholly inclosing the achenes.
  Hypanthium in fruit becoming more or less fleshy, not supplied with hooked bristles; carpels numerous; shrubs........................... 1. ROSA (p. 306).
  Hypanthium dry, the top surrounded by numerous hooked bristles; carpels few; herbs.......... 2. AGRIMONIA (p. 309).
Hypanthium not constricted at the throat, neither fleshy nor prickly, at most loosely investing the fruits.
  Fruit consisting of 1 to 5 dehiscent follicles.
    Seeds winged; leaves persistent.................21. VAUQUELINIA (p. 321).
    Seeds not winged; leaves deciduous.
      Follicles more or less united at the base; leaves broadly ovate, lobed, 2 to 4 cm. long.......................... 3. OPULASTER (p. 309).
      Follicles distinct (usually 5); leaves very small, spatulate, 1 cm. long or less.. 4. PETROPHYTON (p. 310).
  Fruit usually consisting of numerous indehiscent carpels, these becoming either achenes or drupelets.
    Carpels becoming more or less fleshy drupelets.
      Styles filiform; stigmas capitate; leaves compound; stems spiny............18. RUBUS (p. 319).
      Styles club-shaped; stigmas 2-lobed; leaves simple; stems unarmed.
        Drupelets capped by hard hairy cushions; stems suffrutescent, dying back most of their length each winter; leaves large; fruit pleasantly acid, pulpy...............19. RUBACER (p. 320).
        Drupelets without cushions; stems nearly all woody; leaves small; fruit usually dry...............20. OREOBATUS (p. 320).
    Carpels becoming dry achenes.
      Styles articulated to the ovary, deciduous; herbs (except Dasiphora).
        Styles terminal or nearly so; ovules pendulous and anatropous...... 6. POTENTILLA (p. 310).
        Styles lateral or basal; ovules not pendulous.
          Styles nearly basal; ovules ascending or erect, orthotropous...11. DRYMOCALLIS (p. 316).

Styles lateral; ovules ascending, amphitropous.
Achenes hairy; shrubs........10. DASIPHORA (p. 316).
Achenes glabrous; herbs.
Achenes 10 to 15; stamens 5. (Leaves trifoliolate; petals yellow.)............ 9. SIBBALDIA (p. 315).
Achenes numerous; stamens about 20.
Receptacles not enlarged in fruit; leaves pinnate; petals yellow.. 7. ARGENTINA (p. 314).
Receptacles fleshy, much enlarged in fruit, red; leaves trifoliolate; petals white............ 8. FRAGARIA (p. 315).
Styles not articulated to the ovary; shrubs, except Geum and Sieversia.
Styles geniculate above, the upper portion deciduous, the basal portion forming a hook.................13. GEUM (p. 317).
Styles not geniculate, wholly persistent.
Herbs with woody rootstocks and pinnate leaves...............12. SIEVERSIA (p. 316).
Shrubs with simple, at most only deeply lobed, leaves.
Hypanthium bearing bracts; achenes numerous..........14. FALLUGIA (p. 317).
Hypanthium bractless; achenes 1 to 5.
Hypanthium flat, saucer-shaped, or hemispheric; flowers in a large panicle; (carpels 5)............... 5. SERICOTHECA (p. 310).
Hypanthium funnelform or tubular; flowers mostly solitary.
Petals wanting; hypanthium long-tubular; calyx deciduous from the hypanthium..............17. CERCOCARPUS (p. 318).
Petals 5; hypanthium turbinate; calyx persistent.
Carpels about 5; achenes with long plumose tails.................15. COWANIA (p. 318).
Carpels 1; achene not plumose-tailed...........16. PURSHIA (p. 318).

1. ROSA L. Rose.

More or less spiny shrubs, 2 meters high or less, with mostly slender branches and odd-pinnate, 3 to 7-foliolate leaves; stipules conspicuous, adnate to the petioles; flowers solitary or in few-flowered corymbs terminating the branches, large and showy, 3 to 6 cm. in diameter, pink or rose purple, fading lighter; hypanthium spherical or ellipsoidal, not bracteolate; sepals 5, with more or less foliaceous tips; petals broadly obovate to rotund; stamens numerous like the pistils; fruit mostly globose, dry, or somewhat succulent by the softening of the hypanthium.

KEY TO THE SPECIES.

Hypanthium and fruit densely spiny; sepals all or nearly all lobed.
  Young branches densely lepidote-stellate; leaves usually with 3 leaflets........ 1. *R. stellata.*
  Young branches with a dense covering of short, mostly gland-tipped spines, not lepidote-stellate; leaves mostly with 5 leaflets........ 2. *R. mirifica.*
Hypanthium not spiny; sepals not lobed.
  Infrastipular spines wanting.
    Flowers corymbose at the ends of the branches........ 3. *R. suffulta.*
    Flowers solitary at the ends of the branches........ 4. *R. sayi.*
  Infrastipular spines present.
    Sepals bristly glandular........ 5. *R. adenosepala.*
    Sepals not bristly.
      Petioles not glandular, the bracts often glandular-toothed.
        Spines few, stout, strongly curved; leaflets pale and shining, small, thick........ 10. *R. neomexicana.*
        Spines numerous, slender, straight or nearly so; leaflets bright green, not shining, large, thin.
          Leaflets short-villous beneath........ 11. *R. pecosensis.*
          Leaflets not short-villous beneath........ 12. *R. maximiliani.*
      Petioles more or less glandular.
        Leaves finely pubescent beneath, often strongly glandular.
          Spines straight; leaflets oblong to oval........ 8. *R. fendleri.*
          Spines recurved; leaflets obovate........ 9. *R. helleri.*
        Leaves glabrous beneath.
          Spines numerous, slender, straight........ 6. *R. hypoleuca.*
          Spines few, stout, curved........ 7. *R. melina.*

1. Rosa stellata Wooton, Bull. Torrey Club **25:** 152. 1898.

TYPE LOCALITY: Near the Cueva in the Organ Mountains, New Mexico. Type collected by Wooton (no. 126).

RANGE: Dry slopes, Organ and San Andreas Mountains, New Mexico, in the Upper Sonoran Zone.

The stellate pubescence of the stems and the large rose purple flowers are very characteristic. This species is to be expected in some of the drier and hotter mountains of the southern part of the State or in northern Chihuahua, but as yet it has not been found. It lends itself to cultivation tolerably well, though increased moisture and a richer soil tend to make it grow slightly taller and cause the flowers to become paler.

2. Rosa mirifica Greene, Leaflets **2:** 62. 1910.

TYPE LOCALITY: Near the Mescalero Agency, White Mountains, New Mexico. Type collected by Wooton (no. 193).

RANGE: White and Sacramento mountains of New Mexico, in the Upper Sonoran Zone.

Dr. Greene was no doubt correct in his judgment that this should be separated from *R. stellata*, from which it is farther separated than are most of the nearly related species of the genus from each other. He has pointed out the differences very accurately in his description.

3. Rosa suffulta Greene, Pittonia **4:** 12. 1899.

Type locality: Near Las Vegas, New Mexico. Type collected by G. R. Vasey, June, 1881.

Range: Northern and eastern New Mexico.

New Mexico: Las Vegas; Ute Park; Raton Mountains; Sierra Grande; Santa Fe; Perico Creek; White Mountains. Transition Zone.

This is probably the same as *R. pratincola* Greene; the name *suffulta*, however, has priority.

This is perhaps the stoutest of all the roses of the State, though it commonly does not grow very tall. The young stems are stout, greenish, strictly erect, very spiny, and very leafy. The leaves consist of 4 or 5 pairs of leaflets and a terminal one, all elliptic and sharply serrate; they are rather thick, dull green, and somewhat pubescent. The flowers are bright pink, about 5 or 6 cm. in diameter, and produced in crowded corymbs.

4. Rosa sayi Schwein. Narr. Exp. St. Peter's Riv. **2:** 388. 1824.

Type locality: Canada.

Range: Quebec and Alberta to Michigan and northern New Mexico.

New Mexico: Johnsons Mesa (*Wooton*).

5. Rosa adenosepala Woot. & Standl. Contr. U. S. Nat. Herb. **16:** 131. 1913.

Type locality: Along the Pecos River 8 miles east of Glorieta, New Mexico. Type collected by Heller (no. 3674).

Range: Known only from type locality.

Readily distinguished from our other species by the densely glandular-bristly calyx lobes.

This is probably a plant of the Transition Zone. Although the region along the Pecos in this locality supports a characteristic Upper Sonoran vegetation, the low land bordering the river, as the result of the greater amount of moisture with which it is supplied, is covered with plants common in the Transition Zone.

6. Rosa hypoleuca Woot. & Standl. Contr. U. S. Nat. Herb. **16:** 131. 1913.

Type locality: Near Kingston, Sierra County, New Mexico. Type collected by Metcalfe (no. 940).

Range: New Mexico.

New Mexico: Near Kingston; Winsor Creek. Mountains, in the Transition Zone.

Related to *Rosa fendleri*, but differing in having glabrous leaflets which are glaucous underneath.

7. Rosa melina Greene, Pittonia **4:** 10. 1899.

Type locality: Cerro Summit, above Cimarron, Colorado.

Range: Wyoming to northern New Mexico.

New Mexico: Santa Fe (*Wooton*).

8. Rosa fendleri Crép. Bull. Soc. Bot. Belg. **15:** 91. 1876.

Type locality: New Mexico. Type collected by Fendler (no. 210).

Range: Montana and South Dakota to Arizona and New Mexico.

New Mexico: Chama; Santa Fe and Las Vegas mountains; Sandia Mountains; Raton; Sierra Grande; Carrizo Mountains; Tunitcha Mountains; Copper Canyon; Fort Bayard; White Mountains. Transition Zone.

Probably this is the most common rose of the State, occuring at middle elevations in the mountains. It is usually a slender branching bush about 1 meter high with reddish stems and thin dark green leaves, and in these respects it resembles several of the nearly related species. The character of the spines and the pubescence are distinctive, however.

9. Rosa helleri Greene, Leaflets **2**: 259. 1912.
TYPE LOCALITY: Lake Waha, Nez Perces County, Idaho.
RANGE: Idaho to New Mexico.
NEW MEXICO: Coolidge; Las Vegas; Magdalena Mountains. Transition Zone.
The specimens are referred here doubtfully by Dr. Rydberg.

10. Rosa neomexicana Cockerell, Ent. News **1901**: 41. 1901.
TYPE LOCALITY: Cloudcroft, New Mexico.
RANGE: Southern New Mexico.
NEW MEXICO: Mesilla; near Mesilla Park; Sapello Creek; Cloudcroft.

11. Rosa maximiliani Nees in Wied-Neuw. Reis. Nord Amer. **2**: 434. 1841.
TYPE LOCALITY: On the plains along the Missouri River above Fort Pierre, South Dakota.
RANGE: Washington and Saskatchewan to Utah and New Mexico.
NEW MEXICO: Chama; Farmington; Winsors Ranch; Las Vegas; Pecos; Joseph; Raton Mountains; White Mountains; Indian Canyon; Animas Mountains; Organ Mountains. Upper Sonoran and Transition zones.

12. Rosa pecosensis Cockerell, Proc. Acad. Phila. **1904**: 110. 1904.
TYPE LOCALITY: Near Pecos, New Mexico. Type collected by Cockerell.
RANGE: Known only from the vicinity of the type locality.
Just what *Rosa praetincta* Cockerell [1] is, we are unable to determine. It also was described from the vicinity of Pecos.

2. AGRIMONIA L. AGRIMONY.

Herbaceous perennial, 60 to 80 cm. high or less, with interruptedly pinnate 5 or 7-foliolate leaves, and small yellow flowers in elongated slender racemes; hypanthium turbinate, bearing a ring of hooked prickles; petals small; stamens 5 to 12; carpels 1 to 3, becoming achenes and inclosed in the persistent hooked hypanthium.

1. Agrimonia striata Michx. Fl. Bor. Amer. **1**: 287. 1803.
Agrimonia brittoniana Bicknell, Bull. Torrey Club **23**: 517. 1896.
Agrimonia brittoniana occidentalis Rydb. Colo. Agr. Exp. Sta. Bull. **100**: 189. 1906.
TYPE LOCALITY: "In Canada."
RANGE: Quebec and New York to West Virginia, and in the Rocky Mountains.
NEW MEXICO: Santa Fe and Las Vegas mountains; Johnsons Mesa; Sandia Mountains; Chama; Mogollon Mountains; White Mountains. Woods, in the Transition Zone.

3. OPULASTER Medic. NINEBARK.

A low shrub, 1 meter high or less, with exfoliating bark and white flowers in terminal corymbs; leaves simple, rounded-ovate in outline, 3 to 5-lobed, glabrous or nearly so, doubly incised-serrate; flowers small; hypanthium about 3 mm. broad, stellate-pubescent; petals orbicular, about 3 mm. long; follicles 2 or rarely 3, united to above the middle, densely stellate-pubescent, with spreading beaks.

1. Opulaster monogynus (Torr.) Kuntze, Rev. Gen. Pl. **2**: 949. 1891.
Spiraea monogyna Torr. Ann. Lyc. N. Y. **2**: 194. 1827.
Physocarpus monogynus Coulter, Contr. U. S. Nat. Herb. **2**: 104. 1891.
TYPE LOCALITY: "On the Rocky Mountains," Colorado.
RANGE: Wyoming and South Dakota to New Mexico and Texas.

---

[1] Proc. Acad. Phila. **1904**: 110. 1904.

New Mexico: Santa Fe and Las Vegas mountains; Sierra Grande; Sandia Mountains; Magdalena Mountains; Mogollon Mountains; Black Range; Sacramento Mountains. Transition Zone.

A low shrub, not uncommon in the mountains at middle elevations. It should be valuable in cultivation because of its clusters of small white flowers, profusely produced.

4. PETROPHYTON (Nutt.) Rydb.

A densely cespitose depressed undershrub with prostrate branches; leaves spatulate, 5 to 12 mm. long, 2 to 4 mm. wide, densely silky; peduncles 3 to 10 cm. high, with bractlike subulate leaves; inflorescence a dense spike of small whitish flowers; sepals ovate-lanceolate, acute, 1.5 mm. long; petals spatulate, obtuse, of about the same length as the sepals; follicles 3 to 5, 2 mm. long.

1. **Petrophyton caespitosum** (Nutt.) Rydb. Mem. N. Y. Bot. Gard. **1:** 206. 1900.

Spiraea caespitosa Nutt.; Torr. & Gray, Fl. N. Amer. **1:** 418. 1840.

Eriogynia caespitosa S. Wats. Bot. Gaz. **15:** 242. 1890.

Type locality: "On high shelving rocks in the Rocky Mountains, toward the sources of the Platte."

Range: California, Montana, and South Dakota to Arizona and western Texas.

New Mexico: Big Hatchet Mountains; Guadalupe Mountains. Rocky hillsides, in the Upper Sonoran Zone.

A rather rare plant, occurring on rocky cliffs and exposed tops of the mountains, where it spreads flat over the rocks, the roots finding their way into the crevices. It is to be expected in the south end of the Organ Mountains, for it has been collected in the Franklin Range just across the line in Texas.

5. SERICOTHECA Raf.

Shrub 2 or 3 meters high, with few main stems and numerous spreading branches bearing simple leaves and with terminal spreading panicles of small white flowers; leaves obovate-cuneate, decurrent, with a few rounded teeth, densely white-villous beneath; sepals 5, 1.5 mm. long, cream-colored like the hemispheric hypanthium; petals 5, elliptic or oval, about 2 mm. long; stamens about 20; pistils 5.

1. **Sericotheca dumosa** (Nutt.) Rydb. N. Amer. Fl. **22:** 263. 1908.

Spiraea dumosa Nutt.; Torr. & Gray, Fl. N. Amer. **1:** 416. 1840, as synonym; Hook. Lond. Journ. Bot. **6:** 217. 1847.

Holodiscus dumosus Heller, Cat. N. Amer. Pl. 4. 1898.

Holodiscus australis Heller, Bull. Torrey Club **25:** 194. 1898.

Type locality: Rocky Mountains.

Range: Mountains of Wyoming and Utah to New Mexico and Chihuahua.

New Mexico: Baldy; Santa Fe and Las Vegas mountains; Sandia Mountains; Magdalena Mountains; Mogollon Mountains; Black Range; Big Hatchet Mountains; San Luis Mountains; Organ Mountains; White and Sacramento mountains. Upper Sonoran and Transition zones.

A beautiful and graceful shrub, producing an abundance of flowers. It is well worth cultivation and would doubtless grow well at middle elevations in the State.

The type of *Holodiscus australis* was collected in Santa Fe Canyon by Heller (no. 3840).

6. POTENTILLA L. Cinquefoil.

Annual or perennial herbs, when perennial with elongated, scaly, more or less cespitose rootstocks; leaves pinnately or palmately compound; inflorescence usually cymose-paniculate; hypanthium concave, mostly hemispheric; bractlets, sepals, and petals 5; calyx persistent; stamens commonly 20; receptacle hemispheric or conic, bearing numerous (in *P. lemmoni* only 5 to 10) pistils; fruit of small achenes.

KEY TO THE SPECIES.

Annuals or biennials with very leafy many-flowered cymes.
Achenes with a corky gibbosity on the upper suture; leaves pinnate, with 4 or 5 pairs of leaflets .... 1. *P. paradoxa.*
Achenes not gibbous; leaves pinnate or digitate.
Lower leaves pinnate, with 2 approximate pairs of leaflets .... 2. *P. rivalis.*
Leaves all digitate.
Petals about half as long as the sepals; achenes smooth; hypanthium in fruit 5 mm. wide or less; stamens 10 .... 3. *P. millegrana.*
Petals equaling the sepals or nearly so; achenes rugulose; hypanthium about 7 mm. wide; stamens 15 to 20 .... 4. *P. monspeliensis.*
Perennials with few-flowered, not very leafy cymes.
Flowers red.
Leaves green above, white-tomentose beneath .... 5. *P. atrorubens.*
Leaves glabrate or silky beneath, without tomentum.... 6. *P. thurberi.*
Flowers yellow.
Leaves digitate.
Plants glabrous throughout .... 7. *P. sierrae-blancae.*
Plants abundantly pubescent.
Plants tall, more than 20 cm. high ....11. *P. filipes.*
Plants low, less than 20 cm., usually less than 10 cm. high.
Leaves tomentose beneath ....10. *P. bicrenata.*
Leaves not tomentose beneath.
Stems simple, erect; sepals and bractlets not incurved .... 8. *P. diversifolia.*
Stems much branched, spreading; sepals and bractlets incurved... 9. *P. subviscosa.*
Leaves odd-pinnate.
Style not longer than the mature achene, thickened at the base; leaflets deeply lobed into narrow oblong teeth ....12. *P. strigosa.*
Style much longer than the mature achene, filiform; leaflets not lobed, merely coarsely toothed or nearly entire.
Leaves green on both sides, or merely strigose.
Pedicels arcuate-spreading in fruit ....19. *P. plattensis.*
Pedicels erect.
Stems stout, 60 to 70 cm. high; leaflets obovate-oblong, coarsely serrate.13. *P. ambigens.*
Stems 10 to 40 cm. high, slender; leaflets cuneate or linear-oblong, toothed only at the apex.
Stems ascending or decumbent; leaflets conduplicate; pistils numerous ....14. *P. crinita.*
Stems erect; leaves flat; pistils 5 to 12 ....15. *P. lemmoni.*
Leaves tomentose beneath.
Leaves almost equally white-pubescent on both surfaces ....16. *P. hippiana.*

Leaves green and glabrous or merely silky above.
Leaflets usually 9, ascending, the upper pair decurrent, not closely approximate......................17. *P. propinqua.*
Leaflets 5 or 7, the lower spreading or reflexed, the upper pair not decurrent, all closely approximate.18. *P. pulcherrima.*

Potentilla effusa Dougl. and *P. oblanceolata* Rydb. have been reported as occurring in New Mexico by Dr. P. A. Rydberg, but we have seen no specimens of either of these from the State.

1. **Potentilla paradoxa** Nutt.; Torr. & Gray, Fl. N. Amer. **1**: 437. 1840.
Tridophyllum paradoxum Greene, Leaflets **1**: 189. 1905.
TYPE LOCALITY: "Banks of the great western rivers, the Ohio! Mississippi! Missouri! &c. to Oregon."
RANGE: Washington and Ontario to New York and New Mexico, south into Mexico; also in Asia.
NEW MEXICO: Santa Fe; Mesilla Valley. Wet ground.
A common species in the sandy bed of the Rio Grande, often carried into cultivated fields by irrigating water.

2. **Potentilla rivalis** Nutt.; Torr. & Gray, Fl. N. Amer. **1**: 437. 1840.
Tridophyllum rivale Greene, Leaflets **1**: 189. 1905.
TYPE LOCALITY: "In alluvial soil along the Lewis River."
RANGE: British Columbia and Saskatchewan to Mexico.
NEW MEXICO: Santa Fe Creek; above Mimbres. Wet ground.

3. **Potentilla millegrana** Engelm.; Lehm. Delect. Sem. Hort. Hamb. **1849**: 11. 1849.
Potentilla rivalis millegrana S. Wats. Proc. Amer. Acad. **8**: 553. 1873.
TYPE LOCALITY: St. Louis, Missouri.
RANGE: Washington and Manitoba to California, New Mexico, and Illinois.
NEW MEXICO: Gila; Sapello Creek; Grant County; Sandia Mountains; north of Ramah.

4. **Potentilla monspeliensis** L. Sp. Pl. 499. 1753.
Potentilla norvegica L. Sp. Pl. 499. 1755.
Tridophyllum monspeliense Greene, Leaflets **1**: 189. 1905.
Tridophyllum norvegicum Greene, loc. cit.
TYPE LOCALITY: "Habitat Monspelii."
RANGE: British America to Maryland, Kansas, Mexico, and California; also in Europe and Asia.
NEW MEXICO: Tunitcha Mountains; Farmington; Chama; Pajarito Park; mountains near Grants; Santa Fe and Las Vegas mountains; Mogollon Mountains. Wet ground.

5. **Potentilla atrorubens** Rydb. Bull. Torrey Club **24**: 11. 1897.
TYPE LOCALITY: Arizona.
RANGE: Mountains of Arizona, New Mexico, and northern Mexico.
NEW MEXICO: Las Vegas Mountains; Magdalena Mountains; Mogollon Mountains; Santa Rita. Transition Zone.

6. **Potentilla thurberi** A. Gray, Mem. Amer. Acad. n. ser. **5**: 318. 1854.
TYPE LOCALITY: Near Santa Rita del Cobre, New Mexico. Type collected by Thurber in August, 1851.

RANGE: New Mexico to southern California and northern Mexico.

NEW MEXICO: Mogollon Mountains; Black Range; Santa Rita; White Mountains. Damp meadows and along streams in the mountains, Transition Zone.

7. **Potentilla sierrae-blancae** Woot. & Rydb. Mem. Bot. Columb. Coll. **2**: 57. 1898.

TYPE LOCALITY: White Mountain Peak, New Mexico. Type collected by Wooton (no. 469).

RANGE: Known only from the type locality, in the Arctic-Alpine Zone.

8. **Potentilla diversifolia** Lehm. Nov. Stirp. Pugill. **2**: 9. 1830.

TYPE LOCALITY: Summits of Rocky Mountains, British America.

RANGE: British America to California and New Mexico.

NEW MEXICO: Wheeler Peak; Truchas Peak; near the head of the Nambe. Meadows, in the Arctic-Alpine Zone.

9. **Potentilla subviscosa** Greene, Bull. Torrey Club **8**: 97. 1881.

TYPE LOCALITY: On a dry southward slope of the Mogollon Mountains, New Mexico. Type collected by Greene.

RANGE: New Mexico, Arizona, and northern Mexico.

NEW MEXICO: Tierra Amarilla; Mogollon Mountains. Meadows in the mountains, Transition Zone.

10. **Potentilla bicrenata** Rydb. Bull. Torrey Club **23**: 431. 1896.

TYPE LOCALITY: Agua Fria, New Mexico. Type collected by C. D. Walcott (no. 66).

RANGE: New Mexico to Wyoming.

NEW MEXICO: Agua Fria; Stinking Lake; Sandia Mountains; Tierra Amarilla. Meadows, in the Transition Zone.

11. **Potentilla filipes** Rydb. Bull. Torrey Club **28**: 174. 1901.

TYPE LOCALITY: Wahatoya Canyon, Spanish Peaks, Colorado.

RANGE: Manitoba and Alberta to Utah and New Mexico.

NEW MEXICO: Chusca Mountains; Chama; Rio Pueblo; Santa Fe and Las Vegas mountains; Sandia Mountains; White and Sacramento mountains. Meadows, Transition to Arctic-Alpine Zone.

12. **Potentilla strigosa** (Pursh) Pall.; Tratt. Rosac. Monogr. **4**: 31. 1824.

Potentilla pennsylvanica strigosa Pursh, Fl. Amer. Sept. 356. 1814.

Potentilla arachnoidea Dougl.; Rydb. N. Amer. Fl. **22** : 350. 1908.

TYPE LOCALITY: "On the Missouri."

RANGE: Hudson Bay to British Columbia, Kansas, and New Mexico; also in Asia.

NEW MEXICO: North of Ramah; Rio Pueblo; mountains near Grants; Santa Fe and Las Vegas mountains; Sierra Grande; Sandia Mountains; White Mountains. Meadows, in the Transition Zone.

13. **Potentilla ambigens** Greene, Erythea **1**: 5. 1893.

TYPE LOCALITY: Moist meadows along Bear Creek, above Morrison, Colorado.

RANGE: Wyoming to New Mexico.

NEW MEXICO: Santa Fe and Las Vegas mountains; Sierra Grande; White and Sacramento mountains. Meadows, in the Transition Zone.

14. **Potentilla crinita** A. Gray, Mem. Amer. Acad. n. ser. **4**: 41. 1849.

TYPE LOCALITY: Along Santa Fe Creek, New Mexico. Type collected by Fendler (no. 199).

RANGE: Colorado and Utah to New Mexico and Arizona.

NEW MEXICO: Chama; Dulce; Santa Fe. Meadows, in the Transition Zone.

15. **Potentilla lemmoni** (S. Wats.) Greene, Pittonia **1**: 104. 1885.

Ivesia lemmoni S. Wats. Proc. Amer. Acad. **20**: 365. 1885.

TYPE LOCALITY: On vertical rocks bordering Oak Creek, near Flagstaff, Arizona.

RANGE: New Mexico and Arizona.

We have seen no specimens of this from New Mexico, but Doctor Rydberg reports a specimen collected in the State by Lemmon, no locality given.

16. Potentilla hippiana Lehm. Nov. Stirp. Pugill. **2:** 7. 1830.

Potentilla leucophylla Torr. Ann. Lyc. N. Y. **2:** 197. 1827, not Pall. 1773.

Potentilla pennsylvanica hippiana Torr. & Gray, Fl. N. Amer. **1:** 438. 1840.

TYPE LOCALITY: Sources of the Platte, Colorado.

RANGE: British America to Arizona and New Mexico.

NEW MEXICO: Chama; Pajarito Park; Santa Fe and Las Vegas mountains; Magdalena Mountains; White Mountains. Meadows, in the Transition Zone.

17. Potentilla propinqua Rydb. Bull. Torrey Club **28:** 176. 1901.

Potentilla diffusa A. Gray, Mem. Amer. Acad. n. ser. **4:** 41. 1849, not *P. diffusa* Willd. 1809.

Potentilla hippiana propinqua Rydb. Bull. Torrey Club **24:** 3. 1897.

TYPE LOCALITY: Along Santa Fe Creek, New Mexico. Type collected by Fendler (no. 198).

RANGE: Alberta and South Dakota to Arizona and New Mexico.

NEW MEXICO: Mountains near Grants Station; Tunitcha Mountains; Chama; Santa Fe and Las Vegas mountains; Mogollon Mountains; Black Range; Sandia Mountains; White Mountains. Meadows, in the Transition Zone.

18. Potentilla pulcherrima Lehm. Nov. Stirp. Pugill. **2:** 10. 1830.

TYPE LOCALITY: Not stated.

RANGE: British America to Utah and New Mexico.

NEW MEXICO: Gilmores Ranch; Chama; Sierra Grande. Meadows, in the Transition Zone.

19. Potentilla plattensis Nutt.; Torr. & Gray, Fl. N. Amer. **1:** 439. 1840.

TYPE LOCALITY: Plains of the Platte.

RANGE: Saskatchewan to Utah and northern New Mexico.

NEW MEXICO: Costilla Valley (*Wooton*).

7. ARGENTINA Lam.

Prostrate herbaceous perennials with a rosette of interruptedly pinnate leaves, numerous slender runners bearing reduced leaves, and solitary axillary flowers; basal leaves 8 to 30 cm. long, oblong in outline; leaflets 1 to 2 cm. long, elliptic, incised-serrate; flowers 1 to 2 cm. in diameter; hypanthium almost flat; bractlets, sepals, and petals normally 5, often more; stamens 20 to 25; achenes corky, grooved.

KEY TO THE SPECIES.

Leaves silvery on both sides .... 1. *A. argentea.*
Leaves green and glabrate above .... 2. *A. anserina.*

1. Argentina argentea Rydb. Bull. Torrey Club **33:** 143. 1906.

Argentina anserina concolor Rydb. Mem. Bot. Columb. Coll. **2:** 160. 1898.

TYPE LOCALITY: Not stated.

RANGE: British America to Oregon, South Dakota, and New Mexico.

NEW MEXICO: Gallo Spring; Tunitcha Mountains. Wet ground, in the Upper Sonoran Zone.

2. Argentina anserina (L.) Rydb. Mem. Bot. Columb. Coll. **2:** 159. 1898.

Potentilla anserina L. Sp. Pl. 495. 1753.

TYPE LOCALITY: "Habitat in Europae pascuis; in argillosis argentea."

RANGE: Through most of temperate North America, also in Europe and Asia.

New Mexico: Tierra Amarilla; Taos; Santa Fe and Las Vegas mountains; San Domingo; Los Lunas; Socorro; Mogollon Mountains; Mesilla Valley; Sacramento Mountains. Chiefly in the Transition Zone, in wet ground, especially in sandy soil along streams.

8. FRAGARIA L. Strawberry.

Small herbaceous perennials with short scaly rootstocks bearing rosettes of leaves and slender runners rooting to form new plants; leaves trifoliolate; hypanthium almost flat; bractlets, sepals, and petals normally 5; petals white; stamens about 20, in 3 series; receptacle conic or rounded, bearing numerous pistils, becoming enlarged and juicy in fruit.

KEY TO THE SPECIES.

Leaves densely silky beneath, even at maturity .... 1. *F. mexicana.*
Leaves glabrate beneath at maturity.
  Pubescence of the scapes and petioles appressed; leaves thick, cuneate-oblong .... 2. *F. ovalis.*
  Pubescence of scapes and petioles spreading or reflexed; leaves thin, mostly obovate .... 3. *F. bracteata.*

In the North American Flora several other species are reported from New Mexico: *F. californica* Schlecht. & Cham., *F. americana* (Porter) Britton, and *F. glauca* (S. Wats.) Rydb. The reports of *F. americana* and *F. californica* are based, partly at least, upon specimens of *F. bracteata*. We have seen no New Mexican specimens that could be referred to either of these species or to *F. glauca*.

1. **Fragaria mexicana** Schlecht. Linnaea **13**: 265. 1839.

Type locality: Near Jalapa, Mexico.

Range: Central Mexico, Lower California, and southwestern New Mexico.

New Mexico: Mogollon Mountains.

It is doubtful whether our specimens really belong to this species, but they certainly are more closely related to it than to any other. Our material is not in the best of condition, hence it is impossible to make a thorough comparison.

2. **Fragaria ovalis** (Lehm.) Rydb. Bull. Torrey Club **33**: 143. 1906.

Potentilla ovalis Lehm. Delect. Sem. Hort. Hamb. **1849**: 9. 1849.

Fragaria firma Rydb. Mem. Bot. Columb. Coll. **2**: 184. 1898.

Type locality: New Mexico, probably in the mountains east of Santa Fe. Type collected by Fendler (no. 206).

Range: Wyoming to Arizona and New Mexico.

New Mexico: Manzanares Valley; Winsors Ranch; Santa Fe Canyon; near the head of the Nambe. Shaded hillsides, in the Transition Zone.

3. **Fragaria bracteata** Heller, Bull. Torrey Club **25**: 194. 1896.

Type locality: "In a meadow along Santa Fe Creek, 9 miles east of Santa Fe," New Mexico. The labels of the type collection say "four miles" instead of "nine." Type collected by Heller (no. 3615).

Range: British Columbia and Montana to California and New Mexico.

New Mexico: Tunitcha Mountains; Chama; Pajarito Park; Rio Pueblo; Santa Fe and Las Vegas mountains; Sierra Grande; Sandia Mountains; Magdalena Mountains; White and Sacramento mountains. Meadows, in the Transition Zone.

A very common strawberry in the mountains. The fruit, although small, is of good flavor.

9. SIBBALDIA L.

Dwarf tufted alpine perennial with thick trifoliolate stipulate leaves and small yellow flowers on scapelike leafless peduncles; calyx persistent; sepals longer than the petals; achenes 5 to 10, on very short hairy stipes.

1. **Sibbaldia procumbens** L. Sp. Pl. 284. 1753.

TYPE LOCALITY: "Habitat in alpibus Lapponiae, Helvetiae, Scothiae."

RANGE: Arctic and alpine regions of North America, Asia, and Europe.

NEW MEXICO: Santa Fe and Las Vegas mountains. Meadows, in the Arctic-Alpine Zone..

10. DASIPHORA Raf. SHRUBBY CINQUEFOIL.

Low branching shrub with pinnately 3 to 7-foliolate silky leaves and scarious stipules; young branches silky-villous, the older stems brown, with shredded bark; flowers bright yellow, axillary and solitary or in small cymes; hypanthium saucer-shaped; bractlets, sepals, and petals 5; petals nearly orbicular; stamens about 25; achenes densely villous.

1. **Dasiphora fruticosa** (L.) Rydb. Mem. Bot. Columb. Coll. **2**: 188. 1898.

Potentilla fruticosa L. Sp. Pl. 495. 1753.

TYPE LOCALITY: "Habitat in Eboraco, Anglia, Oelandia australi, Sibiria."

RANGE: British America to California, New Mexico, and New Jersey; also in Europe and Asia.

NEW MEXICO: Chama; Catskill; Santa Fe and Las Vegas mountains; Mogollon Mountains. Damp meadows and along streams, Transition to Arctic-Alpine Zone.

A densely branched shrub, often becoming a meter high or more. Above timber line it is very much stunted and gnarled. When covered with its numerous golden yellow flowers it is a handsome plant.

11. DRYMOCALLIS Fourn.

Rather coarse herbaceous perennials with scaly rootstocks, pinnate leaves, and cymose inflorescence; hypanthium saucer-shaped; bractlets, sepals, and petals 5; petals but little exceeding the sepals, whitish or yellow; stamens 20 to 30; pistils very numerous.

KEY TO THE SPECIES.

Petals white or cream-colored; plants slender, low, 30 cm. high or less.. 1. *D. convallaria.*
Petals bright yellow; plants stout, tall, frequently 60 cm. high..... 2. *D. glandulosa.*

1. **Drymocallis convallaria** Rydb. Mem. Bot. Columb. Coll. **2**: 193. 1898.

Potentilla convallaria Rydb. Bull. Torrey Club **24**: 249. 1897.

TYPE LOCALITY: Near Bozeman, Montana.

RANGE: Montana to New Mexico.

NEW MEXICO: Santa Fe and Las Vegas mountains; Sierra Grande. Damp woods, in the Transition and Canadian zones.

2. **Drymocallis glandulosa** (Lindl.) Rydb. Mem. Bot. Columb. Coll. **2**: 198. 1898.

Potentilla glandulosa Lindl. in Edwards's Bot. Reg. **19**: *pl. 1583*. 1833.

TYPE LOCALITY: "California."

RANGE: British Columbia to South Dakota, California, and New Mexico.

NEW MEXICO: Ensenada; Johnsons Mesa; Gallinas Canyon; Chama. Wet ground, in the Transition Zone.

12. SIEVERSIA R. Br.

Low, more or less cespitose, herbaceous perennials with a cluster of twice pinnate leaves and a few scapelike stems 50 cm. high or less; flowers in cymose clusters; hypanthium turbinate, 5-bracted; sepals 5, ovate-lanceolate, acute; petals 5, yellow or purplish; styles not jointed, plumose or appressed-pubescent, sometimes elongating in fruit; achenes pubescent.

KEY TO THE SPECIES.

Petals purplish; styles in fruit much elongated, plumose........... 1. *S. grisea.*
Petals yellow; styles slightly if at all elongated in fruit, appressed-pubescent... 2. *S. turbinata.*

1. Sieversia grisea (Greene) Rydb. N. Amer. Fl. **22**: 409. 1913.
Erythrocoma grisea Greene, Leaflets **1**: 178. 1906.
TYPE LOCALITY: San Francisco Mountains, Arizona.
RANGE: Washington and Montana to Chihuahua.
NEW MEXICO: Tunitcha Mountains; Chama; Tierra Amarilla. Transition Zone.

2. Sieversia turbinata (Rydb.) Greene, Pittonia **4**: 50. 1899.
Geum turbinatum Rydb. Bull. Torrey Club **24**: 91. 1897.
Acomastylis turbinata Greene, Leaflets **1**: 174. 1906.
TYPE LOCALITY: Not stated.
RANGE: Wyoming to New Mexico and Arizona.
NEW MEXICO: Santa Fe and Las Vegas mountains; White Mountain Peak. Meadows, in the Arctic-Alpine Zone.

13. GEUM L. AVENS.

Perennial herbs with large, mostly radical, lyrate or pinnate leaves; stipules adnate to the sheathing petioles; flowers 10 to 25 mm. in diameter, cymose, yellow or purplish tinged; calyx persistent, 5-bracted; petals 5; stamens numerous like the pistils; achenes small, compressed, hispidulous, tipped by the slender, jointed, hooked, more or less plumose, elongated styles.

KEY TO THE SPECIES.

Petals pink or purplish, clawed .... 1. *G. rivale.*
Petals yellow, clawless.
  Upper internode of the style long-hairy, the lower not glandular; petals 5 to 7 mm. long .... 2. *G. strictum.*
  Upper internode of the style sparingly short-hairy, the lower glandular-puberulent; petals 4 to 5 mm. long .... 3. *G. oregonense.*

1. Geum rivale L. Sp. Pl. 501. 1753.
TYPE LOCALITY: "Habitat in Europae pratis subhumidis."
RANGE: British America to New Jersey and New Mexico; also in Europe.
NEW MEXICO: Santa Fe and Las Vegas mountains. Damp woods, in the Transition Zone.

2. Geum strictum Ait. Hort. Kew. **2**: 217. 1789.
TYPE LOCALITY: "North America."
RANGE: British America to Pennsylvania, Missouri, and Mexico.
NEW MEXICO: Chama; Santa Fe and Las Vegas mountains; Rio Pueblo; Mogollon Mountains; Black Range; White and Sacramento mountains. Damp thickets, in the Transition Zone.

3. Geum oregonense Scheutz, Nov. Act. Soc. Sci. Upsal. **7**: 26. 1869.
TYPE LOCALITY: "Habit. in regioni Oregonensi."
RANGE: British America to California and New Mexico.
NEW MEXICO: Tunitcha Mountains; Santa Fe and Las Vegas mountains. Damp ground, in the Transition Zone.

14. FALLUGIA Endl. APACHE PLUME

A much branched evergreen shrub 1 to 2 meters high, with slender white branches and small, fascicled, cuneate-obovate, pinnately divided, hispidulous leaves; divisions of the leaves narrowly oblong, obtuse, revolute; flowers numerous, the inflorescence somewhat corymbosely branched and the floral leaves reduced to bracts; hypanthium hemispheric, with several linear-lanceolate bracts alternating with the ovate, abruptly long-acuminate sepals; petals 5, broadly obovate to rotund, white; stamens numerous; achenes numerous, obovoid-fusiform, long-tailed.

1. **Fallugia paradoxa** (Don) Endl. Gen. Pl. 1246. 1840.
Sieversia paradoxa Don, Trans. Linn. Soc. Bot. **14**: 576. 1825.
Fallugia paradoxa acuminata Wooton, Bull. Torrey Club **25**: 306. 1898.
Fallugia micrantha Cockerell, Ent. News **1901**: 41. 1901.
Fallugia acuminata Cockerell, Proc. Acad. Phila. **1903**: 590. 1903.
TYPE LOCALITY: Mexico.
RANGE: Colorado and Utah to Arizona, Texas, and Mexico.
NEW MEXICO: Espanola; Las Vegas; Albuquerque; Santa Fe; Pajarito Park; Magdalena Mountains; Mogollon Mountains; Black Range; mountains west of San Antonio; Big Hatchet Mountains; Animas and San Luis mountains; Las Cruces; Organ Mountains; White Mountains; Guadalupe Mountains. Chiefly in the Upper Sonoran Zone.

A common shrub in the drier mountains and arroyos, especially in the southern part of the State, where it is a valuable forage plant much browsed by cattle, sheep, and goats. It is well worth cultivation for decorative purposes since it grows rapidly and produces an abundance of white flowers as large as apple blossoms, while the clusters of plumose fruits, first greenish and later reddish tinged, which remain on the plant for some time, are almost as beautiful as the flowers.

The type of *F. paradoxa acuminata* was collected on the mesa near Las Cruces and that of *F. micrantha* is from the same locality.

The native name is "poñil."

15. COWANIA Don.

Spreading shrub, 1 to 2 meters high, with small pinnate cuneate-obovate glandular crowded leaves and solitary flowers; leaves 10 to 15 mm. long, glabrate above, tomentose beneath, the oblong segments with revolute margins; hypanthium turbinate, tomentose and glandular-pubescent; sepals 5, broadly ovate, densely tomentose, glandular on the back; petals broadly obovate, pale yellow; achenes about 5, densely villous, plumose-tailed, the tail sometimes 3 to 5 cm. long.

1. **Cowania stansburiana** Torr. in Stansb. Expl. Great Salt Lake 386. *pl. 3.* 1853.
TYPE LOCALITY: Stansburys Island, Salt Lake, Utah.
RANGE: Utah and Colorado to New Mexico.
NEW MEXICO: Carrizo Mountains; Coolidge; Thoreau; Bear Mountain; Animas Creek; Aragon. Dry hills, in the Upper Sonoran Zone.

16. PURSHIA DC.

A low, intricately branched shrub with small fascicled tomentose cuneate crenate leaves and solitary flowers terminating short branches; hypanthium turbinate; sepals ovate, obtuse; petals small, obovate, yellow; fruit fusiform, pubescent, long-tailed.

1. **Purshia tridentata** (Pursh) DC. Trans. Linn. Soc. Bot. **12**: 158. 1817.
Tigarea tridentata Pursh, Fl. Amer. Sept. 333. 1814.
Kunzia tridentata Spreng. Syst. Veg. **2**: 475. 1825.
TYPE LOCALITY: "In the prairies of the Rocky Mountains, and on the Columbia River."
RANGE: Washington and Montana to California and New Mexico.
NEW MEXICO: From Dulce westward to the Tunitcha and Carrizo mountains. Dry hills, in the Upper Sonoran Zone.

17. CERCOCARPUS H. B. K. MOUNTAIN MAHOGANY.

Shrubs 1 to 4 meters high with stout, rather widely branched stems and hard brittle wood; leaves simple, fascicled, small; flowers solitary or fascicled with the leaves, inconspicuous; hypanthium tubular, 1 cm. long or less, persistent; sepals dull whitish, small; corolla wanting; stamens numerous, in 2 or 3 rows, deciduous with the calyx;

fruit a terete or fusiform, densely villous achene, terminating in a long, slender, variously bent and curved, plumose tail sometimes 5 cm. long.

The species of this genus furnish not a little forage for cattle, sheep, and goats which browse upon them at all seasons of the year. On some of the rockier, drier mountains of the southern part of the State, they are important parts of the scrubby underbrush and furnish much of the firewood. The seasoned wood is very hard and brittle and has a specific gravity near 1. It is so hard that it is difficult to chop with an ax, but so brittle it may be broken easily.

KEY TO THE SPECIES.

Leaves large, 3 or 4 cm. long, coarsely toothed.
- Pubescence of the petioles and flowers appressed, silky..... 1. *C. argenteus.*
- Pubescence of the petioles and flowers spreading, not silky.. 2. *C. montanus.*

Leaves small, 2 cm. long or less, entire, or with a few inconspicuous and very small teeth near the apex.
- Pubescence of the petioles and flowers spreading, loose; upper surface of the leaves mostly soft-pubescent....... 3. *C. paucidentatus.*
- Pubescence appressed, silky; upper surface of the leaves glabrous, or with a few silky, appressed hairs......... 4. *C. breviflorus.*

1. **Cercocarpus argenteus** Rydb. N. Amer. Fl. **22:** 422. 1913.

TYPE LOCALITY: Rocky bluffs on Red River, Randall County, Texas.

RANGE: Colorado to Texas and New Mexico.

NEW MEXICO: Cedar Hill; Santa Fe and Las Vegas mountains; Raton Mountains; Chama; Folsom; Sandia Mountains; White and Sacramento mountains; Capitan Mountains.

A decoction of the root of this shrub, along with juniper ashes and the powdered bark of the alder, was formerly used by the Navahos in dying wool red.

2. **Cercocarpus montanus** Raf. Atl. Journ. 146. 1832.

Cercocarpus fothergilloides H. B. K. err. det. Torr. Ann. Lyc. N. Y. **2:** 198. 1828.

Cercocarpus parvifolius Nutt.; Hook. & Arn. Bot. Beechey Voy. 337. 1840.

TYPE LOCALITY: "On the Rocky Mountains."

RANGE: Montana and South Dakota to Utah and New Mexico.

NEW MEXICO: Tunitcha Mountains; Dulce; Coolidge; Thoreau; Sandia Mountains; Cross L Ranch; Glorieta. Hillsides, in the Upper Sonoran and Transition zones.

3. **Cercocarpus paucidentatus** (S. Wats.) Britton, Trans. N. Y. Acad. **14:** 31. 1894.

Cercocarpus parvifolius paucidentatus S. Wats. Proc. Amer. Acad. **17:** 353. 1882.

TYPE LOCALITY: San Miguelito, San Luis Potosi, Mexico.

RANGE: Western Texas to Arizona, south into Mexico.

NEW MEXICO: Common from the Mogollon Mountains to the Capitan Mountains and south to the Mexican border. Dry hillsides, in the Upper Sonoran Zone.

4. **Cercocarpus breviflorus** A. Gray, Pl. Wright. **2:** 54. 1853.

TYPE LOCALITY: "Sides of mountains near Frontera," Texas. Type collected by Wright (no. 1057).

RANGE: Western Texas to southern Arizona, south into Mexico.

NEW MEXICO: Lake Valley; Kingston; San Luis Mountains; San Andreas Mountains; Queen; Organ Mountains. Dry hills, in the Upper Sonoran Zone.

18. RUBUS L. RASPBERRY.

Prickly shrubs, 1 meter high or less, with 5 to 7-foliolate leaves and white flowers; stems of the first season erect, armed with straight prickles; leaves of the flowering branches with fewer leaflets; leaflets ovate to rhombic-lanceolate, serrate or crenate, the terminal sometimes lobed, green above, white-tomentose beneath; fruits red or black, juicy, with a pleasant taste and odor.

KEY TO THE SPECIES.

Fruit black; leaflets crenate; achenes keeled on the back; stems glaucous .... 1. *R. bernardinus.*
Fruit red; leaflets incised-serrate; stems not glaucous .... 2. *R. arizonicus.*

1. **Rubus bernardinus** (Greene) Rydb. N. Amer. Fl. **22:** 444. 1913.

BLACK RASPBERRY.

Melanobatus bernardinus Greene, Leaflets **1:** 244. 1906.

TYPE LOCALITY: Mill Creek Falls, San Bernardino Mountains, California.

RANGE: Southwestern New Mexico to southern California.

NEW MEXICO: Mogollon Mountains (*Rusby* 123).

This specimen is referred here with some doubt by Doctor Rydberg.

2. **Rubus arizonicus** (Greene) Rydb. N. Amer. Fl. **22:** 446. 1913.

RED RASPBERRY.

Batidaea arizonica Greene, Leaflets **1:** 243. 1906.

TYPE LOCALITY: San Francisco Mountains, Arizona.

RANGE: Mountains of Arizona, New Mexico, and southern Colorado.

NEW MEXICO: Common in all the higher mountains from the Black Range and White Mountains northward. Transition Zone.

The common wild raspberry of middle elevations in the mountains, growing in large patches on the hillsides among the pines. The fruit is abundantly produced and much appreciated by the people of the region, who gather the berries in quantity for table use. It is also one of the favorite foods of the bears.

19. RUBACER Rydb. THIMBLE-BERRY.

A low unarmed perennial, 30 to 60 cm. high, with mostly herbaceous stems arising from a woody base, bearing few large 3 to 5-lobed leaves; flowers white, 3 to 5 cm. broad; calyx densely tomentose; sepals long-acuminate; fruit large, red, pleasantly flavored.

1. **Rubacer parviflorus** (Nutt.) Rydb. Bull. Torrey Club **30:** 275. 1903.

Rubus parviflorus Nutt. Gen. Pl. **1:** 308. 1818.

Rubus nutkanus Moc.; DC. Prodr. **2:** 566. 1825.

Bossekia parviflora Greene, Leaflets **1:** 211. 1906.

TYPE LOCALITY: "Island of the Michilimackinack, Lake Huron."

RANGE: Alaska to California, New Mexico, and Lake Superior.

NEW MEXICO: Common in the higher mountains from the Mogollon and Sacramento mountains northward. Woods, in the Transition and Canadian zones.

A common and conspicuous plant in the higher mountains, rather handsome with its large white flowers. It often completely covers the ground in the deep woods. The fruits are of good quality, but so few are borne on a single plant that picking them is a tedious task.

20. OREOBATUS Rydb.

Unarmed branching shrubs, 1 meter high or less, with 3 to 5-lobed stipulate leaves and brownish shredded bark; hypanthium flat, not bracteolate; sepals broadly ovate, with elongated tips, accrescent, loosely inclosing the fruit; flowers white, conspicuous; fruit fleshy or soon dry.

KEY TO THE SPECIES.

Petals 20 to 35 mm. long; leaves not lobed or with mostly 5 shallow lobes, the teeth very acute.............................. 1. *O. deliciosus.*
Petals less than 20 mm. long; leaves conspicuously 3-lobed, the teeth mostly obtuse.
  Leaves glabrous on the upper surface, small, usually 30 to 40 mm. long, not conspicuously reticulate-veined; pubescence of the petioles short and close........... 2. *O. rubicundus.*
  Leaves soft-pubescent on both surfaces, large, 40 to 65 mm. long, conspicuously reticulate-veined; pubescence of the petioles loose and spreading.................... 3. *O. neomexicanus.*

1. **Oreobatus deliciosus** (Torr.) Rydb. Bull. Torrey Club **30:** 275. 1903.
Rubus deliciosus Torr. Ann. Lyc. N. Y. **2:** 196. 1828.
Bossekia deliciosa A. Nels. in Coulter, New Man. Rocky Mount. 250. 1909.
Type locality: "On the Rocky Mountains," Colorado.
Range: Mountains of Colorado and New Mexico.
New Mexico: Sierra Grande (*Howell* 208, *Standley* 6078). Upper Sonoran and Transition zones.

2. **Oreobatus rubicundus** Woot. & Standl. Contr. U. S. Nat. Herb. **16:** 130. 1913.
Type locality: Van Pattens Camp in the Organ Mountains, New Mexico. Type collected by Standley, June 16, 1906.
Range: Canyons in the Organ Mountains of southern New Mexico, in the Upper Sonoran Zone.

3. **Oreobatus neomexicanus** (A. Gray) Rydb. Bull. Torrey Club **30:** 275. 1903.
Rubus neomexicanus A. Gray, Pl. Wright. **2:** 55. 1853.
Type locality: Mountain sides at the Copper Mines, New Mexico. Type collected by Wright (no. 1061).
Range: Southwestern New Mexico and southern Arizona.
New Mexico: Mogollon Mountains; Black Range; Animas Mountains; San Luis Mountains. Canyons in the mountains, Upper Sonoran Zone.

21. VAUQUELINIA Correa.

Large shrub or small tree with coriaceous persistent serrate leaves and corymbose flowers; stipules small, deciduous; hypanthium short-turbinate; sepals 5, persistent; petals 5; stamens 15 to 25; capsule woody, of 5 follicles coherent at the base; seeds 2 in each cell, winged.

1. **Vauquelinia californica** (Torr.) Sarg. Gard. & For. **2:** 400. 1889.
Spiraea californica Torr. in Emory, Mil. Reconn. 140. 1847.
Vauquelinia corymbosa Torr. U. S. & Mex. Bound. Bot. 64. 1859.
Type locality: High mountains near the Gila, Arizona.
Range: Southwestern New Mexico to southern California, south into Mexico.
We have seen no specimens of this from New Mexico, but Dr. E. A. Mearns states that it is found rather sparsely in Guadalupe Canyon at the southwest corner of the State.

65. MALACEAE. Apple Family.

Trees or shrubs with alternate simple or pinnately compound leaves having fugacious stipules; flowers regular, in racemes or cymes; hypanthium mostly spheroidal, adnate to the 1 to 5-celled ovaries; petals and sepals 5; stamens usually many, distinct; fruit a pome with papery, bony, or leathery carpels.

52576°—15——21

KEY TO THE GENERA.

Cavities of the ovary becoming twice as many as the styles by a partial or complete false partition; flowers racemose or corymbose.
- Styles 5; flowers racemose.......................... 1. AMELANCHIER (p. 322).
- Styles 2; flowers solitary, or sessile in 2 or 3-flowered corymbs.................................. 4. PERAPHYLLUM (p. 324).

Cavities of the ovary not divided, as many as the styles; flowers in corymbiform cymes.
- Leaves simple, lobed; ovules 1 in each carpel........ 2. CRATAEGUS (p. 323).
- Leaves pinnate; ovules 2 in each carpel............. 3. SORBUS (p. 324).

1. AMELANCHIER L. SERVICE BERRY.

Shrubs or small trees, 1 to 2 meters high, with alternate, simple, mostly rather coarsely serrate, small leaves and white flowers in racemes terminating short branches of the year; stamens numerous, all borne on the hypanthium, the latter adnate to the inferior ovary; fruit berry-like.

The fruits of the native service berries were a favorite food among the Indians in earlier days. They were eaten fresh or were dried and preserved for winter use. They are insipid in all the species. Those of the species which grow at lower levels are nearly dry and consequently useless for food.

KEY TO THE SPECIES.

Leaves acutish, oblong-ovate.. 1. *A. rubescens.*
Leaves obtuse to truncate, broader.
- Mature leaves finely pubescent, at least beneath.
  - Leaves crenate, pubescent on both surfaces............... 2. *A. crenata.*
  - Leaves sharply serrate, pubescent beneath............... 3. *A. bakeri.*
- Mature leaves glabrous or loosely villous, never finely pubescent.
  - Whole plant perfectly glabrous.............................. 4. *A. polycarpa.*
  - Bud scales, and usually the young leaves, villous.
    - Mature leaves glabrous, conspicuously cordate, crenate to the base.............................. 5. *A. goldmanii.*
    - Mature leaves with loose pubescence beneath and often above, not cordate or but slightly so, often cuneate, usually entire below the middle.
      - Petals 10 to 15 mm. long.......................... 7. *A. mormonica.*
      - Petals 8 mm. long or less.
        - Leaves thin, bright green; calyx lobes shorter than the fruit, not foliaceous.......... 6. *A. oreophila.*
        - Leaves thick, coriaceous, pale green or glaucescent; calyx lobes longer than the fruit, foliaceous...................... 8. *A. australis.*

1. Amelanchier rubescens Greene, Pittonia 4: 128. 1900.

TYPE LOCALITY: In arroyos and among the hills about Aztec, New Mexico. Type collected by Baker (nos. 380, 381).

RANGE: Southwestern Colorado and northwestern New Mexico.

NEW MEXICO: Western San Juan County; Kingston. Dry hills, in the Upper Sonoran Zone.

2. Amelanchier crenata Greene, Pittonia 4: 127. 1900.

Type locality: On rock declivities near Aztec, New Mexico. Type collected by Baker (no. 377).

Range: Northwestern New Mexico.

New Mexico: Northern San Juan County. Dry hills, in the Upper Sonoran Zone.

3. Amelanchier bakeri Greene, Pittonia 4: 128. 1900.

Type locality: Los Pinos, southern Colorado.

Range: Southwestern Colorado and western New Mexico, probably in eastern Arizona.

New Mexico: Magdalena Mountains; Silver City; Bear Mountain; Canjilon. Mountains, in the Transition Zone.

4. Amelanchier polycarpa Greene, Pittonia 4: 127. 1900.

Type locality: Piedra, southern Colorado.

Range: Wyoming to Colorado and New Mexico.

New Mexico: Zuni Mountains; Chama; Stinking Lake. Damp woods, in the Transition Zone.

5. Amelanchier goldmanii Woot. & Standl. Contr. U. S. Nat. Herb. **16**: 131. 1913.

Type locality: Copper Canyon, Magdalena Mountains, New Mexico. Type collected by E. A. Goldman in 1909.

Range: Mountains of western New Mexico.

New Mexico: Copper Canyon; Mogollon Mountains.

6. Amelanchier oreophila A. Nels. Bot. Gaz. **40**: 65. 1905.

Type locality: Evanston, Wyoming.

Range: Northern New Mexico to Colorado and Wyoming.

New Mexico: Santa Fe and Las Vegas mountains. Mountains, in the Transition Zone.

7. Amelanchier mormonica C. Schneid. Handb. Laubh. **1**: 740. 1906.

Type locality: Mormon Lake, Arizona.

Range: Arizona and New Mexico.

New Mexico: Dulce; Sandia Mountains; Chama; Tunitcha Mountains. Hillsides, in the Transition Zone.

8. Amelanchier australis Standley, Proc. Biol. Soc. Washington **26**: 116. 1913.

Type locality: Ropes Spring, San Andreas Mountains, New Mexico. Type collected by Wooton, September 23, 1912.

Range: Known only from type locality, in the Upper Sonoran Zone.

2. CRATAEGUS L. Hawthorn.

Shrubs or small trees with stout spiny stems, simple, alternate, toothed or lobed leaves, and white flowers in corymbs; hypanthium urceolate, adnate to the ovary; sepals 5, persistent; petals 5, spreading; stamens 5 to 10; fruit small, drupaceous, containing 2 to 5 bony 1-seeded carpels.

KEY TO THE SPECIES.

Spines short, 2 cm. long or less, not very numerous; leaves mostly elliptic, not noticeably lobed, the blades about twice as long as wide........ 1. *C. rivularis.*

Spines longer, 4 cm. or more, commonly numerous; leaves broader, at least some of them more or less lobed.

Leaves elliptic-ovate, rather coarsely serrate, some of them with a few larger lobelike teeth, the smaller teeth gland-tipped; leaves cuneate at the base........ 2. *C. erythropoda.*

Leaves broadly ovate, with 3 or 4 pairs of broad lobes, finely serrate or doubly serrate with straight teeth, not gland-tipped; base of leaves truncate........ 3. *C. wootoniana.*

1. **Crataegus rivularis** Nutt.; Torr. & Gray, Fl. N. Amer. **1**: 464. 1840.

TYPE LOCALITY: "Oregon, along rivulets in the Rocky Mountains."

RANGE: Western Wyoming to Utah and Idaho, south to New Mexico.

NEW MEXICO: Upper Negrito Creek (*Wooton*). Stream banks, in the Transition Zone.

2. **Crataegus erythropoda** Ashe, N. C. Agr. Exp. Sta. Bull. **175**: 113. 1900.

MANZANA DE PUYA LARGA.

Crataegus cerronis A. Nels. Bot. Gaz. **34**: 370. 1902.

TYPE LOCALITY: Foothills of the Cache le Poudre Mountains, northern Colorado.

RANGE: Wyoming to northern New Mexico.

NEW MEXICO: Sandia Mountains; Ponchuelo Creek; El Rito Creek; Chama. Stream banks and canyons, in the Transition Zone.

3. **Crataegus wootoniana** Eggleston, Torreya **7**: 236. 1907.

TYPE LOCALITY: Head of Little Creek, Mogollon Mountains, New Mexico. Type collected by Metcalfe (no. 584).

RANGE: Mountains of central and southern New Mexico.

NEW MEXICO: Mogollon Mountains; White Mountains.

3. SORBUS L. MOUNTAIN ASH.

Shrub 1 to 3 meters high with pinnate leaves and white flowers in compound cymes; hypanthium urceolate or turbinate; leaflets 11 to 15, 3 to 4 cm. long, oblong-lanceolate, serrate, glabrous; sepals 5; petals 5, spreading, short-clawed; stamens 20; styles 3 to 5, distinct, woolly at the base; fruit a berry-like pome.

1. **Sorbus scopulina** Greene, Pittonia **4**: 130. 1900.

TYPE LOCALITY: Santa Fe Canyon, New Mexico. Type collected by Heller (no. 3711).

RANGE: British America to Washington, Utah, Colorado, and New Mexico.

NEW MEXICO: Zuni Mountains; Manzano Mountains; Tunitcha Mountains; Santa Fe Mountains. Damp woods, in the Transition and Canadian zones.

4. PERAPHYLLUM Nutt.

Shrub 1 to 2 meters high, the small, narrowly oblanceolate, serrulate or entire, short-petiolate leaves fascicled at the ends of the branchlets; flowers solitary or in 2 or 3-flowered umbels, pale rose color; fruit globose, crowned with the persistent calyx lobes.

1. **Peraphyllum ramosissimum** Nutt.; Torr. & Gray, Fl. N. Amer. **1**: 474. 1838.

TYPE LOCALITY: "Dry hillsides near the Blue Mountains of the Oregon."

RANGE: Dry hillsides, Oregon and California to Colorado and northwestern New Mexico, in the Upper Sonoran Zone.

We have seen no specimens of this from New Mexico, but it occurs abundantly along the railroad below Durango, Colorado, just above the New Mexico line. Dr. David Griffiths states that he has collected fruit of this plant near Farmington. The fruit is shaped like a small crab apple, and is yellow tinged with purple. The juice is very bitter.

66. AMYGDALACEAE. Almond Family.

Trees or shrubs with alternate, petiolate, simple, mostly serrate leaves and fugacious stipules; bark, leaves, and seeds bitter with prussic acid; flowers perfect, solitary, fascicled, corymbose, or racemose; hypanthium mostly spheroidal, free from the simple solitary ovary; sepals and petals 5; stamens mostly numerous; fruit a drupe.

KEY TO THE GENERA.

Flowers in long racemes, on short leafy branches of the year.. 1. PADUS (p. 325).
Flowers in corymbs or umbels, on short stems of the previous year, preceding the leaves.
Stone of the fruit flattened, with more or less acute edges.. 2. PRUNUS (p. 327).
Stone of the fruit spheroidal, little or not at all flattened.. 3. CERASUS (p. 327).

1. PADUS Borckh. CHOKECHERRY.

Large shrubs or small trees with smooth dark-colored bark; flowers numerous, in elongated racemes terminating short leafy branches of the year; hypanthium spheroidal, sometimes campanulate; sepals 5, short, persistent or deciduous with a part of the hypanthium; petals white, with the numerous stamens on the throat of the hypanthium; carpels solitary; ovary 1-celled, 2-ovuled; drupe small, usually 1 cm. in diameter or less, astringent, not glaucous.

The fruits of these trees and shrubs were eaten by the Indians. They lack the astringent flavor of the eastern chokecherries.

KEY TO THE SPECIES.

Calyx persistent in fruit.
Young branches densely tawny-pubescent; young fruit pubescent.......... 1. *P. rufula.*
Young branches and fruit glabrous.......... 2. *P. virens.*
Calyx deciduous soon after anthesis.
Plants glabrous throughout.......... 3. *P. melanocarpa.*
Plants pubescent on the peduncles, petioles, and lower surface of the leaves.
Leaves not glaucous beneath at maturity, of about the same color on both surfaces.......... 4. *P. pumicea.*
Leaves pale beneath at maturity.
Pedicels longer than the fruit, slender; seeds 8 to 10 mm. in diameter.......... 5. *P. mescaleria.*
Pedicels shorter than the fruit, stout; seeds 7 mm. in diameter or less.
Pedicels glabrous; racemes slender; leaves elliptic, narrowed at the base; buds narrowly lanceolate in outline.......... 6. *P. calophylla.*
Pedicels pubescent; racemes stout; leaves oblong to ovate or obovate, rounded to subcordate at the base; buds ovoid.......... 7. *P. valida.*

1. Padus rufula Woot. & Standl. Contr. U. S. Nat. Herb. **16:** 132. 1913.

TYPE LOCALITY: On the West Fork of the Rio Gila, New Mexico. Type collected by Wooton, August 6, 1900.

RANGE: Mountains of southwestern New Mexico, southeastern Arizona, and adjacent Mexico.

NEW MEXICO: Mogollon Mountains; Black Range.

2. Padus virens Woot. & Standl. Contr. U. S. Nat. Herb. **16:** 132. 1913.

TYPE LOCALITY: Van Pattens Camp in the Organ Mountains, New Mexico. Type collected by Standley, June 6, 1906.

RANGE: Southern New Mexico and Arizona.

NEW MEXICO: Organ Mountains; Bear Mountains; San Francisco Mountains; Mogollon Mountains; Burro Mountains; Kingston; White and Sacramento mountains. Canyons and hills, in the Upper Sonoran Zone.

In the Organ Mountains this species occurs in abundance in the opening of the cañyon in which Van Pattens Camp is located, growing along with *Quercus grisea* and *Q. arizonica.* Attempts have been made to use the trees as stocks for grafting various fruits, but they have been unsuccessful.

The material from the western part of the State may represent a different species. This plant is usually much smaller, only a tall shrub, and its leaves are narrower, thicker, not so bright a green, and on shorter petioles.

3. **Padus melanocarpa** (A. Nels.) Shafer in Britt. & Shaf. N. Amer. Trees 504. 1908.

Cerasus demissa melanocarpa A. Nels. Bot. Gaz. **34:** 25. 1902.

Prunus melanocarpa Rydb. Bull. Torrey Club **33:** 143. 1906.

TYPE LOCALITY: Rocky Mountains.

RANGE: British Columbia and Alberta to California and New Mexico.

NEW MEXICO: Zuni Mountains; Chama; Santa Fe and Las Vegas mountains; Raton Mountains; Magdalena Mountains; Sierra Grande; Sandia Mountains; White and Sacramento mountains. Damp woods, especially along streams, chiefly in the Transition Zone.

The species, if all our material represents a single one, shows considerable variation, possibly because of altitude. At the lower levels in the northern part of the State it is a shrub 2 or 3 meters high with very large fruit. Higher up, at Winsors Ranch, it is a very low shrub, usually with only a single stem in a place, not more than 50 to 60 cm. high. Mature fruit could not be secured at the higher levels, so that it is impossible to tell whether there is any substantial difference between the two plants in that feature.

4. **Padus pumicea** Woot. & Standl. Contr. U. S. Nat. Herb. **16:** 133. 1913.

TYPE LOCALITY: Craters, Valencia County, New Mexico. Type collected by Wooton, July 28, 1906.

RANGE: Mountains of western New Mexico.

NEW MEXICO: Craters; mountains south of Canjilon.

From all our pubescent species this differs in having the leaves of about the same color on both surfaces. The fruits, too, are very few and the pedicels remarkably short. The branches are densely furnished with leaves, so that in general appearance this is unlike any of our other choke cherries.

5. **Padus mescaleria** Woot. & Standl. Contr. U. S. Nat. Herb. **16:** 134. 1913.

TYPE LOCALITY: On Tularosa Creek near the Mescalero Agency, New Mexico. Type collected by Wooton, August 6, 1901.

RANGE: White Mountains of New Mexico, in the Transition Zone.

The most distinctive features of this are the long pedicels, glabrous racemes, large seeds, and rather narrow deep green leaves, strongly glaucous beneath.

6. **Padus calophylla** Woot. & Standl. Contr. U. S. Nat. Herb. **16:** 134. 1913.

TYPE LOCALITY: Five miles west of Chloride, New Mexico. Type collected by E. A. Goldman (no. 1768).

RANGE: Mountains of northwestern New Mexico.

NEW MEXICO: West of Chloride; Tunitcha Mountains; Santa Fe Canyon. Transition Zone.

7. **Padus valida** Woot. & Standl. Contr. U. S. Nat. Herb. **16:** 134. 1913.

TYPE LOCALITY: Canyons near Kingston, Sierra County, New Mexico. Type collected by Metcalfe (no. 1243).

RANGE: Mountains of western New Mexico.

NEW MEXICO: Kingston; Magdalena Mountains. Transition Zone.

2. PRUNUS L. Plum.

Low, treelike or spreading shrubs, 3 meters high or less, forming thickets; branches stout, rigid, somewhat spiny; bark grayish; leaves sharply serrate; flowers white, produced before the leaves; fruit ellipsoidal, red, the stone flattened, acute on both edges.

KEY TO THE SPECIES.

Leaves pubescent beneath; petals 4 to 6 mm. long.................. 1. *P. watsoni.*
Leaves glabrous; petals 8 to 16 mm. long .... 2. *P. americana.*

1. **Prunus watsoni** Sarg. Gard. & For. **7:** 134. *f. 25.* 1894. SAND PLUM.
Prunus angustifolia watsoni Waugh, Rep. Vt. Agr. Exp. Sta. **12:** 239. 1899.
TYPE LOCALITY: Ellis, Kansas.
RANGE: Nebraska to Texas and eastern New Mexico.
NEW MEXICO: Nara Visa (*Fisher* 205). Plains, in the Upper Sonoran Zone.

2. **Prunus americana** Marsh. Arb. Amer. 111. 1785. WILD PLUM.
TYPE LOCALITY: Not definitely stated.
RANGE: Montana and New York to Florida and New Mexico.
NEW MEXICO: Taos; Pecos; Farmington; White Mountains.
In some parts of the State this plum is almost certainly native; in other places it may have been introduced. At Taos the trees are abundant and the fruit is gathered by the Indians. At Farmington the small trees are very numerous along some of the irrigating ditches.
Some similar species has escaped rather abundantly near Mesilla.

3. CERASUS L. Cherry.

A small slender tree 3 to 4 meters high, with smooth purplish or reddish brown bark, slender virgate branches, and corymbose white flowers; leaves 3 to 5 cm. long, oblong-elliptic, slightly attenuate to the base, acute or abruptly short-acuminate, crenulate, on petioles 1 cm. long or less; corymbs about 4-flowered; hypanthium campanulate, glabrous; petals small, white; fruit ovoid, red; stone ovoid.

1. **Cerasus crenulata** Greene, Proc. Biol. Soc. Washington **18:** 56. 1905.
TYPE LOCALITY: West Fork of the Gila, Mogollon Mountains, New Mexico. Type collected by Metcalfe (no. 587).
RANGE: Mountains of southwestern New Mexico.
NEW MEXICO: Mogollon Mountains; Hillsboro Peak. Transition Zone.
The fruits when ripe are a bright cherry red, ellipsoidal, 1 cm. long or less, and decidedly acid as well as somewhat astringent.

67. MIMOSACEAE. Mimosa Family.

Shrubs or suffrutescent perennials with usually spiny stems and bipinnate leaves with usually numerous small leaflets; flowers regular, small, in axillary pedunculate heads or spikes; calyx 4 or 5-parted (sometimes wanting in Acuan); corolla of 4 or 5 distinct or united petals; stamens 5 to 10 or numerous, distinct or united; fruit a more or less flattened, dehiscent or indehiscent legume.

KEY TO THE GENERA.

Stamens numerous, always more than 10, distinct or monadelphous.
Corolla gamopetalous, tubular; stamens monadelphous; low plants with unarmed stems..... 1. CALLIANDRA (p. 328).
Corolla polypetalous; stamens distinct; plants more or less shrubby and spiny (except in *A. cuspidata*)................................ 2. ACACIA (p. 328).

Stamens 5 or 10, distinct.
  Flowers 5-merous; anthers tipped with a gland; pods indehiscent; large shrubs.
    Pods spirally coiled; flowers yellow........... 3. STROMBOCARPA (p. 329).
    Pods elongated, not coiled; flowers greenish... 4. PROSOPIS (p. 330).
  Flowers 4 or 5-merous; anthers not gland-tipped; pods dehiscent; shrubs or herbs.
    Leaves sensitive; pods more or less 4-sided; plants decumbent...................... 5. MORONGIA (p. 330).
    Leaves not sensitive or at most very tardily so; pods flat; plants erect or spreading, not decumbent.
      Plants without spines; stems mostly herbaceous............................ 6. ACUAN (p. 330).
      Plants armed with numerous short recurved triangular spines; shrubs with woody stems................ 7. MIMOSA (p. 331).

1. CALLIANDRA Benth.

Low herbaceous or woody perennials without spines, 30 cm. high or less; flowers in globose heads; corolla gamopetalous, elongate-tubular; stamens numerous, monadelphous, long-exserted; pods flattened, straight or slightly curved, the valves elastically revolute from apex to base.

KEY TO THE SPECIES.

Stems woody throughout.. 1. *C. eriophylla.*
Stems herbaceous, sometimes woody at the base.
  Leaflets 2 or 3 mm. long, pilose; pinnæ 4 or more pairs........ 2. *C. humilis.*
  Leaflets 5 to 10 mm. long, glabrous or nearly so; pinnæ 1 or 2 pairs.. 3. *C. reticulata.*

1. Calliandra eriophylla Benth. Lond. Journ. Bot. **3**: 105. 1844.

Calliandra chamaedrys Engelm. in A. Gray, Mem. Amer. Acad. n. ser. **4**: 39. 1849.

TYPE LOCALITY: "Mexico; Chila in the district of Pueblo."

RANGE: Western Texas to Arizona, south into Mexico.

NEW MEXICO: Southern Grant County.

2. Calliandra humilis Benth. Lond. Journ. Bot. **5**: 103. 1846.

Calliandra ? herbacea Engelm. in A. Gray, Mem. Amer. Acad. n. ser. **4**: 39. 1849.

TYPE LOCALITY: Zacatecas, Mexico.

RANGE: New Mexico and Arizona to Mexico.

NEW MEXICO: Mogollon Mountains; Kingston; San Luis Mountains; Santa Rita; Sandia Mountains. Dry hills, in the Lower and Upper Sonoran zones.

The type of *C. herbacea* was collected by Fendler between San Miguel and Las Vegas.

3. Calliandra reticulata A. Gray, Pl. Wright. **2**: 53. 1853.

TYPE LOCALITY: Stony hills at the Copper Mines, New Mexico. Type collected by Wright (no. 1045).

RANGE: Southern New Mexico and Arizona to Mexico.

NEW MEXICO: West Fork of the Gila; Santa Rita. Dry hillsides.

2. ACACIA L. ACACIA.

Shrubs or low trees with armed or smooth stems and numerous very small leaflets; flowers small, regular, in spikes or heads on axillary peduncles; corolla valvate, of 4 or 5 similar petals; stamens numerous, distinct, exserted; pods flattened or terete, 2-valved, dehiscent.

KEY TO THE SPECIES.

Flowers in elongated spikes; pods flat, 15 to 20 mm. wide, curved; spines short and hooked .... 1. *A. greggii.*
Flowers in globose heads; pods terete, or if flat less than 10 mm. wide, straight; spines straight and slender or none.
Spiny shrub 1 meter high or more; flowers bright yellow, sweet-scented; pods terete, constricted between the seeds .... 2. *A. constricta.*
Unarmed shrubs less than 1 meter high; flowers whitish, odorless; pods flat and thin.
Leaflets 8 to 13 pairs, obtuse; inflorescence nearly always axillary .... 3. *A. cuspidata.*
Leaflets 18 pairs or more, acute; inflorescence becoming paniculate, sometimes axillary .... 4. *A. filicioides.*

1. Acacia greggii A. Gray, Pl. Wright. **1:** 65. 1852.

TYPE LOCALITY: Western Texas.

RANGE: Western Texas to southern Arizona and adjacent Mexico.

NEW MEXICO: Redrock; Lone Pine; Carlisle; west of Roswell. Lower Sonoran Zone.

2. Acacia constricta Benth.; A. Gray, Pl. Wright. **1:** 66. 1852.

TYPE LOCALITY: Prairies near the source of the San Felipe, western Texas.

RANGE: Southwestern Texas to southern New Mexico and Arizona, south into Mexico.

NEW MEXICO: Redrock; Big Hatchet Mountains; mesa west of Organ Mountains; Carlsbad; La Luz Canyon; Lakewood. Dry hills and plains, in the Lower Sonoran Zone.

2a. Acacia constricta paucispina Woot. & Standl. Bull. Torrey Club **36:** 105. 1909.

TYPE LOCALITY: On Animas Creek, in the Black Range, New Mexico. Type collected by Metcalfe (no. 1123).

NEW MEXICO: Animas Creek; Organ Mountains; Carlisle; Burro Mountains; Hillsboro. Upper Sonoran Zone.

This is a larger plant with fewer spines and larger, much less glandular, more pubescent leaves and young stems, occurring usually at slightly higher levels than the species. It is found in western Texas and southern Arizona as well.

3. Acacia cuspidata Schlecht. Linnaea **12:** 573. 1838.

TYPE LOCALITY: "Prope Mexico."

RANGE: Western Texas to southern Arizona, south into Mexico.

NEW MEXICO: Black Range; Mangas Springs; Organ Mountains; Dog Spring; Hanover Mountain. Dry hills, in the Upper Sonoran Zone.

4. Acacia filicioides (Cav.) Trel. Rep. Ark. Geol. Surv. **4:** 178. 1891.

Mimosa filicioides Cav. Icon. Pl. **1:** 55. *pl. 78.* 1791.

TYPE LOCALITY: "Habitat in Mexico."

RANGE: Missouri and Kansas to Texas and Arizona, south into Mexico.

NEW MEXICO: San Luis Mountains; near White Water. Dry hills, in the Lower Sonoran Zone.

3. STROMBOCARPA A. Gray. SCREW BEAN. TORNILLO.

A tall, gracefully spreading shrub 5 meters high or less, branching from the base, with dense dark-colored wood; leaves small, with 1 or 2 pairs of pinnæ; leaflets 5 to 8 pairs, short-oblong; stipular spines rigid, 2 cm. long or less, whitish; flowers yellow, in crowded spikes; pod an indehiscent, spirally coiled legume.

1. **Strombocarpa pubescens** (Benth.) A. Gray, Pl. Wright. **1:** 60. 1852.

Prosopis pubescens Benth. Lond. Journ. Bot. **5:** 82. 1846.

TYPE LOCALITY: "California between San Miguel and Monterey."

RANGE: Western Texas to Arizona and California.

NEW MEXICO: Socorro; Animas Creek; Mesilla Valley. River valleys, in the Lower Sonoran Zone.

This is one of the common large shrubs of the river valleys of the southern part of the State, where it is everywhere known under its Spanish name of "tornillo." The larger stems or trunks serve very well for fence posts and the wood is also extensively used for fuel, being the best for this purpose found at the lower levels. The pods contain a large amount of sugar and are very sweet when chewed.

4. **PROSOPIS** L. MESQUITE.

A much branched shrub 3 meters high or less, seldom larger, with rigid tough stems bearing large stipular spines; leaves with 1 or 2 pairs of pinnæ and numerous oblong entire leaflets; flowers small, greenish yellow, in axillary spikes; fruit an indehiscent, slightly compressed, straight or falcate legume.

1. **Prosopis glandulosa** Torr. Ann. Lyc. N. Y. **2:** 192. 1828.

TYPE LOCALITY: "On the Canadian?," New Mexico.

RANGE: Arizona and New Mexico to Oklahoma and Texas.

NEW MEXICO: Common from the Black Range to Socorro, and Tucumcari and southward across the State. Plains and river valleys, in the Lower and Upper Sonoran zones.

This is one of the best known plants of the arid southwest and in southern New Mexico is of great economic importance. The flowers furnish the best of nectar for honey making. The leaves and pods are eaten by all kinds of grazing animals. The pods, too, on account of their sugar content, are often eaten when ripe by the native people. The large roots and thickened bases of the stems furnish the best fuel of the region. The legumes and seeds were collected by the Indians, who ground them and formed the meal into a sort of bread.

5. **MORONGIA** Britton. SENSITIVE BRIER.

Decumbent perennial with recurved prickles on leaves and stems; flowers pink, in a globose head; pod narrow, 4-sided, 4-valved, spiny.

1. **Morongia occidentalis** Woot. & Standl. Contr. U. S. Nat. Herb. **16:** 135. 1913.

TYPE LOCALITY: Sandy soil at Logan, New Mexico. Type collected by G. L. Fisher (no. 93).

RANGE: Known only from type locality.

6. **ACUAN** Medic.

Suffrutescent or herbaceous perennials with unarmed herbaceous stems and numerous small leaflets; flowers in axillary pedunculate heads, greenish white; calyx sometimes pappiform or wanting; stamens 5 or 10, the anthers not gland-bearing; fruit a flattened dehiscent legume, straight or arcuate.

KEY TO THE SPECIES.

Stamens 5; plants tall, erect; pinnæ 10 to 14 pairs; pods arcuate, in a crowded headlike cluster .... 1. *A. illinoensis.*

Stamens 10; plants low, spreading; pinnæ 3 to 6 pairs; pods straight, fewer, divaricate.. 2. *A. jamesii.*

Acuan velutina was reported from Santa Rita in the Botany of the Mexican Boundary, but the specimens upon which the report is based seem to be *A. jamesii.*

1. **Acuan illinoensis** (Michx.) Kuntze, Rev. Gen. Pl. 1: 158. 1891.
Mimosa illinoensis Michx. Fl. Bor. Amer. 2: 254. 1803.
Acacia brachyloba Willd. Sp. Pl. 4: 1071. 1806.
Desmanthus brachylobus Benth. Lond. Journ. Bot. 4: 358. 1842.
TYPE LOCALITY: "Hab. in pratensibus regionis Illinoensis."
RANGE: Minnesota to Florida, Colorado, and New Mexico.
NEW MEXICO: Albuquerque; Socorro; Sabinal; Mesilla Valley; Roswell; Lakewood; Dayton; Perico. River valleys, in the Lower and Upper Sonoran zones.

2. **Acuan jamesii** (Torr. & Gray) Kuntze, Rev. Gen. Pl. 1: 158. 1891.
Desmanthus jamesii Torr. & Gray, Fl. N. Amer. 1: 402. 1840.
TYPE LOCALITY: "Sources of the Canadian River," Colorado or New Mexico. Type collected by James.
RANGE: Oklahoma and Texas to Arizona.
NEW MEXICO: Rio Zuni; Mogollon Mountains; Kingston; Silver City; Organ Pass; west of Roswell; Gray; Nara Visa, Redlands; Knowles; Buchanan. Dry hills and plains, in the Upper Sonoran Zone.

7. MIMOSA L. CAT-CLAW.

Low shrubs, the stems armed with hooked spines; leaflets small; flowers in spikes or heads, small; sepals and petals 5; stamens 10, distinct; fruit a flattened pod, armed or unarmed, sometimes constricted between the seeds.

KEY TO THE SPECIES.

Flowers in spikes, pink .... 1. *M. dysocarpa.*
Flowers in spherical heads, yellow or pink.
  Pinnæ 4 to 7 pairs, pubescent.
    Young stems not flexuous, somewhat virgate; pods usually not constricted between the seeds, straight .... 2. *M. lemmoni.*
    Young stems flexuous; pods more or less constricted between the seeds, conspicuously arcuate .... 3. *M. biuncifera.*
  Pinnæ 1 to 3 pairs, glabrous.
    Pods more or less spiny .... 4. *M. borealis.*
    Pods not spiny .... 5. *M. fragrans.*

1. **Mimosa dysocarpa** Benth.; A. Gray, Pl. Wright. 1: 62. 1852.
TYPE LOCALITY: "Mountain valleys in the Pass of the Limpia, and beyond," Texas.
RANGE: Western Texas to southern Arizona and adjacent Mexico.
NEW MEXICO: Fort Bayard; Animas Mountains; San Luis Mountains; Little Burro Mountains. Upper Sonoran Zone.
A rather uncommon species in the southwestern part of the State in the lower mountains. The spikes of pink flowers and the yellow young stems are characteristic. The young pods are densely velutinous.

2. **Mimosa lemmoni** A. Gray, Proc. Amer. Acad. 19: 76. 1883.
TYPE LOCALITY: Near Fort Huachuca, southern Arizona.
RANGE: Southern Arizona and New Mexico, south into Mexico.
NEW MEXICO: San Luis Mountains. Upper Sonoran Zone.

3. **Mimosa biuncifera** Benth. Pl. Hartw. 12. 1839.
TYPE LOCALITY: Mexico.
RANGE: Western Texas to southern Arizona and adjacent Mexico.
NEW MEXICO: From the Black Range to the Organ and Guadalupe mountains and southward. Dry plains and hills, in the Upper Sonoran Zone.
A common shrub about 1 meter high, occurring in the foothills and canyons of the drier and rockier mountains of the southern part of the State. The young stems are

browsed to a small extent by cattle, but their thorns are so sharp and strong that they are mostly avoided.

4. Mimosa borealis A. Gray, Mem. Amer. Acad. n. ser. **4**: 39. 1849.

TYPE LOCALITY: "Hillsides, Upper Spring, on the Cimarron." Type collected by Fendler (no. 181).

RANGE: Northern and eastern New Mexico and western Texas.

NEW MEXICO: Upper Cimarron; Logan; Lincoln; Nara Visa. Mountains, in the Upper Sonoran Zone.

5. Mimosa fragrans A. Gray, Bost. Journ. Nat. Hist. **6**: 182. 1850.

TYPE LOCALITY: "Rocky soil, on the Pierdenales," western Texas.

RANGE: Western Texas and eastern New Mexico.

NEW MEXICO: Guadalupe Mountains (*Wooton*). Dry hills and canyons, in the Upper Sonoran Zone.

A not uncommon shrub, 1 to 1.5 meters high, in the mountains of the southeastern part of the State, where it is browsed by cattle, sheep, and goats. The trunk is stout, and the branches rigidly divaricate. The bark is gray and the leaves very small. In herbarium specimens it closely resembles the preceding species, but it is a much more rigid and grayer plant, and the pods are spiny on the margins.

68. CASSIACEAE. Senna Family.

Herbaceous or shrubby annuals or perennials with pinnate or bipinnate, alternate, usually stipulate leaves; flowers perfect, slightly irregular; calyx of 5 more or less united sepals; petals 5, yellow, imbricated, the upper one innermost in bud; stamens 10 or fewer, distinct; pistil simple, the ovary 1-celled, becoming a legume in fruit; seeds usually several.

KEY TO THE GENERA.

Leaves bipinnate.
- Low herbaceous plants 30 cm. high or less........ 1. HOFFMANSEGGIA (p. 332).
- Large shrub over 1 meter high..................... 2. POINCIANA (p. 334).

Leaves once pinnate.
- Corolla almost regular; some of the stamens abortive; calyx lobes obtuse.................. 3. CASSIA (p. 334).
- Corolla irregular, the lower petals noticeably longest; all 10 stamens functional; calyx lobes acuminate.............................. 4. CHAMAECRISTA (p. 335).

1. HOFFMANSEGGIA Cav.

Low herbaceous perennials from tuberous roots or a thickened woody base, the bipinnate leaves with very small leaflets; plants more or less glandular, especially on the flowers and fruit (one species without glands); flowers yellow or the stamens red, in naked racemes, terminal, or opposite the leaves; sepals and petals 5; stamens 10; pods flattened, with few or several seeds.

KEY TO THE SPECIES.

Calyx not oblique at the base; sepals all alike; plants without sessile black glands.
- Plants stipitate-glandular, the inflorescence densely so; pods straight or but slightly curved....................... 1. *H. densiflora.*
- Plants not glandular; pods strongly falcate................. 2. *H. drepanocarpa.*

Calyx oblique at the base; lower sepals broadest; plants covered with sessile black glands.
- Pods rhombic-ovate.. 3. *H. brachycarpa.*
- Pods sublunate... 4. *H. jamesii.*

1. **Hoffmanseggia densiflora** Benth.; A. Gray, Pl. Wright. 1: 55. 1852.

CAMOTE DE RATÓN.

Hoffmanseggia stricta Benth.; A. Gray, op. cit. 56. 1852.

Hoffmanseggia stricta demissa Benth.; A. Gray, loc. cit.

Hoffmanseggia stricta rusbyi Fisher, Contr. U. S. Nat. Herb. 1: 144. 1892.

TYPE LOCALITY: Valley of the Pecos, Texas. Type, Wright's no. 148.

RANGE: Southwestern Texas, southern New Mexico and Arizona, and adjacent Mexico.

NEW MEXICO: Tucumcari; Los Lunas; Socorro; Deming; Tularosa; Alamogordo; Hillsboro; Roswell; Carlsbad; Mangas Springs; Animas Valley; Hachita; Mesilla Valley; Nogal. Plains and river valleys, in the Lower and Upper Sonoran zones.

This species is common in hard alkaline soils in the lower valleys, especially in locations which are flooded occasionally. The small, ellipsoidal or spheroidal tuberous roots 2 to 4 cm. long, which give it its common name, are produced 15 to 30 cm. below the surface, on slender tough roots. They are rather sweet and of not unpleasant flavor, although tough. They are commonly eaten by the Indians.

The various subspecies which have been proposed are probably forms caused by varying quantities of water received by the plants at different times of the year. They can all be found in a small patch of the species by careful search. Bentham himself observed that *H. densiflora* was perhaps too near *H. demissa*, and Doctor Gray in publishing it reduced *H. demissa* of Bentham's manuscript to *H. stricta demissa*. It is unfortunate that he should have published these two forms before the *H. stricta*, which is really the typical form of the plant, whose name is distinctive of its most characteristic difference from *H. falcaria* Cav. But according to the rules of priority the name *densiflora* must stand. The type of *H. falcaria rusbyi* was collected by Rusby at Mangas Springs.

2. **Hoffmanseggia drepanocarpa** A. Gray, Pl. Wright. 1: 58. 1852.

TYPE LOCALITY: "New Mexico, or between Texas and El Paso."

RANGE: New Mexico to southern California.

NEW MEXICO: Acoma; Socorro; Mangas Springs; Organ Mountains; Guadalupe Mountains; Lake Valley; Knowles. Dry hills, in the Upper and Lower Sonoran zones.

3. **Hoffmanseggia brachycarpa** A. Gray, Pl. Wright. 1: 55. 1852.

TYPE LOCALITY: "New Mexico." Type collected by Wright in 1851.

RANGE: Southern New Mexico, southwestern Texas, and probably adjacent Mexico.

We have seen no specimens from New Mexico, but it is included here on the strength of the citation of the type, for which no number is given. Wright's later collections of the same species are from Texas east of the Pecos. It is possible that the original citation is incorrect, as Doctor Gray was indefinite in this citation, while he was usually very particular about this point.

4. **Hoffmanseggia jamesii** Torr. & Gray, Fl. N. Amer. 1: 393. 1840.

Pomaria glandulosa Cav. err. det. Torr. Ann. Lyc. N. Y. 2: 193. 1828.

TYPE LOCALITY: "Sources of the Canadian River," probably in New Mexico. Type collected by James.

RANGE: Colorado to Texas, Arizona, and Mexico.

NEW MEXICO: Farmington; Zuni Reservation; Kennedy; Tijeras Canyon; Sabinal; mesa near Las Cruces; Gage; Clayton; Buchanan; Redlands. Sandy hills and plains, in the Upper and Lower Sonoran zones.

This is the common black-glandular species, occurring mostly in sandy soil, the branching leafy stems arising from a thickened woody root often 10 to 20 cm. long and 3 cm. thick. The dull yellow flowers and sublunate pods are distinctive.

2. POINCIANA L. Bird-of-paradise flower.

1. **Poinciana gilliesii** Hook. Bot. Misc. Hook. **1:** 129. *pl. 34.* 1830.

Caesalpinia gilliesii Wall. Bot. Misc. Hook. **1:** 129. 1830.

Type locality: "Prope Rio Quarto et Rio Quinto, et apud La Punta de San Luis. Abundat circa Mendozam, Americae Meridionalis."

An ill-smelling, erect, sparingly branched shrub with green stems 2 to 3 meters high; leaves large, bipinnate, with very numerous small leaflets; inflorescence of terminal racemes of large yellow flowers with long-exserted, bright red stamens and pistil. It is one of the commonest ornamental plants in gardens, especially in the southern part of the State and is frequently escaped.

3. CASSIA L. Senna.

Shrubs or herbaceous perennials with abruptly pinnate leaves and rather large yellow flowers; calyx lobes 5, obtuse; petals 5, nearly equal; stamens 10, the upper 3 abortive, the anthers opening by terminal pores; pods slightly compressed, elongated, several-seeded.

KEY TO THE SPECIES.

Shrub with leaflets 5 mm. long or less.......................... 1. *C. wislizeni.*
Herbs with leaflets 20 mm. long or more.
  Leaves 2-foliolate.
    Leaflets lanceolate, 30 to 50 mm. long, bright green.... 4. *C. roemeriana.*
    Leaflets oblong, 25 mm. long or less, grayish........... 5. *C. bauhinioides.*
  Leaves 6 to many-foliolate.
    Plants glabrous; leaflets lanceolate, acute............. 2. *C. leptocarpa.*
    Plants pubescent throughout; leaflets oblong, obtuse.
      Leaflets more than 3 pairs, oblong to ovate; pubescence coarse, spreading; pods 5 cm. long or more, obtuse.............................. 3. *C. lindheimeriana.*
      Leaflets 2 or 3 pairs, mostly oblong-obovate; pubescence fine and close; pods 3 cm. long or less, acute.................................... 6. *C. covesii.*

1. **Cassia wislizeni** A. Gray, Pl. Wright. **1:** 60. 1852.

Type locality: "Carrizal and Ojo Caliente, south of El Paso," Chihuahua.

Range: New Mexico and Arizona to Mexico.

New Mexico: Southern Grant and Luna counties. Dry hillsides and plains, in the Lower and Upper Sonoran zones.

2. **Cassia leptocarpa** Benth. Linnaea **22:** 528. 1849.

Type locality: "Ad Zapativa legit Pohl et prope Rio Janeiro."

Range: Texas to Arizona, south into Mexico and South America.

New Mexico: Near White Water (*Mearns* 343).

3. **Cassia lindheimeriana** Scheele, Linnaea **21:** 457. 1848.

Type locality: New Braunfels, Texas.

Range: Western Texas to Arizona.

New Mexico: Florida Mountains; Organ Mountains; Guadalupe Mountains. Dry rocky hills, in the Upper Sonoran Zone.

4. **Cassia roemeriana** Scheele, Linnaea **21:** 457. 1848.

Type locality: "Auf felsigem Boden am obern Guadeloupe," western Texas.

Range: Western Texas and southern New Mexico to Mexico.

New Mexico: Hurrah Creek; Tierra Blanca; Ruidoso Creek; Roswell; Knowles; Queen; Lincoln; Causey. Lower and Upper Sonoran zones.

5. **Cassia bauhinioides** A. Gray, Bost. Journ. Nat. Hist. 6: 180. 1850.

TYPE LOCALITY: On the Rio Grande, Texas.

RANGE: Western Texas to Arizona.

NEW MEXICO: Kingston; Mangas Springs; Dog Spring; mesa west of Organ Mountains; west of Roswell. Mesas, in the Lower Sonoran Zone.

6. **Cassia covesii** A. Gray, Proc. Amer. Acad. 7: 399. 1868.

TYPE LOCALITY: "Camp Grant, and south of Prescott, Arizona."

RANGE: Southwestern New Mexico to Arizona and southward.

NEW MEXICO: Telegraph Mountains (*Wooton*). Dry soil.

4. CHAMAECRISTA Moench. PARTRIDGE PEA.

Herbaceous annuals or perennials with abruptly pinnate leaves and bright yellow flowers in few-flowered extra-axillary clusters; rachis of the leaf bearing one or two glands near the base; calyx lobes 5, acuminate; petals 5, unequal, the lowest little or much the largest; stamens 10, all perfect, the anthers opening by terminal pores; pods narrowly oblong-linear, flattened, the valves elastic; seeds compressed, ovoid or quadrate.

KEY TO THE SPECIES.

Perennial; peduncles 30 to 40 mm. long.............................. 1. *C. wrightii.*
Annuals; peduncles 20 mm. long or less.
  Flowers small, the petals about 5 mm. long, one much longer than the other four.................................... 2. *C. leptadenia.*
  Flowers larger, the petals 12 mm. long or more, all of about the same size.
    Leaflets 16 to 28; pod not beaked or obscurely so.......... 3. *C. fasciculata.*
    Leaflets 10 or 12; pod with a beak 2 or 3 mm. long......... 4. *C. rostrata.*

1. **Chamaecrista wrightii** (A. Gray) Woot. & Standl.

Cassia wrightii A. Gray, Pl. Wright. 2: 50. 1853.

TYPE LOCALITY: "Hillsides, on the Sonoita, near Deserted Rancho, Sonora."

RANGE: New Mexico, Arizona, and Sonora.

NEW MEXICO: On the Mimbres (*Mexican Boundary Survey* 297).

2. **Chamaecrista leptadenia** (Greenm.) Cockerell, Muhlenbergia 4: 68. 1908.

Cassia leptadenia Greenm. Proc. Amer. Acad. 41: 238. 1905.

TYPE LOCALITY: Texas.

RANGE: Western Texas to Arizona.

NEW MEXICO: Organ Mountains. Dry hillsides, in the Upper Sonoran Zone.

3. **Chamaecrista fasciculata** (Michx.) Greene; Small, Fl. Southeast. U. S. 587. 1903.

Cassia fasciculata Michx. Fl. Bor. Amer. 1: 262. 1803.

TYPE LOCALITY: "Hab. in Pennsylvania et Virginia."

RANGE: Maine to South Dakota, Florida, Colorado, and New Mexico.

NEW MEXICO: Organ Mountains (*Wooton*). Dry fields and hillsides, in the Upper Sonoran Zone.

4. **Chamaecrista rostrata** Woot. & Standl. Contr. U. S. Nat. Herb. 16: 135. 1913.

TYPE LOCALITY: Sandy soil at Logan, New Mexico. Type collected by G. L. Fisher (no. 93).

RANGE: Known only from type locality, on sandy plains of the Upper Sonoran Zone.

69. KRAMERIACEAE. Krameria Family.

1. KRAMERIA Loefl.

Low herbaceous or woody perennials with prostrate or widely spreading stems and small silky-pubescent leaves; leaves alternate, exstipulate, entire; flowers perfect, crimson, irregular; calyx of 4 or 5 unequal petaloid sepals, deciduous; corolla of 4 or 5 petals shorter than the sepals, irregular, the posterior petal clawed, sometimes adnate, the anterior thick, sessile; stamens 3 or 4, the filaments united at the base; pistil simple; fruit an indehiscent spiny globose 1-seeded pod.

KEY TO THE SPECIES.

Herb with prostrate branches.................................... 1. *K. secundiflora.*
Shrub with diffuse branches..................................... 2. *K. glandulosa.*

1. **Krameria secundiflora** DC. Prodr. **1**: 341. 1824.
Krameria lanceolata Torr. Ann. Lyc. N. Y. **2**: 168. 1827.
TYPE LOCALITY: Mexico.
RANGE: Kansas and Florida to New Mexico, south into Mexico.
NEW MEXICO: Mangas Springs; Tucumcari; Carrizozo; Perico Creek; Pajarito Valley; Roswell; San Andreas Mountains. Plains, in the Upper Sonoran Zone.

2. **Krameria glandulosa** Rose & Painter, Contr. U. S. Nat. Herb. **10**: 108. 1906.
Krameria parvifolia Benth. err. det. various authors.
TYPE LOCALITY: Near El Paso, Texas.
RANGE: California and Utah to western Texas, southward into Mexico.
NEW MEXICO: Mesa west of Organ Mountains; Buchanan. Dry, sandy hills and mesas, in the Lower Sonoran Zone.

A very common and rather handsome plant on the dry mesas of southern New Mexico. It is a low, densely branched shrub 30 cm. high or less, blooming in early spring.

70. FABACEAE. Pea Family.

Herbs or shrubs, sometimes trees, with simply compound or rarely simple, alternate, stipulate leaves; flowers papilionaceous; calyx of 5 more or less united sepals; petals 5 or fewer, irregular, the upper petal larger than the others and inclosing them in bud, the two lateral ones (wings) oblique, the lower two more or less coherent by their anterior edges and forming the keel; stamens mostly 10, monadelphous, diadelphous, or distinct; fruit a legume, 1-celled (2-celled in some Astragali), containing 1 to many seeds.

KEY TO THE GENERA.

Stamens distinct.
  Leaves palmately trifoliolate; flowers yellow....... 1. THERMOPSIS (p. 338).
  Leaves odd-pinnate; flowers not yellow.
    Herbs; seeds not red.......................... 2. SOPHORA (p. 339).
    Shrub; seeds bright red....................... 3. BROUSSONETIA (p. 339).
Stamens monadelphous or diadelphous.
  Anthers of 2 kinds; stamens monadelphous; leaves palmately compound.
    Stipules not decurrent; pods flattened......... 5. LUPINUS (p. 340).
    Stipules, at least the upper ones, decurrent; pods inflated.......................... 4. CROTALARIA (p. 339).
  Anthers all alike; stamens diadelphous (9 and 1), or sometimes only 5; leaves usually pinnately compound, rarely palmate.

Leaves with an even number of leaflets, terminated by a tendril or a bristle-like appendage.
Styles filiform, hairy all around and below the apex; stamen tube usually oblique at the summit .... 25. VICIA (p. 373).
Styles flattened, hairy on the inner side; stamen tube truncate or nearly so ..... 26. LATHYRUS (p. 375).
Leaves odd-pinnate, rarely palmate, without tendrils.
Pod a loment, breaking transversely into 1-seeded indehiscent segments.
Pods 1 or 2-seeded, more or less spiny or toothed .... 22. ONOBRYCHIS (p. 371).
Pods several seeded, not spiny or toothed.
Leaves 3-foliolate .... 23. MEIBOMIA (p. 371).
Leaves with numerous leaflets .... 24. HEDYSARUM (p. 373).
Pod not a loment, 2-valved or indehiscent, sometimes 2-celled by intrusion of the sutures.
Stems slender, twining; plants herbaceous, annuals or perennials; leaves 3-foliolate.
Keel of the corolla not coiled nor incurved.
Flowers yellow .... 27. DOLICHOLUS (p. 376).
Flowers purplish .... 29. GALACTIA (p. 376).
Keel of the corolla coiled or incurved.
Keel of the corolla spirally twisted; inflorescence racemose .... 30. PHASEOLUS (p. 377).
Keel merely incurved; inflorescence various.
Annual; calyx 5-toothed; inflorescence capitate .... 31. STROPHOSTYLES (p. 378).
Perennials; calyx 4-toothed; flowers solitary or in 2's in the axils .... 32. COLOGANIA (p. 378).
Stems not twining; herbs or shrubs; leaves 3 to many-foliolate.
Foliage glandular-punctate (except in a few species of Parosela).
Pods covered with hooked prickles. 21. GLYCYRRHIZA (p. 371).
Pods not prickly.
Leaves palmately 3 or 5-foliolate or pinnately 3-foliate ...... 11. PSORALEA (p. 348).
Leaves mostly pinnately 5 to many-foliolate.
Stamens only 5 .... 16. PETALOSTEMUM (p. 355).

Stamens 9 or 10.
Corolla of 5 petals.
Pods not falcate; petals very unequal; herbs or low shrubs........ 15. PAROSELA (p. 350).
Pods falcate; petals nearly equal; tall shrub.... 12. VIBORQUIA (p. 349).
Corolla of less than 5 petals.
Petals 1, the banner... 13. AMORPHA (p. 349).
Petals wanting........ 14. PARRYELLA (p. 350).
Foliage not glandular-punctate.
Leaves 3-foliolate.
Leaflets entire or merely undulate.
Shrub; flowers scarlet; stems spiny.................... 28. ERYTHRINA (p. 376).
Herbs; flowers not scarlet; stems not spiny.
Leaflets 4 or more.......... 9. ANISOLOTUS (p. 346).
Leaflets 3.................. 10. ACMISPON (p. 348).
Leaflets more or less toothed.
Pods curved or coiled; flowers racemose................ 6. MEDICAGO (p. 343).
Pods straight; flowers racemose or capitate.
Leaves pinnate; valves of the pod leathery; flowers in racemes.............. 7. MELILOTUS (p. 343).
Leaves digitate; pod valves thin; flowers in heads.. 8. TRIFOLIUM (p. 344).
Leaves several to many-foliolate.
Shrubs or small trees; stems spiny; flowers pink......... 18. ROBINIA (p. 356).
Herbs, at most woody only at the base; stems sometimes spiny; variously colored.
Stipules spiny; leaflets wanting or flowers early deciduous............... 17. PETERIA (p. 356).
Stipules not spiny; leaflets persistent.
Keel prolonged into a beak 20. OXYTROPIS (p. 370).
Keel not prolonged into a beak................ 19. ASTRAGALUS (p. 356).

1. THERMOPSIS R. Br.

Coarse perennial herbs with palmately 3-foliolate leaves; flowers 1 cm. long or more bright yellow, racemose; pods flat, several-seeded; stems 30 to 50 cm. high, sparingly appressed-pubescent.

KEY TO THE SPECIES.

Pods erect, straight.. 1. *T. pinetorum.*
Pods spreading, arcuate.. 2. *T. divaricarpa.*

1. **Thermopsis pinetorum** Greene, Pittonia **4**: 138. 1900.

TYPE LOCALITY: Below Marshall Pass, Colorado.

RANGE: Colorado and New Mexico.

NEW MEXICO: Chama; Santa Fe and Las Vegas mountains; Agua Fria; Sandia Mountains; Mogollon Mountains. Woods, in the Transition Zone.

2. **Thermopsis divaricarpa** A. Nels. Bot. Gaz. **25**: 275. *pl. 18. f. 3.* 1898.

TYPE LOCALITY: Pole Creek, Wyoming.

RANGE: Wyoming to northern New Mexico.

NEW MEXICO: Sierra Grande (*Standley* 6140). Meadows, in the Transition Zone.

2. SOPHORA L.

Low pubescent perennial herbs with pinnate many-foliolate leaves and dense racemes of white or blue flowers; stamens 10, the filaments distinct or nearly so; pods thick, torulose, tardily dehiscent.

KEY TO THE SPECIES.

Leaflets linear; flowers blue.................................... 1. *S. stenophylla.*
Leaflets oblong or oblong-obovate; flowers white or nearly so...... 2. *S. sericea.*

1. **Sophora stenophylla** A. Gray in Ives, Rep. Colo. Riv. **4**: 10. 1861.

TYPE LOCALITY: "Oryabe," Arizona.

RANGE: Southern Utah to northern Arizona and New Mexico.

NEW MEXICO: Sia; San Andreas Mountains. Dry hills, in the Upper Sonoran Zone.

2. **Sophora sericea** Nutt. Gen. Pl. **1**: 280. 1818.

TYPE LOCALITY: "On the elevated plains of the Missouri, near the confluence of White River."

RANGE: Wyoming and South Dakota to Arizona and Texas.

NEW MEXICO: Espanola; Coolidge; Cross L Ranch; Las Vegas; Hebron; Zuni; Clayton; Raton; Nara Visa; San Marcial; Mangas Springs; Dog Spring; Mesilla Valley; Gray; Gavilan Canyon. Dry fields and plains, in the Lower and Upper Sonoran zones.

3. BROUSSONETIA Orteg.

A shrub or small tree, the leaves with 7 to 13 leathery oblong leaflets; flowers in dense racemes; pods 5 to 10 cm. long, 3 or 4-seeded, the seeds scarlet.

1. **Broussonetia secundiflora** Orteg. Hort. Matr. Dec. 61. *pl. 7.* 1798.

Sophora secundiflora Lag.; DC. Cat. Hort. Monsp. 148. 1813.

TYPE LOCALITY: "Habitat in Nova Hispania."

RANGE: Western Texas and southern New Mexico, south to Mexico.

NEW MEXICO: Dark Canyon, Guadalupe Mountains (*Wooton*). Dry hills.

This beautiful evergreen shrub with glossy dark green leaves is well worth cultivation. If once established it would probably endure the very trying conditions of low altitudes in the southern part of the State. It grows naturally in crevices of limestone cliffs and probably would need an open soil containing considerable disintegrated limestone. Its large scarlet beans are said to be poisonous, which would be a drawback to its use as a decorative plant.

4. CROTALARIA L. RATTLEBOX.

A diffuse annual, nearly glabrous, with trifoliolate petioled leaves and few-flowered racemes of yellow flowers opposite the leaves; banner large, cordate; stamens monadelphous, the anthers of two kinds; pods short, oblong, inflated, puberulent.

1. **Crotalaria lupulina** H. B. K. Nov. Gen. & Sp. **6**: 402. *pl. 590.* 1823.

TYPE LOCALITY: "Crescit in monte ignivomo Jorullo, alt. 570 hexap.," Mexico.

RANGE: New Mexico and Arizona, south into Mexico.

NEW MEXICO: Southern Grant County. Lower and Upper Sonoran zones.

5. LUPINUS L. Lupine.

Annual or perennial herbs, 10 to 100 cm. high, with palmate leaves and few or numerous rather large flowers in long or short terminal racemes; leaves alternate, the petioles dilated and somewhat clasping, bearing stipules on the expanded portion; calyx bilabiate, the upper lip more or less lobed, the lower entire or minutely toothed; corolla of various colors, the standard with recurved margins, more or less grooved in the median line; keel and wings various; stamens 10, monadelphous, the anthers alternately of two sizes; pistil simple, becoming a 2 to several-seeded flattened legume.

A widely dispersed genus in the United States, although the species are much more numerous in the western part. Some of the species are of considerable economic importance because they are more or less poisonous to stock.

KEY TO THE SPECIES.

Annuals, small, usually not more than 30 cm. high.
- Pods more than 2-seeded; cotyledons petiolate............ 1. *L. micensis.*
- Pods 2-seeded; cotyledons sessile, clasping.
  - Acaulescent plants, tufted; racemes pedunculate, headlike.
    - Upper lobe of the calyx obsolete................ 2. *L. brevicaulis.*
    - Upper lobe of the calyx evident, deeply 2-lobed... 3. *L. dispersus.*
  - Stems evident, although sometimes short; racemes various.
    - Racemes more or less elongated, several to many-flowered, shorter than the leaves; peduncles very short and stout; lower lobe of the calyx entire.................................... 4. *L. pusillus.*
    - Racemes headlike, on peduncles equaling or exceeding the petioles of the adjacent leaves, few-flowered; lower lobe of the calyx 2 or 3-toothed.
      - Peduncles elongated, some of the flower clusters overtopping the leaves; branches ascending, not widely divaricate; plants conspicuously villous-hirsutulous....... 5. *L. kingii.*
      - Penducles shorter, never much longer than the adjacent petioles, the flower clusters not overtopping the leaves; branches widely divaricate-spreading; pubescence softer, less copious.
        - Upper surface of the leaves pubescent like the lower.......................... 6. *L. argillaceus.*
        - Upper surface of leaves glabrous......... 7. *L. sileri.*

Perennials.
- Leaves permanently silky on the upper surface.
  - Calyx distinctly saccate at the base................... 8. *L. aduncus.*
  - Calyx not saccate.................................. 9. *L. palmeri.*
- Leaves glabrous on the upper surface.
  - Stems hirsute.
    - Calyx strongly gibbous; leaflets acute or obtuse.
      - Leaflets obtuse; banner with a dark spot; plants low..................................10. *L. ammophilus.*
      - Leaflets mostly acute; banner not with a dark spot; plants tall.......................16. *L. amplus.*

Calyx scarcely at all gibbous; leaflets acute or acutish.......................................11. *L. neomexicanus.*

Stems not hirsute.

Flowers 8 to 9 mm. long; banner not with a dark spot......................................12. *L. ingratus.*

Flowers 12 mm. long or more; banner with a dark spot (except in no. 13).

Flowers bright blue; leaves bright green......13. *L. laetus.*

Flowers pale bluish; leaves dull grayish or yellowish green.

Leaflets sharply acute, 60 to 70 mm. long; flowers in an elongated raceme; pubescence loose.................14. *L. sierrae-blancae.*

Leaflets obtuse, 30 to 45 mm. long; raceme short, few-flowered; pubescence closely appressed..................15. *L. aquilinus.*

1. **Lupinus micensis** Jones, Proc. Calif. Acad. II. **5:** 630. 1895.

TYPE LOCALITY: Mica, Utah.

RANGE: Nevada and Utah to Arizona and New Mexico.

NEW MEXICO: Organ Mountains; Mangas Springs; Florida Mountains. Dry hills and mesas, in the Lower Sonoran Zone.

2. **Lupinus brevicaulis** S. Wats. in King, Geol. Expl. 40th Par. **5:** 53. 1871.

TYPE LOCALITY: "In the valleys and lower cañons of Western Nevada to the East Humboldt Mountains."

RANGE: Western New Mexico and Colorado to Nevada and California.

NEW MEXICO: West of Patterson; Silver City; Organ Mountains. Dry hills and plains, in the Upper Sonoran and Transition zones.

3. **Lupinus dispersus** Heller, Muhlenbergia **5:** 141. 1909.

TYPE LOCALITY: Rhyolite, Nye County, Nevada.

RANGE: Nevada to Colorado and New Mexico.

NEW MEXICO: Silver City; Organ Mountains. Dry hillsides, in the Lower and Upper Sonoran zones.

4. **Lupinus pusillus** Pursh, Fl. Amer. Sept. 468. 1814.

TYPE LOCALITY: "On the banks of the Missouri."

RANGE: From the Missouri to the Columbia, southward to Arizona and New Mexico.

NEW MEXICO: Coolidge; Mogollon Mountains; Chama; near Santa Rita. Dry hills, in the Upper Sonoran Zone.

5. **Lupinus kingii** S. Wats. Proc. Amer. Acad. **8:** 534. 1873.

TYPE LOCALITY: "Heber Valley in the Wasatch," Utah.

RANGE: Mountains of Utah, Colorado, Arizona, and New Mexico.

NEW MEXICO: Coolidge; Mogollon Mountains. Transition Zone.

6. **Lupinus argillaceus** Woot. & Standl. Contr. U. S. Nat. Herb. **16:** 137. 1913.

TYPE LOCALITY: Near Pecos, San Miguel County, New Mexico. Type collected by Standley (no. 4974).

RANGE: Northern New Mexico.

NEW MEXICO: Near Pecos; El Rito. Open hills, in the Upper Sonoran Zone.

7. **Lupinus sileri** S. Wats. Proc. Amer. Acad. **10:** 345. 1875.

TYPE LOCALITY: "Southern Utah and on the Rio Grande in Southern Colorado." Type collected by A. L. Siler.

RANGE: Utah, Colorado, and New Mexico.

NEW MEXICO: Rio Zuni; Tunitcha Mountains; Dulce. Dry hillsides, in the Upper Sonoran Zone.

One of our New Mexican plants exactly matches Wolf's no. 195 from the Rio Grande at Loma, Colorado, and the two are certainly distinct from *L. kingii.* We have not seen Siler's plant, which would be the type, but on the assumption that there is no confusion in the specimens of Wolf's collection and that Doctor Watson was correct in referring these plants there, in our judgment this name should not be reduced to synonymy.

8. Lupinus aduncus Greene, Pittonia **4:** 132. 1900.

Lupinus decumbens argophyllus A. Gray, Mem. Amer. Acad. n. ser. **4:** 37. 1849.

Lupinus helleri Greene, Pittonia **4:** 134. 1900.

Lupinus argophyllus Cockerell, Torreya **2:** 42. 1902.

Type locality: Dry ravines among the sandy hills at Aztec, New Mexico. Type collected by Baker (no. 433).

Range: Nebraska to Colorado and New Mexico.

New Mexico: Aztec; Santa Fe; Ramah; Glorieta; Tesuque; Pajarito Park; Johnsons Mesa; Chama; Tunitcha Mountains; Nara Visa. Hills and mesas, in the Upper Sonoran and Transition zones.

The types of *L. decumbens argophyllus* and *L. helleri* came from near Santa Fe.

9. Lupinus palmeri S. Wats. Proc. Amer. Acad. **8:** 530. 1873.

Type locality: San Francisco Mountains, Arizona.

Range: Arizona and New Mexico.

New Mexico: Burro Mountains; Luna; Blue Creek Canyon. Mountains, in the Transition Zone.

The pubescence of fresh specimens is silvery white, but in the herbarium it soon becomes tawny.

10. Lupinus ammophilus Greene, Pittonia **4:** 136. 1900.

Type locality: Sandy bottoms of dry streams at Aztec, New Mexico. Type collected by Baker (no. 434).

Range: Southern Colorado and northwestern New Mexico.

New Mexico: Aztec; Carrizo Mountains; Dulce; Tierra Amarilla. Dry hills, in the Upper Sonoran and Transition zones.

11. Lupinus neomexicanus Greene, Pittonia **4:** 133. 1900.

Type locality: About Silver City and in foothills of the Pinos Altos Mountains, New Mexico. Type collected by Greene.

Range: Southwestern New Mexico.

New Mexico: Silver City; Santa Rita; Mogollon Mountains. Transition Zone.

12. Lupinus ingratus Greene, Pittonia **4:** 133. 1900.

Type locality: Low grassy lands at Chama, New Mexico. Type collected by Baker.

Range: Northern New Mexico and southern Colorado.

New Mexico: Chama; Tunitcha Mountains; Carrizo Mountains; Santa Fe Canyon; Grosstedt Place. Meadows, in the Transition Zone.

13. Lupinus laetus Woot. & Standl. Contr. U. S. Nat. Herb. **16:** 137. 1913.

Type locality: Winter Folly, in the Sacramento Mountains north of Cloudcroft, New Mexico. Type collected by Wooton, August 13, 1899.

Range: Known only from type locality, in the Transition Zone.

14. Lupinus sierrae-blancae Woot. & Standl. Contr. U. S. Nat. Herb. **16:** 138. 1913.

Type locality: On the lower part of White Mountain Peak, New Mexico. Type collected by Wooton, July 6, 1895.

Range: Meadows in the White Mountains of New Mexico, in the Transition Zone.

15. Lupinus aquilinus Woot. & Standl. Contr. U. S. Nat. Herb. **16:** 138. 1913.

TYPE LOCALITY: Gilmores Ranch on Eagle Creek in the White Mountains, New Mexico. Type collected by Wooton & Standley (no. 3613).

RANGE: New Mexico.

NEW MEXICO: Sierra Grande; Raton; White Mountains. Mountains, in the Transition Zone.

16. Lupinus amplus Greene, Pl. Baker. **3:** 36. 1901.

TYPE LOCALITY: Cerro Summit above Cimarron, Colorado.

RANGE: Mountains of Colorado and northern New Mexico.

NEW MEXICO: Chama (*Standley* 6827). Transition Zone.

6. MEDICAGO L.

Annual or perennial herbs, not glandular-dotted, with pinnate 3-foliolate toothed leaves, and small flowers in spikelike racemes; pods spirally coiled, few-seeded.

KEY TO THE SPECIES.

Flowers bluish purple; plants erect.................................. 1. *M. sativa.*
Flowers yellow; plants prostrate..................................... 2. *M. lupulina.*

1. Medicago sativa L. Sp. Pl. 778. 1753. ALFALFA.

TYPE LOCALITY: "Habitat in Hispaniae, Galliae apricis."

NEW MEXICO: Escaped in cultivated and waste ground in nearly all parts of the State.

2. Medicago lupulina L. Sp. Pl. 779. 1753. BLACK MEDIC.

TYPE LOCALITY: "Habitat in Europae pratis."

NEW MEXICO: Tesuque; Taos; Santa Fe; Pecos; Mangas Springs.

An introduction from Europe, occasional along irrigating ditches and in wet fields.

7. MELILOTUS Juss. SWEET CLOVER.

Erect annual or perennial herbs, sometimes 1.5 meters high, with pinnately 3-foliolate toothed leaves, small yellow or white flowers in axillary pedunculate racemes, and small ovoid 1 or 2-seeded coriaceous wrinkled pods.

KEY TO THE SPECIES.

Annual; corolla 2 to 2.5 mm. long, yellow........................ 1. *M. indica.*
Perennials; corolla 5 or 6 mm. long, yellow or white.
Corolla yellow; standard and wing petals about equal........ 2. *M. officinalis.*
Corolla white; standard longer than the wings..... . ... 3. *M. alba.*

1. Melilotus indica (L.) All. Fl. Pedem. **1:** 308. 1785.

Trifolium melilotus indica L. Sp. Pl. 765. 1753.

Melilotus parviflora Desf. Fl. Atlant. **2:** 192. 1800.

TYPE LOCALITY: "Habitat in India, Africa."

NEW MEXICO: Albuquerque; Santa Fe; Pecos; Kingston; Mesilla Valley.

No weed is more common in alfalfa fields. Its seed is a common adulterant of alfalfa seed, and frequently the sweet clover seedlings are more numerous than the alfalfa plants. Because of their bitter taste, probably, the plants are invariably refused by cattle and horses.

2. Melilotus officinalis (L.) Lam. in Lam. & DC. Fl. Franç. **2:** 594. 1778. YELLOW SWEET CLOVER.

Trifolium melilotus officinalis L. Sp. Pl. 765. 1753.

TYPE LOCALITY: "Habitat in Europae campestribus."

NEW MEXICO: Mesilla Valley; Farmington; Cedar Hill.

The plant is well established in orchards in the Mesilla Valley.

3. Melilotus alba Desr. in Lam. Encycl. **4**: 63. 1797. SWEET CLOVER.

TYPE LOCALITY: Siberia.

NEW MEXICO: Frisco; Farmington; Pecos; Gila Hot Springs; Mesilla Valley; Santa Fe; Albuquerque; Las Vegas.

A common and troublesome weed in many parts of the State, especially in alkaline soil. The plant is persistent and spreads rapidly. It is not infrequent in alfalfa fields. It is said to be an excellent bee plant, and for this reason has been introduced in several places, probably on the recommendation of not overscrupulous seedsmen.

8. TRIFOLIUM L. CLOVER.

Low perennial herbs, often tufted or diffuse, with palmately 3-foliolate leaves (occasionally pinnately 3 to 5-foliolate); flowers usually in pedunculate heads, occasionally elongated-spiciform; calyx with slender subulate teeth; corolla attached to the stamen tube; pods small, membranous, indehiscent, often included in the persistent calyx.

KEY TO THE SPECIES.

Heads involucrate.
- Plants low, cespitose; stems scapiform.......... 1. *T. parryi.*
- Plants with elongated leafy stems.
  - Corolla 12 to 15 mm. long; stems not much elongated, mostly erect; peduncles glabrous.......... 2. *T. fendleri.*
  - Corolla 8 to 11 mm. long; stems much elongated, reclining; peduncles glabrous or pubescent.
    - Peduncles glabrous; involucre united well above the base, the divisions broad.......... 3. *T. lacerum.*
    - Peduncles pubescent below the involucre; involucre cleft almost to the base, the divisions very narrow.......... 4. *T. longicaule.*

Heads not involucrate.
- Stems leafy; plants tall or the stems long.
  - Calyx glabrous; flowers white or nearly so.
    - Stems creeping, stoloniferous.......... 6. *T. repens.*
    - Stems erect, tufted.......... 7. *T. hybridum.*
  - Calyx pubescent; flowers mostly purplish.
    - Heads sessile.......... 8. *T. pratense.*
    - Heads long-peduncled.
      - Flowers purplish; stems permanently pubescent.......... 9. *T. neurophyllum.*
      - Flowers white; stems glabrous in age..........12. *T. rydbergii.*
- Stems scapiform; plants low and cespitose.
  - Calyx glabrous; heads 1 to 3-flowered.......... 5. *T. nanum.*
  - Calyx pubescent; heads several to many-flowered.
    - Leaflets obovate, strongly veined, sharply dentate..10. *T. subacaulescens.*
    - Leaflets oblong to lanceolate, entire, not strongly veined..........11. *T. stenolobum.*

1. Trifolium parryi A. Gray, Amer. Journ. Sci. II. **33**: 409. 1862.

TYPE LOCALITY: Rocky Mountains of Colorado.

RANGE: Wyoming and Utah to Colorado and New Mexico.

NEW MEXICO: Pecos Baldy (*Standley*). Meadows, in the Arctic-Alpine Zone.

2. Trifolium fendleri Greene, Pittonia **3**: 221. 1897.

TYPE LOCALITY: Near Santa Fe, New Mexico. Type collected by Fendler.

RANGE: Colorado to New Mexico and Arizona.

NEW MEXICO: Santa Fe and Las Vegas mountains; Rio Pueblo; Albuquerque; Mogollon Mountains; White and Sacramento mountains. Wet meadows, in the Upper Sonoran and Transition zones.

3. **Trifolium lacerum** Greene, Erythea **2:** 182. 1894.

TYPE LOCALITY: "Valley of the Sierra de Las Animas," New Mexico. This locality is not in "southern Colorado or northern New Mexico," as stated by Doctor Greene in the place of publication, but in the southwestern corner of New Mexico. Type collected by Wright (no. 997).

RANGE: Western New Mexico and adjacent Arizona.

NEW MEXICO: Coolidge; Ramah; Silver City; Mogollon Mountains; Animas Mountains; Pescado Spring. Wet ground.

The species is remarkable because of the unusual prolongation of the lateral veins, especially in the uppermost leaves.

4. **Trifolium longicaule** Woot. & Standl. Contr. U. S. Nat. Herb. **16:** 141. 1913.

TYPE LOCALITY: Along Eagle Creek at Gilmores Ranch in the White Mountains, New Mexico. Type collected by Wooton and Standley in 1907.

RANGE: Along small streams in the White and Sacramento mountains of New Mexico, in the Transition Zone.

5. **Trifolium nanum** Torr. Ann. Lyc. N. Y. **1:** 35. *pl. 3. f. 4.* 1824.

TYPE LOCALITY: James Peak, Colorado.

RANGE: Montana to New Mexico.

NEW MEXICO: Pecos Baldy (*Bailey* 618). Meadows, in the Arctic-Alpine Zone.

6. **Trifolium repens** L. Sp. Pl. 767. 1753. WHITE CLOVER.

TYPE LOCALITY: "Habitat in Europae pascuis."

NEW MEXICO: Chama; Winsors Ranch; Santa Fe; Pecos; Farmington; Shiprock.

7. **Trifolium hybridum** L. Sp. Pl. 766. 1753. ALSIKE.

TYPE LOCALITY: "Habitat in Europae cultis."

NEW MEXICO: Pecos (*Standley* 5012).

Abundant in meadows and along irrigating ditches in this locality.

8. **Trifolium pratense** L. Sp. Pl. 768. 1753. RED CLOVER.

TYPE LOCALITY: "Habitat in Europae graminosis."

NEW MEXICO: Harveys Upper Ranch; Wingfields Ranch; Winsors Ranch; Mesilla Valley; Pecos; Raton; Farmington.

Red clover has been noticed in only a few places in the State. Occasionally it appears in alfalfa, but it does not seem to survive long. It has been tried in cultivation at various places, but can not compete with alfalfa on the market and has been cultivated but little.

9. **Trifolium neurophyllum** Greene, Leaflets **1:** 154. 1905.

TYPE LOCALITY: Mogollon Mountains, New Mexico. Type collected by Metcalfe (no. 532).

RANGE: Known only from the type locality.

10. **Trifolium subacaulescens** A. Gray in Ives, Rep. Colo. Riv. **4:** 10. 1860.

TYPE LOCALITY: Pine forests of high table-lands near Fort Defiance, Arizona or New Mexico. Type collected by Palmer.

RANGE: Colorado to Arizona and New Mexico.

NEW MEXICO: Mountains west of Grants Station; Stinking Lake; Tierra Amarilla. Meadows, in the Transition Zone.

11. **Trifolium stenolobum** Rydb. Bull. Torrey Club **28:** 499. 1901.

TYPE LOCALITY: La Plata Mountains, Colorado.

RANGE: Colorado and northern New Mexico.

NEW MEXICO: Sandia Mountains; Santa Fe and Las Vegas mountains. High meadows, Canadian to Arctic-Alpine Zone.

12. **Trifolium rydbergii** Greene, Pittonia 3: 222. 1897.

TYPE LOCALITY: Wind River Mountains, Wyoming.

RANGE: Idaho and Montana to Utah and northern New Mexico.

NEW MEXICO: Chama (*Standley* 6510, *Eggleston* 6647). Moist meadows, in the Transition Zone.

9. ANISOLOTUS Bernh. BIRD'S-FOOT TREFOIL.

Herbaceous annuals or perennials, 50 cm. high or less, generally with numerous rigid and ascending or weak and decumbent or prostrate stems; leaves numerous, small, with black-glandular stipules, pinnate, sometimes appearing palmate by reduction of the rachis, 4 to 7-foliolate, the leaflets small, short-obovate to oblong-linear; flowers axillary and sessile or in few-flowered pedunculate clusters, yellow or reddish orange; calyx lobes mostly very narrow, about the length of the tube; legume straight, slightly or not at all flattened.

KEY TO THE SPECIES.

Annual; plants loosely villous throughout........................ 1. *A. trispermus.*
Perennials; plants more or less puberulent, one species with spreading hirsutulous pubescence.
  Leaves without appreciable rachis, the leaflets crowded on the end of a very short petiole (1 mm. long), or sessile and appearing as a fascicle of simple leaves; flowers pedunculate or sessile.
    Flowers almost all solitary and axillary, with no peduncle, or those of the upper part of the stem occasionally short-peduncled.................... 2. *A. wrightii.*
    Flowers in 1 to 3-flowered clusters, the peduncles usually longer than the leaves........................... 3. *A. rigidus.*
  Leaves with a rachis, although usually a short one, and an appreciable petiole; flowers pedunculate.
    Plants low, decumbent to prostrate; leaflets short and small, obtuse, 8 mm. long or less................. 4. *A. neomexicanus.*
    Plants taller; upper leaflets acute, 10 mm. long or more.
      Branches ascending, stout; leaflets all narrowly oblong-lanceolate or oblanceolate........... 5. *A. puberulus.*
      Branches weak, decumbent, only the ends ascending; leaflets various.
        Pubescence appressed, basal leaflets short and rounded, elliptic-obovate................ 6. *A. nummularius.*
        Pubescence spreading; all leaflets elliptic-lanceolate, about 10 mm. long.......... 7. *A. mollis.*

1. **Anisolotus trispermus** (Greene) Woot. & Standl. Contr. U. S. Nat. Herb. **16**: 135. 1913.

Lotus trispermus Greene, Erythea **1**: 258. 1893.

TYPE LOCALITY: Hills bordering the Mohave Desert, California.

RANGE: California to western New Mexico.

NEW MEXICO: Silver City; Mangas Springs.

2. **Anisolotus wrightii** (A. Gray) Rydb. Bull. Torrey Club **33**: 144. 1906.

Hosackia wrightii A. Gray, Pl. Wright. **2**: 42. 1853.

Lotus wrightii Greene, Pittonia **2**: 143. 1890.

TYPE LOCALITY: Stony hills at the Copper Mines, New Mexico. Type collected by Wright (no. 1000).

RANGE: Colorado to New Mexico and Arizona, south into Mexico.

New Mexico: Chama; Fort Wingate; Ramah; Sandia Mountains; Datil; Mogollon Mountains; Fort Bayard; Burro Mountains. In the mountains at middle elevations, Transition Zone.

3. **Anisolotus rigidus** (Benth.) Rydb. Bull. Torrey Club **33:** 144. 1906.

Hosackia rigida Benth. Pl. Hartw. 305. 1849.

Lotus rigidus Greene, Pittonia **2:** 142. 1890.

Type locality: Monterey, California.

Range: California and Utah to New Mexico.

New Mexico: Hanover Mountain; Mangas Springs; Burro Mountains; Gallinas Planting Station; Pino Canyon. Open slopes, in the Upper Sonoran and Transition zones.

The specimens here listed are doubtfully referred to this apparently little known species, but they fit the original description more nearly than any other material we have seen. Most of the material passing under this name in herbaria belongs elsewhere.

4. **Anisolotus neomexicanus** (Greene) Heller, Muhlenbergia **8:** 60. 1912.

Lotus neomexicanus Greene, Pittonia **2:** 141. 1890.

Type locality: Near Silver City, New Mexico. Type collected by Greene.

Range: Southwestern New Mexico.

New Mexico: Florida Mountains; Tres Hermanas; Silver City. Dry hillsides and mesas, in the Upper Sonoran Zone.

5. **Anisolotus puberulus** (Benth.) Woot. & Standl. Contr. U. S. Nat. Herb. **16:** 135. 1913.

Hosackia puberula Benth. Pl. Hartw. 305. 1849.

Lotus puberulus Greene, Pittonia **2:** 142. 1890.

Type locality: Near Zacatecas, Mexico.

Range: Western Texas to Arizona, south into Mexico.

New Mexico: Las Vegas; Capitan Mountains; White Mountains; Organ Mountains; San Luis Mountains; Fort Bayard. Dry hillsides, in the Upper Sonoran and Transition zones.

This species is generally confused with *A. wrightii*, which it resembles in habit; but its leaflets are arranged pinnately upon a short rachis and have a short petiole, and the flowers have long peduncles.

6. **Anisolotus nummularius** (Jones) Woot. & Standl. Contr. U. S. Nat. Herb. **16:** 135. 1913.

Hosackia rigida nummularia Jones, Bull. Calif. Acad. II. **5:** 633. 1895.

Type locality: Rockville, Utah.

Range: Utah, Arizona, and New Mexico.

New Mexico: Organ Mountains; South Percha Creek. Dry hills and plains, in the Upper Sonoran Zone.

This species has been confused with *A. puberulus* and *A. mollis*. The leaf characters will separate it from the former and the pubescence distinguishes it at once from the latter.

7. **Anisolotus mollis** (Greene) Heller, Muhlenbergia **8:** 60. 1912.

Hosackia mollis Greene, Bull. Calif. Acad. **1:** 185. 1885.

Lotus mollis Greene, Pittonia **2:** 143. 1890.

Type locality: Huachuca Mountains, Arizona.

Range: Southern New Mexico and Arizona to northern Mexico.

New Mexico: Mangas Springs; Juniper Spring; San Luis Pass. Dry plains and hills.

10. ACMISPON Raf.

Annual, similar to Anisolotus; leaves 3-foliolate, rarely 1-foliolate; stipules reduced to mere traces of glands; pod straight, readily dehiscent.

1. **Acmispon americanum** (Nutt.) Rydb. Bull. Torrey Club **40**: 45. 1913.
Lotus sericeus Pursh, Fl. Amer. Sept. 489. 1814, not DC. 1813.
Trigonella americana Nutt. Gen. Pl. **2**: 120. 1818.
Type locality: "On the dry and open alluvial soils of the Missouri, from the river Platte to the Mountains."
Range: Washington and Minnesota to California and New Mexico.
New Mexico: Mule Creek (*Wooton*).

11. PSORALEA L.

Perennial, more or less gland-dotted herbs (glands inconspicuous in some species), with palmately 3 to 5-foliolate leaves; flowers in capitate or racemose clusters; stamens diadelphous; pods small, 1-seeded, indehiscent.

KEY TO THE SPECIES.

Flowers more than 12 mm. long, in dense spikes; plants low, with short stems; roots long, tuberous.
  Stems hirsute .... 6. *P. esculenta.*
  Stems not hirsute, the pubescence sericeous or canescent.
    Leaflets linear to oblanceolate, acutish; spikes elongate .... 1. *P. hypogaea.*
    Leaflets obovate, rounded; inflorescence subcapitate .... 2. *P. megalantha.*
Flowers small, less than 8 mm. long, in racemes or interrupted spikes; plants tall and branching; roots not tuberous.
  Flowers in interrupted spikes; leaves silvery .... 3. *P. argophylla.*
  Flowers in racemes; leaves not silvery.
    Racemes short and dense; fruit globose; upper leaflets linear .... 4. *P. micrantha.*
    Racemes loose, elongate; fruit ovoid; upper leaflets oblanceolate or elliptic .... 5. *P. tenuiflora.*

1. **Psoralea hypogaea** Nutt.; Torr. & Gray, Fl. N. Amer. **1**: 302. 1838.
Type locality: "Plains of the Platte."
Range: Nebraska to New Mexico and Texas.
New Mexico: Taos (*Stevenson*).

2. **Psoralea megalantha** Woot. & Standl. Contr. U. S. Nat. Herb. **16**: 140. 1913.
Type locality: Aztec, New Mexico. Type collected by Baker (no. 440).
Range: Known only from type locality, in the Upper Sonoran Zone.

3. **Psoralea argophylla** Pursh, Fl. Amer. Sept. 475. 1814.
Type locality: "On the banks of the Missouri."
Range: Wisconsin and Saskatchewan to Missouri and New Mexico.
New Mexico: Sierra Grande; Raton Mountains. Open slopes.

4. **Psoralea micrantha** A. Gray, U. S. Rep. Expl. Miss. Pacif. **4**: 77. 1856.
Type locality: "Sand hills, near the last camp on the upper Canadian."
Range: Oklahoma and western Texas to Arizona.
New Mexico: West of Santa Fe; Coolidge; Tesuque; San Augustine Plains; Zuni; Mogollon Mountains. Upper Sonoran Zone.

5. **Psoralea tenuiflora** Pursh, Fl. Amer. Sept. 475. 1814.

TYPE LOCALITY: "On the banks of the Missouri."

RANGE: Montana and South Dakota to Arizona and Arkansas.

NEW MEXICO: Common nearly throughout the State. Plains and hills, in the Upper Sonoran and Transition zones.

6. **Psoralea esculenta** Pursh, Fl. Amer. Sept. 475. 1814.

TYPE LOCALITY: "On the banks of the Missouri."

RANGE: Manitoba and North Dakota to Texas.

NEW MEXICO: Sierra Grande (*Standley* 6156).

12. VIBORQUIA Orteg.

Small tree or large shrub with glandular-punctate odd-pinnate leaves with 10 to 23 pairs of small leaflets, and small white flowers in terminal racemes; corolla only slightly irregular; stamens 10, diadelphous; fruit a falcate pod 10 to 16 mm. long.

1. **Viborquia orthocarpa** (A. Gray) Cockerell, Bull. Amer. Mus. Nat. Hist. **24:** 97. 1908.

Eysenhardtia amorphoides orthocarpa A. Gray, Pl. Wright. **2:** 37. 1853.

Eysenhardtia orthocarpa S. Wats. Proc. Amer. Acad. **17:** 339. 1882.

TYPE LOCALITY: "Mountains at Guadalupe Pass, between San Bernardino, Sonora, and the copper mines."

RANGE: Dry hills and plains, Arizona and New Mexico to Mexico, in the Upper Sonoran Zone.

We have seen no further specimens from New Mexico, but the type locality is on the line between New Mexico and Sonora, perhaps in New Mexico.

13. AMORPHA L.

Shrubs or undershrubs with glandular-punctate odd-pinnate many-foliolate leaves; flowers in terminal, more or less elongated spikes; pods small, 1-seeded; calyx teeth 5, about equal; petal 1, the banner; stamens monadelphous.

KEY TO THE SPECIES.

Tall shrub, 2 meters high or more; pods 2-seeded................ 1. *A. californica.*
Undershrubs, more or less herbaceous, less than 1 meter high; pods 1-seeded.
  Plants green and glabrate.................................... 2. *A. microphylla.*
  Plants white-canescent....................................... 3. *A. canescens.*

1. **Amorpha californica** Nutt.; Torr. & Gray, Fl. N. Amer. **1:** 306. 1838.

FALSE INDIGO.

TYPE LOCALITY: Santa Barbara, California.

RANGE: Southern California to New Mexico; also in Mexico.

NEW MEXICO: Chiz; Albuquerque; Gila; Animas Mountains; Socorro; Kingston; Organ Mountains; Mesilla Valley. River valleys and along streams, Lower Sonoran to Transition Zone.

2. **Amorpha microphylla** Pursh, Fl. Amer. Sept. 466. 1814.

TYPE LOCALITY: "On the banks of the Missouri."

RANGE: Manitoba to Iowa, Nebraska, and New Mexico.

NEW MEXICO: White Mountains; Upper Canadian. Transition Zone.

3. **Amorpha canescens** Pursh, Fl. Amer. Sept. 467. 1814. SHOESTRINGS.

TYPE LOCALITY: "On the banks of the Missouri and Mississippi."

RANGE: Manitoba and Indiana to Louisiana, Texas, and New Mexico.

NEW MEXICO: Ute Park; Sierra Grande; Capitan Mountains; Las Vegas; Beulah. Dry plains and hills, in the Upper Sonoran Zone.

14. PARRYELLA Torr. & Gray.

Low much branched shrub with alternate glandular-dotted many-foliolate odd-pinnate leaves and terminal compound spikes of very small, dull yellowish-green flowers; calyx 5-toothed, with very short teeth; petals wanting; fruit a short, very glandular, 1-seeded pod.

1. **Parryella filifolia** Torr. & Gray, Proc. Amer. Acad. **7:** 397. 1867.

TYPE LOCALITY: Along the Rio Grande below Albuquerque, New Mexico. Type collected by Parry.

RANGE: New Mexico and Arizona.

NEW MEXICO: Zuni Reservation; Rio Puerco; Black Rock; Belen; south of Fruitland; Shiprock. Sand hills, in the Upper Sonoran Zone.

15. PAROSELA Cav.

Annuals or perennials with herbaceous or woody stems 60 cm. high or less; leaves mostly odd-pinnate, in one species palmately trifoliolate; flowers small, in crowded terminal spikes, bracteate; leaflets usually small, 1 to 20 pairs, mostly glandular-punctate; petals 5, 4 attached to the column of the monadelphous stamens, the banner free; stamens 9 or 10; pods 1 or 2-seeded, usually indehiscent, included in the persistent calyx.

KEY TO THE SPECIES.

Calyx deeply lobed, the lobes broadly lanceolate, foliaceous..... 1. *P. calycosa.*
Calyx not deeply lobed, the teeth setaceous or subulate or short-triangular.
  Calyx glabrous, at least outside, short-pubescent on the margins of the teeth or in the throat; calyx lobes short.
    Annual; bracts inconspicuous.......................... 7. *P. urceolata.*
    Perennials; bracts conspicuous.
      Stems prostrate, herbaceous.........................17. *P. glaberrima.*
      Stems erect, more or less woody...................... 2. *P. frutescens.*
  Calyx variously pubescent, the lobes mostly setaceous or subulate.
    Plants glabrous, except the inflorescence; annuals or perennials.
      Stems woody; low shrub.......................... 3. *P. formosa.*
      Stems herbaceous, woody at the base in one or two species.
        Perennials, from a thick woody base.
          Bracts sericeous, sparingly glandular; leaflets numerous.......................11. *P. grayi.*
          Bracts glandular, the margins hyaline; leaflets 5 or 6 pairs.
            Spikes lax; corolla yellow; stamens 9.. 4. *P. enneandra.*
            Spikes crowded; corolla purple; stamens 10.......................... 5. *P. pogonanthera.*
        Annuals; stems slender, mostly erect.
          Leaflets very numerous; spikes elongated; plants generally more than 30 cm. high.......................... 6. *P. dalea.*
          Leaflets 5 or 6 pairs; spikes short, almost capitate; plants generally less than 30 cm. high.

Bracts caducous; leaflets filiform; peduncles very slender; corolla purple .... 10. *P. filiformis.*
Bracts persistent; leaflets oblong, peduncles short and stout; corolla yellow or white.
Corolla yellow; spikes globose ..... 9. *P. brachystachys.*
Corolla white; spikes cylindric .... 8. *P. polygonoides.*
Plants more or less pubescent throughout; perennials.
Spikes lax; plants prostrate, densely pubescent; flowers purple.
Calyx tube villous, the lobes lanceolate .... 15. *P. lanata.*
Calyx tube glabrous, the lobes short-triangular. 16. *P. terminalis.*
Spikes densely flowered, mostly erect; plants at least ascending; pubescence various; flowers variously colored.
Flowers white, in strict terminal spikes; plants finely appressed-pubescent throughout.
Leaflets very small, about 3 mm. long, 15 to 20 pairs .... 12. *P. ordiae.*
Leaflets larger, 5 to 7 mm. long, 10 to 12 pairs .... 13. *P. albiflora.*
Flowers blue, yellow, or purple; pubescence various.
Spikes comparatively few-flowered, globose; flowers blue; calyx lobes short; plants canescent .... 14. *P. scoparia.*
Spikes at least oblong, usually considerably elongated, 1 cm. thick or more; flowers yellow or purple; calyx lobes setaceous; plants sericeous or villous.
Leaves glandular-punctate; spikes very thick and much elongated.
Stems sericeous, not glandular .... 18. *P. aurea.*
Stems tuberculate, black-glandular, minutely pubescent .... 19. *P. lachnostachys.*
Leaves not glandular-punctate; spikes smaller.
Leaves palmately trifoliolate ...... 20. *P. jamesii.*
Leaves pinnate.
Spikes sessile; leaflets acute ... 21. *P. wrightii.*
Spikes short, pedunculate; leaflets obtuse .... 22. *P. nana.*

1. **Parosela calycosa** (A. Gray) Heller, Cat. N. Amer. Pl. ed. 2. 5. 1900.
Dalea calycosa A. Gray, Pl. Wright. **2**: 40. 1853.
TYPE LOCALITY: "Hills near the Deserted Rancho, on the San Pedro, Sonora."
RANGE: Southwestern New Mexico, southern Arizona, and adjacent Mexico.
NEW MEXICO: Cactus Flat; Mangas Springs. Dry hillsides, in the Upper Sonoran Zone.

2. **Parosela frutescens** (A. Gray) Vail, Contr. U. S. Nat. Herb. **8**: 303. 1905.
Dalea frutescens A. Gray, Bost. Journ. Nat. Hist. **6**: 175. 1850.
TYPE LOCALITY: "Rocky hills and high plains, along the margins of thickets, on the Guadelupe, Sabinas, and Pedernales," Texas.

RANGE: Western Texas and southeastern New Mexico.

NEW MEXICO: White and Capitan mountains; Guadalupe Mountains. Open slopes, in the Transition Zone.

3. **Parosela formosa** (Torr.) Vail, Bull. Torrey Club **24:** 16. 1897.

Dalea formosa Torr. Ann. Lyc. N. Y. **2:** 177. 1828.

TYPE LOCALITY: On the Platte, Colorado.

RANGE: Colorado and Utah to Arizona and Texas.

NEW MEXICO: Albuquerque; Logan; west of Santa Fe; Bear Mountain; Florida Mountains; Hillsboro; mesa west of Organ Mountains; Cross L Ranch; Mangas Springs; Socorro Mountain; Jarilla Mountains; Sandia Mountains; Roswell. Dry plains and low hills, in the Lower and Upper Sonoran zones.

4. **Parosela enneandra** (Nutt.) Britton, Mem. Torrey Club **5:** 196. 1894.

Dalea enneandra Nutt. Fraser's Cat. no. 30. 1813.

Dalea laxiflora Pursh, Fl. Amer. Sept. 741. 1814.

TYPE LOCALITY: "Upper Louisiana."

RANGE: Texas to Colorado and the Upper Missouri.

NEW MEXICO: Tucumcari; Nara Visa. Dry plains, in the Upper Sonoran Zone.

5. **Parosela pogonanthera** (A. Gray) Vail, Trans. N. Y. Acad. **14:** 34. 1894.

Dalea pogonanthera A. Gray, Mem. Amer. Acad. n. ser. **4:** 31. 1849.

TYPE LOCALITY: Around Monterey, Mexico.

RANGE: Texas to southern Arizona, south into Mexico.

NEW MEXICO: Zuni; Hillsboro; Carrizalillo Mountains; Organ Mountains; near White Water; Mangas Springs. Plains and low hills, in the Lower and Upper Sonoran zones.

6. **Parosela dalea** (L.) Britton, Mem. Torrey Club **5:** 196. 1894.

Psoralea dalea L. Sp. Pl. 764. 1753.

Dalea alopecuroides Willd. Sp. Pl. **3:** 1336. 1803.

TYPE LOCALITY: "Habitat in America."

RANGE: Illinois to Minnesota and Nebraska, south to Texas and Mexico.

NEW MEXICO: Farmington; Raton; Pecos; Sandia Mountains; Santa Fe; Albuquerque; Socorro; White Mountains; Black Range; San Luis Mountains; Mesilla Valley. Moist fields, Lower Sonoran to Transition Zone.

A common weed in fields, beside ditches, and on cultivated lands, especially after grain has been harvested.

7. **Parosela urceolata** (Greene) Standley, Contr. U. S. Nat. Herb. **13:** 194. 1910.

Dalea urceolata Greene, Leaflets **1:** 199. 1906.

TYPE LOCALITY: Mogollon Mountains, New Mexico. Type collected by Metcalfe (no. 553).

RANGE: Mogollon Mountains of New Mexico, in the Transition Zone.

8. **Parosela polygonoides** (A. Gray) Heller, Cat. N. Amer. Pl. ed. 2. 6. 1900.

Dalea polygonoides A. Gray, Pl. Wright. **2:** 39. 1853.

TYPE LOCALITY: "Pebbly bed of mountain torrents, near the copper mines, New Mexico." Type collected by Wright (no. 991).

RANGE: Southern New Mexico.

NEW MEXICO: West Fork of the Gila; Hillsboro; White Mountains; Santa Rita. Open hillsides, in the Upper Sonoran and Transition zones.

9. **Parosela brachystachys** (A. Gray) Heller, Cat. N. Amer. Pl. ed. 2. 113. 1900.

Dalea brachystachys A. Gray, Pl. Wright. **2:** 39. 1853.

TYPE LOCALITY: "Valleys, in alluvial soil, between the San Pedro and the Sonoita, Sonora." Type collected by Wright (no. 990).

RANGE: New Mexico and Arizona and adjacent Mexico.

NEW MEXICO: Hillsboro; Cliff; Dog Spring; Organ Mountains; Las Vegas Hot Springs. Dry hills, in the Upper Sonoran Zone.

10. Parosela filiformis (A. Gray) Heller, Cat. N. Amer. Pl. ed. 2. 6. 1900.

Dalea filiformis A. Gray, Pl. Wright. **2**: 39. 1853.

TYPE LOCALITY: Hillsides near the Copper Mines, New Mexico. Type collected by Wright (no. 992).

RANGE: Southwestern New Mexico and southeastern Arizona.

NEW MEXICO: Santa Rita; Mogollon Mountains. Exposed hillsides, in the Upper Sonoran and Transition zones.

11. Parosela grayi Vail, Bull. Torrey Club **24**: 14. 1897.

Dalea laevigata A. Gray, Pl. Wright. **2**: 38. 1853, not Moc. & Sessé, 1832.

TYPE LOCALITY: "On the Chiricahui Mountains, and on the Barbocomori, Sonora."

RANGE: Southwestern New Mexico, southern Arizona, and adjacent Mexico.

NEW MEXICO: Mangas Springs (*Metcalfe* 684). Lower and Upper Sonoran zones.

12. Parosela ordiae (A. Gray) Heller, Cat. N. Amer. Pl. ed. 2. 6. 1900.

Dalea ordiae A. Gray, Proc. Amer. Acad. **17**: 200. 1882.

TYPE LOCALITY: Plains near Bowie and Rucker Valley, southern Arizona.

RANGE: Southwestern New Mexico and southern Arizona to Mexico.

NEW MEXICO: Near Fort Bayard; Bear Mountain; Dog Spring. Lower and Upper Sonoran zones.

13. Parosela albiflora (A. Gray) Vail, Trans. N. Y. Acad. **14**: 34. 1894.

Dalea albiflora A. Gray, Pl. Wright. **2**: 38. 1853.

TYPE LOCALITY: "Hill-sides on the San Pedro and Barbocomori, Sonora."

RANGE: New Mexico, southern Arizona, and northern Mexico.

NEW MEXICO: Mogollon Mountains; Mangas Springs; Fort Bayard; Santa Rita; Kingston; headwaters of the Pecos. Lower and Upper Sonoran zones.

14. Parosela scoparia (A. Gray) Heller, Cat. N. Amer. Pl. ed. 2. 7. 1900.

Dalea scoparia A. Gray, Mem. Amer. Acad. n. ser. **4**: 32. 1849.

Dalea scoparia subrosea Cockerell, Science n. ser. **7**: 625. 1898.

TYPE LOCALITY: Jornada del Muerto, New Mexico. Type collected by Wislizenus.

RANGE: Western Texas to southern Arizona and adjacent Mexico.

NEW MEXICO: Albuquerque; Socorro; Sabinal; Jornada del Muerto; Deming; Mesilla Valley. Sand hills, in the Lower Sonoran Zone.

A much branched canescent shrub 1 meter high, growing on sand dunes in the southern part of the State along the Rio Grande Valley. It could profitably be used as a sand binder in such localities.

The type of *Dalea scoparia subrosea* was collected near Mesilla.

15. Parosela lanata (Spreng.) Britton, Mem. Torrey Club **5**: 196. 1894.

Dalea lanata Spreng. Syst. Veg. **3**: 327. 1826.

Dalea lanuginosa Nutt.; Torr. & Gray, Fl. N. Amer. **1**: 307. 1838.

TYPE LOCALITY: "Ad fl. Arkansa Amer. bor."

RANGE: Kansas to Colorado, Texas, and New Mexico.

NEW MEXICO: Nara Visa; Roswell; south of Melrose. Prairies and plains, in the Lower and Upper Sonoran zones.

16. Parosela terminalis (Jones) Heller, Muhlenbergia **6**: 96. 1910.

Dalea terminalis Jones, Contr. West. Bot. **12**: 8. 1908.

TYPE LOCALITY: El Paso, Texas.

RANGE: Utah to northern Mexico.

NEW MEXICO: Albuquerque; Chamita; Socorro; Sabinal; Mesilla Valley. Mesas and dry fields, in the Lower and Upper Sonoran zones.

52576°—15——23

17. Parosela glaberrima (S. Wats.) Rose, Contr. U. S. Nat. Herb. **10:** 103. 1906.

Dalea glaberrima S. Wats. Proc. Amer. Acad. **22:** 470. 1887.

TYPE LOCALITY: "On sand hills 30 or 40 miles south of Paso del Norte," Chihuahua.

RANGE: Southern New Mexico to Chihuahua.

NEW MEXICO: Mesilla (*Wooton* 35). Sandy fields, in the Lower Sonoran Zone.

In the original description Doctor Watson says the plant is erect and branching, a statement in which he is probably wrong. *Dalea arenaria* Jones,[1] is doubtless very near this species, as the author suggests. Our plant has the prostrate, matlike habit of *D. arenaria* and grows on the sands, but is certainly perennial and the bracts are very different.

18. Parosela aurea (Nutt.) Britton, Mem. Torrey Club **5:** 196. 1894.

Dalea aurea Nutt. Gen. Pl. **2:** 101. 1818.

TYPE LOCALITY: "On gravelly hills, near White River, Missouri."

RANGE: From the Upper Missouri to the eastern borders of the Rocky Mountains, south to western Texas.

NEW MEXICO: Cross L Ranch; Las Vegas; Nara Visa; Portales; Redlands. Open hills and plains, in the Upper Sonoran and Transition zones.

19. Parosela lachnostachys (A. Gray) Heller, Cat. N. Amer. Pl. ed. 2. 6. 1900.

Dalea lachnostachys A. Gray, Pl. Wright. **1:** 46. 1852.

TYPE LOCALITY: Hills about 80 miles west of the Pecos, Texas.

RANGE: Western Texas and southwestern New Mexico to southern Arizona and adjacent Mexico.

NEW MEXICO: Dog Spring (*Mearns* 2387). Lower Sonoran Zone.

Characterized by the very densely flowered hispidulous thick spikes 5 to 8 cm. long and the peculiar glands. The stems are thickly beset with black tuberculate conical glands and the leaflets bear a single row of glands near the margin, with occasionally a few scattered over the back.

20. Parosela jamesii (Torr.) Vail, Bull. Torrey Club **24:** 16. 1897.

Psoralea jamesii Torr. Ann. Lyc. N. Y. **2:** 175. 1828.

Dalea jamesii Torr. & Gray, Fl. N. Amer. **1:** 308. 1838.

TYPE LOCALITY: "Sandy plains of the Canadian." (Oklahoma ?).

RANGE: Western Texas and New Mexico to Colorado.

NEW MEXICO: Pecos; Las Vegas; west of Santa Fe; Willard; Carrizozo; Hillsboro; Apache Teju; Silver City; Mangas Springs; Knowles; Queen; Torrance. Dry plains and hills, in the Upper Sonoran Zone.

21. Parosela wrightii (A. Gray) Vail, Bull. Torrey Club **24:** 16. 1897.

Dalea wrightii A. Gray, Pl. Wright. **1:** 49. 1852.

TYPE LOCALITY: "Dry hills 80 miles west of the Pecos, and on the mountains near El Paso," Texas.

RANGE: Western Texas to southern Arizona, and adjacent Mexico.

NEW MEXICO: Hillsboro; Tortugas Mountain; Bishops Cap. Dry hills, in the Lower Sonoran Zone.

22. Parosela nana (Torr.) Heller, Bot. Expl. Texas 49. 1895.

Dalea nana Torr.; A. Gray, Mem. Amer. Acad. n. ser. **4:** 31. 1849.

TYPE LOCALITY: "Sandy soil, Willow Bar, on the Cimarron," New Mexico?

RANGE: Kansas to Texas and New Mexico.

NEW MEXICO: Nara Visa; Organ Mountains; south of Roswell. Dry plains and hills, in the Lower and Upper Sonoran zones.

Parosela neomexicana (A. Gray) Heller,[2] notwithstanding its name, probably does not come into New Mexico, although it may occur in the extreme southwest corner along with other species of similar distribution.

---

[1] Contr. West. Bot. **12:** 8. 1908.

[2] *Dalea mollis neomexicana* A. Gray, Pl. Wright. **1:** 47. 1852.

16. PETALOSTEMUM Michx. PRAIRIE CLOVER.

Herbaceous glandular-punctate annuals or perennials with odd-pinnate leaves, the flowers in dense pedunculate terminal spikes or heads; calyx with 5 connivent teeth; petals with filiform claws, 4 of them nearly alike and adnate by their claws to the stamen tube, the banner free; stamens 5, monadelphous; pods membranous in the persistent calyx, indehiscent, 1-seeded.

KEY TO THE SPECIES.

Stems prostrate; flowers purplish............................... 1. *P. prostratum.*
Stems erect; flowers variously colored.
  Calyx glabrous; corolla white............................... 2. *P. oligophyllum.*
  Calyx pubescent; corolla variously colored.
    Annual.. 3. *P. exile.*
    Perennials.
      Corolla white or yellow............................ 4. *P. compactum.*
      Corolla purple.
        Bracts glabrous; leaflets mucronulate......... 5. *P. purpureum.*
        Bracts sericeous; leaflets obtuse............... 6. *P. tenuifolium.*

1. Petalostemum prostratum Woot. & Standl. Contr. U. S. Nat. Herb. **16:** 138. 1913.

TYPE LOCALITY: Near Albuquerque, New Mexico. Type collected by Winnie Harward (no. 17).

RANGE: Rio Grande Valley of central New Mexico.

NEW MEXICO: Near Albuquerque; near Belen. Lower Sonoran Zone.

A remarkable species, distinguished from all our others by its prostrate habit. In general appearance it resembles some species of Parosela.

2. Petalostemum oligophyllum (Torr.) Rydb. Mem. N. Y. Bot. Gard. **1:** 237. 1900.

Petalostemum gracile oligophyllum Torr. in Emory, Mil. Reconn. 139. 1848.

TYPE LOCALITY: "Valley of the del Norte," New Mexico. Type collected by Emory in 1847.

RANGE: British America to Iowa, Colorado, and Arizona.

NEW MEXICO: Throughout the State. Open slopes, in the Upper Sonoran and Transition zones.

3. Petalostemum exile A. Gray, Pl. Wright. **2:** 41. 1853.

TYPE LOCALITY: Hillsides, near Santa Cruz, Sonora.

RANGE: Southern New Mexico and Arizona and adjacent Mexico.

NEW MEXICO: Mogollon Mountains.

4. Petalostemum compactum (Spreng.) Swezey, Nebr. Pl. Doane Coll. 6. 1891.

Dalea compacta Spreng. Syst. Veg. **3:** 327. 1826.

Petalostemum macrostachyum Torr. Ann. Lyc. N. Y. **2:** 176. 1828.

TYPE LOCALITY: "Ad fl. Rio Roxo in ditione Arkansa Amer. bor."

RANGE: Wyoming and Nebraska to Colorado and New Mexico.

NEW MEXICO: Albuquerque; Rito de los Frijoles; Dona Ana County.

5. Petalostemum purpureum (Vent.) Rydb. Mem. N. Y. Bot. Gard. **1:** 238. **1900.**

Dalea purpurea Vent. Pl. Jard. Cels. *pl. 40.* 1800.

Petalostemum violaceum Michx. Fl. Bor. Amer. **2:** 50. 1803.

TYPE LOCALITY: "Illinois."

RANGE: British America to Missouri and New Mexico.

NEW MEXICO: Nara Visa. Pecos. Open hillsides and plains, in the Upper Sonoran Zone.

6. **Petalostemum tenuifolium** A. Gray, Proc. Amer. Acad. **11:** 73. 1876.

TYPE LOCALITY: "Arkansas, at the crossing of Red River."

RANGE: Kansas and Arkansas to New Mexico.

NEW MEXICO: Las Vegas; Raton Mountains; Causey; Algodones; Sandia Mountains; Buchanan; Sierra Grande. Plains and low hills, in the Upper Sonoran and Transition zones.

17. PETERIA A. Gray.

Low herbaceous perennial with smooth glaucous junciform stems and pinnate many-foliolate leaves; stipules small, spiny; leaflets very small, acute, deciduous from the rachis; flowers rather large, pale greenish, tinged with pink, widely scattered on long peduncles; pods linear, pendulous, 5 cm. long or more.

1. **Peteria scoparia** A. Gray, Pl. Wright. **1:** 50. 1852.

TYPE LOCALITY: Mountain valleys beyond the pass of the Limpio, Texas.

RANGE: Western Texas and adjacent Mexico.

NEW MEXICO: Telegraph Mountains; Tortugas Mountain. Dry hills, in the Lower and Upper Sonoran zones.

Coulter states [1] that the plant has a small edible tuberous rootstock and is known in western Texas as "camote de monte."

18. ROBINIA L. LOCUST.

Spiny shrubs or small trees with odd-pinnate leaves, rather large pink flowers in crowded axillary short-peduncled racemes, and flat pods 6 to 12 cm. long, with prominent sutures and numerous seeds; leaflets 1 to 2 cm. long, oblong-elliptic to oval.

KEY TO THE SPECIES.

Fruit and peduncles densely glandular-hispid.................. 1. *R. neomexicana.*
Fruit glabrous; peduncles not hispid, merely glandular-pubescent
    or puberulent, the glands few and small................ 2. *R. rusbyi.*

Robinia pseudacacia L., the black locust, a tree with white flowers, is often planted as a shade tree. It seems to do better than almost any other introduced shade tree in the drier portions of the State.

1. **Robinia neomexicana** A. Gray, Mem. Amer. Acad. n. ser. **5:** 314. 1854.

NEW MEXICAN LOCUST.

TYPE LOCALITY: Dry hills on the Mimbres, New Mexico. Type collected in May, 1851, by Thurber.

RANGE: Colorado to Arizona and western Texas.

NEW MEXICO: Raton; Sandia Mountains; Cross L Ranch; Magdalena Mountains, Mangas Springs; Burro Mountains; Black Range; Fort Bayard; Organ Mountains; White and Sacramento mountains. Transition Zone.

2. **Robinia rusbyi** Woot. & Standl. Contr. U. S. Nat. Herb. **16:** 140. 1913.

TYPE LOCALITY: On the Mogollon road 15 miles east of Mogollon, New Mexico. Type collected by Wooton, August 8, 1900.

RANGE: Mountains of southern New Mexico.

NEW MEXICO: Mogollon Mountains; Burro Mountains; Mescalero Reservation. Transition Zone.

19. ASTRAGALUS L.

Herbaceous perennials, rarely annuals; leaves odd-pinnate; flowers racemose, sometimes pseudocapitate, whitish, yellow, or purple; stipules either free, adnate to the base of the petiole or connate opposite the petioles forming a partial sheath; calyx

[1] Contr. U. S. Nat. Herb. **2:** 81. 1891.

campanulate or tubular, sometimes gibbous, the teeth mostly subulate or triangular; corolla longer than the calyx, the banner and wings usually exceeding the obtuse keel; banner usually with its side reflexed, occasionally spreading; stamens diadelphous; fruit a few to many-seeded pod, variously inflated, curved, flattened, or grooved, with one or both sutures more or less inflexed, becoming in some cases completely 2-celled.

Mature pods are usually necessary for the determination of the species.

In the consideration of this genus we have been unable to follow those who have separated it into several genera, because the lines of separation do not appear to us to be sufficiently distinct. Acting upon this judgment, we have considered all the species under one generic name. We fully appreciate the merits of Dr. P. A. Rydberg's treatment of the species here referred to Astragalus, in the Flora of Colorado,[1] to which in our own work we are much indebted. While we believe thoroughly in the principle of the segregation of species into small genera wherever possible, the small groups must, in our judgment, possess resemblances in more than one set of characters and their members should be readily recognized as close relatives. The characters of the fruit are especially unsuited for the purpose of generic distinction in this family. As a means of making Doctor Rydberg's valuable work easily usable in connection with our key, we insert his generic names in the key.

By strict rules of priority, if all the plants here referred to Astragalus are placed in one genus it should bear the name Phaca, that having precedence in Linnaeus's Species Plantarum over Astragalus. However, it seems to us that there are extreme cases where such a rule may be disregarded and we have preferred to leave the transference of the species to Phaca to some other writer, if it should ever be considered necessary.

KEY TO THE SPECIES.

Pods completely 2-celled.
  Pod fleshy. (GEOPRUMNON.)[2] .... 1. *A. crassicarpus.*
  Pods not fleshy.
    Pods inflated-membranaceous. (CYSTIUM) .... 2. *A. diphysus.*
    Pods not inflated.
      Pods linear-oblong, somewhat flattened laterally, membranous. (HAMOSA.)
        Stems very short and stout; plants appressed-sericeous. 3. *A. calycosus.*
        Stems slender, elongated, spreading; plants villous, merely canescent, or sometimes glabrate .... 4. *A. nuttallianus.*
      Pods ovoid, never flattened laterally, sometimes compressed dorso-ventrally, coriaceous. (ASTRAGALUS.)
        Mature pods glabrous.
          Flowers purple; stems very short, almost wanting. 5. *A. mollissimus.*
          Flowers yellow or greenish; stems stout, 30 to 60 cm. high.
            Leaves glabrous above, merely strigose beneath; flowers greenish yellow .... 6. *A. oreophilus.*
            Leaves densely villous, almost tomentose, on both surfaces; flowers bright yellow .... 7. *A. yaquianus.*

---

[1] Colo. Agr. Exp. Sta. Bull. **100:** 202–212. 1906.

[2] This and the following names inserted in the key are considered as representing separate genera by Doctor Rydberg, in the Flora of Colorado. By us they are regarded as sections.

Mature pods pubescent.
Plants sparingly appressed-pubescent with short hairs; pods small, not inflated, sparingly villous......8. *A. goniatus.*
Plants densely pubescent; pods much larger, more or less inflated, densely hairy.
Plants small; leaves appressed silky-villous throughout; pods inflated, sulcate along both sutures.
9. *A. matthewsii.*
Plants large and coarse; leaves densely tomentulose; pods not so much inflated, usually not sulcate.
Leaflets large, more than 15 mm. long, oval.
10. *A. bigelovii.*
Leaflets small, mostly 10 mm. long or less, broadly obovate or suborbicular.......12. *A. thompsonae.*

Pods not completely 2-celled.
Dorsal suture of pod more or less introverted in various ways (or often not so in *A. missouriensis*).
Lower (dorsal) suture strongly intruded, making the pod obcordate or inverted V-shaped in cross section; pods mostly membranous. (TIUM.)
Stipes very short (1 mm. long) or none.
Flowers very small, barely 5 mm. long; pods sessile, only slightly longer than the flowers, puberulent..........11. *A. vaccarum.*
Flowers more than 5 mm. long; pods short-stipitate and much larger.
Leaflets broadly obovate, mostly retuse or obcordate; pods not curved, the sulcus narrow...............13. *A. cobrensis.*
Leaflets oblong-lanceolate, acute; pods strongly curved; sulcus broad and shallow ....14. *A. humistratus.*
Stipes conspicuous, as long as the calyx or longer.
Flowers purple, never yellow; plants slender, weak.15. *A. alpinus.*
Flowers yellowish; plants erect, though sometimes slender.
Plants loosely villous.............................52. *A. drummondii.*
Plants appressed-pubescent.
Racemes elongated, lax.......................16. *A. rusbyi.*
Racemes short, flowers numerous and more or less crowded.
Flowers small, hardly 10 mm. long; plants slender.
17. *A. altus.*
Flowers larger, 15 mm. long or more; plants stout.
18. *A. scopulorum.*
Lower suture not intruded as in the previous section, but intruded either as an interior growth forming a partial partition or as the result of dorsoventral flattening; pods various.
Lower suture slightly intruded inside the pod, forming a partial septum; pods not flattened exteriorly.
Plants low, pubescent; pods rather thin (on slender peduncles, scattered) stipitate. (ATELOPHRAGMA.).........19. *A. brandegei.*
Plants tall, stout, almost glabrous (malodorous); pods thick and fleshy, becoming woody, not stipitate. (PHACOPSIS.)
Plants early glabrate, only the very young parts ever pubescent.
20. *A. praelongus.*
Plants sparingly hispidulous-strigose on the stems and the lower surface of the leaves...................21. *A. pattersonii.*

Lower suture intruded as the result of dorsoventral flattening; pods coriaceous or woody.
Corolla yellow; flowers small. (CNEMIDOPHACOS.)
22. *A. flaviflorus.*
Corolla at least partly purple; flowers large. (XYLOPHACOS.)
Pods densely long-hairy; only the keel of the corolla purple.
23. *A. newberryi.*
Pods short-hairy or glabrous; corolla entirely purple.
Flowers small, about 8 mm. long (subcapitate).
24. *A. accumbens.*
Flowers large, 15 mm. long or more.
Pods short and nearly straight (dorsal suture not introverted; mature pod sometimes laterally flattened).
25. *A. missouriensis.*
Pods longer, strongly curved upward.
Pods obtuse at the base, the dorsal suture strongly introverted........................26. *A. shortianus.*
Pods acute at both ends, the dorsal suture usually not much inflexed ....27. *A. amphioxys.*
Neither suture of the pod introverted (except sometimes in *A. subcinereus* and *A. sonorae*).
Pods strongly bisulcate ventrally, parallel to the suture.
Flowers purple or violet, at least on the keel; pods 10 to 20 mm. long.
28. *A. bisulcatus.*
Flowers whitish; pods less than 10 mm. long........29. *A. haydenianus.*
Pods not bisulcate.
Leaflets spinose-tipped. (KENTROPHYTA.).............53. *A. impensus.*
Leaflets not spinose-tipped.
Pods membranous or chartaceous, terete or flattened laterally, not inflated.
Pods long-stipitate; stipe several times the length of the calyx.
Flowers purple; leaflets linear, about 1 cm. long.
36. *A. coltoni.*
Flowers white; leaflets 2 cm. long or more.
37. *A. lonchocarpus.*
Pods sessile or short-stipitate; stipe seldom exceeding the calyx.
Pods more or less flattened laterally.
Pods sessile (linear-oblong, about 3 mm. long).
30. *A. diversifolius.*
Pods stipitate.
Stipe as long as or longer than the calyx; flowers whitish.
31. *A. tenellus.*
Stipe very short, barely 1 mm. long; flowers purple.
32. *A. wingatanus.*
Pods subterete.
Pods sessile...............................30. *A. diversifolius.*
Pods stipitate, the stipe sometimes very short.
Stipe as long as the calyx; plants erect. 33. *A. proximus.*
Stipe very short, almost obsolete; plants erect or decumbent.
Plants erect; pods 5 mm. in diameter at the upper end, tapering to the small stipe, 2 cm. long or more.
34. *A. fendleri.*

Plants decumbent to prostrate; pods 3 mm. in diameter, oblong, about 1 cm. long.........35. *A. flexuosus.*

Pods membranous to woody, more or less inflated, not terete nor flattened laterally.

Pods membranous, strongly inflated.

Plants with only a few leaves; pods mottled. 38. *A. ceramicus.*

Plants with many leaves.

Pods small, about 1 cm. in diameter, spherical; plants about 40 cm. high..............................39. *A. thurberi.*

Pods larger, more than 1 cm. in diameter, elliptic to lunate in vertical section; plants variable in size.

Annual; flowers reddish purple, small....40. *A. wootoni.*

Perennials; flowers various.

Pods almost lunate in cross section, the upper suture nearly straight or slightly recurved. 41. *A. allochrous.*

Pods elliptic, the upper suture curved about as much as the lower, slightly inflexed.....42. *A. subcinercus.*

Pods only slightly inflated, the walls thick, subcoriaceous to woody.

Pods subcoriaceous to chartaceous.

Plants acaulescent, small.

Flowers capitate, on conspicuous peduncles; pods about 5 mm. long............................43. *A. gilensis.*

Flowers (few) racemose, on very short peduncles; pods 20 mm. long or more..............44. *A. elatiocarpus.*

Plants with stems over 10 cm. long.

Plants prostrate, trailing. (Lower suture sometimes introverted.)

Leaflets pubescent on both surfaces.....45. *A. sonorae.*

Leaflets glabrous on the upper surface. 54. *A. hosackiae.*

Plants erect or spreading.

Pod short-stipitate, ellipsoidal, the upper suture curved downward..........................46. *A. greenei.*

Pods sessile, sublunate (acuminate), the upper suture straight or slightly recurved.......47. *A. gertrudis.*

Pods coriaceous to woody when mature, thick-walled or fleshy when young.

Leaflets ovate-lanceolate, acute, numerous; peduncles about 20 cm. long.................48. *A. neomexicanus.*

Leaflets oblong to obovate, obtuse, few; peduncles shorter, less than 20 cm. long.

Flowers small, 10 mm. long or less; pods of about the same length; stems usually short....49. *A. tephrodes.*

Flowers larger, 12 to 15 mm. long; pods 15 to 20 mm. long; stems frequently 20 cm. long.......50. *A. remulcus.*

No. 51, *Astragalus albulus*, is omitted from the key, because without mature fruit it is impossible to place it in any of the sections, all of which depend upon fruit characters. The habit of the plant, the yellow flowers, the rudimentary stipe, the thin pod, partially flattened when little more than an ovary, suggest that the plant may be related to *A. scopulorum*.

1. **Astragalus crassicarpus** Nutt. Fraser's Cat. no. 6. 1813.

Geoprumnon crassicarpum Rydb. Colo. Agr. Exp. Sta. Bull. **100:** 203. 1906.

TYPE LOCALITY: "Above the River Platte."

RANGE: Manitoba and Montana to Missouri, Texas, and New Mexico.

NEW MEXICO: Highest point of the Llano Estacado; near Horse Spring. Upper Sonoran Zone.

2. **Astragalus diphysus** A. Gray, Mem. Amer. Acad. n. ser. **4:** 34. 1849.

Cystium diphysum Rydb. Colo. Agr. Exp. Sta. Bull. **100:** 204. 1906.

TYPE LOCALITY: Plains, around Santa Fe, New Mexico. Type collected by Fendler (no. 146).

RANGE: Colorado and Utah to Arizona and New Mexico.

NEW MEXICO: Santa Fe; Albuquerque; Zuni; Salt Lake; Cerrillos; Lemitar. On the hills and higher plains, in the Upper Sonoran Zone.

3. **Astragalus calycosus** S. Wats. in King, Geol. Expl. 40th Par. **5:** 66. 1871.

TYPE LOCALITY: "In the West Humboldt, East Humboldt, and Clover mountains, Nevada."

RANGE: Nevada and Utah to Arizona and New Mexico.

NEW MEXICO: Aztec (*Baker* 409). Mountains and dry hills, in the Upper Sonoran Zone.

The specimen cited was named by Doctor Greene on the sheet as a new species and does not exactly agree with the other material of *A. calycosus*, the calyx teeth and the pubescence being noticeably different. The material we have seen is without fruit and we prefer to wait for more complete specimens before accepting it as a new species.

4. **Astragalus nuttallianus** DC. Prodr. **2:** 289. 1825.

Hamosa nuttalliana Rydb. Colo. Agr. Exp. Sta. Bull. **100:** 204. 1906.

TYPE LOCALITY: "In planitiebus Amer. Bor. ad Red-river."

RANGE: Mountains of Colorado and New Mexico to Arkansas and Texas.

NEW MEXICO: Farmington; Carrizo Mountains; Lemitar; Florida Mountains; Deming; Mangas Springs; Carrizalillo Mountains; Organ Mountains; Star Peak; mountains west of San Antonio; Roswell. Upper Sonoran Zone.

Our plant is much more pubescent than the eastern form, which is the typical one, being always at least grayish from the abundant appressed pubescence.

5. **Astragalus mollissimus** Torr. Ann. Lyc. N. Y. **2:** 178. 1828. LOCO WEED.

Astragalus simulans Cockerell, Torreya **2:** 154. 1902.

TYPE LOCALITY: On the Platte, Colorado.

RANGE: Wyoming and Nebraska to New Mexico and western Nebraska.

NEW MEXICO: Las Vegas; Santa Rosa; Gray; Roswell; White Mountains; Sierra Grande; Nara Visa. Dry hills and plains, in the Upper Sonoran and Transition zones.

This is one of the commonest and best known loco weeds. As we have seen it growing, it is nowhere very common in New Mexico and so does relatively little damage to stock interests. These remarks apply only to this species, not to all loco weeds. The type of *A. simulans* was collected near Las Vegas by Cockerell.

6. **Astragalus oreophilus** Rydb. Bull. Torrey Club **31:** 561. 1904.

TYPE LOCALITY: Pagosa Springs, Colorado.

RANGE: Colorado and northern New Mexico.

NEW MEXICO: Taos; Pecos; Santa Fe; Tierra Amarilla; Chama. Damp meadows, in the Upper Sonoran and Transition zones.

7. **Astragalus yaquianus** S. Wats. Proc. Amer. Acad. **23:** 270. 1888.

TYPE LOCALITY: Moist banks and gravelly bars of the upper Yaqui River at Guerrero, Chihuahua.

RANGE: Western Texas and southern New Mexico to Mexico.

New Mexico: White Mountains. Pine woods, in the Transition Zone.

Mr. M. E. Jones is of the opinion that this is the same species that Doctor Watson described as *A. giganteus* from the Davis Mountains of western Texas. He is probably correct, but the material of *A. giganteus* which we have seen is not of such a character as to make it possible to decide the question, and the original description is not complete. In case Mr. Jones is correct, the name of the species is *A. giganteus* S. Wats. with *A. yaquianus* S. Wats. and *A. texanus* Sheld.[1] as synonyms.

8. **Astragalus goniatus** Nutt.; Torr. & Gray, Fl. N. Amer. **1**: 330. 1838.

Type locality: "Rocky Mountains, near the sources of the Platte," Colorado.

Range: Saskatchewan and Washington to California and New Mexico.

New Mexico: Ensenada; Las Vegas; Sierra Grande; Chama; Tierra Amarilla. In meadows and river valleys, in the Upper Sonoran and Transition zones.

9. **Astragalus matthewsii** S. Wats. Proc. Amer. Acad. **18**: 192. 1883.

Type locality: Fort Wingate, New Mexico. Type collected by Matthews.

Range: Northwestern New Mexico, southwestern Colorado, and probably adjacent Arizona and Utah.

New Mexico: Fort Wingate; Aztec. Plains and low hills, in the Upper Sonoran Zone.

This is probably too close to *A. thompsonae* S. Wats., and field study will be necessary to determine the status of the two.

10. **Astragalus bigelovii** A. Gray, Pl. Wright. **2**: 42. 1853.

Type locality: Organ Mountains, New Mexico. Type collected by Wright (no. 1358).

Range: Southern New Mexico, western Texas, and northern Chihuahua.

New Mexico: Datil; Magdalena Mountains; mountains west of San Antonio; Lava; Kingston; Burro Mountains; Tres Hermanas; Organ Mountains; Dog Spring; Roswell. Low mountains and foothills, in the Upper Sonoran Zone.

This is one of the common loco weeds of the southern part of the State, occasionally causing considerable loss to stockmen.

Astragalus mogollonicus Greene[2] we have not seen, but Doctor Greene assures us that he is well acquainted with *A. bigelovii*, its nearest relative, and that the two are distinct. It should be found when the region from which it comes ("bleak grassy summits of the Mogollon Mountains") has been more thoroughly explored. Doctor Greene originally published the following comparison: "As compared with its nearest ally, *A. bigelovii*, the plant is a dwarf, being barely a span high. Its still smaller pods are much more densely woolly, and nearly straight at maturity, in which latter character, however, the species is at variance with the rest of the Mollissimi."

11. **Astragalus vaccarum** A. Gray, Pl. Wright. **2**: 43. 1853.

Type locality: Ojo de Vaca, west of the Copper Mines, New Mexico. Type collected by Wright (no. 1002).

Range: Known only from the type locality.

12. **Astragalus thompsonae** S. Wats. Proc. Amer. Acad. **10**: 345. 1875.

Astragalus bigelovii thompsonae Jones, Contr. West. Bot. **8**: 23. 1898.

Type locality: Southern Utah.

Range: Utah to Arizona and northwestern New Mexico.

New Mexico: Aztec (*Baker* 405). Dry hills, in the Upper Sonoran Zone.

13. **Astragalus cobrensis** A. Gray, Pl. Wright. **2**: 43. 1853.

Type locality: Santa Rita, New Mexico. Type collected by Bigelow.

Range: Southwestern New Mexico.

New Mexico: Santa Rita; Burro Mountains; Kingston. In the drier mountains, Upper Sonoran and Transition zones.

---

[1] Minn. Bot. Stud. **9**: 65. 1894.

[2] Bull. Torrey Club **8**: 97. 1881.

14. Astragalus humistratus A. Gray, Pl. Wright. **2:** 43. 1853.

Tium humistratum Rydb. Colo. Agr. Exp. Sta. Bull. **100:** 205. 1906.

TYPE LOCALITY: "Pebbly bed of a stream, and on hills under pine trees, near the Copper Mines," New Mexico. Type collected by Wright (no. 1003).

RANGE: Mountains of Colorado and Utah to New Mexico and Arizona.

NEW MEXICO: Dulce; Pecos; Santa Rita; White Mountains; Black Range. Transition Zone.

This species is fairly common in the southern part of the State and in Arizona. It is associated with a nearly related species in New Mexico and with two close relatives in Arizona. As here treated, *A. humistratus* is the form having deeply sulcate, somewhat recurved pods, thinly villous to glabrate leaflets, and not very conspicuous stipules. *Astragalus sonorae* is the close relative common in southern New Mexico. It has herbage persistently white-silky throughout, with conspicuous white, connate stipules, and a somewhat shorter pod than *A. humistratus*, described as not sulcate, but not infrequently dorsally sulcate and recurved, though generally not so pronouncedly so as that of *A. humistratus*. On account of the pod characters originally given, these two species are usually widely separated in the lists of species, but they are very closely related. The other nearly related species is *A. hosackiae* Greene.

15. Astragalus alpinus L. Sp. Pl. 760. 1753.

Tium alpinum Rydb. Colo. Agr. Exp. Sta. Bull. **100:** 205. 1906.

TYPE LOCALITY: "Habitat in Alpibus Lapponicis, Helveticis."

RANGE: British America to Vermont and northern New Mexico; also in Europe and Asia.

NEW MEXICO: Winsors Ranch; Rio Pueblo; Chama. Damp meadows and open woodlands, in the Transition and Canadian zones.

16. Astragalus rusbyi Greene, Bull. Calif. Acad. **1:** 8. 1884.

TYPE LOCALITY: "On Mt. Humphreys, in the northern part of Arizona."

RANGE: Mountains of western New Mexico and Arizona.

NEW MEXICO: Mountains west of Grant; Magdalena Mountains; Socorro; Craters; Mogollon Mountains. Transition Zone.

17. Astragalus altus Woot. & Standl. Contr. U. S. Nat. Herb. **16:** 136. 1913.

TYPE LOCALITY: Toboggan, in the Sacramento Mountains, Otero County, New Mexico. Type collected by Wooton, July 31, 1899.

RANGE: Sacramento Mountains of New Mexico.

Similar to *Astragalus rusbyi*, but differing in its shorter and broader leaflets and crowded flowers on much shorter peduncles, and in the longer stipes and much less inflexed sutures of the pods.

18. Astragalus scopulorum Porter in Port. & Coult. Syn. Fl. Colo. 24. 1874.

Tium scopulorum Rydb. Colo. Agr. Exp. Sta. Bull. **100:** 205. 1906.

TYPE LOCALITY: South Park, Colorado.

RANGE: Mountains of Colorado and northern New Mexico.

NEW MEXICO: Glorieta; Placitas. Transition Zone.

19. Astragalus brandegei Porter in Port. & Coult. Syn. Fl. Colo. 24. 1874.

Atelophragma brandegei Rydb. Colo. Agr. Exp. Sta. Bull. **100:** 205. 1906.

TYPE LOCALITY: Banks of the Arkansas, near Canyon City, Colorado.

RANGE: Mountains of Colorado and northern New Mexico.

NEW MEXICO: Santa Fe; near Cliff. Transition Zone.

20. Astragalus praelongus Sheld. Minn. Bot. Stud. **9:** 23. 1894.

Astragalus procerus A. Gray, Proc. Amer. Acad. **13:** 369. 1878, not Boiss. & Haussk. 1867.

Astragalus rothrockii Sheld. Minn. Bot. Stud. **9:** 174. 1894.

Astragalus pattersoni procerus Jones, Proc. Calif. Acad. II. **5:** 636. 1895.

Phacopsis praelongus Rydb. Colo. Agr. Exp. Sta. Bull. **100:** 206. 1906.

Type locality: "Near St. Thomas, S. E. Nevada, at the confluence of the Muddy River with the Virgen."

Range: Colorado and Utah to Nevada and New Mexico.

New Mexico: Fort Wingate; Carrizo Mountains; Acoma; hills near Santa Fe; Rito Quemado; San Augustine Plains; Lemitar; Roswell. Upper Sonoran Zone.

The three following specimens are doubtfully placed here; they are not sufficiently distinct to be separated by name. The pods are slightly smaller as are the leaves, the plants probably being starvelings: Albuquerque, June, 1881, *Vasey;* near Carrizozo, July 22, 1901, *Wooton;* near Camp City, April, 1910, *Wooton.*

This and the following species are coarse ill-smelling plants of the open plains and valleys at middle and lower elevations. The separation of the species rests more on geographical distribution than on characters, though the plant which lives in the drier and hotter area is glabrous. Mr. Jones may be right when he recognizes *A. praelongus* as merely a subspecies of *A. pattersoni.* The two are said to be poisonous, but we have never seen any evidence of their being eaten by stock.

21. **Astragalus pattersoni** A. Gray; T. S. Brandeg. Bull. U. S. Geol. Geogr. Surv. Terr. **2:** 235. 1876.

Phacopsis pattersoni Rydb. Colo. Agr. Exp. Sta. Bull. **100:** 206. 1906.

Type locality: Foothills of Gore Mountains, Colorado.

Range: Utah and Colorado to New Mexico.

New Mexico: Cross L Ranch; Shiprock. Plains and hills, in the Upper Sonoran Zone.

22. **Astragalus flaviflorus** (Kuntze) Sheld. Minn. Bot. Stud. **9:** 158. 1895.

Astragalus flavus Nutt.; Torr. & Gray, Fl. N. Amer. **1:** 335. 1838, not *Phaca flava* Hook. & Arn. 1833.

Tragacantha flaviflora Kuntze, Rev. Gen. Pl. **2:** 941. 1891.

Cnemidophacos flaviflorus Rydb. Colo. Agr. Exp. Sta. Bull. **100:** 297. 1906.

Type locality: "Hills of the central chain of the Rocky Mountains, toward the Oregon."

Range: Mountains of New Mexico, northward to Wyoming.

New Mexico: Western San Juan and McKinley counties. Transition Zone.

23. **Astragalus newberryi** A. Gray, Proc. Amer. Acad. **12:** 55. 1877.

Xylophacos newberryi Rydb. Colo. Agr. Exp. Sta. Bull. **100:** 207. 1906.

Type locality: "On the frontiers of Utah and Arizona."

Range: Southwestern Colorado and northwestern New Mexico to Arizona and Utah.

New Mexico: Aztec (*Baker* 420). Upper Sonoran Zone.

24. **Astragalus accumbens** Sheld. Minn. Bot. Stud. **9:** 157. 1894.

Astragalus procumbens S. Wats. Proc. Amer. Acad. **20:** 361. 1885, not Hook. & Arn. 1833.

Type locality: Near Fort Wingate, New Mexico. Type collected by Matthews.

Range: Northwestern New Mexico and adjacent Arizona.

New Mexico: Western McKinley County. Upper Sonoran Zone.

25. **Astragalus missouriensis** Nutt. Gen. Pl. **2:** 99. 1818.

Xylophacos missouriensis Rydb. Colo. Agr. Exp. Sta. Bull. **100:** 206. 1906.

Type locality: "On the hills throughout Upper Louisiana."

Range: From the Upper Missouri through the Rocky Mountains to New Mexico and Texas.

New Mexico: Pecos; Carrizo Mountains; Aztec; Stanley; mountains west of San Antonio; Raton; Sierra Grande; Round Mountain above Tularosa. Open hills and plains, in the Upper Sonoran Zone.

26. Astragalus shortianus Nutt.; Torr. & Gray, Fl. N. Amer. **1:** 331. 1838.
Astragalus cyaneus A. Gray, Mem. Amer. Acad. n. ser. **4:** 34. 1849.
Astragalus shortianus cyaneus Jones, Contr. West. Bot. **8:** 5. 1898.
Xylophacos shortianus Rydb. Colo. Agr. Exp. Sta. Bull. **100:** 297. 1906.
TYPE LOCALITY: "Rocky Mountains, toward the plains of the Oregon."
RANGE: Northern New Mexico to Nebraska and Wyoming.
NEW MEXICO: Santa Fe; Shiprock; Albuquerque. Plains and hills, in the Upper Sonoran Zone.
The type of *A. cyaneus* was collected near Santa Fe by Fendler (no. 148).

27. Astragalus amphioxys A. Gray, Proc. Amer. Acad. **13:** 366. 1878.
Xylophacos amphioxys Rydb. Colo. Agr. Exp. Sta. Bull. **100:** 297. 1906.
TYPE LOCALITY: "Southern Utah and New Mexico and Northern Arizona."
RANGE: Colorado to Arizona and Texas.
NEW MEXICO: Albuquerque; Carrizo Mountains. Dry hills and plains, in the Upper Sonoran Zone.

28. Astragalus bisulcatus (Hook.) A. Gray, U. S. Rep. Expl. Miss. Pacif. **12:** 42. 1860.
Phaca bisulcata Hook. Fl. Bor. Amer. **1:** 145. 1833.
Diholcos bisulcatus Rydb. Colo. Agr. Exp. Sta. Bull. **100:** 297. 1906.
TYPE LOCALITY: "Plains of the Saskatchewan."
RANGE: Saskatchewan and Nebraska to New Mexico.
NEW MEXICO: Springer; Raton; Nutritas Creek. Plains and low hills, in the Upper Sonoran and Transition zones.

29. Astragalus haydenianus A. Gray, Proc. Amer. Acad. **12:** 56. 1876.
Diholcos haydenianus Rydb. Colo. Agr. Exp. Sta. Bull. **100:** 207. 1906.
TYPE LOCALITY: Southwestern Colorado.
RANGE: Wyoming to New Mexico.
NEW MEXICO: Nutria; Ramah; Dulce; Chama. Hillsides, in the Transition Zone.

30. Astragalus diversifolius A. Gray, Proc. Amer. Acad. **6:** 230. 1866.
Homalobus orthocarpus Nutt.; Torr. & Gray, Fl. N. Amer. **1:** 351. 1838, not *Astragalus orthocarpus* Boiss. 1849.
Astragalus junciformis A. Nels. Bull. Torrey Club **26:** 9. 1900.
Homalobus junciformis Rydb. Colo. Agr. Exp. Sta. Bull. **100:** 210. 1906.
TYPE LOCALITY: "Sandy plains of the Colorado of the West, near the sources of the Platte."
RANGE: Idaho to Utah and New Mexico.
NEW MEXICO: Gallup (*Herrick* 812). Plains, in the Upper Sonoran Zone.
Characterized by the peculiar rushlike, apparently leafless stems and slender straight pods.

31. Astragalus tenellus Pursh, Fl. Amer. Sept. 473. 1814.
Astragalus multiflorus A. Gray, Proc. Amer. Acad. **6:** 226. 1864.
Homalobus tenellus Rydb. Colo. Agr. Exp. Sta. Bull. **100:** 209. 1906.
TYPE LOCALITY: "On the banks of the Missouri."
RANGE: Mountains of New Mexico, northward throughout the Rocky Mountains.
NEW MEXICO: Santa Fe; Ponchuelo Creek. Transition Zone.

32. Astragalus wingatanus S. Wats. Proc. Amer. Acad. **18:** 192. 1883.
Homalobus wingatanus Rydb. Colo. Agr. Exp. Sta. Bull. **100:** 209. 1906.
TYPE LOCALITY: Fort Wingate, New Mexico. Type collected by Matthews.
RANGE: Western New Mexico and eastern Arizona.
NEW MEXICO: Fort Wingate; Carrizo Mountains. Hillsides, in the Transition Zone.

33. Astragalus proximus (Rydb.) Woot. & Standl.

Homalobus proximus Rydb. Bull. Torrey Club **32:** 667. 1905.

Type locality: Arboles, Colorado.

Range: Southwestern Colorado and northwestern New Mexico.

New Mexico: Aztec (*Baker* 427). Dry hills, in the Upper Sonoran Zone.

34. Astragalus fendleri A. Gray, Pl. Wright. **2:** 44. 1853.

Phaca fendleri A. Gray, Mem. Amer. Acad. n. ser. **4:** 36. 1849.

Homalobus fendleri Rydb. Colo. Agr. Exp. Sta. Bull. **100:** 210. 1906.

Type locality: Mountains between Santa Fe and Pecos, New Mexico. Type collected by Fendler (no. 157).

Range: Southern Colorado and northern New Mexico.

New Mexico: Santa Fe; Raton. Sandy plains and low hills, in the Upper Sonoran Zone.

There has been much uncertainty about this species, largely because Doctor Gray confused it originally with what he afterwards described as *A. greenei*. The true *Astragalus fendleri* is most closely related to *A. flexuosus* and *A. hallii*. Its pods are larger at the distal end and taper at the base, being almost intermediate between the small terete pod of *A. flexuosus* and the considerably larger terete pod of *A. hallii*, but not stipitate nor inflated as in *A. greenei*. The plant was originally described as being "a foot high."

35. Astragalus flexuosus Dougl.; Hook. Fl. Bor. Amer. **1:** 141. 1833.

Phaca flexuosa Hook. Fl. Bor. Amer. **1:** 141. 1833.

Homalobus flexuosus Rydb. Colo. Agr. Exp. Sta. Bull. **100:** 210. 1906.

Type locality: "Abundant on elevated and dry fertile soils of the Red River and Assinaboin, lat. 50°."

Range: Alberta and Saskatchewan to Kansas and New Mexico.

New Mexico: Sierra Grande; El Rito; Glorieta; White Mountains; Capitan Mountains; Sacramento Mountains. Plains, in the Upper Sonoran and Transition zones.

36. Astragalus coltoni Jones, Zoe **2:** 237. 1891.

Type locality: Canyons of the Coal Range at Castle Gate, Utah.

Range: Utah and northwestern New Mexico.

New Mexico: Carrizo Mountains (*Matthews*).

Our specimens and the material collected later at the type locality by Mr. Jones have as many as 5 pairs of leaflets.

37. Astragalus lonchocarpus Torr. U. S. Rep. Expl. Miss. Pacif. **4:** 80. 1857.

Phaca macrocarpa DC. err. det. A. Gray, Mem. Amer. Acad. n. ser. **4:** 36. 1849.

Homalobus macrocarpus Rydb. Colo. Agr. Exp. Sta. Bull. **100:** 210. 1906.

Type locality: Rocky declivities, near Santa Fe, New Mexico. Type collected by Fendler (no. 160).

Range: Utah and Colorado to northern New Mexico.

New Mexico: Fort Wingate; Santa Fe; Pecos; Llano Estacado; Chama. Open hillsides, in the Upper Sonoran and Transition zones.

38. Astragalus ceramicus Sheld. Minn. Bot. Stud. **9:** 19. 1894.

Phaca picta A. Gray, Mem. Amer. Acad. n. ser. **4:** 37. 1849.

Astragalus pictus A. Gray, Proc. Amer. Acad. **6:** 214. 1866, not Steud. 1840.

Astragalus pictus foliosus A. Gray, loc. cit.

Type locality: Loose, sandy soil on the banks of the Rio Grande del Norte, New Mexico. Type collected by Fendler (no. 161).

Range: Colorado and Utah to New Mexico and Arizona.

New Mexico: Aztec; hills west of Santa Fe. Upper Sonoran Zone.

The number of leaflets and the degree of prolongation of the rachis vary considerably, even on the same plant. It is doubtful if any of the New Mexican material belongs to the typical *Phaca picta* as originally described by Gray, and some of the specimens from the western part of the State approach *Phaca longifolia* (Pursh) Nutt.

39. Astragalus thurberi A. Gray, Mem. Amer. Acad. n. ser. **5**: 312. 1854.

TYPE LOCALITY: "Near Fronteras, &c, Sonora; on dry plains."

RANGE: Southern New Mexico and Arizona and adjacent Mexico.

NEW MEXICO: Below Silver City; Dog Spring. Upper Sonoran Zone.

40. Astragalus wootoni Sheld. Minn. Bot. Stud. **9**: 138. 1894.

Astragalus triflorus A. Gray, Pl. Wright. **2**: 45. 1853, at least in part, not *Phaca triflora* DC.

Astragalus playanus Jones, Contr. West. Bot. **8**: 6. 1898.

TYPE LOCALITY: Near Las Cruces, New Mexico. Type collected by Wooton.

RANGE: Western Texas to southern Arizona and adjacent Mexico.

NEW MEXICO: Mountains west of San Antonio; Cactus Flat; Paraje; Mesilla Valley; Carrizalillo Mountains. Valleys and low flats, in the Lower Sonoran Zone.

The original description of this plant is incorrect in calling it a perennial, for it is an annual. It is similar in general appearance (in pressed specimens) to *A. allochrous*, which occurs in the same region, but the latter is a short-lived perennial and a larger plant with larger bluish purple flowers and somewhat larger pods. The flowers of *A. wootoni* are reddish purple and generally small. It is a member of the rather scanty spring flora of the region and is usually gone by the middle of the summer. It would seem to be closely related to *A. geyeri*, a plant with a more northerly distribution and with paler flowers.

41. Astragalus allochrous A. Gray, Proc. Amer. Acad. **13**: 366. 1878.

Astragalus triflorus A. Gray, Pl. Wright. **2**: 45. 1853, in part.

TYPE LOCALITY: Near Wickenburg, Arizona.

RANGE: Utah and Colorado to Arizona, western Texas, and adjacent Mexico.

NEW MEXICO: Horse Spring; Zuni Reservation; Cactus Flat; Kingston; Mangas Springs; plains south of the White Sands; foothills of Organ Mountains; Rio Gila near Redrock; east of Hachita; Albuquerque. Dry hills, in the Upper Sonoran Zone.

This is evidently the plant Doctor Rydberg cites as *Phaca candolleana* H. B. K. in the Flora of Colorado,[1] and that may be the proper name for the species. We are inclined to doubt such an extended range, however, and retain the later name for the plant of the United States.

42. Astragalus subcinereus A. Gray, Proc. Amer. Acad. **13**: 366. 1878.

TYPE LOCALITY: Mokiak Pass, northwestern Arizona.

RANGE: Arizona and Utah to western New Mexico.

NEW MEXICO: Northwestern corner of the State.

The fruit is strikingly like that figured for *Phaca candolleana* by the authors of that species. We are not acquainted with it in the field.

43. Astragalus gilensis Greene, Bull. Torrey Club **8**: 97. 1881.

TYPE LOCALITY: On a high summit at the mouth of the canyon of the Gila River, New Mexico. Type collected by Greene.

RANGE: Mountains of western New Mexico.

NEW MEXICO: Mogollon Mountains; Hillsboro. Transition Zone.

Mr. Jones's statements[2] that the pods are "thin-chartaceous" and "flattened laterally" do not agree with Doctor Greene's original description nor with the New Mexican specimens.

---

[1] Colo. Agr. Exp. Sta. Bull. **100**: 211. 1906.

[2] Zoe **4**: 27. 1893.

44. Astragalus elatiocarpus Sheld. Minn. Bot. Stud. **9**: 20. 1894.

Astragalus lotiflorus brachypus A. Gray, Proc. Amer. Acad. **6**: 209. 1866, not *A. brachypus* Schrenk. 1841.

Astragalus ammolotus Greene, Erythea **3**: 76. 1895.

Phaca elatiocarpa Rydb. Colo. Agr. Exp. Sta. Bull. **100**: 211. 1906.

TYPE LOCALITY: Silver Lake, Ottertail County, Minnesota.

RANGE: Minnesota and Saskatchewan to Missouri and Texas, in the Rocky Mountains south to New Mexico.

NEW MEXICO: Near San Juan (*Heller* 3768). Plains, in the Upper Sonoran Zone.

45. Astragalus sonorae A. Gray, Pl. Wright. **2**: 44. 1853.

TYPE LOCALITY: Mountain valleys, between the San Pedro and the Sonoita, Sonora.

RANGE: Southern New Mexico and Arizona and northern Mexico.

NEW MEXICO: Mangas Springs; west of Patterson; near Grant; Organ Mountains. Plains and low hills, in the Upper Sonoran Zone.

This species is very near *A. humistratus* A. Gray, possibly too near to be kept distinct. As originally described, the pods are not compressed and neither suture is introverted, but the plants not infrequently have pods almost identical with those of *A. humistratus*, while the other characters (pubescence, stipules, etc.) are those of *A. sonorae*.

46. Astragalus greenei A. Gray, Proc. Amer. Acad. **16**: 105. 1880.

Astragalus fendleri A. Gray, Pl. Wright. **2**: 44. 1853, not *Phaca fendleri* A. Gray, 1849.

Astragalus fàllax S. Wats. Proc. Amer. Acad. **20**: 362. 1885, not Fisher, 1853.

Astragalus famelicus Sheld. Minn. Bot. Stud. **9**: 23. 1894.

Astragalus gracilentus greenei Jones, Contr. West. Bot. **8**: 14. 1898.

Astragalus gracilentus fallax Jones, loc. cit.

TYPE LOCALITY: Foothills of the Mogollon Mountains, New Mexico. Type collected by E. L. Greene, April 20, 1880.

RANGE: Mountains of western New Mexico, and in Arizona.

NEW MEXICO: Santa Rita; Mimbres; Mogollon Mountains. Transition Zone.

The synonymy of this species is considerably involved. Doctor Gray named *Phaca fendleri* from material collected at Santa Fe by Fendler. The species, he thought, might be "too near *P. flexuosa*." Later he gave additional notes on the species, based upon Wright's specimens from near the Copper Mines, and transferred it to the genus Astragalus. Unfortunately the Wright plants were not the same species as those of Fendler. Years afterwards Doctor Gray received material from the region of Santa Rita through Doctor Greene, which he named *Astragalus greenei*, without recognizing it as the species Wright had collected. In 1885 Doctor Watson discovered the mistake made originally by Doctor Gray and named the Wright plant *Astragalus fallax*, apparently without discovering its identity with *A. greenei*. In 1894 Mr. E. P. Sheldon found an older *Astragalus fallax* and renamed the species *A. famelicus*. Mr. Jones has recognized the similarity of *A. greenei* and *A. fallax* and reduced them to subspecies of *A. gracilentus* A. Gray, a plant named from Santa Fe which is, in all probability, *A. flexuosus*, although we have no means of being absolutely sure at this time. At any rate that plant is described as having sessile, oblong pods 5 mm. in diameter, while the plants from the Santa Rita region have short-stipitate pods about 10 mm. in diameter. Assuming that they are not the same as the Santa Fe plant, the Santa Rita plants will take the name *Astragalus greenei*.

47. Astragalus gertrudis Greene, Leaflets **2**: 43. 1910.

TYPE LOCALITY: Taos County, New Mexico. Type collected by Heller (no. 3598).

RANGE: Hills west and northwest of Santa Fe.

NEW MEXICO: Between Barranca and Embudo; hills west of Santa Fe. Upper Sonoran Zone.

48. Astragalus neomexicanus Woot. & Standl. Contr. U. S. Nat. Herb. **16**: 136. 1913.

TYPE LOCALITY: James Canyon in the Sacramento Mountains near Cloudcroft, New Mexico. Type collected by Wooton, July 23, 1899.

RANGE: Known only from the type locality, in the Transition Zone.

49. Astragalus tephrodes A. Gray, Pl. Wright. **2**: 45. 1853.

TYPE LOCALITY: Plains at the base of the Organ Mountains, New Mexico. Type collected by Wright.

RANGE: Southern New Mexico and probably in northern Mexico.

NEW MEXICO: Hillsboro; Florida Mountains; mountains west of San Antonio; Organ Mountains; Tres Hermanas. Low mountains, in the Lower and Upper Sonoran zones.

50. Astragalus remulcus Jones, Proc. Calif. Acad. II. **5**: 658. 1895.

TYPE LOCALITY: Banghartes Ranch, Arizona.

RANGE: Mountains of central Arizona and New Mexico.

NEW MEXICO: Kingman; Socorro; mesa south of Atarque de Garcia; Mimbres. Transition Zone.

The species does not seem to be sufficiently distinct (judging merely from dried material and descriptions) from *A. pephragmenus* M. E. Jones, but we hesitate to change the name without a field knowledge of the two plants. In all the New Mexican material the pods are not glabrous but puberulent.

51. Astragalus albulus Woot. & Standl. Contr. U. S. Nat. Herb. **16**: 136. 1913.

TYPE LOCALITY: In a canyon on the road to Zuni some distance south of Gallup, New Mexico. Type collected by Wooton (no. 2649).

RANGE: Northwestern New Mexico.

NEW MEXICO: Canyon south of Gallup; Shiprock. Dry hills and plains, in the Upper Sonoran Zone.

52. Astragalus drummondii Dougl.; Hook. Fl. Bor. Amer. **1**: 153. *pl. 57*. 1833.

Tium drummondii Rydb. Bull. Torrey Club **32**: 659. 1906.

TYPE LOCALITY: "Eagle and Red-Deer Hills of the Saskatchewan."

RANGE: Saskatchewan and Alberta to Colorado and northern New Mexico.

NEW MEXICO: Sierra Grande (*Standley* 6091). Plains, in the Upper Sonoran Zone.

53. Astragalus impensus (Sheld.) Woot. & Standl.

Astragalus kentrophyta elata S. Wats. in King, Geol. Expl. 40th Par. **5**: 77. 1871.

Astragalus viridis impensus Sheld. Minn. Bot. Stud. **9**: 118. 1894.

Astragalus kentrophyta impensus Jones, Contr. West. Bot. **10**: 63. 1902.

Kentrophyta impensa Rydb. Bull. Torrey Club **32**: 665. 1906.

TYPE LOCALITY: Holmes Creek Valley, Nevada.

RANGE: Nevada to Colorado and northwestern New Mexico.

NEW MEXICO: Shiprock (*Standley* 7850). Sandy plains, in the Upper Sonoran Zone.

54. Astragalus hosackiae Greene, Bull. Calif. Acad. **1**: 157. 1885.

TYPE LOCALITY: Northern Arizona.

RANGE: Arizona and New Mexico.

NEW MEXICO: Hanover Mountain (*Holzinger*). Low hills, in the Upper Sonoran Zone.

20. OXYTROPIS DC. Loco weed.

Like Astragalus in most particulars, but distinguished by the short subulate prolongation or beak of the keel.

Several of the species are noxious loco weeds and do as much damage or more than the species of Astragalus that pass under the same name.

KEY TO THE SPECIES.

Stipules free from the petiole; pods pendulous.................... 1. *O. deflexa*.
Stipules adnate to the petiole; pods erect.
  Leaflets many, subverticillate; flowers purple................. 2. *O. richardsoni*.
  Leaflets few, not verticillate; flowers purple to white.
    Flowers 1 to 3; plants 10 cm. high or less................. 3. *O. parryi*.
    Flowers numerous; plants more than 10 cm. high.
      Flowers purple................................. 4. *O. lambertii*.
      Flowers white, the keel often spotted with purple.
        Plants tall, 30 cm. or more; inflorescence elongated................................. 5. *O. pinetorum*.
        Plants low, 15 cm. or less; inflorescence short and dense................................. 6. *O. vegana*.

1. Oxytropis deflexa (Pall.) A. DC. Astrag. 96. 1802.

Astragalus deflexus Pall. Act. Acad. Petrop. **3**[2]: 268. *pl. 15.* 1779.

Aragallus deflexus Heller, Cat. N. Amer. Pl. 4. 1898.

Type locality: Southern Siberia.

Range: British America to Utah and New Mexico.

New Mexico: Rio Pueblo; Chama. Damp woods, in the Canadian Zone.

2. Oxytropis richardsoni (Hook.) Woot. & Standl.

Oxytropis splendens richardsoni Hook. Fl. Bor. Amer. **1**: 148. 1833.

Aragallus richardsoni Greene, Pittonia **4**: 69. 1899.

Type locality: "From Cumberland-House on the Saskatchewan, north to Fort-Franklin and the Bear Lake, and west to the dry Prairies of the Rocky Mountains."

Range: British America to northern New Mexico.

New Mexico: Santa Fe and Las Vegas mountains; Costilla Valley. High mountain meadows, in the Canadian Zone.

A very handsome plant, occurring in great abundance in parks in the Santa Fe and Las Vegas ranges.

3. Oxytropis parryi A. Gray, Proc. Amer. Acad. **20**: 4. 1885.

Aragallus parryi Greene, Pittonia **3**: 212. 1897.

Type locality: "Rocky Mountains of Northern New Mexico and Colorado, near the limit of trees."

Range: Colorado and northern New Mexico.

New Mexico: Pecos Baldy (*Bailey* 616). High mountain meadows, in the Arctic-Alpine Zone.

4. Oyxtropis lambertii Pursh, Fl. Amer. Sept 740. 1814.

Aragallus lambertii Greene, Pittonia **3**: 212. 1897.

Aragallus metcalfei Greene, Proc. Biol. Soc. Washington **18**: 12. 1905.

Type locality: "On the Missouri."

Range: British America to Arizona and Texas.

New Mexico: Mountains and high plains nearly throughout the State. Meadows, Upper Sonoran to Canadian Zone.

Probably the most abundant and widely scattered loco weed in the State. It closely resembles *Astragalus mollissimus*. The type of *Aragallus metcalfei* was collected in the Black Range (*Metcalfe* in 1904).

5. **Oxytropis pinetorum** (Heller) Woot. & Standl.

Aragallus pinetorum Heller, Bull. Torrey Club **26**: 548. 1899.

TYPE LOCALITY: "On gravelly hills thinly clothed with pine trees, at a point 11 miles southeast of Santa Fe," New Mexico. Type collected by Heller (no. 3751).

RANGE: New Mexico.

NEW MEXICO: Between Santa Fe and Canyoncito; near Clark; Burro Mountains; Gray; White Mountains; Sierra Grande; Knowles; Redlands; Nara Visa; Pecos; Sawyers Peak. Mountains, in the Upper Sonoran and Transition zones.

6. **Oxytropis vegana** (Cockerell) Woot. & Standl.

Aragallus pinetorum veganus Cockerell, Torreya **2**: 155. 1902.

Aragallus veganus Woot. & Standl. Contr. U. S. Nat. Herb. **16**: 136. 1913.

TYPE LOCALITY: Top of the Las Vegas Range, above Sapello Canyon, New Mexico. Type collected by Fabián García, June 26, 1901.

RANGE: Known only from the vicinity of the type locality.

21. GLYCYRRHIZA L. WILD LICORICE.

Erect glandular-punctate herbaceous perennial with more or less resinous, odd-pinnate leaves and short axillary racemes of greenish white flowers; pods short, few-seeded, covered with short hooked prickles.

1. **Glycyrrhiza lepidota** Nutt. Gen. Pl. **2**: 106. 1818.

TYPE LOCALITY: St. Louis, Missouri.

RANGE: British America to Arkansas, New Mexico, and California.

NEW MEXICO: Zuni; San Juan; McCarthy Station; Ojo Caliente; Mogollon Mountains; Mesilla Valley; Farmington; Chama; Raton; Pecos; Albert. Wet ground, in the Lower and Upper Sonoran zones.

A common weed in cultivated ground and along ditch banks.

22. ONOBRYCHIS L.

Slender annual; leaves odd-pinnate, the leaflets cuneate-lanceolate, entire, glabrous; flowers purplish red, in elongate spikes; calyx 5-toothed; stamens diadelphous; fruit a loment, 1 or 2-seeded, the joints spiny-toothed, pubescent.

1. **Onobrychis onobrychis** (L.) Rydb. Mem. N. Y. Bot. Gard. **1**: 256. 1900.

SANFOIN.

Hedysarum onobrychis L. Sp. Pl. 751. 1753.

Onobrychis sativa Lam. in Lam. & DC. Franç. **2**: 652. 1778.

TYPE LOCALITY: European.

NEW MEXICO: Mangas Springs (*J. K. Metcalfe* 5).

Often cultivated and possibly escaped here.

23. MEIBOMIA Heist. TICK TREFOIL.

Annual or perennial, erect or spreading herbs, usually slender, with pinnate 3-foliolate stipulate leaves and elongated, sparsely flowered, terminal racemes; flowers small, pale pink or darker, subtended by bracts; calyx 5-toothed; stamens diadelphous; fruit a loment with flattened, reticulately veined, orbicular to elliptic segments.

The plants of this genus are much prized by sheepmen because of their feeding value. They occur mainly in the mountains in the timbered areas.

KEY TO THE SPECIES.

Annuals.
- Pods twisted; all joints except the terminal one pubescent; leaflets broad .... 1. *M. bigelovii.*
- Pods straight; all joints glabrous or nearly so; leaflets narrow .. 2. *M. neomexicana.*

Perennials.
- Loments glabrous .... 3. *M. batocaulis.*
- Loments pubescent.
  - Leaflets ovate to orbicular, obtuse .... 4. *M. grahami.*
  - Leaflets lanceolate to linear, acute or acutish.
    - Leaflets glabrous and shining above, lanceolate; stems glabrous .... 5. *M. metcalfei.*
    - Leaflets densely pubescent above, linear to linear-oblong; stems pubescent .... 6. *M. arizonica.*

1. **Meibomia bigelovii** (A. Gray) Kuntze, Rev. Gen. Pl. **1:** 197. 1891.
Desmodium bigelovii A. Gray, Pl. Wright. **2:** 47. 1853.
Desmodium spirale bigelovii Robins. & Greenm. Proc. Amer. Acad. **29:** 385. 1894.
TYPE LOCALITY: Valley on the San Pedro, Sonora.
RANGE: Southern New Mexico and Arizona to Mexico.
NEW MEXICO: Mangas Springs; Black Range; San Luis Mountains; Organ Mountains; Gila Hot Springs. Dry hillsides, in the Upper Sonoran Zone.

2. **Meibomia neomexicana** (A. Gray) Kuntze, Rev. Gen. Pl. **1:** 198. 1891.
Desmodium neomexicanum A. Gray, Pl. Wright. **1:** 53. 1852.
Desmodium exiguum A. Gray, Pl. Wright. **2:** 46. 1853.
TYPE LOCALITY: Mountain valley 30 miles east of El Paso, Texas.
RANGE: Western Texas and southern Arizona to Mexico.
NEW MEXICO: Pinos Altos Mountains; Burro Mountains; Mogollon Mountains; San Luis Mountains; Organ Mountains. Dry rocky hills, in the Upper Sonoran Zone.

3. **Meibomia batocaulis** (A. Gray) Kuntze, Rev. Gen. Pl. **1:** 197. 1891.
Desmodium batocaulon A. Gray, Pl. Wright. **2:** 47. 1853.
TYPE LOCALITY: Stony banks of small streams, on the San Pedro, Sonora.
RANGE: New Mexico and Arizona to Mexico.
NEW MEXICO: San Luis Mountains (*Mearns* 2138).

4. **Meibomia grahami** (A. Gray) Kuntze, Rev. Gen. Pl. **1:** 198. 1891.
Desmodium grahami A. Gray, Pl. Wright. **2:** 48. 1853.
TYPE LOCALITY: On mountains, near the Copper Mines, New Mexico. Type collected by Wright (no. 1015).
RANGE: Western Texas to southern Arizona, south into Mexico.
NEW MEXICO: Mangas Springs; Fort Bayard; Mogollon Mountains; San Luis Mountains; Kingston; Organ Mountains. Dry hills and canyons, in the Upper Sonoran Zone.

5. **Meibomia metcalfei** Rose & Painter, Bot. Gaz. **40:** 144. 1905.
TYPE LOCALITY: Animas Creek in the Black Range, Grant County, New Mexico. Type collected by Metcalfe (no. 1137).
RANGE: Southwestern New Mexico.
NEW MEXICO: Animas Creek; Pine Cienaga.

6. **Meibomia arizonica** (S. Wats.) Vail, Bull. Torrey Club **19:** 117. 1892.
Desmodium arizonicum S. Wats. Proc. Amer. Acad. **20:** 363. 1885.
TYPE LOCALITY: Arizona.
RANGE: Arizona and New Mexico to Mexico.
NEW MEXICO: Mogollon Mountains.

24. HEDYSARUM L.

Perennial herb with odd-pinnate leaves and terminal racemes of purple flowers; calyx 5-toothed, the teeth subulate; keel straight, unappendaged, longer than the wings; fruit a loment with 3 to 5 flat, conspicuously nerved, elliptic segments.

1. **Hedysarum pabulare** A. Nels. Proc. Biol. Soc. Washington **15:** 185. 1902.

TYPE LOCALITY: Not stated.

RANGE: Colorado and New Mexico.

NEW MEXICO: Canyoncito; near Fort Defiance; Raton. Open hills, in the Upper Sonoran Zone.

25. VICIA L. WILD VETCH.

Slender herbs climbing by tendrils borne at the ends of the pinnate leaves; stipules semisagittate; flowers in axillary racemes or few-flowered clusters; calyx 5-toothed, the upper divisions sometimes shorter; wings of the corolla adnate to the keel; stamens mostly diadelphous; style filiform, hairy all around or only on the back at the apex; pods flat, 2 to several-seeded, 2-valved.

The plants are never very abundant, but are found almost all over the State, commonest at middle elevations in the mountains. They are all eaten by stock and considered good feed. *Vicia cracca*, the common vetch, is not infrequently cultivated in various parts of the State as a fodder or soiling crop.

KEY TO THE SPECIES.

Flowers sessile or nearly so .... 7. *V. angustifolia.*
Flowers long-pedicellate.
  Flowers 15 mm. long or more.
    Leaflets abundantly pubescent .... 8. *V. caespitosa.*
    Leaflets glabrous, at least in age.
      Leaflets thin, not strongly veined, usually oval, often linear-oblong .... 1. *V. americana.*
      Leaflets thick, strongly veined, linear or linear-oblong.
        Leaflets elongated-linear; plants low .... 9. *V. sparsifolia.*
        Leaflets, at least the upper ones, oblong to linear-oblong; plants tall, climbing .... 10. *V. dissitifolia.*
  Flowers less than 9 mm. long.
    Flowers 6 to 20 in each raceme.
      Flowers pale purple, drying blue; racemes with usually less than 10 flowers .... 2. *V. pulchella.*
      Flowers white; racemes 15 to 20-flowered .... 3. *V. melilotoides.*
    Peduncles with 1 to 4 flowers.
      Peduncles 3 to 4-flowered; flowers bright blue .... 4. *V. leavenworthii.*
      Peduncles 1 or 2-flowered; flowers pale blue or whitish.
        Pods glabrous; calyx not pilose .... 5. *V. exigua.*
        Pods pubescent; calyx pilose .... 6. *V. leucophaea.*

1. **Vicia americana** Muhl.; Willd. Sp. Pl. **3:** 1096. 1801.

Vicia truncata Nutt.; Torr. & Gray, Fl. N. Amer. **1:** 270. 1838.

Vicia linearis Nutt. op. cit. 276.

TYPE LOCALITY: "Pennsylvania."

RANGE: British America to New York and New Jersey, west to Kansas, New Mexico and California.

NEW MEXICO: Common in all the higher mountains of the State. Thickets, in the Transition Zone.

This is the common bright blue flowered vetch of the timbered mountains everywhere in the State. The leaf form varies greatly, the leaflets being from broadly elliptic to narrowly linear-oblong. The name *V. linearis* is often applied to the plants with very narrow leaflets, but the form does not appear to be constant, and all intermediates are found, even in the same locality.

2. Vicia pulchella H. B. K. Nov. Gen. & Sp. **6:** 499. *pl. 583*. 1823.

TYPE LOCALITY: "Crescit in declivitate occidentali montium Mexicanorum prope Mescala, alt. 265 hex."

RANGE: New Mexico and western Texas to Arizona and Mexico.

NEW MEXICO: White and Sacramento mountains. Thickets, in the Transition Zone.

3. Vicia melilotoides Woot. & Standl. Contr. U. S. Nat. Herb. **16:** 141. 1913.

TYPE LOCALITY: Winsors Ranch in the Pecos River National Forest, New Mexico. Type collected by Standley (no. 4364).

RANGE: New Mexico and Arizona.

NEW MEXICO: Santa Fe and Las Vegas mountains; Coolidge; Mogollon Mountains; Black Range; White and Sacramento mountains. Moist meadows and thickets, in the Transition Zone.

This has long been confused with *V. pulchella*, which it closely resembles in general appearance. The flowers, however, are white instead of blue as in that species, and much more numerous, while the peduncles are shorter and the calyx less pubescent. Both species are found in the same region in the White Mountains, where they are at once distinguishable in the field.

4. Vicia leavenworthii Torr. & Gray, Fl. N. Amer. **1:** 271. 1838.

TYPE LOCALITY: "Arkansas."

RANGE: Arkansas and Oklahoma to Texas and New Mexico, in the Upper Sonoran Zone.

NEW MEXICO: Gray (*Skehan* 88).

5. Vicia exigua Nutt.; Torr. & Gray, Fl. N. Amer. **1:** 272. 1838.

TYPE LOCALITY: "Plains of the Oregon and Upper California."

RANGE: Oregon and California to New Mexico and western Texas.

NEW MEXICO: Carrizalillo Mountains; Mangas Springs; Lemitar; Organ Mountains; Mesilla; Gray; Sierra Grande; Star Peak. Hills and canyons, in the Lower and Upper Sonoran zones.

6. Vicia leucophaea Greene, Bot. Gaz. **6:** 217. 1881.

TYPE LOCALITY: "Along streams in the higher mountains of southwestern New Mexico." Type collected by Greene.

RANGE: Western New Mexico and adjacent Arizona.

NEW MEXICO: Mogollon Mountains; Hanover Mountain. Transition Zone.

7. Vicia angustifolia Reich. Fl. Moen. Franc. **2:** 44. 1778.

TYPE LOCALITY: European.

NEW MEXICO: Chama (*Standley* 6696).

Introduced from Europe, but the plants seemed to be at home in a wet meadow near Chama.

8. Vicia caespitosa A. Nels. Bull. Torrey Club **25:** 373. 1908.

TYPE LOCALITY: Laramie Plains, Wyoming.

RANGE: Wyoming to New Mexico.

NEW MEXICO: Clayton; Kingston. Open slopes, in the Upper Sonoran and Transition zones.

9. **Vicia sparsifolia** Nutt.; Torr. & Gray, Fl. N. Amer. **1**: 270. 1838.

TYPE LOCALITY: "Plains of the Oregon."

RANGE: Alberta and Montana to California and Kansas.

NEW MEXICO: Raton (*Standley* 6304). Open hills and plains, in the Upper Sonoran Zone.

10. **Vicia dissitifolia** (Nutt.) Rydb. Bull. Torrey Club **33**: 144. 1906.

Lathyrus dissitifolius Nutt.; Torr. & Gray, Fl. N. Amer. **1**: 277. 1838.

TYPE LOCALITY: Plains of the Platte.

RANGE: Nebraska to northern New Mexico.

NEW MEXICO: Winsors Ranch; near Pecos. Mountains and hills, in the Transition Zone.

26. LATHYRUS L. WILD PEA.

Low, mostly slender and short-stemmed herbs with pinnate leaves terminating in tendrils, or the tendrils much reduced or wanting; stems erect or climbing; similar to Vicia, but flowers usually larger, banner more recurved, and the style flattened near the apex and hairy only along the inner side.

KEY TO THE SPECIES.

Plants erect; tendrils none or much reduced.
  Plants conspicuously pubescent............................ 6. *L. incanus.*
  Plants glabrous.
    Flowers purple, 15 mm. long or more; leaflets thick.... 1. *L. decaphyllus.*
    Flowers ochroleucous, 10 mm. long or less; leaflets thin.
      Leaflets oval to oblong.......................... 2. *L. leucanthus.*
      Leaflets linear or nearly so....................... 7. *L. arizonicus.*
Plants climbing by well-developed tendrils.
  Leaflets linear or nearly so.................................. 3. *L. graminifolius.*
  Leaflets elliptic to oval.
    Plants pubescent throughout; leaflets 35 to 45 mm. long. 4. *L. oreophilus.*
    Plants glabrous or nearly so; leaflets 20 mm. long or less. 5. *L. parvifolius.*

1. **Lathyrus decaphyllus** Pursh, Fl. Amer. Sept. 471. 1814.

Lathyrus polymorphus Nutt. Gen. Pl. **2**: 96. 1818.

TYPE LOCALITY: "On the banks of the Missouri."

RANGE: Idaho to Arizona and New Mexico.

NEW MEXICO: Farmington; Raton; Pecos; Las Vegas; Albuquerque; Sierra Grande; Chama River; Santa Fe; Chiz; Cliff; Gray; Zuni. Plains and open hills, in the Upper Sonoran Zone.

The plant is a rather handsome one with much larger flowers than most of our species. In the northern part of the State it often becomes a weed in cultivated fields.

2. **Lathyrus leucanthus** Rydb. Bull. Torrey Club **28**: 37. 1901.

TYPE LOCALITY: Ojo, Colorado.

RANGE: Colorado and New Mexico.

NEW MEXICO: Chama; Santa Fe and Las Vegas mountains; Sandia Mountains; Organ Mountains; White and Sacramento mountains. Meadows, in the Transition Zone.

An inconspicuous little plant, 10 to 20 cm. high, with few thin leaves without tendrils and few-flowered racemes of small white flowers.

3. **Lathyrus graminifolius** (S. Wats.) White, Bull. Torrey Club **21**: 454. 1894.

Lathyrus palustris graminifolius S. Wats. Proc. Amer. Acad. **23**: 263. 1888.

TYPE LOCALITY: "Frequent from New Mexico to Arizona and northern Mexico."

RANGE: California to New Mexico and Mexico.

NEW MEXICO: Mogollon Mountains; Magdalena Mountains; Ramah; Hanover Mountain; Rio Apache; Hillsboro Peak; White Mountains. Transition Zone.

4. **Lathyrus oreophilus** Woot. & Standl. Muhlenbergia **5:** 87. 1909.

TYPE LOCALITY: James Canyon, about 4 miles east of Cloudcroft, in the Sacramento Mountains, New Mexico. Type collected by Wooton, June 26, 1899.

RANGE: Known only from the type locality, in the Transition Zone.

5. **Lathyrus parvifolius** S. Wats. Proc. Amer. Acad. **17:** 345. 1882.

TYPE LOCALITY: In the San Miguelito Mountains, San Luis Potosi, Mexico.

RANGE: Mexico to Utah and Washington.

NEW MEXICO: Carrizo Mountains. Thickets, in the Transition Zone.

6. **Lathyrus incanus** (Rydb. & Smyth) Rydb. Colo. Agr. Exp. Sta. Bull. **100:** 217. 1906.

Lathyrus ornatus incanus Rydb. & Smyth, Rep. Univ. Nebr. Bot. Surv. **21:** 64. 1895.

TYPE LOCALITY: Nebraska.

RANGE: Wyoming and Nebraska to Utah and northern New Mexico.

NEW MEXICO: Nara Visa (*Fisher* 124, 125). Plains, in the Upper Sonoran Zone.

The flowers are usually purple, but occasionally cream-colored and purplish only on the standard.

7. **Lathyrus arizonicus** Britton, Trans. N. Y. Acad. **8:** 65. 1889.

TYPE LOCALITY: Arizona.

RANGE: Colorado to Arizona and northern New Mexico.

NEW MEXICO: Chama (*Eggleston* 6660a). Open slopes, in the Transition Zone.

27. DOLICHOLUS Medic.

Perennial twining herb with pinnately 3-foliolate leaves and short-peduncled few-flowered axillary racemes of small yellow flowers; pods oblong, flattened, pubescent.

1. **Dolicholus texanus** (Torr. & Gray) Vail, Bull. Torrey Club **26:** 108. 1899.

Rhynchosia texana Torr. & Gray, Fl. N. Amer. **1:** 687. 1838.

TYPE LOCALITY: Texas.

RANGE: Western Texas to Arizona, south into Mexico.

NEW MEXICO: Mogollon Mountains; Fort Bayard; Kingston; Organ Mountains; west of Roswell; Queen. Dry hills, in the Lower Sonoran Zone.

28. ERYTHRINA L. CORAL BEAN.

A tall thick-stemmed shrub, with scattering hooked prickles on stems and leaves; leaves pinnately 3-foliolate, with large fan-shaped leaflets; flowers in short terminal racemes, bright scarlet; calyx campanulate, truncate, white-tomentose; corolla about 4 cm. long, with an elongated banner; stamens 10, monadelphous; legume linear, torulose, long-stipitate; seeds large, bright scarlet.

1. **Erythrina flabelliformis** Kearney, Trans. N. Y. Acad. **14:** 321. 1894.

TYPE LOCALITY: Near Fort Huachuca, Arizona.

RANGE: Southwestern New Mexico, southeastern Arizona, and adjacent Mexico.

NEW MEXICO: San Luis Mountains.

29. GALACTIA P. Br. MILK PEA.

Twining suberect plants with slender stems and pinnately 3-foliolate leaves; flowers purplish, in strict many-flowered racemes; calyx 4-toothed, the teeth nearly twice the length of the tube; pod linear-oblong, flat.

1. **Galactia wrightii** A. Gray, Pl. Wright. **1:** 44. 1852.

Galactia tephrodes A. Gray, Pl. Wright. **2:** 34. 1853.

TYPE LOCALITY: Hills near the Limpio, western Texas.

RANGE: Western Texas to southern Arizona.

New Mexico: Mangas Springs; Condes Camp; Florida Mountains. Dry hills, in the Upper Sonoran Zone.

The type of *G. tephrodes* came from Condes Camp (*Wright* 956).

30. PHASEOLUS L. Bean.

Prostrate or twining annual or perennial herbs with pinnately 3-foliolate leaves, the flowers in axillary, mostly long-peduncled racemes; calyx 5-toothed; standard recurved or spreading, the keel strongly incurved or coiled; pods 2-valved, straight or falcate, with usually many seeds.

Some of the species are of considerable value as forage plants and one of them, introduced into cultivation, promises well for this purpose.

KEY TO THE SPECIES.

Annuals.
- Leaflets triangular-lanceolate or ovate, not much longer than the peduncles .... 1. *P. acutifolius.*
- Leaflets elongated-linear or linear-oblong, about twice as long as the peduncles .... 2. *P. tenuifolius.*

Perennials.
- Plants erect, 15 cm. high or less; peduncles 1 or 2-flowered .. 3. *P. parvulus.*
- Plants twining, stems much more than 15 cm. long; peduncles several-flowered.
  - Stems and leaves pilose .... 4. *P. macropoides.*
  - Stems and leaves glabrous or puberulent, never pilose.
    - Pods 12 mm. wide or more; bracts large, persistent; leaflets rhombic-ovate .... 5. *P. metcalfei.*
    - Pods less than 9 mm. wide; bracts small, deciduous; leaflets linear to rhombic.
      - Leaflets elongated-linear, entire; style slender, 2 mm. long .... 6. *P. angustissimus.*
      - Leaflets broader than linear, lobed; style slender or stout.
        - Leaflets deeply 3-lobed; pods densely pubescent, with a short stout style.... 7. *P. grayanus.*
        - Leaflets with small lobes near the base; pods nearly glabrous, with a long slender style .... 8. *P. dilatatus.*

1. Phaseolus acutifolius A. Gray, Pl. Wright. **1:** 43. 1852.

Type locality: Mountain valley, thirty miles east of El Paso, Texas.

Range: Western Texas to southern Arizona.

New Mexico: Salt Lake; Organ Mountains; Zuni. Dry hills, in the Upper Sonoran Zone.

2. Phaseolus tenuifolius (A. Gray) Woot. & Standl. Contr. U. S. Nat. Herb. **16:** 140. 1913.

Phaseolus acutifolius tenuifolius A. Gray, Pl. Wright. **2:** 33. 1853.

Type locality: Mountain sides near the Copper Mines, New Mexico. Type collected by Wright (no. 950).

Range: Mountains of southwestern New Mexico.

New Mexico: Mangas Springs; Santa Rita.

3. Phaseolus parvulus Greene, Bot. Gaz. **6:** 217. 1881.

Type locality: "Abundant in deep woods of *Pinus ponderosa*, in the Pinos Altos Mountains," New Mexico. Type collected by E. L. Greene in 1880 or 1881.

Range: Southwestern New Mexico and southern Arizona.

We have seen no specimens of this from New Mexico.

4. Phaseolus macropoides A. Gray, Pl. Wright. **2**: 33. 1853.

TYPE LOCALITY: Stony hills at the Copper Mines, New Mexico. Type collected by Wright (no. 953).

RANGE: Western Texas to Arizona and adjacent Mexico.

NEW MEXICO: Mogollon Mountains; Organ Mountains; Dog Spring. Dry hills and canyons, in the Upper Sonoran Zone.

5. Phaseolus metcalfei Woot. & Standl. Contr. U. S. Nat. Herb. **16**: 140. 1913.

METCALFE BEAN.

Phaseolus retusus Benth. Pl. Hartw. 11. 1839, not Moench, 1794.

TYPE LOCALITY: Mexico.

RANGE: Southern New Mexico and Arizona to Mexico.

NEW MEXICO: Mangas Springs; West Fork of the Gila; Kingston; San Luis Mountains; near White Water; Crawfords Ranch. Transition Zone.

Mr. J. K. Metcalfe introduced this plant into cultivation and demonstrated its usefulness as a forage plant. It has a large thickened root and produces prostrate stems frequently 3 to 4 meters long. The crop which can be produced on an acre of ground is a large one. The pods and seeds are large, a fact which adds to the feeding value. The plant was called "Metcalfe bean" in honor of the introducer.

6. Phaseolus angustissimus A. Gray, Pl. Wright. **2**: 33. 1853.

TYPE LOCALITY: Stony hillsides at the crossing of the Rio Grande above Dona Ana, New Mexico. Type collected by Wright (no. 951).

RANGE: New Mexico and Arizona.

NEW MEXICO: Craters; Grand Canyon of the Gila; Organ Mountains; Florida Mountains; Mangas Springs. Dry hills and plains, in the Lower and Upper Sonoran zones.

7. Phaseolus grayanus Woot. & Standl. Contr. U. S. Nat. Herb. **16**: 139. 1913.

Phaseolus wrightii A. Gray, Pl. Wright. **2**: 33. 1853, not op. cit. **1**: 43. 1852.

TYPE LOCALITY: San Luis Mountains, New Mexico. Type collected by E. A. Mearns (no. 2124).

RANGE: Southern New Mexico and Arizona and adjacent Mexico.

NEW MEXICO: San Luis Mountains; Fort Bayard; Mogollon Creek; Mangas Springs; Hanover Mountain.

8. Phaseolus dilatatus Woot. & Standl. Contr. U. S. Nat. Herb. **16**: 139. 1913.

TYPE LOCALITY: Mogollon Mountains, New Mexico. Type collected by Rusby.

RANGE: Mountains of southwestern New Mexico, probably also in adjacent Arizona.

NEW MEXICO: Mogollon Mountains; Burro Mountains.

31. STROPHOSTYLES Ell. WILD BEAN.

Annual with slender trailing or climbing stems; leaves 3-foliolate, with thickish, linear to lanceolate leaflets; flowers in axillary long-pedunculate racemes, purplish; pods linear, flattish, strigose, 2 to 3 cm. long.

1. Strophostyles pauciflora (Benth.) S. Wats. in A. Gray, Man. ed. 6. 145. 1890.

Phaseolus pauciflorus Benth. Ann. Naturhist. Hofmus. Wien **1**: 140. 1837.

TYPE LOCALITY: Texas.

RANGE: Minnesota and Indiana to Nebraska, New Mexico, Texas, and Louisiana.

NEW MEXICO: Mesilla Valley; Chavez; Queen; Organ Mountains. Lower and Upper Sonoran zones.

32. COLOGANIA Kunth.

Perennial trailing or twining herbs with 3-foliolate leaves, prominent bracts, and axillary, 1 to 3-flowered clusters of reddish purple flowers; peduncles short or none; calyx tubular, 4-toothed; keel somewhat incurved, not coiled; style beardless; pods linear to oblong, flattened, straight or slightly curved.

KEY TO THE SPECIES.

Leaflets linear or linear-oblong, twice as long as the petioles or more.. 1. *C. longifolia.*
Leaflets oblong-ovate, about equaling the petioles.................... 2. *C. pulchella.*

1. Cologania longifolia A. Gray, Pl. Wright. **2:** 35. 1853.

TYPE LOCALITY: Hills near the Copper Mines, New Mexico. Type collected by Wright (no. 961).

RANGE: New Mexico, Arizona, and adjacent Mexico.

NEW MEXICO: Hanover Mountain; Mogollon Mountains. Dry hills, in the Upper Sonoran Zone.

2. Cologania pulchella H. B. K. Nov. Gen. & Sp. **6:** 413. 1823.

TYPE LOCALITY: "Crescit in Regno Novae Hispaniae, prope Pazcuaro, alt. 1130 hex."

RANGE: Western Texas to Arizona, south into Mexico.

NEW MEXICO: Mangas Springs; Organ Mountains; Cliff; Hondo Hill; Cactus Flat; Queen; Organ Mountains. Dry hillsides.

Order 28. GERANIALES.

KEY TO THE FAMILIES.

Plants with secreting glands in the leaves or bark.
  Filaments united into a cup or tube, wholly or in part............................79. **MELIACEAE** (p. 390).
  Filaments distinct nearly or quite to the base.
    Leaf blades punctate with oil glands....77. **RUTACEAE** (p. 388).
    Leaves not punctate....................78. **SIMARUBACEAE** (p. 390).
Plants destitute of secreting glands or cells.
  Sepals bearing 1 or 2 dorsal glands..........76. **MALPIGHIACEAE** (p. 388).
  Sepals without dorsal glands.
    Styles united around a central column, breaking away from this at maturity..........................71. **GERANIACEAE** (p. 379).
    Styles distinct or permanently united.
      Styles distinct or partially united, the tips and the stigmas distinct.
        Leaves simple; stamens 5.......72. **LINACEAE** (p. 381).
        Leaves compound; stamens 10 to 15....................73. **OXALIDACEAE** (p. 383).
      Styles and stigmas permanently united.
        Filaments normally appendaged; seeds straight or nearly so.................74. **ZYGOPHYLLACEAE** (p. 385).
        Filaments not appendaged; seeds strongly bent.......75. **KOEBERLINIACEAE** (p. 387).

71. GERANIACEAE. Cranesbill Family.

Annual or perennial herbs, often glandular-pubescent, with lobed or dissected leaves; flowers regular, complete, symmetrically pentamerous, in few-flowered axillary pedunculate clusters; sepals persistent; petals mostly conspicuous; stamens of the same number as or 2 or 3 times the number of the petals; pistil of 5 united carpels, the united styles forming a persistent column; fruit a capsule, each carpel breaking away from the column.

KEY TO THE GENERA.

Stamens 10; carpel tails naked on the inner side............ 1. GERANIUM (p. 380).
Stamens 5; carpel tails hairy on the inner side............... 2. ERODIUM (p. 381).

1. GERANIUM L. CRANESBILL.

Herbaceous perennials or annuals, with often glandular branching stems 10 to 60 cm. long; leaves long-petioled, palmately lobed or dissected, pentagonal to rotund in outline; petals alternating with 5 glands; stamens mostly 10, of two lengths, alternating; capsule 5-lobed, with a long beak; carpels breaking from the column and curling upward, remaining attached at the tip for a time.

KEY TO THE SPECIES.

Annual... 1. *G. langloisii.*
Perennials.
  Flowers white or pale pinkish.
    Petals 10 mm. long or less; leaf segments obtuse; plants densely glandular........................ 2. *G. lentum.*
    Petals 12 to 20 mm. long; leaf segments acute; plants glandular only on the pedicels.................. 3. *G. richardsonii.*
  Flowers purple or rose purple.
    Petals narrowly obovate or oblong, dark purple.
      Plants glandular throughout; tips of the sepals 0.5 to 1 mm. long......................... 4. *G. furcatum.*
      Plants not glandular; tips of sepals 1.5 to 2 mm. long................................. 5. *G. atropurpureum.*
    Petals obcordate or broadly obovate, light purple.
      Sepals glandular like the upper part of the stems; leaves somewhat canescent; plants stout.... 6. *G. fremontii.*
      Sepals not glandular, the plants glandular only on the pedicels, slender; leaves nearly glabrous................................ 7. *G. eremophilum.*

1. **Geranium langloisii** Greene, Pittonia **3:** 171. 1897.

TYPE LOCALITY: St. Martinsville, Louisiana.

RANGE: Oklahoma and Louisiana to Texas and southern New Mexico.

NEW MEXICO: Organ Mountains (*Wooton*). Dry hillsides, in the upper Sonoran Zone.

2. **Geranium lentum** Woot. & Standl. Contr. U. S. Nat. Herb. **16:** 142. 1913.

TYPE LOCALITY: West Fork of the Gila, New Mexico. Type collected by Wooton, August 7, 1900.

RANGE: Mountains of western New Mexico.

NEW MEXICO: Mogollon Mountains; Craters, Valencia County. Transition Zone.

3. **Geranium richardsonii** Fisch. & Trautv. Ind. Sem. Hort. Petrop. **4:** 37. 1837.

Geranium pentagynum Engelm. in Wisliz. Mem. North. Mex. 90. 1848.

Geranium gracilentum Greene; Rydb. Colo. Agr. Exp. Sta. Bull. **100:** 218. 1906.

TYPE LOCALITY: Valleys of the Rocky Mountains.

RANGE: California and South Dakota to Colorado and New Mexico.

NEW MEXICO: Common in all the higher mountains of the State. Damp woods, in the Transition and Canadian zones.

The type of *Geranium pentagynum* was collected by Wislizenus on Wolf Creek, in northern New Mexico.

4. **Geranium furcatum** Hanks, N. Amer. Fl. **25:** 16. 1907.

TYPE LOCALITY: Grand Canyon of the Colorado, Arizona.

RANGE: Mountains of New Mexico and Arizona.

NEW MEXICO: Water Canyon; Hop Canyon; Cebolla; Tunitcha Mountains. Transition Zone.

5. **Geranium atropurpureum** Heller, Bull. Torrey Club **28:** 195. 1898.

Geranium gracile Engelm. in A. Gray, Mem. Amer. Acad. n. ser. **4:** 27. 1849, not Ledeb. 1837.

TYPE LOCALITY: Along Santa Fe Creek, New Mexico. Type collected by Heller (no. 3723).

RANGE: Colorado to northern Mexico.

NEW MEXICO: Higher mountains throughout the State. Moist slopes, Transition to the Hudsonian Zone.

6. **Geranium fremontii** Torr.; A. Gray, Mem. Amer. Acad. n. ser. **4:** 26. 1849.

TYPE LOCALITY: Bottom lands of the Mora River, New Mexico. Type collected by Fendler (no. 90).

RANGE: Colorado and Utah to New Mexico and Arizona.

NEW MEXICO: Johnsons Mesa; Coolidge; Sierra Grande.

7. **Geranium eremophilum** Woot. & Standl. Contr. U. S. Nat. Herb. **16:** 142. 1913.

TYPE LOCALITY: San Luis Mountains, New Mexico. Type collected by E. A. Mearns (no. 2142).

RANGE: Mountains of southwestern New Mexico.

NEW MEXICO: San Luis Mountains; Organ Mountains.

2. ERODIUM L'Her.

Very similar to Geranium in general appearance, but stamens only 5 and the carpel tails long-bearded and becoming spirally twisted; distal peduncles appearing terminal, but really axillary.

KEY TO THE SPECIES.

Leaf blades pinnately divided; petals small, 6 mm. long or less, pale purplish.......... 1. *E. cicutarium.*
Leaf blades merely lobed; petals large, 10 mm. long or more, purple. 2. *E. texanum.*

1. **Erodium cicutarium** (L.) L'Her.; Ait. Hort. Kew. **2:** 414. 1789.

ALFILERIA. FILAREE.

Geranium cicutarium L. Sp. Pl. 680. 1753.

TYPE LOCALITY: "Habitat in Europae sterilibus cultis."

NEW MEXICO: Introduced into nearly all parts of the State.

An inconspicuous prostrate plant, widely introduced into North America. In certain parts of the Southwest it has been found to be a valuable forage plant but it has never been utilized in New Mexico. Nowhere is it very abundant.

2. **Erodium texanum** A. Gray, Gen. Fl. Amer. **2:** 130. *pl. 151.* 1849.

TYPE LOCALITY: Texas.

RANGE: Texas to California.

NEW MEXICO: Tortugas Mountain; Tres Hermanas; near Roswell; Organ Mountains. Dry plains and hills, in the Lower and Upper Sonoran zones.

72. LINACEAE. Flax Family.

Annual or perennial herbs, with mostly low slender stems; leaves simple, alternate, scattered or crowded; stipules wanting or mere glands; flowers complete and regular, mostly axillary to bracts similar to the leaves but smaller; calyx persistent; petals early deciduous, yellow or blue; stamens 5, united at the base and sometimes bearing 5 alternating staminodia; pistil of 5 united carpels, each more or less 2-celled by a partial or complete membranous partition; seeds oily.

KEY TO THE GENERA.

Flowers blue; stigmas introrse, more or less elongated; sepals glandless .... 1. LINUM (p. 382.)
Flowers yellow; stigmas capitate; sepals, at least the inner ones, glandular on the margins............ 2. CATHARTOLINUM (p. 382).

1. LINUM L. WILD FLAX.

Slender perennial with several erect stems from a ligneous root, glabrous throughout, glaucescent, branching above; leaves small, 2 cm. long or less, sessile, oblong-lanceolate; flowers large, on slender pedicels, bright blue; sepals broadly ovate, acute or obtuse, some of them with narrow membranous margins; petals obovate, early deciduous; capsule spheroidal, with incomplete false septa; seeds flat.

1. **Linum lewisii** Pursh, Fl. Amer. Sept. 210. 1814.
Linum perenne lewisii Eat. & Wright, N. Amer. Bot. 302. 1840.
TYPE LOCALITY: "In the valleys of the Rocky-mountains and on the banks of the Missouri."
RANGE: Alaska and British America to California, Texas, and Mexico.
NEW MEXICO: Higher mountains throughout the State. Meadows, in the Upper Sonoran and Transition zones.

2. CATHARTOLINUM Reichenb. YELLOW FLAX.

Annual or short-lived perennial herbs, with slender angled stems and small sessile leaves; flowers yellow, with persistent sepals and caducous petals; sepals acute to aristate, with fine gland-tipped teeth along the margins of at least some of them; capsules ovoid to spherical, the partitions mostly complete.

KEY TO THE SPECIES.

Styles distinct; flowers in virgate racemes .... 1. *C. neomexicanum.*
Styles more or less united; flowers corymbose.
  Stems puberulent.
    Plants stout, branched from the base; petals 10 mm. long or less.................................. 2. *C. puberulum.*
    Plants slender, simple below; petals 13 to 15 mm. long 3. *C. vestitum.*
  Stems glabrous.
    Petals 6 to 9 mm. long .... 4. *C. australe.*
    Petals 13 to 22 mm. long.
      Sepals not long-aristate; leaves numerous and imbricated on the lower part of the stem...... 5. *C. vernale.*
      Sepals long-aristate; leaves scattered along the stem.
        Plants slender, not leafy, the leaves narrow, mostly involute, thick .... 6. *C. aristatum.*
        Plants stout, densely leafy, the leaves broad and thin .... 7. *C. berlandieri.*

1. **Cathartolinum neomexicanum** (Greene) Small, N. Amer. Fl. **25:** 73. 1907.
Linum neomexicanum Greene, Bot. Gaz. **6:** 183. 1881.
TYPE LOCALITY: "In woods of *Pinus ponderosa* on the Pinos Altos Mountains," New Mexico. Type collected in 1880 by E. L. Greene.

RANGE: New Mexico and Arizona to northern Mexico.

NEW MEXICO: Hanover Mountain; Mogollon Mountains; Burro Mountains; San Luis Mountains. Transition Zone.

2. **Cathartolinum puberulum** (Engelm.) Small, N. Amer. Fl. **25**: 80. 1907.

Linum rigidum puberulum Engelm. in A. Gray, Pl. Wright. **1**: 25. 1852.

Linum puberulum Heller, Pl. World **1**: 22. 1897.

TYPE LOCALITY: "Santa Fe to the Cimarron River," New Mexico. Type collected by Fendler (no. 85).

RANGE: Colorado and Utah to New Mexico and Arizona.

NEW MEXICO: Santa Fe; White Mountains; Carrizo Mountains; Farmington; Santa Rita; Grass Mountain; Pecos; Pajarito Park; White and Sacramento mountains. Dry hills and plains, in the Upper Sonoran Zone.

3. **Cathartolinum vestitum** Woot. & Standl. Contr. U. S. Nat. Herb. **16**: 142. 1913.

TYPE LOCALITY: Mangas Springs, New Mexico. Type collected by Metcalfe in 1901.

RANGE: Southwestern New Mexico.

NEW MEXICO: Mangas Springs; Carrizalillo Mountains; Kingston; Middle Fork of the Gila; San Andreas Mountains. Dry hills.

4. **Cathartolinum australe** (Heller) Small, N. Amer. Fl. **25**: 81. 1907.

Linum australe Heller, Bull. Torrey Club **25**: 627. 1898.

TYPE LOCALITY: "On an open slope in dry ground, at the head of the reservoir, 4 miles east of Santa Fe," New Mexico. Type collected by Heller (no. 3724).

RANGE: Colorado and Arizona to northern Mexico.

NEW MEXICO: Santa Fe and Las Vegas mountains; Sandia Mountains; Mogollon Mountains; White and Sacramento mountains; Tunitcha Mountains; Raton; Roswell. Dry hills, in the Upper Sonoran and Transition zones.

5. **Cathartolinum vernale** (Wooton) Small, N. Amer. Fl. **25**: 80. 1907.

Linum vernale Wooton, Bull. Torrey Club **25**: 452. 1898.

TYPE LOCALITY: Tortugas Mountain, New Mexico. Type collected by Wooton (no. 589).

RANGE: Western Texas and southern New Mexico.

NEW MEXICO: Mesa west of Organ Mountains; La Luz Canyon. Mesas, in the Lower Sonoran Zone.

6. **Cathartolinum aristatum** (Engelm.) Small, N. Amer. Fl. **25**: 83. 1907.

Linum aristatum Engelm. in Wisliz. Mem. North. Mex. 101. 1848.

TYPE LOCALITY: "In sandy soil near Carizal, south of El Paso," Chihuahua.

RANGE: Western Texas and southern New Mexico to northern Mexico.

NEW MEXICO: West of Cambray; mesa west of Organ Mountains; Nara Visa. Lower Sonoran Zone.

7. **Cathartolinum berlandieri** (Hook.) Small, N. Amer. Fl. **25**: 82. 1907.

Linum berlandieri Hook. Curtis's Bot. Mag. **63**: *pl. 3480*. 1836.

Linum rigidum berlandieri Torr. & Gray, Fl. N. Amer. **1**: 204. 1838.

TYPE LOCALITY: Near San Antonio, Texas.

RANGE: Kansas and Colorado to New Mexico and Arizona.

NEW MEXICO: Guadalupe Mountains; Roswell. Dry plains and hills, in the Upper Sonoran Zone.

73. OXALIDACEAE. Wood-sorrel Family.

Low perennial herbs from small bulblike or elongated rootstocks, acaulescent or caulescent; leaves alternate or all basal, palmately 3 to several-foliolate, the leaflets mostly obcordate; flowers in scapose cymes or in few-flowered axillary clusters; calyx of 5 herbaceous sepals; petals rose or yellow; stamens 10, the filaments united at the base; pistil of 5 united carpels; fruit a capsule, the seeds transversely wrinkled.

KEY TO THE GENERA.

Plants acaulescent, with tuberous roots; flowers rose to almost violet.. 1. IONOXALIS (p. 384).
Plants caulescent, rhizomatous, sometimes with tubers; flowers yellow.. 2. XANTHOXALIS (p. 385).

1. IONOXALIS Small. VIOLET WOOD-SORREL.

Perennial acaulescent herbs with leaves and scapes rising from scaly bulbs; petioles elongated; leaf blades palmately 3 to several-foliolate, the leaflets narrowly to broadly obcordate, with cuneate bases; flowers in bracted cymose clusters terminating long slender weak scapelike peduncles, usually not very numerous; sepals 5, each with apical tubercles; stamens 10, the filaments united at the base; capsules erect.

KEY TO THE SPECIES.

Leaves with 4 to several leaflets.
  Leaflets as broad as long or broader; plants 8 cm. high or less.... 1. *I. caerulea.*
  Leaflets much longer than broad; plants more than 8 cm. high.. 2. *I. grayi.*
Leaflets 3.
  Filaments appendaged on the back.
    Longer filaments pubescent, the shorter ones glabrous; leaflets shallowly notched........................ 3. *I. metcalfei.*
    All filaments pubescent; leaflets deeply notched.......... 4. *I. amplifolia.*
  Filaments not appendaged.
    Tubercles of the sepals distinct.......................... 5. *I. monticola.*
    Tubercles confluent at the tips.......................... 6. *I. violacea.*

1. Ionoxalis caerulea Small, N. Amer. Fl. **25**: 33. 1907.

TYPE LOCALITY: Lincoln County, New Mexico. Type collected by Miss Josephine Skehan (no. 112).

RANGE: Known only from the type locality, in the Transition Zone.

2. Ionoxalis grayi Rose, Contr. U. S. Nat. Herb. **10**: 112. 1906.

Oxalis decapetala H. B. K. err. det. A. Gray, Pl. Wright. **2**: 25. 1853.

TYPE LOCALITY: Copper Mines, New Mexico. Type collected by Wright in 1851.

RANGE: New Mexico and northern Mexico.

NEW MEXICO: West Fork of the Gila; James Canyon. Damp woods and canyons, in the Transition Zone.

3. Ionoxalis metcalfei Small, N. Amer. Fl. **25**: 39. 1907.

TYPE LOCALITY: Mogollon Mountains, Socorro County, New Mexico. Type collected by Metcalfe (no. 299).

RANGE: Southwestern New Mexico.

NEW MEXICO: Mogollon Creek; Burro Mountains. Transition Zone.

4. Ionoxalis amplifolia (Trel.) Rose, Contr. U. S. Nat. Herb. **10**: 110. 1906.

Oxalis latifolia H. B. K. err. det. Trel. Mem. Bost. Soc. Nat. Hist. **4**: 91. *pl. 11. f. 12.* 1888.

Oxalis divergens amplifolia Trel. in A. Gray, Syn. Fl. **1**[1]: 368. 1897.

TYPE LOCALITY: Arizona.

RANGE: Western Texas to Arizona.

NEW MEXICO: San Luis Mountains; White and Sacramento mountains. Damp slopes, in the Transition Zone.

5. Ionoxalis monticola Small, N. Amer. Fl. **25:** 42. 1907.

TYPE LOCALITY: Iron Creek, Grant County, New Mexico. Type collected by Metcalfe (no. 1220).

RANGE: Western Texas and southern New Mexico.

NEW MEXICO: Iron Creek; Burro Mountains; Organ Mountains. Shaded slopes, in the Transition Zone.

6. Ionoxalis violacea (L.) Small, Fl. Southeast. U. S. 665. 1903.

Oxalis violacea L. Sp. Pl. 434. 1753.

TYPE LOCALITY: "Habitat in Virginia, Canada."

RANGE: Maine and Florida to the Rocky Mountains.

NEW MEXICO: Chama; Santa Fe and Las Vegas mountains. Open slopes, in the Transition Zone.

Our specimens are not typical *I. violacea*, but seem nearer to that species than to any other.

2. XANTHOXALIS Small. YELLOW WOOD-SORREL.

Low herbs with horizontal rootstocks sometimes bearing fusiform tubers; stems erect or decumbent, 10 to 30 cm. long, more or less pubescent or glabrate; leaflets obcordate; flowers rather small, in axillary few-flowered cymes; petals yellow; capsules columnar.

KEY TO THE SPECIES.

Rootstocks arising from an elongated tuberous root; leaves pubescent on both surfaces........ 1. *X. albicans.*
Rootstocks with fibrous roots, or plant with a slender taproot; leaves glabrous or nearly so........ 2. *X. stricta.*

1. Xanthoxalis albicans (H. B. K.) Small, N. Amer. Fl. **25:** 54. 1907.

Oxalis albicans H. B. K. Nov. Gen. & Sp. **5:** 244. 1821.

Oxalis wrightii A. Gray, Pl. Wright. **1:** 27. 1852.

Xanthoxalis wrightii Abrams, Bull. Torrey Club **34:** 264. 1907.

TYPE LOCALITY: "Crescit prope Moran Mexicanorum et Llactacunga Quitensium, alt. 1340 et 1480 hex."

RANGE: Western Texas to Arizona, south into Mexico and South America.

NEW MEXICO: Middle Fork of the Gila; Ash Canyon.

2. Xanthoxalis stricta (L.) Small, Fl. Southeast. U. S. 667. 1903.

Oxalis stricta L. Sp. Pl. 435. 1753.

Oxalis corniculata stricta Sav. in Lam. Encycl. **4:** 683. 1797.

TYPE LOCALITY: "Habitat in Virginia."

RANGE: Nova Scotia and Wyoming to Florida and Mexico.

NEW MEXICO: Agua Fria; Las Vegas; Cloverdale; West Fork of the Gila; Chama; Sierra Grande. Open slopes and fields, in the Upper Sonoran and Transition zones.

74. ZYGOPHYLLACEAE. Caltrop Family.

Trailing herbs or spreading shrubs with compound, 2 to several-foliolate, abruptly pinnate leaves; leaflets small, entire, sometimes inequilateral; flowers perfect, yellow or orange, mostly rather small, regular or nearly so; calyx of 5 sepals; petals of the same number; stamens twice as many, in two whorls; pistil of 5 united carpels (sometimes more or fewer); fruit a capsule of various forms, 3 to 12-celled, separating at maturity into nutlets containing the seeds.

KEY TO THE GENERA.

Spreading heavy-scented shrub with 2-foliolate evergreen resinous leaves........................ 1. COVILLEA (p. 386).
Trailing prostrate herbs with several pairs of leaflets, neither resinous nor evergreen.
  Fruit with several strong sharp spines............ 2. TRIBULUS (p. 386).
  Fruit merely tuberculate, not spiny.............. 3. KALLSTROEMIA (p. 386).

1. COVILLEA Vail. CREOSOTE BUSH.

Erect, spreading, strongly scented, evergreen, resinous shrub 2 meters high or less with 2-foliolate leaves and numerous rather small, bright yellow flowers; leaflets small, 5 to 10 mm. long, inequilateral; flowers solitary on short axillary peduncles; sepals unequal, caducous, yellow; petals obovate to spatulate, twisted like the blades of a propeller; stamens inserted on the 10-lobed disk, the filaments winged below; fruit a densely hairy spheroidal capsule breaking into 5 indehiscent nutlets.

1. **Covillea glutinosa** (Engelm.) Rydb. N. Amer. Fl. **25**: 108. 1910.

Larrea glutinosa Engelm. in Wisliz. Mem. North. Mex. 93. 1848.

TYPE LOCALITY: "Olla and Fray Cristobal," New Mexico. Type collected by Wislizenus in 1846.

RANGE: Western Texas to southern California and southward.

NEW MEXICO: From the south side of the Black Range to Socorro and Tularosa, and southward. Dry plains, in the Lower Sonoran Zone.

A characteristic shrub of the mesas in the southern part of the State, where the people most often call it "greasewood." The native people call it "hediondilla."

2. TRIBULUS L. BURNUT.

Trailing annual with odd-pinnate leaves and small yellow flowers followed by a spiny fruit somewhat resembling a sandbur; leaflets about 5 mm. long, oblong, acute, 4 to 6 pairs; flowers solitary on axillary peduncles, pale or bright yellow; stamens 10, hypogynous, the filaments not winged; pistil of 5 carpels, surrounded by an urceolate disk; fruit 5-angled, separating at maturity into 5 bony, reticulately veined carpels, each bearing 1 to 3 stout spines and divided interiorly into 3 to 5 1-seeded cells.

1. **Tribulus terrestris** L. Sp. Pl. 387. 1753.

TYPE LOCALITY: "Habitat in Europa australi ad semitas."

RANGE: A native of Europe, introduced into many parts of North America.

NEW MEXICO: Glorieta; Kingston; Mesilla Valley; Deming; Nara Visa; Filmore Canyon; Hillsboro.

A pernicious weed, introduced into many parts of North America from Europe. The spines of the fruit are almost as sharp as tacks and will penetrate the soles of shoes. They are a constant danger to rubber tires, too, growing by roads as they do.

3. KALLSTROEMIA Scop.

Trailing annuals or perennials, closely resembling the last, but with usually larger flowers with orange or bright orange petals, and never spiny, merely tuberculate-roughened fruit; sepals 5 or 6, mostly persistent and accrescent in age; petals 4 to 6, caducous; stamens 10 or 12, the filaments not winged; carpels 8 to 12, consisting of usually 1-seeded, bony nutlets falling from a persistent central column.

KEY TO THE SPECIES.

Petals 20 to 30 mm. long.................................. 1. *K. grandiflora.*
Petals 12 mm. long or less.
  Beak not longer than the body of the fruit, usually shorter.. 2. *K. brachystylis.*
  Beak much longer than the body of the fruit.
    Leaves strongly hirsute and paler beneath; beak glabrous or nearly so.................................. 4. *K. hirsutissima.*
    Leaves glabrous beneath or nearly so, of about the same color on both surfaces; beak finely puberulent.
      Plants prostrate; petals 5 to 7 mm. long, barely exceeding the sepals; carpels bluntly tuberculate..... 3. *K. parviflora.*
      Plants erect; petals 7 to 12 mm. long; carpels sharply tuberculate.................................. 5. *K. laetevirens.*

1. **Kallstroemia grandiflora** Torr.; A. Gray, Pl. Wright. **1:** 28. 1852.

Type locality: "Borders of the Gila," Arizona or New Mexico.

Range: Arizona and New Mexico.

New Mexico: Carlisle; Mangas Springs; Florida Mountains; Organ Mountains; Little Burro Mountains. Dry hills, in the Upper and Lower Sonoran zones.

2. **Kallstroemia brachystylis** Vail, Bull. Torrey Club **24:** 206. 1897.

Type locality: Mesa near Las Cruces, New Mexico. Type collected by Wooton.

Range: Arizona to Western Texas and southward.

New Mexico: Carrizo Mountains; Chama River; San Juan; Santa Fe; Mangas Springs; Albuquerque; Organ Mountains; Mesilla Valley; Pecos; Canada Valley. Dry plains, in the Lower and Upper Sonoran zones.

3. **Kallstroemia parviflora** Norton, Rep. Mo. Bot. Gard. **9:** 153. *pl. 46.* 1898.

Type locality: Agricultural College, Mississippi.

Range: Mississippi to New Mexico and Arizona, and southward.

New Mexico: Santa Fe; Socorro; Kingston; Organ Mountains; San Luis Mountains; Gray; Nara Visa; Raton; Rio Frisco; Tularosa; Dayton; Tucumcari. Dry hills and plains, in the Upper Sonoran Zone.

4. **Kallstroemia hirsutissima** Vail in Small, Fl. Southeast. U. S. 670. 1903.

Type locality: Plains south of the White Sands, New Mexico. Type collected by Wooton (no. 564).

Range: Kansas and Colorado to Texas and Mexico.

New Mexico: Plains south of White Sands; Socorro; Albuquerque; Santa Fe; Redrock; south of Roswell; Lakewood. Dry plains, in the Lower and Upper Sonoran zones.

5. **Kallstroemia laetevirens** Thornber, Contr. U. S. Nat. Herb. **16:** 143. 1913.

Type locality: Hanover Mountain, New Mexico. Type collected by J. M. Holzinger in 1911.

Range: Arizona and New Mexico.

New Mexico: Farmington; Hanover Mountain. Dry hills and plains, in the Upper Sonoran Zone.

75. KOEBERLINIACEAE. Junco Family.

1. KOEBERLINIA Zucc. Junco.

Much branched leafless shrub, 1 meter high or less, rarely 2 meters; stems of hard wood, green, the oldest blackish, each branch ending in a sharp thorn; leaves reduced to small scales; flowers slender-pediceled, in small lateral racemes on short peduncles; sepals 4, 1 mm. long; petals 4, twice to 3 times as long, greenish white; stamens 8, shorter than the petals, the filaments enlarged in the middle; fruit a spherical black berry about 6 mm. in diameter.

1. **Koeberlinia spinosa** Zucc. Abh. Akad. Wiss. München 1: 358. 1832.

TYPE LOCALITY: Mexico.

RANGE: Southern New Mexico and Arizona, western Texas, and adjacent Mexico.

NEW MEXICO: Warm Spring; plains south of the White Sands; west of Dona Ana Mountains; Black Range; Hachita; Organ Mountains; north of Deming. Dry plains and hills, in the Lower Sonoran Zone.

A very spiny, much branched shrub, usually branched from the base, but sometimes with a definite trunk. It is sometimes called "crown of thorns."

76. MALPIGHIACEAE. Malpighia Family.

1. JANUSIA A. Juss.

Low twining perennial with woody stems; leaves opposite, narrowly lanceolate, 1 to 3 cm. long, pubescent on both surfaces; sepals 5; petals 5, yellow, turning reddish brown; stamens 5; styles united; fruit a samara, 9 to 12 mm. long.

1. **Janusia gracilis** A. Gray, Pl. Wright. 1: 37. 1852.

TYPE LOCALITY: Mountains east of El Paso, Texas.

RANGE: Western Texas to southern Arizona and adjacent Mexico.

NEW MEXICO: Parkers Well; Tortugas Mountain. Dry hills, in the Lower Sonoran Zone.

77. RUTACEAE. Rue Family.

Aromatic shrubs or low herbaceous perennials; leaves alternate, simple or compound, glandular-punctate; flowers perfect or by abortion polygamous, in cymes or short raceme-like clusters, not conspicuous; sepals 4 or 5, small; petals of the same number, dull-colored, small; stamens of the same or twice the same number, inserted on a hypogynous disk; pistil of 2 or 3 united carpels; fruit a capsule or samara.

KEY TO THE GENERA.

Low herbaceous plants; leaves small, simple........... 1. RUTOSMA (p. 388).
Shrubs; leaves 3 to 10-foliolate, with large leaflets.
  Fruit a circular samara; leaves 3-foliolate......... 2. PTELEA (p. 388).
  Fruit a 2-celled pod without wings; leaves palmately 5 to 10-foliolate..................... 3. ASTROPHYLLUM (p. 390).

1. RUTOSMA A. Gray.

Perennial herb, 30 cm. high or less, with small linear sessile leaves and inconspicuous flowers; fruit of 2 divergent carpels.

1. **Rutosma purpurea** Woot. & Standl. Contr. U. S. Nat. Herb. 16: 143. 1913.

TYPE LOCALITY: On an arid rocky slope at Bishops Cap at the south end of the Organ Mountains, New Mexico. Type collected by Wooton, March 30, 1905.

RANGE: Western Texas to southern Arizona.

NEW MEXICO: San Andreas Mountains; Organ Mountains; Carrizalillo Mountains; south of Hillsboro; Mangas Springs. Dry rocky hills, in the Lower and Upper Sonoran zones.

2. PTELEA L. SHRUBBY TREFOIL.

Branching shrubs, 2 to 3 meters high, with smooth dark-colored bark on the old stems and greenish or yellow or reddish brown bark on the young stems, strongly scented; leaves 3-foliolate, the leaflets oblong-lanceolate or rhombic, the terminal one attenuate at the base, the lateral ones inequilateral, pellucid-punctate; flowers polygamous, greenish yellow, small, cymose; sepals, petals, and stamens 4 or 5, the last abortive in the pistillate flowers; ovary 2 or 3-celled; fruit a flattened, 2 or 3-seeded, disk-shaped, reticulate samara.

KEY TO THE SPECIES.

Younger branches pale yellow or straw-colored; plants sweet-scented; leaves turning bright yellow in autumn.......... 1. *P. angustifolia.*
Younger branches reddish brown; plants ill-scented; leaves green when shed.. 2. *P. tomentosa.*

1. Ptelea angustifolia Benth. Pl. Hartw. 9. 1839.

Ptelea verrucosa Greene, Contr. U. S. Nat. Herb. **10:** 69. 1906.

Ptelea confinis Greene, op. cit. 72.

TYPE LOCALITY: Mexico.

RANGE: Western Texas to southern California, south into Mexico.

NEW MEXICO: Bishops Cap; Dona Ana Mountains; Florida Mountains. Dry hills, in the Lower and Upper Sonoran zones.

For further synonymy of this species see the North American Flora.[1] *Ptelea angustifolia* is there considered to be a synonym of *P. baldwinii* Torr. & Gray, but the southwestern plants seem different from those of Florida, the type locality of *P. baldwinii.* The type of *P. verrucosa* was collected by the Mexican Boundary Survey, possibly in New Mexico. The type of *P. confinis* was collected near El Paso, Texas, by G. R. Vasey.

2. Ptelea tomentosa Raf. Fl. Ludov. 108. 1817.

Ptelea formosa Greene, Contr. U. S. Nat. Herb. **10:** 59. 1906.

Ptelea villosula Greene, op. cit. 60.

Ptelea undulata Greene, op. cit. 62.

Ptelea cognata Greene, loc. cit.

Ptelea jucunda Greene, op. cit. 63.

Ptelea parvula Greene, op. cit. 64.

Ptelea monticola Greene, loc. cit.

Ptelea similis Greene, op. cit. 65.

Ptelea polyadenia Greene, loc. cit.

Ptelea subvestita Greene, op. cit. 67.

Ptelea neomexicana Greene, op. cit. 68.

TYPE LOCALITY: Louisiana.

RANGE: Arizona and New Mexico, eastward to the Atlantic coast.

NEW MEXICO: Pajarito Park; Sandia Mountains; Coolidge; Barranca; Black Range; Burro Mountains; west of Chloride; Big Hatchet Mountains; Organ Mountains; White Mountains; San Luis Mountains. Canyons, in the Upper Sonoran and Transition zones.

For further synonomy of the species see the North American Flora.[2] A great many segregates of this group have been described recently, but there seems to be no means of separating the named forms definitely, at least without ampler material than is now available. The type of *P. formosa* came from the White Mountains of New Mexico (*Wooton* 657); that of *P. villosula* from the Organ Mountains (*Wooton* 134); that of *P. undulata* from the Burro Mountains (*Rusby* 111); that of *P. cognata* from Fort Huachuca, Arizona; the type of *P. jucunda* from the San Luis Mountains (*Mearns* 383); of *P. parvula* from the White Mountains (*Wooton* 658); of *P. monticola* from the Guadalupe Mountains of western Texas; of *P. similis* from near Clifton, Arizona, near the New Mexico line; of *P. polyadenia* from the Canadian River, possibly in New Mexico (*Bigelow*); of *P. subvestita* from about Silver City and Fort Bayard (*Greene*); of *P. neomexicana* from the Black Range (*Metcalfe* 1479).

---

[1] **25:** 210. 1911.

[2] **25:** 209. 1911.

3. ASTROPHYLLUM Torr.

Low shrub with rather thick rough stems bearing opposite, palmately 5 to 10-foliolate leaves crowded near the ends; leaflets linear, thick, bright green, coarsely dentate, conspicuously glandular-punctate; flowers large, 10 to 20 mm. in diameter, solitary or in 2 to 4-flowered clusters, axillary, white; stamens 8 to 10; ovary 5-lobed, hairy, becoming a 2-celled capsule by the abortion of some of the cells.

1. **Astrophyllum dumosum** Torr. U. S. Rep. Expl. Miss. Pacif. **2**: 161. 1855.

SORILLA.

TYPE LOCALITY: On the Organ Mountains, New Mexico. Type collected by Pope.

RANGE: Western Texas to southern Arizona.

NEW MEXICO: San Andreas Mountains. Dry hills, in the Upper Sonoran Zone.

Probably the type came from the San Andreas Range rather than from the Organs, for the shrub has not been found in the latter range in recent years, although it occurs in the San Andreas just to the north. Pope's expedition crossed over the pass lying between the two ranges.

78. SIMARUBACEAE. Quassia Family.

1. AILANTHUS Desf.

1. **Ailanthus glandulosa** Desf. Mém. Acad. Sci. Paris **1786**: 265. 1789.

TREE-OF-HEAVEN.

This has been somewhat extensively introduced into the southern part of the State, where it is of value as a shade tree. It frequently is 10 meters high or more. It has smooth bark and leaves resembling in shape those of the sumac or walnut, 30 to 60 cm. long. The small, dull white, polygamous, very malodorous flowers are borne in large terminal panicles; they are succeeded by the cluster of winged, 1-seeded, reddish samaras. It is a native of Eastern Asia.

79. MELIACEAE. China-berry Family.

1. MELIA L.

1. **Melia azederach** L. Sp. Pl. 384. 1753. CHINA-BERRY.

An introduced tree, often escaped in the southern part of the State, where it is one of the most common shade trees. It reaches a height of 8 to 10 meters and is widely branching and umbraculiform. The leaves are large, 30 to 90 cm. long, twice pinnate, with large, glossy, green leaflets. The abundant, pale lavender, sweet-scented flowers are complete, 5 or 6-merous, with 10 to 12 monadelphous stamens, the filaments being produced beyond the anthers. The "berries" are at first pulpy but at last a spherical, dry, several-seeded fruit 10 to 15 mm. in diameter. It is native in the warm parts of the Old World.

Order 29. POLYGALALES.

80. POLYGALACEAE. Milkwort Family.

Herbs or low shrubs with simple entire leaves and no stipules; flowers mostly small, papilionaceous in appearance; sepals 2; petals 3; stamens monadelphous or diadelphous, with 1-celled anthers opening by a terminal pore; pods flat, 1 or 2-celled.

KEY TO THE GENERA.

Pods 1-celled .... 1. MONNINA (p. 391).
Pods 2-celled .... 2. POLYGALA (p. 391).

1. MONNINA Ruiz & Pav.

Slender annual 30 cm. high or less, with terminal racemes of small blue flowers; leaves linear-lanceolate, 2 to 5 cm. long, acute; flowers about 3 mm. long, on deflexed pedicels; stamens 6, in two groups; fruit a small, circular, winged, minutely reticulate pod, the sides carinately 1-nerved.

1. **Monnina wrightii** A. Gray, Pl. Wright. **2**: 31. 1853.

TYPE LOCALITY: Crevices of rocks, mountain sides, near the Copper Mines, New Mexico. Type collected by Wright (no. 938).

RANGE: Southern New Mexico and Arizona and adjacent Mexico.

NEW MEXICO: Mogollon Creek; Kingston; Mangas Springs. Transition Zone.

2. POLYGALA L. MILKWORT.

Low herbs or shrubs with solitary or racemose flowers and small simple leaves; sepals 5, the 2 lateral ones large and petaloid; petals 3, united to each other and to the stamen tube, the middle one (keel) often crested or appendaged; stamens 6 or 8; pods 2-celled, flattened contrary to the partition, sometimes winged.

KEY TO THE SPECIES.

Annual.......... 9. *P. viridescens.*
Perennials.
  Keel of the corolla with a fimbriate crest; flowers white.
    Fruit not winged.......... 1. *P. alba.*
    Fruit winged.
      Mature capsule obscurely winged; inflorescence sparingly puberulent; leaves rigid, erect, linear.......... 2. *P. scoparia.*
      Mature capsule with a broad half-wing; inflorescence glabrous; leaves spreading, thin, broader.......... 3. *P. hemipterocarpa.*
  Keel not crested, sometimes with a solitary beaklike process; flowers variously colored.
    Flowers solitary.......... 4. *P. macradenia.*
    Flowers racemose.
      Stems woody, with spinose tips.......... 5. *P. subspinosa.*
      Stems herbaceous, not spinose-tipped.
        Keel furnished with a beaklike process; leaves glabrous, shining.......... 6. *P. parvifolia.*
        Keel not beaked; leaves puberulent, never shining.
          Faces of the fruit puberulent; leaves lanceolate, acute, thin.......... 7. *P. neomexicana.*
          Faces of the fruit glabrous; leaves linear or oblong-linear, mostly obtuse, thick. 8. *P. puberula.*

1. **Polygala alba** Nutt. Gen. Pl. **2**: 87. 1818.

TYPE LOCALITY: "On the plains of the Missouri."

RANGE: Washington and North Dakota to Arizona and Texas.

NEW MEXICO: Coolidge; Bear Mountain; Magdalena Mountains; White Mountains; Capitan Mountains; Torrance; Redlands; Nara Visa; Organ Mountains; Clayton; Sandia Mountains; Queen; Knowles. Dry hills, in the Upper Sonoran and Transition zones.

2. **Polygala scoparia** H. B. K. Nov. Gen. & Sp. **5**: 399. 1821.

Polygala scoparia multicaulis A. Gray, Pl. Wright. **1**: 38. 1852.

TYPE LOCALITY: "Crescit prope Mexico, alt. 1170 hex."

RANGE: Western Texas to southern Arizona, south into Mexico.

NEW MEXICO: Dog Spring (*Mearns* 41).

3. Polygala hemipterocarpa A. Gray, Pl. Wright. **2**: 31. 1853.

TYPE LOCALITY: "Stony hills of the Sonoita, near Deserted Rancho, on the borders of Sonora."

RANGE: Western Texas to southern Arizona and adjacent Mexico.

NEW MEXICO: Carrizalillo Mountains (*Mearns* 42).

4. Polygala macradenia A. Gray, Pl. Wright. **1**: 39. 1852.

TYPE LOCALITY: Hills at the head of the San Felipe, western Texas.

RANGE: Western Texas to southern Arizona.

NEW MEXICO: Upper Corner Monument; Guadalupe Mountains; San Andreas Mountains.

5. Polygala subspinosa S. Wats. Amer. Nat. **7**: 299. 1873.

TYPE LOCALITY: Silver City, Nevada.

RANGE: Colorado and Nevada to Arizona and New Mexico.

NEW MEXICO: Aztec (*Baker*). Dry hills, in the Upper Sonoran Zone.

6. Polygala parvifolia (Wheelock) Woot. & Standl.

Polygala lindheimeri parvifolia Wheelock, Mem. Torrey Club **2**: 143. 1891.

TYPE LOCALITY: Foothills of the Santa Rita Mountains, Arizona.

RANGE: Southwestern New Mexico and southeastern Arizona.

NEW MEXICO: Big Hatchet Mountains (*Mearns* 43).

From *P. lindheimeri* this may at once be distinguished by the smaller size, puberulent instead of pubescent stems, smaller leaves, and smaller glabrous fruit.

7. Polygala neomexicana Woot. & Standl. Contr. U. S. Nat. Herb. **16**: 144. 1913.

TYPE LOCALITY: Miller Hill, Grant County, New Mexico. Type collected by Metcalfe in 1897.

RANGE: Southwestern New Mexico and southeastern Arizona.

NEW MEXICO: Grant and Sierra counties. Dry hills, in the Upper Sonoran Zone.

A species related to *Polygala puberula*, but differing in its taller, more slender stems, larger, broader, thinner, nearly glabrous leaves, larger flowers, and larger puberulent fruit.

8. Polygala puberula A. Gray, Pl. Wright. **1**: 40. 1852.

TYPE LOCALITY: Valley of the Limpio, western Texas.

RANGE: Western Texas to southern Arizona.

NEW MEXICO: Organ Mountains; Hillsboro; Queen; Mangas Springs; White and Sacramento mountains. Dry hillsides, in the Upper Sonoran Zone.

9. Polygala viridescens L. Sp. Pl. 705. 1753.

Polygala sanguinea L. loc. cit.

TYPE LOCALITY: "Habitat in Virginia."

RANGE: Open slopes, New England to North Carolina, west to Minnesota and New Mexico, in the Upper Sonoran Zone.

We have seen no New Mexican specimens of this species, but Doctor Gray states[1] that it was collected by Fendler on "low prairies, near Las Vegas" (no. 109). It is to be expected on the plains of northeastern New Mexico.

Order 30. EUPHORBIALES.

KEY TO THE FAMILIES.

Styles and stigmas distinct or nearly so, cleft or foliaceous; ovary 3-celled; land plants... 81. **EUPHORBIACEAE** (p. 393).

Styles united by pairs; ovary 4-celled; aquatic plants.................................. 82. **CALLITRICHACEAE** (p. 405).

[1] Mem. Amer. Acad. n. ser. **4**: 30. 1849.

81. EUPHORBIACEAE. Spurge Family.

Monœcious or diœcious herbs or shrubs with acrid or milky sap; leaves simple, sessile or petiolate, alternate or opposite; stipules present or wanting; inflorescence various, the flowers involucrate in one section, the involucre resembling a calyx and the true calyx much reduced; number of parts of the perianth varying in the staminate and pistillate flowers in the same species; corolla often wanting, especially in pistillate flowers; stamens few or numerous, variously united or distinct; ovary usually 3-celled, with 1 or 2 ovules in each cell; fruit mostly a 3-celled capsule, separating at maturity into three 2-valved carpels each containing 2 or mostly 1 large seed.

KEY TO THE GENERA.

Flowers involucrate; calyx represented by a minute scale at the base of a filament-like pedicel.
- Glands of the involucre without petaloid appendages, naked, sometimes with crescent-shaped horns.
  - Stems not terminated by an umbel; stipules glandlike; involucres cymose-clustered, each with a single gland or rarely with 4 glands and fimbriate lobes.............................. 1. POINSETTIA (p. 394).
  - Stems terminating in an umbel; stipules none; involucres in open cymes, each with 4 glands and entire or toothed lobes........................ 2. TITHYMALUS (p. 395).
- Glands of the involucres with petaloid appendages, these sometimes much reduced.
  - Leaves alternate or scattered, at least below the inflorescence.
    - Perennials; stipules wanting; bracts not petaloid; leaves small, narrow.... 3. TITHYMALOPSIS (p. 396).
    - Annuals or biennials; stipules narrow; bracts petaloid; leaves large and broad (those of the inflorescence variegated)...................... 4. DICHROPHYLLUM (p. 396).
  - Leaves all opposite.
    - Leaf blades not oblique at the base; leaves mostly rather large........ 5. ZYGOPHYLLIDIUM (p. 397).
    - Leaf blades oblique at the base; leaves mostly small...................... 6. CHAMAESYCE (p. 397).

Flowers not involucrate; calyx of several sepals.
- Ovules and seeds 2 in each cell.
  - Annual; stamens 2; filaments distinct...... 7. REVERCHONIA (p. 401).
  - Perennial; stamens 3; filaments partly united................................ 8. PHYLLANTHUS (p. 401).
- Ovules and seeds solitary in each cell.
  - Petals present, at least in the staminate flowers.
    - Flowers in terminal cymes; petals conspicuous........................ 9. JATROPHA (p. 401).

Flowers in crowded spikelike clusters, terminal or axillary, the pistillate usually below; petals small and inconspicuous, often wanting in the pistillate flowers.
Stamens 10, monadelphous.........10. DITAXIS (p. 401).
Stamens 6, the filaments distinct...11. CROTON (p. 402).
Petals wanting in all flowers.
Stamens numerous, 8 to 20.
Shrubs; stigmas 2-cleft.............15. BERNARDIA (p. 405).
Herbs; stigmas dissected...........12. ACALYPHA (p. 403).
Stamens few, 2 to 5.
Plants slender, small, covered with stinging hairs; capsules pediceled.....................13. TRAGIA (p. 404).
Plants stout, much larger, glabrous; capsules sessile...............14. STILLINGIA (p. 404).

1. POINSETTIA Graham. POINSETTIA.

Annuals, 30 to 60 cm. high, the stems simple or branched from the base; upper leaves often colored, alternate below, opposite above; stipules glandlike; involucres in axillary or terminal cymes or solitary, the lobes fimbriate; glands fleshy, solitary or rarely 3 or 4, sessile or short-stalked, without appendages, the missing ones represented by narrow lobes; capsules exserted; seeds narrowed above, tuberculate or roughened.

KEY TO THE SPECIES.

Glands of the involucre sessile or nearly so; leaves discolored at the base; plants glabrous.............................. 1. *P. havanensis.*
Glands of the involucre stalked; bracts and leaves not discolored; plants more or less pubescent.
Leaf blades linear to linear-lanceolate; seeds not prominently tuberculate; glands of the involucre 3 or 4... 2. *P. cuphosperma.*
Leaf blades ovate to ovate-lanceolate; seeds prominently tuberculate; glands of the involucre solitary........ 3. *P. dentata.*

1. Poinsettia havanensis (Willd.) Small, Fl. Southeast. U. S. 722. 1903.
Euphorbia havanensis Willd.; Boiss. in DC. Prodr. **15**^{2}: 73. 1862.
TYPE LOCALITY: Cuba.
RANGE: Southeastern United States to Arizona and in tropical America.
NEW MEXICO: Organ Mountains.

2. Poinsettia cuphosperma (Engelm.) Small, Fl. Southeast. U. S. 721. 1903.
Euphorbia dentata cuphosperma Engelm. in Torr. U. S. & Mex. Bound. Bot. 190. 1859.
Euphorbia cuphosperma Boiss. in DC. Prodr. **15**^{2}: 73. 1862.
TYPE LOCALITY: Copper Mines, New Mexico. Type collected by Wright (no. 1834).
RANGE: Wyoming and South Dakota to Texas.
NEW MEXICO: Upper Pecos River; Santa Rita; Organ Mountains; Roswell; White Mountains. Dry, open slopes, in the Upper Sonoran Zone.

3. Poinsettia dentata (Michx.) Klotzsch & Garcke, Monatsb. Preuss. Akad. Wiss. Berlin **1859:** 253. 1859.
Euphorbia dentata Michx. Fl. Bor. Amer. **2:** 211. 1803.
TYPE LOCALITY: "Hab. in Tennassee."

RANGE: South Dakota and Pennsylvania to Utah and New Mexico.

NEW MEXICO: Santa Fe Creek; Pecos; Organ Mountains; Gila Hot Springs; White Mountains; Hanover Mountain; Taos; Las Vegas; Sacramento Mountains. Dry hills and plains, in the Upper Sonoran Zone.

2. TITHYMALUS Klotzsch & Garcke.

Annual or perennial herbs, light green, glabrous, with erect stems umbellately branching above; leaves alternate below, the upper opposite, crowded, sessile, exstipulate, mostly entire; involucres sessile or pedunculate, in terminal cymes, the lobes often toothed; glands 4, transversely oblong, reniform, or crescent-shaped by the hornlike appendages, the missing one represented by a thin, often ciliate lobe; capsules exserted, smooth or tuberculate; seeds variously pitted, sometimes carunculate.

KEY TO THE SPECIES.

Capsules tuberculate.
- Stems cymosely branched below the umbel; capsules short-warty .... 1. *T. missouriensis.*
- Stems racemosely branched; capsules long-warty.
  - Leaves coarsely serrate; stems few .... 3. *T. altus.*
  - Leaves obscurely serrulate; stems very numerous .... 2. *T. mexicanus.*

Capsules smooth.
- Cauline leaves broadest above the middle .... 4. *T. luridus.*
- Cauline leaves broadest near the base.
  - Seeds broadly truncate at the base; capsules 5 mm. long. 5. *T. chamaesula.*
  - Seeds rounded at the base; capsules 3 to 4 mm. long .... 6. *T. montanus.*

1. Tithymalus missouriensis (Norton) Small, Fl. Southeast. U. S. 721. 1903.

Euphorbia arkansana missouriensis Norton, Rep. Mo. Bot. Gard. **11:** 103. 1899.

TYPE LOCALITY: "In the Missouri River Valley, usually in open prairie or waste places, from Missouri to South Dakota and west to Colorado and Idaho, and extending into eastern Washington."

RANGE: As under type locality.

NEW MEXICO: Las Vegas; Farmington. Upper Sonoran Zone.

2. Tithymalus mexicanus (Engelm.) Woot. & Standl. Contr. U. S. Nat. Herb. **16:** 145. 1913.

Euphorbia dictyosperma mexicana Engelm. in Torr. U. S. & Mex. Bound. Bot. 191. 1859.

Euphorbia mexicana Norton, Rep. Mo. Bot. Gard. **11:** 105. 1899.

TYPE LOCALITY: "Valley of the Nagas, Bolson de Mapimi," Mexico.

RANGE: Western Texas to southern Arizona and Mexico.

We have seen no specimens of this from New Mexico, but Norton refers here one collected by Thurber at Mule Spring (no. 282).

3. Tithymalus altus (Norton) Woot. & Standl. Contr. U. S. Nat. Herb. **16:** 145. 1913.

Euphorbia alta Norton, Rep. Mo. Bot. Gard. **11:** 108. 1899.

TYPE LOCALITY: "In the mountains of southern Arizona and New Mexico, and in Mexico."

RANGE: Arizona and New Mexico to Mexico.

NEW MEXICO: Mountains west of San Antonio; Mogollon Mountains; White Mountains; Sacramento Mountains; Nutt. Transition Zone.

4. Tithymalus luridus (Engelm.) Woot. & Standl. Contr. U. S. Nat. Herb. **16:** 145. 1913.

Euphorbia lurida Engelm. Proc. Amer. Acad. **5:** 173. 1861.

TYPE LOCALITY: Base of the San Francisco Mountains, Arizona.

RANGE: Utah to Arizona and New Mexico.

NEW MEXICO: Ramah; San Lorenzo; Magdalena Mountains; Horse Camp; Tunitcha Mountains; Willow Creek, Rio Arriba County. Meadows.

5. Tithymalus chamaesula (Boiss.) Woot. & Standl. Contr. U. S. Nat. Herb. **16:** 145. 1913.

Euphorbia chamaesula Boiss. Cent. Euphorb. 38. 1860.

TYPE LOCALITY: Near the Copper Mines, New Mexico. Type collected by Wright (no. 1820).

RANGE: Arizona and New Mexico and adjacent Mexico.

NEW MEXICO: Mogollon Mountains, and south to the Mexican boundary. Transition Zone.

6. Tithymalus montanus (Engelm.) Small; Rydb. Colo. Agr. Exp. Sta. Bull. **100:** 224. 1906.

Euphorbia montana Engelm. in Torr. U. S. & Mex. Bound. Bot. 192. 1859.

Euphorbia montana gracilior Engelm. loc. cit.

TYPE LOCALITY: Near Santa Fe, New Mexico. Type collected by Fendler (no. 786).

RANGE: Utah and Colorado to Arizona and western Texas.

NEW MEXICO: Farmington; Sandia Mountains; Chama; Santa Fe and Las Vegas mountains; Black Range; San Luis Mountains; White and Sacramento mountains. Meadows, in the Transition Zone.

The type of *E. montana gracilior* was collected near Santa Fe (*Fendler* 786, in part).

3. TITHYMALOPSIS Klotzsch & Garcke.

Slender glabrous perennial herb with erect stems and oblong-linear alternate leaves; involucres on pubescent pedicels 3 to 5 mm. long, clustered; glands sessile or stalked, with yellowish white appendages; capsule exserted; seeds narrowed upward, more or less punctate, without caruncles.

1. Tithymalopsis strictior (Holzinger) Woot. & Standl.

Euphorbia strictior Holzinger, Contr. U. S. Nat. Herb. **1:** 214. *pl. 18.* 1892.

TYPE LOCALITY: Oldham County, Texas.

RANGE: Texas to eastern New Mexico.

NEW MEXICO: Tucumcari Mountain; Nara Visa. Plains and hills, in the Upper Sonoran Zone.

4. DICHROPHYLLUM Klotzsch & Garcke. SNOW-ON-THE-MOUNTAIN.

Annual with erect stems umbellately branched above, 40 to 80 cm. high; leaves alternate or opposite, sessile, ovate, entire, acute, about 5 cm. long, those of the inflorescence broadly white-margined, attenuate below into a short petiole; flowers crowded at the summit of the stem; involucres campanulate, in rather dense cymes, the lobes fimbriate, the 5 glands peltate, somewhat concave, with pink and white petal-like appendages; capsule exserted, pubescent; seeds narrowed upward, carunculate.

1. Dichrophyllum marginatum (Pursh) Klotzsch & Garcke, Monatsb. Preuss. Akad. Wiss. Berlin **1859:** 249. 1859.

Euphorbia marginata Pursh, Fl. Amer. Sept. 607. 1814.

Tithymalus marginatus Cockerell, Univ. Mo. Stud. Sci. 2^2: 165. 1911.

TYPE LOCALITY: "On the Yellow-stone river."

RANGE: Montana and Minnesota to New Mexico and Texas.

NEW MEXICO: Frio Draw; Tucumcari; Red Lake east of Elida; Albert; Roswell. Plains, in the Upper Sonoran Zone.

A showy plant, not rare in cultivation. It is easily recognized by the white-margined floral leaves.

5. ZYGOPHYLLIDIUM Small.

Annuals with erect branching stems; leaves opposite, or rarely alternate below, not oblique at the base, entire; stipules glandlike, often obsolete; involucres delicate, short-pedunculate in the upper forks of the stems; glands 5, broader than long, subtended by petal-like appendages; capsules long-pediceled, 3-lobed; seeds terete, usually narrowed upward, more or less papillose, the caruncle sometimes wanting.

KEY TO THE SPECIES.

Plants glabrous; upper leaves ovate, long-petioled.............. 1. *Z. delicatulum.*
Plants more or less pubescent; upper leaves linear or oblong-linear.
  Glands bilobate; seeds not carunculate.................... 2. *Z. bilobatum.*
  Glands entire; seeds carunculate.......................... 3. *Z. exstipulatum.*

1. **Zygophyllidium delicatulum** Woot. & Standl. Contr. U. S. Nat. Herb. **16:** 145. 1913.

TYPE LOCALITY: Mineral Creek, Sierra County, New Mexico. Type collected by Metcalfe (no. 1414).

RANGE: Mountains of southern New Mexico.

NEW MEXICO: Black Range; White Mountains. Transition Zone.

A very different plant from any of the other species of the genus, differing most noticeably in the width and shape of the leaf blades and in the glabrous involucres.

2. **Zygophyllidium bilobatum** (Engelm.) Standley, Contr. U. S. Nat. Herb. **13:** 199. 1910.

Euphorbia bilobata Engelm. in Torr. U. S. & Mex. Bound. Bot. 190. 1859.

TYPE LOCALITY: Near the Copper Mines, New Mexico. Type collected by Bigelow.

RANGE: New Mexico and Arizona.

NEW MEXICO: Mogollon Mountains to the Organ Mountains and southward. Hills and canyons, in the Upper Sonoran Zone.

3. **Zygophyllidium exstipulatum** (Engelm.) Woot. & Standl. Contr. U. S. Nat. Herb. **16:** 146. 1913.

Euphorbia exstipulata Engelm. in Torr. U. S. & Mex. Bound. Bot. 189. 1859.

TYPE LOCALITY: Western Texas.

RANGE: Western Texas to southern Arizona and adjacent Mexico.

NEW MEXICO: Stanley; Tortugas Mountain; Kingston; plains between Fort Wingate and Belen; Carrizozo; Apache Teju. Lower and Upper Sonoran zones.

6. CHAMAESYCE S. F. Gray. SPURGE.

Annual or perennial herbs, mostly branching from the base; branches erect, ascending, or prostrate; leaves opposite, the blades entire or toothed, more or less oblique at the base; stipules delicate, entire or fimbriate; involucres axillary, solitary or in cymes; glands 4, sessile or stalked, naked or usually appendaged, one sinus glandless; capsules smooth, sometimes pubescent, the angles sharp or rounded; seeds smooth or transversely wrinkled, with minute caruncles.

The different species are known among the Mexicans as "golondrina." The plants are reputed, everywhere in the Southwest, to be a remedy for rattlesnake bites.

KEY TO THE SPECIES.

Leaf blades toothed, at least near the apex.
  Capsules pubescent.
    Glands of the involucre with conspicuous petal-like appendages..................................... 1. *C. indivisa.*
    Glands of the involucre with small and inconspicuous appendages..................................... 2. *C. stictospora.*

Capsules glabrous.
Plants more or less pilose on the stems and leaves.
Stems erect; leaves 2 cm. long or more............ 3. *C. nutans.*
Stems prostrate; leaves usually less than 1 cm. long. 4. *C. serrula.*
Plants glabrous, or at least never pilose.
Stems erect or strongly ascending; plants yellowish green.................................... 6. *C. neomexicana.*
Stems prostrate; plants not yellowish.
Seeds conspicuously wrinkled horizontally....17. *C. glyptosperma.*
Seeds not wrinkled, pitted or smooth.
Seeds pitted..............................18. *C. rugulosa.*
Seeds smooth or nearly so................ 5. *C. serpyllifolia.*
Leaf blades entire.
Perennials.
Leaves strongly pubescent........................... 7. *C. lata.*
Leaves glabrous.
Stems prostrate; involucres corolla-like .... 8. *C. albomarginata.*
Stems erect or ascending; involucres inconspicuous.
Leaves scarcely if at all longer than broad, rounded-obtuse......................... 9. *C. fendleri.*
Leaves twice as long as broad, acute..........10. *C. chaetocalyx.*
Annuals.
Stems prostrate; leaves usually very oblique at the base.
Stipules ciliate......................................13. *C. micromera.*
Stipules not cilitate.
Leaves about as long as broad; stipules triangular-subulate...........................11. *C. serpens.*
Leaves twice as long as broad; stipules filiform.12. *C. geyeri.*
Stems erect; leaves slightly if at all unequal at the base.
Capsules less than 1.5 mm. broad; plants 15 cm. high or less; branches divaricate............14. *C. revoluta.*
Capsules more than 1.5 mm. broad; plants 30 cm. high or more; branches ascending.
Appendages of glands conspicuous, white.....15. *C. petaloidea.*
Appendages inconspicuous, greenish white, or obsolete......................................16. *C. flagelliformis.*

1. Chamaescyce indivisa (Engelm.) Millsp. Field Mus. Bot. **2**: 387. 1914.

Euphorbia dioica indivisa Engelm. in Torr. U. S. & Mex. Bound. Bot. 187. 1859.

TYPE LOCALITY: Near the Copper Mines, New Mexico. Type collected by Wright (no. 1845).

RANGE: Western Texas and southern Arizona to Mexico.

NEW MEXICO: Hillsboro; Organ Mountains. Dry plains and low hills, in the Upper Sonoran Zone.

2. Chamaesyce stictospora (Engelm.) Small, Fl. Southeast. U. S. 714. 1903.

Euphorbia stictospora Engelm. in Torr. U. S. & Mex. Bound. Bot. 187. 1859.

TYPE LOCALITY: "From Kansas (Fendler, 798) to Santa Fe (Fendler, 797) and Doña Ana (Wright, 59), New Mexico, and Corallitas, Chihuahua."

RANGE: Kansas and Colorado to Mexico.

NEW MEXICO: Pecos; Fort Cummings; Kingston; Sandia Mountains; Mesilla Valley; Carrizozo; south of Roswell; Gray. Hills and plains, in the Lower and Upper Sonoran zones.

3. **Chamaesyce nutans** (Lag.) Small, Fl. Southeast. U. S. 712. 1903.
Euphorbia nutans Lag. Gen. & Sp. Nov. 17. 1816.
Euphorbia preslii Guss. Fl. Sic. Prodr. 539. 1827.
TYPE LOCALITY: "Habitat in N. [ova] H. [ispania]."
RANGE: Eastern North America, west to the Rocky Mountains, south into Mexico.
NEW MEXICO: Mangas Springs; Organ Mountains; Lower Plaza; Roswell. Damp fields and canyons, in the Upper Sonoran Zone.

4. **Chamaesyce serrula** (Engelm.) Woot. & Standl. Contr. U. S. Nat. Herb. **16:** 144. 1913.
Euphorbia serrula Engelm. in Torr. U. S. & Mex. Bound. Bot. 188. 1859.
TYPE LOCALITY: "Western Texas and New Mexico."
RANGE: Western Texas to Arizona and adjacent Mexico.
NEW MEXICO: Farmington; Tesuque; mesa west of Organ Mountains; White Sands; south of Roswell; Carlsbad. Dry plains and hills, in the Lower and Upper Sonoran zones.

5. **Chamaesyce serpyllifolia** (Pers.) Small, Fl. Southeast. U. S. 712. 1903
Euphorbia serpyllifolia Pers. Syn. Pl. **2:** 14. 1807.
TYPE LOCALITY: "In Amer. [ica] calidiore."
RANGE: Wisconsin to California, south to Mexico.
NEW MEXICO: Common throughout the State except along the eastern border. Plains and low hills, Lower Sonoran to Transition Zone.

6. **Chamaesyce neomexicana** (Greene) Standley, Contr. U. S. Nat. Herb. **13:** 199. 1910.
Euphorbia neomexicana Greene, Bull. Calif. Acad. **2:** 55. 1886.
TYPE LOCALITY: Plains of the upper Gila in western New Mexico.
RANGE: Southwestern New Mexico and adjacent Arizona.
NEW MEXICO: Hillsboro; Sycamore Creek; Organ Mountains; Gray; south of Roswell. Upper Sonoran Zone.

7. **Chamaesyce lata** (Engelm.) Small, Fl. Southeast. U. S. 710. 1903.
Euphorbia lata Engelm. in Torr. U. S. & Mex. Bound. Bot. 188. 1859.
TYPE LOCALITY: Western Texas.
RANGE: Kansas to Texas and New Mexico.
NEW MEXICO: Organ Mountains; Roswell; Nara Visa; Carrizozo; south of the White Sands; Guadalupe Mountains; Buchanan; Redlands; Carlsbad. Plains, Lower and Upper Sonoran zones.

8. **Chamaesyce albomarginata** (Torr. & Gray) Small, Fl. Southeast. U. S. 710. 1903.
Euphorbia albomarginata Torr. & Gray, U. S. Rep. Expl. Miss. Pacif. **2**^{2}**:** 174. 1855.
TYPE LOCALITY: "Head-waters of the Colorado."
RANGE: California to Texas and southward.
NEW MEXICO: Fort Wingate; Socorro; Mangas Springs; Kingston; Deming; Organ Mountains; mesa near Las Cruces; Tucumcari; Gray; south of Roswell; Tularosa; Carrizalillo Mountains; Zuni; Water Canyon. Dry plains and hills, in the Lower and Upper Sonoran zones.

9. **Chamaesyce fendleri** (Torr. & Gray) Small, Fl. Southeast. U. S. 710. 1903.
Euphorbia fendleri Torr. & Gray, U. S. Rep. Expl. Miss. Pacif. **2**^{2}**:** 175. 1855.
TYPE LOCALITY: New Mexico. Type collected by Fendler (no. 800).
RANGE: Wyoming and Nebraska to Texas and Arizona.
NEW MEXICO: Nearly throughout the State. Dry hills and plains, in the Lower and Upper Sonoran zones.

10. **Chamaesyce chaetocalyx** (Boiss.) Woot. & Standl. Contr. U. S. Nat. Herb. **16**: 144. 1913.

Euphorbia fendleri chaetocalyx Boiss. in DC. Prodr. **15**[2]: 39. 1862.

TYPE LOCALITY: "In Novo-Mexico." Type collected by Wright (no. 1847).

RANGE: Western Texas to Arizona.

NEW MEXICO: Canyoncito; Los Lunas; Albuquerque; San Augustine Plains; Kingston; Organ Mountains; Tortugas Mountain; White Mountains; Queen. Dry mesas and hills, in the Lower Sonoran Zone.

11. **Chamaesyce serpens** (H. B. K.) Small, Fl. Southeast. U. S. 709. 1903.

Euphorbia serpens H. B. K. Nov. Gen. & Sp. **2**: 52. 1817.

TYPE LOCALITY: "Crescit in umbrosis Cumanae prope Borcones et Punta Araya," Venzuela.

RANGE: Iowa to New Mexico, western Texas, and Mexico.

NEW MEXICO: South of Roswell; Carlsbad; mesa west of Organ Mountains. Plains and mesas, in the Lower Sonoran Zone.

12. **Chamaesyce geyeri** (Engelm.) Small, Fl. Southeast. U. S. 709. 1903.

Euphorbia geyeri Engelm. Bost. Journ. Nat. Hist. **5**: 260. 1845.

TYPE LOCALITY: "Beardstown, Illinois, and Upper Missouri."

RANGE: Minnesota and Illinois to Oklahoma and New Mexico.

NEW MEXICO: Zuni; Estancia; Pecos. Dry plains and hills, in the Upper Sonoran Zone.

13. **Chamaesyce micromera** (Boiss.) Woot. & Standl. Contr. U. S. Nat. Herb. **16**: 144. 1913.

Euphorbia micromera Boiss. in DC. Prodr. **15**[2]: 44. 1862.

TYPE LOCALITY: Sonora.

RANGE: New Mexico and Arizona and adjacent Mexico.

NEW MEXICO: Mesa west of Organ Mountains. Mesas, in the Lower Sonoran Zone.

14. **Chamaesyce revoluta** (Engelm.) Small, Fl. Southeast. U. S. 711. 1903.

Euphorbia revoluta Engelm. in Torr. U. S. & Mex. Bound. Bot. 186. 1859.

TYPE LOCALITY: Gravelly hills near Rock Creek, western Texas.

RANGE: Western Texas to Arizona, south into Mexico.

NEW MEXICO: Mesa west of Organ Mountains; Gray; Hillsboro; Mangas Springs. Mesas and dry hills, in the Lower Sonoran Zone.

15. **Chamaesyce petaloidea** (Engelm.) Small, Fl. Southeast. U. S. 711. 1903.

Euphorbia petaloidea Engelm. in Torr. U. S. & Mex. Bound. Bot. 185. 1859.

TYPE LOCALITY: Not stated.

RANGE: Wyoming and Idaho to New Mexico and Texas.

NEW MEXICO: San Juan Valley; lower Pecos Valley. Dry hills, in the Upper and Lower Sonoran zones.

16. **Chamaesyce flagelliformis** (Engelm.) Rydb. Colo. Agr. Exp. Sta. Bull. **100**: 223. 1906.

Euphorbia petaloidea flagelliformis Engelm. in Torr. U. S. & Mex. Bound. Bot. 185. 1859.

Euphorbia flagelliformis Engelm.; T. S. Brandeg. Bull. U. S. Geol. Geogr. Surv. Terr. **2**: 243. 1876.

TYPE LOCALITY: New Mexico.

RANGE: Colorado to Texas and New Mexico.

NEW MEXICO: Sabinal; Lava; Mesilla Valley; south of Melrose; Farmington. Dry hills and plains, in the Lower and Upper Sonoran zones.

17. **Chamaesyce glyptosperma** (Engelm.) Small, Fl. Southeast. U. S. 712. 1903.

Euphorbia glyptosperma Engelm. in Torr. U. S. & Mex. Bound. Bot. 187. 1859.

TYPE LOCALITY: On the Rio Grande, Texas.

RANGE: British America to Texas and Mexico.

NEW MEXICO: Shiprock; Dulce; Tunitcha Mountains. Sandy soil, in the Upper Sonoran Zone.

18. **Chamaesyce rugulosa** (Engelm.) Rydb. Bull. Torrey Club **33:** 145. 1906.
Euphorbia serpyllifolia rugulosa Engelm.; Millsp. Pittonia **2:** 85. 1890.

TYPE LOCALITY: San Bernardino, California.

RANGE: California to Colorado and New Mexico.

NEW MEXICO: Dulce (*Standley* 8135). Open hills, in the Upper Sonoran Zone.

7. REVERCHONIA A. Gray.

Slender, smooth, divaricately branched annual, 30 cm. high or less, with glaucous stems, narrow entire leaves, and small axillary clusters of dark purple flowers; staminate flowers with 4 sepals and 2 short stamens; pistillate flowers with 5 sepals and a 6-lobed disk; ovary 3-celled; styles 3, distinct; fruit a dry capsule.

1. **Reverchonia arenaria** A. Gray, Proc. Amer. Acad. **16:** 107. 1881.

TYPE LOCALITY: Sandhills of the Brazos, Baylor County, Texas.

RANGE: Texas and southern New Mexico, south into Mexico.

NEW MEXICO: Lava; sands northeast of Jornada Range, Dona Ana County; Roswell. Dry hills and plains, in the Lower Sonoran Zone.

8. PHYLLANTHUS L.

Low herbaceous perennial from a woody base, 10 to 20 cm. high, with slender smooth stems, small leaves, and very small greenish flowers; leaves narrowly oblong-cuneate, with lanceolate stipules at the base of the short petiole; flowers in axillary few-flowered clusters, monœcious; staminate flowers about 0.5 mm. long, the pistillate about 3 times as large, on very slender pedicels; sepals 5 or 6, green, white-margined, persistent at the base of the 3-celled 6-valved 3-seeded capsule.

1. **Phyllanthus polygonoides** Nutt.; Spreng. Syst. Veg. **3:** 23. 1826.

TYPE LOCALITY: "Ditio Arcansa Amer. bor."

RANGE: New Mexico to Texas and Oklahoma.

NEW MEXICO: Organ Mountains; Queen; Hatchet Ranch. Dry, rocky hillsides, in the Upper Sonoran Zone.

9. JATROPHA L.

Herbaceous perennial from a thickened tuberous root; stems thick and succulent, 30 to 50 cm. high, erect; leaves petiolate, palmately 3 to 5-lobed, 10 cm. in diameter or less, the lobes triangular-lanceolate, with rather numerous coarse aristate teeth; stipules, bracts, and sepals laciniately lobed into linear segments; flowers large, in a terminal cyme; sepals 5, united below; petals 5, bright pink, about 1 cm. long; filaments united below, unequal; fruit a 3-celled capsule, each 2-valved cell containing a single large caruncúlate seed.

1. **Jatropha macrorhiza** Benth. Pl. Hartw. 8. 1839.

TYPE LOCALITY: Mexico.

RANGE: Southern New Mexico and Arizona to Mexico.

NEW MEXICO: Southern Grant and Luna counties. Sandy plains, in the Lower Sonoran Zone.

10. DITAXIS Vahl.

Monœcious herbaceous perennials with simple alternate leaves and axillary clusters of flowers; staminate flowers with 4 or 5 sepals, the petals of the same number and alternate with them; stamens once, twice, or thrice the number of the petals, with united filaments; petals of the pistillate flowers smaller or rudimentary; styles 3, 2-cleft; fruit a 3-seeded capsule; seeds globose.

52576°—15——26

KEY TO THE SPECIES.

Plants hirsute.. 1. *D. neomexicana.*
Plants glabrous.
  Stems slender, much branched; leaves petiolate............ 2. *D. laevis.*
  Stems stout, simple; leaves sessile......................... 3. *D. cyanophylla.*

1. **Ditaxis neomexicana** (Muell. Arg.) Heller, Cat. N. Amer. Pl. 5. 1898.
Argyrothamnia neomexicana Muell. Arg. Linnaea **34:** 19. 1865.
TYPE LOCALITY: "In Novo-Mexico." The types are Wright's 643 and 1797. The first of these is certainly Texan, but the second may have come from New Mexico.
RANGE: Western Texas to Arizona and adjacent Mexico.
NEW MEXICO: South of Hillsboro (*Metcalfe* 1287).

2. **Ditaxis laevis** (Torr.) Heller, Cat. N. Amer. Pl. 5. 1898.
Aphora laevis Torr. U. S. & Mex. Bound. Bot. 196. 1859.
Argyrothamnia laevis Muell. Arg. Linnaea **34:** 147. 1865.
TYPE LOCALITY: Western Texas.
RANGE: Western Texas and southern New Mexican.
NEW MEXICO: Roswell and vicinity. Dry plains, in the Lower Sonoran Zone.

3. **Ditaxis cyanophylla** Woot. & Standl. Bull. Torrey Club **36:** 106. 1909.
TYPE LOCALITY: Kingston, Sierra County,. New Mexico. Type collected by Metcalfe, May 25, 1904.
RANGE: Western New Mexico and adjacent Arizona.
NEW MEXICO: Kingston; south of Rito Quemado.

11. CROTON L.

Herbaceous or woody annuals or perennials, more or less stellate-pubescent, with alternate simple entire leaves and inconspicuous monœcious or diœcious flowers, these in axillary or terminal, spicate or racemose clusters, sometimes crowded; staminate flowers uppermost; sepals 4 to 6, usually 5; petals mostly present but small, alternating with glands; stamens 5 or more; pistillate flowers usually loosely clustered; their sepals 5 to 10, the petals usually wanting; stigmas much divided; capsule 3-celled, splitting into 2-valved carpels each containing 1 seed.

KEY TO THE SPECIES

Annuals.
  Plants stellate-pubescent throughout, grayish............... 1. *C. texensis.*
  Plants glabrous, bright yellowish green...................... 2. *C. luteovirens.*
Perennials.
  Low shrub with cordate or subcordate leaves............... 3. *C. fruticulosus.*
  Herbs, sometimes suffrutescent near the base, the leaves never cordate, mostly oval or oblong, rounded at the base.
    Staminate flowers petaliferous.
      Leaves gray on both surfaces, the upper ones obtuse. 4. *C. corymbulosus.*
      Leaves green above, the upper ones acute.......... 5. *C. eremophilus.*
    Staminate flowers apetalous.
      Plants loosely stellate-pubescent, not silvery...... 6. *C. tenuis.*
      Plants densely lepidote-stellate, silvery............ 7. *C. neomexicanus.*

1. **Croton texensis** (Klotzsch) Muell. Arg. in DC. Prodr. **15**^{2}: 692. 1866.
Hendecandra texensis Klotzsch in Wiegmann, Archiv Naturg. **7:** 252. 1841.
Hendecandra multiflora Torr. in Frém. Rep. Exped. Rocky Mount. 96. 1845.
TYPE LOCALITY: Texas.
RANGE: Wyoming and Illinois to Arizona and Texas, south into Mexico.

New Mexico: Santa Fe; Pecos; Zuni Reservation; Sandia Mountains; Socorro; Mangas Springs; Gila; Nara Visa; Malaga; Tucumcari; south of Melrose. Plains and low hills, in the Upper Sonoran Zone.

A common weed in many parts of the State, especially abundant in draws or flats on overstocked ranges.

2. Croton luteovirens Woot. & Standl. Contr. U. S. Nat. Herb. **16:** 145. 1913.

Type locality: On the Rio Gila, New Mexico. Type collected by Wooton, August 15, 1902.

Range: Known only from the type locality.

The plant is very abundant in this region, growing with the related *C. texensis*. Patches of the two are distinguishable at a distance because of their different color.

3. Croton fruticulosus Engelm. in Torr. U. S. & Mex. Bound. Bot. 194. 1859.

Type locality: "Mountain sides and rocky ravines, western Texas."

Range: Western Texas and southern New Mexico to Mexico.

New Mexico: Organ Mountains. Rocky hills and canyons, in the Upper Sonoran Zone.

A low shrub, 1 meter high or less.

4. Croton corymbulosus Engelm. in Wheeler, Rep. U. S. Surv. 100th Merid. **5:** 242. 1878.

Type locality: Camp Bowie, Arizona.

Range: Western Texas to Arizona, and southward.

New Mexico: Silver City Draw; Organ Mountains; Tortugas Mountain; near White Water; Tucumcari; Jarilla; between Fierro and Hanover; Pecos Valley. Sandy mesas and barren rocky hills, in the Lower Sonoran Zone.

5. Croton eremophilus Woot. & Standl. Contr. U. S. Nat. Herb. **16:** 144. 1913.

Type locality: Dog Spring in the Dog Mountains, New Mexico. Type collected by E. A. Mearns (no. 2336).

Range: Southwestern New Mexico.

New Mexico: Dog Spring; Parkers Well. Dry hills and plains, in the Lower Sonoran Zone.

6. Croton tenuis S. Wats. Proc. Amer. Acad. **14:** 297. 1879.

Croton californicus tenuis Ferguson, Rep. Mo. Bot. Gard. **12:** 64. 1901.

Type locality: Southern California.

Range: Southwestern New Mexico to California and adjacent Mexico

New Mexico: Near White Water (*Mearns* 2269).

7. Croton neomexicanus Muell. Arg. Linnaea **34:** 141. 1865.

Type locality: Western Texas.

Range: Western Texas, southern New Mexico, and adjacent Mexico.

New Mexico: Grant County; mesa west of Organ Mountains; Guadalupe Mountains. Mesas and low, dry hills, in the Lower Sonoran Zone.

12. ACALYPHA L. Three-seeded mercury.

Annual or perennial herbs with simple petiolate leaves and monœcious flowers in axillary or terminal spikes; leaves thin, punctate, serrate; staminate flowers with 4 sepals and 8 to 16 united stamens; pistillate flowers subtended by foliaceous bracts, the sepals 3 to 5, the stigmas fringed; fruit a 3-celled 3-seeded capsule.

KEY TO THE SPECIES.

Annual; stigmas greenish, inconspicuous; inflorescence mostly axillary.. 1. *A. neomexicana.*

Perennial; stigmas bright red, showy; inflorescence terminal..... 2. *A. lindheimeri.*

1. **Acalypha neomexicana** Muell. Arg. Linnaea **34**: 19. 1865.

TYPE LOCALITY: New Mexico. Type collected by Wright (no. 1817).

RANGE: Western Texas to Arizona and adjacent Mexico.

NEW MEXICO: West Fork of the Gila; Kingston; Mangas Springs; Organ Mountains; White Mountains; Queen; Carlsbad. Upper Sonoran Zone.

2 **Acalypha lindheimeri** Muell. Arg. Linnaea **34**: 47. 1865.

TYPE LOCALITY: Texas.

RANGE: Western Texas to southern Arizona, south into Mexico.

NEW MEXICO: Santa Rita; Deer Creek; Mangas Springs; Kingston; San Luis Mountains; Animas Valley; west of Hope; Guadalupe Mountains.

13. TRAGIA L.

Low herbaceous much-branched perennials with slender wiry stems armed with stinging hairs; leaves simple, alternate, small, coarsely toothed, short-petiolate; flowers small, in small clusters near the ends of the stems, monœcious, apetalous; stamens 3 or 5; sepals 3 to 5 in the staminate flowers, 5 in the pistillate flowers; fruit a 3-celled 3-seeded capsule.

KEY TO THE SPECIES.

Stems appressed-pubescent; staminate calyx with 3 sepals; stamens 3........ 1. *T. nepetaefolia.*
Stems hirsute; staminate calyx with 4 or 5 sepals; stamens 4 or 5.. 2. *T. ramosa.*

1. **Tragia nepetaefolia** Cav. Icon. Pl. **6**: 37. *pl. 557. f. 1.* 1801.

TYPE LOCALITY: "Habitat inter Ixmiquilpan et Cimapan," Mexico.

RANGE: Kansas and Arizona, south into Mexico.

NEW MEXICO: Organ Mountains; White Mountains; south of Roswell. Dry hills and plains, in the Upper Sonoran and Transition zones.

2. **Tragia ramosa** Torr. Ann. Lyc. N. Y. **2**: 245. 1828.

Tragia stylaris Muell. Arg. Linnaea **34**: 180. 1865.

TYPE LOCALITY: "Sources of the Canadian ?," New Mexico ?

RANGE: Colorado and Missouri to Arizona and Texas.

NEW MEXICO: Pecos; Raton; Sierra Grande; Albuquerque; Mangas Springs; White Water; Tortugas Mountain; Organ Mountains; Socorro Mountain; Tularosa Creek; Queen; mountains west of San Antonio. Dry hills, among rocks, in the Upper Sonoran and Transition zones.

14. STILLINGIA L. QUEEN'S DELIGHT.

Monœcious herbaceous perennials 30 to 60 cm. high, with several stems umbellately branched above; leaves narrow, glabrous, shining, 3 to 8 cm. long, serrulate; spikes terminal, bracteate; staminate flowers with a 2 or 3-lobed calyx and 2 or 3 exserted stamens; capsule 2 or 3-celled, with a single large globose seed in each cell.

KEY TO THE SPECIES.

Leaves linear or nearly so; capsules less than 10 mm. wide....... 1. *S. linearifolia.*
Leaves lanceolate to elliptic; capsules more than 10 mm. wide.... 2. *S. smallii.*

1. **Stillingia linearifolia** (Torr.) Small, Fl. Southeast. U. S. 704. 1903.

Sapium sylvaticum linearifolium Torr. U. S. & Mex. Bound. Bot. 201. 1859.

Stillingia sylvatica linearifolia Muell. Arg. in DC. Prodr. **15**[2]: 1158. 1866.

TYPE LOCALITY: "Ravines on the San Pedro River and on limestone rocks higher up on the Rio Grande," western Texas.

RANGE: Texas and southeastern New Mexico and adjacent Mexico.

NEW MEXICO: Sands south of Melrose (*Wooton*). Plains and low hills, in the Upper Sonoran Zone.

2. **Stillingia smallii** Woot. & Standl.

Stillingia sylvatica salicifolia Torr.; Small, Fl. Southeast. U. S. 704. 1903.

Stillingia salicifolia Small, loc. cit., not Baill. 1865.

Type locality: "In sandy soil, Kansas to Arkansas and Texas."

Range: Kansas and Arkansas to eastern New Mexico.

New Mexico: Nara Visa; Roswell. Plains and low hills, in the Upper Sonoran Zone.

15. BERNARDIA P. Br.

Low, much branched shrub with alternate stipulate stellate-pubescent ovate-oblong repand-dentate short-petiolate leaves and small diœcious flowers in axillary racemes; staminate calyx 3-parted; stamens 3 to 20; ovary 3-celled, 3-ovuled; seeds not carunculate.

1. **Bernardia myricaefolia** (Scheele) Benth. & Hook. Gen. Pl. **3**: 308. 1883.

Tyria myricaefolia Scheele, Linnaea **25**: 581. 1852.

Type locality: New Braunfels, Texas.

Range: Western Texas and southeastern New Mexico to Mexico.

New Mexico: Queen; San Andreas Mountains. Dry hills, in the Lower Sonoran Zone.

82. CALLITRICHACEAE. Water starwort Family.

1. CALLITRICHE L. Water starwort.

Small aquatic herbs with opposite entire leaves and minute solitary polygamous flowers in the axils; flowers without calyx or corolla, the staminate ones with a single stamen subtended by 2 bracts, the pistillate ones bearing a single pistil with a 4-celled ovary; styles united in pairs; fruit in ours globose, sessile, 1.5 mm. in diameter.

With us a single species with linear, sessile, submersed leaves and spatulate, rounded or retuse emersed ones.

1. **Callitriche palustris** L. Sp. Pl. 969. 1753.

Callitriche autumnalis L. Fl. Suec. ed. 2. 2. 1755.

Type locality: "Habitat in Europae fossis paludibus."

Range: British America to New Mexico; also in Europe.

New Mexico: Horsethief Canyon; Taos; Costilla Valley; Brazos Canyon. In water.

Order 31. SAPINDALES.

KEY TO THE FAMILIES.

Fruit a double samara; stamens alternate with the sepals; leaves opposite....................85. **ACERACEAE** (p. 410).

Fruit not a double samara; stamens opposite the sepals; leaves opposite or alternate.

Plants with resiniferous tissue; leaves compound................................83. **ANACARDIACEAE** (p. 405).

Plants without resiniferous tissue; leaves compound or simple.

Leaf blades simple....................84. **CELASTRACEAE** (p. 409).

Leaf blades pinnate....................86. **SAPINDACEAE** (p. 412).

83. ANACARDIACEAE. Cashew Family.

Shrubs, sometimes small, usually large, with acrid, sometimes poisonous sap and polygamous or diœcious flowers; leaves pinnately 3 to many-foliolate, exstipulate; flowers small, usually inconspicuous, in crowded clusters, these sometimes large; calyx 3 to 7-cleft; petals of the same number as the calyx lobes; stamens as many or twice as many, inserted at the base of a disk; ovary superior, 1-celled, 1-ovuled, the styles often 3, ultimately becoming a small dry drupelike fruit.

Our species have all been recognized as belonging to one genus, Rhus, but they are so different in general appearance that it seems best to follow Doctor Greene in separating them into several genera, though the characters upon which the separation is based are mostly vegetative rather than floral.

KEY TO THE GENERA.

Flowers appearing before the leaves, in small crowded clusters.
- Leaves 1 or 3-foliolate; flowers yellow, tinged with red. 1. SCHMALTZIA (p. 406).
- Leaves 5 to 9-foliolate; flowers white.................. 2. RHOEIDIUM (p. 408).

Flowers appearing after the leaves, in their axils or in panicles terminating the stems.
- Leaves 3-foliolate, poisonous; generally undershrubs with slender stems.................. 3. TOXICODENDRON (p. 408).
- Leaves pinnately several to many-foliolate, not poisonous; shrubby plants with thick stems, one species with hard wood................ 4. RHUS (p. 408).

1. SCHMALTZIA Desv. LEMITA.

Widely branching shrubs, 2 meters high or less; leaves unifoliolate or trifoliolate, the leaflets mostly cuneate-obovate, crenately and coarsely few-toothed, the terminal ones often 3-lobed; flowers yellow or reddish yellow, in crowded clusters on very short peduncles on the branches of the previous season, appearing before the leaves; fruit orange scarlet, globose, 4 to 6 mm. in diameter.

The roots of these plants are used by the Indians in forming patterns for their basketry. The bark is of a dark reddish brown color. The plants are also used in setting dyes. Mexicans sometimes use the stems in making baskets, mixing them with willow branches. The berries were used as food by some of the Indians.

KEY TO THE SPECIES.

Leaves unifoliolate.... .... 1. *S. affinis.*

Leaves trifoliolate.
- Young twigs densely velvety-pubescent with long yellowish hairs.. 2. *S. emoryi.*
- Young twigs merely puberulent or soft-pubescent, the hairs not yellowish.
  - Bracts of the aments tomentose all over.
    - Fruit densely long-hirsute; leaflets small, thick, nearly glabrous, at least on the upper surface......... 6. *S. quercifolia.*
    - Fruit sparingly short-hirsute; leaflets large, thin, pubescent on both surfaces .... 7. *S. bakeri.*
  - Bracts of the aments glabrous or glabrate, at least on the upper half.
    - Terminal leaflet abruptly contracted at the base, deeply 3-lobed.................................. 3. *S. pulchella.*
    - Terminal leaflet gradually cuneate at the base, shallowly lobed.
      - Teeth of the leaves rounded; blades densely pubescent; all the leaflets toothed........... 4. *S. leiocarpa.*
      - Teeth of the leaves acutish; blades mostly glabrous on the upper surface; some of the leaflets entire.................................. 5. *S. cognata.*

There is room for considerable doubt concerning the validity of the numerous segregates of this group recently proposed. The differences between the species are mostly variations in leaf outline. Upon a single shrub one finds great variation in this respect, so that it is questionable whether it is wise to maintain more than a single species of the group, *S. trilobata*. There is some variation in pubescence, but this is only in quantity. Doctor Greene speaks of certain glabrous plants, but we have been unable to find any that are truly glabrous. Field study is necessary to determine with any degree of precision the exact relationship between the numerous forms.

1. **Schmaltzia affinis** Greene, Leaflets **1**: 135. 1905.

TYPE LOCALITY: "Shrub of southern Utah deserts, collected at Kanab, Springdale, and Silver Reef."

RANGE: Southern Utah to northern Arizona and northwestern New Mexico.

NEW MEXICO: Carrizo Mountains (*Standley* 7338). Dry hills among rocks, in the Upper Sonoran Zone.

This is probably an extreme variant of *S. trilobata*, influenced largely by the very arid situations in which it grows. The leaves are commonly unifoliolate, but sometimes they are cleft nearly to the base, and frequently the plants bear a few trifoliolate leaves.

2. **Schmaltzia emoryi** Greene, Leaflets **1**: 133. 1905.

Rhus trilobata mollis A. Gray; Patterson, Check List 21. 1892.

Rhus emoryi Wooton; Greene, op. cit. 134, as synonym.

TYPE LOCALITY: "Hills and low mountains of eastern and southern New Mexico." Type collected by Wooton (no. 584).

RANGE: Southern New Mexico and Arizona.

NEW MEXICO: Horse Camp; Mangas Springs; Burro Mountains; Dog Spring; mountains west of San Antonio; Organ Mountains; Mesilla Park; White Sands; Roswell; Perico; Queen; south of Torrance. Dry hills and valleys, in the Lower and Upper Sonoran zones.

3. **Schmaltzia pulchella** Greene, Leaflets **1**: 134. 1905.

TYPE LOCALITY: "Toward the Rio Limpio, western Texas.

RANGE: Western Texas and southeastern New Mexico.

NEW MEXICO: Queen (*Wooton*). Dry hills, in the Upper Sonoran Zone.

4. **Schmaltzia leiocarpa** Greene, Leaflets **1**: 133. 1905.

TYPE LOCALITY: Valley of the Rio Grande at Mesilla, New Mexico. Type collected by Wooton (no. 48).

RANGE: Southern New Mexico and southeastern Arizona.

NEW MEXICO: Reserve; Black Range; San Luis Mountains; Mesilla; Dona Ana Mountains; above Tularosa; Cloudcroft; Lincoln National Forest. Mesas and river valleys, in the Lower Sonoran Zone.

5. **Schmaltzia cognata** Greene, Leaflets **1**: 141. 1905.

TYPE LOCALITY: Durango, Colorado.

RANGE: Southern Colorado and northwestern New Mexico.

NEW MEXICO: Tunitcha Mountains; Cedar Hill. Dry hills, in the Upper Sonoran Zone.

6. **Schmaltzia quercifolia** Greene, Leaflets **1**: 141. 1905.

TYPE LOCALITY: Canyons in Seward County, southwestern Kansas.

RANGE: Kansas to northeastern New Mexico.

NEW MEXICO: Glorieta; Folsom; Nara Visa; Raton. Low hills and open plains, in the Upper Sonoran Zone.

7. **Schmaltzia bakeri** Greene, Leaflets **1**: 132. 1905.

TYPE LOCALITY: Near Fort Collins, Colorado.

RANGE: Colorado to New Mexico and Arizona.

NEW MEXICO: West of Santa Fe; Tierra Amarilla; Sierra Grande; Farmington; Carrizo Mountains; Chama; Reserve; Fort Bayard; Silver City. Low hills, in the Upper Sonoran Zone.

2. RHOEIDIUM Greene.

Stiff, woody, widely branching desert shrub, often 2 meters high and of equal diameter, with stems intricately interlaced, the short ones sometimes spinescent; leaves generally about 2 cm. long, with about 7 elliptic leaflets borne on a winged rachis; leaflets acute, mostly entire; flowers small, in crowded clusters on the naked branches of the previous season in the axils above the leaf scars; calyx lobes orbicular, concave, entire; petals white, finely ciliate; fruit globose, about 6 mm. in diameter, hispidulous, viscid.

This genus is very close to Schmaltzia, as here understood, the differences in the leaves and the color of the flowers being hardly sufficient for separation. The description of the fruit of Rhoeidium given by Doctor Greene is not correct for the fruit of the species in New Mexico. It is always orange scarlet, nearly like that of Schmaltzia. The plant is also strikingly like the species of that genus in general habit, instead of being very diverse, as has been suggested, and the two grow side by side.

1. **Rhoeidium microphyllum** (Engelm.) Greene, Leaflets **1**: 143. 1905.

Rhus microphylla Engelm. in A. Gray, Pl. Wright. **1**: 31. 1852.

TYPE LOCALITY: "Margins of thickets, on the top of hills, in the large prairie between New Braunfels and San Antonio," Texas.

RANGE: Western Texas to southern Arizona, south into Mexico.

NEW MEXICO: Socorro; Berendo Creek; Hachita; Tortugas Mountain; Hopkins Mill; Organ Mountains. Dry hills, in the Lower Sonoran Zone.

3. TOXICODENDRON Mill. POISON OAK.

Low shrubs, usually less than 1 meter high, with 3-foliolate poisonous leaves having large, broadly ovate to rhombic, acuminate, coarsely few-toothed or entire leaflets; flowers inconspicuous, greenish yellow, in small several-flowered axillary panicles; fruit depressed-globose, glabrous, thin-walled, white and shining when mature.

1. **Toxicodendron rydbergii** (Small) Greene, Leaflets **1**: 117. 1905.

Rhus rydbergii Small in Rydb. Mem. N. Y. Bot. Gard. **1**: 268. 1900.

Toxicodendron punctatum Greene, Leaflets **1**: 125. 1905.

TYPE LOCALITY: Not definitely stated, apparently Montana.

RANGE: British Columbia and Montana to Nebraska and New Mexico.

NEW MEXICO: Winsors Ranch; Black Range; Sandia Mountains; Mogollon Mountains; White Mountains. In woods, in the Transition Zone.

In New Mexico the plant goes under the name of "poison oak," but in other parts of the United States the name given is more often "poison ivy," which would seem much more appropriate. The type of *T. punctatum* came from the Black Range (*Metcalfe* 1088).

4. RHUS L. SUMAC.

Erect spreading shrubs 1 to 2 meters high or more, with pinnately 5 to many-foliolate leaves and axillary or terminal panicles of small, dull whitish or yellowish flowers; leaves persistent or deciduous, the leaflets large, 3 to 8 cm. long; flowers and fruit as described under the family.

KEY TO THE SPECIES.

Leaves evergreen, thick; flowers axillary in small clusters; wood of stems very hard.. 1. *R. choriophylla.*
Leaves deciduous, thin; flowers in dense terminal panicles; wood of stems soft, with large pith.
  Rachis winged; leaflets densely pubescent beneath, of the same color on both surfaces........................... 2. *R. lanceolata.*
  Rachis not winged; leaflets glabrous and paler beneath....... 3. *R. cismontana.*

1. **Rhus choriophylla** Woot. & Standl. Contr. U. S. Nat. Herb. **16:** 146. 1913.

TYPE LOCALITY: Guadalupe Canyon, New Mexico. Type collected by E. A. Mearns (no. 699).

RANGE: Southern New Mexico and Arizona and adjacent Mexico.

NEW MEXICO: Guadalupe Canyon; San Andreas Mountains; Organ Mountains. Dry hills and canyons, in the Upper Sonoran Zone.

2. **Rhus lanceolata** (A. Gray) Britton in Britt. & Shaf. N. Amer. Trees 606. 1908.

Rhus copallina lanceolata A. Gray, Bost. Journ. Nat. Hist. **6:** 158. 1850.

Schmaltzia lanceolata Small, Fl. Southeast. U. S. 728. 1903.

TYPE LOCALITY: Rocky soil and high prairies, New Braunfels, Texas.

RANGE: Central Texas to southeastern New Mexico.

NEW MEXICO: Queen; San Andreas Mountains. Dry hills and canyons.

Our specimens are not altogether typical, the pubescence being more abundant and spreading than is usual, while the leaflets are smaller and the inflorescence shorter.

3. **Rhus cismontana** Greene, Proc. Washington Acad. Sci. **8:** 189. 1906.

Rhus sorbifolia Greene, op. cit. 195. 1906.

TYPE LOCALITY: Thomas County, Nebraska.

RANGE: Utah and North Dakota to New Mexico and Arizona.

NEW MEXICO: Sandia Mountains; mountains west of Las Vegas; Kingston; Mogollon Creek; Guadalupe Canyon; White Mountains. Along streams, in the Transition Zone.

The type of *Rhus sorbifolia* was collected in the mountains near Las Vegas.

84. CELASTRACEAE. Staff-tree Family.

Low shrubs, sometimes spiny; leaves simple, small, alternate or opposite; flowers normally cymose, small and inconspicuous, perfect; calyx and corolla 4 or 5-merous; stamens 4 to 10, inserted on a disk lining the hypanthium; fruit a capsule, drupe, or berry, the seeds often arillate.

KEY TO THE GENERA.

Stamens 10; plants spiny; stems green.................... 1. FORSELLESIA (p. 409).
Stamens 4 or 5; plants not spiny; stems yellowish or brown.
  Flowers 4-merous; fruit a 2-valved capsule; leaves opposite, smooth.............................. 2. PACHISTIMA (p. 410).
  Flowers 5-merous; fruit indehiscent; leaves alternate, scurfy.................................. 3. MORTONIA (p. 410).

1. FORSELLESIA Greene.

A spiny green-stemmed shrub 30 to 40 cm. high or less, with small obovate acute leaves 1 cm. long or smaller, these nearly smooth, short-petioled, entire; flowers small, pentamerous, white.

1. **Forsellesia spinescens** (A. Gray) Greene, Erythea 1: 206. 1893.

Glossopetalon spinescens A. Gray, Pl. Wright. 2: 29. *pl. 12*. 1853.

TYPE LOCALITY: "In a mountain ravine near Frontera," Texas. The type came from within a few miles of the southern boundary of New Mexico. It may even have come from the south end of the Organ Mountains.

RANGE: Washington and California to Colorado and western Texas.

NEW MEXICO: Upper Corner Monument; San Andreas Mountains; Llano Estacado; Organ Mountains. Dry hills, in the Upper Sonoran Zone.

A rare shrub of dry rocky slopes, mostly on limestone soil.

2. PACHISTIMA Raf.

Prostrate evergreen shrub with glabrous opposite short-petioled serrulate leaves; flowers solitary or in few-flowered axillary cymes; calyx 4-lobed, with a short tube; petals 4; ovary 2-celled; capsule small, 1 or 2-seeded.

1. **Pachistima myrsinites** (Pursh) Raf. Fl. Tellur. 42. 1838.

Ilex ? myrsinites Pursh, Fl. Amer. Sept. 119. 1814.

TYPE LOCALITY: "On the Rocky Mountains and near the Pacific Ocean."

RANGE: British Columbia and California to New Mexico.

NEW MEXICO: Zuni Mountains; Carrizo Mountains; Sandia Mountains; Bear Mountains; Cloudcroft; Mogollon Road; Tunitcha Mountains; Chama; Lookout Mines; Santa Fe Canyon. Woods, in the Transition and Canadian zones.

3. MORTONIA A. Gray.

A low shrub; leaves elliptic, thick, entire, acute, contracted into a very short petiole, crowded, 1 cm. long or less; stems yellowish like the leaves; flowers in short terminal bracteate racemes; whole plant densely scurfy.

1. **Mortonia scabrella** A. Gray, Pl. Wright. 2: 28. 1853.

TYPE LOCALITY: "Mountain-sides, near the San Pedro, Sonora," and "Mountains near El Paso."

RANGE: Western Texas and southern Arizona, south into Mexico.

NEW MEXICO: Upper Corner Monument (*Mearns* 64, 247). Dry hills, in the Lower Sonoran Zone.

85. ACERACEAE. Maple Family.

Small or large trees with smooth exfoliating bark; leaves opposite, simple and palmately lobed or pinnately compound; flowers polygamous or diœcious, in axillary racemes or corymbs; sepals 4 or 5; petals as many or mostly wanting; stamens as many as the sepals, rarely 8, inserted on a disk, or the disk wanting; pistil of 2 or more united carpels, becoming 2 laterally winged samaras.

KEY TO THE GENERA.

Leaves simple or palmately 3-foliolate; young branches reddish or gray; flowers polygamous .... 1. ACER (p. 410).

Leaves pinnately 3 or 5-foliolate; young branches green; flowers diœcious .... 2. NEGUNDO (p. 411).

1. ACER L. MAPLE.

Trees with reddish, brownish, or grayish twigs, rather smooth bark, and palmately 5-lobed or 3-foliolate leaves; flowers polygamous, preceding the leaves, inconspicuous, on slender pendent pedicels; petals sometimes present; fruit as described for the family.

KEY TO THE SPECIES.

Corymbs long-peduncled; teeth of the leaves acute.
  Leaves, at least most of them, 3-parted .... 1. *A. neomexicanum.*
  Leaves merely 3 or 5-lobed, never parted .... 2. *A. glabrum.*
Corymbs nearly sessile; teeth of the leaves obtuse.
  Lobes of the leaves broadly oblong, with several teeth, broadest near the apex; wing of the fruit 30 mm. long or more .... 3. *A. grandidentatum.*
  Lobes of the leaves triangular-lanceolate, mostly entire, broadest at the base; wings 15 mm. long .... 4. *A. brachypterum.*

1. **Acer neomexicanum** Greene, Pittonia **5**: 3. 1902.

TYPE LOCALITY: Mountains near Las Vegas, New Mexico.

RANGE: New Mexico and Arizona to southern Colorado.

NEW MEXICO: Santa Fe and Las Vegas Mountains; Sandia Mountains; Copper Canyon; Mogollon Mountains; Lookout Mines; Cloudcroft; White Mountains. Damp woods and along streams, in the Transition and Canadian zones.

This may be the same as *A. tripartitum* Nutt., but it seems different. *Acer neomexicanum* and the following are slender shrubs, usually from 2 to 4 meters high.

2. **Acer glabrum** Torr. Ann. Lyc. N. Y. **2**: 172. 1828.

TYPE LOCALITY: "On the Rocky Mountains."

RANGE: Wyoming and Nebraska to Utah and New Mexico.

NEW MEXICO: Zuni Mountains; Tunitcha Mountains; Dulce; Sierra Grande. Damp woods, in the Transition and Canadian zones.

The specimens show none of the divided leaves of *A. neomexicanum.*

3. **Acer grandidentatum** Nutt.; Torr. & Gray, Fl. N. Amer. **1**: 247. 1838.

TYPE LOCALITY: "Rocky Mountains, on Bear River of Timpanagos."

RANGE: Montana to Arizona and western Texas.

NEW MEXICO: Holts Ranch; Lookout Mines; Organ Mountains; White and Sacramento mountains. Woods, in the Transition Zone.

A medium-sized tree with spreading branches.

4. **Acer brachypterum** Woot. & Standl. Contr. U. S. Nat. Herb. **16**: 146. 1913.

TYPE LOCALITY: San Luis Mountains, New Mexico. Type collected by E. A. Mearns (no. 535).

RANGE: Southwestern New Mexico, southeastern Arizona, and adjacent Mexico.

NEW MEXICO: San Luis Mountains.

A species near *A. grandidentatum*, but the leaves more densely pubescent and with very different lobes and the wings of the fruit much shorter.

NEGUNDO Boehmer. BOX ELDER.

Medium-sized tree with pinnate leaves; young twigs smooth and glaucous, green; leaflets 3, sometimes 5, ovate, with a few coarse teeth near the apex, or sometimes somewhat lobed.

1. **Negundo interius** (Britton) Rydb. Bull. Torrey Club **40**: 56. 1913.

Acer interior Britton in Britt. & Shaf. N. Amer. Trees 655. *f. 608.* 1908.

Rulac interior Nieuwland, Amer. Mid. Nat. **2**: 139. 1911.

TYPE LOCALITY: Chaparral-covered hills southeast of Ouray, Colorado.

RANGE: New Mexico and Arizona to Saskatchewan and Manitoba.

NEW MEXICO: Pecos; Hurrah Creek; west of Chloride; Black Range; Cliff; Gila; Animas Peak; Organ Mountains; Trinchera Pass; Tunitcha Mountains; White Mountains. Along streams, in the Transition Zone.

86. SAPINDACEAE. Soapberry Family.

Shrubs or trees with alternate pinnate leaves; inflorescence lateral or terminal, mostly paniculate; flowers white or pink, polygamous, usually conspicuous; sepals 4 or 5; petals 4 or 5, regular or irregular; stamens 7 to 10, inserted on a disk; ovary 2 to 4-celled; fruit a capsule or berry-like.

KEY TO THE GENERA.

Trees with small white flowers; fruit berry-like, with a single seed.. 1. SAPINDUS (p. 412).
Shrubs with large pink flowers; fruit a 3-celled capsule with 3 seeds.. 2. UNGNADIA (p. 412)

1. SAPINDUS L. SOAPBERRY.

Tree 7 or 8 meters high or less, with rather smooth yellowish gray bark and thick foliage; leaves with 8 to 19 narrowly lanceolate leaflets 4 to 8 cm. long, somewhat falcate, acuminate, glabrous above, soft-pubescent beneath; flowers white, small, numerous, in terminal panicles; sepals and petals 4 or 5, the latter twice as long as the former and more or less lacerate; fruit consisting of a globose, yellow, fleshy to leathery pericarp about 1 cm. in diameter, containing a single globose seed, drying black.

1. **Sapindus drummondii** Hook. & Arn. Bot. Beechey Voy. 281. 1840.

TYPE LOCALITY: Texas.

RANGE: Kansas, Arkansas, and Louisiana to Arizona.

NEW MEXICO: Mangas Springs; Fairview; Fort Bayard; Black Range; Carrizalillo Mountains; Dog Spring; east of Deming; Organ Mountains; Roswell; Albert. Canyons, in the Lower and Upper Sonoran zones.

A single species occurring in the mountains and foothills of the southern part of the State, sometimes cultivated. Young plants are commonly bushy, with several stems from the root, and will be recognized only by the leaves, since they do not bloom. The Spanish name for this is "jaboncillo."

2. UNGNADIA Endl. NEW MEXICAN BUCKEYE.

Branched shrubs 2 meters high or less, with reddish twigs and large leaves with 3 to 9 leaflets; leaflets usually 7, broadly lanceolate, acuminate, irregularly serrate; flowers rather large, bright pink, numerous, appearing before the leaves, irregular, polygamous; sepals 5; petals 4 or 5; stamens 5 to 10, exserted; capsule long-stipitate, coriaceous to woody, 3 to 5 cm. in diameter, 3-celled; seeds globose, brown, smooth, shining, about 10 mm. in diameter.

1. **Ungnadia speciosa** Endl. Atact. Bot. *pl. 36.* 1833.

TYPE LOCALITY: Not ascertained.

RANGE: Central Texas to southern New Mexico.

NEW MEXICO: Organ Mountains; Guadalupe Mountains. Rocky hills, in the Upper Sonoran Zone.

Order 32. RHAMNALES.

KEY TO THE FAMILIES.

Sepals evident; petals involute; fruit capsular or drupaceous; shrubs or trees.................... **87. RHAMNACEAE** (p. 413).
Sepals minute or obsolete; petals valvate; fruit a berry; vines with tendrils.................... **88. VITACEAE** (p. 415).

87. RHAMNACEAE. Buckthorn Family.

More or less spiny shrubs 2 meters high or less, with simple leaves and small stipules; flowers perfect or polygamo-diœcious, mostly small and inconspicuous; calyx of 4 or 5 valvate sepals, with a disk lining the hypanthium; petals 4 or 5 or wanting; stamens 4 or 5, opposite the petals on the throat of the hypanthium or on the disk; pistil of 2 or 3 united carpels; ovaries united with the disk and hypanthium to form the berry-like fruit.

KEY TO THE GENERA.

Fruit fleshy, black, with a 1 to 3-celled stone.
- Petals present; young stems glaucous .... 1. ZIZYPHUS (p. 413).
- Petals wanting; young stems not glaucous .... 2. CONDALIA (p. 413).

Fruit dry or somewhat berry-like, 2 or 3-seeded.
- Plants low; petals hooded or long-clawed; stigmas 3.... 3. CEANOTHUS (p. 413).
- Tall shrubs 1 meter high or more; petals not clawed nor hooded; stigmas 2 .... 4. RHAMNUS (p. 414).

1. ZIZYPHUS Juss. LOTE BUSH.

Rigid spiny shrub 1 to 2 meters high, with glaucous young branches and small glaucous leaves, these 15 mm. long or less, ovate to oblong-elliptic, acute or obtuse; flowers small, in axillary corymbs; sepals 5, triangular, keeled within; petals and stamens 5, opposite each other on the disk; ovary 2 or 3-celled; fruit a pulpy black berry, green within.

1. **Zizyphus lycioides** A. Gray, Bost. Journ. Nat. Hist. 6: 168. 1850.

TYPE LOCALITY: Between Matamoros and Mapimi, Mexico.

RANGE: Western Texas to southern New Mexico and northeastern Mexico.

NEW MEXICO: Mangas Springs; Berendo Creek; west of Cambray; Florida Mountains; Hachita; Dog Spring; mesa west of Organ Mountains; Organ Mountains. Dry mesas and hills, in the Lower Sonoran Zone.

2. CONDALIA Cav.

Very similar to the preceding, but the leaves spatulate and finely pubescent, and the petals wanting.

1. **Condalia spathulata** A. Gray, Pl. Wright. 1: 32. 1852.

TYPE LOCALITY: "On the Rio Grande, Texas; and prairies on the San Felipe."

RANGE: Western Texas and New Mexico to northeastern Mexico.

NEW MEXICO: East of Hachita; Las Palomas Hot Springs; mesa near Las Cruces; Guadalupe Mountains. Mesas, in the Lower Sonoran Zone.

This and the last occur together on the mesas of the southern part of the State. This plant is nearly always leafy, the leaves being more or less persistent, while *Zizyphus lycioides* is most frequently without leaves. Much of the time it appears to be merely a bush composed of spines. It may be recognized by its younger branches, which are always bluish green or glaucous, even after the leaves have fallen. The fruits of this and the preceding are sometimes eaten, but their seeds are very large and the amount of pulp small.

3. CEANOTHUS L. BUCKTHORN.

Low shrubs, more or less spinescent, mostly less than 1 meter high; leaves simple, alternate, with minute caducous stipules; flowers small, in crowded terminal racemes or corymbs; sepals 5, white, petaloid; disk filling the hypanthium; petals 5, white, long-clawed, hooded; stamens 5, exserted; ovary immersed in the disk; fruit at last dry, 3-celled, berry-like.

KEY TO THE SPECIES.

Leaf blades thin, bright green, nearly or quite glabrous, 25 to 35 mm. long; inflorescence much exceeding the leaves........ 1. *C. mogollonicus.*

Leaf blades thick, grayish green, densely pubescent, at least beneath, 20 mm. long or less; inflorescence usually not exceeding the leaves.

Branches spinescent; leaves sericeous beneath, elliptic to lanceolate, acutish.................................. 2. *C. fendleri.*

Branches not spinescent; leaves never sericeous, hirtellous-puberulent, mostly obovate, rounded or retuse at the apex.. 3. *C. greggii.*

1. Ceanothus mogollonicus Greene, Leaflets **1:** 67. 1904.

TYPE LOCALITY: Mogollon Creek in the Mogollon Mountains, New Mexico. Type collected by Metcalfe (no. 239).

RANGE: Known only from type locality.

2. Ceanothus fendleri A. Gray, Mem. Amer. Acad. n. ser. **4:** 29. 1849.

TYPE LOCALITY: Mountains east of Santa Fe, New Mexico. Type collected by Fendler (no. 105).

RANGE: Wyoming and South Dakota to Arizona and New Mexico.

NEW MEXICO: Santa Fe and Las Vegas mountains; Sandia Mountains; Capitan Mountains; Chama; Tunitcha Mountains; Sawyers Peak; Mogollon Mountains; White Mountains. Open slopes and thickets, in the Transition Zone.

3. Ceanothus greggii A. Gray, Pl. Wright. **2:** 28. 1853.

TYPE LOCALITY: Buena Vista, Mexico.

RANGE: Western Texas to southern Arizona, south into Mexico.

NEW MEXICO: Bear Mountain; Organ Mountains; Queen; San Andreas Mountains.

4. RHAMNUS L. BUCKTHORN.

Unarmed shrubs over 1 meter high, with rather large alternate leaves; flowers perfect or polygamo-diœcious, in small axillary clusters; sepals 4 or 5; disks lining the hypanthium; petals 4 or 5, sometimes wanting, clawless, on the margin of the hypanthium; stamens 4 or 5, inserted on the edge of the disk; ovary 2 to 4-celled; fruit a 2 to 4-seeded, rather dry berry.

KEY TO THE SPECIES.

Flowers fascicled, 2 or 3 in each axil; leaves small, 35 mm. long or less, yellowish beneath; seeds 2.

Mature leaves obtuse, finely pubescent on the upper surface.. 1. *R. fasciculata.*

Mature leaves acutish, glabrous on the upper surface........ 4. *R. smithii.*

Flowers in peduncled cymes, numerous; leaves usually more than 35 mm. long, not yellowish beneath; seeds 2 or 3.

Seeds 2; leaves pale beneath with a dense tomentulose pubescence.. 2. *R. ursina.*

Seeds 3; leaves green on both surfaces, sparingly pubescent beneath.. 3. *R. betulaefolia.*

1. Rhamnus fasciculata Greene, Leaflets **1:** 63. 1904.

TYPE LOCALITY: South Fork of Tularosa Creek, 3 miles east of the Mescalero Agency, New Mexico. Type collected by Wooton (no. 203).

RANGE: Southern New Mexico.

NEW MEXICO: White and Sacramento mountains; Guadalupe Mountains. Mountains, in the Transition Zone.

This is doubtfully distinct from *R. smithii.*

2. Rhamnus ursina Greene, Leaflets 1: 63. 1904.

TYPE LOCALITY: Bear Mountain near Silver City, New Mexico. Type collected by Metcalfe (no. 172).

RANGE: Southwestern New Mexico.

NEW MEXICO: Sycamore Creek; Bear Mountain; Mangas Springs; Gila; Berendo Creek; San Andreas Mountains. Mountains, in the Upper Sonoran Zone.

3. Rhamnus betulaefolia Greene, Pittonia 3: 16. 1896.

TYPE LOCALITY: Banks of streams in the Mogollon Mountains, New Mexico. Type collected by Rusby in 1881.

RANGE: Mountains of southern New Mexico and adjacent Arizona.

NEW MEXICO: Mogollon Mountains; San Francisco Mountains; Kingston; Animas Peak; Tularosa Creek. In the Transition Zone.

4. Rhamnus smithii Greene, Pittonia 3: 17. 1896.

TYPE LOCALITY: Pagosa Springs, southwestern Colorado.

RANGE: Southern Colorado and northern New Mexico.

NEW MEXICO: Chama; between Tierra Amarilla and Park View. Open hillsides, in the Transition Zone.

88. VITACEAE. Grape Family.

Woody vines, trailing or climbing by tendrils; leaves large, simple or compound, petiolate, the blades flat and mostly thin; inflorescence axillary, cymose or paniculate; flowers small and inconspicuous, greenish or yellowish, sometimes delicately perfumed, perfect, polygamous, or diœcious, regular; calyx and corolla 4 or 5-merous, a disk present or wanting; stamens of the same number as the petals and opposite them; pistil compound; fruit a berry.

KEY TO THE GENERA.

Leaves simple.................................... 1. VITIS (p. 415).
Leaves compound.
  Leaves 5-foliolate, thin.......................... 2. PARTHENOCISSUS (p. 415).
  Leaves 3-foliolate, fleshy........................ 3. CISSUS (p. 416).

1. VITIS L. GRAPE.

Trailing or climbing vines with shreddy bark and forking tendrils; leaves simple, more or less palmately lobed or angled, with small caducous stipules; flowers in axillary panicles, diœcious, polygamo-diœcious, or rarely perfect; calyx minute; corolla caducous, the petals coherent; stamens exserted, alternate with the lobes of the disk; fruit a few-seeded globose berry; seeds hard and bony, pear-shaped, relatively large.

1. Vitis arizonica Engelm. Amer. Nat. 2: 321. 1868.

TYPE LOCALITY: Arizona.

RANGE: Western Texas to Arizona.

NEW MEXICO: McCarthy Station; Sandia Mountains; Magdalena Mountains; Mangas Springs; Mogollon Mountains; Fort Bayard; Bear Mountain; Black Range; Organ Mountains; Roswell; Gray; Queen; Cloverdale; Animas Mountains; White Mountains. Canyons and thickets, in the Upper Sonoran and Transition zones.

The berries of this grape are not very palatable, but they were used for food by the Indians.

2. PARTHENOCISSUS Planch. VIRGINIA CREEPER.

Trailing or climbing woody vines with forking tendrils and alternate, palmately 5-foliolate leaves; leaflets 4 to 10 cm. long, coarsely toothed; flowers small, greenish, in axillary cymes; calyx and corolla 5-merous, disk wanting; stamens 5; fruit a depressed-globose berry, blackish, not edible.

1. **Parthenocissus vitacea** (Knerr) Hitchc. Spr. Fl. Manhattan 26. 1894.
Ampelopsis quinquefolia vitacea Knerr, Bot. Gaz. **18:** 71. 1893.
Psedera vitacea Greene, Leaflets 1: 220. 1906.
TYPE LOCALITY: Not stated.
RANGE: Wyoming and Michigan to Ohio and Arizona.
NEW MEXICO: Pecos; Sandia Mountains; Magdalena Mountains; Gila Hot Springs; Burro Mountains; Guadalupe Canyon; Gray; Organ Mountains; Cedar hill; Raton. Canyons, in the Transition Zone.

3. CISSUS L.

A succulent vine 1 to 10 meters long, with warty bark and forking tendrils; leaves 3-foliolate, fleshy, the leaflets 3 to 10 cm. long, coarsely toothed, the terminal one sometimes 3-lobed; flowers in trichotomous umbel-like cymes; berries obovoid to globose, 10 to 12 mm. long, blackish, on recurved pedicels.

1. **Cissus incisa** (Nutt.) Desmoul.; S. Wats. Bibl. Ind. 173. 1878.
Vitis incisa Nutt.; Torr. & Gray, Fl. N. Amer. 1: 243. 1840.
TYPE LOCALITY: "Arkansas."
RANGE: Florida to Arkansas, Texas, and southern New Mexico.
NEW MEXICO: Guadalupe Canyon (*Mearns* 691).

Order 33. MALVALES.

89. MALVACEAE. Mallow Family.

Annual or perennial herbs (one species suffruticose) with simple, alternate, petiolate, variously lobed or dissected leaves, and rather large and conspicuous flowers; plants mostly pubescent, frequently with stellate hairs; inflorescence axillary or by reduction of the uppermost leaves becoming racemose or paniculate; calyx of 5 sepals more or less united at the base, sometimes subtended by few to several bracts forming an involucre; petals 5, more or less united at the base and with the base of the tube of the numerous monadelphous stamens; pistil of 5 to many carpels with united styles and separate stigmas; fruit a 5 to many-celled capsule of dehiscent or indehiscent, 1 to several-seeded carpels.

KEY TO THE GENERA.

Fruit a loculicidal capsule; stamen column anther-bearing for a considerable part of its length.
  Annuals; calyx inflated, conspicuously nerved..... 1. TRIONUM (p. 417).
  Perennials; calyx neither inflated nor conspicuously nerved.................................... 2. HIBISCUS (p. 417).
Fruit of several radially disposed carpels, these separating at maturity; stamen column anther-bearing mostly at the summit.
  Carpels indehiscent; ovules solitary; styles stigmatic on the inner side.
    Bractlets wanting; carpels 5 to 9................ 3. SIDALCEA (p. 418).
    Bractlets 1 to 3; carpels more numerous.
      Carpels beaked; bractlets 1 to 3; flowers bright purplish red.................. 4. CALLIRRHOE (p. 418).
      Carpels not beaked; bractlets 3; flowers white to rose-colored................ 5. MALVA (p. 419).
  Carpels dehiscent; ovules 1 to several in each cell; stigmas capitate.
    Seeds 2 or more in each carpel.
      Calyx without bracts; seeds several in each carpel.

Carpels membranous, rounded at the apex .... 6. GAYOIDES (p. 419).
Carpels leathery, usually acute or cuspidate .... 7. ABUTILON (p. 419).
Calyx bracted; seeds 1 to 3 in each carpel.
Capsules hirsute as well as stellate-pubescent, 2 or 3-seeded, smooth on the sides at the base .... 8. PHYMOSIA (p. 420).
Capsules stellate-pubescent but not hirsute, 2-seeded, more or less reticulate at the base .... 9. SPHAERALCEA (p. 420).
Seeds solitary in each carpel.
Calyx bracteate.
Flowers white; stems prostrate.........10. DISELLA (p. 424).
Flowers orange to pink; stems erect....11. MALVASTRUM (p. 425).
Calyx not bracted.
Perennials; carpels erect, with connivent or erect tips..............12. SIDA (p. 425).
Annuals; carpels depressed, the tips spreading.
Carpels 9 to 20, hirsute; flowers axillary.......................13. ANODA (p. 426).
Carpels 5 to 9, not hirsute; inflorescence racemose or paniculate.14. SIDANODA (p. 427).

1. TRIONUM Medic. FLOWER-OF-AN-HOUR.

More or less hispid annual, branching from the base, with palmately 3 to 5-lobed or parted leaves and dull white flowers with an inflated nerved calyx; capsules ovoid, about 15 mm. high, inclosed in the persistent calyx.

1. Trionum trionum (L.) Woot. & Standl.
Hibiscus trionum L. Sp. Pl. 697. 1753.
TYPE LOCALITY: "Habitat in Italia, Africa."
NEW MEXICO: Ramah; north of Kennedy; Las Vegas; Raton.
An introduced weed, occasionally found about gardens.

2. HIBISCUS L. ROSE MALLOW.

Herbaceous perennials with large or small pubescent ovate leaves, and axillary pink or purplish flowers; calyx of 5 more or less united sepals, subtended by several bracts of about the same length.

KEY TO THE SPECIES.

Plants large, 1 meter high or more; flowers 6 to 8 cm. long..... 1. *H. lasiocarpus.*
Plants small, 40 cm. high or less; flowers 3 cm. long or less...... 2. *H. involucellatus.*

1. Hibiscus lasiocarpus Cav. Monad. Diss. 159. *pl. 70. f. 1.* 1787.
TYPE LOCALITY: Not stated.
RANGE: Low ground, New Mexico to Illinois, Georgia, and Louisiana.
NEW MEXICO: Roswell (*Earle* 357).

2. Hibiscus involucellatus (A. Gray) Woot. & Standl.
Hibiscus denudatus involucellatus A. Gray, Pl. Wright. 1: 22. 1852.
TYPE LOCALITY: "Sides of hills near El Paso."
RANGE: Western Texas to southern Arizona, south into Mexico.
NEW MEXICO: Tortugas Mountain; Grant County. Dry hills, in the Lower Sonoran Zone.

52576°—15——27

3. SIDALCEA A. Gray.

Tall perennial herbs, 1 meter high or less, with rounded, palmately lobed or dissected leaves and showy flowers in elongated, terminal, simple or paniculate, bracted racemes; basal leaves often merely coarsely crenate, the upper deeply lobed; flowers large, 2 cm. long or more, purple or white; calyx 5-cleft, without bracts; carpels 5 to 9, 1-seeded, not beaked, indehiscent.

KEY TO THE SPECIES.

Flowers cream-colored; inflorescence and calyx densely stellate-pubescent.. 1. *S. candida.*
Flowers purple; inflorescence and calyx hirsute................. 2. *S. neomexicana.*

1. **Sidalcea candida** A. Gray, Mem. Amer. Acad. n. ser. **4**: 24. 1849.
Sidalcea candida tincta Cockerell, Bot. Gaz. **29**: 280. 1900.
TYPE LOCALITY: Along Santa Fe Creek, New Mexico. Type collected by Fendler (no. 80).
RANGE: Wyoming and Utah to New Mexico.
NEW MEXICO: Santa Fe and Las Vegas mountains; Sandia Mountains; White and Sacramento mountains; Chama. Wet ground, in the Transition and Canadian zones.
The type of *S. candida tincta* was collected by Cockerell at Harveys Ranch. It is a common form in which the corolla is tinged with pink.

2. **Sidalcea neomexicana** A. Gray, Mem. Amer. Acad. n. ser. **4**: 23. 1849.
TYPE LOCALITY: Moist meadows, Santa Fe, New Mexico. Type collected by Fendler (no. 79).
RANGE: Wyoming and Utah to New Mexico and southern California.
NEW MEXICO: Chama; Santa Fe and Las Vegas mountains; Kingston; Mangas Springs; Mogollon Mountains; Las Huertas Canyon; Ramah; White Mountains. Wet ground, in the Transition Zone.

4. CALLIRRHOE Nutt.

Perennial herbs with long thick roots; leaves alternate, petiolate, lobed or cleft; flowers showy, axillary; petals purple; carpels 10 to 20, 1-celled, 1-seeded, beaked when mature, forming a disklike fruit about the axis.

KEY TO THE SPECIES.

Calyx subtended by 3 bracts.. 1. *C. involucrata.*
Calyx naked.. 2. *C. alcaeoides.*

1. **Callirrhoe involucrata** (Torr. & Gray) A. Gray, Mem. Amer. Acad. n. ser. **4**: 16. 1849.
Malva involucrata Torr. & Gray, Fl. N. Amer. **1**: 226. 1838.
A common weed in gardens and cultivated ground.
TYPE LOCALITY: "Valley of the Loup Fork of the Platte."
RANGE: Utah and Minnesota to Texas and northeastern New Mexico.
NEW MEXICO: Sierra Grande; Perico Creek. Plains, in the Upper Sonoran Zone.

2. **Callirrhoe alcaeoides** (Michx.) A. Gray, Mem. Amer. Acad. n. ser. **4**: 18. 1849.
Sida alcaeoides Michx. Fl. Bor. Amer. **2**: 44. 1803.
TYPE LOCALITY: "Hab. in glareosis Kentucky et Tennessee."
RANGE: Nebraska and Kansas to New Mexico and Texas.
NEW MEXICO: Las Vegas (*Porter*). Plains, in the Upper Sonoran Zone.

5. MALVA L. Mallow.

Annual or perennial herbs with orbicular or reniform, sometimes obscurely lobed or crenate leaves; flowers axillary, solitary or in small clusters; calyx with 3 or 2 distinct bracts; carpels numerous, 1-celled, reniform when mature, beakless, disposed around the axis in a disklike fruit.

KEY TO THE SPECIES.

Margins of the leaves crisped; plants erect........................ 3. *M. crispa.*
Margins of the leaves not crisped; plants erect or prostrate.
  Stems erect; calyx reflexed in fruit........................... 1. *M. parviflora.*
  Stems prostrate; calyx not reflexed in fruit.................. 2. *M. rotundifolia.*

1. Malva parviflora L. Amoen. Acad. 3: 416. 1756.
Type locality: Not stated.
Range: A native of the Old World, introduced in the southern and western United States.
New Mexico: Santa Fe; Pecos; Kingston; Las Cruces; Teel; Capitan; Taos; Las Vegas; White Mountains. Upper Sonoran Zone.

2. Malva rotundifolia L. Sp. Pl. 688. 1753. Common mallow.
Type locality: "Habitat in Europae ruderatis, viis, plateis."
New Mexico: Mangas Springs; Kingston; Raton; Santa Fe; Gallinas Planting Station.
A weed in waste ground.

3. Malva crispa L. Sp. Pl. ed. 2. 970. 1763. Curled mallow.
Malva verticillata crispa L. Sp. Pl. 689. 1753.
Type locality: Not stated.
New Mexico: Tularosa; Shiprock.

6. GAYOIDES Small.

Slender perennial herb with the aspect of Abutilon, but the carpels with very thin membranous walls and rounded at the apex rather than acute or beaked.

1. Gayoides crispum (L.) Small, Fl. Southeast. U. S. 764. 1903.
Sida crispa L. Sp. Pl. 685. 1753.
Abutilon crispum Sweet, Hort. Brit. ed. 1. 53. 1827.
Type locality: "Habitat in Carolina, Providentia, Bahama."
Range: Florida to New Mexico and Arizona, and throughout the tropics.
New Mexico: Bishops Cap (*Wooton*). Dry hills.

7. ABUTILON Gaertn. Indian mallow.

Erect or decumbent, stout or slender, densely pubescent, herbaceous perennials, with simple cordate leaves and axillary flowers; calyx not bracteate; corolla red or yellow; carpels 5 to 10, leathery, beaked or rounded, dehiscent, 2 to several-seeded, 6 mm. high or more.

KEY TO THE SPECIES.

Carpels short-beaked or at least acute; stems slender, prostrate or ascending... 1. *A. parvulum.*
Carpels rounded at the apex; stems stout and erect.
  Leaves not lobed, as broad as long; sepals 8 mm. long........... 2. *A. malacum.*
  Leaves more or less 3-lobed, longer than broad; sepals not more than 5 mm. long.. 3. *A. texense.*

1. **Abutilon parvulum** A. Gray, Pl. Wright. **1:** 21. 1852.

TYPE LOCALITY: "Calcareous hills of the San Felipe and the San Pedro Rivers," Texas.

RANGE: Western Texas to Arizona, south into Mexico.

NEW MEXICO: Mangas Springs; Anton Chico; west of Roswell; Hondo Hill; Filmore Canyon. Dry hills, in the Upper Sonoran Zone.

2. **Abutilon malacum** S. Wats. Proc. Amer. Acad. **21:** 446. 1886.

TYPE LOCALITY: Wilson County, western Texas.

RANGE: Western Texas to southern New Mexico.

NEW MEXICO: Tortugas Mountain. Dry hills, in the Lower Sonoran Zone.

A coarse, yellow-flowered plant, 1 meter high or less, with large velvety leaves. In New Mexico it is known from but a single locality and is certainly rare, although it may be expected in the mountains of the southwestern corner.

3. **Abutilon texense** Torr. & Gray, Fl. N. Amer. **1:** 231. 1838.

TYPE LOCALITY: Texas.

RANGE: Central Texas to southern New Mexico.

NEW MEXICO: Hillsboro; Dog Spring. Upper Sonoran Zone.

8. PHYMOSIA Desv.

Tall herbaceous perennials, appearing glabrous, but pubescent; leaves large, 3 to 7-cleft; flowers in interrupted spikes terminating the branches, large, rose or white, 3 to 5 cm. in diameter, carpels hispid or hirsute and with fine stellate pubescence, usually 3-seeded, not reticulated on the sides.

1. **Phymosia grandiflora** Rydb. Bull. Torrey Club **40:** 60. 1913.

Sphaeralcea grandiflora Rydb. Bull. Torrey Club **31:** 565. 1904.

TYPE LOCALITY: Mesa Verde, Colorado.

RANGE: Colorado and northern New Mexico.

NEW MEXICO: Sandia Mountains.

9. SPHAERALCEA St. Hil.

Low or tall, coarse, perennial herbs with stellate pubescence; leaves petioled, various in outline, simple or dissected; flowers in small axillary clusters or by reduction of the leaves forming narrow crowded panicles; pedicels usually short; calyx subtended by 2 or 3 bracts; fruit a capsule, consisting of numerous 2-ovuled, 1 or 2-seeded, 1-celled carpels.

KEY TO THE SPECIES.

Leaves digitately 5-parted.
- Flowers solitary, on long slender pedicels........................ 1. *S. tenuipes.*
- Flowers fascicled, on short stout pedicels........................ 2. *S. pedata.*

Leaves 3-parted or simple, never 5-parted.
- Fruit depressed-globose; upper ovule usually not maturing; mature carpels mostly reniform, completely deciduous from the axis.
  - Leaves round-ovate, simple, or with 3 rounded lobes, obtuse........................ 3. *S. marginata.*
  - Leaves subhastate, lanceolate, or pinnatifid, acute.
    - Flowers 15 to 20 mm. long; leaves subhastate, silvery-stellate........................ 4. *S. martii.*
    - Flowers 10 to 12 mm. long; leaves various.
      - Pubescence very dense, fine, silvery, giving the plants a whitish appearance; leaves pinnatifid........................ 5. *S. glabrescens.*

Pubescence loose and coarse, yellowish or at least not white; plants not whitish; leaves pinnatifid or merely toothed or lobed.

Leaves pinnatifid, the terminal portion conspicuously lobed; pedicels long, slender .... 6. *S. pumila.*

Leaves subhastate, the terminal lobe not lobed; pedicels various.

Pedicels very short and stout; pubescence coarse, tawny; leaves more or less lobed .... 7. *S. subhastata.*

Pedicels long and slender; pubescence fine and close, grayish; leaves often lanceolate and not at all lobed .... 8. *S. arenaria.*

Fruit little or not at all depressed, the carpels with 2 or 3 ovules and 1 or 2 seeds, usually oblong, after separation from the axis cohering with each other by their sides and held by a short thread.

Carpels smooth at the base or nearly so .... 9. *S. leiocarpa.*

Carpels strongly reticulated at the base.

Leaves narrowly oblong, or lanceolate and subhastate; plants tall and stout.

Leaves narrowly oblong, not lobed ....10. *S. cuspidata.*

Leaves lanceolate, subhastate, with 2 low rounded lobes at the base, broader ....11. *S. lobata.*

Leaves round-ovate or rhombic-ovate in outline, often pinnatifid or variously lobed; plants mostly lower and more slender.

Pubescence fine and very dense, almost velvety (yellowish); lobes of the leaves usually round-ovate; flowers usually very numerous and dense ....12. *S. incana.*

Pubescence coarse and loose; lobes of the leaves usually narrower, mostly acutish; flowers few or numerous.

Leaves divided into 3 almost equal rounded-oblong entire lobes, the lateral lobes divergent ....13. *S. tripartita.*

Leaves not equally 3-parted, the lobes toothed and the lateral ones not divergent.

Pubescence yellowish, the whole plant green, not grayish ....14. *S. fendleri.*

Pubescence white, giving the plants a grayish appearance.

Leaves broadly cordate-ovate, mostly simple; petioles usually longer than the blades; pedicels short and comparatively stout ....15. *S. ribifolia.*

Leaves deeply 3-lobed; petioles shorter than the blades; pedicels very long and slender .....16. *S. laxa.*

1. **Sphaeralcea tenuipes** Woot. & Standl. Contr. U. S. Nat. Herb. **16**: 148. 1913.

TYPE LOCALITY: Tortugas Mountain southeast of Las Cruces, New Mexico. Type collected by Standley, May 6, 1906.

RANGE: Southwestern New Mexico, western Texas, and adjacent Mexico.

NEW MEXICO: Tortugas Mountain; between El Paso and Monument 40; Bishops Cap. Dry hills, in the Lower and Upper Sonoran zones.

2. **Sphaeralcea pedata** Torr. in A. Gray, Mem. Amer. Acad. n. ser. **4**: 23. 1849.

Malvastrum digitatum Greene, Leaflets **1**: 154. 1905.

TYPE LOCALITY: None given, but the type collected on Frémont's Third Expedition.

RANGE: Western Texas to Arizona.

NEW MEXICO: Chiz; mountains southeast of Patterson; mountains west of San Antonio; Organ Mountains; Kingston; Tortugas Mountain; Puertecito; Reserve; Socorro Mountain; Berendo Creek. Dry, open hills, in the Upper Sonoran Zone.

The type of *Malvastrum digitatum* was collected near Kingston (*Metcalfe* 941).

3. **Sphaeralcea marginata** York, Bull. Torrey Club **33**: 145. 1906.

TYPE LOCALITY: Grand Junction, Colorado.

RANGE: Southwestern Colorado to northern New Mexico.

NEW MEXICO: Ojo Caliente; Tiznitzin; Zuni Reservation; Farmington; Tunitcha Mountains; Carrizo Mountains. Sandy plains, in the Upper Sonoran Zone.

4. **Sphaeralcea martii** Cockerell, Bot. Gaz. **32**: 60. 1901.

TYPE LOCALITY: Picacho Mountain, Mesilla Valley, New Mexico.

RANGE: Known only from the type locality, on dry, rocky hills, in the Lower Sonoran Zone.

5. **Sphaeralcea glabrescens** Woot. & Standl. Bull. Torrey Club **36**: 107. 1909.

TYPE LOCALITY: Providencia Lake, about 30 miles west of Las Cruces, New Mexico. Type collected by Wooton, July 3, 1900.

RANGE: Known only from the type locality, in the Lower Sonoran Zone.

6. **Sphaeralcea pumila** Woot. & Standl. Bull. Torrey Club **36**: 110. 1909.

TYPE LOCALITY: Diamond A Wells in the Silver City Draw, Grant County, New Mexico. Type collected by Wooton, July 1, 1906.

RANGE: Western New Mexico.

NEW MEXICO: Diamond A Wells; Bear Mountain; north of Ramah. Dry plains and low hills, in the Lower and Upper Sonoran zones.

7. **Sphaeralcea subhastata** Coulter, Contr. U. S. Nat. Herb. **2**: 38. 1891.

Sphaeralcea simulans Woot. & Standl. Bull. Torrey Club **36**: 109. 1909.

TYPE LOCALITY: "In southwestern Texas and adjacent New Mexico and Mexico."

RANGE: Western Texas and southwestern New Mexico.

NEW MEXICO: Plains near Deming; Mangas Springs. Plains, in the Lower Sonoran Zone.

The type of *S. simulans* was collected on the plains near Deming (*Wooton* in 1906).

8. **Sphaeralcea arenaria** Woot. & Standl. Contr. U. S. Nat. Herb. **16**: 147. 1913.

TYPE LOCALITY: White Sands, Otero County, New Mexico. Type collected by Wooton (no. 165).

RANGE: New Mexico.

NEW MEXICO: White Sands; Albuquerque; Providencia Lake; mesa west of Organ Mountains; Suwanee; between Tularosa and Mescalero Agency; lake east of Dona Ana Mountains; Zuni Reservation; Mangas Springs. Sandy plains, in the Lower and Upper Sonoran zones.

9. **Sphaeralcea leiocarpa** Woot. & Standl. Bull. Torrey Club **36**: 107. 1909.

TYPE LOCALITY: Mangas Springs, Grant County, New Mexico. Type collected by Metcalfe (no. 721).

RANGE: Western New Mexico and adjacent Arizona.

NEW MEXICO: Mangas Springs; Mogollon Mountains; Fierro; Fort Bayard; Hatchet Ranch. Mountains, in the Upper Sonoran and Transition zones.

10. Sphaeralcea cuspidata (A. Gray) Britton in Britt. & Brown, Illustr. Fl. **2:** 519. 1898.

Sida stellata Torr. Ann. Lyc. N. Y. **2:** 171. 1828, not Cav. 1790.

Sphaeralcea stellata Torr. & Gray, Fl. N. Amer. **1:** 228. 1838.

Sphaeralcea angustifolia cuspidata A. Gray, Proc. Amer. Acad. **22:** 293. 1887.

TYPE LOCALITY: "Sources of the Arkansa."

RANGE: Colorado and Kansas to Arizona, Texas, and adjacent Mexico.

NEW MEXICO: Cross L Ranch; Clayton; Hopkins Mill; Silver City; Mangas Springs; Dog Spring; mesa west of Organ Mountains; White Sands; Alamogordo; Gray; White Mountains; Las Vegas; Roswell; Albert; Deming; Carlsbad. Plains and low hills, often in cultivated ground, in the Upper Sonoran Zone.

11. Sphaeralcea lobata Wooton, Bull. Torrey Club **25:** 306. 1898.

NIGGER WEED. YERBA DEL NEGRO.

Sphaeralcea incana ? oblongifolia A. Gray, Pl. Wright. **2:** 21. 1853.

Sphaeralcea lobata perpallida Cockerell, Bull. Torrey Club **27:** 87. 1900.

Sphaeralcea fendleri lobata Cockerell, Entomologist **1900:** 217. 1900.

TYPE LOCALITY: Mesilla, New Mexico. Type collected by Wooton (no. 2).

RANGE: Western Texas and New Mexico.

NEW MEXICO: Ojo Caliente; Santa Fe; Las Vegas; Albuquerque; Silver City; Kingston; Mesilla Valley; Hillsboro; White and Sacramento mountains. Open hills and in river valleys, in the Lower and Upper Sonoran zones.

A common weed in the lower Rio Grande Valley in irrigated fields. It does not commonly exhibit much variation, but occasionally aberrant forms occur. The usual color of the flowers is orange or orange scarlet, but sometimes they are pale, almost white. On most plants the leaves are oblong with an inconspicuous rounded lobe on each side at the base, but we have abnormal forms in which the lobes are more numerous. In one plant noticed, the lateral lobes were extremely narrow, reduced almost to the midveins, with an enlarged portion near the apex.

The typical form becomes almost a meter high and is erect, strict, and sparingly branched below, differing from the related *S. fendleri* which is much smaller and more branched.

12. Sphaeralcea incana Torr. in A. Gray, Mem. Amer. Acad. n. ser. **4:** 23. 1849.

TYPE LOCALITY: "In New Mexico." Type collected by Abert.

RANGE: New Mexico and Arizona to Chihuahua.

NEW MEXICO: Albuquerque; Laguna; Big Hatchet Mountains; San Luis Mountains; Organ Mountains; White Sands; west of Roswell; White Mountains; Guadalupe Mountains. Dry hills and plains, in the Upper Sonoran Zone.

13. Sphaeralcea tripartita Woot. & Standl. Bull. Torrey Club **36:** 108. 1909.

TYPE LOCALITY: Kingston, Sierra County, New Mexico. Type collected by Metcalfe (no. 1103).

RANGE: Known only from type locality.

14. Sphaeralcea fendleri A. Gray, Mem. Amer. Acad. n. ser. **4:** 29. 1849.

TYPE LOCALITY: Fields and wet meadows, Santa Fe, New Mexico. Type collected by Fendler (no. 78).

RANGE: New Mexico and Arizona.

NEW MEXICO: From the Las Vegas Mountains to the White Mountains and westward across the State, in the mountains and foothills. Open slopes, in the Upper Sonoran and Transition zones.

15. Sphaeralcea ribifolia Woot. & Standl. Bull. Torrey Club **36:** 109. 1909.

TYPE LOCALITY: Martin and Sloan Ranch, Grant County, New Mexico. Type collected by Wooton, August 13, 1902.

RANGE: Known only from the type locality.

16. Sphaeralcea laxa Woot. & Standl. Bull. Torrey Club **36:** 108. 1909.

TYPE LOCALITY: Frisco, Socorro County, New Mexico. Type collected by Wooton, July 25, 1900.

RANGE: Southwestern New Mexico.

NEW MEXICO: Frisco; Graham. Upper Sonoran Zone.

10. DISELLA Greene.

Prostrate or ascending herbaceous perennials, stellate-scurfy or lepidote, with rather stout short stems and simple leaves oblique at the base; flowers solitary or few in the axils, usually pale yellowish within, pink-tinged without; calyx more or less 5-angled, with 2 or 3 deciduous bractlets.

KEY TO THE SPECIES.

Plants merely loosely stellate-pubescent; leaves rounded at the apex .... 1. *D. hederacea.*

Plants densely lepidote-pubescent; leaves acute.

  Leaves obliquely ovate or deltoid-lanceolate, seldom or never with lobes at the base .... 2. *D. lepidota.*

  Leaves linear-lanceolate or narrowly oblong, with conspicuous narrow lobes at the base .... 3. *D. sagittaefolia.*

1. Disella hederacea (Dougl.) Greene, Leaflets **1:** 209. 1906. MELONCILLA.

Malva hederacea Dougl.; Hook. Fl. Bor. Amer. **1:** 107. 1830.

Sida hederacea Torr.; A. Gray, Mem. Amer. Acad. n. ser. **4:** 23. 1849.

TYPE LOCALITY: "Sides of streams, upon their low projecting banks, in the interior districts of the Columbia."

RANGE: Western Texas to Washington and California.

NEW MEXICO: Albuquerque; Magdalena; Mesilla Valley. River valleys and plains, mostly in alkaline soil, in the Lower Sonoran Zone.

This and the next are common weeds, in irrigated lands of the Rio Grande Valley in particular, though not restricted to this region. They are usually abundant on rather compact and sometimes slightly alkaline soils which get occasional irrigation, where they carpet the ground with their spreading, decumbent stems. The peculiar oblique, truncate leaves are characteristic.

2. Disella lepidota (A. Gray) Greene, Leaflets **1:** 209. 1906.

Sida lepidota A. Gray, Pl. Wright. **1:** 18. 1852.

TYPE LOCALITY: "New Mexico." Probably this should be western Texas.

RANGE: Western Texas to southern Arizona.

NEW MEXICO: Cactus Flat; Mangas Springs; Mesilla Valley; White Sands; Deming; Roswell; Hanover Mountain. Dry fields, in the Lower Sonoran Zone.

3. Disella sagittaefolia (A. Gray) Greene, Leaflets **1:** 209. 1906.

Sida lepidota sagittaefolia A. Gray, Pl. Wright. **1:** 18. 1852.

TYPE LOCALITY: Mountain valley, sixty miles west of the Pecos, Texas.

RANGE: Western Texas to Arizona and southern Colorado.

NEW MEXICO: Laguna Colorado; Socorro; near White Water; lake east of Dona Ana Mountains. Dry plains, in the Lower and Upper Sonoran zones.

11. MALVASTRUM A. Gray.

Low perennial herbs with branched stems, simple or lobed leaves, and small axillary clusters of flowers, these by the reduction of the leaves forming narrow panicles; calyx subtended by 2 or 3 small bracts; petals reddish; fruit a capsule consisting of numerous 1 or 2-ovuled, 1-seeded, 1-celled carpels.

KEY TO THE SPECIES.

Plants densely silvery-lepidote with peltate scales; divisions of the leaves linear, mostly entire........................ 1. *M. leptophyllum.*
Plants loosely canescent with stellate hairs; divisions of the leaves broader than linear, usually lobed.
  Flowers 8 mm. long or less; leaves 12 mm. long or shorter, the divisions very narrow............................... 2. *M. micranthum.*
  Flowers more than 10 mm. long; leaves 20 mm. long or more, the divisions broader.
    Plants 30 to 40 cm. high, slender; racemes loose, elongated; terminal lobe of the leaves much longer than the others................................. 3. *M. elatum.*
    Plants 10 to 20 cm. high, stout; racemes crowded; terminal segments of the leaves only slightly longer than the others................................. 4. *M. coccineum.*

1. **Malvastrum leptophyllum** A. Gray, Pl. Wright. **1:** 17. 1852.
TYPE LOCALITY: Western Texas.
RANGE: Southern Utah and Colorado to Texas and Mexico.
NEW MEXICO: Fort Cummings; Socorro Mountain; Magdalena Mountains; Farmington; Carrizozo; west of Roswell; Carrizo Mountains; San Andreas Mountains. Dry hills and plains, in the Upper Sonoran Zone.

2. **Malvastrum micranthum** Woot. & Standl. Contr. U. S. Nat. Herb. **16:** 147. 1913.
TYPE LOCALITY: Near Tiznitzin, New Mexico. Type collected by Wooton (no. 2673).
RANGE: Northwestern New Mexico.
NEW MEXICO: Near Tiznitzin; mountains southeast of Patterson. Upper Sonoran Zone.

3. **Malvastrum elatum** (Baker) A. Nels. Bot. Gaz. **34:** 25. 1902.
Malvastrum coccineum elatum Baker, Journ. Bot. Brit. & For. **29:** 171. 1891.
TYPE LOCALITY: Bed of the Limpio River, western Texas. Type, Wright's no. 41.
RANGE: Arizona and Colorado to western Texas.
NEW MEXICO: Ojo Caliente; Gallup; Zuni; Las Vegas; Pecos; Reserve; Water Canyon; Carrizalillo Mountains; Mangas Springs; San Augustine Ranch; Horse Camp; White Mountains. Open slopes, Upper Sonoran and Transition zones.

4. **Malvastrum coccineum** (Pursh) A. Gray, Mem. Amer. Acad. n. ser. **4:** 21. 1849.
Cristaria coccinea Pursh, Fl. Amer. Sept. 453. 1814.
Sida dissecta Nutt.; Torr. & Gray, Fl. N. Amer. **1:** 235. 1840.
Malvastrum cockerellii A. Nels. Bot. Gaz. **34:** 24. 1902.
TYPE LOCALITY: "On the dry prairies and extensive plains of the Missouri."
RANGE: Oregon and Saskatchewan to Arizona, Texas, and Iowa.
NEW MEXICO: Common throughout the State. Open hills and plains, in the Upper Sonoran Zone.

12. SIDA L.

Prostrate or erect herbaceous perennials, with simple, alternate, mostly narrow leave; and slender stems; flowers solitary or in small axillary clusters, yellow or oranges calyx more or less 5-angled, sometimes accrescent in fruit, not bracteate; carpels 5 or numerous, 1-celled, 1-ovuled, indehiscent, or dehiscent at the apex.

KEY TO THE SPECIES.

Calyx strongly accrescent at maturity.......................... 1. *S. physocalyx.*
Calyx not accrescent.
  Plants erect, not hirsute.................................... 2. *S. neomexicana.*
  Plants prostrate, hirsute.................................... 3. *S. diffusa.*

1. Sida physocalyx A. Gray, Bost. Journn. Nat. Hist. **6:** 163. 1850.

TYPE LOCALITY: On the Liano, western Texas.

RANGE: Texas to Arizona, south into Mexico.

NEW MEXICO: Organ Mountains; south of Roswell; Lakewood; Tortugas Mountain. Dry hills, in the Upper Sonoran Zone.

2. Sida neomexicana A. Gray, Proc. Amer. Acad. **22:** 296. 1887.

TYPE LOCALITY: Mountains at the Copper Mines, New Mexico. Type collected by Wright.

RANGE: Western Texas to Arizona and adjacent Mexico.

NEW MEXICO: Santa Rita; Fort Bayard; San Luis Mountains; Organ Mountains. Dry hills, in the Upper Sonoran Zone.

3. Sida diffusa H. B. K. Nov. Gen. & Sp. **5:** 257. 1821.

TYPE LOCALITY: "Crescit prope Zelaya Mexicanorum, alt. 950 hex."

RANGE: Texas and New Mexico to Mexico.

NEW MEXICO: North Percha Creek; Mangas Springs; Organ Mountains; Tortugas Mountain; between Santa Rita and Mimbres. Dry hills, in the Lower and Upper Sonoran zones.

13. ANODA Cav.

Erect annuals 60 to 150 cm. high, with alternate, simple, hastate or deltoid-cordate leaves and solitary axillary flowers, or these becoming somewhat paniculate above by the reduction of the leaves; calyx lobes triangular, spreading, thin; capsules depressed and radiate, of 9 to 20 long-beaked carpels, the flat summit hirsute.

KEY TO THE SPECIES.

Corolla lavender; sepals much exceeding the hispid carpels.... 1. *A. lavaterioides.*
Corolla yellow; sepals slightly exceeding the stellate-hirsute
    carpels... 2. *A. wrightii.*

1. Anoda lavaterioides Medic. Malvenfam. 19. 1787.

TYPE LOCALITY: Not stated.

RANGE: Western Texas to southern Arizona, south into Mexico and South America.

NEW MEXICO: Mangas Springs; Kingston; Fort Bayard; San Luis Mountains; Mesilla Valley; Organ Mountains; White Mountains; Gray; Albuquerque; Belen; Bernalillo; Capitan; Dayton. Open hills, often in cultivated fields, in the Lower and Upper Sonoran zones.

A common weed in the southern part of the State in fields and orchards, especially in summer after grain crops have been harvested.

2. Anoda wrightii A. Gray, Pl. Wright. **2:** 22. 1853.

TYPE LOCALITY: Summit of mountains near the Copper Mines, New Mexico. Type collected by Wright (no. 894).

RANGE: Southwestern New Mexico.

NEW MEXICO: Hillsboro; Santa Rita.

14. SIDANODA (Robinson) Woot. & Standl.

Sidanoda Woot. & Standl. gen. nov.

Anoda section *Sidanoda* Robinson in A. Gray, Syn. Fl. 1[1]: 320. 1897.

Erect, much branched annuals, with short stellate pubescence; leaves various in outline, alternate, long-petiolate; flowers small, long-pediceled, in open racemes or panicles; petals yellow or blue; carpels 5 to 9, depressed or ascending, dorsally umbonate or short-cuspidate, puberulent, never hirsute; seeds resupinate-pendulous.

TYPE SPECIES: *Anoda pentaschista* A. Gray.

1. **Sidanoda pentaschista** (A. Gray) Woot. & Standl.

Anoda pentaschista A. Gray, Pl. Wright. 2: 22. 1853.

TYPE LOCALITY: Valley between Ojo de Gavilan and Condes Camp, beyond the Copper Mines, New Mexico. Type collected by Wright (no. 893).

RANGE: Western Texas to southern Arizona and adjacent Mexico.

NEW MEXICO: Between Ojo de Gavilan and Condes Camp; Mesilla Valley. Lower Sonoran Zone.

Order 34. HYPERICALES.

KEY TO THE FAMILIES.

Styles wanting.
- Herbs; placentæ axial........................ 90. **ELATINACEAE** (p. 427).
- Shrubs; placentæ basal........................ 91. **TAMARICACEAE** (p. 427).

Styles present.
- Petals united to above the middle.......... 92. **FOUQUIERIACEAE** (p. 428).
- Petals distinct, or merely coherent at the base.
  - Styles united........................ 95. **VIOLACEAE** (p. 428).
  - Styles distinct.
    - Sepals united into a tube; leaves not pellucid-dotted.......... 93. **FRANKENIACEAE** (p. 428).
    - Sepals distinct; leaves pellucid-dotted........................ 94. **HYPERICACEAE** (p. 428).

90. ELATINACEAE. Waterwort Family.

1. ELATINE L. WATERWORT.

Small, fragile, often aquatic, glabrous herbs with opposite or whorled leaves; flowers minute, usually solitary in the axils; sepals 2; petals and stamens 2 or 3; capsules subglobose, rarely 1 mm. in diameter; seeds small, striate.

1. **Elatine americana** (Pursh) Arnold, Edinburgh Journ. Sci. 1: 430. 1830.

Peplis americana Pursh, Fl. Amer. Sept. 238. 1814.

TYPE LOCALITY: Pennsylvania.

RANGE: British America to Oregon, New Mexico, and Virginia.

We have seen no specimens of this from New Mexico, but in the Botany of the Mexican Boundary it is reported from "hills near the Copper Mines," collected by Bigelow.

91. TAMARICACEAE. Tamarix Family.

1. TAMARIX L.

1. **Tamarix gallica** L. Sp. Pl. 270. 1753. SALT CEDAR.

TYPE LOCALITY: "Habitat in Gallia, Hispania, Italia."

A cultivated plant, used very effectively for hedges in many places, often escaped. It may be recognized by its habit, which suggests the name of cedar (though it is not evergreen), and by its large panicles of small pink flowers borne profusely in the spring or early summer. It grows rapidly from cuttings and withstands continued drought very well, nor is it easily hurt by alkali in the soil, characteristics which make it especially valuable for cultivation in an arid climate.

92. FOUQUIERIACEAE. Ocotillo Family.

1. FOUQUIERIA H. B. K. OCOTILLO.

Spiny shrubs with several erect or ascending virgate stems 3 meters long or less, bearing leaves for but a short time in the summer, the spines formed by the indurated mid-ribs of the leaves of previous seasons; leaves oblanceolate-spatulate, entire; flowers perfect, in thyrsoid terminal panicles, bright scarlet, appearing usually before the leaves; sepals 5; corolla 5-merous, gamopetalous, broadly tubular, with a spreading limb; stamens 10, epipetalous; fruit an ovoid capsule with many seeds.

1. **Fouquieria splendens** Engelm. in Wisliz. Mem. North. Mex. 98. 1848.

TYPE LOCALITY: Jornada del Muerto, New Mexico. Type collected by Wislizenus in 1846.

RANGE: Western Texas to Arizona, south into Mexico.

NEW MEXICO: Black Range; Upper Corner Monument; Big Hatchet Mountains; Hachita; mesa west of Organ Mountains. Mesas, in the Lower Sonoran Zone.

A form with white flowers was collected by Metcalfe near Kingston in 1904.

93. FRANKENIACEAE. Frankenia Family.

1. FRANKENIA L.

Branching shrubs 1 meter high or less, with small crowded leaves on numerous fascicled short branches and small white flowers, these solitary, axillary, sessile; sepals 5, united into a persistent tube; petals 5, white, clawed; stamens 6; fruit a few-seeded capsule included in the calyx.

1. **Frankenia jamesii** Torr. in A. Gray, Proc. Amer. Acad. **8**: 622. 1873.

TYPE LOCALITY: Colorado.

RANGE: Southern Colorado to New Mexico and western Texas.

NEW MEXICO: White Sands; Alamogordo; Los Mitos; Organ Mountains. In alkaline soil, in the Lower Sonoran Zone.

94. HYPERICACEAE. St. Johnswort Family.

1. HYPERICUM L. ST. JOHNSWORT.

Herbs 30 to 60 cm. high, with yellow, loosely cymose flowers and opposite sessile leaves, these usually black-dotted along the margins; sepals 5; petals 5, bright yellow, with a few black glands; stamens numerous, united at the base into 3 or 5 clusters; styles 3, distinct; fruit a 3-lobed capsule with numerous seeds.

1. **Hypericum formosum** H. B. K. Nov. Gen. & Sp. **5**: 196. *pl. 460.* 1821.

TYPE LOCALITY: "Crescit prope Pazcuaro Mexicanorum, alt. 1130 hex."

RANGE: Colorado and Utah to California and Mexico.

NEW MEXICO: Higher mountains throughout the State. Damp meadows, in the Transition Zone.

95. VIOLACEAE. Violet Family.

Low herbs, often acaulescent, with simple alternate stipulate leaves and complete irregular flowers; sepals 5; petals 5, irregular, one of them often spurred; stamens 5, the anthers erect or connivent; pistil of 3 carpels, with a single style, becoming a 1-celled capsule with several seeds.

KEY TO THE GENERA.

Sepals auriculate at the base; flowers mostly large and showy; lower petal spurred........................ 1. VIOLA (p. 429).
Sepals not auriculate; flowers small, greenish; upper and lateral petals markedly unequal.................... 2. CALCEOLARIA (p. 431).

1. VIOLA L. VIOLET.

Low perennial herbs, acaulescent or with short stems, with alternate stipulate leaves of various shapes; flowers solitary, scapose, on axillary peduncles, often of two kinds, the later ones cleistogamous; petals irregular, the lowermost spurred or saccate at the base; capsules elastically dehiscent.

The writers wish to acknowledge their appreciation of the assistance of Dr. Ezra Brainerd in the preparation of the account of this genus.

KEY TO THE SPECIES.

Plants acaulescent.
  Leaf blades lobed.
    Lobes of leaves linear or nearly so, numerous, extending nearly to the base.............................. 1. *V. pedatifida.*
    Lobes oblong, few, separate only about half way to the base... 2. *V. wilmattae.*
  Leaf blades not lobed.
    Flowers white.. 3. *V. pallens.*
    Flowers blue.
      Leaves broadly ovate, obtuse; capsules 5 to 10 mm. long... 4. *V. nephrophylla.*
      Leaves deltoid, acutish; capsules 10 to 15 mm. long.. 5. *V. missouriensis.*
Plants caulescent.
  Flowers yellow or brownish................................ 6. *V. pinetorum.*
  Flowers blue or white.
    Flowers blue.
      Leaves deeply cordate; stems much elongated, slender, not cespitose........................ 7. *V. montanensis.*
      Leaves rounded to acutish at the base; stems stout, thick, cespitose.
        Leaves glabrous or nearly so.................... 8. *V. adunca.*
        Leaves puberulent............................... 9. *V. puberula.*
    Flowers white or nearly so.
      Stipules fimbriate ....10. *V. reptans.*
      Stipules entire.
        Leaves nearly glabrous beneath; petals not retuse...................................11. *V. canadensis.*
        Leaves muriculate-scabrous on both surfaces; petals retuse.............................12. *V. muriculata.*

1. Viola pedatifida Don, Hist. Dichl. Pl. **1**: 320. 1831.

TYPE LOCALITY: North America.

RANGE: Colorado and New Mexico to Saskatchewan and Illinois.

NEW MEXICO: Sierra Grande; between Park View and Tierra Amarilla. Plains and low hills, in the Upper Sonoran Zone.

2. Viola wilmattae Pollard & Cockerell, Proc. Biol. Soc. Washington **15**: 18. 1902.

TYPE LOCALITY: Sapello Canyon, Beulah, New Mexico. Type collected by Mrs. W. P. Cockerell in 1901.

RANGE: Northern New Mexico.

NEW MEXICO: Beulah; Sierra Grande.

This appears to be a hybrid between *V. nephrophylla* and *V. pedatifida*, and has recently been described as such by Dr. Ezra Brainerd.[1]

3. **Viola pallens** (Banks) Brainerd, Rhodora **7**: 247. 1905.

Viola rotundifolia pallens Banks; DC. Prodr. **1**: 295. 1824.

TYPE LOCALITY: "In Labrador et Kamtschatka."

RANGE: New Mexico to Tennessee, New England, and Labrador; also in Siberia.

NEW MEXICO: East Canyon (*Holzinger*).

The plant appears to be of this species, although it is not in such a condition as to make certain determination possible.

4. **Viola nephrophylla** Greene, Pittonia **3**: 144. 1896.

TYPE LOCALITY: "In dry thickets of scrubby willows and *Potentilla fruticosa*, the valley of the Cimarron River, western Colorado."

RANGE: Idaho and Wyoming to Nevada and New Mexico.

NEW MEXICO: Tunitcha Mountains; Chama; Rio Pueblo; Santa Fe and Las Vegas mountains; Holts Ranch; Middle Fork of the Gila; Iron Creek. Moist shaded slopes, in the Transition Zone.

5. **Viola missouriensis** Greene, Pittonia **4**: 141. 1900.

TYPE LOCALITY: Leeds, Missouri.

RANGE: Missouri to Oklahoma and northeastern New Mexico.

NEW MEXICO: Gallinas River near Las Vegas (*Cockerell*).

6. **Viola pinetorum** Greene, Pittonia **2**: 14. 1889.

TYPE LOCALITY: "Pine woods of the higher mountains south of Tehachapi, Kern Co., California."

RANGE: California and Oregon to northern New Mexico.

NEW MEXICO: Cross L Ranch (*Griffiths* 4308).

7. **Viola montanensis** Rydb. Mem. N. Y. Bot. Gard. **1**: 263. 1900.

TYPE LOCALITY: Jack Creek Canyon, Montana.

RANGE: Montana to northern New Mexico.

NEW MEXICO: Chama (*Standley* 6823). Damp woods, in the Canadian Zone.

8. **Viola adunca** J. E. Smith, Rees's Cycl. **37**: no. 63. 1817.

TYPE LOCALITY: West coast of North America.

RANGE: British America to California and New Mexico.

NEW MEXICO: Winsor Creek; Pecos Baldy. Damp woods, Transition to Hudsonian Zone.

9. **Viola puberula** (S. Wats.) Howell, Fl. Northw. Amer. **1**: 72. 1897.

Viola canina puberula S. Wats.; A. Gray, Man. ed. 6. 81. 1890.

Viola retroscabra Greene, Pittonia **4**: 290. 1901.

TYPE LOCALITY: "Sandy or stony shores and islands of Lakes Huron and Superior."

RANGE: Washington and California to Colorado and New Mexico.

NEW MEXICO: Pecos Baldy; Nutritas Creek below Tierra Amarilla; top of Las Vegas Range. Transition Zone.

10. **Viola reptans** Robinson, Proc. Amer. Acad. **27**: 165. 1892.

Viola pringlei Rose & House, Proc. U. S. Nat. Mus. **29**: 444. 1905.

TYPE LOCALITY: Hills of Patzcuaro, Michoacán, Mexico.

RANGE: Mexico to northern New Mexico.

NEW MEXICO: Near Dulce; San Luis Mountains; Animas Valley. Mountains, in the Transition Zone.

---

[1] Bull. Torrey Club **40**: 259, 1913.

This is a remarkable extension of range for the species, described from specimens from southwestern Mexico. The plants collected at Dulce were growing on a bank under pine trees.

11. Viola canadensis L. Sp. Pl. 936. 1753.

Viola neomexicana Greene, Pittonia **5**: 28. 1902.

Viola canadensis neomexicana House; Rydb. Colo. Agr. Exp. Sta. Bull. **100**: 233. 1906.

TYPE LOCALITY: "Habitat in Canada."

RANGE: British America southward to New Mexico.

NEW MEXICO: Chama; Tunitcha Mountains; Sandia Mountains; Santa Fe and Las Vegas mountains; Kelly; Holts Ranch; Iron Creek; White and Sacramento mountains. Transition and Canadian zones.

12. Viola muriculata Greene, Pittonia **5**: 28. 1902.

TYPE LOCALITY: "In subalpine woods of Mt. San Francisco, near Flagstaff, Arizona."

RANGE: Mountains of Arizona and New Mexico.

NEW MEXICO: Mogollon Creek; Iron Creek; Magdalena Mountains.

2. CALCEOLARIA Loefl.

Low perennial herb with branched stems about 10 cm. high, small simple narrow leaves, and very small pale flowers; sepals equal, not auricled; petals unequal, the two upper ones smallest, the lower largest, gibbous at the base; anthers connivent, the filaments distinct, the two lower ones glandular at the base; capsules elastically 3-valved.

1. Calceolaria verticillata (Orteg.) Kuntze, Rev. Gen. Pl. **1**: 41. 1891.

Viola verticillata Orteg. Hort. Matr. Dec. **4**: 50. 1797.

Ionidium lineare Torr. Ann. Lyc. N. Y. **2**: 168. 1827.

Hybanthus verticillatus A. Nels. in Coulter, New Man. Rocky Mount. 323. 1909.

TYPE LOCALITY: "Nova Hispania."

RANGE: Colorado and Kansas to Texas, Arizona, and Mexico.

NEW MEXICO: Sierra Grande; Mangas Springs; Black Range; Tortugas Mountain; Organ Mountains; Roswell; Florida Mountains; Queen. Dry hills, in the Upper Sonoran Zone.

Order 35. OPUNTIALES.

KEY TO THE FAMILIES.

Sepals and petals very unlike, 4 or 5; leaves ample; plants not succulent, not armed with spines............ **96. LOASACEAE** (p. 431).

Sepals and petals nearly alike, numerous; leaves reduced to mere scales or wanting; plants succulent, armed with spines.................................... **97. CACTACEAE** (p. 436).

96. LOASACEAE. Loasa Family.

Herbaceous annuals or perennials with whitish stems; leaves simple, entire to deeply pinnatifid, covered with coarse barbed or stinging hairs; hypanthium more or less tubular; sepals 5, persistent; petals 5, often with 5 petal-like staminodia, white, yellow, or orange; stamens 5 to many, the filaments often petaloid; capsules 1-celled, with 1 to 3 parietal placentæ; seeds 1 to many.

KEY TO THE GENERA.

Stamens 5; plants covered with stinging hairs............. 1. CEVALLIA (p. 432).
Stamens numerous; plants rough-hispid with barbed or hooked but never stinging hairs.
Placentæ with horizontal lamellæ between the seeds; seeds flat, winged.............................. 2. NUTTALLIA (p. 432).
Placentæ without lamellæ; seeds not winged.
Seeds prismatic, muricate; leaves sessile........... 3. ACROLASIA (p. 435).
Seeds pyriform or ellipsoid, striate with parallel curved lines; leaves petiolate (petioles sometimes very short)......................... 4. MENTZELIA (p. 436).

1. CEVALLIA Lag.

Canescent branched perennial herbs armed with stinging hairs; leaves alternate, sessile, sinuate-pinnatifid; flowers in terminal heads; tube of the calyx short, with erect linear lobes; petals 5, plumose; stamens 5, erect, with very short filaments; fruit dry, indehiscent, 1-seeded.

1. Cevallia sinuata Lag. Var. Cienc. **21**: 35. 1805.

TYPE LOCALITY: Mexico.

RANGE: Western Texas to southern Arizona and southward.

NEW MEXICO: Rincon; mesa west of Organ Mountains; Dog Spring; Lake Valley; Lordsburg; Hatchet Ranch. Dry mesas, especially along arroyos, in the Lower Sonoran Zone.

2. NUTTALLIA Raf.

Herbaceous short-lived perennials (sometimes annuals) with white stems, at first rough, becoming smooth and shining below, or the epidermis exfoliating; leaves simple, alternate, more or less densely rough-hirsute with stiff, barbed or hooked, white hairs; hypanthium mostly campanulate, becoming hemispheric to cylindric in fruit; sepals 5, persistent; petals 5 or with 5 additional petal-like staminodia, of some shade of yellow (often described as white); stamens numerous, the outer rows of filaments often petaloid; capsules 1-celled with parietal placentæ lamellated between the flattened, winged, finely tuberculate seeds.

KEY TO THE SPECIES.

Leaves pinnately toothed or lobed.
Lobes of the leaves linear, several times as long as broad, 1 to 2 mm. wide.
Plants small, 30 to 50 cm. high; flowers pale yellow, small. 1. *N. gypsea.*
Plants taller, spreading, 60 to 80 cm. high; flowers deep golden yellow, larger.............................. 2. *N. laciniata.*
Lobes of the leaves oblong, hardly more than twice as long as broad, 2 to 3 mm. wide.
Petals bright yellow; some of the leaves entire.
Some of the leaves oblanceolate; plants stout; capsules nearly as broad as long........................12. *N. integra.*
All leaves linear; plants slender; capsules twice as long as broad....................................13. *N. springeri.*
Petals whitish or pale yellow; leaves all lobed or toothed.
Lobes of the leaves few, 1 to 3 on each side, often entirely wanting; plants cespitose............... 3. *N. perennis.*

Lobes of the leaves more numerous, 4 to many on each side, the leaves never entire; plants not cespitose.

Plants stout, widely spreading; flowers of medium size for the genus, numerous................ 4. *N. multiflora.*

Plants tall and slender, not spreading; flowers small for the genus, few.................. 5. *N. procera.*

Leaves sinuate-dentate, not pinnatifid.

Flowers very large, petals about 6 cm. long; filaments not dilated.. 6. *N. decapetala.*

Flowers small, petals 2 to 3 cm. long; outer filaments usually dilated.

Cauline leaves all sessile, not cuneate at the base.

Petals 2 cm. long; flowers crowded at the top of the stem; leaves large, 5 to 10 cm. long, long-acuminate.................................. 7. *N. rusbyi.*

Petals about 3 cm. long; flowers mostly solitary; cauline leaves short and small, acute................... 8. *N. strictissima*

Cauline leaves, at least the lower ones, cuneate at the base or tapering into a petiole.

Flowers bright yellow................................ 9. *N. speciosa.*

Flowers pale yellow.

Involucral bracts narrow, entire..................10. *N. nuda.*

Involucral bracts laciniately lobed, one sepal occasionally so...............................11. *N. stricta.*

1. **Nuttallia gypsea** Woot. & Standl. Contr. U. S. Nat. Herb. **16:** 149. 1913.

TYPE LOCALITY: On pure gypsum, near Lakewood, New Mexico. Type collected by Wooton, August 6, 1909.

RANGE: Southeastern New Mexico.

NEW MEXICO: Near Lakewood; 35 miles south of Torrance; Roswell. Gypsum soil.

2. **Nuttallia laciniata** (Rydb.) Woot. & Standl. Contr. U. S. Nat. Herb. **16:** 150. 1913.

Touterea laciniata Rydb. Bull. Torrey Club **31:** 565. 1904.

TYPE LOCALITY: Pagosa Springs, Colorado.

RANGE: Southwestern Colorado, western New Mexico, and probably Utah and Arizona.

NEW MEXICO: Gallup; McCarthy Station; above Chamita; Zuni. Upper Sonoran and Transition zones.

3. **Nuttallia perennis** (Wooton) Cockerell, Trans. Amer. Ent. Soc. **32:** 300. 1906.

Mentzelia perennis Wooton, Bull. Torrey Club **25:** 260. 1898.

Hesperaster perennis Cockerell, Torreya **1:** 143. 1901.

Touterea perennis Rydb. Bull. Torrey Club **30:** 277. 1903.

TYPE LOCALITY: In gypseous soil at Round Mountain, half way between Tularosa and the Mescalero Agency, White Mountains, New Mexico. Type collected by Wooton (no. 184).

RANGE: Known only from the type locality, in the Upper Sonoran Zone.

4. **Nuttallia multiflora** (Nutt.) Greene, Leaflets **1:** 210. 1906.

Bartonia multiflora Nutt. Journ. Acad. Phila. II. **1:** 180. 1848.

Mentzelia multiflora A. Gray, Mem. Amer. Acad. n. ser. **4:** 48. 1849.

Touterea multiflora Rydb. Bull. Torrey Club **30:** 277. 1903.

TYPE LOCALITY: Sandy hills along the borders of the Rio del Norte, Santa Fe, New Mexico. Type collected by Gambel.

52576°—15——28

RANGE: Colorado to western Texas and northeastern Mexico.

NEW MEXICO: Common nearly throughout the State except in the higher mountains. Dry hills and plains, in the Lower and Upper Sonoran zones.

So far as we are able to judge from field observations, the plants of the western and southwestern sections of the State agree very well with the plants about Santa Fe. The plant found in the south-central part of the State appears somewhat different when growing, but we are unable to find characters to separate the two. The Santa Fe plant has rather bright yellow flowers which open only just before sundown, and the plant is rather widely spreading and, when full grown, almost 1 meter high; while the plant of the Rio Grande Valley, at the south end of the State, is generally lower, about 60 cm. high or less, its flowers pale yellow, fading almost white, and opening early in the afternoon, not infrequently by 2 or 3 o'clock.

5. **Nuttallia procera** Woot. & Standl. Contr. U. S. Nat. Herb. **16**: 150. 1913.

TYPE LOCALITY: White Sands, New Mexico. Type collected by Wooton & Standley, August 18, 1907.

RANGE: Southern New Mexico.

NEW MEXICO: White Sands; above Tularosa. Gypseous soil, in the Lower Sonoran Zone.

6. **Nuttallia decapetala** (Pursh) Greene, Leaflets **1**: 210. 1906.

Bartonia decapetala Pursh in Curtis's Bot. Mag. **18**: *pl. 1487*. 1812.

Bartonia ornata Pursh, Fl. Amer. Sept. 327. 1814.

Mentzelia ornata Torr. & Gray, Fl. N. Amer. **1**: 534. 1840.

Touterea decapetala Rydb. Bull. Torrey Club **30**: 276. 1903.

TYPE LOCALITY: "Banks of the Missouri."

RANGE: South Dakota and Alberta to Texas and Nevada.

NEW MEXICO: Colfax; Raton Mountains; Albert. Plains, in the Upper Sonoran and Transition zones.

7. **Nuttallia rusbyi** (Wooton) Cockerell, Trans. Amer. Ent. Soc. **32**: 300. 1906.

Mentzelia rusbyi Wooton, Bull. Torrey Club **25**: 261. 1898.

Hesperaster rusbyi Cockerell, Torreya **1**: 143. 1901.

Touterea rusbyi Rydb. Bull. Torrey Club **30**: 276. 1903.

TYPE LOCALITY: Bellmont, Arizona.

RANGE: New Mexico to Wyoming and Montana.

NEW MEXICO: Farmington; Mora; Santa Fe and Las Vegas mountains; Mogollon Mountains; White and Sacramento mountains; Gray; Dulce. Damp slopes in the mountains, in the Transition Zone.

8. **Nuttallia strictissima** Woot. & Standl. Contr. U. S. Nat. Herb. **16**: 150. 1913.

TYPE LOCALITY: Arroyo Ranch near Roswell, New Mexico. Type collected by David Griffiths (no. 5701).

RANGE: Southern New Mexico.

NEW MEXICO: Arroyo Ranch; twenty miles south of Roswell. Plains, in the Lower Sonoran Zone.

9. **Nuttallia speciosa** (Osterhout) Greene, Leaflets **1**: 210. 1906.

Mentzelia aurea Osterhout, Bull. Torrey Club **28**: 644. 1901, not Nutt. 1818.

Mentzelia speciosa Osterhout, op. cit. 689. 1901.

Touterea speciosa Osterhout, Bull. Torrey Club **30**: 276. 1903.

TYPE LOCALITY: Estes Park, Larimer County, Colorado.

RANGE: Wyoming to northern New Mexico.

NEW MEXICO: Colfax (*Wooton*). Hills and dry valleys, in the Upper Sonoran Zone.

10. **Nuttallia nuda** (Pursh) Greene, Leaflets **1**: 210. 1906.

Mentzelia nuda Torr. & Gray, Fl. N. Amer. **1**: 535. 1840.

Hesperaster nuda Cockerell, Torreya **1**: 143. 1901.

TYPE LOCALITY: "On the bank of the Missouri."

RANGE: Nebraska and Wyoming to northern New Mexico.

NEW MEXICO: Clayton (*Howell* 153). Plains, in the Upper Sonoran Zone.

The original description of this species might with perfect accuracy be applied to any one of a half dozen of the species now recognized in the genus, in all but one particular. The statement that the ovary is naked is to be taken in relation to the description of the species immediately preceding (*N. decapetala*), in which the ovary is hidden by conspicuous laciniate bracts.

The plant here accepted as *Nuttallia nuda* is the only one we have seen "germine nudo," even by comparison, and it has one or two inconspicuous bracts at the bases of the capsules. Its leaves are oblong, sessile, attenuate at the base, and repand-dentate, as described by Pursh, but not "somewhat lanceolate, interruptedly pinnatifid" as described by Torrey and Gray. It is evidently a much rarer plant than *Nuttallia stricta*, which has passed as *Nuttallia nuda* in herbaria for a long time.

11. **Nuttallia stricta** (Osterhout) Greene, Leaflets **1**: 210. 1906.

Hesperaster stricta Osterhout, Bull. Torrey Club **29**: 174. 1902.

Touterea stricta Rydb. Bull. Torrey Club **30**: 276. 1903.

TYPE LOCALITY: New Windsor, Weld County, Colorado.

RANGE: Western Colorado to northern New Mexico and western Texas.

NEW MEXICO: Clayton; Nara Visa; Cross L Ranch; Perico. Plains, in the Upper Sonoran Zone.

12. **Nuttallia integra** (Jones) Rydb. Bull. Torrey Club **40**: 61. 1913.

Mentzelia multiflora integra Jones, Proc. Calif. Acad. II. **5**: 689. 1895.

Touterea integra Rydb. Colo. Agr. Exp. Sta. Bull. **100**: 235. 1906.

TYPE LOCALITY: Rockville, Utah.

RANGE: Southern Utah to northern Arizona and northwestern New Mexico.

NEW MEXICO: Tunitcha Mountains; Carrizo Mountains. Dry hills and mesas, in the Upper Sonoran Zone.

13. **Nuttallia springeri** Standley, Proc. Biol. Soc. Washington **26**: 115. 1913.

TYPE LOCALITY: Mesa above the Abbott Ranch, Rito de los Frijoles, northwest of Santa Fe, New Mexico. Type collected by Frank Springer (no. 4).

RANGE: Known only from type locality, in the Upper Sonoran Zone.

3. ACROLASIA Presl.

Annuals with slender, branching, rather weak, white stems, long internodes, and narrowly lanceolate sessile cauline leaves, with a rosette of basal leaves, the whole plant hispidulous with barbed hairs; leaves entire, coarsely toothed, or pinnatifid; flowers small, axillary or congested at the ends of the branches; sepals 2 to 3 mm. long; petals 5, yellow, obovate, 3 to 4 mm. long; filaments not petaloid; capsules elongated, cylindric or clavate, over 1 cm. long, with filiform placentæ, not lamellated between the seeds; seeds several, irregularly prismatic, finely tuberculate.

KEY TO THE SPECIES.

Cauline leaves pinnatifid.. 1. *A. albicaulis*.
Cauline leaves mostly entire, linear-lanceolate...................... 2. *A. parviflora*.

1. **Acrolasia albicaulis** (Dougl.) Rydb. Bull. Torrey Club **30**: 277. 1903.

Mentzelia albicaulis Dougl.; Hook. Fl. Bor. Amer. **1**: 222. 1833.

TYPE LOCALITY: "On the arid sandy plains of the river Columbia, under the shade of *Purshia tridentata*."

RANGE: New Mexico to Montana and British Columbia.

NEW MEXICO: Aztec; Cliff; Organ Mountains; Carrizo Mountains. Dry hills, in the Upper Sonoran Zone.

2. Acrolasia parviflora Heller, Muhlenbergia **1:** 138. 1906.

Mentzelia parviflora Heller, Bull. Torrey Club **25:** 199. 1898.

TYPE LOCALITY: Eleven miles southwest of Santa Fe, New Mexico, on the road leading to Canyoncito. Type collected by Heller (no. 3750).

RANGE: New Mexico and Arizona.

NEW MEXICO: Near Santa Fe; Organ Mountains; Cliff; Hillsboro; near Rio Apache; Wheelers Ranch; Sandia Mountains. Plains and foothills, in the Upper Sonoran Zone.

The plants from the southwestern part of the State have the cauline leaves mostly linear-lanceolate and entire, with only occasionally a toothed one. The species is close to *Acrolasia albicaulis*.

4. MENTZELIA L.

Herbaceous annuals or perennials with alternate, simple, coarsely toothed or lobed leaves having short petioles, and solitary sessile axillary orange-colored flowers; younger stems, leaves, and capsules covered with stiff barbed white hairs; flowers of medium size; petals and sepals 5; filaments not dilated; fruit clavate-cylindric, with a few ellipsoid or pyriform seeds, these finely striate in curved lines.

KEY TO THE SPECIES

Annual; leaves acute or acuminate; seeds several; capsules thin-walled .... 1. *M. asperula*.

Perennial with tuberous root; leaves acute or obtuse; seeds solitary; capsules thick-walled and woody .... 2. *M. monosperma*.

1. Mentzelia asperula Woot. & Standl. Contr. U. S. Nat. Herb. **16:** 148. 1913.

TYPE LOCALITY: Trujillo Creek, Sierra County, New Mexico. Type collected by Metcalfe (no. 1364).

RANGE: Western Texas to southern Arizona, south into Mexico.

NEW MEXICO: Trujillo Creek; Organ Mountains. Canyons, in the Upper Sonoran Zone.

2. Mentzelia monosperma Woot. & Standl. Contr. U. S. Nat. Herb. **16:** 149. 1913.

TYPE LOCALITY: Organ Mountains, New Mexico. Type collected by Wooton, August 29, 1894.

RANGE: Southern New Mexico.

NEW MEXICO: Organ Mountains; 35 miles west of Roswell. Dry hills, in the Upper Sonoran Zone.

97. CACTACEAE. Cactus Family.

Green fleshy-stemmed spiny perennials, mostly leafless xerophytes of peculiar aspect; stems globose, cylindric, or flattened, tuberculate or ridged, often jointed, the spines and spicules borne on restricted areas known as areoles; flowers mostly large and handsome; sepals numerous, in several series, gradually becoming petaloid; petals numerous, of delicate texture and handsome colors; stamens very numerous; ovary inferior, with a thick style and several stigmas; fruit a dry or pulpy berry with thin or thickened rind and numerous seeds in the single cell.

KEY TO THE GENERA.

Plants with small terete caducous leaves; stems jointed; spines often barbed, accompanied by glochids; tube of the flowers short .... 1. OPUNTIA (p. 437).

Plants without leaves; stems not jointed; spines not barbed, without glochids; tube of flowers more or less elongated.

Stems short, mostly ovoid, globose, or short-cylindric, tuberculate; ovary and fruit smooth, neither scaly nor spiny .... 2. MAMILLARIA (p. 447).
Stems mostly short or long-cylindric, ovoid or occasionally globose, mostly larger than in the preceding genus; tubercles confluent into longitudinal ridges; ovary and fruit not smooth.
  Ovary and fruit scaly, not spiny; flowers borne in the center of the stem at the apex.... 3. ECHINOCACTUS (p. 451).
  Ovary and fruit spiny; flowers borne laterally on the stem some distance from the apex.
    Flowers usually brightly colored, red, yellow, or greenish, open during the day; stems thick and very spiny, with 6 or more ribs........................ 4. ECHINOCEREUS (p. 454).
    Flowers white, open at night; stems slender, 4 or 5-ribbed; spines very short and inconspicuous..................... 5. PENIOCEREUS (p. 458).

1. OPUNTIA Mill. PRICKLY PEAR.

Perennials with jointed stems, bearing small, terete or conic, fleshy, caducous leaves; joints of the stems flattened ("prickly pears" or "nopales"), cylindric ("chollas" or "cane cacti," inpart), or clavate or tumid, smooth to strongly tuberculate; leaves usually 1 cm. long or less, to be seen only on the young joints or the young ovary; areoles with numerous retrorsely barbed glochids 3 to 15 mm. long and 1 to several slender or stout, long or short spines (in one section the spines covered by a papery sheath); flowers mostly large, with numerous sepals and petals, very numerous stamens, and a single thick style with several stigmas; fruit tuberculate or smooth, with several to many areoles, these bristle-bearing or sometimes spine-bearing, occasionally proliferous, dry or berry-like, with a thick rind (berry-like fruits known as "tunas").

KEY TO THE SPECIES.

Joints clavate, tumid, or cylindric, not conspicuously flattened.
  Joints clavate or tumid, smooth or tuberculate; plants low, 30 cm. high or less, spreading; spines without sheaths.
    Joints tumid when fresh and growing, simulating some forms of the Platyopuntiae when dry, very small, 2 to 5 cm. long, 2 to 3 cm. wide and nearly as thick; tubercles not conspicuous; spines of the fruit merely spreading.
      Joints elliptic-ovate, 3 to 5 cm. long and 2 to 3 cm. wide; bristles and spines very numerous, the latter white and small..................... 3. *O. arenaria.*
      Joints circular to short-obovate, 2 or 3 cm. long; bristles few.
        Spines white or whitish; joints short-obovate.. 1. *O. brachyarthra.*
        Spines yellow or brownish; joints nearly orbicular.................................. 2. *O. fragilis.*
    Joints clavate (in one species almost cylindric); tubercles conspicuous, especially in dried specimens; spines of the fruit in radiating clusters.

Joints large, 10 cm. long or more, nearly cylindric, 3 to 4 cm. in diameter (strongly tuberculate); spines 3 to 5 cm. long, yellowish brown (flattened, stout) .... 4. *O. stanlyi.*

Joints smaller, rarely over 5 cm. long, clavate; spines various but not yellowish.

Spines 3 to 5 cm. long, straight, slender, terete, ashy gray; joints rather slender, about 15 mm. in diameter .... 5. *O. grahami.*

Spines shorter, 2 cm. long or less, one stout and flattened, white, strongly recurved .... 6. *O. clavata.*

Joints cylindric, more or less strongly tuberculate; spines inclosed in a chartaceous sheath; plants mostly taller, 60 cm. to 3 meters or more.

Plants diffuse, about 60 cm. high, seldom more; stems 10 to 15 mm. in diameter; flowers yellow.

Sheaths pale yellowish to white; plant of western New Mexico and Arizona .... 7. *O. whipplei.*

Sheaths yellowish brown; plant of eastern New Mexico and Texas .... 8. *O. davisii.*

Plants taller, branching from a simple or divided main stem, 1 to 3 meters high or more; flowers mostly purple, yellow in *O. leptocaulis.*

Stems slender, 1 cm. in diameter or less; plants about 1 meter high when full grown; spines long and slender, 2 to 4 cm. long, 1 to 3 in each areole; flowers yellow or purple; fruit fleshy, smooth or tuberculate, scarlet or reddish.

Fruit small, about 1 cm. long, a scarlet berry, smooth; flowers small, yellow .... 9. *O. leptocaulis.*

Fruit larger, 2 cm. long, green, more or less tinged with red, somewhat tuberculate; flowers dull greenish purple .... 10. *O. kleiniae.*

Stems stout, 2 cm. in diameter or more; plants more or less arborescent when fully grown, 3 meters high or more; spines short, 1 to 2 cm. long, numerous in the areole; flowers bright purple; fruit dry, tuberculate, yellow when mature.

Spines 2 cm. long, 4 to 10 in the areole; tubercles large and conspicuous; sheaths loose, shining and white at the base, brown-tipped, giving the stem a whitish tinge .... 11. *O. arborescens.*

Spines mostly less than 1 cm. long, 12 to 20 in an areole; tubercles mostly not conspicuous; spines giving a decided pinkish tinge to the stems .... 12. *O. spinosior.*

Joints flattened, becoming orbicular, obovate, or elliptic, several times as wide as thick.

Fruit dry, not succulent; spines light-colored, numerous (except in *O. sphaerocarpa*); joints small; plants prostrate, spreading.

Spines 35 to 100 mm. long, very numerous, 5 to 8 large ones in each areole .... 13. *O. hystricina.*

Spines 2 to 5 cm. long, less numerous, 1 to 5 large ones in each areole.
Joints bright green, rather strongly tuberculate, wrinkled; spines few, in the upper areoles, about 2 cm. long, slender, white; lower part of the joints almost naked; fruit spherical...14. *O. sphaerocarpa.*
Joints paler, mostly obscured by the spines, not noticeably tuberculate nor wrinkled when dry; spines mostly covering the joints, the largest 4 to 5 cm. long, stouter, yellowish or brown-tinged; fruit longer than broad.
Lower spines hairlike, especially on the older joints....................................15. *O. trichophora.*
Lower spines not hairlike......................16. *O. polyacantha.*
Fruit succulent, a thick, rather tough rind surrounding the pulpy interior, red or purple when ripe; joints mostly larger (small in two species); spines and habit various.
Plants erect or suberect, often 1 meter high or more, the branches often decumbent at the base, but at least 3 or 4 joints high.
Spines pale yellow, uniform in color throughout, 1 to 2 cm. long, 3 to 5 in each areole, appressed-reflexed; stems with a noticeable though short trunk..17. *O. chlorotica.*
Spines not as described above; trunklike stem always wanting (except in very young plants).
Joints mostly reddish or purplish, or bluish green and glaucous, very thin, at least some of them armed with long, dark brown or black spines 4 to 8 cm. long..............................18. *O. macrocentra.*
Joints never reddish, mostly yellowish green, thick; spines brown, yellow, white, or variegated.
Joints prevailingly obovate, of medium size, the younger 10 to 18 cm. long, the older sometimes larger.
Spines few, short, mostly white or yellow, somewhat reflexed; seeds small, about 3 mm. in diameter.
Spines white................................19. *O. dulcis.*
Spines clear bright yellow....................20. *O. lindheimeri.*
Spines more numerous, darker colored in the typical form, the young spines dark brown, 3 to 5 cm. long, porrect, spreading; seeds 5 mm. in diameter..........................21. *O. phaeacantha.*
Joints prevailingly orbicular or ovate, the younger ones sometimes very broadly obovate, large, 20 to 30 cm. long, or even larger.
Joints distinctly narrowed at the apex, ovate or elliptic; spines large, divaricate, 3 to 5 cm. long, stout, yellow at the apex, reddish brown at the base..........................22. *O. wootoni.*
Joints mostly orbicular, sometimes broadly obovate; spines various.
Spines few, on some joints none, yellow or brownish at the base, 1 to 2 cm. long....23. *O. dillei.*

Spines more numerous, 2 to 5 cm. long, divaricate, or the two lowest appressed, somewhat flattened, whitish, or at least pale.
Fruit oblong to clavate, large............24. *O. engelmanni.*
Fruit spheroidal, small..................32. *O. cyclodes.*
Plants prostrate and mainly widely spreading, at most ascending, usually only the one or two terminal joints erect, mostly under 30 cm. high, sometimes forming irregular beds of an area of several square meters.
Spines mostly white or pale, at most yellow or yellowish brown at base or tip; joints mostly small, 15 cm. long or less, of various shapes.
Plants small, usually consisting of only a few joints, these 6 or 7 cm. long; spines slender, 3 to 5 cm. long; glochids yellow, very long on the edges of young joints; flowers red..................31. *O. filipendula.*
Plants larger, often of numerous joints 8 to 15 cm. long; spines various; glochids usually not prominent, except on old joints, then very numerous; flowers yellow.
Joints spiny only on the upper marginal areoles..25. *O. stenochila.*
Joints spiny almost throughout.
Spines mostly 2 to 4 cm. long (plants resembling *O. polyacantha*, but with pulpy fruit); some of the joints nearly orbicular, 8 to 10 cm. long; seeds larger, about 6 mm. in diameter; roots fibrous....................26. *O. cymochila.*
Spines longer, 3 to 6 cm. long, slender; joints narrowly obovate, 10 to 15 cm. long or even larger; seeds smaller, 3.5 to 4 mm. in diameter; roots sometimes tuberous.............27. *O. tenuispina.*
Spines darker, stouter, light brown or dark brown, sometimes lighter at the tip or with age; joints mostly about 15 cm. long, sometimes larger, obovate, often broadly so, or the older joints almost orbicular.
Spines yellowish brown at the base, lighter toward the tip, the lateral spines usually lighter colored and more or less appressed................28. *O. toumeyi.*
Spines darker colored, mostly dark brown, widely spreading, 5 to 6 cm. long, or even more, rarely if ever appressed.
Glochids on old joints very abundant and large, 10 to 12 mm. long; plant of the southern part of the State...............................29. *O. chihuahuensis.*
Glochids less numerous and shorter; plant of the north-central part of the State...............30. *O. camanchica.*

In the treatment here given there is some uncertainty about several of the species, the degree and character of the doubt being expressed in the notes following the separate species, which should be consulted.

Just what *Opuntia phaeacantha brunnea* Engelm. is we are unable to determine. It is probably a form very similar to that here referred to *O. toumeyi* Rose, but should be an erect, spreading plant, similar to the species.

Further critical study of the genus will undoubtedly result in the recognition of many more species in this distribution area and in the better limitation of the species now recognized.

A specimen from Gallup, collected October 20, 1896, by Ashmun we are unable to determine. Its spines suggest *Opuntia cymochila*, but the joints are orbicular or even broader than long.

1. **Opuntia brachyarthra** Engelm. & Bigel. Proc. Amer. Acad. 3: 302. 1856.

Opuntia fragilis brachyarthra Coulter, Contr. U. S. Nat. Herb. 3: 440. 1896.

TYPE LOCALITY: Inscription Rock, near Zuni, New Mexico. Type collected by Bigelow in 1853.

RANGE: Known only from the type locality.

As described, this is a small tumid-jointed plant, possibly related to *O. arenaria* or *O. fragilis*. It has not been collected in recent years. A specimen from Santa Fe, collected by Bigelow and referred to this species, may be *Opuntia fragilis*, as that species comes into the mountains a short distance north of Santa Fe. Possibly Doctor Coulter may have been right in considering *Opuntia brachyarthra* a subspecies of *O. fragilis*. Until further material of *O. brachyarthra* from the type locality can be studied, it is probably better to retain it as a species.

2. **Opuntia fragilis** (Nutt.) Haw. Syn. Pl. Succ. Suppl. 82. 1819.

Cactus fragilis Nutt. Gen. Pl. 1: 296. 1818.

TYPE LOCALITY: "From the Mandans to the mountains, in sterile but moist situations."

RANGE: Wisconsin and British Columbia to Kansas and northern New Mexico.

NEW MEXICO: Lake La Jara; Tunitcha Mountains. Upper Sonoran Zone.

3. **Opuntia arenaria** Engelm. Proc. Amer. Acad. 3: 301. 1856.

TYPE LOCALITY: Sandy bottoms of the Rio Grande near El Paso, Texas or Chihuahua.

RANGE: Southern New Mexico and western Texas.

NEW MEXICO: Mesquite Lake (*Standley*). Sandy soil, in the Lower Sonoran Zone.

Coulter refers Fendler's 7, 150, and 153 to this species and to New Mexico. Where they were obtained we have been unable to ascertain. *Opuntia arenaria* has such a limited distribution in the type locality and occurs in a habitat so different from the region about Santa Fe (where Fendler did most of his collecting) that we are inclined to doubt the accuracy of the reference. It is possible the plant may occur in the Rio Grande Valley west of Santa Fe, but even this is doubtful. The collection at Mesquite Lake is the only one made since the type was gathered. The plants are not at all abundant here. They grow on the dunes of pure sand about one of the lakes or oxbows formed in the old bed of the Rio Grande.

4. **Opuntia stanlyi** Engelm. in Emory, Mil. Reconn. 157. *f. 9.* 1848.

TYPE LOCALITY: On the Del Norte and Gila, New Mexico. Type collected by Emory.

RANGE: Southern New Mexico and Arizona and adjacent Mexico.

NEW MEXICO: Near Carlisle (*Wooton*). Lower Sonoran Zone.

Opuntia stanlyi is a very distinct species with thick joints 10 to 12 cm. long and 3 to 4 cm. in diameter, with large tubercles and various spines. It forms beds often 2 or 3 meters across and only 30 cm. high or less. It grows on sandy mesas in the southwestern part of the State, but is more common in Arizona. Our specimens are from very near the type locality.

5. **Opuntia grahami** Engelm. Proc. Amer. Acad. 3: 304. 1856.

TYPE LOCALITY: Sandy bottoms of the Rio Grande near El Paso, Texas or Chihuahua.

RANGE: Rio Grande Valley about El Paso and southward, in the Lower Sonoran Zone.

This is probably to be found in southern New Mexico, although we have seen no specimens. It is very common on the mesas about El Paso.

6. Opuntia clavata Engelm. in Wisliz. Mem. North. Mex. 95. 1848.

TYPE LOCALITY: About Albuquerque, New Mexico. Type collected by Wislizenus.

RANGE: New Mexico.

NEW MEXICO: Santa Fe; El Rito; Las Vegas; Carrizozo; Stanley; Tesuque; Socorro; Albuquerque; Laguna; Los Lunas; Cubero. Dry plains, in the Upper Sonoran Zone.

A common plant on the high mesas and plains of the central and northern parts of the State. It is rarely over 10 cm. high and forms irregular beds sometimes 1 or 2 meters across.

7. Opuntia whipplei Engelm. & Bigel. Proc. Amer. Acad. **3:** 307. 1856.

Opuntia whipplei laevior Engelm. & Bigel. loc. cit.

TYPE LOCALITY: "From Zuñi westward to Williams River," Arizona or New Mexico.

RANGE: Western New Mexico to Arizona.

NEW MEXICO: Gallup; Aztec; Puertecito; south of Ojo Caliente; Farmington; Cedar Hill. Upper Sonoran Zone.

This is a characteristic low, cylindric-stemmed plant, reaching a height of 60 cm. in western New Mexico in the region about Zuni. It often forms dense beds 1 meter or more in diameter. The New Mexican plant, so far as we know, is always of this form and size, and has yellow flowers. Farther west, in Arizona, it becomes much larger and often has purple flowers.

The species is most like the next, *Opuntia davisii*, which is fairly common in the eastern part of the State, on plains south and east of Portales. Both are low and very spiny and have medium-sized yellow flowers and tuberculate fruit. *Opuntia davisii* is stouter, with shorter joints and more numerous spines. The two are most easily distinguished by the sheaths of the spines. *Opuntia whipplei* always looks whitish or very pale yellow, while *Opuntia davisii* is a golden brown, these colors being due to the sheaths.

8. Opuntia davisii Engelm. & Bigel. Proc. Amer. Acad. **3:** 305. 1856.

TYPE LOCALITY: On the Llano Estacado, near the upper Canadian River, New Mexico or Texas.

RANGE: Eastern New Mexico and western Texas.

NEW MEXICO: Red Lake (*Wooton*). Sandy plains, in the Upper Sonoran Zone.

9. Opuntia leptocaulis DC. Mém. Mus. Hist. Nat. **17:** 118. 1829.

TYPE LOCALITY: Mexico.

RANGE: Southern New Mexico and Arizona to western Texas, south into Mexico.

NEW MEXICO: Mesa near Agricultural College; ten miles east of Hillsboro; Orogrande; Guadalupe Mountains; Kingston; Upper Corner Monument; Tularosa; Socorro; Hachita. Lower Sonoran Zone.

The slender-stemmed species of Opuntia of this type usually pass under the name of "tasajilla" among the Mexicans, although this species is also called "garrambullo," a name applied to almost any shrub with red berries.

10. Opuntia kleiniae DC. Mém. Mus. Hist. Nat. **17:** 118. 1829.

Opuntia wrightii Engelm. Proc. Amer. Acad. **3:** 308. 1856.

TYPE LOCALITY: Mexico.

RANGE: Western Texas and southern New Mexico to Mexico.

NEW MEXICO: Low hills west of San Antonio; mesa near Agricultural College. Lower Sonoran Zone.

The specimens here listed are referred to the above species tentatively and with considerable doubt. They agree in most of their characters with the description of *Opuntia wrightii*, but the flowers are a dull purple, more or less streaked with green.

11. Opuntia arborescens Engelm. in Wisliz. Mem. North. Mex. 90. 1848.

CANE CACTUS.

TYPE LOCALITY: Northern New Mexico.

RANGE: Colorado to Arizona and western Texas, southward into Mexico.

NEW MEXICO: Near El Rito; Organ Mountains; Mangas Springs; Logan; San Rafael; Raton; Rincon; Cubero; Santa Fe; Socorro; Fairview; Queen; Sierra Grande. Plains and hills, in the Lower and Upper Sonoran zones.

This is the common species found almost throughout the State on the mesas and in the foothills of the mountains. It often stands 3 meters high or more. The stems are used to some extent for the manufacture of canes, the reticulated woody part of the stem giving them a peculiar appearance. A number of names are applied to the plant besides the one given above. "Candelabrum cactus" is used, "tree cactus" is fairly common in southern New Mexico, and "velas de coyote" (coyote candles) is often used by the Mexicans. It is sometimes utilized as stock feed, though rarely.

A low form, less than 2 meters high, occurs on the mesas of southern New Mexico near the Agricultural College, which may be different from the more common form, but it has not yet been separated.

12. Opuntia spinosior (Engelm. & Bigel.) Toumey, Bot. Gaz. **25:** 119. 1898.

Opuntia whipplei spinosior Engelm. & Bigel. Proc. Amer. Acad. **3:** 307. 1856.

TYPE LOCALITY: "South of the Gila," Arizona.

RANGE: Western New Mexico to Arizona and Sonora.

NEW MEXICO: Hermanas; Silver City; Deming; Steins Pass; White Water. Upper Sonoran Zone.

The species is easily confused with the preceding if one has only the descriptions to work with. Professor Toumey confused them in one of his earlier publications, but later corrected his mistake and pointed out very clearly the mistake made by Engelmann and Bigelow in associating the plant with *Opuntia whipplei*.

Opuntia spinosior deserves its name. It is more spiny than most of our Cylindropuntiae; though its spines are short they are numerous and close together. The peculiar pinkish tinge given the stem by the color of the spines and sheaths is easily recognized when once seen and is very characteristic. This species is known in New Mexico only from the southwestern part of the State, where it is about the size of *O. arborescens*. In Arizona it is frequently larger.

13. Opuntia hystricina Engelm. Proc. Amer. Acad. **3:** 299. 1856.

TYPE LOCALITY: "West of the Rio Grande, to the San Francisco Mountains," New Mexico and Arizona.

RANGE: Western New Mexico to Arizona and California.

NEW MEXICO: Aztec (*Baker* 481).

The specimens upon which the species was founded were collected "at the Colorado Chiquito and on the San Francisco mountains," both of the localities being in Arizona. Doctor Coulter does not report any New Mexican material and we have seen only one doubtful specimen. It is reported here also on the authority of the first collector, Doctor Bigelow. The original description says nothing about the flower, but Doctor Coulter says it is yellow or purple, a statement we are much inclined to doubt. Collectors should look for the species in the region between Albuquerque and Zuni, keeping in mind its strong resemblance to *Opuntia polyacantha*, from which it differs in the longer and more numerous spines.

14. Opuntia sphaerocarpa Engelm. & Bigel. Proc. Amer. Acad. **3:** 300. 1856.

TYPE LOCALITY: Mountains near Albuquerque, New Mexico. Type collected by Bigelow.

RANGE: Known only from the Sandia Mountains, New Mexico.

The species suggests some forms of *O. polyacantha*, but seems abundantly distinct.

15. Opuntia trichophora (Engelm.) Britt. & Rose, Smiths. Misc. Coll. **50:** 535. 1908.

Opuntia missouriensis trichophora Engelm. Proc. Amer. Acad. **3:** 300. 1856.

TYPE LOCALITY: Mountains near Albuquerque, New Mexico. Type collected by Bigelow.

RANGE: Central New Mexico.

NEW MEXICO: Arroyo Hondo; Mesa Redonda twelve miles south of Tucumcari; south edge of San Augustine Plains; sandhills near the Chincherita Mountains. Upper Sonoran Zone.

The species is similar to *Opuntia polyacantha.* The slender, hairlike lower spines give it its specific name and are the distinguishing character. The condition is more noticeable in dried than in growing specimens.

16. Opuntia polyacantha Haw. Syn. Pl. Succ. Suppl. 82. 1819.

Cactus ferox Nutt. Gen. Pl. **1:** 296. 1818, not Willd. 1813.

Opuntia missouriensis DC. Prodr. **3:** 472. 1828.

TYPE LOCALITY: "In arid situations on the plains of the Missouri."

RANGE: From near the northern boundary of the United States to western Texas and eastern New Mexico.

NEW MEXICO: Laguna; near Fort Defiance; near Albuquerque; 20 miles north of Gallup; south of Tierra Amarilla; west of Magdalena; west of Tiznitzin; Arroyo Hondo; Farmington; Raton. Plains and sometimes in the mountains, Upper Sonoran Zone.

A variable and widely distributed species.

The subspecies *albispina* Engelm. & Bigel., from the "sandy bottoms and dry beds of streamlets on the upper Canadian, 250 miles east of the Pecos" and "on the Sandia Mountains near Albuquerque," is the form most common in New Mexico. It is scarcely to be distinguished from subspecies *rufispina,* collected in "rocky places on the Pecos," (probably in the region of Santa Rosa or Anton Chico), which is the type form. The color variations in the spines indicated by the names are hardly constant.

17. Opuntia chlorotica Engelm. & Bigel. Proc. Amer. Acad. **3:** 291. 1856.

TYPE LOCALITY: "Western Colorado country, between New Mexico and California from the San Francisco Mountains to Mojave Creek."

RANGE: Southern New Mexico, Arizona, and California and adjacent Mexico.

NEW MEXICO: Hatchet Mountains; Dona Ana Mountains; Steins Pass; Lake Valley; Red Rock. Lower Sonoran Zone.

This is the one Opuntia we have so far failed to transplant successfully, and we have tried several times. It is very rare, occurring only in the southwestern part of the State. The appressed, short, clear yellow spines are very characteristic, as is the short trunk which is usually present.

18. Opuntia macrocentra Engelm. Proc. Amer. Acad. **3:** 292. 1856.

TYPE LOCALITY: Sandhills on the Rio Grande, near El Paso, Texas or Chihuahua.

RANGE: Western Texas, southern New Mexico, Arizona, and adjacent Mexico.

NEW MEXICO: Mesa near Las Cruces; Garfield; Deming; Mangas Springs; Steins Pass; Lordsburg; White Water; Dog Spring. Sandy plains and hills, in the Lower Sonoran Zone.

This is probably *Opuntia violacea* Engelm. of Emory's Reconnaissance, although the drawing does not closely resemble the plant nor is there any means of determining what the artist had to make his picture from. The only reason for believing that this name might apply is that *Opuntia macrocentra* is the common red-jointed Platyopuntia with long spines in the region mentioned and is "suberect."

It seems useless in most cases to try to attach the names proposed by Doctor Engelmann in this report to any species, because he himself did not try to use them, although he worked with the cacti of that region more than almost any other person until very recently, and knew American species better than anyone else of his day.

Well grown plants of this are suberect, about 1 meter high or even more. Many times the plants are smaller and sometimes, though not usually, they are spreading. It may be recognized readily by its thin, mostly circular, reddish joints and its long, nearly black spines. Occasional joints have no spines but abundant brown bristles. If grown where it gets plenty of water the joints grow much thicker, and lose their red color, becoming bluish green and glaucous.

19. Opuntia dulcis Engelm. Proc. Amer. Acad. **3:** 291. 1856.

Type locality: Near Presido del Norte, Texas.

Range: Western Texas to southern New Mexico.

New Mexico: Mesilla Valley; Guadalupe Mountains; Mangas Springs. Lower Sonoran Zone.

This is the plant referred to as *Opuntia laevis?* in Griffiths and Hare's bulletins on cacti. The Mangas Springs specimens may be true *O. laevis*. The plant is introduced in the Mesilla Valley and is widely cultivated for hedges. It is said to have come from Chihuahua.

20. Opuntia lindheimeri Engelm. Bost. Journ. Nat. Hist. **5:** 207. 1845.

Type locality: About New Braunfels, Texas.

Range: Western Texas to southern New Mexico.

New Mexico: Guadalupe Mountains (*Wooton* 5505). Dry hills.

21. Opuntia phaeacantha Engelm. Mem. Amer. Acad. n. ser. **4:** 352. 1849.

Type locality: "On rocky hills about Santa Fe, and on the Rio Grande," New Mexico. Type collected by Fendler.

Range: Colorado and New Mexico.

New Mexico: Santa Fe; Puertecito; Gallup; Flora Vista; Chamita; near Magdalena; Rio Hondo; Gallinas Mountains. Dry hills, in the Upper Sonoran Zone.

This is the common suberect plant of the mountains and plains of the northern part of the State. When well grown it is often a meter high and where it gets abundance of water its joints are rather bluish green and thick and have dark spines. Where it is drier, the joints are often yellowish green and the spines lighter in color. Young plants are much smaller than the key requires and seem to be procumbent.

22. Opuntia wootoni Griffiths, Rep. Mo. Bot. Gard. **21:** 171. 1910.

Type locality: Organ Mountains, New Mexico.

Range: Known so far only from the Organ Mountains and Tortugas Mountain, southern New Mexico, in the Lower and Upper Sonoran zones.

This is perhaps most closely related to *Opuntia engelmanni*, having its spines arranged much as in that species, but the joints are narrowed at the top, being ovate or oval, and the long stout spines are yellow at the tip and reddish or brownish at the base.

23. Opuntia dillei Griffiths, Rep. Mo. Bot. Gard. **20:** 83. 1909.

Type locality: San Andreas Canyon of the Sacramento Mountains, New Mexico, about 5 miles south of Alamogordo.

Range: Known only from type locality, in the Upper Sonoran Zone.

The large, circular, thick joints with few or no spines are characteristic.

24. Opuntia engelmanni Salm-Dyck; Engelm. Proc. Amer. Acad. **3:** 291. 1856.

Type locality: Near Chihuahua, Mexico.

Range: Western Texas and southern New Mexico to Mexico.

New Mexico: Mesa west of Organ Mountains; Organ Mountains; Dog Spring; Lordsburg; Hatchet Mountains; east of Hillsboro; Red Rock; Deming. Lower and Upper Sonoran zones.

The plant here called *Opuntia engelmanni* has been referred to under the name of *Opuntia engelmanni cyclodes* Engelm. & Bigel. in one or two publications of recent years. That plant was first collected near Anton Chico, New Mexico, and Doctor Engelmann separated it on the characters of small, globose fruit with larger seeds than the species and fewer and shorter spines on the joints. The plants from southern New Mexico match almost exactly, so far as spine characters go, material from near the city of Chihuahua, and fruit of the New Mexican plant is never globose but ellipsoid to slightly obovoid, about twice as long as broad.

25. Opuntia stenochila Engelm. Proc. Amer. Acad. **3**: 296. 1856.

Opuntia mesacantha stenochila Coulter, Contr. U. S. Nat. Herb. **3**: 430. 1896.

TYPE LOCALITY: Canyon near Zuni, New Mexico.

RANGE: Known only from the original collection by Bigelow. We have seen no material of this species.

26. Opuntia cymochila Engelm. & Bigel. Proc. Amer. Acad. **3**: 295. 1856.

TYPE LOCALITY: "Along the Canadian River east of the Llano Estacado, and on that plain," Texas.

RANGE: Eastern New Mexico and the Panhandle region of Texas.

NEW MEXICO: Nara Visa; Lakewood; Knowles. Lower and Upper Sonoran zones.

27. Opuntia tenuispina Engelm. Proc. Amer. Acad. **3**: 294. 1856.

TYPE LOCALITY: Sandhills near El Paso, Texas or Chihuahua.

RANGE: Western Texas, southern New Mexico, and adjacent Mexico.

NEW MEXICO: Rio Mimbres; Deming; Mesilla Valley. Lower Sonoran Zone.

This is the most common species in the lower Rio Grande Valley on the heavier soils.

28. Opuntia toumeyi Rose, Contr. U. S. Nat. Herb. **12**: 402. 1909.

TYPE LOCALITY: Tucson, Arizona.

RANGE: Southern New Mexico and Arizona, probably in adjacent Mexico.

NEW MEXICO: North of Kellys Ranch, west of Frisco and north of Alma; Lordsburg. Low mountains, in the Lower Sonoran Zone.

29. Opuntia chihuahuensis Rose, Contr. U. S. Nat. Herb. **12**: 291. 1909.

Opuntia mesacantha sphaerocarpa Coulter, Contr. U. S. Nat. Herb. **3**: 431. 1896.

TYPE LOCALITY: Santa Eulalia, near Chihuahua, Mexico.

RANGE: Western Texas and central and southern New Mexico to Chihuahua.

NEW MEXICO: Organ Mountains; Ancho. Dry hills, in the Lower Sonoran Zone.

The New Mexican specimens exactly match the type specimen of *Opuntia chihuahuensis* and also match a specimen determined as *O. mesacantha oplocarpa* Coulter by Doctor Coulter himself. We have cultivated the plant in the garden at the Agricultural College for years and have generally called it *Opuntia camanchica* Engelm., although with some doubt in our minds, because we were not certain as to what that species really is. Recently in the type locality of *O. camanchica* it was possible to see the plant growing in its native habitat all the way from the "Llano Estacado, at the base of the hills * * * * to the Tucumcari hills," where it is everywhere the common species. The plant from that region is darker green than that of the mesas about the Agricultural College; its joints are slightly smaller, but the spines are much the same; the habit of the plant is the same. It will be possible to determine the differences now, since we have the two growing side by side in the garden.

30. Opuntia camanchica Engelm. & Bigel. Proc. Amer. Acad. **3**: 293. 1856.

TYPE LOCALITY: Llano Estacado, on the upper Canadian River, Texas.

RANGE: Eastern New Mexico and northwestern Texas.

NEW MEXICO: Foot of Tucumcari Mountain; hills near Tucumcari. Upper Sonoran Zone.

31. Opuntia filipendula Engelm. Proc. Amer. Acad. **3:** 294. 1856.

Opuntia ballii Rose, Contr. U. S. Nat. Herb. **13:** 309. 1911.

Type locality: "Alluvial bottoms of the Rio Grande near El Paso, and eastward on the Pecos," Texas.

Range: Western Texas and southern New Mexico.

New Mexico: Knowles; Redlands; Queen; Lakewood; Jornada del Muerto; Lower and Upper Sonoran zones.

Doctor Griffiths reports having seen this about Alamogordo.

32. Opuntia cyclodes (Engelm. & Bigel.) Rose, Contr. U. S. Nat. Herb. **13:** 309. 1911.

Opuntia engelmanni var. ? *cyclodes* Engelm. & Bigel. Proc. Amer. Acad. **3:** 291. 1856.

Type locality: "On the upper Pecos, in New Mexico." The type was collected about the mouth of the Gallinas River, near Anton Chico.

Range: Known only from the vicinity of the type locality, in the Upper Sonoran Zone.

What *Opuntia angustata* Engelm.[1] from Zuni may be, we are unable to determine. Britton and Rose are of the opinion that there is some mixture passing under this name, and Doctor Engelmann's description and illustrations suggest this possibility, for a tuberculate fruit of the character drawn is an anomaly when attached to a flat-jointed Opuntia, although of course such a thing might exist. So far, however, we have never seen such a combination. The region contains two, and possibly three, species of Cylindropuntiae with such fruit. It seems best to omit the name from our list until it is better known. As figured and described, the joints are narrowly obovate, 15 to 25 cm. long and half to two-thirds as wide, with yellowish or whitish spines much like those of *O. engelmanni*. The plant is prostrate; the flower is not known. One of the spiny fruits figured suggests those of *Opuntia polyacantha*, somewhat enlarged; the other looks like that of *Opuntia arborescens*.

Opuntia cymochila montana Engelm.[2] and *Opuntia microcarpa* Engelm.,[3] both of which came from New Mexico, are uncertain. The latter was described only from a pencil drawing.

2. MAMILLARIA Haw. Pincushion cactus.

Mostly small, solitary, proliferous or cespitose, globose to short-cylindric plants with spines borne on the ends of conic teatlike tubercles; flowers borne in the axils of the tubercles; ovary smooth; fruit neither scaly nor spiny; seeds smooth or pitted.

KEY TO THE SPECIES.

Tubercles not grooved on the upper side; spines sometimes hooked.
- Central spines wanting; plants small, 3 to 5 cm. high, with very numerous small white spines.
  - Spines glabrous; plants depressed to umbilicate at the apex; spines of the upper tubercles often elongated .... 1. *M. micromeris.*
  - Spines pubescent; plants oval, rarely cespitose; upper spines not elongated .... 2. *M. lasiacantha.*
- Central spines present, 1 or more; plants larger.
  - At least one of the central spines hooked; plants small, globose or oval, not flat-topped; spines all slender, the radials white, numerous, the centrals dark-colored, brown or black.

[1] Proc. Amer. Acad. **3:** 292. 1856.

[2] Proc. Amer. Acad. **3:** 296. 1856.

[3] In Emory, Mil. Reconn. 157. *f.* 7. 1848.

Only 1 central spine in each areole hooked; radials 15 to 30 .... 3. *M. grahami.*

More than 1 central hooked in some of the areoles; radials 8 to 12 .... 4. *M. wrightii.*

None of the spines hooked, the centrals usually 1, sometimes wanting; plants flat-topped, with turbinate root and milky juice.

Radial spines few, 5 to 9, stout, dull-colored .... 5. *M. meiacantha.*

Radials more numerous, 10 to 20, slender, white ..... 6. *M. heyderi.*

Tubercles grooved on the upper side (in *M. macromeris* the groove wanting in young plants and never reaching the axil); none of the spines hooked.

Central spines none or 1; plants small, subglobose, 2 to 5 cm. high; radials very numerous, 30 to 50, 5 to 10 mm. long .... 7. *M. dasyacantha.*

Centrals several, generally 3 or more; plants larger, of various shapes; radials various.

Tubercles large, 12 to 35 mm. long (mostly about 20 mm.); plants large, with long spines.

Flowers rose purple; central spines mostly 4, dark, slender but strong, 2 to 5 cm. long; plants cespitose .... 8. *M. macromeris.*

Flowers brownish yellow; centrals 2 to 5, light-colored, stout, 2 to 3 cm. long (one usually curved downward at the tip); plants mostly solitary.. 9. *M. scheerii.*

Tubercles smaller, usually less than 12 mm. long; plants small, with relatively short and numerous spines.

Fruit bright red; lower spines deciduous, leaving the base of the plant tuberculate with dry corky protuberances; spines numerous, white; centrals 5 to 9, glaucous, purple-tipped ....10. *M. tuberculosa.*

Fruit green; lower spines rarely deciduous; base of the plant little or not at all tuberculate; spines numerous, but the centrals mostly darker and not quite so numerous (except in *M. radiosa neomexicana*).

Stigmas short-mucronate; plants proliferous and cespitose; seeds yellowish brown ....11. *M. vivipara.*

Stigmas obtuse; plants sparingly proliferous, usually solitary; seeds reddish brown, slightly larger ....12. *M. radiosa.*

The references by Coulter[1] of *Cactus scolymoides* and *C. echinus* to New Mexico are probably dependent upon an incorrect determination of *Mamillaria scheerii*, for the former at least; and he does not cite any specimens of the latter species, although including it in our range. We have seen no specimens of either species from New Mexico.

1. **Mamillaria micromeris** Engelm. Proc. Amer. Acad. **3**: 260. 1856.

TYPE LOCALITY: "From El Paso eastward to the San Pedro River," Texas.

RANGE: Western Texas and southern New Mexico.

NEW MEXICO: Sacramento Mountains; Guadalupe Mountains; Capitan Mountains. Dry limestone mountains, in the Lower and Upper Sonoran zones.

---

[1] Contr. U. S. Nat. Herb. **3**: 115, 116. 1896.

This little plant is rather interesting and is prized as somewhat of a rarity by cactus growers. The flat or sunken top, the numerous fine, smooth, white spines, at the top much longer and spirally arranged, and the small size make the plant easily recognizable. Its habit of producing the clavate, few-seeded, red fruit several months after flowering is a striking peculiarity.

2. **Mamillaria lasiacantha** Engelm. Proc. Amer. Acad. 3: 261. 1856.

TYPE LOCALITY: On the Pecos River, western Texas.

RANGE: Western Texas and southern New Mexico and Arizona; also in adjacent Mexico.

NEW MEXICO: Mouth of Dark Canyon, Guadalupe Mountains (*Wooton*). Dry limestone hills, in the Upper Sonoran Zone.

This superficially resembles the preceding to such an extent that a careless observer may mistake it for that species. But it is never flat or sunken at the top, and may always be recognized by its pubescent, fine, white spines.

3. **Mamillaria grahami** Engelm. Proc. Amer. Acad. 3: 262. 1856.

TYPE LOCALITY: "Mountainous regions from El Paso, southward and westward," Chihuahua.

RANGE: Utah to western Texas, southern California, and northern Mexico.

NEW MEXICO: Tortugas Mountain; mountains east of Dona Ana; Mangas Springs; Burro Mountains. Lower and Upper Sonoran zones.

4. **Mamillaria wrightii** Engelm. Proc. Amer. Acad. 3: 262. 1856.

TYPE LOCALITY: Near the Copper Mines, New Mexico. Type collected by Wright.

RANGE: Western Texas, New Mexico, and adjacent Mexico.

NEW MEXICO: Mangas Springs; Burro Mountains; White Oaks. Upper Sonoran Zone.

Pressed material of this species is hard to distinguish from the preceding, but the characters given in the key will hold. Growing plants are more easily distinguishable. *Mamillaria grahami* is usually so thickly covered with fine white radials that it is difficult to see the tubercles, and the hooked central spines are reddish brown; while in *M. wrightii* the plants appear green because of the fewer radials and the almost black, hooked centrals are noticeably more numerous.

5. **Mamillaria meiacantha** Engelm. Proc. Amer. Acad. 3: 263. 1856.

TYPE LOCALITY: 'Western Texas and New Mexico."

RANGE: Mountains of western Texas, southern New Mexico, and northern Chihuahua.

NEW MEXICO: Queen. Upper Sonoran Zone.

This species and the next are easily separated from all other Mamillarias by the shape of the plant, this appearing as a flat-topped disk of spiny tubercles at most only a few centimeters above the surface of the soil and often about flush with it. Often the plants occur in crevices of the rocks and surrounded by grasses and other plants in such a way as to be easily overlooked. They are difficult to dig up because they have large, thickened, turbinate, sometimes branching roots. The tubercles are rather wide apart in well-grown plants and stand erect, with the short stout spines surmounting them. The flowers are small and inconspicuous, but the fruits are bright red. The spine characters given in the key will separate the two species.

6. **Mamillaria heyderi** Mühlenpf. Allg. Gartenz. 16: 20. 1848.

TYPE LOCALITY: Texas.

RANGE: Western Texas to southern Arizona and adjacent Mexico.

NEW MEXICO: Organ Mountains; Tortugas Mountain; Cooks Peak; Hillsboro; Steins Pass; Mangas Springs Upper Sonoran Zone.

All of the New Mexican material that we have seen is to be referred to *Mamillaria heyderi* rather than the subspecies *hemisphaerica*, if the character which gives rise to the name is considered. The New Mexican plant is always flat-topped, with more or less turbinate thickened root. It is not infrequently even larger than described

7. Mamillaria dasyacantha Engelm. Proc. Amer. Acad. **3**: 268. 1856.

TYPE LOCALITY: "El Paso and eastward," Texas.

RANGE: Western Texas to southern Arizona and adjacent Mexico.

NEW MEXICO: Big Hatchet Mountains; Kingston; Lake Valley; Mogollon Creek. Dry mountains, in the Lower and Upper Sonoran zones.

8. Mamillaria macromeris Engelm. in Wisliz. Mem. North. Mex. 97. 1848.

TYPE LOCALITY: Sandy soil near Dona Ana, New Mexico. Type collected by Wislizenus.

RANGE: Southern New Mexico to western Texas and Chihuahua.

NEW MEXICO: Dona Ana; Parkers Well; plains south of White Sands; Tortugas Mountain. Mesas and sandy soil, in the Lower Sonoran Zone.

One of the commonest Mamillarias of the southern part of the State, growing on sandy mesas, forming rounded clumps sometimes almost a meter in diameter. The individual plants are frequently 20 cm. long, fully half of the length being under ground. They are rather dark green; the tubercles are large, the groove never reaching the summit and sometimes wanting in young plants; the spines are long, the radials dull-colored and often bent, the centrals dark, almost black, slender but stiff. The flowers are a bright rose purple, sometimes lighter, often turning lavender; they are large, 5 cm. long or more and opening as wide, and usually are produced in profusion in the middle of the summer. The species is a very desirable one for cultivation

9. Mamillaria scheerii Mühlenpf. Allg. Gartenz. **15**: 97. 1847.

TYPE LOCALITY: Mexico.

RANGE: Southern New Mexico, trans-Pecos Texas, and adjacent Mexico.

NEW MEXICO: Lordsburg; mesa near Agricultural College. Gravelly mesas and in the mountains, in the Lower Sonoran Zone.

Only 4 or 5 plants have been found around the Agricultural College.

The flowers are 5 or 6 cm. long and of a peculiar bronze or brownish yellow, different from most of our other Cactaceae. The single plants sometimes reach a height of 15 cm. and almost as great a diameter, being the largest single Mamillaria plants found in New Mexico. The tubercles in such plants are 25 mm. long, and are distant, spreading, and conic; the central spines are stout, one of them more or less curved downward at the tip but not hooked. The fruit is green, pulpy, irregularly clavate or obovate, with numerous brownish red seeds. It is one of the most interesting of the New Mexican Mamillarias for pot culture.

10. Mamillaria tuberculosa Engelm. Proc. Amer. Acad. **3**: 268. 1856.

Mamillaria strobiliformis Scheer in Salm-Dyck, Cact. Hort. Dyck. 104. 1850, not Mühlenpf. 1848, nor Engelm. 1848.

TYPE LOCALITY: "From the Pecos to Leon Springs, Eagle Springs, and El Paso, on the higher mountains," Texas.

RANGE: Southern New Mexico, trans-Pecos Texas, and adjacent Mexico.

NEW MEXICO: Tortugas Mountain; Van Pattens; near Hillsboro. Low, dry mountains, in the Lower and Upper Sonoran zones, mostly on limestone soil.

This little plant, with its dense coat of white radial spines and dusky-tipped centrals, its proliferous habit, its tuberculate base, its small pink flowers, and its bright red, tart fruit, is one of the commonest species of the southern part of the State, where it is found growing in the crevices of limestone rocks.

11. Mamillaria vivipara (Nutt.) Haw. Syn. Pl. Succ. Suppl. 72. 1819.

Cactus viviparus Nutt. Fraser's Cat. no. 22. 1813.

TYPE LOCALITY: "Near the Mandan towns on the Missourie: lat. near 49°."

RANGE: British America to Montana, Nebraska, Utah, and northern New Mexico.

NEW MEXICO: Chusca Mountains; Tierra Amarilla. Plains.

It is difficult to separate this species from *Mamillaria radiosa* and its subspecies, but, generally speaking, the plants referred to *M. vivipara* have the more northerly range, are smaller and cespitose, and have fewer, shorter, and more slender spines, though the differences seem to be of degree rather than kind.

Mamillaria vivipara is very rare in the State, only two collections of it being on record, but it is to be expected in the mountains or on the high plains in the northern part. *Mamillaria radiosa* and its subspecies *neomexicana* are common almost everywhere above 1,500 meters throughout the State and it is altogether probable that the subspecies *arizonica* occurs along the southwestern border.

12. Mamillaria radiosa Engelm. Bost. Journ. Nat. Hist. **6:** 196. 1850.

TYPE LOCALITY: "Sterile, sandy soil on the Pierdenales," Texas.

RANGE: Colorado and New Mexico to western Texas and northeastern Mexico.

NEW MEXICO: Pecos; head of the Rio Mimbres; Lake Valley; Nara Visa; Tierra Amarilla; Mogollon Creek; Sierra Grande; Farmington; Santa Fe; Raton. Upper Sonoran and Transition zones.

12a. Mamillaria radiosa neomexicana Engelm. U. S. & Mex. Bound. Bot. Cact. 64. 1859.

Mamillaria vivipara radiosa neomexicana Engelm. Proc. Amer. Acad. **3:** 269. 1856.

Mamillaria neomexicana A. Nels. in Coulter, New Man. Rocky Mount. 327. 1909.

TYPE LOCALITY: "From western Texas to New Mexico."

RANGE: New Mexico to western Texas and Mexico.

NEW MEXICO: Stinking Lake; Tucumcari; Mule Creek; near Black Rock; foot of Eagle Peak; San Antonio; Burro Mountains; Gallup; Mangas Springs; Inscription Rock; Santa Fe; Magdalena Mountains; Sandia Mountains; Cooks Peak. Upper Sonoran Zone.

This is with difficulty distinguishable from the species or from subspecies *arizonica*. Those forms with 20 to 30 radials 6 to 8 mm. long and 4 or 5 centrals of about the same length not pronouncedly purplish or reddish brown, on tubercles 8 to 12 mm. long, are referred to *M. radiosa*. *Neomexicana* has more numerous radials, 14 to 40, and centrals mostly 6 to 9 (3 to 12), slightly larger tubercles, and longer spines, the centrals purplish, especially with age. *Arizonica* is somewhat stouter, with fewer but longer spines, radials 15 to 20, up to 20 mm. long, centrals 3 to 6, reddish brown a bove, and tubercles 12 to 25 mm. long; its flowers, also, are considerably larger. In the present state of our knowledge, it is most convenient to recognize these as forms of a single species, perhaps the commonest Mamillaria in the State.

3. ECHINOCACTUS Link & Otto.

Globose or short-cylindric plants, mostly solitary, with tubercles coalescing into vertical or spirally twisted ridges bearing clusters of mostly stout, more or less flattened, curved or sometimes hooked spines; flower-bearing areolæ above the young spine-bearing ones, the plants thus blooming in the center at the top; ovary scaly or woolly, not spiny; fruit dry or succulent, scaly or smooth.

KEY TO THE SPECIES.

Some of the spines hooked.
- Central spines 1, hooked, some of the laterals also hooked, slender, terete, about 10 cm. long; plants small, about 10 cm. high .... 1. *E. uncinatus wrightii.*
- Centrals 4, stout, flattened or quadrangular, 5 to 7.5 cm. long (rarely 10 cm.); plants often large, up to 90 cm. high and 30 cm. in diameter.
  - Plants small, 10 cm. high or less, 8 cm. or less in diameter .... 2. *E. glaucus.*
  - Plants large, 50 cm. high or over, more than 30 cm. in diameter .... 3. *E. wislizeni.*

None of the spines hooked.
- Centrals 2 to 4, not stout; plants small, 10 cm. high or less.
  - Spines flat, flexible, chartaceous; ridges broken into tubercles .... 4. *E. papyracanthus.*
  - Spines terete, stiff though small; ridges continuous.. 5. *E. intertextus.*
- Centrals 1 or none, very stout and horny; plants larger, 10 to 25 cm. in diameter.
  - Ribs few, 8 to 10, rounded; spines mostly terete; plants about 10 cm. in diameter, spheroidal... 6. *E. horizonthalonius.*
  - Ribs more numerous, 13 to 21, more acute; spines compressed; plants 20 to 30 cm. in diameter, depressed .... 7. *E. texensis.*

1. Echinocactus uncinatus wrightii Engelm. Proc. Amer. Acad. **3**: 277. 1856.

TYPE LOCALITY: "Near El Paso and on the river below."

RANGE: Southern New Mexico, trans-Pecos Texas, and adjacent Mexico.

NEW MEXICO: Pena Blanca; Bishops Cap. Lower and Upper Sonoran zones.

A rare species, known in New Mexico only from the dry foothills in the southern part of the State. It is subcylindric to almost hemispheric, about 10 cm. high and nearly as great in diameter, not including the long, slender, hooked, yellow spines, which are sometimes twice as long as the body of the plant. The ribs are rather prominent, the radials about 8, slender, light-colored; the flowers are small for the genus, about 25 mm. long, of a dull brownish purple and glabrous.

2. Echinocactus glaucus Schum. Gesamtb. Kakt. 438. 1903.

TYPE LOCALITY: Dry Creek, Mesa Grande, Colorado.

RANGE: Colorado and Utah to Arizona and northwestern New Mexico.

NEW MEXICO: Carrizo Mountains; Shiprock. Dry hills, in the Upper Sonoran Zone.

3. Echinocatus wislizeni Engelm. in Wisliz. Mem. North. Mex. 96. 1848.

VIZNAGA.

TYPE LOCALITY: Near Dona Ana, New Mexico. Type collected by Wislizenus in 1846.

RANGE: Utah and Arizona to western Texas and neighboring Mexico.

NEW MEXICO: Pena Blanca; mesa west of Organ Mountains; Filmore Canyon; Little Florida Mountains. Lower Sonoran Zone.

The largest cylindric-stemmed cactus found in our range. It is sometimes 70 to 90 cm. high and mostly about 40 to 50 cm. in diameter. The spines are numerous; the lowest radials are slender and whitish, the others stouter and rigid, and the 4 centrals stout, reddish, banded, the lowest ones sometimes 5 or 6 cm. long, flattened and bony, strongly hooked downward.

The plants of this species are used considerably in southern New Mexico and about El Paso, Texas, as a decorative plant in dooryards, especially where water is scanty, and they lend themselves readily to such treatment. They bloom rather late in summer, the blossoms being followed by the scaly yellow fruit which often persists for several years.

The pulpy interior tissue of these plants is used by Mexican candy makers, who cut it into irregular pieces and candy it by boiling in a saturated sugar solution, making what they call "cubiertas," or "dulce de viznaga," a most palatable sweetmeat.

4. Echinocactus papyracanthus Engelm. Trans. Acad. St. Louis **2**: 202. 1863.

Mamillaria papyracantha Engelm. Mem. Amer. Acad. n. ser. **4**: 49. 1849.

TYPE LOCALITY: In a valley between the lower hills, near Santa Fe, New Mexico.

RANGE: Known only from the type locality.

The type was collected by Fendler in 1847 (no. 279). Coulter also reports a specimen collected near Santa Fe in 1882 by Bandelier.

5. Echinocactus intertextus Engelm. Proc. Amer. Acad. **3**: 277. 1856.

TYPE LOCALITY: "From El Paso to the Limpio," Texas.

RANGE: Southern New Mexico, trans-Pecos Texas, and adjacent Mexico.

NEW MEXICO: Mesa west of Organ Mountains; Socorro; Organ Mountains; Rincon; Cooks Peak. Dry hills, in the Upper Sonoran Zone.

This species is a small plant, short-cylindric or globose, 10 cm. high or less and usually about 5 to 7 cm. in diameter, suggesting some of the species of Mamillaria more than Echinocactus. Its spines are white, reddish above, short, about 1 cm. long, numerous, and very closely set, densely covering the plant. The flowers are small, 15 to 20 mm. long, with numerous pale pink, acute petals, followed by a small dry fruit.

The subspecies *dasyacantha* Engelm.[1] has longer and more erect upper spines. It seems to be merely a growth form. We have seen a specimen of this collected at Rincon by Evans in 1891.

6. Echinocactus horizonthalonius Lem. Cact. Hort. Monv. 19. 1839.

Echinocactus horizonthalonius centrispinus Engelm. Proc. Amer. Acad. **3**: 276. 1856.

TYPE LOCALITY: Not stated.

RANGE: Southern New Mexico, trans-Pecos Texas, and Mexico.

NEW MEXICO: Tortugas Mountain; Guadalupe Mountains; Bishops Cap; Guadalupe Canyon. Limestone soil, in the Lower Sonoran Zone.

This plant is about spherical, 20 cm. or less in diameter, with 8 to 10 rounded ribs, suggesting a canteloupe in general form. The spines are few, 6 to 9 in each areole, stout, compressed, horny, reddish or ashy, recurved, forming a coarse network which sometimes persists and maintains the form of the plant even after the soft parts have decayed; the single central is not hooked.

The large, bright pink flowers, imbedded in dense white wool at the base, open in bright sunshine and persist for two or three days, partly closing at night and opening again in the sunlight, like those of many other cacti. They generally darken as they age. The plant is hard to transplant, unless the soil in which it is placed contains considerable lime. The plants usually occur in crevices in limestone rocks.

7. Echinocactus texensis Hopf. Allg. Gartenz. **10**: 297. 1842.

DEVIL'S PINCUSHION.

TYPE LOCALITY: Western Texas.

RANGE: Southeastern New Mexico and western Texas and northeastern Mexico.

NEW MEXICO: Knowles (*Wooton*). Lower and Upper Sonoran zones.

Depressed-hemispheric plants, about 30 cm. in diameter and less than half as high, frequently only a little above the level of the ground, rather dark green, and with 20

[1] Loc. cit.

or more acute ridges. The spines are few, 6 to 9 in the areole, stout, straight or somewhat curved, compressed, bony or horny, reddish, with a single recurved central, this not hooked. The flowers are moderately large, about 5 cm. long, bright pink, fading as they grow older, white-woolly outside. They are followed by a pulpy, bright red fruit which persists for some time, unless eaten by birds or small rodents.

4. ECHINOCEREUS Engelm.

Plants globose to cylindric (ours all erect and stout), solitary, proliferous or cespitose, ribbed, 5 to 60 cm. high, usually less than 30 cm.; areoles usually approximate, often with the spines overlapping and almost concealing the stem; flowers borne close above old spine-bearing areolæ, thus lateral on the stem; ovary spiny but the spines deciduous from the ripe fruit; fruit succulent, with thin rind, edible, with a pleasant flavor in most species; seeds small, numerous.

KEY TO THE SPECIES.

Flowers small, about 2 cm. long, green; spines reddish and white.
- Radial spines long, 5 to 10 mm.; centrals 3 to 6, the lower one about 25 mm. long, somewhat reflexed; plants conic at the apex .... 1. *E. chloranthus.*
- Radials short, 2 to 6 mm. long, rigid, pectinate; centrals mostly wanting, occasionally a few about 25 mm. long; plants depressed at the apex .... 2. *E. viridiflorus.*

Flowers larger, 3 to 10 cm. long, not green; spines variously colored.
- Flowers bright yellow, large, about 10 cm. long, closing at night; spines short, pectinate, more or less tinged with pink .... 3. *E. dasyacanthus.*
- Flowers never yellow; flowers and spines various.
  - Flowers large, 7 to 10 cm. long, open only in daylight, purple or rose, never scarlet; petals mostly acute.
    - Spines short, rigid, pectinate; centrals mostly wanting; flowers purple to rose.
      - Flowers purple; spines white; plants small, 5 to 8 cm. high .... 4. *E. pectinatus.*
      - Flowers rose to red; spines variegated red and white; plants larger, 10 to 20 cm. high .. 5. *E. rigidissimus.*
    - Spines longer, not pectinate; centrals long and conspicuous; flowers purple.
      - Spines dark, comparatively few, the upper centrals connivent-curved; stems only a few together or solitary .... 6. *E. fendleri.*
      - Spines pale yellow to straw-colored, very numerous and long; young spines straight, dusky; plants usually forming large mounds, often a meter across or more ... 7. *E. stramineus.*
  - Flowers of medium size, 3 to 7 cm. long (rarely a little larger), open day and night, bright scarlet, orange scarlet, or cardinal, never purple; petals mostly obtuse.
    - Spines very stout, strongly angled, relatively few in the areole.
      - Spines 6 to 8, mostly 6, twisted and curved; radials 2 to 3 cm. long; centrals 4 cm. long or more .... 8. *E. gonacanthus.*

Spines 3 to 6, mostly 3, shorter and not so stout. 9. *E. triglochidiatus.*
Spines more slender though rigid, mostly terete, more numerous in the areole (except in *E. paucispinus*).
Centrals none or sometimes 1; radials 3 to 6; spines all terete or but slightly flattened, stout for the group....................10. *E. paucispinus.*
Centrals 1 to several, mostly 3 to 5 or 6; radials 8 to 16, mostly 10 to 13; spines often flattened or angled.
Centrals mostly 6; flowers small, the petals acute..............................11. *E. neomexicanus.*
Centrals 3 to 5, mostly 4; flowers larger; petals obtuse.
Centrals stout, terete, usually gray or pinkish gray when young, dark gray in age, 15 to 20 mm. long.. 12. *E. rosei.*
Centrals more slender, yellowish to gray or darker.
Spines short, mostly yellowish; centrals 14 to 40 mm. long, mostly about 25 mm.; plants rounded or somewhat depressed at the apex........13. *E. coccineus.*
Spines much longer, mostly dark, the centrals 25 to 80 mm. long, commonly about 50 mm.; plants conic at the apex......................14. *E. conoideus.*

Echinocereus hexaedrus (Engelm.) Rümpl. from near Zuni, known only from the type locality, is probably only a form of *E. gonacanthus*, orginally from the same region, where it is fairly common. What the plant may be which Doctor Coulter refers to *E. octacanthus* (Fendler's 272, in part, from Santa Fe) we are unable to say, but it is likely to be either a form of *E. paucispinus* or an aberrant *E. coccineus.* Schumann is certainly incorrect in reducing *E. gonacanthus* and *E. triglochidiatus* to *E. paucispinus;* the two angular-spined species may be the same, although this is doubtful, but *E. paucispinus* is more closely related to the *E. polyacanthus* group, notwithstanding its few spines.

1. **Echinocereus chloranthus** (Engelm.) Rümpl. in Först. Handb. Cact. ed. 2. 814. 1886.

Cereus chloranthus Engelm. Proc. Amer. Acad. **3:** 278. 1856.

TYPE LOCALITY: "Stony hills and mountain sides near El Paso."

RANGE: Southern New Mexico, trans-Pecos Texas, and adjacent Mexico.

NEW MEXICO: Tortugas Mountain; Organ Mountains; San Mateo Peak; Queen; Cooks Peak; Rincon; Lake Valley. Limestone hills, in the Lower Sonoran Zone.

2. **Echinocereus viridiflorus** Engelm. in Wisliz. Mem. North. Mex. 7. 1848.

Cereus viridiflorus Engelm. in A. Gray, Mem. Amer. Acad. n. ser. **4:** 50. 1349.

TYPE LOCALITY: Prairies on Wolf Creek, New Mexico. Type collected by Wislizenus in 1846.

RANGE: Southern Wyoming to New Mexico and western Texas.

NEW MEXICO: Pecos; Colfax; west of Santa Fe; Organ Mountains; White Mountains; Sierra Grande; Nara Visa. Upper Sonoran Zone.

The Organ Mountains plants are probably the subspecies *cylindricus* Engelm. and subspecies *tubulosus* Coulter, which are nothing but growth forms. The northern plants are usually much smaller.

3. **Echinocereus dasyacanthus** Engelm. in Wisliz. Mem. North. Mex. 100. 1848.

Cereus dasyacanthus Engelm. Mem. Amer. Acad. n. ser. **4**: 50. 1849.

Type locality: "El Paso del Norte [now Ciudad Juárez]," Chihuahua.

Range: Southern Arizona to trans-Pecos Texas and adjacent Mexico.

New Mexico: Queen; mesa 10 miles west of Carlsbad. Lower and Upper Sonoran zones.

4. **Echinocereus pectinatus** (Scheidw.) Engelm. in Wisliz. Mem. North. Mex. 110. 1848.

Echinocactus pectinatus Scheidw. Bull. Acad. Sci. Brux. **5**: 492. 1838.

Cereus pectinatus Engelm. in A. Gray, Mem. Amer. Acad. n. ser. **4**: 50. 1849.

Type locality: "Habitat prope l'ila del Pennaso in locis temperatis."

Range: Western Texas and southeastern New Mexico to Mexico.

New Mexico: Knowles (*Wooton*). Lower Sonoran Zone.

Our plant is not typical and may be incorrectly determined. It is smaller than the common forms of this species, being only 5 or 6 cm. high; the areoles are not so numerous nor so closely approximated as is common in that species; the flower is larger, opens only in daylight, and is purple. Further study of the species will be necessary for more accurate determination.

5. **Echinocereus rigidissimus** (Engelm.) Rose, Contr. U. S. Nat. Herb. **12**: 293. 1909. Rainbow cactus.

Cereus pectinatus rigidissimus Engelm. Proc. Amer. Acad. **3**: 279. 1856.

Type locality: "In the Sierras of Pimeria Alta in Sonora."

Range: Western Texas to southern Arizona and adjacent Mexico.

New Mexico: Deer Creek; Hatchet Ranch; near White Water; Dog Spring. Lower and Upper Sonoran zones.

6. **Echinocereus fendleri** (Engelm.) Rümpl. in Först. Handb. Cact. ed. 2. 801. 1886.

Cereus fendleri Engelm. in A. Gray, Mem. Amer. Acad. n. ser. **4**: 50. 1849.

Cereus fendleri pauperculus Engelm. op. cit. 51.

Type locality: Santa Fe, New Mexico. Type collected by Fendler.

Range: Utah and Colorado to western Texas and northern Mexico.

New Mexico: Albuquerque; Fort Defiance; San Mateo Peak; Apache Mountains; Socorro; Mangas Springs; Steins Pass; Hermosa; Hillsboro; Graham; Cooks Peak; Organ Mountains; Farmington; Carrizo Mountains. Plains and mesas, chiefly in the Upper Sonoran Zone.

7. **Echinocereus stramineus** (Engelm.) Rümpl. in Först. Handb. Cact. ed. 2. 797. 1886.

Cereus stramineus Engelm. Proc. Amer. Acad. **3**: 282. 1856.

Type locality: "Mountain slopes, from El Paso to the Pecos and Gila Rivers," Texas and New Mexico.

Range: Trans-Pecos Texas to southern Arizona and adjacent Mexico.

New Mexico: Tortugas Mountain. Lower Sonoran Zone, usually on limestone soil.

8. **Echinocereus gonacanthus** (Engelm. & Bigel.) Lem. Cact. Hort. Monv. 57. 1868.

Cereus gonacanthus Engelm. & Bigel. Proc. Amer. Acad. **3**: 283. 1856.

Type locality: Near Zuni, New Mexico. Type collected by Bigelow.

Range: Southern Colorado and northern New Mexico.

New Mexico: Twenty-five miles north of Gallup; south of Zuni Reservation; White Sands. Upper Sonoran Zone.

9. Echinocereus triglochidiatus Engelm. in Wisliz. Mem. North. Mex. 93. 1848.

Cereus triglochidiatus Engelm. in A. Gray, Mem. Amer. Acad. n. ser. **4:** 50. 1849.

TYPE LOCALITY: On Wolf Creek, New Mexico. Type collected by Wislizenus in 1846.

RANGE: Trans-Pecos Texas to New Mexico.

NEW MEXICO: Near Santa Fe. Upper Sonoran Zone.

10. Echinocereus paucispinus (Engelm.) Rümpl. in Först. Handb. Cact. ed. 2. 794. 1886.

Cereus paucispinus Engelm. Proc. Amer. Acad. **3:** 285. 1856.

TYPE LOCALITY: "Western Texas, from the San Pedro to the mouth of the Pecos."

RANGE: Colorado and New Mexico to western Texas.

NEW MEXICO: Santa Fe; mountains near Albuquerque; Cedar Hill; Tunitcha Mountains; Farmington. Upper Sonoran Zone.

11. Echinocereus neomexicanus Standley, Bull. Torrey Club **35:** 87. 1908.

TYPE LOCALITY: Mesa west of the Organ Mountains, New Mexico. Type collected by Standley (no. 383).

RANGE: Known only from the type locality.

12. Echinocereus rosei Woot. & Standl.

Echinocereus polyacanthus Engelm. err. det. Standley, Bull. Torrey Club **35:** 85 *f. 1.* 1908.

TYPE LOCALITY: Agricultural College, New Mexico. Type in the U. S. National Herbarium, no. 535093, collected by Paul C. Standley in 1907 (no. 1235).

RANGE: Southern New Mexico, western Texas, and adjacent Mexico.

NEW MEXICO: Agricultural College; Big Hatchet Mountains; Socorro; San Mateo Peak. Dry plains and low hills, in the Lower Sonoran Zone.

This has always passed as *E. polyacanthus*, and specimens from the vicinity of El Paso were referred here by Doctor Engelmann. That species, however, is amply separated by the presence of long, white wool in the areolæ of the ovary and fruit.

13. Echinocereus coccineus Engelm. in Wisliz. Mem. North. Mex. 93. 1848.

Cereus coccineus Engelm. in A. Gray, Mem. Amer. Acad. n. ser. **4:** 50. 1849.

Cereus phoeniceus Engelm. Proc. Amer. Acad. **3:** 284. 1856.

Cereus aggregatus Coulter; Contr. U. S. Nat. Herb. **3:** 306. 1896, possibly *Mamillaria aggregata* Engelm. 1848.

TYPE LOCALITY: About Santa Fe, New Mexico. Type collected by Wislizenus.

RANGE: Colorado to western Texas and Arizona.

NEW MEXICO: Santa Fe; Burro Mountains; Upper Pecos; Anton Chico; Zuni Mountains; Burro Mountains; head of the Rio Mimbres; Raton; Carrizo Mountains; Tunitcha Mountains. Open rocky slopes, mountains, in the Upper Sonoran and Transition zones.

14. Echinocereus conoideus (Engelm. & Bigel.) Rümpl. in Först. Handb. Cact. ed. 2. 807. 1886.

Cereus conoideus Engelm. & Bigel. Proc. Amer. Acad. **3:** 284. 1856.

TYPE LOCALITY: Rocky places on the Upper Pecos, New Mexico. Type collected by Bigelow.

RANGE: New Mexico, Arizona, and western Texas.

NEW MEXICO: Anton Chico; Fort Wingate; between Barranca and Embudo; east of Hillsboro; Las Cruces; Cubero; Lordsburg. Dry hills and mesas, in the Lower and Upper Sonoran zones.

5. **PENIOCEREUS** Britt. & Rose.

1. **Peniocereus greggii** (Engelm.) Britt. & Rose, Contr. U. S. Nat. Herb. **12:** 428. 1909.

Cereus greggii Engelm. in Wisliz. Mem. North. Mex. 102. 1848.

Type locality: "North and south of Chihuahua."

Range: Southern New Mexico and Arizona to Chihuahua and Sonora.

New Mexico: Tortugas Mountain (*Mrs. E. O. Wooton*). Dry plains and hills, in the Lower Sonoran Zone.

Reported as collected somewhere in New Mexico in 1891 by W. H. Evans.

This species is apparently rare. The stems are slender, 20 to 40 cm. high or more, about 2 cm. in diameter, mostly 4-angled, with very small, stout spines 1 to 2 mm. long from swollen bases. The plant usually has one or two ascending stems from a large tuberous root. The flowers are about 15 cm. long, with a slender tube and white funnelform perianth: they open at dusk and remain open only during the night.

Order 36. THYMELAEALES.

98. ELAEAGNACEAE. Oleaster Family.

Shrubs or trees with silvery lepidote or stellate pubescence; leaves opposite or alternate, the blades entire; flowers perfect, polygamous, or diœcious, usually clustered in the axils of the branches of the present or previous year; calyx of 4 or sometimes 2 sepals surmounting the hypanthium; petals wanting; stamens 4 or 8 on the tube of the hypanthium; pistil simple, becoming a drupelike fruit.

Elaeagnus angustifolia, the oleaster, is cultivated in a number of places in the State. It is especially attractive because of the sweet odor of its flowers.

1. LEPARGYREA Raf. Buffalo berry

KEY TO THE SPECIES.

Leaves ovate or oval, green above; stems not spiny; low shrub.... 1. *L. canadensis.*
Leaves oblong, silvery on both surfaces; stems spiny; tall treelike shrub.. 2. *L. argentea.*

1. **Lepargyrea canadensis** (L.) Greene, Pittonia **2:** 122. 1890.

Hippophae canadensis L. Sp. Pl. 1024. 1753.

Shepherdia canadensis Nutt. Gen. Pl. **2:** 240. 1818.

Type locality: "Habitat in Canada."

Range: Alaska and Newfoundland to Oregon, New Mexico, and New York.

New Mexico: Brazos Canyon; Santa Fe Mountains. Damp woods, in the Canadian Zone.

A low shrub less than a meter high, growing in deep damp woods.

2. **Lepargyrea argentea** (Pursh) Greene, Pittonia **2:** 122. 1890.

Hippophae argentea Pursh, Fl. Amer. Sept. 115. 1814.

Shepherdia argentea Nutt. Gen. Pl. **2:** 240. 1818.

Type locality: "On the banks of the Missouri."

Range: British America to Kansas and New Mexico.

New Mexico: San Juan Valley. Along streams, in the Upper Sonoran Zone.

A tall shrub, usually about 3 meters high, with small, silvery leaves. The small, bright red berries are borne in great profusion. They have a pleasant acid flavor, similar to that of red currants, and are gathered for making jellies.

Order 37. MYRTALES.

KEY TO THE FAMILIES.

Styles wanting; aquatics........................... 101. **GUNNERACEAE** (p. 473).
Styles present; land plants.
  Hypanthium merely inclosing the ovary........ 99. **LYTHRACEAE** (p. 459).
  Hypanthium adnate to the ovary................ 100. **EPILOBIACEAE** (p. 459).

99. LYTHRACEAE. Loosestrife Family.

1. LYTHRUM L. LOOSESTRIFE.

Herbs with simple sessile entire leaves and slender angled stems; flowers perfect, dimorphous, mostly solitary in the axils; hypanthium short, tubular or narrowly funnelform, about 5 mm. long; petals 4 to 6, rose purple; stamens 8 to 12; ovary 2-celled, superior, becoming 1-celled by the breaking down of the septum; seeds several.

1. **Lythrum linearifolium** (A. Gray) Small, Fl. Southeast. U. S. 828. 1903.
Lythrum alatum linearifolium A. Gray, Bost. Journ. Nat. Hist. **6**: 188. 1850.
TYPE LOCALITY: "Rocks in the Cibolo River," Texas.
RANGE: Texas to Arizona, south into Mexico.
NEW MEXICO: Black Range; Dog Spring; Organ Mountains; Mogollon Mountains; Fresnal; Roswell. Wet ground, in the Upper Sonoran and Transition zones.

100. EPILOBIACEAE. Evening primrose Family.

Annual or perennial herbs, rarely somewhat shrubby; leaves simple, alternate or opposite, exstipulate; flowers 2 or 4-merous, mostly conspicuous, small in two or three genera, axillary, spicate, or racemose, regular or slightly irregular; hypanthium mostly tubular and more or less elongated, sometimes spreading; sepals 2 or 4; petals of the same number; stamens as many as the sepals or twice as many; pistil usually of 4 carpels, with one style and 1 to 4 more or less united stigmas; fruit a 4-valved capsule, sometimes indehiscent and nutlike.

KEY TO THE GENERA.

Flowers 2-merous; fruit indehiscent, obovoid, bristly with black hairs................................ 1. CIRCAEA (p. 460).
Flowers 4-merous; fruit various.
  Fruit indehiscent, nutlike......................... 2. GAURA (p. 461).
  Fruit a dehiscent capsule.
    Seeds with a tuft of silky hairs.
      Flowers bright deep red.................. 5. ZAUSCHNERIA (p. 464).
      Flowers purple to white, never bright red.
        Hypanthium not prolonged beyond the ovary; flowers about 2 cm. in diameter...................... 3. CHAMAENERION (p. 463).
        Hypanthium prolonged beyond the ovary; flowers less than 1 cm. in diameter...................... 4. EPILOBIUM (p. 463).
    Seeds without a tuft of silky hairs.
      Hypanthium not prolonged beyond the ovary; flowers minute............. 6. GAYOPHYTUM (p. 464).
      Hypanthium prolonged beyond the ovary; flowers much larger.
        Stigmas discoid or capitate.
          Stigmas capitate; annuals........ 7. SPHAEROSTIGMA (p. 465).

Stigmas discoid; plants perennial, woody at the base.
Hypanthium tube longer than the ovary; stigma entire.................... 8. GALPINSIA (p. 465).
Hypanthium tube shorter than the ovary; stigma 4-lobed .... 9. MERIOLIX (p. 466).
Stigmas divided into 4 linear lobes.
Stamens equal; capsules without appendages.
Petals white at first, sometimes pink-tinged later; seeds in 1 row; buds drooping.................10. ANOGRA (p. 467).
Petals yellow; seeds in 2 or more rows; buds usually erect.
Seeds prismatic-angled....11. OENOTHERA (p. 469).
Seeds not angled..........12. RAIMANNIA (p. 470).
Stamens unequal, the alternate ones longer; capsules winged or angled.
Ovules and seeds numerous, clustered on slender funiculi; plants with branching stems................13. HARTMANNIA (p. 470).
Ovules and seeds few, sessile in 1 or 2 rows; plants acaulescent, or caulescent.
Plants caulescent, with long wiry stems........16. GAURELLA (p. 473).
Plants acaulescent or nearly so, with thick stems.
Capsules distinctly double-crested on the angles; petals white...............14. PACHYLOPHUS (p. 471).
Capsules winged or sharply angled; petals yellow or white........15. LAVAUXIA (p. 472).

1. CIRCAEA L. ENCHANTER'S NIGHTSHADE.

Low weak glabrous herb with opposite petiolate leaves and small white reddish-tinged flowers in racemes; hypanthium slightly prolonged beyond the ovary; sepals, petals, and stamens 2; fruit 1 or 2-seeded, indehiscent.

1. Circaea alpina L. Sp. Pl. 9. 1753.

TYPE LOCALITY: "Habitat ad radices montium in frigidis Europae."

RANGE: British America and Washington to New Mexico and Georgia.

NEW MEXICO: West Fork of the Gila (*Metcalfe* 516). Damp woods, in the Canadian Zone.

2. GAURA L.

Annual, biennial, or perennial herbs with alternate, mostly narrow and small leaves, and small, red or pink flowers in terminal, sometimes elongated, racemes; hypanthium prolonged beyond the ovary; sepals and petals 4, the latter clawed and unequal; stamens usually 8, declined; ovary 4-celled; style declined; stigmas 4-lobed, surrounded by a cuplike border; fruit nutlike, ribbed or angled, indehiscent.

KEY TO THE SPECIES.

Anthers oval, attached near the middle........................ 1. *G. parviflora.*
Anthers linear or narrowly oblong, attached near the base.
  Fruit glabrous.
    Branches of the inflorescence, bracts, and calyx glandular........................ 2. *G. glandulosa.*
    Plants not glandular.
      Leaves deeply lobed or sinuate-dentate.
        Buds glabrous; bracts broadly obovate, acuminate........................ 3. *G. brassicacea.*
        Buds strigillose; bracts ovate or ovate-lanceolate, acute........................ 4. *G. strigillosa.*
      Leaves entire or merely shallowly repand-dentate.
        Buds strigillose; fruit not stipitate........................ 5. *G. gracilis.*
        Buds glabrous; fruit short-stipitate........................ 6. *G. podocarpa.*
  Fruit pubescent.
    Fruit on long slender stipes.
      Branches of the inflorescence glabrous........................ 7. *G. villosa.*
      Branches of the inflorescence cinereous........................ 8. *G. cinerea.*
    Fruit not on long slender stipes, sessile or with a short thick angled stipe.
      Fruit not constricted below the middle, with spreading pubescence........................ 9. *G. neomexicana.*
      Fruit constricted below the middle, appressed-pubescent.
        Stems strigose to hirsute........................10. *G. coccinea.*
        Stems glabrous.
          Bracts linear, acute, much exceeding the ovaries........................11. *G. induta.*
          Bracts lanceolate or ovate, acuminate, much shorter than the ovaries........................12. *G. linearis.*

1. **Gaura parviflora** Dougl.; Hook. Fl. Bor. Amer. **1:** 208. 1830.
TYPE LOCALITY: "Sandy banks of the Wallahwallah River."
RANGE: Washington and North Dakota to Louisiana and Mexico.
NEW MEXICO: Throughout the State. Lower and Upper Sonoran zones.

2. **Gaura glandulosa** Woot. & Standl. Contr. U. S. Nat. Herb. **16:** 153. 1913.
TYPE LOCALITY: Reserve, New Mexico. Type collected by Wooton, July 9, 1906.
RANGE: Mogollon Mountains of New Mexico, in the Transition Zone.

3. **Gaura brassicacea** Woot. & Standl. Contr. U. S. Nat. Herb. **16:** 152. 1913.
TYPE LOCALITY: Socorro, New Mexico. Type collected by G. R. Vasey in 1881.
RANGE: Known only from type locality.

4. **Gaura strigillosa** Woot. & Standl. Contr. U. S. Nat. Herb. **16:** 152. 1913.
TYPE LOCALITY: Wingfields Ranch on Ruidoso Creek, White Mountains, New Mexico. Type collected by Wooton, July 8, 1895.
RANGE: Known only from type locality, in the Transition Zone.

5. **Gaura gracilis** Woot. & Standl. Contr. U. S. Nat. Herb. **16:** 153. 1913.

TYPE LOCALITY: Forest Nursery, Fort Bayard, New Mexico. Type collected by J. C. Blumer (no. 44).

RANGE: Southern New Mexico.

NEW MEXICO: Fort Bayard; Hanover Mountain; Organ Mountains. Mountains, in the Upper Sonoran Zone.

6. **Gaura podocarpa** Woot. & Standl. Contr. U. S. Nat. Herb. **16:** 154. 1913.

TYPE LOCALITY: Bear Mountains near Silver City, Grant County, New Mexico. Type collected by Metcalfe (no. 166).

RANGE: Southwestern New Mexico and adjacent Arizona.

NEW MEXICO: Bear Mountains; West Fork of the Gila; Santa Rita; Organ Mountains. Mountains, in the Upper Sonoran Zone.

This and the three preceding species have passed as *Gaura suffulta* Engelm., a plant originally described from Lindheimer's collections. All four of our plants have much narrower leaves, broader and much shorter bracts, smaller flowers, and larger fruit; while each, in addition, differs from that species in other particulars.

7. **Gaura villosa** Torr. Ann. Lyc. N. Y. **2:** 200. 1828.

TYPE LOCALITY: "Sources of the Canadian," New Mexico. Type collected by James.

RANGE: New Mexico to Kansas and Texas.

NEW MEXICO: Near Portales (*Wooton*). Plains, in the Upper Sonoran Zone.

8. **Gaura cinerea** Woot. & Standl. Contr. U. S. Nat. Herb. **16:** 152. 1913.

TYPE LOCALITY: Twenty miles south of Roswell, Chaves County, New Mexico. Type collected by Earle (no. 533).

RANGE: Known only from type locality.

9. **Gaura neomexicana** Wooton, Bull. Torrey Club **25:** 307. 1898.

TYPE LOCALITY: South Fork of Tularosa Creek 3 miles east of the Mescalero Agency, White Mountains, New Mexico. Type collected by Wooton (no. 204).

RANGE: Southern Colorado to New Mexico.

NEW MEXICO: White and Sacramento mountains; Chama. Moist meadows, in the Transition Zone.

10. **Gaura coccinea** Pursh, Fl. Amer. Sept. 733. 1814.

TYPE LOCALITY: "In Upper Louisiana."

RANGE: Montana to Arizona and Texas.

NEW MEXICO: Common throughout the State. Plains and low hills, in the Upper Sonoran Zone.

11. **Gaura induta** Woot. & Standl. Contr. U. S. Nat. Herb. **16:** 153. 1913.

TYPE LOCALITY: Dry, clay hills near Pecos, New Mexico. Type collected by Standley (no. 4933).

RANGE: Central New Mexico and Arizona to Utah, Wyoming, and South Dakota.

NEW MEXICO: Pecos; Santa Fe; Las Vegas; Sandia Mountains; Tesuque; Patterson; Farmington; Cedar Hill; Dulce; Nutritas Creek; Raton; Ramah; Estancia; Hebron. Dry hills, in the Upper Sonoran Zone.

12. **Gaura linearis** Woot. & Standl. Contr. U. S. Nat. Herb. **16:** 154. 1913.

TYPE LOCALITY: On gypsum soil near Lakewood, New Mexico. Type collected by Wooton, August 6, 1909.

RANGE: Known only from type locality.

3. CHAMAENERION Adans. FIREWEED.

Perennial herb with several stems, 80 cm. high or less; leaves mostly lanceolate, nearly entire, short-petioled; flowers in terminal racemes; hypanthium tube little or not at all prolonged; petals slightly irregular, rose purple; stamens in a single series, the filaments dilated below; stigma with 4 divergent lobes; capsules linear-fusiform, many-seeded, the seeds fusiform, comose.

1. **Chamaenerion angustifolium** (L.) Scop. Fl. Carn. ed. 2. **1**: 271. 1772.
Epilobium angustifolium L. Sp. Pl. 347. 1753.
TYPE LOCALITY: "Habitat in Europa boreali."
RANGE: British America to North Carolina and California.
NEW MEXICO: Santa Fe and Las Vegas mountains; Mogollon Mountains; Black Range; White and Sacramento mountains. Damp woods and open clearings, in the Transition and Canadian zones.

A common and rather showy plant in the higher mountains. It receives its common name from the fact that it is one of the first plants to spring up where forests have been swept by fire, persisting in the "burns" until they are reforested. This habit of the plant results largely from the structure of its seeds, which are peculiarly adapted to dispersal by wind, being furnished with tufts of down.

4. EPILOBIUM L. WILLOW-HERB.

Herbs, 80 cm. high or less, with alternate or opposite, narrow leaves and small, axillary or racemose flowers; hypanthium tube produced beyond the ovary; sepals 4, deciduous; petals 4, obovate to obcordate; stamens 8; ovary 4-celled; fruit an elongated linear-oblong 4-sided 4-celled capsule; seeds small, comose.

KEY TO THE SPECIES.

Annual; stigmas 4-cleft .... 1. *E. adenocladon.*
Perennials; stigmas entire or merely notched.
  Leaves linear, entire, cinereous .... 2. *E. lineare.*
  Leaves lanceolate to ovate, toothed, glabrous or nearly so.
    Stoloniferous, low, 10 to 30 cm. high, simple or sparingly branched .... 3. *E. alpinum.*
    Not stoloniferous, tall, 30 to 60 cm. high, much branched.
      Leaves narrowly lanceolate .... 4. *E. fendleri.*
      Leaves elliptic or ovate-lanceolate .... 5. *E. novomexicanum.*

1. **Epilobium adenocladon** (Hausskn.) Rydb. Bull. Torrey Club **33**: 146. 1906.
Epilobium paniculatum adenocladon Hausskn. Monogr. Epilob. 247. 1884.
TYPE LOCALITY: Mountains of Colorado.
RANGE: Wyoming and South Dakota to Utah and New Mexico.
NEW MEXICO: North of Chama (*Wooton* 2736). Transition Zone.

2. **Epilobium lineare** Muhl. Cat. Pl. 39. 1813.
Epilobium palustre lineare A. Gray, Man. ed. 2. 130. 1856.
TYPE LOCALITY: New England.
RANGE: British Columbia and New Brunswick to Delaware, Oklahoma, and New Mexico.
NEW MEXICO: White Mountains (*Wooton* 661). Transition Zone.

3. **Epilobium alpinum** L. Sp. Pl. 348. 1753.
TYPE LOCALITY: "Habitat in Alpibus Helveticis, Lapponicis."
RANGE: Arctic regions to Oregon, New Mexico, and New Hampshire; also in Europe and Asia.

NEW MEXICO: Winsors Ranch; Mogollon Creek; Eagle Peak. Transition to Arctic-Alpine Zone.

Our plants are larger than typical *alpinum* and occur at rather low levels, but they seem to belong here rather than with any other species.

4. Epilobium fendleri Hausskn. Monogr. Epilob. 261. 1884.

TYPE LOCALITY: New Mexico. Type collected by Fendler (no. 217, in part).

RANGE: New Mexico.

We have seen no collections besides the original one, but Doctor Trelease reports a few others.

5. Epilobium novomexicanum Hausskn. Monogr. Epilob. 260. 1884.

TYPE LOCALITY: Santa Fe, New Mexico. Type collected by Fendler (no. 217, in part).

RANGE: New Mexico.

NEW MEXICO: Farmington; Santa Fe; Taos; Upper Pecos River; Trinchera Pass; Sandia Mountains; Mangas Springs; Mogollon Mountains; Guadalupe Canyon; Organ Mountains; White Mountains; Tunitcha Mountains; Chama; Carrizo Mountains. Mountains, in the Transition Zone.

5. ZAUSCHNERIA Presl.

Ascending suffrutescent plant about 40 cm. high, with large bright red flowers somewhat resembling those of the cultivated Fuchsias, but erect rather than pendent; hypanthium elongated, narrowly funnelform above the ovary; tube bearing 8 small scales within near the base; petals 4, obcordate; stamens 8, in two series, unequally inserted; capsules narrowly fusiform, many-seeded; seeds comose.

1. Zauschneria arizonica Davidson, Bull. South. Calif. Acad. **1:** 4. 1902.

TYPE LOCALITY: Chase Creek, Metcalf, Arizona.

RANGE: Southern New Mexico and Arizona and adjacent Mexico.

NEW MEXICO: Bear Mountains; San Luis Mountains.

6. GAYOPHYTUM Juss.

Slender annuals 30 to 40 cm. high, with linear entire leaves and with very small axillary flowers near the ends of the branches; hypanthium tube not prolonged beyond the ovary; capsules narrowly oblong, 5 to 8 mm. long; seeds not comose.

KEY TO THE SPECIES.

Capsules not exceeding the stipes, usually shorter; petals about 1 mm. long .... 1. *G. ramosissimum.*
Capsules about twice as long as the stipes; petals 1.5 to 2.5 mm. long .... 2. *G. intermedium.*

1. Gayophytum ramosissimum Torr. & Gray, Fl. N. Amer. **1:** 513. 1840.

TYPE LOCALITY: "Rocky Mountains."

RANGE: Washington and Montana to California and New Mexico.

NEW MEXICO: North of Ramah; Tierra Amarilla; Dulce. Open slopes, in the Transition Zone.

2. Gayophytum intermedium Rydb. Bull. Torrey Club **31:** 569. 1904.

TYPE LOCALITY: Ouray, Colorado.

RANGE: Washington and Montana to California and New Mexico.

NEW MEXICO: Tunitcha Mountains; Chama. Open fields in the mountains, in the Transition Zone.

7. SPHAEROSTIGMA Fisch. & Mey.

Small annual with the general appearance of Epilobium, but the seeds naked and the capsule sessile.

1. **Sphaerostigma chamaenerioides** (A. Gray) Small, Bull. Torrey Club **23:** 189. 1896.

Oenothera chamaenerioides A. Gray, Pl. Wright. **2:** 58. 1853.

TYPE LOCALITY: "Stony hills along the Rio Grande near El Paso."

RANGE: Utah to Texas and southern California.

NEW MEXICO: Bishops Cap (*Wooton*). Lower Sonoran Zone.

8. GALPINSIA Britton.

Low perennials, more or less woody at the base, usually cespitose, 30 cm. high or less, with mostly small crowded leaves and large showy yellow flowers; hypanthium prolonged beyond the ovary, dilated upward; sepals green in bud, becoming yellowish and blotched with red in anthesis; petals broad, bright yellow, turning red on drying; stigma discoid, entire; capsules elongated, narrowed at the base, somewhat curved.

KEY TO THE SPECIES.

Free portion of the hypanthium 9 to 13 mm. long, the lower part very slender, plants low and spreading, densely glandular .... 1. *G. tubicula.*
Free portion of the hypanthium more than 30 mm. long, the lower part relatively stout; plants erect or spreading, with or without glandular pubescence.
  Capsules with spreading pubescence.
    Corolla 35 mm. long .... 2. *G. lampsana.*
    Corolla 12 to 18 mm. long .... 3. *G. camporum.*
  Capsules with appressed pubescence or merely glandular.
    Leaves densely grayish-strigose; plants low and tufted. 4. *G. lavandulaefolia.*
    Leaves merely glandular-pubescent to glabrous; plants taller and usually not tufted.
      Free tips of the calyx lobes 3.5 to 4 mm. long; stems wiry, erect .... 5. *G. toumeyi.*
      Free tips of the calyx lobes 2.5 mm. long or less; stems stouter, lower, more or less spreading.
        Leaves glabrous or nearly so, linear-oblong; petals 25 to 30 mm. long .... 6. *G. fendleri.*
        Leaves pubescent or glandular, linear or filiform; petals less than 20 mm. long.
          Leaves puberulent, linear; petals 15 to 20 mm. long .... 7. *G. hartwegi.*
          Leaves densely glandular, nearly filiform; petals 12 to 15 mm. long .... 8. *G. filifolia.*

1. **Galpinsia tubicula** (A. Gray) Small, Bull. Torrey Club **23:** 186. 1896.

Oenothera tubicula A. Gray, Pl. Wright. **1:** 71. 1852.

TYPE LOCALITY: Prairies beyond the Pecos, western Texas.

RANGE: Western Texas and southeastern New Mexico.

NEW MEXICO: Hondo Hill (*Wooton*). Dry hills, in the Upper Sonoran Zone.

52576°—15——30

2. **Galpinsia lampsana** (Buckl.) Woot. & Standl. Contr. U. S. Nat. Herb. **16:** 152. 1913.

Oenothera lampsana Buckl. Proc. Acad. Phila. **1861:** 454. 1861.

TYPE LOCALITY: Prairies, Lampasas County, Texas.

RANGE: Western Texas and eastern New Mexico.

NEW MEXICO: Cabra Springs; Organ Mountains. Upper Sonoran Zone.

3. **Galpinsia camporum** Woot. & Standl. Contr. U. S. Nat. Herb. **16:** 152. 1913.

TYPE LOCALITY: Knowles, New Mexico. Type collected by Wooton, July 29, 1909.

RANGE: Plains and low hills of eastern New Mexico.

NEW MEXICO: Knowles, Nara Visa; Causey; Buchanan; Hondo Hill. Upper Sonoran Zone.

4. **Galpinsia lavandulaefolia** (Torr. & Gray) Small, Fl. Southeast. U. S. 845. 1903.

Oenothera lavandulaefolia Torr. & Gray, Fl. N. Amer. **1:** 501. 1840.

Oenothera hartwegi lavandulaefolia S. Wats. Proc. Amer. Acad. **8:** 590. 1873.

TYPE LOCALITY: "Plains of the Platte."

RANGE: Kansas and Wyoming to New Mexico and Texas.

NEW MEXICO: Hell Canyon; Sierra Grande; Nara Visa. Dry hills and plains, in the Upper Sonoran Zone.

5. **Galpinsia toumeyi** Small, Bull. Torrey Club **25:** 317. 1898.

TYPE LOCALITY: Chiricahua Mountains, Arizona.

RANGE: Southwestern New Mexico and southern Arizona to northern Mexico.

NEW MEXICO: West Fork of the Gila (*Metcalfe* 555). Transition Zone.

6. **Galpinsia fendleri** (A. Gray) Heller, Cat. N. Amer. Pl. ed. 2. 8. 1900.

Oenothera fendleri A. Gray, Mem. Amer. Acad. n. ser. **4:** 45. 1849.

Oenothera hartwegi fendleri A. Gray, Pl. Wright. **2:** 56. 1853.

TYPE LOCALITY: "Sunny hillsides at Santa Fe, and on the Rio del Norte," New Mexico. Type collected by Fendler (no. 230).

RANGE: New Mexico and Texas.

NEW MEXICO: Banks of the Nutria; Gallinas Mountains; Pecos; Ramah; west of Santa Fe; Cactus Flat; Bear Mountains; Sacramento Mountains; Ojo Caliente. Open hills, in the Upper Sonoran Zone.

7. **Galpinsia hartwegi** (Benth.) Britton, Mem. Torrey Club **5:** 236. 1894.

Oenothera hartwegi Benth. Pl. Hartw. 5. 1839.

Salpingia hartwegi Raim. in Engl. & Prantl, Pflanzenfam. 3^{7}: 217. 1893.

TYPE LOCALITY: Mexico.

RANGE: Mexico to Colorado and Nebraska.

NEW MEXICO: Willard; White Mountains. Plains and low hills, in the Upper Sonoran Zone.

8. **Galpinsia filifolia** (Eastw.) Wooton; Heller, Cat. N. Amer. Pl. ed. 2. 8. 1900.

Oenothera tubicula filifolia Eastw. Proc. Calif. Acad. II. **1:** 72. 1897.

TYPE LOCALITY: White Sands, New Mexico. Type collected by T. D. A. Cockerell.

RANGE: Southeastern New Mexico.

NEW MEXICO: White Sands; twenty miles south of Roswell. Dry plains, in the Lower Sonoran Zone.

9. MERIOLIX Raf.

Low perennial herb; flowers smaller than in the last genus, pale yellow; hypanthium tube expanded only immediately below the sepals; stigma discoid, 4-lobed; capsules sessile, not tapering at the base; seeds longitudinally grooved.

1. **Meriolix serrulata** (Nutt.) Walp. Repert. Bot. **2:** 79. 1843.

Oenothera serrulata Nutt. Gen. Pl. **1:** 246. 1818.

TYPE LOCALITY: "From the river Platte to the mountains, on dry hills."

RANGE: Manitoba and Minnesota to New Mexico and Texas.

NEW MEXICO: Nara Visa; Knowles; Lakewood; Redlands. Dry plains, in the Upper Sonoran Zone.

10. ANOGRA Spach.

Spreading biennials or perennials, 30 to 40 cm. high or less, somewhat woody at the base, with simple, more or less sinuate-toothed leaves and showy white flowers opening in the evening and at night, turning rose pink the second day; cortex often papery and exfoliating, the stems often white and shining; buds drooping; ovules numerous, in a single row in the cell; capsules sessile, sometimes enlarged at the base, often woody, 4-celled, the seeds terete.

KEY TO THE SPECIES.

Tips of the calyx segments not free in bud, the buds merely acute .... 1. *A. albicaulis.*
Tips of the calyx segments free, the buds abruptly acuminate.
  Throat of the calyx villous within .... 2. *A. coronopifolia.*
  Throat of the calyx not villous.
    Plants glabrous throughout .... 3. *A. pallida.*
    Plants not glabrous.
      Calyx appressed-pubescent, not at all hirsute.
        Plants grayish; leaves densely appressed-pubescent, slightly pinnatifid or entire. 4. *A. gypsophila.*
        Plants green; leaves sparingly pubescent, deeply pinnatifid .... 5. *A. runcinata.*
      Calyx more or less villous or hirsute.
        Calyx merely villous, without appressed pubescence.
          All leaves petioled, nearly glabrous; capsules 20 to 25 mm. long; petals 18 to 20 mm. long .... 10. *A. neomexicana.*
          Upper leaves clasping, abundantly pubescent; capsules 35 mm. long or more; petals about 15 mm. long .... 11. *A. amplexicaulis.*
        Calyx with appressed pubescence as well as spreading hairs.
          Whole plant densely villous or hirsute.
            Leaves nearly entire, sessile .... 8. *A. engelmanni.*
            Leaves deeply pinnatifid, petioled .... 9. *A. leucotricha.*
          Plants with only a few spreading hairs, most of the pubescence appressed.
            Leaves with only a few shallow teeth, not hirsute .... 6. *A. latifolia.*
            Leaves deeply pinnatifid, more or less hirsute .... 7. *A. ctenophylla.*

1. **Anogra albicaulis** (Pursh) Britton, Mem. Torrey Club **5:** 234. 1894.

Oenothera albicaulis Pursh, Fl. Amer. Sept. 733. 1814.

Oenothera pinnatifida Nutt. Gen. Pl. **1:** 245. 1818.

TYPE LOCALITY: "In Upper Louisiana."

RANGE: Montana and North Dakota to Mexico and Texas.

NEW MEXICO: West of Santa Fe; Magdalena Mountains; Socorro; Mangas Springs; Kingston; Farmington; Carrizo Mountains; Las Vegas; Raton; Nara Visa; mountains west of San Antonio; Atarque de Garcia; Tres Hermanas; Agricultural College. Lower and Upper Sonoran zones.

2. **Anogra coronopifolia** (Torr. & Gray) Britton, Bull. Torrey Club **23:** 174. 1896.

Oenothera coronopifolia Torr. & Gray, Fl. N. Amer. **1:** 495. 1840.

TYPE LOCALITY: "Forks of the Platte."

RANGE: Wyoming and South Dakota to Utah, New Mexico, and Kansas.

NEW MEXICO: Santa Fe and Las Vegas mountains; Johnsons Mesa; San Augustine Plains; Gray; Shiprock; Chama; Raton; Sierra Grande; Socorro Mountain; Mogollon Mountains; White Mountains. Open slopes, in the Upper Sonoran and Transition zones.

3. **Anogra pallida** (Lindl.) Britton, Bull. Torrey Club **23:** 175. 1896.

Oenothera pallida Lindl. Edwards's Bot. Reg. **14:** *pl. 1142*. 1828.

Oenothera pinnatifida integrifolia A. Gray, Mem. Amer. Acad. n. ser. **4:** 44. 1849.

TYPE LOCALITY: "In the Northwest of North America."

RANGE: British Columbia to Mexico.

NEW MEXICO: Farmington; Shiprock; Mesilla Valley; Tunitcha Mountains; Aztec. Sandy fields, in the Lower and Upper Sonoran zones.

4. **Anogra gypsophila** (Eastw.) Heller, Cat. N. Amer. Pl. ed. 2. 8. 1900.

Oenothera albicaulis gypsophila Eastw. Proc. Calif. Acad. III. **1:** 73. 1897.

TYPE LOCALITY: White Sands, New Mexico. Type collected by T. D. A. Cockerell.

RANGE: Known only from the type locality, in the Lower Sonoran Zone.

5. **Anogra runcinata** (Engelm.) Woot. & Standl. Contr. U. S. Nat. Herb. **16:** 151. 1913.

Oenothera albicaulis runcinata Engelm. Amer. Journ. Sci. II. **34:** 334. 1862.

Anogra pallida runcinata Small, Bull. Torrey Club **23:** 175. 1896.

TYPE LOCALITY: Near Sante Fe, New Mexico. Type collected by Fendler (no. 223).

RANGE: New Mexico and Arizona.

NEW MEXICO: Santa Fe; Pecos; Coolidge; mountains west of San Antonio; Zuni Reservation; Albuquerque; Gallup; Reserve; Chavez; Mesilla Valley. Dry hills, in the Lower and Upper Sonoran zones.

6. **Anogra latifolia** Rydb. Bull. Torrey Club **31:** 570. 1904.

Oenothera pallida latifolia Rydb. Contr. U. S. Nat. Herb. **3:** 159. 1895.

TYPE LOCALITY: Mullen, Nebraska.

RANGE: Nebraska and Kansas to Colorado and northern New Mexico.

NEW MEXICO: Near Salt Lake; Zuni Reservation. Dry hills, in the Upper Sonoran Zone.

7. **Anogra ctenophylla** Woot. & Standl. Contr. U. S. Nat. Herb. **16:** 151. 1913.

TYPE LOCALITY: Near Zuni, New Mexico. Type collected by Mrs. M. C. Stevenson (no. 99).

RANGE: Western and southern New Mexico.

NEW MEXICO: Zuni; Crawfords Ranch; Reserve; Defiance; Burro Mountains; Ruidoso Creek. Hillsides, in the Upper Sonoran and Transition zones.

8. **Anogra engelmanni** (Small) Woot. & Standl. Contr. U. S. Nat. Herb. **16:** 151. 1913.

Oenothera albicaulis trichocalyx Engelm. Amer. Journ. Sci. II. **34:** 335. 1862, not *O. trichocalyx* Nutt. 1840.

Anogra pallida engelmanni Small, Bull. Torrey Club **23:** 176. 1896.

TYPE LOCALITY: Las Vegas, New Mexico. Type collected by Wislizenus in 1846 (no. 473).

RANGE: Plains of northern New Mexico.

NEW MEXICO: Nara Visa; Las Vegas. Upper Sonoran Zone.

9. Anogra leucotricha Woot. & Standl. Contr. U. S. Nat. Herb. **16:** 151. 1913.

TYPE LOCALITY: San Augustine Plains, New Mexico. Type collected by Wooton (no. 2735).

RANGE: Plains of central New Mexico.

NEW MEXICO: San Augustine Plains; Willard. Upper Sonoran Zone.

10. Anogra neomexicana Small, Bull. Torrey Club **23:** 176. 1896.

TYPE LOCALITY: Sandy bed of a creek near the Copper Mines, New Mexico. Type collected by Wright (no. 1068).

RANGE: Western New Mexico and adjacent Arizona.

NEW MEXICO: Magdalena; Santa Rita; Hop Canyon; Mogollon Mountains; Organ Mountains. Canyons, in the Upper Sonoran and Transition zones.

11. Anogra amplexicaulis Woot. & Standl. Contr. U. S. Nat. Herb. **16:** 150. 1913.

TYPE LOCALITY: Sandbar along the Mimbres River, New Mexico. Type collected by Metcalfe (no. 1054).

RANGE: Known only from the type locality.

11. OENOTHERA L. EVENING PRIMROSE.

Biennial or perennial herbs, 50 to 200 cm. high or more, with erect or spreading branching stems, alternate, mostly sessile, sometimes short-petioled leaves, and large yellow flowers; leaves mostly undulate-toothed, sometimes entire; hypanthium tube prolonged above the ovary, in one species very long; petals broad; ovary 4-celled; seeds horizontal, prismatic, angled, in 2 or more rows in each cell.

KEY TO THE SPECIES.

Hypanthium tube 15 to 19 cm. long; stems spreading............ 1. *O. macrosiphon.*
Hypanthium tube 5 cm. long or less; stems erect.
  Petals 12 to 14 mm. long; stems simple; pubescence spreading.. 2. *O. procera.*
  Petals 30 mm. long or more; stems simple or branched; pubescence spreading or appressed.
    Pubescence cinereous or strigose, dense, grayish; plants 1 to 2 meters high, much branched..................... 3. *O. irrigua.*
    Pubescence mostly hirsute, loose, not grayish; plants usually less than 1 meter high, rarely branched...... 4. *O. hookeri.*

1. Oenothera macrosiphon Woot. & Standl. Contr. U. S. Nat. Herb. **16:** 155. 1913.

TYPE LOCALITY: Organ Mountains, New Mexico. Type collected by Wooton, August 29, 1894.

RANGE: Canyons in the Organ Mountains, New Mexico, in the Upper Sonoran Zone.

A beautiful plant with larger flowers than any other species of the genus. It occurs in the Organs in deep rocky canyons, principally about the edges of pools. It has been called *O. jamesii*, but that species has much smaller flowers and abundant appressed pubescence.

2. Oenothera procera Woot. & Standl. Contr. U. S. Nat. Herb. **16:** 156. 1913.

TYPE LOCALITY: Along Winsor Creek in the Pecos National Forest, New Mexico. Type collected by Standley (no. 4212).

RANGE: New Mexico and Arizona to southern Colorado.

NEW MEXICO: Santa Fe and Las Vegas mountains; White Mountains; West Fork of the Gila; James Canyon; Chama. Damp meadows and along streams, in the Transition Zone.

A common plant in the mountains in the Transition Zone. It grows usually on moist open slopes, but sometimes along streams. Seldom or never does it exceed a meter in height, and the stems are almost invariably simple. The type collection was distributed as *O. strigosa* Rydb., but that is a plant with much larger flowers and different pubescence.

3. **Oenothera irrigua** Woot. & Standl. Contr. U. S. Nat. Herb. **16:** 155. 1913.

TYPE LOCALITY: Mesilla Valley, Dona Ana County, New Mexico. Type collected by Wooton and Standley, June, 1906.

RANGE: New Mexico.

NEW MEXICO: Mesilla Valley; Farmington; Albuquerque; Shiprock. River valleys, usually in wet ground, in the Lower and Upper Sonoran zones.

4. **Oenothera hookeri** Torr. & Gray, Fl. N. Amer. **1:** 493. 1840.

Oenothera biennis hirsutissima A. Gray, Mem. Amer. Acad. n. ser. **4:** 43. 1849.

Onagra hookeri Small, Bull. Torrey Club **23:** 171. 1896.

TYPE LOCALITY: "California."

RANGE: Rocky mountains, west to the Pacific coast, south into Mexico.

NEW MEXICO: Pecos; Chama; Farmington; Santa Fe Canyon; Ramah; Belen; Kingston; Fort Bayard; Mesilla Valley; Roswell; Capitan Mountains; White and Sacramento mountains. Transition Zone.

12. RAIMANNIA Rose.

Diffusely branched biennial 20 to 30 cm. high, with alternate oblanceolate sinuate-toothed leaves, the lowest almost lyrate; flowers of medium size, yellow; seeds not prismatic-angled.

1. **Raimannia mexicana** (Spach) Woot. & Standl.

Oenothera mexicana Spach, Nouv. Ann. Mus. Hist. Nat. **4:** 347. 1835.

Oenothera sinuata hirsuta Torr. & Gray, Fl. N. Amer. **1:** 494. 1840.

Oenothera laciniata mexicana Small, Bull. Torrey Club **23:** 173. 1896.

TYPE LOCALITY: Texas.

RANGE: Nebraska to New Mexico, Texas, and Mexico.

NEW MEXICO: Mouth of Mora River; Burro Mountains; Mogollon Mountains; White Mountains. Open slopes, in the Transition Zone.

13. HARTMANNIA Spach.

Slender branching herbs 20 to 40 cm. high, with medium-sized rose purple or large white flowers and sinuate or almost lyrate leaves; resembling superficially species of Anogra, but with stamens of different lengths and the seeds clustered on slender funicles.

KEY TO THE SPECIES.

Petals less than 20 mm. long, rose purple or pink; body of the capsule shorter than the pedicel-like base .... 1. *H. rosea.*

Petals more than 20 mm. long, white turning pink; capsule without a pedicel-like base .... 2. *H. speciosa.*

1. **Hartmannia rosea** (Ait.) Don in Sweet, Hort. Brit. ed. 3. 236. 1839.

Oenothera rosea Ait. Hort. Kew. **2:** 3. 1789.

TYPE LOCALITY: Peru.

RANGE: Texas to New Mexico and southward; also in South America.

NEW MEXICO: Clayton (*Bartlett*). Plains and prairies, in the Lower and Upper Sonoran zones.

2. **Hartmannia speciosa** (Nutt.) Small, Bull. Torrey Club **23**: 181. 1896.
Oenothera speciosa Nutt. Journ. Acad. Phila. **2**: 119. 1821.
TYPE LOCALITY: "On the plains of Red River."
RANGE: Missouri and Kansas to New Mexico and Mexico.
NEW MEXICO: Kingston (*Metcalfe* 1017).

14. PACHYLOPHUS Spach.

Cespitose, almost acaulescent, herbaceous perennials with rosettes of narrowly oblanceolate, entire or pinnatifid, acute leaves, and large white flowers with long hypanthium tubes; capsules basal, woody, doubly crested on the angles.

KEY TO THE SPECIES.

Pubescence fine, short, appressed .... 1. *P. australis.*
Pubescence, coarse, long, spreading.
Hypanthium, calyx, and fruit densely hirsute.
Plants acaulescent; ridges of the fruit slightly tuberculate. 2. *P. hirsutus.*
Plants caulescent; ridges of the fruit with lobed, somewhat foliaceous crests .... 3. *P. eximius.*
Hypanthium, calyx, and fruit with only a few scattered hairs or glabrous.
Plants caulescent; capsules elongated conic-ovoid, with low ridges .... 4. *P. caulescens.*
Plants acaulescent; capsules shortly conic-ovoid, with very thick ridges .... 5. *P. macroglottis.*

1. **Pachylophus australis** Woot. & Standl. Contr. U. S. Nat. Herb. **16**: 156. 1913.
TYPE LOCALITY: On the South Fork of Tularosa Creek, New Mexico. Type collected by Wooton, July 31, 1897.
RANGE: Known only from type locality.

2. **Pachylophus hirsutus** Rydb. Bull. Torrey Club **31**: 571. 1904.
TYPE LOCALITY: Georgetown, Colorado.
RANGE: Colorado and Utah to New Mexico.
NEW MEXICO: Carrizo Mountains; Tunitcha Mountains; Shiprock; west of Santa Fe; Santa Fe Canyon; Burro Mountains; White Mountains; Organ Mountains; San Luis Mountains. Mountains and in river valleys, Upper Sonoran and Transition zones.

3. **Pachylophus eximius** (A. Gray) Woot. & Standl. Contr. U. S. Nat. Herb. **16**: 157. 1913.
Oenothera eximia A. Gray, Mem. Amer. Acad. n. ser. **4**: 45. 1849.
Pachylophus exiguus Rydb. Colo. Agr. Exp. Sta. Bull. **100**: 246. 1906.
TYPE LOCALITY: Along Santa Fe Creek, New Mexico. Type collected by Fendler (no. 228).
RANGE: Colorado and New Mexico.
NEW MEXICO: Winsor Creek; Santa Fe Creek. Meadows, in the Transition Zone.

4. **Pachylophus caulescens** Rydb. Bull. Torrey Club **31**: 571. 1901.
TYPE LOCALITY: Palisades, Colorado.
RANGE: Colorado and northern New Mexico.
NEW MEXICO: Mouth of Holy Ghost Creek (*Standley*). Mountains, in the Transition Zone.

5. **Pachylophus macroglottis** Rydb. Bull. Torrey Club **30**: 259. 1903.
TYPE LOCALITY: Tributaries of Turkey Creek, Colorado.
RANGE: Colorado and northern New Mexico.
NEW MEXICO: Cedar Hill; Chama; Dulce; Tierra Amarilla. Open hills, in the Upper Sonoran and Transition zones.

15. LAVAUXIA Spach.

Low, acaulescent, annual or perennial herbs, with simple or lyrate leaves and large yellow or white flowers; hypanthium tube often much elongated; capsules stout, the angles variously appendaged or smooth; seeds few.

KEY TO THE SPECIES.

Tips of the calyx segments not free in bud; leaves copiously hirsute........ 1. *L. primiveris.*
Tips of the calyx segments free in bud; leaves not hirsute, or if so only on the petioles.
  Leaves thick, grayish, densely cinereous, entire or but little lobed.
    Petals 4 cm. long or less; capsules ovoid or ellipsoid..... 2. *L. brachycarpa.*
    Petals 5 to 6 cm. long; capsules attenuate to the apex.... 3. *L. wrightii.*
  Leaves thin, bright green, not cinereous, deeply pinnatifid or lyrate.
    Capsules with divaricate hooked beaks on the angles above the middle........ 4. *L. hamata.*
    Capsules with smooth angles.
      Petals 20 mm. long; leaves with very numerous crowded lobes........ 5. *L. flava.*
      Petals 35 to 40 mm. long; leaves with few distant lobes........ 6. *L. taraxacoides.*

1. **Lavauxia primiveris** (A. Gray) Small, Bull. Torrey Club **23:** 182. 1896.
Oenothera primiveris A. Gray, Pl. Wright. **2:** 58. 1853.
TYPE LOCALITY: "Dry hills near El Paso." Type collected by Wright (no. 1376).
RANGE: Western Texas to southern New Mexico.
NEW MEXICO: Rincon; Organ foothills; Mesilla Valley; Tres Hermanas. Low hills and mesas, in the Lower Sonoran Zone.

2. **Lavauxia brachycarpa** (A. Gray) Britton, Mem. Torrey Club **5:** 235. 1894.
Oenothera brachycarpa A. Gray, Pl. Wright. **1:** 70. 1852.
TYPE LOCALITY: "Between western Texas and El Paso."
RANGE: Western Texas and southern New Mexico.
NEW MEXICO: Hillsboro; Big Hatchet Mountains; Kingston.

3. **Lavauxia wrightii** (A. Gray) Small, Bull. Torrey Club **23:** 183. 1896.
Oenothera wrightii A. Gray, Pl. Wright. **2:** 57. 1853.
TYPE LOCALITY: Stony hills near the Copper Mines, New Mexico. Type collected by Wright (no. 1072).
RANGE: Southern New Mexico.
NEW MEXICO: Santa Rita; San Andreas Mountains.

4. **Lavauxia hamata** Woot. & Standl. Contr. U. S. Nat. Herb. **16:** 154. 1913.
TYPE LOCALITY: Socorro, New Mexico. Type collected by G. R. Vasey, May, 1881.
RANGE: Known only from type locality.

5. **Lavauxia flava** A. Nels. Bull. Torrey Club **31:** 243. 1904.
TYPE LOCALITY: Laramie, Wyoming.
RANGE: Wyoming to New Mexico.
NEW MEXICO: Rio Pueblo; Winsors Ranch; Pecos; Tunitcha Mountains; Raton; Chama; Farmington; Gallo Spring. Moist meadows, in the Transition Zone.

6. **Lavauxia taraxacoides** Woot. & Standl. Contr. U. S. Nat. Herb. **16**: 155. 1913.
TYPE LOCALITY: James Canyon, Sacramento Mountains, New Mexico. Type collected by Wooton, July 6, 1899.
RANGE: Meadows, White and Sacramento mountains of New Mexico, in the Transition Zone.

16. GAURELLA Small.

Low perennial canescent herb with slender wiry branched stems; leaves alternate, sessile, linear-lanceolate, entire or sparingly denticulate; flowers axillary, sessile; hypanthium tube slender, canescent, only slightly longer than the calyx lobes; petals white or pink, striped with red; capsule sessile, ovoid, sharply 4-angled, 8 to 10 mm. long.

1. **Gaurella canescens** (Torr.) A. Nels. in Coulter, New Man. Rocky Mount. 341. 1909.
Oenothera canescens Torr. in Frém. Rep. Exped. Rocky Mount. 315. 1845.
Oenothera guttulata Geyer; Hook. Lond. Journ. Bot. **6**: 222. 1847.
Gaurella guttulata Small, Fl. Southeast. U. S. 844. 1903.
TYPE LOCALITY: "On the upper waters of the Platte."
RANGE: Colorado and New Mexico to Nebraska.
NEW MEXICO: Sierra Grande (*Howell* 216). Plains and low hills, in the Upper Sonoran Zone.

101. GUNNERACEAE. Water milfoil Family.

Aquatic perennial herbs with alternate or whorled leaves, these dissected when immersed; flowers inconspicuous, sessile in the axils, perfect, monoecious, or diœcious; calyx entire or 4-lobed; corolla small or wanting, of 2 to 4 petals; stamens 1 to 8; ovary 1 to 4-celled, adnate to the calyx; fruit indehiscent, nutlike, of 2 to 4-seeded carpels.

KEY TO THE GENERA.

Leaves entire; stamen 1; ovary 1-celled.................. 1. HIPPURIS (p. 473).
At least some of the leaves dissected; stamens 4 to 8;
ovary 4-celled.................................. 2. MYRIOPHYLLUM (p. 473).

1. HIPPURIS L. MARESTAIL.

Smooth erect herb with running rootstocks, growing in shallow water; leaves simple, entire, linear, in whorls of 8 to 12; flowers solitary, perfect; petals wanting; fruit 1-seeded, drupelike.

1. **Hippuris vulgaris** L. Sp. Pl. 4. 1755.
TYPE LOCALITY: "Habitat in Europae fontibus."
RANGE: From the arctic regions to California and New Mexico; also in Europe and Asia.
NEW MEXICO: Near Santa Fe; Dulce Lake; near Aztec In shallow water.

2. MYRIOPHYLLUM L. WATER MILFOIL.

Smooth leafy aquatic herb with leaves of two sorts, the immersed ones pinnatifid, in whorls of 3 or 4; flowers monœcious, the upper ones usually staminate; petals 2 to 4, wanting in the pistillate flowers; stamens 8; ovary 4-celled; stigmas recurved and plumose.

1. **Myriophyllum spicatum** L. Sp. Pl. 992. 1753.
TYPE LOCALITY: "Habitat in Europae aquis quietis."
RANGE: Canada to California and Florida.
NEW MEXICO: La Jara (*Standley* 8273). In quiet water.

Order 38. UMBELLALES.

KEY TO THE FAMILIES.

Fruit dry, a cremocarp; gynœcium 2-carpellary; stigmas terminal.................................104. **APIACEAE** (p. 475).
Fruit drupaceous or baccate; gynœcium 1 to several-carpellary, if 2-carpellary the stigmas introrse.
  Ovule with a dorsal raphe; leaves mostly opposite, the blades entire or merely toothed..102. **CORNACEAE** (p. 474).
  Ovule with a ventral raphe; leaves mostly alternate, the blades compound...............103. **HEDERACEAE** (p. 475).

102. CORNACEAE. Dogwood Family.

Trees or shrubs with simple, entire, mainly opposite, exstipulate leaves and perfect or unisexual flowers in spikes or cymes; calyx lobes minute; petals and stamens 4, epigynous; ovary inferior, becoming a 1 or 2-seeded drupe or berry, sometimes dry at maturity.

KEY TO THE GENERA.

Flowers perfect, cymose; leaves deciduous.................... 1. CORNUS (p. 474).
Flowers diœcious, spicate; leaves evergreen................. 2. GARRYA (p. 474).

1. CORNUS L. CORNEL.

Shrub 1 to 2 meters high with reddish branches; leaves elliptic-ovate, entire, short-petioled, thin, deciduous; flowers white, perfect, in flat-topped cymes, without involucres; fruit white.

1. **Cornus instolonea** A. Nels. Bot. Gaz. **53:** 224. 1912.
Svida stolonifera riparia Rydb. Bull. Torrey Club **31:** 573. 1904.
TYPE LOCALITY: Crystal Creek, Colorado.
RANGE: Alaska and Montana to Arizona and Nebraska.
NEW MEXICO: Cedar Hill; Tunitcha Mountains; Chama; Santa Fe and Las Vegas mountains; Zuni; Sandia Mountains; Mogollon Mountains; White and Sacramento mountains. Along streams and in wet ground, in the Transition Zone.
The leaves of the shrub vary in outline from broadly ovate and abruptly short-acuminate to lanceolate and long-acuminate, and the base may be either broadly rounded or somewhat narrowed.

2. GARRYA Dougl.

Evergreen shrubs from 50 cm. to over 3 meters high, with elliptic to ovate, entire, short-petioled, coriaceous leaves and diœcious flowers in loose drooping axillary spikes; petals wanting; calyx 4-merous in the staminate flowers, with 4 stamens, 2-lobed or obsolete in the pistillate flowers; ovary 1-celled, with 2 persistent styles; fruit a blue black berry 5 mm. in diameter or less, becoming dry.

KEY TO THE SPECIES.

Mature leaves glabrous; plants mostly 2 to 3 meters high.......... 1. *G. wrightii.*
Mature leaves densely pubescent; plants low, usually less than 1 meter high.. 2. *G. goldmanii.*

1. **Garrya wrightii** Torr. U. S. Rep. Expl. Miss. Pacif. **4:** 136. 1856.
TYPE LOCALITY: Copper Mines, New Mexico. Type collected by Wright (no. 1789).
RANGE: Western Texas to Arizona.
NEW MEXICO: Mogollon and Magdalena mountains, to the Organ and White mountains and southward. Dry hills and canyons, in the Upper Sonoran Zone.

2. **Garrya goldmanii** Woot. & Standl. Contr. U. S. Nat. Herb. **16:** 157. 1913.

TYPE LOCALITY: Limestone ledges near Queen, New Mexico. Type collected by Wooton, July 31, 1909.

RANGE: Western Texas and southern New Mexico.

NEW MEXICO: Big Hatchet Mountains; Queen; San Andreas Mountains. Dry hills, in the Upper Sonoran Zone.

103. HEDERACEAE. Ivy Family.

1. ARALIA L.

Perennial herbs having the habit of some of the Apiaceae, about 1 meter high, the leaves ternately or pinnately twice or thrice compound with large leaflets; flowers small, whitish, in large compound umbels, 5-merous, regular; fruit a berry.

1. **Aralia bicrenata** Woot. & Standl. Contr. U. S. Nat. Herb. **16:** 157. 1913.

TYPE LOCALITY: Near Holts Ranch in the Mogollon Mountains, New Mexico. Type collected by Wooton, July 20, 1900.

RANGE: Mountains of New Mexico.

NEW MEXICO: Mogollon Mountains; Las Vegas Hot Springs; Sierra Grande; White Mountains; Brazos Canyon. Transition Zone.

104. APIACEAE. Parsley Family.

Annual, biennial, or perennial herbs with mostly hollow stems; leaves usually alternate, often all basal, variously compound, rarely simple; petioles expanded or sheathing at the base; flowers small, in umbels (in one genus in heads); umbels simple or more frequently 1 to several times compound; bracts forming involucres and involucels or sometimes wanting; calyx represented by 5 teeth or wanting; petals and stamens 5; ovary inferior, the two filiform styles often borne on a stylopodium; fruit sometimes flattened laterally or dorsally; carpels 2, ribbed or winged, the wings either thin or corky-thickened; oil tubes present in the carpel wall in most species.

KEY TO THE GENERA.

Flowers in dense heads.......................... 2. ERYNGIUM (p. 477).
Flowers not in heads, evidently umbellate.
  Fruit conspicuously bristly.
    Bristles not hooked....................... 3. WASHINGTONIA (p. 478).
    Bristles hooked or barbed at the tips.
      Whole fruit covered with hooked bristles. (See also *Spermolepis echinatus.*).................... 1. SANICULA (p. 477).
      Only the ribs of the fruit bristly, the bristles barbed.................21. DAUCUS (p. 484).
  Fruit not bristly.
    Fruit strongly flattened dorsally, with lateral ribs, more or less prominently winged.
      Oil tubes solitary in the intervals between the ribs.
        Stylopodium conical.
          Plants slender, glabrous......18. OXYPOLIS (p. 483).
          Plants stout, pubescent......20. HERACLEUM (p. 484).
        Stylopodium flat or wanting.
          Plants acaulescent or nearly so.

Stylopodium wanting; calyx teeth minute or obsolete; dorsal ribs filiform.......19. COGSWELLIA (p. 484).

Stylopodium present; calyx teeth evident; dorsal ribs sharp or winged...25. CYNOMARATHRUM (p. 485).

Plants caulescent, branching.

Flowers white; dorsal ribs prominent....13. ANGELICA (p. 481).

Flowers yellow; dorsal ribs filiform.......23. PASTINACA (p. 485).

Oil tubes more than one in the intervals between the ribs. (See also Aulospermum.)

Plants caulescent, branching.

Leaves ternately or pinnately compound.............13. ANGELICA (p. 481).

Leaves several times compound.................12. CONIOSELINUM (p. 480).

Plants acaulescent or nearly so.

Lateral wings of the fruit thin...................19. COGSWELLIA (p. 484).

Lateral wings of the fruit thickened.

Dorsal ribs very prominent or slightly winged...........17. PSEUDOCYMOPTERUS (p. 482).

Dorsal ribs filiform.......16. CYMOPTERUS (p. 482).

Fruit not strongly flattened dorsally, usually more or less laterally compressed.

Oil tubes more than one in the intervals between the ribs.

Stylopodium conical.

Fruit spherical, with globose carpels and very slender inconspicuous ribs. 9. BERULA (p. 479).

Fruit ovoid or oblong, with prominent equal ribs..10. LIGUSTICUM (p. 480).

Stylopodium flat or wanting.

Seed face sulcate or concave..15. AULOSPERMUM (p. 481).

Seed face plane or but slightly concave.

Leaves pinnate, with short crowded segments; flowers purplish..............14. PHELLOPTERUS (p. 481).

Leaves ternate-pinnate, with distant segments; flowers yellow................26. PTERYXIA (p. 485).

Oil tubes solitary in the intervals between the ribs.
Stylopodium flat or wanting.
Flowers white; an annual.... 4. APIUM (p. 478).
Flowers yellow; perennials.
Ribs equal, broad, corky. 11. OREOXIS (p. 480).
Ribs filiform............. 8. ALETES (p. 479).
Stylopodium conical.
Leaflets broad, large......... 6. CICUTA (p. 479).
Leaflets (at least the upper) linear to filiform.
Involucres wanting.
Flowers white.......22. CORIANDRUM (p. 484).
Flowers yellow......24. FOENICULUM (p. 485).
Involucres present.
Fruit smooth (ribs filiform); leaves with few leaflets... 7. CARUM (p. 479).
Fruit tuberculate or bristly; leaves finely dissected........ 5. SPERMOLEPIS (p. 478).

1. SANICULA L. Self-heal.

Smooth perennial with almost naked or sparsely leafy stems and palmately cleft leaves with pinnatifid or incised lobes; involucres present; flowers greenish yellow, in irregularly compound few-rayed umbels; calyx teeth somewhat foliaceous, persistent; fruit subglobose, densely covered with hooked bristles.

1. Sanicula marilandica L. Sp. Pl. 235. 1753.
Type locality: "Habitat in Marilandia, Virginia."
Range: New England and Nebraska to New Mexico and Alabama.
New Mexico: Gallinas Planting Station (*Bartlett*); Brazos Canyon (*Standley & Bollman*). Damp woods, in the Transition Zone.

2. ERYNGIUM L.

Coarse glabrous perennials 30 cm. high or more, with rigid, coriaceous, sometimes spinulose leaves, and small flowers in involucrate heads; sepals prominent, rigid, persistent; fruit ovoid, laterally flattened, scaly or tuberculate; ribs obsolete; stylopodium wanting; oil tubes mostly 5; seed face plane.

KEY TO THE SPECIES.

Leaves pectinate-dentate or pinnatifid, the lobes spinose-tipped, not parallel-veined........................... 1. *E. wrightii*.
Leaves elongate-linear, mostly entire, parallel-veined...... 2. *E. sparganophyllum*.

1. Eryngium wrightii A. Gray, Pl. Wright. 1: 78. 1852.
Type locality: "Bed of the Limpia or Wild Rose Creek," Texas.
Range: Western Texas to southern Arizona.
New Mexico: Animas Valley; near White Water.

2. **Eryngium sparganophyllum** Hemsl. in Hook. Icon. Pl. IV. **6:** *pl. 2508.* 1897.
Eryngium longifolium A. Gray, Pl. Wright. **2:** 65. 1853, not Cav. 1793.
TYPE LOCALITY: Las Playas Springs, near the Sierra de las Animas, New Mexico. Type collected by Wright (no. 1103).
RANGE: Known only from type locality.

3. WASHINGTONIA Raf. SWEET CICELY.

Plants glabrate or pubescent, from thick aromatic roots; leaves ternately decompound, the leaflets ovate to lanceolate, coarsely toothed or cleft; calyx teeth obsolete; fruit linear-oblong or clavate, bristly on the ribs; carpels hardly flattened; oil tubes obsolete in mature fruit, often numerous in the young fruit.

KEY TO THE SPECIES.

Involucels of several bracts; fruit beaked........................ 1. *W. longistylis.*
Involucels wanting; fruit obtuse.................................. 2. *W. obtusa.*

1. **Washingtonia longistylis** (Torr.) Britton in Britt. & Brown, Illustr. Fl. **2:** 530. 1897.
Myrrhis longistylis Torr. Fl. North. & Mid. U. S. 310. 1824.
Osmorrhiza longistylis DC. Prodr. **4:** 232. 1830.
TYPE LOCALITY: Wet meadows near Albany, New York.
RANGE: Nova Scotia and North Carolina to Montana and New Mexico.
NEW MEXICO: Johnsons Mesa (*Wooton*). Damp woods, in the Transition Zone.

2. **Washingtonia obtusa** Coult. & Rose, Contr. U. S. Nat. Herb. **7:** 64. 1900.
TYPE LOCALITY: Ishowood Creek, northwestern Wyoming.
RANGE: British Columbia and Wyoming to California and New Mexico.
NEW MEXICO: Tunitcha Mountains; Chama; Santa Fe and Las Vegas mountains; Hillsboro Peak; Cloudcroft. Damp woods, in the Transition and Canadian zones.

4. APIUM L.

Erect glabrous herb with pinnately or ternately divided leaves with thick, strongly scented petioles and umbels of white flowers opposite the leaves; calyx teeth obsolete; fruit laterally flattened, ovoid or broader than long, the ribs prominent, obtuse, corky; oil tubes solitary in the intervals.

1. **Apium graveolens** L. Sp. Pl. 264. 1753. CELERY.
TYPE LOCALITY: "Habitat in Europae humectis, praesertim maritimis."
NEW MEXICO: Kingston; above Tularosa.
Escaped from cultivation.

5. SPERMOLEPIS Raf.

Slender smooth branched annuals with finely dissected leaves having filiform or linear segments, the small flowers in involucellate, very unequally few-rayed, pedunculate umbels; calyx teeth obsolete; fruit ovoid, flattened laterally, bristly or tuberculate; seed face plane.

KEY TO THE SPECIES.

Fruit tuberculate.. 1. *S. divaricatus.*
Fruit with hooked bristles.. 2. *S. echinatus.*

1. **Spermolepis divaricatus** (Walt.) Britton, Mem. Torrey Club **5:** 244. 1894.
Daucus divaricatus Walt. Fl. Carol. 114. 1788.
Leptocaulis divaricatus DC. Mém. Ombel. 39. 1829.
TYPE LOCALITY: North or South Carolina.
RANGE: North Carolina and Florida to Kansas and New Mexico.
NEW MEXICO: Las Vegas (*Dewey*). Open slopes, in the Upper Sonoran Zone.

2. **Spermolepis echinatus** (Nutt.) Heller, Bot. Expl. Texas 3. 1895.
Leptocaulis echinatus Nutt.; DC. Prodr. **4:** 107. 1830.
TYPE LOCALITY: Red River, Arkansas.
RANGE: Alabama and southern California to Mexico.
NEW MEXICO: Carrizalillo Mountains; Florida Mountains; Tres Hermanas; Mesilla Valley; Organ Mountains. Mesas and low hills, in the Upper Sonoran Zone.

6. CICUTA L. WATER HEMLOCK.

Large coarse glabrous perennial from a thickened root, with twice pinnately compound leaves having large, ovate to lanceolate, serrate leaflets, and with large umbels of white flowers; calyx teeth prominent; fruit oblong to orbicular, glabrous; carpels with strongly flattened corky ribs, the lateral ones largest; oil tubes large.

1. **Cicuta occidentalis** Greene, Pittonia **2:** 7. 1889.
TYPE LOCALITY: Trinidad, Colorado.
RANGE: Idaho and South Dakota to Nevada and New Mexico.
NEW MEXICO: San Juan Valley; Santa Fe and Las Vegas mountains; Black Range; White and Sacramento mountains. In swamps and along streams and ditches, in the Upper Sonoran and Transition zones.

7. CARUM L.

Smooth erect slender herb with tuberous or fusiform roots, pinnate leaves with few linear leaflets, involucels of several narrow bracts, and white flowers; calyx teeth prominent; fruit orbicular to oblong, glabrous; carpels with filiform inconspicuous ribs; oil tubes 2 to 6 on the commissural side, solitary in the intervals.

1. **Carum gairdneri** (Hook. & Arn.) A. Gray, Proc. Amer. Acad. **7:** 344. 1867.
Atenia gairdneri Hook. & Arn. Bot. Beechey Voy. 349. 1840.
TYPE LOCALITY: Near San Francisco or Monterey, California.
RANGE: British Columbia and California to South Dakota and New Mexico.
NEW MEXICO: West Fork of the Gila (*Metcalfe* 497).

8. ALETES Coult. & Rose.

Acaulescent glabrous perennial with pinnate leaves, broad, sharply toothed, rather distant leaflets, and yellow flowers; calyx teeth prominent; fruit oblong, glabrous, the ribs prominent, equal; oil tubes large, solitary.

1. **Aletes acaulis** (Torr.) Coult. & Rose, Rev. Umbell. 126. 1888.
Deweya? acaulis Torr. U. S. Rep. Expl. Miss. Pacif. **4^{1}:** 94. 1856.
Oreosciadium acaule A. Gray, Proc. Amer. Acad. **7:** 343. 1867.
TYPE LOCALITY: Crevices of rocks near Santa Antonita, New Mexico. Type collected by Bigelow in 1853.
RANGE: Mountains of New Mexico and Colorado.
NEW MEXICO: Socorro; Sandia Mountains. Transition Zone.

9. BERULA Hoffm.

Smooth aquatic perennial, 20 to 40 cm. high or more, with once pinnate leaves, variously cut leaflets, conspicuous involucres, and medium-sized umbels of white flowers; calyx teeth minute; fruit rotund, emarginate at the base; carpels nearly globose, with slender inconspicuous ribs and thick corky pericarp; oil tubes numerous, contiguous, closely surrounding the seed cavity; seeds terete.

1. **Berula erecta** (Huds.) Coville, Contr. U. S. Nat. Herb. **4:** 115. 1893.
Sium erectum Huds. Fl. Angl. 103. 1762.
Sium angustifolium L. Sp. Pl. ed. 2. 1672. 1763.

TYPE LOCALITY: Not stated.

RANGE: British America to California and Mexico; also in Europe.

NEW MEXICO: Zuni; Farmington; Silver City; Mogollon Mountains; Fort Bayard; Dog Spring; White Mountains; Carrizozo; Roswell. Edge of streams, in the Transition and Canadian zones.

10. LIGUSTICUM L.

Glabrous perennial from a thickened cormlike base; stems a meter high or less, with large, decompound, dark green leaves and a large terminal umbel of white flowers; calyx teeth small or obsolete; fruit oblong-ovoid, 6 to 7 mm. long, with rather prominent winged ribs; oil tubes 4 to 6 in the intervals, 8 to 10 on the commissural side.

1. **Ligusticum porteri** Coult. & Rose, Rev. Umbell. 86. 1888. CHUCHUPATE.

TYPE LOCALITY: "Head waters of the Platte," Colorado.

RANGE: Wyoming to Arizona and New Mexico.

NEW MEXICO: Sandia Mountains; Tunitcha Mountains; Sierra Grande; Santa Fe and Las Vegas mountains; Hillsboro Peak; Sawyers Peak; Mogollon Mountains; White and Sacramento mountains. Damp woods, Transition to Hudsonian Zone.

11. OREOXIS Raf.

Cespitose alpine perennials with pinnate basal leaves having narrow segments; umbels compact, more or less headlike; calyx teeth prominent; fruit globose, slightly flattened laterally; carpels with thick equal corky prominent ribs; oil tubes 1 to 3 in the narrow intervals.

KEY TO THE SPECIES.

Bracts of involucels linear, entire .... 1. *O. humilis.*
Bracts of involucels cuneate, toothed .... 2. *O. bakeri.*

1. **Oreoxis humilis** Raf. Bull. Bot. Seringe 217. 1830.

TYPE LOCALITY: "Rocky Mountains."

RANGE: Colorado and northern New Mexico.

NEW MEXICO: Top of Las Vegas Range (*Cockerell*). High mountain meadows, in the Arctic-Alpine Zone.

2. **Oreoxis bakeri** Coult. & Rose, Contr. U. S. Nat. Herb. 7: 144. 1900.

TYPE LOCALITY: Mountains near Pagosa Peak, southern Colorado.

RANGE: Southern Colorado and northern New Mexico.

NEW MEXICO: Pecos Baldy (*Standley* 4318). Meadows in the mountains, Arctic-Alpine Zone.

12. CONIOSELINUM Hoffm.

Glabrous perennial 60 to 90 cm. high, with often large leaves, these twice or thrice ternate, then once or twice pinnate, or twice pinnate, or simply pinnately compound; leaflets laciniately pinnatifid; umbels with numerous rays, involucellate with linear elongated bractlets; calyx teeth obsolete; fruit oblong, glabrous, about 6 mm. long; oil tubes usually 1 in the dorsal interval and 2 in the lateral, 2 to 4 in the commissural side; ribs prominent, the laterals winged.

1. **Conioselinum scopulorum** (A. Gray) Coult. & Rose, Contr. U. S. Nat. Herb. 7: 151. 1900.

Ligusticum scopulorum A. Gray, Proc. Amer. Acad. 7: 347. 1868.

TYPE LOCALITY: Santa Antonita, New Mexico.

RANGE: Oregon to Arizona and New Mexico.

NEW MEXICO: Santa Antonita; Chusca Mountains; Mogollon Mountains; Copper Canyon. Mountains, in the Transition Zone.

13. ANGELICA L.

Stout glabrous perennial 60 to 90 cm. high, with simply pinnate leaves having 2 to 4 pairs of ovate to narrowly lanceolate, sharply serrate to entire leaflets; flowers white or purplish, in large spreading umbels; calyx teeth mostly obsolete; fruit oblong, glabrous, 4 to 6 mm. long, the dorsal and intermediate ribs prominent, thick, the lateral ones winged; oil tubes solitary in the intervals.

1. **Angelica pinnata** S. Wats. in King, Geol. Expl. 40th Par. **5**: 126. 1871.

TYPE LOCALITY: Uinta Mountains, Utah.

RANGE: Wyoming and Montana to New Mexico.

NEW MEXICO: Tunitcha Mountains; Chama. Wet meadows and in bogs, in the Transition Zone.

14. PHELLOPTERUS Nutt.

Acaulescent perennials with pale, once to thrice pinnate leaves having mostly short broad crowded leaflets and with purplish flowers; involucre hyaline, conspicuous; calyx teeth evident; fruit orbicular, with broad wings and an oblong body 10 to 15 mm. long; oil tubes 2 to 4 in the intervals, 4 to 8 on the commissural side.

KEY TO THE SPECIES.

Bracts of involucels variously toothed at the apex.............. 1. *P. macrorhizus*.
Bracts of involucels entire.
  Involucre conspicuous; bracts and bractlets 1 to 3-nerved.. 2. *P. utahensis*.
  Involucre mostly a low hyaline sheath; bracts and bractlets many-nerved.. 3. *P. multinervatus*.

1. **Phellopterus macrorhizus** (Buckl.) Coult. & Rose, Contr. U. S. Nat. Herb. **7**: 167. 1900.

Cymopterus macrorhizus Buckl. Proc. Acad. Phila. **1861**: 455. 1862.

TYPE LOCALITY: Prairies north of Austin, Texas.

RANGE: Texas to eastern New Mexico.

NEW MEXICO: Nara Visa (*Fisher* 109). Plains, in the Upper Sonoran Zone.

2. **Phellopterus utahensis** (Jones) Woot. & Standl. Contr. U. S. Nat. Herb. **16**: 158. 1913.

Cymopterus montanus purpurascens A. Gray in Ives, Rep. Colo. Riv. 15. 1860.

Cymopterus utahensis Jones, Proc. Calif. Acad. II. **5**: 684. 1895.

Phellopterus purpurascens Coult. & Rose, Contr. U. S. Nat. Herb. **7**: 168. 1900.

TYPE LOCALITY: Above Pagumpa, Arizona.

RANGE: Idaho and Nevada to New Mexico.

NEW MEXICO: Barranca; Fort Wingate; Aztec; Sandia Mountains; Tierra Amarilla. Open slopes, in the Upper Sonoran Zone.

3. **Phellopterus multinervatus** Coult. & Rose, Contr. U. S. Nat. Herb. **7**: 169. 1900.

TYPE LOCALITY: Peach Springs, northern Arizona.

RANGE: Utah to Arizona and New Mexico.

NEW MEXICO: Mangas Springs; Carrizalillo Mountains. Upper Sonoran Zone.

15. AULOSPERMUM Coult. & Rose.

Purplish, nearly acaulescent, glabrous herb, with broadly triangular leaf blades, twice or thrice pinnately compound, and unequally 8 to 12-rayed umbels exceeding the leaves; flowers yellowish purple; calyx teeth evident; fruit orbicular, glabrous, 8 to 10 mm. long; carpels with 3 to 5 broad wings scarcely thickened at the base; carpels somewhat flattened, with broadly concave faces; oil tubes 4 or 5 in the intervals, 8 on the commissural side.

52576°—15——31

1. **Aulospermum purpureum** (S. Wats.) Coult. & Rose, Contr. U. S. Nat. Herb. 7: 178. 1900.

Cymopterus purpureus S. Wats. Amer. Nat. 7: 300. 1873.

TYPE LOCALITY: New Mexico. Type collected by Palmer.

RANGE: Colorado and Utah to New Mexico and Arizona.

NEW MEXICO: Aztec; Fort Wingate; Stinking Lake. Upper Sonoran Zone.

16. CYMOPTERUS Raf.

Low (5 to 10 cm.) glabrous perennials from thick elongated roots; leaves usually exceeding the peduncles, twice or thrice pinnate, the leaflets oblong, incised; umbels with few unequal rays; flowers yellow; involucels conspicuous, exceeding the flowers; fruit oblong; carpels each with 3 or 4 wings, these thin at the margin and near the body; oil tubes several in the intervals, 4 to 8 on the commissural side.

1. **Cymopterus fendleri** A. Gray, Mem. Amer. Acad. n. ser. 4: 56. 1849.

TYPE LOCALITY: Gravelly hills, Santa Fe, New Mexico. Type collected by Fendler (no. 274).

RANGE: New Mexico to Colorado.

NEW MEXICO: Santa Fe; Sandia Mountains. Upper Sonoran Zone.

17. PSEUDOCYMOPTERUS Coult. & Rose.

Caulescent or almost acaulescent perennials, 30 or 40 cm. high or less, from perennial roots, with bipinnate leaves, mostly no involucre, and yellow or purple flowers in usually small or medium-sized umbels; calyx teeth evident; fruit oblong, glabrous; carpels with very prominent and acute (sometimes narrowly winged) dorsal and intermediate ribs and broad thickish lateral wings; oil tubes 1 to 4 in the intervals, 2 to 8 on the commissural side.

KEY TO THE SPECIES.

Flowers purple .... 1. *P. purpureus.*
Flowers yellow.
  Ultimate divisions of the leaves short, ovate or lanceolate; plants tall, stems usually solitary .... 2. *P. montanus.*
  Ultimate divisions of the leaves linear or nearly so; plants tall or low; stems solitary or cespitose.
    Divisions of the leaves 18 to 60 mm. long, few; leaves twice compound; flowers pale yellow .... 3. *P. tenuifolius.*
    Divisions of the leaves short, usually less than 15 mm. long, very numerous, crowded; leaves mostly thrice compound; flowers bright yellow.
      Basal leaves long, 20 to 25 cm., very numerous; segments usually sessile, with a pair of lobes at the base; umbels 15 mm. wide or less .... 4. *P. filicinus.*
      Basal leaves short, usually less than 15 cm. long, few; segments long-petioled; umbels usually more than 20 mm. broad .... 5. *P. multifidus.*

1. **Pseudocymopterus purpureus** (Coult. & Rose) Rydb. Bull. Torrey Club 33: 147. 1906.

Pseudocymopterus montanus purpureus Coult. & Rose, Rev. Umbell. 75. 1888.

TYPE LOCALITY: Fort Humphreys, Arizona.

RANGE: Utah to Arizona and New Mexico.

NEW MEXICO: Burro Mountains; Bullards Peak; Mogollon Peak; Organ Mountains. Meadows and on cliffs, in the Transition Zone.

This form seems to us distinct enough to rank as a species. It certainly is much more easily recognized than most of the species of the family. We have never seen it occurring with the other species nor have we ever seen intergradient forms.

2. **Pseudocymopterus montanus** (A. Gray) Coult. & Rose, Rev. Umbell. 74. 1888.

Thaspium? montanum A. Gray, Mem. Amer. Acad. n. ser. **4:** 57. 1849.

TYPE LOCALITY: Sunny declivities at the foot of mountains, along Santa Fe Creek, New Mexico. Type collected by Fendler (no. 276).

RANGE: Wyoming to Arizona and New Mexico.

NEW MEXICO: Mountains west of Grants Station; Santa Fe and Las Vegas mountains; Black Range; White Mountains. Meadows and damp woods, chiefly in the Transition Zone.

3. **Pseudocymopterus tenuifolius** (A. Gray) Rydb. Bull. Torrey Club **33:** 147. 1906.

Thaspium? montanum tenuifolium A. Gray, Pl. Wright. **2:** 65. 1853.

Pseudocymopterus montanus tenuifolius Coult. & Rose, Rev. Umbell. 75. 1888.

TYPE LOCALITY: "Hillsides of Coppermine Creek, New Mexico." Type collected by Wright (no. 1107).

RANGE: New Mexico and Arizona.

NEW MEXICO: Mogollon Mountains and Black Range to the Organ Mountains and southward. Canyons and faces of cliffs, in the Transition Zone.

Apparently this is a very good species, distinguished from *P. montanus* by its tufted habit, much elongated, very narrow leaf segments, and pale flowers. It is found in different situations, too, preferring crevices of cliffs in the deep canyons, always growing in shade.

4. **Pseudocymopterus filicinus** Woot. & Standl. Contr. U. S. Nat. Herb. **16:** 158. 1913.

TYPE LOCALITY: Bear Mountain near Silver City, Grant County, New Mexico. Type collected by Metcalfe (no. 165).

RANGE: Mountains of western New Mexico.

NEW MEXICO: Bear Mountain; Mangas Springs; Holts Ranch; Pinos Altos.

A very handsome plant, for the family, its leaves strongly suggesting some of the ferns. It is distinguished from our other species by the very numerous leaves of peculiar form and by the small umbels which usually but slightly exceed the leaves.

5. **Pseudocymopterus multifidus** Rydb. Colo. Agr. Exp. Sta. Bull. **100:** 257. 1906.

Pseudocymopterus montanus multifidus Rydb. Bull. Torrey Club **31:** 574. 1904.

TYPE LOCALITY: Range between Sapello and Pecos rivers, New Mexico. Type collected by Cockerell in 1900.

RANGE: Colorado and New Mexico.

NEW MEXICO: Tunitcha Mountains; Chama; Jemez Mountains; Santa Fe and Las Vegas mountains; Rio Pueblo; Sandia Mountains; White and Sacramento mountains; Organ Mountains. Meadows in the mountains, Transition to the Arctic-Alpine Zone.

This becomes much larger than is suggested in the original description or in Coulter & Nelson's flora, being often 30 or 40 cm. high.

18. OXYPOLIS Raf.

Smooth erect herb 30 to 60 cm. high, from fascicled tubers; leaves simply pinnate, with 5 to 9 leaflets; flowers white; involucre and involucels wanting; calyx teeth evident; fruit ovoid, scarcely 4 mm. long, with prominent dorsal and intermediate and narrower lateral wings; oil tubes solitary in the intervals, 2 to 4 on the commissural side.

1. **Oxypolis fendleri** (A. Gray) Heller, Bull. Torrey Club **24:** 478. 1897.
Archemora fendleri A. Gray, Mem. Amer. Acad. n. ser. **4:** 56. 1849.
Tiedemannia fendleri Coult. & Rose, Rev. Umbell. 48. 1888.

TYPE LOCALITY: Margins of Santa Fe Creek, New Mexico. Type collected by Fendler (no. 272).

RANGE: Wyoming to New Mexico.

NEW MEXICO: Chama; Santa Fe and Las Vegas mountains; Rio Pueblo; West Fork of the Gila. Edges of streams and in bogs, in the Transition and Canadian zones.

19. COGSWELLIA Spreng.

Acaulescent or short-caulescent perennial with thickened roots, bipinnate leaves with small, entire or toothed, oblong segments and unequally 5 to 8-rayed umbels of whitish or purplish flowers; calyx teeth obsolete; fruit almost orbicular, emarginate at the base, glabrous, the dorsal ribs filiform, the lateral winged, coherent till maturity; oil tubes solitary in the intervals (rarely 2 in the lateral intervals), 4 on the commissural side.

1. **Cogswellia orientalis** (Coult. & Rose) Jones, Contr. West. Bot. **12:** 33. 1908.
Lomatium orientale Coult. & Rose, Contr. U. S. Nat. Herb. **7:** 220. 1900.

TYPE LOCALITY: Plains around Denver, Colorado.

RANGE: Washington and North Dakota to Kansas and Arizona.

NEW MEXICO: Mangas Springs; Organ Mountains. Dry hills, in the Upper Sonoran Zone.

20. HERACLEUM L. COW PARSNIP.

Tall stout perennial 1 to 2 meters high, more or less pubescent or woolly, with large ternately compound leaves, deciduous involucres, involucels of many bractlets, and large umbels of white flowers having obcordate petals; calyx teeth small or obsolete; fruit broadly obovate, somewhat pubescent, 8 to 12 mm. long; dorsal and intermediate ribs filiform, the lateral ones broadly winged; oil tubes about half as long as the carpels, conspicuous, 2 to 4 on the commissural side; seed much flattened dorsally.

1. **Heracleum lanatum** Michx. Fl. Bor. Amer. **1:** 166. 1803.

TYPE LOCALITY: "Canada."

RANGE: Canada to North Carolina, New Mexico, and California.

NEW MEXICO: Chama; Santa Fe and Las Vegas mountains. Along streams and in bogs, in the Transition and Canadian zones.

21. DAUCUS L. CARROT.

Bristly annuals or biennials with pinnately decompound leaves, foliaceous cleft involucral bracts, involucels of entire or toothed bractlets, and usually white flowers in concave umbels; calyx teeth obsolete; fruit oblong, flattened dorsally; carpels with 5 slender primary ribs and 4 secondary ones, each bearing a single row of prominent barbed prickles; stylopodium depressed or wanting.

1. **Daucus pusillus** Michx. Fl. Bor. Amer. **1:** 164. 1803.

TYPE LOCALITY: "In campestribus Carolinae."

RANGE: North Carolina and Florida to California and British Columbia.

NEW MEXICO: Upper Corner Monument; Nutt Mountain.

22. CORIANDRUM L. CORIANDER.

Slender glabrous annual with pinnately dissected leaves and compound umbels of white flowers; involucres none, the involucels few-parted; fruit nearly globose, the ribs filiform or acutish.

1. **Coriandrum sativum** L. Sp. Pl. 257. 1753.

TYPE LOCALITY: "Habitat in Italiae agris."

NEW MEXICO: Zuni; Mesilla Valley.

The common coriander of the gardens, whose fruit is extensively used in flavoring, occurs occasionally in waste ground, where it has escaped from cultivation.

23. PASTINACA L. PARSNIP.

Tall glabrous biennial with pinnately compound leaves and yellow flowers; calyx teeth obsolete; involucre and involucels none; fruit oval, flattened dorsally, the dorsal ribs filiform, the lateral ones expanded into broad wings; stylopodium depressed.

1. **Pastinaca sativa** L. Sp. Pl. 262. 1753.

TYPE LOCALITY: "Habitat in Europae australioris ruderalis et pascuis."

NEW MEXICO: Farmington; Santa Fe and Las Vegas mountains.

The parsnip is a common weed in many parts of the United States, but so far it is not common in New Mexico and nowhere is it a troublesome weed.

24. FOENICULUM Adans. FENNEL.

A stout glabrous aromatic herb with large leaves dissected into numerous filiform segments, large umbels of yellow flowers, and oblong glabrous fruit, terete or nearly so, with prominent ribs and solitary oil tubes.

1. **Foeniculum foeniculum** (L.) Karst. Deutsch. Fl. 837. 1880.

Anethum foeniculum L. Sp. Pl. 263. 1753.

Foeniculum vulgare Hill, Brit. Herb. 413. 1756.

TYPE LOCALITY: "Habitat in Narbonae, Armoriae, Maderae rupibus cretaccis."

NEW MEXICO: Sabinal; above Rincon.

Escaped from cultivation.

25. CYNOMARATHRUM Nutt.

Acaulescent perennial with long thick caudices and very thick long roots, the caudices thickly covered with the leaves and their persistent bases; calyx teeth evident; flowers yellow; fruit oblong, strongly flattened dorsally, with sharp or winged dorsal and intermediate ribs, and winged laterals; oil tubes mostly several in the intervals, rarely obscure.

1. **Cynomarathrum nuttallii** (A. Gray) Coult. & Rose, Contr. U. S. Nat. Herb. **7**: 245. 1900.

Seseli nuttallii A. Gray, Proc. Amer. Acad. **8**: 287. 1870.

TYPE LOCALITY: "Rocky Mountains."

RANGE: Wyoming and Nebraska to Utah and northwestern New Mexico.

NEW MEXICO: Cedar Hill (*Standley* 8025). Dry hills, in the Upper Sonoran Zone.

Probably our material represents an undescribed form, for it does not altogether agree with other material of this species. However, it is not complete enough for a thorough comparison.

26. PTERYXIA. Nutt.

A plant, apparently of this genus, but the material too poor for satisfactory determination, was collected by Standley on sandstone hills at the north end of the Carrizo Mountains, in 1911 (no. 7352). It is probably an undescribed species.

Order 39. ERICALES.

KEY TO THE FAMILIES.

Gynœcium inferior; fruit baccate or drupaceous........108. **VACCINIACEAE** (p. 489).
Gynœcium superior; fruit usually capsular.
  Herbaceous saprophytes without green leaves.....105. **MONOTROPACEAE** (p. 486).
  Herbs or shrubs with green leaves.
    Corolla of essentially distinct petals; herbs....106. **PYROLACEAE** (p. 486).
    Corolla of more or less united petals; shrubs..107. **ERICACEAE** (p. 488).

105. MONOTROPACEAE. Indian pipe Family.

Fleshy herbs with pale reddish stems destitute of chlorophyll, parasitic or saprophytic, the leaves reduced to scales; flowers similar to those of the Pyrolaceae, in enlarged racemes.

KEY TO THE GENERA.

Petals united, persistent; plants glandular, dark purplish red when growing.................................. 1. PTEROSPORA (p. 486).
Petals distinct, deciduous; plants glabrous or slightly pubescent above, not glandular, bright red when growing. 2. HYPOPITYS (p. 486).

1. PTEROSPORA Nutt. PINEDROPS.

Stems tall, 30 to 50 cm. high, slightly woody on drying, solitary, from a thick base; calyx deeply 5-parted; corolla globular, urceolate; stamens 10; disk none; stigma 5-lobed; capsule depressed-globose, 5-lobed; seeds numerous, broadly winged at the apex.

1. Pterospora andromedea Nutt. Gen. Pl. **1:** 269. 1818.
TYPE LOCALITY: "In Upper Canada near the Falls of Niagara."
RANGE: British America to California, New Mexico, and Pennsylvania.
NEW MEXICO: Tunitcha Mountains; Winsor Creek; Sandia Mountains; Gallinas Planting Station; Mogollon Creek; Sawyers Peak. Transition Zone.

2. HYPOPITYS Hill. PINESAP.

Stems thick and fleshy, 20 cm. high or less, usually several in a cluster; sepals and petals 3 to 5, the latter saccate at the base; anthers reniform, the cells completely confluent into one, opening by very unequal valves; stigma glandular on the margins.

1. Hypopitys latisquama Rydb. Bull. Torrey Club **40:** 461. 1913.
TYPE LOCALITY: Bridger Mountains, Montana.
RANGE: British Columbia and Montana to California and Mexico.
NEW MEXICO: Tunitcha Mountains; Sandia Mountains; Santa Fe and Las Vegas mountains; Middle Fork of the Gila; Black Range; White and Sacramento mountains. Damp woods, in the Canadian and Hudsonian zones.

106. PYROLACEAE. Wintergreen Family.

Low perennial herbs with simple petiolate leaves and perfect solitary, racemose, or corymbose flowers; calyx 4 or 5-merous; corolla of 5 distinct or slightly united petals; stamens twice as many as the petals; anthers opening introrsely by pores or short slits, inverted in anthesis; ovary superior; style often declined; fruit a capsule with numerous seeds.

KEY TO THE GENERA.

Stems leafy; flowers corymbose .... 1. CHIMAPHILA (p. 487).
Stems scapiform, naked; leaves basal; proper stems short; flowers not corymbose.
  Flowers racemose .... 2. PYROLA (p. 487).
  Flowers solitary .... 3. MONESES (p. 488).

1. CHIMAPHILA Pursh. PIPSISSEWA.

Low perennial herb with thick evergreen shining leaves rather crowded on the short suffruticose stems; flowers purplish, fragrant, in a terminal corymb; stamens 10; filaments enlarged and hairy at the middle; style short, inversely conical, immersed in the depressed summit of the globular ovary; stigma broad, orbicular, 5-crenate.

1. **Chimaphila umbellata** (L.) Nutt. Gen. Pl. **1**: 274. 1818.

Pyrola umbellata L. Sp. Pl. 396. 1753.

TYPE LOCALITY: "Habitat in Europae, Asiae et Americae septentrionalis sylvis."

RANGE: British America to Georgia and Mexico; also in Europe and Asia.

NEW MEXICO: Elizabethtown; Harveys Upper Ranch; Gallinas Planting Station; Mogollon Mountains. Deep woods, in the Canadian and Hudsonian zones.

2. PYROLA L. WINTERGREEN.

Small perennial evergreen herbs with basal leaves on a short scaly stem arising from a slender rhizome; flowers white, greenish, or purplish, in terminal racemes, nodding; stamens 10, the anthers emarginate or 2-beaked at the base; disk usually obsolete; ovary 5-valved from the base, the valves with cobwebby margins.

KEY TO THE SPECIES.

Style straight; stamens connivent .... 1. *P. secunda.*
Style and stamens declined.
  Leaves mottled .... 2. *P. picta.*
  Leaves green, not mottled.
    Leaves reniform, cordate; flowers pink .... 3. *P. asarifolia.*
    Leaves rounded or narrowed at the base; flowers greenish white.
      Leaf blades orbicular, coriaceous, usually shorter than the petioles .... 4. *P. chlorantha.*
      Leaf blades oval or obovate, thin, longer than the petioles .... 5. *P. elliptica.*

1. **Pyrola secunda** L. Sp. Pl. 396. 1753.

TYPE LOCALITY: "Habitat in Europa frigidiore."

RANGE: British America to California and Virginia; also in Europe and Asia.

NEW MEXICO: Tunitcha Mountains; Rio Pueblo; Winsor Creek; Harveys Upper Ranch; Mogollon Mountains. Deep woods, in the Canadian and Transition zones.

2. **Pyrola picta** J. E. Smith, Rees's Cycl. **29**: no. 8. 1814.

TYPE LOCALITY: "Found on the west coast of North America."

RANGE: British Columbia and Wyoming to California and New Mexico.

NEW MEXICO: Harveys Upper Ranch; Mogollon Mountains. Deep woods, in the Canadian and Hudsonian zones.

3. **Pyrola asarifolia** Michx. Fl. Bor. Amer. **1**: 251. 1803.

Pyrola rotundifolia asarifolia Hook. Fl. Bor. Amer. **2**: 46. 1834.

TYPE LOCALITY: Not stated.

RANGE: British America to New Mexico, Minnesota, and Massachusetts.

NEW MEXICO: Santa Fe and Las Vegas mountains. Cold woods, in the Canadian and Hudsonian zones.

4. **Pyrola chlorantha** Swartz, Svensk. Vet. Akad. Handl. **1810:** 190. *pl. 5.* 1810.
TYPE LOCALITY: Carlsberg near Stockholm, Sweden.
RANGE: British America to Oregon, New Mexico, Nebraska, and Virginia.
NEW MEXICO: Tunitcha Mountains; Sandia Mountains; Santa Fe and Las Vegas mountains. Deep woods, in the Canadian and Hudsonian zones.

5. **Pyrola elliptica** Nutt. Gen. Pl. **1:** 273. 1818. SHINLEAF.
TYPE LOCALITY: "Common around Philadelphia, and in the woods of New Jersey."
RANGE: British America to New Mexico, Illinois, and Maryland.
NEW MEXICO: Gallinas Planting Station; West Fork of the Gila. Damp woods, in the Transition and Canadian zones.

3. MONESES Salisb. ONE-FLOWERED WINTERGREEN.

Similar to Pyrola, but the 5 pure white petals widely spreading, orbicular; filaments subulate, naked, the anthers conspicuously 2-horned; style straight, exserted; stigma large, peltate, with 5 radiating lobes; valves of the capsule naked; scapes 1-flowered.

1. **Moneses uniflora** (L.) A. Gray, Man. 273. 1848.
Pyrola uniflora L. Sp. Pl. 397. 1753.
TYPE LOCALITY: "Habitat in Europae borealis silvis."
RANGE: British America to Oregon, New Mexico, and Pennsylvania.
NEW MEXICO: Top of Las Vegas Range; Rio Pecos near Truchas Peak. Deep woods, in the Hudsonian Zone.

107. ERICACEAE. Heath Family.

Shrubs or trees with scaly buds and simple alternate exstipulate leaves, usually evergreen; flowers perfect, 4 or 5-merous, in small axillary or terminal clusters; corolla urceolate or globular, 4 or 5-toothed, deciduous; stamens twice as many as the corolla lobes, included, the anthers dehiscent by terminal pores or chinks; fruit fleshy, drupaceous or berry-like.

KEY TO THE GENERA.

Ovary 5-celled, ripening into a granular-coated berry with many seeds and a firm endocarp; trees... 1. ARBUTUS (p. 488).
Ovary 4 to 10-celled, with solitary ovules, in fruit becoming a drupe with as many seedlike nutlets, or a solid stone; low shrubs, one of them prostrate.. 2. ARCTOSTAPHYLOS (p. 489).

1. ARBUTUS L.

Good-sized trees with exfoliating bark; leaves evergreen, coriaceous, alternate, petiolate; flowers small, white or flesh-colored, in a terminal cluster of racemes or panicles; calyx small, 5-parted; corolla globular to ovoid; ovules crowded on a fleshy placenta projecting from the inner angle of each cell; styles long, the stigmas obtuse.

KEY TO THE SPECIES.

Leaves elliptic-lanceolate, acute, glabrous.......................... 1. *A. arizonica.*
Leaves oblong or ovate, obtuse, permanently pubescent beneath..... 2. *A. texana.*

1. **Arbutus arizonica** (A. Gray) Sarg. Gard. & For. **4:** 317. 1891.
Arbutus xalapensis arizonica A. Gray, Syn. Fl. ed. 2. **2**[1]: 396. 1886.
TYPE LOCALITY: Mountains of southern Arizona.
RANGE: Southern New Mexico and Arizona and adjacent Mexico.
NEW MEXICO: San Luis Mountains; Animas Peak. Low mountains.

Doctor Mearns says of one of his collections: "Three large trunks sprang from a common stump, perhaps measuring 9 meters in circumference. The individual trunks measured, respectively, 1 meter above the common bole, 318, 251, and 107 cm. in circumference, and were about 20 meters in height."

2. Arbutus texana Buckl. Proc. Acad. Phila. **1861:** 460. 1862.

TYPE LOCALITY: Hays County, Texas.

RANGE: Western Texas and southeastern New Mexico.

NEW MEXICO: Queen (*Wooton*). Low hills, in the Upper Sonoran Zone.

Our specimens have broader and more pubescent leaves than the presumably typical form.

2. ARCTOSTAPHYLOS Adans.

Low erect or trailing shrubs with alternate evergreen entire leaves and small white or rose-colored flowers; flowers in structure similar to those of the preceding genus, but the ovules solitary instead of numerous in the cells.

KEY TO THE SPECIES.

Erect shrub, 1 to 2 meters high; leaves oblong, acute or acutish; branches not glandular .... 1. *A. pungens.*
Creeping shrub; leaves oblanceolate or obovate, obtuse or retuse; branches glandular .... 2. *A. uva-ursi.*

1. Arctostaphylos pungens H. B. K. Nov. Gen. & Sp. **3:** 278. *pl. 259*. 1818.

MANZANITA.

TYPE LOCALITY: "Crescit in Regno Mexicano, locis alsis, juxta Moran et Villapando, alt. 1300–1400 hex."

RANGE: Utah to California and Mexico.

NEW MEXICO: Mogollon Mountains; San Luis Mountains; Hillsboro. Lower mountains, in the Upper Sonoran Zone.

2. Arctostaphylos uva-ursi (L.) Spreng. Syst. Veg. **2:** 287. 1825. BEARBERRY.

Arbutus uva-ursi L. Sp. Pl. 395. 1753.

Daphnidostylis fendleriana Klotzsch, Linnaea **24:** 80. 1851.

TYPE LOCALITY: "Habitat in Europa frigida, Canada."

RANGE: Arctic regions to California, New Mexico, and Arizona; also in Europe and Asia.

NEW MEXICO: Tunitcha Mountains; Santa Fe and Las Vegas mountains. Moist woods, Transition to Hudsonian Zone.

The type of *Daphnidostylis fendleriana* was collected by Fendler near Santa Fe.

108. VACCINIACEAE. Huckleberry Family.

1. VACCINIUM L. BLUEBERRY.

Low shrubs, 40 cm. high or less, with slender branches and small alternate leaves; flower small, solitary, axillary; calyx lobes nearly obsolete; corolla globular, the limb 5-lobed; stamens 10; anthers prolonged upward into tubes, opening by terminal pores; fruit a reddish or black berry.

KEY TO THE SEPCIES.

Fruit purplish black; leaves usually over 15 mm. long, obtuse or rounded .... 1. *V. oreophilum.*
Fruit red; leaves usually less than 15 mm. long, acute or acutish.. 2. *V. scoparium.*

1. Vaccinium oreophilum Rydb. Bull. Torrey Club **33:** 148. 1906.

TYPE LOCALITY: Uinta Mountain, Utah.

RANGE: British Columbia and Alberta to northern New Mexico.

NEW MEXICO: Costilla Pass (*Howell* 202).

2. **Vaccinium scoparium** Leiberg, Mazama **1**: 196. 1897.

Vaccinium erythrococcum Rydb. Mem. N. Y. Bot. Gard. **1**: 301. 1900.

TYPE LOCALITY: "Alpine woods near the Height of Land and Columbia Portage."

RANGE: British Columbia and Alberta to California and northern New Mexico.

NEW MEXICO: Winsor Creek; Baldy; Harveys Upper Ranch. Deep woods, in the Canadian and Hudsonian zones.

This is one of the few plants found in the dense spruce forests. The berries are of good flavor and are frequently gathered for food.

Order 40. PRIMULALES.

KEY TO THE FAMILIES.

Styles distinct; fruit an achene or utricle; ovule 1. **109. PLUMBAGINACEAE** (p. 490).
Styles united; fruit capsular or drupelike; ovules several........................... **110. PRIMULACEAE** (p. 490).

109. PLUMBAGINACEAE. Plumbago Family.

1. LIMONIUM Adans.

Acaulescent herbaceous perennial, 30 cm. high or less, with leathery entire basal leaves and a corymbosely branched inflorescence of small white flowers; leaves oblanceolate-spatulate; flowers sessile; calyx funnelform, the tube 10-ribbed, the limb scarious, plicate; petals 5, clawed, the claws united at the base; ovary 1-celled, 5-angled; styles 5, distinct; fruit a utricle, the seed filling the cavity.

1. **Limonium limbatum** Small, Bull. Torrey Club **25**: 317. 1898.

Statice limbata Schum. Bot. Jahrb. Engler **26**: 390. 1900.

TYPE LOCALITY: "In alkaline soil, Texas and New Mexico." Type, Wright's 1435.

RANGE: Texas and southeastern New Mexico.

NEW MEXICO: Tularosa; Belen; plains near the White Sands; Roswell. Alkaline soil, in the Lower Sonoran Zone.

110. PRIMULACEAE. Primrose Family.

Annual or perennial herbs, most of them low, some of them very small; leaves mostly simple, alternate, opposite, or whorled, basal or cauline, exstipulate; inflorescence various; flowers perfect, usually regular; calyx of 4 to 9 partially united sepals, usually persistent; corolla gamopetalous, hypogynous (except in Samolus); stamens as many as the corolla lobes and opposite them, attached to the tube; ovary free (except in Samolus), 1-celled, with a central placenta; fruit a capsule with 2 to 8 valves.

KEY TO THE GENERA.

Ovary inferior, attached to the base of the calyx and corolla tubes.................................... 1. SAMOLUS (p. 491).
Ovary superior, free from all other flower parts.
    Leafy-stemmed plants.
        Capsule opening lengthwise; flowers yellow.... 2. STEIRONEMA (p. 491).
        Capsule circumscissile; flowers reddish........ 3. ANAGALLIS (p. 492).
    Scapose or very short-stemmed plants, the leaves all basal.
        Corolla lobes reflexed; stamens exserted, connivent in a cone or somewhat monadelphous................................ 4. DODECATHEON (p. 492).

Corolla lobes erect or spreading; stamens included, distinct.
Corolla tube equaling or exceeding the calyx, bright pink or rose purple to crimson; flowers showy............ 5. PRIMULA (p. 492).
Corolla tube shorter than the calyx, white or yellow; flowers very small.
Perennials, the stems branched at the base; capsules few-seeded....... 7. DROSACE (p. 494).
Annuals, the stems simple; capsules many-seeded.................... 6. ANDROSACE (p. 493).

1. SAMOLUS L. BROOKWEED.

Annual or perennial caulescent herbs with alternate entire fleshy leaves and small white flowers in racemes or panicles; sepals united at the base, persistent, adherent to the ovary below; corolla perigynous, the tube very short; stamens 5, adnate to the corolla, sometimes alternating with 5 staminodia; ovary 1-celled; capsule 5-valved, hardly elongated; seeds numerous.

KEY TO THE SPECIES.

Racemes sessile or nearly so, numerous, glabrous; calyx 2.5 mm. broad.. 1. *S. floribundus.*
Racemes long-pedunculate, few, glandular; calyx 3.5 to 5 mm. broad.. 2. *S. cuneatus.*

1. Samolus floribundus H. B. K. Nov. Gen. & Sp. **2:** 181. 1817.
TYPE LOCALITY: "Crescit in maritimis Peruviae juxta portum Callao de Lima."
RANGE: British America to California, Florida, Mexico, and South America.
NEW MEXICO: Guadalupe Canyon; Roswell. Wet ground.

2. Samolus cuneatus Small, Bull. Torrey Club **24:** 491. 1897.
TYPE LOCALITY: "On limestone rocks or soil, Texas."
RANGE: Texas and southeastern New Mexico.
NEW MEXICO: Roswell; Carlsbad; Dona Ana Mountains. Wet ground, in the Upper Sonoran Zone.

2. STEIRONEMA Raf. LOOSESTRIFE.

Leafy-stemmed perennial herbs with opposite petiolate simple entire leaves and axillary yellow flowers on slender pedicels; flowers 5-merous; corolla rotate, 5-lobed, each lobe convolute or involute about its stamen; stamens 5, alternating with 5 staminodia; capsules rounded, naked; seeds 10 to 20.

KEY TO THE SPECIES.

Leaf blades ovate, ciliate.. 1. *S. ciliatum.*
Leaf blades lanceolate, not ciliate.................................. 2. *S. validulum.*

1. Steironema ciliatum (L.) Raf. Ann. Gén. Phys. **7:** 192. 1820.
Lysimachia ciliata L. Sp. Pl. 147. 1753.
TYPE LOCALITY: "Habitat in Virginia, Canada."
RANGE: British America to Arizona, Alabama, and Georgia.
NEW MEXICO: Pecos; Las Vegas; Beulah; Gallinas Planting Station. Wet ground, in the Transition Zone.

2. Steironema validulum Greene, Contr. U. S. Nat. Herb. **16:** 158. 1913.
TYPE LOCALITY: Along Oak Creek, near Flagstaff, Arizona.
RANGE: New Mexico and Arizona.
NEW MEXICO: Mogollon Mountains. Wet soil.

3. ANAGALLIS L. PIMPERNEL.

Slender branched leafy-stemmed annual with entire, mostly opposite leaves and small, axillary, scarlet to white flowers on slender peduncles 1 to 2 cm. long.

1. **Anagallis arvensis** L. Sp. Pl. 148. 1753.
TYPE LOCALITY: "Habitat in Europae arvis."
NEW MEXICO: Kingston (*Metcalfe* 1339).
Widely introduced into North America from Europe.

4. DODECATHEON L. SHOOTING STAR.

Showy perennial herbs with short rootstocks, smooth entire leaves forming a rosette, and rose-colored, violet, or white flowers on an umbellate scape; calyx 5-merous, narrow, the lobes reflexed in flower, longer than the tube; corolla tube short, the lobes reflexed; stamens 5, exserted; anthers large, the filaments short or obsolete, stout, united at the base; ovary free; capsule partially 5-valved.

KEY TO THE SPECIES.

Anthers on conspicuous filaments; petals rose-colored............... 1. *D. radicatum.*
Anthers sessile; petals white... 2. *D. ellisiae.*

1. **Dodecatheon radicatum** Greene, Erythea **3**: 37. 1895.
TYPE LOCALITY: Near Santa Fe, New Mexico. Type collected by Fendler (no. 549).
RANGE: Wyoming and South Dakota to Kansas and New Mexico.
NEW MEXICO: Chama; Santa Fe and Las Vegas mountains; White Mountains. In wet meadows, from the Transition to the Arctic-Alpine Zone.

2. **Dodecatheon ellisiae** Standley, Proc. Biol. Soc. Washington **26**: 195. 1913.
TYPE LOCALITY: Capulin Canyon, Sandia Mountains.
NEW MEXICO: Type collected by Miss Charlotte C. Ellis (no. 330).
RANGE: Known only from type locality.

5. PRIMULA L. PRIMROSE.

Perennial scapose herbs; leaves radical, forming a thick tuft, mostly glabrous; flowers in umbels surmounting the usually stout scapes (these sometimes only 1 or 2-flowered); calyx oblong to campanulate, farinaceous, accrescent and persistently surrounding the fruit; corolla narrowly funnelform or salverform, the tube longer than the calyx, the limb of various shades of pink and rose purple or lilac purple; stamens 5, distinct, epipetalous; capsules 5-valved at the summit, many-seeded.

KEY TO THE SPECIES.

Scapes with 1 or 2 flowers; plants 5 cm. high or less............. 1. *P. angustifolia.*
Scapes with 3 to many flowers; plants 10 cm. high or more.
  Plants 25 to 40 cm. high, stout; leaves 3 to 5 cm. wide, usually entire.. 2. *P. parryi.*
  Plants less than 25 cm. high, slender; leaves less than 2 cm. wide, evidently denticulate.
    Scapes about equaling the leaves; calyx 7 mm. high..... 3. *P. ellisiae.*
    Scapes twice as long as the leaves; calyx 4 to 5 mm. high.. 4. *P. rusbyi.*

1. **Primula angustifolia** Torr. Ann. Lyc. N. Y. **1**: 34. *pl. 3.* 1824.
Primula angustifolia helenae Pollard & Cockerell, Proc. Biol. Soc. Washington **15**: 179. 1902.
TYPE LOCALITY: James Peak, Colorado.
RANGE: Colorado and northern New Mexico.

New Mexico: Santa Fe and Las Vegas mountains. High mountain meadows, in the Arctic-Alpine Zone.

The type of *P. angustifolia helenae* came from the Las Vegas Mountains.

2. **Primula parryi** A. Gray, Amer. Journ. Sci. II. **34:** 257. 1862.

Type locality: Rocky Mountains of Colorado.

Range: Colorado to Nevada and New Mexico.

New Mexico: Santa Fe and Las Vegas mountains. Bogs in the mountains, in the Hudsonian and Arctic-Alpine zones.

This is one of our most beautiful native plants, with its many-flowered clusters of large bright reddish purple flowers. It grows in wet soil near the snow banks and in bogs lower down.

3. **Primula ellisiae** Pollard & Cockerell, Proc. Biol. Soc. Washington **15:** 178. 1902.

Type locality: Sandia Mountains, New Mexico. Type collected by Miss Charlotte C. Ellis (no. 3).

Range: High mountain meadows in the Sandia and White mountains of New Mexico, in the Hudsonian Zone.

4. **Primula rusbyi** Greene, Bull. Torrey Club **8:** 122. 1881.

Primula serra Small, Bull. Torrey Club **25:** 319. 1898.

Type locality: On rich moist slopes near the summits of the Mogollon Mountains, New Mexico. Type collected by Rusby (no. 252).

Range: Mountains of New Mexico and Arizona.

New Mexico: Black Range; Mogollon Mountains. Canadian and Hudsonian zones.

6. ANDROSACE L.

Small annuals with a rosette of basal leaves and scapose umbels of very small white flowers, sometimes tinged with pink; calyx 5-lobed, persistent; corolla short-salverform or funnelform, the tube varying in length, the limb 5-lobed; stamens 5, distinct, included; capsule short, 5-valved, with many seeds.

KEY TO THE SPECIES.

Bracts of the involucre ovate or oblong.
- Calyx lobes triangular, acute .... 1. *A. occidentalis.*
- Calyx lobes broadly oblong to ovate, obtuse .... 2. *A. platysepala.*

Bracts of the involucre lanceolate to subulate.
- Corolla longer than the calyx.
  - Peduncles 10 to 20 cm. high, much longer than the ascending or erect pedicels........................ 3. *A. pinetorum.*
  - Peduncles 3 cm. high or less, often equaled or exceeded by the spreading pedicels...................... 4. *A. subumbellata.*
- Corolla shorter than the calyx.
  - Plants abundantly glandular............................ 5. *A. glandulosa.*
  - Plants not glandular.
    - Pedicels and calyx lobes glabrous or nearly so..... 6. *A. diffusa.*
    - Pedicels and calyx lobes puberulent.
      - Calyx longer than the mature capsule, the lobes spreading, equaling the tube .... 7. *A. subulifera.*
      - Calyx shorter than the mature capsule, the lobes erect, shorter than the tube............. 8. *A. puberulenta.*

1. **Androsace occidentalis** Pursh, Fl. Amer. Sept. 137. 1814.

Type locality: "On the banks of the Missouri."

Range: Montana and Manitoba to California, New Mexico, and Missouri.

New Mexico: Chama; Tierra Amarilla; Organ Mountains. Open slopes, in the Upper Sonoran Zone.

2. **Androsace platysepala** Woot. & Standl. Bull. Torrey Club **34:** 519. 1907.

TYPE LOCALITY: Kingston, Sierra County, New Mexico. Type collected by Metcalfe (no. 1547).

RANGE: Southwestern New Mexico.

NEW MEXICO: Kingston; Bear Mountain. Upper Sonoran Zone.

3. **Androsace pinetorum** Greene, Pittonia **4:** 149. 1900.

Androsace septentrionalis pinetorum Knuth in Engl. Pflanzenreich **22:** 215. 1905.

TYPE LOCALITY: "In pine woods of Graham's Park, Rio de los Pinos, at 7,800 feet, southern Colorado."

RANGE: British America to Arizona and New Mexico.

NEW MEXICO: Stinking Lake; Tierra Amarilla; Sandia Mountains; Dulce; Santa Fe Canyon; Wheelers Ranch; South Percha Creek. Shaded slopes in the mountains, in the Transition Zone.

4. **Androsace subumbellata** (A. Nels.) Small, Bull. Torrey Club **25:** 319. 1898.

Androsace septentrionalis subumbellata A. Nels. Wyo. Agr. Exp. Sta. Bull. **28:** 149. 1896.

TYPE LOCALITY: On a grassy hillside near the summit of Union Peak, Wyoming.

RANGE: Montana to Arizona and New Mexico.

NEW MEXICO: Santa Fe and Las Vegas mountains. Wooded slopes and canyons, Transition to Arctic-Alpine Zone.

5. **Androsace glandulosa** Woot. & Standl. Bull. Torrey Club **34:** 519. 1907.

TYPE LOCALITY: Middle Fork of the Rio Gila, New Mexico. Type collected by Wooton, August 5, 1900.

RANGE: Mountains of southwestern New Mexico and southeastern Arizona.

NEW MEXICO: Middle Fork of the Gila; East Canyon.

6. **Androsace diffusa** Small, Bull. Torrey Club **25:** 318. 1898.

Androsace septentrionalis diffusa Knuth in Engl. Pflanzenreich **22:** 215. 1905.

TYPE LOCALITY: "In rocky soil, western Arctic America to the Dakotas, New Mexico, and Arizona."

RANGE: British America to New Mexico and Arizona.

NEW MEXICO: Chama; Tierra Amarilla; Santa Fe and Las Vegas mountains; Albuquerque; Black Range; Sacramento Mountains. Shaded slopes and canyons, in the Transition and Canadian zones.

7. **Androsace subulifera** (A. Gray) Rydb. Bull. Torrey Club **33:** 148. 1906.

Androsace septentrionalis subulifera A. Gray, Syn. Fl. **2**[1]**:** 60. 1878.

TYPE LOCALITY: Mountains near Boulder, Colorado.

RANGE: Montana to New Mexico.

NEW MEXICO: Rio Pueblo; Santa Fe and Las Vegas mountains; Mogollon Mountains; White and Sacramento mountains. Shaded slopes, Transition to Arctic-Alpine Zone.

8. **Androsace puberulenta** Rydb. Bull. Torrey Club **30:** 260. 1903.

TYPE LOCALITY: Southern Colorado, near Veta Pass.

RANGE: Manitoba and Alberta to New Mexico.

NEW MEXICO: Grass Mountain; Chama; Tierra Amarilla; James Canyon. Open slopes, in the Transition and Canadian zones.

7. DROSACE A. Nels.

Low herb very similar to the preceding genus, but perennial, with cespitose branched stems and few-seeded capsules; umbels subcapitate.

1. **Drosace carinata** (Torr.) A. Nels. in Coulter, New Man. Rocky Mount. 374. 1909.

Androsace carinata Torr. Ann. Lyc. N. Y. **1:** 30. *pl. 3. f. 1.* 1824.

Androsace chamaejasme carinata Knuth in Engl. Pflanzenreich **22:** 190. 1905.

TYPE LOCALITY: James Peak, Colorado.
RANGE: Alberta to New Mexico.
NEW MEXICO: Taos Mountains; Santa Fe and Las Vegas mountains. Summits of mountains, in the Arctic-Alpine Zone.

Order 41. EBENALES.

111. SAPOTACEAE. Sapodilla Family.

1. BUMELIA Swartz. BUCKTHORN.

Low tree with rigid spreading branches; leaves alternate, simple, entire; flowers in few-flowered axillary fascicles, perfect, 5-merous; calyx persistent; corolla white, deciduous, the lobes longer than the tube; stamens 5, epipetalous; staminodia 5, petaloid; ovary 5-celled; fruit drupelike.

1. **Bumelia rigida** (A. Gray) Small, Bull. N. Y. Bot. Gard. **1**: 444. 1900.
Bumelia lanuginosa rigida A. Gray, Syn. Fl. **2**[1]: 68. 1878.
TYPE LOCALITY: "S. Texas to S. Arizona."
RANGE: Western Texas to southern Arizona.
NEW MEXICO: Dog Spring; Deer Creek. Dry hills, in the Upper Sonoran Zone.

Order 42. OLEALES.

112. OLEACEAE. Olive Family.

Trees, shrubs, or herbaceous perennials, with opposite (rarely alternate), simple or pinnate, exstipulate leaves and regular, 2 to 4-parted, perfect, polygamous, or diœcious flowers in panicles, corymbs, or fascicles; calyx usually small (sometimes wanting), of 4 or more sepals; corolla of 2 to 6 distinct petals or gamopetalous; stamens 2 to 4, adnate to the base of the corolla; ovary superior, 2-celled; fruit a capsule, samara, berry, or drupe.

KEY TO THE GENERA.

Fruit fleshy, a small, bluish black drupe; flowers apetalous, polygamo-diœcious; good sized shrubs...... 1. FORESTIERA (p. 495).
Fruit dry, a capsule or samara; flowers various; herbs, trees, or shrubs.
  Fruit a samara; flowers diœcious; trees with mostly pinnate leaves.............................. 2. FRAXINUS (p. 496).
  Fruit a thin-walled capsule; flowers perfect; low, herbaceous or suffrutescent plants with simple leaves and bright yellow flowers.
    Corolla rotate or campanulate; stamens exserted; filaments filiform.......................... 3. MENODORA (p. 497).
    Corolla salverform, with a long tube; stamens included, the anthers nearly sessile..... 4. MENODOROPSIS (p. 497).

1. FORESTIERA Poir. IRONWOOD.

Rather large shrubs with divaricately branching stems bearing broad simple leaves and inconspicuous polygamo-diœcious flowers in lateral clusters; flowers appearing before the leaves on stems of the previous year; calyx usually present but small; corolla mostly wanting; stamens 2 to 4; fruit a blue black drupe.

KEY TO THE SPECIES.

Leaves glabrous.. 1. *F. neomexicana.*
Leaves pubescent.. 2. *F. pubescens.*

1. **Forestiera neomexicana** A. Gray, Proc. Amer. Acad. **12:** 63. 1876.

Adelia neomexicana Kuntze, Rev. Gen. Pl. **1:** 410. 1891.

TYPE LOCALITY: New Mexico. Type collected by Fendler.

RANGE: Colorado to Arizona and western Texas.

NEW MEXICO: San Juan Valley; Carrizo Mountains; Pajarito Park; Craters; Puertecito; Mangas Springs; Sapello Creek; Rio Negrito. River valleys, in the Upper Sonoran Zone.

2. **Forestiera pubescens** Nutt. Trans. Amer. Phil. Soc. n. ser. **5:** 177. 1837.

Adelia pubescens Kuntze, Rev. Gen. Pl. **1:** 410. 1891.

TYPE LOCALITY: "In the prairies of Red River."

RANGE: Oklahoma and Texas to eastern New Mexico.

NEW MEXICO: Queen; Lincoln National Forest. Upper Sonoran Zone.

2. FRAXINUS L. ASH.

Large or small trees with opposite pinnate leaves and inconspicuous diœcious flowers in clusters or panicles; calyx with a short tube and 4 unequal lobes or wanting; stamens 2 to 4, the filaments short or elongated; ovary 2-celled, the styles united, the stigma 2-cleft; fruit a samara, with flat or terete body and a single wing partly surrounding it.

KEY TO THE SPECIES.

Leaves usually simple.. 4. *F. anomala.*

Leaves pinnate.

Flowers with a 4-parted corolla; leaflets small, 35 mm. long or less. 1. *F. cuspidata.*

Flowers apetalous; leaflets more than 40 mm. long.

Leaflets sessile or nearly so.............................. 2. *F. velutina.*

Leaflets distinctly stalked.............................. 3. *F. attenuata.*

1. **Fraxinus cuspidata** Torr. U. S. & Mex. Bound. Bot. 166. 1859. FLOWERING ASH.

TYPE LOCALITY: "Eagle Mountains and Great Cañon of the Rio Grande," Texas.

RANGE: Western Texas to Arizona and adjacent Mexico.

NEW MEXICO: Grant; McCarthy Station; San Andreas Mountains; Big Hatchet Mountains. Upper Sonoran Zone.

2. **Fraxinus velutina** Torr. in Emory, Mil. Reconn. 149. 1848.

Fraxinus pistaciaefolia Torr. U. S. Rep. Expl. Miss. Pacif. **4:** 128. 1856.

TYPE LOCALITY: "Between the waters of the Del Norte and the Gila," New Mexico. Type collected by Emory in 1847.

RANGE: Western Texas to Arizona.

NEW MEXICO: San Luis Mountains, Florida Mountains; Animas Mountains; Organ Mountains; White Mountains. Low mountains, Upper Sonoran and Transition zones.

3. **Fraxinus attenuata** Jones, Contr. West. Bot. **12:** 59. 1908.

Fraxinus toumeyi Britt. & Shaf. N. Amer. Trees 803. 1908.

TYPE LOCALITY: Valley of Palms, Lower California, and Santa Catalina Mountains, Arizona.

RANGE: New Mexico, Arizona, and adjacent Mexico.

NEW MEXICO: Mogollon Mountains; Black Range; Dog Mountains; San Luis Mountains; Organ Mountains; Las Palomas. Upper Sonoran and Transition zones.

4. **Fraxinus anomala** Torr.; S. Wats. in King, Geol. Expl. 40th Par. **5:** 283. 1871.

TYPE LOCALITY: "In Labyrinth Cañon on the Colorado River, Utah."

RANGE: Utah and Nevada to Colorado and northwestern New Mexico.

NEW MEXICO: Carrizo Mountains (*Standley* 7316). Dry rocky hills, in the Upper Sonoran Zone.

3. MENODORA Humb. & Bonpl.

Low herbs, suffrutescent at the base, with simple entire leaves and bright yellow flowers; calyx persistent, with a short tube and 5 to 15 narrow lobes; corolla rotate or short-campanulate, with 5 or 6 lobes; stamens 2 or 3, exserted, on slender filiform filaments; ovary 2-celled; stigma capitate; fruit didymous, circumscissile near the middle.

KEY TO THE SPECIES.

Plants glabrous throughout.. 1. *M. laevis.*
Plants scabrous-puberulent, rough to the touch........................ 2. *M. scabra.*

1. Menodora laevis Woot. & Standl. Contr. U. S. Nat. Herb. **16:** 158. 1913.

TYPE LOCALITY: Organ Mountains, New Mexico. Type collected by G. R. Vasey in 1881.

RANGE: Low mountains of southern New Mexico.

NEW MEXICO: Organ Mountains; Duck Creek Flats; La Luz Canyon. Upper Sonoran Zone.

2. Menodora scabra A. Gray, Amer. Journ. Sci. II. **14:** 43. 1852.

TYPE LOCALITY: New Mexico. Type collected by Wislizenus in 1846.

RANGE: Western Texas to Arizona and southward.

NEW MEXICO: Santa Fe; Las Vegas; Zuni Reservation; Socorro; Pinos Altos; Albuquerque; Silver City; Rincon; Tortugas Mountain; Organ Mountains. Dry hills, in the Upper and Lower Sonoran zones.

4. MENODOROPSIS Small.

Low suffrutescent herb, about 30 cm. high, with tufted stems, simple, mostly opposite leaves, and conspicuous bright yellow flowers with long-salverform corollas; calyx pediceled, ribbed, 10-lobed; stamens included, the anthers nearly sessile on the throat of the corolla; capsule didymous, circumscissile near the middle.

1. Menodoropsis longiflora (A. Gray) Small, Fl. Southeast. U. S. 917. 1903.

Menodora longiflora A. Gray, Amer. Journ. Sci. II. **14:** 45. 1852.

TYPE LOCALITY: Texas.

RANGE: Western Texas to southeastern New Mexico.

NEW MEXICO: Queen (*Wooton*). Dry hills, in the Upper Sonoran Zone.

Order 43. GENTIANALES.

113. GENTIANACEAE. Gentian Family.

Smooth herbs with bitter colorless juice; leaves opposite, rarely alternate or verticillate, exstipulate; flowers perfect, regular; calyx 4 to 12-lobed or toothed, often marcescent; corolla gamopetalous, 4 to 12-lobed or toothed; stamens as many as the corolla lobes and alternate with them, epipetalous; ovary superior, 1-celled, rarely 2-celled, with parietal placentæ or the whole wall ovuliferous; capsule dehiscent through the placentæ; seeds numerous.

KEY TO THE GENERA.

Styles filiform, mostly deciduous; anthers recurved or twisted at maturity.
  Corolla small, red, rose, or yellowish, the tube surpassing the calyx; anthers spirally twisted.................................. 1. CENTAURIUM (p. 498).

Corolla large, blue, purple, or white, the tube much shorter than the calyx; stamens recurved .... 2. EUSTOMA (p. 499).

Styles stout and short or none; stamens straight.

Corolla with nectariferous glands, pits, or scales.

Corolla campanulate; flowers yellow; low annual .... 3. HALENIA (p. 499).

Corolla rotate; flowers yellow or blue; perennials.

Style manifest; leaves verticillate; corolla with a crown at the base..... 4. FRASERA (p. 499).

Style wanting; leaves opposite, occasionally alternate; corolla without a crown .... 5. SWERTIA (p. 500).

Corolla without nectariferous glands, pits, or scales.

Calyx parted to near the base; corolla rotate or nearly so; stamens inserted on the base of the corolla .... 10. PLEUROGYNA (p. 503).

Calyx merely lobed; corolla not rotate; stamens inserted in the corolla tube.

Corolla without plaits or lobes at the sinuses; calyx without an intercalycine membrane; sepals imbricated.

Flowers 4-merous, rather large, usually over 3 cm. long; corolla lobes fringed or toothed; inner sepals broader, membranous-margined .... 6. ANTHOPOGON (p. 501).

Flowers 5-merous (seldom 4-merous), small, less than 2 cm. long; outer sepals broader; corolla lobes never fringed, rarely toothed .... 7. AMARELLA (p. 501).

Corolla plicate in the sinuses, the plaits more or less extended into membranous lobes or teeth; calyx with an intercalycine membrane, its lobes valvate.

Dwarf annuals or biennials; flowers solitary, terminal; anthers cordate, versatile .... 8. CHONDROPHYLLA (p. 502).

Perennials; flowers short-pedunculate, at least some of them axillary; anthers linear or oblong, extrorse .... 9. DASYSTEPHANA (p. 502).

1. CENTAURIUM Hill. CENTAURY.

Low branched annuals with bright rose pink flowers; calyx 4 or 5-parted, with slender keeled divisions; corolla salverform, with 4 or 5 lobes; stamens 4 or 5, partly adnate to the corolla tube; anthers twisted at maturity; ovary 1-celled; stigmas 2; capsule 2-valved, oblong.

KEY TO THE SPECIES.

Plants low, less than 15 cm. high; flowers small, the corolla lobes much shorter than the tube; anthers oblong.................. 1. *C. texense.*
Plants larger, usually more than 20 cm. high; flowers large, the corolla lobes only slightly shorter than the tube; anthers linear.. 2. *C. calycosum.*

1. Centaurium texense (Griseb.) Fernald, Rhodora **10:** 54. 1908.
Erythraea texensis Griseb. in DC. Prodr. **9:** 58. 1845.
TYPE LOCALITY: "In rep. Texas pr. S. Felipe."
RANGE: Texas to southeastern New Mexico.
NEW MEXICO: White Sands (*Wooton & Standley*). Lower Sonoran Zone.

2. Centaurium calycosum (Buckl.) Fernald, Rhodora **10:** 54. 1908.
Erythraea calycosa Buckl. Proc. Acad. Phila **1862:** 7. 1863.
TYPE LOCALITY: North of Fort Mason, Texas.
RANGE: Texas to Arizona.
NEW MEXICO: Carrizo Mountains; Farmington; Gila Hot Springs; Mesilla Valley. Wet ground, in the Lower and Upper Sonoran zones.
A plant with white flowers was collected at Mesquite Lake (*Wooton & Standley* 3933).

2. EUSTOMA Salisb.

Glaucous perennial, often 60 cm. high, with opposite entire sessile clasping thick leaves and large bluish flowers; calyx 5-parted (rarely 6-parted), with narrow keeled lobes; corolla campanulate-funnelform, with deeply 5 or 6-lobed limb; anthers oblong, versatile, recurved in age; ovary 1-celled, the ovules numerous.

1. Eustoma russellianum (Hook.) Griseb. in DC. Prodr. **9:** 51. 1845.
Lisianthus glaucifolius Jacq. err. det. Nutt. Trans. Amer. Phil. Soc. n. ser. **5:** 197. 1837.
Lisianthus russellianus Hook. Curtis's Bot. Mag. **65:** *pl. 3626.* 1839.
TYPE LOCALITY: "On the sandy banks of the Great Salt River of Arkansas."
RANGE: Nebraska and Colorado to New Mexico and Louisiana; also in Mexico.
NEW MEXICO: Sabinal; Shalam; White Sands; Tularosa Creek; Roswell. Alkaline soil, in the Lower Sonoran Zone.
A form with white flowers is common about the White Sands.

3. HALENIA Borkh.

Low annual, 10 to 20 cm. high, with linear opposite leaves and loose cymes of small yellow 4-merous flowers; sepals linear-lanceolate; corolla 8 to 10 mm. long, the lobes ovate; spurs divaricate-ascending, shorter than the corolla.

1. Halenia rothrockii A. Gray, Proc. Amer. Acad. **11:** 84. 1876.
TYPE LOCALITY: Mount Graham, Arizona.
RANGE: Southern Arizona and New Mexico to Mexico.
NEW MEXICO: Mogollon Mountains (*Rusby* 264).

4. FRASERA Walt. DEER'S EARS.[1]

Biennial or perennial herbs with tall erect hollow stems and entire, opposite or verticillate leaves; flowers numerous, in paniculate or thyrsoid cymes; calyx 4-lobed, the lobes narrow; corolla rotate, dull whitish or yellowish, with 4 lobes and 1 or 2 fringed nectariferous glands; stamens 4, adnate to the throat of the corolla; filaments distinct, or united at the base, the anthers versatile; ovary 1-celled, 2-valved; capsule ovoid, leathery, often flattened; seeds flattened, margined or winged.

---

[1] A translation of the Navaho name.

KEY TO THE SPECIES.

Inflorescence loosely paniculate; leaves with cartilaginous white margins; petals 12 mm. long or less........................ 1. *F. paniculata.*
Inflorescence a dense leafy thyrse; leaves not margined; petals more than 12 mm. long.
  Leaves scabrous-puberulent, conspicuously veined............ 2. *F. venosa.*
  Leaves glabrous, not conspicuously veined.
    Sepals linear, much exceeding the petals.................. 3. *F. stenosepala.*
    Sepals linear-lanceolate, scarcely equaling the petals...... 4. *F. speciosa.*

1. **Frasera paniculata** Torr. U. S. Rep. Expl. Miss. Pacif. **4**: 126. 1856.

TYPE LOCALITY: Sand bluffs, Inscription Rock, New Mexico.

RANGE: Arizona and New Mexico.

We have seen no further specimens of this plant from New Mexico and only a few from Arizona; apparently it is very rare. The type was collected by Bigelow in 1853.

2. **Frasera venosa** Greene, Pittonia **4**: 185. 1900.

Frasera speciosa scabra Jones, Zoe **4**: 277. 1893.

Frasera scabra Rydb. Bull. Torrey Club **33**: 149. 1906.

TYPE LOCALITY: Hills near Santa Rita del Cobre, New Mexico. Type collected by Greene in 1880.

RANGE: Colorado to Arizona and New Mexico.

NEW MEXICO: Tunitcha Mountains; Sierra Grande; Kingston; West Fork of the Gila; Santa Rita; White Mountains. Mountains, in the Transition Zone.

The Sierra Grande specimen has longer sepals than our others, and comes from well outside the usual range of the species.

3. **Frasera stenosepala** Rydb. Bull. Torrey Club **33**: 149. 1906.

Frasera speciosa stenosepala Rydb. Bull. Torrey Club **31**: 632. 1904.

TYPE LOCALITY: Foothills, Larimer County, Colorado.

RANGE: Wyoming to northern New Mexico.

NEW MEXICO: Santa Fe and Las Vegas mountains. Wet meadows and along streams, in the Transition and Canadian zones.

4. **Frasera speciosa** Dougl.; Griseb. in Hook. Fl. Bor. Amer. **2**: 66. *pl. 153*. 1838.

TYPE LOCALITY: "On the low hills near Spokan and Salmon Rivers and subalpine parts of the Blue Mountains, near the Kooskooka River."

RANGE: Oregon and South Dakota to California and New Mexico.

NEW MEXICO: Chama; White and Sacramento mountains. Wet ground in the mountains, in the Transition and Canadian zones.

5. SWERTIA L.

Simple-stemmed herbaceous perennial 20 to 30 cm. high, with opposite or sometimes alternate leaves, at least the lower tapering into petioles; flowers 5-merous or 4-merous, dark blue; sepals subulate-lanceolate, about half as long as the petals; corolla rotate, the lobes about 1 cm. long; glands orbicular, the appendages 10 or fewer; capsules ovoid, the seeds lenticular, winged.

1. **Swertia palustris** A. Nels. Bull. Torrey Club **28**: 227. 1901.

TYPE LOCALITY: Nashs Fork, Wyoming.

RANGE: Wyoming to northern New Mexico.

NEW MEXICO: Taos Mountains; Santa Fe and Las Vegas mountains; Costilla Valley. Bogs in the high mountains, in the Hudsonian and Arctic-Alpine zones.

6. **ANTHOPOGON** Necker. Fringed gentian.

Annual or biennial herbs with opposite entire sessile leaves and large, terminal, mostly 4-parted flowers; calyx relatively large, the lobes unequal, the inner broader than the outer, scarious or hyaline-margined; corolla blue, showy, the tube campanulate-funnelform, the lobes large, rounded, fimbriate or lacerate, the sinuses not plaited; stamens usually accompanied by a row of glands at the base of the filaments; capsule stipitate, the seeds numerous.

KEY TO THE SPECIES.

Annual; flowers long-pedunculate.................................. 1. *A. elegans.*
Perennial; flowers short-pedunculate, nearly sessile................ 2. *A. barbellatus.*

1. **Anthopogon elegans** (A. Nels.) Rydb. Bull. Torrey Club **33:** 148. 1906.
Gentiana elegans A. Nels. Bull. Torrey Club **25:** 276. 1898.
Type locality: Cummins, Wyoming.
Range: British America to Arizona and New Mexico.
New Mexico: Rio Pueblo; Taos Mountains; Santa Fe and Las Vegas mountains. Meadows in the mountains, from the Transition to the Arctic-Alpine Zone.

2. **Anthopogon barbellatus** (Engelm.) Rydb. Bull. Torrey Club **33:** 148. 1906.
Gentiana barbellata Engelm. Trans. Acad. St. Louis **2:** 26. 1863.
Type locality: "On the alpine summit of Mount Flora, in the Snowy Range, Colorado."
Range: Colorado and northern New Mexico.
New Mexico: Baldy; Costilla Pass; Brazos Canyon. Meadows in the mountains, Canadian to Arctic-Alpine Zone.

7. **AMARELLA** Gilib. Gentian.

Annuals, biennials, or perennials with opposite, entire, mainly sessile leaves and rather small flowers, solitary or in cymes; flowers 4 or 5-merous; calyx usually small, with imbricated equal or unequal lobes; corolla funnelform or salverform, the lobes entire or sparingly toothed, not plicate in the sinuses, often filamentous at the base; ovary 1-celled; capsule usually sessile, the seeds numerous.

KEY TO THE SPECIES.

Calyx lobes very unequal, two of them large, foliaceous, ovate or oval, much broader than the rest and covering them...... 1. *A. heterosepala.*
Calyx lobes slightly unequal, all alike in general form, lanceolate or linear.
Flowers yellowish, numerous, crowded, short-pediceled; leaves usually equaling or exceeding the internodes... 2. *A. strictiflora.*
Flowers bluish, few, distinctly pediceled; middle internodes elongated, usually longer than the leaves............... 3. *A. scopulorum.*

1. **Amarella heterosepala** (Engelm.) Greene, Leaflets **1:** 53. 1904.
Gentiana heterosepala Engelm. Trans. Acad. St. Louis **2:** 215. 1863.
Gentiana distegia Greene, Pittonia **4:** 182. 1900.
Amarella distegia Greene, Leaflets **1:** 53. 1904.
Type locality: "Northern slope of the Uintah Mountains, east of the Great Salt Lake, Utah."
Range: Utah and Colorado to New Mexico.
New Mexico: Tunitcha Mountains; Sandia Mountains; Hillsboro Peak; Cox Canyon; Cloudcroft. Mountains, in the Transition Zone.
This species has broader leaves and is more slender than our others, while the peduncles are much longer.

2. **Amarella strictiflora** (Rydb.) Greene, Leaflets **1**: 53. 1904.
Gentiana acuta stricta Hook. Fl. Bor. Amer. **2**: 63. 1838.
Gentiana acuta strictiflora Rydb. Mem. N. Y. Bot. Gard. **1**: 309. 1900.
Amarella cobrensis Greene, Leaflets **1**: 56. 1904.
TYPE LOCALITY: "Canada to the Rocky Mountains and Slave Lake."
RANGE: British America to California and New Mexico.
NEW MEXICO: Jemez Mountains; Santa Fe and Las Vegas mountains; Taos Mountains; Carrizo Mountains; Baldy; Santa Rita; Capitan Mountains; White Mountains. Meadows in the mountains, Transition to the Arctic-Alpine Zone.
The type of *Amarella cobrensis* was collected at Santa Rita (*Greene* in 1880).

3. **Amarella scopulorum** Greene, Leaflets **1**: 55. 1904.
Amarella revoluta Greene, loc. cit.
TYPE LOCALITY: "Rocky Mountain region from Colorado to Montana."
RANGE: Montana and South Dakota to Arizona and New Mexico.
NEW MEXICO: Rio Pueblo; Santa Fe and Las Vegas mountains; West Fork of the Gila; Jemez Mountains; White and Sacramento mountains. Mountain meadows, in the Transition Zone.
The type of *A. revoluta* was collected in the White Mountains (*Wooton* 552).

8. CHONDROPHYLLA (Bunge) A. Nels.

Small annual or biennial, less than 10 cm. high, with single or several stout stems; leaves numerous, small, opposite, with white scarious margins; flowers solitary, terminal; calyx narrow, 4 or 5-toothed; corolla salverform when expanded, plicate at the sinuses with broad emarginate lobes or plaits, without crown or glands; anthers cordate, versatile.

1. **Chondrophylla fremontii** (Torr.) A. Nels. Bull. Torrey Club **31**: 245. 1904.
Gentiana fremontii Torr. in Frém. Rep. Exped. Rocky Mount. 94. 1845.
TYPE LOCALITY: "Wind River Mountains."
RANGE: Wyoming to northern New Mexico.
NEW MEXICO: Winsors Ranch; Costilla Valley. Moist meadows, in the Transition Zone.

9. DASYSTEPHANA Adans. CLOSED GENTIAN.

Annuals, biennials, or usually perennials, with opposite entire leaves, sometimes scabrous-ciliolate or erose, and usually 5-parted flowers variously arranged; calyx persistent, the lobes minute or foliaceous, smooth and glabrous or scabrous-ciliolate; corolla salverform, funnelform, or clavate, without glands at the base of the tube or filaments at the base of the lobes, the lobes erect, converging, plaited in the sinuses; stamens with converging or cohering anthers; capsules stipitate.

KEY TO THE SPECIES.

Floral leaves broadened, more or less scarious; flowers deep purple. 2. *D. parryi.*
Floral leaves narrow, green; flowers not deep purple.
  Calyx glabrous; corolla white with purplish dots............... 1. *D. romanzovii.*
  Calyx scabrous; corolla not dotted, white or purplish.
    Corolla white, 3 cm. long or more............................ 3. *D. rusbyi.*
    Corolla purple, 2 cm. long or less.
      Flowers in a short dense cluster; plants less than 15 cm. high... 4. *D. bigelovii.*
      Flowers in an elongated raceme; plants more than 20 cm. high... 5. *D. interrupta.*

1. **Dasystephana romanzovii** (Ledeb.) Rydb. Bull. Torrey Club **33:** 148. 1906.
Gentiana romanzovii Ledeb.; Bunge, Nouv. Mém. Soc. Nat. Moscou **1:** 215. *pl. 11. f. 1.* 1829.
Gentiana frigida of authors, not Haenke, 1788.
Type locality: Gulf of St. Lawrence, Siberia.
Range: Alaska to Montana, Utah, and New Mexico; also in Siberia.
New Mexico: Santa Fe and Las Vegas Mountains. High mountain meadows, in the Arctic-Alpine Zone.

2. **Dasystephana parryi** (Engelm.) Rydb. Bull. Torrey Club **33:** 149. 1906.
Gentiana parryi Engelm. Trans. Acad. St. Louis **2:** 218. *pl. 10.* 1863.
Type locality: "Near the base of Alpine slopes on the Snowy Range, Colorado."
Range: Wyoming to Utah and New Mexico.
New Mexico: Baldy; Taos Mountains; Costilla Pass; Santa Fe and Las Vegas mountains. Mountain meadows, in the Hudsonian and Arctic-Alpine zones.

3. **Dasystephana rusbyi** (Greene) Woot. & Standl. Contr. U. S. Nat. Herb. **16:** 159. 1913.
Gentiana rusbyi Greene; A. Gray, Syn. Fl. **2**[1]: 406. 1878.
Type locality: Mogollon Mountains, New Mexico. Type collected by Rusby in 1881.
Range: Known only from type locality.

4. **Dasystephana bigelovii** (A. Gray) Rydb. Bull. Torrey Club **33:** 149. 1906.
Gentiana bigelovii A. Gray, Proc. Amer. Acad. **19:** 87. 1883.
Type locality: Sandia Mountains, New Mexico. Type collected by Bigelow in 1853.
Range: Colorado and New Mexico.
New Mexico: Sandia Mountains; Upper Pecos; Beulah; Lincoln National Forest; Castle Rock. Mountains, in the Transition Zone.

5. **Dasystephana interrupta** (Greene) Rydb. Bull. Torrey Club **33:** 149. 1906.
Gentiana interrupta Greene, Pittonia **4:** 182. 1900.
Type locality: Meadows along streams at Pagosa Springs, southern Colorado.
Range: Colorado to Nevada and New Mexico.
New Mexico: Dulce; Upper Pecos; Chicorico Canyon; Sandia Mountains; Carrizo Mountains; West Fork of the Gila; White Mountains. Meadows in the mountains, in the Transition Zone.

10. PLEUROGYNA Eschsch.

Slender glabrous annual with linear leaves and whitish flowers in a narrow panicle; corolla rotate or campanulate, pentamerous; pedicels slender, erect; sepals linear, green; stamens inserted on the corolla tube near its base; style none, the stigmas decurrent along the sutures of the ovary.

1. **Pleurogyna fontana** A. Nels. Proc. Biol. Soc. Washington **17:** 177. 1904.
Type locality: Crow Creek, Colorado.
Range: Wyoming to northern New Mexico.
New Mexico: Costilla Valley (*Wooton*). Bogs.

Order 44. ASCLEPIADALES.

KEY TO THE FAMILIES.

Styles united; stamens distinct or gynandrous; pollen loosely granular .... 114. **APOCYNACEAE** (p. 504).
Styles distinct; stamens monadelphous; pollen united into waxy masses, or the grains in groups of 4.............................. 115. **ASCLEPIADACEAE** (p 506).

114. APOCYNACEAE. Dogbane Family.

Perennial herbs with milky sap; leaves opposite or alternate, sessile or short-petiolate, simple, usually entire, exstipulate; flowers perfect, regular, in terminal cymes; calyx 5-parted, persistent; corolla 5-parted; stamens as many as the corolla lobes and alternate with them, inserted in the tube or throat of the corolla; anthers linear-oblong, sagittate, 2-celled; ovary superior or partly inferior, of 2 distinct carpels; fruit of paired follicles; seeds often comose.

KEY TO THE GENERA.

Leaves alternate; anthers free from the stigma; corolla 10 mm. long or more, tubular or funnelform.................... 1. AMSONIA (p. 504).
Leaves opposite; anthers converging about the stigma and slightly adherent to it; corolla 3 to 5 mm. long, short-campanulate or urceolate.............................. 2. APOCYNUM (p. 505).

1. AMSONIA Walt.

Herbaceous perennials, sometimes woody at the base, with alternate entire narrow leaves and tubular or salverform flowers in terminal cymes; calyx lobes 5, acuminate; disk wanting; corolla 1 cm. long or more, the tube narrow below, slightly enlarged above, villous within; stamens included, not appendaged; carpels 2, connected by the slender styles; stigma appendaged by a reflexed membrane; follicles erect, several-seeded, the seeds not appendaged.

KEY TO THE SPECIES.

Stems and leaves glabrous.
  Pods constricted between the seeds; corolla tube about 1 cm. long.. 1. *A. brevifolia.*
  Pods continuous, not constricted; corolla tube 3 to 4 cm. long.... 4. *A. longiflora.*
Stems and leaves variously pubescent.
  Pods constricted between the seeds; plants tomentose........... 2. *A. arenaria.*
  Pods continuous, not constricted; plants hirtellous.............. 3. *A. hirtella.*

1. Amsonia brevifolia A. Gray, Proc. Amer. Acad. **12**: 64. 1876.

TYPE LOCALITY: "S. Utah and W. Arizona, to the border of California."

RANGE: Southern California to Utah and northwestern New Mexico.

NEW MEXICO: Zuni (*Stevenson* 53). Dry soil, in the Upper Sonoran Zone.

2. Amsonia arenaria Standley, Proc. Biol. Soc. Washington **26**: 117. 1913.

TYPE LOCALITY: Sandhills between Strauss and Anapra, near the southeast corner of Dona Ana County, New Mexico. Type collected by Elmer Stearns (no. 372).

RANGE: Southern New Mexico to northern Chihuahua.

NEW MEXICO: Between Strauss and Anapra; San Andreas Mountains. Sandhills, in the Lower Sonoran Zone.

3. Amsonia hirtella Standley, Proc. Biol. Soc. Washington **26**: 118. 1913.

TYPE LOCALITY: Near the Upper Corner Monument, southern Grant County, New Mexico. Type collected by E. A. Mearns (no. 117).

RANGE: Western Texas and southern New Mexico to Chihuahua.

NEW MEXICO: Grant County (*Mearns* 117). Dry hills, in the Lower Sonoran Zone.

4. Amsonia longiflora Torr. U. S. & Mex. Bound. Bot. 159. 1859.

TYPE LOCALITY: "Rocky ravines near El Paso," Chihuahua or Texas.

RANGE: Southern New Mexico and western Texas to northern Mexico.

NEW MEXICO: Rio Gila (*Wooton*).

The specimen is in fruit only, but it has the leaves and the glabrous stems of this species.

2. APOCYNUM L. Dogbane.

Perennial herbs with reddish or greenish stems, 1.5 meters high or less, with tough fibrous bark and opposite nucronate entire leaves; flowers small, pinkish or greenish white, in terminal cymes; calyx small, deeply 5-parted, adnate to the ovary by a thickish disk; corolla short-campanulate or urceolate, the limb erect or spreading, 5-lobed, bearing appendages within alternate with the 5 included stamens; filaments short, the anthers sagittate; fruit a pair of follicles, terete or long-fusiform; seeds numerous, with a coma at the apex.

KEY TO THE SPECIES.

Corolla 5 mm. long or more, its lobes spreading; leaves ovate; low plants with spreading branches and bright pink flowers.
- Leaves glabrous beneath; sepals narrowly lanceolate...... 1. *A. ambigens.*
- Leaves pubescent beneath; sepals narrowly or broadly lanceolate.
  - Sepals broadly lanceolate; corolla open-campanulate; leaves thick, dark green, decidedly pubescent beneath.. 2. *A. scopulorum.*
  - Sepals narrowly lanceolate; corolla narrowly campanulate; leaves pale green, pubescent only on the petioles and veins............................... 3. *A. lividum.*

Corolla 3 mm. long or less, with erect lobes; leaves oblong to narrowly elliptic-oblong; tall plants with erect or strongly ascending branches and pale flowers.
- Leaves pubescent beneath................................. 4. *A. laurinum.*
- Leaves glabrous.
  - Cauline leaves broadly oblong, clasping, obtuse........ 5. *A. hypericifolium.*
  - Cauline leaves narrower, petioled, acute, bright green.
    - Leaves rounded or obtuse at the base, oblong to elliptic-oblong.............................. 6. *A. viride.*
    - Leaves acute at the base (on longer petioles), elliptic-lanceolate................................ 7. *A. angustifolium.*

1. Apocynum ambigens Greene, Pl. Baker. **3:** 17. 1901.

Type locality: "In the Black Cañon," Colorado.

Range: Washington and Montana to California and New Mexico.

New Mexico: Raton; White Mountains. Mountains, in the Transition Zone.

2. Apocynum scopulorum Greene; Rydb. Colo. Agr. Exp. Sta. Bull. **100:** 269. 1906.

Type locality: Colorado.

Range: British America to Colorado and New Mexico.

New Mexico: Chama; Santa Fe and Las Vegas mountains; Sierra Grande; Sandia Mountains; Lookout Mines; Turkey Creek; White and Sacramento mountains. Open slopes in the mountains, in the Transition Zone.

3. Apocynum lividum Greene, Pl. Baker. **3:** 17. 1901.

Apocynum cannabinum lividum A. Nels. in Coulter, New Man. Rocky Mount. 386. 1909.

Type locality: "Common on railway embankments in Black Cañon," Colorado.

Range: Mountains of Colorado and northern New Mexico.

New Mexico: Sierra Grande (*Standley* 6195). Transition Zone.

Professor Nelson's reduction of this species to rank as a subspecies of *A. cannabinum* is peculiarly unfortunate, since the two plants are not closely related.

4. Apocynum laurinum Greene, Pittonia **5:** 64. 1902.

TYPE LOCALITY: Organ Mountains, New Mexico. Type collected by Wooton (no. 113).

RANGE: Mountains of southern New Mexico.

NEW MEXICO: Organ Mountains; Ruidoso Creek.

5. Apocynum hypericifolium Ait. Hort. Kew. **1:** 304. 1789.

TYPE LOCALITY: North America.

RANGE: British America to Ohio and New Mexico.

NEW MEXICO: Mesilla Valley. River valleys, in the Lower and Upper Sonoran zones.

6. Apocynum viride Woot. & Standl. Contr. U. S. Nat. Herb. **16:** 159. 1913.

TYPE LOCALITY: Gilmores Ranch on Eagle Creek, White Mountains, Lincoln County, New Mexico. Type collected by Wooton and Standley (no. 3451).

RANGE: New Mexico.

NEW MEXICO: Farmington; Cedar Hill; Las Vegas; Pecos; Reserve; White and Sacramento mountains. Wet ground, in the Upper Sonoran and Transition zones.

7. Apocynum angustifolium Wooton, Contr. U. S. Nat. Herb. **16:** 159. 1913.

TYPE LOCALITY: Gila River bottom near Cliff, Grant County, New Mexico. Type collected by Metcalfe (no. 132).

RANGE: Southwestern New Mexico.

NEW MEXICO: Cliff; Mimbres; Lower Plaza; Eagle Creek. Along streams, in the Upper Sonoran Zone.

A very distinct species because of its very narrow bright green leaves, nearly all of them acute at the base.

115. ASCLEPIADACEAE. Milkweed Family.

Herbaceous perennials, erect or twining, mostly with milky juice; leaves simple, alternate, opposite, or whorled, exstipulate, mostly entire; flowers perfect, regular, mostly umbellate; calyx hypogynous, the tube very short or none, imbricated in bud; corolla 5-merous, campanulate, urceolate, rotate, or funnelform, the lobes commonly reflexed; a 5-lobed crown (corona) borne between the corolla and stamens and adnate to one or both of them; stamens 5, adnate to the corolla usually near its base, monadelphous or distinct; anthers converging around the stigmas, sometimes united to each other, the sacs tipped with an inflexed or erect scarious membrane or not appendaged at the top, sometimes appendaged at the base; pollen in waxy masses (pollinia) connected with the stigma in pairs; pistil of 2 carpels; styles 2, with a single peltate discoid stigma; fruit a pair of several to many-seeded follicles; seeds comose.

KEY TO THE GENERA.

Pollinia horizontal or nearly so; stigma sharply 5-angled, depressed; twining vine........................ 1. VINCETOXICUM (p. 507).

Pollinia pendulous; stigmas various; stems twining or straight.

  Crown double, the outer a shallow ring, the inner consisting of 5 fleshy hoodlike scales; twining vines.................................... 2. PHILIBERTELLA (p. 507).

  Crown single; erect, ascending, or procumbent herbs.

    Corolla lobes erect-spreading during anthesis; hoods of the crown pendulous or saccate at the base, curved upward, obtuse, crested within, at least in the upper part. 3. ASCLEPIODORA (p. 508).

    Corolla lobes reflexed during anthesis; hoods various.

Hoods of the crown not corniculate, or very obscurely so.
  Leaves mostly scattered, not densely tomentose beneath; flowers greenish; anther wings narrowed below.................... 4. ACERATES (p. 508).
  Leaves opposite, densely tomentose beneath; flowers dark purple; anther wings not narrowed below........................... 7. GOMPHOCARPUS (p. 513).
Hoods of the crown aristate or corniculate-appendaged within.
  Hoods short, not surpassing the stamens, or if long much broader below, never constricted at the base... 5. ASCLEPIAS (p. 509).
  Hoods long, erect, much surpassing the stamens, laminately expanded below, narrow at the summit, mostly bicorniculate........... 6. PODOSTEMMA (p. 512).

1. VINCETOXICUM Walt. ANGLE-POD.

Herbaceous twining pubescent perennial from a slender rootstock; leaves opposite, sagittate-cordate; flowers greenish, in small clusters; sepals lance-elliptic, green; corolla narrowly campanulate, the lobes as long as the tube, oblong, obtuse; crown nearly equaling the column, entire or barely undulate, inserted at the base of the short column and connected with it by 5 adnate crests, free at the 2-toothed or entire apex; pollinia obliquely inserted on broad winged caudicles; follicles ovoid, 3 cm. long or more, puberulent.

1. Vincetoxicum productum (Torr.) Vail, Bull. Torrey Club **26:** 431. 1899.

Gonolobus productus Torr. U. S. & Mex. Bound. Bot. 165. 1859.

TYPE LOCALITY: "Banks of Rock Creek," "Valley of the Limpio, and along the Rio Grande," Texas.

RANGE: Western Texas to southern Arizona and southward.

NEW MEXICO: Organ Mountains; Mesilla Valley; Gray. Moist ground, in the Lower and Upper Sonoran zones.

2. PHILIBERTELLA Vail.

Herbaceous twining perennials with opposite leaves and mostly fragrant flowers in axillary umbels; calyx small, 5-lobed, the lobes acute; corolla campanulate or rotate, a shallow entire or undulate ring forming an outer crown in its throat, the lobes 5, longer than the tube, the inner crown of 5 turgid, fleshy or hard scales or flattish appendages attached in a circle at the base of the sessile or slightly stalked column, forming a hollow, entire or undulate, spreading surface near the level of the conical stigmas; follicles naked, slender.

KEY TO THE SPECIES.

Column conspicuous; peduncles much shorter than the crispate leaves; stems cinereous.................................. 1. *P. crispa.*
Column inconspicuous or none; peduncles equaling or surpassing the leaves, these not crispate; stems glabrous or pubescent.
  Stems pubescent; flowers numerous in each umbel, white; leaves cordate-ovate to subsagittate.................. 2. *P. cynanchoides.*
  Stems glabrous or nearly so; flowers few, reddish; leaves linear.. 3. *P. heterophylla.*

1. **Philibertella crispa** (Benth.) Vail, Bull. Torrey Club **24**: 306. 1897.
Sarcostemma crispum Benth. Pl. Hartw. 291. 1841.
Sarcostemma undulata Torr. U. S. & Mex. Bound. Bot. 161. 1859.
Philibertia undulata A. Gray, Proc. Amer. Acad. **12**: 95. 1876.
TYPE LOCALITY: Aguas Calientes, Mexico.
RANGE: Western Texas to New Mexico and southward.
NEW MEXICO: Lone Pine; Deming; Organ Mountains; Queen. Rocky hillsides and canyons, in the Lower and Upper Sonoran zones.

2. **Philibertella cynanchoides** (Decaisne) Vail, Bull. Torrey Club **24**: 307. 1897.
Sarcostemma cynanchoides Decaisne in DC. Prodr. **8**: 540. 1844.
Philibertia cynanchoides A. Gray, Proc. Amer. Acad. **12**: 64. 1876.
TYPE LOCALITY: Mexico.
RANGE: Utah and Arizona to western Texas.
NEW MEXICO: Mountains west of San Antonio; Kingston; Mesilla Valley; Tularosa; Roswell. Mesas and low hills, in the Lower Sonoran Zone.

3. **Philibertella heterophylla** (Engelm.) Cockerell, Bot. Gaz. **26**: 279. 1890.
Sarcostemma heterophylla Engelm. in Torr. U. S. Rep. Expl. Miss. Pacif. **5**: 362. 1856.
Philibertia linearis heterophylla A. Gray, Syn. Fl. $\mathbf{2^1}$: 88. 1878.
TYPE LOCALITY: Near Fort Yuma, Arizona.
RANGE: Southern California to western Texas and adjacent Mexico.
NEW MEXICO: Mesa west of Organ Mountains; Organ Mountains; Knowles. Dry hills and mesas, in the Lower Sonoran Zone.

3. ASCLEPIODORA A. Gray. ANTELOPE HORNS.[1]

Erect or spreading perennial herb, sparingly pubescent, with several stout ascending stems from a woody root; leaves alternate, elongate-lanceolate, short-petiolate; flowers rather large; calyx lobes narrowed, acute; corolla rotate, the lobes broadly ovate, green, spreading in anthesis; hoods inserted over the whorl of the short column, dark purple, obtuse, crested within; anther wings salient above the middle; follicles large, about 8 cm. long, erect, on recurved pedicels.

1. **Asclepiodora decumbens** (Nutt.) A. Gray, Proc. Amer. Acad. **12**: 66. 1876.
Anantherix decumbens Nutt. Trans. Amer. Phil. Soc. n. ser. **5**: 201. 1837.
Acerates decumbens Decaisne in DC. Prodr. **8**: 521. 1844.
TYPE LOCALITY: "On dry hills near the confluence of Kiamesha and Red river."
RANGE: Arkansas and Texas to Utah and Arizona.
NEW MEXICO: Tunitcha Mountains; Carrizo Mountains; Dulce; Sierra Grande; west of Santa Fe; Las Vegas; Magdalena Mountains; Kingston; Organ Mountains, Gray; Cloudcroft; White Mountains. Plains and hills, in the Lower and Upper Sonoran zones.

4. ACERATES Ell.

Erect perennial herbs 80 cm. high or less, with opposite or alternate leaves, the flowers in lateral axillary umbels on short peduncles; horns of the hoods wanting or very obscure; anther wings angled near the middle; otherwise like Asclepias.

KEY TO THE SPECIES.

Leaves linear.
Hoods truncate at the apex, with an obscure horn within..... 3. *A. rusbyi.*
Hoods emarginate, crestless within.............................. 4. *A. auriculata.*
Leaves not linear.
Leaves oblong; anther wings gradually narrowed below........ 1. *A. viridiflora.*
Leaves elongated-lanceolate; anther wings abruptly narrowed below... 2. *A. ivesii.*

[1] A translation of the very appropriate Navaho name, referring to the form of the pods.

1. **Aceràtes viridiflora** (Raf.) Eaton, Man. Bot. ed. 5. 90. 1829.
Asclepias viridiflora Raf. Med. Repos. N. Y. **5**: 360. 1808.
TYPE LOCALITY: "In several parts of Maryland and Pennsylvania, mostly in fields."
RANGE: Saskatchewan and New England to Florida and Mexico.
NEW MEXICO: Sierra Grande (*Standley* 6060). Plains and dry hills, in the Upper Sonora Zone.

2. **Acerates ivesii** (Britton) Woot. & Standl.
Asclepias lanceolata Ives, Amer. Journ. Sci. **1**: 252. 1819, not Walt. 1788.
Acerates viridiflora ivesii Britton, Mem. Torrey Club **5**: 265. 1894.
TYPE LOCALITY: "On the sandy plains east of Cedar Hill, in New Haven."
RANGE: Nebraska and South Dakota to Arizona and New Mexico.
NEW MEXICO: Santa Fe Canyon; Kingman. Plains and low hills, in the Upper Sonoran Zone.

3. **Acerates rusbyi** Vail, Bull. Torrey Club **25**: 37. 1898.
TYPE LOCALITY: Oak Creek, Arizona.
RANGE: New Mexico and Arizona.
NEW MEXICO: Near Tesuque; south of Roswell. Upper Sonoran Zone.

4. **Acerates auriculata** Engelm. in Torr. U. S. & Mex. Bound. Bot. 160. 1859.
Asclepias auriculata Holzinger, Bot. Gaz. **17**: 125. 1892.
TYPE LOCALITY: Dry ravines near the Copper Mines and along the Mimbres, New Mexico. Type collected by Bigelow.
RANGE: Nebraska and Kansas to New Mexico and Texas.
NEW MEXICO: Lower Plaza; Organ Mountains; White Mountains. Dry hills, in the Upper Sonoran and Transition zones.

5. ASCLEPIAS L. MILKWEED.

Perennial herbs of various habit, with opposite, alternate, or verticillate leaves, the flowers in pedunculate umbels, borne mostly near the top of the stem, pseudoterminal; calyx small, green, often with minute glands at the base of the lobes; corolla rotate, deeply 5-parted, the lobes reflexed in anthesis; hoods of the crown of various shapes, not narrowed below, bearing hornlike processes within the hood; anther wings broader below the middle; stigma 5-angled, flat-topped; follicles mostly smooth; seeds comose.

KEY TO THE SPECIES.

Leaves linear or filiform.
  Hoods 3 times as long as the anthers, acute, recurved; stems much branched, puberulent throughout; leaves nearly filiform.................................. 1. *A. macrotis.*
  Hoods slightly if at all longer than the anthers, toothed or obtuse, erect; stems simple or sparingly branched, glabrous or nearly so; leaves broader.
    Hoods 5-toothed................................. 2. *A. quinquedentata.*
    Hoods entire.
      Leaves scattered, rigid; plants 10 to 20 cm. high... 3. *A. pumila.*
      Leaves opposite or verticillate, weak; plants 40 to 60 cm. high or more.................... 4. *A. galioides.*
Leaves lanceolate or broader.
  Leaves narrowly lanceolate to narrowly oblong, acute.
    Stems hirsute; corolla orange........................ 5. *A. tuberosa.*
    Stems not hirsute; corolla never orange.
      Pedicels erect in fruit; flowers bright purple; plants tall, erect, 60 to 100 cm. high, nearly simple.................................. 6. *A. incarnata.*

Pedicels deflexed in fruit; flowers dull purplish or whitish; plants lower, 50 cm. high or less, spreading, much branched near the base.
Hoods longer than the anthers; flowers greenish white .... 7. *A. involucrata.*
Hoods shorter than the anthers; flowers purplish.
Hoods about half as long as the anthers; umbels pedunculate.............. 8. *A. brachystephana.*
Hoods only slightly shorter than the anthers; umbels subsessile.......... 9. *A. uncialis.*
Leaves broadly oblong to orbicular, mostly obtuse or retuse.
Fruit echinate; leaves acute or acutish...............10. *A. speciosa.*
Fruit smooth; leaves obtuse or retuse.
Stems very short, 5 cm. long or less; leaves as broad as long or broader.....................11. *A. nummularia.*
Stems 20 cm. long or more; leaves usually much longer than broad.
Leaves thin, glabrous, more or less glaucous, sessile or clasping......................12. *A. elata.*
Leaves very thick, more or less tomentose, at least when young, never glaucous, short-petiolate.
Stems, pedicels, and young leaves densely tomentose..........................13. *A. arenaria.*
Stems, pedicels, and leaves sparingly tomentose or glabrous.............14. *A. latifolia.*

1. Asclepias macrotis Torr. U. S. & Mex. Bound. Bot. 164. *pl. 45.* 1859.

Type locality: "Rocky hills near El Paso, and on the mountains below San Elceario," Chihuahua.

Range: Western Texas and southern New Mexico and southward.

New Mexico: Mangas Springs; Kingston; Organ Mountains; Buchanan; Sandia Mountains. Dry, rocky hills, in the Upper Sonoran Zone.

2. Asclepias quinquedentata A. Gray, Proc. Amer. Acad. **12:** 71. 1876.

Asclepias quinquedentata neomexicana Greene, Proc. Amer. Acad. **16:** 103. 1880.

Type locality: San Pedro River, western Texas.

Range: Western Texas to Arizona.

New Mexico: Coolidge; Mogollon Creek; Gilmores Ranch. Transition Zone.

3. Asclepias pumila (A. Gray) Vail in Britt. & Brown, Illustr. Fl. **3:** 12. 1898.

Asclepias verticillata pumila A. Gray, Proc. Amer. Acad. **12:** 71. 1876.

Type locality: "The western dry plains from Nebraska to New Mexico."

Range: South Dakota and Arkansas to Colorado and New Mexico.

New Mexico: Cabra Springs; Nara Visa; Gavilan Creek; Leachs. Plains and low hills, in the Upper Sonoran Zone.

4. Asclepias galioides H. B. K. Nov. Gen. & Sp. **3:** 188. 1818.

Type locality: "Crescit in temperatis Regni Novae Hispaniae, inter Valladolid de Mechoacan et locum Cuiseo."

Range: Kansas and Colorado to Arizona, western Texas, and Mexico.

New Mexico: Common throughout the State. River valleys and wet grounds, in the Lower and Upper Sonoran zones.

A common weed in cultivated fields and along irrigating ditches. Our specimens may include *A. verticillata*, but we have been unable to separate them definitely. They also include specimens cited by various authors as *A. subverticillata*. In our opinion there is only one species of this type in New Mexico.

5. Asclepias tuberosa L. Sp. Pl. 217. 1753. BUTTERFLY WEED.

TYPE LOCALITY: "Habitat in America boreali."

RANGE: British America to Florida, Texas, and Arizona.

NEW MEXICO: Santa Fe and Las Vegas mountains; Sandia Mountains; Kingston; Mogollon Mountains; Burro Mountains; San Luis Mountains; Capitan Mountains; White and Sacramento mountains. Transition Zone.

The western form is variable and somewhat different from the typical eastern one. It has usually narrower, more crowded leaves and often paler flowers. In the higher mountains the flowers are darker and as brightly colored as in the eastern plants.

6. Asclepias incarnata L. Sp. Pl. 215. 1753. SWAMP MILKWEED.

Asclepias incarnata longifolia A. Gray, Syn. Fl. **2**[1]: 91. 1878.

TYPE LOCALITY: "Habitat in Canada, Virginia."

RANGE: British America to New Mexico and Florida.

NEW MEXICO: White and Sacramento mountains; Roswell. Wet ground, in the Upper Sonoran and Transition zones.

The subspecies proposed by Doctor Gray does not seem to deserve a name. Our plants have the leaves narrow, but not more so than some of the eastern ones.

7. Asclepias involucrata Engelm. in Torr. U. S. & Mex. Bound. Bot. 163. 1859.

TYPE LOCALITY: Sandy soil, on the Mimbres and near the Copper Mines, New Mexico. Type, Mexican Boundary Survey no. 1074.

RANGE: Southern Utah to Arizona and New Mexico.

NEW MEXICO: Farmington; Sierra Grande; Nara Visa; Santa Fe Creek; Las Vegas; Mogollon Creek; San Augustine Plains; Santa Rita. Dry hills and plains, in the Upper Sonoran Zone.

8. Asclepias brachystephana Engelm. in Torr. U. S. & Mex. Bound. Bot. 163. 1859.

TYPE LOCALITY: "Sandy soils, valley of the upper Rio Grande, Chihuahua, and Sonora."

RANGE: Wyoming to Arizona and Texas and southward.

NEW MEXICO: Socorro; Hillsboro; Organ Mountains; Chosa Spring; Roswell; Carlsbad. Sandy plains, in the Lower Sonoran Zone.

9. Asclepias uncialis Greene, Bot. Gaz. **5**: 64. 1880.

TYPE LOCALITY: Open hilltops in southwestern New Mexico, about Silver City. Type collected by E. L. Greene.

RANGE: Wyoming to Arizona and New Mexico.

NEW MEXICO: Silver City.

10. Asclepias speciosa Torr. Ann. Lyc. N. Y. **2**: 218. 1828.

TYPE LOCALITY: "On the Canadian?," Colorado or New Mexico.

RANGE: British Columbia and Manitoba to California and New Mexico.

NEW MEXICO: Farmington; Shiprock; Chama; Perico Creek; Pecos; Las Vegas; Joseph; Middle Fork of the Gila; White and Sacramento mountains. River valleys and wet ground, in the Upper Sonoran and Transition zones.

11. Asclepias nummularia Torr. U. S. & Mex. Bound. Bot. 163. *pl. 45*. 1859.

TYPE LOCALITY: Copper Mines, New Mexico. Type, Mexican Boundary Survey no. 1073.

RANGE: Southern New Mexico and Arizona and southward.

NEW MEXICO: San Luis Mountains; Dog Spring; Santa Rita. Dry hills and mesas, in the Lower Sonoran Zone.

12. Asclepias elata Benth. Pl. Hartw. 290. 1848.

TYPE LOCALITY: Aguas Calientes, Mexico.

RANGE: New Mexico and Arizona and southward.

NEW MEXICO: Black Range; Van Pattens; Queen.

This may be the same as *A. glaucescens* H. B. K., as stated by Doctor Gray, but the two seem different.

13. Asclepias arenaria Torr. U. S. & Mex. Bound. Bot. 162. 1859.

TYPE LOCALITY: Sandy banks, Jornada del Muerto, and on the upper Rio Gila, New Mexico.

RANGE: New Mexico and Arizona.

NEW MEXICO: Between El Paso and Monument 53; White Sands; Roswell. Mesas and low hills, in the Lower Sonoran Zone.

14. Asclepias latifolia (Torr.) Raf. Atl. Journ. 146. 1833.

Asclepias obtusifolia latifolia Torr. Ann. Lyc. N. Y. **2**: 217. 1828.

Asclepias jamesii Torr. U. S. & Mex. Bound. Bot. 162. 1859.

TYPE LOCALITY: "On the Canadian?" New Mexico or Colorado.

RANGE: Colorado to Arizona and Texas.

NEW MEXICO: Cross L Ranch; Sierra Grande; Nara Visa; lower Gila above Duncan; Hillsboro; Burro Mountains; Emory Peak; Ruidoso Creek; Redlands. Plains and low hills, in the Lower and Upper Sonoran zones.

ASCLEPIAS SCAPOSA Vail, Bull. Torrey Club **25**: 171. 1897.

TYPE LOCALITY: Near Santa Rita, New Mexico. Type collected by Wright in 1851.

This we have not seen. It was described without flowers, hence it is impossible to tell whether it is an Asclepias or a Podostemma. Its general appearance, as described, suggests the latter genus.

6. PODOSTEMMA Greene.

Coarse perennial herbs, resembling Asclepias; flowers rather large, in short-pedunculate umbels, the blooming ones well down the stem, thus appearing lateral; hoods large, erect, much longer than the anthers and surpassing the crown by half their length, narrow below (pseudostipitate), expanded above and mostly bicorniculate; anther wings broadest in the middle, notched.

KEY TO THE SPECIES.

Hoods 3.5 mm. long or less; leaves lanceolate .... 1. *P. emoryi.*

Hoods 7 mm. long; leaves oblong .... 2. *P. lindheimeri.*

1. Podostemma emoryi Greene, Pittonia **3**: 237. 1897.

TYPE LOCALITY: Texas or New Mexico.

RANGE: Western Texas to southern New Mexico.

NEW MEXICO: Mangas Springs. Lower Sonoran Zone.

It is impossible to tell where the type was collected, for part of the type number came from Texas and part from New Mexico.

2. Podostemma lindheimeri (Engelm. & Gray) Greene, Pittonia **3**: 236. 1897.

Asclepias lindheimeri Engelm. & Gray, Bost. Journ. Nat. Hist. **5**: 250. 1845.

Asclepias wrightii Greene; A. Gray, Proc. Amer. Acad. **16**: 102. 1880.

TYPE LOCALITY: Near Industry, Texas.

RANGE: Western Texas and southern New Mexico.

NEW MEXICO: Santa Rita; Trujillo Creek; mesa west of Organ Mountains; Organ Mountains; south of Roswell. Mesas, in the Lower and Upper Sonoran zones.

7. GOMPHOCARPUS R. Br.

Tomentose perennial, similar to Asclepias, but differing in the absence of horns or crests to the hoods.

1. **Gomphocarpus hypoleucus** A. Gray, Proc. Amer. Acad. **17**: 222. 1882.
TYPE LOCALITY: Santa Rita Mountains, Arizona.
RANGE: Southern New Mexico and Arizona and southward.
NEW MEXICO: Turkey Creek, Mogollon Mountains (*Metcalfe* 566).

Order 45. POLEMONIALES.

KEY TO THE FAMILIES.

Stamens 5.
Gynœcium of 2 distinct carpels ....116. **DICHONDRACEAE** (p. 514).
Gynœcium of 2 or more partially or wholly united carpels.
Fruit drupaceous or of 2 or 4 nutlets.
Style or stigmas furnished with a glandular ring.
122. **HELIOTROPACEAE** (p. 537).
Style or stigmas without a glandular ring.
Fruit a group of 2 or 4 nutlets; style arising from between the lobes of the ovary ....123. **BORAGINACEAE** (p. 538).
Fruit drupaceous; style terminating the lobeless ovary.
121. **EHRETIACEAE** (p. 535).
Fruit capsuiar or baccate, the ovary never 4-lobed.
Styles or stigmas wholly united.
Median axis of the gynœcium in the same axis as the stem; seeds mostly pitted; corolla regular...126. **SOLANACEAE** (p. 566).
Median axis of the gynœcium not in the axis of the stem; seeds mostly tuberculate; corolla usually irregular.
127. **SCROPHULARIACEAE** (p. 575).
Styles or stigmas distinct.
Ovary 3-celled; stigmas 3.
Calyx lobes imbricated; corolla mostly plaited in the bud.
118. **CONVOLVULACEAE** (p. 515).
Calyx lobes valvate; corolla merely convolute in the bud.
119. **POLEMONIACEAE** (p. 519).
Ovary 1 or 2-celled, rarely 4-celled; stigmas 2.
Ovary 1-celled..........120. **HYDROPHYLLACEAE** (p. 530).
Ovary 2-celled or 4-celled.
Corolla unappendaged within; plants with normal leaves.
118. **CONVOLVULACEAE** (p. 515).
Corolla appendaged within; parasitic twining plants with scalelike leaves......117. **CUSCUTACEAE** (p. 514).
Stamens 4 and didynamous, or 1 or 2.
Carpels ripening into nutlets (grouped in 4's) or into achenes or drupes.
Style apical on the entire ovary.............124. **VERBENACEAE** (p. 548).
Style rising from between the 4 lobes of the ovary.
125. **MENTHACEAE** (p. 551).
Carpels ripening into a capsule.
Placentæ of the ovary axile.
Ovary 1-celled ....129. **PINGUICULACEAE** (p. 599).
Ovary 2-celled, rarely 3 or 5-celled.
Corolla lobes imbricated; capsules not elastically dehiscent.
127. **SCROPHULARIACEAE** (p. 575).

Corolla lobes convolute; capsules elastically dehiscent.
128. ACANTHACEAE (p. 597).

Placentæ of the ovary parietal.

Plants parasitic; leaves scalelike, without green coloring matter.
130. OROBANCHACEAE (p. 599).

Plants not parasitic; leaves green.

Ovary and capsule 2-celled; shrubs; seeds winged.
131. BIGNONIACEAE (p. 600).

Ovary and capsule 1-celled; herbs; seeds wingless.
132. MARTYNIACEAE (p. 601).

116. DICHONDRACEAE. Dichondra Family.

1. DICHONDRA Forst.

Annual or perennial creeping herbs with small petiolate entire leaves having orbicular or reniform blades, the inconspicuous solitary flowers on short peduncles; calyx of 5 distinct or nearly distinct sepals; corolla about 5 mm. in diameter, rotate or campanulate, 5-lobed; stamens 5, shorter than the corolla; pistil of 2 carpels; styles 2, distinct; capsules 2-celled, the carpels more or less united, indehiscent; seeds solitary, smooth.

KEY TO THE SPECIES.

Leaves silvery, densely sericeous.................................... 1. *D. argentea.*
Leaves green, sparingly villous.................................... 2. *D. brachypoda.*

1. Dichondra argentea Willd. Hort. Berol. *pl. 81.* 1816.

TYPE LOCALITY: "Habitat in America meridionali."

RANGE: Western Texas to Arizona and southward.

NEW MEXICO: Organ Mountains. Shaded slopes, in the Upper Sonoran and Transition zones.

2. Dichondra brachypoda Woot. & Standl. Contr. U. S. Nat. Herb. **16**: 160. 1913.

TYPE LOCALITY: Filmore Canyon in the Organ Mountains, New Mexico. Type collected by Wooton and Standley, September 23, 1906.

RANGE: Western Texas and southern New Mexico.

NEW MEXICO: Kingston; Organ Mountains; Queen. Canyons and in woods.

117. CUSCUTACEAE. Dodder Family.

1. CUSCUTA L. DODDER.

Herbaceous parasites with twining yellow or orange stems, the leaves reduced to minute scales; flowers perfect, waxy white, cymose; calyx of 5 or 4 imbricated lobes, accompanied by as many alternating, crenulate or appendaged scales, or these obsolete; stamens 5 or 4, attached to the corolla in the throat or near the sinuses above the scales; ovary 2-celled; styles 2, mostly distinct; stigmas capitate or elongated; capsules subglobose, depressed or elongated, circumscissile or indehiscent, 1 to 4-seeded.

It is said that the Navahos used the parched seeds of dodder as food.

KEY TO THE SPECIES.

Calyx of 5 almost distinct overlapping sepals, subtended by bracts. 1. *C. squamata.*
Calyx gamosepalous, bractless.
  Styles equal, with elongated stigmas.............................. 4. *C. epithymum.*
  Styles unequal, with capitate stigmas.
    Petals acute; styles longer than the capsule.............. 2. *C. umbellata*
    Petals obtuse; styles shorter than the capsule.............. 3. *C. curta.*

1. Cuscuta squamata Engelm. Trans. Acad. St. Louis. 1: 510. 1859.

TYPE LOCALITY: "El Paso."

RANGE: Western Texas and southern New Mexico and southward.

NEW MEXICO: Mesilla Valley; south of Roswell. Lower Sonoran Zone.

Common on *Helianthus ciliaris* and other plants of the valleys.

2. Cuscuta umbellata H. B. K. Nov. Gen. & Sp. **3:** 121. 1818.

TYPE LOCALITY: "Crescit in Nova Hispania, inter Querétaro et Salmanca, alt. 900 hex."

RANGE: Colorado and Texas to Arizona and Mexico.

NEW MEXICO: Santa Fe; Pajarito Park; Clayton; Torrevios; Hillsboro; Tortugas Mountain; Chavez; Duck Creek Flats. Lower and Upper Sonoran zones.

On many small herbs, such as *Trianthema portulacastrum, Kallstroemia brachystylis, Cladothrix lanuginosa, Chamaesyce* spp., *Eriogonum rotundifolium, Bahia dealbata, Wedeliella glabra, Boerhaavia torreyana, Cassia bauhinioides, Chamaesaracha conioides,* and many others; also on cultivated beets.

3. Cuscuta curta Engelm.; Rydb. Colo. Agr. Exp. Sta. Bull. **100:** 273. 1906.

Cuscuta gronovii curta Engelm. Trans. Acad. St. Louis **1:** 508. 1859.

TYPE LOCALITY: "Northwest America."

RANGE: Utah and Colorado to Arizona and New Mexico.

NEW MEXICO: Shiprock; Farmington; Albuquerque; Santa Fe; Sandia Mountains; Nara Visa; Ojo Caliente; Chiz; Mogollon Mountains; Kingston; Mesilla Valley; La Luz; White Mountains. Chiefly in the Upper Sonoran and Transition zones.

On various plants, such as *Gaertneria acanthicarpa, Peritoma serrulatum,* Salix, *Salsola pestifer, Xanthium commune, Helianthus annuus, Rumex mexicanus,* and *Aster hesperius,* and on cultivated plants such as beets and chile.

4. Cuscuta epithymum L. Sp. Pl. 124. 1753. CLOVER DODDER.

TYPE LOCALITY: "Habitat in Plantis Europae parasitica."

NEW MEXICO: Cedar Hill (*Standley* 8058).

On alfalfa; introduced from Europe.

118. CONVOLVULACEAE. Morning-glory Family.

Annual or perennial herbs, often twining; leaves alternate, exstipulate; flowers perfect and regular, axillary, solitary or cymose; calyx of 5 more or less united imbricated sepals, persistent; corolla hypogynous, convolute in bud, the limb often entire; stamens 5, alternate with the divisions of the corolla, often epipetalous; pistil of 2 more or less united carpels, the ovary 2 to 5-celled, on a fleshy disk; styles often united; fruit a capsule, 1 to 5-celled, the seeds large; endosperm mucilaginous.

KEY TO THE GENERA.

Styles distinct or at least partly so; decumbent or ascending herbs, not twining or trailing.
- Styles partially united, entire; limb of corolla deeply 5-lobed, the lobes ovate-lanceolate... 1. CRESSA (p. 516).
- Styles distinct, each 2-cleft; corolla limb not lobed. 2. EVOLVULUS (p. 516).

Styles united up to the stigma; climbing or trailing vines (except *Ipomoea leptophylla*).
- Corolla narrowly funnelform, nearly salverform, bright scarlet; stamens and style exserted.... 3. QUAMOCLIT (p. 517).
- Corolla broadly funnelform, never scarlet; stamens and style included.
  - Stigmas 1 to 3, ovoid or subglobose............ 4. IPOMOEA (p. 517).
  - Stigmas usually 2, filiform to oblong-cylindric.. 5. COLVOLVULUS (p. 519).

1. CRESSA L.

Low herbaceous perennial with slender ascending stems and very numerous small lanceolate entire sericeous leaves; flowers small, axillary, solitary, white; calyx of 5 nearly equal distinct lobes; corolla persistent, the limb 5-lobed; stamens exserted; capsule little longer than broad, the seeds smooth and shining, often solitary.

1. **Cressa truxillensis** H. B. K. Nov. Gen. & Sp. **3:** 119. 1818.

Cressa cretica truxillensis Choisy in DC. Prodr. **9:** 440. 1845.

TYPE LOCALITY: "In arenosis salsis Oceani Pacifici, prope Truxillo Peruvianorum."

RANGE: California and Utah to Texas and Mexico.

NEW MEXICO: Mesilla Valley. In heavy soil of river valleys, in the Lower Sonoran Zone.

2. EVOLVULUS L.

Small prostrate or diffuse perennials with more or less densely pubescent stems and small simple entire leaves; flowers solitary in the axils; sepals nearly equal; corolla funnelform or nearly rotate, white or blue, the limb entire; ovary 2-celled, the capsules subglobose, 2 to 4-valved, the seeds 1 to 4.

KEY TO THE SPECIES.

Flowers white; upper surface of leaves glabrous.................. 1. *E. wilcoxianus.*
Flowers blue; leaves pubescent on both surfaces.
  Pedicels 3 mm. long or less; flowers 10 to 12 mm. in diameter.. 2. *E. pilosus.*
  Pedicels 25 to 40 mm. long; flowers not more than 8 mm. in diameter.. 3. *E. linifolius.*

Evolvulus arizonicus should be found in New Mexico about the southwestern corner. It is similar to *E. linifolius* but has flowers about 15 mm. broad.

1. **Evolvulus wilcoxianus** House, Bull. Torrey Club **33:** 315. 1906.

TYPE LOCALITY: Near Fort Huachuca, Arizona.

RANGE: New Mexico and Arizona and adjacent Mexico.

NEW MEXICO: Las Vegas; Mangas Springs; Kingston; Water Canyon; San Luis Pass; Organ Mountains. Dry hills and plains, in the Upper Sonoran Zone.

2. **Evolvulus pilosus** Nutt. Gen. Pl. **1:** 174. 1818.

Evolvulus argenteus Pursh, Fl. Amer. Sept. 187. 1814, not R. Br. 1810.

Evolvulus nutallianus Roem. & Schult. Syst. Veg. **6:** 198. 1820.

Evolvulus oreophilus Greene, Leaflets **1:** 151. 1905.

TYPE LOCALITY: "On the banks of the Missouri."

RANGE: South Dakota and Colorado to Arizona and Mexico.

NEW MEXICO: Farmington; Raton; Nara Visa; Perico Creek; Albuquerque; Portales; Torrance; Socorro Mountain; Hillsboro; Organ Mountains; Leachs Ranch; Redlands. Plains and low hills, in the Upper Sonoran Zone.

The type of *E. oreophilus* was collected near Hillsboro (*Metcalfe* 1228).

3. **Evolvulus linifolius** L. Sp. Pl. ed. 2. 392. 1762.

TYPE LOCALITY: Jamaica.

RANGE: Southern New Mexico and Arizona to Mexico and the West Indies.

NEW MEXICO: Carrizalillo Mountains; Organ Mountains.

Our specimens are scarcely typical, having more abundant pubescence and larger flowers than the southern forms. They probably belong to an undescribed species, but it seems inadvisable to attempt to separate them until the difficult genus can be carefully revised.

3. QUAMOCLIT Moench.

Annual or perennial twining vines, glabrous throughout, with entire or lobed leaves and axillary few-flowered cymes of bright scarlet flowers on peduncles exceeding the leaves; sepals mostly equal; corolla about 25 mm. long, narrowly funnelform to salverform, the limb short and spreading, not lobed, more or less pentagonal, expanding in daylight; stamens exserted; ovary 2-celled or falsely 4-celled; capsules subglobose, 7 to 8 mm. in diameter; seeds 4.

KEY TO THE SPECIES.

Leaves entire or nearly so.. 1. *Q. coccinea.*
Leaves deeply 3-lobed, the lateral lobes usually again lobed........ 2. *Q. hederifolia.*

1. Quamoclit coccinea (L.) Moench, Meth. Pl. 453. 1794.
Ipomoea coccinea L. Sp. Pl. 160. 1753.
TYPE LOCALITY: "Habitat in Domingo."
RANGE: Tropical regions of America, frequently escaped in the United States.
NEW MEXICO: Santa Fe; Pecos; Animas Creek; Fort Bayard.

2. Quamoclit hederifolia (L.) Choisy in DC. Prodr. **9**: 336. 1845.
Ipomoea hederifolia L. Syst. Veg. ed. 10. **2**: 925. 1759.
TYPE LOCALITY: Not stated.
RANGE: Western Texas to southern Arizona and southward.
NEW MEXICO: Las Vegas; Fort Bayard; Kingston; Organ Mountains; Gray; White Mountains. Moist ground, in the Upper Sonoran and Transition zones.

4. IPOMOEA L. MORNING-GLORY.

Annual or perennial herbs with twining or erect stems; leaves entire or lobed; flowers axillary, solitary or in few-flowered cymes; sepals membranous or rather fleshy, sometimes becoming leathery, closely imbricated, not elongated, persistent; corolla white, pink, blue, or purple, funnelform, the limb usually spreading and relatively ample, pentagonal or circular; stamens included; ovary 2 to 4-celled; capsules 2 to 4-valved; seeds often pubescent.

KEY TO THE SPECIES.

Stems erect, stout, much branched, forming dense clumps; leaves linear.. 1. *I. leptophylla.*
Stems twining or decumbent, slender; leaves not linear.
  Sepals with green foliaceous tips, acute or attenuate, hirsute.
    Leaves entire or nearly so................................ 6. *I. purpurea.*
    Leaves 3-lobed or 3-parted.
      Tips of the sepals merely acute......................... 7. *I. hirsutula.*
      Tips of the sepals with long attenuate tips.
        Corolla 3 cm. long or less; sepals scarcely dilated at the base, the tips spreading............. 8. *I. desertorum.*
        Corolla 8 cm. long; sepals much dilated at the base, the tips erect......................... 9. *I. lindheimeri.*
  Sepals scarious, at least on the margins, not with foliaceous tips, rounded to acuminate at the apex, glabrous or nearly so.
    Leaves cordate-ovate, not cleft............................ 5. *I. cardiophylla.*
    Leaves 5 to 7-cleft.
      Corolla more than 5 cm. long........................... 2. *I. tenuiloba.*
      Corolla less than 3 cm. long.
        Sepals muricate; perennial from a tuberous root.. 3. *I. muricata.*
        Sepals smooth; annual.............................. 4. *I. costellata.*

1. Ipomoea leptophylla Torr. Frém. Rep. Exped. Rocky Mount. 94. 1845.
Convolvulus canadensis Buckl. Proc. Acad. Phila. **1862**: 6. 1862.
TYPE LOCALITY: "Forks of the Platte to Laramie river."
RANGE: Montana and South Dakota to Texas and northeastern New Mexico.
NEW MEXICO: Las Vegas; Sierra Grande; Cross L Ranch; Nara Visa; Clayton; south of Melrose. Plains and low hills, in the Upper Sonoran Zone.

2. Ipomoea tenuiloba Torr. U. S. & Mex. Bound. Bot. 148. 1859.
TYPE LOCALITY: "Hills and rocky places near Puerto de Paysano, western Texas."
RANGE: Western Texas and southern New Mexico.
NEW MEXICO: Guadalupe Mountains (*Bailey* 720).

3. Ipomoea muricata Cav. Icon. Pl. **5**: 52. *pl. 478. f. 2.* 1794.
Convolvulus capillaceus H. B. K. Nov. Gen. & Sp. **3**: 97. 1819.
Ipomoea capillacea Don, Hist. Dichl. Pl. **4**: 267. 1838.
Ipomoea capillacea patens A. Gray, Syn. Fl. ed. 2. **2**[1]: 434. 1886.
Ipomoea patens House, Ann. N. Y. Acad. **18**: 237. 1908.
TYPE LOCALITY: "Habitat in Huanajuato," Mexico.
RANGE: New Mexico and Arizona to Mexico.
NEW MEXICO: Las Vegas; Mogollon Mountains; Hanover Mountain; White and Sacramento mountains. Open slopes in the mountains, in the Transition Zone.

The form named *patens* certainly does not deserve nomenclatorial recognition. It is a mere seasonal variation and even at any time in a given spot one may collect both forms. The smaller, more erect plant later becomes spreading and has longer stems and leaf segments.

4. Ipomoea costellata Torr. U. S. & Mex. Bound. Bot. 149. 1859.
TYPE LOCALITY: "On the Rio Grande, from the mouth of Pecos to El Paso, and near the Copper Mines of New Mexico."
RANGE: Western Texas to Arizona and southward.
NEW MEXICO: Mangas Springs; Mogollon Mountains; Organ Mountains; Roswell; Queen. Open slopes, from the Lower Sonoran to the lower part of the Transition Zone.

5. Ipomoea cardiophylla A. Gray, Syn. Fl. **2**[1]: 213. 1878.
TYPE LOCALITY: "In the mountains near El Paso."
RANGE: Western Texas and southern New Mexico to Mexico.
NEW MEXICO: Organ Mountains; White Mountains. Canyons, in the Upper Sonoran Zone.

6. Ipomoea purpurea (L.) Lam. Tabl. Encycl. **1**: 466. 1791.
Convolvulus purpureus L. Sp. Pl. ed. 2. 219. 1762.
Pharbitis purpurea Voigt, Hort. Calcutt. 354. 1845.
TYPE LOCALITY: "Habitat in America."
RANGE: Throughout tropical America, frequently introduced elsewhere.
NEW MEXICO: Manzanares Valley; Hillsboro; Mesilla Valley; Gilmores Ranch.

7. Ipomoea hirsutula Jacq. Eclog. Pl. Rar. **1**: 63. 1811.
Ipomoea mexicana A. Gray, Syn. Fl. **2**[1]: 218. 1878.
TYPE LOCALITY: Mexico.
RANGE: Western Texas to Arizona and southward.
NEW MEXICO: Santa Fe; Pecos; Fort Bayard; Grand Canyon of the Gila; San Luis Mountains; Mesilla Valley; Organ Mountains; White Sands; White Mountains; Gray; Roswell. Waste ground, in the Lower and Upper Sonoran zones.

8. Ipomoea desertorum House, Ann. N. Y. Acad. **18**: 203. 1908.
TYPE LOCALITY: Tucson, Arizona.
RANGE: Southern Arizona and New Mexico.
NEW MEXICO: Mesa west of Organ Mountains; Florida Mountains. Dry hills and mesas, in the Lower Sonoran Zone.

9. Ipomoea lindheimeri A. Gray, Syn. Fl. **2**[1]: 210. 1878.

Pharbitis lindheimeri Small, Fl. Southeast. U. S. 964. 1903.

TYPE LOCALITY: Western Texas.

RANGE: Western Texas and southern New Mexico.

NEW MEXICO: Guadalupe Mountains; west of Hope. Dry hills, in the Upper Sonoran Zone.

5. CONVOLVULUS L. BINDWEED.

Annual or perennial twining herbs with petiolate, hastate or cordate leaves and solitary or clustered, axillary, white or pink flowers; calyx naked or subtended by bracts; sepals nearly equal or the outer one larger; corolla funnelform, the limb entire or somewhat 5-angled; stamens included; ovary 1 or 2-celled; capsules globose or nearly so, 2 to 4-valved; seeds glabrous.

KEY TO THE SPECIES.

Bracts large, near the calyx and inclosing it; flowers white............ 1. *C. sepium.*
Bracts small, remote from the calyx; flowers pinkish.
  Plants nearly glabrous, sparingly pilose; leaf blades hastate, otherwise entire.. 2. *C. arvensis.*
  Plants canescent; blades linear or narrowly oblong, with deeply cleft basal lobes.. 3. *C. incanus.*

1. Convolvulus sepium L. Sp. Pl. 153. 1753. BINDWEED.

TYPE LOCALITY: "Habitat in Europae sepibus."

RANGE: British America to New Mexico and North Carolina; in New Mexico apparently introduced.

NEW MEXICO: Farmington; Abiquiu; Mesilla Valley.

Not uncommon as a weed in cultivated fields.

2. Convolvulus arvensis L. Sp. Pl. 153. 1753.

Convolvulus ambigens House, Bull. Torrey Club **32**: 139. 1905.

TYPE LOCALITY: "Habitat in Europae agris."

RANGE: Of wide occurrence in North America, in New Mexico introduced from the east or from Europe.

NEW MEXICO: Farmington; Cedar Hill; Raton; Santa Fe; Chama; Clovis; Kingston; Silver City; Rio Gila; Mesilla Valley.

This is a very variable species, but it seems ill advised to attempt to separate any of the forms. In New Mexico it is common in some localities in cultivated fields, where it is evidently introduced, as it doubtless is everywhere in the Rocky Mountain region. The amount of variation among the different forms is very slight, and every possible intermediate can be found between them. *Convolvulus ambigens* seems to differ in no way from numerous European specimens of *C. arvensis* in the U. S. National Herbarium.

3. Convolvulus incanus Vahl, Symb. Bot. **3**: 23. 1794.

TYPE LOCALITY: "America."

RANGE: Colorado and Kansas to Texas and Mexico.

NEW MEXICO: Albert; Tucumcari; Las Vegas; Clayton; Frisco; Socorro; Mangas Springs; Kingston; San Luis Mountains; Carrizalillo Mountains; Organ Mountains; Gray; Roswell; Queen. Dry hills and plains, in the Upper Sonoran Zone.

119. POLEMONIACEAE. Phlox Family.

Annual or perennial herbs or low shrubs, never twining, with opposite or alternate, simple or compound leaves, and regular 5-merous flowers; calyx gamosepalous, 5-lobed, persistent, imbricated; corolla convolute in the bud; stamens 5, equally or unequally inserted, epipetalous; ovary 3-celled, with a thick axis; styles united; stigmas 3; fruit a 3-celled loculicidal capsule; seeds several or solitary in each cell, the coats sometimes mucilaginous.

KEY TO THE GENERA.

Calyx not ruptured by the capsule; leaves alternate; seeds with mucilage and spiracles (spirally twisted threads) when wetted.

- Calyx teeth spinulose-tipped; leaves pinnatifid with linear segments; flowers blue, in few-flowered woolly heads .... 1. ERIASTRUM (p. 520).
- Calyx teeth herbaceous, not spinulose-tipped; leaves and flowers various.
  - Leaves simple, entire; flowers small, capitate-crowded; stamens straight .... 2. COLLOMIA (p. 521).
  - Leaves pinnatifid; flowers large, variously arranged; stamens declined .... 3. POLEMONIUM (p. 521).

Calyx at length ruptured by the maturing capsule; leaves alternate or opposite; seeds with or without mucilage and spiracles when wet.

- Calyx tube not at all or but slightly scarious, early splitting; leaves sessile, divided into several linear spinulose segments; seeds without mucilage and spiracles; plants strongly scented .... 4. LEPTODACTYLON (p. 522).
- Calyx tube more or less scarious between the lobes, distended, then ruptured by the capsule; leaves sessile or petiolate; seeds with or without mucilage when wet.
  - Corolla salverform, with a very narrow throat.
    - Seeds mucilaginous when wetted; annual with alternate floral leaves .... 5. MICROSTERIS (p. 523).
    - Seeds not altered when wetted; perennials with all the leaves opposite .... 6. PHLOX (p. 523).
  - Corolla funnelform or tubular, with an open throat.
    - Leaves alternate; inflorescence paniculate, thyrsiform, or capitate ..... 7. GILIA (p. 525).
    - Leaves opposite, spinulose-tipped; inflorescence various.
      - Flowers bluish, sessile; leaves with a few narrow unequal segments or simple .... 8. LINANTHUS (p. 529).
      - Flowers bright yellow, long-pediceled; leaves with several nearly equal segments appearing as a verticel of linear leaves.... 9. DACTYLOPHYLLUM (p. 529).

1. ERIASTRUM Woot. & Standl.

Low, wiry-stemmed, widely spreading annual, at first white-woolly throughout, the stems and lower leaves later glabrate; leaves alternate, sessile; flowers small, blue or white, in few-flowered glomerate heads surrounded by several crowded woolly leaflike bracts; leaves with 3 narrowly linear spinulose lobes, the segments unequal; corolla salverform, the tube longer than the sepals; capsules not distending the calyx, several-seeded.

1. **Eriastrum filifolium** (Nutt.) Woot. & Standl. Contr. U. S. Nat. Herb. **16:** 160. 1913.

Gilia filifolia Nutt. Journ. Acad. Phila. II. **1:** 156. 1848.

Navarretia filifolia Brand in Engl. Pflanzenreich **27:** 167. 1907.

TYPE LOCALITY: Near Santa Barbara, California.

RANGE: Washington and California to New Mexico and western Texas.

NEW MEXICO: Nutt Mountain; Crawfords Ranch; Mesilla Valley; Organ Mountains. Dry plains, in the Lower Sonoran Zone.

2. COLLOMIA Nutt.

Erect annual with alternate entire leaves and small flowers crowded at the top of the stem; calyx scarious between the lobes, not distended by the capsule; corolla tubular-funnelform with open throat and short obtuse lobes; stamens unequal, unequally inserted on the corolla tube; capsules narrowed at the base; seeds mucilaginous when wetted, emitting spiracles.

1. **Collomia linearis** Nutt. Gen. Pl. **1:** 126. 1818.

Gilia linearis A. Gray, Proc. Amer. Acad. **17:** 223. 1882.

TYPE LOCALITY: "Near the banks of the Missouri about the confluence of Shian River, and in the vicinity of the Arikaree village."

RANGE: British Columbia and North Dakota to California, New Mexico, and Nebraska.

NEW MEXICO: Tunitcha Mountains; Chama; Santa Fe and Las Vegas mountains. Meadows in the mountains, Transition Zone.

3. POLEMONIUM L.

Perennial herbs 1 meter high or less, with pinnate leaves and blue, purplish, white, or yellow flowers in panicles or glomerate terminal clusters; calyx rotate-campanulate to tubular, not scarious between the herbaceous lobes; corolla narrowly funnelform to campanulate or almost rotate; stamens equally inserted, the filaments more or less declined, pilose-appendaged at the base; capsules oblong to globose, not bursting from the persistent calyx; seeds black or brown, oblong, often angled or winged.

KEY TO THE SPECIES.

Corolla funnelform.
- Corolla purple, the tube thick .... 1. *P. confertum.*
- Corolla ochroleucous, the tube slender .... 2. *P. mellitum.*

Corolla campanulate to nearly rotate.
- Corolla yellow .... 3. *P. flavum.*
- Corolla blue, rarely white, never yellow.
  - Plants low, 10 to 20 cm.; flowers pale blue.
    - Seeds not winged .... 4. *P. scopulinum.*
    - Seeds narrowly winged .... 5. *P. pterospermum.*
  - Plants tall, 40 to 100 cm.; flowers deep blue.
    - Inflorescence narrow, thyrsiform; leaflets lanceolate, attenuate, rather distant .... 6. *P. filicinum.*
    - Inflorescence broad and open; leaflets elliptic, less acute.
      - Corolla about 20 mm. long; leaves and lower part of the stem glabrous .... 7. *P. grande.*
      - Corolla 10 to 12 mm. long; leaves and stem viscid-pubescent .... 8. *P. molle.*

1. **Polemonium confertum** A. Gray, Proc. Acad. Phila. **1863:** 73. 1864.

TYPE LOCALITY: Rocky Mountains of Colorado.

RANGE: Wyoming to northern New Mexico.

NEW MEXICO: Wheelers Peak; Upper Pecos River. High mountain meadows, in the Arctic-Alpine Zone.

2. **Polemonium mellitum** (A. Gray) A. Nels. Bull. Torrey Club **26:** 354. 1899.

Polemonium confertum mellitum A. Gray, Proc. Acad. Phila. **1863:** 73. 1864.

TYPE LOCALITY: Rocky Mountains of Colorado.

RANGE: Wyoming and Nevada to Colorado and New Mexico.

NEW MEXICO: Sandia Mountains. High mountains, in the Hudsonian Zone.

3. **Polemonium flavum** Greene, Bot. Gaz. **6:** 217. 1881.

TYPE LOCALITY: Cold northward slopes of the highest Pinos Altos Mountains, New Mexico. Type collected by E. L. Greene in 1880.

RANGE: Mountains of southwestern New Mexico.

NEW MEXICO: Willow Creek; Eagle Peak; Hillsboro Peak; West Fork of the Gila; near East View. Transition Zone.

4. **Polemonium scopulinum** Greene; Rydb. Colo. Agr. Exp. Sta. Bull. **100:** 280. 1906.

TYPE LOCALITY: Colorado.

RANGE: Colorado and northern New Mexico.

NEW MEXICO: Santa Fe and Las Vegas mountains. Deep woods, Canadian to Arctic-Alpine Zone.

5. **Polemonium pterospermum** Nels. & Cockerell, Proc. Biol. Soc. Washington **16:** 45. 1903.

TYPE LOCALITY: Cloudcroft, Sacramento Mountains, New Mexico. Type collected by Cockerell, September, 1900.

RANGE: Known only from type locality, in the Canadian Zone.

We have seen no specimens of this species.

6. **Polemonium filicinum** Greene, Pittonia **1:** 124. 1887.

TYPE LOCALITY: Pinos Altos Mountains, New Mexico. Type collected by E. L. Greene in 1880.

RANGE: Mountains of southern New Mexico and Arizona.

NEW MEXICO: Lookout Mines; White and Sacramento mountains. In the Transition and Canadian zones.

7. **Polemonium grande** Greene, Leaflets **1:** 153. 1905.

TYPE LOCALITY: Near Pagosa Peak, southern Colorado.

RANGE: Colorado and New Mexico.

NEW MEXICO: Santa Fe and Las Vegas mountains; Sacramento Mountains. Damp woods and along streams, from the Transition to the Hudsonian Zone.

8. **Polemonium molle** Greene, Leaflets **1:** 153. 1905.

TYPE LOCALITY: Piedra, southern Colorado.

RANGE: Colorado and northern New Mexico.

NEW MEXICO: Chama; Sandia Mountains; Pecos Baldy. Mountains, in the Canadian Zone.

4. LEPTODACTYLON Hook. & Arn.

Erect, tufted, more or less woody plants, 50 cm. high or less, with opposite or alternate, palmately 3 to 7-parted leaves having linear spinulose segments, the white or yellowish flowers with tubular-funnelform corollas; calyx tube not at all or but slightly scarious between the lobes, at length ruptured by the swelling capsule; seeds not mucilaginous, without spiracles.

KEY TO THE SPECIES.

Leaves alternate, rigid, spinescent.................................. 1. *L. brevifolium.*
Leaves opposite, neither rigid nor spinescent........................ 2. *L. nuttallii.*

1. **Leptodactylon brevifolium** Rydb. Bull. Torrey Club **40:** 474. 1913.

TYPE LOCALITY: Juniper Range, Utah.

RANGE: Washington and Nevada to Colorado and New Mexico.

NEW MEXICO: Cedar Hill; near Ojo Caliente. Dry hills and plains, in the Upper Sonoran Zone.

2. **Leptodactylon nuttallii** (A. Gray) Rydb. Colo. Agr. Exp. Sta. Bull. **100:** 279. 1906.

Gilia nuttallii A. Gray, Proc. Amer. Acad. **8:** 267. 1870.

TYPE LOCALITY: "Rocky Mountains of Colorado and Utah to the Sierra Nevada in California."

RANGE: Washington and California to Wyoming and New Mexico.

NEW MEXICO: Ramah; Burro Mountains; Mogollon Mountains; White Mountains. Mountains, in the Transition Zone.

Most of our specimens are considerably taller and more slender than the typical form, and are inclined to be woody throughout.

5. MICROSTERIS Greene.

Slender annual with mostly alternate leaves and small, loosely cymose or scattered, purplish flowers; calyx at length ruptured by the capsule; corolla salverform, with a narrow throat; seeds mucilaginous when wetted, without spiracles.

1. **Microsteris micrantha** (Kellogg) Greene, Pittonia **3:** 303. 1898.

Collomia micrantha Kellogg, Proc. Calif. Acad. **3:** 18. 1863.

TYPE LOCALITY: "Vicinity of Silver City, Nevada Territory."

RANGE: Nebraska and Wyoming to California and New Mexico.

NEW MEXICO: Sandia Mountains; Tierra Amarilla. Open slopes, in the Transition Zone.

6. PHLOX L. PHLOX.

Perennial herbs, 30 or 40 cm. high or less, the base of the stem often woody, with opposite sessile leaves and cymose flowers with showy corollas; calyx narrow, scarious between the lobes; corolla salverform, constricted in the throat; stamens unequally inserted, included; capsules ovoid, ultimately rupturing the persistent calyx.

KEY TO THE SPECIES.

Plants densely cespitose, forming thick mats; leaves more or less fascicled.
  Leaves with many cobwebby hairs, not glandular-ciliate; plant of low dry hills.. 1. *P. canescens.*
  Leaves glandular-ciliate, without cobwebby hairs; plant of alpine meadows.. 2. *P. caespitosa.*
Plants not cespitose, loose, erect; leaves not fascicled.
  Tube of the corolla fully twice as long as the calyx.
    Calyx 13 to 14 mm. long; leaves long, 20 to 40 mm., linear or nearly so.. 3. *P. stansburyi.*
    Calyx less than 10 mm. long; leaves short, 10 to 20 mm., linear-lanceolate.. 4. *P. grayi.*
  Tube of the corolla considerably less than twice as long as the calyx.

Plants not glandular.
Corolla lobes rounded-obovate, 12 mm. wide; leaves spreading.. 5. *P. triovulata.*
Corolla lobes narrowly obovate, 5 mm. wide; leaves erect.. 6. *P. longifolia.*
Plants glandular, at least on the inflorescence.
Leaves ascending or erect, linear-lanceolate; stems densely glandular.. 7. *P. nana.*
Leaves divaricate, linear; stems sparingly glandular.
Corolla lobes orbicular-obovate, 10 to 15 mm. broad. 8. *P. mesoleuca.*
Corolla lobes cuneate-oblanceolate, 5 mm. broad... 9. *P. tenuis.*

1. **Phlox canescens** Torr. & Gray, U. S. Rep. Expl. Miss. Pacif. **2**[2]: 122. *pl. 6.* 1855.

TYPE LOCALITY: "On the Cedar Mountains, south of Great Salt Lake, Utah."

RANGE: Washington and California to Colorado and northwestern New Mexico.

NEW MEXICO: Farmington; Aztec. Dry hills, in the Upper Sonoran Zone.

2. **Phlox caespitosa** Nutt. Journ Acad. Phila. **7**: 41. 1834.

TYPE LOCALITY: "Flat-Head River on the sides of dry hills."

RANGE: Washington and Montana to Colorado and northern New Mexico.

NEW MEXICO: Pecos Baldy; Truchas Peak. Mountain meadows, in the Arctic-Alpine Zone.

3. **Phlox stansburyi** (Torr.) Heller, Bull. Torrey Club **24**: 478. 1897.

Phlox speciosa stansburyi Torr. U. S. & Mex. Bound. Bot. 145. 1859.

Phlox longifolia stansburyi A. Gray, Proc. Amer. Acad. **8**: 255. 1870.

TYPE LOCALITY: Gravelly hills near the Organ Mountains, New Mexico. Type collected by Bigelow.

RANGE: New Mexico.

NEW MEXICO: Barranca; mountains west of San Antonio; Organ Mountains. Dry foothills, in the Upper Sonoran Zone.

We have seen little material that agrees with the typical form. That from Utah, Nevada, and Arizona referred here certainly represents a different species.

4. **Phlox grayi** Woot. & Standl. Contr. U. S. Nat. Herb. **16**: 161. 1913.

Phlox longifolia stansburyi forma *brevifolia* A. Gray, Proc. Amer. Acad. **8**: 255. 1870, not *P. brevifolia* Baum. 1824.

Phlox longifolia brevifolia A. Gray, Syn. Fl. **2**[1]: 133. 1878.

TYPE LOCALITY: Not definitely stated.

RANGE: Utah and Nevada to Arizona and northwestern New Mexico.

NEW MEXICO: A single specimen seen, probably from the Navaho Reservation (*Marsh* 4). Upper Sonoran Zone.

5. **Phlox triovulata** Thurb.; Torr. U. S. & Mex. Bound. Bot. 145. 1859.

? *Phlox nana glabella* A. Gray, Proc. Amer. Acad. **8**: 256. 1870.

TYPE LOCALITY: Ravines, Mule Spring, New Mexico. Type collected by Thurber.

RANGE: Southern New Mexico and Arizona and adjacent Mexico.

NEW MEXICO: Berendo Creek; Hanover Mountain; Organ Mountains; Queen. Low hills, in the Upper Sonoran Zone.

6. **Phlox longifolia** Nutt. Journ. Acad. Phila. **7**: 41. 1834.

TYPE LOCALITY: "Valleys of the Rocky Mountains generally."

RANGE: Washington and Montana to Colorado and northern New Mexico.

NEW MEXICO: Nacimiento Mountain; Aztec; Chama; Tierra Amarilla. Open slopes and low hills, in the Upper Sonoran Zone.

7. **Phlox nana** Nutt. Journ. Acad. Phila. II. **1:** 153. 1848.

Phlox nana oculata Cockerell, Amer. Nat. **36:** 813. 1902.

Phlox nana lilacina Cockerell, loc. cit.

TYPE LOCALITY: "Rocky Mountains near Santa Fe," New Mexico. Type collected by Gambel.

RANGE: New Mexico and Arizona.

NEW MEXICO: Santa Fe; Gallinas Canyon; Glorieta; Bernal; Gallinas Mountains; Magdalena Mountains; Jicarilla Mountains. Hills and mountains, in the Upper Sonoran and Transition zones.

8. **Phlox mesoleuca** Greene, Leaflets **1:** 152. 1905.

TYPE LOCALITY: Kingston, New Mexico. Type collected by Metcalfe (no. 1272).

RANGE: Southern New Mexico.

NEW MEXICO: Kingston; White and Sacramento mountains. Mountains, in the Transition Zone.

9. **Phlox tenuis** Woot. & Standl. Contr. U. S. Nat. Herb. **16:** 161. 1913.

TYPE LOCALITY: Barranca, Taos County, New Mexico. Type collected by Heller (no. 3589).

RANGE: Northern New Mexico and southern Colorado.

NEW MEXICO: Barranca (*Heller* 3589). Dry hills, in the Upper Sonoran Zone.

7. **GILIA** Ruiz & Pavon.

Annuals or perennials with alternate, mostly pinnately divided leaves, the flowers in thyrsiform, paniculate, or glomerate clusters; calyx tubular or campanulate, scarious between the lobes, burst by the enlarging capsule; corolla tubular-funnelform, with open throat and usually a short limb, red, white, pink, or blue; stamens inserted equally or unequally; seeds mucilaginous when wetted, producing spiracles.

KEY TO THE SPECIES.

Corolla scarlet, usually spotted with bright yellow.
- Lobes of the calyx about half as long as the tube; corolla yellowish red .... 1. *G. greeneana.*
- Lobes of the calyx equaling the tube; corolla seldom yellowish.
  - Stems glabrate in age; calyx lobes usually spreading, attenuate .... 2. *G. formosissima.*
  - Stems permanently tomentulose; calyx lobes erect, acuminate .... 3. *G. texana.*

Corolla never red, sometimes pinkish.
- Flowers capitately glomerate; annuals.
  - Leaves pinnatifid; corolla bluish .... 18. *G. pumila.*
  - Leaves entire; corolla white .... 19. *G. gunnisonii.*
- Flowers openly paniculate or thyrsiform-paniculate; annuals or perennials.
  - Lobes of the corolla twice as long as the tube or more; corolla bright blue; cauline leaves with rigid acerose segments .... 20. *G. acerosa.*
  - Lobes of the corolla not exceeding the tube; corolla of various colors; leaves various.
    - Inflorescence thyrsiform.
      - Corolla lobes about equaling the tube, the tube not exserted from the calyx .... 5. *G. brachysiphon.*

Corolla lobes much shorter than the tube, the latter much exserted.
Tube of the corolla strongly bent downward; flowers white.................. 6. *G. campylantha.*
Tube of the corolla straight, erect; flowers white or blue.
Flowers white; calyx lobes obtuse, one-third as long as the tube or less.... 4. *G. candida.*
Flowers blue; calyx lobes about equaling the tube, never obtuse.
Tube of the corolla less than 15 mm. long .... 7. *G. multiflora.*
Tube of the corolla 25 mm. long or more.
Corolla tube 25 to 30 mm. long; lobes long-caudate........ 8. *G. pringlei.*
Corolla tube 30 to 35 mm. long; lobes obtuse or short-apiculate................. 9. *G. thurberi.*
Inflorescence openly paniculate.
Leaves entire or the lower ones toothed or lobed.
Leaves all entire.........................10. *G. formosa.*
At least part of the leaves toothed or lobed.
Corolla less than 1 cm. long............11. *G. leptomeria.*
Corolla more than 1 cm. long.
Cauline leaves oblanceolate, often toothed; stems stout; capsules 5 mm. long............12. *G. crandallii.*
Cauline leaves lance-linear, entire; stems slender; capsules 3.5 mm. long...................13. *G. haydeni.*
Leaves all except the uppermost pinnately divided.
Corolla tube over 1 cm. long.
Corolla tube 30 to 40 mm. long, the lobes obtuse.....................16. *G. longiflora.*
Corolla tube 15 to 25 mm. long, the lobes acute.....................17. *G. laxiflora.*
Corolla tube less than 1 cm. long.
Stamens exserted; corolla salverform...14. *G. viscida.*
Stamens included; corolla funnelform or nearly so.....................15. *G. inconspicua.*

1. **Gilia greeneana** Woot. & Standl. Contr. U. S. Nat. Herb. **16**: 161. 1913.
Callisteris collina Greene, Leaflets **1**: 159. 1905, not *Gilia collina* Eastw. 1904.
Batanthes collina Greene, op. cit. 224. 1906.
Gilia attenuata collina Cockerell, Univ. Mo. Stud. Sci. **2**[2]: 197. 1911.

Type locality: Bluffs of Clear Creek on the plains not far from Denver, Colorado.

Range: Colorado and New Mexico.

New Mexico: Rio Pueblo; Raton Mountains; Trinchera Pass; Santa Clara Canyon; Santa Fe; Winsors Ranch; Chama; Raton; Beulah; Glorieta; Sandia Mountains. Meadows in the mountains, in the Transition and Canadian zones.

We can not agree with Doctor Greene in separating the genus Batanthes from Gilia. At first glance these species seem to form a group distinct enough, but after one studies them carefully in their relation to other groups of the genus, it seems impossible to separate them on any logical ground. About the only distinction is the color of the flowers, scarcely a sufficient basis for generic segregation; and even this is not constant, for among the red-flowered plants white-flowered individuals are common. If the group is accorded generic rank, it should receive the name Ipomopsis, typified by *Gilia coronopifolia*. It is scarcely possible to conceive of that species and *G. aggregata* as belonging to different genera.

This and the next species were segregated from *G. aggregata*. That species certainly does not occur in New Mexico, at least among the specimens we have examined.

2. **Gilia formosissima** (Greene) Woot. & Standl. Contr. U. S. Nat. Herb. **16:** 161. 1913.

Callisteris formosissima Greene, Leaflets **1:** 160. 1905.

Batanthes formosissima Greene, op. cit. 224. 1906.

TYPE LOCALITY: Black Range, southern New Mexico. Type collected by Metcalfe.

RANGE: New Mexico and Arizona.

NEW MEXICO: Burro Mountains; Tunitcha Mountains; Carrizo Mountains; Farmington; Dulce; West Fork of the Gila; Hillsboro Peak; Organ Mountains. Mountains, in the Transition and Canadian zones.

3. **Gilia texana** (Greene) Woot. & Standl. Contr. U. S. Nat. Herb. **16:** 161. 1913.

Callisteris texana Greene, Leaflets **1:** 160. 1905.

Batanthes texana Greene, op. cit. 224. 1906.

TYPE LOCALITY: Guadalupe Mountains, western Texas.

RANGE: Mountains of southern New Mexico and western Texas.

NEW MEXICO: White and Sacramento mountains; Capitan Mountains; Queen. Transition Zone.

According to Mr. Hightower, the plant is known among the Mexican population as "Vara de San José."

4. **Gilia candida** Rydb. Bull. Torrey Club **28:** 29. 1901.

TYPE LOCALITY: Mesa near La Veta, Colorado.

RANGE: Mountains of Colorado and northeastern New Mexico.

NEW MEXICO: Raton; Sierra Grande. Transition and Canadian zones.

5. **Gilia brachysiphon** Woot. & Standl. Contr. U. S. Nat. Herb. **16:** 160. 1913.

TYPE LOCALITY: Van Pattens Camp in the Organ Mountains, New Mexico. Type collected by Wooton, August 29, 1894.

RANGE: Southern and western New Mexico.

NEW MEXICO: Kingston; Carlisle; mountains southeast of Patterson; Organ Mountains. Open woods, in the Upper Sonoran and Transition zones.

6. **Gilia campylantha** Woot. & Standl. Contr. U. S. Nat. Herb. **16:** 160. 1913.

TYPE LOCALITY: San Luis Mountains, New Mexico. Type collected by E. A. Mearns (no. 2242).

RANGE: Known only from the San Luis Mountains of New Mexico and Mexico.

A remarkable species, distinguished from all the related ones by its small white flowers and the peculiarly formed corolla tube. Otherwise it suggests *G. glomeriflora* Benth., but that has a very different calyx.

7. **Gilia multiflora** Nutt. Journ. Acad. Phila. II. **1:** 154. 1848.

? *Gilia macombii* Torr.; A. Gray, Proc. Amer. Acad. **20:** 301. 1885.

TYPE LOCALITY: "Sandy hills along the borders of the Rio del Norte," New Mexico. Type collected by Gambel.

RANGE: New Mexico and Arizona.

NEW MEXICO: Gallup; Santa Fe; Canyon Largo; Ramah; Zuni; Mogollon Mountains; Santa Rita; Bear Mountains. Plains and dry hills, in the Upper Sonoran Zone.

8. **Gilia pringlei** A. Gray, Proc. Amer. Acad. **21**: 401. 1886.
Collomia pringlei Peter in Engl. & Prantl, Pflanzenfam. **14**[3a]: 48. 1891.
Gilia macombii pringlei Brand in Engl. Pflanzenreich **27**: 114. 1907.
TYPE LOCALITY: Hillsides west of Chihuahua, Mexico.
RANGE: Southern New Mexico and Arizona to northern Mexico.
NEW MEXICO: Santa Rita; San Luis Mountains; Dog Spring.

9. **Gilia thurberi** Torr.; A. Gray, Proc. Amer. Acad. **8**: 261. 1870.
Collomia thurberi A. Gray, loc. cit.
TYPE LOCALITY: Near Santa Rita, New Mexico. Type collected by Thurber.
RANGE: Southern Arizona and New Mexico and adjacent Mexico.
We have seen no further specimens of this from New Mexico.

10. **Gilia formosa** Greene; Brand in Engl. Pflanzenreich **27**: 119. 1907.
TYPE LOCALITY: Aztec, New Mexico.
RANGE: Known only from type locality in the Upper Sonoran Zone.

11. **Gilia leptomeria** A. Gray, Proc. Amer. Acad. **8**: 278. 1870.
TYPE LOCALITY: "Mountain valleys of Nevada and Utah."
RANGE: Colorado and New Mexico to California.
NEW MEXICO: Carrizo Mountains; Shiprock; Farmington. Dry hills and mesas, in the Upper Sonoran Zone.

12. **Gilia crandallii** Rydb. Bull. Torrey Club **31**: 634. 1904.
Gilia bakeri Greene; Brand in Engl. Pflanzenreich **27**: 119. 1907, as synonym.
TYPE LOCALITY: Durango, Colorado.
RANGE: Colorado and New Mexico to Nevada.
NEW MEXICO: San Juan Valley. Dry hills, in the Upper Sonoran Zone.

13. **Gilia haydeni** A. Gray, Proc. Amer. Acad. **11**: 85. 1876.
Gilia subnuda haydeni Brand in Engl. Pflanzenreich **27**: 119. 1907.
TYPE LOCALITY: "Mesa San Juan, southern border of Colorado or adjacent part of Utah."
RANGE: Colorado and northwestern New Mexico.
NEW MEXICO: Twenty miles south of Fruitland (*Wooton* 2853). Dry plains, in the Upper Sonoran Zone.

14. **Gilia viscida** Woot. & Standl. Contr. U. S. Nat. Herb. **16**: 161. 1913.
Gilia pinnatifida Nutt.; A. Gray, Proc. Amer. Acad. **8**: 276. 1870, not Moc. & Sessé, 1837.
TYPE LOCALITY: "N. New Mexico and Colorado to Snake River."
RANGE: Wyoming and Nebraska to northern New Mexico.
NEW MEXICO: Santa Fe; Sandia Mountains; Canyoncito. Mesas and low hills, in the Upper Sonoran Zone.

15. **Gilia inconspicua** (Smith) Dougl.; Hook. in Curtis's Bot. Mag. **56**: *pl. 2883*. 1829.
Ipomopsis inconspicua J. E. Smith, Exot. Bot. **1**: *pl. 14*. 1804.
TYPE LOCALITY: North America.
RANGE: British Columbia and California to Colorado and Texas.
NEW MEXICO: Aztec; west of Santa Fe; Mangas Springs; Kingston; Sandia Mountains; mountains west of San Antonio; Carrizalillo Mountains; Florida Mountains; Organ Mountains. Dry hills and mesas, in the Upper Sonoran Zone.

16. **Gilia longiflora** (Torr.) Don, Hist. Dichl. Pl. **4**: 245. 1838.
Cantua longiflora Torr. Ann. Lyc. N. Y. **2**: 221. 1828.
Collomia longiflora A. Gray, Proc. Amer. Acad. **8**: 261. 1870.
TYPE LOCALITY: "On the Canadian," New Mexico or Colorado.

RANGE: Colorado and Nebraska to Texas and Mexico.

NEW MEXICO: Common throughout the State. Dry mesas and plains, in the Lower and Upper Sonoran zones.

17. Gilia laxiflora (Coulter) Osterhout, Bull. Torrey Club **24:** 51. 1897.

Gilia macombii laxiflora Coulter, Contr. U. S. Nat. Herb. **1:** 44. 1889.

TYPE LOCALITY: Camp Charlotte, Ixion County, Texas.

RANGE: Colorado and Utah to New Mexico and western Texas.

NEW MEXICO: Mountainair; Stanley; Cabra Springs; Santa Fe; Nara Visa; Buchanan. Plains, in the Upper Sonoran Zone.

18. Gilia pumila Nutt. Journ. Acad. Phila. II. **1:** 156. 1848.

Navarretia pumila Smyth, Check List Pl. Kans. 18. 1892.

TYPE LOCALITY: "Near the first range of the Rocky Mountains of the Platte," Colorado.

RANGE: Wyoming and Kansas to Arizona and western Texas.

NEW MEXICO: Carrizo Mountains; Farmington; Glorieta; west of Santa Fe; San Andreas Mountains; San Marcial; White Sands; above Tularosa. Dry hills and plains, in the Upper Sonoran Zone.

19. Gilia gunnisonii Torr. & Gray, U. S. Rep. Expl. Miss. Pacif. **2**[2]**:** 128. *pl. 9.* 1855.

TYPE LOCALITY: "Sand-banks of Green River, Utah."

RANGE: Utah to Arizona and northwestern New Mexico.

NEW MEXICO: Carrizo Mountains; Shiprock. Dry hills and mesas, in the Upper Sonoran Zone.

20. Gilia acerosa (A. Gray) Britton, Man. 761. 1901.

Gilia rigidula acerosa A. Gray, Proc. Amer. Acad. **8:** 280. 1870.

TYPE LOCALITY: "North New Mexico to Arizona."

RANGE: Western Texas to southern Arizona.

NEW MEXICO: Pojoaque; west of Santa Fe; Berendo Creek; Placitas; Torrance; Carrizalillo Mountains; Lakewood; south of Roswell; Dayton. Dry hills, in the Upper Sonoran Zone.

8. LINANTHUS Benth.

Annual, 20 cm. high or less, divaricately branched, with slender stems, opposite, simple or 3-lobed leaves with linear spinulose segments, and solitary subsessile flowers between the equal branches; calyx tubular-funnelform, scarious between the equal linear spinulose lobes; corolla tubular-funnelform, blue, fading whitish, the tube not as long as the calyx; capsules oblong, bursting the persistent calyx.

1. Linanthus bigelovii (A. Gray) Greene, Pittonia **2:** 253. 1892.

Gilia dichotoma parviflora Torr. U. S. & Mex. Bound. Bot. 147. 1859.

Gilia bigelovii A. Gray, Proc. Amer. Acad. **8:** 265. 1870.

TYPE LOCALITY: Cooks Spring, New Mexico. Type collected by Bigelow.

RANGE: Western Texas to southern California.

NEW MEXICO: Foothills of the Organ Mountains (*Wooton*). Dry hills.

9. DACTYLOPHYLLUM Spach.

Low, divaricately branched annual, about 10 cm. high, with opposite, palmately divided leaves having linear spinulose segments, and with conspicuous yellow axillary flowers on long slender pedicels; calyx funnelform-campanulate, scarious between the lobes, hirtellous like the leaves; corolla funnelform, 10 to 12 mm. long, with a broad spreading limb, bright yellow; stamens slightly exserted; capsules splitting the persistent calyx.

1. **Dactylophyllum aureum** (Nutt.) Heller, Muhlenbergia **2**: 231. 1906.
Gilia aurea Nutt. Journ. Acad. Phila. II. **1**: 155. *pl. 22*. 1848.
Linanthus aureus Greene, Pittonia **2**: 257. 1892.
TYPE LOCALITY: Santa Barbara, California.
RANGE: Southern California to western Texas.
NEW MEXICO: Mangas Springs (*Metcalfe* 54).

120. HYDROPHYLLACEAE. Waterleaf Family.

Annual or perennial herbs 1 meter high or less, mostly small plants, with usually opposite, exstipulate, simple or compound leaves and perfect flowers solitary in the axils or in terminal helicoid cymes or 1-sided racemes; calyx of 5 more or less united sepals, the sinuses sometimes appendaged; corolla regular, 5-lobed, mostly funnelform, often appendaged within at the base of the tube; stamens 5, the filaments adnate to the base of the corolla; ovary superior, 1-celled or rarely 2-celled; fruit a capsule with 1 or 2 incomplete cells; seeds usually few.

KEY TO THE GENERA.

Leaf blades entire; ovary more or less 2-celled; styles 2, distinct.
  Corolla urceolate; flowers cymose; leaves long-linear; plants fleshy .... 1. ANDROPUS (p. 530).
  Corolla funnelform; flowers solitary; leaves linear-oblong or broader; plants not fleshy .... 2. MARILAUNIDIUM (p. 531).
Leaf blades more or less toothed, lobed, or dissected; ovary 1-celled; style 2-cleft.
  Corolla lobes imbricated in the bud; placentæ narrow .... 3. PHACELIA (p. 532).
  Corolla lobes convolute in the bud; placentæ dilated.
    Stamens included; calyx enlarged in fruit; leaves opposite .... 4. NYCTELEA (p. 535).
    Stamens exserted; calyx not enlarged in fruit; leaves alternate .... 5. HYDROPHYLLUM (p. 535).

Doctor Gray in the Synoptical Flora[1] states that *Eriodictyon angustifolium* Nutt. occurs in New Mexico, but we have seen no specimens nor does it seem probable that the plant comes within our borders.

1. ANDROPUS Brand.

Fleshy cespitose herbaceous perennial, hispid throughout; leaves narrowly linear; flowers in terminal crowded cymes, not conspicuously helicoid; calyx of 5 linear sepals coalescent for a short distance at the base, not appendaged; corolla long-urceolate, not appendaged within, the limb with 5 short reflexed lobes; stamens 5, included, the filaments slightly expanded at the base; ovary 2-celled; styles 2, with capitate stigmas; seeds numerous, horizontal, angled.

1. **Andropus carnosus** (Wooton) Brand, Repert. Nov. Sp. Fedde **10**: 281. 1912.
Conanthus ? carnosus Wooton, Bull. Torrey Club **25**: 262. 1898.
TYPE LOCALITY: White Sands New Mexico. Type collected by Wooton (no. 164).
RANGE: Southern New Mexico and western Texas
NEW MEXICO: White Sands; Lakewood. Gypsum soil, in the Lower Sonoran Zone.

---

[1] **2**[1]: 176.

2. MARILAUNIDIUM Kuntze.

Low annual or perennial herbs with pubescent, mostly diffusely branched stems, small simple entire alternate nearly sessile leaves, and solitary axillary flowers; corolla whitish to blue or purple, funnelform, the lobes broad, imbricated in bud; stamens mostly included; ovary 1-celled, sometimes imperfectly 2-celled; styles 2, distinct; capsules oblong to subglobose, 2-valved; seeds numerous, small, rugose.

KEY TO THE SPECIES.

Perennial, from a thick woody base.......................... 1. *M. xylopodum.*
Annuals.
  Corolla not at all or but slightly exceeding the calyx...... 2. *M. angustifolium.*
  Corolla much exceeding the calyx.
    Corolla pale blue, 5 mm. long; plants slender, with few erect branches.............................. 3. *M. tenue.*
    Corolla deep purple, more than 7 mm. long; plants stout, with very numerous dense spreading branches.
      Leaves of the inflorescence obovate to broadly oblong, flat, strongly hispid or hirsute; tube of corolla shorter than the calyx............ 4. *M. foliosum.*
      Leaves of the inflorescence linear to narrowly oblong, revolute; tube of corolla exceeding the calyx.................................. 5. *M. hispidum.*

1. **Marilaunidium xylopodum** Woot. & Standl. Contr. U. S. Nat. Herb. **16:** 162. 1913.

TYPE LOCALITY: Crevices of limestone rocks near Queen, New Mexico. Type collected by Wooton, July 31, 1909.

RANGE: Southern New Mexico and western Texas.

NEW MEXICO: Queen (*Wooton*). Dry hills.

2. **Marilaunidium angustifolium** (A. Gray) Kuntze, Rev. Gen. Pl. **2:** 384. 1891.

Nama dichotomum angustifolium A. Gray, Proc. Amer. Acad. **8:** 284. 1870.

Conanthus angustifolius Heller, Bull. Torrey Club **24:** 479. 1897.

TYPE LOCALITY: "New Mexico." Type collected by Fendler in 1847, near Santa Fe (no. 644).

RANGE: Colorado to New Mexico and western Texas.

NEW MEXICO: Tunitcha Mountains; Pecos; Santa Fe Canyon; Santa Rita; Mogollon Mountains; Organ Mountains; White and Sacramento mountains. Open slopes, in the Upper Sonoran and Transition zones.

3. **Marilaunidium tenue** Woot. & Standl. Contr. U. S. Nat. Herb. **16:** 162. 1913.

TYPE LOCALITY: Limestone hills 3 miles south of Hillsboro, Sierra County, New Mexico. Type collected by Metcalfe (no. 1291).

RANGE: Known only from type locality.

4. **Marilaunidium foliosum** Woot. & Standl. Contr. U. S. Nat. Herb. **16:** 162. 1912.

TYPE LOCALITY: On saltgrass flats near Roswell, New Mexico. Type collected by Earle (no. 531).

RANGE: Southeastern New Mexico.

NEW MEXICO: Roswell; Lake Arthur; Fort Stanton. Dry plains and hills, in the Lower and Upper Sonoran zones.

5. **Marilaunidium hispidum** (A. Gray) Kuntze, Rev. Gen. Pl. **2:** 434. 1891.

Nama hispida A. Gray, Proc. Amer. Acad. **5:** 339. 1862.

Conanthus hispidus Heller, Bull. Torrey Club **24:** 479. 1897.

TYPE LOCALITY: Not definitely stated.

RANGE: Western Texas to southern California.

NEW MEXICO: Shiprock; Farmington; headwaters of the Pecos; Sandia Mountains; Stanley; west of Santa Fe; Socorro; Mangas Springs; Black Range; Belen; Mesilla Valley; Florida Mountains. Dry hills and plains, in the Lower and Upper Sonoran zones.

3. PHACELIA Juss.

Annual or perennial herbs, mostly pubescent or glandular, with alternate (sometimes opposite below) lobed or dissected or entire leaves and helicoid cymes of flowers; calyx slightly accrescent, not appendaged; corolla white, blue, or purple, mostly funnelform, the tube often appendaged within, the lobes imbricated, spreading; stamens included or exserted; ovary 1-celled, the style 2-cleft; capsules 1-celled or almost 2-celled by the dilation of the placentæ, 2-valved, the seeds reticulate or roughened.

KEY TO THE SPECIES.

Perennials.
- Leaves simple and entire or some of the lower pinnate but with entire divisions .... 1. *P. heterophylla.*
- Leaves pinnately parted into linear or linear-oblong divisions. 2. *P. sericea.*

Annuals or rarely biennials.
- Leaves entire..17. *P. demissa.*
- Leaves not entire.
  - Corolla lobes dentate or erose.
    - Stamens twice as long as the corolla; leaves with narrow segments............................ 3. *P. alba.*
    - Stamens only slightly exceeding the corolla; leaves with broad segments........................ 4. *P. neomexicana.*
  - Corolla lobes entire or sinuate-crenate.
    - Leaves bipinnate.................................. 5. *P. popei.*
    - Leaves not bipinnate.
      - Leaves sinuate-crenate halfway to the midrib.... 7. *P. corrugata.*
      - Leaves, at least part of them, pinnately divided to the midrib.
        - Sepals much longer than the capsule; stems slender, weak, decumbent.................... 6. *P. rupestris.*
        - Sepals slightly if at all exceeding the capsule; stems stout; usually erect.
          - Branches very numerous from the base, spreading.
            - Corolla blue, 6 to 8 mm. long.............. 8. *P. similis.*
            - Corolla white, 3 to 4 mm. long............ 9. *P. arizonica.*
          - Branches few, most of them from above the base, erect.
            - Stamens included.........................10. *P. caerulea.*
            - Stamens exserted.
              - Flowers 10 mm. long or more...........11. *P. glandulosa.*
              - Flowers 8 mm. long or less.
                - Stems and leaves merely glandular-viscid.
                  - Calyx densely hirsute; flowers 7 to 8 mm. long.......................12. *P. crenulata.*
                  - Calyx sparingly hirsute or merely viscid; flowers 5 mm. long or less.13. *P. intermedia.*
                - Stems densely villous or hirsute as well as glandular.

Calyx lobes rounded-obovate; leaves with numerous lobed divisions...14. *P. depauperata.*
Calyx lobes oblanceolate or oblong, acutish; leaves with few nearly entire divisions.
Flowers nearly sessile; leaves densely silky-strigose, almost all pinnatifid; calyx sparingly hirsute.......................15. *P. bombycina.*
Flowers on conspicuous slender pedicels; leaves sparingly strigose, only the lower ones pinnatifid; calyx densely hirsute....16. *P. tenuipes.*

Phacelia infundibuliformis Torr. was reported by Doctor Gray [1] from New Mexico but we have seen no specimens. Probably the report was due to a wrong determination.

1. **Phacelia heterophylla** Pursh, Fl. Amer. Sept. 140. 1814.
TYPE LOCALITY: "On dry hills on the banks of the Kooskooskee," Idaho.
RANGE: British Columbia and Montana to California and New Mexico.
NEW MEXICO: Tunitcha Mountains; Chama; Sierra Grande; Santa Fe and Las Vegas mountains; Middle Fork of the Gila; White and Sacramento mountains. Meadows, normally in the Transition Zone.

2. **Phacelia sericea** (Graham) A. Gray, Proc. Amer. Acad. **10:** 323. 1875.
Eutoca sericea Graham; Hook. Curtis's Bot. Mag. **57:** *pl. 3003.* 1830.
TYPE LOCALITY: "Rocky Mountains, North America."
RANGE: British Columbia and Saskatchewan to Nevada and New Mexico.
NEW MEXICO: Wheeler Peak (*Bailey*). Meadows, in the Arctic-Alpine Zone.

3. **Phacelia alba** Rydb. Bull. Torrey Club **28:** 30. 1901.
TYPE LOCALITY: Sangre de Cristo Creek, Colorado.
RANGE: Colorado to New Mexico.
NEW MEXICO: Dulce; Coolidge; Inscription Rock; Gallup; Santa Fe and Las Vegas mountains; Gila; Reserve; White and Sacramento mountains. Open slopes in the mountains, in the Transition Zone.

4. **Phacelia neomexicana** Thurb.; Torr. U. S. & Mex. Bound. Bot. 143. 1859.
TYPE LOCALITY: Pine woods near the Copper Mines, New Mexico.
RANGE: Southern New Mexico and Arizona.
NEW MEXICO: Hillsboro Peak; Capitan Mountains. Transition Zone.

5. **Phacelia popei** Torr. & Gray, U. S. Rep. Expl. Miss. Pacif. **2:** 172. *pl. 10.* 1855.
TYPE LOCALITY: "On the Llano Estacado and Pecos," New Mexico or Texas. Type collected in 1854 by Pope.
RANGE: Eastern New Mexico and adjacent Texas.
NEW MEXICO: Cabra Spring; near head of Little Creek; San Andreas Mountains.

6. **Phacelia rupestris** Greene, Leaflets **1:** 152. 1905.
TYPE LOCALITY: Foothills of the Black Range, New Mexico. Type collected by Metcalfe (no. 1012).
RANGE: Southern New Mexico and Arizona.
NEW MEXICO: Hillsboro; Mogollon Mountains; Florida Mountains; Organ Mountains; Dona Ana Mountains; White Mountains. Canyons, in the Upper Sonoran Zone.

---

[1] U. S. & Mex. Bound. Bot. 144. 1859.

7. Phacelia corrugata A. Nels. Bot. Gaz. **34:** 26. 1902.

TYPE LOCALITY: Rifle, Garfield County, Colorado.

RANGE: Southern Utah and Colorado to western Texas and Mexico.

NEW MEXICO: Tunitcha Mountains; Albuquerque; Pajarito Park; Espanola; San Augustine Plains; mountains west of San Antonio; Mesilla Valley; Organ Mountains; White Sands; Blazers Mill; south of Torrance. Plains and sandhills, in the Lower and Upper Sonoran zones.

8. Phacelia similis Woot. & Standl. Bull. Torrey Club **36:** 111. 1909.

TYPE LOCALITY: Plains near Nutt Station, Sierra County, New Mexico. Type collected by Metcalfe (no. 1665).

RANGE: Known only from type locality, in the Lower Sonoran Zone.

9. Phacelia arizonica A. Gray, Syn. Fl. **2**[1]: 394. 1878.

TYPE LOCALITY: Southern Arizona.

RANGE: Southern Arizona and New Mexico.

NEW MEXICO: Near Cliff; Nutt Flats. Dry plains and foothills, in the Lower Sonoran Zone.

10. Phacelia caerulea Greene, Bull. Torrey Club **8:** 122. 1881.

TYPE LOCALITY: "Southern New Mexico and Arizona." Type collected by E. L. Greene.

RANGE: Southern Arizona to western Texas and adjacent Mexico.

NEW MEXICO: Tres Hermanas; Hillsboro; Florida Mountains; Mesilla Valley; Organ Mountains; mountains west of San Antonio. Dry plains and foothills, in the Lower Sonoran Zone.

11. Phacelia glandulosa Nutt. Journ. Acad. Phila. II. **1:** 160. 1848.

TYPE LOCALITY: "About Ham's Fork of the Colorado of the West, on dry, bare hills."

RANGE: Montana to Arizona and Texas.

NEW MEXICO: Baldy (*Wooton*). Hudsonian Zone.

12. Phacelia crenulata Torr.; S. Wats. in King, Geol. Expl. 40th Par. **5:** 251. 1871.

TYPE LOCALITY: Trinity Mountains, Nevada.

RANGE: Nevada and Utah to Arizona and New Mexico.

NEW MEXICO: Aztec; Berendo Creek; headwaters of the Pecos. Dry plains and hills, in the Upper Sonoran Zone.

13. Phacelia intermedia Wooton, Bull. Torrey Club **25:** 457. 1898.

TYPE LOCALITY: Mesa near Las Cruces, New Mexico. Type collected by Wooton, April 10, 1893.

RANGE: Southern New Mexico.

NEW MEXICO: Mesa near Agricultural College; Mesilla Valley; San Andreas Mountains. Sandy mesas and valleys, in the Lower Sonoran Zone.

14. Phacelia depauperata Woot. & Standl. Contr. U. S. Nat. Herb. **16:** 163. 1913.

TYPE LOCALITY: Arroyo Ranch near Roswell, New Mexico. Type collected by David Griffiths (no. 4249).

RANGE: Known only from type locality.

15. Phacelia bombycina Woot. & Standl. Contr. U. S. Nat. Herb. **16:** 163. 1913.

TYPE LOCALITY: Mangas Springs, New Mexico. Type collected by Rusby (no. 276).

RANGE: Southwestern New Mexico.

NEW MEXICO: Mangas Springs; Bear Mountains. Dry hills.

16. Phacelia tenuipes Woot. & Standl. Contr. U. S. Nat. Herb. **16:** 163. 1913.

TYPE LOCALITY: Carrizalillo Spring, New Mexico. Type collected by E. A. Mearns (no. 91).

RANGE: Known only from type locality.

17. **Phacelia demissa** A. Gray, Proc. Amer. Acad. **10**: 326. 1875.

TYPE LOCALITY: "New Mexico." Type collected by Dr. E. Palmer in 1869, doubtless near Fort Defiance.

RANGE: Utah, Arizona, and northwestern New Mexico.

We have seen no further specimens of this from New Mexico.

4. NYCTELEA Scop.

Small annual herbs with pinnate leaves and small blue or whitish flowers in terminal cymes; calyx lobes relatively large, obovate, accrescent, not appendaged; corolla convolute in bud, about equaling the calyx; stamens included; ovary 1-celied, with fleshy placentæ, styles 2-cleft.

KEY TO THE SPECIES.

Calyx in fruit less than 4 mm. long, the lobes rounded-oblong...... 1. *N. micrantha.*
Calyx in fruit 8 mm. long or more, the lobes narrowly triangular, acute.. 2. *N. nyctelea.*

1. **Nyctelea micrantha** (Torr.) Woot. & Standl.

Phacelia micrantha Torr. U. S. & Mex. Bound. Bot. 144. 1859.

Macrocalyx micrantha Coville, Contr. U. S. Nat. Herb. **4**: 157. 1893.

TYPE LOCALITY: Stony hills near El Paso.

RANGE: California to Arizona and western Texas.

NEW MEXICO: Organ Mountains; Tortugas Mountain. Low hills, in the Lower and Upper Sonoran zones.

2. **Nyctelea nyctelea** (L.) Britton in Britt. & Brown, Illustr. Fl. ed. 2. **3**: 67. 1913.

Polemonium nyctelea L. Sp. Pl. 231. 1753.

Ellisia nyctelea L. Sp. Pl. ed. 2. 1662. 1763.

Macrocalyx nyctelea Kuntze, Rev. Gen. Pl. **2**: 434. 1891.

TYPE LOCALITY: "Habitat in Virginia."

RANGE: Montana and Saskatchewan to New Mexico and Virginia.

NEW MEXICO: Sierra Grande (*Standley* 6162). Damp ground, in the Transition Zone.

5. HYDROPHYLLUM L. WATERLEAF.

Perennial herb, 30 cm. high or more, from a horizontal rootstock, with large radical leaves or alternate, pinnately parted, long-petioled cauline leaves and terminal long-peduncled cymes of pale bluish flowers; corolla with short funnelform tube and 5 rounded lobes; stamens with slender filaments, exserted; ovary 1-celled, pubescent; style 2-cleft; ovules 4, inclosed in the fleshy placenta; capsule 2-valved; seeds 1 to 4.

1. **Hydrophyllum fendleri** (A. Gray) Heller, Pl. World **1**: 23. 1897.

Hydrophyllum occidentale fendleri A. Gray, Proc. Amer. Acad. **10**: 314. 1875.

TYPE LOCALITY: Near Santa Fe, New Mexico. Type collected by Fendler.

RANGE: Idaho and Wyoming to New Mexico.

NEW MEXICO: Chama; Santa Fe and Las Vegas mountains; West Fork of the Gila; James Canyon. Wet ground in the mountains, in the Transition and Canadian zones.

121. EHRETIACEAE.

Low spreading or prostrate herbs or undershrubs with branched stems; leaves simple, alternate, entire; flowers solitary, axillary; calyx of 5 narrow sepals united at the base; corolla gamopetalous, narrowly funnelform to campanulate, not appendaged in the throat, the limb of 5 rounded lobes; stamens 5, epipetalous; anthers 2-celled, opening by longitudinal pores; ovary 4-celled, the fruit various.

KEY TO THE GENERA.

Nutlet 1, the other cells abortive; inflorescence capitate; sepals linear-subulate.............................. 1. **Ptilocalyx** (p. 536).
Nutlets 4; inflorescence axillary; sepals linear-lanceolate.
Nutlets cohering by their inner faces so as to form a smooth depressed-globose fruit; stamens equally inserted; plants woody almost throughout, canescent.................................... 2. **Stegnocarpus** (p. 536).
Nutlets roughened, cohering by the inner angle to form a 4-parted fruit; stamens unequally inserted; plants woody only at the base if at all, hispid.. 3. **Eddya** (p. 536).

1. PTILOCALYX Torr.

Low, much branched shrub with small ovate leaves and white flowers in short-capitate clusters; calyx lobes filiform-subulate, densely hairy; corolla campanulate, the lobes rotund to obovate, crenulate; stamens included, equally inserted near the base of the corolla; ovary 4-lobed, 4-celled, with an obscure glandular ring at the base; styles 2-parted above; fruit 1-seeded by abortion.

1. **Ptilocalyx greggii** Torr. U. S. Rep. Expl. Miss. Pacif. **2:** 110. *pl. 8.* 1855.
Coldenia greggii A. Gray, Syn. Fl. 2^1: 182. 1878.
Type locality: Near Buena Vista, Mexico.
Range: Western Texas and southern New Mexico and southward.
New Mexico: Bishops Cap (*Wooton & Standley*). Dry hills, in the Lower Sonoran Zone.

2. STEGNOCARPUS Torr.

Prostrate perennial with slender, much branched, woody stems 10 cm. long or less; leaves appressed-canescent; flowers pale pink, axillary; calyx of 5 lanceolate sepals; corolla campanulate-funnelform, the lobes orbicular-obovate, crenulate; ovary ovoid, slightly 4-lobed; fruit of 4 smooth rounded triangular nutlets.

1. **Stegnocarpus canescens** (DC.) Torr. U. S. Rep. Expl. Miss. Pacif. **2:** 169. *pl. 7.* 1855.
Coldenia canescens DC. Prodr. **9:** 559. 1845.
Type locality: "In Mexico inter Santander et Victoria."
Range: Western Texas to Arizona.
New Mexico: Berendo Creek; Tortugas Mountain; Organ Mountains; White Mountains; Guadalupe Mountains; Lakewood. Dry hills and plains, in the Lower and Upper Sonoran zones.

3. EDDYA Torr.

Small, prostrate, densely hispid perennials with much branched slender stems about 10 cm. long, from thick woody roots; leaves small, crowded; flowers small, axillary, solitary; corolla salverform or funnelform, with a broad limb; ovary 4-lobed; nutlets 4, cohering by the inner angle, muricate-scabrous.

KEY TO THE SPECIES.

Plants merely hispid; leaves linear, revolute; corolla 5 mm. long.. 1. *E. hispidissima.*
Plants white with villous or canescent pubescence; leaves spatulate, flat; corolla 10 mm. long.............................. 2. *E. gossypina.*

1. **Eddya hispidissima** Torr. U. S. Rep. Expl. Miss. Pacif. **2:** 170. *pl. 9.* 1855.
Coldenia hispidissima A. Gray, Proc. Amer. Acad. **5:** 340. 1862.
Type locality: "Common on the Rio Grande about El Paso."

RANGE: Western Texas to Arizona and Utah.

NEW MEXICO: El Rito; south of Torrance; White Sands; Suwanee; Lakewood; Guadalupe Mountains. Dry hills and plains, in the Lower and Upper Sonoran zones.

2. **Eddya gossypina** Woot. & Standl. Contr. U. S. Nat. Herb. **16:** 164. 1913.

TYPE LOCALITY: Tortugas Mountain, New Mexico. Type collected by Wooton, September 2, 1894.

RANGE: Known only from type locality, in the Lower Sonoran Zone.

122. HELIOTROPACEAE. Heliotrope Family.

Low annual or perennial herbs; leaves alternate, exstipulate, entire; flowers perfect, mostly in helicoid cymes; calyx of 5 partially united sepals; corolla gamopetalous, funnelform to salverform, 5-lobed; stamens 5, adnate to the corolla; ovary 2 to 4-celled; styles united; stigma annular, surmounted by a 2-lobed appendage; fruit drupaceous or separating into 2 or 4 nutlets.

KEY TO THE GENERA.

Fruit of 2 nutlets; flowers large, axillary to leaflike bracts, appearing as if truly axillary............. 1. EUPLOCA (p. 537).
Fruit of 4 nutlets; flowers small, in terminal helicoid cymes.. 2. HELIOTROPIUM (p. 537).

1. EUPLOCA Nutt.

Spreading strigose annual with entire ovate-lanceolate thin leaves and conspicuous white flowers in the axils of leaflike bracts; calyx lobes narrow; corolla salverform, with a limb 1 to 2 cm. wide, strongly plicate; stamens included, the anthers cohering by their minutely bearded tips; ovary 4-celled; fruit of 2 1-seeded nutlets.

1. **Euploca convolvulacea** Nutt. Trans. Amer. Phil. Soc. n. ser. **5:** 181. 1837.

Heliotropium convolvulaceum A. Gray, Proc. Amer. Acad. **6:** 403. 1857.

Euploca grandiflora Torr. in Emory, Mil. Reconn. 147. 1848.

TYPE LOCALITY: "On the sandy banks of the Arkansas."

RANGE: Nebraska and Colorado to Arizona and Mexico.

NEW MEXICO: Carrizo Mountains; Sabinal; Albuquerque; Nara Visa; Socorro; Melrose; mesa near Las Cruces; Roswell. Dry plains and hills, in the Lower and Upper Sonoran zones.

The type of *Euploca grandiflora* was collected on the Rio Grande below Santa Fe by Emory in 1847.

2. HELIOTROPIUM L.

Annual or perennial herbs with alternate narrow leaves and small white flowers in terminal helicoid cymes; calyx lobes narrow; corolla narrowly funnelform, with a short 5-lobed limb; stamens included; ovary 4-celled or 2-celled; fruit of 4 1-seeded nutlets.

KEY TO THE SPECIES.

Stems prostrate; leaves fleshy, glabrous.......................... 1. *H. xerophilum.*
Stems erect or nearly so; leaves thin, hispid..................... 2. *H. greggii.*

1. **Heliotropium xerophilum** Cockerell, Bot. Gaz. **33:** 379. 1902.

Heliotropium spathulatum Rydb. Bull. Torrey Club **30:** 262. 1903.

TYPE LOCALITY: Albuquerque, New Mexico. Type collected by Cockerell in 1901.

RANGE: Washington and Saskatchewan to California and New Mexico.

NEW MEXICO: San Augustine Plains; Albuquerque; Berendo Creek; Mesilla Valley; Roswell. Alkaline soil, in the Lower Sonoran Zone.

2. **Heliotropium greggii** Torr. U. S. & Mex. Bound. Bot. 137. 1859.

TYPE LOCALITY: Valley of the Conchos near Santa Rosalia, Mexico.

RANGE: Western Texas to New Mexico and Mexico.

NEW MEXICO: South of White Sands; mesa near Agricultural College; Artesia; Roswell. Sandy plains, in the Lower Sonoran Zone.

123. BORAGINACEAE. Borage Family.

More or less rough-hairy annual or perennial herbs, with simple, alternate, exstipulate leaves, the flowers in cymes, these often helicoid; leaves mostly entire and small; flowers perfect, usually regular; calyx of 5 united persistent sepals; corolla tubular, funnelform, or salverform, deciduous; stamens 5, adnate to the corolla tube; filaments sometimes appendaged; pistil with a single style; ovary deeply 4-lobed; stigmas simple or 2-lobed; fruit of 2 or 4 seedlike nutlets.

KEY TO THE GENERA.

Corolla oblique, the lobes unequal......................13. ECHIUM (p. 547).
Corolla regular, with equal lobes.
  Nutlets with hooked prickles, at least on the margins.
    Nutlets spreading, covered with prickles...... 1. CYNOGLOSSUM (p. 539).
    Nutlets erect or incurved, with prickles on the back or angles........................ 2. LAPPULA (p. 539).
  Nutlets unarmed or if prickly the prickles not curved.
    Receptacle (gynobase) flat or slightly convex; nutlets attached laterally or by the base.
      Nutlets obliquely attached; corolla blue or pink, tubular-funnelform....... 3. MERTENSIA (p. 541).
      Nutlets attached at the very base; corolla never blue or pink, of various shapes.
        Corolla salverform, mostly yellow (in one species greenish).......... 4. LITHOSPERMUM (p. 542).
        Corolla tubular, slightly open at the throat, with erect lobes, mostly dull greenish yellow.
          Stamens included; corolla about 1 cm. long.................... 5. ONOSMODIUM (p. 543).
          Stamens exserted; corolla 3 cm. long or more.................... 6. MACROMERIA (p. 543).
    Receptacle (gynobase) elongated or conic; nutlets attached laterally.
      Nutlets attached below the middle, mostly near the base.
        Perennial, cespitose; flowers bright blue............................ 7. ERITRICHIUM (p. 544).
        Annual; flowers white................12. ALLOCARYA (p. 547).
      Nutlets attached at the middle or from the base to or above the middle.
        Perennials............................10. OREOCARYA (p. 544).
        Annuals.
          Nutlets attached by the middle.. 9. PLAGIOBOTHRYS (p. 544).

Nutlets attached from the base to or above the middle.
  Calyx lobes strongly nerved; plants mostly bristly-hispid; roots without coloring matter.......11. CRYPTANTHE (p. 546).
  Calyx lobes nerveless or faintly nerved; plants not bristly-hispid; roots with purplish coloring matter.................. 8. EREMOCARYA (p. 544).

1. CYNOGLOSSUM L. HOUND'S-TONGUE.

Coarse biennial herb, 20 to 60 cm. high, with rather large flat leaves and with reddish purple flowers in terminal racemes; stamens included; ovary of 4 nearly distinct lobes; nutlets flat or convex, covered with short barbed prickles.

1. **Cynoglossum officinale** L. Sp. Pl. 134. 1753.

TYPE LOCALITY: "Habitat in Europae ruderatis."

NEW MEXICO: Raton Mountains.

A common weed in many parts of the United States, introduced from Europe.

2. LAPPULA Moench. STICKSEED.

Hispid or canescent annual, biennial, or perennial herbs, 1 meter high or less, with alternate leaves and small white or blue flowers in spikes or racemes; calyx 5-lobed; corolla salverform or funnelform, with a short tube; stamens included; ovary 4-lobed; fruit burlike, of 4 nutlets armed on the back or margins with barbed prickles.

KEY TO THE SPECIES.

Inflorescence leafy throughout, the floral leaves merely smaller than those of the stem; annuals.
  Annular margin connecting the bases of the prickles inconspicuous in all 4 nutlets................................ 1. *L. occidentalis.*
  Annular margin connecting the bases of the prickles, in at least 3 of the nutlets, broadened and forming a cup...... 2. *L. texana.*
Inflorescence leafy-bracted only at the base, the bracts minute above; biennials or perennials.
  Prickles of the fruit united for about half their length.
    Flowers white; cauline leaves narrowly oblong, sessile or nearly so; sepals obtuse........................... 3. *L. leucantha.*
    Flowers blue; cauline leaves lanceolate or lance-ovate, petiolate; sepals acute.............................. 4. *L. ursina.*
  Prickles distinct to the base.
    Stems hirsute; margins of the leaves long-ciliate.......... 5. *L. hirsuta.*
    Stems not hirsute; leaves not long-ciliate.
      Plants densely grayish-strigose throughout, the bases of the hairs white and much enlarged; inflorescence loose, few-flowered....................... 6. *L. grisea.*
      Plants not densely grayish-strigose, the hairs short and comparatively soft; bases of the hairs small and inconspicuous; inflorescence dense and many-flowered.
        Cauline leaves linear-oblong, sessile or nearly so.. 7. *L. floribunda.*
        Cauline leaves lanceolate, conspicuously petiolate. 8. *L. pinetorum.*

1. **Lappula occidentalis** (S. Wats.) Greene, Pittonia **4:** 97. 1899.

Echinospermum redowskii occidentale S. Wats. in King, Geol. Expl. 40th Par. **5:** 246. 1871.

Type locality: "From Western Texas to Arizona and northward to the Saskatchewan, Bear Lake and Fort Youkon."

Range: Washington and Saskatchewan to Arizona and Missouri.

New Mexico: Tunitcha Mountains; Tierra Amarilla; Dulce; Chama; Santa Fe; Raton; Sierra Grande; Carrizo Mountains; Rio Apache; Rio Pueblo; Las Vegas; Winsors Ranch; Cross L Ranch; mountains west of San Antonio; Mangas Springs; Mogollon Creek; Carrizalillo Mountains; White and Sacramento mountains. Open slopes and dry hills, in the Upper Sonoran and Transition zones.

2. **Lappula texana** (Scheele) Greene, Pittonia **4:** 94. 1899.

Echinospermum texanum Scheele, Linnaea **25:** 260. 1852.

Echinospermum redowskii cupulatum A. Gray in Brewer & Wats. Bot. Calif. **1:** 530. 1876.

Lappula cupulata Rydb. Bull. Torrey Club **28:** 31. 1901.

Type locality: San Antonio, Texas.

Range: Idaho and South Dakota to New Mexico and Texas.

New Mexico: Aztec; Mogollon Mountains; Tortugas Mountain; Organ Mountains. Dry plains and hills, in the Lower and Upper Sonoran zones.

3. **Lappula leucantha** Greene, Leaflets **1:** 152. 1905.

Type locality: Shady canyons, Iron Creek, Black Range, New Mexico. Type collected by Metcalfe (no. 1475).

Range: Mountains of southwestern New Mexico.

New Mexico: Mogollon Mountains; Iron Creek; East Canyon. Transition Zone.

4. **Lappula ursina** Greene, Pittonia **2:** 182. 1891.

Echinospermum ursinum Greene; A. Gray, Proc. Amer. Acad. **17:** 224. 1882.

Type locality: On gravel beds of Bear Canyon in the Bear Mountains, New Mexico. Type collected by E. L. Greene in 1880.

Range: Mountains of southwestern New Mexico.

New Mexico: Mogollon Mountains; Organ Mountains. Transition Zone.

5. **Lappula hirsuta** Woot. & Standl. Contr. U. S. Nat. Herb. **16:** 164. 1913.

Type locality: Santa Fe Canyon 9 miles east of Santa Fe, New Mexico. Type collected by Heller (no. 3793).

Range: Mountains of New Mexico.

New Mexico: Santa Fe and Las Vegas mountains; Water Canyon; Mogollon Creek. Transition Zone.

6. **Lappula grisea** Woot. & Standl. Contr. U. S. Nat. Herb. **16:** 164. 1913.

Type locality: James Canyon, Sacramento Mountains, New Mexico. Type collected by Wooton, August 6, 1905.

Range: White and Sacramento mountains of New Mexico, in the Transition Zone.

7. **Lappula floribunda** (Lehm.) Greene, Pittonia **2:** 182. 1891.

Echinospermum floribundum Lehm. Nov. Stirp. Pugill. **2:** 24. 1830.

Type locality: "Lake Pentanguishene to the Rocky Mountains."

Range: Washington and Saskatchewan to California and New Mexico.

New Mexico: Santa Fe and Las Vegas mountains; Chama; White and Sacramento mountains. Damp thickets in the mountains, in the Transition Zone.

8. **Lappula pinetorum** Greene, Pittonia **2:** 182. 1891.

Echinospermum pinetorum Greene; A. Gray, Proc. Amer. Acad. **17:** 224. 1872.

Type locality: Pinos Altos Mountains, New Mexico. Type collected by Greene in 1880.

RANGE: Mountains of New Mexico and Arizona.

NEW MEXICO: Pinos Altos Mountains; Lookout Mines; Holts Ranch; Hanover Mountain; White Mountains. Transition Zone.

This is evidently near *L. floribunda*, but differs in foliage characters. Piper[1] separates the two by differences in the corolla appendages, but the distinctions used by him in his key seem to us not to exist in the plants.

3. MERTENSIA Roth. LUNGWORT.

Perennial herbs, 1 meter high or less, with alternate simple leaves and blue, pink, or white flowers in paniculate terminal clusters; calyx lobes narrow, usually somewhat enlarged in fruit; corolla tubular to funnelform-campanulate, the limb 5-lobed, with small crests in the throat; stamens included, at least not surpassing the limb; ovary 4-parted, the nutlets attached laterally near the base to a slightly elevated receptacle.

KEY TO THE SPECIES.

Leaves glabrous on both surfaces, glaucous, thick and fleshy... 1. *M. caelestina.*
Leaves pubescent on one or both surfaces, or at least pustulate, not glaucous above, thin.
  Leaves densely pubescent on both surfaces.
    Stems in dense clumps from a woody base.............. 2. *M. amoena.*
    Stems mostly solitary, never in clumps from a woody base.
      Calyx lobes densely pubescent, not hispid-ciliate; basal leaves numerous...................... 3. *M. amplifolia.*
      Calyx lobes glabrate on the back, hispid-ciliate; basal leaves usually wanting............... 4. *M. lateriflora.*
  Leaves glabrous beneath or nearly so.
    Calyx divided to about the middle; plants low, 30 cm. high or less................................ 5. *M. fendleri.*
    Calyx divided almost to the base; plants 40 to 100 cm. high.
      Leaves merely pustulate on the upper surface, somewhat glaucous beneath; calyx lobes hispid-ciliate................................ 6. *M. cynoglossoides.*
      Leaves pubescent on the upper surface, green on both sides; calyx lobes not hispid-ciliate.
        Calyx lobes lanceolate, acute; most of the cauline leaves petioled or narrowed at the base; tube and limb of corolla about equal. 7. *M. pratensis.*
        Calyx lobes oblong, mostly obtuse; cauline leaves broad and sessile at the base; limb of the corolla much shorter than the tube.............................. 8. *M. grandis.*

1. **Mertensia caelestina** Nels. & Cockerell, Proc. Biol. Soc. Washington **16:** 46. 1903.

TYPE LOCALITY: Truchas Peaks, above timber line, New Mexico. Type collected by Mrs. W. P. Cockerell.

RANGE: Northern New Mexico.

NEW MEXICO: Truchas Peak; Taos Mountains. Mountain peaks, in the Arctic-Alpine Zone.

---

[1] Bull. Torrey Club **29:** 536. 1902.

2. **Mertensia amoena** A. Nels. Bot. Gaz. **30:** 95. 1900.

Mertensia bakeri amoena A. Nels. in Coulter, New Man. Rocky Mount. 422. 1909.

TYPE LOCALITY: Monida, Montana.

RANGE: Wyoming to northern New Mexico.

NEW MEXICO: Sierra Grande (*Standley* 6153). Mountains, in the Canadian Zone.

3. **Mertensia amplifolia** Woot. & Standl. Contr. U. S. Nat. Herb. **16:** 165. 1913.

TYPE LOCALITY: Glorieta, New Mexico. Type collected by G. R. Vasey, June, 1881.

RANGE: Known only from type locality, in the Transition Zone.

4. **Mertensia lateriflora** Greene, Pl. Baker. **3:** 18. 1901.

Mertensia bakeri lateriflora A. Nels. in Coulter, New Man. Rocky Mount. 423. 1909.

TYPE LOCALITY: Carson, Colorado.

RANGE: Mountains of Colorado and northern New Mexico.

NEW MEXICO: Sierra Grande (*Standley* 6151, 6152). Canadian Zone.

5. **Mertensia fendleri** A. Gray, Amer. Journ. Sci. II. **34:** 339. 1862.

Mertensia lanceolata fendleri A. Gray, Proc. Amer. Acad. **10:** 53. 1874.

TYPE LOCALITY: Santa Fe Creek, New Mexico. Type collected by Fendler (no. 625).

RANGE: Northern New Mexico.

NEW MEXICO: Chama; Sierra Grande; Sandia Mountains; Santa Fe and Las Vegas mountains; Jemez Mountains. Mountains, in the Transition Zone.

6. **Mertensia cynoglossoides** Greene, Pl. Baker. **3:** 19. 1901.

TYPE LOCALITY: "On moist ledges in the Black Cañon," Colorado.

RANGE: Colorado to northern New Mexico.

NEW MEXICO: Chama (*Standley* 6523). Shaded damp thickets, in the Transition Zone.

7. **Mertensia pratensis** Heller, Bull. Torrey Club **26:** 550. 1899.

TYPE LOCALITY: Meadow in Santa Fe Canyon, 9 miles east of Santa Fe, New Mexico. Type collected by Heller (no. 3641).

RANGE: Colorado and New Mexico.

NEW MEXICO: Tunitcha Mountains; Chama; Taos Mountains; Santa Fe and Las Vegas mountains; White and Sacramento mountains. Meadows and thickets, in the Transition and Hudsonian zones.

8. **Mertensia grandis** Woot. & Standl. Contr. U. S. Nat. Herb. **16:** 165. 1913.

TYPE LOCALITY: Hillsboro Peak, Black Range, New Mexico. Type collected by Metcalfe (no. 1319).

RANGE: Mountains of southwestern New Mexico.

NEW MEXICO: Hillsboro Peak; Mogollon Mountains. Transition Zone.

4. LITHOSPERMUM L. PUCCOON.

Hispid or hirsute perennial herbs with alternate sessile leaves and leafy-bracted spikes or helicoid cymes of flowers; calyx of 5 narrow sepals united only at the base; corolla salverform or funnelform, the tube mostly longer than the sepals, yellow or greenish, sometimes crested or pubescent in the throat; stamens short, included; nutlets hard and bony, ovoid, smooth and shining, attached by the base to the nearly flat receptacle.

KEY TO THE SPECIES.

Flowers dull greenish; cauline leaves ovate to lanceolate......... 1. *L. viride.*
Flowers bright yellow; cauline leaves linear, oblong, or narrowly lanceolate.
  Corolla lobes fimbriate, the tube 2 to 3 cm. long; late flowers cleistogamous.. 2. *L. linearifolium.*
  Corolla lobes entire, the tube usually less than 10 mm. long; flowers all alike.
    Corolla tube much exceeding the calyx, usually twice as long; limb about 5 mm. broad..................... 3. *L. multiflorum.*
    Corolla tube about equaling the calyx; limb 1 cm. broad. 4. *L. cobrense.*

1. **Lithospermum viride** Greene, Bot. Gaz. **6**: 158. 1881.

TYPE LOCALITY: Mimbres Mountains, near Georgetown, New Mexico. Type collected by E. L. Greene.

RANGE: Mountains of southern New Mexico and Arizona.

NEW MEXICO: Bear Mountain; Queen; White Mountains. Transition Zone.

2. **Lithospermum linearifolium** Goldie, Edinburgh Phil. Journ. **1822**: 319. 1822.

Lithospermum angustifolium Michx. Fl. Bor. Amer. **1**: 130. 1803, not Forsk. 1775.

Lithospermum oblongum Greene, Pittonia **4**: 92. 1899.

TYPE LOCALITY: North America.

RANGE: British Columbia, Manitoba, and Illinois, to Arizona and Texas.

NEW MEXICO: Dulce; Farmington; Raton; Sierra Grande; Clayton; Bear Canyon; Aztec; Santa Fe; Kingston; Mangas Springs; Filmore Canyon; Roswell; Buchanan; Gilmores Ranch. Plains and hills, chiefly in the Upper Sonoran Zone.

3. **Lithospermum multiflorum** Torr.; S. Wats. in King, Geol. Expl. 40th Par. **5**: 238. 1871.

TYPE LOCALITY: Not stated.

RANGE: Wyoming to Mexico.

NEW MEXICO: Common in all the higher mountains. Mountains and hills, in the Transition Zone.

4. **Lithospermum cobrense** Greene, Bot. Gaz. **6**: 157. 1881.

TYPE LOCALITY: Santa Rita, New Mexico.

RANGE: Mountains of southern New Mexico and Arizona and adjacent Mexico.

NEW MEXICO: Burro Mountains; Mogollon Mountains; mountains west of Grants; San Luis Mountains; Animas Valley; White and Sacramento mountains. Transition Zone.

5. ONOSMODIUM Michx.

Coarse erect rough-hairy leafy perennial herb with broad, strongly veined leaves and numerous dull yellowish green flowers in terminal helicoid cymes; calyx deeply 5-parted; corolla a little longer than the calyx, tubular-funnelform, the lobes erect; stamens included, with very short filaments; ovary 4-parted, the style long-exserted, persistent; nutlets ovoid to globular, usually but 1 or 2 maturing, attached by the base to the nearly flat receptacle.

1. **Onosmodium occidentale** Mackenz. Bull. Torrey Club **32**: 502. 1905.

TYPE LOCALITY: Not definitely stated.

RANGE: British America to Montana, Texas, and Illinois.

NEW MEXICO: Sierra Grande; Clayton. Upper Sonoran Zone.

6. MACROMERIA Don.

Very similar to the preceding genus, but with a much more elongated corolla, exserted stamens, versatile anthers, enlarged and persistent style base, and usually more numerous nutlets.

1. **Macromeria thurberi** (A. Gray) Mackenz. Bull. Torrey Club **32:** 496. 1905.
Onosmodium thurberi A. Gray, Syn. Fl. **2**[1]: 205. 1878.
TYPE LOCALITY: New Mexico. Type collected by Thurber.
RANGE: New Mexico, Arizona, and southward.
NEW MEXICO: Gallinas Planting Station; Hermits Peak; Mogollon Mountains; Black Range; White and Sacramento mountains. Transition Zone

7. ERITRICHIUM Schrad. MOUNTAIN FORGET-ME-NOT.

Dwarf, densely cespitose, white-hairy, perennial herb, with narrow leaves and small bright blue flowers; corolla rotate, the throat with 5 crests; stamens included; nutlets divergent, with sharply toothed margins.

1. **Eritrichium elongatum** (Rydb.) W. F. Wight, Bull. Torrey Club **29:** 408. 1902.
Eritrichium aretioides elongatum Rydb. Mem. N. Y. Bot. Gard. **1:** 327. 1900.
TYPE LOCALITY: Spanish Basin, Montana.
RANGE: Oregon and Montana to northern New Mexico.
NEW MEXICO: Lake Peak; Truchas Peak; Wheeler Peak; Costilla Range. On rocky slopes of mountain tops, in the Arctic-Alpine Zone.

8. EREMOCARYA Greene.

Low canescent annual, 10 cm. high or less, with much branched stems; leaves in a basal rosette, usually wanting in herbarium specimens; bracts linear-oblong; flowers very small, white; sepals united only at the very base; corolla short-funnelform; ovary 4-lobed, the 4 nutlets attached for their whole length, erect, not margined.

1. **Eremocarya micrantha** (Torr.) Greene, Pittonia **1:** 58. 1887.
Eritrichium micranthum Torr. U. S. & Mex. Bound. Bot. 141. 1859.
TYPE LOCALITY: "Sand hills, Frontera, Texas, and in other places along the Rio Grande."
RANGE: Utah and California to western Texas.
NEW MEXICO: Mesa west of Organ Mountains. Dry mesas and low hills, in the Lower Sonoran Zone.

9. PLAGIOBOTHRYS Fisch. & Mey.

Coarse annual, the roots, leaves, and stems producing a purplish coloring matter; stems hispid almost throughout, 30 to 40 cm. high; leaves narrowly oblong, entire, sessile, acute; flowers small, in elongated bracted racemes; calyx campanulate, the lobes about as long as the tube, narrow; corolla 2 to 3 mm. long, narrowly funnelform; stamens included; nutlets broadly ovoid or somewhat 3-angled, often incurved, rough or smooth, 2 or 3 of them sometimes abortive.

1. **Plagiobothrys arizonicus** (A. Gray) Greene; A. Gray, Proc. Amer. Acad. **20:** 284. 1885.
Eritrichium canescens arizonicum A. Gray, Proc. Amer. Acad. **17:** 227. 1882.
TYPE LOCALITY: Arizona.
RANGE: Southwestern New Mexico to California.
NEW MEXICO: Silver City; Bear Mountain; Gila. Dry plains and low hills.

10. OREOCARYA Greene.

Coarse, usually erect, branched or tufted, rough-hispid biennials or perennials, 40 cm. high or less, with white or yellow flowers in crowded paniculate or thyrsoid clusters; calyx deeply 5-parted, open in fruit, persistent; corolla salverform, the tube usually but little surpassing the calyx, the throat crested; stamens included, sometimes of 2 kinds; ovary deeply 4-lobed, the nutlets 4, sharply angled or wing-margined, attached laterally to the pyramidal receptacle.

KEY TO THE SPECIES.

Fruit conic or ovoid, the nutlets touching each other; flowers large, the calyx 8 to 10 mm. long.
  Corolla yellow .... 1. *O. lutescens.*
  Corolla white.
    Cauline leaves rounded-obtuse, scarcely reduced; lateral divisions of the inflorescence long-pedunculate even in flower .... 2. *O. urticacea.*
    Cauline leaves acute, much reduced; lateral divisions of the inflorescence pedunculate only in age, if at all.
      Inflorescence and stems yellow-hispid; corolla tube much exserted .... 3. *O. fulvocanescens.*
      Inflorescence and stems mostly grayish, never bright yellow; corolla tube very slightly, if at all, exserted .... 4. *O. hispidissima.*
Fruit depressed; nutlets separated by an open space; flowers smaller, the calyx seldom more than 5 mm. long.
  Plants tall, the inflorescence much exceeding the leaves, not grayish, densely yellowish-hispid above; racemes elongated .... 5. *O. multicaulis.*
  Plants low, the inflorescence only slightly exceeding the leaves, grayish, not yellowish-hispid above; racemes short and dense .... 6. *O. suffruticosa.*

1. Oreocarya lutescens Greene, Pittonia **4:** 93. 1899.

TYPE LOCALITY: Hills above Aztec, New Mexico. Type collected by Baker (no. 562).

RANGE: Known only from the type locality, in the Upper Sonoran Zone.

2. Oreocarya urticacea Woot. & Standl. Contr. U. S. Nat. Herb. **16:** 166. 1913.

TYPE LOCALITY: Canyoncito, Santa Fe County, New Mexico. Type collected by Heller (no. 3731).

RANGE: Northern New Mexico.

NEW MEXICO: Canyoncito; Glorieta; Sierra Grande. Low hills, in the Upper Sonoran Zone.

3. Oreocarya fulvocanescens (S. Wats.) Greene, Pittonia **1:** 58. 1887.

Eritrichium glomeratum fulvocanescens S. Wats. in King, Geol. Expl. 40th Par. **5:** 243. 1871.

Eritrichium fulvocanescens A. Gray, Proc. Amer. Acad. **10:** 91. 1875.

Krynitzkia fulvocanescens A. Gray, Proc. Amer. Acad. **20:** 280. 1885.

TYPE LOCALITY: Near Santa Fe, New Mexico. Type collected by Fendler (no. 632).

RANGE: Northern New Mexico.

NEW MEXICO: Aztec; Santa Fe. Low hills and plains, in the Upper Sonoran Zone.

4. Oreocarya hispidissima (Torr.) Rydb. Bull. Torrey Club **33:** 150. 1906.

Eritrichium glomeratum hispidissimum Torr. U. S. & Mex. Bound. Bot. 140. 1859.

TYPE LOCALITY: New Mexico.

RANGE: Western Texas and southern New Mexico.

NEW MEXICO: Mesa west of Organ Mountains; Pena Blanca. Dry mesas and hills, in the Lower Sonoran Zone.

5. Oreocarya multicaulis (Torr.) Greene, Pittonia **3:** 114. 1896.

Eritrichium multicaule Torr. in Marcy, Expl. Red Riv. 62. 1854.

TYPE LOCALITY: New Mexico, probably near Santa Fe. Type collected by Fendler (no. 636).

RANGE: Colorado to Arizona and western Texas.

NEW MEXICO: San Juan Valley; Tunitcha Mountains; Willard; Pajarito Park; Belen; Horse Camp; Bernal; Pecos; Santa Fe; Mangas Springs; Mimbres River; Mogollon Mountains; Carrizalillo Mountains; Animas Valley; San Luis Mountains; Organ Mountains. Dry plains and hills, in the Lower and Upper Sonoran zones.

6. **Oreocarya suffruticosa** (Torr.) Greene, Pittonia **1:** 57. 1887.

Myosotis suffruticosa Torr. Ann. Lyc. N. Y. **2:** 225. 1827.

Eritrichium jamesii Torr. in Marcy, Expl. Red Riv. 262. 1854.

Krynitzkia jamesii A. Gray, Proc. Amer. Acad. **20:** 277. 1884.

TYPE LOCALITY: "Barren desert along the Platte."

RANGE: Wyoming and South Dakota to Arizona and western Texas.

NEW MEXICO: Round Mountain; Carrizozo; Nara Visa; south of Melrose. Plains and low hills, in the Upper Sonoran Zone.

11. CRYPTANTHE Lehm.

Low, branched, rough-hispid or canescent annuals or perennials with narrow leaves and small flowers in terminal spikes or racemes; calyx closely embracing the fruit and deciduous with it; corolla white, salverform or funnelform; nutlets 4 or by abortion fewer, smooth or tuberculate, attached laterally almost to the apex, the margins sometimes acute or even winged.

KEY TO THE SPECIES.

Nutlets wing-margined .... 1. *C. pterocarya.*
Nutlets not winged.
  Nutlets smooth, shining; stems simple or branched at the base.
    Leaves narrowly linear; plants tall, usually 20 cm. or more, simple at the base .... 2. *C. fendleri.*
    Leaves mostly spatulate or oblanceolate; plants low, seldom more than 10 cm. high, branched from the base .... 3. *C. pattersonii.*
  Nutlets not smooth; stems branched from the base.
    Racemes permanently leafy-bracted throughout .... 4. *C. ramosa.*
    Racemes not bracted except sometimes at the base.
      Nutlets dissimilar; calyx 5 mm. long; lower leaves narrowly spatulate .... 5. *C. crassisepala.*
      Nutlets alike; calyx smaller, less than 4 mm. long; all leaves linear.
        Calyx lobes linear-filiform; plants stout, strongly hispid, leafy .... 6. *C. angustifolia.*
        Calyx lobes ovate-lanceolate; plants very slender, slightly hispid, not very leafy .... 7. *C. pusilla.*

1. **Cryptanthe pterocarya** (Torr.) Greene, Pittonia **1:** 120. 1887.

Eritrichium pterocaryum Torr. U. S. & Mex. Bound. Bot. 142. 1859.

Krynitzkia pterocarya A. Gray, Proc. Amer. Acad. **20:** 276. 1885.

TYPE LOCALITY: "Near El Paso."

RANGE: Washington to California and western Texas.

NEW MEXICO: Organ Mountains; mountains west of San Antonio. Dry hills and plains, in the Upper Sonoran Zone.

2. **Cryptanthe fendleri** (A. Gray) Greene, Pittonia **1:** 120. 1887.

Krynitzkia fendleri A. Gray, Proc. Amer. Acad. **20:** 268. 1885.

TYPE LOCALITY: "New Mexico." Type collected by Fendler in 1847, probably near Santa Fe.

RANGE: Wyoming to New Mexico and Arizona.

NEW MEXICO: Tunitcha Mountains; Stinking Lake; Chama; Santa Fe; Negrito Creek; Aragon. Open slopes in the mountains, chiefly in the Upper Sonoran Zone.

3. **Cryptanthe pattersonii** (A. Gray) Greene, Pittonia **1**: 120. 1887.

Krynitzkia pattersonii A. Gray, Proc. Amer. Acad. **20**: 268. 1885.

TYPE LOCALITY: "At the base of the Rocky Mountains in Colorado."

RANGE: Wyoming to northern New Mexico.

NEW MEXICO: Zuni; Chama. Open slopes, in the Upper Sonoran and Transition zones.

4. **Cryptanthe ramosa** (Lehm.) Greene, Pittonia **1**: 115. 1887.

Lithospermum ramosum Lehm. Pl. Asper. 328. 1818.

Krynitzkia ramosa A. Gray, Proc. Amer. Acad. **20**: 274. 1885.

TYPE LOCALITY: Mexico.

RANGE: Southeastern New Mexico and western Texas to Mexico.

NEW MEXICO: Near Roswell (*Wooton*). Lower Sonoran Zone.

5. **Cryptanthe crassisepala** (Torr. & Gray) Greene, Pittonia **1**: 112. 1887.

Eritrichium crassisepalum Torr. & Gray, U. S. Rep. Expl. Miss. Pacif. **2**[2]: 171. 1854.

Krynitzkia crassisepala A. Gray, Proc. Amer. Acad. **20**: 268. 1885.

Cryptanthe dicarpa A. Nels. Proc. Biol. Soc. Washington **16**: 30. 1903.

TYPE LOCALITY: "On the Pecos, Llano Estacado, etc."

RANGE: Montana and Saskatchewan to Arizona, western Texas, and Mexico.

NEW MEXICO: Aztec; Albuquerque; Estancia; Santa Fe; Hillsboro; Cliff; Florida Mountains; Mesilla Valley; Organ Mountains; Queen. Dry hills and plains, in the Lower and Upper Sonoran zones.

6. **Cryptanthe angustifolia** (Torr.) Greene, Pittonia **1**: 112. 1887.

Eritrichium angustifolium Torr. U. S. Rep. Expl. Miss. Pacif. **5**: 363. 1857.

Krynitzkia angustifolia A. Gray, Proc. Amer. Acad. **20**: 272. 1885.

TYPE LOCALITY: "On the Colorado and Lower Gila, westward to the mountains."

RANGE: Southern California to western Texas.

NEW MEXICO: Mesilla Valley. Dry mesas and sandhills, in the Lower Sonoran Zone.

7. **Cryptanthe pusilla** (A. Gray) Greene, Pittonia **1**: 115. 1887.

Eritrichium pusillum Torr. & Gray, U. S. Rep. Expl. Miss. Pacif. **2**[2]: 171. 1854.

Krynitzkia pusilla A. Gray, Proc. Amer. Acad. **20**: 274. 1885.

TYPE LOCALITY: "Rio Pecos to Llano Estacado," New Mexico or Texas.

RANGE: Southern New Mexico and western Texas.

NEW MEXICO: Hillsboro; mesa near Agricultural College. Dry plains and hills, in the Lower Sonoran Zone.

12. ALLOCARYA Greene.

Slender hirsute annual, branched from the base, with nearly linear opposite leaves and very small flowers on thickened pedicels in slender racemes; corolla salverform, white with a yellow throat; nutlets 4, rugulose or muriculate.

1. **Allocarya scopulorum** Greene, Pittonia **1**: 16. 1887.

TYPE LOCALITY: "Colorado, Wyoming, and Montana."

RANGE: Washington and Montana to Nevada and New Mexico.

NEW MEXICO: Chama (*Standley* 6800). Wet ground, in the Transition Zone.

13. ECHIUM L. VIPER'S BUGLOSS.

Rough-bristly biennial with linear-lanceolate sessile cauline leaves and showy blue flowers in an open panicle; stamens mostly exserted, unequal; nutlets roughened or wrinkled, attached by a flat base.

1. Echium vulgare L. Sp. Pl. 139. 1753.

TYPE LOCALITY: "Habitat in Europa ad vias & agros."

NEW MEXICO: Mesilla (*Cockerell*).

A native of Europe, frequently established as a weed in the United States.

124. VERBENACEAE. Vervain Family.

Herbs or shrubs with opposite exstipulate leaves (sometimes apparently fascicled); flowers in elongated or contracted bracted spikes, these somewhat elongated in fruit; calyx limb 2 to 5-lobed, sometimes unequally so; corolla mostly 2-lipped, tubular, 5-merous; stamens didynamous, epipetalous, alternate with the corolla lobes; style single, with 1 or 2 stigmas; ovary 2 to 4-celled; fruit dry, separating at maturity into 2 or 4 nutlets.

KEY TO THE GENERA.

Fruit of 4 nutlets.................................... 1. VERBENA (p. 548).
Fruit of 2 nutlets.
  Shrubs, aromatic; spikes elongated.................. 2. LIPPIA (p. 550).
  Herbs, odorless; spikes contracted, headlike........ 3. PHYLA (p. 550).

1. VERBENA L. VERVAIN.

Annual or perennial herbs, erect or prostrate, with simple or pinnately lobed leaves and terminal bracted spikes of flowers; calyx 5-toothed, sometimes unequally so; corolla usually bluish or purple, with 5 unequal lobes; stamens 4, didynamous, the connective sometimes with a gland; ovary 4-celled, the stigma 2-lobed; fruit of 4 nutlets separating at maturity, without a stylopodium.

KEY TO THE SPECIES.

Anthers of the upper stamens glandular-appendaged; flowers in depressed spikes; corollas conspicuous.
  Calyx lobes shortly setaceous-tipped, nearly equal; bracts merely acute; plants erect; nutlets nearly smooth on the lower fourth; flowers rose purple.................. 1. *V. wrightii.*
  Calyx lobes with long setaceous tips, conspicuously unequal; bracts acute or long-acuminate; plants prostrate; nutlets reticulate almost to the base; flowers rose or bluish purple.
    Segments of the leaves oblong-lanceolate, 3 to 4 mm. wide; flowers bluish purple; plants stout, occasionally with some suberect stems................ 2. *V. ambrosiaefolia.*
    Segments of the leaves narrowly oblong-linear, 1 to 2 mm. wide; flowers rose purple; plants lower and with slender stems.............................. 3. *V. pubera.*
Anthers not glandular-appendaged; flowers in elongated spikes; corollas usually small; leaves simple or once pinnate.
  Bracts longer than the fruiting calyx; stems prostrate or erect.
    Plants prostrate, only the ends of the stems ascending. 4. *V. bracteosa.*
    Plants erect, all the stems rigidly upright............. 5. *V. imbricata.*
  Bracts shorter than the fruiting calyx; stems erect.
    Leaves not lobed; spikes dense.
      Fruiting calyx almost glabrous; plants sparingly hispid.................................... 6. *V. hastata.*
      Fruiting calyx appressed-hirsutulous; plants hispid-villous throughout......................... 7. *V. macdougalii.*

Leaves pinnatifid or pinnate; spikes slender, interrupted.
Leaves, at least the basal ones, obovate-oblanceolate, the segments 5 mm. broad or more..... 8. *V. neomexicana.*
Leaves mostly linear-oblong, the lower ones with a few salient teeth........................ 9. *V. perennis.*

1. **Verbena wrightii** A. Gray, Syn. Fl. **2**[1]: 337. 1878.

TYPE LOCALITY: "Near Frontera, on the borders of Texas, and adjacent New Mexico and Chihuahua."

RANGE: Western Texas to southern Arizona and southward.

NEW MEXICO: San Rafael; Sandia Mountains; Mangas Springs; Datil; Magdalena Mountains; Socorro; Black Range; Burro Mountains; Organ Mountains; White Mountains; Mesilla Valley. Mountains and foothills, in the Upper Sonoran and Transition zones.

The plants here listed are the common erect plant of the region from which *V. wrightii* was described, but they do not agree with Doctor Gray's original description in the characters "stems simple below" and "from an annual root." We know of no single-stemmed Verbena from the region, and *V. wrightii* is probably a short-lived perennial, flowering the first year. Its erect habit and rose purple flowers are very distinctive.

2. **Verbena ambrosiaefolia** Rydb. in Small, Fl. Southeast. U. S. 1011. 1903.

TYPE LOCALITY: Rocky Ford, Colorado.

RANGE: Arkansas and Texas to Colorado and New Mexico.

NEW MEXICO: Common throughout the State. Plains, in the Upper Sonoran Zone.

3. **Verbena pubera** Greene, Pittonia **5**: 136. 1903.

TYPE LOCALITY: Davis Mountains, western Texas.

RANGE: Mountains of western Texas and southern New Mexico.

NEW MEXICO: Albuquerque; Horse Camp; Magdalena; Fairview; San Antonio; Carrizalillo Mountains; Clemow; Lordsburg. Upper Sonoran Zone

4. **Verbena bracteosa** Michx. Fl. Bor. Amer. **2**: 13. 1803.

Verbena rudis Greene, Pittonia **4**: 152. 1900.

Verbena confinis Greene, loc. cit.

TYPE LOCALITY: "Hab. in regione Illinoensi et in urbe Nash-ville."

RANGE: Across the United States.

NEW MEXICO: Common throughout the State. Plains and waste ground, in the Upper Sonoran and Transition zones.

The type of *V. confinis* was collected in the Organ Mountains (*Wooton* 409).

5. **Verbena imbricata** Woot. & Standl. Contr. U. S. Nat. Herb. **16**: 166. 1913.

TYPE LOCALITY: Farmington, New Mexico. Type collected by Wooton (no. 2831).

RANGE: Known only from type locality.

6. **Verbena hastata** L. Sp. Pl. 20. 1753.

TYPE LOCALITY: "Habitat in Canadae humidis."

RANGE: British America to California and Florida.

NEW MEXICO: Mule Creek; near the Copper Mines. Wet ground, in the Upper Sonoran Zone.

7. **Verbena macdougalii** Heller, Bull. Torrey Club **26**: 588. 1899.

TYPE LOCALITY: Moist soil in valley near Flagstaff, Arizona.

RANGE: Colorado to Arizona and New Mexico.

NEW MEXICO: Dulce; Sierra Grande; Las Huertas Canyon; Santa Fe and Las Vegas mountains; mountains west of Grant; Ramah; Agua Fria Spring; White and Sacramento mountains. Meadows in the mountains, in the Transition Zone.

Plants of this species sometimes have rose-colored flowers and in a specimen from Fresnal the corolla is white.

8. **Verbena neomexicana** (A. Gray) Small, Fl. Southeast. U. S. 1010. 1903.

Verbena officinalis hirsuta Torr. U. S. & Mex. Bound. Bot. 128. 1859.

Verbena canescens neomexicana A. Gray, Syn. Fl. 2[1]: 337. 1878.

TYPE LOCALITY: Santa Rita, New Mexico.

RANGE: Western Texas to southern Arizona and Chihuahua.

NEW MEXICO: Socorro Mountain; mountains west of San Antonio; Mogollon Mountains; Kingston; Gray; White Mountains. Low mountains, in the Upper Sonoran and Transition zones.

9. **Verbena perennis** Wooton, Bull. Torrey Club **25**: 363. 1898.

TYPE LOCALITY: Crevices of rocks along the road about 2 miles west of the Mescalero Agency, in the White Mountains, New Mexico. Type collected by Wooton in 1897.

RANGE: Mountains of southern New Mexico.

NEW MEXICO: White Mountains; Capitan Mountains; Queen; plains south of Torrance. Upper Sonoran Zone.

2. LIPPIA L.

Branched shrub, about 1 meter high, with slender stems, small, ovate, crenate-serrate, strongly scented leaves, and terminal spikes of small white flowers; bracts ovate-lanceolate, acuminate, about the length of the calyx; calyx about 2 mm. long, with 4 acute equal lobes, densely white-hirsute; corolla about twice the length of the calyx, glabrous within; nutlets thin-walled.

1. **Lippia wrightii** A. Gray, Amer. Journ. Sci. II. **16**: 98. 1853.

TYPE LOCALITY: Not stated; probably western Texas.

RANGE: Western Texas to southern Arizona and adjacent Mexico.

NEW MEXICO: Rio Alamosa; Magdalena Mountains; Socorro; Mangas Springs; Florida Mountains; Dona Ana and Organ mountains; Capitan Mountains; Burro Mountains; Orogrande; White Mountains. Rocky hills, in the Lower Sonoran Zone.

3. PHYLA Lour.

Prostrate herbs, green and glabrate or strigillose, with simple leaves, the small flowers in bracted heads or very short spikes; calyx 2-toothed; corolla 2-lipped, the upper lip notched, the lower 3-lobed; stamens 4, didynamous; style short and slender, the stigma oblique; fruit of 2 nutlets inclosed in a persistent calyx.

KEY TO THE SPECIES.

Leaves broadly lanceolate or elliptic-lanceolate, acute, decurrent into a short petiole, bright green, with about 15 serrate teeth.. 1. *P. lanceolata.*

Leaves narrowly to broadly cuneate, sometimes oblanceolate, with no proper petiole, cinereous, with a few coarse teeth above the middle.

Peduncles little or not at all exceeding the leaves............ 2. *P. cuneifolia.*

Peduncles 2 to 5 times the length of the leaves................. 3. *P. incisa.*

1. **Phyla lanceolata** (Michx.) Greene, Pittonia **4**: 47. 1899.

Lippia lanceolata Michx. Fl. Bor. Amer. **2**: 15. 1803.

TYPE LOCALITY: "Hab. in Carolina, juxta amniculum."

RANGE: Colorado southward and across the continent.

NEW MEXICO: Roswell (*Earle* 355). Lower Sonoran Zone.

2. **Phyla cuneifolia** (Torr.) Greene, Pittonia 4: 47. 1899.
Zapania cuneifolia Torr. Ann. Lyc. N. Y. 2: 234. 1827.
Lippia cuneifolia Steud. in Marcy, Expl. Red Riv. 293. 1854.
TYPE LOCALITY: "On the Platte."
RANGE: Wyoming and South Dakota to Texas and northern Mexico.
NEW MEXICO: Rio Zuni; Red Lake; Nara Visa; Sierra Grande. Upper Sonoran Zone.

3. **Phyla incisa** Small, Fl. Southeast. U. S. 1012. 1903.
TYPE LOCALITY: Near Corpus Christi, Texas.
RANGE: Western Texas to southern California.
NEW MEXICO: Socorro; Deming; Mesilla Valley; White Sands. Chiefly in adobe soil, in the Lower Sonoran Zone.

125. MENTHACEAE. Mint Family.

Aromatic herbs or shrubs with 4-sided stems and opposite leaves; leaves simple, entire or toothed; inflorescence of small cymose clusters in the axils of the normal or reduced leaves or in terminal spikes or heads; flowers perfect, irregular, rarely nearly regular; calyx free, persistent, tubular or campanulate, regular or irregular, 5-toothed; corolla usually bilabiate; stamens 4 or by abortion 2; anthers 2-celled; ovary superior, deeply 4-lobed, 4-celled; fruit of 4 small nutlets in the persistent calyx.

KEY TO THE GENERA.

Ovary of 4 united carpels, 4-lobed; style not basal; nutlets attached to each other along the inner angle.
  Corolla very irregular, the upper lip much reduced.
    Flowers in a terminal spikelike inflorescence; leaves merely toothed, not pinnatifid..... 1. TEUCRIUM (p. 553).
    Flowers solitary, axillary to leaflike bracts; leaves deeply pinnatifid.................. 2. MELOSMON (p. 553).
  Corolla nearly regular, the lobes almost alike.
    Flowers in small axillary cymose clusters....... 3. TETRACLEA (p. 553).
    Flowers in thyrsoid panicles at the ends of the stems.................................... 4. TRICHOSTEMA (p. 554).
Ovary of 4 distinct or nearly distinct carpels; styles basal; nutlets attached at or near the base.
  Corolla nearly regular, 4 or 5-toothed.
    Anther-bearing stamens 4........................ 5. MENTHA (p. 554).
    Anther-bearing stamens 2........................ 6. LYCOPUS (p. 555).
  Corolla conspicuously bilabiate.
    Calyx 2-lipped, the lips entire, the upper ones crested; stamens 4....................... 7. SCUTELLARIA (p. 555).
    Calyx with more than 2 divisions, not crested; stamens 4, or by abortion 2.
      Stamens included in the corolla tube....... 8. MARRUBIUM (p. 556).
      Stamens more or less exserted.
        Upper lip of the corolla flat, not concave.
          Stamens straight, distant and diverging; calyx almost regularly 5-toothed; anther-bearing stamens 4; flowers in terminal bracted heads............... 9. MADRONELLA (p. 556).

- Stamens curved, often converging; calyx various; anther-bearing stamens 2 or 4; inflorescence various.
  - Shrub; calyx regularly 5-toothed; anther-bearing stamens 2 .... 10. POLIOMINTHA (p. 556).
  - Herbs, calyx 2-lipped; anther-bearing stamens 2 or 4.
    - Flowers in terminal heads; leaves large; anther-bearing stamens 4 .... 11. CLINOPODIUM (p. 556).
    - Flowers in axillary clusters or in interrupted spikelike inflorescences; leaves small; anther-bearing stamens 2 .... 12. HEDEOMA (p. 557).

Upper lip of the corolla more or less concave.

- Anther-bearing stamens 2.
  - Calyx equally 5-toothed; connective short, the anther sacs confluent; flowers in crowded, bracteate, verticillate or terminal heads, the bracts often colored .... 13. MONARDA (p. 559).
  - Calyx 2-lipped, the upper lip obscurely 3-toothed or entire, the lower 2-toothed; connective long, with a perfect anther sac at one end and a rudimentary one at the other; flowers never in headlike clusters .... 14. SALVIA (p. 560).
- Anther-bearing stamens 4.
  - Upper stamens longer than the lower.
    - Calyx distinctly 2-lipped; flowers very small, obscured by the numerous bracts .... 15. MOLDAVICA (p. 562
    - Calyx about equally 5-toothed; flowers usually larger and the bracts of the inflorescence not conspicuous.
      - Anther sacs parallel or nearly so; stamens divergent .... 16. AGASTACHE (p. 562).
      - Anther sacs divaricate; stamens approximate in pairs .... 17. NEPETA (p. 564).

Upper stamens shorter than the lower.
  Calyx distinctly 2-lipped, closed in fruit..........18. PRUNELLA (p. 565).
  Calyx not 2-lipped, 5-toothed, open in fruit.
    Upper leaves clasping; nutlets 3-sided, truncate above..................19. LAMIUM (p. 565).
    Upper leaves not clasping; nutlets nearly terete, rounded above........20. STACHYS (p. 565).

1. TEUCRIUM L. GERMANDER.

Erect branched perennial herb with villous-hirsute stems and ovate-oblong, sharply serrate, short-petioled leaves; flowers in long terminal panicles, on short pedicels; calyx campanulate, the lobes shorter than the tube; corolla pinkish purple, 8 to 12 mm. long, tomentulose, glandular, the upper lip short, the lower with 2 short lateral lobes and a central elongated one; stamens 4, exserted.

1. **Teucrium occidentale** A. Gray, Syn. Fl. **2**[1]**:** 349. 1878.

TYPE LOCALITY: Nebraska.

RANGE: British America to California, east to the Atlantic Coast.

NEW MEXICO: Cedar Hill; Farmington; Ojo Caliente; Inscription Rock; Frisco; Mesilla Valley; Roswell. Moist ground, in the Upper Sonoran and Transition zones.

2. MELOSMON Raf.

Low spreading perennial herb about 10 cm. high, with laciniately parted leaves and solitary white flowers crowded among leaflike bracts at the ends of the stems; calyx turbinate, strongly 10-ribbed, the lobes linear-subulate, several times longer than the tube; corolla 15 to 20 mm. long, the lower lip much surpassing the calyx; nutlets obscurely reticulate, granular.

1. **Melosmon laciniatum** (Torr.) Small, Fl. Southeast. U. S. 1019. 1903.

Teucrium laciniatum Torr. Ann. Lyc. N. Y. **2:** 231. 1828.

TYPE LOCALITY: "On the Rocky Mountains," Colorado.

RANGE: Colorado to Texas and Arizona.

NEW MEXICO: Perico Creek; Las Vegas; Santa Fe; Sierra Grande; Nara Visa; White Mountains; Gallinas Mountains; Gray; Buchanan. Plains and meadows, in the Upper Sonoran and Transition zones.

3. TETRACLEA A. Gray.

Perennial herbs with toothed leaves and rather few flowers in bracted few-flowered axillary clusters on short peduncles and pedicels; calyx campanulate, the limb 5-parted, accrescent, persistent; the 5 lobes of the corolla entire, elliptic-obovate, cream-colored, tinged with red outside; stamens exserted; fruit of 4 pyriform nutlets, very strongly and coarsely reticulated and finely pubescent, the commisural scar somewhat rugose.

KEY TO THE SPECIES.

Leaves broadly ovate, at least the lower ones entire............... 1. *T. coulteri*.
Leaves narrowly oblong, conspicuously toothed.................... 2. *T. angustifolia*.

1. **Tetraclea coulteri** A. Gray, Amer. Journ. Sci. II. **16:** 98. *pl. 41.* 1853.

TYPE LOCALITY: Mexico.

RANGE: Western Texas to southern Arizona and southward.

NEW MEXICO: Mangas Springs; south of Hillsboro; near White Water; Dog Spring; Organ Mountains; Tortugas Mountain; south of Roswell; Dayton. Rocky hills and canyons, in the Lower and Upper Sonoran zones.

2. **Tetraclea angustifolia** Woot. & Standl. Contr. U. S. Nat. Herb. **16:** 170. 1913.

TYPE LOCALITY: Plains south of the White Sands, Otero County, New Mexico. Type collected by Wooton (no. 403).

RANGE: Known only from type locality, in the Lower Sonoran Zone.

4. TRICHOSTEMA L. BLUE CURLS.

Low shrub, 30 to 60 cm. high, sparingly puberulent or glabrate; leaves small, ovate, petiolate, entire; flowers in small axillary cymes on the upper part of the stems, becoming thyrsoid-paniculate; calyx campanulate, with 5 equal acute lobes; corolla with a short tube not equaling the calyx, the lobes longer than the tube; stamens with long capillary curved filaments 2 cm. long or more, much surpassing the corolla; style still longer; fruit of 4 coarsely reticulate-roughened nutlets.

1. **Trichostema arizonicum** A. Gray, Proc. Amer. Acad. **8:** 371. 1872.

TYPE LOCALITY: Chiricahua Mountains, southern Arizona.

RANGE: Southwestern New Mexico and adjacent Arizona and Mexico.

NEW MEXICO: Berendo Creek; Guadalupe Canyon. Mountains, in the Upper Sonoran Zone.

5. MENTHA L. MINT.

Aromatic perennial herbs with toothed leaves, the small flowers in clusters in the axils of the leaves or forming terminal spikes or verticillate axillary clusters; calyx 5-toothed, tubular or campanulate, 10-ribbed, the teeth equal or nearly so; corolla nearly regular, with short included tube and 5 lobes; stamens 4, erect, the anthers 2-celled, the cells parallel; nutlets smooth.

KEY TO THE SPECIES.

Whorls of flowers all axillary.. 1. *M. penardi.*
Whorls of flowers in terminal spikes.
  Plants glabrous; leaves acute.................................. 2. *M. spicata.*
  Plants canescent or tomentose; leaves obtuse................. 3. *M. rotundifolia.*

1. **Mentha penardi** (Briq.) Rydb. Bull. Torrey Club **33:** 150. 1906.

Mentha arvensis penardi Briq. Bull. Herb. Boiss. **3:** 215. 1895.

TYPE LOCALITY: Boulder, Colorado.

RANGE: British Columbia and Nebraska to Arizona and New Mexico.

NEW MEXICO: Farmington; Tunitcha Mountains; Chama; Ramah; mountains west of Grants; Santa Fe and Las Vegas mountains; Albuquerque; Mimbres; Mangas Springs; Middle Fork of the Gila; Mesilla Valley; White Mountains; Roswell. Wet ground, from the Lower Sonoran to the Transition Zone.

2. **Mentha spicata** L. Sp. Pl. 576. 1753. SPEARMINT.

Mentha spicata viridis L. loc. cit.

Mentha viridis L. Sp. Pl. ed. 2. 804. 1763.

TYPE LOCALITY: European.

NEW MEXICO: Santa Fe; Mesilla Valley; Anton Chico.

A native of Europe, frequently cultivated and often escaped.

3. **Mentha rotundifolia** (L.) Huds. Fl. Angl. 221. 1762. ROUND-LEAVED MINT.
Mentha spicata rotundifolia L. Sp. Pl. 576. 1753.
TYPE LOCALITY: European.
NEW MEXICO: Mesilla Valley; Tularosa.
Common along ditch banks, introduced from Europe.

6. LYCOPUS L. BUGLEWEED.

Perennial herbs from slender branching rootstocks, with lanceolate or narrowly oblong, toothed leaves, the small flowers sessile in crowded clusters in the axils; calyx 2 to 3 mm. long, with equal triangular-subulate teeth; corolla little longer than the calyx, whitish; nutlets triangular, with a thickened border along the edges, shorter than the calyx.

KEY TO THE SPECIES.

Leaves narrowly oblong, merely serrate, sessile.................... 1. *L. lucidus*.
Leaves lanceolate to ovate in outline, sinuate-pinnatifid, petiolate. 2. *L. americanus*.

1. **Lycopus lucidus** Turcz.; Benth. in DC. Prodr. **12**: 178. 1848.
Lycopus lucidus americanus A. Gray, Proc. Amer. Acad. **8**: 286. 1870.
TYPE LOCALITY: "In montibus Ircutiae."
RANGE: British Columbia and Nebraska to California and New Mexico; also in Eurasia.
NEW MEXICO: Farmington (*Standley* 7019). Wet ground, in the Upper Sonoran Zone.

2. **Lycopus americanus** Muhl.; Barton, Fl. Phila. Prodr. 15. 1815.
Lycopus sinuatus Ell. Bot. S. C. & Ga. **1**: 26. 1817.
TYPE LOCALITY: Philadelphia, Pennsylvania.
RANGE: Nearly across North America.
NEW MEXICO: Farmington; Cedar Hill; Pecos; Chavez; Sandia Mountains; Mangas Springs. Wet ground and in meadows, in the Upper Sonoran Zone.

7. SCUTELLARIA L. SKULLCAP.

Annual or perennial herbs with small, entire or toothed, short-petioled leaves and solitary flowers axillary to foliar leaves or leaflike bracts; calyx campanulate, 2-lipped, the lips entire, the upper crested, persistent, slightly accrescent; corolla blue or violet, with a recurved tube dilated at the throat, the upper lip arched, the lower with 2 small lateral lobes and a large middle one; stamens 4; nutlets papillose-tuberculate.

KEY TO THE SPECIES.

Annual; plants villous, somewhat glandular...................... 1. *S. drummondii*.
Perennials; plants cinereous-puberulent.
Woody at the base, not stoloniferous, 20 cm. high or less...... 2. *S. wrightii*.
Not woody at the base, stoloniferous, 50 to 100 cm. high...... 3. *S. galericulata*.

1. **Scutellaria drummondii** Benth. Labiat. Gen. Sp. 441. 1834.
TYPE LOCALITY: "Hab. in America boreali: ad Rio Brazos a provinciae Texas Mexicanorum."
RANGE: Texas and southern New Mexico.
NEW MEXICO: Sixteen Spring Canyon; Roswell. Dry hills, in the Lower and Upper Sonoran zones.

2. **Scutellaria wrightii** A. Gray, Proc. Amer. Acad. **8**: 370. 1872.
TYPE LOCALITY: Texas.
RANGE: Texas to southern Arizona.
NEW MEXICO: Black Range; Dog Mountains. Lower and Upper Sonoran zones.

3. Scutellaria galericulata L. Sp. Pl. 599. 1753.

TYPE LOCALITY: "Habitat in Europae littoribus."

RANGE: Alaska and British America to Arizona and North Carolina; also in Europe.

NEW MEXICO: Farmington (*Standley* 7153). Along streams and in swamps, in the Upper Sonoran Zone.

Common in this locality in the cattail swamps.

8. MARRUBIUM L. HOREHOUND.

Erect, branched, densely white-tomentose perennial herb, with rugose crenate short-petioled leaves and numerous small flowers in crowded axillary clusters; calyx tubular, 10-ribbed, the lobes 10, equal, rigid, uncinate; corolla pale purplish, the tube twice as long as the calyx; stamens 4, included; nutlets granular.

1. Marrubium vulgare L. Sp. Pl. 583. 1753.

TYPE LOCALITY: "Habitat in Europae borealioris ruderatis."

RANGE: A native of Eurasia, introduced in waste places in most parts of North America.

NEW MEXICO: In nearly all the moister parts of the State; often abundant, especially where sheep range.

9. MADRONELLA Greene.

Perennial herb 30 to 40 cm. high, with many slender erect stems from a woody root; leaves small, about 2 cm. long, oblong-lanceolate, obtuse, entire; flowers in a terminal bracted head; calyx tubular, 5-toothed, 10 to 13-nerved; corolla somewhat bilabiate, the upper lip 2-cleft, the lower 3-lobed into narrow similar lobes; stamens 4, sometimes unequal.

1. Madronella parvifolia (Greene) Rydb. Bull. Torrey Club **33:** 150. 1906.

Monardella parvifolia Greene, Pl. Baker. **3:** 22. 1901.

TYPE LOCALITY: Canyon of the Gunnison near Cimarron, Colorado.

RANGE: Mountains of Colorado, Arizona, and New Mexico.

NEW MEXICO: Near Mogollon; West Fork of the Gila. Transition Zone.

10. POLIOMINTHA A. Gray.

Hoary-canescent shrub about 1 meter high, with entire, linear-oblong to lanceolate, nearly sessile leaves about 2 cm. long; flowers in small axillary clusters toward the ends of the branches; calyx broadly tubular, with 5 narrow equal teeth, densely villous with spreading white hairs; corolla pale purplish, the tube surpassing the calyx; fertile stamens 2.

1. Poliomintha incana (Torr.) A. Gray, Proc. Amer. Acad. **8:** 296. 1870.

Hedeoma incana Torr. U. S. & Mex. Bound. Bot. 130. 1859.

TYPE LOCALITY: Sandy places near El Paso.

RANGE: Western Texas to Utah and southern Arizona.

NEW MEXICO: White Sands. Sandy plains, in the Lower Sonoran Zone.

11. CLINOPODIUM L. WILD BASIL.

Slender perennial herb, 30 to 50 cm. high, with villous-hirsute stems, petiolate, thin, oval or ovate-lanceolate, obscurely toothed leaves, and terminal headlike clusters of small flowers; calyx 13-ribbed, tubular, a little inflated below, the lobes linear, subulate, unequal; corolla longer than the calyx, 2-lipped, the upper lip notched, the lower 3-lobed; stamens 4; nutlets smooth.

1. **Clinopodium vulgare** L. Sp. Pl. 587. 1753.
Melissa clinopodium Benth. Labiat. Gen. Sp. 393. 1834.
Calamintha clinopodium Benth. in DC. Prodr. **12:** 233. 1848.
TYPE LOCALITY: "Habitat in rupestribus Europae, Canadae."
RANGE: British America to North Carolina and New Mexico; also in Europe.
NEW MEXICO: Gallinas Canyon; West Fork of the Gila; White and Sacramento mountains; Brazos Canyon. In woods, in the Transition Zone.

12. HEDEOMA Pers. PENNYROYAL.

Annual or perennial herbs, 40 cm. high or mostly less, with small entire or toothed leaves and small flowers in axillary clusters among the reduced upper leaves; calyx tubular, 2-lipped, the 3 upper teeth shorter than the 2 lower; corolla narrowly tubular to tubular-funnelform, purple or paler, 2-lipped, the upper lip entire or emarginate, the lower 3-lobed; stamens 2, sometimes with 2 sterile filaments; nutlets smooth.

KEY TO THE SPECIES.

Upper calyx lobes triangular-lanceolate, noticeably broader than the lower ones; calyx sparingly hispidulous; flowers conspicuous, with gaping corolla; leaves entire.
  Stems rather stout; leaves elliptic-oblong, obtuse, spreading, not conspicuously veined .... 1. *H. pulcherrima.*
  Stems slender, wiry; leaves narrowly lanceolate, erect or ascending, very conspicuously veined .... 2. *H. hyssopifolia.*
Upper calyx lobes shorter but scarcely narrower than the lower, subulate, villous-hirsute; flowers small, the corolla tube narrow; leaves entire or toothed.
  At least some of the leaves toothed.
    Leaves mostly entire, only a few with occasional inconspicuous teeth; upper leaves lanceolate .... 3. *H. oblongifolia.*
    Leaves all conspicuously toothed, ovate.
      Corolla conspicuously long-tubular; leaves not plicate, their veins inconspicuous .... 4. *H. pulchella.*
      Corolla very short, scarcely exceeding the calyx; leaves plicate, with very conspicuous veins... 5. *H. plicata.*
  All the leaves entire.
    Leaves shorter than the subtended calyces .... 6. *H. nana.*
    Leaves longer than the subtended calyces.
      Calyx 5 to 6 mm. long, the corolla but little longer, barely exserted .... 7. *H. ciliata.*
      Calyx 7 to 8 mm. long, the corolla almost twice as long, much exserted .... 8. *H. lata.*

1. **Hedeoma pulcherrima** Woot. & Standl. Contr. U. S. Nat. Herb. **16:** 168. 1913.
TYPE LOCALITY: White Mountains, Lincoln County, New Mexico.
RANGE: White and Sacramento mountains of New Mexico, in the Transition Zone.

2. **Hedeoma hyssopifolia** A. Gray, Proc. Amer. Acad. **11:** 96. 1876.
TYPE LOCALITY: Mount Graham, Arizona.
RANGE: Mountains of Arizona and New Mexico and adjacent Mexico.
NEW MEXICO: Mogollon Mountains; Pinos Altos Mountains; San Luis Mountains. Upper Sonoran Zone.

3. **Hedeoma oblongifolia** (A. Gray) Heller, Muhlenbergia **1**: 4. 1900.

Hedeoma piperita oblongifolia A. Gray, Proc. Amer. Acad. **8**: 367. 1872.

Hedeoma thymoides oblongifolia A. Gray, Syn. Fl. **2**^{1}: 362. 1878.

TYPE LOCALITY: "New Mexico and Arizona."

RANGE: Southern New Mexico and Arizona and adjacent Mexico.

NEW MEXICO: Fort Tularosa; Hanover Mountain; Middle Fork of the Gila; Dog Spring; Organ Mountains; Shalam Hills. Low mountains, in the Upper Sonoran Zone.

There is very little doubt that this is the plant Doctor Gray had before him, notwithstanding that the name applies much better to some of the other species. There is also little doubt that this is a plant which Doctor Torrey included in his *H. dentata* in the Botany of the Mexican Boundary Survey, a reference in which others have followed him, although the two plants are very unlike. The leaves of *H. oblongifolia* are not oblong; the lower cauline leaves are ovate to elliptic, acute and decurrent, entire, with occasionally a few inconspicuous teeth; the upper ones are lanceolate and acute. Doctor Gray calls attention to these facts in the descriptions cited above.

4. **Hedeoma pulchella** Greene, Leaflets **1**: 213. 1906.

TYPE LOCALITY: Limestone hills near Kingston, New Mexico. Type collected by Metcalfe (no. 1599).

RANGE: Known only from type locality.

5. **Hedeoma plicata** Torr. U. S. & Mex. Bound. Bot. 130. 1859.

TYPE LOCALITY: Dry ravines near the Limpio Mountains, Texas.

RANGE: Western Texas to southern Arizona and adjacent Mexico.

NEW MEXICO: Organ Mountains; Queen; White Mountains. Dry hills, in the Upper Sonoran Zone.

6. **Hedeoma nana** (Torr.) Greene, Pittonia **3**: 339. 1898.

Hedeoma dentata nana Torr. U. S. & Mex. Bound. Bot. 130. 1859.

TYPE LOCALITY: "Rocky hills of the Rio Grande, near El Paso."

RANGE: Western Texas to southern Arizona and southward.

NEW MEXICO: Hillsboro; Organ Mountains; Tortugas Mountain. Dry hills, in the Upper Sonoran Zone.

7. **Hedeoma ciliata** Nutt. Journ. Acad. Phila. II. **1**: 183. 1848.

Hedeoma sancta Small, Bull. N. Y. Bot. Gard. **1**: 287. 1899.

TYPE LOCALITY: Santa Fe, New Mexico. Type collected by Gambel.

RANGE: Colorado to Texas and New Mexico.

NEW MEXICO: Santa Fe; Gallinas Canyon; Sandia Mountains; Tesuque; Pecos; Magdalena Mountains. Open hills, in the Upper Sonoran Zone.

We have not seen types of either *H. sancta* Small or *H. ciliata* Nutt. and can not be sure that they are the same plant. What we take to be *H. sancta* seems to be the common plant of this genus in the region about Santa Fe, the type locality of the other species, and is reported by Doctor Rydberg from southern Colorado. Through the kindness of Dr. J. H. Barnhart we have learned that there is every reason to believe that *H. ciliata* Nutt. was published a short time before *H. ciliata* Benth. Thus *H. ciliata* Nutt. would seem to be the proper name for the Santa Fe plant, with *H. sancta* as a synonym, providing Doctor Rydberg and we have been correct in referring this southern Colorado and northern New Mexico plant to the latter species.

8. **Hedeoma lata** Small, Fl. Southeast. U. S. 1040. 1903.

TYPE LOCALITY: "On rocky prairies, Texas and New Mexico."

RANGE: Western Texas and New Mexico.

NEW MEXICO: Carrizo Mountains; Dulce; Aztec; Ramah; Magdalena; Santa Rita; Mangas Springs; mountains west of San Antonio; White Mountains; Organ Mountains; Capitan Mountains; Torrance; south of Roswell. Dry plains and low hills, in the Upper Sonoran Zone.

13. MONARDA L. Horsemint.

Annual or usually perennial herbs, 30 to 60 cm. high, with petioled leaves, the flowers in crowded verticillate headlike clusters, these forming either terminal heads or a series of clusters in the axils of the upper leaves and surrounded by conspicuous, often colored bracts; calyx tubular, 15-ribbed, 5-toothed, the teeth about equal; corolla flesh-colored, rose, or purplish, 2-lipped, the upper lip arched, the lower 3-lobed; anther-bearing stamens 2, 2 rudimentary filaments present or wanting; nutlets smooth.

KEY TO THE SPECIES.

Flower clusters terminal and solitary; flowers rose purple.
  Stems and petioles villous-hirsute, the former especially so below the nodes........ 1. *M. comata.*
  Stems and petioles finely strigose or puberulent.
    Plants pale green; leaves finely puberulent, velvety to the touch, especially beneath........ 2. *M. menthaefolia.*
    Plants bright green; leaves glabrous, in the dried specimens feeling papery to the touch........ 3. *M. stricta.*
Flowers in several verticillate glomerules in the axils of the upper leaves, mostly pale.
  Calyx lobes triangular, acute, about as long as the width of the tube (long-ciliate); bracts conspicuously velvety, whitish or rose purple above........ 4. *M. lasiodonta.*
  Calyx lobes narrowly subulate-aristate, several times longer than the width of the tube; bracts mostly greenish, not conspicuously velvety.
    Bracts lanceolate, tapering into the aristate apex, green; calyx teeth, petioles, and bases of leaves sparingly or not at all ciliate; plants stout........ 5. *M. tenuiaristata.*
    Bracts ovate to oblong, abruptly aristate, sometimes purplish; calyx teeth densely long villous hirsute, the petioles and bases of the leaves sparingly so; plants rather slender........ 6. *M. pectinata.*

1. **Monarda comata** Rydb. Bull. Torrey Club **28:** 502. 1901.

Type locality: Wahatoya Creek, Colorado.

Range: Colorado and northern New Mexico.

New Mexico: Gallinas Planting Station; Sierra Grande. Open slopes, in the Transition Zone.

2. **Monarda menthaefolia** Graham, Edinburgh Phil. Journ. **1829:** 347. 1829.

Type locality: "Between Norway House and Canada," British America.

Range: Arizona and New Mexico and northward.

New Mexico: Rito de Frijoles; Chama; Santa Fe Canyon; Middle Fork of the Gila. Mountains, in the Transition Zone.

The distinctions between this and the next are so slight that it is doubtful whether they should be kept separate, but the various species of this group are all so closely related that they are separated with difficulty, and then only with large series of specimens.

3. **Monarda stricta** Wooton, Bull. Torrey Club **25:** 263. 1898.

Type locality: On the divide 9 miles northeast of the Mescalero Agency, White Mountains, New Mexico. Type collected by Wooton.

Range: Colorado and New Mexico.

New Mexico: Carrizo Mountains; Tunitcha Mountains; Pajarito Park; Santa Fe and Las Vegas mountains; Hop Canyon; Mogollon Mountains; Copper Mines; White and Sacramento mountains. Meadows in the mountains, in the Transition Zone.

4. **Monarda lasiodonta** (A. Gray) Small, Fl. Southeast. U. S. 1038. 1903.

Monarda punctata lasiodonta A. Gray, Proc. Amer. Acad. **8**: 369. 1872, name only; Syn. Fl. **2**[1]: 375. 1878.

Type locality: Texas.

Range: Oklahoma and Texas to Arizona.

New Mexico: Mountains west of Grants; San Lorenzo; Inscription Rock; Nara Visa; Organ Mountains. Low hills and plains, in the Upper Sonoran Zone.

This is probably *M. punctata humilis* Torr. in Sitgreaves Report,[1] from near Zuni, although Doctor Torrey does not mention the very characteristic pubescence of the calyx teeth, the distinguishing peculiarity of the species.

5. **Monarda tenuiaristata** (A. Gray) Small, Fl. Southeast. U. S. 1038. 1903.

Monarda citriodora tenuiaristata A. Gray, Proc. Amer. Acad. **8**: 369. 1872, name only; Syn. Fl. **2**[1]: 375. 1878, as synonym.

Type locality: Not stated.

Range: Arkansas and Kansas to New Mexico.

New Mexico: Middle Fork of the Gila; Santa Rita; Animas Valley; San Luis Mountains; Organ Mountains. Low hills, in the Upper Sonoran Zone.

6. **Monarda pectinata** Nutt. Journ. Acad. Phila. II. **1**: 182. 1848.

Type locality: Near Santa Fe, New Mexico. Type collected by Gambel.

Range: Colorado and New Mexico.

New Mexico: Dulce; Gallinas Planting Station; Pajarito Park; Barranca; Glorieta; Hermits Peak; Inscription Rock; Laguna Blanca; Luna; Water Canyon; Cactus Flat; Burro Mountains; Gila Hot Springs; Middle Fork of the Gila; Gray; Nara Visa. Dry plains and low hills, in the Upper Sonoran and Transition zones.

14. SALVIA L. Sage.

Annual or perennial herbs or shrubs with petiolate glandular leaves, the flowers in terminal, paniculate or crowded, verticillate clusters; calyx tubular or campanulate, bilabiate, sometimes obscurely so, the upper lip entire or trifid, the lower bifid, the throat smooth; corolla usually brightly colored, blue or red, bilabiate, the upper lip erect, the lower 3-lobed; stamens 2; nutlets ovoid or 3-sided, smooth.

KEY TO THE SPECIES.

Flowers red; leaves pinnatifid; stems white-villous................ 1. *S. henryi.*
Flowers blue or white; leaves not pinnatifid; stems variously pubescent or glabrous.
  Shrubs; leaves oblong to ovate.
    Leaves oblong to elliptic, acute or obtuse, entire or the uppermost obscurely dentate, nearly glabrous; calyx conspicuously veined.......................... 2. *S. ramosissima.*
    Leaves ovate or deltoid-ovate, acute, crenate, finely but densely canescent or finely puberulent; calyx not conspicuously veined.
      Calyx limb wine-colored; leaves at most puberulent.. 3. *S. vinacea.*
      Calyx limb green; leaves densely white-canescent beneath.................................... 4. *S. pinguifolia.*
  Herbs; leaves linear, lanceolate, or narrowly oblong.

[1] Sitgreaves, Rep. Zuñi & Colo. 166. 1854.

Annuals, sparingly puberulent; calyx conspicuously bilabiate and ribbed.
Corolla pale blue, barely exceeding the calyx; leaves entire or with a few inconspicuous teeth; stems and calyx green.............................. 5. *S. lanceaefolia.*
Corolla dark blue, one and one-half times the length of the calyx; leaves conspicuously sinuate-dentate; stems and calyx often deep purple....... 6. *S. subincisa.*
Perennials, the inflorescence canescent or tomentulose; calyx truncate, the lips very short.
Calyx densely white or blue-tomentulose, 6 to 8 mm. long, the lobes hardly distinguishable.......... 7. *S. earlei.*
Calyx sparingly canescent, whitish or bluish, slightly more than 8 mm. long; lobes conspicuous...... 8. *S. pitcheri.*

1. **Salvia henryi** A. Gray, Proc. Amer. Acad. **8:** 368. 1872.

Type locality: New Mexico, on the Mimbres.

Range: Southern New Mexico and Arizona and southward.

New Mexico: Bear Mountains; Socorro; San Andreas Mountains; Kingston; Upper Corner Monument; Soledad Canyon; Shalam Hills; Tortugas Mountain. Arid hills, in the Lower and Upper Sonoran zones.

2. **Salvia ramosissima** Fernald, Proc. Amer. Acad. **35:** 521. 1900.

Type locality: "Cañons of the Rio Grande," Texas.

Range: Western Texas and southern New Mexico to Mexico.

New Mexico: Organ Mountains. Dry hills and canyons, in the Upper Sonoran Zone.

3. **Salvia vinacea** Woot. & Standl. Contr. U. S. Nat. Herb. **16:** 170. 1913.

Type locality: Florida Mountains, New Mexico. Type collected by E. A. Goldman (no. 1501).

Range: Southern New Mexico.

New Mexico: Florida Mountains; San Andreas Mountains. Canyons in the mountains, Upper Sonoran Zone.

4. **Salvia pinguifolia** (Fernald) Woot. & Standl. Contr. U. S. Nat. Herb. **16:** 169. 1913.

Salvia ballotaeflora pinguifolia Fernald, Proc. Amer. Acad. **35:** 523. 1900.

Type locality: Probably in New Mexico. Type collected by Wright (no. 1524).

Range: Southern Arizona and New Mexico.

New Mexico: Burro Mountains; Mangas Springs; Hatchet Ranch; Organ Mountains. Low hills, in the Upper Sonoran Zone.

5. **Salvia lanceaefolia** Poir. in Lam. Encycl. Suppl. **5:** 49. 1817.

Type locality: Described from cultivated plants thought to have come from Peru.

Range: Colorado and Kansas to Arizona and Mexico.

New Mexico: Common throughout the State. Waste ground and damp fields, in the Upper Sonoran and Transition zones.

A common but not troublesome weed in gardens and grain fields.

6. **Salvia subincisa** Benth. Pl. Hartw. 20. 1839.

Type locality: Mexico.

Range: Western Texas to southern Arizona and southward.

New Mexico: Pajarito Park; Gallinas Mountains; south of Santa Fe; Las Vegas; Grant; Pecos; Black Range; Mogollon Mountains; Magdalena; San Luis Mountains; Organ Mountains; White Mountains; Gray; Queen. Open slopes, in the Upper Sonoran and Transition zones.

7. Salvia earlei Woot. & Standl. Contr. U. S. Nat. Herb. **16:** 169. 1913.

TYPE LOCALITY: Thirty-five miles west of Roswell, New Mexico. Type collected by Earle (no. 375).

RANGE: Western Texas and southern New Mexico.

NEW MEXICO: West of Roswell; Roswell; Sixteen Spring Canyon. Plains and low hills, in the Upper Sonoran Zone.

8. Salvia pitcheri Torr.; Benth. Labiat. Gen. Sp. 251. 1833.

TYPE LOCALITY: "Hab. in America septentrionali ad Red River."

RANGE: Nebraska to New Mexico and Texas.

NEW MEXICO: Canadian River (*Bigelow*). Plains, in the Upper Sonoran Zone.

The specimen probably came from some locality near Tucumcari.

15. MOLDAVICA Adans. DRAGON-HEAD.

Coarse herb, 20 to 70 cm. high, with branched stems, lanceolate to ovate-lanceolate, coarsely serrate, petiolate leaves, and flowers in large crowded bracted headlike spikes terminating the branches; calyx tubular, 15-nerved, 5-lobed, the lobes foliaceous, the upper broadest; corolla small, blue, shorter than the calyx; nutlets ovoid, smooth.

1. Moldavica parviflora (Nutt.) Britton in Britt. & Brown, Illustr. Fl. ed. 2. **3:** 114. 1913.

Dracocephalum parviflorum Nutt. Gen. Pl. **2:** 35. 1818.

TYPE LOCALITY: "Around Fort Mandan, on the Missouri."

RANGE: British America to New Mexico.

NEW MEXICO: Tunitcha Mountains; Farmington; Chama; Sierra Grande; Raton; Santa Fe and Las Vegas mountains; Santa Antonita; Ramah; Mogollon Mountains; Bear Mountain; Organ Mountains; White and Sacramento mountains. Open slopes and plains, in the Upper Sonoran and Transition zones.

16. AGASTACHE Clayt. GIANT HYSSOP.

Aromatic perennial herbs, 40 to 80 cm. high, more or less puberulent throughout, with petiolate leaves, the flowers in terminal, dense or interrupted, spikelike panicles; calyx tubular, often colored, 15-nerved, 5-toothed, the teeth triangular to subulate, erect; corolla greenish to red purple, narrowly tubular-funnelform, more or less arched, from barely exceeding the calyx to 2 or 3 times as long, the limb bilabiate; stamens 4, nearly equal; nutlets small, brown, smooth, or granular at the apex.

KEY TO THE SPECIES.

Calyx 6 mm. long or less; flowers small and inconspicuous.
  Calyx purplish-tinged, 5 to 6 mm. long........................ 1. *A. verticillata.*
  Calyx green or whitish, 3 to 4 mm. long.
    Panicles dense; corolla white, scarcely exceeding the calyx; calyx mostly green, the lobes acute......... 2. *A. micrantha.*
    Panicles interrupted, verticillate; corolla colored, twice as long as the calyx; calyx teeth subulate, white..... 3. *A. wrightii.*
Calyx 8 mm. long or more; flowers large and mostly conspicuous.
  Corolla not over twice the length of the calyx; whole flower 20 mm. long or less.
    Calyx green, at most with whitish teeth; corolla pale, half as long again as the calyx.......................... 4. *A. pallidiflora.*
    Calyx colored, at least the teeth; corolla various.
      Corolla fully twice as long as the calyx, arched, pale; calyx teeth pinkish.............................. 5. *A. greenei.*

Corolla less than twice as long as the calyx, nearly straight, deep red purple; calyx purple throughout .... 6. *A. neomexicana.*

Corolla over twice as long as the calyx; whole flower more than 20 mm. long.

Leaves large, 2 to 3 cm. broad and a third longer, all coarsely toothed; panicles dense; calyx lobes triangular-subulate .... 7. *A. mearnsii.*

Leaves smaller, less than 2 cm. broad, some of them entire; panicles loose; calyx lobes merely triangular, acute.

Leaves lanceolate or linear-lanceolate, entire; plants cinereous .... 8. *A. rupestris.*

Leaves broader, some of them ovate, most of them crenate-dentate, at least on the sides, entire at the apex; plants not cinereous .... 9. *A. cana.*

1. **Agastache verticillata** Woot. & Standl. Contr. U. S. Nat. Herb. **16:** 168. 1913.

TYPE LOCALITY: Organ Mountains, New Mexico. Type collected by Wooton & Standley, September 23, 1906.

RANGE: Southern New Mexico.

NEW MEXICO: Organ Mountains; Mogollon Mountains. Mountains, in the Upper Sonoran Zone.

2. **Agastache micrantha** (A. Gray) Woot. & Standl. Contr. U. S. Nat. Herb. **16:** 168. 1913.

Cedronella micrantha A. Gray, Proc. Amer. Acad. **8:** 369. 1872.

TYPE LOCALITY: "S. W. Texas near the borders of New Mexico."

RANGE: Western Texas and southern New Mexico, south into Mexico.

NEW MEXICO: Carpenter Creek; Middle Fork of the Gila; Organ Mountains. Canyons, in the Upper Sonoran Zone.

3. **Agastache wrightii** (Greenman) Woot. & Standl. Contr. U. S. Nat. Herb. **16:** 168. 1913.

Cedronella wrightii Greenman, Proc. Amer. Acad. **41:** 244. 1905.

Brittonastrum wrightii Robinson, Proc. Amer. Acad. **43:** 26. 1907.

TYPE LOCALITY: Mountains near Santa Cruz, Sonora.

RANGE: Southern New Mexico and Arizona and northern Sonora.

NEW MEXICO: Mangas Springs; near Graham. Upper Sonoran Zone.

4. **Agastache pallidiflora** (Heller) Rydb. Bull. Torrey Club **33:** 150. 1906.

Brittonastrum pallidiflorum Heller, Bull. Torrey Club **26:** 621. 1899.

TYPE LOCALITY: Canyon near the eastern base of Bill Williams Mountain, Arizona.

RANGE: Mountains of Arizona and New Mexico.

NEW MEXICO: Chama; La Jara; Middle Fork of the Gila; N Bar Ranch; Gilmores Ranch. Transition Zone.

5. **Agastache greenei** (Briq.) Woot. & Standl. Contr. U. S. Nat. Herb. **16:** 167. 1913.

Brittonastrum greenei Briq. Ann. Cons. Jard. Genève **6:** 157. 1902.

TYPE LOCALITY: Chama, New Mexico. Type collected by Baker (no. 567).

RANGE: Northern New Mexico.

NEW MEXICO: Chama; near Defiance. Along cliffs in the mountains, in the Transition Zone.

We do not agree with Doctor Rydberg in considering this a synonym of *A. pallidiflora*. A rather extended series of the latter plant, from different localities in Arizona and western New Mexico, shows that it has a green calyx and pale whitish flowers, while *A. greenei* has the calyx teeth and upper part of the tube pink or purplish and the corolla of a deeper color. Besides these more conspicuous differences, the corollas in *A. greenei* are noticeably longer, more arched, and wider at the throat. These differences in color and size of corolla and calyx, and evident differences in the calyx teeth, seem to be the most important diagnostic characters in a group of closely related but distinct species, which, until recently, have been taken to belong to two or three very variable ones.

A character which is indescribable with our present vocabulary is to be found in the peculiar odors of these plants, because of the volatile oils they contain. These differ very perceptibly in the different species, some of them being pronounced.

6. **Agastache neomexicana** (Briq.) Standley, Contr. U. S. Nat. Herb. **13:** 211. 1910.

Brittonastrum neomexicanum Briq. Ann. Cons. Jard. Genève **6:** 158. 1902.

TYPE LOCALITY: White Mountains, New Mexico. Type collected by Wooton (no. 266).

RANGE: Mountains of New Mexico.

NEW MEXICO: Tunitcha Mountains; Chama; Sandia Mountains; Ramah; Mogollon Mountains; White and Sacramento mountains; Organ Mountains. Transition Zone.

7. **Agastache mearnsii** Woot. & Standl. Contr. U. S. Nat. Herb. **16:** 167. 1913.

TYPE LOCALITY: San Luis Mountains, New Mexico. Type collected by E. A. Mearns (no. 2251).

RANGE: Mountains of southwestern New Mexico and adjacent Mexico.

NEW MEXICO: San Luis Mountains; Animas Valley; Burro Mountains; Pinos Altos Mountains. Upper Sonoran Zone.

8. **Agastache rupestris** (Greene) Standley, Contr. U. S. Nat. Herb. **13:** 212. 1910.

Cedronella cana lanceolata A. Gray, Syn. Fl. **2**[1]: 462. 1878, in part.

Cedronella rupestris Greene, Pittonia **1:** 164. 1888.

Brittonastrum lanceolatum Heller, Muhlenbergia **1:** 4. 1900.

Brittonastrum rupestre Heller, loc. cit.

Agastache lanceolata Standley, Contr. U. S. Nat. Herb. **13:** 212. 1910.

TYPE LOCALITY: Mangas Springs, New Mexico.

RANGE: Mountains of southwestern New Mexico.

NEW MEXICO: Mangas Springs; Mogollon Mountains; Burro Mountains.

The type of *Cedronella cana lanceolata* was collected in New Mexico.

9. **Agastache cana** (Hook.) Woot. & Standl. Contr. U. S. Nat. Herb. **16:** 166. 1913.

Cedronella cana Hook. in Curtis's Bot. Mag. **77:** *pl. 4618*. 1851.

Cedronella cana lanceolata A. Gray, Syn. Fl. **2**[1]: 462. 1878, in part.

TYPE LOCALITY: Western Texas.

RANGE: Mountains of western Texas and southern New Mexico.

NEW MEXICO: Headwaters of the Pecos; Santa Rita; Hillsboro; Organ and Dona Ana mountains. Upper Sonoran Zone.

17. NEPETA L. CATNIP.

Perennial tomentulose-canescent herb with petiolate, ovate to oblong, subcordate, dentate leaves and pale whitish flowers in crowded verticillate clusters; calyx tubular, slightly oblique, 15-nerved, 5-toothed, obscurely if at all bilabiate; corolla tube enlarged above, the limb strongly bilabiate, the upper lip erect, entire or emarginate, the lower spreading, 3-lobed; nutlets ovoid, smooth.

1. **Nepeta cataria** L. Sp. Pl. 570. 1753.

TYPE LOCALITY: "Habitat in Europa."

NEW MEXICO: Farmington; Cedar Hill.

A native of Europe, frequent in cultivation and widely established as a weed.

18. PRUNELLA L. SELF-HEAL.

Perennial herb with ovate-lanceolate to oblong-lanceolate, obscurely toothed, petiolate leaves and stout crowded bracted terminal spikes of purple flowers; calyx reddish purple, 2-lipped, the tube 10-ribbed; corolla 2-lipped, the tube longer than the calyx, the upper lip arched, the lower 3-lobed; stamens 4, 2 of the filaments sterile; nutlets smooth.

1. **Prunella vulgaris** L. Sp. Pl. 600. 1753.

TYPE LOCALITY: "Habitat in Europae pascuis."

RANGE: Throughout temperate North America, Asia, and Europe.

NEW MEXICO: Chama; Tunitcha Mountains; Santa Fe and Las Vegas mountains; East Canyon; Ruidoso Creek; James Canyon. Transition Zone.

Standley's 6787 from Chama is a form with white corollas.

19. LAMIUM L. HENBIT.

Low annual with diffusely branched stems and incised or toothed leaves, the lower ones mostly petioled; flowers in axillary clusters; calyx campanulate or tubular-campanulate, the tube 5-nerved, the limb 5-toothed; corolla slender tubular-funnelform, the upper lip erect, concave, usually entire, the lower spreading, 3-lobed; stamens 4, all fertile; nutlets smooth or tuberculate.

1. **Lamium amplexicaule** L. Sp. Pl. 579. 1753.

TYPE LOCALITY: "Habitat in Europae cultis."

RANGE: A native of the Old World, introduced in many parts of North America.

NEW MEXICO: Santa Fe (*Cockerell*).

20. STACHYS L. HEDGE NETTLE.

Perennial herbs, 10 to 60 cm. high, with freely branched stems and toothed leaves, the flowers in axillary clusters at the ends of the stems, the upper leaves gradually reduced and the flower clusters appearing as verticels, gradually approximated into an interrupted spike; calyx campanulate-tubular, 5 to 10-ribbed, the 5 lobes equal or nearly so; corolla pink, purple, or bright red, 2-lipped, the upper lip entire or notched, erect, the lower 3-lobed, the middle lobe largest and entire or 2-lobed; stamens 4; nutlets smooth, obtuse.

KEY TO THE SPECIES.

Corolla much exserted, bright red, about 25 mm. long; leaves long-petioled; plants puberulent, sparingly hirsute .... 1. *S. coccinea.*

Corollas barely exceeding the calyx, purplish or pink, about 15 mm. long; leaves short-petioled or sessile; plants hirsute or woolly throughout, especially on the younger parts.

  Plants woolly, especially on the inflorescence, low, 10 to 20 cm. high, branched from the base; leaves narrowly oblong-lanceolate, obtuse .... 2. *S. rothrockii.*

  Plants hirsute throughout, densely so in the inflorescence, 30 to 60 cm. high, branched above; leaves oblong to elliptic-lanceolate, acute .... 3. *S. scopulorum.*

1. **Stachys coccinea** Jacq. Pl. Hort. Schönbr. **3:** 18. *pl. 284.* 1798.

SCARLET HEDGE NETTLE.

TYPE LOCALITY: Not known.

RANGE: Western Texas to southern Arizona and southward.

NEW MEXICO: Kingston; Mangas Springs; San Luis Mountains; Dog Mountains; Organ Mountains. Canyons, in the Upper Sonoran Zone.

2. Stachys rothrockii A. Gray, Proc. Amer. Acad. **12:** 82. 1876.

TYPE LOCALITY: Zuni, New Mexico. Type collected by Rothrock (no. 177).

RANGE: Western New Mexico and adjacent Arizona.

NEW MEXICO: Zuni; Ojo Caliente; mesa south of Atarque de Garcia. Upper Sonoran Zone.

3. Stachys scopulorum Greene, Pittonia **3:** 342. 1898.

TYPE LOCALITY: "Colorado Rocky Mountains."

RANGE: Alberta and Minnesota to Arizona and New Mexico.

NEW MEXICO: Dulce; mountains west of Grants; Middle Fork of the Gila; Santa Fe and Las Vegas mountains; White and Sacramento mountains. Damp slopes and along streams, in the Transition Zone.

126. SOLANACEAE. Nightshade Family.

Annual or perennial herbs or shrubs with alternate or opposite, exstipulate, simple or compound leaves; flowers perfect, mostly regular; inflorescence various; calyx commonly of 5 more or less united sepals; corolla gamopetalous, more or less 5-lobed, mostly plicate in bud; stamens 5, inserted on the tube and alternate with the lobes of the corolla; style and stigma single; ovary mostly 2-celled, many-ovuled, with a central placenta; fruit a berry or capsule.

KEY TO THE GENERA.

Fruit a capsule.
- Capsules spiny in fruit; seeds flattened........... 1. DATURA (p. 567).
- Capsules not spiny; seeds not flattened.
  - Flowers racemose; sepals united; stems erect; flowers white or yellow................ 2. NICOTIANA (p. 567).
  - Flowers solitary in the axils; sepals nearly distinct; stems spreading; flowers purplish red............................ 3. PETUNIA (p. 568).

Fruit a berry, sometimes dry at maturity.
- Shrubs; corolla lobes little if at all plicate, valvate.................................... 4. LYCIUM (p. 568).
- Herbs; corolla plicate.
  - Fruiting calyx bladdery-inflated, inclosing the fruit, more or less 5-angled.
    - Corolla rotate, violet or purple.......... 5. QUINCULA (p. 569).
    - Corolla funnelform, open campanulate, or urceolate, yellowish or greenish, sometimes merely tinged with violet.
      - Corolla minutely toothed on the constricted orifice, urceolate...... 6. MARGARANTHUS (p. 569).
      - Corolla not toothed, mostly open-campanulate or funnelform.... 7. PHYSALIS (p. 570).
  - Fruiting calyx not bladdery-inflated.
    - Calyx not inclosing the berry.......... 8. SOLANUM (p. 572).
    - Calyx closely investing the berry.
      - Stamens dissimilar, declined; prickly annuals...................... 9. ANDROCERA (p. 574).
      - Stamens alike, not declined; low unarmed perennials...........10. CHAMAESARACHA (p. 574).

The tomato, *Lycopersicum esculentum*, frequently escapes from cultivation and persists for some time.

1. DATURA L. Thorn-apple.

Rank, ill-smelling, annual or perennial herbs with coarse stems, alternate simple leaves, and large axillary solitary flowers; calyx with a long angled tube, often circumscissile near the base; corolla funnelform, white to violet, with a plaited 5-lobed limb, the lobes abruptly acuminate; ovary 2-celled or falsely 4-celled; capsules large, prickly, 4-valved or opening irregularly.

KEY TO THE SPECIES.

Corolla 15 to 20 cm. long; capsules rather fleshy, bursting irregularly .... 1. *D. meteloides.*
Corolla 10 cm. long or less; capsule dry, 4-valved.
  Leaves sinuate-pinnatifid; spines of the capsule long and stout, few .... 2. *D. quercifolia.*
  Leaves sinuately angled or toothed; spines short, slender, numerous.
    Corolla white; lower spines of capsules shorter .... 3. *D. stramonium.*
    Corolla violet; spines of the capsules all alike .... 4. *D. tatula.*

1. Datura meteloides DC.; Dunal in DC. Prodr. **13**[1]**:** 544. 1852.

Type locality: "In calidis Novae Hispaniae regionibus."

Range: Colorado and western Texas to California and southward.

New Mexico: Cedar Hill; San Ildefonso; Zuni; Albuquerque; Lake Valley; Mangas Springs; Kingston; Rio Frisco; Dog Spring; Organ Mountains; Mesilla; south of Roswell; White Mountains. Dry hills and mesas, in the Lower and Upper Sonoran zones.

A very handsome plant, bearing numerous large, pure white, rather heavily scented flowers. It is abundant on sandy mesas, especially along arroyos.

2. Datura quercifolia H. B. K. Nov. Gen. & Sp. **3:** 7. 1818.

Type locality: "Crescit locis temperatis Regni Mexicani prope Zelaya et Molino de Sarabia, alt. 930 hex."

Range: Texas and Arizona to Mexico.

New Mexico: Santa Fe; Albuquerque; Glorieta; Fort Bayard; Mangas Springs; Mimbres; Mesilla Valley. Waste and cultivated ground.

A not uncommon weed along ditch banks.

3. Datura stramonium L. Sp. Pl. 179. 1753.

Type locality: "Habitat in America, nunc vulgaris per Europam."

Range: Throughout eastern and southern North America; introduced in New Mexico.

New Mexico: Fresnal (*Wooton*). Waste ground.

4. Datura tatula L. Sp. Pl. ed. 2. 256. 1762.

Type locality: Not stated.

Range: Throughout the warmer parts of North America, widely introduced elsewhere.

New Mexico: Farmington; Cedar Hill. Waste ground.

Introduced into New Mexico from the East.

2. NICOTIANA L. Tobacco.

Clammy-pubescent or glabrous herbs or shrubs, with ample, alternate, entire or undulate leaves and terminal panicles of white or yellow flowers; calyx campanulate to tubular, 5-lobed; corolla funnelform or salverform, with an elongated tube; stamens included; capsule 2-celled, 2 to 4-valved from the apex; seeds small, very numerous.

KEY TO THE SPECIES.

Plants glaucous, more than a meter high, woody below......... 1. *N. glauca.*
Plants densely viscid, green, less than a meter high, herbaceous throughout.
  Leaves clasping at the base; flowers diurnal............... 2. *N. trigonophylla.*
  Leaves petiolate, not clasping; flowers nocturnal.......... 3. *N. attenuata.*

1. Nicotiana glauca Graham, Edinburgh Phil. Journ. **1828:** 174. 1828.

TYPE LOCALITY: Cultivated from seeds received from Buenos Aires.

RANGE: South America; abundantly introduced into southern North America from Texas to southern California and Mexico.

NEW MEXICO: One mile south of Kingston (*Metcalfe* 1009).

2. Nicotiana trigonophylla Dunal in DC. Prodr. **13**[1]**:** 562. 1852.

TYPE LOCALITY: Aguas Calientes, Mexico.

RANGE: Western Texas to southern California and southward.

NEW MEXICO: Fairview; Carlisle; Mangas Springs; Kingston; Dog Mountains; mountains west of San Antonio; Tortugas Mountain; Organ Mountains; Tularosa; Lincoln; Lakewood. Canyons, in the Lower and Upper Sonoran zones.

3. Nicotiana attenuata Torr.; S. Wats. in King, Geol. Expl. 40th Par. **5:** 276. *pl. 27. f. 1.* 1871.

TYPE LOCALITY: Not definitely stated.

RANGE: California to Utah and Texas.

NEW MEXICO: Aztec; Carrizo Mountains; Silver City; Mangas Springs; Berendo Creek. Sandy plains, Upper Sonoran Zone.

3. PETUNIA Juss.

Diffuse prostrate annual with numerous small entire leaves and inconspicuous solitary axillary flowers; calyx of 5 narrow sepals united only at the base; corolla funnelform, pale purplish red, about 5 mm. long; capsules ovoid, 3 to 4 mm. long, surpassed by the calyx lobes.

1. Petunia parviflora Juss. Ann. Mus. Paris **2:** 216. *pl. 47.* 1803.

TYPE LOCALITY: "De l'embouchure de la Plata."

RANGE: Southern Florida to Texas and southern California; also in tropical America.

NEW MEXICO: Albuquerque; Mesilla Valley. Waste ground and along stream beds, in the Lower Sonoran Zone.

4. LYCIUM L. TOMATILLA.

Shrubs with divaricate branches, many of them ending in spines; leaves alternate or often fascicled, thickish, entire or undulate; flowers mostly in few-flowered axillary cymes; calyx of 5 sepals united at the base, persistent in fruit; corolla funnelform to almost salverform, greenish or purplish, 5-lobed; stamens mostly exserted; ovary 2-celled; berries globose, fleshy, scarlet.

KEY TO THE SPECIES.

Stems slender, recurved or climbing; introduced plant.......... 4. *L. halimifolium.*
Stems stout; native plant.
  Flowers greenish, 20 mm. long; older branches dark reddish brown...................................... 1. *L. pallidum.*
  Flowers purplish, 12 mm. long or less; branches grayish.
    Corolla 8 to 12 mm. long; leaves large, numerous....... 2. *L. torreyi.*
    Corolla 5 mm. long, leaves small...................... 3. *L. parviflorum.*

1. **Lycium pallidum** Miers, Illustr. S. Amer. Pl. **2:** 108. 1849–57.

TYPE LOCALITY: "In Nova Mexico." Type collected by Fendler (no. 670), probably near Santa Fe.

RANGE: Utah and Colorado to New Mexico and Arizona.

NEW MEXICO: Carrizo Mountains; Cedar Hill; Barranca; Zuni Reservation; Thoreau; Tiznitzin; Fort Bayard; Mangas Springs; Bear Mountain; Dog Spring; Organ Mountains; Nogal. Dry hills and plains, in the Upper Sonoran Zone.

2. **Lycium torreyi** A. Gray, Proc. Amer. Acad. **6:** 47. 1862.

TYPE LOCALITY: "Texas, on the Rio Grande."

RANGE: Western Texas to southern California.

NEW MEXICO: Black Range; Playas Valley; north of Deming; Las Palomas Hot Springs; Socorro; Mesilla Valley; Organ Mountains. Dry valleys and plains, in the Lower Sonoran Zone.

The fruit of this and other species is eaten by the native people. The flavor is rather insipid.

3. **Lycium parviflorum** A. Gray, Proc. Amer. Acad. **6:** 48. 1862.

TYPE LOCALITY: Arizona.

RANGE: Southern New Mexico and Arizona.

NEW MEXICO: Florida Mountains; White Sands; White Mountains; east of Deming. Dry plains, in the Lower Sonoran Zone.

4. **Lycium halimifolium** Mill. Gard. Dict. ed. 8. no. 6. 1768. MATRIMONY VINE.

Lycium barbarum vulgare Ait. Hort. Kew. ed. 2. **2:** 3. 1811.

Lycium vulgare Dunal in DC. Prodr. **13**[1]**:** 509. 1852.

TYPE LOCALITY: China.

A native of Europe and Asia, common in cultivation in the United States and frequently escaped. It is said to be established in the vicinity of Las Vegas. It is cultivated at Raton and in the Mesilla Valley.

5. QUINCULA Raf.

Low diffuse scurfy perennial herb with sinuately toothed or lobed leaves and violet flowers in small axillary clusters; calyx campanulate, inflated at maturity, sharply 5-angled, reticulated; corolla rotate, pentagonal; seeds few, reniform, somewhat flattened, thick-margined, rugose-tuberculate.

1. **Quincula lobata** (Torr.) Raf. Atl. Journ. 145. 1832.

Physalis lobata Torr. Ann. Lyc. N. Y. **2:** 226. 1827.

TYPE LOCALITY: "On the Canadian ?" Colorado or New Mexico.

RANGE: Kansas and New Mexico to California and Mexico.

NEW MEXICO: Mora; Raton; Roswell; Ruidoso Creek; La Lande; Buchanan; Lincoln. Plains, in the Upper Sonoran Zone.

6. MARGARANTHUS Schlecht.

Smooth annuals with the appearance of Physalis, but with an urceolate corolla more or less constricted and minutely 5-toothed at the throat and with a rather dry fruit.

KEY TO THE SPECIES.

Calyx half as long as the corolla, in fruit about 8 mm. in diameter.. 1. *M. solanaceus*.
Calyx fully two-thirds as long as the corolla, in fruit 12 to 15 mm. in diameter........ 2. *M. purpurascens*.

1. Margaranthus solanaceus Schlecht. Ind. Sem. Hort. Hal. 1838; Linnaea **13**: Litt. 99. 1839.

TYPE LOCALITY: "Nascitur in terris Mexicanis locis calidioribus." Described from plants grown from seeds sent by Ehrenberg.

RANGE: New Mexico and Arizona to Mexico.

NEW MEXICO: Santa Rita; Van Pattens; Mangas Springs.

2. Margaranthus purpurascens Rydb. Mem. Torrey Club **4**: 317. 1896.

TYPE LOCALITY: New Mexico. Type collected by Rusby (no. 307).

RANGE: Southwestern New Mexico.

NEW MEXICO: Mogollon Mountains; Kingston.

7. PHYSALIS L. GROUND-CHERRY.

Annual or perennial herbs, 50 cm. high or less, glabrous or pubescent; leaves entire or toothed; flowers axillary, usually solitary, nodding in anthesis; calyx much inflated, completely covering the fruit, usually 5-angled, the lobes connivent; corolla open campanulate-funnelform, yellowish, often with a darker center, with 5 short lobes; fruit a globose pulpy berry with numerous seeds.

KEY TO THE SPECIES.

Annuals.
 Plants pubescent nearly throughout.
  Plants stout, erect; leaves sinuate-crenate .... 1. *P. neomexicana.*
  Plants slender, diffuse; leaves nearly entire .... 2. *P. pubescens.*
 Plants glabrous, or with a few hairs on the young branches.
  Corolla narrowly campanulate, less than 3 mm. broad .... 3. *P. lanceifolia.*
  Corolla rotate-campanulate, more than 10 mm. broad.
   Pedicels much longer than the fruiting calyx; corolla yellow .... 4. *P. wrightii.*
   Pedicels usually shorter than the fruiting calyx; corolla with a purplish blotch in the center .... 5. *P. ixocarpa.*
Perennials.
 Leaves glabrous, the rest of the plant mostly so.
  Leaves broadly ovate, sinuate; calyx pyramidal, much inflated .... 6. *P. macrophysa.*
  Leaves lanceolate, entire or nearly so; calyx ovoid, little inflated .... 7. *P. longifolia.*
 Leaves, as well as the rest of the plant, pubescent.
  Pubescence at least in part stellate.
   Plants low, densely and conspicuously stellate-pubescent; leaves as broad as long, many of them orbicular, obtuse .... 10. *P. cinerascens.*
   Plants tall, finely and sparsely stellate-pubescent, the branching of the hairs noticeable only under a lens; leaves ovate, acute .... 11. *P. fendleri.*
  Pubescence not at all stellate.
   Pubescence fine and dense, appressed, often grayish .. 12. *P. hederaefolia.*
   Pubescence, at least in part, of long spreading hairs.
    Plants very viscid, densely pubescent, spreading; leaves broadly ovate, obtuse .... 13. *P. comata.*
    Plants with a few long hairs, scarcely viscid, erect; leaves lanceolate or nearly so, acute.
     Pubescence long and soft, spreading .... 8. *P. lanceolata.*
     Pubescence of fewer, short, scarcely spreading hairs .... 9. *P. polyphylla.*

1. **Physalis neomexicana** Rydb. Mem. Torrey Club **4**: 325. 1896.

TYPE LOCALITY: New Mexico. Type collected by Fendler (no. 679), probably near Santa Fe.

RANGE: Colorado and New Mexico.

NEW MEXICO: Sierra Grande; Albert; Gallinas Mountains; Pecos; Santa Fe; Mogollon Mountains; Kingston; San Luis Mountains; Organ Mountains; White Mountains; Gray. Fields in the mountains, in the Upper Sonoran and Transition zones.

A common weed in cultivated and waste ground. The berries are very large and are sometimes gathered for making preserves.

2. **Physalis pubescens** L. Sp. Pl. 183. 1753.

TYPE LOCALITY: "Habitat in India utraque."

RANGE: Pennsylvania and Florida to California and tropical America.

NEW MEXICO: Mangas Springs (*Rusby* 310).

3. **Physalis lanceifolia** Nees, Linnaea **6**: 473. 1831.

TYPE LOCALITY: Peru.

RANGE: Western Texas to Arizona, south to Mexico and South America.

NEW MEXICO: Mesilla Valley (*Wooton & Standley* 3149).

4. **Physalis wrightii** A. Gray, Proc. Amer. Acad. **10**: 63. 1874.

Chamaesaracha physaloides Greene, Bull. Torrey Club **9**: 122. 1882.

TYPE LOCALITY: Prairies along the San Pedro River, western Texas.

RANGE: Western Texas to southern California and Mexico.

NEW MEXICO: Two specimens from the southwestern part of the State seen, both without definite locality.

5. **Physalis ixocarpa** Brot.; Hornem. Hort. Hafn. Suppl. 26. 1819.

Physalis aequata Jacq. f.; Nees, Linnaea **6**: 470. 1831.

TYPE LOCALITY: Not stated.

RANGE: Colorado and Texas to California and Mexico, and in tropical America; also introduced eastward.

NEW MEXICO: Ojo Caliente (*Wooton* 2697).

Rydberg [1] reports Fendler's 680 as this species.

6. **Physalis macrophysa** Rydb. Bull. Torrey Club **22**: 308. 1895.

TYPE LOCALITY: Not definitely stated.

RANGE: New Mexico and Texas to Kansas and Arkansas.

NEW MEXICO: Filmore Canyon (*Wooton*).

7. **Physalis longifolia** Nutt. Trans. Amer. Phil. Soc. n. ser. **5**: 193. 1837.

TYPE LOCALITY: "On sandy banks of the Arkansas near Belle Point."

RANGE: Wyoming and Arizona to Iowa and Arkansas, and in Mexico.

NEW MEXICO: Farmington; Aztec; Perico Creek; Chama; Pecos; Albuquerque; Garfield; Zuni; Kingston; Mesilla Valley; Roswell. Moist ground, in the Lower and Upper Sonoran zones.

8. **Physalis lanceolata** Michx. Fl. Bor. Amer. **1**: 149. 1803.

Physalis pennsylvanica lanceolata A. Gray, Man. ed. 5. 382. 1867.

TYPE LOCALITY: "Habitat in Carolina."

RANGE: North and South Carolina to Wyoming and Arizona.

NEW MEXICO: Near Santa Fe; West Fork of the Gila.

9. **Physalis polyphylla** Greene, Pittonia **4**: 150. 1900.

TYPE LOCALITY: Piedra, southern Colorado.

RANGE: Colorado and New Mexico.

---

[1] Mem. Torrey Club **4**: 335. 1896.

New Mexico: Mouth of Indian Creek; Pecos; Nara Visa. Open slopes, in the Upper Sonoran Zone.

Our specimens have broader leaves than the type and are larger and stouter, but the pubescence, flowers, and fruit are the same.

10. Physalis cinerascens (Dunal) Hitchc. Spr. Fl. Manhattan 32. 1894.

Physalis pennsylvanica cinerascens Dunal in DC. Prodr. **13**[1]: 435. 1852.

Physalis mollis cinerascens A. Gray, Proc. Amer. Acad. **10**: 66. 1874.

Type locality: Texas or Mexico.

Range: Oklahoma and Texas to California and Mexico.

New Mexico: Knowles; 20 miles south of Roswell. Plains, in the Upper Sonoran Zone.

11. Physalis fendleri A. Gray, Proc. Amer. Acad. **10**: 66. 1874.

Type locality: "In the northern part of New Mexico." Type collected by Fendler.

Range: Colorado to Arizona and Mexico.

New Mexico: Las Vegas; Cedar Hill; Carrizo Mountains; Dulce; Raton; Pecos; Zuni; Magdalena Mountains; Mangas Springs; Fort Bayard; Dog Spring; Apache Spring. Dry hills, in the Upper Sonoran Zone.

12. Physalis hederaefolia A. Gray, Proc. Amer. Acad. **10**: 65. 1874.

Physalis digitalifolia Britton, Mem. Torrey Club **5**: 288. 1895.

Physalis palmeri A. Gray, Syn. Fl. **2**[1]: 235. 1878.

Type locality: Western Texas.

Range: Colorado and Texas to southern California and Mexico.

New Mexico: Mangas Springs; Big Hatchet Mountains; Organ Mountains; Tortugas Mountain; White Mountains; Guadalupe Mountains. Dry hills, in the Lower and Upper Sonoran zones.

13. Physalis comata Rydb. Bull. Torrey Club **22**: 306. 1895.

Type locality: Nebraska.

Range: Nebraska and Kansas to New Mexico and Texas.

New Mexico: Stanley; Las Vegas; Raton; Sierra Grande; Las Cruces; Gray. Hills and plains, in the Upper Sonoran Zone.

8. SOLANUM L.

Annual or perennial herbs, sometimes prickly, with simple or pinnate leaves, the inflorescence of terminal or axillary cymes; calyx mostly rotate, 5-lobed; corolla white, purplish, or violet, rotate, the limb 5-angled or lobed, plicate; stamens adnate near the throat, the anthers narrowed upward, opening by terminal pores or slits; ovary smooth, 2-celled; berries subglobose in the persistent calyx, pulpy or dry, with numerous flattened seeds.

KEY TO THE SPECIES.

Perennials.
- Plants stellate-pubescent, prickly; roots without tubers... 1. *S. elaeagnifolium*.
- Plants neither stellate nor prickly; roots with tubers.
  - Flowers white; leaflets mostly lanceolate............. 2. *S. jamesii*.
  - Flowers purplish; leaflets elliptic to ovate............ 3. *S. fendleri*.

Annuals.
- Leaves pinnatifid.. 4. *S. triflorum*.
- Leaves sinuate-dentate or entire.
  - Plants viscid-villous; fruit greenish or yellowish ...... 7. *S. villosum*.
  - Plants not viscid-villous; fruit black.
    - Leaves sparingly strigose beneath; corolla 3 to 4 mm. long; fruiting calyx spreading........ 5. *S. interius*.
    - Leaves densely strigose beneath; corolla 6 to 8 mm. long; fruiting calyx erect.................. 6. *S. douglasii*.

1. **Solanum elaeagnifolium** Cav. Icon. Pl. **3**: 22. *pl. 243*. 1794. TROMPILLO.

Solanum flavidum Torr. Ann. Lyc. N. Y. **2**: 227. 1827.

TYPE LOCALITY: "Habitat in America calidiore."

RANGE: Kansas and Colorado to Arizona and Mexico; also in South America.

NEW MEXICO: Abundant except in the San Juan Valley and the higher mountains. Plains and valleys, in the Lower and Upper Sonoran zones.

A very abundant and troublesome weed in cultivated fields of the valleys. It is remarkable for the length of its roots. Ordinarily the flowers are violet, but they vary to blue and very frequently white. The berries are used by the native people as a substitute for rennet in curdling milk.

2. **Solanum jamesii** Torr. Ann. Lyc. N. Y. **2**: 227. 1827.

TYPE LOCALITY: "The station was not recorded but is probably on the Arkansa."

RANGE: Colorado to Arizona and New Mexico.

NEW MEXICO: Las Vegas; Santa Fe; Pecos; Pajarito Park; west of Grants Station; Zuni Reservation; Magdalena Mountains; Mangas Springs; Mogollon Mountains; Animas Creek; White Mountains; Alamogordo. Plains and low hills, Lower Sonoran to the Transition Zone.

3. **Solanum fendleri** A. Gray, Amer. Journ. Sci. II. **22**: 285. 1856. WILD POTATO.

Solanum tuberosum boreale A. Gray, Syn. Fl. **2**[1]: 227. 1878.

TYPE LOCALITY: "In the northern part of New Mexico." Type collected by Fendler in 1847.

RANGE: New Mexico and Arizona to Mexico.

NEW MEXICO: Gallinas Canyon; Mogollon Mountains; Organ Mountains; White and Sacramento mountains; Hanover Mountain. Damp shaded slopes, in the Transition Zone.

4. **Solanum triflorum** Nutt. Gen. Pl. **1**: 128. 1818.

TYPE LOCALITY: "As a weed in and about the gardens of the Mandans and Minitaries, and in no other situations. Near Fort Mandan."

RANGE: British America to New Mexico.

NEW MEXICO: Common, except along the lower Pecos Valley. Plains, especially about prairie dog towns, in the Lower and Upper Sonoran zones.

5. **Solanum interius** Rydb. Bull. Torrey Club **31**: 641. 1905. BLACK NIGHTSHADE.

TYPE LOCALITY: Middle Loup River, near Mullen, Nebraska.

RANGE: Nebraska and Colorado to Texas and California.

NEW MEXICO: Raton; Chiz; Santa Rita; Middle Fork of the Gila; Cliff; Kingston; Mesilla Valley; Dog Spring; Organ Mountains; White Mountains. Shaded slopes, in the Upper Sonoran and Transition zones.

A common weed in cultivated and waste ground, especially along irrigating ditches.

6. **Solanum douglasii** Dunal in DC. Prodr. **13**[1]: 48. 1852.

Solanum nigrum douglasii A. Gray in Brewer & Wats. Bot. Calif. **1**: 538. 1876.

TYPE LOCALITY: "In Nova California."

RANGE: Southern California to western New Mexico and southward.

NEW MEXICO: Santa Rita (*Holzinger*).

7. **Solanum villosum** Mill. Gard. Dict. ed. 8. no. 2. 1768.

TYPE LOCALITY: Barbados.

RANGE: British Columbia and Wyoming to Mexico and the West Indies; introduced in the eastern United States.

NEW MEXICO: Las Vegas Hot Springs (*Rose & Fitch* 17585).

9. ANDROCERA Nutt. BUFFALO BUR.

Prickly herb 30 cm. high or less, with spreading branches; leaves once or twice pinnatifid, with broad undulate or sinuate lobes; calyx spreading, 5-lobed, closely investing the fruit; corolla rotate, 5-angled; stamens 5, anthers unequal, tapering upward, opening by terminal pores; berry dry, the seeds flattened.

KEY TO THE SPECIES.

Flowers yellow; pubescence stellate............................ 1. *A. rostrata.*
Flowers purple; pubescence mostly glandular.................. 2. *A. novomexicana.*

1. **Androcera rostrata** (Dunal) Rydb. Bull. Torrey Club **33**: 150. 1906.

Solanum rostratum Dunal, Sol. Syn. 234. *pl. 24.* 1813.

Solanum heterandrum Pursh, Fl. Amer. Sept. 156. *pl.* 7. 1814.

Androcera lobata Nutt. Gen. Pl. **1**: 129. 1818.

TYPE LOCALITY: Described from cultivated plants.

RANGE: Wyoming and North Dakota to Texas and Mexico.

NEW MEXICO: Farmington; Santa Fe; Las Vegas; Pecos; Rio Frisco; Kingston; Cliff; Cloverdale; Angus; Roswell; Albert; Elida; Nara Visa. Plains, in the Upper Sonoran Zone.

2. **Androcera novomexicana** (Bartlett) Woot. & Standl. Contr. U. S. Nat. Herb. **16**: 170. 1913.

Solanum heterodoxum novomexicanum Bartlett, Proc. Amer. Acad. **44**: 628. 1909.

TYPE LOCALITY: New Mexico. Type collected by Fendler (no. 673).

RANGE: New Mexico.

NEW MEXICO: San Juan; Las Vegas; Pecos; Santa Fe; Fort Bayard; Santa Rita. Plains and low hills, in the Upper Sonoran Zone.

10. CHAMAESARACHA A. Gray.

Low, perennial, diffusely spreading herbs with entire or pinnatifid leaves and flowers in axillary few-flowered clusters; calyx campanulate, 5-lobed, somewhat enlarged at maturity, closely investing the fruit, open at the mouth, neither ribbed nor angled; corolla rotate, ochroleucous, often purple-tinged; anthers oblong, opening by longitudinal slits; seeds reniform, flattened, rugose-favose or punctate; berry pulpy.

KEY TO THE SPECIES.

Plants villous, densely viscid.. 1. *C. conioides.*
Plants sparingly stellate-pubescent, scarcely if at all viscid......... 2. *C. coronopus.*

1. **Chamaesaracha conioides** (Moric.) Britton, Mem. Torrey Club **5**: 287. 1895.

Solanum conioides Moric.; Dunal in DC. Prodr. **13**[1]: 64. 1852.

Withania ? sordida Dunal, op. cit. 64.

Chamaesaracha sordida A. Gray in Brewer & Wats. Bot. Calif. **1**: 540. 1876.

TYPE LOCALITY: Between Laredo and San Antonio, Texas.

RANGE: California and Kansas to Mexico and Texas.

NEW MEXICO: Tucumcari; Hatchet Ranch; Kingston; Dog Spring; Organ Mountains; Tortugas Mountain; Gray; Knowles; Lincoln; Buchanan; mountains west of San Antonio; Lakewood. Plains and low hills, in the Lower and Upper Sonoran zones.

2. **Chamaesaracha coronopus** (Dunal) A. Gray in Brewer & Wats. Bot. Calif. **1**: 540. 1876.

Solanum coronopus Dunal in DC. Prodr. **13**[1]: 64. 1852.

TYPE LOCALITY: Between Laredo and San Antonio, Texas.

RANGE: California, Utah, and Kansas to Mexico.

NEW MEXICO: Nearly throughout the State. Dry plains and low hills, in the Lower and Upper Sonoran zones.

127. SCROPHULARIACEAE. Figwort Family.

Herbs or shrubs with usually terete stems and opposite, alternate, or whorled, exstipulate, simple, lobed, or parted leaves; flowers perfect, irregular; calyx of 4 or 5 more or less united sepals, persistent; corolla gamopetalous, commonly 2-lipped; stamens 4 or 5, usually of 2 kinds, more or less adnate to the corolla, alternate with its lobes; ovary 2-celled, superior, the styles united or nearly distinct; fruit a 2-celled 2-valved capsule, or rarely baccate; seeds usually numerous.

KEY TO THE GENERA.

Anther-bearing stamens 5 .... 1. VERBASCUM (p. 577).
Anther-bearing stamens 4 or 2.
  Low shrubs; leaves silvery-canescent; corolla funnelform .... 2. LEUCOPHYLLUM (p. 577).
  Herbs (a few suffrutescent at the base); leaves and corolla various.
    Corolla tube spurred or saccate on the lower side.
      Corolla tube spurred; flowers in terminal spikes, dark blue; stems erect .... 3. LINARIA (p. 577).
      Corolla tube saccate at the base; flowers mostly solitary, axillary, light blue; stems twining.
        Capsules subglobose, opening by 2 or 3 pores; seeds with irregular corky ridges .... 4. ANTIRRHINUM (p. 578).
        Capsules ovoid, the base of the style becoming hardened and broadened, the lateral valves adherent at the base; seeds winged. 5. MAURANDIA (p. 578).
    Corolla not spurred nor saccate at the base.
      Stamens 5, 4 anther-bearing, the fifth not antheriferous, often rudimentary.
        Staminodium merely a scale adnate to the upper side of the corolla tube.
          Perennials, 50 cm. high or more; seeds numerous; corolla not gibbous at the base .... 6. SCROPHULARIA (p. 578).
          Annuals; 10 cm. high or less; seeds few or solitary; corolla gibbous at the base .... 7. COLLINSIA (p. 579).
        Staminodium an elongated filament, usually longer than the anther-bearing ones.
          Corolla strongly bilabiate, more or less oblique .... 8. PENTSTEMON (p. 579).
          Corolla nearly salverform, not strongly bilabiate, the lobes nearly alike .... 9. LEIOSTEMON (p. 586).
      Stamens 4 or 2.
        Upper lip or lobes of the corolla external in the bud.
          Corolla nearly regular; leaves in a basal rosette .... 13. LIMOSELLA (p. 588).

Corolla bilabiate; stems leafy.
Anther-bearing stamens 2....10. GRATIOLA (p. 586).
Anther-bearing stamens 4.
Sepals united into an angled tube; leaves merely toothed, or entire.11. MIMULUS (p. 586).
Sepals distinct or nearly so; leaves pinnatifid................12. CONOBEA (p. 588).
Lower lip or lobes of the corolla external in the bud.
Stamens 2.
Corolla almost regularly 4-lobed; leaves mostly cauline; inflorescence not scapelike..........14. VERONICA (p. 588).
Corolla none, or 2-lipped; leaves basal; inflorescence scapelike.......15. BESSEYA (p. 590).
Stamens 4.
Corolla slightly 2-lipped, yellow; stamens not ascending under the upper lip.......16. DASYSTOMA (p. 590).
Corolla distinctly 2-lipped, of various colors; stamens ascending under the upper lip.
Anther sacs dissimilar, the inner one pendulous by its apex; leaves mostly alternate.
Calyx 2-phyllous, or by absence of the lower part 1-phyllous............17. ADENOSTEGIA (p. 590).
Calyx gamosepalous, united into a tube below.
Calyx deeply cleft before and behind, less deeply so (or not at all) on the sides; upper lip of the corolla much longer than the 3-lobed lower one; mostly perennials...18. CASTILLEJA (p. 591).
Calyx almost equally 4-cleft; upper lip of corolla slightly if at all longer than the 1 to 3-saccate lower one; annuals........19. ORTHOCARPUS (p. 594).
Anther sacs alike, parallel; leaves mostly opposite.

Calyx 4-toothed, inflated and veiny in fruit; capsules orbicular.....20. RHINANTHUS (p. 595).
Calyx cleft below or both above and below, not inflated; capsules ovoid or oblong, oblique.
Galea prolonged into a filiform recurved beak; throat with a tooth on each side...........21. ELEPHANTELLA (p. 595).
Galea not prolonged into a beak, or this not filiform, straight; throat without teeth.......22. PEDICULARIS (p. 595).

1. VERBASCUM L. MULLEIN.

Coarse, densely woolly, biennial herb 1 meter high or more, with large, thick, spatulate or elliptic-spatulate, decurrent leaves on a thick stem, and a rosette of similar basal ones; flowers in a crowded thick terminal spike; sepals 5, partly united; corolla rotate, yellow; stamens 5, exserted; seeds rugose, wingless.

1. **Verbascum thapsus** L. Sp. Pl. 177. 1753.

TYPE LOCALITY: "Habitat in Europæ glareosis sterilibus."

RANGE: A native of the Old World, widely introduced into North America, especially in pastures and along roadsides.

NEW MEXICO: Cedar Hill; Pecos; Mogollon; Ruidoso Creek.

The specimens collected along Ruidoso Creek were found at an altitude of at least 2,700 meters. They had the appearance of being a native plant, but had been carried in by cattle. The plant is also well established at Mogollon as a roadside and garden weed.

2. LEUCOPHYLLUM Humb. & Bonpl.

Low spreading shrub with small silvery-canescent obovate-spatulate leaves and solitary axillary pink flowers; calyx lobes valvate, the outer ones linear; corolla funnelform-campanulate, with 5 rounded lobes; stamens 4, included; styles united; seeds numerous, strongly rugose.

1. **Leucophyllum minus** A. Gray in Torr. U. S. & Mex. Bound. Bot. 115. 1859.

TYPE LOCALITY: "Hills on and near the Pecos," western Texas.

RANGE: Western Texas and southern New Mexico.

NEW MEXICO: Guadalupe Mountains (*Wooton*). Dry hills.

3. LINARIA L.

Slender glabrous annual, 30 cm. high or less, with erect scapelike flowering stems and a cluster of weak leafy ones at the base; leaves entire, oblong to elliptic, acute, alternate or verticillate; flower racemose, deep blue; sepals 5, partly united; corolla irregular, the tube spurred at the base; stamens 4, included; capsules subglobose.

1. **Linaria canadensis** (L.) DuM. de Cours. Bot. Cult. 2: 96. 1802.

BLUE TOADFLAX.

Antirrhinum canadense L. Sp. Pl. 618. 1753.

TYPE LOCALITY: "Habitat in Virginia, Canada."

RANGE: British America to California and Florida.

NEW MEXICO: Glorieta; Florida Mountains; Hillsboro; mesa west of Organ Mountains. Sandy plains, in the Upper Sonoran Zone.

4. ANTIRRHINUM L.

Slender vine with thin triangular-hastate glabrous leaves and light blue solitary axillary flowers; sepals partially united, narrowly lanceolate, persistent and slightly accrescent, not carinate at the base in fruit; corolla 2-lipped, the tube more or less saccate at the base, the limb short; stamens 4, included; capsule subglobose, 5 to 8 mm. in diameter, opening by chinks near the persistent base of the style; seeds numerous, corky-thickened on the angles.

1. **Antirrhinum antirrhiniflorum** (Willd.) Hitchc. Rep. Mo. Bot. Gard. **4**: 113. 1893.

Maurandia antirrhiniflora Willd. Hort. Berol. *pl. 83.* 1816.

Antirrhinum maurandioides A. Gray, Proc. Amer. Acad. **6**: 376. 1868.

TYPE LOCALITY: Mexico.

RANGE: Western Texas and southern Arizona to Mexico.

NEW MEXICO: Clemow; Lower Plaza; Silver City; Kingston; Grand Canyon of the Gila; Guadalupe Canyon; Dog Spring; Organ Mountains. Dry hills and sandy fields, in the Lower and Upper Sonoran zones.

5. MAURANDIA Ortega.

Very similar in most respects to the preceding, but a larger coarser plant, the leaves thicker, somewhat narrower; flowers larger, light blue, the corolla scarcely gibbous at the base; calyx lobes accrescent and persistent, strongly carinate outside at the base, closely investing the capsule; this ovoid, surmounted by the hardened and flattened base of the style, opening by 2 long slits at the top; seeds numerous, dark brown, winged.

1. **Maurandia wislizeni** A. Gray in Torr. U. S. & Mex. Bound. Bot. 111. 1859.

TYPE LOCALITY: Along the Rio Grande below Dona Ana, New Mexico. Type collected by Wislizenus in 1846.

RANGE: Western Texas and southern New Mexico to Mexico.

NEW MEXICO: San Marcial; Deming; Mesilla Valley. Sandy mesas and river valleys, in the Lower Sonoran Zone.

6. SCROPHULARIA L. FIGWORT.

Large coarse branched leafy green perennial herbs, mostly over 1 meter high, with opposite or whorled leaves and terminal panicles; leaves mostly large, flat, variously toothed; calyx lobes 5, relatively broad; corolla dull reddish or greenish, the tube open-campanulate, 2-lipped, the lips short; anther-bearing stamens 4, the upper filament reduced to a scale; capsules ovoid, the seeds numerous, wingless, wrinkled.

KEY TO THE SPECIES.

Corolla 15 to 20 mm. long, reddish .... 1. *S. coccinea.*
Corolla 12 mm. long or mostly less, dull greenish.
  Branches of the inflorescence ascending, forming a dense panicle; leaves finely and evenly toothed .... 2. *S. montana.*
  Branches of the inflorescence spreading, forming a loose few-branched panicle; leaves coarsely and unevenly laciniate-serrate.
    Plants glabrous throughout or nearly so; petioles long, usually half as long as the blades, these light green .. 3. *S. laevis.*
    Plants densely and finely puberulent on the stems, variously pubescent elsewhere; petioles less than one-third as long as the blades, these dull dark green .... 4. *S. parviflora.*

1. **Scrophularia coccinea** A. Gray in Torr. U. S. & Mex. Bound. Bot. 111. 1859.

TYPE LOCALITY: "At the base of a rocky ledge near the summit of a mountain, Santa Rita del Cobre," New Mexico. Type collected by Wright (no. 1470).

RANGE: Known only from type locality, in the Transition Zone.

An extremely rare plant, apparently, collected only twice.

2. **Scrophularia montana** Wooton, Bull. Torrey Club **25**: 308. 1898.

TYPE LOCALITY: Eagle Creek near Gilmores Ranch in the White Mountains, New Mexico. Type collected by Wooton in 1897.

RANGE: Mountains of New Mexico.

NEW MEXICO: Santa Fe and Las Vegas mountains; White and Sacramento mountains; Brazos Canyon. Transition Zone.

3. **Scrophularia laevis** Woot. & Standl. Contr. U. S. Nat. Herb. **16**: 173. 1913.

TYPE LOCALITY: Organ Peak, New Mexico. Type collected by Wooton & Standley, September 23, 1906.

RANGE: Moist canyons of the Organ Mountains, New Mexico, in the Transition Zone.

4. **Scrophularia parviflora** Woot. & Standl. Contr. U. S. Nat. Herb. **16**: 173. 1913.

TYPE LOCALITY: In the Mogollon Mountains on the West Fork of the Gila, New Mexico. Type collected by Metcalfe (no. 345).

RANGE: Mountains of western New Mexico, probably in adjacent Arizona.

NEW MEXICO: West Fork of the Gila; Graham. Transition Zone.

7. COLLINSIA Nutt.

Slender low annual with obtuse, entire, oblong or lanceolate, sessile, opposite leaves and solitary long-pediceled flowers in the axils of the leaves; corolla blue or blue and white, deeply 2-lipped, the upper lip 2-cleft, the lower 3-lobed, the middle lobe a keel-shaped sac inclosing the 4 declined stamens and style; anther cells confluent; fifth stamen represented by a gland near the base of the corolla; capsules ovoid or globose.

1. **Collinsia tenella** (Pursh) Piper, Contr. U. S. Nat. Herb. **11**: 496. 1906.

Antirrhinum tenellum Pursh, Fl. Amer. Sept. 421. 1814.

Collinsia parviflora Dougl.; Lindl. in Edwards's Bot. Reg. **13**: *pl. 1082*. 1827.

TYPE LOCALITY: "On the banks of the Missouri."

RANGE: British Columbia and Lake Superior to California and northern New Mexico.

NEW MEXICO: Hills southwest of Tierra Amarilla (*Eggleston* 6504). Open slopes, in the Transition Zone.

8. PENTSTEMON Soland. BEARD-TONGUE.

Perennial caulescent herbs with opposite, entire or toothed leaves, these sometimes clasping or perfoliate; flowers in terminal racemes or panicles; calyx lobes 5, entire or toothed; corolla usually showy, mostly elongated tubular-funnelform, white to purplish or scarlet, distinctly 2-lipped; anther-bearing stamens 4, the fifth filament sterile, more or less bearded or glabrous; capsules ovoid; seeds numerous, wingless, angled or rounded.

KEY TO THE SPECIES.

Anthers horseshoe-shaped or sagittate, opening only on the proximal part.
- Inflorescence glandular; stems glabrous; tube of corolla only slightly dilated .... 1. *P. bridgesii.*
- Inflorescence glabrous; stems puberulent; tube much dilated .... 2. *P. spinulosus.*

Anthers variously shaped, dehiscent for nearly their whole length.
- Flowers scarlet or cardinal red.
  - Plants strongly glaucous.................................. 3. *P. superbus.*
  - Plants green, seldom or never glaucescent.
    - Corolla contracted at the mouth, the lobes 3 mm. long or less.
      - Sepals broadly ovate, obtuse; cauline leaves triangular-lanceolate, thick............. 4. *P. crassulus.*
      - Sepals lance-ovate or narrower, acute; cauline leaves triangular-ovate, thin............ 5. *P. cardinalis.*
    - Corolla not contracted at the mouth, the lobes much more than 3 mm. long.
      - Corolla inflated in the throat; leaves elongated-linear.................................. 6. *P. lanceolatus.*
      - Corolla nearly tubular; leaves not elongated-linear.
        - Leaves filiform, crowded.................. 7. *P. pinifolius.*
        - Leaves broad, lanceolate or wider, not crowded.
          - Corolla obscurely bilabiate, the lobes scarcely spreading; pedicels short, stout...............................12. *P. eatoni.*
          - Corolla strongly bilabiate, the lobes spreading; pedicels long, slender.
            - Lower lip of corolla bearded within. 8. *P. barbatus.*
            - Lower lip not bearded.
              - Anthers glabrous................. 9. *P. torreyi.*
              - Anthers long-bearded...........10. *P. trichander.*
- Flowers never scarlet nor cardinal red, crimson to blue or white.
  - Upper leaves perfoliate.................................11. *P. spectabilis.*
  - None of the leaves perfoliate.
    - Anthers bearded with long villous hairs.
      - Stems and leaves puberulent...................15. *P. comarrhenus.*
      - Stems and leaves glabrous.
        - Calyx lobes acute or obtuse...............13. *P. strictus.*
        - Calyx lobes long-acuminate...............14. *P. strictiformis.*
    - Anthers not bearded, sometimes short-hirsute.
      - Plants suffruticose at the base; leaves linear or nearly so.
        - Calyx lobes scarious-margined, dentate or erose; floral leaves reduced.........16. *P. linarioides.*
        - Calyx lobes scarcely scarious-margined, entire; floral leaves not reduced.
          - Leaves green, glabrate................17. *P. crandallii.*
          - Leaves grayish, canescent-puberulent.18. *P. teucrioides.*
      - Plants not suffruticose at the base; leaves various.
        - Corolla tube almost cylindric or very slightly widened upward.
          - Leaves more or less dentate...............19. *P. humilis.*
          - Leaves entire.
            - Calyx lobes merely acute; inflorescence glandular, loose.....................21. *P. oliganthus.*

Calyx lobes attenuate or abruptly acuminate; inflorescence glandular or glabrous, open or dense.
Inflorescence glandular; flowers few, remote..........................20. *P. gracilis.*
Inflorescence not glandular; flowers numerous, dense..................22. *P. rydbergii.*
Corolla decidedly funnelform, the throat much wider than the tube.
Sterile stamen glabrous.
Sepals long-attenuate, villous; most of the leaves denticulate...............23. *P. whippleanus.*
Sepals merely acute to truncate, not attenuate, nearly glabrous; leaves dentate or entire.
Stems puberulent; sepals acute......24. *P. virgatus.*
Stems glabrous; sepals acute to truncate.
Sepals obtuse or truncate..........25. *P. neomexicanus.*
Sepals attenuate or abruptly acuminate.
Anthers short-hirsute; pedicels pubescent; inflorescence very dense; upper leaves ovate or cordate.......................26. *P. brandegei.*
Anthers glabrous; pedicels glabrous; inflorescence open; upper leaves lanceolate......27. *P. unilateralis.*
Sterile stamen bearded.
Stems glandular, at least about the inflorescence; plants seldom glaucous.
Stems glabrous below..................28. *P. stenosepalus.*
Stems pubescent throughout.
Stems densely glandular throughout..............................30. *P. similis.*
Stems glandular only above.
Stems pruinose-puberulent; leaves entire.................29. *P. metcalfei.*
Stems hirsute; leaves evidently serrate........................31. *P. pulchellus.*
Stems glabrous throughout; plants glaucous.
Bracts of the inflorescence mostly longer than the flowers; basal leaves linear or nearly so.........32. *P. caudatus.*
Bracts shorter than the flowers; basal leaves spatulate or oblanceolate.
Bracts ovate to nearly orbicular....33. *P. cyathophorus.*
Bracts lanceolate or linear-lanceolate.
Calyx lobes lanceolate............34. *P. secundiflorus.*
Calyx lobes ovate................35. *P. fendleri.*

1. **Pentstemon bridgesii** A. Gray, Proc. Amer. Acad. **7:** 379. 1868.

TYPE LOCALITY: California.

RANGE: Colorado and New Mexico to California and Nevada.

NEW MEXICO: Trujillos Ranch on the Rio Frisco (*Wooton*). Upper Sonoran Zone.

2. **Pentstemon spinulosus** Woot. & Standl. Contr. U. S. Nat. Herb. **16:** 173. 1913.

TYPE LOCALITY: Magdalena Mountains, New Mexico. Type collected by G. R. Vasey, June, 1881.

RANGE: Known only from the type locality.

3. **Pentstemon superbus** A. Nels. Proc. Biol. Soc. Washington **17:** 100. 1904.

Pentstemon puniceus A. Gray in Torr. U. S. & Mex. Bound. Bot. 113. 1859, not Lilja, 1843.

TYPE LOCALITY: Guadalupe Canyon, Sonora.

RANGE: Southwestern New Mexico and adjacent Arizona and Mexico.

NEW MEXICO: Mangas Springs (*Metcalfe* 68). Upper Sonoran Zone.

4. **Pentstemon crassulus** Woot. & Standl. Contr. U. S. Nat. Herb. **16:** 172. 1913.

TYPE LOCALITY: Lincoln National Forest, New Mexico. Type collected by Fred G. Plummer in 1903.

RANGE: Known only from type locality.

5. **Pentstemon cardinalis** Woot. & Standl. Contr. U. S. Nat. Herb. **16:** 171. 1913.

TYPE LOCALITY: White Mountain Peak just above the forks of Ruidoso Creek, New Mexico. Type collected by Wooton, July 6, 1895.

RANGE: Known only from type locality, in the Canadian Zone.

6. **Pentstemon lanceolatus** Benth. Pl. Hartw. 22. 1839.

Pentstemon pauciflorus Greene, Bot. Gaz. **6:** 218. 1881.

TYPE LOCALITY: Mexico.

RANGE: Southwestern New Mexico and adjacent Arizona and Mexico.

NEW MEXICO: Dog Spring (*Mearns* 102, 2353).

The type of *P. pauciflorus* was collected along the Gila River by Greene, in 1880.

7. **Pentstemon pinifolius** Greene, Bot. Gaz. **6:** 218. 1881.

TYPE LOCALITY: "Summits of the San Francisco range, back of Clifton, in southeastern Arizona."

RANGE: Southern New Mexico and Arizona.

NEW MEXICO: Mogollon Mountains; San Luis Mountains; Animas Mountains; Hillsboro Peak. Transition Zone.

8. **Pentstemon barbatus** (Cav.) Nutt. Gen. Pl. **2:** 53. 1818.

Chelone barbata Cav. Icon. Pl. **3:** 22. *pl. 242*. 1794.

Pentstemon coccineus Engelm. in Wisliz. Mem. North. Mex. 107. 1848.

TYPE LOCALITY: Mexico.

RANGE: Colorado to Mexico.

NEW MEXICO: Bear Mountain; West Fork of the Gila; Reserve; Graham. Mountains, in the Transition Zone.

9. **Pentstemon torreyi** Benth. in DC. Prodr. **10:** 324. 1846.

Pentstemon barbatus torreyi A. Gray, Proc. Amer. Acad. **6:** 59. 1862.

TYPE LOCALITY: "Versus montes Scopulosos," Colorado.

RANGE: Colorado to Mexico.

NEW MEXICO: Common in all the higher mountains. Meadows in the mountains, Transition and Canadian zones.

Standley's 6790 from Chama is a form with salmon-colored corollas. It is common in this locality.

10. Pentstemon trichander (A. Gray) Rydb. Bull. Torrey Club **33:** 151. 1906.

Pentstemon barbatus trichander A. Gray, Proc. Amer. Acad. **11:** 94. 1875.

TYPE LOCALITY: Southwestern Colorado.

RANGE: Mountains of Colorado and New Mexico.

NEW MEXICO: Carrizo Mountains; Tunitcha Mountains; Dulce; near Fort Defiance; Coolidge. Upper Sonoran and Transition zones.

Standley's no. 7775 from the Tunitcha Mountains is a form with purplish red flowers. A considerable number of such plants were noticed in the Tunitcha Mountains, but the prevailing color is scarlet.

11. Pentstemon spectabilis Thurb.; Torr. & Gray, U. S. Rep. Expl. Miss. Pacif. **4:** 119. 1856.

TYPE LOCALITY: Southern California.

RANGE: Southern New Mexico to California.

NEW MEXICO: Organ Mountains. Rocky canyons, in the Upper Sonoran Zone.

12. Pentstemon eatoni A. Gray, Proc. Amer. Acad. **8:** 395. 1873.

TYPE LOCALITY: Provo Canyon, Wahsatch Mountains, Utah.

RANGE: Utah and Nevada to northwestern New Mexico.

NEW MEXICO: Carrizo Mountains (*Standley* 7315). Rocky canyons, in the Upper Sonoran Zone.

13. Pentstemon strictus Benth. in DC. Prodr. **10:** 324. 1846.

TYPE LOCALITY: "In montibus Scopulosis ad fontes fl. Sweetwater."

RANGE: Wyoming to New Mexico and Utah.

NEW MEXICO: Chusca Mountains; Tunitcha Mountains. Mountains, in the Transition Zone.

14. Pentstemon strictiformis Rydb. Bull. Torrey Club **31:** 642. 1905.

TYPE LOCALITY: Mancos, Colorado.

RANGE: Colorado and northern New Mexico.

NEW MEXICO: Ford of the Chama; Sandia Mountains; Fort Wingate.

15. Pentstemon comarrhenus A. Gray, Proc. Amer. Acad. **12:** 81. 1887.

TYPE LOCALITY: Utah.

RANGE: Utah and Colorado to northern New Mexico.

NEW MEXICO: Dulce (*Standley* 8118). Open hills, in the Transition Zone.

16. Penstemon linarioides A. Gray in Torr. U. S. & Mex. Bound. Bot. 112. 1859.

TYPE LOCALITY: Organ Mountains, New Mexico.

RANGE: Western Texas to Utah and Arizona and adjacent Mexico.

NEW MEXICO: Tunitcha Mountains; Ramah; Carrizo Mountains; Mangas Springs; Luna; Crawfords Ranch; west of Silver City; San Luis Mountains; Las Animas; Organ Mountains. Dry hillsides, in the Upper Sonoran Zone.

17. Pentstemon crandallii A. Nels. Bull. Torrey Club **26:** 354. 1899.

Pentstemon xylus A. Nels. Bot. Gaz. **34:** 32. 1902.

TYPE LOCALITY: Near Como, Park County, Colorado.

RANGE: Colorado to northern New Mexico.

NEW MEXICO: Dulce; Stinking Lake; Tierra Amarilla. Low mountains, in the Transition Zone.

18. Pentstemon teucrioides Greene, Pl. Baker. **3:** 23. 1901.

TYPE LOCALITY: Sapinero, Colorado.

RANGE: Southern Colorado and northern New Mexico.

NEW MEXICO: Hills south of Tierra Amarilla (*Eggleston* 6549). Open hills, in the Transition Zone.

19. Pentstemon humilis Nutt.; A. Gray, Proc. Amer. Acad. **6:** 69. 1862.

TYPE LOCALITY: "Rocky Mountains."

RANGE: Montana and Alberta to Nevada and New Mexico.

NEW MEXICO: Sierra Grande; Hermits Peak; Cross L Ranch. Open meadows, in the Transition Zone.

20. Pentstemon gracilis Nutt. Gen. Pl. **2:** 52. 1818.

Pentstemon pubescens gracilis A. Gray, Proc. Amer. Acad. **6:** 69. 1862.

TYPE LOCALITY: "From the Arikarees to Fort Mandan, in depressed soils."

RANGE: Wyoming and Nebraska to Colorado and New Mexico.

NEW MEXICO: Glorieta; Sandia Mountains; Santa Fe; Beulah. Open hills and plains, in the Upper Sonoran and Transition zones.

21. Pentstemon oliganthus Woot. & Standl. Contr. U. S. Nat. Herb. **16:** 172. 1913.

TYPE LOCALITY: Mountains west of Grants Station, New Mexico. Type collected by Wooton, August 1, 1892.

RANGE: Known only from type locality.

22. Pentstemon rydbergii A. Nels. Bull. Torrey Club **25:** 281. 1898.

Pentstemon erosus Rydb. Bull. Torrey Club **28:** 28. 1901.

TYPE LOCALITY: Indian Creek Pass, Colorado.

RANGE: Wyoming to northern New Mexico.

NEW MEXICO: Tunitcha Mountains; Chama. Moist meadows, in the Transition Zone.

23. Pentstemon whippleanus A. Gray, Proc. Amer. Acad. **6:** 73. 1862.

TYPE LOCALITY: Arroyos in the Sandia Mountains, New Mexico. Type collected by Bigelow.

RANGE: Mountains of New Mexico.

NEW MEXICO: Sandia Mountains; Beulah; Jemez Mountains; White Mountain Peak. Transition Zone.

24. Pentstemon virgatus A. Gray in Torr. U. S. & Mex. Bound. Bot. 113. 1859.

TYPE LOCALITY: Santa Rita del Cobre, New Mexico. Type collected by Wright (no. 1476).

RANGE: Mountains of New Mexico and Arizona.

NEW MEXICO: Pajarito Park; Inscription Rock; Sandia Mountains; Glorieta; Ramah; mountains west of Grants; Fort Wingate; Mogollon Mountains; Burro Mountains; Santa Rita; Hanover Mountain. Transition Zone.

This has generally been confounded with *P. secundiflorus* A. Gray, not Benth. (*P. unilateralis* Rydb.), but it is a very different plant.

25. Pentstemon neomexicanus Woot. & Standl. Contr. U. S. Nat. Herb. **16:** 172. 1913.

TYPE LOCALITY: In pine woods near Gilmores Ranch on Eagle Creek, White Mountains, New Mexico. Type collected by Wooton & Standley (no. 3507).

RANGE: Mountains of southern New Mexico.

NEW MEXICO: White and Sacramento mountains; Capitan Mountains. Transition Zone.

26. Pentstemon brandegei Porter, Mem. N. Y. Bot. Gard. **1:** 343. 1900.

Pentstemon cyananthus brandegei Porter in Port. & Coult. Syn. Fl. Colo. 91. 1874.

TYPE LOCALITY: Sierra Mojada, Colorado.

RANGE: Colorado and northern New Mexico.

NEW MEXICO: Raton; Sierra Grande. Plains and low hills, in the Upper Sonoran Zone.

This is one of the most showy and probably the handsomest of all our species.

27. Pentstemon unilateralis Rydb. Bull. Torrey Club **33**: 150. 1906.
Pentstemon secundiflorus Benth. err. det. A. Gray, Syn. Fl. **2**[1]: 263. 1878.
TYPE LOCALITY: "Mountains of Colorado, common at 8 or 9,000 feet."
RANGE: Wyoming to northern New Mexico.
NEW MEXICO: Near Chama (*Wooton* 2796, *Standley* 6550). Mountain meadows, in the Transition Zone.

28. Pentstemon stenosepalus (A. Gray) Howell, Fl. Northw. Amer. 514. 1903.
Pentstemon glaucus stenosepalus A. Gray, Proc. Amer. Acad. **6**: 70. 1862.
TYPE LOCALITY: Rocky Mountains of Colorado.
RANGE: Colorado and Utah to New Mexico.
NEW MEXICO: Baldy; Santa Fe and Las Vegas mountains; Mogollon Mountains. Mountain meadows, in the Canadian and Hudsonian zones.

29. Pentstemon metcalfei Woot. & Standl. Torreya **9**: 145. 1909.
Pentstemon puberulus Woot. & Standl. Bull. Torrey Club **36**: 112. 1909, not M. E. Jones, 1908.
TYPE LOCALITY: On sandy slopes at the Lookout Mine, Sierra County, New Mexico. Type collected by Metcalfe (no. 1605).
RANGE: Known only from type locality, in the Transition Zone.

30. Pentstemon similis A. Nels. Bull. Torrey Club **25**: 548. 1898.
TYPE LOCALITY: Near Santa Fe, New Mexico. Type collected by Fendler (no. 575).
RANGE: Colorado to western Texas.
NEW MEXICO: Pecos; Clayton; Chusca Mountains; Sierra Grande; Nara Visa; Magdalena Mountains; Santa Fe; mountains west of Patterson. Hills and plains, in the Upper Sonoran Zone.

31. Pentstemon pulchellus Lindl. in Edwards's Bot. Reg. **14**: *pl. 1138*. 1828.
TYPE LOCALITY: Mexico.
RANGE: Southwestern New Mexico and northern Mexico.
NEW MEXICO: San Luis Mountains (*Mearns* 2112, 2222).

32. Pentstemon caudatus Heller, Minn. Bot. Stud. **2**: 34. 1898.
Pentstemon angustifolius caudatus Rydb. Bull. Torrey Club **33**: 151. 1906.
TYPE LOCALITY: Barranca, Taos County, New Mexico. Type collected by Heller (no. 3581).
RANGE: Colorado and New Mexico.
NEW MEXICO: Barranca; Aztec. Dry hills, in the Upper Sonoran Zone.

33. Pentstemon cyathophorus Rydb. Bull. Torrey Club **31**: 643. 1904.
TYPE LOCALITY: Pearl, Colorado.
RANGE: Colorado and New Mexico.
NEW MEXICO: Arroyo Ranch; Hillsboro; Santa Fe. Dry hills, in the Upper Sonoran Zone.

34. Pentstemon secundiflorus Benth. in DC. Prodr. **10**: 325. 1846.
TYPE LOCALITY: Rocky Mountains.
RANGE: Wyoming to northern New Mexico.
NEW MEXICO: Cedar Hill; Farmington; Raton. Dry hills, in the Upper Sonoran Zone.

35. Pentstemon fendleri A. Gray, U. S. Rep. Expl. Miss. Pacif. **2**: 168. *pl. 5*. 1855.
TYPE LOCALITY: "On the Pecos and Llano Estacado," New Mexico.
RANGE: Colorado and New Mexico.
NEW MEXICO: Santa Fe; Water Canyon; Sierra Grande; Tierra Amarilla; Nara Visa; mountains west of San Antonio; Carrizalillo Mountains; Organ Mountains. Hills and plains, in the Upper Sonoran and Transition zones.

9. LEIOSTEMON Raf.

Slender, much branched perennial herbs, about 80 cm. high or less, with narrow leaves and showy paniculate flowers; corolla hardly bilabiate, obliquely salverform, white or tinged with red outside; fifth stamen glabrous, occasionally with a rudimentary anther.

KEY TO THE SPECIES.

Tube of corolla 10 to 15 mm. long; throat little dilated .... 1. *L. ambiguus.*
Tube of corolla 5 mm. long or less; throat much dilated .... 2. *L. thurberi.*

1. **Leiostemon ambiguus** (Torr.) Greene, Leaflets **1**: 223. 1906.
Pentstemon ambiguus Torr. Ann. Lyc. N. Y. **2**: 228. 1828.
Leiostemon purpureus Raf. Atl. Journ. 145. 1833.
TYPE LOCALITY: "Near the Rocky Mountains," Colorado or New Mexico.
RANGE: Utah and Colorado to western Texas and southern California.
NEW MEXICO: Clayton; Nara Visa; Albuquerque; Aden; Hop Canyon; Jarilla; south of Melrose; Buchanan. Dry hills and plains, in the Lower and Upper Sonoran zones.

2. **Leiostemon thurberi** (Torr.) Greene, Leaflets **1**: 223. 1906.
Pentstemon thurberi Torr. U. S. Rep. Expl. Miss. Pacif. **7**: 15. 1857.
Pentstemon ambiguus thurberi A. Gray, Proc. Amer. Acad. **6**: 64. 1862.
TYPE LOCALITY: Burro Mountains, New Mexico. Type collected by Antisell in August, 1854.
RANGE: New Mexico and Arizona to northern Mexico.
NEW MEXICO: Magdalena Mountains; Albuquerque; near Juniper Spring; Carrizozo; Gage. Dry hills and plains, in the Lower and Upper Sonoran zones.

10. GRATIOLA L. HEDGE HYSSOP.

Low annual, 30 cm. high or mostly less, with narrow sessile leaves and small axillary flowers on peduncles nearly equaling the leaves; calyx lobes nearly equal; corolla deeply bilabiate, the upper lip entire or 2-cleft, the lower 3-cleft; capsules 4-valved, many-seeded.

1. **Gratiola virginiana** L. Sp. Pl. 17. 1753.
TYPE LOCALITY: "Habitat in Virginia."
RANGE: British America to California and Florida.
NEW MEXICO: Chama (*Standley* 6659); Brazos Canyon (*Standley & Bollman*). Wet ground, in the Transition Zone.

11. MIMULUS L. MONKEY FLOWER.

Annual or perennial herbs, usually growing in wet soil, 50 cm. high or less, with solitary, axillary, or by the reduction of the leaves racemose flowers; sepals united into a 5-angled tube with short oblique limb; corolla strongly 2-lipped, yellow or purple, the upper lip spreading or reflexed, the lower erect; stamens 4; capsules loculicidally dehiscent, inclosed in the persistent papery calyx.

KEY TO THE SPECIES.

Corolla purple or reddish.
  Corolla 8 mm. long or less; annual .... 1. *M. rubellus.*
  Corolla 35 mm. long or more; perennial .... 8. *M. cardinalis.*
Corolla yellow.
  Corolla tube twice as long as the calyx .... 2. *M. parvulus.*
  Corolla tube little or not at all exceeding the calyx.
    Calyx teeth equal; leaves elliptic or narrowly oblong ...... 3. *M. gratioloides.*

Calyx teeth unequal; leaves broader, mostly orbicular-ovate.
  Plants pubescent.................................... 4. *M. puberulus.*
  Plants glabrous or nearly so.
    Stems usually floating, rooting at the nodes; corolla about 8 mm. long.................. 5. *M. geyeri.*
    Stems neither floating nor rooting at the nodes; corolla more than 10 mm. long.
      Corolla 2 cm. long or less; slender annual; leaves usually as broad as long........ 6. *M. cordatus.*
      Corolla 2 to 3 cm. long; stout perennial; leaves longer than broad.............. 7. *M. langsdorfii.*

1. **Mimulus rubellus** A. Gray, in Torr. U. S. & Mex. Bound. Bot. 116. 1859.

TYPE LOCALITY: Wet ravines of the Organ Mountains and Copper Mines, New Mexico.

RANGE: Southern New Mexico to California.

NEW MEXICO: Hillsboro; Organ Mountains. Wet ground, in the Upper Sonoran Zone.

2. **Mimulus parvulus** Woot. & Standl. Contr, U. S. Nat. Herb. **16:** 171. 1913.

TYPE LOCALITY: Rocky Canyon, Grant County, New Mexico. Type collected by J. M. Holzinger in 1911.

RANGE: Known only from type locality.

3. **Mimulus gratioloides** Rydb. Bull. Torrey Club **28:** 27. 1901.

TYPE LOCALITY: Butte, 5 miles southwest of La Veta, Colorado.

RANGE: Southern Colorado and northern New Mexico.

NEW MEXICO: Near Tierra Amarilla (*Eggleston* 6473, 6479, 6505). Open hills, in the Upper Sonoran and Transition zones.

4. **Mimulus puberulus** Greene, Leaflets **2:** 4. 1909.

TYPE LOCALITY: Pagosa Springs, Colorado.

RANGE: Colorado and northern New Mexico.

NEW MEXICO: Tunitcha Mountains; Chama; Santa Fe and Las Vegas mountains. Along streams in the mountains, Transition Zone.

5. **Mimulus geyeri** Torr. in Nicoll. Rep. Miss. 157. 1843.

Mimulus jamesii Torr. & Gray; Benth. in DC. Prodr. **10:** 371. 1846.

TYPE LOCALITY: "Fresh water springs, Devil's Lake."

RANGE: North Dakota and Michigan to Illinois and New Mexico.

NEW MEXICO: Farmington; Apache Spring; Mangas Springs; White Mountains. In water.

6. **Mimulus cordatus** Greene, Leaflets **2:** 5. 1909.

TYPE LOCALITY: Bear Mountain, near Silver City, New Mexico. Type collected by Metcalfe (no. 28).

RANGE: Mountains of New Mexico.

NEW MEXICO: Sandia Mountains; Bear Mountains; Dog Spring; San Luis Mountains; Organ Mountains; White Mountains. Along streams, in the Upper Sonoran and Transition zones.

Probably this is *M. nasutus* Greene, and not constantly different from *M. langsdorfii.*

7. **Mimulus langsdorfii** Don; Sims in Curtis's Bot. Mag. **36:** *pl. 1501.* 1812.

Mimulus guttatus DC. Cat. Hort. Monsp. 127. 1813.

TYPE LOCALITY: Unalaska.

RANGE: Alaska and California to Colorado and New Mexico.

NEW MEXICO: Manguitas Spring; mountains southeast of Patterson; Fort Bayard; Lower Plaza; Hillsboro; Cloverdale; Organ Mountains; White Mountains. Along streams, in the Transition Zone.

8. Mimulus cardinalis Dougl.; Benth. Scroph. Ind. 28. 1835.

TYPE LOCALITY: California.

RANGE: Wet ground, Oregon and California to Mexico and western New Mexico.

NEW MEXICO: Gila Canyon (*P. F. Mohr*).

12. CONOBEA Aublet.

Low slender annual with pinnatifid leaves and small axillary flowers; calyx lobes unequal, longer than the tube; corolla 2-lipped; stamens 4, included, the anther sacs parallel, contiguous; styles united, incurved; capsules ovoid-conic, septicidally dehiscent, the valves entire or 2-cleft; seeds striate.

1. Conobea intermedia A. Gray in Torr. U. S. & Mex. Bound. Bot. 117. 1859.

TYPE LOCALITY: Dry hills around the Copper Mines, New Mexico. Type collected by Wright (no. 1485).

RANGE: Southern New Mexico and Arizona and adjacent Mexico.

NEW MEXICO: Mogollon Mountains; Santa Rita.

13. LIMOSELLA L. MUDWORT.

Small glabrous aquatic plants with fibrous roots, a cluster of entire fleshy leaves at the nodes of the stolons, and short scapelike naked pedicels from the axils, each bearing a small white flower; calyx campanulate; corolla rotate-campanulate, 5-lobed, nearly regular; stamens 4, the anther cells confluent.

1. Limosella aquatica L. Sp. Pl. 631. 1753.

TYPE LOCALITY: "Habitat in Europae septentrionalis inundatis."

RANGE: British America to California and New Mexico; also in Europe, Asia, and South America.

NEW MEXICO: Tunitcha Mountains; Chama; West Fork of the Gila; Bartlett Ranch. In mud and shallow water, in the Transition Zone.

The plants from southern New Mexico have remarkably narrow leaves but not so narrow as in *L. tenuifolia* Hoffm.

14. VERONICA L. SPEEDWELL.

Low annual or perennial caulescent herbs with opposite or sometimes alternate, entire or toothed leaves and axillary racemose or spicate flowers; sepals 4, slightly united at the base; corolla whitish or blue, rotate, slightly irregular, the lower lobe usually narrowest; stamens 2, on either side of the upper corolla lobe; capsules flattened, notched or 2-lobed at the apex.

KEY TO THE SPECIES.

Flowers in axillary racemes.
 Leaves all short-petioled, the blades ovate to oblong...... 1. *V. americana.*
 Leaves mostly sessile, the blades of various shapes.
  Leaves oblong or oblong-lanceolate, conspicuously serrate, 35 mm. long or more, not narrowed at the base; sepals not exceeding the capsule........... 2. *V. anagallis-aquatica.*
  Leaves oval or obovate, entire or nearly so, less than 30 cm. long, narrowed at the base; sepals conspicuously exceeding the capsule................ 3. *V. micromera.*

Flowers in terminal spikes or racemes or solitary in the axils of the leaves.
- Annuals; flowers solitary in the axils of the little reduced leaves.
  - Pedicels shorter than the oblong to linear cauline leaves.. 4. *V. xalapensis.*
  - Pedicels longer than the ovate cauline leaves........... 5. *V. tournefortii.*
- Perennials; flowers in terminal spikes or racemes; leaves of the inflorescence much reduced.
  - All leaves sessile, ovate to oblong; capsules merely emarginate.................................... 6. *V. wormskjoldii.*
  - Lower leaves petioled, the blades rounded-oval to oblong; capsules obcordate............................... 7. *V. serpyllifolia.*

1. Veronica americana Schwein.; Benth. in DC. Prodr. **10:** 468. 1846.

AMERICAN BROOKLIME.

TYPE LOCALITY: "In America boreali a Canada et Carolina usque ad flum. Oregon et in ins. Sitcha."

RANGE: British America to California, New Mexico, and Pennsylvania.

NEW MEXICO: Carrizo Mountains; Tunitcha Mountains; Farmington; Chama; Santa Fe and Las Vegas mountains; Gallo Spring; West Fork of the Gila; Reserve; White and Sacramento mountains. Along streams, in the Upper Sonoran and Transition zones.

2. Veronica anagallis-aquatica L. Sp. Pl. 12. 1753. WATER SPEEDWELL.

TYPE LOCALITY: "Habitat in Europa ad fossas."

RANGE: British America to Arizona and North Carolina.

NEW MEXICO: Rio Mimbres (*Thurber* 217). Along streams.

3. Veronica micromera Woot. & Standl. Contr. U. S. Nat. Herb. **16:** 174. 1913.

TYPE LOCALITY: Along ditches in the vicinity of Shiprock, on the Navajo Reservation, New Mexico. Type collected by Standley (no. 7283).

RANGE: Known only from type locality, in the Upper Sonoran Zone.

4. Veronica xalapensis H. B. K. Nov. Gen. & Sp. **2:** 389. 1817.

PURSLANE SPEEDWELL.

Veronica peregrina of many authors, not L. 1753.

TYPE LOCALITY: "Crescit in Regno Mexicano prope Xalapa (alt. 630 hex.), in nemoribus Liquidambaris Styracifluae."

RANGE: British America to California and Mexico.

NEW MEXICO: Santa Fe and Las Vegas mountains; Sandia Mountains; Shiprock; Chama; Mogollon Mountains; Kingston; Organ Mountains; White and Sacramento mountains. Chiefly in the Transition Zone.

5. Veronica tournefortii K. Gmel. Fl. Badens. **1:** 39. 1805.

Veronica buxbaumii Ten. Fl. Napol. **1:** 7. *pl. 1.* 1811.

TYPE LOCALITY: "Prope Carlsruhe in agris am Holzhof, ante aliquot annos ex horto botanico emigrata et nunc quasi spontanea," Germany.

NEW MEXICO: Beulah (*Cockerell*).

A native of Europe, introduced in many places in the United States.

6. Veronica wormskjoldii Roem. & Schult. Syst. Veg. **1:** 101. 1817.

ALPINE SPEEDWELL.

TYPE LOCALITY: Greenland.

RANGE: British America to Arizona and New Hampshire.

NEW MEXICO: Santa Fe and Las Vegas mountains. Wet ground, from the Canadian to the Arctic-Alpine Zone.

7. Veronica serpyllifolia L. Sp. Pl. 12. 1753. THYME-LEAVED SPEEDWELL.

Veronica serpyllifolia neomexicana Cockerell, Amer. Nat. **40:** 872. 1906.

TYPE LOCALITY: "Habitat in Europa & America septentrionali, ad vias, agros."

RANGE: British America to California, New Mexico, and Georgia; also in Europe.

NEW MEXICO: Tunitcha Mountains; Chama; Santa Fe and Las Vegas mountains. Wet meadows, in the Transition Zone.

The type of *V. serpyllifolia neomexicana* was collected on the top of the Las Vegas Range.

15. BESSEYA Rydb.

Perennial herbs with simple scapose stems and toothed leaves, these mostly basal, the cauline ones reduced; flowers small, the inflorescence spicate; calyx 4-parted, with oblong divisions; corolla short-campanulate, 4-parted, somewhat irregular; stamens 2, exserted, the anther cells not confluent, mostly parallel; capsules compressed, obtuse or emarginate.

KEY TO THE SPECIES.

Upper lip of corolla twice as long as the calyx; plants 10 to 15 cm. high .... 1. *B. alpina.*

Upper lip of corolla only slightly exceeding the calyx; plants 15 to 30 cm. high .... 2. *B. plantaginea.*

1. Besseya alpina (A. Gray) Rydb. Bull. Torrey Club **30:** 280. 1903.

Synthyris alpina A. Gray, Amer. Journ. Sci. II. **34:** 251. 1862.

TYPE LOCALITY: Rocky Mountains of Colorado.

RANGE: Wyoming to New Mexico.

NEW MEXICO: Pecos Baldy; Wheeler Peak. Meadows, in the Arctic-Alpine Zone.

2. Besseya plantaginea (Benth.) Rydb. Bull. Torrey Club **30:** 280. 1903.

Synthyris plantaginea Benth. in DC. Prodr. **10:** 455. 1846.

TYPE LOCALITY: "Ad origines flum. Platte."

RANGE: Wyoming to Arizona and New Mexico.

NEW MEXICO: Santa Fe and Las Vegas mountains; White Mountain Peak. In woods, Transition to Hudsonian Zone.

16. DASYSTOMA Raf. FALSE FOXGLOVE.

Perennial herb, 30 to 60 cm. high, with slender erect stems, linear leaves, and long-pediceled flowers from the axils of the reduced upper leaves, the inflorescence becoming somewhat racemose; sepals 5, united below; corolla yellow, open-funnelform, slightly irregular, the lobes spreading; stamens 4, included, the filaments pubescent; capsules acute, beaked, loculicidal.

1. Dasystoma wrightii (A. Gray) Woot. & Standl. Contr. U. S. Nat. Herb. **16:** 171. 1913.

Gerardia wrightii A. Gray in Torr. U. S. & Mex. Bound. Bot. 118. 1859.

TYPE LOCALITY: "Hill sides between Babocomori and Santa Cruz, Sonora."

RANGE: Arizona and New Mexico and adjacent Mexico.

NEW MEXICO: San Luis Mountains (*Mearns* 2435).

17. ADENOSTEGIA Benth.

Annual, 30 to 40 cm. high, freely branched, scabrous-puberulent, with crowded terminal spikes of dull yellow flowers, and with alternate pinnatifid leaves with linear segments; calyx spathelike, of 1 or 2 lobes, not colored; corolla tubular, the lips subequal; stamens 4, the anther cells dissimilar, ciliate or bearded; style hooked at the apex.

1. **Adenostegia wrightii** (A. Gray) Greene, Pittonia 2: 180. 1891.

Cordylanthus wrightii A. Gray in Torr. U. S. & Mex. Bound. Bot. 120. 1859.

TYPE LOCALITY: "Prairies, from 6 to 30 miles east of El Paso, Western Texas."

RANGE: Western Texas to Arizona.

NEW MEXICO: Tunitcha Mountains; Farmington; Gallup; Zuni Reservation; San Lorenzo; Ramah; Mogollon Mountains; Bear Mountain. Dry plains and hills, in the Upper Sonoran Zone.

A decoction of this plant is used by the Navahos as a cure for syphilis.

18. CASTILLEJA Mutis. PAINTED CUP.

Annual or perennial herbs, 80 cm. high or mostly less, often parasitic on the roots of other plants; leaves alternate, simple or lobed; flowers in terminal spikes, subtended by conspicuous, often bright-colored bracts; calyx of almost wholly united sepals, the tube laterally compressed; corolla often highly colored, strongly 2-lipped, the upper lip arched, the lower very short; stamens 4, inclosed in the upper lip of the corolla, the anther sacs unequal, the outer attached at the middle, the inner pendulous by its apex; stigma entire or 2-lobed; capsules loculicidal; seeds reticulated.

KEY TO THE SPECIES.

Annuals.
- Upper leaves and bracts linear; plants slender............ 1. *C. minor.*
- Upper leaves and bracts lanceolate; plants stout........... 2. *C. exilis.*

Perennials.
- Galea less than 3 times as long as the lip, rarely half as long as the corolla tube; bracts mostly yellow or brownish.
  - Leaves pinnately divided, at least the upper ones; bracts reddish, entire or lobed.
    - Plants lanuginous throughout; corolla scarcely surpassing the calyx........................ 3. *C. lineata.*
    - Plants not lanuginous; corolla much exceeding the calyx.................................... 4. *C. sessiliflora.*
  - Leaves entire; bracts entire or slightly lobed, tinged with yellow.
    - Plants low, 5 to 15 cm. high; corolla less than 2 cm. long.................................... 5. *C. occidentalis.*
    - Plants taller, 20 to 40 cm. high; corolla usually more than 2 cm. long.
      - Plants darkening in drying or turning bright bluish green; leaves merely acute, lanceolate or elliptic................... 6. *C. luteovirens.*
      - Plants usually not darkening in drying; leaves long-attenuate or acuminate, linear-lanceolate............................ 7. *C. sulphurea.*
- Galea several times longer than the lip, usually at least two-thirds as long as the corolla tube; bracts scarlet, crimson, or rose.
  - Calyx cleft much deeper in front than behind.
    - Upper leaves lanceolate, 3-nerved, usually entire. 8. *C. wootoni.*
    - All the leaves linear, 1-nerved, pinnatifid........ 9. *C. linariaefolia.*
  - Calyx about equally cleft before and behind.
    - Stems villous-canescent or densely woolly.
      - Plants densely white-woolly throughout; leaves lanuginous on both surfaces.....10. *C. lanata.*
      - Plants villous-canescent; leaves glabrous on the upper surface....................11. *C. integra.*

Stems neither villous-canescent nor lanuginous.
- Leaves, at least the upper ones, pinnately cleft like the bracts.
  - Galea much longer than the tube of the corolla........................ 12. *C. eremophila.*
  - Galea shorter than the tube of the corolla.
    - Plants glabrous up to the inflorescence; bracts rose-colored..... 13. *C. haydeni.*
    - Plants hirsute or villous throughout; bracts scarlet to crimson...... 14. *C. angustifolia.*
- Leaves entire or the uppermost slightly 3-lobed.
  - Leaves mostly linear, conspicuously pubescent........................ 15. *C. organorum.*
  - Leaves linear-lanceolate or broader, minutely puberulent or glabrous.
    - Bracts crimson or purplish.
      - Bracts mostly entire, if lobed with an obtuse middle lobe; stems glabrous or nearly so.................. 16. *C. lauta.*
      - Bracts 3-cleft with lanceolate lobes; stems hirsute or villous throughout........... 17. *C. trinervis.*
    - Bracts scarlet or nearly so.
      - Bracts usually entire, rarely with 2 linear lobes, obtuse; stems with spreading pubescence.... 18. *C. austromontana.*
      - Bracts all lobed, the lobes often narrow, acute or obtuse; stems with spreading or appressed pubescence.
        - Pubescence spreading, rough; bracts much divided; plants branched above........... 19. *C. inconstans.*
        - Pubescence weak or appressed, not spreading; bracts sparingly lobed; stems simple... 20. *C. confusa.*

1. **Castilleja minor** A. Gray, in Brewer & Wats. Bot. Calif. **1**: 573. 1876.

Castilleja affinis minor A. Gray in Torr. U. S. & Mex. Bound. Bot. 119. 1859.

Type locality: Near the Copper Mines, New Mexico.

Range: Southern New Mexico and Arizona and adjacent Mexico.

New Mexico: Mogollon Mountains; Animas Creek; Santa Rita. Upper Sonoran Zone.

2. **Castilleja exilis** A. Nels. Proc. Biol. Soc. Washington **17**: 100. 1904.

Castilleja stricta Rydb. Mem. N. Y. Bot. Gard. **1**: 354. 1900, not DC. 1846.

Type locality: Ruby Valley, Nevada.

Range: Washington and Montana to Nevada and northwestern New Mexico.

New Mexico: Aztec; Farmington. Wet ground, in the Upper Sonoran Zone.

3. **Castilleja lineata** Greene, Pittonia **4**: 151. 1900.

Type locality: Moist slopes near Pagosa Springs, southern Colorado.

Range: Colorado and northern New Mexico.

New Mexico: Chama; Dulce; Canjilon; Vermejo Park. Wet meadows, in the Transition Zone.

4. **Castilleja sessiliflora** Pursh, Fl. Amer. Sept. 738. 1814.

TYPE LOCALITY: "In Upper Louisiana."

RANGE: Saskatchewan and Illinois to Missouri, Texas, and Arizona.

NEW MEXICO: Sierra Grande; Cross L Ranch; mountains west of San Antonio; Carrizalillo Mountains; San Andreas Mountains. Dry plains and foothills, in the Upper Sonoran Zone.

5. **Castilleja occidentalis** Torr. Ann. Lyc. N. Y. **2**: 230. 1827.

Castilleja pallida occidentalis A. Gray in Brewer & Wats. Bot. Calif. **1**: 575. 1876.

TYPE LOCALITY: "On the Rocky Mountains," Colorado.

RANGE: British America to northern New Mexico.

NEW MEXICO: Pecos Baldy; Truchas Peak. Meadows, in the Arctic-Alpine Zone.

6. **Castilleja luteovirens** Rydb. Bull. Torrey Club **28**: 26. 1901.

TYPE LOCALITY: Sangre de Cristo Creek, Colorado.

RANGE: Wyoming and Utah to northern New Mexico.

NEW MEXICO: Chama; Santa Fe and Las Vegas mountains; Manzanares Valley; Ensenada. Mountains, in the Transition Zone.

7. **Castilleja sulphurea** Rydb. Mem. N. Y. Bot. Gard. **1**: 359. 1900.

TYPE LOCALITY: Electric Peak, Montana.

RANGE: Wyoming and South Dakota to Utah and northern New Mexico.

NEW MEXICO: Catskill; Jemez Mountains; Santa Fe and Las Vegas mountains. Wet meadows, Transition to Arctic-Alpine Zone.

8. **Castilleja wootoni** Standley, Muhlenbergia **5**: 84. 1909.

TYPE LOCALITY: Gilmores Ranch on Eagle Creek, White Mountains, New Mexico. Type collected by Wooton & Standley (no. 3411).

RANGE: White and Sacramento mountains of New Mexico, in the Transition and Canadian zones.

9. **Castilleja linariaefolia** Benth. in DC. Prodr. **10**: 532. 1846.

TYPE LOCALITY: Rocky Mountains.

RANGE: Open woods, Wyoming to California and Mexico.

NEW MEXICO: Farmington; Chama; Dulce; Rosa; Ramah; Winsor Creek; Aztec; Gallinas Canyon; San Ignacio. Transition Zone.

10. **Castilleja lanata** A. Gray in Torr. U. S. & Mex. Bound. Bot. 118. 1859.

TYPE LOCALITY: "Along and near the Rio Grande, from Eagle Pass, etc., to El Paso," Texas.

RANGE: Western Texas to southern Arizona and adjacent Mexico.

NEW MEXICO: Dog Mountains; Soledad Canyon. Dry hills, in the Upper Sonoran Zone.

11. **Castilleja integra** A. Gray in Torr. U. S. & Mex. Bound. Bot. 119. 1859.

TYPE LOCALITY: Organ Mountains, New Mexico.

RANGE: Colorado to Arizona and Mexico.

NEW MEXICO: Common nearly throughout the State. Dry hills and plains, in the Upper Sonoran Zone.

12. **Castilleja eremophila** Woot. & Standl. Contr. U. S. Nat. Herb. **16**: 171. 1913.

TYPE LOCALITY: Arid sandy mesas about the north end of the Carrizo Mountains, northeast corner of Arizona. Type collected by Standley (no. 7464).

RANGE: Known only from the type locality, in the Upper Sonoran Zone.

13. **Castilleja haydeni** (A. Gray) Cockerell, Bull. Torrey Club **17**: 37. 1898.

Castilleja pallida haydeni A. Gray, Syn. Fl. **2**[1]: 297. 1878.

TYPE LOCALITY: "Alpine regions of the Sierra Blanca, S. Colorado."

RANGE: Colorado and northern New Mexico.

NEW MEXICO: Pecos Baldy; Truchas Peak; Baldy. Meadows in the mountains, Arctic-Alpine Zone.

14. Castilleja angustifolia (Nutt.) Don, Hist. Dichl. Pl. 4: 616. 1837.

Euchroma angustifolia Nutt. Journ. Acad. Phila. 7: 46. 1834.

TYPE LOCALITY: "Native in dry prairies on the borders of the Little Godding River, near the source of the Columbia."

RANGE: British Columbia to California and northern New Mexico.

NEW MEXICO: Aztec; southeast of Tierra Amarilla. Open hills, in the Upper Sonoran Zone.

15. Castilleja organorum Standley, Muhlenbergia 5: 86. 1909.

TYPE LOCALITY: Rocky sides of the Organ Mountains not far from Van Pattens Camp, New Mexico. Type collected by Standley, June 6, 1906.

RANGE: Organ Mountains of New Mexico, in the Upper Sonoran Zone.

16. Castilleja lauta A. Nels. Bull. Torrey Club 27: 269. 1900.

TYPE LOCALITY: Dunraven Peak, Yellowstone Park.

RANGE: Montana to northern New Mexico.

NEW MEXICO: Near Chama (*Standley* 6842). Marshes in the mountains, in the Canadian Zone.

17. Castilleja trinervis Rydb. Bull. Torrey Club 28: 26. 1901.

TYPE LOCALITY: Headquarters of Sangre de Cristo Creek, Colorado.

RANGE: Colorado and northern New Mexico.

NEW MEXICO: Near Chama (*Standley* 6841). Marshes in the mountains, in the Canadian Zone.

18. Castilleja austromontana Standl. & Blumer, Muhlenbergia 7: 44. 1911.

TYPE LOCALITY: Pine woods at Mannings Camp, Rincon Mountains, Arizona.

RANGE: Mountains of southern New Mexico and Arizona.

NEW MEXICO: Mogollon Mountains; East Canyon. Transition Zone.

19. Castilleja inconstans Standley, Muhlenbergia 5: 83. 1909.

TYPE LOCALITY: Winsors Ranch, on the headquarters of the Pecos River, New Mexico. Type collected by Standley (no. 4000).

RANGE: Mountains of northern New Mexico.

NEW MEXICO: Santa Fe and Las Vegas mountains; Tunitcha Mountains; Chama. Transition and Canadian zones.

What appears to be a hybrid between either this or *C. confusa* and one of the yellow-bracted species was discovered by Prof. T. D. A. Cockerell at Harveys Ranch near Las Vegas and described as *C. confusa*×*acuminata*.[1] This was later renamed *C.* × *porterae*.[2] A number of plants with orange instead of scarlet bracts have been noticed at the type locality of the species.

20. Castilleja confusa Greene, Pittonia 4: 1. 1899.

TYPE LOCALITY: "Of the more southerly or southwesterly Colorado Rocky Mountains, and those of adjacent New Mexico."

RANGE: Mountains of Colorado and northern New Mexico.

NEW MEXICO: Chama; Jemez Mountains; Santa Fe and Las Vegas mountains; Sandia Mountains; Baldy. From the Transition to the Hudsonian Zone.

19. ORTHOCARPUS Nutt.

Erect annuals, 40 cm. high or less, with alternate simple or pinnatifid leaves, and yellow, purple, or white flowers in terminal crowded bracted spikes, the bracts often colored; calyx tubular or tubular-campanulate, 4-cleft, the lobes about equal; corolla

---

[1] Bot. Gaz. 29: 280. 1900. [2] Cockerell, Nature 70: 319. 1904.

very irregular, the tube slender, the limb 2-lipped, the upper lip little if at all longer than the 3-lobed lower one; stamens 4, ascending under the upper lip; anther sacs dissimilar, the outer fixed by the middle, the other pendulous by its upper end; style filiform, the stigma entire; capsules oblong, many-seeded.

KEY TO THE SPECIES.

Corolla yellow; spikes densely flowered........................ 1. *O. luteus.*
Corolla purple and white; spikes lax.......................... 2. *O. purpureo-albus.*

1. **Orthocarpus luteus** Nutt. Gen. Pl. **2**: 57. 1818.
TYPE LOCALITY: "In humid situations on the plains of the Missouri, near Fort Mandan."
RANGE: Washington and Saskatchewan to Nevada and New Mexico.
NEW MEXICO: Raton; mountains west of Grants Station; Chama; Taos; Johnsons Mesa; Pajarito Park; Santa Fe and Las Vegas mountains; Mogollon Mountains. Plains and hillsides, in the Upper Sonoran and Transition zones.

2. **Orthocarpus purpureo-albus** A. Gray in King, Geol. Expl. 40th Par. **5**: 458. 1871.
TYPE LOCALITY: "New Mexico."
RANGE: Utah and Colorado to Arizona and New Mexico.
NEW MEXICO: Tunitcha Mountains; Dulce; mountains west of Grants; Coolidge; Ramah; Datil Mountains; Mogollon Mountains; Silver City; Rito de los Frijoles. Dry plains and hills, in the Transition and Upper Sonoran zones.

20. RHINANTHUS L. YELLOW RATTLE.

Erect annual about 30 cm. high, with simple opposite leaves and yellow axillary flowers crowded in a bracted terminal spike; calyx ventricose-compressed, 4-toothed, inflated in fruit; corolla tube cylindric, the upper lip galeate, ovate, obtuse, compressed, entire at the apex, with a minute tooth on each side; lower lip shorter, with 3 spreading lobes; capsules orbicular, compressed, the seeds suborbicular, winged.

1. **Rhinanthus crista-galli** L. Sp. Pl. 603. 1753.
TYPE LOCALITY: "Habitat in Europae pratis."
RANGE: British America to New Mexico and New York; also in Europe and Asia.
NEW MEXICO: Rio Pueblo; Bartlett Ranch; Brazos Canyon. Woods, in the Transition Zone.

21. ELEPHANTELLA Rydb. LITTLE RED ELEPHANT.

Similar in general appearance to the next genus, but distinguished by having the galea prolonged into a filiform recurved beak, and by the presence of teeth on each side of the throat of the corolla.

1. **Elephantella groenlandica** (Retz.) Rydb. Mem. N. Y. Bot. Gard. **1**: 363. 1900.
Pedicularis groenlandica Retz. Fl. Scand. Prodr. ed. 2. 145. 1795.
TYPE LOCALITY: Greenland.
RANGE: Greenland and British America to California and New Mexico.
NEW MEXICO: Santa Fe and Las Vegas mountains. Bogs, Candian to the Arctic-Alpine Zone.

22. PEDICULARIS L. LOUSEWORT.

Perennial herbs, mostly low, sometimes as much as a meter high, with opposite or alternate leaves and terminal crowded spikes or racemes of rather conspicuous flowers; leaves more or less dissected; calyx of 5 mostly united sepals, sometimes cleft on the lower or upper side; corolla of various colors, strongly 2-lipped, the upper lip concave or conduplicate, laterally flattened, the lower erect or spreading; stamens 4, of 2 lengths, ascending under the upper lip of the corolla, the anthers alike; capsules flattened, oblique or curved, beaked, loculicidal; seeds numerous, reticulated, striate, pitted, or ribbed.

KEY TO THE SPECIES.

Leaves merely crenate, linear to linear-lanceolate.
- Galea with a long beak; lip 10 to 12 mm. long .... 1. *P. racemosa.*
- Galea not beaked; lip about 7 mm. long .... 2. *P. angustissima.*

Leaves pinnatifid.
- Galea produced into a distinct beak.
  - Plants 20 cm. high or less; beak not incurved .... 3. *P. parryi.*
  - Plants 50 cm. high or more; beak strongly incurved..... 4. *P. mogollonica.*
- Galea not produced into a beak, often with several small teeth at the apex.
  - Inflorescence shorter than the leaves; plants glabrous.... 5. *P. centranthera.*
  - Inflorescence much longer than the leaves; plants more or less pubescent.
    - Leaves divided to the midrib or nearly so, the segments acute; plants tall, mostly about a meter high .... 6. *P. grayi.*
    - Leaves divided about halfway to the base, the segments obtuse; plants lower, 35 cm. high or less. 7. *P. fluviatilis.*

1. **Pedicularis racemosa** Dougl.; Hook. Fl. Bor. Amer. **2**: 108. 1838.

TYPE LOCALITY: "Abundant on the summit of the high mountains of the Grand Rapids of the Columbia."

RANGE: British Columbia and California.

NEW MEXICO: Santa Fe and Las Vegas mountains. Damp woods, in the Canadian and Hudsonian zones.

2. **Pedicularis angustissima** Greene, Leaflets **1**: 151. 1905.

TYPE LOCALITY: Mogollon Mountains, New Mexico. Type collected by Metcalfe (no. 534).

RANGE: Mogollon Mountains of New Mexico.

3. **Pedicularis parryi** A. Gray, Amer. Journ. Sci. II. **34**: 250. 1862.

TYPE LOCALITY: Rocky Mountains of Colorado.

RANGE: Wyoming to Utah and northern New Mexico.

NEW MEXICO: Truchas Peak; Baldy; Pecos Baldy. Meadows, in the Arctic-Alpine Zone.

4. **Pedicularis mogollonica** Greene, Leaflets **1**: 151. 1905.

TYPE LOCALITY: Mogollon Mountains, New Mexico. Type collected by Metcalfe (no. 496).

RANGE: Mogollon Mountains, New Mexico.

5. **Pedicularis centranthera** A. Gray in Torr. U. S. & Mex. Bound. Bot. 120. 1859.

TYPE LOCALITY: Ben More, New Mexico. Type collected by Bigelow.

RANGE: Utah to New Mexico and Arizona.

NEW MEXICO: Sandia Mountains.

6. **Pedicularis grayi** A. Nels. Proc. Biol. Soc. Washington **17**: 100. 1904.

Pedicularis procera A. Gray, Amer. Journ. Sci. II. **34**: 251. 1862, not Adams, 1823.

TYPE LOCALITY: Colorado Rocky Mountains.

RANGE: Wyoming to New Mexico.

NEW MEXICO: Tunitcha Mountains; Chama; Santa Fe and Las Vegas mountains; Sandia Mountains; Hillsboro Peak; Mogollon Mountains; White and Sacramento mountains. Damp woods and along streams, in the Transition and Canadian zones.

7. **Pedicularis fluviatilis** Heller, Minn. Bot. Stud. **2**: 33. 1898.

TYPE LOCALITY: Meadow 9 miles east of Santa Fe, New Mexico. Type collected by Heller (no. 3639).

RANGE: Northern New Mexico and southern Colorado.

NEW MEXICO: Santa Fe and Las Vegas mountains; Sierra Grande. Mountains, in the Transition Zone.

This is closely related to the eastern *P. canadensis* L., but appears to be fairly distinct.

128. ACANTHACEAE. Acanthus Family.

Annual or perennial herbs or shrubs, with alternate or opposite leaves; flowers perfect, irregular, sometimes solitary, often subtended by large bracts; calyx of 5 variously united or distinct sepals; corolla of 5 partially united petals, 2-lipped; stamens 2 and equal or 4 and didynamous; styles terminal, united; fruit a capsule, usually with 2 cavities, opening with an elastic longitudinal dehiscence.

KEY TO THE GENERA.

Cauline leaves reduced to imbricated scales............ 1. TUBIFLORA (p. 597).
Cauline leaves not reduced to scales.
  Shrubs.
    Corolla purplish red; tall shrub 50 cm. high or more; corolla deeply bilabiate.......... 2. ANISACANTHUS (p. 597).
    Corolla white; low shrub, 30 cm. high or less; corolla nearly regular.................... 6. RUELLIA (p. 598).
  Low herbs.
    Corolla convolute in bud; flowers covered by large bracts............................ 3. DIAPEDIUM (p. 598).
    Corolla imbricated in bud; flowers not covered by bracts.
      Stamens 4; plants hirsute................. 4. CARLOWRIGHTIA (p. 598).
      Stamens 2; plants glabrous................ 5. STENANDRIUM (p. 598).

1. TUBIFLORA J. F. Gmel.

Perennial caulescent herb with numerous large basal leaves, those of the stems reduced to imbricated scales; flowers in dense spikes; corolla white or blue, with a slender tube; stamens 2.

1. **Tubiflora squamosa** (Jacq.) Kuntze, Rev. Gen. Pl. **2**: 500. 1891.

Verbena squamosa Jacq. Pl. Hort. Schönbr. **1**: 3. *pl. 5*. 1797.

Elytraria tridentata Vahl, Enum. Pl. **1**: 107. 1804.

TYPE LOCALITY: Not known.

RANGE: Western Texas to southern Arizona, southward through tropical America; also in Africa.

NEW MEXICO: San Luis Mountains (*Mearns* 539). Dry, rocky hills.

2. ANISACANTHUS Nees.

Shrub with opposite entire leaves; flowers solitary, axillary; corolla purplish red, with a slender elongated tube; upper lip entire or 2-cleft, the lower 3-lobed; stamens included; capsules contracted into a stipelike base; seeds 4 or fewer.

1. **Anisacanthus thurberi** (Torr.) A. Gray, Syn. Fl. **2**[1]: 328. 1878.

Drejera thurberi Torr. U. S. & Mex. Bound. Bot. 124. 1859.

TYPE LOCALITY: "Along water-courses, Las Animas, Sonora." Type collected by Thurber.

RANGE: Southwestern New Mexico to Arizona and Mexico.

NEW MEXICO: Mangas Springs; Lordsburg; Big Hatchet Mountains. Dry hills and mesas.

3. DIAPEDIUM König.

Glabrous perennial herb with opposite petiolate lanceolate leaves and axillary few-flowered peduncles; flowers subtended by large cordate bracts; stamens 2, barely equaling the lips; ovules 2 in each cavity.

1. **Diapedium torreyi** (A. Gray) Woot. & Standl.

Dicliptera torreyi A. Gray, Proc. Amer. Acad. **20:** 309. 1885.

TYPE LOCALITY: Arizona.

RANGE: Southwestern New Mexico and adjacent Arizona and Mexico.

NEW MEXICO: Guadalupe Canyon (*Mearns* 2037).

4. CARLOWRIGHTIA A. Gray.

Slender branched glabrous perennial herb with opposite linear leaves; flowers axillary; corolla deeply 4-parted, with a short tube; stamens 2, nearly equaling the corolla lobes; capsules ovoid, acuminate, on slender stipes; seeds few.

1. **Carlowrightia linearifolia** (Torr.) A. Gray, Proc. Amer. Acad. **13:** 364. 1877.

Schaueria linearifolia Torr. U. S. & Mex. Bound. Bot. 123. 1859.

TYPE LOCALITY: "Rocks at the mouth of the Great cañon of the Rio Grande, and on the Burro mountains," Texas and New Mexico.

RANGE: Western Texas to southern Arizona and southward.

NEW MEXICO: Twenty miles north of Rincon; Dog Spring; mesa west of the Rio Grande, near Mesilla; Organ Mountains. Dry, rocky hills and mesas, in the Lower and Upper Sonoran zones.

5. STENANDRIUM Nees.

Perennial short-stemmed herb with hirsute foliage; leaves approximate, entire; flowers in terminal bracted spikes, rose purple; corolla with a slender tube and oblique 5-lobed limb; stamens 4, included; ovules 2 in each cavity.

1. **Stenandrium barbatum** Torr. & Gray, U. S. Rep. Expl. Miss. Pacif. **2**[2]**:** 168. *pl. 4.* 1855.

TYPE LOCALITY: "On the Pecos," Texas or New Mexico.

RANGE: Southern New Mexico, western Texas, and northeastern Mexico.

NEW MEXICO: Guadalupe Mountains (*Wooton*). Dry hills, in the Lower Sonoran Zone.

6. RUELLIA L.

Low shrub, 30 cm. high or less, with thick obovate petiolate ciliate leaves and few solitary axillary flowers; calyx lobes linear; corolla white, with a long tube and a nearly regular limb; stamens 4, included; ovules 3 to 10 in each cavity.

1. **Ruellia parryi** A. Gray, Syn. Fl. **2**[1]**:** 326. 1878.

Dipteracanthus suffruticosus Torr. U. S. & Mex. Bound. Bot. 122. 1859, not *Ruellia suffruticosa* Roxb. 1814.

TYPE LOCALITY: Presidio del Norte, Texas.

RANGE: Western Texas and southern New Mexico to northeastern Mexico.

NEW MEXICO: Dark Canyon, Guadalupe Mountains (*Wooton*). Dry hills.

129. PINGUICULACEAE. Bladderwort Family.

1. UTRICULARIA L. Bladderwort.

Small slender aquatic herb with capillary-dissected leaves bearing small bladder-like appendages, and with short, 1 to few-flowered scapes; calyx 2-lipped, the lips entire or nearly so; corolla deeply bilabiate, yellow, the lower lip larger and 3-lobed, spurred at the base in front; ovary free; style very short or none; stigma 2-cleft.

1. **Utricularia vulgaris** L. Sp. Pl. 18. 1753.

Type locality: "Habitat in Europae fossis paludibus profundioribus."

Range: Throughout most of North America and in Europe, Asia, and Africa.

New Mexico: Tunitcha Mountains (*Standley* 7559). In quiet water.

130. OROBANCHACEAE. Broomrape Family.

Perennial parasitic or saprophytic herbs, less than 40 cm. high; leaves reduced to scales, without chlorophyll; flowers perfect, rarely diœcious, sometimes cleistogamous; calyx of 4 or 5 more or less united sepals, persistent; corolla irregular, bilabiate, persistent; stamens 4, didynamous, mostly included; anthers 2-celled, rarely 1-celled; ovary 1 or 2-celled; style 1; stigma capitate or 2-lobed; fruit a 1 or 2-celled capsule; seeds many, minute.

KEY TO THE GENERA.

Calyx irregular, split on the lower side, the upper part with 3 or 4 toothlike lobes.............................. 1. Conopholis (p. 599).

Calyx regular or nearly so, with 2 to 5 equal or unequal lobes.

Calyx with a deep sinus above and below, the lateral lobes often 2-cleft.......................... 2. Myzorrhiza (p. 599).

Calyx nearly equally 5-lobed.......................... 3. Thalesia (p. 600).

1. CONOPHOLIS Wallr.

Low herb, about 20 cm. high, with very thick glabrous yellow stems, appressed or erect scalelike leaves, and perfect flowers in a dense scaly-bracted terminal spike; calyx accompanied by 2 bractlets, spathelike, split on the lower side; corolla with a curved tube and strongly 2-lipped limb, the upper lip arching, notched, the lower shorter and 3-lobed; ovary 1-celled, with 4 placentæ.

1. **Conopholis mexicana** A. Gray; S. Wats. Proc. Amer. Acad. **18:** 131. 1883.

Squaw root.

Type locality: "In the Sierra Madre, south of Saltillo, and at Soledad, Coahuila, growing at the foot of oaks," Mexico.

Range: New Mexico and Arizona to Mexico.

New Mexico: Gallinas Planting Station; Sandia Mountains; Santa Fe Canyon; Magdalena Mountains; Kingston; Mogollon Creek; San Luis Mountains; Organ Mountains; Gilmores Ranch.

2. MYZORRHIZA Phil. Broomrape.

Herbs, 20 to 40 cm. high, more or less glandular-pubescent, purplish or brownish, with scalelike leaves, the flowers in a terminal spike or panicle; calyx nearly equally 5-lobed; corolla purplish, the tube slightly curved, the limb 2-lipped, the lips often nearly erect; ovary 1-celled.

A decoction of these plants is used by the Navahos in the treatment of sores.

KEY TO THE SPECIES.

Corolla 20 to 25 mm. long; anthers woolly........................ 1. *M. multiflora.*
Corolla 15 to 18 mm. long; anthers glabrous (before dehiscence).. 2. *M. ludoviciana.*

1. **Myzorrhiza multiflora** (Nutt.) Rydb. Bull. Torrey Club **33**: 151. 1906.
Orobanche multiflora Nutt. Journ. Acad. Phila. II. **1**: 179. 1848.
Aphyllon multiflorum A. Gray in Brewer & Wats. Bot. Calif. **1**: 585. 1876.
TYPE LOCALITY: Sandy ground along the borders of the Rio del Norte, New Mexico. Type collected by Gambel.
RANGE: Utah and Colorado to California and Texas.
NEW MEXICO: Upper Pecos; Zuni; Socorro; San Marcial; Mesilla Valley; Gilmores Ranch; East View; Organ Mountains. Parasitic on the roots of various plants.

2. **Myzorrhiza ludoviciana** (Nutt.) Rydb. in Small, Fl. Southeast. U. S. 1093. 1903.
Orobanche ludoviciana Nutt. Gen. Pl. **2**: 58. 1818.
Aphyllon ludovicianum A. Gray in Brewer & Wats. Bot. Calif. **1**: 585. 1876.
? *Orobanche xanthochroa* Nels. & Cockerell, Bot. Gaz. **37**: 278. 1904.
TYPE LOCALITY: "In sandy alluvial soils, around Fort Mandan," North Dakota.
RANGE: Washington and Illinois to California and Texas.
NEW MEXICO: Zuni; Albuquerque; Carrizo Mountains; Cedar Hill; Organ Mountains; Dog Spring; Guadalupe Canyon; Cactus Flat.
The type of *Orobanche xanthochroa* was collected near Pecos.

3. THALESIA Raf. CANCER ROOT.

Low herb, 10 cm. high or less, with pale yellowish or pinkish stems, solitary or few together, bearing a few terminal flowers and scalelike leaves, the latter mostly at the base of the stem; flowers on slender pedicels; calyx lobes nearly equal, acute or acuminate; corolla sometimes more deeply colored than the rest of the plant, the tube curved, the limb slightly 2-lipped, the upper lip often 2-lobed, the lower spreading, with 3 more or less unequal lobes; stamens included; ovary 1-celled, with 4 placentæ.

1. **Thalesia fasciculata** (Nutt.) Britton, Mem. Torrey Club **5**: 298. 1894.
Orobanche fasciculata Nutt. Gen. Pl. **2**: 59. 1818.
Aphyllon fasciculatum Torr. & Gray in A. Gray, Man. ed. 2. 281. 1856.
TYPE LOCALITY: "In sandy alluvial soil about Fort Mandan," North Dakota.
RANGE: British Columbia and California to Saskatchewan and Texas.
NEW MEXICO: Mogollon Creek; Barranca; Carrizo Mountains; Organ Mountains; Gilmores Ranch; Kingston; Winsors Ranch. Parasitic on the roots of various plants.

131. BIGNONIACEAE. Bignonia Family.

Shrubs or low trees with simple or pinnate exstipulate leaves and large perfect flowers in terminal racemes; calyx hypogynous, of 2 more or less united sepals; corolla irregular, large, funnelform, 2-lipped, deciduous; stamens 5, 1 or 3 reduced to sterile filaments; ovary 1-celled with 2 parietal placentæ or 2-celled by a false partition; style 1; stigmas 2; fruit a slender terete capsule, with numerous winged seeds.

The two species of Catalpa, natives of the Central and Southern States, are sometimes cultivated as shade trees in New Mexico.

KEY TO THE GENERA.

Leaves simple; flowers purplish........................ 1. CHILOPSIS (p. 601).
Leaves pinnate; flowers bright yellow................... 2. STENOLOBIUM (p. 601).

1. CHILOPSIS D. Don. DESERT WILLOW.

Large shrub, sometimes treelike, 2 to 5 meters high; leaves narrowly lanceolate, light green; flowers in terminal racemes, showy, purplish; calyx splitting into 2 concave lobes; corolla about 25 mm. long, obscurely 2-lipped, the limb narrow; capsules tapering at both ends, terete, 10 cm. long or more; seeds small, numerous.

1. **Chilopsis linearis** (Cav.) Sweet, Hort. Brit. 283. 1827.

Bignonia ? linearis Cav. Icon. Pl. **3**: 35. *pl. 269*. 1794.

Chilopsis saligna Don, Edinburgh Phil. Journ. **9**: 261. 1823.

TYPE LOCALITY: Origin of type unknown.

RANGE: Western Texas to southern California and southward.

NEW MEXICO: Albuquerque; Clemow; Berendo Creek; Mangas Springs; Dog Spring; Apache Mountains; mesa near Las Cruces; west of Roswell; Organ Mountains. Low hills and sandy mesas, especially along arroyos, in the Lower Sonoran Zone.

A form of this is not infrequent in cultivation as an ornamental shrub. Some of the cultivated plants have white flowers.

2. STENOLOBIUM Don.

Low shrub, about 1 meter high, with pinnate, incisely serrate, bright green leaves; flowers bright yellow, showy.

1. **Stenolobium incisum** Rose & Standl. Contr. U. S. Nat. Herb. **16**: 174. 1913.

TYPE LOCALITY: Hills near Chihuahua, Mexico.

RANGE: Southern New Mexico, southward into northern Mexico.

NEW MEXICO: Dona Ana Mountains (*Wooton & Standley*). Dry hills, in the Lower Sonoran Zone.

132. MARTYNIACEAE. Unicorn plant Family.

1. PROBOSCIDEA Moench. UNICORN PLANT.

Coarse clammy herbs with thick stems, long-petioled, usually large, opposite or alternate leaves, and axillary few-flowered racemes of large yellowish purple flowers; calyx lobes 4 or 5, more or less unequal; corolla campanulate to broadly funnelform, obscurely 2-lipped; stamens 2 or 4, their filaments filiform, the anthers divergent; ovary 1-celled, with 2 parietal placentæ; style 1, stigmas 2; fruit a beaked curved capsule, becoming hard and woody, the beak splitting and forming 2 large opposed hooklike appendages upon drying; seeds irregularly angled or flattened.

The plants are often known as devil horns or devil claws, because of the form of the fruit. Some of the Arizona Indians use the black fiber of the pods in forming the patterns of their basketry.

KEY TO THE SPECIES.

Leaves small, 6 cm. wide or less, deeply lobed; plants low, 30 cm. high and 60 cm. wide or usually less........................ 1. *P. altheaefolia*.

Leaves large, about 10 cm. wide or more, shallowly lobed or entire; plants much larger.

Flowers 4 cm. long or more; leaves entire.................... 2. *P. louisiana*.

Flowers 3 cm. long or less; leaves shallowly lobed or angled.. 3. *P. parviflora*.

1. **Proboscidea altheaefolia** (Benth.) Decaisne, Ann. Sci. Nat. V. Bot. **3**: 324. 1865.

Martynia altheaefolia Benth. Bot. Voy. Sulph. 37. 1844.

TYPE LOCALITY: Bay of Magdalena, Lower California.

RANGE: Western Texas to Lower California.

NEW MEXICO: Las Cruces; Deming; Tortugas Mountain. Dry mesas, in the Lower Sonoran Zone.

2. Proboscidea louisiana (Mill.) Woot. & Standl.

Martynia louisiana Mill. Gard. Dict. ed. 8. no. 3. 1768.

Martynia proboscidea Glox. Obs. Bot. 14. 1785.

TYPE LOCALITY: Vera Cruz, Mexico.

RANGE: Iowa and Indiana to Mexico.

NEW MEXICO: South of Roswell; Lake Arthur; Albert; Buchanan. Plains, in the Upper Sonoran Zone.

3. Proboscidea parviflora (Wooton) Woot. & Standl.

Martynia parviflora Wooton, Bull. Torrey Club **25:** 453. 1898.

TYPE LOCALITY: San Augustine Ranch, at the base of the Organ Mountains, New Mexico. Type collected by Wooton (no. 580).

RANGE: Western Texas and southern New Mexico.

NEW MEXICO: Las Cruces; San Augustine Ranch; Crain Brothers Ranch; Carlisle; Gila River; Engle; Socorro. Dry mesas and low hills, in the Lower Sonoran Zone.

Order 46. PLANTAGINALES.

133. PLANTAGINACEAE. Plantain Family.

1. PLANTAGO L. PLANTAIN.

Annual or perennial acaulescent herbs, with usually numerous basal leaves; inflorescence spicate, on scapes; flowers perfect, monœcious, or diœcious, sessile, bracteate; calyx of 4 persistent, often scarious-margined sepals; corolla hypogynous, scarious or membranous, nerveless, usually persistent, tubular-salverform, with 4 erect or spreading lobes; stamens 4 or 2, adnate to the throat of the corolla; ovary superior, 1 or 2-celled or apparently 3 or 4-celled; fruit a circumscissile capsule or pyxis; seeds 1 to several in each cell.

KEY TO THE SPECIES.

Leaves linear.
  Pubescence loose and spreading; bracts much longer than the flowers .... 1. *P. purshii.*
  Pubescence sericeous, appressed; bracts shorter than the flowers .... 2. *P. argyrea.*
Leaves lanceolate to ovate.
  Spikes short, oblong; seeds concave on the faces .... 3. *P. lanceolata.*
  Spikes elongated, cylindric; seeds not concave on the faces.
    Leaves ovate, abruptly contracted at the base; seeds more than 2 in each cell .... 4. *P. major.*
    Leaves lanceolate, gradually tapering to the petiole; seeds not more than 2 in each cell.
      Plants with copious brown wool at the base .... 5. *P. eriopoda.*
      Plants not woolly at the base .... 6. *P. tweedyi.*

1. Plantago purshii Roem. & Schult. Syst. Veg. **3:** 120. 1818.

Plantago gnaphalioides Nutt. Gen. Pl. **1:** 100. 1818.

Plantago patagonica gnaphalioides A. Gray, Syn. Fl. **2**[1]: 391. 1878.

TYPE LOCALITY: "In dry situations on the banks of the Missouri."

RANGE: British America to Arizona, Texas, and Missouri.

NEW MEXICO: Carrizo Mountains; Farmington; Sierra Grande; Nara Visa; Mountainair; Pajarito Park; Clayton; Las Vegas; Springer; Santa Fe; Socorro Mountain; Cliff; Carrizalillo Mountains; Aden; Las Cruces; Organ Mountains. Dry plains and hills, in the Lower and Upper Sonoran zones.

2. Plantago argyrea Morris, Bull. Torrey Club **27:** 111. 1900.

TYPE LOCALITY: Castle Creek, Arizona.

RANGE: Arizona and New Mexico.

NEW MEXICO: Tunitcha Mountains; Middle Fork of the Gila; Ramah. Mountains and low hills, in the Upper Sonoran and Transition zones.

3. Plantago lanceolata L. Sp. Pl. 113. 1753. ENGLISH RIBGRASS.

TYPE LOCALITY: "Habitat in Europae campis sterilibus."

NEW MEXICO: Farmington; Mesilla Valley; Lake Valley.

A native of Europe, widely introduced into North America. It grows in New Mexico chiefly in alfalfa fields, where it spreads rapidly.

4. Plantago major L. Sp. Pl. 112. 1753. COMMON PLANTAIN.

TYPE LOCALITY: "Habitat in Europa ad vias."

RANGE: Nearly cosmopolitan; a common weed throughout the United States.

NEW MEXICO: Nearly throughout the State.

Our plants are very variable in their pubescence; in some it is appressed, in others spreading, and in some nearly wanting.

5. Plantago eriopoda Torr. Ann. Lyc. N. Y. **2:** 237. 1827.

Plantago retrorsa Greene, Pl. Baker. **3:** 32. 1901.

TYPE LOCALITY: "Depressed and moist situations along the Platte."

RANGE: British America to Nevada and northern New Mexico.

NEW MEXICO: North of Ramah (*Wooton*). Transition Zone.

6. Plantago tweedyi A. Gray, Syn. Fl. 2^1: 390. 1878.

TYPE LOCALITY: "N. W. Wyoming, on grassy slopes of the East Fork of the Yellowstone River."

RANGE: Montana to Utah and northern New Mexico.

NEW MEXICO: Pecos Baldy (*Standley* 4329). Meadows in the mountains, in the Arctic-Alpine Zone.

Order 47. RUBIALES.

KEY TO THE FAMILIES.

Stamens twice as many as the corolla lobes; low herbs with ternately dissected leaves......136. **ADOXACEAE** (p. 612).

Stamens as many as the corolla lobes; herbs or shrubs with simple or pinnate leaves.

- Leaves with stipules, these often leaflike, adnate to the stems between the leaf bases.................................134. **RUBIACEAE** (p. 603).
- Leaves without stipules, or, if present, these adnate to the petioles.................135. **CAPRIFOLIACEAE** (p. 608).

134. RUBIACEAE. Madder Family.

Low herbs or sometimes shrubs, annuals or usually perennials, with opposite or whorled simple leaves, and mostly small flowers in axillary or terminal cymes or panicles; flowers perfect or polygamous, regular or nearly so; hypanthium adnate to the ovary; sepals deciduous or persistent; corolla inserted near the top of the hypanthium, of 3 to 6, rarely 10, more or less united petals; stamens as many as the lobes of the corolla and alternate with them, adnate to the tube; ovary partly inferior, 2 to 5-celled, the styles united, the stigmas 2 to several; fruit a drupe, a capsule, or a berry; seeds 1 to several, sometimes flattened on one side.

KEY TO THE GENERA.

Carpels 1-seeded.
- Stipules foliaceous, resembling the leaves, the leaves thus appearing whorled; stems 4-sided; corolla rotate .... 1. GALIUM (p. 604).
- Stipules more or less connate-sheathing, lacerate, the leaves distinctly opposite; stems not conspicuously 4-sided; corolla salverform.......... 2. CRUSEA (p. 606).

Carpels several to many-seeded.
- Seeds winged; corolla long-tubular, bright red; leaves broadly lanceolate, 2 to 5 cm. long; ovules and seeds numerous.............................. 3. BOUVARDIA (p. 606).
- Seeds not winged; corolla short, salverform, never bright red; leaves narrower, mostly less than 2 cm. long; ovules and seeds several.
  - Top of the capsule extending beyond the hypanthium................................ 4. HOUSTONIA (p. 607).
  - Top of the capsule not exceeding the hypanthium................................ 5. OLDENLANDIA (p. 608).

1. GALIUM L. BEDSTRAW.

Low annual or perennial herbs with 4-angled stems and small whorled leaves, the stipules often as large as the leaves; flowers small, perfect, white, yellowish, or purple, in axillary or terminal cymes or panicles; sepals often obsolete; corolla rotate, with 3 or usually 4 lobes; ovary 2-celled, with one ovule in each cell; styles 2; stigmas capitate; fruit of 2 nearly separate, 1-seeded carpels, either glabrous or hispid, dry or fleshy.

KEY TO THE SPECIES.

Fruit covered with long straight hairs.
- Corolla purplish, the lobes abruptly long-acuminate.
  - Leaves linear or linear-oblong, nearly glabrous; stems puberulent.. 1. *G. rothrockii.*
  - Leaves oblanceolate or elliptic, strongly pubescent; stems hirtellous.................................. 2. *G. wrightii.*
- Corolla white or yellowish; petals obtuse or merely acute.
  - Leaves ovate to elliptic-lanceolate, thick, fleshy....... 3. *G. acutissimum.*
  - Leaves linear-oblong, thin............................... 4. *G. fendleri.*

Fruit glabrous or covered with hooked hairs.
- Fruit scaberulous to glabrous, never with hooked hairs.
  - Leaves acute; bracts persistent at the base of the fruit.. 5. *G. microphyllum.*
  - Leaves obtuse; fruit not bracteate...................... 6. *G. brandegei.*
- Fruit covered with hooked hairs.
  - Stems stout, erect; leaves 3-nerved.................... 7. *G. boreale.*
  - Stems slender, weak, usually reclining; leaves mostly 1-nerved.
    - Leaves not cuspidate-pointed; slender annual.... 8. *G. proliferum.*
    - Leaves cuspidate-pointed; annuals or perennials.
      - Pedicels scarcely exceeding the bracts; stout plants, usually annuals.
        - Leaves linear or oblanceolate.............. 9. *G. aparine.*
        - Leaves elliptic............................10. *G. flaviflorum.*
      - Pedicels much exceeding the bracts; perennials.
        - Leaves broadly elliptic; corolla yellowish; stems nearly smooth.................11. *G. triflorum.*
        - Leaves narrowly oblong; corolla purplish; stems very rough....................12. *G. asperrimum.*

1. **Galium rothrockii** A. Gray, Proc. Amer. Acad. **17**: 203. 1882.
TYPE LOCALITY: Southern Arizona.
RANGE: Southern Arizona and New Mexico and adjacent Mexico.
NEW MEXICO: Carrizo Mountains; Mangas Springs; Mogollon Mountains. Dry hills, in the Upper Sonoran and Transition zones.

2. **Galium wrightii** A. Gray, Pl. Wright. **1**: 80. 1852.
TYPE LOCALITY: "Crevices of rocks, on mountains, in the Pass of the Limpia," Texas.
RANGE: Western Texas to southern Arizona and adjacent Mexico.
NEW MEXICO: Mogollon Mountains; Hillsboro; Organ Mountains; Dona Ana Mountains; Gilmores Ranch. Dry hills, in the Upper Sonoran Zone.

3. **Galium acutissimum** A. Gray, Proc. Amer. Acad. **7**: 350. 1867.
TYPE LOCALITY: "Between the Rio del Norte and New Mexico." Type collected by Newberry in 1859.
RANGE: Utah to Arizona and New Mexico.
NEW MEXICO: Carrizo Mountains (*Standley* 7353). Dry hills, in the Upper Sonoran Zone.

4. **Galium fendleri** A. Gray, Mem. Amer. Acad. n. ser. **4**: 60. 1849.
TYPE LOCALITY: Sunny side of the high mountains, valley of Santa Fe Creek, New Mexico. Type collected by Fendler (no. 288).
RANGE: New Mexico and Arizona.
NEW MEXICO: Santa Fe and Las Vegas mountains; Mogollon Mountains; Mangas Springs; White Mountains; Capitan Mountains. Damp woods, in the Transition Zone.

5. **Galium microphyllum** A. Gray, Pl. Wright. **1**: 80. 1852.
Relbunium microphyllum Hemsl. Biol. Centr. Amer. Bot. **2**: 63. 1881.
TYPE LOCALITY: "Mountains at the Pass of the Limpia, in crevices of rocks, and in the valley of the Limpia," Texas.
RANGE: Western Texas to southern Arizona and southward.
NEW MEXICO: Graham; Fort Bayard; Guadalupe Canyon; Organ Mountains; Capitan Mountains; White Mountains; Queen. Dry hills and rocky canyons, in the Upper Sonoran Zone.

6. **Galium brandegei** A. Gray, Proc. Amer. Acad. **12**: 58. 1876.
TYPE LOCALITY: "Valley of the Rio Grande, New Mexico, on the Los Pinos trail." Type collected by T. S. Brandegee, September, 1875.
RANGE: Wyoming to New Mexico and California.
NEW MEXICO: Ponchuelo Creek (*Standley* 4194). Transition Zone.

7. **Galium boreale** L. Sp. Pl. 108. 1853. NORTHERN BEDSTRAW.
TYPE LOCALITY: "Habitat in Europae borealis pratis."
RANGE: British America to California, Texas, and Pennsylvania; also in Europe and Asia.
NEW MEXICO: Tunitcha Mountains; Chama; Sierra Grande; Santa Fe and Las Vegas mountains; Sandia Mountains; White and Sacramento mountains. Meadows and damp slopes, in the Transition Zone.

8. **Galium proliferum** A. Gray, Pl. Wright. **2**: 67. 1853.
Galium virgatum diffusum A. Gray, op. cit. **1**: 80. 1852.
TYPE LOCALITY: "High, rocky hills of the Pecos," Texas.
RANGE: Western Texas to southern Arizona and adjacent Mexico.
NEW MEXICO: Tortugas Mountain; Bishops Cap. Dry hills, in the Lower Sonoran Zone.

9. Galium aparine L. Sp. Pl. 108. 1753. GOOSEGRASS.

TYPE LOCALITY: "Habitat in Europae cultis & ruderatis."

RANGE: British America to California and Texas; also in Europe and Asia.

NEW MEXICO: Sierra Grande; Organ Mountains; Ruidoso Creek; Gilmores Ranch. Upper Sonoran and Transition zones.

10. Galium flaviflorum Heller; Rydb. Colo. Agr. Exp. Sta. Bull. **100:** 322. 1906.

TYPE LOCALITY: Santa Fe Canyon 9 miles east of Santa Fe, New Mexico. Type collected by Heller (no. 3823).

RANGE: Colorado and New Mexico.

NEW MEXICO: Santa Fe and Las Vegas mountains; Sandia Mountains. Moist thickets, in the Transition Zone.

11. Galium triflorum Michx. Fl. Bor. Amer. **1:** 80. 1803. SWEET-SCENTED BEDSTRAW.

TYPE LOCALITY: "In umbrosis Canadae sylvis."

RANGE: British America to California, Texas, and Alabama.

NEW MEXICO: West Fork of the Gila; Tunitcha Mountains; Chama; Las Huertas Canyon. Transition Zone.

12. Galium asperrimum A. Gray, Mem. Amer. Acad. n. ser. **4:** 60. 1849.

TYPE LOCALITY: Wet places near irrigating ditches, Santa Fe, New Mexico. Type collected by Fendler (no. 289).

RANGE: Mountains of New Mexico and Arizona.

NEW MEXICO: Santa Fe and Las Vegas mountains; Sandia Mountains; Rio Pueblo; San Luis Mountains; White and Sacramento mountains. Transition Zone.

2. CRUSEA Cham.

Low glabrous annual, 10 to 20 cm. high, with linear leaves and sheathing stipules; flowers small, white, in few-flowered clusters; calyx teeth 2 or 3, lanceolate, foliaceous, with 1 or 2 much smaller and partly scarious ones; corolla salverform; stamens 4, exserted; ovary 2-celled, the styles wholly or partly united, capillary; capsules didymous, the carpels separating, indehiscent, thin-walled; seeds mostly oblong.

1. Crusea subulata (Pavon) A. Gray, Proc. Amer. Acad. **19:** 78. 1883.

Spermacoce subulata Pavon; DC. Prodr. **4:** 543. 1830.

Borreria subulata DC. loc. cit.

TYPE LOCALITY: Mexico.

RANGE: Southern New Mexico and Arizona and southward.

NEW MEXICO: Mogollon Mountains; Kingston; White Mountains. Open slopes, in the Transition Zone.

3. BOUVARDIA Salisb. BOUVARDIA.

Low shrub, the upper part of the stems herbaceous, with ovate-lanceolate short-petiolate leaves 25 to 50 mm. long, mostly in whorls of 4, and with terminal cymes of conspicuous heterogone-dimorphous red flowers; corolla slender-tubular; hypanthium turbinate or campanulate; sepals 4, persistent; corolla lobes short, valvate in bud; styles slender, more or less exserted in some of the flowers; stigmas 2, obtuse; ovary 2-celled; capsule didymous-globose, coriaceous; seeds numerous, flat, winged, imbricated on the placenta.

1. Bouvardia ovata A. Gray, Pl. Wright. **2:** 67. 1853.

TYPE LOCALITY: "Mountain valleys from San Pedro to Santa Cruz, Sonora."

RANGE: Southern New Mexico and Arizona and southward.

NEW MEXICO: Animas Peak; Dog Mountains; San Luis Mountains. Dry hills.

4. HOUSTONIA L. Bluets.

Annual or perennial herbs, sometimes woody at the base or throughout, 30 cm. high or less; leaves opposite, entire, often ciliate, mostly narrow; flowers perfect, often dimorphous, solitary or cymose; hypanthium subglobose or obovoid; sepals 4; corolla white or blue or bright rose pink, narrowly funnelform or salverform; stamens 4; ovary 2-celled, the styles united to the narrow stigmas; capsules partly inferior, more or less distinctly 2-lobed; seeds few or several in each cell.

KEY TO THE SPECIES.

Annual.. 1. *H. humifusa.*
Perennials.
  Leaves fasciculate; stems woody throughout................ 2. *H. fasciculata.*
  Leaves opposite; plants woody only near the base, if at all.
    Corolla tube 2 cm. long; plants 6 cm. high or less...... 5. *H. rubra.*
    Corolla tube less than 1 cm. long; plants more than 6 cm. high.
      Pedicels reflexed after anthesis.................. 6. *H. wrightii.*
      Pedicels erect.
        Stems glabrous, herbaceous; leaves soft, not acerose-tipped........................ 4. *H. rigidiuscula.*
        Stems puberulent, woody below, leaves rigid, acerose-tipped........................ 3. *H. polypremoides.*

1. Houstonia humifusa A. Gray, Proc. Amer. Acad. **4:** 314. 1859.
Hedyotis humifusa A. Gray, Bost. Journ. Nat. Hist. **6:** 216. 1850.
Type locality: "Open gravelly banks of streamlets, near Fredericksburg," Texas.
Range: Western Texas to eastern New Mexico.
New Mexico: Twenty miles south of Roswell (*Earle* 278).

2. Houstonia fasciculata A. Gray, Proc. Amer. Acad. **17:** 203. 1882.
Type locality: "Southwestern borders of Texas, at Presidio."
Range: Western Texas and southern New Mexico; also in Mexico.
New Mexico: Organ Mountains (*Vasey*). Upper Sonoran Zone.

3. Houstonia polypremoides A. Gray, Proc. Amer. Acad. **21:** 379. 1886.
Type locality: Santa Eulalia Mountains, near Chihuahua, Mexico.
Range: Southern New Mexico and western Texas to Mexico.
New Mexico: Clayton; Bishops Cap; White Mountains; Queen; Gray; Buchanan; Lakewood; Torrance. Lower and Upper Sonoran zones.

4. Houstonia rigidiuscula (A. Gray) Woot. & Standl. Contr. U. S. Nat. Herb. **16:** 175. 1913.
Houstonia angustifolia rigidiuscula A. Gray, Syn. Fl. 2^1: 27. 1878.
Type locality: Texas.
Range: Western Texas and southern New Mexico to Mexico.
New Mexico: Tucumcari; Lincoln; Queen; James Canyon. Upper Sonoran Zone.

5. Houstonia rubra Cav. Icon. Pl. **5:** 48. *pl. 474.* 1799.
Hedyotis rubra A. Gray, Mem. Amer. Acad. n. ser. **4:** 61. 1849.
Oldenlandia rubra A. Gray, Pl. Wright. **2:** 68. 1853.
Type locality: "Habitat prope oppidum mexicanum Ixmiquilpan."
Range: New Mexico and Arizona to Mexico.
New Mexico: Puertecito; Zuni; Albuquerque; Santa Fe; Fort Wingate; Socorro Mountain; Apache Teju; Round Mountain; East View; Buchanan. Dry hills and plains, in the Upper Sonoran Zone.

6. Houstonia wrightii A. Gray, Proc. Amer. Acad. **17**: 202. 1882.

TYPE LOCALITY: On the Limpio, western Texas.

RANGE: Western Texas to southern Arizona and adjacent Mexico.

NEW MEXICO: Ramah; Coolidge; Kingston; Burro Mountains; Mogollon Mountains; Hanover Hills; Magdalena Mountains; Pinos Altos. Transition Zone.

5. OLDENLANDIA L.

Low erect annual with diffusely branched stems and narrow opposite leaves; flowers small, white, sessile in the forks of the branches and in the axils; calyx teeth triangular-subulate, about the length of the hypanthium; corolla 3 to 4 mm. long, salverform, the tube slightly surpassing the calyx lobes; capsules quadrangular-hemispheric, at first somewhat turbinate; seeds moderately angled.

1. Oldenlandia greenei A. Gray, Proc. Amer. Acad. **19**: 77. 1883.

TYPE LOCALITY: Pinos Altos Mountains, New Mexico. Type collected by E. L. Greene in 1880 (no. 149).

RANGE: Southern New Mexico and Arizona.

We have seen no further specimens of this from New Mexico.

135. CAPRIFOLIACEAE. Honeysuckle Family.

Shrubs, trees, woody vines, or rarely low herbs, with opposite, exstipulate, often perfoliate leaves and perfect flowers variously arranged in axillary pairs or in axillary or terminal cymes; hypanthium adnate to the 2 to 5-celled ovary; calyx of 4 or 5 sepals; corolla rotate, tubular, or funnelform, sometimes bilabiate, the lobes 4 or 5, imbricated; stamens 4 or 5, sometimes partly adnate to the corolla, alternate with its lobes; ovary 2 to 5-celled, inferior; fruit a berry or drupe.

KEY TO THE GENERA.

Corolla rotate; styles deeply 3 to 5-cleft, short; inflorescence compound-cymose; fruit drupaceous; erect shrubs; leaves pinnate .... 1. SAMBUCUS (p. 609.)

Corolla tubular, campanulate, funnelform, or salverform, sometimes 2-lipped, never rotate; styles slender, not divided; inflorescence simple, few-flowered; fruit dry or berry-like; erect or prostrate shrubs or undershrubs; leaves simple.

 Trailing evergreen plant; stamens 4, didynamous; flowers in pairs, pedicellate and on long peduncles .... 2. LINNAEA (p. 610).

 Erect or vinelike shrubs with woody stems; stamens mostly 5; flowers variously disposed.

  Corolla regular, tubular-funnelform; fruit 2-seeded .... 3. SYMPHORICARPOS (p. 610).

  Corolla more or less bilabiate, broadly funnelform, or nearly regular; fruit few to many-seeded.

   Upper leaves connate-perfoliate; flowers mostly in terminal clusters, pseudoverticillate or crowded; corolla not saccate at the base .... 4. LONICERA (p. 611).

   Upper leaves not connate; flowers axillary in pairs, sessile on the end of the common peduncle, subtended by bracts; corolla conspicuously saccate at the base.

Bracts and bractlets very small, inconspicuous, green, not accrescent...................... 5. XYLOSTEON (p. 611).
Bracts foliaceous and bractlets accrescent, reddish brown...... 6. DISTEGIA (p. 612).

1. SAMBUCUS L. ELDERBERRY.

Shrubs or trees with soft wood, large pith, and opposite, pinnately compound leaves with large leaflets, the small white or ochroleucous flowers in terminal compound cymes; hypanthium turbinate or ovoid; sepals 3 to 5, equal; corolla rotate, with 3 to 5 equal, imbricated, rarely valvate, lobes; stamens 5, adnate to the base of the corolla; anthers opening extrorsely by clefts; ovary 3 to 5-celled, becoming a 1-seeded drupe-like fruit.

KEY TO THE SPECIES.

Cymes not flat-topped, thyrsoid-paniculate, the axis continuous.
  Fruit red; cymes, in flower, seldom more than 4 cm. broad... 1. *S. microbotrys.*
  Fruit black; cymes larger, 6 cm. wide or more.............. 2. *S. melanocarpa.*
Cymes flat-topped, with several compound rays, the axis not continuous.
  Leaflets less than 6 cm. long, ovate to oblong, abruptly short-acuminate; a good-sized tree........................... 3. *S. mexicana.*
  Leaflets larger, 8 to 15 cm. long, lanceolate, long-attenuate; small trees or shrubs.
    Branches and leaflets pubescent; flowers less than 4 mm. broad; shrub with several nearly simple shoots from the root.................................... 4. *S. vestita.*
    Branches and leaves glabrous; flowers 5 or 6 mm. broad; small tree with well-defined trunk............... 5. *S. neomexicana.*

1. **Sambucus microbotrys** Rydb. Bull. Torrey Club **28:** 503. 1901.

TYPE LOCALITY: Bottomless Pit and below Halfway House, Pikes Peak, Colorado.

RANGE: Wyoming to Arizona and New Mexico.

NEW MEXICO: Tunitcha Mountains; Chama; Santa Fe and Las Vegas mountains; Eagle Peak; White Mountains. Mountains, Transition to Hudsonian Zone.

2. **Sambucus melanocarpa** A. Gray, Proc. Amer. Acad. **19:** 76. 1883.

TYPE LOCALITY: "New Mexico." Type collected by Fendler, probably east of Santa Fe.

RANGE: Oregon and Alberta to Colorado and New Mexico.

NEW MEXICO: Pecos Baldy; Santa Fe Creek. Mountains, in the Transition and Canadian zones.

3. **Sambucus mexicana** Presl in DC. Prodr. **4:** 322. 1830.

Sambucus canadensis mexicana Sarg. Silv. N. Amer. **5:** 88. 1893.

TYPE LOCALITY: Mexico.

RANGE: New Mexico to California and southward.

NEW MEXICO: Burro Mountains; Silver City; Mesilla Valley. River valleys, in the Lower Sonoran Zone.

This is often cultivated as a shade tree in southern New Mexico and where it receives plenty of water sometimes reaches a large size. The leaves remain green all winter and young ones are continually unfolding, while the flowers open almost any month of the year.

52576°—15——39

4. **Sambucus vestita** Woot. & Standl. Contr. U. S. Nat. Herb. **16:** 175. 1913.

TYPE LOCALITY: Ice Canyon above Van Pattens Camp, Organ Mountains, New Mexico. Type collected by Standley, June 11, 1906.

RANGE: Southern New Mexico and Arizona.

NEW MEXICO: Mogollon Mountains; San Mateo Peak; Black Range; Organ Mountains. Canyons and along streams, in the Transition Zone.

The plant is common in the canyons of the southwestern mountains. It is related to *S. neomexicana*, but has smaller flowers and pubescent instead of glabrous branches. In habit the two are dissimilar, for *S. neomexicana* has usually a well-developed trunk with branches, while *S. vestita* consists of a clump of mostly simple shoots.

5. **Sambucus neomexicana** Wooton, Bull. Torrey Club **25:** 309. 1898.

Sambucus intermedia neomexicana Schwerin, Mitt. Deutsch. Dendr. Ges. **1909:** 38. 1909.

Sambucus glauca neomexicana A. Nels. in Coulter, New Man. Rocky Mount. 469. 1909.

TYPE LOCALITY: Ruidoso Crossing in the White Mountains, New Mexico. Type collected by Wooton (no. 648).

RANGE: White and Sacramento mountains of New Mexico, in the Transition and Canadian zones.

2. LINNAEA Gron. TWIN FLOWER.

Slender creeping perennial herb with small rounded opposite few-toothed short-petiolate leaves and long slender peduncles forking into 2 slender pedicels, each bearing a nodding flower; bracts at the base of the calyx very glandular; calyx lobes lanceolate; corolla about 1 cm. long, funnelform, about equally 5-lobed, pink; stamens 4, didynamous; ovary and small dry pod 3-celled, the latter 1-seeded.

1. **Linnaea americana** Forbes, Hort. Woburn. 135. 1833.

Linnaea borealis longiflora Torr. in Wilkes, U. S. Expl. Exped. **15:** 327. 1874.

Linnaea longiflora Howell, Fl. Northw. Amer. 280. 1900.

TYPE LOCALITY: "America."

RANGE: British America to Oregon, New Mexico, and Maryland.

NEW MEXICO: Horsethief Canyon (*Standley* 4883). Damp mountain woods, in the Canadian and Hudsonian zones.

3. SYMPHORICARPOS L. SNOWBERRY.

Branching shrubs 1.5 meters high or less, with opposite simple short-petiolate leaves and small flowers in few-flowered axillary clusters; hypanthium cup-shaped to subglobose; sepals 4 or 5, unequal; corolla white or reddish tinged, sometimes campanulate, mostly tubular-funnelform, the 4 or 5 lobes almost equal; stamens 4 or 5, adnate to the corolla; ovary 4-celled; berry fleshy, 2-seeded, white.

KEY TO THE SPECIES.

Corolla 5 mm. long or less, open-campanulate.................... 1. *S. pauciflorus.*

Corolla 8 to 12 mm. long, salverform or narrowly funnelform.

  Corolla 6 to 8 mm. long; leaves thick, densely pubescent; seeds rounded at both ends........................... 2. *S. rotundifolius.*

  Corolla 8 to 12 mm. long; leaves thin, sparingly pubescent; seeds acute at one end................................ 3. *S. oreophilus.*

1. **Symphoricarpos pauciflorus** (Robbins) Britton, Mem. Torrey Club **5:** 305. 1894.

Symphoricarpos racemosus pauciflorus Robbins; A. Gray, Man. ed. 5. 203. 1867.

TYPE LOCALITY: "Rocky woods of L. Superior."

RANGE: Mountains from British America to California, Colorado, and Pennsylvania.

NEW MEXICO: White Mountains. Transition Zone.

2. **Symphoricarpos rotundifolius** A. Gray, Pl. Wright. **2**: 66. 1853.

TYPE LOCALITY: Sides of mountains around the Copper Mines, New Mexico. Type collected by Wright (no. 1388).

RANGE: Idaho and Wyoming to New Mexico.

NEW MEXICO: Santa Fe and Las Vegas mountains; Santa Antonita; Mount Sedgwick; Magdalena Mountains; Bear Mountains; Black Range; Animas Mountains; Organ Mountains; White Mountains. Mountains, in the Transition Zone.

3. **Symphoricarpos oreophilus** A. Gray, Journ. Linn. Soc. Bot. **14**: 12. 1875.

TYPE LOCALITY: "Rocky Mountains, Colorado Territory and New Mexico to the eastern side of the Sierra Nevada, California."

RANGE: Colorado and New Mexico to Utah and Arizona.

NEW MEXICO: Carrizo Mountains; Tunitcha Mountains; Chama; Santa Fe and Las Vegas mountains; Barranca; Zuni Mountains; Magdalena Mountains; Mogollon Creek; Burro Mountains; Lookout Mines; Animas Peak; Organ Mountains. Mountains, in the Transition Zone.

4. LONICERA L. HONEYSUCKLE.

Woody vines with trailing, rather stiff stems and shredded bark; leaves opposite, entire, short-petiolate or the upper connate-perfoliate; flowers mostly sessile and whorled at the ends of the stems; hypanthium subglobose or ovoid; sepals 5; corolla tubular-funnelform or broader, more or less 2-lipped; stamens 5, adnate to the corolla tube; ovary 2 or 3-celled, the ovules numerous; fruit a fleshy few-seeded berry.

KEY TO THE SPECIES.

Limb of corolla nearly regular; inflorescence pedunculate; leaves conspicuously ciliate.. 1. *L. arizonica.*
Limb of corolla deeply bilabiate; inflorescence sessile; leaves not ciliate.. 2. *L. dumosa.*

1. **Lonicera arizonica** Rehder, Trees and Shrubs **1**: 45. *pl. 23.* 1902.

TYPE LOCALITY: No type is cited, but the first specimen listed is one collected in the Rincon Mountains of Arizona by Pringle.

RANGE: Utah to Arizona and New Mexico.

NEW MEXICO: Carrizo Mountains; Zuni Mountains; San Mateo Peak; Magdalena Mountains; Lookout Mines; San Luis Mountains. Mountains, in the Transition Zone.

2. **Lonicera dumosa** A. Gray, Pl. Wright. **2**: 66. 1853.

TYPE LOCALITY: "Banks of a torrent between Rock Creek and the Limpio," Texas.

RANGE: Western Texas to southern Arizona.

NEW MEXICO: West Fork of the Gila; Kingston; Animas Mountains; Capitan Mountains; San Andreas Mountains; Organ Mountains; Craters; Gilmores Ranch; Queen. Mountains, in the Transition Zone.

5. XYLOSTEON B. Juss. FLY HONEYSUCKLE.

Erect branching shrub with opposite simple leaves, these entire, sessile or short-petiolate, not connate above; flowers sessile in pairs on the ends of solitary axillary peduncles, subtended by 2 minute bracts and bractlets; calyx minute or obsolete; corolla broadly funnelform, 1 cm. long or more, saccate at the base, the limb 5-lobed, the lobes nearly equal; ovary usually 2-celled, the red berries distinct or didymous.

1. **Xylosteon utahense** (S. Wats.) Howell, Fl. Northw. Amer. 282. 1900.

Lonicera utahensis S. Wats. in King, Geol. Expl. 40th Par. **5**: 133. 1871.

TYPE LOCALITY: Cottonwood Canyon, Wasatch Mountains, Utah.

RANGE: British Columbia and Montana to Utah and New Mexico.

New Mexico: Mogollon Road (*Wooton*). Mountains, in the Transition Zone.

Our specimens are very close to this species, but appear somewhat different. They probably represent an undescribed species, but our material is not sufficient for thorough study.

6. DISTEGIA Raf.

Stout shrub similar to the last, but the flowers sometimes in 3's and not coherent, the bracts leaflike, and the bractlets strongly accrescent and purplish, surrounding the large black berries.

1. **Distegia involucrata** (Richards.) Raf. New Fl. N. Amer. 3: 21. 1836.

Xylosteon involucratum Richards. Bot. App. Frankl. Journ. 733. 1823.

Lonicera involucrata Banks; Richards. loc. cit.

Type locality: "Wooded country from 54° to 64° north," British America.

Range: British America to California and New Mexico.

New Mexico: Tunitcha Mountains; Chama; Copper Canyon; Santa Fe and Las Vegas mountains. Damp deep woods, in the Transition Zone.

136. ADOXACEAE. Moschatel Family.

1. ADOXA L. Musk-root. Moschatel.

Low glabrous herb with scaly or bulbiferous rootstocks, basal and opposite, ternately compound leaves, and small green flowers in terminal capitate clusters; hypanthium hemispheric, adnate to the ovary; sepals 2 or 3; corolla rotate, regular, 4 to 6-lobed; stamens twice as many as the corolla lobes, inserted in pairs on the tube; anthers peltate, 1-celled; ovary 3 to 5-celled; style 3 to 5-parted; fruit a small drupe with 3 to 5 nutlets.

1. **Adoxa moschatellina** L. Sp. Pl. 367. 1753.

Type locality: "Habitat in Europae nemoribus."

Range: Arctic America to Wisconsin and northern New Mexico; also in Europe.

New Mexico: Pecos Baldy (*Standley* 4330). Cold woods and high meadows, in the Arctic-Alpine Zone.

Order 48. CAMPANULALES.

KEY TO THE FAMILIES.

Endosperm wanting; flowers monœcious or diœcious; vines with tendrils.................137. **CUCURBITACEAE** (p. 612).

Endosperm present; flowers perfect; plants not vines.

Corolla regular..............................138. **CAMPANULACEAE** (p. 616).

Corolla split on one side, more or less irregular..................................139. **LOBELIACEAE** (p. 617).

137. CUCURBITACEAE. Gourd Family.

Annual or perennial herbaceous vines, mostly tendril-bearing, prostrate or climbing, some of them from enlarged tuberous roots; leaves alternate, simple, palmately veined or lobed, without stipules; flowers monœcious or diœcious; calyx of 4 or 5 united sepals, adherent to the ovary in the pistillate flower; petals as many as the sepals, united and adherent to them; stamens usually 3, 2 of them with 2-celled anthers, the third with a 1-celled anther; filaments distinct or variously united; staminodia sometimes present in pistillate flowers; ovary 1 to 3-celled, with one, several, or many ovules; fruit a berry, pepo, or thin-walled, more or less inflated, dry fruit, dehiscent or indehiscent; seeds usually rather large, flattened, numerous or sometimes solitary.

KEY TO THE GENERA.

Ovary 2 or 3-celled; ovules solitary or few in the cell.
Plants perennial; fruit ovoid or globose, with 1 to 4 seeds in each cell; seeds turgid, rounded at both ends, smooth.......................... 1. MARAH (p. 613).
Plants annual; fruit oblong, attenuate at each end; with 2 to 6 seeds in each cell; seeds small, flattened, rugose.......................... 2. ECHINOPEPON (p. 613).
Ovary 1-celled; ovules solitary or numerous.
Ovules solitary, pendulous; fruit dry.............. 3. SICYOS (p. 614).
Ovules numerous, borne on 3 to 5 placentæ, mostly horizontal; fruit various.
Anthers straight or merely curved; slender or coarse vines.
Slender climbing vine; flowers diœcious; fruit a berry...................... 4. IBERVILLEA (p. 614).
Coarse prostrate vine; flowers monœcious; fruit a leathery pepo................ 5. APODANTHERA (p. 615).
Anthers much contorted; coarse prostrate vines.
Anthers cohering in a head; plants ill-scented.......................... 6. CUCURBITA (p. 615).
Anthers distinct or but slightly cohering; plants not ill-scented.
Tendrils branched; leaves deeply lobed; connective not produced beyond the anthers.................. 7. CITRULLUS (p. 615).
Tendrils not branched; leaves not deeply lobed; connective produced beyond the anthers.............. 8. CUCUMIS (p. 616).

1. MARAH Kellogg.

Herbaceous vine climbing by tendrils, with thin lobed leaves and small monœcious flowers; staminate flowers in racemes or panicles; hypanthium broadly campanulate; sepals 5 or 6; corolla white or greenish, rotate, 5 or 6-lobed; stamens 2 or 3, the filaments united, the anthers nearly horizontal; pistillate flowers usually solitary, sometimes clustered in the axils, their calyx and corolla similar to those of the staminate flowers; staminodia more or less prominent; ovary echinate, 1 to 4-celled; ovules 1 to 4 in each cell; fruit echinate, fibrous within.

1. **Marah gilensis** Greene, Leaflets **2**: 36. 1910.
Megarrhiza gilensis Greene, Bull. Torrey Club **8**: 97. 1881.
Echinocystis gilensis Greene, Bull. Calif. Acad. **1**: 189. 1885.
Micrampelis gilensis Britton, Trans. N. Y. Acad. **8**: 67. 1889.
TYPE LOCALITY: In deep sand on the banks of the upper Gila River and its tributaries, New Mexico. Type collected by E. L. Greene.
RANGE: Southwestern New Mexico and adjacent Arizona.
NEW MEXICO: Burro Mountains (*Rusby* 141).

2. ECHINOPEPON Naud.

Annual herbs with cordate, entire or parted leaves; flowers monœcious, the staminate in long racemes, with 5-lobed limbs; pistillate flowers solitary; ovary ovoid, beaked, hispid or echinate, usually 3-celled.

KEY TO THE SPECIES.

Racemes 10 to 20 cm. long; staminate flowers large, not punctate-glandular .... 1. *E. confusus.*
Racemes shorter, mostly less than 10 cm. long; staminate flowers smaller, conspicuously punctate-glandular .... 2. *E. wrightii.*

1. Echinopepon confusus Rose, Contr. U. S. Nat. Herb. **5:** 115. 1897.

TYPE LOCALITY: Pinos Altos Mountains, New Mexico. Type collected by Greene.

RANGE: Mountains of southwestern New Mexico.

NEW MEXICO: Copper Mines.

2. Echinopepon wrightii (A. Gray) S. Wats. Bull. Torrey Club **13:** 158. 1887.

Elaterium wrightii A. Gray, Pl. Wright. **2:** 61. 1853.

Echinocystis wrightii Cogn. Mém. Acad. Sci. Belg. **28:** 88. 1878.

TYPE LOCALITY: Mountains near Guadalupe Pass, New Mexico. Type collected by Wright (no. 1090).

RANGE: Southern New Mexico and Arizona and adjacent Mexico.

NEW MEXICO: Guadalupe Pass.

3. SICYOS L. ONE-SEEDED BUR CUCUMBER.

Slender climbing vines with lobed leaves and branched tendrils; flowers monœcious, the staminate in racemes or corymbs, the hypanthium broadly campanulate or nearly flat, the corolla whitish or pale yellow, rotate, 5-lobed; stamens with their filaments united into a column, the anthers 2 to 5, distinct or united; pistillate flowers usually clustered at the end of a peduncle arising from the same node as the longer staminate peduncle; ovary 1-celled, bristly, glandular, or glabrous; ovule solitary, pendulous; fruit not inflated, thin-walled, indehiscent.

KEY TO THE SPECIES.

Fruit glabrous .... 1. *S. glaber.*
Fruit hispid.
Lobes of the leaves triangular, attenuate, the basal sinus usually broad and open .... 2. *S. parviflorus.*
Lobes of the leaves rounded, obtuse, the sinus usually closed. 3. *S. ampelophyllus.*

1. Sicyos glaber Wooton, Bull. Torrey Club **25:** 310. 1898.

TYPE LOCALITY: Organ Mountains, south of San Augustine Ranch, New Mexico. Type collected by Wooton (no. 606).

RANGE: Organ Mountains of New Mexico, in the Upper Sonoran Zone.

2. Sicyos parviflorus Willd. Sp. Pl. **4:** 626. 1805.

TYPE LOCALITY: Mexico.

RANGE: Western Texas and southern Arizona to Mexico.

NEW MEXICO: Fort Bayard; Teel; White Mountains; Gray. Canyons, in the Upper Sonoran and Transition zones

3. Sicyos ampelophyllus Woot. & Standl. Bull. Torrey Club **36:** 111. 1909.

TYPE LOCALITY: Kingston, Sierra County, New Mexico. Type collected by Metcalfe (no. 1195).

RANGE: Southwestern New Mexico and adjacent Arizona.

NEW MEXICO: Kingston; Fort Bayard; Burro Mountains; Sapello Creek; Gila; Santa Rita.

4. IBERVILLEA Greene.

Climbing herbaceous vines from thickened roots; leaves deeply 3 to 5-lobed; flowers small, diœcious, the staminate ones in racemes, the pistillate ones solitary in the axils; hypanthium cylindric or cylindric-campanulate; corolla salverform, yellow;

stamens 3, the connective not produced beyond the anthers; ovary 1-celled, with 2 or 3 placentæ, the stigma 3-lobed; berry globose, red, the seeds somewhat swollen.

1. **Ibervillea tenuisecta** (A. Gray) Small, Fl. Southeast. U. S. 1136. 1903.
Sicydium lindheimeri tenuisecta A. Gray, Pl. Wright. **1**: 75. 1852.
TYPE LOCALITY: "Dry sandy soil, near the Rio Grande, Texas, and New Mexico."
RANGE: Western Texas to New Mexico and Chihuahua.
NEW MEXICO: Dog Spring; above Rincon; Tortugas Mountain; Organ Mountains; Guadalupe Mountains. Sandy mesas and low hills, in the Lower Sonoran Zone.

5. APODANTHERA Arn. MELÓN LOCO.

Rough prostrate coarse vines having a very disagreeable odor, from thick perennial roots; leaves round-reniform, entire or lobed; flowers large, yellow, monœcious, the staminate racemose or corymbose from the lower axils, the pistillate solitary in the upper axils; calyx tube subcylindric; anthers distinct, sessile, dorsally fixed; ovary 1-celled, with 3 placentæ; seeds horizontal, numerous; fruit 7 to 10 cm. in diameter, nearly spherical, ridged, with a tough or somewhat woody rind.

1. **Apodanthera undulata** A. Gray, Pl. Wright. **2**: 60. 1853.
TYPE LOCALITY: "In valleys from Eagle Springs to the Limpio," Texas.
RANGE: Western Texas to southern Arizona and southward.
NEW MEXICO: Near White Water; mesa west of Organ Mountains; Mangas Valley. Dry sandy mesas, in the Lower Sonoran Zone.

6. CUCURBITA L. GOURD.

Coarse, rough, usually ill-scented, prostrate, tendril-bearing, herbaceous vines from thickened roots; leaves large, nearly entire or lobed; flowers large, showy, yellow, monœcious, solitary in the axils; hypanthium of staminate flowers campanulate or rarely tubular, that of the pistillate flowers subglobose; stamens 3, the filaments distinct, the anthers linear, coherent, contorted; staminodia in pistillate flowers 3; ovary 1-celled, with 3 to 5 placentæ; fruit a pepo, usually large (in ours 8 cm. in diameter or less), woody.

KEY TO THE SPECIES.

Leaves deltoid-ovate, entire or angled.......................... 1. *C. foetidissima.*
Leaves 5-lobed to the base.. 2. *C. digitata.*

1. **Cucurbita foetidissima** H. B. K. Nov. Gen. & Sp. **2**: 123. 1817.
Cucurbita perennis A. Gray, Bost. Journ. Nat. Hist. **6**: 193. 1850.
TYPE LOCALITY: "Prope Guanaxuato Mexicanorum, altit. 1080 hexap."
RANGE: Nebraska and Missouri to California and Texas and southward.
NEW MEXICO: Black Range; Socorro; Dog Spring; Albert; Nara Visa; mesa west of Organ Mountains; Eagle Creek. Plains, in the Lower and Upper Sonoran zones.

2. **Cucurbita digitata** A. Gray, Pl. Wright. **2**: 60. 1853.
TYPE LOCALITY: Between the Copper Mines and Condes Camp, New Mexico. Type collected by Wright (no. 1088).
RANGE: Southwestern New Mexico and adjacent Arizona.
NEW MEXICO: Lordsburg; mesa south of Gila; near White Water. Dry sandy plains, in the Lower Sonoran Zone.

7. CITRULLUS Forsk.

1. **Citrullus citrullus** (L.) Small, Bull. Torrey Club **25**: 606. 1898. WATERMELON.
Cucurbita citrullus L. Sp. Pl. 1010. 1753.
Citrullus vulgaris Schrad.; Eckl. & Zeyh. Enum. Pl. Afr. Austr. 279. 1834–7.

TYPE LOCALITY: Given as "Apulia, Cabria, Siculia," but the original home of the plant is probably Africa.

NEW MEXICO: Mesilla Valley (*Wooton & Standley*).

This is a common escape in the Rio Grande Valley, especially along ditches.

8. CUCUMIS L.

1. **Cucumis melo** L. Sp. Pl. 1011. 1753. CANTELOUPE.

TYPE LOCALITY: Not stated.

NEW MEXICO: Mesilla Valley (*Wooton & Standley* 3225).

This, like the watermelon, is often found growing in the waste ground of the Rio Grande Valley and elsewhere.

138. CAMPANULACEAE. Bluebell Family.

Herbs with alternate exstipulate leaves; flowers racemose or solitary, blue, usually showy; calyx tube adnate to the ovary; corolla gamopetalous, campanulate or rotate; stamens commonly 5, distinct, the filaments broad and membranaceous at the base; stigmas 3; ovary 3-celled; capsule short, opening by 3 valves or pores.

KEY TO THE GENERA.

Corolla campanulate; flowers terminal or axillary, long-pediceled; perennials.................................. 1. CAMPANULA (p. 616).
Corolla rotate; flowers axillary, sessile; annuals............. 2. SPECULARIA (p. 616).

1. CAMPANULA L. BLUEBELL.

Perennial herbs with narrow leaves and terminal or axillary, pedicellate, showy, blue flowers; corolla campanulate, 5-lobed; capsules short, opening on the sides by valves or pores.

KEY TO THE SPECIES.

Capsules erect; flowers solitary; sepals nearly equaling the corolla... 1. *C. parryi*.
Capsules nodding; flowers usually several; sepals about half as long as the corolla.. 2. *C. petiolata*.

1. **Campanula parryi** A. Gray, Syn. Fl. ed. 2. **2**[1]: 395. 1886.

TYPE LOCALITY: Rocky Mountains of Colorado.

RANGE: Wyoming to New Mexico and Arizona.

NEW MEXICO: Winsors Ranch; Chusca Canyon; Zuni; Abiquiu Peak; Rio Pueblo; Embudo. Meadows, in the Transition and Canadian zones.

2. **Campanula petiolata** A. DC. Monogr. Campan. 278. 1830.

TYPE LOCALITY: "Habitat in America boreali prope lacum dictum *Slave Lake*."

RANGE: British America to Utah and New Mexico.

NEW MEXICO: Sandia Mountains; Santa Fe and Las Vegas mountains; Tunitcha Mountains; Jemez Mountains; Magdalena Mountains; Mogollon Mountains; Black Range; White and Sacramento mountains; Capitan Mountains. Meadows, from the Transition to the Arctic-Alpine Zone.

2. SPECULARIA Heist. VENUS'S LOOKING-GLASS.

Annual with mostly simple angled pubescent stems and broadly ovate crenate sessile leaves; corolla rotate, blue or violet, the flowers sessile in the axils of the leaves; capsules oblong or turbinate, opening about the middle.

1. **Specularia perfoliata** (L.) A. DC. Monogr. Campan. 351. 1830.

Campanula perfoliata L. Sp. Pl. 169. 1753.

Legouzia perfoliata Britton, Mem. Torrey Club **5**: 309. 1894.

TYPE LOCALITY: "Habitat in Virginia."
RANGE: British America to Mexico and Florida.
NEW MEXICO: Mountains west of Grants Station; Sandia Mountains; Middle Fork of the Gila; Organ Mountains. Dry slopes, in the Upper Sonoran Zone.

139. LOBELIACEAE. Lobelia Family.

1. LOBELIA L. LOBELIA.

Herbs with alternate, exstipulate, linear to spatulate, toothed leaves and red or blue flowers in terminal racemes; hypanthium campanulate to subglobose, ribbed, adnate to the ovary; sepals 5, linear-lanceolate; corolla conspicuously bilabiate, the upper lip 2-lobed, the lower 3-lobed; stamens 5, the filaments united above; ovary 2 to 5-celled, with numerous ovules; fruit a many-seeded capsule.

KEY TO THE SPECIES.

Corolla red; cauline leaves denticulate.............................. 1. *L. splendens.*
Corolla blue; cauline leaves usually entire.......................... 2. *L. gruina.*

1. **Lobelia splendens** Willd. Hort. Berol. *pl. 86.* 1816. CARDINAL FLOWER.
TYPE LOCALITY: Mexico.
RANGE: Texas to California and southward.
NEW MEXICO: Pajarito Park; headwaters of the Pecos; West Fork of the Gila; Lower Plaza; Kingston; White Mountains; Roswell; Zuni. Wet ground, in the Upper Sonoran and Transition zones.

2. **Lobelia gruina** Cav. Icon. Pl. **6:** 8. *pl. 511.* 1801.
TYPE LOCALITY: Mexico.
RANGE: Arizona and western New Mexico to central Mexico.
NEW MEXICO: Mogollon Mountains. Transition Zone.

Order 49. VALERIANALES.

140. VALERIANACEAE. Valerian Family.

Perennial herbs, sometimes 1.5 meters tall, from thickened cormlike roots, with opposite leaves, and small flowers in terminal panicles; leaves more or less pinnately divided, some of the basal ones entire; flowers perfect, monœcious, or diœcious, small; calyx of 3 to 5 sepals or pappus-like or obsolete; corolla tube narrowly funnelform or salverform; stamens 1 to 4, adnate to the corolla tube; ovary inferior, 3-celled, 2 of the cells abortive; fruit a 1-seeded nutlet crowned with the calyx or naked.
With us a single genus with the characters of the family.

1. VALERIANA L. VALERIAN.

KEY TO THE SPECIES.

Leaves thick, entire or with long linear divisions; venation almost parallel; tall plant, about 1 meter high..................... 1. *V. trachycarpa.*
Leaves thin, the cauline ones pinnate, the lobes not linear; venation distinctly pinnate; plants lower, seldom more than 30 cm. high.
Basal leaves ovate-cordate.................................... 2. *V. ovata.*
Basal leaves spatulate or lanceolate, tapering at the base..... 3. *V. acutiloba.*

1. **Valeriana trachycarpa** Rydb. Bull. Torrey Club **31:** 645. 1904.
TYPE LOCALITY: Red Mountain, Colorado.
RANGE: Colorado and New Mexico.

New Mexico: Tunitcha Mountains; Chama; Sierra Grande; Santa Fe and Las Vegas mountains; West Fork of the Gila; Hillsboro Peak; Luna; White and Sacramento mountains. Meadows and thickets in the mountains, Transition to Hudsonian Zone.

2. Valeriana ovata Rydb. Bull. Torrey Club **31:** 645. 1904.

Valeriana acutiloba ovata A. Nels. in Coulter, New Man. Rocky Mount. 476. 1909.

Type locality: "Cameron's Cove," Colorado.

Range: Mountains of Colorado and New Mexico.

New Mexico: San Mateo Mountains; Santa Fe and Las Vegas mountains; Sandia Mountains; Magdalena Mountains; Holts Ranch; Kingston; Organ Peak; White Mountains. Transition and Canadian zones.

3. Valeriana acutiloba Rydb. Bull. Torrey Club **28:** 24. 1901.

Type locality: Near Gray-Back Mining Camp, Sangre de Cristo Range, Colorado.

Range: Wyoming to Arizona and New Mexico.

New Mexico: Santa Fe and Las Vegas mountains. Damp woods, in the Transition and Canadian zones.

Order 50. ASTERALES.

KEY TO THE FAMILIES.

Flowers all with tubular corollas or none, or only the imperfect ray flowers with ligulate corollas.
- Stamens distinct; flowers unisexual........... 143. **AMBROSIACEAE** (p. 631).
- Stamens united by the anthers or if distinct the flowers perfect........................... 144. **ASTERACEAE** (p. 637).

Corollas of all or only of the perfect flowers bilabiate or ligulate.
- Corollas all ligulate; herbage usually with milky juice; style branches filiform........... 141. **CICHORIACEAE** (p. 618).
- Corollas of all or only the perfect flowers bilabiate; herbage without milky juice; style branches short, not filiform........ 142. **MUTISIACEAE** (p. 630).

141. CICHORIACEAE. Chicory Family.

Herbs, mostly with milky bitter juice; leaves alternate; heads homogamous and ligulate, the flowers all perfect and with ligulate corollas; ligules 5-toothed at the apex; anthers auriculate at the base, not caudate; style branches filiform, minutely papillose, not appendaged.

KEY TO THE GENERA.

Pappus of plumose bristles, often paleaceous at the base.
- Achenes truncate at the apex, not beaked; flowers pink.. 1. Ptiloria (p. 620).
- Achenes beaked; flowers white to purple.
  - Flowers white or pinkish; leaves runcinate; involucres with a few calyculate outer bracts.................................... 2. Nemoseris (p. 621).
  - Flowers purple; leaves entire; involucres without outer calyculate bracts................ 3. Tragopogon (p. 621).

Pappus not plumose.
  Pappus, at least in part, of scales, or these reduced and united into a crown.
    Involucre simple, that is, with no short calyculate outer bracts; outer pappus of minute scales scarcely visible except under a strong lens; flowers yellow................ 4. CYNTHIA (p. 621).
    Involucre with few or many calyculate outer bracts; outer series of pappus conspicuous; flowers yellow or blue.
      Flowers blue; plants caulescent............ 5. CICHORIUM (p. 622).
      Flowers yellow; plants acaulescent......... 6. UROPAPPUS (p. 622).
  Pappus of capillary bristles.
    Achenes flattened.
      Achenes narrowed at the top or beaked; pappus bristles falling separately; involucres cylindric...................... 7. LACTUCA (p. 622).
      Achenes truncate; pappus bristles not falling separately; involucres campanulate.. 8. SONCHUS (p. 623).
    Achenes not flattened.
      Pappus bristles promptly deciduous, usually together............................ 9. MALACOTHRIX (p. 624).
      Pappus persistent or tardily deciduous.
        Achenes with distinct slender beaks.
          Plant caulescent; pappus tawny....10. SITILIAS (p. 624).
          Plants scapose; pappus white or nearly so.
            Achenes 10-ribbed or nerved; not spinose-muricate; involucres more or less imbricated................11. AGOSERIS (p. 624).
            Achenes 4 or 5-ribbed; muricate-spinulose, at least near the apex; involucre of a single series of equal inner bracts with some calyculate outer ones......12. TARAXACUM (p. 626).
        Achenes not beaked.
          Flowers rose-colored, never yellow.
            Receptacle bearing capillary bristles.................13. CALYCOSERIS (p. 627).
            Receptacle naked.
              Pappus scabrous; heads 15 to 20-flowered; cauline leaves not scale-like. One species of..................14. HIERACIUM (p. 627).
              Pappus not scabrous; heads 3 to 12-flowered; cauline leaves scalelike or none............15. LYGODESMIA (p. 628).

Flowers yellow or white.
Achenes tapering upward; pappus white; bracts in fruit more or less thickened at the base or on the midrib....................16. CREPIS (p. 629).
Achenes not tapering upward; pappus usually sordid or reddish, rarely white; bracts not thickened.....14. HIERACIUM (p. 627).

1. PTILORIA Raf.

Nearly or quite glabrous annuals or perennials with branched stems, a tuft of usually pinnatifid basal and small, often scalelike cauline leaves; heads small, 3 to 20-flowered, paniculate or corymbose, with pink flowers; involucre cylindric, of several narrow appressed bracts and a few short calyculate outer ones; achenes 5-angled, often with intermediate ribs; pappus a single series of soft plumose bristles.

KEY TO THE SPECIES.

Annuals; pappus dilated at the base.
Plants stout, simple at the base; stems very glaucous or nearly white .... 1. *P. bigelovii.*
Plants slender, much branched from the base; stems green... 2. *P. exigua.*
Perennials; pappus not dilated at the base.
Heads large, 10 to 20-flowered, 11 to 12 mm. high........... 3. *P. thurberi.*
Heads smaller, 5 to 8-flowered, 9 mm. high or less.
Plants low, seldom more than 20 cm. high, diffusely and densely branched.............................. 4. *P. neomexicana.*
Plants taller, usually more than 30 cm., sparingly and not diffusely branched.
Pappus white, plumose to the base.................. 5. *P. ramosa.*
Pappus tawny, only scabrous at the base............ 6. *P. pauciflora.*

1. Ptiloria bigelovii (A. Gray) Woot. & Standl. Contr. U. S. Nat. Herb. **16:** 176. 1913.

Hemiptilium bigelovii A. Gray in Torr. U. S. & Mex. Bound. Bot. 105. 1859.

TYPE LOCALITY: Frontera, western Texas.

RANGE: Western Texas to Arizona.

NEW MEXICO: Farmington; mesa west of Organ Mountains. Dry hills and mesas, in the Lower and Upper Sonoran zones.

2. Ptiloria exigua (Nutt.) Greene, Pittonia **2:** 132. 1890.

Stephanomeria exigua Nutt. Trans. Amer. Phil. Soc. n. ser. **7:** 428. 1841.

TYPE LOCALITY: "On the Rocky Mountain plains, toward the Colorado."

RANGE: California and Nevada to Wyoming and New Mexico.

NEW MEXICO: Carrizo Mountains; Mesilla Valley. Dry hills and plains, in the Lower and Upper Sonoran zones.

3. Ptiloria thurberi (A. Gray) Greene, Pittonia **2:** 133. 1890.

Stephanomeria thurberi A. Gray, Mem. Amer. Acad. n. ser. **5:** 325. 1854.

TYPE LOCALITY: Sierra de las Animas, Sonora or New Mexico.

RANGE: Southwestern New Mexico to southern Arizona and southward.

NEW MEXICO: Mangas Springs; Kingston. Lower and Upper Sonoran zones.

4. Ptiloria neomexicana Greene, Bull. Torrey Club **25:** 123. 1898.

TYPE LOCALITY: Mesas near Las Cruces, New Mexico. Type collected by Wooton (no. 482).

RANGE: Western Texas and southern New Mexico.

NEW MEXICO: Frisco; near White Water; mesa west of Organ Mountains; Gray; Parkers Well. Sandy mesas, in the Lower Sonoran Zone.

5. Ptiloria ramosa Rydb. Mem. N. Y. Bot. Gard. **1:** 453. 1900.

TYPE LOCALITY: Scotts Bluff, Nebraska.

RANGE: Montana and Nebraska to Colorado and New Mexico.

NEW MEXICO: Carrizo Mountains; Pecos; Sierra Grande; Las Vegas Canyon; Sandia Mountains; Silver City; Capitan Mountains. Dry plains and hills, in the Upper Sonoran Zone.

6. Ptiloria pauciflora (Torr.) Raf. Atl. Journ. 145. 1832.

Prenanthes ? pauciflora Torr. Ann. Lyc. N. Y. **2:** 210. 1828.

TYPE LOCALITY: "Near the Rocky Mountains."

RANGE: Nevada and Arizona to Colorado and Texas.

NEW MEXICO: Cedar Hill; Chiz; Rosa; Kingston; Dog Spring; Dona Ana Mountains; Organ Mountains. Plains and hills, in the Lower and Upper Sonoran zones.

2. NEMOSERIS Greene.

Glabrous succulent winter annual with pinnatifid alternate leaves and large heads of white or pinkish flowers; involucre cylindric, of 7 to 15 narrow equal bracts and numerous short calyculate outer ones; achenes terete, fusiform, few-ribbed, attenuate to a slender beak; pappus of 10 to 15 slender long-plumose white bristles.

1. Nemoseris neomexicana (A. Gray) Greene, Pittonia **2:** 193. 1891.

Rafinesquia neomexicana A. Gray, Pl. Wright. **2:** 103. 1853.

TYPE LOCALITY: "Stony hills along the Rio Grande near El Paso," Texas or Chihuahua.

RANGE: Western Texas to Utah and southern California.

NEW MEXICO: Mangas Springs; Glorieta; Nutt Flats; mesa near Las Cruces; Organ Mountains. Dry mesas and low hills, in the Lower Sonoran Zone.

3. TRAGOPOGON L. SALSIFY.

Tall glabrous biennial or perennial with fleshy tap-root, alternate entire linear-lanceolate long-acuminate leaves, and large long-pedunculate heads of purplish flowers; involucre narrowly campanulate, the few bracts in a single series; achenes linear, terete, slender-beaked; pappus a single series of plumose bristles connate at the base.

1. Tragopogon porrifolius L. Sp. Pl. 789. 1753.

TYPE LOCALITY: Not stated.

NEW MEXICO: Pecos; Santa Fe; Ramah; Mesilla.

The plant is frequently cultivated in gardens and often escapes.

4. CYNTHIA Don.

Nearly glabrous branched perennial with a rosette of basal leaves and a few sessile alternate cauline ones; heads medium-sized, the flowers orange-colored; involucre campanulate, of 9 to 15 lanceolate nerveless bracts; achenes cylindric, striate; pappus of 10 to 15 minute linear scales and as many or more inner bristles, the outer scales visible only under a strong lens.

1. Cynthia viridis Standley, Contr. U. S. Nat. Herb. **13:** 357. 1911.

TYPE LOCALITY: Near Cowles, Pecos River National Forest, San Miguel County, New Mexico. Type collected by Standley (no. 4418).

RANGE: Mountains of Colorado, New Mexico, and Arizona.

NEW MEXICO: Cowles; Gallinas Planting Station; Las Vegas Hot Springs; West Fork of the Gila; Fresnal. Transition Zone.

5. CICHORIUM L. CHICORY.

Erect branching herbaceous perennial with basal and alternate cauline leaves and large heads of blue flowers in sessile clusters along the branches; involucre of 2 series of herbaceous bracts, the outer spreading, the inner erect; achenes 5-angled, truncate; pappus of 2 or 3 series of short blunt scales.

1. **Cichorium intybus** L. Sp. Pl. 811. 1753.

TYPE LOCALITY: "Habitat in Europa ad margines agrorum viarumque."

NEW MEXICO: Near Albuquerque (*Munson & Hopkins*).

A common weed in many parts of the United States but, so far, rare in New Mexico.

6. UROPAPPUS Nutt.

Acaulescent annual, with narrow, entire or laciniate-toothed leaves and medium-sized heads of yellow flowers; involucre of a series of narrow inner bracts and a few calyculate outer ones; achenes attenuate to a short beak; pappus white, of 5 scarious awn-tipped scales.

1. **Uropappus pruinosus** Greene, Leaflets 1: 213. 1906.

TYPE LOCALITY: "Southwestern New Mexico and adjacent Arizona."

RANGE: Southern New Mexico and Arizona.

NEW MEXICO: Nutt Mountain; Organ Mountains. Lower Sonoran Zone.

7. LACTUCA L. LETTUCE.

Tall or low perennial or biennial herbs with milky juice, leafy stems, and paniculate inflorescence of small heads of blue or yellow flowers; involucre glabrous, cylindric, of few subequal bracts in a single series and numerous calyculate outer ones; achenes compressed, beaked; pappus of usually white, slender bristles falling separately.

KEY TO THE SPECIES.

Achenes not transversely rugulose.
  Flowers yellow; leaves spinulose .... 1. *L. integrata.*
  Flowers blue; leaves not spinulose .... 2. *L. pulchella.*
Achenes transversely rugulose.
  Leaves spinulose on the margins and midribs .... 3. *L. ludoviciana.*
  Leaves never spinulose.
    Involucre about 10 mm. high; leaves mostly pinnatifid, with broad lobes .... 4. *L. canadensis.*
    Involucre 15 to 20 mm. high; leaves mostly entire, linear, or the lowest with narrow lobes .... 5. *L. graminifolia.*

1. **Lactuca integrata** (Gren. & Godr.) A. Nels. in Coulter, New Man. Rocky Mount. 596. 1909. PRICKLY LETTUCE.

Lactuca scariola integrata Gren. & Godr. Fl. France 2: 320. 1850.

TYPE LOCALITY: France.

NEW MEXICO: Farmington; Shiprock; Mesilla Valley; Gilmores Ranch; Ruidoso.

Introduced from the eastern States and originally from Europe, this has become a troublesome weed in some of the river valleys of New Mexico.

2. **Lactuca pulchella** (Pursh) DC. Prodr. 7: 134. 1838.

Sonchus pulchellus Pursh, Fl. Amer. Sept. 502. 1814.

Lactuca integrifolia Nutt. Gen. Pl. 2: 124. 1818.

TYPE LOCALITY: "On the banks of the Missouri."

RANGE: British America to Michigan, Kansas, and New Mexico.

NEW MEXICO: Farmington; Tunitcha Mountains; Dulce; Pescado Spring; Johnsons Mesa; Pecos; Perico Creek; Santa Fe; Silver City; Mangas Springs; Mogollon Mountains; Mesilla Valley; White and Sacramento mountains. Open slopes and in waste ground, in the Upper Sonoran and Transition zones.

3. **Lactuca ludoviciana** (Nutt.) DC. Prodr. **7**: 141. 1838.

Sonchus ludovicianus Nutt. Gen. Pl. **2**: 25. 1818.

TYPE LOCALITY: "In humid places, in the open plains, and Fort Mandan on the Missouri."

RANGE: Montana and Minnesota to New Mexico and Texas.

NEW MEXICO: Farmington (*Wooton* 2592). River valleys, in the Upper Sonoran Zone.

4. **Lactuca canadensis** L. Sp. Pl. 796. 1753.

TYPE LOCALITY: "Habitat in Canada."

RANGE: British America to New Mexico and Florida.

NEW MEXICO: Tunitcha Mountains; Gallinas Canyon; Sandia Mountains; Mimbres River; Ruidoso Creek. Upper Sonoran and Transition zones.

5. **Lactuca graminifolia** Michx. Fl. Bor. Amer. **2**: 85. 1803.

TYPE LOCALITY: "In Carolina inferiore."

RANGE: Colorado and Arizona to North Carolina and Florida.

NEW MEXICO: Winsors Ranch; Las Vegas; South Percha Creek; Middle Fork of the Gila; Parkers Well; Cloudcroft; Tularosa Creek; Ruidoso Creek. Open slopes and meadows, in the Upper Sonoran and Transition zones.

8. SONCHUS L. Sow THISTLE.

Succulent annuals or biennials with alternate, auriculate-clasping, dentate or pinnatifid, prickly leaves, and corymbs of medium-sized heads of yellow flowers; involucre ovoid or campanulate, becoming thickened at the base, the bracts imbricated in several series, the outer successively smaller; receptacle flat, naked; achenes oval, flattened, ribbed, truncate; pappus of soft smooth capillary bristles.

KEY TO THE SPECIES.

Heads 25 mm. high; involucres glandular-pubescent.................. 3. *S. arvensis*.

Heads about 15 mm. high; involucres not glandular-pubescent.

  Auricles of leaves acute; achenes not transversely wrinkled...... 1. *S. asper*.

  Auricles obtuse; achenes transversely wrinkled................... 2. *S. oleraceus*.

1. **Sonchus asper** (L.) All. Fl. Pedem. **1**: 222. 1785.

Sonchus oleraceus asper L. Sp. Pl. 794. 1753.

TYPE LOCALITY: European.

NEW MEXICO: Farmington; Carrizo Mountains; Pecos; Santa Fe; Sandia Mountains; Mangas Springs; Berendo Creek; Cloverdale; Mesilla Valley; Fresnal; Round Mountain.

A common weed in gardens and cultivated fields, widely introduced into North America from Europe.

2. **Sonchus oleraceus** L. Sp. Pl. 794. 1753.

TYPE LOCALITY: "Habitat in Europae cultis."

NEW MEXICO: Kingston; Patterson.

Introduced from Europe.

3. **Sonchus arvensis** L. Sp. Pl. 793. 1753.

TYPE LOCALITY: "Habitat in Europae agris argillosis."

NEW MEXICO: Shiprock (*Standley* 7825).

Abundant along irrigating ditches in this one locality; introduced from Europe.

9. MALACOTHRIX DC.

Low branching annual with numerous stems terminated by medium-sized pedunculate yellow heads; leaves mostly basal, pinnatifid; involucre campanulate, of numerous narrow, equal, somewhat imbricated bracts and a few short outer ones; achenes truncate at both ends; pappus a very shallow entire cup.

1. **Malacothrix fendleri** A. Gray, Pl. Wright. **2:** 104. 1853.

TYPE LOCALITY: Low sandy banks of the Rio del Norte, New Mexico. Type collected by Fendler.

RANGE: Western Texas to southern California.

NEW MEXICO: Santa Fe; Sandia Mountains; mesa west of Organ Mountains; Mesilla Valley; Organ Mountains. Dry mesas, in the Lower and Upper Sonoran zones.

10. SITILIAS Raf. FALSE DANDELION.

Nearly glabrous perennial with a rosette of entire to pinnatifid basal leaves, alternate divided cauline ones, nearly simple or much branched stems, and large heads of yellow flowers; involucre a series of equal appressed narrow bracts with a few linear short outer ones; achenes fusiform, with a long slender beak; pappus reddish, simple, capillary, surrounded at the base by a soft-villous ring.

1. **Sitilias multicaulis** (DC.) Greene, Pittonia **2:** 179. 1891.

Pyrrhopappus multicaulis DC. Prodr. **7:** 144. 1838.

TYPE LOCALITY: "In Mexico ad Tamaulipas et S. Fernando de Bexar." The first locality is Mexican, the second Texan.

RANGE: Texas to Arizona and southward.

NEW MEXICO: San Juan; Pecos; Albuquerque; Los Lunas; Socorro; Mangas Springs; Kingston; Mesilla Valley; Roswell. Wet ground, in the Lower and Upper Sonoran zones.

This is a very variable plant. There seems to be no means of distinguishing *Pyrrhopappus rothrockii* A. Gray. That is a mere form and not worthy of nomenclatural recognition.

11. AGOSERIS Raf.

Acaulescent perennial herbs with mostly narrow, entire, toothed, or pinnatifid basal leaves, and large scapose heads of yellow, orange, or purplish flowers; involucre narrowly campanulate or cylindric, of numerous narrow imbricated bracts, the outer successively shorter, herbaceous, not thickened; achenes oblong or linear, terete, 10-ribbed, with a long or short beak; pappus of numerous capillary white bristles.

KEY TO THE SPECIES.

Beak of achenes slender, about as long as the body, nearly smooth at the middle.
- Flowers yellow; leaves linear or nearly so .... 1. *A. graminifolia.*
- Flowers purplish or orange; leaves not linear.
  - Bracts conspicuously blotched with purple .... 2. *A. purpurea.*
  - Bracts not blotched with purple, sometimes slightly purplish along the midrib.
    - Leaves glaucous, glabrous, pinnatifid with linear lobes .... 3. *A. greenei.*
    - Leaves green, somewhat pubescent, entire or with broad lobes .... 4. *A. aurantiaca.*

Beak of achenes stout, short, about half as long as the body, striate throughout.
Bracts usually villous-ciliate.
Outer bracts much broader than the inner, usually somewhat obtuse; plants low; leaves obtuse, not deeply pinnatifid.. 5. *A. pumila.*
Outer bracts not much broader than the inner, acute; plants tall; leaves acute, deeply pinnatifid....... 6. *A. laciniata.*
Bracts glabrous, sometimes slightly tomentose when young.
Leaves thick, glaucous, oblanceolate, toothed............ 7. *A. glauca.*
Leaves thin, green, nearly linear, entire.................. 8. *A. parviflora.*

1. **Agoseris graminifolia** Greene, Bull. Torrey Club **25:** 124. 1898.

TYPE LOCALITY: White Mountains, New Mexico. Type collected by Wooton (no. 513).

RANGE: Southern New Mexico.

NEW MEXICO: Hillsboro Peak; Middle Fork of the Gila; White and Sacramento mountains. Damp meadows in the mountains, Transition Zone.

2. **Agoseris purpurea** (A. Gray) Greene, Pittonia **2:** 177. 1891.

Macrorhynchus purpureus A. Gray, Mem. Amer. Acad. n. ser. **4:** 114. 1849.

Troximon purpureum A. Nels. in Coulter, New Man. Rocky Mount. 599. 1909.

TYPE LOCALITY: Santa Fe Creek, New Mexico. Type collected by Fendler (no. 487).

RANGE: Mountains of Colorado and New Mexico.

NEW MEXICO: Rio Pueblo; Santa Fe; Beattys Cabin; Beulah; Baldy. Transition Zone.

3. **Agoseris greenei** (A. Gray) Rydb. Mem. N. Y. Bot. Gard. **1:** 459. 1900.

Troximon gracilens greenei A. Gray, Proc. Amer. Acad. **19:** 71. 1883.

TYPE LOCALITY: Scott Mountains, Siskiyou County, California.

RANGE: British America to California and New Mexico.

NEW MEXICO: Winsor Creek; Dulce; Chama. Transition Zone.

4. **Agoseris aurantiaca** (Hook.) Greene, Pittonia **2:** 177. 1891.

Troximon aurantiacum Hook. Fl. Bor. Amer. **1:** 300. *pl. 104.* 1833.

TYPE LOCALITY: "Alpine prairies of the Rocky Mountains."

RANGE: British Columbia and Montana to Colorado and New Mexico.

NEW MEXICO: Tunitcha Mountains; Chama; Winsor Creek; Harveys Upper Ranch. Meadows in the mountains, Transition Zone.

5. **Agoseris pumila** (Nutt.) Rydb. Mem. N. Y. Bot. Gard. **1:** 457. 1900.

Troximon pumilum Nutt. Trans. Amer. Phil. Soc. n. ser. **7:** 434. 1841.

Troximon glaucum pumilum A. Nels. in Coulter, New Man. Rocky Mount. 599. 1909.

TYPE LOCALITY: "Plains of the Rocky Mountains, in Oregon."

RANGE: Montana to northern New Mexico.

NEW MEXICO: Truchas Peak (*Standley* 4819). Mountains, in the Arctic-Alpine Zone.

6. **Agoseris laciniata** (Nutt.) Greene, Pittonia **2:** 178. 1891.

Troximon laciniatum A. Gray, Proc. Amer. Acad. **19:** 71. 1883.

TYPE LOCALITY: Not stated.

RANGE: Idaho and Wyoming to California and New Mexico.

NEW MEXICO: Tunitcha Mountains; Santa Fe Canyon; Chama; Tierra Amarilla. Meadows, in the Transition Zone.

52576°—15——40

7. **Agoseris glauca** (Pursh) Steud. Nom. Bot. ed. 2. 1: 37. 1840.

Troximon glaucum Nutt. Gen. Pl. 2: 128. 1818.

TYPE LOCALITY: "On the banks of the Missouri."

RANGE: Washington and Saskatchewan to Utah and northwestern New Mexico.

NEW MEXICO: Near Fort Defiance; Carrizo Mountains. Meadows in the mountains, Transition Zone.

8. **Agoseris parviflora** (Nutt.) D. Dietr. Syn. Pl. 4: 1332. 1847.

Troximon parviflorum Nutt. Trans. Amer. Phil. Soc. n. ser. 7: 434. 1841.

TYPE LOCALITY: "On the plains of the Platte to the Rocky Mountains."

RANGE: Alberta and North Dakota to Colorado and New Mexico.

NEW MEXICO: Tunitcha Mountains; Sandia Mountains; Grosstedt Place. Meadows, in the Transition Zone.

12. TARAXACUM Hall. DANDELION.

Perennial acaulescent herbs with pinnatifid, toothed, or rarely entire leaves and large heads of yellow flowers terminating the naked hollow scapes; bracts of the involucre mostly equal, with a number of shorter calyculate outer ones; receptacle flat, naked; achenes oblong to fusiform, angled and nerved, spinulose at the summit, beaked; pappus copious, of unequal persistent slender bristles.

KEY TO THE SPECIES.

Outer bracts reflexed from the base; leaves deeply pinnatifid.
- Terminal segments of the leaves broadly triangular; achenes greenish brown .... 1. *T. taraxacum.*
- Terminal segments of the leaves triangular or oblong; achenes red .... 2. *T. mexicanum.*

Outer bracts not reflexed from the base; leaves pinnatifid or denticulate.
- Outer bracts nearly as long as the inner, reflexed above the middle; leaves denticulate, yellowish green .... 3. *T. dumetorum.*
- Outer bracts less than half as long as the inner, usually appressed; leaves pinnatifid, bright green .... 4. *T. montanum.*

1. **Taraxacum taraxacum** (L.) Karst. Deutsch. Fl. 1138. 1880–83.

Leontodon taraxacum L. Sp. Pl. 798. 1753.

Taraxacum officinale Web. Prim. Fl. Hols. 56. 1780.

TYPE LOCALITY: European.

RANGE: Nearly throughout the United States, in some places introduced from Europe, in others apparently native.

NEW MEXICO: Santa Fe; Las Vegas; Farmington; Chama; Raton. Open fields and waste ground.

2. **Taraxacum mexicanum** DC. Prodr. 7: 146. 1838.

TYPE LOCALITY: Near the City of Mexico.

RANGE: Colorado to Mexico.

NEW MEXICO: Chama; Nutritas Creek; Tierra Amarilla. Mountains, in the Transition Zone.

3. **Taraxacum dumetorum** Greene, Pittonia 4: 230. 1901.

Taraxacum oblanceolatum Rydb. Colo. Agr. Exp. Sta. Bull. 100: 410. 1906.

TYPE LOCALITY: Dale Creek, Wyoming.

RANGE: Alberta to Colorado and northern New Mexico.

NEW MEXICO: Upper Pecos River; top of Las Vegas Range. High mountains, in the Arctic-Alpine Zone.

4. **Taraxacum montanum** Nutt. Trans. Amer. Phil. Soc. n. ser. **7**: 430. 1841.

TYPE LOCALITY: "On the banks of the Platte, in subsaline situations toward the Rocky Mountains, and in the highest valleys of the Colorado of the West."

RANGE: Montana to New Mexico.

NEW MEXICO: Winsors Ranch; Santa Fe Canyon; Sacramento Mountains. Mountains, in the Transition Zone.

A specimen from Cloudcroft has entire leaves. It is not improbable that the material from the Sacramento Mountains represents an undescribed species.

13. CALYCOSERIS A. Gray.

Slender glaucous winter annual with branching stems, the leaves pinnately parted into linear lobes, the heads rather large, pedunculate, terminating the branches; flowers white or pinkish; heads and peduncles glandular-hispid; involucre of numerous erect narrow bracts in a single series and of a loose calyculate outer series; receptacle bristle-bearing; achenes fusiform or oblong, 5-ribbed, attenuate to a short beak terminating in a shallow scarious crown; pappus of soft capillary bristles, deciduous.

1. **Calycoseris wrightii** A. Gray, Pl. Wright. **2**: 104. *pl. 14*. 1853.

TYPE LOCALITY: "Stony hills around El Paso," Texas or Chihuahua.

RANGE: Western Texas to Arizona and Utah.

NEW MEXICO: Mesa west of Organ Mountains; Picacho Mountain. Mesas, in the Lower Sonoran Zone.

14. HIERACIUM L. HAWKWEED.

Hirsute, lanate, or glandular perennial herbs with mostly entire leaves and paniculate heads of yellowish, whitish, or pink flowers; heads 12 to many-flowered; involucre of narrow, somewhat imbricated bracts, a few short ones at the base; achenes short, striate, not beaked; pappus a single row of slender white or tawny bristles.

KEY TO THE SPECIES.

All leaves glabrous or nearly so, none long-hairy.................. 1. *H. gracile*.
Basal leaves hirsute or lanate.
  Flowers whitish or flesh-colored; pappus bright white; stems leafy.
    Plants stout; cauline leaves and stem hirsute; flowers whitish.. 2. *H. lemmoni*.
    Plants slender; cauline leaves and stems glabrous; flowers flesh-colored.................................. 3. *H. carneum*.
  Flowers yellow; pappus tawny; stems leafy or nearly naked.
    Leaves lanate.. 4. *H. pringlei*.
    Leaves hirsute, never lanate.
      Stems bearing numerous leaves; involucres 7 mm. high or less.................................. 5. *H. rusbyi*.
      Stems bearing only 1 or 2 reduced leaves; involucres 9 to 15 mm. high.
        Basal leaves oblanceolate-spatulate to obovate, long-hirsute; involucres 12 to 15 mm. high. 6. *H. fendleri*.
        Basal leaves linear-oblong, short-hairy; involucres not more than 10 mm. high............... 7. *H. brevipilum*.

1. **Hieracium gracile** Hook. Fl. Bor. Amer. **1**: 298. 1833.

Hieracium triste gracile A. Gray in Brewer & Wats. Bot. Calif. **1**: 441. 1876.

TYPE LOCALITY: Rocky Mountains.

RANGE: Mountains, Alaska and Montana to California and New Mexico.

NEW MEXICO: Truchas Peak (*Standley* 4810). Arctic-Alpine Zone.

2. Hieracium lemmoni A. Gray, Proc. Amer. Acad. **19:** 70. 1883.

TYPE LOCALITY: Cave Canyon, near Fort Huachuca, southern Arizona.

RANGE: Mountains of southern Arizona and southwestern New Mexico.

NEW MEXICO: Hillsboro Peak (*Metcalfe* 1507). Transition Zone.

3. Hieracium carneum Greene, Bot. Gaz. **6:** 184. 1881.

TYPE LOCALITY: South base of the Pinos Altos Mountains, New Mexico. Type collected by Greene in 1880.

RANGE: Mountains of Arizona, southwestern New Mexico, and adjacent Mexico.

NEW MEXICO: Eagle Creek; Pinos Altos Mountains.

4. Hieracium pringlei A. Gray, Proc. Amer. Acad. **19:** 69. 1883.

TYPE LOCALITY: Santa Rita Mountains, southern Arizona.

RANGE: Mountains of southwestern New Mexico and southeastern Arizona.

NEW MEXICO: San Luis Mountains (*Mearns* 404).

5. Hieracium rusbyi Greene, Bull. Torrey Club **9:** 64. 1882.

TYPE LOCALITY: Mogollon Mountains, New Mexico. Type collected by Rusby (no. 177).

RANGE: Mountains of southwestern New Mexico.

NEW MEXICO: Mogollon Mountains.

6. Hieracium fendleri Schultz Bip. Bonplandia **9:** 173. 1861.

Crepis ambigua A. Gray, Mem. Amer. Acad. n. ser. **4:** 114. 1849, not *Hieracium ambiguum* Schult. 1809.

TYPE LOCALITY: Along Santa Fe Creek, New Mexico. Type collected by Fendler.

RANGE: New Mexico and Arizona to South Dakota; also in Mexico.

NEW MEXICO: Tunitcha Mountains; Agua Fria; Gallinas Planting Station; Tierra Amarilla; Santa Fe Canyon; Sandia Mountains; Pinos Altos Mountains; White Mountains; Cloudcroft. Wooded hills, in the Transition Zone.

7. Hieracium brevipilum Greene, Bull. Torrey Club **9:** 64. 1882.

Hieracium fendleri mogollonense A. Gray, Proc. Amer. Acad. **19:** 69. 1883.

TYPE LOCALITY: Mogollon Mountains, New Mexico. Type collected by Rusby (no. 178).

RANGE: Known only from type locality.

15. LYGODESMIA Don.

Glabrous and often glaucous perennials with rigid stems, linear or scalelike leaves, and rather large terminal heads of pink flowers; involucre cylindric, of few equal bracts and a few short calyculate outer ones; achenes terete, linear or fusiform, obscurely striate or angled or smooth; pappus of numerous soft, white or tawny bristles.

KEY TO THE SPECIES.

Involucres about 10 mm. high, usually 5-flowered................. 1. *L. juncea.*

Involucres 15 to 25 mm. high, 6 to 10-flowered.

Stems leafy, only the uppermost leaves reduced; stems branched from the base.............................. 2. *L. grandiflora.*

Stems nearly naked, most of the leaves reduced to scales; stems usually simple at the base...................... 3. *L. texana.*

1. Lygodesmia juncea (Pursh) Don, Edinburgh Phil. Journ. **6:** 311. 1829.

Prenanthes juncea Pursh, Fl. Amer. Sept. 498. 1814.

TYPE LOCALITY: "On the banks of the Missouri."

RANGE: British America to New Mexico and Missouri.

NEW MEXICO: Dulce; Clayton; Raton; Kennedy; Colfax; Las Vegas; Pecos; Ramah; Sierra Grande; Nara Visa. Plains and low hills, in the Upper Sonoran Zone.

2. **Lygodesmia grandiflora** (Nutt.) Torr. & Gray, Fl. N. Amer. **2**: 485. 1842.

Erythremia grandiflora Nutt. Trans. Amer. Phil. Soc. n. ser. **7**: 445. 1841.

TYPE LOCALITY: "In the Rocky Mountain range, on the borders of the Platte."

RANGE: Wyoming and Idaho to Arizona and northwestern New Mexico.

NEW MEXICO: Farmington (*Standley* 6936). Plains and low hills, in the Upper Sonoran Zone.

3. **Lygodesmia texana** (Torr. & Gray) Greene; Small, Fl. Southeast. U. S. 1315. 1903.

Lygodesmia aphylla texana Torr. & Gray, Fl. N. Amer. **2**: 485. 1842.

TYPE LOCALITY: Texas.

RANGE: Plains of New Mexico and Texas.

NEW MEXICO: Near Fort Cummings; Hondo Hill; 40 miles west of Roswell. Upper Sonoran Zone.

Prenanthella exigua (A. Gray) Rydb. probably occurs in New Mexico. It was described from plants collected on hills above El Paso. We have seen no New Mexican specimens, although the range of the plant extends well to the north.

16. CREPIS L.

Perennial herbs with mostly basal, toothed, pinnatifid, or nearly entire leaves and rather small corymbose yellow heads; involucre cylindric or campanulate, swollen at the base, of few narrow bracts with several short basal ones; achenes columnar, ribbed or nerved; pappus of copious soft whitish capillary bristles.

KEY TO THE SPECIES.

Leaves canescent.
- Involucres glabrous.. 1. *C. acuminata.*
- Involucres canescent.. 2. *C. occidentalis.*

Leaves glabrous.
- Involucres glabrous.
  - Leaves nearly linear, toothed, erect, green.............. 3. *C. mogollonica.*
  - Leaves oblanceolate, entire, flattened upon the ground, glaucous.. 4. *C. chamaephylla.*
- Involucres glandular or hirsute.
  - Leaves long-petioled, the petioles half as long as the blades or more.. 5. *C. petiolata.*
  - Leaves sessile or on short, broadly winged petioles.
    - Leaves thick, lobed with deep obtuse lobes; involucres about 9 mm. high .... 6. *C. neomexicana.*
    - Leaves thin, shallowly dentate with acute teeth; involucres about 10 mm. high............... 7. *C. perplexans.*

1. **Crepis acuminata** Nutt. Trans. Amer. Phil. Soc. n. ser. **7**: 437. 1841.

TYPE LOCALITY: "Plains of the Platte."

RANGE: Washington and Montana to California and New Mexico.

NEW MEXICO: A single specimen seen, without definite locality but probably from western McKinley County. Mountains, in the Transition Zone.

2. **Crepis occidentalis** Nutt. Journ. Acad. Phila. **7**: 29. 1834.

TYPE LOCALITY: "Columbia River."

RANGE: Washington and Montana to California and northern New Mexico.

NEW MEXICO: Pass southeast of Tierra Amarilla (*Eggleston* 6596). Open hills, in the Upper Sonoran and Transition zones.

3. **Crepis mogollonica** Greene, Contr. U. S. Nat. Herb. **16**: 176. 1913.

TYPE LOCALITY: West Fork of the Gila, Mogollon Mountains, Socorro County, New Mexico. Type collected by Metcalfe (no. 576).

RANGE: Kn-wn only from type locality, in the Transition Zone.

4. Crepis chamaephylla Woot. & Standl. Contr. U. S. Nat. Herb. **16:** 175. 1913.

TYPE LOCALITY: North end of the Carrizo Mountains, northeast corner of Arizona. Type collected by Standley (no. 7419).

RANGE: Known only from the type locality, in the Upper Sonoran Zone.

5. Crepis petiolata Rydb. Bull. Torrey Club **32:** 134. 1905.

TYPE LOCALITY: Along Bear River, five miles east of Hayden, Colcrado.

RANGE: Wyoming to New Mexico.

NEW MEXICO: Chama; Cienaga Ranch; Tularosa Creek; Costilla Valley; Castle Rock. Mountains, in the Transition Zone.

6. Crepis neomexicana Woot. & Standl. Contr. U. S. Nat. Herb. **16:** 176. 1913.

TYPE LOCALITY: Tularosa Creek, Socorro County, New Mexico. Type collected by Wooton, July 14, 1906.

RANGE: Known only from type locality.

7. Crepis perplexans Rydb. Bull. Torrey Club **34:** 134. 1905.

TYPE LOCALITY: Encampment, Carbon County, Wyoming.

RANGE: British America to New Mexico and Nebraska.

NEW MEXICO: Upper Pecos (*Bartlett*). Mountains, in the Transition Zone.

142. MUTISIACEAE.

Herbs or shrubs with alternate, simple, sessile or petiolate leaves and solitary, corymbose, or paniculate heads of flowers surrounded by an involucre of more or less imbricated bracts; heads homagamous or heterogamous; corollas either regularly 5-cleft or bilabiate in the perfect flowers and simply ligulate in the fertile ray flowers; anthers long-caudate; receptacle naked; style branches of the perfect flowers not appendaged, mostly very short.

KEY TO THE GENERA.

Heads radiate; plants scapose .... 1. CHAPTALIA (p. 630).
Heads not radiate; plants caulescent.
  Herbs; flowers white or pink .... 2. PEREZIA (p. 630).
  Shrub; flowers yellow .... 3. TRIXIS (p. 631).

1. CHAPTALIA Vent.

Scapose perennial herb with obovate or oblong leaves, pinnatifid at the base, and solitary heads; involucre turbinate, of narrow appressed imbricated bracts, the outer successively shorter; heads heterogamous, radiate; achenes oblong to fusiform, 5-nerved, with a filiform beak; pappus of soft capillary bristles.

1. Chaptalia alsophila Greene, Leaflets **1:** 158. 1905.

TYPE LOCALITY: Black Range, New Mexico. Type collected by Metcalfe (no. 1454).

RANGE: Mountains of southwestern New Mexico and southeastern Arizona.

NEW MEXICO: South Percha Creek; East Canyon. Transition Zone.

It is questionable whether our plant might not well be referred to one of the Mexican species.

2. PEREZIA Lag.

Perennial herbs with thick reticulate-veined spinulose alternate sessile leaves and solitary or corymbose heads of white or rose-colored flowers; involucre of thin bracts imbricated in several series, the outer successively shorter; heads homogamous, all the flowers with bilabiate corollas; receptacle flat, naked; achenes puberulent, elongated, not rostrate; pappus of copious scabrous bristles.

KEY TO THE SPECIES.

Low, 20 cm. high or less; heads solitary, 20 to 30-flowered, the flowers purplish.. 1. *P. nana*.
Tall, 60 cm. or more high; heads corymbose or paniculate, 5 to 12-flowered, the flowers nearly white.
  Heads 5 or 6-flowered; bracts obtuse or acutish, glabrous on the back.. 2. *P. wrightii*.
  Heads 8 to 12-flowered; bracts abruptly acuminate, scaberulous on the back.. 3. *P. thurberi*.

1. **Perezia nana** A. Gray, Mem. Amer. Acad. n. ser. **4**: 111. 1849.

Type locality: Near Chihuahua, Mexico.

Range: Western Texas to southern Arizona and southward.

New Mexico: Cliff; Laguna Colorado; Carrizalillo Mountains; Dog Spring; Deming; San Marcial; Organ Mountains; White Sands; Knowles; Nogal; La Luz Canyon. Dry plains and hills, in the Lower Sonoran Zone.

2. **Perezia wrightii** A. Gray, Pl. Wright. **1**: 127. 1852.

Type locality: "On the Rio Seco and westward; also on the Rio Grande, Texas."

Range: Western Texas to southern Arizona and southward.

New Mexico: Mangas Springs; Florida Mountains; Dona Ana Mountains; Organ Mountains; Tortugas Mountain. Dry hills and ravines, in the Lower and Upper Sonoran zones.

3. **Perezia thurberi** A. Gray, Mem. Amer. Acad. n. ser. **5**: 324. 1854.

Type locality: Rocky hills, near Santa Cruz, Sonora.

Range: Southern New Mexico and Arizona, southward into Mexico.

New Mexico: Dog Spring; San Luis Mountains.

3. TRIXIS P. Br.

Low woody perennial with entire or denticulate alternate lanceolate leaves and corymbose medium-sized heads of yellow flowers; involucres many-flowered, the bracts 8 to 12, equal, in a single series, subtended by a few bractlike leaves; achenes slender, with a tapering summit; pappus yellowish, of capillary bristles.

1. **Trixis californica** Kellogg, Proc. Calif. Acad. **2**: 182. *f. 53*. 1862.

Type locality: Cedros Island, Lower California.

Range: Western Texas to Arizona and Mexico.

New Mexico: Black Range; Mangas Springs; Bear Mountains; Dog Spring; near White Water; Dona Ana Mountains; Organ Mountains; Tortugas Mountain; south of Roswell. Rocky canyons and hills, in the Lower and Upper Sonoran zones.

143. AMBROSIACEAE. Ragweed Family.

Annual or perennial herbs or shrubs with alternate leaves; flowers small, aggregated on a receptacle, surrounded by an involucre, the staminate and pistillate in the same or separate heads; involucral bracts few, distinct or united, those of the pistillate flowers often nutlike or burlike at maturity or winged; stamens usually 5, distinct; corollas all tubular; ovary 1-celled; stigmas 2, hairy or brushlike at the apex.

KEY TO THE GENERA.

Staminate and pistillate flowers in the same heads, the latter few (rarely solitary or none) in the margins.
  Achenes flattened, wing-margined; involucre with 1 or 2 inner enlarged scarious bracts......... 1. Dicoria (p. 632).

Achenes turgid, ovoid or pear-shaped; bracts of the involucre alike, herbaceous.................... 2. Iva (p. 633).
Staminate and pistillate flowers in separate heads, the latter 1 to 4, without corollas, inclosed in a nutlike or burlike involucre.
Shrub; pistillate involucre with several scarious wings.. 3. Hymenoclea (p. 634).
Herbs; pistillate involucre not winged.
Involucres of staminate heads with distinct bracts; pistillate involucres large, with hooked spines............................ 4. Xanthium (p. 634).
Involucres of staminate heads with united bracts; pistillate involucres small, the spines not hooked.
Spines or tubercles of the 1-flowered pistillate heads in a single row............ 5. Ambrosia (p. 635).
Spines of the 1 to 4-flowered pistillate heads in more than one row................. 6. Gaertneria (p. 636).

1. DICORIA Torr. & Gray.

Low branched canescent annuals with alternate petiolate leaves; heads heterogamous, of 1 or 2 fertile and several staminate flowers; pistillate flowers without corollas; involucre of 5 short, oval or oblong, herbaceous bracts and 1 or 2 inner enlarged scarious ones; achenes flat on the inner surface, convex on the outer, with dentate or thin and scarious pectinate edges; pappus rudimentary, of several small scales.

KEY TO THE SPECIES.

Achenes 1 in each head; teeth of achenes conspicuous, often connected at the base.. 1. *D. brandegei.*
Achenes 2 in each head; teeth of achenes inconspicuous, few, distinct... 2. *D. paniculata.*

1. Dicoria brandegei A. Gray, Proc. Amer. Acad. **11**: 76. 1876.

Type locality: Along the San Juan, between McElmo and Recapture Creeks, Utah.

Range: Colorado and Utah to northern Arizona and northwestern New Mexico.

New Mexico: Shiprock (*Standley* 7188). Sandy soil in valleys, in the Upper Sonoran Zone.

2. Dicoria paniculata Eastw. Proc. Calif. Acad. II. **6**: 298. *pl. 45.* 1896.

Type locality: Sandy flats along the San Juan River near the junction with McElmo Creek, Utah.

Range: Southeastern Utah and northwestern New Mexico.

New Mexico: Shiprock (*Standley* 7188a). Sandhills in valleys, in the Upper Sonoran Zone.

The two species were growing together on sandbars along the San Juan at the Shiprock Agency, and in the field were taken to be the same species. *D. paniculata,* as stated by Miss Eastwood, blooms earlier than *D. brandegei,* and the plants collected at Shiprock were in mature fruit and had lost many of their leaves, while those of the latter species were only flowering.

2. IVA L. MARSH ELDER.

Coarse herbaceous perennials or annuals with entire or dissected leaves, at least some of them opposite; heads numerous, small, axillary or loosely paniculate; involucre hemispheric, of few rounded bracts; receptacle with linear or spatulate chaff; marginal flowers pistillate, 1 to 5, their corollas tubular or wanting, the disk flowers perfect, with 5-lobed funnelform corollas; achenes flattened, glabrous; pappus none.

KEY TO THE SPECIES.

Perennial; leaves small, entire, sessile; heads axillary.......... 1. *I. axillaris.*
Annuals; leaves large, not entire, petiolate; heads not axillary.
  Heads in terminal bracteate spikes; fertile flowers with evident corollas.................................... 2. *I. ciliata.*
  Heads naked-paniculate; corolla of fertile flowers rudimentary or none.
    Heads conspicuously pedicellate; leaves twice or thrice pinnately parted.............................. 3. *I. ambrosiaefolia.*
    Heads nearly sessile; leaves toothed or laciniate-pinnatifid.
      Stems 1 meter high or more; plants bright green; leaves serrate................................ 4. *I. xanthiifolia.*
      Stems 60 cm. high or less; plants densely tomentose; leaves mostly laciniate-pinnatifid............ 5. *I. dealbata.*

1. **Iva axillaris** Pursh, Fl. Amer. Sept. 743. 1814.
TYPE LOCALITY: "In Upper Louisiana."
RANGE: British Columbia and Saskatchewan to California, New Mexico, and Oklahoma.
NEW MEXICO: Farmington; San Juan; Anniston. Alkaline soil, in the Upper Sonoran Zone.

2. **Iva ciliata** Willd. Sp. Pl. 3: 2386. 1803.
TYPE LOCALITY: "Habitat in America boreali."
RANGE: Nebraska and Illinois to Louisiana and New Mexico.
In Plantae Fendlerianae this is said to occur "From Sand Creek, New Mexico, to Fort Leavenworth, in low prairies." We have seen no New Mexican specimens, but the plant is to be expected in the northeast corner of the State. Plains and dry fields, in the Upper Sonoran Zone.

3. **Iva ambrosiaefolia** A. Gray, Syn. Fl. 1^2: 246. 1884.
Euphrosyne ambrosiaefolia A. Gray, Pl. Wright. 1: 102. 1852.
TYPE LOCALITY: "Mountains near El Paso," Texas or Chihuahua.
RANGE: Western Texas to southern New Mexico and southward.
NEW MEXICO: Trujillo Creek; Mangas Springs; mesa west of Organ Mountains; Organ Mountains; Florida Mountains. Dry mesas and sandhills, in the Lower and Upper Sonoran zones.

4. **Iva xanthiifolia** Nutt. Gen. Pl. 2: 185. 1818.
Euphrosyne xanthiifolia A. Gray, Pl. Wright. 2: 85. 1853.
TYPE LOCALITY: "In arid soils, near Fort Mandan, &c., on the banks of the Missouri."
RANGE: Saskatchewan and Nebraska to Washington and New Mexico.
NEW MEXICO: Cedar Hill; Shiprock; Chama; Pecos; Santa Fe; Sandia Mountains; Mountainair; Taos; Hebron; Las Vegas; Belen. Along streams and in waste ground, in the Upper Sonoran and Transition zones.
A common weed in cultivated fields in some parts of the State.

5. **Iva dealbata** A. Gray, Pl. Wright. **1:** 104. 1852.

TYPE LOCALITY: "In a mountain valley, between the Limpia and the Rio Grande," western Texas.

RANGE: Western Texas and southern New Mexico.

NEW MEXICO: Kingston; Lake Valley; near White Water; north of Parkers Well; White Mountains; Lakewood; Carlsbad; Artesia; Carrizozo. Lower and Upper Sonoran zones.

3. HYMENOCLEA Torr. & Gray.

Much branched slender shrub, 1 to 2 meters high, with alternate linear-filiform leaves, the lower irregularly pinnately parted; heads small, unisexual, very numerous; involucre of staminate flowers saucer-shaped, 4 to 6-lobed; bracts of the receptacle subtending the outer flowers obovate or spatulate; involucre of the solitary fertile flower ovoid or fusiform, beaked at the apex, furnished below with 9 to 12 dilated scarious transverse wings.

1. **Hymenoclea monogyra** Torr. & Gray, Mem. Amer. Acad. n. ser. **4:** 79. 1849.

TYPE LOCALITY: Valley of the Gila, New Mexico.

RANGE: Western Texas to southern California and adjacent Mexico.

NEW MEXICO: Burro Mountains; Socorro; Deming; Rincon; Rio Alamosa; Dona Ana Mountains; mesa west of Organ Mountains; Van Pattens. Along arroyos, in the Lower Sonoran Zone.

4. XANTHIUM L. COCKLEBUR.

Coarse annuals with branched stems and alternate toothed or lobed leaves; sterile and fertile flowers in different heads, the latter clustered below the short spikes or racemes of the staminate ones; fertile involucre coriaceous, ellipsoid or ovoid, covered with hooked prickles so as to form a bur, 2-celled; achenes oblong, flat.

KEY TO THE SPECIES.

Leaves attenuate to both ends, short-petiolate, armed with triple spines in the axils.. 1. *X. spinosum.*

Leaves cordate-ovate, long-petiolate, unarmed.

Fruit densely covered with spines, the body 5 to 10 mm. in diameter.. 2. *X. commune.*

Fruit with few spines, the body 5 mm. in diameter or less.... 2a. *X. commune wootoni.*

1. **Xanthium spinosum** L. Sp. Pl. 987. 1753.

TYPE LOCALITY: "Habitat in Lusitania."

RANGE: Waste ground in many parts of the United States; native of tropical America.

NEW MEXICO: Pecos; Silver City; Santa Rita; Las Vegas.

A noxious weed, in appearance very unlike the common cocklebur.

2. **Xanthium commune** Britton, Man. 912. 1901.

TYPE LOCALITY: Westport, New York.

RANGE: New York and Quebec to Utah and Arizona.

NEW MEXICO: Shiprock; Carrizo Mountains; Tunitcha Mountains; Nara Visa; Zuni; Pecos; Las Vegas; Dog Spring; Mesilla Valley. Waste ground and cultivated fields.

A very common and troublesome weed in cultivated fields in many parts of the State. Probably it has been introduced into New Mexico, but in some localities it seems to be at home.

2a. Xanthium commune wootoni Cockerell, Proc. Biol. Soc. Washington **16:** 9. 1903.

TYPE LOCALITY: "At Española, N. M., and Las Vegas, N. M."

RANGE: New Mexico.

NEW MEXICO: Las Vegas; Albuquerque. Valleys and cultivated fields, in the Lower and Upper Sonoran zones.

This seems distinct enough from *X. commune* to be regarded as a species. It certainly is more easily separable from that than are most of the eastern species from each other. Professor Cockerell states, however, that he has found both forms of fruit on the same plant, hence we hesitate to raise the subspecies to specific rank. The occurrence of both forms of fruit on a single plant would not necessarily invalidate either as a species but would rather seem to be a result of hybridization. Ordinarily the two plants are distinct enough.

5. AMBROSIA L. RAGWEED.

Coarse annual or perennial herbs with lobed or dissected, opposite or alternate leaves and small inconspicuous flowers; sterile heads racemose, bractless; fertile flowers mostly glomerate in the lower axils; involucre of staminate flowers hemispheric to turbinate, 5 to 12-lobed or truncate; receptacle flat, with filiform chaff among the outer flowers; involucre of the solitary pistillate flower nutlike, beaked at the apex, usually with 4 to 8 tubercles or stout spines in a row below the beak.

KEY TO THE SPECIES.

Leaves mostly 3 or 5-cleft; involucre of staminate heads 3 or 4-ribbed .... 1. *A. aptera.*
Leaves once to thrice pinnatifid; involucre of staminate flowers not ribbed.
  Annual; fruit with sharp tubercles; leaves mostly twice parted .... 2. *A. artemisiaefolia.*
  Perennial; fruit with blunt tubercles or unarmed; leaves mostly once pinnatifid .... 3. *A. psilostachya.*

1. Ambrosia aptera DC. Prodr. **5:** 527. 1836. GREAT RAGWEED.

Ambrosia trifida texana Scheele, Linnaea **22:** 156. 1849.

TYPE LOCALITY: Near San Antonio, Texas.

RANGE: Texas to southern Arizona.

NEW MEXICO: Vermejo Peak; Mangas Springs; Brockmans Ranch; Cliff; Crains Ranch. Wet ground.

2. Ambrosia artemisiaefolia L. Sp. Pl. 987. 1753. COMMON RAGWEED.

TYPE LOCALITY: "Habitat in Virginia, Pennsylvania."

RANGE: British America to Mexico and South America.

NEW MEXICO: Santa Fe; Ogle; Agricultural College. Waste ground.

The common ragweed of the Eastern States is, so far, a rare introduction into New Mexico.

3. Ambrosia psilostachya DC. Prodr. **5:** 526. 1836. WESTERN RAGWEED.

TYPE LOCALITY: "In Mexico inter San-Fernando et Matamoros."

RANGE: Illinois and Saskatchewan to Arizona and California, south into Mexico.

NEW MEXICO: Chama; Shiprock; Pecos; Santa Fe; Clayton; Nara Visa; Malaga; Albuquerque; Brockmans Ranch; Kingston; Dog Spring; Mesilla Valley. Plains, in the Sonoran and Transition zones.

6. **GAERTNERIA** Medic.

Herbaceous annuals or perennials with mostly alternate, variously parted leaves; heads of staminate flowers as in Ambrosia or sometimes mixed with the pistillate; fertile involucre 1 to 4-flowered, 1 to 4-celled, with 1 to 4 beaks, armed with numerous sharp spines in several series.

KEY TO THE SPECIES.

Leaves simply pinnate or simple .... 1. *G. grayi.*
Leaves twice or thrice pinnately dissected.
  Leaves interruptedly pinnate, with ovate or triangular divisions, tomentose beneath .... 2. *G. tomentosa.*
  Leaves regularly pinnate with linear to oblong divisions, not tomentose beneath.
    Annual; staminate involucres cleft below the middle.. 3. *G. acanthicarpa.*
    Perennial; staminate involucres not cleft to the middle.. 4. *G. tenuifolia.*

1. **Gaertneria grayi** A. Nels. Bot. Gaz. **34:** 35. 1902.

Franseria tomentosa A. Gray, Mem. Amer. Acad. n. ser. **4:** 80. 1849, not *Ambrosia tomentosa* Nutt. 1818.

TYPE LOCALITY: "High banks of Walnut Creek, between Council Grove and Fort Mann, of the Arkansas."

RANGE: Kansas and Colorado to New Mexico and Texas.

We have seen no specimens of this from New Mexico, but it was collected by Griffiths at Texline, Texas, so no doubt occurs in eastern and northeastern New Mexico. Valleys, in the Upper Sonoran Zone.

2. **Gaertneria tomentosa** (Nutt.) A. Nels. Bot. Gaz. **34:** 34. 1902.

Ambrosia tomentosa Nutt. Gen. Pl. **2:** 186. 1818.

Franseria discolor Nutt. Trans. Amer. Phil. Soc. n. ser. **7:** 507. 1841.

Franseria tomentosa A. Nels. in Coulter, New Man. Rocky Mount. 542. 1909.

TYPE LOCALITY: "In Upper Louisiana on the banks of the Missouri."

RANGE: Montana to New Mexico.

NEW MEXICO: Winsors Ranch; Cleveland; Maxwell City; Mora. Waste and cultivated ground, in the Upper Sonoran and Transition zones.

3. **Gaertneria acanthicarpa** (Hook.) Britton, Mem. Torrey Club **5:** 332. 1894.

Ambrosia acanthicarpa Hook. Fl. Bor. Amer. **1:** 309. 1830.

Franseria hookeriana Nutt.; Torr. & Gray, Fl. N. Amer. **2:** 294. 1842.

TYPE LOCALITY: "Banks of the Saskatchewan and Red River."

RANGE: British America to Texas and California.

NEW MEXICO: Common throughout the State. Plains and valleys, especially in cultivated and waste ground, in the Lower and Upper Sonoran zones.

One of the commonest weeds in cultivated fields almost everywhere in the State. It is often called "ragweed," but of course is very different from the true ragweed of the Eastern States.

4. **Gaertneria tenuifolia** (A. Gray) Kuntze, Rev. Gen. Pl. **1:** 339. 1891.

Franseria tenuifolia A. Gray, Mem. Amer. Acad. n. ser. **4:** 80. 1849.

TYPE LOCALITY: "Poñi Creek, between Bent's Fort and Santa Fe; also at Santa Fe," New Mexico. Type collected by Fendler.

RANGE: Kansas and Colorado to Texas and California, also in Mexico.

NEW MEXICO: Santa Fe; Raton Mountains; Clayton; Maxwell City; Albert; Kingston; Organ Mountains; Gray; Eagle Creek. Plains and moist fields, in the Upper Sonoran Zone.

144. ASTERACEAE. Aster Family.

Annual or perennial herbs or low shrubs with opposite or alternate leaves; flowers aggregated on a naked or scaly receptacle, surrounded by an involucre; involucre of distinct or partly united bracts in one or more series; calyx of bristles, awns, or scales or cuplike, forming pappus at maturity, sometimes wanting; corolla 5-lobed, that of the marginal flowers often produced into a ray; ovary 1-celled; stigmas 2; fruit an achene.

KEY TO THE TRIBES.

Stigmatic lines at the base of the stigmas or below the middle; heads discoid, never yellow nor brown; anthers not caudate at the base.
  Stigmas filiform or subulate, hispidulous; coarse perennial herbs with corymbose purple heads....................................**I. VERNONIEAE** (p. 637).
  Stigmas more or less clavate, papillose-puberulent; habit various.
II. EUPATORIEAE (p. 637).
Stigmatic lines extending to the tips of the stigmas or their appendages; heads discoid or radiate, variously colored; anthers sometimes caudate at the base.
  Anther sacs caudate at the base; heads never radiate; corollas yellow only in a few species of Cirsium.
    Anthers not appendaged at the top; heads heterogamous or diœcious; pistillate flowers with filiform corollas; mostly small whitish plants with very small heads..................................**IV. GNAPHALIEAE** (p. 640).
    Anthers with elongate, cartilaginous, mostly caudate appendages at the top; flowers all perfect or the marginal neutral; corolla not filiform; coarse plants with large heads and often spiny leaves.
IX. CYNAREAE (p. 645).
  Anther sacs not caudate at the base; heads commonly radiate and with yellow or brown disk flowers.
    Stigmas of the perfect flowers with more or less distinct appendages, these usually strongly hairy outside, glabrous inside, but never with a ring of longer hairs...**III. ASTEREAE** (p. 638).
    Stigmas of the perfect flowers without appendages or these, if present, hairy on both sides and with a ring of longer hairs.
      Pappus capillary; stigmas often appendaged.
VIII. SENECIONEAE (p. 645).
      Pappus never capillary; stigmas rarely appendaged.
        Bracts of the involucres dry and scarious; rays mostly white or inconspicuous.....................**VII. ANTHEMIDEAE** (p. 645).
        Bracts of the involucres herbaceous or foliaceous; rays various.
          Receptacle with chaffy scales subtending the flowers.
V. HELIANTHEAE (p. 641).
          Receptacle naked or in Gaillardia with bristles, never chaffy-bracted..........................**VI. HELENIEAE** (p. 643).

KEY TO THE GENERA.

Tribe I. VERNONIEAE.

A single genus.. 1. VERNONIA (p. 645).

Tribe II. EUPATORIEAE.

Achenes 5-angled, destitute of intervening ribs.
  Pappus wholly of capillary bristles.
    Annuals; pappus bristles plumose..................2. CARMINATIA (p. 646).

Perennials; pappus merely scabrous.
Receptacle flat............................3. EUPATORIUM (p. 646).
Receptacle conic...........................4. CONOCLINIUM (p. 647).
Pappus at least partly of paleæ.
Flowers bluish; involucres many-flowered, campanulate.
5. COELESTINA (p. 648).
Flowers pink or white; involucres 3 to 5-flowered, cylindric.
6. STEVIA (p. 648).
Achenes 8 to 10-striate or costate.
Bracts of the involucres herbaceous, not striate nor nerved.
Shrub; leaves opposite; pappus paleaceous-aristiform.
7. CARPOCHAETE (p. 648).
Herbs; leaves alternate; pappus of plumose or barbellate bristles.
8. LACINIARIA (p. 649).
Bracts of the involucres not herbaceous, conspicuously striate-nerved.
Pappus plumose; leaves mostly alternate.............9. KUHNIA (p. 649).
Pappus barbellate or scabrous; leaves alternate or opposite.
10. COLEOSANTHUS (p. 650).

Tribe III. ASTEREAE.

Heads unisexual, discoid; plants diœcious..................30. BACCHARIS (p. 671).
Heads not unisexual, discoid or radiate; plants not diœcious.
Marginal pistillate flowers not ligulate, reduced to a filiform or short tube.
31. ESCHENBACHIA (p. 673).
Marginal pistillate flowers, if present, ligulate.
Ray flowers blue, pink, or white, never yellow.
Pappus coroniform; rays white.................42. APHANOSTEPHUS (p. 691).
Pappus at least in part of awns or bristles; rays of various colors.
Pappus of a few long awns or coarse bristles or in ray flowers reduced to paleæ.
43. TOWNSENDIA (p. 691).
Pappus of numerous capillary bristles, at least in the disk achenes.
Rays only slightly if at all exceeding the pappus; annuals.
Bracts in 2 or 3 rows, the outer foliaceous; stigma tips acute.
34. BRACHYACTIS (p. 682).
Bracts in 1 or 2 series, narrow, not foliaceous; stigma tips obtuse.
32. LEPTILON (p. 673).
Rays conspicuously exceeding the pappus, usually equaling or exceeding the width of the disk (rarely wanting); annuals or perennials.
Pappus wanting or a mere trace in the ray flowers.
41. PSILACTIS (p. 690).
Pappus present and usually similar in both ray and disk flowers.
Stigma tips triangular or ovate, obtuse or rarely acutish; bracts not foliaceous.
Involucres turbinate; bracts imbricated in several rows; rays white.
40. LEUCELENE (p. 690).
Involucres hemispheric or broader; bracts in 1 to 3 rows; rays white to purple..............................33. ERIGERON (p. 674).
Stigma tips lanceolate or oblong to filiform; bracts often foliaceous.
Annuals or biennials, without rootstocks; bracts in many series, with herbaceous spreading or filiform tips.
39. MACHAERANTHERA (p. 687).
Perennials, with rootstocks or caudices; bracts various.
Plants with axillary spines; leaves of stems scalelike or wanting.
38. LEUCOSYRIS (p. 686).

Plants not spiny; leaves not scalelike.
Leaves very thick and rigid, spinose-toothed.
37. HERRICKIA (p. 686).
Leaves thin, not spinose-toothed.
Bracts broad, with a distinct keel or midvein, not at all foliaceous; plants more or less glaucous.
36. EUCEPHALUS (p. 686).
Bracts mostly narrow, when broad neither keeled nor with a prominent midvein; plants not glaucous.
35. ASTER (p. 682).
Rays yellow or wanting.
Pappus none to coroniform, paleaceous, or of few rigid awns, never of numerous capillary bristles.
Pappus of few deciduous stout awns; heads large, usually viscid.
11. GRINDELIA (p. 653).
Pappus not of awns, rarely a few in some of the disk flowers, paleaceous or wanting; heads various.
Pappus altogether wanting; rays short, not surpassing the disk corollas; perennial plant, woody at the base.......12. GYMNOSPERMA (p. 656).
Pappus present but often obscure; rays much surpassing the disk corollas; annuals or perennials.
Rays 12, 1 cm. long; pappus coroniform; plant annual; heads large, hemispheric........................13. XANTHOCEPHALUM (p. 656).
Rays 10 or less, less than 5 mm. long; pappus not coroniform; annuals or perennials; heads small.
Only the ray flowers fertile; pappus in disk flowers of aristiform paleæ dilated at the base..................14. AMPHIACHYRIS (p. 656).
All or most of the disk flowers as well as those of the rays fertile; pappus of numerous paleæ, those of the ray small.
15. GUTIERREZIA (p. 656).
Pappus, at least in part, of numerous capillary bristles.
Ray achenes usually without pappus, never with numerous bristles.
16. HETEROTHECA (p. 658).
All achenes with pappus of numerous capillary bristles.
Pappus of 2 series, the inner of capillary bristles, the outer of scales or short bristles...................................17. CHRYSOPSIS (p. 658).
Pappus wholly of capillary bristles.
Rays wanting.
Involucres narrowly turbinate; bracts arranged in vertical ranks; usually shrubs..................18. CHRYSOTHAMNUS (p. 660).
Involucres broadly turbinate or hemispheric; bracts not in vertical ranks; herbs.
Stigma tips obtuse; involucral bracts narrow, slightly imbricated.
(Rayless species of) 33. ERIGERON (p. 674).
Stigma tips acute; bracts broad or well imbricated or both.
Leaves with spine-tipped teeth; corolla tube slender.
(Rayless species of) 19. SIDERANTHUS (p. 663).
Leaves not with spine-tipped teeth; corolla dilated above.
20. ISOCOMA (p. 665).
Rays present.
Shrub....................................21. CHRYSOMA (p. 666).
Herbs.
Leaves pinnately cleft, or the teeth spine-tipped, or both.
19. SIDERANTHUS (p. 663).

Leaves entire or toothed, the teeth never spine-tipped.
Bracts of the involucre longitudinally striate.
22. OLIGONEURON (p. 666).
Bracts not longitudinally striate.
Plants low, cespitose, with short woody stems; leaves evergreen; heads solitary, long-pedunculate....23. STENOTUS (p. 666).
Stems wholly herbaceous; leaves not evergreen; inflorescence various.
Bracts, at least the outer, foliaceous or with prominent foliaceous tips.
Disk flowers tubular; plants with tap-roots.
24. PYRROCOMA (p. 667).
Disk flowers widened upward; plants not with tap-roots.
Heads corymbose; rays few; plants tall.
25. OREOCHRYSUM (p. 667).
Heads solitary; rays numerous; plants low.
26. TONESTUS (p. 667).
Bracts not foliaceous, or merely with green tips.
Rays more numerous than the disk flowers; receptacle fimbriolate; heads corymbose.....27. EUTHAMIA (p. 667).
Rays not more numerous than the disk flowers; receptacle alveolate; inflorescence various.
Inflorescence racemose or paniculate; bracts not in vertical rows; basal leaves not rigid nor sharp-pointed.
28. SOLIDAGO (p. 668).
Inflorescence corymbose; bracts in distinct vertical rows; basal leaves rigid, sharp-pointed.
29. PETRADORIA (p. 671).

Tribe IV. GNAPHALIEAE.

Plants not woolly, tall, usually 1 meter or more.
Stems woody; leaves sericeous; pappus bristles of sterile flowers with thickened tips.......................................44. BERTHELOTIA (p. 693).
Stems herbaceous; leaves not sericeous; pappus bristles not thickened at the tips.
45. PLUCHEA (p. 693).
Plants woolly, much less than 1 meter high, seldom more than 40 cm.
Receptacle chaffy.
Bracts of the receptacle inclosing the achenes and falling with them, hyaline-appendaged................................46. STYLOCLINE (p. 693).
Bracts merely subtending the achenes, persistent, not hyaline-appendaged.
47. EVAX (p. 693).
Receptacle naked.
Plants not diœcious; flowers fertile throughout the heads.
48. GNAPHALIUM (p. 694).
Plants diœcious or the pistillate heads with a few perfect flowers in the center.
Pappus bristles of pistillate flowers falling separately, those of the staminate flowers scarcely clavellate; perfect flowers present in the centers of pistillate heads................49. ANAPHALIS (p. 695).
Pappus bristles of pistillate flowers falling in a ring, those of the staminate flowers clavellate; central perfect flowers none.
50. ANTENNARIA (p. 695).

Tribe V. HELIANTHEAE.

Involucral bracts (uniserial) partly or wholly inclosing the achenes of the ray flowers; pappus none in the ray achenes.
 Rays inconspicuous, scarcely exserted; plants viscid; achenes laterally compressed .... 51. MADIA (p. 697).
 Rays showy; plants not viscid; achenes not laterally compressed.
 52. BLEPHARIPAPPUS (p. 698).
Involucral bracts not inclosing the ray achenes; pappus various.
 Disk flowers sterile.
 Involucres cylindric or fusiform, few-flowered.
 Leaves, at least some of them, connate, simple; achenes oblong, without divergent awns or horns .... 53. GUARDIOLA (p. 698).
 Leaves never connate, dissected into linear-filiform segments; most of the achenes bearing a pair of divergent awns or horns.
 54. DICRANOCARPUS (p. 698).
 Involucres campanulate or hemispheric, many-flowered.
 Rays white.
 Rays long and showy; leaves entire .... 55. MELAMPODIUM (p. 698).
 Rays scarcely if at all exceeding the disk; leaves lobed or pinnatifid.
 56. PARTHENIUM (p. 699).
 Rays yellow.
 Achenes strongly flattened, unicostate on the inner face; pappus nearly obsolete or else evanescent; plants finely canescent.
 57. BERLANDIERA (p. 699).
 Achenes carinate on both surfaces; pappus conspicuous, persistent; plants rough-hirsute .... 58. ENGELMANNIA (p. 700).
 Disk flowers fertile.
 Rays persistent on the achenes and becoming papery.
 Annuals; rays short and inconspicuous, greenish white; leaves petiolate, entire .... 59. SANVITALIA (p. 700).
 Perennials; rays showy, white or yellow; leaves mostly sessile, sometimes toothed.
 Disk achenes compressed; leaves entire; plants low, less than 20 cm. high.
 60. CRASSINA (p. 700).
 Disk achenes obtusely 4-angled; leaves toothed; plants tall, usually 50 cm. high or more .... 61. HELIOPSIS (p. 701).
 Rays deciduous, thin, or wanting.
 Pappus of several to many hyaline paleæ (rarely wanting).
 Rays white; heads very small; leaves opposite ..... 62. GALINSOGA (p. 701).
 Rays yellow or brown; heads large; leaves alternate.
 97. GAILLARDIA (p. 719).
 Pappus various but never of conspicuous hyaline scales.
 Achenes obcompressed; involucre of 2 distinct series.
 Rays pink; achenes slender, beaked; slender annual.
 63. COSMOS (p. 701).
 Rays yellow, rarely wanting; achenes various; annuals or perennials.
 Bracts of inner involucre united at least to the middle.
 64. THELESPERMA (p. 702).
 Bracts distinct or nearly so.
 Achenes not awned; pappus obscure or wanting; rays neutral.
 65. COREOPSIS (p. 703).

Achenes with retrorsely barbed awns; rays sterile or fertile.
Rays sterile; all achenes awned................66. BIDENS (p. 703).
Rays fertile; outer achenes usually not awned.
67. HETEROSPERMUM (p. 705).
Achenes, at least those of the disk, never obcompressed; involucre various.
Rays white; stems prostrate, spreading; bracts of receptacle reduced to awnlike bristles subtending the naked achenes.
68. ECLIPTA (p. 705).
Rays yellow or brown; stems erect; bracts of receptacle concave, loosely subtending the disk achenes.
Receptacle conic or columnar.
Achenes flattened, distinctly margined or winged.
69. RATIBIDA (p. 705).
Achenes not flattened, terete or 4-angled, not margined nor winged.
Achenes terete or nearly so; leaves somewhat glaucous.
70. DRACOPIS (p. 706).
Achenes 4-angled; leaves never glaucous.
Achenes quadrangular-compressed, the apex covered by the base of the corolla tube..................72. GYMNOLOMIA (p. 707).
Achenes nearly equally 4-angled, the apex not covered by the base of the corolla...................71. RUDBECKIA (p. 707).
Receptacle from flat to convex.
Achenes of the disk not winged nor very flat, when flat not margined nor sharp-edged; pappus sometimes deciduous, or often wanting.
Rays fertile.
Pappus of conspicuous awns; leaves not 3-lobed.
74. WYETHIA (p. 709).
Pappus none in the disk achenes; leaves 3-lobed.
73. ZALUZANIA (p. 708).
Rays sterile (rarely wanting).
Pappus none or a minute ring..........72. GYMNOLOMIA (p. 707).
Pappus of 2 awns or paleæ.
Achenes pubescent; pappus persistent or deciduous.
75. VIGUIERA (p. 709).
Achenes glabrous; pappus deciduous..76. HELIANTHUS (p. 710).
Achenes of the disk winged, or flat-compressed and margined, or thin-edged; pappus persistent.
Shrub; rays wanting.......................77. FLOURENSIA (p. 711).
Herbs; rays present.
Achenes not winged nor margined.
Leaves linear; pappus of 2 paleæ without intermediate squamellæ...................................78. ENCELIA (p. 712).
Leaves not linear; paleæ with intermediate squamellæ.
79. HELIANTHELLA (p. 712).
Achenes winged or margined.
Only the ray flowers fertile; perennial.
81. WOOTONELLA (p. 713).
All the flowers fertile; annuals or perennials.
Annual; involucral bracts foliaceous, spreading.
80. XIMENESIA (p. 713).
Perennials; bracts usually not foliaceous, all, or all except the outermost, appressed............82. VERBESINA (p. 713).

Tribe VI. HELENIEAE.

Plant tissues, especially the leaves and involucres, with oil glands.
  Rays wanting; plants tall, usually about a meter high. 83. POROPHYLLUM (p. 714).
  Rays present; plants mostly less than 30 cm. high.
    Pappus wholly of capillary bristles.
      Stems woody almost throughout.......... 84. CHRYSACTINIA (p. 714).
      Stems herbaceous; annuals........................ 85. PECTIS (p. 714).
    Pappus at least in part of paleæ.
      Bracts of the involucre distinct.
        Perennials; leaves alternate; flowers flesh-colored.
          86. NICOLLETIA (p. 715).
        Annuals; leaves opposite; flowers yellow...... 85. PECTIS (p. 714).
      Bracts more or less united.
        Bracts united only at the base.............. 87. BOEBERA (p. 715).
        Bracts united for half their length or more.
          Involucres cylindric; achenes compressed, not striate; annual.
            88. TAGETES (p. 716).
          Involucres campanulate; achenes terete, striate; annuals or perennials.
            Low shrub; leaves all entire; heads sessile or nearly so.
              89. ACIPHYLLAEA (p. 716).
            Herbs; leaves mostly pinnatifid; heads slender-peduncled.
              90. THYMOPHYLLA (p. 716).

Plant tissues without oil-glands.
  Rays persistent; plants more or less woolly.
    Pappus present; rays 3 to 4; heads small, corymbose.
      91. PSILOSTROPHE (p. 718).
    Pappus none; rays 5 to many; heads large, solitary at the ends of the branches.
      92. BAILEYA (p. 718).
  Rays not persistent; plants with various pubescence.
    Achenes flat.
      Pappus of about 20 bristles; plants scapose..... 93. LAPHAMIA (p. 719).
      Pappus a low crown or of 1 or 2 awns; plants not scapose.
        Pappus a lacerate-ciliate crown with sometimes a pair of awns; tall plant 50 cm. high or more with triangular-hastate leaves.
          94. PERICOME (p. 719).
        Pappus of 2 awns; low plants seldom more than 10 cm. high, with pinnately parted leaves.............. 95. PERITYLE (p. 719).
    Achenes angled or terete, never flat.
      Receptacle chaffy.
        Stems woody; leaves fleshy; pappus of 20 to 25 bristles.
          96. CLAPPIA (p. 719).
        Stems herbaceous; leaves not fleshy; pappus of 5 to 10 scarious paleæ.
          97. GAILLARDIA (p. 719).
      Receptacle naked.
        Bracts of the involucre pale or colored, at least the margins and tips scarious.
          Involucres 3 to 9-flowered............ 98. SCHKUHRIA (p. 720).
          Involucres 12 to many-flowered.
            Rays wanting.................... 99. HYMENOPAPPUS (p. 720).

Rays present.
Involucral bracts spatulate or linear-oblanceolate; rays deeply cleft, purple; heads turbinate.
100. OTHAKE (p. 722).
Involucral bracts obovate or broadly oblong; rays not deeply cleft, not purple; heads campanulate.
Rays inconspicuous, scarcely exceeding the disk; heads small; plants glabrous...101. HYMENOTHRIX (p. 722).
Rays large, white; heads large; plants tomentose.
Pappus of numerous prominent paleæ.
99. HYMENOPAPPUS (p. 720).
Pappus an obscure crown ...102. LEUCAMPYX (p. 723).
Bracts neither colored nor scarious.
Flowers pink..........................103. CHAENACTIS (p. 723).
Flowers never pink, mostly yellow.
Achenes 4-angled.
Leaves alternate.
Leaves entire; pappus of 10 paleæ.
107. PLATYSCHKUHRIA (p. 725).
Leaves dissected; pappus wanting or of 12 or more scales.
108. VILLANOVA (p. 725).
Leaves, at least the lower, opposite.
Foliage not impressed-punctate; receptacle conic; leaves entire..........................104. BAERIA (p. 723).
Foliage impressed-punctate; receptacle flat or convex; leaves dissected.
Annuals; leaf segments filiform or nearly so.
106. ACHYROPAPPUS (p. 724).
Perennials; leaf segments oblong to linear.
Leaf blades impressed-punctate; involucral bracts carinate............105. PICRADENIOPSIS (p. 724).
Leaf blades not impressed-punctate; bracts not carinate ....105a. BAHIA (p. 724).
Achenes 5 to 10-ribbed.
Bracts of the involucre spreading or reflexed.
Leaves decurrent; tubes of disk corollas very short or reduced to a ring......................109. HELENIUM (p. 726).
Leaves not decurrent; tube of the disk corollas long.
110. DUGALDEA (p. 726).
Bracts of the involucre erect, never spreading nor reflexed.
Pappus wanting; involucres few-flowered.
111. FLAVERIA (p. 726).
Pappus present; involucres many-flowered.
Bracts of the involucre unequal, the outer united at the base.
Leaves opposite; plant glabrous.
112. SARTWELLIA (p. 727).
Leaves alternate or basal; plants more or less pubescent.
Stems monocephalous; plants woody throughout.
113. MACDOUGALIA (p. 727).
Stems branched and bearing numerous heads; plants sparingly if at all woolly.
114. HYMENOXYS (p. 727).

Bracts of the involucre nearly equal and alike; all distinct.
Leaves dissected; stems leafy throughout.
115. RYDBERGIA (p. 730).
Leaves entire; stems naked or with only a few leaves.
116. TETRANEURIS (p. 730).

Tribe VII. ANTHEMIDEAE.

Receptacle chaffy; rays white.
Achenes flattened; heads small..........................117. ACHILLEA (p. 733).
Achenes terete; heads large............................118. ANTHEMIS (p. 733).
Receptacle naked; rays white or usually wanting.
Heads radiate; rays white.....................119. CHRYSANTHEMUM (p. 734).
Heads discoid.
Heads corymbose, bright yellow, showy...........120. TANACETUM (p. 734).
Heads not corymbose, not bright yellow, inconspicuous.
Plants spiny; achenes cobwebby..........121. PICROTHAMNUS (p. 734).
Plants not spiny; achenes not cobwebby.......122. ARTEMISIA (p. 734).

Tribe VIII. SENECIONEAE.

Involucre of 4 to 6 bracts; stems woody throughout.......123. TETRADYMIA (p. 739).
Involucre of more than 6 bracts; stems woody at the base or entirely herbaceous.
Pappus of plumose bristles; leaves mostly wanting or reduced to scales.
124. BEBBIA (p. 740).
Pappus not of plumose bristles; leaves evident.
Involucral bracts erect-connivent, equal, sometimes with a few short outer ones.
Leaves opposite; pappus a single series of rigid bristles.
125. ARNICA (p. 740).
Leaves alternate; pappus of numerous soft bristles.
Heads usually radiate; corollas yellow; leaves bipinnatifid to entire.
126. SENECIO (p. 740).
Heads discoid; corollas white; leaves thrice pinnatifid into narrow segments..........................130. MESADENIA (p. 749).
Involucral bracts not erect-connivent, usually unequal and overlapping.
Leaves opposite, linear-filiform.............127. HAPLOESTHES (p. 748).
Leaves alternate, not linear-filiform.
Heads homogamous; achenes terete.....128. PSATHYROTES (p. 748).
Heads heterogamous; achenes compressed.
129. BARTLETTIA (p. 749).

Tribe IX. CYNAREAE.

Annual; achenes attached obliquely.......................131. CENTAUREA (p. 749).
Perennials or biennials; achenes not attached obliquely.
Leaves prickly; involucral bracts not slender-subulate, the tips not hooked.................................132. CIRSIUM (p. 749).
Leaves not prickly; involucral bracts slender-subulate, hooked...133. ARCTIUM (p. 753).

1. VERNONIA Schreb. IRONWEED.

Tall herbaceous perennials with leafy stems and corymbose inflorescence; heads many-flowered, campanulate, the bracts closely imbricated in several series; flowers purple; achenes ribbed, the pappus in two series, the outer of short scales or bristles, the inner long and capillary.

KEY TO THE SPECIES.

Leaves lanceolate, tomentose beneath .... 1. *V. missurica.*
Leaves narrowly linear-lanceolate, glabrous .... 2. *V. marginata.*

1. Vernonia missurica Raf. Herb. Raf. 28. 1833.
Vernonia drummondii Werner, Journ. Cincinnati Soc. Nat. Hist. **16:** 171. 1894.
TYPE LOCALITY: "In Missouri, barrens."
RANGE: Kansas and Illinois to Oklahoma, Texas, and eastern New Mexico.
NEW MEXICO: Capitan Mountains (*Earle* 540).

2. Vernonia marginata (Torr.) Raf. Atl. Journ. **1:** 146. 1832.
Vernonia altissima marginata Torr. Ann. Lyc. N. Y. **2:** 210. 1828.
Vernonia jamesii Torr. & Gray, Fl. N. Amer. **2:** 58. 1841.
TYPE LOCALITY: "On the Arkansa ?"
RANGE: Kansas and Oklahoma to Texas and southeastern New Mexico.
NEW MEXICO: Roswell; Buchanan; Quay. Prairies, in the Upper Sonoran Zone.

2. CARMINATIA Mocino.

Low slender annual with opposite or alternate thin long-petiolate ovate sinuate-serrate leaves; heads racemosely paniculate, cylindric, about 12 mm. high, several-flowered; bracts linear-lanceolate, thin, imbricated, striate; flowers whitish; achenes slender, with pappus of 10 to 18 plumose bristles slightly coherent at the base.

1. Carminatia tenuiflora DC. Prodr. **7:** 267. 1838.
TYPE LOCALITY: Near Guanajuato, Mexico.
RANGE: New Mexico and Arizona to Mexico.
NEW MEXICO: Kingston; Mogollon Mountains; Florida Mountains; San Luis Mountains; Organ Mountains. Canyons, in the Upper Sonoran and Transition zones.

3. EUPATORIUM L.

Erect coarse perennial herbs with verticillate, opposite, or alternate, entire or toothed leaves and corymbose heads of whitish or purplish flowers; heads discoid, 3 to many-flowered; involucre cylindric to campanulate, of numerous imbricated bracts; receptacle flat, naked; achenes 5-angled, the pappus a single row of capillary slightly roughened bristles.

KEY TO THE SPECIES.

Leaves verticillate in 3's or 4's; flowers purplish .... 1. *E. bruneri.*
Leaves opposite; flowers white.
  Bracts glabrous; leaves truncate or tapering at the base .. 2. *E. rothrockii.*
  Bracts pubescent; leaves usually cordate.
    Heads 3 to 5-flowered; leaves ovate-lanceolate .... 3. *E. solidaginifolium.*
    Heads 12 to 25-flowered; leaves ovate-cordate.
      Stems woody; leaves less than 15 mm. long; pappus and bracts purplish .... 4. *E. wrightii.*
      Stems herbaceous; leaves 30 mm. long or more; heads not purplish.
        Bracts conspicuously nerved, villous and ciliate; heads about 7 mm. high ...... 5. *E. fendleri.*
        Bracts not conspicuously nerved, minutely puberulent; heads mostly 5 mm. high. 6. *E. arizonicum.*

1. Eupatorium bruneri A. Gray, Syn. Fl. 1^2: 96. 1884. JOE PYE WEED.
Eupatorium rydbergii Britton, Man. 921. 1901.
TYPE LOCALITY: Fort Collins, Colorado.
RANGE: British Columbia and Minnesota to Utah and New Mexico.
NEW MEXICO: South Fork of Tularosa Creek (*Wooton*). Transition Zone.

2. **Eupatorium rothrockii** A. Gray, Syn. Fl. 1[2]: 102. 1884.

TYPE LOCALITY: Mount Graham, Arizona.

RANGE: Mountains of Arizona and southern New Mexico.

NEW MEXICO: White Mountains; Cloudcroft; Capitan Mountains. Transition Zone.

3. **Eupatorium solidaginifolium** A. Gray, Pl. Wright. 1: 87. 1852.

TYPE LOCALITY: Mountains between the Limpio and the Rio Grande, western Texas.

RANGE: Western Texas to southern Arizona.

NEW MEXICO: Guadalupe Pass (*Wright* 1146).

Guadalupe Pass is on the southern boundary of the State, and Wright's specimens may have come from either Mexico or New Mexico.

4. **Eupatorium wrightii** A. Gray, Pl. Wright. 1: 87. 1852.

TYPE LOCALITY: Sides of the Guadalupe Mountains, 40 miles east of El Paso, Texas.

RANGE: Western Texas to southern Arizona.

NEW MEXICO: Bishops Cap (*Wooton*). Dry hills, in the Upper Sonoran Zone.

This should be found at other places in the southern part of the State. It is a low shrub with small, thick, scabrous, ovate leaves and numerous small heads conspicuously tinged with purple on the bracts and at the base of the pappus.

5. **Eupatorium fendleri** A. Gray, Proc. Amer. Acad. 17: 205. 1882.

Brickellia fendleri A. Gray, Mem. Amer. Acad. n. ser. 4: 63. 1849.

TYPE LOCALITY: "Foot of mountains, on the sunny side along the creek, 11 miles above Santa Fe," New Mexico. Type collected by Fendler (no. 347).

RANGE: Mountains of New Mexico and Arizona.

NEW MEXICO: Santa Fe and Las Vegas mountains; Sandia Mountains; Black Range; Mogollon Mountains. Transition Zone.

6. **Eupatorium arizonicum** (A. Gray) Greene, Pittonia 4: 280. 1901.

Eupatorium ageratifolium var.? *herbaceum* A. Gray, Pl. Wright. 2: 74. 1853.

Eupatorium occidentale arizonicum A. Gray, Syn. Fl. 1[2]: 101. 1884.

TYPE LOCALITY: "Mountains, east of Santa Cruz, Sonora (a small-leaved form); also at Guadalupe Pass, and at the Copper Mines, under trees." The last two localities are in New Mexico.

RANGE: Mountains of Arizona and New Mexico.

NEW MEXICO: Las Vegas Canyon; Santa Rita; Burro Mountains; Mangas Springs; Organ Mountains; Cloudcroft. Transition Zone.

4. CONOCLINIUM DC.

Branched perennial herb with opposite, palmately cleft or parted leaves and corymbose clusters of heads on naked peduncle-like branches; involucre campanulate, 4 to 6 mm. high, the bracts linear; corolla bluish purple; achenes narrow, 5-angled, truncate; pappus of few slender bristles in a single series; receptacle conic, naked.

1. **Conoclinium dissectum** A. Gray, Pl. Wright. 1: 88. 1852.

Eupatorium dissectum A. Gray, Proc. Amer. Acad. 18: 100. 1853, not Benth. 1844.

Eupatorium greggii A. Gray, Syn. Fl. 1[2]: 102. 1884.

Conoclinium greggii Small, Fl. Southeast. U. S. 1169. 1903.

TYPE LOCALITY: "Damp place, Rio Seco, and on the Rio Grande, Texas." Type collected by Wright (no. 258).

RANGE: Western Texas to southeastern Arizona, south to northern Mexico.

NEW MEXICO: San Andreas Mountains (*Wooton*). Upper Sonoran Zone.

Our specimens have unusually large heads. The plants from Arizona have commonly larger heads than those from Texas and Mexico.

5. COELESTINA Cass.

Perennial herb, suffrutescent at the base, puberulent; leaves opposite, petiolate, the blades ovate or deltoid, crenate, thick, gland-dotted beneath; heads few, campanulate, many-flowered, glomerate at the ends of the long naked branches; bracts linear, appressed, striate; flowers bluish; achenes 5-angled, the pappus a short dentate crown.

1. Coelestina sclerophylla Woot. & Standl. Contr. U. S. Nat. Herb. **16**:176. 1913.

TYPE LOCALITY: In Guadalupe Canyon, Sonora. Type collected by E. C. Merton (no. 2031).

RANGE: Northern Sonora to southwestern New Mexico and southeastern Arizona.

NEW MEXICO: Guadalupe Canyon.

6. STEVIA Cav.

Annual or perennial herbs with opposite or alternate, mostly narrow, entire or toothed, sessile leaves and small narrow heads in panicles or corymbs; heads cylindric, the flowers white or purplish; achenes linear, sometimes compressed; pappus paleaceous or aristiform, or of both awns and short scales.

KEY TO THE SPECIES.

Annual; heads loosely paniculate .... 1. *S. micrantha.*
Perennials; heads corymbose.
  Upper leaves mostly alternate, linear to linear-lanceolate; stems rather densely leafy, pubescent or hirsute .... 2. *S. serrata.*
  All leaves opposite, lanceolate (conspicuously veined); stems less leafy, puberulent .... 3. *S. plummerae.*

1. Stevia micrantha Lag. Gen. & Sp. Nov. 27. 1816.

Stevia macella A. Gray, Pl. Wright. **2**: 70. 1853.

TYPE LOCALITY: Mexico.

RANGE: New Mexico and Arizona to Mexico.

NEW MEXICO: Fort Bayard (*Blumer* 128). Low mountains.

2. Stevia serrata Cav. Icon. Pl. **4**: 33. *pl. 35.* 1797.

TYPE LOCALITY: "Habitat in Nova-Hispania."

RANGE: Southern New Mexico and Arizona to Mexico.

NEW MEXICO: Mogollon Mountains; Hanover Mountain; Sacramento Mountains. Hills and mountains, in the Upper Sonoran and Transition zones.

3. Stevia plummerae A. Gray, Proc. Amer. Acad. **17**: 204. 1882.

TYPE LOCALITY: Rucker Valley, Chiricahua Mountains, southern Arizona.

RANGE: Mountains of southern Arizona and southwestern New Mexico.

NEW MEXICO: Mogollon Mountains; Sawyers Peak. Transition Zone.

7. CARPOCHAETE A. Gray.

Low shrub, 40 cm. high or less, with slender brittle branches; leaves opposite, entire, sessile, spatulate-oblong, bearing fascicles of leaves in their axils; heads solitary or clustered; flowers rose-colored; involucre cylindric, of few acuminate bracts; achenes puberulent, the pappus paleaceous-aristiform.

1. Carpochaete bigelovii A. Gray, Pl. Wright. **1**: 89. 1852.

TYPE LOCALITY: "On the boundary between Mexico and New Mexico." Type collected by Bigelow.

RANGE: Southern New Mexico and Arizona and southward.

NEW MEXICO: Emory Peak; Organ Mountains. Dry hills and canyons, in the Upper Sonoran Zone.

8. LACINIARIA Hill. Blazing star.

Handsome perennial herbs with thick globose rootstocks; stems simple, leafy, bearing large rose-purple heads in racemes or spikes; leaves alternate, narrow, entire; heads 4 to many-flowered; involucral bracts spirally imbricated; receptacle naked; achenes slender, pubescent; pappus a single series of plumose or merely barbellate bristles.

KEY TO THE SPECIES.

Pappus plumose; bracts abruptly acuminate .... 1. *L. punctata.*
Pappus merely barbellate; bracts rounded-obtuse.
  Heads many, nearly sessile, 1 cm. broad or less; some of the leaves trinervate .... 2. *L. lancifolia.*
  Heads few, pedunculate, 2 cm. broad; none of the leaves trinervate .... 3. *L. ligulistylis.*

1. Laciniaria punctata (Hook.) Kuntze, Rev. Gen. Pl. **2**: 349. 1891.

Liatris punctata Hook. Fl. Bor. Amer. **1**: 306. *pl. 105.* 1833.

Type locality: "Plains of the Saskatchewan, *Drummond;* and on the Red Deer and Eagle hills, in dry soils."

Range: Montana and Saskatchewan to Iowa, Arizona, and Texas.

New Mexico: Gallinas Planting Station; Clovis; Pecos; Folsom; Logan; Capitan Mountains; Colfax; Johnsons Mesa; Raton Mountains; Nara Visa; Melrose. Dry plains, in the Upper Sonoran Zone.

2. Laciniaria lancifolia Greene, Bull. Torrey Club **25**: 118. 1898.

Type locality: White Mountains, New Mexico. Type collected by Wooton (no. 254).

Range: Southeastern New Mexico.

New Mexico: White Mountains; Roswell. Upper Sonoran Zone.

3. Laciniaria ligulistylis A. Nels. Bot. Gaz. **31**: 405. 1901.

Liatris ligulistylis A. Nels. in Coulter, New Man. Rocky Mount. 488. 1909.

Type locality: Laramie Peak, Wyoming.

Range: Wyoming and Black Hills of South Dakota to northeastern New Mexico.

New Mexico: Sierra Grande (*Howell* 212).

9. KUHNIA L.

Low, much branched, perennial herbs with narrow entire alternate leaves and paniculate-corymbose discoid heads of whitish flowers; heads rather few-flowered, the flowers perfect; involucral bracts thin, striate-nerved, narrow, loosely imbricated; achenes cylindric, 10-striate; pappus a single row of plumose bristles.

KEY TO THE SPECIES.

Leaves linear; bracts narrow, thin, straw-colored, in 2 evident series, pubescent only on the margins, strongly glandular.. 1. *K. rosmarinifolia.*
Leaves mostly linear-lanceolate; bracts broad, thick, green, not in 2 evident series, finely pubescent, sparingly if at all glandular .... 2. *K. chlorolepis.*

1. Kuhnia rosmarinifolia Vent. Pl. Jard. Cels *pl. 91.* 1800.

Eupatorium canescens Orteg. Hort. Matr. Dec. 34. 1797–1800, not Vahl, 1793.

Kuhnia leptophylla Scheele, Linnaea **21**: 598. 1849.

Type locality: Given as Cuba, but this is probably incorrect and should be Mexico.

Range: Texas and Arizona to Mexico.

New Mexico: Dulce; Pajarito Park; Cleveland; Pecos; Laguna; Anton Chico; Socorro; Kingston; Mogollon Mountains; Dona Ana and Organ mountains; White and Sacramento mountains; Artesia; Carlsbad; Nara Visa. Dry hills, in the Upper Sonoran and lower part of the Transition zones.

2. Kuhnia chlorolepis Woot. & Standl. Contr. U. S. Nat. Herb. **16**: 177. 1913.

TYPE LOCALITY: Mangas Springs, New Mexico. Type collected by Metcalfe (no. 104).

RANGE: Southwestern New Mexico and adjacent Arizona and Mexico.

NEW MEXICO: Mangas Springs; Cliff; Alamo Viejo. Open hillsides.

10. COLEOSANTHUS Cass.

Large herbs or low shrubs with opposite or alternate leaves and variously arranged discoid heads of white or greenish flowers; bracts thin, striate, regularly imbricated, the outer shorter; receptacle naked; pappus a single series of capillary barbellate bristles.

KEY TO THE SPECIES.

Heads 25 to 50-flowered; plants herbaceous to the base.
  Peduncles densely glandular-viscid; heads rather small.... 1. *C. modestus.*
  Peduncles not viscid; heads large.
    Peduncles mostly shorter than the few congested heads.................................... 2. *C. umbellatus.*
    Peduncles equaling or exceeding the numerous heads.
      Outer bracts merely acute, the inner obtuse; leaves acute, subcordate........................ 3. *C. ambigens.*
      Outer bracts long-acuminate, the inner acute; leaves various.
        Heads few, on slender erect branches; leaves attenuate, truncate or cuneate at the base.................................. 4. *C. grandiflorus.*
        Heads numerous, on stout spreading branches; leaves merely acute, subcordate........ 5. *C. petiolaris.*
Heads 9 to 25-flowered; plants mostly woody at the base or nearly throughout.
  Leaves sessile, or the lower short-petiolate.
    Leaves alternate, merely scaberulous or glabrous.
      Leaves serrate; heads axillary...................... 6. *C. brachyphyllus.*
      Leaves entire; heads in terminal corymbs........ 7. *C. linifolius.*
    Leaves opposite, abundantly pubescent.
      Stems hirsute; leaves crenate...................... 8. *C. betonicaefolius.*
      Stems canescent or puberulent; leaves not crenate.
        Leaves narrowly oblong or linear; heads long-pedunculate............................ 9. *C. venosus.*
        Leaves lanceolate; heads on very short stout peduncles ....10. *C. wootoni.*
  Leaves all conspicuously petiolate.
    Bracts and peduncles glandular-viscid.
      Tips of bracts spreading; heads terminating slender branches covered with reduced bractlike leaves.....................................11. *C. scaber.*
      Tips of bracts appressed; heads on short naked peduncles.
        Heads nearly sessile; leaves very small, laciniate-toothed.
          Leaves coriaceous, very viscid; heads 15 to 18-flowered........................13. *C. baccharideus.*
          Leaves thin, scabrous; heads 9 to 12-flowered...............................12. *C. laciniatus.*

Heads conspicuously pedunculate; leaves larger, not deeply toothed.
Peduncles 2 to 4 cm. long; leaves rounded or cuneate at the base.............14. *C. chenopodinus.*
Peduncles 1 cm. long or less; leaves truncate or subcordate at the base......15. *C. floribundus.*
Bracts and peduncles not viscid.
Leaves longer than broad, mostly attenuate.
Leaves long-attenuate, seldom toothed above the middle, thin, sparingly pubescent..16. *C. rusbyi.*
Leaves usually merely acute, toothed almost to the apex, thick, densely pubescent..17. *C. wrightii.*
Leaves about as broad as long or broader, acutish or obtuse.
Leaves 3 to 5 cm. wide, thin, bright green.....18. *C. axillaris.*
Leaves 25 mm. wide or narrower, thick, grayish................................19. *C. reniformis.*

1. Coleosanthus modestus Greene, Pittonia **4:** 230. 1900.

Type locality: Grays Peak, Lincoln County, New Mexico. Type collected by Earle (no. 161).

Range: Vicinity of the type locality.

2. Coleosanthus umbellatus Greene, Pittonia **4:** 238. 1901.

Brickellia grandiflora minor A. Gray, Proc. Acad. Phila. **1863:** 67. 1864.

Coleosanthus congestus A. Nels. Bot. Gaz. **31:** 401. 1901.

Type locality: "Rather common in the mountains of northern Arizona."

Range: Wyoming to New Mexico and Arizona.

New Mexico: Tunitcha Mountains; Dulce; Sandia Mountains; Raton Mountains; Copper Canyon; Eagle Peak; Kingston; Organ Mountains; Capitan Mountains. Shaded mountain slopes, in the Transition Zone.

3. Coleosanthus ambigens Greene, Bull. Torrey Club **25:** 118. *pl. 330.* 1898.

Type locality: White Mountains, New Mexico. Type collected by Wooton (no. 335).

Range: White Mountains of New Mexico, in the Transition Zone.

4. Coleosanthus grandiflorus (Hook.) Kuntze, Rev. Gen. Pl. **1:** 328. 1891.

Eupatorium ? grandiflorum Hook. Fl. Bor. Amer. **2:** 26. 1834.

Brickellia grandiflora Nutt. Trans. Amer. Phil. Soc. n. ser. **7:** 287. 1841.

Type locality: "In the Rocky Mountain range by streams in gravelly places, and west to the lower falls of the Columbia."

Range: Washington and Montana to Arizona and New Mexico.

New Mexico: Santa Fe and Las Vegas mountains; Mogollon Mountains; Mangas Springs; Fort Bayard; San Luis Mountains; White and Sacramento mountains. Shaded mountain slopes, in the Transition and Canadian zones.

5. Coleosanthus petiolaris (A. Gray) Greene, Bull. Torrey Club **25:** 117. 1898.

Brickellia grandiflora petiolaris A. Gray, Proc. Amer. Acad. **17:** 207. 1882.

Type locality: Mountains of southern Arizona.

Range: Mountains of southern Arizona and New Mexico.

New Mexico: Organ Mountains. Transition Zone.

6. Coleosanthus brachyphyllus (A. Gray) Kuntze, Rev. Gen. Pl. **1:** 328. 1891.

Clavigera brachyphylla A. Gray, Mem. Amer. Acad. n. ser. **4:** 63. 1849.

Brickellia brachyphylla A. Gray, Pl. Wright. **1:** 84. 1852.

Type locality: "Foot of high rocks, 2 miles east of the Mora River," New Mexico.

Type collected by Fendler (no. 339).

RANGE: Texas and New Mexico to Arizona

NEW MEXICO: Dulce; Sierra Grande; Cerrillos; Sandia Mountains; Santa Rita; Mogollon Mountains; Mangas Springs; Black Range; White and Capitan mountains. Dry hills, in the Upper Sonoran Zone.

7. Coleosanthus linifolius (D. C. Eaton) Kuntze, Rev. Gen. Pl. **1:** 328. 1891.

Brickellia linifolia D. C. Eaton in King, Geol. Expl. 40th Par. **5:** 137. *pl. 15.* 1871.

Coleosanthus humilis Greene, Pittonia **4:** 124. 1900.

Brickellia humilis A. Nels. in Coulter, New Man. Rocky Mount. 487. 1909.

TYPE LOCALITY: Sandy bottoms of American Fork, Jordan Valley, Utah.

RANGE: Colorado and New Mexico to Nevada and California.

NEW MEXICO: Carrizo Mountains; Farmington. Dry hills and plains, in the Upper Sonoran Zone.

A low plant, only about 20 cm. high, with simple erect stems in dense clumps.

8. Coleosanthus betonicaefolius (A. Gray) Kuntze, Rev. Gen. Pl. **1:** 328. 1891.

Brickellia betonicaefolia A. Gray, Pl. Wright. **2:** 72. 1853.

TYPE LOCALITY: Hills near the Copper Mines, New Mexico. Type collected by Wright (no. 1137).

RANGE: Southern New Mexico and Arizona and southward.

NEW MEXICO: Kingston; Santa Rita; West Fork of the Gila; Capitan Mountains. Dry hills.

9. Coleosanthus venosus Woot. & Standl. Contr. U. S. Nat. Herb. **16:** 177. 1913.

TYPE LOCALITY: Mangas Springs, New Mexico. Type collected by Metcalfe (no. 653).

RANGE: Southern New Mexico and Arizona and adjacent Mexico.

NEW MEXICO: Mangas Springs; Burro Mountains; San Luis Mountains. Dry hills, in the Upper Sonoran Zone.

10. Coleosanthus wootoni Greene, Bull. Torrey Club **24:** 511. 1897.

TYPE LOCALITY: Organ Mountains, New Mexico. Type collected by Wooton.

RANGE: Mountains of southern New Mexico.

NEW MEXICO: Hillsboro; Organ Mountains. Upper Sonoran Zone.

11. Coleosanthus scaber Greene, Pittonia **3:** 100. 1896.

Brickellia scabra A. Nels. in Coulter, New Man. Rocky Mount. 487. 1909.

TYPE LOCALITY: Mountains near Grand Junction, Colorado.

RANGE: Utah and Colorado to Arizona and New Mexico.

NEW MEXICO: Carrizo Mountains; Farmington. Dry hills among rocks, in the Upper Sonoran Zone.

In general appearance this is very unlike our other species. It grows usually about the edges of cliffs, in large clumps, with its slender wiry branches densely interlaced.

12. Coleosanthus laciniatus (A. Gray) Kuntze, Rev. Gen. Pl. **1:** 328. 1891.

Brickellia laciniata A. Gray, Pl. Wright. **1:** 87. 1852.

TYPE LOCALITY: "Mountain valley, 40 miles east of El Paso," Texas.

RANGE: Western Texas to southern New Mexico and Mexico.

NEW MEXICO: Mesa west of Organ Mountains. Along arroyos and in dry foothills, Lower Sonoran Zone.

13. Coleosanthus baccharideus (A. Gray) Kuntze, Rev. Gen. Pl. **1:** 328. 1891.

Brickellia baccharidea A. Gray, Pl. Wright. **1:** 87. 1852.

TYPE LOCALITY: "Mountains near El Paso," Texas or Chihuahua.

RANGE: Western Texas to southern Arizona.

We have seen no specimens of this, but Doctor Gray reported [1] that it was collected by Bigelow near Santa Rita.

---

[1] Torr. U. S. & Mex. Bound. Bot. 75. 1859.

14. Coleosanthus chenopodinus Greene, Contr. U. S. Nat. Herb. **16:** 177. 1913.

TYPE LOCALITY: Gila River bottoms near Cliff, Grant County, New Mexico. Type collected by Metcalfe (no. 776).

RANGE: Known only from type locality.

15. Coleosanthus floribundus (A. Gray) Kuntze, Rev. Gen. Pl. **1:** 328. 1891.

Brickellia floribunda A. Gray, Pl. Wright. **2:** 73. 1853.

TYPE LOCALITY: Ravines near Santa Cruz, and on rocky banks of the San Pedro, Sonora.

RANGE: Southern Arizona and New Mexico to northern Mexico.

NEW MEXICO: Kingston; Burro Mountains; Dog Spring. Lower and Upper Sonoran zones.

16. Coleosanthus rusbyi (A. Gray) Kuntze, Rev. Gen. Pl. **1:** 328. 1891.

Brickellia rusbyi A. Gray, Syn. Fl. 1^2: 106. 1884.

TYPE LOCALITY: Mountains of New Mexico. Type collected by Rusby.

RANGE: Southern New Mexico and Arizona.

NEW MEXICO: South Percha Creek; Organ Mountains. Upper Sonoran Zone.

17. Coleosanthus wrightii (A. Gray) Britton, Trans. N. Y. Acad. **14:** 43. 1894.

Brickellia wrightii A. Gray, Pl. Wright. **2:** 72. 1853.

TYPE LOCALITY: Hills near the Copper Mines, New Mexico. Type collected by Wright (no. 1139).

RANGE: Hills of southwestern New Mexico.

NEW MEXICO: Sapello Creek; Santa Rita; Capitan Mountains. Upper Sonoran Zone.

It is a question whether the preceding and the two succeeding species should not be united with this. The four certainly differ very little from each other, probably no more than is to be expected from the various environmental conditions under which they grow.

18. Coleosanthus axillaris Greene, Leaflets **1:** 149. 1905.

TYPE LOCALITY: Southward slopes of the Black Range, New Mexico. Type collected by Metcalfe (no. 1446).

RANGE: Mountains of southwestern New Mexico.

NEW MEXICO: Near Monument 26; west of Hillsboro. Upper Sonoran Zone.

19. Coleosanthus reniformis (A. Gray) Rydb. Bull. Torrey Club **31:** 646. 1905.

Brickellia reniformis A. Gray, Pl. Wright. **1:** 86. 1852.

Coleosanthus melissaefolius Greene, Leaflets **1:** 150. 1905.

Coleosanthus albicaulis Rydb. loc. cit.

TYPE LOCALITY: Mountain valley 35 miles east of El Paso, Texas.

RANGE: Utah and Colorado to Arizona and western Texas.

NEW MEXICO: Cedar Hill; Carrizo Mountains; San Domingo; Sandia Mountains; Mangas Springs; San Mateo Mountains; Fort Bayard; Florida Mountains; Dona Ana and Organ mountains. Dry hills, in the Lower and Upper Sonoran zones.

The type of *C. melissaefolius* was collected in the Organ Mountains (*Wooton* in 1897).

11. GRINDELIA Willd. GUM PLANT.

Coarse biennial or perennial, resinous herbs with thick, rigid, entire or serrate leaves and numerous discoid or radiate, rather large heads of yellow flowers; involucre hemispheric or globose, of numerous much imbricated, narrow, erect or recurved, often strongly viscid bracts; achenes short and thick, mostly compressed, glabrous, the pappus of 2 to 8 caducous, smooth or barbellate, stout awns.

KEY TO THE SPECIES.

Heads discoid.
- Leaves merely dentate .... 1. *G. aphanactis.*
- Leaves laciniately toothed, the lower ones pinnatifid with dentate segments .... 2. *G. pinnatifida.*

Heads radiate.
- Peduncles pubescent .... 3. *G. scabra.*
- Peduncles glabrous.
  - Pappus awns conspicuously barbellate .... 4. *G. subalpina.*
  - Pappus awns smooth except under a compound microscope.
    - Tips of the outer bracts spreading, none reflexed.
      - Disk about 20 mm. broad; cauline leaves ovate to obovate .... 5. *G. arizonica.*
      - Disk 12 mm. broad or less; cauline leaves oblanceolate.
        - Leaves conspicuously spinulose-toothed; bracts scarcely at all viscid .... 6. *G. setulifera.*
        - Leaves not spinulose-toothed; bracts strongly viscid .... 7. *G. decumbens.*
    - Tips of the outer bracts squarrose, those of the outermost strongly reflexed.
      - Cauline leaves ovate or oblong, broadest at the base; tips of the bracts subulate.
        - Leaves bluish green, spinulose-dentate; heads very broad and flat, the bracts broad, with flattened tips .... 8. *G. texana.*
        - Leaves yellowish green, merely dentate; heads hemispheric, the bracts narrow, with terete tips .... 9. *G. squarrosa.*
      - Cauline leaves narrowly oblong or oblanceolate, not broadest at the base; tips of the bracts flat, never subulate.
        - Leaves coarsely and irregularly incised-toothed; heads densely viscid, 15 mm. wide or smaller .... 10. *G. subincisa.*
        - Leaves evenly and finely serrate; heads various.
          - Bracts abundantly viscid, the tips narrow and thick, strongly reflexed .... 11. *G. serrulata.*
          - Bracts scarcely at all viscid, the tips broad and flat, slightly reflexed .... 12. *G. neomexicana.*

1. Grindelia aphanactis Rydb. Bull. Torrey Club **31:** 647. 1904.

TYPE LOCALITY: Durango, Colorado.

RANGE: Southern Colorado to Arizona and New Mexico.

NEW MEXICO: Farmington; Dulce; Santa Fe; Glorieta; Ramah; Watrous; Pajarito Park; Gallinas Planting Station; Raton Mountains; Gallup; Belen; Socorro; Laguna; Albuquerque; Mesilla Valley; Organ Mountains; White Oaks; Roswell; Lake Valley. Plains and low hills, in the Lower and Upper Sonoran zones.

2. Grindelia pinnatifida Woot. & Standl. Contr. U. S. Nat. Herb. **16:** 178. 1913.

TYPE LOCALITY: Open slopes about Chama, New Mexico. Type collected by Standley (no. 6606).

RANGE: Vicinity of the type locality, in the Upper Sonoran and Transition zones.

3. **Grindelia scabra** Greene, Bull. Torrey Club **25**: 120. 1898.

TYPE LOCALITY: White Mountains, New Mexico. Type collected by Wooton (no. 224).

RANGE: New Mexico.

NEW MEXICO: Sandia Mountains; White and Sacramento mountains. Meadows in the mountains, Transition Zone.

A common and handsome plant of the open parks of the mountains.

4. **Grindelia subalpina** Greene, Pittonia **3**: 297. 1898.

TYPE LOCALITY: "High plains of southern Wyoming, and at subalpine elevations on the mountains of northern Colorado."

RANGE: Montana and Utah to Colorado and northern New Mexico.

NEW MEXICO: Johnsons Mesa; Raton; Sierra Grande. Mountains, in the Upper Sonoran and Transition zones.

5. **Grindelia arizonica** A. Gray, Proc. Amer. Acad. **17**: 208. 1882.

TYPE LOCALITY: Arizona.

RANGE: Southern Arizona and New Mexico.

NEW MEXICO: Bear Mountains; Santa Rita. Transition Zone.

6. **Grindelia setulifera** Woot. & Standl. Contr. U. S. Nat. Herb. **16**: 179. 1913.

TYPE LOCALITY: High summits of the Mogollon Mountains, New Mexico. Type collected by Rusby (no. 206).

RANGE: Known only from type locality.

7. **Grindelia decumbens** Greene, Pittonia **4**: 102. 1896.

TYPE LOCALITY: Mountains about Cimarron, Colorado.

RANGE: Kansas and Colorado to northern New Mexico.

NEW MEXICO: Aztec; Farmington; Dulce. Plains and low hills, in the Upper Sonoran and lower part of the Transition Zone.

8. **Grindelia texana** Scheele, Linnaea **21**: 601. 1849.

TYPE LOCALITY: New Braunfels, Texas.

RANGE: Colorado and New Mexico to Texas.

NEW MEXICO: Raton. Plains and low hills, in the Upper Sonoran Zone.

9. **Grindelia squarrosa** (Pursh) Dunal in DC. Prodr. **5**: 315. 1836.

Donia squarrosa Pursh, Fl. Amer. Sept. 559. 1814.

TYPE LOCALITY: "In open prairies, on the banks of the Missouri."

RANGE: Wyoming and Iowa to Arizona and Texas.

NEW MEXICO: West of Roswell. Plains, in the Upper Sonoran Zone.

10. **Grindelia subincisa** Greene, Pittonia **4**: 154. 1900.

TYPE LOCALITY: Chama, New Mexico. Type collected by Baker (no. 683).

RANGE: Northern New Mexico and southern Colorado.

NEW MEXICO: Chama. Meadows in the mountains, in the Transition Zone.

11. **Grindelia serrulata** Rydb. Bull. Torrey Club **31**: 646. 1905.

TYPE LOCALITY: Fort Collins, Colorado.

RANGE: Wyoming to northern New Mexico.

NEW MEXICO: Cedar Hill (*Standley* 8044). Plains and low hills, in the Upper Sonoran Zone.

Our plant is more slender and has thinner leaves than the typical form of the species, but otherwise seems to be the same.

12. **Grindelia neomexicana** Woot. & Standl. Contr. U. S. Nat. Herb. **16**: 178. 1913.

TYPE LOCALITY: Mountains north of Santa Rita, New Mexico. Type collected by Wooton, August 23, 1900.

RANGE: Mountains of western New Mexico.

NEW MEXICO: Mountains north of Santa Rita; mountains southeast of Patterson; G O S Ranch.

12. GYMNOSPERMA Less.

Perennial herb, often woody at the base, glabrous, resinous-viscid; leaves alternate, entire, linear or linear-lanceolate; heads small, yellow-flowered, in fastigiately corymbose cymes; bracts obtuse; rays very small; achenes oblong, slightly compressed, 4 or 5-nerved, glabrous; pappus wanting.

1. **Gymnosperma corymbosum** DC. Prodr. **5:** 312. 1836.

TYPE LOCALITY: "In Mexico circa Matamoros."

RANGE: Western Texas and southern Arizona to Mexico.

NEW MEXICO: Bear Mountains; Gila Hot Springs; Florida Mountains; Animas Mountains; San Luis Mountains; Organ Mountains; White Mountains. Dry hills and rocky canyons, in the Upper Sonoran Zone.

13. XANTHOCEPHALUM Willd.

Slender branched annual with alternate, entire, linear or linear-oblong leaves; heads loosely cymose, about 10 mm. broad; rays about 12, bright yellow, oblong, as long as the disk; achenes truncate, with an obscure coroniform border.

1. **Xanthocephalum wrightii** A. Gray, Proc. Amer. Acad. **8:** 632. 1873.

Gutierrezia wrightii A. Gray, Pl. Wright. **2:** 78. 1853.

TYPE LOCALITY: "Margin of dried-up streams, between Barbocomori and Santa Cruz, Sonora."

RANGE: Southern Arizona and New Mexico to Mexico.

NEW MEXICO: Black Range; Mogollon Mountains. Upper Sonoran and Transition zones.

14. AMPHIACHYRIS Nutt.

Slender annual, effusely corymbose-branched; leaves linear to filiform; heads very numerous, long-pedunculate, hemispheric, with 10 to 12 firm, coriaceous, ovate to oval bracts; rays 5 to 10, oval or oblong; disk flowers 10 to 20, sterile; achenes with minute coroniform pappus.

1. **Amphiachyris dracunculoides** (DC.) Nutt. Trans. Amer. Phil. Soc. n. ser. **7:** 313. 1841.

Brachyris dracunculoides DC. Mém. Soc. Phys. Hist. Nat. Genève **7:** 265. *pl. 1.* 1836.

TYPE LOCALITY: "Arkansas."

RANGE: Kansas and Oklahoma to Texas and eastern New Mexico.

NEW MEXICO: Roswell (*Earle* 347). Plains, in the Upper Sonoran Zone.

15. GUTIERREZIA Lag. SNAKEWEED.

Slender, viscid, much branched herbs or shrubby perennials, with slender alternate linear leaves and numerous small heads of yellow flowers, these either solitary or clustered; rays 1 to 8; achenes short, obovate or oblong, terete or 5-angled; pappus of numerous paleæ, these often minute.

It is said that the Navahos chew these plants and apply them to the stings of bees, wasps, or ants.

KEY TO THE SPECIES.

Annual; heads about 5 mm. broad .... 1. *G. sphaerocephala.*
Perennials; heads less than 4 mm. broad.
  Disk and ray flowers in each head 1 or 2 each............ 2. *G. glomerella.*
  Disk and ray flowers in each head 3 to 7 each.
    Branches, leaves, and outer bracts densely lepidote-scurfy .... 3. *G. furfuracea.*

Plants not lepidote-scurfy.
Heads all pedunculate; plants very slender...... 4. *G. filifolia.*
Heads mostly or at least partly sessile; plants stouter.
Leaves linear, 2 to 6 cm. long.
Plants 30 cm. high or less, woody only at the base; involucres campanulate.. 5. *G. diversifolia.*
Plants 50 to 100 cm. high, shrubby; involucres elongated-turbinate...... 6. *G. longifolia.*
Leaves linear-filiform, mostly short.
Plants low, 15 cm. high or less, densely branched; heads very numerous, dense; stems very slender......... 7. *G. juncea.*
Plants more than 20 cm. high, sparingly branched; heads fewer, loosely arranged; stems stout.............. 8. *G. tenuis.*

1. **Gutierrezia sphaerocephala** A. Gray, Mem. Amer. Acad. n. ser. **4:** 73. 1849.
Type locality: "Low prairie, from the Upper to the Middle spring of the Cimarron."
Range: New Mexico and Arkansas to Texas.
New Mexico: Magdalena; Nara Visa; Palomas; Socorro; mesa west of Organ Mountains; Roswell; Artesia; Dayton; Carlsbad; Lake Valley. Plains, in the Lower and Upper Sonoran zones.

2. **Gutierrezia glomerella** Greene, Pittonia **4:** 54. 1899.
Type locality: Organ Mountains, New Mexico. Type collected by Wooton (no. 449).
Range: Colorado and New Mexico to western Texas.
New Mexico: Tunitcha Mountains; Gallup; Magdalena; Redrock; Magdalena Mountains; San Marcial; Deming; Mangas Springs; Organ Mountains; Guadalupe Mountains; Florida Mountains; Carlsbad; Fort Bayard; Orogrande; Albuquerque; Dayton. Dry plains and hills, in the Lower and Upper Sonoran zones.

3. **Gutierrezia furfuracea** Greene, Repert. Nov. Sp. Fedde **7:** 195. 1909.
Type locality: Cactus Flat, New Mexico. Type collected by E. A. Goldman (no. 1568).
Range: Southern New Mexico.
New Mexico: Cactus Flat; Bishops Cap; Tortugas Mountain; Lake Valley. Dry plains and hills, in the Lower and Upper Sonoran zones.

4. **Gutierrezia filifolia** Greene, Pittonia **4:** 55. 1899.
Type locality: Round Mountain, New Mexico. Type collected by Wooton in 1897.
Range: Plains and foothills of New Mexico and western Texas.
New Mexico: Santa Fe; Albert; Nara Visa; Dog Spring; near White Water; Rincon; Round Mountain; Roswell; Carlsbad. Upper Sonoran Zone.

5. **Gutierrezia diversifolia** Greene, Pittonia **4:** 53. 1899.
Type locality: Laramie, Wyoming.
Range: Montana and Saskatchewan to Utah and New Mexico.
New Mexico: Tunitcha Mountains; Park View; Johnsons Mesa; Mora Creek; Folsom; Raton; Hebron; Maxwell City. Plains and low hills, in the Upper Sonoran and Transition zones.

6. **Gutierrezia longifolia** Greene, Pittonia **4:** 54. 1899.
Type locality: White Mountains, New Mexico. Type collected by Wooton (no. 377).
Range: Colorado to New Mexico.

New Mexico: Cross L Ranch; Raton; San Rafael; Animas Valley; Organ Mountains; White Mountains. Plains and low hills, in the Upper Sonoran and Transition zones.

7. **Gutierrezia juncea** Greene, Pittonia **4**: 56. 1899.

Type locality: Near Gray, New Mexico. Type collected by Miss Josephine Skehan (no. 78).

Range: Oklahoma and Colorado to New Mexico and Arizona.

New Mexico: Clayton; McIntosh; Carrizozo; Gray; Nara Visa; Estancia; White Oaks; Endee; Ogle. Plains, in the Lower and Upper Sonoran zones.

8. **Gutierrezia tenuis** Greene, Pittonia **4**: 55. 1899.

Gutierrezia linearis Rydb. Bull. Torrey Club **31**: 647. 1904.

Gutierrezia goldmanii Greene, Repert. Nov. Sp. Fedde **7**: 195. 1909.

Type locality: Foothills of the mountains back of Silver City, New Mexico.

Range: Colorado and Arizona to western Texas.

New Mexico: Common at lower altitudes nearly throughout the State. Plains and hills, in the Lower and Upper Sonoran zones.

This is by far our commonest species. It is found almost everywhere at low and middle elevations, all over the State except in the extreme southwestern corner. In most localities where it occurs it is very abundant and one of the most characteristic plants. It is especially prominent upon overstocked ranges and spreads rapidly where overstocking takes place. This and the other species are known variously as "yellow weed," "brownweed," "sheepweed," "snakeweed," and "yerba de víbora."

The type of *Gutierrezia linearis* came from Gray (*Earle* 474), and that of *G. goldmanii* from the Florida Mountains (*Goldman* in 1908).

16. HETEROTHECA Cass.

Annual or biennial herb, glandular and hirsute, the leaves alternate, oblong or oblong-ovate, serrate, those of the stem clasping; involucre 7 to 8 mm. high, the bracts linear-lanceolate to linear, acuminate; rays yellow; achenes flattened, those of the rays without pappus, those of the disk with pappus in 2 series, the inner of numerous long slender tawny bristles, the outer of very short ones.

1. **Heterotheca subaxillaris** (Lam.) Britt. & Rusby, Trans. N. Y. Acad. **7**: 10. 1887.

Inula subaxillaris Lam. Encycl. **3**: 259. 1789.

Type locality: "Dans le Caroline, le Maryland."

Range: Delaware and Kansas to Florida and eastern New Mexico.

New Mexico: Nara Visa (*Fisher* 49). Open fields and plains, in the Upper Sonoran Zone.

17. CHRYSOPSIS Nutt.

Perennial herbs with usually numerous stems and large, solitary or corymbose heads; leaves entire, mostly sessile; heads many-flowered, with numerous bright yellow rays; involucre campanulate or hemispheric, of narrow, much imbricated bracts; achenes compressed, obovate; pappus in 2 series, the inner of numerous capillary scabrous bristles, the outer of minute short bristles.

KEY TO THE SPECIES.

Heads subtended by few to many, thin, broad, leaflike bracts.
  Plants appressed-sericeous throughout .... 1. *C. nitidula.*
  Plants with spreading pubescence, never sericeous.
    Floral leaves ovate, acute; leaves merely scaberulous,
      very glandular; stems slender .... 2. *C. cryptocephala.*

Floral leaves oblong to obovate or oblanceolate, obtuse; leaves abundantly villous or hirsute, scarcely if at all glandular; stems stout.
Heads several, clustered at the ends of the branches; pubescence of peduncles and floral leaves long-villous, white .... 3. *C. senilis.*
Heads mostly solitary, never clustered at the ends of the branches; pubescence nowhere long-villous or very dense, not white............. 4. *C. fulcrata.*
Heads on naked or nearly naked peduncles.
Pubescence of the stems mostly appressed.
Heads small, about 7 mm. wide, nearly sessile, clustered at the ends of the branches; leaves crowded, all white-sericeous.................................. 5. *C. berlandieri.*
Heads larger, 10 mm. wide or more, long-pedunculate, not clustered at the ends of the branches, usually corymbose; leaves not crowded, usually only the uppermost whitish.............................. 6. *C. villosa.*
Pubescence of the stems spreading, none appressed.
Leaves more or less silvery-sericeous, abundantly pubescent, sparingly glandular................... 7. *C. hirsutissima.*
Leaves all green, sparingly pubescent, densely glandular. 8. *C. hispida.*

1. **Chrysopsis nitidula** Woot. & Standl. Contr. U. S. Nat. Herb. **16**: 179. 1913.

TYPE LOCALITY: Mogollon Mountains, on the West Fork of the Gila, New Mexico. Type collected by Metcalfe (no. 552).

RANGE: Mountains of western New Mexico.

NEW MEXICO: West Fork of the Gila; north of Ramah; Middle Fork of the Gila.

2. **Chrysopsis cryptocephala** Woot. & Standl. Contr. U. S. Nat. Herb. **16**: 179. 1913.

TYPE LOCALITY: In Section 23 of the V Pasture, in the White Mountains, New Mexico. Type collected by Wooton, July 23, 1905.

RANGE: White Mountains, New Mexico, in the Transition Zone.

3. **Chrysopsis senilis** Woot. & Standl. Contr. U. S. Nat. Herb. **16**: 179. 1913.

TYPE LOCALITY: Organ Mountains, New Mexico. Type collected by Wooton (no. 509).

RANGE: Organ Mountains, New Mexico, in the Upper Sonoran Zone.

4. **Chrysopsis fulcrata** Greene, Bull. Torrey Club **25**: 119. 1898.

TYPE LOCALITY: Organ Mountains, New Mexico. Type collected by Wooton (no. 510).

RANGE: Mountains of southern New Mexico.

NEW MEXICO: San Luis Mountains; Animas Valley; Organ Mountains; White Mountains. Upper Sonoran Zone.

5. **Chrysopsis berlandieri** Greene, Erythea **2**: 96. 1894.

Aplopappus canescens DC. Prodr. **5**: 349. 1836.

Chrysopsis canescens Torr. & Gray, Fl. N. Amer. **2**: 256. 1842, not DC. 1836.

TYPE LOCALITY: Texas.

RANGE: New Mexico to western Texas.

NEW MEXICO: Colfax; Nara Visa; Clayton; Bishops Cap. Dry plains and low hills, in the Upper Sonoran Zone.

6. **Chrysopsis villosa** (Pursh) Hook. Fl. Bor. Amer. **2**: 22. 1834.

Amellus villosus Pursh, Fl. Amer. Sept. 564. 1814.

TYPE LOCALITY: "On the Missouri."

RANGE: Idaho and Minnesota to New Mexico and Texas.

NEW MEXICO: Tunitcha Mountains; Aztec; Carrizo Mountains; Chama; south of Gallup; Santa Fe; Las Vegas; Hermits Peak; Organ Mountains; Dona Ana Mountains; Tortugas Mountain; west of Roswell; Gilmores Ranch. Dry hills and plains, in the Upper Sonoran and Transition zones.

7. Chrysopsis hirsutissima Greene, Pittonia 4: 153. 1900.

TYPE LOCALITY: Arboles, southern Colorado.

RANGE: Colorado and New Mexico.

NEW MEXICO: Santa Fe; Ensenada; Tesuque; Laguna; James Canyon; Wingfields Ranch. Dry hillsides, in the Upper Sonoran and Transition zones.

8. Chrysopsis hispida (Hook.) Nutt. Trans. Amer. Phil. Soc. n. ser. 7: 316. 1841.

Diplopappus hispidus Hook. Fl. Bor. Amer. 2: 22. 1834.

Chrysopsis villosa hispida A. Gray, Proc. Acad. Phila. **1863**: 65. 1864.

TYPE LOCALITY: "Carlton-House Fort."

RANGE: Saskatchewan and Alberta to Arizona and New Mexico.

NEW MEXICO: Farmington; Tunitcha Mountains; Dulce; mountains west of Grants Station; Upper Pecos; Pajarito Park; Santa Fe; Las Vegas; Sandia Mountains; Laguna; Sierra Grande; Water Canyon; Tortugas Mountain; White Mountains; Gray; mountains west of San Antonio. Plains and dry hills, in the Upper Sonoran and Transition zones.

18. CHRYSOTHAMNUS Nutt. RABBIT BRUSH.

Coarse plants, usually shrubby, sometimes woody only at the base, 30 cm. to 2 meters high, with entire narrow leaves, and usually corymbose small heads of yellow flowers; heads narrow, cylindric or turbinate, mostly 5-flowered; involucre of narrow keeled dry bracts, these often with spreading tips, arranged in 5 vertical ranks; achenes slender, glabrous or pubescent; pappus of nearly equal bristles.

A decoction of the heads of various species of Chrysothamnus was formerly used by the Navahos in dyeing wool yellow.

KEY TO THE SPECIES.

Achenes glabrous.
- Stems, leaves, and involucres more or less floccose .... 1. *C. bigelovii.*
- Plants without floccose pubescence.
  - Involucres about 5 mm. high, the bracts not acuminate .. 2. *C. vaseyi.*
  - Involucres 10 mm. high or more, the bracts usually abruptly acuminate.
    - Leaves oblanceolate or nearly so, puberulent .... 3. *C. depressus.*
    - Leaves mostly linear, usually glabrous except on the margins.
      - Leaves ciliolate; heads, including the pappus, less than 15 mm. long .... 4. *C. baileyi.*
      - Leaves not ciliolate; heads 20 mm. long or more.
        - Leaves densely puberulent on the faces; tall shrub 75 cm. high; bracts acuminate ....17. *C. elatior.*
        - Leaves glabrous on the faces; low shrub 30 cm. high or less; bracts not acuminate .... 5. *C. pulchellus.*

Achenes pubescent.
- Plants entirely glabrous or at least never tomentose.
  - Bracts obtuse or acutish.
    - Leaves linear, twisted .... 6. *C. elegans.*
    - Leaves linear-lanceolate, not twisted .... 7. *C. glaucus.*

Bracts abruptly acuminate.
Tall shrub, 1 meter high or more; leaves narrowly elliptic.................................. 8. *C. linifolius.*
Low plants, 20 to 40 cm. high; leaves linear or filiform.
Leaves linear; heads long-pedunculate, subtended by very short bracts............. 9. *C. greenei.*
Leaves filiform, heads short-pedunculate or sessile, subtended by long bracts........10. *C. filifolius.*
Plants tomentose, at least on the young branches and in the axils of the leaves.
Bracts long-acuminate, with spreading tips..............11. *C. newberryi.*
Bracts obtuse or acute, not acuminate, appressed.
Branches permanently white-tomentose.
Plants tall, about 1 meter high, with erect branches; leaves permanently tomentose.12. *C. latisquameus.*
Plants low, 30 cm. or less, with spreading branches; leaves soon glabrate..........13. *C. formosus.*
Branches and leaves soon glabrate or at least the branches never permanently tomentose.
Leaves permanently tomentose on both surfaces...................................14. *C. pulcherrimus.*
Leaves glabrate in age.
Bracts glabrous; lobes of the corolla spreading in age; plants much branched...15. *C. graveolens.*
Bracts villous-ciliate; lobes of the corolla erect; plants sparingly branched....16. *C. confinis.*

1. **Chrysothamnus bigelovii** (A. Gray) Greene, Erythea **3**: 102. 1904.
Linosyris bigelovii A. Gray, U. S. Rep. Expl. Miss. Pacif. **4**: 98. 1856.
Bigelovia bigelovii A. Gray, Proc. Amer. Acad. **8**: 642. 1873.
Type locality: "Hills and arroyos, Cienegella, above Albuquerque," New Mexico. Type collected by Bigelow in 1853.
Range: Southern Colorado and northern New Mexico.
New Mexico: Carrizo Mountains; Tunitcha Mountains; Farmington; Stinking Lake; Laguna; Zuni Valley; Moreno Valley; Pecos; Horse Camp; Clark; Embudo; Albuquerque; Santa Fe; San Augustine Plains. Dry plains and hills, in the Upper Sonoran Zone.

2. **Chrysothamnus vaseyi** (A. Gray) Greene, Erythea **3**: 96. 1894.
Bigelovia vaseyi A. Gray, Proc. Amer. Acad. **12**: 58. 1876.
Type locality: Middle Park, Colorado.
Range: Wyoming and Utah to Colorado and New Mexico.
New Mexico: Dulce; Stinking Lake; Moreno Valley; Chama. Low hills and valleys, in the Upper Sonoran and lower part of the Transition zones.

3. **Chrysothamnus depressus** Nutt. Journ. Acad. Phila. II. **1**: 171. 1847.
Bigelovia depressa A. Gray, Proc. Amer. Acad. **8**: 643. 1873.
Type locality: "In the Sierra and Upper California."
Range: Colorado to northern New Mexico.
New Mexico: Tunitcha Mountains; Sandia Mountains; Dulce. Dry hills and plains, in the Upper Sonoran and lower part of the Transition zones.
A low shrub, 15 to 20 cm. high, forming dense clumps.

4. **Chrysothamnus baileyi** Woot. & Standl. Contr. U. S. Nat. Herb. **16:** 181. 1913.
TYPE LOCALITY: North end of the Guadalupe Mountains, New Mexico. Type collected by Vernon Bailey (no. 498).
RANGE: Southern New Mexico.
NEW MEXICO: Guadalupe Mountains; White Mountains; Buchanan. Dry hills, in the Upper Sonoran Zone.

5. **Chrysothamnus pulchellus** (A. Gray) Greene, Erythea **3:** 107. 1895.
Linosyris pulchella A. Gray, Pl. Wright. **1:** 96. 1852.
Bigelovia pulchella A. Gray, Proc. Amer. Acad. **17:** 209. 1882.
TYPE LOCALITY: Western Texas.
RANGE: Colorado and Utah to Arizona and western Texas.
NEW MEXICO: White Sands; north of Deming; Nara Visa. Dry hills and plains, in the Lower and Upper Sonoran zones.

6. **Chrysothamnus elegans** Greene, Erythea **3:** 94. 1894.
TYPE LOCALITY: Gunnison Valley, Colorado.
RANGE: Plains of Colorado and northern New Mexico.
NEW MEXICO: Tunitcha Mountains; south of Gallup; Zuni Valley. Upper Sonoran Zone.

7. **Chrysothamnus glaucus** A. Nels. Bull. Torrey Club **25:** 377. 1898.
Bigelovia douglasii serrulata A. Gray, Proc. Amer. Acad. **8:** 644. 1873.
Chrysothamnus serrulatus Rydb. Bull. Torrey Club **33:** 152. 1906.
TYPE LOCALITY: Dry slopes in the foothills of the Medicine Bow Mountains, Chimney Rock, Wyoming.
RANGE: Wyoming and Utah to Colorado and northern New Mexico.
NEW MEXICO: Moreno Valley (*Bailey* 3652).

8. **Chrysothamnus linifolius** Greene, Pittonia **3:** 24. 1896.
TYPE LOCALITY: Near Rock Springs, Wyoming.
RANGE: Wyoming to northern New Mexico.
NEW MEXICO: Aztec; Shiprock; Farmington; Dulce. Valleys, especially in alkaline soil, in the Upper Sonoran Zone.

9. **Chrysothamnus greenei** (A. Gray) Greene, Erythea **3:** 94. 1894.
Bigelovia greenei A. Gray, Proc. Amer. Acad. **11:** 75. 1876.
Chrysothamnus scoparius Rydb. Bull. Torrey Club **28:** 503. 1901.
TYPE LOCALITY: Huerfano Plains, southern Colorado.
RANGE: Utah and Colorado to northwestern New Mexico.
NEW MEXICO: Cedar Hill (*Standley* 7977). Dry hills, in the Upper Sonoran Zone.

10. **Chrysothamnus filifolius** Rydb. Bull. Torrey Club **28:** 503. 1901.
TYPE LOCALITY: Granite, Colorado.
RANGE: Plains of southern Colorado and northern New Mexico.
NEW MEXICO: Mesa near Atarque de Garcia; Black Rock; Tiznitzin; Carrizo Mountains; Farmington; Shiprock. Upper Sonoran Zone.

11. **Chrysothamnus newberryi** Rydb. Bull. Torrey Club **31:** 652. 1904.
TYPE LOCALITY: Canyon Largo, New Mexico. Type collected by Newberry.
RANGE: Southern Colorado and northwestern New Mexico.
NEW MEXICO: Canyon Largo; Cedar Hill; Dulce. Dry hills, in the Upper Sonoran Zone.

12. **Chrysothamnus latisquameus** (A. Gray) Greene, Pittonia **4:** 42. 1899.
Bigelovia graveolens latisquamea A. Gray, Proc. Amer. Acad. **8:** 645. 1873.
Bigelovia graveolens appendiculata Eastw. Proc. Calif. Acad. III. **1:** 74. ***pl. 6.*** 1897.
Chrysothamnus appendiculatus Heller, Muhlenbergia **1:** 6. 1900.

TYPE LOCALITY: New Mexico. Type collected by Bigelow.

RANGE: New Mexico.

NEW MEXICO: Sandia Mountains; Santa Fe; Dulce; San Lorenzo; Burro Mountains; Gila; White Sands. Dry plains and foothills, in the Lower and Upper Sonoran zones.

The type of *Bigelovia graveolens appendiculata* was collected on the White Sands by Cockerell.

13. Chrysothamnus formosus Greene, Pittonia **4:** 41. 1899.

TYPE LOCALITY: In the neighborhood of a mineral spring among the hills a few miles southwest from Grand Junction, Colorado.

RANGE: Southwestern Colorado to northwestern New Mexico and adjacent Arizona.

NEW MEXICO: Carrizo Mountains (*Standley* 7343). Dry hills, in the Upper Sonoran Zone.

14. Chrysothamnus pulcherrimus A. Nels. Bot. Gaz. **28:** 370. 1899.

TYPE LOCALITY: Woods Landing, Wyoming.

RANGE: Montana to northwestern New Mexico.

NEW MEXICO: Carrizo Mountains (*Standley* 7467). Open hills, in the Upper Sonoran Zone.

A densely branched shrub about 1 meter high.

15. Chrysothamnus graveolens (Nutt.) Greene, Erythea **3:** 108. 1894.

Chrysocoma graveolens Nutt. Gen. Pl. **2:** 136. 1818.

Bigelovia graveolens A. Gray, Proc. Amer. Acad. **8:** 644. 1873.

TYPE LOCALITY: "On the banks of the Missouri in denudated soils."

RANGE: Montana and Nebraska to Utah and New Mexico.

NEW MEXICO: Tunitcha Mountains; Farmington; Dulce; Zuni Valley; Grants Station; Embudo; Taos; Cross L Ranch; Chama; Moreno Valley; Cebolla Spring; Datil; Patterson. Plains and low hills, in the Upper Sonoran Zone.

16. Chrysothamnus confinis Greene, Pittonia **5:** 62. 1902.

TYPE LOCALITY: White Mountains, New Mexico. Type collected by Wooton (no. 379).

RANGE: Along streams, White Mountains of New Mexico, in the Upper Sonoran Zone.

17. Chrysothamnus elatior Standley, Proc. Biol. Soc. Washington **26:** 118. 1913.

TYPE LOCALITY: Sandhills north of Goldenbergs, New Mexico Range Reserve, Dona Ana County, New Mexico. Type collected by Wooton, October 12, 1912.

RANGE: Known only from type locality.

19. SIDERANTHUS Fraser.

Annual or perennial herbs with alternate, simple to pinnatifid leaves and numerous rather large heads of yellow flowers with yellow rays; leaves small, toothed or pinnatifid, sessile; involucre campanulate, many-flowered, the bracts with green tips, in several series; receptacle naked; achenes obtuse, compressed, sericeous, 8 to 10-nerved; persistent pappus of 1 or more series of unequal, smooth or barbellate bristles.

KEY TO THE SPECIES.

Teeth of leaves not bristle-tipped; whole plant densely glandular.. 1. *S. viscidus*.
Teeth bristle-tipped; plants not densely glandular.
  Heads discoid.......... 2. *S. grindelioides*.
  Heads radiate.
    Leaves merely toothed; heads 15 mm. broad............ 3. *S. serratus*.
    Leaves, at least the lower ones, pinnatifid; heads seldom more than 10 mm. broad.

Stems and leaves conspicuously floccose, at least when young.
Pubescence all or nearly all floccose, dense and persistent .... 4. *S. wootoni.*
Pubescence mostly glandular or scabrous, the floccose hairs soon deciduous .... 5. *S. spinulosus.*
Stems and leaves never floccose.
Plants glabrous throughout, or puberulent only on the bracts.
Stems slender, branched, sparingly leafy; leaves with few shallow teeth, bright green; heads few, solitary .... 6. *S. laevis.*
Stems stout, simple up to the inflorescence, densely leafy; leaves deeply toothed or pinnatifid, somewhat glaucous; heads numerous, clustered at the ends of the branches .... 7. *S. glaberrimus.*
Plants abundantly pubescent.
Annual; stems strigose .... 8. *S. gracilis.*
Perennial; stems glandular-puberulent ..... 9. *S. australis.*

1. Sideranthus viscidus Woot. & Standl. Contr. U. S. Nat. Herb. **16:** 180. 1913.

TYPE LOCALITY: Near Hope, New Mexico. Type collected by Wooton, August 3, 1905.

RANGE: Plains of southeastern New Mexico.

NEW MEXICO: Near Hope; Dayton.

2. Sideranthus grindelioides (Nutt.) Britton, Bull. Torrey Club **27:** 620. 1900.

Eriocarpum grindelioides Nutt. Trans. Amer. Phil. Soc. n. ser. **7:** 321. 1841.

TYPE LOCALITY: "On shelving rocks in the Rocky Mountain range, Oregon."

RANGE: Manitoba and Nebraska to Arizona and New Mexico.

NEW MEXICO: Carrizo Mountains; Tunitcha Mountains; Farmington. Dry plains and hills, in the Upper Sonoran Zone.

3. Sideranthus serratus (Greene) Standley, Contr. U. S. Nat. Herb. **13:** 222. 1910.

Eriocarpum serratum Greene, Bull. Torrey Club **25:** 119. 1898.

TYPE LOCALITY: White Mountains, New Mexico. Type collected by Wooton (no. 251).

RANGE: Southeastern New Mexico.

NEW MEXICO: White and Sacramento mountains; Capitan Mountains; west of Roswell; Queen. Transition Zone.

4. Sideranthus wootoni (Greene) Standley, Contr. U. S. Nat. Herb. **13:** 222. 1910.

Eriocarpum wootoni Greene, Bull. Torrey Club **25:** 120. 1898.

TYPE LOCALITY: White Mountains, New Mexico. Type collected by Wooton (no. 518).

RANGE: New Mexico.

NEW MEXICO: Farmington; Laguna; Albuquerque; Roswell; White Mountains. Hills and plains, in the Upper Sonoran Zone.

5. Sideranthus spinulosus (Pursh) Sweet, Hort. Brit. 227. 1826.

Amellus spinulosus Pursh, Fl. Amer. Sept. 564. 1814.

Aplopappus spinulosus DC. Prodr. **5:** 347. 1836.

TYPE LOCALITY: "In open prairies on the Missouri."

RANGE: Montana and Minnesota to Arizona and Texas.

New Mexico: Raton; Sierra Grande; Las Vegas; Estancia; Santa Fe; Socorro; Bernal; Suwanee; Pecos; Torrance; White Sands; Gray; Pajarito Park. Open slopes and on plains, in the Lower and Upper Sonoran zones.

6. **Sideranthus laevis** Woot. & Standl. Contr. U. S. Nat. Herb. **16:** 180. 1913.

Type locality: Gypsum hills near Lakewood, New Mexico. Type collected by Wooton, August 6, 1909.

Range: Known only from type locality.

7. **Sideranthus glaberrimus** Rydb. Bull. Torrey Club **27:** 621. 1900.

Sideranthus spinulosus glaberrimus A. Nels. in Coulter, New Man. Rocky Mount. 489. 1909.

Type locality: Nebraska.

Range: Nebraska and Kansas to Colorado and northeastern New Mexico.

New Mexico: Raton; Nara Visa; highest point of the Llano Estacado. Plains and low hills, in the Upper Sonoran Zone.

8. **Sideranthus gracilis** (Nutt.) Rydb. Colo. Agr. Exp. Sta. Bull. **100:** 344. 1906.

Dieteria gracilis Nutt. Journ. Acad. Phila. II. **1:** 177. 1848.

Aplopappus gracilis A. Gray, Mem. Amer. Acad. n. ser. **4:** 76. 1849.

Eriocarpum gracile Greene, Erythea **2:** 189. 1894.

Type locality: Near Santa Fe, New Mexico. Type collected by Gambel.

Range: Colorado and Utah to Arizona, western Texas, and Mexico.

New Mexico: Stinking Lake; Dulce; Fort Bayard; Mangas Springs; Socorro Mountain; Gila Hot Springs; San Luis Mountains; Organ Mountains. Plains and low hills, in the Lower and Upper Sonoran zones.

9. **Sideranthus australis** (Greene) Rydb. Bull. Torrey Club **27:** 621. 1900.

Eriocarpum australe Greene, Erythea **2:** 108. 1894.

Type locality: "Texas, New Mexico, Arizona, and adjacent Mexico."

Range: Western Texas to Arizona and southward.

New Mexico: Carrizo Mountains; Santa Fe; Kennedy; Socorro Mountain; Magdalena; Burro Mountains; Carrizalillo Mountains; Sabinal; Kingston; Mesilla Valley; Organ Mountains; Redlands. Plains and dry hills, in the Lower and Upper Sonoran zones.

20. ISOCOMA Nutt. Rayless goldenrod.

Stout perennial herbs, 50 cm. high or less, with linear to oblanceolate, entire or toothed leaves; heads very numerous, corymbose, yellow-flowered; involucres turbinate or narrowly campanulate, the bracts thick, coriaceous, appressed in several series; corollas inflated; achenes short, sericeous.

KEY TO THE SPECIES.

Bracts all acute .... 1. *I. oxylepis.*
Bracts obtuse or only the innermost acute.
    Leaves linear; heads slender-pedunculate .... 2. *I. wrightii.*
    Leaves linear-oblanceolate, usually narrowly so; heads mostly sessile or subsessile .... 3. *I. heterophylla.*

1. **Isocoma oxylepis** Woot. & Standl. Contr. U. S. Nat. Herb. **16:** 180. 1913.

Type locality: Near White Water, New Mexico or Chihuahua. Type collected by E. A. Mearns (no. 2288).

Range: Mountains of southwestern New Mexico and adjacent Mexico.

New Mexico: Near White Water; Dog Spring.

2. **Isocoma wrightii** (A. Gray) Woot. & Standl. Contr. U. S. Nat. Herb. **16:** 181. 1913.

Linosyris wrightii A. Gray, Pl. Wright. **1:** 95. 1852.

Bigelovia wrightii A. Gray, Proc. Amer. Acad. **8:** 639. 1873.

TYPE LOCALITY: Valley of the Rio Grande, 60 or 70 miles below El Paso, Texas.

RANGE: Western Texas to southern New Mexico.

NEW MEXICO: Along the Rio Grande; Carlsbad; Tucumcari. Dry hills and plains, in the Lower Sonoran Zone.

3. **Isocoma heterophylla** (A. Gray) Greene, Erythea **2**: 111. 1894.

Linosyris heterophylla A. Gray, Pl. Wright. **1**: 95. 1852.

Linosyris hirtella A. Gray, loc. cit.

TYPE LOCALITY: Valley of the Pecos, western Texas.

RANGE: Western Texas to Arizona and southward.

NEW MEXICO: Carrizo Mountains; Bueyeros; Socorro; Sabinal; Laguna; Horace; Albuquerque; Mesilla Valley; White Sands; White Mountains; Roswell. Dry plains and low hills, in the Lower and Upper Sonoran zones.

21. CHRYSOMA Nutt.

Low densely branched shrub, 60 cm. high or less, with linear-acerose rigid resinous-punctate crowded leaves and very numerous small heads of bright yellow flowers in dense cymose clusters; bracts of the involucre appressed, in 2 or 3 series, subulate-linear, acute; rays 3 to 6; disk flowers 10 or 12; achenes slender, villous, with fine and soft capillary pappus.

1. **Chrysoma laricifolia** (A. Gray) Greene, Erythea **3**: 11. 1895.

Aplopappus laricifolius A. Gray, Pl. Wright. **2**: 80. 1853.

TYPE LOCALITY: Mountains at Guadalupe Pass, New Mexico.

RANGE: Western Texas to southern Arizona.

NEW MEXICO: Burro Mountains; Bear Mountains; Redrock; Florida Mountains; Animas Mountains; Organ Mountains; Dona Ana Mountains. Dry, rocky hills and canyons, in the Lower and Upper Sonoran zones.

This is one of the handsomest plants of the southwestern foothills. When in full flower it is a solid mass of golden yellow. It is especially effective because of its densely branched crown and the very numerous dark green leaves.

22. OLIGONEURON Small.

Coarse perennial with numerous basal and many broad, serrate or entire, thick cauline leaves; heads comparatively large, compactly corymbose or cymose; involucral bracts broad, longitudinally striate; achenes turgid, 12 to 15-nerved, glabrous.

1. **Oligoneuron canescens** Rydb. Bull. Torrey Club **31**: 652. 1905.

TYPE LOCALITY: Buffalo, Wyoming.

RANGE: Montana and Nebraska to Colorado and New Mexico.

NEW MEXICO: Beulah; Chama; Sierra Grande; Hermits Peak; Raton Mountains; Johnsons Mesa; Baldy; Sacramento Mountains. Mountains, in the Transition Zone.

23. STENOTUS Nutt.

A low scapose cespitose perennial from a thick woody caudex; leaves mostly basal, linear-oblanceolate; heads large, 12 to 15 mm. broad, radiate; bracts thin, oblong, obtuse; achenes oblong-turbinate, villous; pappus white, of numerous unequal scabrous bristles.

1. **Stenotus armerioides** Nutt. Trans. Amer. Phil. Soc. n. ser. **7**: 335. 1841.

Aplopappus armerioides A. Gray, Syn. Fl. 1^2: 132. 1884.

TYPE LOCALITY: "Toward the sources of the Platte, in the Rocky Mountain range, on shelving rocks."

RANGE: British America to Colorado and New Mexico.

NEW MEXICO: Fort Wingate; hills 10 miles north of Santa Fe. Dry hills and plains, in the Upper Sonoran Zone.

24. PYRROCOMA Nutt.

Perennial herb with usually simple stems, alternate leaves, and large showy heads of yellow flowers with bright yellow rays; bracts foliaceous, oblong, mostly obtuse, numerous, appressed; achenes linear, 3-angled, striate, glabrous; pappus of slender tawny bristles.

1. **Pyrrocoma crocea** (A. Gray) Greene, Erythea **2**: 69. 1894.
Aplopappus croceus A. Gray, Proc. Acad. Phila. **1863**: 65. 1864.
Pyrrocoma amplectens Greene, Leaflets **2**: 10. 1909.
Type locality: Middle Park, Colorado.
Range: Wyoming to Arizona and New Mexico.
New Mexico: Chama; El Rito; Baldy; Santa Fe and Las Vegas mountains; Mogollon Mountains. Meadows in the mountains, in the Transition and Canadian zones.

A common and very handsome plant in the open meadows of the higher mountains. When growing it suggests the more common *Dugaldea hoopesii* and is likely to be taken for that by careless observers. The type of *P. amplectens* was collected on the Middle Fork of the Gila by Metcalfe (no. 540).

25. OREOCHRYSUM Rydb.

Nearly glabrous perennial herb, in aspect like the Solidagos, with numerous basal leaves, a low leafy stem, and numerous rather large corymbose heads of yellow flowers; involucre campanulate, the broad, foliaceous or chartaceous, oblong, obtuse bracts in 2 or 3 unequal series; rays numerous, small, narrow, pale yellow; achenes short, glabrous or nearly so.

1. **Oreochrysum parryi** (A. Gray) Rydb. Bull. Torrey Club **33**: 153. 1906.
Aplopappus parryi A. Gray, Amer. Journ. Sci. II. **33**: 239. 1862.
Solidago parryi Greene, Erythea **2**: 57. 1894.
Type locality: Upper Clear Creek, Colorado.
Range: Wyoming to Arizona and New Mexico.
New Mexico: Tunitcha Mountains; Sandia Mountains; Jemez Mountains; Santa Fe and Las Vegas mountains; Hillsboro Peak; Mogollon Mountains; White and Sacramento mountains. Deep woods, in the Canadian and Hudsonian zones.

The plant of the White and Sacramento mountains has narrower bracts and smaller heads than the typical form found farther north and west.

26. TONESTUS A. Nels.

Low herbaceous perennial from a thick woody root; stems simple, monocephalous; leaves linear-spatulate; bracts oblong, obtuse, the outer foliaceous; rays conspicuous, numerous; achenes pubescent, the pappus white, capillary.

1. **Tonestus pygmaeus** (Torr. & Gray) A. Nels. Bot. Gaz. **37**: 262. 1904.
Stenotus pygmaeus Torr. & Gray, Fl. N. Amer. **2**: 237. 1841.
Aplopappus pygmaeus A. Gray, Amer. Journ. Sci. II. **33**: 239. 1862.
Macronema pygmaeum Greene, Erythea **2**: 73. 1894.
Type locality: "Rocky Mountains, probably in about lat. 41°."
Range: Wyoming to northern New Mexico.
New Mexico: Pecos Baldy; Baldy; Truchas Peak. High mountains, in the Arctic-Alpine Zone.

27. EUTHAMIA Nutt.

Tall, paniculately branched perennial with glabrous stems and alternate linear leaves; heads small, glomerately cymose, each with numerous flowers; rays small and inconspicuous, more numerous than the disk flowers; achenes villous, short, turbinate; receptacle fimbriolate.

1. Euthamia occidentalis Nutt. Trans. Amer. Phil. Soc. n. ser. **7:** 326. 1841.
Solidago occidentalis Torr. & Gray, Fl. N. Amer. **2:** 226. 1841.
TYPE LOCALITY: "Banks of the Oregon and Wahlamet, and Lewis River."
RANGE: Washington and Montana to Colorado and New Mexico.
NEW MEXICO: Mesilla Valley. Lower Sonoran Zone.

28. SOLIDAGO L. GOLDENROD.

Perennial herbs with sessile or nearly sessile, alternate leaves and very numerous racemose or clustered, small heads of yellow flowers; heads few to many-flowered, the rays 1 to 16, pistillate; bracts appressed, without herbaceous tips; receptacle naked; achenes terete, many-ribbed, with simple pappus of nearly equal capillary bristles.

KEY TO THE SPECIES.

Leaves glabrous, or slightly pubescent along the veins and margins.
  Leaves not triple-veined; branches of the inflorescence short, not recurved-spreading; heads not secund.
    Plants low, 10 to 15 cm. high; inflorescence with few heads, short, congested........................ 1. *S. decumbens.*
    Plants taller, 20 to 40 cm.; inflorescence with many heads, elongated.
      Branches of the inflorescence villous; leaves ciliate at the base.................................. 2. *S. scopulorum.*
      Branches of the inflorescence not villous; leaves not ciliate.
        Heads about 7 mm. high; bracts acute or abruptly acuminate.................... 5. *S. neomexicana.*
        Heads about 5 mm. high; bracts obtuse, thick.
          Rays pale yellow.......................... 4. *S. oreophila.*
          Rays deep golden yellow.................. 3. *S. aureola.*
  Leaves triple-veined; heads mostly secund on longer recurved branches.
    Cauline leaves lanceolate; stems tall, usually a meter high or more, deep purple at the base........... 9. *S. pitcheri.*
    Cauline leaves oblanceolate or narrower; stems usually lower, not deep purple at the base.
      Heads fully 5 mm. high; stems stout; branches of the inflorescence long and widely spreading...... 6. *S. marshallii.*
      Heads 4 mm. high or less; stems more slender; branches of the inflorescence narrower and less spreading.
        Plants nearly 1 meter high, very slender; cauline leaves linear or linear-oblanceolate; inflorescence very narrow................ 7. *S. tenuissima.*
        Plants lower, stout; cauline leaves lanceolate; inflorescence broader, the branches stouter, spreading...................... 8. *S. glaberrima.*
Leaves canescent, usually on both surfaces.
  Leaves lanceolate to ovate; plants mostly 1 meter high or more (lower in *S. bigelovii*).
    Leaves broadly lanceolate or ovate; branches of the inflorescence erect; heads not secund............14. *S. bigelovii.*
    Leaves lanceolate or narrowly so; branches of the inflorescence spreading; heads secund.

Heads 5 mm. high or more; panicles very broad....10. *S. arizonica.*
Heads 4 mm. high or less; panicles narrow.
Leaves densely canescent; whole plant yellowish green............................11. *S. gilvocanescens.*
Leaves scarcely at all pubescent except on the veins and margins; plants bright green..12. *S. canadensis.*
Leaves, at least the lower, oblanceolate, spatulate, or elliptic.
Heads large, 8 to 10 mm. high, few....................13. *S. wrightii.*
Heads less than 7 mm. high, very numerous.
Leaves thick, rigid; bracts narrowly oblong, obtuse..15. *S. howellii.*
Leaves thin, soft; bracts lanceolate, acute..........16. *S. trinervata.*

1. **Solidago decumbens** Greene, Pittonia **3:** 161. 1897.
Solidago humilis nana A. Gray, Syn. Fl. **1**^{2}: 148. 1884.
Type locality: "Rocky Mountains of Colorado and northward, in subalpine and alpine situations."
Range: Wyoming to northern New Mexico.
New Mexico: Jemez Mountains; Santa Fe and Las Vegas mountains. High mountain meadows, in the Arctic-Alpine Zone.

2. **Solidago scopulorum** (A. Gray) A. Nels. Bot. Gaz. **37:** 264. 1904.
Solidago multiradiata scopulorum A. Gray, Proc. Amer. Acad. **17:** 191. 1882.
Type locality: "Higher Rocky Mountains to New Mexico, Utah, &c."
Range: British America to Utah and northern New Mexico.
New Mexico: Tunitcha Mountains (*Standley* 7543). Hills and mountains, in the Transition Zone.

3. **Solidago aureola** Greene, Pittonia **4:** 236. 1900.
Type locality: Capitan Mountains, New Mexico. Type collected by Earle, July 28, 1900.
Range: Known only from type locality.
We have seen no specimens of this. Probably it is the same as *S. oreophila.*

4. **Solidago oreophila** Rydb. Mem. N. Y. Bot. Gard. **1:** 387. 1900.
Type locality: "Gap in the Belt Mountains above White's Gulch," Montana.
Range: British America to northern New Mexico.
New Mexico: Tunitcha Mountains; Santa Fe and Las Vegas mountains; Sandia Mountains; Chama. Mountains, in the Transition Zone.

5. **Solidago neomexicana** (A. Gray) Woot. & Standl. Contr. U. S. Nat. Herb. **16:** 182. 1913.
Solidago multiradiata neomexicana A. Gray, Proc. Amer. Acad. **17:** 191. 1882.
Type locality: High summits of one of the Mogollon Mountains, New Mexico. Type collected by Rusby (no. 228½).
Range: Known only from type locality.

6. **Solidago marshallii** Rothr. in Wheeler, Rep. U. S. Surv. 100th Merid. **6:** 146. 1879.
Type locality: Chiricahua Agency, southern Arizona.
Range: Mountains of southern Arizona and New Mexico.
New Mexico: Luna; Hop Canyon; Middle Fork of the Gila. Transition Zone.

7. **Solidago tenuissima** Woot. & Standl. Contr. U. S. Nat. Herb. **16:** 182. 1913.
Type locality: Guadalupe Canyon near Cloverdale, New Mexico. Type collected by E. A. Mearns (no. 466).
Range: Mountains of southwestern New Mexico and adjacent Arizona and Mexico.
New Mexico: Guadalupe Canyon; Mogollon Mountains.

8. **Solidago glaberrima** Martens, Bull. Acad. Sci. Brux. **8:** 67. 1841.

Type locality: Probably near St. Louis, Missouri.

Range: Idaho and Michigan to Missouri, Texas, and Arizona.

New Mexico: Mouth of Holy Ghost Creek; mountains southeast of Patterson. Mountains, in the Transition Zone.

9. **Solidago pitcheri** Nutt. Journ. Acad. Phila. **7:** 101. 1834.

Type locality: Arkansas.

Range: Washington and Minnesota to New Mexico and Arkansas.

New Mexico: White and Sacramento mountains. Along streams, in the Transition Zone.

10. **Solidago arizonica** (A. Gray) Woot. & Standl. Contr. U. S. Nat. Herb. **16:** 181. 1913.

Solidago canadensis arizonica A. Gray, Proc. Amer. Acad. **17:** 197. 1882.

Type locality: Arizona.

Range: Mountains of southern Arizona and New Mexico.

New Mexico: Beulah; Zuni; Upper Pecos; Middle Fork of the Gila; Black Range; Fort Bayard; Dog Spring; Roswell. Upper Sonoran and Transition zones.

11. **Solidago gilvocanescens** (Rydb.) Smyth, Trans. Kans. Acad. **16:** 161. 1899.

Solidago canadensis gilvocanescens Rydb. Contr. U. S. Nat. Herb. **3:** 162. 1895.

Type locality: Codys Lakes, Hooker County, Nebraska.

Range: North Dakota and Minnesota to Colorado and New Mexico.

New Mexico: Farmington; Albuquerque; Mesilla Valley; Chavez; Round Mountain. River valleys, in the Lower and Upper Sonoran zones.

12. **Solidago canadensis** L. Sp. Pl. 878. 1753.

Type locality: "Habitat in Virginia, Canada."

Range: British America to New Mexico and eastward.

New Mexico: Chama; Dulce; Pecos; Cedar Hill. Moist ground, in the Upper Sonoran and Transition zones.

13. **Solidago wrightii** A. Gray, Proc. Amer. Acad. **16:** 80. 1880.

Type locality: "W. Texas to Arizona."

Range: Southern Arizona and New Mexico.

New Mexico: Capitan Mountains. Transition Zone.

14. **Solidago bigelovii** A. Gray, Proc. Amer. Acad. **16:** 80. 1880.

Type locality: New Mexico. Type collected by Bigelow.

Range: Western Texas to Arizona and southward.

New Mexico: Sandia Mountains; Copper Canyon; San Luis Mountains; Mogollon Mountains; Organ Mountains; White and Sacramento mountains; Capitan Mountains; Hillsboro Peak; Santa Rita; Bear Mountains. Mountains, in the Upper Sonoran and Transition zones.

15. **Solidago howellii** Woot. & Standl. Contr. U. S. Nat. Herb. **16:** 181. 1913.

Type locality: On the Sierra Grande, New Mexico. Type collected by A. H. Howell (no. 219).

Range: Northeastern New Mexico.

New Mexico: Sierra Grande; Trinchera Pass; Clayton; Folsom; Capitan Mountains; Nara Visa. Plains and low hills, in the Upper Sonoran Zone.

16. **Solidago trinervata** Greene, Pittonia **3:** 100. 1896.

Type locality: "Common along the foothills of the mountains in southern and western Colorado."

Range: Wyoming and South Dakota to Arizona and New Mexico.

New Mexico: Tunitcha Mountains; Cedar Hill; Dulce; Chama; Sandia Mountains; Santa Fe and Las Vegas mountains; Johnsons Mesa; Middle Fork of the Gila; Kingston; Burro Mountains; Water Canyon; San Luis Mountains; Organ Mountains; White and Sacramento mountains. Open slopes and in thickets, chiefly in the Transition Zone.

29. PETRADORIA Greene.

Low tufted perennials with mostly basal narrow rigid sharp-pointed leaves and small heads of yellow flowers in corymbs; heads narrowly oblong, 5 to 8-flowered, with 1 to 3 short rays; bracts much imbricated, firm, broad, slightly carinate, with small green tips; achenes compressed, 5-nerved, with short rigid pappus.

KEY TO THE SPECIES.

Basal leaves linear-oblanceolate, with 3 or more nerves, long, usually more than half as long as the stems; plants usually 20 cm. high or more .... 1. *P. pumila.*
Leaves all linear, 1-nerved, short, less than half as long as the stems, more numerous; plants lower .... 2. *P. graminea.*

1. Petradoria pumila (Torr. & Gray) Greene, Erythea **3:** 13. 1895.
Solidago pumila Torr. & Gray, Fl. N. Amer. **2:** 210. 1840.

Type locality: "In open situations, on shelving rocks toward the western declivity of the Rocky Mountains."

Range: Wyoming and Nevada to Colorado and New Mexico.

New Mexico: Abiquiu; Pecos River; Rio Zuni; Farmington; Carrizo Mountains. Dry hills and plains, in the Upper Sonoran Zone.

2. Petradoria graminea Woot. & Standl. Contr. U. S. Nat. Herb. **16:** 183. 1913.

Type locality: Northwestern New Mexico. Type collected by C. C. Marsh (no. 209).

Range: Northwestern New Mexico.

New Mexico: Gallup; Tunitcha Mountains. Plains and low hills, in the Upper Sonoran Zone.

30. BACCHARIS L.

Often viscid shrubs, rarely perennial herbs, with alternate, simple, entire or toothed leaves, the branches commonly striate or angled; heads usually small, whitish or yellowish, diœcious; involucre of small, much imbricated, mostly acute bracts; receptacle usually flat and naked; pappus of fertile flowers of numerous bristles in 1 or several series, often elongated in fruit.

KEY TO THE SPECIES.

Pappus scant, little if at all elongated in fruit, not exceeding the styles.
  Bracts with green midribs; leaves linear or oblong, rather obtuse, 1-nerved, 4 cm. long or less .... 1. *B. bigelovii.*
  Bracts yellowish throughout; leaves elongated-lanceolate, acute, 3-nerved at the base, 4 to 10 cm. long .... 2. *B. glutinosa.*
Pappus abundant, much elongated in fruit and exceeding the styles.
  Pappus in several series; plants 50 cm. high or less, herbaceous to the base .... 7. *B. wrightii.*
  Pappus in a single series; plants usually 1 meter high or more, shrubby.

Branches smooth or nearly so, scabrous; leaves crowded or fasciculate; heads racemose or virgate............ 3. *B. pteronioides.*
Branches striate-angled, glabrous; leaves not crowded; heads paniculate or corymbose.
Leaves linear, entire, 25 mm. long or usually less; plants broomlike in general appearance....... 4. *B. sarothroides.*
Leaves oblong to obovate, mostly toothed, usually more than 30 mm. long; plants not broomlike.
Fertile heads hemispheric, 5 to 10 mm. broad... 5. *B. salicina.*
Fertile heads cylindric or long-campanulate, 3 to 5 mm. broad............................ 6. *B. emoryi.*

1. **Baccharis bigelovii** A. Gray in Torr. U. S. & Mex. Bound. Bot. 84. 1859.

TYPE LOCALITY: "Puerto de Paysano," Sonora.

RANGE: Southern Arizona and New Mexico to northern Mexico.

NEW MEXICO: Carpenter Creek; West Fork of the Gila. Transition Zone.

2. **Baccharis glutinosa** Pers. Syn. Pl. **2**: 425. 1807.

TYPE LOCALITY: Chile.

RANGE: Colorado to western Texas and California, south to Mexico, Central America, and South America.

NEW MEXICO: Mangas Springs; North Percha Creek; Socorro; Redrock; Deming; Mesilla Valley; Organ Mountains; between Tularosa and Mescalero. River valleys and wet ground, in the Lower Sonoran Zone.

One of the most common shrubs of the river valleys, preferring land that is sometimes inundated. It becomes a meter high or even more. Along the banks of the Rio Grande it covers large areas.

3. **Baccharis pteronioides** DC. Prodr. **5**: 410. 1836.

TYPE LOCALITY: "In Mexico inter Tampico et Real del Monte."

RANGE: Southern New Mexico and Arizona to Mexico.

NEW MEXICO: Mangas Springs; Kingston; Dog Spring; Organ Mountains; Queen. Dry hills, in the Upper Sonoran Zone.

4. **Baccharis sarothroides** A. Gray, Proc. Amer. Acad. **17**: 211. 1882.

TYPE LOCALITY: "Southern borders of California, San Diego County, near the old Mission station, the boundary monument, etc."

RANGE: Southern California to southwestern New Mexico.

NEW MEXICO: Near Carlisle (*Wooton*). Along streams.

5. **Baccharis salicina** Torr. & Gray, Fl. N. Amer. **2**: 258. 1842.

Baccharis salicifolia Nutt. Trans. Amer. Phil. Soc. n. ser. **7**: 337. 1841.

TYPE LOCALITY: "Banks of the Arkansas."

RANGE: Colorado and western Kansas to New Mexico and Texas.

NEW MEXICO: Fruitland; Laguna; Albuquerque; San Juan. River valleys and banks of streams, in the Upper Sonoran Zone.

6. **Baccharis emoryi** A. Gray in Torr. U. S. & Mex. Bound. Bot. 83. 1859.

TYPE LOCALITY: On the Gila River, Arizona.

RANGE: Colorado and New Mexico to Arizona and California.

NEW MEXICO: Roswell; Garfield. Along streams, in the Lower Sonoran Zone.

7. **Baccharis wrightii** A. Gray, Pl. Wright. **1**: 101. 1852.

TYPE LOCALITY: Valley of the Limpio, western Texas.

RANGE: Kansas and Colorado to Arizona, western Texas, and Mexico.

NEW MEXICO: McCarthys Ranch; Sierra Grande; Nara Visa; Las Vegas; Raton; Albuquerque; Socorro Mountain; San Augustine Plains; Aden; between Tularosa and Mescalero; Gray. Plains and dry hills, in the Upper Sonoran Zone.

31. ESCHENBACHIA Moench.

Coarse annuals with branched, very leafy stems, small, narrow, toothed, pinnatifid or entire, sessile, alternate leaves, and very numerous heads in leafy panicles; heads small, campanulate, many-flowered, the flowers whitish, the corollas of the pistillate flowers reduced to a short-filiform tube; rays wanting; bracts narrow, in 1 to 3 series; achenes small, compressed; pappus a single series of soft capillary bristles.

KEY TO THE SPECIES.

Leaves merely serrate or laciniate; stems very glandular, densely leafy .... 1. *E. coulteri.*
Leaves twice pinnatifid; stems sparingly glandular, not densely leafy .... 2. *E. tenuisecta.*

1. **Eschenbachia coulteri** (A. Gray) Rydb. Bull. Torrey Club **33**: 154. 1906.
Conyza coulteri A. Gray, Proc. Amer. Acad. **7**: 355. 1868.
Conyzella coulteri Greene, Fl. Franc. 386. 1897.
TYPE LOCALITY: Mexico.
RANGE: Colorado and western Texas to California and southward.
NEW MEXICO: Garfield; Cienaga Ranch; Plaza Larga; Mangas Springs; Organ Mountains; south of Tularosa; south of Roswell; Alamogordo; Carlsbad; Dayton; Guadalupe Mountains. Plains and hills, in the Lower and Upper Sonoran zones.

2. **Eschenbachia tenuisecta** (A. Gray) Woot. & Standl. Contr. U. S. Nat. Herb. **16**: 186. 1913.
Conyza coulteri tenuisecta A. Gray, Syn. Fl. **1**^{2}: 221. 1884.
TYPE LOCALITY: Near Fort Huachuca, southern Arizona.
RANGE: Southern New Mexico and Arizona.
NEW MEXICO: Bear Mountains; Mineral Creek; Fort Bayard; Organ Mountains. Canyons, in the Upper Sonoran Zone.

32. LEPTILON Raf.

Coarse annuals or biennials with simple or branched stems, alternate entire or toothed leaves, and numerous paniculate or racemose, small heads of whitish flowers; involucre cylindric or campanulate, of numerous small narrow green bracts; rays very short, scarcely if at all surpassing the pappus, in several rows; pappus a single series of capillary bristles.

KEY TO THE SPECIES.

Plants usually a meter high, often taller; leaves mostly linear, ciliate; heads less than 5 mm. wide, numerous .... 1. *L. canadense.*
Plants 50 cm. high or less; leaves oblong to lanceolate, not ciliate; heads 6 or 7 mm. wide, few .... 2. *L. integrifolium.*

1. **Leptilon canadense** (L.) Britton in Britt. & Brown, Illustr. Fl. **3**: 391. 1898. HORSEWEED.
Erigeron canadensis L. Sp. Pl. 863. 1753.
TYPE LOCALITY: "Canada, Virginia."
RANGE: Throughout temperate North America, a common weed in cultivated fields and waste ground; also in Asia.
NEW MEXICO: Common throughout the State.

2. **Leptilon integrifolium** Woot. & Standl. Contr. U. S. Nat. Herb. **16**: 183. 1913.
TYPE LOCALITY: West Fork of the Gila, Mogollon Mountains, New Mexico. Type collected by Metcalfe (no. 610).
RANGE: Mountains of New Mexico.
NEW MEXICO: West Fork of the Gila; Mineral Creek; White Mountains; East Las Vegas. Transition Zone.

33. **ERIGERON** L. Fleabane.

Annual or perennial herbs, sometimes woody at the base, with entire or toothed, mostly sessile leaves, these sometimes all basal; heads solitary or corymbose, often showy, the disk flowers yellow, the rays white to purple, sometimes short and inconspicuous, rarely wanting; involucral bracts narrow, imbricated in 2 to 4 series, the outer sometimes shorter; receptacle flat or convex, naked; achenes flattened or nearly terete, pubescent or glabrous, smooth or striate; pappus a single series of bristles or with an outer series of short bristles or inconspicuous paleæ.

KEY TO THE SPECIES.

Bracts of the involucre in 3 or 4 series, imbricated, thickened on the back, the outer successively shorter.
- Pubescence spreading; basal leaves oblanceolate, obtuse.... 1. *E. caespitosus.*
- Pubescence appressed; basal leaves linear or linear-oblanceolate, mostly acute.
  - Achenes glabrous.......... 2. *E. canus.*
  - Achenes pubescent.
    - Heads large, 10 to 12 mm. broad; stems sparingly strigose; bracts thin.......... 3. *E. pulcherrimus.*
    - Heads small, 6 to 8 mm. broad; stems densely grayish-strigose; bracts very thick.......... 4. *E. utahensis.*

Bracts in 1 or 2 series, about equal, not thickened on the back.
- Rays inconspicuous, slightly if at all exceeding the disk, usually with a row of rayless pistillate flowers inside; heads racemose.......... 5. *E. minor.*
- Rays conspicuous (in one species wanting), much longer than the disk, without rayless pistillate flowers inside; heads not racemose.
  - Plants with runners, at least at maturity.
    - Leaves and stem with spreading pubescence.......... 6. *E. commixtus.*
    - Pubescence all appressed.
      - Whole plant nearly glabrous, bright green; heads about 5 mm. broad.......... 7. *E. tonsus.*
      - Plants abundantly strigose, grayish; heads more than 5 mm. broad.
        - Stems branched at the base; cauline leaves narrowly oblanceolate to linear, acute.......... 8. *E. flagellaris.*
        - Stems branched above the base; cauline leaves obovate or oblanceolate, obtuse.......... 9. *E. senilis.*
  - Plants without runners.
    - Annuals or biennials.
      - Cauline leaves broad, cordate-clasping; rays white or pink.......... 10. *E. philadelphicus.*
      - Cauline leaves narrower, not cordate-clasping; rays variously colored.
        - Pubescence more or less appressed, not always closely.
          - Rays about 100; disk 7 to 8 mm. broad.......... 11. *E. gilensis.*
          - Rays 40 to 50 or fewer; disk less than 7 mm. broad.
            - Pappus in 2 series; rays narrow; achenes slightly compressed.......... 12. *E. modestus.*
            - Pappus in a single series; rays broader; achenes conspicuously compressed.

Plants 30 cm. high or more, corymbosely branched at the middle or above, stout; stems coarsely striate....................13. *E. bellidiastrum.*
Plants low, 10 to 14 cm. high, diffusely branched from or near the base, slender; stems smooth..........................14. *E. eastwoodiae.*
Pubescence spreading.
Stems simple below..............................15. *E. wootoni.*
Stems branched from the base.
Rays white; stems spreading....................16. *E. arenarius.*
Rays purplish; stems erect or ascending........17. *E. divergens.*
Perennials.
Leaves dissected or deeply cleft.
Rays purple; stems low, less than 10 cm. high, scapiform..18. *E. pinnatisectus.*
Rays white or pinkish; stems tall, not scapiform...19. *E. neomexicanus.*
Leaves entire or slightly toothed.
Stems low, less than 20 cm. high, scapiform, usually bearing a single head.
Involucres and peduncles long-villous.
Involucres and peduncles with black purple hairs......................................20. *E. melanocephalus*
Involucres and peduncles white-hairy.
Plants 5 to 8 cm. high; disk 10 to 12 mm. broad..................................21. *E. uniflorus.*
Plants 10 cm. high or more; disk 13 to 15 mm. wide....................................22. *E. leucotrichus.*
Involucres and peduncles hirsute to glandular-puberulent or glabrate, never long-villous.
Leaves and stems glabrous or nearly so; leaves spatulate.................................23. *E. leiomerus.*
Leaves and stems pubescent; leaves much narrower.
Heads 12 to 13 mm. high......................24. *E. grayi.*
Heads 6 mm. high or less.
Leaves linear-filiform; pubescence of the stems appressed........................25. *E. nematophyllus.*
Leaves linear; pubscence of stems spreading.26. *E. vetensis.*
Stems taller, 20 to 100 cm. high, usually with several to numerous heads, not scapiform.
Stems densely cespitose from a thick tap-root; heads small, the disk usually less than 1 cm. wide.
Rays wanting.................................27. *E. aphanactis.*
Rays present.
Pappus double; stems hirsute.................28. *E. concinnus.*
Pappus uniseriate; stems strigose............29. *E. eatoni.*
Stems usually solitary from the ends of simple or branched rootstocks; heads larger, the disk usually more than 1 cm. wide.
Bracts loose, with reflexed tips; rays broad; pappus uniseriate.
Bracts villous; rays white....................30. *E. coulteri.*
Bracts glandular-puberulent; rays purplish..31. *E. salsuginosus.*

Bracts appressed except at the very tips; rays narrow; pappus usually double.
Rays white; pappus uniseriate .... 32. *E. rusbyi.*
Rays lavender to deep purple; pappus double.
Upper cauline leaves reduced, linear-lanceolate; none of the leaves 3-nerved; peduncles long, erect.
Leaves canescent .... 33. *E. deminutus.*
Leaves glabrous or nearly so, or merely ciliate, never canescent.
Involucres glandular-puberulent, sparingly if at all hairy.
Stem with long white hairs throughout; lower leaves hairy .... 34. *E. viscidus.*
Stem glabrous below, glandular-puberulent above; leaves glabrous . 35. *E. smithii.*
Involucres densely hirsute, villous, or strigose.
Stems villous throughout; plants low, usually 20 cm. high or less ...... 36. *E. pecosensis.*
Stems glabrous below, usually much more than 20 cm. high.
Disk more than 15 mm. broad; heads solitary or very few ..... 37. *E. formosissimus.*
Disk less than 12 mm. broad; heads more numerous .... 38. *E. glabellus.*
Upper cauline leaves ample, not much smaller than the lower, lanceolate to ovate; lower leaves 3-nerved; peduncles short, ascending.
Leaves glabrous, smooth to the touch, sometimes ciliate.
Basal leaves broadly obovate-spatulate, often denticulate; cauline leaves distant, shorter than the internodes, not ciliate .... 39. *E. superbus.*
Basal leaves oblanceolate, entire; cauline leaves more numerous, longer than the internodes, usually ciliate.
Bracts glandular-puberulent, not at all hirsute .... 40. *E. macranthus.*
Bracts hirsute or villous .... 41. *E. speciosus.*
Leaves more or less pubescent, rough to the touch.
Leaves densely canescent, grayish .... 42. *E. subtrinervis.*
Leaves not canescent, green.
Stems hirsute throughout.
Upper leaves ovate to oblong, mostly obtuse .... 43. *E. platyphyllus.*
Upper leaves lanceolate, very acute .... 44. *E. rudis.*
Stems not hirsute, at least below.
Upper leaves broadly lanceolate, scaberulous; stems puberulent . 45. *E. semirasus.*

Upper leaves narrowly lanceolate, sparingly soft-pubescent; stems abundantly soft-pubescent........................46. *E. bakeri.*

1. **Erigeron caespitosus** Nutt. Trans. Amer. Phil. Soc. n. ser. **7:** 307. 1841.

TYPE LOCALITY: "On the summits of dry hills in the Rocky Mountain range, on the Colorado of the West."

RANGE: British America to Colorado and New Mexico.

NEW MEXICO: Puertecito (*Wooton*). Plains and low hills, in the Upper Sonoran Zone.

2. **Erigeron canus** A. Gray, Mem. Amer. Acad. n. ser. **4:** 67. 1849.

Wyomingia cana A. Nels. in Coulter, New Man. Rocky Mount. 531. 1909.

TYPE LOCALITY: Dry places on gravelly hills at the foot of mountains, Santa Fe, New Mexico. Type collected by Fendler (no. 375).

RANGE: Wyoming and South Dakota to Colorado and northern New Mexico.

NEW MEXICO: West of Santa Fe; between Santa Fe and Canyoncito; Sierra Grande; Raton. Plains and low hills, in the Upper Sonoran Zone.

3. **Erigeron pulcherrimus** Heller, Bull. Torrey Club **25:** 200. 1898.

Wyomingia pulcherrima A. Nels. Bull. Torrey Club **26:** 249. 1899.

TYPE LOCALITY: Sandy hills 10 miles north of Santa Fe, New Mexico. Type collected by Heller (no. 3664).

RANGE: Known only from the type locality.

4. **Erigeron utahensis** A. Gray, Proc. Amer. Acad. **16:** 89. 1880.

Erigeron stenophyllus tetrapleurus A. Gray, Proc. Amer. Acad. **8:** 650. 1873.

TYPE LOCALITY: Southern Utah.

RANGE: Southern Utah to northern New Mexico and Arizona.

NEW MEXICO: Carrizo Mountains (*Standley* 7351). Dry hills among rocks, in the Upper Sonoran Zone.

5. **Erigeron minor** (Hook.) Rydb. Bull. Torrey Club **24:** 295. 1897.

Erigeron glabratus minor Hook. Fl. Bor. Amer. **2:** 18. 1834.

TYPE LOCALITY: "Rocky Mountains."

RANGE: British Columbia and Saskatchewan to Utah and northern New Mexico.

NEW MEXICO: Rio Pueblo: Taos; Costilla Valley. Wet ground, in the Transition Zone.

6. **Erigeron commixtus** Greene, Pittonia **5:** 58. 1902.

Erigeron cinereus A. Gray, Mem. Amer. Acad. n. ser. **4:** 68. 1849, not Hook. & Arn. 1836.

Erigeron colomexicanus A. Nels. in Coulter, New Man. Rocky Mount. 529. 1909.

TYPE LOCALITY: Canyon of the Limpio, western Texas.

RANGE: Utah and Colorado to Arizona and western Texas.

NEW MEXICO: Common in the higher mountains throughout the State. Open slopes and in canyons, in the Upper Sonoran and Transition zones.

The type of *E. cinereus* was collected near Santa Fe by Fendler (no. 374).

7. **Erigeron tonsus** Woot. & Standl. Contr. U. S. Nat. Herb. **16:** 186. 1913.

TYPE LOCALITY: Near the N Bar Ranch, New Mexico. Type collected by Wooton, August 2, 1900.

RANGE: Western New Mexico.

NEW MEXICO: N Bar Ranch; Luna Valley; north of Ramah.

8. **Erigeron flagellaris** A. Gray, Mem. Amer. Acad. n. ser. **4:** 68. 1849.

TYPE LOCALITY: Low, moist places along Santa Fe Creek, New Mexico. Type collected by Fendler (no. 381).

RANGE: Wyoming and South Dakota to Utah and New Mexico.

NEW MEXICO: Tunitcha Mountains; Chama; Tierra Amarilla; Espanola; Pajarito Park; Santa Fe and Las Vegas mountains; Raton; Hanover Mountains; Ruidoso Creek. Open slopes and meadows, chiefly in the Transition Zone, sometimes in the Upper Sonoran.

9. **Erigeron senilis** Woot. & Standl. Contr. U. S. Nat. Herb. **16:** 185. 1913.

TYPE LOCALITY: Canyon above Van Pattens Camp in the Organ Mountains, New Mexico. Type collected by Standley, June 9, 1906.

RANGE: Organ Mountains, New Mexico.

10. **Erigeron philadelphicus** L. Sp. Pl. 863. 1753.

TYPE LOCALITY: "Habitat in Canada."

RANGE: British America to California, New Mexico, and Florida.

NEW MEXICO: Aztec; San Juan; Pecos; Cienaga Ranch; Socorro; near Mesilla; White and Sacramento mountains. Meadows, in the Upper Sonoran and Transition zones.

11. **Erigeron gilensis** Woot. & Standl. Contr. U. S. Nat. Herb. **16:** 184. 1913.

TYPE LOCALITY: North Fork of the Rio Gila, New Mexico. Type collected by Wooton, August 4, 1900.

RANGE: Known only from type locality.

12. **Erigeron modestus** (DC.) A. Gray, Mem. Amer. Acad. n. ser. **4:** 68. 1849.

Distasis modesta DC. Prodr. **5:** 279. 1836.

TYPE LOCALITY: Laredo, Texas.

RANGE: Western Texas to eastern New Mexico.

NEW MEXICO: Llano Estacado (*Bigelow*). Plains, in the Upper Sonoran Zone.

13. **Erigeron bellidiastrum** Nutt. Trans. Amer. Phil. Soc. n. ser. **7:** 307. 1841.

TYPE LOCALITY: "On the borders of the Platte, within the Rocky Mountains."

RANGE: Nebraska and Colorado to New Mexico and western Texas.

NEW MEXICO: Nara Visa (*Fisher* 15, 81). Plains, in the Upper Sonoran Zone.

14. **Erigeron eastwoodiae** Woot. & Standl. Contr. U. S. Nat. Herb. **16:** 183. 1913.

TYPE LOCALITY: Dry hills at the north end of the Carrizo Mountains, northeast corner of Arizona. Type collected by Standley (no. 7433).

RANGE: Southern Utah and Colorado to northern Arizona and New Mexico.

NEW MEXICO: Carrizo Mountains; Shiprock. Sandy hills and plains, in the Upper Sonoran Zone.

15. **Erigeron wootoni** Rydb. Bull. Torrey Club **33:** 153. 1906.

Erigeron cinereus var. γ A. Gray, Mem. Amer. Acad. n. ser. **4:** 68. 1849.

TYPE LOCALITY: Valley of Santa Fe Creek, New Mexico. Type collected by Fendler (no. 385).

RANGE: Colorado to Arizona and New Mexico.

NEW MEXICO: Carrizo Mountains; Tunitcha Mountains; Chama; Santa Fe and Las Vegas mountains; head of Canada Alamosa; White Mountains. Hills and meadows, in the Upper Sonoran and Transition zones.

It is doubtful whether this is in any essential respect different from *Erigeron divergens*. The two are almost invariably found growing together.

16. **Erigeron arenarius** Greene, Bull. Torrey Club **25:** 121. 1898.

TYPE LOCALITY: Sandhills near Mesilla, New Mexico. Type collected by Wooton (no. 23).

RANGE: Southwestern New Mexico.

NEW MEXICO: Mangas Springs; Mesilla Valley; northeast corner of Turney Range, Dona Ana County. Sandy soil, in the Lower Sonoran Zone.

The species is very abundant about the type locality, growing on dunes of nearly pure sand. It occurs at a much lower level than most of the species and comes into flower earlier in the season, continuing, however, until late fall. Its earlier flowering probably results from the warmer climate of the region in which it grows.

17. Erigeron divergens Torr. & Gray, Fl. N. Amer. **2:** 175. 1841.

Erigeron deustus Greene, Leaflets **1:** 211. 1906.

Type locality: "In the Rocky Mountains and the plains of Oregon."

Range: Washington and Montana to California and New Mexico.

New Mexico: Common at higher altitudes except along the eastern side of the State. Plains and hills, in the Upper Sonoran and Transition zones.

The type of *E. deustus* was collected on the West Fork of the Gila by Metcalfe.

18. Erigeron pinnatisectus (A. Gray) A. Nels. Bull. Torrey Club **26:** 246. 1899.

Erigeron compositus pinnatisectus A. Gray, Proc. Amer. Acad. **16:** 90. 1880.

Type locality: "High mountains of Colorado."

Range: Wyoming to northern New Mexico.

New Mexico: Truchas Peak (*Standley* 4820). Mountain meadows, in the Arctic-Alpine Zone.

19. Erigeron neomexicanus A. Gray, Proc. Amer. Acad. **19:** 2. 1883.

Type locality: Mountains at the Copper Mines, New Mexico. Type collected by Wright (no. 1170).

Range: Mountains of New Mexico and Arizona.

New Mexico: Mangas Springs; Hillsboro Peak; Middle Fork of the Gila; Bullards Peak; Santa Rita; San Luis Mountains. Upper Sonoran and Transition zones.

20. Erigeron melanocephalus A. Nels. Bull. Torrey Club **26:** 246. 1899.

Type locality: La Plata Mines, Medicine Bow Mountains, Wyoming.

Range: Wyoming to northern New Mexico.

New Mexico: Truchas Peak; Pecos Baldy. High mountain meadows, in the Arctic-Alpine Zone.

21. Erigeron uniflorus L. Sp. Pl. 864. 1753.

Erigeron simplex Greene, Fl. Franc. 387. 1897.

Type locality: "Habitat in Alpibus Lapponiae, Helvetiae."

Range: Arctic regions and on alpine peaks, Montana to California and northern New Mexico; also in the Old World.

New Mexico: Taos Mountains (*Bailey*). High mountain meadows, in the Arctic-Alpine Zone.

22. Erigeron leucotrichus Rydb. Bull. Torrey Club **28:** 23. 1901.

Type locality: Big Horn Mountains, Wyoming.

Range: Wyoming to northern New Mexico.

New Mexico: Truchas Peak; Baldy. High mountain meadows, in the Arctic-Alpine Zone.

23. Erigeron leiomerus A. Gray, Syn. Fl. **1**^{2}**:** 211. 1884.

Erigeron spathulifolius Rydb. Bull. Torrey Club **26:** 545. 1899.

Type locality: "Rocky Mountains of Colorado, Utah, and Nevada, in the alpine region."

Range: Wyoming to Utah and New Mexico.

New Mexico: Las Vegas Range. High mountain meadows, in the Hudsonian and Arctic-Alpine zones.

24. Erigeron grayi Woot. & Standl. nom. nov.

Erigeron stenophyllus A. Gray, U. S. Rep. Expl. Miss. Pacif. **4:** 98. 1856, not Nutt. 1847.

Type locality: On hillsides and steep banks of the Pecos, New Mexico. Type collected by Bigelow.

Range: Known only from the type locality, in the Upper Sonoran Zone.

25. Erigeron nematophyllus Rydb. Bull. Torrey Club **32:** 124. 1905.
TYPE LOCALITY: Dale Creek, Colorado.
RANGE: Wyoming to northern New Mexico.
NEW MEXICO: Fort Wingate (*Marsh*). Hills and plains, in the Upper Sonoran Zone.

26. Erigeron vetensis Rydb. Bull. Torrey Club **32:** 126. 1905.
TYPE LOCALITY: Mountains near Veta Pass, Colorado.
RANGE: Mountains of southern Colorado and northern New Mexico.
NEW MEXICO: Pass southeast of Tierra Amarilla (*Eggleston* 6535, 6603). Upper Sonoran and Transition zones.

27. Erigeron aphanactis (A. Gray) Greene, Fl. Franc. 389. 1897.
Erigeron concinnus aphanactis A. Gray, Proc. Amer. Acad. **6:** 540. 1865.
TYPE LOCALITY: Near Carson City, Nevada.
RANGE: Colorado and New Mexico to Nevada and California.
NEW MEXICO: Carrizo Mountains (*Standley* 7465). Dry plains and hills, in the Upper Sonoran Zone.

28. Erigeron concinnus (Hook. & Arn.) Torr. & Gray, Fl. N. Amer. **2:** 174. 1841.
Distasis ? concinnus Hook. & Arn. Bot. Beechey Voy. 350. 1840.
Erigeron setulosus Greene, Pittonia **4:** 319. 1901.
TYPE LOCALITY: "Snake River, below the Salmon Falls, Snake Country."
RANGE: British Columbia and Montana to California and New Mexico.
NEW MEXICO: Farmington; Cedar Hill; Stinking Lake; west of Santa Fe. Dry plains and hills, in the Upper Sonoran Zone.
The type of *E. setulosus* was collected at Aztec by Baker.

29. Erigeron eatoni A. Gray, Proc. Amer. Acad. **16:** 91. 1880.
TYPE LOCALITY: Uinta and Wahsatch mountains, Utah.
RANGE: Colorado and Wyoming to Utah and New Mexico.
NEW MEXICO: Las Vegas (*Dewey*). Upper Sonoran Zone.

30. Erigeron coulteri Porter in Port. & Coult. Syn. Fl. Colo. 61. 1874.
TYPE LOCALITY: "Wetson's Pass, at 10,000 feet altitude," Colorado.
RANGE: Colorado and Utah to northern New Mexico.
NEW MEXICO: Chama; Santa Fe and Las Vegas mountains. Meadows and moist slopes, in the Transition and Canadian zones.

31. Erigeron salsuginosus (Richards.) A. Gray, Proc. Amer. Acad. **16:** 93. 1881.
Aster salsuginosus Richards. Bot. App. Frankl. Journ. **2:** 748. 1823.
TYPE LOCALITY: "On the Salt Plains in the Athabasca."
RANGE: Alaska and Alberta to California and northern New Mexico.
NEW MEXICO: Truchas Peak (*Standley* 4827). Mountain meadows and bogs, in the Hudsonian and Arctic-Alpine zones.

32. Erigeron rusbyi A. Gray, Syn. Fl. 1^2: 217. 1884.
TYPE LOCALITY: Mogollon Mountains, New Mexico. Type collected by Rusby (no. 197).
RANGE: Mountains of southern New Mexico and Arizona.
NEW MEXICO: Mogollon Mountains; White and Sacramento mountains. Transition Zone.

33. Erigeron deminutus Woot. & Standl. Contr. U. S. Nat. Herb. **16:** 183. 1913.
TYPE LOCALITY: North of Ramah, New Mexico. Type collected by Wooton, July 25, 1906.
RANGE: Mountains of northwestern New Mexico.
NEW MEXICO: North of Ramah; mountains west of Grants Station.

34. Erigeron viscidus Rydb. Bull. Torrey Club **28:** 24. 1901.

TYPE LOCALITY: Near the Gray-Back Mining Camps, Colorado.

RANGE: Mountains of Colorado and New Mexico.

NEW MEXICO: Hermits Peak (*Snow*).

35. Erigeron smithii Rydb. Bull. Torrey Club **32:** 125. 1905.

TYPE LOCALITY: Parlin, Gunnison County, Colorado.

RANGE: Colorado and New Mexico.

NEW MEXICO: Tunitcha Mountains; Santa Fe Canyon; Chama; Trout Springs; Harveys Upper Ranch; Sierra Grande; Pajarito Park. Meadows, in the Transition Zone.

36. Erigeron pecosensis Standley, Muhlenbergia **5:** 29. 1909.

TYPE LOCALITY: Wet meadow along the Pecos River near Winsors Ranch, New Mexico. Type collected by Standley (no. 4358).

RANGE: Mountains of northern New Mexico.

NEW MEXICO: Santa Fe and Las Vegas mountains; Jemez Mountains. Transition to the Hudsonian Zone.

37. Erigeron formosissimus Greene, Bull. Torrey Club **25:** 121. 1898.

TYPE LOCALITY: White Mountain Peak, New Mexico. Type collected by Wooton (no. 352).

RANGE: Mountains of southern New Mexico.

NEW MEXICO: Mogollon Mountains; Hillsboro Peak; Capitan Mountains; White and Sacramento mountains. Canadian and Hudsonian zones.

38. Erigeron glabellus Nutt. Gen. Pl. **2:** 147. 1818.

TYPE LOCALITY: "On the plains of the Missouri (around Fort Mandan abundant)."

RANGE: British America to Wisconsin and New Mexico.

NEW MEXICO: Banks of the Nutria; Middle Fork of the Gila.

39. Erigeron superbus Greene; Rydb. Colo. Agr. Exp. Sta. Bull. **100:** 361, 364. 1906.

TYPE LOCALITY: Colorado.

RANGE: Mountains of Colorado and northern New Mexico.

NEW MEXICO: Santa Fe and Las Vegas mountains; Baldy; Chama; Willow Creek. Canadian and Hudsonian zones.

40. Erigeron macranthus Nutt. Trans. Amer. Phil. Soc. n. ser. **7:** 310. 1841.

TYPE LOCALITY: "Sources of the Missouri and the plains of the Platte."

RANGE: British Columbia and Montana to Oregon, Utah, and New Mexico.

NEW MEXICO: Carrizo Mountains; Tunitcha Mountains; Pecos; Middle Fork of the Gila; White Mountains; Castle Rock. Meadows in the mountains, in the Transition Zone.

41. Erigeron speciosus (Lindl.) DC. Prodr. **5:** 284. 1836.

Stenactis speciosa Lindl. in Edwards's Bot. Reg. **17:** *pl. 1577*. 1833.

TYPE LOCALITY: "Native of California."

RANGE: British Columbia and California to Colorado and New Mexico.

NEW MEXICO: Eagle Creek (*Wooton*). Transition Zone.

It is doubtful whether this specimen really belongs here, but we have been unable to place it elsewhere.

42. Erigeron subtrinervis Rydb. Mem. Torrey Club **5:** 328. 1894.

Erigeron glabellus mollis A. Gray, Proc. Acad. Phila. **1863:** 64. 1864.

TYPE LOCALITY: Colorado.

RANGE: Wyoming and South Dakota to New Mexico.

NEW MEXICO: Chama; Sandia Mountains; Baldy; below Winsors Ranch; Upper Rio Tesuque; Costilla Valley. Meadows in the mountains, in the Transition Zone.

43. Erigeron platyphyllus Greene, Leaflets **1:** 145. 1905.

Type locality: Santa Rita Mountain, New Mexico. Type collected by Metcalfe (no. 1469).

Range: Known only from the type locality, in the Transition Zone.

44. Erigeron rudis Woot. & Standl. Contr. U. S. Nat. Herb. **16:** 184. 1913.

Type locality: White Mountains, Lincoln County, New Mexico. Type collected by Wooton (no. 270).

Range: Mountains of southern New Mexico.

New Mexico: White Mountains; Burro Mountains; Mogollon Mountains; Capitan Mountains. Damp ground, in the Transition Zone.

45. Erigeron semirasus Woot. & Standl. Contr. U. S. Nat. Herb. **16:** 185. 1913.

Type locality: Mogollon Creek, Mogollon Mountains, New Mexico. Type collected by Metcalfe (no. 320).

Range: Mountains of New Mexico.

New Mexico: Mogollon Mountains; Santa Fe and Las Vegas mountains. Damp woods, in the Transition and Canadian zones.

46. Erigeron bakeri Woot. & Standl. Contr. U. S. Nat. Herb. **16:** 185. 1913.

Type locality: Near Chama, New Mexico. Type collected by Baker (no. 678).

Range: Northern New Mexico.

New Mexico: Chama; Dulce. Meadows, in the Transition Zone.

34. BRACHYACTIS Ledeb.

Low branched leafy-stemmed annual with alternate entire leaves and small racemose-paniculate heads; involucres campanulate, rather small, of 2 or 3 series of narrow, nearly equal bracts, the outer foliaceous and resembling the leaves, the inner membranaceous or scarious; achenes narrow, not compressed, 2 or 3-nerved, pubescent; pappus simple, of capillary bristles; rays very short.

1. Brachyactis woodhousei (Wooton) Woot. & Standl.

Aster woodhousei Wooton, Bull. Torrey Club **25:** 458. 1898.

Type locality: Near Zuni, New Mexico. Type collected by Woodhouse.

Range: Northwestern New Mexico.

New Mexico: Zuni; Albuquerque.

35. ASTER L. Aster.

Perennial branched herbs with leafy stems, sessile or petiolate leaves, and corymbose, paniculate, or racemose heads of yellow flowers; rays purple, blue, or white; ray flowers fertile; involucral bracts more or less imbricated, appressed or spreading, with green foliaceous tips; receptacle flat, alveolate; achenes more or less flattened; pappus of capillary bristles in a single series.

KEY TO THE SPECIES.

Involucres and often the stems glandular.
  Leaves scabrous or variously pubescent.
    Bracts linear, about equal; plants stout.............. 1. *A. novae-angliae.*
    Bracts oblong or lanceolate, very unequal; plants comparatively slender.......................... 2. *A. oblongifolius.*
  Leaves glabrous, except sometimes on the margins.
    Leaves conspicuously ciliate; stems clustered, very numerous.................................. 3. *A. fendleri.*
    Leaves not ciliate; stems few or solitary.
      Heads large, about 12 mm. wide................. 4. *A. hydrophilus.*
      Heads smaller, less than 10 mm. wide.
        Leaves long, nearly linear, acute; heads few. 5. *A. pauciflorus.*

Leaves short, narrowly oblong, obtuse; heads numerous .... 6. *A. boltoniae.*
Plants nowhere glandular.
Outer bracts foliaceous, equaling or surpassing the inner.
Plants low, 10 to 20 cm. high; leaves broad .... 8. *A. apricus.*
Plants tall, 40 to 60 cm.; leaves narrow .... 9. *A. canbyi.*
Outer bracts not conspicuously foliaceous, shorter than the inner.
Bracts pubescent on the back, usually ciliate.
Bracts not bristle-pointed; rays purple .... 10. *A. vallicola.*
Bracts bristle-pointed; rays usually white.
Pubescence of the stems spreading or reflexed.
Heads small, 5 mm. high or less, very numerous .... 11. *A. hebecladus.*
Heads larger, 7 mm. long or more, few .. 12. *A. crassulus.*
Pubescence of the stems appressed.
Heads small, about 5 mm. high, numerous; bracts much imbricated, often reflexed .... 13. *A. multiflorus.*
Heads larger, 7 mm. high or more, few; bracts nearly equal, erect .... 14. *A. commutatus.*
Bracts glabrous, sometimes ciliate.
Leaves conspicuously ciliate .... 15. *A. blepharophyllus.*
Leaves not ciliate.
Plants glabrous throughout.
Rays 2 mm. wide and 8 mm. long, showy; cauline leaves clasping .... 16. *A. laevis.*
Rays 1 mm. wide or less, not more than 3 mm. long; cauline leaves not clasping.
Stems much branched; leaves thick and fleshy, short; heads long-pedunculate .... 7. *A. neomexicanus.*
Stems branched only above; leaves thin, long; heads short-pedunculate .... 17. *A. exilis.*
Stems pubescent, at least with pubescent lines above.
Pubescence covering the peduncles ..... 18. *A. lonchophyllus.*
Pubescence in lines on the peduncles.
Heads short-pedunculate or nearly sessile; leaves thick .... 19. *A. salicifolius.*
Heads long-pedunculate; leaves thin.
Heads large, 10 to 15 mm. wide, few; stems little branched; leaves of the inflorescence little reduced .... 20. *A. wootonii.*
Heads smaller, usually less than 10 mm. wide, very numerous; stems tall, much branched; leaves of the inflorescence much re-reduced .... 21. *A. hesperius.*

1. **Aster novae-angliae** L. Sp. Pl. 875. 1753.
Type locality: "Habitat in Nova Anglia."
Range: Canada to New Mexico, Colorado, and South Carolina.
New Mexico: Near Pecos; Gallinas Planting Station. Open fields, in the Transition Zone.

2. **Aster oblongifolius** Nutt. Gen. Pl. **2:** 156. 1818.
Type locality: "On the banks of the Missouri."
Range: Minnesota and Kansas to northeastern New Mexico, Texas, and Maryland.
New Mexico: Raton Mountains (*Wooton*). Upper Sonoran Zone.

3. **Aster fendleri** A. Gray, Mem. Amer. Acad. n. ser. **4:** 66. 1849.
Type locality: "On the Ocate Creek," New Mexico. Type collected by Fendler (no. 372).
Range: Colorado and northern New Mexico to Kansas and Oklahoma.
New Mexico: Llano Estacado; Ocate Creek. Upper Sonoran Zone.

4. **Aster hydrophilus** Greene, Contr. U. S. Nat. Herb. **16:** 187. 1913.
Type locality: Along the edge of Berendo Creek at the south end of the Black Range, Sierra County, New Mexico. Type collected by Metcalfe (no. 1393).
Range: Known only from type locality.

5. **Aster pauciflorus** Nutt. Gen. Pl. **2:** 154. 1818.
Type locality: "On the margins of saline springs, near Fort Mandan, on the Missouri."
Range: Saskatchewan and South Dakota to Arizona and New Mexico.
New Mexico: Shiprock; Albuquerque; Barranca; Horace; Ojo Caliente; Santo Domingo; Chavez. Wet ground, in the Lower and Upper Sonoran zones.

6. **Aster boltoniae** Greene, Pittonia **3:** 248. 1897.
Type locality: "In irrigated fields and along ditches in western Texas and southern New Mexico."
Range: Western Texas and southern New Mexico.
New Mexico: Mesilla Valley. River valleys, in the Lower Sonoran Zone.

7. **Aster neomexicanus** Woot. & Standl. Contr. U. S. Nat. Herb. **16:** 187. 1913.
Type locality: Roswell, New Mexico. Type collected by Earle (no. 327).
Range: Southeastern New Mexico.
New Mexico: Roswell; Lake Arthur. Wet ground, in the Lower Sonoran Zone.

8. **Aster apricus** (A. Gray) Rydb. Mem. N. Y. Bot. Gard. **1:** 396. 1900.
Aster foliaceus apricus A. Gray, Syn. Fl. **1²:** 193. 1884.
Type locality: "High mountains of Colorado, at Union Pass, Rothrock, and near Gray's Peak."
Range: British Columbia and Montana to Colorado and northern New Mexico.
New Mexico: Ponchuelo Creek (*Standley* 4583); Chama (*Standley* 6574). Meadows in the mountains, in the Transition and Canadian zones.
The first specimen cited may be only a depauperate form of *A. canbyi*. The second has very narrow leaves, but otherwise seems to belong here.

9. **Aster canbyi** Vasey; A. Gray, Syn. Fl. **1²:** 193. 1884.
Aster foliaceus canbyi A. Gray, loc. cit.
Type locality: White River, western Colorado.
Range: Mountains of Colorado and northern New Mexico.
New Mexico: Rio Pueblo; Ponchuelo Creek. Transition and Canadian zones.

10. **Aster vallicola** Greene, Pittonia **4:** 221. 1900.
Type locality: Moist meadows of Pine Valley, above Palisade, Nevada.
Range: Wyoming and Nevada to Colorado and northern New Mexico.

NEW MEXICO: Carrizo Mountains; Ensenada; Taos; Baldy; Santa Fe Creek; Chama; Dulce. Wet ground, especially along streams, in the Upper Sonoran and Transition zones.

11. Aster hebecladus DC. Prodr. **5**: 242. 1836.

TYPE LOCALITY: Texas.

RANGE: Western Texas to southern Arizona and southward.

NEW MEXICO: Gallinas Canyon; Mangas Springs; near White Water; Mesilla Valley; Fort Bayard; Silver City; Dayton. Chiefly in river valleys, often along ditches, Lower Sonoran to Transition Zone.

12. Aster crassulus Rydb. Bull. Torrey Club **28**: 504. 1901.

TYPE LOCALITY: Mesas, La Veta, Colorado.

RANGE: Idaho and North Dakota to Colorado and New Mexico.

NEW MEXICO: Santa Fe; Zuni; Raton. Upper Sonoran Zone.

13. Aster multiflorus Ait. Hort. Kew. **3**: 203. 1789.

TYPE LOCALITY: North America.

RANGE: Montana and Maine to Georgia and Mexico.

NEW MEXICO: Albuquerque; Aztec; Farmington; Las Huertas Canyon. Lower and Upper Sonoran zones.

14. Aster commutatus (Torr. & Gray) A. Gray, Syn. Fl. **1**^{2}: 185. 1884.

Aster multiflorus commutatus Torr. & Gray, Fl. N. Amer. **2**: 125. 1841.

TYPE LOCALITY: "Upper Missouri."

RANGE: Wyoming and Minnesota to Nevada and New Mexico.

NEW MEXICO: Dulce; Santa Fe; Beulah; Taos; Ensenada; Pecos; Folsom; Belen; Gilmores Ranch; north of Capitan Mountains. Open fields, in the Upper Sonoran and lower part of the Transition zones.

15. Aster blepharophyllus A. Gray, Pl. Wright. **2**: 77. 1853.

TYPE LOCALITY: Las Playas Springs, New Mexico. Type collected by Wright (no. 1164).

RANGE: Known only from type locality.

16. Aster laevis L. Sp. Pl. 876. 1753.

TYPE LOCALITY: "Habitat in America septentrionali."

RANGE: Saskatchewan and New England to New Mexico and Louisiana.

NEW MEXICO: Dulce; Raton Mountains; Santa Fe and Las Vegas mountains; Santa Antonita; Folsom; Eagle Creek. Open slopes, in the Upper Sonoran and Transition zones.

17. Aster exilis Ell. Bot. S. C. & Ga. **2**: 344. 1821.

TYPE LOCALITY: "Grows in damp soils in the western districts of Georgia."

RANGE: California and New Mexico to Texas and Georgia.

NEW MEXICO: Mogollon Mountains; Berendo Creek; Organ Mountains. Transition Zone.

18. Aster lonchophyllus Greene, Leaflets **1**: 146. 1905.

TYPE LOCALITY: Stony slopes, Crested Butte, southern Colorado.

RANGE: Mountains of Colorado and northern New Mexico.

NEW MEXICO: Chama (*Standley* 6573). Meadows, in the Transition Zone.

Our specimens are very small as compared with the type, but this is probably because of the rather unfavorable environment in which they grew.

19. Aster salicifolius Lam. Encycl. **1**: 306. 1783.

TYPE LOCALITY: Canada.

RANGE: British America to New Mexico, Texas, and Florida.

NEW MEXICO: Mesilla Valley (*Wooton & Standley*).

20. Aster wootonii Greene, Leaflets **1:** 146. 1905.

Aster hesperius wootonii Greene, Bull. Torrey Club **25:** 119. 1898.

TYPE LOCALITY: Eagle Creek, White Mountains, New Mexico. Type collected by Wooton (no. 329).

RANGE: Mountains of New Mexico and Arizona.

NEW MEXICO: Tunitcha Mountains; Farmington; Dulce; Santa Fe; Pecos; Catskill; Taos; Ensenada; Chama; Grand Canyon of the Gila; Kingston; White Mountains. Wet ground, especially along streams, in the Transition Zone.

21. Aster hesperius A. Gray, Syn. Fl. 1²: 192. 1884.

TYPE LOCALITY: "Damp soil and along streams, S. Colorado and New Mexico to Arizona and S. California."

RANGE: Colorado and western Texas to southern California.

NEW MEXICO: Albuquerque; Mesilla Valley; Round Mountain. Wet ground, in the Lower and Upper Sonoran zones.

36. EUCEPHALUS Nutt.

Glabrous, usually glaucous, perennial herbs with alternate sessile entire leaves and numerous showy heads in terminal corymbs; bracts broad, imbricated, the outer successively shorter, ciliate, dry and chartaceous, with prominent midribs, usually purplish; rays purple or blue; achenes strigose.

KEY TO THE SPECIES.

Inner bracts acute; leaves not narrowed at the base................. 1. *E. glaucus.*
All bracts obtuse or short-mucronate; leaves conspicuously narrowed at the base.. 2. *E. formosus.*

1. Eucephalus glaucus Nutt. Trans. Amer. Phil. Soc. n. ser. **7:** 299. 1841.

Aster glaucus Torr. & Gray, Fl. N. Amer. **2:** 159. 1842.

TYPE LOCALITY: "Towards the sources of the Platte and in the Rocky Mountains."

RANGE: Wyoming to Utah and northern New Mexico.

NEW MEXICO: Dulce (*Standley* 8224). Open slopes in the mountains, in the Transition Zone.

2. Eucephalus formosus Greene, Pittonia **4:** 156. 1900.

Aster glaucus formosus A. Nels. in Coulter, New Man. Rocky Mount. 513. 1909.

TYPE LOCALITY: Mountains near Pagosa Peak, southern Colorado.

RANGE: Mountains of southern Colorado and northern New Mexico.

NEW MEXICO: Chama (*Standley* 6832). Transition Zone.

37. HERRICKIA Woot. & Standl.

Perennial herb with alternate thick rigid toothed sessile leaves; heads solitary at the ends of the slender leafy branches; rays purple; disk flowers yellow, drying purplish; bracts of the involucre in several series, about equal, conspicuously keeled, with green foliaceous tips and spinescent points, the outer bracts foliaceous and changing gradually into the proper leaves; achenes compressed, striate, glabrous; pappus simple, of numerous stout, simple, nearly equal, strongly barbellate bristles.

1. Herrickia horrida Woot. & Standl. Contr. U. S. Nat. Herb. **16:** 186. 1913.

TYPE LOCALITY: Baldy, New Mexico. Type collected by Wooton, August 14, 1910.

RANGE: Mountains of northern New Mexico, probably in adjacent Colorado.

NEW MEXICO: Baldy; Raton; Spring Canyon.

38. LEUCOSYRIS Greene. SPINY ASTER.

Nearly leafless perennial, often woody at the base, with slender striate green branches; cauline leaves small, fugacious, with stout spines in or above the axils; involucre hemispheric, small, of thin lanceolate bracts imbricated in about 3 series; rays white; achenes glabrous.

1. **Leucosyris spinosus** (Benth.) Greene, Pittonia 3: 244. 1897.

Aster spinosus Benth. Pl. Hartw. 20. 1839.

TYPE LOCALITY: Mexico.

RANGE: Utah and California to Texas and New Mexico and southward.

NEW MEXICO: Albuquerque; Socorro; Cliff; near White Water; Mesilla Valley; San Antonio. In heavy, usually adobe soil, in the Lower Sonoran Zone.

39. MACHAERANTHERA Nees.

Stout, rather coarse annuals or biennials with slender or thick taproots; stems usually much branched; leaves alternate, from simple and dentate to bipinnately parted into linear segments; heads rather large, campanulate, terminating the branches, with purple rays; involucre much imbricated, the bracts linear, coriaceous below, with foliaceous, often spreading tips; achenes narrowed downward, compressed, few-nerved, pubescent or glabrous; pappus copious, of rather rigid unequal bristles.

KEY TO THE SPECIES.

Annuals; leaves pinnatifid; achenes mostly terete.
  Leaf segments acute, usually bristle-tipped; heads large; plants abundantly pubescent.
    Heads large, 12 to 15 mm. in diameter; bracts narrow, subulate-tipped; plants spreading, large, 30 cm. high or more .... 1. *M. tanacetifolia.*
    Heads about 10 mm. in diameter; bracts broad, not subulate-tipped; plants erect, small, usually less than 20 cm. high .... 2. *M. tagetina.*
  Leaf segments obtuse, not bristle-tipped; heads small; plants sparingly pubescent.
    Leaves densely glandular; many of the leaves twice pinnatifid, the segments broad .... 3. *M. pygmaea.*
    Leaves not glandular; leaves once pinnatifid, the segments narrow .... 4. *M. parviflora.*
Biennials, sometimes perennials; leaves merely toothed or entire; achenes compressed.
  Tips of the bracts short, lanceolate or rhombic, usually little reflexed.
    Heads small, less than 10 mm. broad; leaves thick, sparingly toothed, little if at all ciliate; peduncles very leafy .... 5. *M. cichoriacea.*
    Heads 12 mm. broad or more; leaves thin, salient-toothed, ciliate; peduncles sparingly leafy .... 6. *M. linearis.*
  Tips of the bracts long, linear or subulate, reflexed.
    Plants nowhere glandular.
      Stems stout, much branched; leaves thick, lanceolate or oblong .... 7. *M. asteroides.*
      Stems slender, simple or nearly so; leaves thin, oblanceolate .... 8. *M. simplex.*
    Plants glandular, at least on the bracts.
      Plants densely and coarsely glandular almost throughout.
        Cauline leaves narrow, mostly 1-veined; leaves stiff, rigid .... 9. *M. viscosa.*
        Cauline leaves broader, triple-veined; leaves thinner.

Stems with few erect branches, green; heads less than 15 mm. broad; leaves with numerous salient teeth..10. *M. centaureoides.*
Stems with numerous, mostly spreading branches; heads more than 15 mm. broad; leaves with few, never salient teeth........................11. *M. bigelovii.*
Plants glandular only on the bracts, or also sparingly on the peduncles, then only finely glandular.
Leaves linear, very acute, entire..............12. *M. angustifolia.*
Leaves oblong to obovate, toothed, mostly obtuse.
Basal leaves rounded-spatulate, the cauline ones obovate, tapering to the base..............................13. *M. amplifolia.*
Basal leaves oblanceolate, acute, the cauline ones oblong to elliptic or lanceolate, not noticeably tapering to the base........................14. *M. aquifolia.*

1. Machaeranthera tanacetifolia (H. B. K.) Nees, Gen. Sp. Aster. 224. 1832.
Aster tanacetifolius H. B. K. Nov. Gen. & Sp. **4**: 95. 1820.
Dieteria coronopifolia Nutt. Trans. Amer. Phil. Soc. n. ser. **7**: 302. 1841.
Machaeranthera coronopifolia A. Nels. Bot. Gaz. **37**: 268. 1904.
TYPE LOCALITY: Mexico.
RANGE: Wyoming and Nebraska to Arizona and western Texas and southward.
NEW MEXICO: Throughout the State at lower altitudes. Sandy soil, in the Lower and Upper Sonoran zones.

2. Machaeranthera tagetina Greene, Pittonia **4**: 71. 1899.
Machaeranthera tanacetifolia humilis A. Gray, Pl. Wright. **2**: 74. 1853.
Machaeranthera humilis Standley, Muhlenbergia **5**: 48. 1909.
TYPE LOCALITY: Arizona.
RANGE: Southwestern New Mexico and southern Arizona.
NEW MEXICO: Near Ojo de Gavilan (*Wright* 1151). Lower Sonoran Zone.

3. Machaeranthera pygmaea (A. Gray) Woot. & Standl. Contr. U. S. Nat. Herb. **16**: 189. 1913.
Macheranthera tanacetifolia pygmaea A. Gray, Pl. Wright. **2**: 74. 1853.
Aster tanacetifolius pygmaeus A. Gray, Syn. Fl. **1**²: 206. 1884.
TYPE LOCALITY: "Dry, stony hills, valley of the Salado, Chihuahua, and near El Paso."
RANGE: Western Texas to southern Arizona and northern Mexico.
NEW MEXICO: A single specimen seen, without definite locality.

4. Machaeranthera parviflora A. Gray, Pl. Wright. **1**: 90. 1852.
Aster parviflorus A. Gray in Brewer & Wats. Bot. Calif. **1**: 322. 1876.
TYPE LOCALITY: Along the Rio Grande, western Texas.
RANGE: Western Texas to southern Arizona.
NEW MEXICO: Farmington; Cedar Hill; Albuquerque; Mesilla Valley; White Sands; Alamogordo. Sandy soil of plains, in the Lower and Upper Sonoran zones.

5. Machaeranthera cichoriacea Greene, Leaflets **1**: 148. 1905.
TYPE LOCALITY: Canyon of Deer Run, southern Colorado.
RANGE: Southern Colorado and northern New Mexico.
NEW MEXICO: Zuni; Shiprock; Carrizo Mountains; Cedar Hill; Farmington; Chama. Dry plains, in the Upper Sonoran Zone.

6. **Machaeranthera linearis** Greene, Bull. Torrey Club **24:** 511. 1897.

TYPE LOCALITY: Sandy fields, Mesilla Valley, New Mexico.

RANGE: Valleys of New Mexico and western Texas.

NEW MEXICO: Albuquerque; Mesilla Valley; White Sands. Lower Sonoran Zone.

7. **Machaeranthera asteroides** (Torr.) Greene, Pittonia **3:** 63. 1896.

Dieteria asteroides Torr. in Emory, Mil. Reconn. 142. 1848.

Machaeranthera canescens latifolia A. Gray, Pl. Wright. **2:** 75. 1853.

TYPE LOCALITY: "Elevated land between the del Norte and the waters of the Gila," New Mexico. Type collected by Emory.

RANGE: Mountains of New Mexico and Arizona.

NEW MEXICO: Mangas Springs; Fort Bayard; Jicarilla Mountains; Dog Spring; Santa Rita. Upper Sonoran and Transition zones.

8. **Machaeranthera simplex** Woot. & Standl. Contr. U. S. Nat. Herb. **16:** 189. 1913.

TYPE LOCALITY: Capitan Mountains, New Mexico. Type collected by Earle (no. 390).

RANGE: Known only from type locality.

9. **Machaeranthera viscosa** (Nutt.) Greene, Pittonia **4:** 22. 1899.

Dieteria viscosa Nutt. Trans. Amer. Phil. Soc. n. ser. **7:** 301. 1841.

Aster canescens viscosus A. Gray, Syn. Fl. **1**[2]: 206. 1884.

TYPE LOCALITY: "With the above (*Dieteria divaricata*) particularly near Scott's Bluff, on the Platte."

RANGE: Washington and Nebraska to northern New Mexico.

NEW MEXICO: Raton (*Cockerell*). Dry hills, in the Upper Sonoran Zone.

10. **Machaeranthera centaureoides** Greene, Contr. U. S. Nat. Herb. **16:** 188. 1913.

TYPE LOCALITY: Mogollon Mountains, on the Middle Fork of the Rio Gila, Socorro County, New Mexico. Type collected by Metcalfe (no. 440).

RANGE: Mountains of western New Mexico.

NEW MEXICO: Middle fork of the Gila; Luna.

11. **Machaeranthera bigelovii** (A. Gray) Greene, Pittonia **3:** 63. 1896.

Aster bigelovii A Gray, U. S. Rep. Expl. Miss. Pacif. **4:** 10. 1856.

Machaeranthera varians Greene, Pittonia **4:** 98. 1899.

TYPE LOCALITY: Arroyos in the Sandia Mountains, New Mexico. Type collected by Bigelow.

RANGE: Southern Colorado and New Mexico.

NEW MEXICO: Sandia Mountains; Johnsons Mesa; Baldy; Chama; Jemez Mountains; Rio Pueblo; Glorieta; Mogollon Mountains; White and Sacramento mountains. Meadows in the mountains, Transition Zone.

12. **Machaeranthera angustifolia** Woot. & Standl. Contr. U. S. Nat. Herb. **16:** 188. 1913.

TYPE LOCALITY: Probably in the Sandia Mountains, New Mexico. Type collected by Bigelow.

RANGE: Mountains of northern New Mexico.

NEW MEXICO: Sandia Mountains; Dulce; Pecos; Glorieta; Tunitcha Mountains. Upper Sonoran and Transition zones.

13. **Machaeranthera amplifolia** Woot. & Standl. Contr. U. S. Nat. Herb. **16:** 187. 1913.

TYPE LOCALITY: Filmore Canyon, Organ Mountains, New Mexico. Type collected by Wooton & Standley, September 23, 1906.

RANGE: Canyons of the Organ Mountains, New Mexico, in the Transition Zone.

14. Machaeranthera aquifolia Greene, Contr. U. S. Nat. Herb. **16**: 188. 1913.

TYPE LOCALITY: Gila Hot Springs, Mogollon Mountains, Socorro County, New Mexico. Type collected by Metcalfe (no. 856).

RANGE: Mountains of southern New Mexico and Arizona.

NEW MEXICO: Gila Hot Springs; Magdalena Mountains; Grand Canyon of the Gila; Mangas Springs; East Canyon; Gilmores Ranch. Upper Sonoran and Transition zones.

The specimens from the Magdalena Mountains are stouter and have more heads than the typical form.

40. LEUCELENE Greene.

Slender perennials with diffusely branched, ascending or prostrate stems and linear-subulate to spatulate, strigose or hispid-ciliate leaves; heads small, terminal, with white or pink rays; involucres campanulate, of narrow, much imbricated bracts with scarious margins; achenes slender, compressed, hirsutulous; pappus a single series of scabrous bristles.

KEY TO THE SPECIES.

Upper leaves strigose, sparingly if at all glandular; not hispid-ciliate.. 1. *L. arenosa.*
Upper leaves strongly glandular, conspicuously hispid-ciliate........ 2. *L. ericoides.*

Other species have been reported from New Mexico, but our material seems to show only two forms. *Leucelene arenosa* may be only the later stage of the vernal *L. ericoides.*

1. Leucelene arenosa Heller, Cat. N. Amer. Pl. 8. 1898.

Aster ericaefolius tenuis A. Gray, Syn. Fl. 1[2]: 198. 1884.

TYPE LOCALITY: "New Mexico."

RANGE: Colorado to western Texas and Arizona and southward.

NEW MEXICO: Horse Camp; Tunitcha Mountains; Carrizo Mountains; Shiprock; Zuni; Dulce; Sierra Grande; Santa Fe; Raton; Pecos; Willard; Albuquerque; Water Canyon; Las Vegas; Jarilla; Mogollon Mountains; San Luis Mountains; Hanover Mountain; Organ Mountains; White Mountains; Gray; Torrance. Dry hills, in the Lower and Upper Sonoran zones.

2. Leucelene ericoides (Torr.) Greene, Pittonia **3**: 148. 1896.

Inula ? ericoides Torr. Ann. Lyc. N. Y. **2**: 212. 1828.

Diplopappus ericoides hirtella A. Gray, Mem. Amer. Acad. n. ser. **4**: 69. 1849.

Leucelene hirtella Rydb. Colo. Agr. Exp. Sta. Bull. **100**: 358. 1906.

TYPE LOCALITY: "On the Canadian?," Colorado or New Mexico.

RANGE: Colorado to Arizona and New Mexico.

NEW MEXICO: Las Vegas; Farmington; Tierra Amarilla; Laguna Colorado; Santa Fe; Nara Visa; Santa Rita; Mangas Springs; San Augustine Plains; Carrizalillo Mountains; Organ Mountains; Roswell; Gray; mountains west of San Antonio; Aden. Open slopes, in the Lower and Upper Sonoran zones.

The type of *Diplopappus ericoides hirtella* was collected near Santa Fe (*Fendler* 348).

41. PSILACTIS A. Gray.

Tall, sparingly branched annual, glandular-puberulent, with narrow, often pinnatifid leaves and small heads terminating the branches; involucre hemispheric, the bracts imbricated in 2 or 3 series; rays purple; achenes pubescent; pappus none in the ray flowers, a single series of capillary bristles in the disk flowers.

1. Psilactis asteroides A. Gray, Mem. Amer. Acad. n. ser. **4**: 72. 1849.

TYPE LOCALITY: "Llanos, in the Sierra Madre, west of Chihuahua," Mexico.

RANGE: Western Texas and southern Arizona to northern Mexico.

NEW MEXICO: Mesilla Valley. Valleys, in the Lower Sonoran Zone.

42. APHANOSTEPHUS DC.

Slender low annuals or perennials, much branched, with entire to pinnatifid leaves and pedunculate white-rayed heads; bracts broadly lanceolate, well imbricated; receptacle conic or hemispheric; achenes terete truncate, striate, with low, laciniate or ciliate pappus.

KEY TO THE SPECIES.

Perennial; leaves linear.. 1. *A. perennis.*
Annuals or biennials; leaves oblanceolate, oblong-lanceolate, or spatulate, often lobed or toothed.
  Corolla much thickened at the base in age; pappus conspicuously dentate or laciniate........................ 2. *A. skirrobasis.*
  Corolla not thickened at the base; pappus merely a ciliate-fringed edge.. 3. *A. ramosissimus.*

1. **Aphanostephus perennis** Woot. & Standl. Contr. U. S. Nat. Herb. **16:** 189. 1913.
TYPE LOCALITY: Knowles, New Mexico. Type collected by Wooton, July 29, 1909.
RANGE: Known only from type locality, in the Upper Sonoran Zone.

2. **Aphanostephus skirrobasis** (DC.) Trel. Rep. Ark. Geol. Surv. **4:** 191. 1891.
Keerlia skirrobasis DC. Prodr. **5:** 310. 1836.
Egletes arkansana Nutt. Trans. Amer. Phil. Soc. n. ser. **7:** 394. 1841.
Aphanostephus arkansanus A. Gray, Pl. Wright. **1:** 93. 1852.
TYPE LOCALITY: "In Mexico inter Bejar et flum. Trinitas." This is now Texas rather than Mexico.
RANGE: Kansas and Arkansas to New Mexico and Mexico.
NEW MEXICO: Arroyo Ranch (*Griffiths* 5677, 5736). Plains and low hills, in the Upper Sonoran Zone.

3. **Aphanostephus ramosissimus** DC. Prodr. **5:** 310. 1836.
Egletes ramosissima A. Gray, Mem. Amer. Acad. n. ser. **4:** 71. 1849.
TYPE LOCALITY: "In Mexico circa Bejar et Laredo." This is now Texas.
RANGE: Western Texas to southern Arizona, south into Mexico.
NEW MEXICO: North of Santa Fe; McCarthys Ranch; Mesilla Valley; Florida Mountains; Dog Spring; mountains west of San Antonio; Rincon; Artesia; Roswell. Dry hills and mesas, in the Lower and Upper Sonoran zones.

43. TOWNSENDIA Hook.

Annual, biennial, or perennial herbs, often cespitose, caulescent or acaulescent; heads large, with purple or white rays; involucre hemispheric, of numerous large, imbricated, scarious-margined, often purplish bracts; pappus of numerous barbellate bristles, that of the ray flowers shorter or reduced to very short bristles or scales.

KEY TO THE SPECIES.

Plants apparently acaulescent.. 1. *T. exscapa.*
Plants with evident stems.
  Bracts not acuminate; rays whitish.
    Pappus of ray achenes equaling that of the disk achenes; heads 10 to 15 mm. high........................ 2. *T. arizonica.*
    Pappus of ray achenes much shorter than that of the disk flowers; heads less than 10 mm. high.
      Perennial, with very short stems; heads usually equaled by the leaves........................ 3. *T. incana.*
      Annual or biennial with long stems; heads much surpassing the leaves........................ 4. *T. fendleri.*

Bracts acuminate; rays purplish.
Stems low, with spreading basal branches.................. 5. *T. grandiflora.*
Stems erect, mostly simple.
Bracts glabrous on the back; heads usually more than one; leaves little reduced above........... 6. *T. eximia.*
Bracts pubescent on the back; heads solitary; leaves much reduced upward......................... 7. *T. formosa.*

1. **Townsendia exscapa** (Richards.) Porter, Mem. Torrey Club **5**: 321. 1894.
Aster ? exscapus Richards. Bot. App. Frankl. Journ. 32. 1832.
TYPE LOCALITY: "Hab. at Carlton House."
RANGE: Montana and Saskatchewan to New Mexico and Texas.
NEW MEXICO: Tierra Amarilla; west of Patterson; Carrizo Mountains; Las Vegas; Agua Fria; Sandia Mountains; Magdalena Mountains; Gilmores Ranch. Plains and low hills, in the Upper Sonoran Zone.

2. **Townsendia arizonica** A. Gray, Proc. Amer. Acad. **16**: 85. 1880.
TYPE LOCALITY: Fort Trumbull, Arizona.
RANGE: Utah and Arizona to northwestern New Mexico.
NEW MEXICO: Carrizo Mountains (*Standley* 7332). Sandhills, in the Upper Sonoran Zone.

3. **Townsendia incana** Nutt. Trans. Amer. Phil. Soc. n. ser. **7**: 305. 1841.
TYPE LOCALITY: "On the Black Hills."
RANGE: Wyoming and Utah to Arizona and New Mexico.
NEW MEXICO: Farmington; Cedar Hill; Aztec; Fort Wingate; San Lorenzo. Sandy soil, in the Upper Sonoran Zone.

4. **Townsendia fendleri** A. Gray, Mem. Amer. Acad. n. ser. **4**: 70. 1849.
TYPE LOCALITY: Gravelly hillsides, Santa Fe, New Mexico. Type collected by Fendler (no. 350).
RANGE: Colorado and New Mexico.
NEW MEXICO: Aztec; Shiprock; Carrizo Mountains; Gallup; Espanola; Fort Wingate; Sandia Mountains; Socorro; North Percha Creek; Mesilla Valley; Organ Mountains; mountains west of San Antonio. Low hills and valleys, in the Lower and Upper Sonoran zones.

This species is very close to *T. strigosa* Nutt. and it is doubtful whether it is possible to separate the two. That species has been reported from New Mexico, but among our material it is impossible to distinguish more than a single species. Our specimens agree very well with those from farther north said to be *T. strigosa.*

5. **Townsendia grandiflora** Nutt. Trans. Amer. Phil. Soc. n. ser. **7**: 306. 1841.
TYPE LOCALITY: "On the Black Hills * * * near the banks of the Platte."
RANGE: Wyoming and South Dakota to New Mexico and Oklahoma.
NEW MEXICO: Raton (*Standley* 6357). Dry hills, in the Upper Sonoran and lower part of the Transition zones.

6. **Townsendia eximia** A. Gray, Mem. Amer. Acad. n. ser. **4**: 70. 1849.
TYPE LOCALITY: "Sides of high mountains, Santa Fe Creek, and prairies on the Mora River," New Mexico. Type collected by Fendler (no. 353).
RANGE: Colorado to Arizona and New Mexico.
NEW MEXICO: Santa Fe and Las Vegas mountains; Sandia Mountains; Laguna Blanca; Rio Pueblo; Pajarito Park. Open slopes and in thickets in the mountains, Transition Zone.

7. **Townsendia formosa** Greene, Leaflets **1**: 213. 1906.
TYPE LOCALITY: Black Range, New Mexico. Type collected by Metcalfe (no. 1434).

RANGE: Mountains of southern New Mexico and Arizona.

NEW MEXICO: Black Range; Mogollon Mountains; White and Sacramento mountains. Transition Zone.

44. BERTHELOTIA DC.

Shrub 1 meter high or more with numerous erect branches; leaves alternate, sericeous, linear-lanceolate, entire; involucre campanulate, the outer bracts obtuse and tomentose, the inner linear and deciduous; pappus of sterile flowers of rigid bristles with thickened tips; flowers pinkish.

1. **Berthelotia sericea** (Nutt.) Rydb. Bull. Torrey Club **33**: 154. 1906.

CACHANILLA. ARROW-WOOD.

Polypappus sericeus Nutt. Journ. Acad. Phila. II. **1**: 178. 1848.

Tessaria borealis Torr. & Gray in Emory, Mil. Reconn. 143. 1848.

Pluchea borealis A. Gray, Proc. Amer. Acad. **17**: 212. 1882.

TYPE LOCALITY: "Rocky Mountains of Upper California."

RANGE: Colorado and Utah to Texas and California.

NEW MEXICO: Rio Grande Valley, from Socorro southward. Along streams and in valleys, in the Lower Sonoran Zone.

45. PLUCHEA Cass. MARSH FLEABANE.

Stems herbaceous, 50 to 100 cm. high; leaves oblong-lanceolate to ovate, dentate, large; bracts ovate or lanceolate, acute, purplish; pappus alike in both kinds of flowers, soft, not thickened at the tips; flowers dull purple.

1. **Pluchea camphorata** (L.) DC. Prodr. **5**: 451. 1836.

Erigeron camphoratum L. Sp. Pl. 864. 1753.

TYPE LOCALITY: "Habitat in Virginia."

RANGE: Massachusetts to Florida, Texas, Arizona, and California.

NEW MEXICO: Roswell. Salt marshes and alkaline soil.

46. STYLOCLINE Nutt.

Densely woolly annual, branched from the base, with entire alternate leaves and small woolly heads clustered at the ends of the branches; bracts thin, inclosing the achenes and falling with them; achenes compressed, obovoid or oblong; pappus of few capillary bristles in the sterile flowers.

1. **Stylocline micropoides** A. Gray, Pl. Wright. **2**: 84. 1853.

TYPE LOCALITY: "Hills near Frontera, New Mexico."

The type locality is doubtless in Texas instead of New Mexico, but very near the State line. The plant has also been collected in the Black Range by Metcalfe.

47. EVAX Gaertn.

Slender woolly annual, much branched from the base, with narrow alternate leaves and small, densely woolly heads clustered at the ends of the branches; bracts of the involucre very thin, loose, deciduous at maturity from the convex receptacle; achenes compressed, smooth or nearly so; pappus none.

1. **Evax multicaulis** DC. Prodr. **5**: 459. 1836.

Filaginopsis multicaulis Torr. & Gray, Fl. N. Amer. **2**: 263. 1842.

TYPE LOCALITY: "In Mexico circa lacum Sancti-Nicolai in sinu Spiritus-Sancti."

RANGE: Texas to southern California, south into Mexico.

NEW MEXICO: Hondo Hill (*Wooton*). Dry plains and hills, in the Upper Sonoran Zone.

48. GNAPHALIUM L. CUDWEED.

Floccose-woolly herbs with narrow sessile entire leaves and cymose or glomerate heads of yellowish or whitish flowers; heads heterogamous, discoid; involucre of numerous thin scarious bracts imbricated in several series; pappus of numerous scabrous bristles in a single series; achenes terete or compressed, mostly nerveless.

KEY TO THE SPECIES.

Heads leafy-bracted; plants low, usually less than 20 cm.
- Plants loosely floccose; leaves mostly oblong or oblanceolate. 1. *G. palustre.*
- Plants appressed-tomentose; leaves linear to linear-oblanceolate.
  - Stems erect, simple or with few erect branches........ 2. *G. strictum.*
  - Stems abundantly divaricate-branched from the base... 3. *G. angustifolium.*

Heads not leafy-bracted; plants mostly 30 cm. high or more.
- Leaves green and glandular-viscid on the upper surface.... 4. *G. decurrens.*
- Leaves tomentose on both surfaces, not apparently viscid.
  - Leaves narrowed at the base, not decurrent; stems slender, weak; heads numerous, loosely corymbose.. 5. *G. wrightii.*
  - Leaves not narrowed at the base, decurrent; stems stout, erect; heads densely clustered at the ends of the branches.
    - Bracts densely arachnoid at the base, nearly white; pubescence of the stems and leaves white; stems densely leafy throughout............ 6. *G. chilense.*
    - Bracts nearly glabrous, yellowish; pubescence of stems and leaves yellowish; stems with only a few distant reduced leaves above......... 7. *G. sulphurescens.*

1. **Gnaphalium palustre** Nutt. Trans. Amer. Phil. Soc. n. ser. **7:** 403. 1841.

TYPE LOCALITY: "Rocky Mountains, Oregon, California and Chili."

RANGE: British Columbia and Montana to California and northwestern New Mexico.

NEW MEXICO: Cedar Hill (*Standley* 7930). Wet ground.

2. **Gnaphalium strictum** A. Gray, U. S. Rep. Expl. Miss. Pacif. **4:** 110. 1856.

TYPE LOCALITY: Banks of the Rio Grande, near Albuquerque, New Mexico. Type collected by Bigelow.

RANGE: Colorado and Wyoming to Arizona and New Mexico.

NEW MEXICO: Albuquerque; Tunitcha Mountains; West Fork of the Gila; Santa Fe Creek. Wet soil, in the Upper Sonoran and Transition zones.

3. **Gnaphalium angustifolium** A. Nels. Bull. Torrey Club **26:** 357. 1899.

TYPE LOCALITY: Head of Woods Creek, Medicine Bow Mountains, Wyoming.

RANGE: Wyoming to northern New Mexico.

NEW MEXICO: Chama; Ensenada. Wet soil, in the Transition Zone.

4. **Gnaphalium decurrens** Ives, Amer. Journ. Sci. **1:** 380. *pl. 1.* 1819.

TYPE LOCALITY: Near New Haven, Connecticut.

RANGE: Idaho and Nova Scotia to New Mexico and Pennsylvania.

NEW MEXICO: Santa Fe and Las Vegas mountains; Santa Rita; Black Range; Fort Tularosa; Capitan Mountains; White Mountains; Organ Mountains. Meadows, in the Transition Zone.

5. **Gnaphalium wrightii** A. Gray, Proc. Amer. Acad. 17: 214. 1882.
TYPE LOCALITY: "Common from S. Arkansas and W. Texas to New Mexico."
RANGE: Colorado and New Mexico to California and Texas.
NEW MEXICO: Hurrah Creek; Bear Mountains; Fort Bayard; Dona Ana Mountains; Organ Mountains; Capitan Mountains. Upper Sonoran and Transition zones.

6. **Gnaphalium chilense** Spreng. Syst. Veg. 3: 480. 1826.
TYPE LOCALITY: California.
RANGE: Oregon and Montana to California and Texas.
NEW MEXICO: Ramah; Ojo Caliente; Gila; Mogollon Creek; Kingston; Rincon; mountains southeast of Patterson; Mesilla Valley; Roswell. Low mountains and hills, in the Lower and Upper Sonoran zones.

7. **Gnaphalium sulphurescens** Rydb. Mem. N. Y. Bot. Gard. 1: 415. 1900.
TYPE LOCALITY: Yellowstone Park.
RANGE: Washington and Wyoming to northern New Mexico.
NEW MEXICO: Stinking Lake (*Standley* 8272). Meadows in the mountains, in the Transition Zone.

49. ANAPHALIS DC. PEARLY EVERLASTING.

White-tomentulose perennial herb with very leafy, usually simple, erect stems; leaves entire, narrowly lanceolate; heads numerous, corymbose, diœcious, usually with a few perfect flowers in the center of the pistillate heads; bristles of the pappus of staminate flowers little or not at all thickened at the tips.

1. **Anaphalis subalpina** (A. Gray) Rydb. Mem. N. Y. Bot. Gard. 1: 415. 1900.
Anaphalis margaritacea subalpina A. Gray, Syn. Fl. 1^2: 233. 1884.
TYPE LOCALITY: Mountains of Colorado.
RANGE: British Columbia and South Dakota to California and New Mexico.
NEW MEXICO: Santa Fe and Las Vegas mountains; West Fork of the Gila; Eagle Creek. Open woods in the mountains, in the Transition and Canadian zones.

50. ANTENNARIA Gaertn. INDIAN TOBACCO.

Perennial white-woolly herbs with mostly basal, broad, entire leaves and corymbose or racemose heads; heads many-flowered, diœcious, the flowers all tubular; involucre dry, scarious, white or colored, imbricated; receptacle naked; achenes terete or flattish; pappus a single row of bristles, in the fertile flowers capillary, in the staminate ones clavellate-thickened at the tips.

KEY TO THE SPECIES.

Leaves glabrous on the upper surface.
  Bracts and peduncles densely viscid; heads subtended by large bracts .... 1. *A. marginata.*
  Bracts and peduncles not viscid or only obscurely so; bracts mostly wanting.
    Stems 12 to 16 cm. high; heads large, about 10 mm. high; all the bracts obtuse .... 2. *A. fendleri.*
    Stems less than 10 cm. high; heads smaller, 6 to 8 mm. high; inner bracts very acute .... 3. *A. peramoena.*
Leaves tomentose on both surfaces.
  Plants acaulescent; heads subsessile among the rosettes of basal leaves .... 4. *A. rosulata.*

Plants caulescent.
Bracts of the involucre with dark umber-colored upper portions.
Inner bracts not rose-colored.......................... 5. *A. umbrinella.*
Inner bracts rose-colored.............................. 6. *A. concinna.*
Bracts with white, pale yellowish, or pink upper portions.
Heads large, 8 to 12 mm. high.
Plants tall, 20 to 30 cm.; heads pedunculate; lower leaves 3-nerved; stolons much elongated.................................. 7. *A. obovata.*
Plants low, usually 10 cm. or less; heads subsessile; lower leaves not 3-nerved; stolons short.................................... 8. *A. aprica.*
Heads small, 5 to 8 mm. high.
Bracts with pink upper portions................. 9. *A. imbricata.*
Bracts with white or pale yellowish upper portions.
Tomentum of leaves closely appressed, silky, shining.............................10. *A. microphylla.*
Tomentum of leaves loose, not silky.
Inflorescence, stems, and usually the leaves glandular; bracts with greenish or brownish spots, the tips pale yellowish..............11. *A. viscidula.*
Plants not glandular; bracts not spotted with green or brown, the tips white.............................12. *A. arida.*

1. Antennaria marginata Greene, Pittonia **3**: 290. 1898.

TYPE LOCALITY: New Mexico, probably about Santa Fe. Type collected by Fendler (no. 523).

RANGE: Mountains of Colorado, New Mexico, and Arizona.

NEW MEXICO: Stinking Lake; Tierra Amarilla; Santa Fe. Transition Zone.

2. Antennaria fendleri Greene, Leaflets **2**: 143. 1911.

TYPE LOCALITY: Near Santa Fe, New Mexico. Type collected by Heller (no. 3612).

RANGE: Mountains of northern New Mexico.

NEW MEXICO: Santa Fe; Hermits Peak. Transition Zone.

3. Antennaria peramoena Greene, Leaflets **2**: 144. 1911.

TYPE LOCALITY: Wheelers Ranch, New Mexico. Type collected by Wooton, July 11, 1906.

RANGE: Mountains of southwestern New Mexico.

NEW MEXICO: Wheelers Ranch; Pinos Altos; Burro Mountains. Transition Zone.

4. Antennaria rosulata Rydb. Bull. Torrey Club **24**: 300. 1897.

TYPE LOCALITY: Arizona.

RANGE: Colorado to Arizona and New Mexico.

NEW MEXICO: Tunitcha Mountains; Tierra Amarilla. Meadows and hills, in the Upper Sonoran and Transition zones.

5. Antennaria umbrinella Rydb. Bull. Torrey Club **24**: 302. 1897.

Antennaria mucronata E. Nels. Bot. Gaz. **27**: 209. 1899.

TYPE LOCALITY: Long Baldy, Little Belt Mountains, Montana.

RANGE: Mountains from Oregon and Wyoming to northern New Mexico.

NEW MEXICO: Truchas Peak; Pecos Baldy. Arctic-Alpine Zone.

6. **Antennaria concinna** E. Nels. Proc. U. S. Nat. Mus. **23**: 705. 1901.
TYPE LOCALITY: Olympic Mountains, Clallam County, Washington.
RANGE: Washington to Utah and northern New Mexico.
NEW MEXICO: Top of Pecos Baldy (*Bailey* 638). Arctic-Alpine Zone.

7. **Antennaria obovata** E. Nels. Bot. Gaz. **27**: 213. 1899.
TYPE LOCALITY: Near Soldier Canyon, Colorado.
RANGE: South Dakota to Colorado and northern New Mexico.
NEW MEXICO: Las Vegas Hot Springs; Sandia Mountains. Hills and plains, in the Transition Zone.

8. **Antennaria aprica** Greene, Pittonia **3**: 282. 1898.
Antennaria holmii Greene, Pittonia **4**: 81. 1899.
Antennaria latisquamea Greene, Leaflets **1**: 145. 1905, not Piper, 1901.
Antennaria anacleta Greene, Leaflets **1**: 200. 1906.
TYPE LOCALITY: "Very common species of the whole Rocky Mountain region."
RANGE: Alberta and South Dakota to Utah and New Mexico.
NEW MEXICO: Tierra Amarilla; Winsors Ranch; Wheelers Ranch; Sawyers Peak; Gilmores Ranch; Vermejo Park. Hills and mountains, in the Upper Sonoran and Transition zones.
The type of *A. latisquamea* was collected in the Black Range (*Metcalfe* 1433).

9. **Antennaria imbricata** E. Nels. Bot. Gaz. **27**: 211. 1899.
Antennaria rosea imbricata E. Nels. Proc. U. S. Nat. Mus. **23**: 707. 1901.
TYPE LOCALITY: Wyoming.
RANGE: Montana to Utah and New Mexico.
NEW MEXICO: Near Tierra Amarilla (*Eggleston* 6461, 6615, 6532). Meadows, in the Upper Sonoran and Transition zones.

10. **Antennaria microphylla** Rydb. Bull. Torrey Club **24**: 303. 1897.
Antennaria formosa Greene, Leaflets **1**: 145. 1905.
TYPE LOCALITY: Manhattan, Montana.
RANGE: British Columbia and Saskatchewan to Nebraska and New Mexico.
NEW MEXICO: Taos; Costilla Valley; Brazos Canyon. Meadows, in the Transition Zone.

11. **Antennaria viscidula** (E. Nels.) Rydb. Colo. Agr. Exp. Sta. Bull. **100**: 369. 1906.
Antennaria arida viscidula E. Nels. Proc. U. S. Nat. Mus. **23**: 710. 1901.
TYPE LOCALITY: Laramie Peak, Wyoming.
RANGE: Wyoming to northern New Mexico.
NEW MEXICO: Chama; hills southwest of Tierra Amarilla. Meadows, in the Transition Zone.

12. **Antennaria arida** E. Nels. Bot. Gaz. **27**: 210. 1899.
TYPE LOCALITY: Tipton, Wyoming.
RANGE: Wyoming to Utah and northern New Mexico.
NEW MEXICO: Chama; top of range between Sapello and Pecos rivers; Jemez Mountains. Meadows, from the Transition to the Hudsonian Zone.

51. MADIA Molina. TARWEED.

Glandular-viscid annual with mostly alternate, linear leaves and small glomerate heads; involucre oblong, angled by the salient backs of the bracts; receptacle flat or convex, bearing a single series of bracts, these inclosing the disk flowers as a kind of inner involucre; ray flowers 2 to 5 or none, the rays inconspicuous; achenes narrow, laterally compressed, angled.

1. **Madia glomerata** Hook. Fl. Bor. Amer. **2**: 24. 1834.

TYPE LOCALITY: "Plains of the Saskatchawan."

RANGE: British America to Colorado and northern New Mexico.

NEW MEXICO: Chama; Brazos Canyon. Moist ground in the mountains, in the Transition Zone.

52. BLEPHARIPAPPUS Hook.

Stout, hirsute, sparingly branched annual with alternate, entire, linear or oblong, sessile leaves, and rather large heads of white flowers; rays large and showy; involucre of a single series of narrow bracts with scarious margins, inclosing the achenes of the ray flowers; receptacle flat, bearing a series of chaffy bracts between the ray and disk flowers; achenes somewhat compressed, those of the ray flowers destitute of pappus, the inner ones with a pappus of numerous bristles.

1. **Blepharipappus glandulosus** Hook. Fl. Bor. Amer. **1**: 316. 1830.

Layia glandulosa Hook. & Arn. Bot. Beechey Voy. 350. 1833.

Layia neomexicana A. Gray, Pl. Wright. **2**: 98. 1853.

TYPE LOCALITY: "Common on the plains of the Columbia, in sandy soils, under the shade of Purshia and Artemisia."

RANGE: British Columbia and Idaho to California and New Mexico.

NEW MEXICO: Mangas Springs.

53. GUARDIOLA Humb. & Bonpl.

Erect annual with sessile, mostly connate, oblong-lanceolate leaves and small 4-flowered turbinate heads, each with a single ray; involucre of 3 concave membranaceous bracts; achenes oblong, slightly compressed, glabrous; pappus wanting.

1. **Guardiola diehlii** Jones, Contr. West. Bot. **12**: 48. 1908.

TYPE LOCALITY: "Albuquerque and Socorro, New Mexico."

RANGE: Known only from the type collections.

We have seen no specimens of this species.

54. DICRANOCARPUS A. Gray.

Low slender annual, the leaves divided into linear segments; heads small, with 3 or 4 ray flowers and 3 or 4 disk flowers, the rays small and inconspicuous; involucre of 3 or 4 narrow bracts and sometimes 1 or 2 small foliaceous outer ones; achenes dimorphous, 1 or 2 elongated, puberulent, smooth, with 2 long divergent awns, the others short, more or less tuberculate, bearing 2 short divaricate horns.

1. **Dicranocarpus dicranocarpus** (A. Gray) Woot. & Standl. Contr. U. S. Nat. Herb. **16**: 189. 1913.

Heterospermum dicranocarpum A. Gray, Pl. Wright. **1**: 109. 1852.

Dicranocarpus parviflorus A. Gray, Mem. Amer. Acad. n. ser. **5**: 322. 1854.

Wootonia parviflora Greene, Bull. Torrey Club **25**: 122. 1898.

TYPE LOCALITY: Plains between the Guadalupe Mountains and the Pecos, western Texas.

RANGE: Southern New Mexico and western Texas to northeastern Mexico.

NEW MEXICO: White Sands. Alkaline soil, in the Lower Sonoran Zone.

55. MELAMPODIUM L.

Perennial herb, 30 cm. high or less, often woody at the base, with opposite, entire, linear to spatulate leaves; heads long-pedunculate, with large white rays; bracts in 2 series, the outer 4 or 5 flat, ovate, partially united, the inner each embracing an achene and deciduous with it; achenes obovate, incurved; pappus none.

1. **Melampodium leucanthum** Torr. & Gray, Fl. N. Amer. **2**: 271. 1842.

TYPE LOCALITY: Texas.

RANGE: Colorado and Kansas to Arizona and Texas and southward.

NEW MEXICO: Bernal; Pecos; Nara Visa; Clayton; Socorro Mountain; Albuquerque; Knowles; Melrose; Fort Bayard; Mangas Springs; Middle Fork of the Gila; Organ Mountains; Roswell; south of Stanley. Dry plains and hills, in the Lower and Upper Sonoran zones.

The ray flowers of this species are normally a clear, bright white, but not infrequently turn pink with age, especially late in the season, and plants with pinkish rays are occasional.

56. PARTHENIUM L.

Herbaceous or woody biennials or perennials with alternate, pubescent, variously lobed or pinnatifid leaves, and numerous corymbs of small heads of white flowers; involucre of few broad appressed bracts in about two series; achenes oval or obovate, usually pubescent, with a narrow callous margin; pappus of 2 chaffy awns or scales.

KEY TO THE SPECIES.

Herb; leaves twice pinnatifid.. 1. *P. lyratum.*

Low shrub; leaves few-lobed.. 2. *P. incanum.*

Statements have been made that *P. argentatum,* the Mexican rubber plant or "guayule," occurs in New Mexico, but so far as we know there is no foundation for these reports, that species ranging much farther south.

1. **Parthenium lyratum** A. Gray, Syn. Fl. 1^2: 244. 1884.

Parthenium hysterophorus lyratum A. Gray, Proc. Amer. Acad. **17**: 216. 1882.

TYPE LOCALITY: Western Texas.

RANGE: Western Texas and southern New Mexico; also in Mexico.

NEW MEXICO: Soledad Canyon; Hondo Hill; Guadalupe Mountains; Carlsbad; Lake Valley. Dry hills, in the Upper Sonoran Zone.

2. **Parthenium incanum** H. B. K. Nov. Gen. & Sp. **4**: 260. *pl. 391.* 1820. MARIOLA.

TYPE LOCALITY: "Colitur in horto botanico Mexicano."

RANGE: Western Texas to southern Arizona and southward.

NEW MEXICO: Socorro; Mangas Springs; Rincon; Tucumcari; Cuchillo; Florida Mountains; Tortugas Mountain; Organ Mountains; White Mountains. Dry hills and along arroyos, in the Lower Sonoran Zone.

This plant has been used for the production of rubber. For this purpose, however, it is far less valuable than the Mexican rubber plant, or "guayule," mentioned above.

57. BERLANDIERA DC.

Coarse canescent perennial herbs with alternate, simple or lyrate-pinnatifid leaves and large pedunculate heads with showy yellow rays; bracts in about 3 series, the outermost small and foliaceous, the inner thin, membranaceous in age; achenes flat, obovate, wingless, unicostate on the inner surface; pappus mostly obsolete.

KEY TO THE SPECIES.

Leaves all pinnatifid; heads numerous.............................. 1. *B. lyrata.*

Leaves merely crenate; heads few.................................... 2. *B. macrophylla.*

1. **Berlandiera lyrata** Benth. Pl. Hartw. 17. 1839.

Berlandiera incisa Torr. & Gray, Fl. N. Amer. **2**: 282. 1842.

TYPE LOCALITY: Mexico.

RANGE: Arkansas and Texas to Arizona and Mexico.

NEW MEXICO: Knowles; Clayton; Albuquerque; Laguna; north of Santa Fe; Socorro; Lake Valley; Mangas Springs; near White Water; Animas Valley; Organ Mountains; between Ruidoso and Eagle creeks; Gray; Roswell; Nara Visa. Plains and hills, in the Lower and Upper Sonoran zones.

2. **Berlandiera macrophylla** (A. Gray) Jones, Contr. West. Bot. **12:** 48. 1908.
Berlandiera lyrata macrophylla A. Gray, Syn. Fl. 1^2: 243. 1884.
TYPE LOCALITY: Southern Arizona.
RANGE: Western Texas to southern Arizona and adjacent Mexico.
NEW MEXICO: Queen (*Wooton*).

58. ENGELMANNIA A. Gray.

Coarse perennial herb, 60 cm. high or less, with stout branched stems and rough pinnatifid leaves; heads about 1 cm. high, with bright yellow rays; bracts in 2 series, the outer linear, foliaceous, the inner coriaceous, oval or obovate, with foliaceous tips; achenes obovate, wingless; pappus a persistent crown cleft into 3 or 4 irregular lobes or into a pair of lanceolate scales.

1. **Engelmannia pinnatifida** Nutt. Trans. Amer. Phil. Soc. n. ser. **7:** 343. 1841.
Engelmannia texana Scheele, Linnaea **22:** 155. 1849.
TYPE LOCALITY: "Plains of Red River."
RANGE: Arkansas and Louisiana to Arizona.
NEW MEXICO: Gallinas Planting Station; Clayton; Las Vegas; Magdalena; Socorro; Organ Mountains; Gray; Fort Stanton; Redlands; Ruidoso Creek; Nara Visa. Plains and low hills, in the Upper Sonoran Zone.

59. SANVITALIA Lam.

Low branched annual with narrowly lanceolate, petiolate, opposite leaves and small heads of greenish yellow flowers; rays short, greenish white, persistent; involucre a single series of dry bracts, these with rigid cuspidate tips; achenes flattened, corky-thickened, those of the ray flowers winged, bearing 3 very short awns or tubercles, those of the disk flowers wingless, usually awnless.

1. **Sanvitalia aberti** A. Gray, Mem. Amer. Acad. n. ser. **4:** 87. 1849.
TYPE LOCALITY: Woodlands, between Santa Fe and Pecos, New Mexico. Type collected by Fendler (no. 538).
RANGE: Western Texas to southern Arizona.
NEW MEXICO: Carrizo Mountains; Pecos; Torrance; Pajarito Park; Black Range; Bear Mountains; Dog Spring; Organ Mountains; Gray; White Mountains; Carlsbad. Low hills and canyons, in the Upper and Lower Sonoran zones.

60. CRASSINA Scepin. ZINNIA.

Low, densely branched, suffruticose or herbaceous plants with entire, mostly linear, sessile leaves and solitary showy heads terminating the branches; involucre campanulate to cylindric, with appressed dry firm broad bracts rounded at the summit and often margined; receptacle conic to cylindric; rays broad, firm, persistent; achenes 2 to 4-aristate.

KEY TO THE SPECIES.

Rays white, 10 mm. long or less .... 1. *C. pumila.*
Rays bright yellow, 12 to 30 mm. long .... 2. *C. grandiflora.*

1. **Crassina pumila** (A. Gray) Kuntze, Rev. Gen. Pl. **1:** 331. 1891.
Zinnia pumila A. Gray, Mem. Amer. Acad. n. ser. **4:** 81. 1849.
TYPE LOCALITY: "High plain near San Juan de la Viqueria, and at Castaniola, in Northern Mexico."

RANGE: Western Texas to southern Arizona and southward.

NEW MEXICO: Grant County; mesa west of Organ Mountains. Mesas and dry hills, in the Lower Sonoran Zone.

A very handsome plant, growing in dense flat-topped clumps 20 cm. high or less. The heads are very showy with their large white rays. It would be well adapted to use as a border plant in cultivation.

2. **Crassina grandiflora** (Nutt.) Kuntze, Rev. Gen. Pl. **1**: 331. 1891.

Zinnia grandiflora Nutt. Trans. Amer. Phil. Soc. n. ser. **7**: 348. 1841.

TYPE LOCALITY: "In the Rocky Mountains, toward Mexico."

RANGE: Colorado and Kansas to Texas and Arizona.

NEW MEXICO: Raton; Sierra Grande; Laguna; north of Santa Fe; Albuquerque; Zuni; Nara Visa; Hillsboro; Socorro; Dog Spring; mesa near Las Cruces; Capitan Mountains; Nogal; south of Roswell; Queen; Redlands; Torrance; Gallinas Mountains; Puertecito; Aden; White Sands. Plains and low hills, in the Lower and Upper Sonoran zones.

This species is equally as handsome as *C. pumila*. It does not bear so many flowers, nor is the plant so compact and densely branched, but the large bright yellow rays are even more showy than those of that species.

61. **HELIOPSIS** Pers. OX-EYE.

Coarse perennial herb with opposite ovate-lanceolate petiolate leaves and large pedunculate terminal heads with yellow rays; heads many-flowered; ray flowers 10 or more, fertile; bracts nearly equal, in 2 or 3 series, the outer foliaceous, spreading; receptacle conic; achenes smooth, 4-angled, truncate; pappus none or a mere border.

1. **Heliopsis scabra** Dunal, Mém. Mus. Hist. Nat. **5**: 56. *pl. 4*. 1819.

Heliopsis laevis scabra Torr. & Gray, Fl. N. Amer. **2**: 303. 1843.

TYPE LOCALITY: "Hab. in America boreali secus amnem Missouri."

RANGE: Saskatchewan and New York, south to New Mexico and Arkansas.

NEW MEXICO: Santa Fe and Las Vegas mountains; White and Sacramento mountains. Meadows, in the Transition Zone.

Our western plant is not altogether like the one found farther east, its leaves being smaller and fewer, with fewer, more appressed, blunter teeth.

62. **GALINSOGA** Ruiz & Pav.

Slender, loosely branched, erect or ascending annual with thin, opposite, petiolate, lanceolate to ovate, serrate leaves and small slender-pedunculate heads of yellow flowers with 4 or 5 barely exserted white rays; involucre campanulate, of ovate, thin, nearly equal bracts in 2 series; achenes turbinate, 4 or 5-angled; pappus of 8 to 16 short paleæ.

1. **Galinsoga parviflora** Cav. Icon. Pl. **3**: 41. *pl. 281*. 1794.

Galinsoga parviflora semicalva A. Gray, Pl. Wright. **2**: 98. 1853.

TYPE LOCALITY: Peru.

RANGE: Moist slopes and canyons, New Mexico and Arizona, southward through tropical America; widely introduced in eastern North America.

NEW MEXICO: Beulah; Mogollon Mountains; Organ Mountains; White Mountains.

63. **COSMOS** Cav. COSMOS.

Slender annual with opposite leaves dissected into linear segments; heads small, on long slender peduncles; involucre of 2 series of bracts, the outer linear, foliaceous, the inner broad, scarious-margined; rays conspicuous, pink; achenes slender, beaked, 4-angled, papillose-roughened.

1. **Cosmos parviflorus** (Jacq.) H. B. K. Nov. Gen. & Sp. **4**: 241. 1820.

Coreopsis parviflora Jacq. Pl. Hort. Schönbr. **3**: 65. *pl. 374*. 1798.

Cosmos bipinnatus parviflorus A. Gray, Pl. Wright. **2**: 90. 1853.

TYPE LOCALITY: Not known.

RANGE: Western Texas to southern Arizona and southward.

NEW MEXICO: Santa Fe and Las Vegas mountains; Las Huertas Canyon; Laguna Blanca; Gallinas Mountains; Fort Bayard; Mogollon Mountains; Hop Canyon; Hanover Mountain; San Luis Mountains; Organ Mountains; Gray; White Mountains. Meadows and along streams, in the Upper Sonoran and Transition zones.

64. THELESPERMA Less.

Slender annual or usually perennial, strong-scented, glabrous herbs, with opposite, finely dissected leaves and long-pedunculate, radiate or discoid heads of yellow flowers; inner bracts united to form a cup, the outer shorter and narrow, connate at the base with the inner; rays, when present, about 8, bright yellow; achenes terete or slightly obcompressed, narrowly oblong to linear, neither margined nor beaked, crowned with a pair of stout persistent awns, or pappus sometimes wanting.

KEY TO THE SPECIES.

Rays present.
- Leaf segments linear-filiform, 1 mm. wide or less; leaves scattered along the stems .... 1. *T. trifidum.*
- Leaf segments linear or broader, 2 mm. wide or more; leaves clustered at the base of the stems .... 2. *T. subnudum.*

Rays wanting.
- Lobes of disk corollas lanceolate; peduncles much shorter than the leafy stems; leaves scattered along the stems; heads about 10 mm. broad .... 3. *T. gracile.*
- Lobes of disk corollas ovate; peduncles much exceeding the leafy stems; leaves crowded at base of stems; heads 6 mm. wide or less .... 4. *T. longipes.*

Thelesperma ambiguum and *T. subsimplicifolium* have been reported from New Mexico, but the specimens apparently are referable to *T. trifidum.*

1. **Thelesperma trifidum** (Poir.) Britton, Trans. N. Y. Acad. **9**: 182. 1890.

Coreopsis trifida Poir. in Lam. Encycl. Suppl. **2**: 253. 1811.

Cosmidium filifolium Torr. & Gray, Fl. N. Amer. **2**: 350. 1842.

Thelesperma filifolium A. Gray, Journ. Bot. Kew Misc. **1**: 252. 1849.

Thelesperma formosum Greene, Pittonia **5**: 56. 1902.

TYPE LOCALITY: North America.

RANGE: Nebraska and Colorado to New Mexico and western Texas.

NEW MEXICO: Pecos; between Santa Fe and Canyoncito; Socorro; Elk Canyon; Gallinas Mountains; Nara Visa. Plains and low hills, in the Upper Sonoran Zone.

The type of *T. formosum* is Heller's 3747, collected between Santa Fe and Canyoncito.

2. **Thelesperma subnudum** A. Gray, Proc. Amer. Acad. **10**: 72. 1875. NAVAHO TEA.

TYPE LOCALITY: St. George, southern Utah.

RANGE: Utah and Colorado to New Mexico and Arizona.

NEW MEXICO: Carrizo Mountains (*Standley* 7307). Dry plains and rocky hills, in the Upper Sonoran Zone.

3. **Thelesperma gracile** (Torr.) A. Gray, Journ. Bot. Kew Misc. **1**: 253. 1849.

Bidens gracilis Torr. Ann. Lyc. N. Y. **2**: 215. 1827.

TYPE LOCALITY: "On the Canadian ?," New Mexico or Colorado.

RANGE: Nebraska and Colorado to western Texas and Arizona.

NEW MEXICO: Common throughout the State. Plains and low hills, in the Upper Sonoran Zone.

4. Thelesperma longipes A. Gray, Pl. Wright. **1:** 109. 1852. COTA.

TYPE LOCALITY: "Hills and dry banks of the San Pedro or Devil's River," western Texas.

RANGE: Western Texas to southern Arizona and southward.

NEW MEXICO: Hillsboro; Tortugas Mountain; Organ Mountains; Capitan Mountains; White Mountains; Guadalupe Mountains; Torrance. Dry hills and mesas, in the Lower and Upper Sonoran zones.

In the southern part of the State this is said to be used as a substitute for tea by the native people. When boiled it gives the water a deep red tinge. The same material may be boiled several times before losing its strength.

65. COREOPSIS L.

Annual or perennial herbs with simple or pinnately divided leaves and solitary or numerous pedunculate heads of yellow or brown flowers; involucre campanulate or hemispheric, the bracts in 2 series, more or less united at the base, those of the outer series narrow and herbaceous, the inner ones broad, colored, thin and scarious or with scarious margins; rays conspicuous, yellow to brown; receptacle flat or slightly convex, chaffy; achenes flat, oblong to orbicular, winged or wingless; pappus wanting or minute.

KEY TO THE SPECIES.

Perennial; leaves simple.. 1. *C. lanceolata.*
Annuals; leaves pinnately divided.
  Achenes winged.. 2. *C. cardaminefolia.*
  Achenes wingless.. 3. *C. tinctoria.*

1. Coreopsis lanceolata L. Sp. Pl. 908. 1753.

TYPE LOCALITY: "Habitat in Carolina."

RANGE: Ontario and Florida to Colorado and New Mexico.

NEW MEXICO: Santa Fe and Las Vegas mountains; East View. Open fields, in the Transition Zone.

2. Coreopsis cardaminefolia (DC.) Torr. & Gray, Fl. N. Amer. **2:** 346. 1842.
Calliopsis cardaminefolia DC. Prodr. **5:** 568. 1836.

TYPE LOCALITY: "In Mexici prov. Texas inter Bejar et flum. Trinitatis, ad Matamoros et ad lacum Sancti-Nicolai in sinu Sancti Spiritus."

RANGE: Louisiana and Texas to Kansas and New Mexico.

NEW MEXICO: North of Ramah; Pescado Spring. Plains, in the Upper Sonoran Zone.

3. Coreopsis tinctoria Nutt. Journ. Acad. Phila. **2:** 114. 1821.
Calliopsis tinctoria DC. Prodr. **5:** 568. 1836.

TYPE LOCALITY: "Throughout the Arkansas territory to the banks of Red River, chiefly in the prairies which are subject to temporary inundation."

RANGE: Saskatchewan and Minnesota to Louisiana, Texas, and Arizona.

NEW MEXICO: Shiprock (*Standley* 7233). Upper Sonoran Zone.

Doctor Gray states [1] that this was collected "East of Mora River, in low places," by Fendler in 1847.

66. BIDENS L. BEGGAR-TICKS.

Slender or coarse annuals with opposite, simple or compound leaves and medium-sized heads of yellow or brownish flowers; involucre of 2 series of bracts, the inner thin and colored, the outer narrow and foliaceous; rays 3 to 8 or none, yellow, neutral; achenes flattened parallel to the bracts of the involucre or very narrow; pappus of 2 awns or short teeth.

[1] Mem. Amer. Acad. n. ser. **4:** 85. 1849.

KEY TO THE SPECIES.

Leaves simple.
  Outer bracts about equaling the inner; achenes not corky on the angles.......... 1. *B. prionophylla.*
  Outer bracts as long as the rays or longer; achenes corky on the angles.......... 2. *B. glaucescens.*
Leaves pinnate or pinnatifid.
  Achenes flat, obovate or cuneate.......... 3. *B. frondosa.*
  Achenes linear, tetragonal.
    Divisions of the leaves lanceolate, oblong, or ovate.
      Leaves once pinnate, the divisions sometimes pinnatifid into large divisions, thick and firm..... 4. *B. anthriscoides.*
      Leaves thrice parted or more into small divisions, thin.......... 5. *B. bigelovii.*
    Divisions of the leaves linear or linear-filiform.
      Divisions of the leaves linear-filiform.......... 6. *B. heterosperma.*
      Divisions of the leaves linear.
        Heads of achenes narrow, 8 mm. wide or less; bracts nearly glabrous, the outer much shorter than the inner.......... 7. *B. cognata.*
        Heads of achenes broad, 10 mm. wide or more; bracts strongly villous, the outer about equaling the inner.......... 8. *B. tenuisecta.*

1. **Bidens prionophylla** Greene, Pittonia 4: 256. 1901.
TYPE LOCALITY: "River Moira, Ontario."
RANGE: British America to New Mexico and Illinois.
NEW MEXICO: Lower Plaza (*Wooton*). Wet ground.

2. **Bidens glaucescens** Greene, Pittonia 4: 258. 1901.
TYPE LOCALITY: "Peculiar to the western mountain districts and the plains adjacent, but beginning in Kansas, perhaps in Missouri."
RANGE: British America to Kansas and New Mexico.
NEW MEXICO: Farmington; west of Clayton. Wet ground, in the Upper Sonoran Zone.

3. **Bidens frondosa** L. Sp. Pl. 832. 1753.
TYPE LOCALITY: "Habitat in America septentrionali."
RANGE: British America to Florida, Texas, and New Mexico.
NEW MEXICO: Farmington; Albuquerque. Wet ground, in the Upper Sonoran Zone.

4. **Bidens anthriscoides** DC. Prodr. 5: 600. 1836.
TYPE LOCALITY: "In Mexici Cordillera de Guachilaqua."
RANGE: Southern New Mexico, south into Mexico.
NEW MEXICO: Mesilla Valley (*Wooton*). Lower Sonoran Zone.
As nearly as we can judge from the description and from the Mexican specimens examined, this collection belongs here. It exactly matches specimens collected by Palmer near Durango. The plant is closely related to *Bidens pilosa* L., a species of Mexico and the West Indies found in California, but has different leaves, pubescence, and achenes.

5. **Bidens bigelovii** A. Gray in Torr. U. S. & Mex. Bound. Bot. 91. 1859.
TYPE LOCALITY: Banks of the Rio Limpio, Texas.
RANGE: Western Texas to southern Arizona and southward.
NEW MEXICO: Mangas Springs; Kingston; West Fork of the Gila; Organ Mountains. Canyons, in the Upper Sonoran Zone.

6. **Bidens heterosperma** A. Gray, Pl. Wright. 2: 90. 1853.

TYPE LOCALITY: Near the Copper Mines, New Mexico. Type collected by Wright in 1851.

RANGE: Mountains of southern Arizona and New Mexico.

We have seen no further collections from New Mexico.

7. **Bidens cognata** Greene, Leaflets 1: 149. 1905.

TYPE LOCALITY: Sawyers Peak, Black Range, New Mexico. Type collected by Metcalfe (no. 1436).

RANGE: Mountains of New Mexico.

NEW MEXICO: Hurrah Creek; Sawyers Peak; West Fork of the Gila. Transition Zone.

8. **Bidens tenuisecta** A. Gray, Mem. Amer. Acad. n. ser. 4: 86. 1849.

TYPE LOCALITY: "Margins of Poñi Creek (between Bent's Fort and Santa Fe)," New Mexico. Type collected by Fendler (no. 449).

RANGE: Idaho and Colorado to Mexico.

NEW MEXICO: Farmington; Chama; Santa Fe and Las Vegas Mountains; Raton; Ensenada; Pajarito Park; Cleveland; Sandia Mountains; Middle Fork of the Gila; White and Sacramento mountains. Wet ground, in the Upper Sonoran and Transition Zones.

67. HETEROSPERMUM Cav.

Small slender glabrous annual with opposite, pinnately or ternately dissected leaves, and small heads of yellow flowers; involucre in 2 series, the outer of 3 to 5 linear foliaceous bracts, the inner of oval striate ones; outer achenes oval, without pappus, the inner usually infertile, subulate, attenuate to a scabrous beak.

1. **Heterospermum pinnatum** Cav. Icon. Pl. 3: 34. *pl. 267*. 1794.

Heterospermum tagetinum A. Gray, Mem. Amer. Acad. n. ser. 4: 87. 1849.

TYPE LOCALITY: "Habitat in Nova-Hispania."

RANGE: Western Texas to Arizona and southward.

NEW MEXICO: Glorieta; Gallinas Mountains; Hurrah Creek; Mogollon Mountains; Hanover Mountain; Kingston; Organ Mountains; White Mountains; Gray. Open hills, in the Upper Sonoran and lower part of the Transition zones.

The type of *H. tagetinum* was collected west of Las Vegas by Fendler (no. 534).

68. ECLIPTA L.

Annual with procumbent or ascending stems and opposite, lanceolate or oblong, sparingly serrate leaves; heads small, solitary, white-flowered; rays short; disk flowers perfect; involucral bracts 10 to 12, in 2 rows, foliaceous, ovate-lanceolate; receptacle flat; achenes short, 3 or 4-angled, roughened on the sides, hairy at the summit; pappus none or an obscure crown.

1. **Eclipta alba** (L.) Hassk. Pl. Jav. Rar. 528. 1848.

Verbesina alba L. Sp. Pl. 902. 1753.

Eclipta erecta L. Mant. Pl. 2: 286. 1771.

TYPE LOCALITY: "Habitat in Virginia, Surinamo."

RANGE: New Jersey and Texas to New Mexico and southward throughout the tropics.

NEW MEXICO: Albuquerque; Organ Mountains; Mesilla Valley; Roswell. Along ditch banks and in wet ground, in the Lower Sonoran Zone.

69. RATIBIDA Raf. CONE FLOWER.

Perennial herbs with pinnately parted alternate leaves and long-pedunculate terminal heads, with showy yellow to brownish purple, drooping rays; disk yellowish, turning darker; achenes short, broad, compressed, sometimes winged on the edges; pappus a chaffy or aristiform tooth over one or both edges, or wanting.

52576°—15——45

KEY TO THE SPECIES.

Disk in fruit oblong, about 1 cm. long .... 1. *R. tagetes.*
Disk in fruit cylindric, 2 to 4 cm. long.
Rays yellow .... 2. *R. columnifera.*
Rays at least in part brownish purple .... 2a. *R. columnifera pulcherrima.*

1. **Ratibida tagetes** (James) Barnhart, Bull. Torrey Club **24**: 410. 1897.
Rudbeckia tagetes James in Long, Exped. **2**: 68. 1823.
Lepachys tagetes A. Gray, U. S. Rep. Expl. Miss. Pacif. **4**: 103. 1856.
TYPE LOCALITY: About 15 miles southwest of La Junta, Colorado.
RANGE: Colorado and Kansas to Texas and Arizona.
NEW MEXICO: Santa Fe; Nara Visa; Las Vegas; Albuquerque; Sandia Mountains; Cross L Ranch; Gallinas Mountains; Estancia; Socorro; Mesilla Valley; Queen; Gray. Plains and river valleys, in the Lower and Upper Sonoran zones.

2. **Ratibida columnifera** (Nutt.) Woot. & Standl.
Rudbeckia columnifera Nutt. Fraser's Cat. no. 75. 1813.
Rudbeckia columnaris Pursh, Fl. Amer. Sept. 575. 1814.
Ratibida columnaris D. Don in Sweet, Brit. Flower Gard. II. **4**: *pl. 361*. 1838.
Lepachys columnaris Torr. & Gray, Fl. N. Amer. **2**: 313. 1842.
TYPE LOCALITY: Upper Louisiana.
RANGE: British Columbia and Saskatchewan to Arizona, Texas, and Tennessee.
NEW MEXICO: Sierra Grande; mountains west of Grants Station; Santa Fe and Las Vegas mountains; Clayton; Lower Plaza; White and Sacramento mountains. Plains and low hills, in the Upper Sonoran and Transition zones.

2a. **Ratibida columnifera pulcherrima** (DC.) Woot. & Standl.
Obeliscaria pulcherrima DC. Prodr. **5**: 559. 1836.
Ratibida columnaris pulcherrima D. Don in Sweet, Brit. Flower Gard. II. **4**: *pl. 361*. 1830.
Lepachys columnaris pulcherrima Torr. & Gray, Fl. N. Amer. **2**: 313. 1842.
TYPE LOCALITY: "In Mexici provinc. Texas ad San-Fernando de Bejar, et in sinu Spiritus-Sancti ad lacum Sancti-Nicolai."
RANGE: With the species, but more common in New Mexico.
NEW MEXICO: Dulce; Chama; Pecos; Santa Antonita; Ramah; near Las Vegas; mountains west of Grants Station; El Cedro; Tucumcari; Mogollon Mountains; White Mountains; Buchanan; Redlands; Queen; Knowles; Artesia.

This is a mere form of the type and hardly deserves a name. Both forms almost invariably occur together, although occasionally they grow alone. It is possible to find in a single patch every possible gradation in the color of the rays from pure bright yellow to solid brown-purple. The same variation in color occurs in *R. tagetes*, but since that has very small and inconspicuous rays no one has yet thought to distinguish the various forms by name.

70. DRACOPIS Cass.

Annual, 30 to 60 cm. high, with somewhat glaucous, entire or serrate, sessile or clasping leaves; involucre of a few small foliaceous bracts; rays oblong, yellow; disk brownish, cylindric in age; achenes small, minutely rugulose, nearly terete, not angled; pappus none.

1. **Dracopis amplexicaulis** (Vahl) Cass. Dict. Sci. Nat. **35**: 273. 1836.
Rudbeckia amplexicaulis Vahl, Skrivt. Naturh.-Selsk. (Kjøbenhavn) **2**[2]: 29. *pl. 4*. 1793.
TYPE LOCALITY: "Habitat in Louisiana?"
RANGE: Missouri and Louisiana to Texas and New Mexico.
NEW MEXICO: Las Cruces (*Wooton*). Low ground.
Collected but once in the Mesilla Valley, where it had probably been introduced,

71. RUDBECKIA L.

Perennial herbs with alternate, simple or divided leaves and large showy heads terminating the stems or branches; ray flowers neutral, those of the disk perfect; bracts foliaceous, spreading, in about 2 series; receptacle conic or elongated; achenes 4-angled, prismatic; pappus a coriaceous, often 4-toothed crown.

KEY TO THE SPECIES.

Leaves entire or sparingly toothed; plants hirsute.................... 1. *R. flava.*
Leaves, except the uppermost, 3 to 5-cleft or pinnatifid; plants glabrous or nearly so.. 2. *R. laciniata.*

1. **Rudbeckia flava** Moore, Pittonia **4**: 179. 1900. BLACK-EYED SUSAN.
TYPE LOCALITY: "Near the Big Muddy, Wyoming."
RANGE: Wyoming and North Dakota to Colorado and New Mexico.
NEW MEXICO: Santa Fe and Las Vegas mountains; Ensenada; Rio Pueblo. Open slopes and meadows, in the Transition Zone.

2. **Rudbeckia laciniata** L. Sp. Pl. 907. 1753. CONE FLOWER.
Rudbeckia ampla A. Nels. Bull. Torrey Club **28**: 234. 1901.
TYPE LOCALITY: "Habitat in Virginia, Canada."
RANGE: Idaho and Arizona to Quebec and Florida.
NEW MEXICO: Tunitcha Mountains; Chama; Santa Fe and Las Vegas mountains; Mogollon Mountains; White Mountains. Along streams and in damp thickets, in the Transition and Canadian zones.

72. GYMNOLOMIA H. B. K.

Annual or perennial herbs or low shrubs with chiefly opposite leaves and medium-sized pedunculate heads with yellow rays; involucre hemispheric, with numerous bracts in 2 to 4 series; receptacle conic, chaffy; rays flowers neutral; disk flowers numerous, perfect; achenes obovoid, thickish, somewhat compressed laterally or 4-angled, rounded at the summit; pappus none, rarely a ring of 2 to 4 laciniate scales.

KEY TO THE SPECIES.

Stems shrubby; disk and receptacle low; leaves pinnately parted.... 1. *G. tenuifolia.*
Stems herbaceous; disk and receptacle high; leaves entire or sparingly toothed.
  Pubescence mostly loose and spreading; leaves (linear or nearly so) conspicuously ciliate for almost their whole length; bracts ciliate; annual.................................. 2. *G. ciliata.*
  Pubescence mostly or all appressed; leaves ciliate only at the base or not at all; bracts canescent, not ciliate; annuals or perennials.
    Annuals; leaves linear or linear-lanceolate.
      Leaves linear, often involute; heads 7 to 10 mm. broad.. 3. *G. annua.*
      Leaves lance-linear or oblong-linear, flat; heads about 12 mm. broad.................................. 4. *G. longifolia.*
    Perennials; leaves narrowly lanceolate or mostly broader.
      Leaves lanceolate to linear-oblong, several times as long as broad.................................. 5. *G. multiflora.*
      Leaves elliptic-ovate to ovate or elliptic, less than twice as long as broad.......................... 6. *G. brevifolia.*

1. **Gymnolomia tenuifolia** (A. Gray) Benth. & Hook. Gen. Pl. **2:** 364. 1873.
Heliomeris tenuifolia A. Gray, Mem. Amer. Acad. n. ser. **4:** 84. 1849.

TYPE LOCALITY: "Dry valleys, at Rinconada, Saltillo, Mapimi, and Andabazo, Northern Mexico."

RANGE: Western Texas to southern New Mexico and southward.

NEW MEXICO: Organ Mountains; Guadalupe Mountains. Dry hills, in the Upper Sonoran Zone.

2. **Gymnolomia ciliata** (Robins. & Greenm.) Rydb. Bull. Torrey Club **37:** 328. 1910.
Gymnolomia hispida ciliata Robins. & Greenm. Proc. Bost. Soc. Nat. Hist. **29:** 93. 1899.

TYPE LOCALITY: Not stated.

RANGE: Utah and New Mexico to southern California and Mexico.

NEW MEXICO: Zuni Mountains; Santa Antonita; near White Water. Lower and Upper Sonoran zones.

3. **Gymnolomia annua** Robins. & Greenm. Proc. Bost. Soc. Nat. Hist. **29:** 93. 1899.

TYPE LOCALITY: Not stated.

RANGE: New Mexico and Arizona and southward.

NEW MEXICO: Mangas Springs; near Defiance; near the Copper Mines; Deming; Bishops Cap. Dry hills and mesas, in the Lower and Upper Sonoran zones.

4. **Gymnolomia longifolia** Robins. & Greenm. Proc. Bost. Soc. Nat. Hist. **29:** 92. 1899.

TYPE LOCALITY: Not stated.

RANGE: Western Texas to southern Arizona and southward.

NEW MEXICO: Burro Mountains; Kingston; Organ Mountains.

5. **Gymnolomia multiflora** (Nutt.) Benth. & Hook.; Rothr. in Wheeler, Rep. U. S. Surv. 100th Merid. **6:** 160. 1878.
Heliomeris multiflora Nutt. Journ. Acad. Phila. II. **1:** 171. 1848.

TYPE LOCALITY: "Mountains of Upper California."

RANGE: Idaho and Wyoming to California and New Mexico, and southward into Mexico.

NEW MEXICO: Tunitcha Mountains; Chama; Sandia Mountains; Baldy; Santa Fe and Las Vegas mountains; Pajarito Park; Magdalena Mountains; Santa Rita; Middle Fork of the Gila; Organ Mountains; White and Sacramento mountains. Open slopes and in canyons, chiefly in the Transition Zone.

6. **Gymnolomia brevifolia** Greene, Contr. U. S. Nat. Herb. **16:** 190. 1913.

TYPE LOCALITY: Mogollon Mountains, on the West Fork of the Rio Gila, New Mexico. Type collected by Metcalfe (no. 511).

RANGE: Known only from type locality.

73. ZALUZANIA Pers.

Coarse perennial herb with opposite petiolate 3-lobed leaves and large pedunculate radiate heads of yellow flowers; involucre hemispheric, the bracts in about 2 series, canescent, the receptacle conic, paleaceous; disk and ray flowers fertile; achenes of the disk flowers somewhat flattened, those of the ray flowers trigonous; pappus none or of a few deciduous scales in the ray flowers.

1. **Zaluzania grayana** Robins. & Greenm. Proc. Amer. Acad. **34:** 531. 1899.
Gymnolomia triloba A. Gray, Proc. Amer. Acad. **17:** 217. 1882, not *Zaluzania triloba* Pers. 1807.

TYPE LOCALITY: Peaks of the Chiricahua Mountains, south of Ruckers Valley, Arizona.

RANGE: Southern Arizona to southwestern New Mexico and Chihuahua.

NEW MEXICO: San Luis Mountains (*Mearns* 2240).

74. WYETHIA Nutt.

Coarse perennial herbs with mostly simple stems, alternate, usually entire leaves, and large heads of yellow flowers; involucre campanulate or hemispheric, the bracts loosely imbricated in 2 or more series, foliaceous; receptacle slightly convex, the chaff lanceolate, equaling and embrácing the flowers; rays large, pistillate; achenes elongated, 4 or 5-angled, with coroniform 5 to 10-toothed or laciniate pappus.

KEY TO THE SPECIES.

Leaves oblong-lanceolate, tapering to both ends, soft-pubescent; bracts equal, in 2 or 3 series; stems not white .... 1. *W. arizonica.*

Leaves linear-oblong, scabrous, not tapering at the ends; bracts very unequal, in 5 to 6 series; stems white .... 2. *W. scabra.*

1. Wyethia arizonica A. Gray, Proc. Amer. Acad. **8:** 655. 1873.

Type locality: Near Bear Springs, northern Arizona.

Range: Colorado and Utah to Arizona and northern New Mexico.

New Mexico: Southeast of Tierra Amarilla (*Eggleston* 6516). Open slopes in the mountains, in the Transition Zone.

2. Wyethia scabra Hook. Lond. Journ. Bot. **6:** 245. 1847.

Type locality: "Clayey argillaceous declivities of the high hills of Upper Colorado River."

Range: New Mexico and Colorado to Utah and Wyoming.

New Mexico: Carrizo Mountains (*Standley* 7439). Dry hills among rocks, in the Upper Sonoran Zone.

75. VIGUIERA H. B. K.

Coarse perennials with chiefly opposite, petioled or sessile, mostly ovate or cordate leaves and large heads on long terminal peduncles; rays bright yellow, showy, sterile; bracts much imbricated, the outer usually foliaceous; achenes pubescent, quadrangular-compressed, not margined nor winged; pappus of 2 awns or paleæ, one at each of the principal angles.

KEY TO THE SPECIES.

Disk stongly convex at maturity; leaves thin, all petioled, rounded or acute at the base, pubescent; stems much branched .... 1. *V. texana.*

Disk flattish; leaves thick, the upper sessile or nearly so, subcordate, scabrous; stems sparingly branched .... 2. *V. cordifolia.*

1. Viguiera texana Torr. & Gray, Fl. N. Amer. **2:** 318. 1842.

Type locality: Texas.

Range: Western Texas to southern Arizona.

New Mexico: Fort Bayard; Bear Mountains; San Luis Mountains; near White Water; White Mountains; Van Pattens. Thickets and open fields, in the Transition and Upper Sonoran zones.

2. Viguiera cordifolia A. Gray, Pl. Wright. **1:** 107. 1852.

Type locality: Plains at the base of the Guadalupe Mountains, Texas.

Range: Western Texas to southern Arizona.

New Mexico: Mangas Springs; Mogollon Mountains; Fairview; Dog Spring; Organ Mountains; White Mountains: Gray; Sandia Mountains. Dry hills and canyons, in the Upper Sonoran Zone.

76. HELIANTHUS L. Sunflower.

Coarse annual or perennial herbs with simple or branched stems, alternate or opposite leaves, and often very large heads; involucre flat to hemispheric, the thick bracts in several series; receptacle flat or convex, chaffy; rays mostly large and showy, yellow, neutral; disk flowers perfect, the corollas brownish, purple, or yellowish; achenes flattened or slightly quadrangular, the pappus of 2 awns or scales, early deciduous.

KEY TO THE SPECIES.

Perennials.
- Leaves glaucous and smooth, ciliate, undulate........... 1. *H. ciliaris.*
- Leaves not glabrous nor glaucous, not ciliate, flat.
  - Leaves soft-villous beneath; stems hispid throughout. 2. *H. neomexicanus.*
  - Leaves scabrous or at least very rough beneath, not soft-villous; stems glabrous, at least above.
    - Disk flowers dark brown or purple; leaves rhombic-ovate; stems pubescent.............. 3. *H. subrhomboideus.*
    - Disk flowers yellow; leaves narrowly lanceolate; stems glabrous or nearly so.
      - Leaves coarsely toothed; bracts hirsute-ciliate............................ 4. *H. grosseserratus.*
      - Leaves sparingly denticulate or entire; bracts not ciliate or ciliate only at the base.. 5. *H. fascicularis.*

Annuals.
- Bracts ciliate, hispid, ovate, abruptly acuminate.
  - Lower leaves, at least, ovate or cordate, conspicuously toothed, dull green......................... 6. *H. annuus.*
  - Leaves lanceolate or narrowly deltoid, obscurely toothed or entire, shining.................... 7. *H. aridus.*
- Bracts canescent-strigose, not ciliate, lanceolate.
  - Leaves green; pubescence of peduncles appressed, short...................................... 8. *H. petiolaris.*
  - Leaves grayish or whitish; pubescence of peduncles long, spreading............................. 9. *H. canus.*

1. Helianthus ciliaris DC. Prodr. **5:** 587. 1836. Blueweed. Yerba parda.

Type locality: "In Mexico prope Reynosa de Tamaulipas."

Range: Western Texas to southern Arizona and southward.

New Mexico: Socorro; Tucumcari; Mesilla Valley; Tularosa; Elida; Carlsbad; Artesia; Roswell. River valleys, usually in alkaline soil, in the Lower Sonoran Zone.

A common and troublesome weed in cultivated fields in the Rio Grande and Pecos valleys. In general appearance this is very unlike our other species.

2. Helianthus neomexicanus Woot. & Standl. Contr. U. S. Nat. Herb. **16:** 190. 1913.

Type locality: Mangas Springs, New Mexico. Type collected by Wooton, August 19, 1902.

Range: Known only from type locality.

3. Helianthus subrhomboideus Rydb. Mem. N. Y. Bot. Gard. **1:** 419. 1900.

Type locality: Whitman, Nebraska.

Range: British America to Nebraska and New Mexico.

New Mexico: Winsors Ranch; Raton Mountains; Gallinas Planting Station; White Mountains; Baldy; Dulce. Plains and hills, in the Transition Zone.

4. **Helianthus grosseserratus** Martens, Sel. Sem. Hort. Loven. 1839.
TYPE LOCALITY: St. Louis, Missouri.
RANGE: New York and Wyoming to Pennsylvania, Texas, and New Mexico.
NEW MEXICO: Ensenada; White Mountains. Plains and low hills, in the Upper Sonoran Zone.

5. **Helianthus fascicularis** Greene, Pl. Baker. **3**: 28. 1901.
Helianthus giganteus utahensis D. C. Eaton, in King, Geol. Expl. 40th Par. **5**: 169. 1871.
Helianthus utahensis A. Nels. Bull. Torrey Club **29**: 405. 1902.
TYPE LOCALITY: Cimarron, Colorado.
RANGE: British America to Colorado and New Mexico.
NEW MEXICO: Pecos; Kingston; Middle Fork of the Gila; Dulce. Mountain valleys, in the Transition Zone.

6. **Helianthus annuus** L. Sp. Pl. 904. 1753. COMMON SUNFLOWER.
Helianthus lenticularis Dougl. in Edwards's Bot. Reg. **15**: *pl. 1225*. 1829.
TYPE LOCALITY: "In Peru, Mexico."
RANGE: British America to California and Texas and southward.
NEW MEXICO: Farmington; Tunitcha Mountains; Carrizo Mountains; Dulce; Winsors Ranch; Pecos; Raton; Zuni; Espanola; Cleveland; Fort Bayard; Mangas Springs; Mesilla Valley; Tularosa; Gilmores Ranch; Gray; Carrizozo. Plains and cultivated ground, from the Lower Sonoran to the Transition Zone.
One of our commonest weeds in cultivated ground.

7. **Helianthus aridus** Rydb. Bull. Torrey Club **32**: 127. 1895.
TYPE LOCALITY: Great Falls, Montana.
RANGE: Montana and Nebraska to New Mexico.
NEW MEXICO: Pajarito Park; Pecos; north of El Vado; mountains southeast of Patterson; Upper Negrito Creek. In dry soil, in the Upper Sonoran Zone.

8. **Helianthus petiolaris** Nutt. Journ. Acad. Phila. **2**: 115. 1821.
TYPE LOCALITY: "On the sandy shores of the Arkansas."
RANGE: Oregon and Saskatchewan to Arizona and Texas.
NEW MEXICO: Shiprock; Farmington; Carrizo Mountains; Gallup; Nara Visa; Cliff; Mangas Springs; Pecos; Roswell. Plains and dry hills, in the Lower and Upper Sonoran Zones.

9. **Helianthus canus** (Britton) Woot. & Standl. Contr. U. S. Nat. Herb. **16**: 190. 1913.
Helianthus petiolaris canescens A. Gray, Pl. Wright. **1**: 108. 1852, not *H. canescens* Michx. 1803.
Helianthus petiolaris canus Britton, Mem. Torrey Club **5**: 334. 1894.
TYPE LOCALITY: Valley of the Rio Grande 60 or 70 miles below El Paso, Texas.
RANGE: Western Texas to southern Arizona.
NEW MEXICO: Zuni; Tesuque; mesa near Las Cruces; Bishops Cap; Mesilla Valley. River valleys and mesas in the Lower and Upper Sonoran Zones.

77. FLOURENSIA DC. TAR-BUSH.

Shrub 1 to 2 meters high, viscid, much branched, with small thick alternate entire leaves and corymbose or paniculate, short-pedunculate heads of yellowish flowers in the upper axils; involucre of 2 or 3 series of lanceolate bracts, part of them foliaceous; heads discoid; receptacle flat, the chaffy bracts conduplicate about the achenes and deciduous with them; achenes compressed, narrowly oblong-cuneate, callous-margined, villous, the pappus a subulate awn from each angle of the summit with occasionally some smaller ones.

1. Flourensia cernua DC. Prodr. **5**: 593. 1836.

TYPE LOCALITY: Monterey, Mexico.

RANGE: Western Texas to southern Arizona and southward.

NEW MEXICO: Las Palomas; Hachita; Lake Valley; mesa west of Organ Mountains; San Andreas Mountains; Tularosa; Pecos Valley near Texas line. Sandy plains and mesas and low hills, in the Lower Sonoran Zone.

78. ENCELIA Adans.

Perennial herb, 30 cm. high or more, with linear entire leaves crowded at the base of the slender scapiform monocephalous stems; rays several, yellow; achenes densely villous, the pappus of 2 chaffy awns.

1. Encelia scaposa A. Gray, Proc. Amer. Acad. **19**: 7. 1883.

Simsia ? scaposa A. Gray, Pl. Wright. **2**: 98. 1853.

TYPE LOCALITY: Stony hills between the Mimbres and the Rio Grande, New Mexico. Type collected by Wright in 1851.

RANGE: Known only from type locality.

We have seen no further specimens of this; apparently it is a very rare plant.

79. HELIANTHELLA Torr. & Gray.

Perennial herbs with simple or sparingly branched stems, entire scattered sessile leaves, and large heads with yellow rays; bracts in about 2 series, loose, foliaceous; paleæ embracing the achenes; ray flowers sterile, those of the disk perfect; achenes compressed, slightly winged on one or both margins.

KEY TO THE SPECIES.

Disk flowers purple; rays few, scarcely surpassing the disk.... 1. *H. microcephala.*
Disk flowers yellow; rays numerous, much surpassing the disk.
  Disk 2 to 3 cm. broad; leaves lanceolate, thin, not strongly reticulate, very acute.............................. 2. *H. quinquenervis.*
  Disk less than 2 cm. wide; leaves oblanceolate or narrower, thick, strongly reticulate-veined, obtuse or nearly so. 3. *H. parryi.*

1. Helianthella microcephala A. Gray, Proc. Amer. Acad. **19**: 10. 1883.

Encelia microcephala A. Gray, Proc. Amer. Acad. **8**: 657. 1873.

TYPE LOCALITY: "Sierra Abayo, New Mexico?" Type collected by Newberry in 1859.

RANGE: Utah and Colorado to northern Arizona and New Mexico.

NEW MEXICO: Carrizo Mountains (*Standley* 7350). Dry rocky hills, in the Upper Sonoran Zone.

2. Helianthella quinquenervis (Hook.) A. Gray, Proc. Amer. Acad. **19**: 10. 1883.

Helianthus quinquenervis Hook. Lond. Journ. Bot. **6**: 247. 1847.

Helianthella majuscula Greene, Leaflets **1**: 148. 1905.

TYPE LOCALITY: "Stony ridges, hills of Upper Platte."

RANGE: Idaho and South Dakota to Colorado and New Mexico.

NEW MEXICO: Tunitcha Mountains; Chama; Santa Fe and Las Vegas mountains; West Fork of the Gila; Black Range; White and Sacramento mountains. Meadows and thickets, in the Transition and Canadian zones.

The type of *H. majuscula* was collected in the Black Range (*Metcalfe* 1435).

3. Helianthella parryi A. Gray, Proc. Acad. Phila. **1863**: 65. 1864.

TYPE LOCALITY: Middle Park, near the foot of Pikes Peak, Colorado.

RANGE: Colorado, New Mexico, and Arizona.

NEW MEXICO: Santa Fe and Las Vegas mountains; Baldy; mountains west of Grants; Vermejo Park. High mountains, in the Canadian and Hudsonian zones.

80. XIMENESIA Cav.

Annual, more or less canescent, with alternate petiolate toothed leaves and large showy heads of yellow flowers; involucre of spreading linear foliaceous equal bracts; disk and receptacle merely convex; rays numerous, large, bright yellow, usually fertile; achenes flat, obovate, broadly winged, with short setiform awns, the awns not hooked.

1. **Ximenesia exauriculata** (Robins. & Greenm.) Rydb. Bull. Torrey Club **33:** 154. 1906.

Verbesina encelioides exauriculata Robins. & Greenm. Proc. Amer. Acad. **34:** 544. 1899.

TYPE LOCALITY: Kansas.

RANGE: Kansas and Colorado to Arizona and western Texas and southward.

NEW MEXICO: Abundant throughout the State. Fields and low hills, from the Lower Sonoran to the Transition Zone.

This is one of the commonest plants of New Mexico, being found in abundance almost everywhere except in the highest parts of the mountains and on the driest plains. It is nearly always to be seen in cultivated fields and waste ground. In the northern part of the State, especially in favorable seasons, it covers large areas of ground to the exclusion of almost everything else, presenting a wide unbroken sheet of rich yellow.

81. WOOTONELLA Standley.

Low perennial, 20 cm. high or less, with slender deep-seated rootstocks; stems simple or branched, ascending, canescent; lower leaves opposite, the upper alternate, irregularly dentate, narrowed into winged petioles, these mostly dilated and dentate at the base; heads 15 to 20 mm. broad, solitary on naked terminal peduncles; bracts foliaceous, canescent; rays rather pale yellow, conspicuously exceeding the involucre; ray flowers fertile, the disk flowers sterile; paleæ very narrow, nearly filiform, persistent; achenes obovate or oblong, villous, broadly winged, the wings corky-thickened near the apex; pappus none.

1. **Wootonella nana** (A. Gray) Standley, Proc. Biol. Soc. Washington **25:** 120. 1912.

Ximenesia encelioides nana A. Gray, Pl. Wright. **2:** 92. 1853.

Verbesina nana Robinson, Proc. Amer. Acad. **34:** 543. 1899.

TYPE LOCALITY: "Around the dwellings of Prairie-dogs, between the Limpio and the Rio Grande," Texas.

RANGE: Southern New Mexico to western Texas and northeastern Mexico.

NEW MEXICO: Artesia; Dayton. Plains, in the Lower Sonoran Zone.

This is said to be a common weed in cultivated fields of the Pecos Valley.

82. VERBESINA L. CROWNBEARD.

Coarse annual or perennial herbs with opposite or alternate, petioled or sessile leaves and few or numerous small or medium-sized heads of yellow flowers; bracts imbricated in 2 or more series, appressed or at least erect, not elongated; receptacle convex to conic; rays several or numerous, large and showy, usually sterile; achenes flat, glabrous or nearly so; awns of the pappus straight, often obsolete or wanting.

KEY TO THE SPECIES.

Leaves elongated-linear.. 1. *V. longifolia.*
Leaves lanceolate to ovate or oblong.
  Leaves thick, sessile, cordate; heads few, 15 mm. in diameter or more.. 2. *V. rothrockii.*
  Leaves thin, petioled, narrowed at the base; heads several to many, 12 mm. in diameter or less......................... 3. *V. oreophila.*

1. **Verbesina longifolia** A. Gray, Proc. Amer. Acad. **19:** 12. 1883.
Actinomeris longifolia A. Gray, Pl. Wright. **2:** 89. 1853.
TYPE LOCALITY: Mountains east of Santa Cruz, Sonora.
RANGE: Southwestern New Mexico, southeastern Arizona, and southward.
NEW MEXICO: Summit of Animas Mountains; Animas Valley; San Luis Mountains.

2. **Verbesina rothrockii** Robins. & Greenm. Proc. Amer. Acad. **34:** 541. 1899.
Verbesina wrightii A. Gray, Proc. Amer. Acad. **19:** 12. 1883, in part, not *V. wrightii* Griseb. 1866.
TYPE LOCALITY: Camp Bowie, Arizona.
RANGE: Southern New Mexico and Arizona and southward.
NEW MEXICO: Between Copper Mines and Condes Camp; Hachita.

3. **Verbesina oreophila** Woot. & Standl. Contr. U. S. Nat. Herb. **16:** 190. 1913.
TYPE LOCALITY: Cloudcroft, Sacramento Mountains, New Mexico. Type collected by Wooton, August 24, 1899.
RANGE: Known only from type locality, in the Transition Zone.

83. POROPHYLLUM Vaill.

Tall perennial, about 1 meter high, woody at the base, with numerous erect, slender, nearly naked, glaucous branches; leaves linear; heads small, long-pedunculate, of yellow flowers; bracts 5 to 10, glaucous, usually gland-dotted; ray flowers none; disk flowers few; achenes linear, the pappus of simple scabrous bristles.

1. **Porophyllum scoparium** A. Gray, Pl. Wright. **1:** 119. 1852.
TYPE LOCALITY: "Rocky banks of the San Pedro River, and mountains east of El Paso," Texas.
RANGE: Western Texas to southern New Mexico and southward.
NEW MEXICO: Arroyo near Bishops Cap; Upper Corner Monument. Dry hills, in the Lower Sonoran Zone.

84. CHRYSACTINIA A. Gray.

Low woody perennial, 30 cm. high or less, branched, with alternate glandular-punctate linear leaves and slender-pedunculate heads terminating the branches; involucral bracts lanceolate, in a single series, each bearing an apical oil gland; rays bright yellow; achenes linear, the simple pappus of scabrous bristles.

1. **Chrysactinia mexicana** A. Gray, Mem. Amer. Acad. n. ser. **4:** 93. 1849.
Pectis taxifolia Greene, Leaflets **1:** 148. 1905.
TYPE LOCALITY: "Dry valley west of Saltillo, * * * and on high grounds near Buena Vista," Mexico.
RANGE: Western Texas and southern New Mexico to Mexico.
NEW MEXICO: Grant County; San Andreas Mountains; Queen. Dry hills, in the Upper Sonoran Zone.
The type of *Pectis taxifolia* is Metcalfe's 1440 from the Black Range.

85. PECTIS L. LIMONCILLO.

Low annuals with erect or prostrate stems, gland-dotted, strong-scented, opposite, narrow, often ciliate leaves and small cymose or solitary heads of yellow flowers; rays yellow; involucre cylindric to campanulate, the narrow distinct bracts in a single series; receptacle naked; achenes linear, striate; pappus of several or numerous scales, awns, or bristles.

KEY TO THE SPECIES.

Stems prostrate; leaves oblanceolate to linear-spatulate.......... 1. *P. prostrata.*
Stems erect; leaves narrowly linear.
  Pappus of numerous barbellate bristles........................ 2. *P. papposa.*

Pappus of a few paleæ or slender awns.
Pappus inconspicuous, a crown of 4 or 5 connate scales; heads in dense clusters, nearly sessile.............. 3. *P. angustifolia.*
Pappus of 2 or 3 rigid awns and a few interposed small scales; heads few, long-pedunculate................. 4. *P. filipes.*

1. **Pectis prostrata** Cav. Icon. Pl. **4**: 12. *pl. 324.* 1797.
Chthonia prostrata Cass. Dict. Sci. Nat. **9**: 173. 1817.
TYPE LOCALITY: "Habitat in Nova-Hispania."
RANGE: Western Texas to southern Arizona, south through tropical America.
NEW MEXICO: Gila Hot Springs; Mangas Springs; Organ Mountains. Plains and low hills, in the Lower and Upper Sonoran zones.

2. **Pectis papposa** A. Gray, Mem. Amer. Acad. n. ser. **4**: 62. 1849.
TYPE LOCALITY: "California."
RANGE: Southern California to Utah and New Mexico.
NEW MEXICO: Mesa near Las Cruces; Mesilla Valley. Dry plains and low hills, in the Lower Sonoran Zone.

3. **Pectis angustifolia** Torr. Ann. Lyc. N. Y. **2**: 214. 1828.
Pectis papposa sessilis Jones, Contr. West. Bot. **12**: 46. 1908.
TYPE LOCALITY: "On the Rocky Mountains."
RANGE: Colorado and Kansas to Arizona and western Texas, south into Mexico.
NEW MEXICO: Zuni; Canyon Largo; Santa Fe; Albuquerque; Cerrillos; Las Vegas; Mountainair; Gallinas Mountains; Clayton; Nara Visa; Bear Mountains; Fort Bayard; San Marcial; Graham; Guadalupe Canyon; Santa Rita; Organ Mountains; Tortugas Mountain; Gray; Roswell. Plains and low dry hills, in the Lower and Upper Sonoran zones.

4. **Pectis filipes** A. Gray, Mem. Amer. Acad. n. ser. **4**: 62. 1849.
TYPE LOCALITY: "California."
RANGE: Western Texas to southern Arizona.
NEW MEXICO: Carlisle; Burro Mountains; Mangas Springs; Gila Hot Springs; Fort Bayard; Dog Spring; Guadalupe Canyon. Dry hills and mesas, in the Lower and Upper Sonoran zones.

86. **NICOLLETIA** A. Gray.

Low perennial herb with alternate, pinnately parted leaves and large heads of purple or flesh-colored flowers; involucre of 8 or 9 distinct bracts, oblong or cylindric; receptacle naked; achenes linear-filiform; pappus double, the outer of numerous capillary bristles, the inner of 5 lanceolate hyaline awned scales.

1. **Nicolletia edwardsii** A. Gray, Pl. Wright. **1**: 119. *pl. 8.* 1852.
TYPE LOCALITY: Near Guajuquilla, Chihuahua, Mexico.
RANGE: Western Texas and southern New Mexico to Mexico.
We have seen no specimens of this from within our limits, but Doctor Gray reported specimens from New Mexico.

87. **BOEBERA** Willd. FETID MARIGOLD.

Diffusely branched, strong-scented annual with opposite, finely dissected leaves and small heads of yellow flowers; rays few, short; involucre campanulate, of 8 to 10 appressed oblong bracts with a few narrow shorter outer ones; achenes narrowly obpyramidal, 3 to 5-angled, pubescent; pappus of about 10 scales, divided to below the middle into stiff capillary bristles.

1. Boebera papposa (Vent.) Rydb. in Britton, Man. 1012. 1901.

Tagetes papposa Vent. Pl. Jard. Cels *pl. 36.* 1800.

Dysodia chrysanthemoides Lag. Gen. & Sp. Nov. 29. 1816.

Dysodia papposa Hitchc. Trans. Acad. St. Louis **5:** 503. 1891.

TYPE LOCALITY: "Illinois."

RANGE: Nebraska and Ohio to Arizona and Louisiana and southward.

NEW MEXICO: Cedar Hill; mouth of Holy Ghost Creek; Pecos; Farmington; Galisteo; Tesuque; El Rito Creek; Folsom; Albuquerque; Raton; Estancia; Beulah; Santa Fe; Cliff; Burro Mountains; Kingston; West Fork of the Gila; Mesilla Valley; White Mountains; Capitan Mountains. Hillsides and roadsides, in the Upper Sonoran and Transition zones.

88. TAGETES L.

Slender, diffusely branched annual, 30 cm. high or smaller, with opposite leaves 3 to 5-parted into linear-filiform divisions; heads small, with 1 to 3 rays; involucre fusiform, the narrow bracts united for nearly their whole length; achenes slender, glabrate; pappus of 2 oval or truncate thin paleæ and 2 longer awns.

1. Tagetes micrantha Cav. Icon. Pl. **4:** 31. *pl. 352.* 1797.

TYPE LOCALITY: "Habitat in Nova-Hispania iuxta urbem Queretaro."

RANGE: New Mexico and Arizona and southward.

NEW MEXICO: Mogollon Creek; near Las Vegas.

89. ACIPHYLLAEA A. Gray.

Low shrubby perennial, 20 cm. high or less, with opposite entire rigid filiform leaves and small, nearly sessile heads of yellow flowers with bright yellow rays; involucre of equal, narrowly oblong, gland-dotted bracts in a single series; achenes linear, striate; pappus of 18 to 20 paleæ, each of these parted above into 3 or 5 capillary bristles.

1. Aciphyllaea acerosa (DC.) A. Gray, Mem. Amer. Acad. n. ser. **4:** 91. 1849.

Dysodia acerosa DC. Prodr. **5:** 641. 1836.

Hymenatherum acerosum A. Gray, Pl. Wright. **1:** 115. 1852.

TYPE LOCALITY: "In Mexici prov. Sancti-Ludovici de Potosi."

RANGE: Western Texas to southern Arizona and southward.

NEW MEXICO: Albuquerque; Plaza Larga; Socorro; Grant County; Tortugas Mounain; Tularosa Creek. Dry hills, in the Lower Sonoran Zone.

90. THYMOPHYLLA Lag.

Annual or perennial herbs with gland-dotted, alternate or opposite, pinnately parted leaves and small pedunculate heads of yellow flowers; involucre campanulate, the bracts united into a cup; receptacle naked or fimbrillate; achenes linear, striate; pappus of several or many scales or bristles.

KEY TO THE SPECIES.

Annuals; divisions of the leaves linear, not rigid.
    Pappus of 5 to 8 oblong erose-truncate scales .... 1. *T. aurea.*
    Pappus of 10 to 20 aristate scales.
        Rays inconspicuous, not surpassing the disk; scales of the pappus thick, firm .... 2. *T. neomexicana.*
        Rays conspicuous, much surpassing the disk; scales of the pappus hyaline .... 3. *T. polychaeta.*

Perennials; divisions of the leaves filiform, rigid.
Scales of the pappus all alike, awned........................ 4. *T. thurberi.*
Scales of pappus in 2 series, the inner awned, the outer obtuse and pointless.
Plants puberulent; bracts ciliate; rays shorter than the involucre.................................... 5. *T. pentachaeta.*
Plants glabrous; bracts not ciliate; rays equaling the involucre.. 6. *T. hartwegi.*

1. **Thymophylla aurea** (A. Gray) Greene; Britt. & Brown, Illustr. Fl. **3:** 453. 1898.
Lowellia aurea A. Gray, Mem. Amer. Acad. n. ser. **4:** 91. 1849.
Hymenatherum aureum A. Gray, Proc. Amer. Acad. **19:** 42. 1883.
Dysodia aurea A. Nels. in Coulter, New Man. Rocky Mount. 563. 1909.
TYPE LOCALITY: "Between Cold Spring and Upper Spring, west of Cimarron Creek," probably in Oklahoma.
RANGE: Kansas and Colorado to New Mexico and Texas.
NEW MEXICO: Plaza Larga (*Bigelow*). Plains, in the Upper Sonoran Zone.

2. **Thymophylla neomexicana** (A. Gray) Woot. & Standl. Contr. U. S. Nat. Herb. **16:** 191. 1913.
Adenophyllum wrightii A. Gray, Pl. Wright. **2:** 92. 1853, not *Hymenatherum wrightii* A. Gray, 1849.
Hymenatherum neomexicanum A. Gray, Proc. Amer. Acad. **19:** 40. 1883.
TYPE LOCALITY: Hillsides near the Copper Mines, New Mexico. Type collected by Wright (no. 1240).
RANGE: Known only from type locality.

3. **Thymophylla polychaeta** (A. Gray) Small, Fl. Southeast. U. S. 1295. 1903.
Hymenatherum polychaetum A. Gray, Pl. Wright. **1:** 116. 1852.
TYPE LOCALITY: "Prairies at the Pass of the Limpio," Texas.
RANGE: Western Texas to southern New Mexico and southward.
NEW MEXICO: Plains east of Fort Cummings; El Paso to Monument 53. Lower Sonoran Zone.

4. **Thymophylla thurberi** (A. Gray) Woot. & Standl. Contr. U. S. Nat. Herb. **16:** 191. 1913.
Hymenatherum thurberi A. Gray, Proc. Amer. Acad. **19:** 41. 1883.
TYPE LOCALITY: "Stony hills near El Paso," Texas or Chihuahua.
RANGE: Western Texas and southern New Mexico to northern Mexico.
NEW MEXICO: Laguna; Tres Hermanas; Tortugas Mountain. Dry hills and mesas, in the Lower and Upper Sonoran zones.

5. **Thymophylla pentachaeta** (DC.) Small, Fl. Southeast. U. S. 1295. 1903.
Hymenatherum pentachaetum DC. Prodr. **5:** 642. 1836.
TYPE LOCALITY: Near Monterey, Mexico.
RANGE: Southern Utah to Arizona and western Texas and southward.
NEW MEXICO: Van Pattens; Tortugas Mountain; Hondo Canyon, 50 miles west of Roswell. Dry hills, in the Lower and Upper Sonoran zones.

6. **Thymophylla hartwegi** (A. Gray) Woot. & Standl. Contr. U. S. Nat. Herb. **16:** 191. 1913.
Hymenatherum berlandieri DC. err. det. Benth. Pl. Hartw. 18. 1839.
Hymenatherum hartwegi A. Gray, Pl. Wright. **1:** 117. 1852.
TYPE LOCALITY: Mexico.
RANGE: Western Texas to southern Arizona, south into Mexico.
NEW MEXICO: Laguna (*Lemmon*). Lower and Upper Sonoran zones.

91. PSILOSTROPHE DC.

Low, corymbosely branched, woolly perennial herbs with alternate, spatulate to linear, often pinnatifid leaves and small heads of yellow flowers with persistent yellow rays; involucre cylindric-campanulate, of 4 to 10 woolly bracts; achenes narrow, terete, obscurely striate; pappus of 4 to 6 hyaline paleæ.

KEY TO THE SPECIES.

Stems covered with a dense matted tomentum; heads long-pedunculate; rays 15 mm. long or more .... 1. *P. cooperi.*
Stems not densely matted-tomentose, the pubescence loose; heads short-pedunculate; rays 10 mm. long or less.
  Stems densely villous or floccose .... 2. *P. tagetina.*
  Stems scantily pubescent, often nearly glabrous, softly hirsute. 3. *P. sparsiflora.*

1. **Psilostrophe cooperi** (A. Gray) Greene, Pittonia **2**: 176. 1891.
Riddellia cooperi A. Gray, Proc. Amer. Acad. **7**: 358. 1868.
TYPE LOCALITY: Gravelly banks at Fort Mohave, California.
RANGE: Southern California to southwestern New Mexico.
NEW MEXICO: Near Duncan (*Davidson* 1032). Dry plains and hills, in the Lower Sonoran Zone.
The handsomest species of the genus by its very large heads whose brilliant rays contrast well with the white stems.

2. **Psilostrophe tagetinae** (Nutt.) Britt. & Brown, Illustr. Fl. **3**: 444. 1898.
Riddellia tagetinae Nutt. Trans. Amer. Phil. Soc. n. ser. **7**: 371. 1841.
TYPE LOCALITY: "The southern range of the Rocky Mountains, toward the sources of the Platte."
RANGE: Colorado to western Texas and southern Arizona.
NEW MEXICO: West of Santa Fe; Cebolla Spring; Pajarito Park; Zuni; Reserve; Sandia Mountains; Mangas Springs; Black Range; Gila Hot Springs; Magdalena; San Luis Mountains; Strauss Station; Mesilla Valley; Organ Mountains; Tucumcari; Tularosa; White Sands; Jarilla; Buchanan; Melrose; Guadalupe Mountains; west of Roswell. Dry plains and low hills, in the Lower and Upper Sonoran zones.

3. **Psilostrophe sparsiflora** (A. Gray) A. Nels. Proc. Biol. Soc. Washington **16**: 23. 1903.
Riddellia tagetina sparsiflora A. Gray, Syn. Fl. **1**[2]: 318. 1884.
TYPE LOCALITY: Southern Utah.
RANGE: Southern Utah to northern Arizona and New Mexico.
NEW MEXICO: Mesa La Vaca (*Marsh*). Dry hills, in the Upper Sonoran Zone.

92. BAILEYA Harv. & Gray.

Densely floccose-woolly biennial or perennial with alternate pinnatifid leaves and long-pedunculate heads of yellow flowers with showy, bright yellow, persistent rays, these reflexed in age; involucre hemispheric, of numerous linear bracts in 2 or 3 series, very woolly; achenes oblong-linear or clavate, angled, striate; pappus none.

1. **Baileya multiradiata** Harv. & Gray in Emory, Mil. Reconn. 144. *pl. 6.* 1848.
Baileya pleniradiata Harv. & Gray, Mem. Amer. Acad. n. ser. **4**: 105. 1849.
TYPE LOCALITY: Along the Rio Grande, New Mexico. Type collected by Emory.
RANGE: Western Texas to southern Utah and California, southward into Mexico.
NEW MEXICO: Albuquerque; Pajarito Park; Socorro; Mangas Springs; Florida Mountains; Dog Spring; Mesilla Valley; mesa west of Organ Mountains; Orogrande; Three Rivers. Dry plains and low hills, in the Lower Sonoran Zone.
A very handsome plant, common on the mesas in early spring. Sometimes the plants continue flowering until late in the fall.

93. LAPHAMIA A. Gray.

Low monocephalous perennial with alternate, petiolate, dentate, broadly ovate leaves; bracts of the hemispheric involucre distinct, imbricated, oblong; rays none; margin of achenes naked or sparingly ciliate; pappus of about 20 bristles.

1. **Laphamia cernua** Greene, Bull. Torrey Club **25**: 122. 1898.

TYPE LOCALITY: Organ Mountains, New Mexico. Type collected by Wooton (no. 476).

RANGE: Known only from the type locality.

94. PERICOME A. Gray.

Tall, much branched perennial herb with bright green opposite long-petiolate triangular-hastate leaves and numerous small heads of yellow flowers in terminal corymbiform cymes; involucre a single series of numerous narrow bracts lightly connate by their edges into a cup; achenes villous-ciliate; pappus a lacerate-ciliate crown with sometimes a pair of short awns.

1. **Pericome caudata** A. Gray, Pl. Wright. **2**: 82. 1853.

TYPE LOCALITY: Sides of the mountains at the Copper Mines, New Mexico. Type collected by Wright (no. 1195).

RANGE: Southern Colorado to New Mexico and Arizona.

NEW MEXICO: Rito de los Frijoles; Sierra Grande; Las Vegas Hot Springs; Clayton; mountains west of Grants Station; Puertecito; Canada Alamosa; Santa Rita; Mogollon Mountains; Hillsboro Peak; Organ Mountains; South Bonito Creek; Gray. Canyons and rocky hills, in the Upper Sonoran and lower part of the Transition zones.

95. PERITYLE Benth.

Low slender perennial, woody at the base, with pedately or pinnately parted leaves, the lowest ones opposite, the upper alternate; heads small, pedunculate, the disk flowers yellow, the rays white; bracts of the involucre distinct, somewhat imbricated; achenes narrowly oblong, glabrate on the faces, densely hirsute, ciliate, the pappus of 2 awns.

1. **Perityle coronopifolia** A. Gray, Pl. Wright. **2**: 82. 1853.

TYPE LOCALITY: Sides of the mountains at the Copper Mines, New Mexico. Type collected by Wright (no. 1196).

RANGE: Southern New Mexico and Arizona.

NEW MEXICO: Mogollon Mountains; Organ Mountains. In crevices of rocks, in the Upper Sonoran Zone.

96. CLAPPIA A. Gray.

Low shrub with alternate fleshy terete leaves, these entire or the lower 3 to 5-parted; heads pedunculate, terminating the branches; involucre hemispheric, of few oval obtuse striate bracts imbricated in 2 or 3 series; rays 12 to 15, linear; achenes oblong-turbinate, terete, 8 to 10-nerved, hirtellous on the nerves; pappus of 20 to 25 rigid scabrous distinct bristles.

1. **Clappia suaedifolia** A. Gray in Torr. U. S. & Mex. Bound. Bot. 93. 1859.

TYPE LOCALITY: Laredo, Texas.

RANGE: Western Texas to southeastern New Mexico.

NEW MEXICO: White Sands; Roswell. Alkaline flats, in the Lower Sonoran Zone.

97. GAILLARDIA Foug. BLANKET FLOWER.

Branched annual or perennial herbs with alternate, entire or pinnatifid leaves and long-pedunculate heads; involucre depressed-hemispheric, the linear or lanceolate bracts imbricated in 2 or 3 series; receptacle convex or globose, fimbrillate; rays cuneate, yellow or party-colored, neutral; achenes turbinate, 5-ribbed, villous; pappus of 6 to 12 awned scales.

KEY TO THE SPECIES.

Lobes of the disk corollas short, obtuse; leaves mostly pinnatifid... 1. *G. pinnatifida.*
Lobes of the disk corollas narrow, acute; leaves mostly entire..... 2. *G. pulchella.*

1. **Gaillardia pinnatifida** Torr. Ann. Lyc. N. Y. **2**: 214. 1828.

TYPE LOCALITY: "On the Canadian ?," Colorado or New Mexico.

RANGE: Colorado to Arizona and Texas.

NEW MEXICO: Nearly throughout the State. Plains and low hills, in the Lower and Upper Sonoran zones.

2. **Gaillardia pulchella** Foug. Mém. Acad. Sci. Paris **1786**: 5. *pl. 1, 2.* 1788.

TYPE LOCALITY: "Louisiane."

RANGE: Arizona to Arkansas and Louisiana.

NEW MEXICO: Pajarito Park; Mogollon Mountains; Nutt Mountain; Burro Mountains; Animas Valley; Organ Mountains; Buchanan; south of Torrance; Nara Visa; Roswell. Plains and hills, in the Upper Sonoran Zone.

98. SCHKUHRIA Roth.

Slender, paniculately much branched annual with alternate leaves pinnately parted into filiform divisions and small pedunculate heads of yellow or purplish flowers; involucre turbinate, of 4 or 5 erect scarious-tipped bracts, 3 to 9-flowered; achenes obpyramidal, the pappus of 8 scarious paleæ.

1. **Schkuhria wrightii** A. Gray, Pl. Wright. **2**: 95. 1853.

TYPE LOCALITY: "On the Sonoita near Deserted Rancho, Sonora."

RANGE: Southern New Mexico and Arizona and adjacent Mexico.

NEW MEXICO: Trujillo Creek; Organ Mountains. Moist canyons, in the Upper Sonoran Zone.

99. HYMENOPAPPUS L'Her.

Perennial or biennial herbs with angled erect stems, alternate, once or twice parted or entire leaves, and corymbose or solitary, pedunculate heads of yellow or whitish flowers; rays wanting except in one species; involucres campanulate, many-flowered, of 6 to 12 appressed bracts with scarious tips; achenes obpyramidal, 4 or 5-angled, the faces 1 to 3-nerved; pappus of 10 to 20 hyaline obtuse scales.

KEY TO THE SPECIES.

Rays present.. 1. *H. radiatus.*
Rays wanting.
 Basal and lower cauline leaves entire.......................... 2. *H. integer.*
 Basal and cauline leaves, at least most of them, pinnate or pinnatifid.
  Divisions of the leaves broadly linear to oblong or lanceolate.
   Corollas whitish; stems nearly naked................. 3. *H. mexicanus.*
   Corollas bright yellow; stems very leafy............. 4. *H. flavescens.*
  Divisions of the leaves linear or filiform.
   Stems very leafy throughout.
    Plants permanently and densely tomentose; lobes of the corolla nearly equaling the throat... 5. *H. robustus.*
    Plants glabrate in age, thinly tomentose when young; lobes of the corolla much shorter than the throat.......................... 6. *H. tenuifolius.*
   Stems scapose or the leaves much reduced and few.
    Heads 12 to 15 mm. in diameter; stems nearly naked, bearing only 1 or 2 much reduced leaves; stems densely arachnoid, tall...... 7. *H. nudatus.*

Heads mostly less than 10 mm. in diameter; stems with more numerous, less reduced leaves; stems densely or sparsely pubescent, low.
Stems densely tomentose; pappus evident, nearly equaling the corolla lobes...... 8. *H. arenosus.*
Stems sparingly tomentose; pappus hidden by the hairs of the achenes.......... 9. *H. filifolius.*

1. **Hymenopappus radiatus** Rose, Contr. U. S. Nat. Herb. **1:** 122. 1891.

TYPE LOCALITY: Willow Springs, Arizona.

RANGE: Mountains of southern Arizona and New Mexico.

NEW MEXICO: Black Range; White and Sacramento mountains. Transition Zone.

This plant is difficult to distinguish from *Leucampyx newberryi*, the two being very similar in general appearance. Leucampyx is usually a little taller and more branched. The two may be definitely distinguished by the presence of chaff on the disk in Leucampyx; in Hymenopappus there is none.

2. **Hymenopappus integer** Greene, Pittonia **3:** 249. 189

TYPE LOCALITY: Mogollon Mountains, New Mexico. Type collected by Rusby (no. 180).

RANGE: Mountains of southwestern New Mexico.

NEW MEXICO: Mogollon Mountains; G O S Ranch.

3. **Hymenopappus mexicanus** A. Gray, Proc. Amer. Acad. **19:** 29. 1883.

TYPE LOCALITY: In the higher mountains near San Luis Potosí, Mexico.

RANGE: Mountains of southern New Mexico and northeastern Mexico.

NEW MEXICO: Mogollon Mountains; Santa Rita; Pinos Altos Mountains. Transition Zone.

4. **Hymenopappus flavescens** A. Gray, Mem. Amer. Acad. n. ser. **4:** 97. 1849.

Hymenoppapus fisheri Woot. & Standl. Contr. U. S. Nat. Herb. **16:** 191. 1913.

TYPE LOCALITY: Between San Miguel and Las Vegas, New Mexico. Type collected by Fendler (no. 464).

RANGE: Western Texas to New Mexico.

NEW MEXICO: Near Las Vegas; Vara Nisa. Plains, Upper Sonoran Zone.

5. **Hymenopappus robustus** Greene, Bull. Torrey Club **9:** 63. 1882.

TYPE LOCALITY: New Mexico.

RANGE: Western Texas to Arizona.

NEW MEXICO: Willard; Alamocitas Canyon; Aden; Silver City; Mesilla Valley; Organ Mountains; Roswell. Sandy plains and valleys, in the Lower and Upper Sonoran zones.

6. **Hymenopappus tenuifolius** Pursh, Fl. Amer. Sept. 742. 1814.

TYPE LOCALITY: "In Upper Louisiana."

RANGE: Nebraska and Arkansas to Texas and New Mexico.

NEW MEXICO: Raton; Sierra Grande; San Lorenzo; Patterson; Hope. Plains, in the Upper Sonoran Zone.

7. **Hymenopappus nudatus** Woot. & Standl. Contr. U. S. Nat. Herb. **16:** 191. 1913.

TYPE LOCALITY: Burro Mountains, Grant County, New Mexico. Type collected by Metcalfe (no. 107).

RANGE: Western New Mexico.

NEW MEXICO: Burro Mountains; west of Patterson; Cactus Flat; Santa Rita; Silver City. Plains and low hills, in the Upper Sonoran Zone.

8. **Hymenopappus arenosus** Heller, Bull. Torrey Club **25**: 200. 1898.

TYPE LOCALITY: Near Espanola, Santa Fe County, New Mexico. Type collected by Heller (no. 3542).

RANGE: Colorado and Utah to New Mexico.

NEW MEXICO: Carrizo Mountains; Tunitcha Mountains; Farmington; Dulce; Santa Fe; Espanola; Ojo Caliente; Chama River; Albuquerque; White Sands. Dry plains and low hills, in the Lower and Upper Sonoran zones.

9. **Hymenopappus filifolius** Hook. Fl. Bor. Amer. **1**: 317. 1833.

TYPE LOCALITY: "On the undulating arid grounds of the Columbia, near Walla Walla, and on the banks of the Spokane and Flat-head Rivers."

RANGE: British America to Nebraska, New Mexico, and Oregon.

NEW MEXICO: Glorieta; Santa Fe; Round Mountain. Plains and dry hills, in the Upper Sonoran Zone.

100. OTHAKE Raf.

Scabrous, more or less viscid herbs with lanceolate to linear, entire, petiolate, alternate leaves and loosely cymose pedunculate heads of rose purple flowers; involucre turbinate, of linear-lanceolate bracts with scarious tips, these in 2 series; rays palmately 3-cleft, fertile; stamens exserted; achenes linear to clavate, quadrangular, minutely pubescent; pappus of 6 to 12 hyaline scales.

KEY TO THE SPECIES.

Heads homogamous; achenes broadened upward.................. 1. *O. texanum.*
Heads heterogamous; achenes not broadened upward............. 2. *O. sphacelatum.*

1. **Othake texanum** (DC.) Bush, Trans. Acad. St. Louis **14**: 176. 1904.

Palafoxia texana DC. Prodr. **5**: 125. 1836.

Polypteris texana A. Gray, Proc. Amer. Acad. **19**: 30. 1883.

TYPE LOCALITY: Texas.

RANGE: Western Texas and eastern New Mexico.

We have seen no New Mexican specimens of this, but it is said by Doctor Gray to have been collected at Los Moros by Bigelow in 1853.

2. **Othake sphacelatum** (Nutt.) Rydb. Bull. Torrey Club **37**: 331. 1910.

Stevia sphacelata Nutt.; Torr. Ann. Lyc. N. Y. **2**: 214. 1828.

Polypteris hookeriana A. Gray, Proc. Amer. Acad. **19**: 30. 1883.

TYPE LOCALITY: Not stated.

RANGE: Nebraska and Colorado to New Mexico and Texas.

NEW MEXICO: Albuquerque; Jemez; Mesilla Valley; Tortugas Mountain; south of Melrose; Carlsbad; Roswell; Nara Visa. Plains and mesas, in the Lower and Upper Sonoran zones.

101. HYMENOTHRIX A. Gray.

Slender, nearly glabrous annuals, having once to thrice parted leaves with linear or filiform divisions, and numerous small corymbose heads with yellowish, whitish, or purplish flowers; involucre turbinate-campanulate, about 30-flowered, the bracts 7 to 10, obovate to oblong, with scarious tips; ray flowers 6 to 10; achenes 4 or 5-angled, obpyramidal; pappus about as long as the achene, of 12 to 20 lanceolate hyaline scales.

KEY TO THE SPECIES.

Bracts of the involucre glabrous, purplish; disk corollas white or purplish.. 1. *H. wrightii.*
Bracts pubescent, yellowish; disk corollas yellow.................... 2. *H. wislizeni.*

1. **Hymenothrix wrightii** A. Gray, Pl. Wright. **2:** 97. 1853.

TYPE LOCALITY: "On hills between the Barbocomori and Santa Cruz, and on the side of the Chiricahui Mountains, Sonora."

RANGE: Southern New Mexico and Arizona and southward.

NEW MEXICO: Hillsboro; Mogollon Creek; Dog Spring; Organ Mountains. Upper Sonoran and Transition zones.

2. **Hymenothrix wislizeni** A. Gray, Mem. Amer. Acad. n. ser. **4:** 102. 1849.

TYPE LOCALITY: "Grassy places, Ojo de Gallejo, between El Paso del Norte and Chihuahua."

RANGE: Southern New Mexico and Arizona and southward.

NEW MEXICO: Mangas Springs; Dog Spring; Organ Mountains; mesa west of Organ Mountains. Low hills and mesas, in the Lower and Upper Sonoran zones.

102. **LEUCAMPYX** A. Gray. WILD COSMOS.

Perennial herb, in general appearance like Hymenopappus; involucre hemispheric, the bracts in 2 or 3 series, imbricate, broadly scarious at the apex; rays large, white; achenes cuneate, compressed, triquetrous; pappus none.

1. **Leucampyx newberryi** A. Gray in Port. & Coult. Syn. Fl. Colo. 77. 1874.

TYPE LOCALITY: "New Mexico."

RANGE: Colorado to Arizona and New Mexico.

NEW MEXICO: Rio Pueblo; Santa Fe and Las Vegas mountains. Open parks and meadows in the mountains, in the Transition and Canadian zones.

On Crews Mesa near Beulah Professor Cockerell found a form with pink rays.

103. **CHAENACTIS** DC.

Low annuals or perennials with alternate, pinnately dissected leaves and pedunculate, solitary or cymose heads of flesh-colored flowers; receptacle flat, naked; heads rayless but the marginal flowers usually enlarged; achenes slender, pubescent; pappus of 4 lanceolate hyaline scales.

KEY TO THE SPECIES.

Perennial; divisions of the leaves pinnatifid or toothed............. 1. *C. douglasii.*
Annual; divisions of the leaves mostly entire........................ 2. *C. stevioides.*

1. **Chaenactis douglasii** (Hook.) Hook. & Arn. Bot. Beechey Voy. 354. 1840.

Hymenopappus douglasii Hook. Fl. Bor. Amer. **1:** 316. 1830.

TYPE LOCALITY: "Common on the barren dry sandy grounds of the Columbia, from the 'Great Falls' to the Rocky Mountains."

RANGE: Washington and Montana to California and New Mexico.

NEW MEXICO: Dulce (*Standley* 8204). Sandy slopes, in the Transition Zone.

2. **Chaenactis stevioides** Hook. & Arn. Bot. Beechey Voy. 353. 1840.

TYPE LOCALITY: "Snake country," Idaho.

RANGE: Idaho and Nevada to New Mexico and Arizona.

NEW MEXICO: Aztec; Mangas Springs; Tortugas Mountain; Organ Mountains. Dry plains and hills, in the Lower and Upper Sonoran zones.

104. **BAERIA** Fisch. & Meyer.

Low slender annual with opposite entire sessile leaves and slender-pedunculate terminal heads of yellow flowers; rays yellow, showy; involucre campanulate, of many narrow bracts, these somewhat carinate, at least below; achenes clavate-linear to linear-cuneate; pappus of 3 or 4 awn-bearing paleæ.

1. **Baeria gracilis** (DC.) A. Gray, Proc. Amer. Acad. **9**: 196. 1874.
Burrielia gracilis DC. Prodr. **5**: 664. 1836.
TYPE LOCALITY: "In Nova-California."
RANGE: California to western New Mexico.
NEW MEXICO: Duck Creek (*Greene*). Dry plains and low hills.

105. PICRADENIOPSIS Rydb.

Perennial herbs with opposite, cleft or dissected leaves and pedunculate heads of yellow flowers; involucre campanulate, of obovate to oblanceolate, carinate, distinct bracts; achenes narrow, quadrangular; pappus of 4 to 8 obovate or spatulate paleæ with scarious tips, the nerves extending to the apex of the paleæ.

KEY TO THE SPECIES.

Pappus scales acute; leaves green.............................. 1. *P. woodhousei.*
Pappus scales obtuse; leaves canescent.......................... 2. *P. oppositifolia.*

1. **Picradeniopsis woodhousei** (A. Gray) Rydb. Bull. Torrey Club **37**: 333. 1910.
Achyropappus woodhousei A. Gray, Proc. Amer. Acad. **6**: 546. 1865.
Bahia woodhousei A. Gray, Proc. Amer. Acad. **19**: 28. 1883.
TYPE LOCALITY: "New Mexico." Type collected by Woodhouse.
RANGE: Plains of northern New Mexico and Arizona.
NEW MEXICO: Hurrah Creek; mesa near Atarque de Garcia; Zuni; Nara Visa. Upper Sonoran Zone.

2. **Picradeniopsis oppositifolia** (Nutt.) Rydb. in Britton, Man. 1008. 1901.
Trichophyllum oppositifolium Nutt. Gen. Pl. **2**: 167. 1818.
Bahia oppositifolia Nutt.; Torr. & Gray, Fl. N. Amer. **2**: 376. 1842.
TYPE LOCALITY: "On denudated sterile hills, near Fort Mandan," North Dakota.
RANGE: Montana and North Dakota to Arizona and western Texas.
NEW MEXICO: Santa Fe; Sierra Grande; Las Vegas; Kennedy; Albuquerque; Clayton; Johnsons Basin; Gray; Taos. Plains and low hills, in the Upper Sonoran Zone.

105a. BAHIA Lag.

Perennial herb with opposite impressed-punctate 3-cleft leaves, or the lower leaves sometimes entire; flowers yellow, the heads pedunculate; involucre campanulate, the bracts distinct, not carinate; achenes pubescent, the pappus of 4 to 8 obovate paleæ, the nerves not extending to the apex of the paleæ.

1. **Bahia dealbata** A. Gray, Mem. Amer. Acad. n. ser. **4**: 99. 1849.
Bahia absinthifolia dealbata A. Gray, Pl. Wright. **1**: 121. 1852.
TYPE LOCALITY: "Valley between Mapimi and Guajuquilla, and at Cadenas, Chihuahua."
RANGE: Western Texas to southern Arizona and southward.
NEW MEXICO: Socorro; Grant County; Dona Ana Mountains; mesa west of Organ Mountains; James Canyon. Sandy plains and low hills, Lower Sonoran Zone.

106. ACHYROPAPPUS H. B. K.

Slender annual with opposite dissected impressed-punctate leaves and small pedunculate heads; involucre hemispheric, of about 10 distinct oblanceolate bracts; rays none; achenes hirsute at the slender base; pappus of short obovate paleæ with scarious tips and somewhat thickened bases.

1. Achyropappus neomexicanus A. Gray; Rydb. Colo. Agr. Exp. Sta. Bull. **100:** 377. 1906.

Schkuhria neomexicana A. Gray, Mem. Amer. Acad. n. ser. **4:** 96. 1849.

Bahia neomexicana A. Gray, Proc. Amer. Acad. **19:** 27. 1883.

TYPE LOCALITY: Margin of fields, Santa Fe, New Mexico. Type collected by Fendler.

RANGE: Colorado to New Mexico and Arizona.

NEW MEXICO: Pecos; Las Vegas; Santa Fe; Canyon Largo; Cubero; Chama River· Willard; Hanover Mountain; Mogollon Mountains; White Mountains. Sandy soil, in the Upper Sonoran Zone.

In Rydberg's Flora of Colorado[1] the combination under Achyropappus is credited to Dr. Gray, but this must be a slip of the pen, as we do not find that Dr. Gray ever assigned the species to this genus.

107. PLATYSCHKUHRIA (A. Gray) Rydb.

Perennial herb, 20 cm. high or less, with a leafy stem bearing few long-pedunculate heads of yellow flowers; leaves alternate, oblanceolate or oblong, nearly glabrous, thick, entire; rays bright yellow, showy; achenes sparingly pubescent; pappus of about 10 linear-lanceolate scales with excurrent costæ.

1. Platyschkuhria oblongifolia (A. Gray) Rydb. Bull. Torrey Club **33:** 155. 1906.

Schkuhria integrifolia oblongifolia A. Gray, Amer. Nat. **8:** 213. 1874.

Bahia oblongifolia A. Gray, Proc. Amer. Acad. **19:** 27. 1883.

TYPE LOCALITY: "San Juan," Utah or New Mexico.

RANGE: Southern Utah and Colorado to northeastern Arizona and northwestern New Mexico.

NEW MEXICO: Carrizo Mountains; Farmington. Dry hills and plains, in the Upper Sonoran Zone.

108 VILLANOVA Benth. & Hook.

Stout, somewhat viscid, strong-scented annuals or biennials with alternate, twice or thrice ternately parted leaves and small, loosely cymose-paniculate heads; involucre hemispheric, yellow-flowered; rays oblong to obovate, yellow; achenes tetragonal-clavate, the paleæ of the pappus oblong to narrowly lanceolate, or wanting.

KEY TO THE SPECIES.

Pappus present.. 1. *V. biternata.*
Pappus wanting.. 2. *V. dissecta.*

1. Villanova biternata (A. Gray) Woot. & Standl.

Bahia biternata A. Gray, Pl. Wright. **2:** 95. 1853.

TYPE LOCALITY: Gravelly hills near Ojo de Gavilan, New Mexico. Type collected by Wright (no. 1256).

RANGE: Western Texas to southern New Mexico.

NEW MEXICO: Tucumcari; Albuquerque; mesa west of Organ Mountains. Dry plains and hills, in the Lower Sonoran Zone.

2. Villanova dissecta (A. Gray) Rydb. Bull. Torrey Club **37:** 333. 1910.

Amauria ? dissecta A. Gray, Mem. Amer. Acad. n. ser. **4:** 104. 1849.

Villanova chrysanthemoides A. Gray, Pl. Wright. **2:** 96. 1853.

Bahia chrysanthemoides A. Gray, Proc. Amer. Acad. **19:** 28. 1883.

Bahia dissecta Britton, Trans. N. Y. Acad. **8:** 68. 1889.

TYPE LOCALITY: A few miles east of the Mora River, New Mexico. Type collected by Fendler (no. 537).

---

[1] Colo. Agr. Exp. Sta. Bull. **100:** 377. 1906.

RANGE: Wyoming and Colorado to New Mexico and Arizona.

NEW MEXICO: Hurrah Creek; Dulce; Tunitcha Mountains; Pajarito Park; Santa Fe and Las Vegas mountains; Raton; Fort Bayard; Hop Canyon; West Fork of the Gila; Kingston; Animas Valley; Organ Mountains; White Mountains; Capitan Mountains. Open hills and meadows, in the Transition Zone.

109. HELENIUM L. SNEEZEWEED.

Erect, corymbosely branched, perennial herb, with alternate decurrent leaves and winged stems; involucre of 1 or 2 series of linear spreading bracts; receptacle subglobose, naked; rays yellow, drooping; achenes turbinate, ribbed; pappus of 5 to 8 ovate, often lacerate or toothed scales.

1. **Helenium montanum** Nutt. Trans. Amer. Phil. Soc. n. ser. **7**: 384. 1841.

Helenium autumnale pubescens Britton, Mem. Torrey Club **5**: 339. 1894.

TYPE LOCALITY: "In the Rocky Mountain range, on the borders of Lewis' River, etc."

RANGE: Washington and Saskatchewan to New Mexico and Mississippi.

NEW MEXICO: Taos; Rociada; Pecos; Tularosa Creek; Roswell. Wet meadows, in the Upper Sonoran Zone.

This species is very near *Helenium autumnale*, and perhaps hardly separable from it.

110. DUGALDEA Cass. OWL'S CLAWS.

Tall stout perennial herb, with mostly basal, oblanceolate, impressed-punctate leaves and several large long-pedunculate heads of bright yellow flowers; involucral bracts in 2 series, numerous, finally reflexed; flowers numerous; achenes villous, the paleæ of the pappus hyaline, lanceolate, with long-attenuate tips.

1. **Dugaldea hoopesii** (A. Gray) Rydb. Mem. N. Y. Bot. Gard. **1**: 425. 1900.

Helenium hoopesii A. Gray, Proc. Acad. Phila. **1863**: 65. 1864.

TYPE LOCALITY: "South Park and west of Pike's Peak," Colorado.

RANGE: Montana and California to New Mexico and Arizona.

NEW MEXICO: Tunitcha Mountains; Chama; Santa Fe and Las Vegas mountains; Baldy; Mogollon Mountains; Iron Creek; White and Sacramento mountains. Meadows in the mountains, in the Transition and Canadian zones.

The common name which we have applied to this very abundant and showy Rocky Mountain plant is a translation of the one used by the Navahos. It refers to the appearance of the involucral bracts, especially in age.

111. FLAVERIA Juss.

Annuals with opposite, sessile or petiolate, entire or dentate leaves and dense cymes of small heads of yellow flowers; involucral bracts 2 to 8, subequal; receptacle naked or setose; rays usually one to each head; achenes linear-oblong, glabrous, 8 to 10-ribbed; pappus none.

KEY TO THE SPECIES.

Heads 10 to 15-flowered; leaves perfoliate, entire .... 1. *F. chloraefolia.*
Heads 2 to 8-flowered; leaves not perfoliate, dentate.
  Receptacle setose; leaves lanceolate .... 2. *F. repanda.*
  Receptacle not setose; leaves linear-lanceolate .... 3. *F. campestris.*

1. **Flaveria chloraefolia** A. Gray, Mem. Amer. Acad. n. ser. **4**: 88. 1849.

TYPE LOCALITY: "Pelayo, northwest of Mapimi, in the State of Chihuahua."

RANGE: Western Texas to southeastern New Mexico and northeastern Mexico.

NEW MEXICO: Roswell (*Cockerell*).

2. **Flaveria repanda** Lag. Gen. & Sp. Nov. 33. 1816.

TYPE LOCALITY: "Nova Hispania."

RANGE: Texas and New Mexico, south to tropical America.

NEW MEXICO: Pajarito Park; Las Vegas; Mesilla Valley; above Tularosa; south of Roswell; Lake Arthur; Carlsbad; Guadalupe Mountains. Valleys, especially in cultivated ground, in the Lower and Upper Sonoran zones.

A common weed in cultivated fields of the Rio Grande and Pecos valleys.

3. **Flaveria campestris** J. R. Johnston, Proc. Amer. Acad. **39**: 287. 1903.

TYPE LOCALITY: Courtney, Missouri. (No type is designated, but the first specimen cited is from this locality.)

RANGE: Missouri and Kansas to Colorado and eastern New Mexico.

NEW MEXICO: Roswell (*Cockerell*).

112. SARTWELLIA A. Gray.

Branched annual, 30 to 60 cm. high, with entire opposite linear leaves and numerous small heads of yellow flowers in corymbiform cymes; bracts 5, oval or oblong; rays mostly entire, obovate; achenes cylindric, striate; pappus a cup with fimbriolate edge.

1. **Sartwellia flaveriae** A. Gray, Pl. Wright. **1**: 122. *pl. 6.* 1852.

TYPE LOCALITY: "Prairies of the Rio Seco, Texas, and mountain valleys and plains of the Pecos, and base of the Guadalupe Mountains."

RANGE: Western Texas to southeastern New Mexico.

NEW MEXICO: White Sands; White Mountains; Roswell; Lake Arthur; Buchanan; mesa west of Organ Mountains. Plains and rocky hills, in the Upper Sonoran Zone.

113. MACDOUGALIA Heller.

Loosely woolly, cespitose perennial with mostly basal linear leaves and scapelike monocephalous stems; involucre hemispheric, the bracts lanceolate, acute, coriaceous, about 12 in each series, distinct, those of the inner series slightly longer and scarious-margined; rays about 12, yellow; paleæ of the pappus about 10, subulate-lanceolate, with evident costæ, attenuate into a bristle-like cusp.

1. **Macdougalia bigelovii** (A. Gray) Heller, Bull. Torrey Club **25**: 629. 1898.

Actinella bigelovii A. Gray, Pl. Wright. **2**: 96. 1853.

TYPE LOCALITY: Mountains near the Copper Mines, New Mexico. Type collected by Bigelow.

RANGE: Mountains of New Mexico and Arizona.

NEW MEXICO: Santa Rita; Agua Fria.

114. HYMENOXYS Cass. COLORADO RUBBER PLANT.

Herbaceous perennials or annuals with gland-dotted alternate divided leaves and few or numerous showy pedunculate heads; involucre of 2 series of bracts, the outer ones thick, united at the base, the inner ones thinner and broader, often with fimbriate margins; ray flowers pistillate or the heads homogamous; rays pale to bright yellow; achenes turbinate, pubescent; pappus of 5 to 12 conspicuous hyaline pointed paleæ.

KEY TO THE SPECIES.

Annuals with often spreading branches; inner bracts not very different from the outer; disk corollas distinctly expanded at the mouth.

Plants tall, 30 to 50 cm. high; heads about 8 mm. in diameter.. 1. *H. cockerellii.*

Plants low, 20 cm. high or less, more branched; heads less than 8 mm. in diameter.

Rays pale yellow; heads about 7 mm. broad; plants dull green, pubescent, usually densely branched........ 2. *H. mearnsii.*

Rays bright yellow; heads 5 or 6 mm. wide; plants bright green, glabrate, sparingly branched................. 3. *H. multiflora.*

Perennials with erect stems and branches; inner bracts usually different from the outer, fimbriate; disk corollas not expanded at the mouth.

Peduncles and other bracts not cinereous or tomentose, glabrous or nearly so; plants 50 cm. high or more.

Leaves entire or with few divisions, the segments linear-oblong or broadly linear; rays mostly equaling the involucre.. 4. *H. rusbyi.*

Leaves all much divided, with narrowly linear or filiform divisions; rays much shorter than the involucre.... 5. *H. brachyactis.*

Peduncles and outer bracts cinereous or tomentose; plants usually less than 35 cm. high.

Basal leaves all entire.................................. 6. *H. olivacea.*

Basal leaves all or mostly divided.

Inner involucral bracts truncate or obtuse; plants loosely branched almost throughout; corollas of the disk flowers long-hairy at the base....... 7. *H. vaseyi.*

Inner involucral bracts acute or acutish, never truncate; plants not loosely branched, usually closely corymbose above; disk corollas not long-hairy.

Paleæ of the pappus abruptly acuminate, with aristiform tips; plants bright green, usually tall; heads large, 7 mm. in diameter....... 8. *H. metcalfei.*

Paleæ of pappus long-attenuate; plants various; heads large or small.

Heads large, 8 mm. high, 8 or 9 mm. in diameter, few; pappus longer than the achenes, nearly equaling the disk corollas.................................. 9. *H. macrantha.*

Heads smaller, 5 or 6 mm. high, numerous; pappus shorter than the achenes, much shorter than the disk corollas.........10. *H. floribunda.*

1. **Hymenoxys cockerellii** Woot. & Standl. Contr. U. S. Nat. Herb. **16:** 192. 1913.

Hymenoxys chrysanthemoides juxta Cockerell, Bull. Torrey Club **31:** 503. 1904.

Type locality: Mangas Springs, New Mexico. Type collected by Metcalfe (no. 118).

Range: Western Texas and southern New Mexico to northern Mexico.

New Mexico: Mangas Springs; Berendo Creek; Rincon; Mesilla Valley. River valleys and low plains, in the Lower Sonoran Zone.

2. **Hymenoxys mearnsii** (Cockerell) Woot. & Standl. Contr. U. S. Nat. Herb. **16:** 192. 1913.

Hymenoxys chrysanthemoides Cockerell, Bull. Torrey Club **31:** 506. 1904.

Type locality: Dog Spring, New Mexico. Type collected by E. A. Mearns (no. 142).

Range: Southern New Mexico and Arizona.

New Mexico: Lake Valley; Mangas Springs; Dog Spring; San Luis Pass; Cliff; Silver City; Mesilla Valley; Roswell. Plains and dry hills, in the Lower Sonoran Zone.

3. **Hymenoxys multiflora** (Buckl.) Rydb. Bull. Torrey Club **33**: 157. 1906.
Phileozera multiflora Buckl. Proc. Acad. Phila. **1861**: 459. 1862.
Picradenia multiflora Greene, Pittonia **3**: 273. 1898.
TYPE LOCALITY: "Prairies north of Fort Belknap," Texas.
RANGE: Oklahoma and Kansas to Texas and eastern New Mexico.
NEW MEXICO: Clayton; Tucumcari. Plains, in the Upper Sonoran Zone.

4. **Hymenoxys rusbyi** (A. Gray) Cockerell, Bull. Torrey Club **31**: 496. 1904.
Actinella rusbyi A. Gray, Proc. Amer. Acad. **19**: 33. 1883.
Picradenia rusbyi Greene, Pittonia **3**: 271. 1898.
TYPE LOCALITY: Grassy slopes of the Mogollon Mountains, New Mexico. Type collected by Rusby (no. 246½).
RANGE: Mountains of southwestern New Mexico and southeastern Arizona.
NEW MEXICO: Mogollon Mountains; north of Santa Rita. Transition Zone.

5. **Hymenoxys brachyactis** Woot. & Standl. Contr. U. S. Nat. Herb. **16**: 192. 1913.
TYPE LOCALITY: Near East View, New Mexico. Type collected by Wooton, August 4, 1906.
RANGE: Known only from type locality.

6. **Hymenoxys olivacea** Cockerell, Bull. Torrey Club **31**: 297. 1904.
TYPE LOCALITY: Hanover Hills, New Mexico. Type collected by Miss A. I. Mulford (no. 807).
RANGE: Southwestern New Mexico.
NEW MEXICO: Hanover Hills; Santa Rita. Low hills and plains.

7. **Hymenoxys vaseyi** (A. Gray) Cockerell, Bull. Torrey Club **31**: 493. 1904.
Actinella vaseyi A. Gray, Proc. Amer. Acad. **17**: 219. 1882.
Picradenia vaseyi Greene, Pittonia **3**: 272. 1898.
TYPE LOCALITY: Organ Mountains, New Mexico. Type collected by G. R. Vasey in 1881.
RANGE: Known only from type locality, in the Upper Sonoran Zone.

8. **Hymenoxys metcalfei** Cockerell, Bull. Torrey Club **31**: 492. 1904.
TYPE LOCALITY: Burro Mountains, New Mexico. Type collected by Metcalfe (no. 179).
RANGE: Mountains and high plains of western New Mexico.
NEW MEXICO: South of Gallup; Burro Mountains; Kingston; G O S Ranch. Upper Sonoran and Transition zones.

9. **Hymenoxys macrantha** (A. Nels.) Rydb. Bull. Torrey Club **33**: 156. 1906.
Picradenia macrantha A. Nels. Bot. Gaz. **28**: 130. 1899.
Hymenoxys richardsoni macrantha Cockerell, Bull. Torrey Club **31**: 475. 1904.
TYPE LOCALITY: Fort Steele, Wyoming.
RANGE: Wyoming to northern New Mexico.
NEW MEXICO: Sandia Mountains (*Wooton*). Transition Zone.

10. **Hymenoxys floribunda** (A. Gray) Cockerell, Bull. Torrey Club **31**: 485. 1904.
Actinella richardsoni floribunda A. Gray, Mem. Amer. Acad. n. ser. **4**: 101. 1849.
Picradenia floribunda Greene, Pittonia **3**: 272. 1898.
TYPE LOCALITY: "Rocky hills, as well as plains and creek bottoms, around Santa Fe," New Mexico. Type collected by Fendler (no. 460).
RANGE: Colorado to New Mexico and Arizona.
NEW MEXICO: Carrizo Mountains; Santa Fe; Chama; Tunitcha Mountains; Glorieta; north of Ramah; between Barranca and Embudo; Holy Ghost Creek; Raton; Cebolla Spring; Defiance; Galisteo; hills west of Magdalena; mountains southeast of Patterson; San Augustine Plains; near Apache Spring. Plains and lower mountains, in the Upper Sonoran and Transition zones.

Attempts have been made to obtain rubber from this and allied species, in southern Colorado. It is possible to secure rubber from this source, but the quantity of the plant available does not seem sufficient to warrant an elaborate equipment for its extraction.

115. RYDBERGIA Greene.

Stout, low, sparingly branched, woolly, alpine perennial with pinnately parted, mostly basal leaves and large heads with spreading, bright yellow rays; bracts of the involucre all alike, distinct, herbaceous, in several series, woolly; receptacle large, hemispheric; rays 15 to 30, linear-cuneiform; paleæ of the pappus 5 or 6, elongated-lanceolate, attenuate to a subulate point.

1. **Rydbergia brandegei** (Porter) Rydb. Bull. Torrey Club **33:** 156. 1906.
Actinella grandiflora glabrata Porter in Port. & Coult. Syn. Fl. Colo. 76. 1874.
Actinella brandegei Porter; A. Gray, Proc. Amer. Acad. **13:** 373. 1878.
TYPE LOCALITY: Sangre de Cristo Pass, Colorado.
RANGE: Southern Colorado to New Mexico.
NEW MEXICO: Taos Mountains; Baldy; Santa Fe and Las Vegas mountains; White Mountain Peak. High mountain peaks, in the Arctic-Alpine Zone.

116. TETRANEURIS Greene.

Usually scapose, annual or perennial herbs with mostly basal, entire, often punctate leaves and long-pedunculate heads; disk flowers and rays bright yellow; involucre hemispheric, of 2 or 3 series of similar appressed imbricated bracts; receptacle convex or conic, naked; achenes turbinate, 5 to 10-ribbed or angled, variously pubescent; pappus of 5 to 12 scarious, aristate, truncate or acuminate paleæ.

KEY TO THE SPECIES.

Annual; stems much branched .... 1. *T. linearifolia.*
Perennials; stems simple or nearly so.
  Flowering stems bearing 2 to 4 leaves, sometimes with 1 or 2 branches.
    Paleæ of pappus long-aristate.
      Leaves and stem densely sericeous; heads 12 mm. in diameter or less .... 2. *T. argentea.*
      Leaves and stems glabrate; heads about 15 mm. broad .... 3. *T. ivesiana.*
    Paleæ of pappus broader, abruptly acuminate, never aristate-tipped.
      Involucres 7 mm. high; rays 15 mm. long .... 4. *T. formosa.*
      Involucres 5 mm. high; rays 10 mm. long or less .... 5. *T. leptoclada.*
  Flowering stems naked, simple.
    Plants less than 10 cm. high; heads little or not at all exceeding the leaves.
      Leaves 5 cm. long or less, nearly linear, 3-nerved at the base .... 6. *T. trinervata.*
      Leaves less than 3 cm. long, oblanceolate, not 3-nerved at the base.
        Leaves loosely villous; heads 12 to 13 mm. broad; rays equaling the involucre .... 7. *T. depressa.*
        Leaves densely sericeous; heads less than 10 mm. broad; rays shorter than the involucre .... 8. *T. pygmaea.*
    Plants more than 10 cm. high; heads much exceeding the leaves.
      Leaves densely sericeous .... 9. *T. acaulis.*

Leaves loosely villous or glabrate.
Stems and involucres copiously lanate or villous..10. *T. lanata.*
Stems and usually the involucres appressed-pubescent.
Branches of the caudex abundantly villous, the hairs at first white, later brown...11. *T. crandallii.*
Branches of the caudex little or not at all villous.
Leaf bases several times broader than the linear blades..................12. *T. linearis.*
Leaf bases not wider than the usually oblanceolate, at least linear-oblanceolate, blades.
Leaves oblanceolate, thick, dull green; bracts slightly shorter than the disk................13. *T. glabriuscula.*
Leaves linear-oblanceolate, thin, bright green; bracts one-fourth shorter than the disk..14. *T. angustifolia.*

1. Tetraneuris linearifolia (Hook.) Greene, Pittonia **3:** 269. 1897.
Hymenoxys linearifolia Hook. Icon. Pl. **2:** *pl. 146.* 1838.
Actinella linearifolia Torr. & Gray, Fl. N. Amer. **2:** 383. 1842.
TYPE LOCALITY: San Felipe, Texas.
RANGE: Louisiana to Texas and southeastern New Mexico.
NEW MEXICO: Artesia (*Wooton*). Plains and low hills, in the Upper Sonoran Zone.

2. Tetraneuris argentea (A. Gray) Greene, Pittonia **3:** 269. 1897.
Actinella argentea A. Gray, Mem. Amer. Acad. n. ser. **4:** 100. 1849.
TYPE LOCALITY: Hills around Santa Fe, New Mexico. Type collected by Fendler (no. 457).
RANGE: New Mexico.
NEW MEXICO: Sandia Mountains; Hop Canyon; Datil; Horse Spring; Santa Fe; Las Vegas. Low plains and hills, in the Upper Sonoran Zone.

3. Tetraneuris ivesiana Greene, Pittonia **3:** 269. 1897.
TYPE LOCALITY: On the Rio Zuni, New Mexico. Type collected by Woodhouse in 1851.
RANGE: Northwestern New Mexico and northeastern Arizona.
NEW MEXICO: Carrizo Mountains; Cedar Hill; Dulce; Gallup; Fort Wingate; Aztec; Tunitcha Mountains. Plains and dry hills, in the Upper Sonoran Zone.

4. Tetraneuris formosa Greene, Contr. U. S. Nat. Herb. **16:** 192. 1913.
TYPE LOCALITY: Dry hills near Kingston, Sierra County, New Mexico. Type collected by Metcalfe (no. 1235).
RANGE: Mountains of western New Mexico.
NEW MEXICO: Kingston; Magdalena Mountains.

5. Tetraneuris leptoclada (A. Gray) Greene, Pittonia **3:** 269. 1897.
Actinella leptoclada A. Gray, U. S. Rep. Expl. Miss. Pacif. **4:** 107. 1856.
TYPE LOCALITY: Mountains and rocky places near Santa Antonita, New Mexico. Type collected by Bigelow.
RANGE: Southern Colorado to northern New Mexico and Arizona.
NEW MEXICO: Santa Antonita; Stanley; Holy Ghost Creek; Pecos; mountains near Grants Station. Dry plains and hills, in the Upper Sonoran Zone.

6. Tetraneuris trinervata Greene, Pittonia **3:** 267. 1897.

TYPE LOCALITY: Sandia Mountains, New Mexico. Type collected by Bigelow.

RANGE: Southern Colorado and northern New Mexico.

NEW MEXICO: Sandia Mountains (*Bigelow*).

7. Tetraneuris depressa (Torr. & Gray) Greene, Pittonia **3:** 266. 1897.

Actinella depressa Torr. & Gray, Mem. Amer. Acad. n. ser. **4:** 100. 1849.

TYPE LOCALITY: "Rocky Mountains, apparently at a great elevation, the locality unknown."

RANGE: Mountains of Utah and Colorado to northern New Mexico.

NEW MEXICO: Taos Mountains (*Bailey* 854–857). Arctic-Alpine Zone.

8. Tetraneuris pygmaea (Torr. & Gray) Woot. & Standl. Contr. U. S. Nat. Herb. **16:** 193. 1913.

Actinella depressa pygmaea Torr. & Gray, Mem. Amer. Acad. n. ser. **4:** 100. 1849.

TYPE LOCALITY: Raton Mountains, New Mexico. Type collected by A. Gordon in 1848.

RANGE: Mountains of northern New Mexico.

NEW MEXICO: Sandia Mountains (*Wooton*).

It is by no means certain that this specimen really is the plant described by Torrey and Gray as *pygmaea*, but it is closer to that than to any published species.

9. Tetraneuris acaulis (Pursh) Greene, Pittonia **3:** 265. 1898.

Gaillardia acaulis Pursh, Fl. Amer. Sept. 743. 1814.

Actinella acaulis Nutt. Gen. Pl. **2:** 173. 1818.

TYPE LOCALITY: "In Upper Louisiana."

RANGE: Idaho and Nebraska to Colorado and New Mexico.

NEW MEXICO: Raton; Sierra Grande; Nara Visa; Las Vegas; Gray. Plains and hills, in the Upper Sonoran Zone.

10. Tetraneuris lanata (Nutt.) Greene, Pittonia **3:** 265. 1898.

Actinella lanata Nutt. Trans. Amer. Phil. Soc. n. ser. **7:** 379. 1841.

TYPE LOCALITY: "On the lofty hills or mountains called the "Three Butes" of the upper Platte, on shelving rocks."

RANGE: Wyoming to northern New Mexico.

NEW MEXICO: Sierra Grande (*Standley* 6145). High mountains, Canadian to Arctic-Alpine Zone.

11. Tetraneuris crandallii Rydb. Bull. Torrey Club **32:** 127. 1905.

TYPE LOCALITY: Grand Junction, Colorado.

RANGE: Southwestern Colorado and northwestern New Mexico.

NEW MEXICO: Carrizo Mountains (*Matthews*). Dry hills, in the Upper Sonoran Zone.

12. Tetraneuris linearis (Nutt.) Greene, Pittonia **3:** 267. 1897.

Actinella scaposa linearis Nutt. Trans. Amer. Phil. Soc. n. ser. **7:** 379. 1841.

Actinella linearis A. Nels. in Coulter, New Man. Rocky Mount. 560. 1909.

TYPE LOCALITY: Texas.

RANGE: New Mexico and Texas.

NEW MEXICO: Cuervo; Las Vegas; above Tularosa; Organ Mountains; Queen. Plains and hills, in the Upper Sonoran Zone.

13. Tetraneuris glabriuscula Rydb. Bull. Torrey Club **33:** 155. 1906.

Tetraneuris glabra Greene, Pittonia **3:** 268. 1897, not *Actinella glabra* Nutt. 1841.

TYPE LOCALITY: "Rocky Mountain plains."

RANGE: Nevada and Colorado to New Mexico.

NEW MEXICO: Highest point of the Llano Estacado; Springer; Nara Visa. Plains and low hills, in the Upper Sonoran Zone.

14. Tetraneuris angustifolia Rydb. Bull. Torrey Club **32:** 128. 1905.

Type locality: White Mountains, New Mexico. Type collected by Wooton (no. 374).

Range: Colorado to New Mexico and Texas.

New Mexico: White Mountains. Hills and mountains, in the Upper Sonoran and Transition zones.

117. ACHILLEA L. Yarrow. Sneezeweed.

Perennial herbs with erect leafy stems, finely dissected, alternate, strong-scented leaves, and small heads in corymbs at the ends of the branches; involucres campanulate, of appressed imbricated bracts, the outer ones shorter; achenes oblong or obovoid, slightly compressed; pappus none.

KEY TO THE SPECIES.

Leaves glabrate, green; heads rather few, loosely corymbose, 6 to 7 mm. high, long-pedunculate.................................. 1. *A. laxiflora.*
Leaves densely arachnoid; heads numerous, in a dense corymb, about 4 mm. high, short-pedunculate.
  Bracts with pale brownish margins; plants 25 to 80 cm. high.. 2. *A. lanulosa.*
  Bracts with dark brown, nearly black margins; plants less than 20 cm. high....................................... 3. *A. subalpina.*

1. Achillea laxiflora Pollard & Cockerell, Proc. Biol. Soc. Washington **15:** 179. 1902.

Type locality: Sandia Mountains, New Mexico. Type collected by Miss Charlotte C. Ellis in 1900.

Range: Mountains of northern New Mexico.

New Mexico: Sandia Mountains; Carrizo Mountains (*Standley* 7379). Transition Zone.

The type of the species is a mere fragment, consisting of the terminal portion of a branch with two small leaves. The plant from the Carrizos seems to be the same, although the inflorescence is composed of more numerous heads in a denser corymb. The latter plant grew in oak thickets among pine trees. It is from 80 to 100 cm. high, with numerous large, broad, bright green leaves, the upper ones but little reduced. The leaf segments are widely spaced on the rachis, not crowded as in *A. lanulosa*, and they are much larger than in that species.

2. Achillea lanulosa Nutt. Journ. Acad. Phila. **7:** 36. 1834.

Type locality: Rocky Mountains.

Range: British Columbia and South Dakota to Kansas and Texas and southward.

New Mexico: Common in all the higher mountains. Meadows, in the Upper Sonoran and Transition zones.

3. Achillea subalpina Greene, Leaflets **1:** 145. 1905.

Achillea lanulosa alpicola Rydb. Mem. N. Y. Bot. Gard. **1:** 426. 1900.

Achillea alpicola Rydb. Bull. Torrey Club **33:** 157. 1906.

Type locality: Subalpine slopes of Mount Ouray, southern Colorado.

Range: Montana to northern New Mexico.

New Mexico: Truchas Peak; Jemez Mountains; Pecos Baldy. High mountain meadows, in the Arctic-Alpine Zone.

118. ANTHEMIS L.

Glabrous or pubescent, branched, ill-scented annual with dissected leaves; heads pedunculate, the receptacle convex, bearing filiform bracts; rays white; achenes 10-ribbed, rugose; pappus none.

1. **Anthemis cotula** L. Sp. Pl. 894. 1753. MAYWEED.

Maruta cotula DC. Prodr. **6**: 13. 1837.

TYPE LOCALITY: "Habitat in Europae ruderatis praecipue in Ucrania."

RANGE: Native of Europe, widely naturalized in North America, Asia, Africa, and Australia.

NEW MEXICO: Balsam Park, Sandia Mountains (*Ellis* 333).

119. CHRYSANTHEMUM L. OX-EYE DAISY.

Perennial herb with alternate, dentate or incised leaves and pedunculate heads of yellow flowers with white rays; involucre hemispheric, the oblong-lanceolate bracts appressed-imbricated in several series; achenes angled or terete, 5 to 10-ribbed; pappus none.

1. **Chrysanthemum leucanthemum pinnatifidum** Lec. & Lam. Cat. Pl. France 227. 1847.

Chrysanthemum leucanthemum subpinnatifidum Fernald, Rhodora **5**: 181. 1903.

TYPE LOCALITY: "Mont Dore; pâturages et pentes herbeuses de Chaudefour, bords du chemin de Sancy à Vassiviére," France.

NEW MEXICO: Near Pecos (*Cockerell*).

A native of Europe, widely introduced into North America. In the East it is a troublesome weed, but it is still very rare in most parts of the West.

120. TANACETUM L. TANSY.

Erect, strongly scented, perennial herb with alternate, pinnately divided leaves and numerous discoid corymbose heads; involucres hemispheric, the oblong-lanceolate bracts appressed in several series; achenes 5-angled, truncate; pappus a short crown.

1. **Tanacetum vulgare** L. Sp. Pl. 844. 1753.

TYPE LOCALITY: "Habitat in Europae aggeribus."

NEW MEXICO: Farmington; Cedar Hill; Aztec.

The plant is well established at these places. It is common in cultivation and has become naturalized in many parts of the United States.

121. PICROTHAMNUS Nutt.

Low shrub, 50 cm. high or less, with numerous spreading spiny branches; leaves small, pedately 5-parted, the divisions 3-lobed; heads globose, racemosely glomerate, on short branches; involucral bracts 5 or 6, broadly obovate; female flowers 1 to 4, sterile ones 4 to 8; achenes and flowers densely covered with long cobwebby hairs.

1. **Picrothamnus desertorum** Nutt. Trans. Amer. Phil. Soc. n. ser. **7**: 417. 1841.

Artemisia spinescens D. C. Eaton in King, Geol. Expl. 40th Par. **5**: 180. *pl. 19*. 1871.

TYPE LOCALITY: "Rocky Mountain plains in arid deserts, toward the north sources of the Platte."

RANGE: Idaho and Montana to California and northwestern New Mexico.

NEW MEXICO: Carrizo Mountains; Farmington; Aztec. Dry hills, in the Upper Sonoran Zone.

122 ARTEMISIA L. SAGEBRUSH. WORMWOOD.

Bitter aromatic herbs or shrubs with alternate, entire to pinnatifid leaves and small rayless heads of inconspicuous, yellow, whitish, or brownish flowers in panicles or rarely in simple racemes; heads few to many-flowered, the flowers homogamous or heterogamous; bracts imbricated in few series; anthers commonly tipped with subulate-acuminate, erect appendages; achenes mostly with a small epigynous disk and no pappus.

KEY TO THE SPECIES.

Annuals or biennials.
  Divisions of the leaves lanceolate or linear, incised or pinnatifid .... 1. *A. biennis.*
  Divisions of the leaves linear-filiform, entire .... 2. *A. caudata.*
Perennials.
  Heads homogamous, i. e., all the flowers fertile.
    All the leaves entire, linear to narrowly oblong .... 3. *A. cana.*
    Most, or at least part, of the leaves 3-toothed at the apex.
      Bracts of the involucre nearly glabrous, resinous; inflorescence lax, racemose .... 4. *A. nova.*
      Bracts tomentose; inflorescence not racemose.
        Heads campanulate, recurved .... 5. *A. petrophila.*
        Heads turbinate, erect.
          Tall shrub; inflorescence much branched. 6. *A. tridentata.*
          Low shrub, 40 cm. high or less; inflorescence rather simple and spikelike.. 7. *A. arbuscula.*
  Heads heterogamous, i. e., the flowers unlike, the marginal ones pistillate, the central ones perfect though sometimes sterile by the abortion of the pistil.
    Central flowers sterile, the style mostly entire and the ovary abortive.
      Plants shrubby, white-tomentose .... 8. *A. filifolia.*
      Plants herbaceous except at the very base, greenish, though sometimes pubescent.
        Upper leaves all linear, only the basal ones pinnatifid; plants pubescent or glabrous.
          Plants glabrous throughout .... 9. *A. dracunculoides.*
          Plants sparingly villous on stems and leaves .... 10. *A. dracunculina.*
        All the leaves pinnatifid; plants somewhat pubescent.
          Stems tall, 50 to 100 cm. high, very leafy, in age nearly glabrous; divisions of the leaves linear-filiform; pubescence scanty .... 11. *A. scouleriana.*
          Stems lower, 20 to 60 cm. high; divisions of the basal leaves linear-oblanceolate, more or less canescently villous .... 12. *A. forwoodii.*
    Central flowers fertile; style 2-cleft.
      Receptacle densely woolly with long hairs.
        Heads large, 6 to 12 mm. in diameter; bracts with dark brown or blackish margins; plants low, 10 to 20 cm. high; spikes scapose; of alpine meadows .... 13. *A. scopulorum.*
        Heads small, 4 to 5 mm. in diameter; bracts light brown; plants taller, 30 to 60 cm. high; inflorescence paniculate; of the low plains .... 14. *A. frigida.*
      Receptacle not woolly, mostly smooth.
        Shrub, 30 cm. high or less; leaves cuneate, tridentate, silky-villous .... 15. *A. bigelovii.*

Herbs, at most woody only at the base; leaves various, more or less tomentose.
Involucres glabrous, at least in age; lower leaves bipinnate .... 16. *A. franserioides.*
Involucres densely tomentose, at least when young; lower leaves toothed or pinnatifid.
Lower leaves merely coarsely toothed or lobed, the teeth or lobes short, the upper leaves entire; plants permanently white-tomentose throughout.
Leaves glabrous or glabrate on the upper surface .... 17. *A. silvicola.*
Leaves permanently tomentose on the upper surface.
Panicles strict; plants stout, strict; heads 3 to 4 mm. wide ...... 18. *A. rhizomata.*
Panicles lax, spreading; plants slender, much branched above; heads about 2 mm. wide .... 19. *A. albula.*
At least the lower leaves pinnatifid with long, filiform to lanceolate segments; pubescence various.
Divisions of the leaves filiform; variously pubescent.
Leaves permanently white-tomentose .... 20. *A. kansana.*
Leaves glabrate above in age .... 21. *A. wrightii.*
Divisions of the leaves broader, oblong-linear to lanceolate, never filiform, glabrate above.
All leaves divided, the segments narrowly oblong-linear ...... 22. *A. redolens.*
Upper leaves lanceolate, undivided, the segments of the lower leaves oblong-lanceolate .... 23. *A. mexicana.*

1. **Artemisia biennis** Willd. Sp. Pl. **3:** 1842. 1804.

Type locality: "Nova Zelandia?"

Range: British America to California, New Mexico, and Pennsylvania.

New Mexico: Shiprock; Dulce; Castle Rock. Wet ground, in the Upper Sonoran and Transition zones.

2. **Artemisia caudata** Michx. Fl. Bor. Amer. **2:** 129. 1803.

Type locality: "Hab. ad ripas sabulosas fluminis Missouri."

Range: Manitoba and Vermont to New Mexico, Texas, and the Atlantic coast.

New Mexico: Nara Visa (*Fisher* 174). Upper Sonoran Zone.

3. **Artemisia cana** Pursh, Fl. Amer. Sept. 521. 1814.

Type locality: "On the Missouri."

Range: Rocky Mountain Region, south to northern New Mexico.

New Mexico: Dulce (*Standley* 8082). Plains and valleys, in the Upper Sonoran and Transition zones.

4. **Artemisia nova** A. Nels. Bull. Torrey Club **27**: 274. 1900.

TYPE LOCALITY: Medicine Bow, Wyoming.

RANGE: Wyoming to northern New Mexico.

NEW MEXICO: Carrizo Mountains; Chusca Mountains; Dulce. Dry hills, in the Upper Sonoran and lower part of the Transition zones.

5. **Artemisia petrophila** Woot. & Standl. Contr. U. S. Nat. Herb. **16**: 193. 1913.

TYPE LOCALITY: On a dry sandstone mesa at the north end of the Carrizo Mountains, northeast corner of Arizona. Type collected by Standley (no. 7355).

RANGE: Northwestern New Mexico and northeastern Arizona.

NEW MEXICO: Carrizo Mountains; Farmington. Dry rocky hills, in the Upper Sonoran Zone.

6. **Artemisia tridentata** Nutt. Trans. Amer. Phil. Soc. n. ser. **7**: 398. 1841.

TYPE LOCALITY: "Plains of the Oregon and Lewis River."

RANGE: British Columbia and Nebraska to California and New Mexico.

NEW MEXICO: Carrizo Mountains; Tunitcha Mountains; Farmington; Nutritas Creek; Dulce; Embudo; Canjilon; Pajarito Park. Plains, in the Upper Sonoran Zone.

This is one of the most characteristic plants of the Upper Sonoran Zone in northern and northwestern New Mexico. In places it covers large areas of the plains to the exclusion of almost everything else. It is usually a shrub about 1 meter high, but where it receives an abundance of water it is often larger.

7. **Artemisia arbuscula** Nutt. Trans. Amer. Phil. Soc. n. ser. **7**: 398. 1841.

TYPE LOCALITY: "On the arid plains of Upper California, on Lewis River."

RANGE: Northern New Mexico and Colorado, west to California.

NEW MEXICO: Chusca Mountains; near Black Rock. Dry plains and low hills, in the Upper Sonoran Zone.

8. **Artemisia filifolia** Torr. Ann. Lyc. N. Y. **2**: 211. 1828.

TYPE LOCALITY: Not stated. Plant collected somewhere on the plains east of the Rocky Mountains, probably in Colorado or New Mexico, by Doctor James.

RANGE: Wyoming to western Texas, Arizona, and northern Mexico.

NEW MEXICO: Farmington; Laguna; Black Rock; Noria; Magdalena; Burro Mountains; Lake Valley; Deming; Mesilla Valley; Jarilla; Orogrande; Roswell. Dry plains and sandhills, in the Lower and Upper Sonoran zones.

9. **Artemisia dracunculoides** Pursh, Fl. Amer. Sept. 521. 1814.

Artemisia cernua Nutt. Gen. Pl. **2**: 143. 1818.

TYPE LOCALITY: "On the Missouri." Exact locality, near the mouth of White River, Lyman County, South Dakota.

RANGE: Washington and Montana to California and Texas.

NEW MEXICO: Tunitcha Mountains; Farmington; Chama; Pajarito Park; Espanola; Hebron; Winsors Ranch; Magdalena; mountains southeast of Patterson; Santa Rita; West Fork of the Gila; Hillsboro; White Mountains; Organ Mountains. Hillsides and meadows, in the Upper Sonoran and Transition zones.

10. **Artemisia dracunculina** S. Wats. Proc. Amer. Acad. **23**: 279. 1888.

TYPE LOCALITY: "At the base of cliffs in the Sierra Madre, Chihuahua."

RANGE: Mountains of southern New Mexico to Chihuahua.

NEW MEXICO: Hillsboro; Organ Mountains; Cox Canyon. Upper Sonoran and Transition zones.

11. **Artemisia scouleriana** (Besser) Rydb. Bull. Torrey Club **33**: 157. 1906.

Artemisia desertorum scouleriana Besser; Hook. Fl. Bor. Amer. **1**: 325. 1833.

TYPE LOCALITY: "North-West coast of America, Fort Vancouver, and Straits of de Fuca."

RANGE: British Columbia to northern New Mexico.

NEW MEXICO: Dulce; Ponchuelo Creek; Pecos. Valleys in the mountains, in the Transition Zone.

12. Artemisia forwoodii S. Wats. Proc. Amer. Acad. **25**: 133. 1890.

TYPE LOCALITY: Black Hills, South Dakota.

RANGE: Montana and Saskatchewan to New Mexico.

NEW MEXICO: Cedar Hill; Tunitcha Mountains; Luna; Fort Tularosa; mountains southeast of Patterson; Santa Rita; Mogollon Mountains; Capitan Mountains. Plains and hills, in the Upper Sonoran and Transition zones.

13. Artemisia scopulorum A. Gray, Proc. Acad. Phila. **1863**: 66. 1864.

TYPE LOCALITY: Middle Park, Colorado.

RANGE: Wyoming to Utah and northern New Mexico.

NEW MEXICO: Truchas Peak; Pecos Baldy; Baldy. High mountain meadows, in the Arctic-Alpine Zone.

14. Artemisia frigida Willd. Sp. Pl. **3**: 1838. 1804. ESTAFIATA.

TYPE LOCALITY: "Dauuriae."

RANGE: Washington and Nevada to Minnesota and Texas; also in the Old World.

NEW MEXICO: Tunitcha Mountains; Farmington; Cedar Hill; Dulce; Zuni; Sierra Grande; Pecos; Cebolla Spring; Ramah; San Augustine Plains; Gallinas Mountains; Sandia Mountains; Torrance; Pajarito Park; Hebron; Raton. Dry hills and plains, chiefly in the Upper Sonoran Zone.

The name "estafiata" is most commonly applied to this species, but it is sometimes extended to include other members of the genus.

15. Artemisia bigelovii A. Gray, U. S. Rep. Expl. Miss. Pacif. **4**: 110. 1856.

TYPE LOCALITY: "Rocks and cañons on the Upper Canadian and Llano Estacado."

RANGE: Colorado to New Mexico and Arizona.

NEW MEXICO: South of Atarque de Garcia; Albuquerque; Cerrillos; Carrizozo; White Mountains; Suwanee; Guadalupe; Torrance; Nara Visa. Plains, in the Upper Sonoran Zone.

16. Artemisia franserioides Greene, Bull. Torrey Club **10**: 42. 1883.

TYPE LOCALITY: Near the summit of the Pinos Altos Mountains, New Mexico. Type collected by Greene.

RANGE: Mountains of Colorado, New Mexico, and Arizona.

NEW MEXICO: Santa Fe and Las Vegas mountains; Black Range; Mogollon Mountains; White and Sacramento Mountains; Organ Peak. Shaded slopes, chiefly in the Canadian Zone.

17. Artemisia silvicola Osterhout, Bull. Torrey Club **28**: 645. 1901.

Artemisia mexicana silvicola A. Nels. in Coulter, New Man. Rocky Mount. 569. 1909.

TYPE LOCALITY: Along MacIntyre Creek, Larimer County, Colorado.

RANGE: Colorado to New Mexico and Arizona.

NEW MEXICO: Tunitcha Mountains; Carrizo Mountains. Meadows and canyons, in the Transition Zone.

18. Artemisia rhizomata A. Nels. Bull. Torrey Club **27**: 34. 1900.

TYPE LOCALITY: Sweetwater, Wyoming.

RANGE: Wyoming and North Dakota to Colorado and New Mexico.

NEW MEXICO: Farmington; Cedar Hill; Dulce; Chama; Las Vegas Hot Springs; Albuquerque; near Condes Camp; Gray. Prairies and river banks, in the Upper Sonoran and lower part of the Transition zones.

19. Artemisia albula Wooton, Contr. U. S. Nat. Herb. **16**: 193. 1913.

Artemisia microcephala Wooton, Bull. Torrey Club **25**: 455. 1898, not Hillebr. 1888.

TYPE LOCALITY: Organ Mountains, New Mexico. Type collected by Wooton (no. 504).

RANGE: Western Texas to southern Arizona, southward into Mexico.

NEW MEXICO: Laguna; Albuquerque; Bear Mountains; Hillsboro Mountains; Mangas Springs; Lordsburg; Dog Spring; Dona Ana Mountains; Tortugas Mountain; Organ Mountains; Eagle Creek. Dry hills, in the Lower and Upper Sonoran zones.

20. Artemisia kansana Britton in Britt. & Brown, Illustr. Fl. **3:** 466. 1898.

TYPE LOCALITY: Plains, Lane County, Kansas.

RANGE: Kansas and Colorado to New Mexico and western Texas.

NEW MEXICO: Clayton; Cross L Ranch; Folsom; Cimarron; Cebolla Spring; Bernal; Gray; Nara Visa. Dry plains, in the Upper Sonoran Zone.

21. Artemisia wrightii A. Gray, Proc. Amer. Acad. **19:** 48. 1883.

TYPE LOCALITY: Mountains around the Copper Mines, New Mexico. Type collected by Wright (no. 1279).

RANGE: Colorado to New Mexico.

NEW MEXICO: Pajarito Park; Santa Fe and Las Vegas mountains; Chama; Grants; Tunitcha Mountains; Zuni; Mogollon Mountains; Santa Rita; Bear Mountains; White Mountains; Gray. Canyons and meadows, in the Upper Sonoran and Transition zones.

22. Artemisia redolens A. Gray, Proc. Amer. Acad. **21:** 393. 1886.

TYPE LOCALITY: "Chihuahua, on cool slopes under cliffs."

RANGE: New Mexico to Chihuahua.

NEW MEXICO: Santa Fe and Las Vegas mountains; Baldy; Mogollon Mountains; Hillsboro Peak; San Luis Mountains; Cloverdale; Organ Mountains; White and Sacramento mountains. Mountain meadows, in the Upper Sonoran and Transition zones.

This plant has been referred to *Artemisia mexicana* and to *A. discolor*, from both of which it is amply distinct. Doctor Gray's remark that it has the appearance of *A. dracunculoides* was certainly unfortunate, if Pringle's 296 in the U. S. National Herbarium is to be taken as typical of it. Our plant is much more closely related to *A. wrightii* and to *A. underwoodii* Rydb. It does not belong in the same group with *A. dracunculoides*. Metcalfe's 1248 was distributed under a manuscript name of Doctor Greene's, but so far as we can learn the name has never been published.

23. Artemisia mexicana Willd.; Spreng. Syst. Veg. **3:** 490. 1825.

TYPE LOCALITY: Mexico.

RANGE: New Mexico and Arizona to western Texas and southward.

NEW MEXICO: Carrizo Mountains; Farmington; Pecos; Winsors Ranch; West Fork of the Gila; White and Sacramento mountains. Plains and hills, in the Upper Sonoran and Transition zones.

123. TETRADYMIA DC.

Low shrubs with alternate entire tomentose leaves and narrow heads of yellow flowers; involucre cylindric, of 4 to 6 imbricated bracts; rays none; receptacle flat; achenes terete, 5-nerved; pappus of numerous soft capillary bristles.

KEY TO THE SPECIES.

Leaves oblong or elliptic, mostly 1 cm. long or less; involucres 6 mm. long or smaller .... 1. *T. inermis*.
Leaves filiform, 2 to 4 cm. long; involucres 8 to 10 mm. long .... 2. *T. filifolia*.

1. Tetradymia inermis Nutt. Trans. Amer. Phil. Soc. n. ser. **7:** 415. 1841.

Tetradymia canescens inermis A. Gray in Brewer & Wats. Bot. Calif. **1:** 408. 1876.

TYPE LOCALITY: "On the dry barren plains of the Rocky Mountains; common, particularly near Lewis' River of the Shoshonee."

RANGE: Montana to Nevada and New Mexico.

NEW MEXICO: Tunitcha Mountains; Farmington; Chupadero; Gallup; Atarque de Garcia; Moreno Valley; San Augustine Plains. Dry hills and plains, in the Upper Sonoran Zone.

A low, densely branched shrub, 50 cm. high or less, rarely larger. The small yellow heads are very numerous, making it a handsome plant when in full flower.

2. **Tetradymia filifolia** Greene, Bull. Torrey Club **25**: 123. 1898.

TYPE LOCALITY: Round Mountain, White Mountain Range, New Mexico. Type collected by Wooton (no. 183).

RANGE: Known only from type locality, in the Upper Sonoran Zone.

124. BEBBIA Greene.

Much branched plant about 1 meter high, shrubby at the base, with slender rushlike erect branches and few alternate linear leaves; heads pedunculate, 20 to 30-flowered; involucre campanulate, the bracts imbricated in 2 or 3 series, oblong, appressed; achenes turbinate, hirsute, faintly 5-nerved; pappus of 15 to 20 plumose bristles in a single series.

1. **Bebbia juncea** (Benth.) Greene, Bull. Calif. Acad. **1**: 179. 1885.

Carphephorus junceus Benth. Bot. Voy. Sulph. 21. 1844.

TYPE LOCALITY: Magdalena Bay, Lower California.

RANGE: Southern New Mexico to California and Mexico.

NEW MEXICO: Parkers Well (*Wooton*). Sandy plains, in the Lower Sonoran Zone.

125. ARNICA L.

Erect simple-stemmed perennials with opposite leaves and long-pedunculate heads of yellow flowers; rays yellow, showy; involucre campanulate, the narrow bracts in 1 or 2 series, nearly equal; receptacle flat; achenes linear, pubescent; pappus a single series of slender barbellate bristles.

KEY TO THE SPECIES.

Leaves cordate-ovate, the cauline ones petiolate.................... 1. *A. cordifolia.*
Leaves lanceolate, oblanceolate, or lance-oblong, the cauline ones sessile.................... 2. *A. foliosa.*

1. **Arnica cordifolia** Hook. Fl. Bor. Amer. **1**: 331. 1834.

TYPE LOCALITY: "Alpine woods of the Rocky Mountains on the east side."

RANGE: British America to California, Nevada, and northern New Mexico.

NEW MEXICO: Chama; Winsor Creek. Shaded hillsides, in the Canadian Zone.

2. **Arnica foliosa** Nutt. Trans. Amer. Phil. Soc. n. ser. **7**: 407. 1841.

TYPE LOCALITY: "On the alluvial flats of the Colorado of the West, particularly near Bear River of the lake Timpanagos."

RANGE: Montana to northern New Mexico.

NEW MEXICO: Tunitcha Mountains; Chama. Wet ground, especially meadows about lakes and streams, in the Transition Zone.

This is very common in the Tunitcha Mountains, where it covers large areas of wet meadow to the exclusion of almost everything else. About Chama only two or three plants were seen.

126. SENECIO L.

Tall or low, annual or perennial herbs with alternate, sessile or petiolate, entire to pinnatifid or subpinnate leaves and solitary, racemose, or corymbose heads of yellow flowers; heads radiate or discoid, many-flowered; involucre cylindric to campanulate, of few or many erect connivent bracts, sometimes with a few smaller ones at the base; receptacle flat, naked; pappus of numerous soft capillary bristles.

KEY TO THE SPECIES.

Annual.. 1. *S. vulgaris.*
Perennials.
 Heads discoid.
  Heads small, 7 to 10 mm. high.
   Glabrous; leaves linear to linear-oblanceolate, entire or nearly so.. 2. *S. pudicus.*
   Slightly tomentose; leaves cordate-ovate to oblong-lanceolate, coarsely toothed........................ 3. *S. sacramentanus.*
  Heads large, 12 to 20 mm. high.
   Leaves long-villous on the midrib beneath; heads usually racemose.. 4. *S. scopulinus.*
   Leaves not long-villous; heads not racemose.
    Plants more or less tomentose; leaves thick and fleshy. 5. *S. chloranthus.*
    Plants glabrous; leaves thin.
     Leaves coarsely and irregularly laciniate-serrate, abruptly contracted at the base, mostly petiolate.. 6. *S. bigelovii.*
     Leaves finely and evenly serrate, attenuate to the very short petiole, or sessile.................. 7. *S. rusbyi.*
 Heads radiate.
  Plants densely viscid.................................. 8. *S. parryi.*
  Plants not viscid.
   Leaves or their divisions narrowly linear or linear-filiform.
    Plants permanently tomentose........................ 9. *S. filifolius.*
    Plants glabrous.
     Leaves entire, rarely with a pair of short filiform lobes at the base..............................10. *S. spartioides.*
     Leaves, except the uppermost, pinnately divided.
      Heads campanulate; bracts 12 to 15...............11. *S. riddellii.*
      Heads cylindric or nearly so; bracts 8 to 10.......12. *S. multicapitatus.*
   Leaves neither linear-filiform nor with linear-filiform divisions.
    Heads nodding.
     Plants low, not more than 10 cm. high; leaves obovate-orbicular, purplish.........................13. *S. taraxacoides.*
     Plants taller, 50 cm. high; leaves lanceolate or oblanceolate, not purplish.......................14. *S. amplectens.*
    Heads not nodding.
     Heads solitary.
      Stems and leaves densely tomentose; stems with only a few much reduced leaves.............15. *S. actinella.*
      Stems and usually the leaves glabrate; stems leafy..16. *S. mogollonicus.*
     Heads not solitary, except in depauperate specimens.
      Plants equally leafy throughout.
       Leaves pinnatifid or pinnate.
        Leaves pinnate, the terminal lobe orbicular or semiorbicular, obtusely toothed........17. *S. sanguisorboides.*
        Leaves pinnatifid, all the segments lanceolate or narrower, with acute teeth.
         Involucre 7 mm. high; leaves shallowly pinnatifid, with broad segments.......18. *S. ambrosioides.*

Involucre 5 mm. high or less; leaves pinnatifid almost to the midrib, the segments very narrow.........................19. *S. macdougalii.*

Leaves merely toothed or entire.

Leaves, at least the lower, triangular or triangular-cordate; plants usually 1 meter high or taller.............................20. *S. triangularis.*

Leaves not triangular, narrowed at the base; plants lower, seldom more than 30 cm. high.

Leaves lanceolate or elliptic, acute, denticulate; stems simple..................22. *S. crassulus.*

Leaves obovate or oval, obtuse, coarsely salient-dentate; stems branched.......21. *S. carthamoides.*

Upper leaves of the stems much reduced.

Rootstock very short, never woody; leaves lanceolate, denticulate; stems stout, 40 cm. high or less................................23. *S. lapathifolius.*

Rootstocks elongated, horizontal or ascending, woody; leaves various; stems low or tall.

Leaves linear, less than 2 mm. wide; plants small, usually less than 15 cm. high......24. *S. thurberi.*

Leaves 5 mm. wide or more; plants mostly taller.

Plants glaucous..........................25. *S. microdontus.*

Plants not glaucous.

Plants tall, stout, 50 cm. high or more; heads very numerous; bracts with black tips..........................26. *S. atratus.*

Plants lower, slender, mostly less than 50 cm. high; heads few; bracts not with black tips.

Leaves and stems permanently floccose.

All leaves pinnatifid, with toothed segments.........................27. *S. fendleri.*

Leaves merely coarsely toothed, the segments entire..................28. *S. canovirens.*

Leaves and stems finally glabrate.

Basal leaves pinnatifid...............29. *S. multilobatus.*

Basal leaves not pinnatifid.

Leaves very thin, yellowish.

Basal leaves very deeply cordate, with a narrow sinus.........30. *S. cardamine.*

Basal leaves not deeply or not at all cordate.

Basal leaves cordate...........31. *S. pseudaureus.*

Basal leaves not cordate.

Basal leaves oblanceolate or elliptic-oblong, attenuate at the base, serrate..32. *S. quaerens.*

Basal leaves oblong or ovate, not attenuate at the base, crenate............33. *S. flavulus.*

Leaves thick and rather fleshy, not yellowish. (SEE CONTINUATION.)

(CONTINUATION.)

Plants comparatively stout, mostly more than 30 cm. high; heads 1 cm. broad..34. *S. neomexicanus.*
Plants slender, lower; heads less than 1 cm. broad.
Basal leaves obovate or broadly oblanceolate.
Bracts 5 to 6 mm. high; plants conspicuously tomentose..35. *S. pentodontus.*
Bracts 8 to 9 mm. high; plants almost entirely glabrous.36. *S. cymbalarioides.*
Basal leaves narrowly oblanceolate or oblong.
Upper cauline leaves pinnatifid or sharply serrate.....37. *S. mutabilis.*
Upper cauline leaves entire.
Basal leaves with numerous coarse, usually sharp teeth..38. *S. oblanceolatus.*
Basal leaves entire, or with a few obscure obtuse teeth.
Stems tall, more than 30 cm., bearing numerous mostly petiolate leaves............39. *S. cynthioides.*
Stems less than 20 cm. high, all the leaves much reduced, sessile.
Leaves mostly entire, sometimes with 2 or 3 inconspicuous teeth; heads 9 mm. high or less; achenes glabrous.40. *S. metcalfei.*
Leaves with several teeth at the apex; heads about 12 mm. high; achenes pubescent on the angles............41. *S. remifolius.*

1. Senecio vulgaris L. Sp. Pl. 867. 1753. COMMON GROUNDSEL.

TYPE LOCALITY: European.

NEW MEXICO: Santa Fe.

Introduced from Europe; a weed in gardens.

2. Senecio pudicus Greene, Pittonia **4:** 118. 1900.

Senecio cernuus A. Gray, Amer. Journ. Sci. **33:** 239. 1862, not L. f. 1781.

TYPE LOCALITY: Dry hillsides, and in the crevices of rocks, upper part of Clear Creek, Colorado.

RANGE: Mountains of Colorado and New Mexico.

NEW MEXICO: Fierro (*Holzinger*). Transition Zone.

Our one specimen comes from a region so far removed from the ordinary range of the species that one would scarcely expect it to be the same. It differs from the form common in Colorado only in having slightly smaller heads. The plant is not known from any of the intervening mountain ranges of New Mexico, this locality being in the southwestern part of the State.

3. Senecio sacramentanus Woot. & Standl. Contr. U. S. Nat. Herb. **16:** 194. 1913.

TYPE LOCALITY: Vicinity of Cloudcroft, near the summit of the Sacramento Mountains, New Mexico. Type collected by Wooton, August 15, 1901.

RANGE: Sacramento Mountains of New Mexico, in the Transition Zone.

4. Senecio scopulinus Greene, Pittonia **4:** 117. 1900.

Senecio bigelovii hallii A. Gray, Proc. Acad. Phila. **1863:** 67. 1864.

TYPE LOCALITY: "Common in the mountains of middle and southern Colorado."

RANGE: Wyoming to New Mexico.

NEW MEXICO: Baldy; Costilla Valley; Chama; Middle Fork of the Gila. Mountains, in the Transition Zone.

5. Senecio chloranthus Greene, Pittonia 4: 118. 1900.

TYPE LOCALITY: Mountains of southern Colorado, near Pagosa Peak.

RANGE: Colorado and northern New Mexico.

NEW MEXICO: Tunitcha Mountains; Rio Pueblo; Baldy; Santa Fe and Las Vegas mountains. Moist shaded slopes in the mountains, in the Canadian and Hudsonian zones.

6. Senecio bigelovii A. Gray, U. S. Rep. Expl. Miss. Pacif. **4**: 111. 1856.

TYPE LOCALITY: Mountain arroyos, near Camp Douglas, New Mexico. Type collected by Bigelow.

RANGE: Mountains of New Mexico.

NEW MEXICO: Sandia Mountains; White Mountains. Transition and Canadian zones.

Typically a tall plant, sometimes 1.5 meters high, with many ascending branches, but sometimes low and scarcely at all branched. Mr. Wooton collected the plant in its type locality in the summer of 1910. His specimens seem to have smaller heads than Bigelow's, but those of the latter collection owe their size to the fact that they are in fruit (they must have been collected about the first of October) and that they were pressed very flat. In some of Wooton's specimens the leaves are truncate at the base, but in most of them they are abruptly contracted.

7. Senecio rusbyi Greene, Bull. Torrey Club **9**: 64. 1882.

TYPE LOCALITY: Mogollon Mountains, New Mexico. Type collected by Rusby (no. 215).

RANGE: Mountains of southern New Mexico and Arizona.

NEW MEXICO: Hillsboro Peak; Mogollon Mountains. Transition and Canadian zones.

The species is certainly very close to *S. bigelovii*, and it is questionable whether it might not better be reduced to synonymy.

8. Senecio parryi A. Gray in Torr. U. S. & Mex. Bound. Bot. 103. 1859.

TYPE LOCALITY: "In live-oak groves, 150 miles above the mouth of the Pecos, on the Mexican side of the Rio Grande."

RANGE: Southwestern New Mexico to southern California and adjacent Mexico.

NEW MEXICO: Mogollon (*Wooton*). Mountains, in the Transition Zone.

A most distinct species, entirely different from all our others by its viscid pubescence.

9. Senecio filifolius Nutt. Trans. Amer. Phil. Soc. n. ser. **7**: 414. 1841.

TYPE LOCALITY: "The banks of the Missouri, toward the Rocky Mountains."

RANGE: Utah and Colorado to Arizona and western Texas, south into Mexico.

NEW MEXICO: Common throughout the State. Plains and low hills, in the Lower and Upper Sonoran, rarely in the Transition, zones.

No plant has a wider distribution in New Mexico than this. It occurs everywhere at low and middle elevations. It is in flower almost every month in the year in the southern part of the State, and is one of the earliest plants to bloom farther north. It is extremely variable in pubescence and in the form of the leaves, and probably when it is studied in a larger series of specimens it will be found to be an aggregate of species. Sometimes almost all the leaves are entire; again even the uppermost are pinnatifid. The plants of the higher elevations are more nearly glabrous than those of the mesas and foothills. On the sandy plains in the southern part of the State the plants are mostly woody at the base and even among the branches.

10. Senecio spartioides Torr. & Gray, Fl. N. Amer. **2**: 438. 1842.

TYPE LOCALITY: "Upper Platte; on a steep sand bank of the Sweet-water River."

RANGE: Nebraska and Wyoming to Arizona and western Texas.

NEW MEXICO: Tunitcha Mountains; Chama; Fruitland; Taos; Ruidoso Creek. Plains and low mountain valleys, in the Upper Sonoran and Transition zones.

11. Senecio riddellii Torr. & Gray, Fl. N. Amer. **2:** 444. 1842.

Type locality: Texas.

Range: Nebraska and Colorado to New Mexico and Texas.

New Mexico: Shiprock; Pecos; Lincoln. Plains and hills, in the Upper Sonoran Zone.

12. Senecio multicapitatus Greenm. Bull. Torrey Club **33:** 160. 1906.

Type locality: "Plains and in mountain valleys of Colorado, New Mexico, and Arizona."

Range: Colorado to Arizona and New Mexico.

New Mexico: Farmington; Dulce; Ramah; Santa Clara Canyon; Raton; Santa Fe; Albuquerque; Arch; Clayton; Folsom; Nara Visa; Mogollon Mountains; Mesilla Valley; Dona Ana Mountains; Capitan Mountains; White Oaks. Plains and open slopes, from the Lower Sonoran to the Transition Zone.

13. Senecio taraxacoides (A. Gray) Greene, Pittonia **4:** 119. 1900.

Senecio amplectens taraxacoides A. Gray, Proc. Acad. Phila. **1863:** 67. 1864.

Type locality: Rocky Mountains of Colorado.

Range: Mountain peaks of Colorado and northern New Mexico.

New Mexico: Truchas Peak; Baldy. Arctic-Alpine Zone.

14. Senecio amplectens A. Gray, Amer. Journ. Sci. II. **33:** 240. 1862.

Type locality: "In the mountains high up, at the foot of the snowy range," Colorado.

Range: Colorado and northern New Mexico.

New Mexico: Truchas Peak; Pecos Baldy; Baldy. High mountain peaks, in the Hudsonian and Arctic-Alpine zones.

Probably this is our handsomest species, by its large heads and the peculiar long, bright yellow rays. It grows only above or at the edge of the timber on the high mountains, often under the low shrubs.

15. Senecio actinella Greene, Bull. Torrey Club **10:** 87. 1883.

Type locality: Near Flagstaff, Arizona.

Range: Mountains of Arizona and New Mexico.

New Mexico: Mogollon Mountains between the Middle and West Forks of the Gila (*Wooton*).

16. Senecio mogollonicus Greene, Leaflets **1:** 212. 1906.

Type locality: Dry Flats on the West Fork of the Gila, Mogollon Mountains, New Mexico. Type collected by Metcalfe (no. 410).

Range: Known only from type locality.

17. Senecio sanguisorboides Rydb. Bull. Torrey Club **27:** 170. 1900.

Type locality: Santa Fe Canyon, New Mexico. Type collected by Heller (no. 3820).

Range: Mountains of New Mexico.

New Mexico: Santa Fe and Las Vegas mountains; Rio Pueblo; White and Sacramento mountains. Transition to Hudsonian Zone.

18. Senecio ambrosioides Rydb. Bull. Torrey Club **37:** 467. 1910.

Type locality: Green Mountain Falls, El Paso County, Colorado.

Range: Wyoming to New Mexico.

New Mexico: Tunitcha Mountains; Carrizo Mountains; Chama; Santa Fe Canyon; Pecos; Las Vegas; Sandia Mountains; Black Range. Mountains, in the Transition Zone.

19. Senecio macdougalii Heller, Bull. Torrey Club **26:** 592. 1899.

Type locality: Near Flagstaff, Arizona.

Range: Mountains of Arizona and New Mexico.

NEW MEXICO: Santa Fe and Las Vegas mountains; Jemez Mountains; G O S Ranch; White and Sacramento mountains. Transition Zone.

Our specimens differ from the type in having larger and narrower leaf segments. The material from Arizona is insufficient to determine whether the New Mexican plant is specifically distinct.

20. Senecio triangularis Hook. Fl. Bor. Amer. **1**: 332. *pl. 115.* 1833.

TYPE LOCALITY: "Moist prairies among the Rocky Mountains."

RANGE: British America to California and northern New Mexico.

NEW MEXICO: Chama; Santa Fe and Las Vegas mountains. Along streams and in bogs, in the Canadian and Hudsonian zones.

21. Senecio carthamoides Greene, Pittonia **4**: 122. 1900.

TYPE LOCALITY: Little Ouray Mountain, at Marshall Pass, Colorado.

RANGE: Mountains of Colorado and northern New Mexico.

NEW MEXICO: Upper Pecos River (*Maltby & Coghill* 198). Arctic-Alpine Zone.

22. Senecio crassulus A. Gray, Proc. Amer. Acad. **19**: 54. 1883.

TYPE LOCALITY: "Colorado Rocky Mountains."

RANGE: Idaho and Montana to Utah and northern New Mexico.

NEW MEXICO: Truchas Peak (*Standley* 4794). High mountains, in the Hudsonian and Arctic-Alpine zones.

23. Senecio lapathifolius Greene; Rydb. Colo. Agr. Exp. Sta. Bull. **100**: 394. 1906.

TYPE LOCALITY: Mountains of Colorado.

RANGE: Southern Colorado and northern New Mexico.

NEW MEXICO: Chama (*Standley* 6733). Wet meadows, in the Transition Zone.

24. Senecio thurberi A. Gray, Proc. Acad. Phila. **1863**: 68. 1864.

TYPE LOCALITY: Santa Rita, New Mexico.

RANGE: Mountains of New Mexico and Arizona.

NEW MEXICO: Santa Rita.

25. Senecio microdontus (A. Gray) Heller, Bull. Torrey Club **24**: 497. 1897.

Senecio toluccanus microdontus A. Gray, Syn. Fl. 1^2: 388. 1884.

Senecio wootonii Greene, Bull. Torrey Club **25**: 122. 1898.

Senecio anacletus Greene, Pittonia **4**: 307. 1901.

TYPE LOCALITY: Pinos Altos Mountains, New Mexico. Type collected by Greene.

RANGE: Colorado to Arizona and New Mexico.

NEW MEXICO: Tierra Amarilla; Santa Fe and Las Vegas mountains; White and Sacramento mountains. Mountains, in the Transition Zone.

The type of *S. wootonii* was collected in the White Mountains (*Wooton* 491).

26. Senecio atratus Greene; Rydb. Colo. Agr. Exp. Sta. Bull. **100**: 395. 1906.

TYPE LOCALITY: Mountains of Colorado.

RANGE: Mountains of Colorado and northern New Mexico.

NEW MEXICO: Baldy (*Wooton*).

27. Senecio fendleri A. Gray, Mem. Amer. Acad. n. ser. **4**: 108. 1849.

TYPE LOCALITY: "Foot of high mountains, along the creek 12 miles above Santa Fe," New Mexico. Type collected by Fendler (no. 478).

RANGE: Colorado and Utah to New Mexico.

NEW MEXICO: Santa Fe and Las Vegas mountains; Sandia Mountains; Jemez Mountains; Rio Pueblo; Baldy. Transition Zone.

28. Senecio canovirens Rydb. Bull. Torrey Club **27**: 187. 1900.

TYPE LOCALITY: White Mountains, New Mexico. Type collected by Wooton (no. 244).

RANGE: Mountains of Colorado and New Mexico.

NEW MEXICO: Black Range; White and Sacramento mountains; Capitan Mountains; Water Canyon; Mogollon Creek; Craters. Transition Zone.

29. Senecio multilobatus Torr. & Gray, Mem. Amer. Acad. n. ser. **4**: 109. 1849.

TYPE LOCALITY: "Abundant on the Uintah River, in the interior of California."

RANGE: Utah and Colorado to Arizona and western Texas.

NEW MEXICO: Carrizo Mountains; Cedar Hill; Farmington; north of Ramah. Plains and low hills, in the Upper Sonoran Zone.

30. Senecio cardamine Greene, Bull. Torrey Club **8**: 98. 1881.

TYPE LOCALITY: Cold northward slopes of the higher Mogollon Mountains, New Mexico. Type collected by Greene in 1881.

RANGE: Mogollon Mountains of southwestern New Mexico.

31. Senecio pseudaureus Rydb. Bull. Torrey Club **24**: 298. 1897.

TYPE LOCALITY: Little Belt Mountains, Montana.

RANGE: British America to Nevada and New Mexico.

NEW MEXICO: Along the Pecos 8 miles east of Glorieta; Chama. Damp meadows in the mountains, Transition Zone.

32. Senecio quaerens Greene, Leaflets **1**: 214. 1906.

Senecio prionophyllus Greene, Leaflets **1**: 212. 1906, not *S. prionophyllus* Greene, 1902.

TYPE LOCALITY: Moist flats on the West Fork of the Gila, Mogollon Mountains, New Mexico. Type collected by Metcalfe (no. 409).

RANGE: Mogollon Mountains of New Mexico, in the Transition Zone.

33. Senecio flavulus Greene, Pittonia **4**: 108. 1900.

Senecio flavovirens Rydb. Bull. Torrey Club **27**: 181. *pl. 5*. 1900.

TYPE LOCALITY: Arboles, southern Colorado.

RANGE: Idaho and Montana to Colorado and New Mexico.

NEW MEXICO: Winsors Ranch; Magdalena Mountains; Kingston; Tierra Amarilla. Wet meadows, in the Transition Zone.

34. Senecio neomexicanus A. Gray, Proc. Amer. Acad. **19**: 55. 1883.

TYPE LOCALITY: New Mexico.

RANGE: Mountains of New Mexico and Arizona.

NEW MEXICO: Burro Mountains; between Santa Fe and Canyoncito; Magdalena Mountains; Organ Mountains. Upper Sonoran Zone.

35. Senecio pentodontus Greene, Pl. Baker. **3**: 26. 1901.

TYPE LOCALITY: Open knolls below the limit of trees, near Carson, Colorado.

RANGE: Mountains of Colorado and New Mexico.

NEW MEXICO: Agua Fria; Tierra Amarilla; Tunitcha Mountains; Chama; Las Vegas; Black Range. Transition Zone.

36. Senecio cymbalarioides Nutt. Trans. Amer. Phil. Soc. n. ser. **7**: 412. 1841.

TYPE LOCALITY: "In Oregon."

RANGE: British America to Utah and New Mexico.

NEW MEXICO: Baldy; Costilla Valley. Meadows.

37. Senecio mutabilis Greene, Pittonia **4**: 113. 1900.

TYPE LOCALITY: Dry lowlands about Arboles and Los Pinos, southern Colorado.

RANGE: Mountains of southern Colorado and northern New Mexico.

NEW MEXICO: Near Tierra Amarilla (*Eggleston* 6546, 6602). Transition Zone.

38. Senecio oblanceolatus Rydb. Bull. Torrey Club **27**: 175. 1900.

TYPE LOCALITY: Como, South Park, Colorado.

RANGE: Wyoming to New Mexico and western Texas.

NEW MEXICO: Chama; Raton; Sierra Grande. Plains and low hills, in the Upper Sonoran and Transition zones.

39. Senecio cynthioides Greene, Leaflets **1:** 212. 1906.

Type locality: Hillsides along Turkey Creek, Mogollon Mountains, New Mexico. Type collected by Metcalfe (no. 574).

Range: Mountains of New Mexico.

New Mexico: Tunitcha Mountains; West Fork of the Gila.

40. Senecio metcalfei Greene, Contr. U. S. Nat. Herb. **16:** 193. 1913.

Type locality: Open slopes on Hillsboro Peak, at the south end of the Black Range, New Mexico. Type collected by Metcalfe (no. 938).

Range: Known only from type locality.

41. Senecio remifolius Woot. & Standl. Contr. U. S. Nat. Herb. **16:** 194. 1913.

Type locality: Willow Creek, New Mexico. Type collected by Wooton, August 8, 1900.

Range: Known only from type locality.

127. HAPLOESTHES A. Gray.

Glabrous perennial with the aspect of Sartwellia, the simple stems bearing a corymbosely branched inflorescence; leaves opposite, linear; heads small, campanulate, of yellow flowers; involucre of 4 or 5 oval, somewhat imbricated bracts; rays small, pale yellow; achenes linear, terete, striate, glabrous; pappus a single series of scabrous whitish bristles.

1. Haploesthes greggii A. Gray, Mem. Amer. Acad. n. ser. **4:** 109. 1849.

Type locality: Valley near Ciénaga Grande, Coahuila.

Range: Colorado and western Texas to New Mexico and southward.

New Mexico: Round Mountain (*Wooton*). Dry hills and plains in alkaline soil, in the Upper Sonoran Zone.

128. PSATHYROTES A. Gray

Low winter annuals with alternate, round-cordate or ovate, petiolate leaves and small heads of yellowish flowers; heads homogamous; involucre of rather numerous bracts in 2 series, the outer somewhat herbaceous; receptacle flat; achenes terete, obscurely striate, villous or hirsute; pappus of unequal bristles.

KEY TO THE SPECIES.

Plants scapose, erect; corollas nearly glabrous; leaves almost entire... 1. *P. scaposa.*
Plants much branched, spreading, leafy; corollas woolly at the top; leaves coarsely toothed.. 2. *P. annua.*

1. Psathyrotes scaposa A. Gray, Pl. Wright. **2:** 100. *pl. 13.* 1853.

Type locality: "Stony hills above El Paso."

Range: Western Texas and southern New Mexico.

We have seen no specimens of this from New Mexico, but the type came from very close to the southern border and perhaps from within our limits.

2. Psathyrotes annua (Nutt.) A. Gray, Pl. Wright. **2:** 100. 1853.

Bulbostylis annua Nutt. Journ. Acad. Phila. II. **1:** 179. 1848.

Type locality: "Rocky Mountains near Santa Fe," New Mexico. Type collected by Gambel.

Range: Plains, California and Nevada to Utah and New Mexico.

We have seen no specimens of this from within our borders. There may have been some mistake made in the citation of the source of the original specimens, but it is not improbable that the plant occurs in the northwest corner of New Mexico.

129. BARTLETTIA A. Gray.

Slender annual, nearly glabrous, with long-petiolate rounded repand-dentate leaves and long-pedunculate, rather large heads, these heterogamous, radiate; involucre campanulate, of 12 to 14 oblong-lanceolate bracts in 2 or 3 series; achenes cuneate-oblong, compressed; pappus of numerous unequal bristles in a single series.

1. **Bartlettia scaposa** A. Gray, Mem. Amer. Acad. n. ser. **5:** 323. 1854.
   Type locality: On a prairie near Corralitas, Chihuahua.
   Range: Southern New Mexico, south into Mexico.
   New Mexico: Lordsburg (*A. Davidson* 1422).

130. MESADENIA Raf.

Perennial herb; leaves thrice pinnatifid into narrow segments; heads 10 to 15 mm. high, numerous, in corymbiform cymes; involucre turbinate, of a single series of 5 or 6 narrow bracts; flowers 5 or 6, with white corollas, none radiate; achenes glabrous.

1. **Mesadenia decomposita** (A. Gray) Standley.
   Cacalia decomposita A. Gray, Pl. Wright. **2:** 99. 1853.
   Type locality: Mountains east of Santa Cruz, Sonora.
   Range: Southern Arizona and southwestern New Mexico to northern Mexico.
   New Mexico: San Luis Mountains (*Mearns* 2219, 527, 531).

131. CENTAUREA L. Star thistle.

Tall, usually simple-stemmed annual with alternate, spatulate to oblong-lanceolate, entire or denticulate leaves and large showy heads of tubular flowers; involucre campanulate, the bracts appressed in many series, pectinate; outer corollas pink or purplish, enlarged and radiant, the inner ones ochroleucous; achenes compressed, smooth; pappus of several series of unequal bristles.

1. **Centaurea americana** Nutt. Journ. Acad. Phila. **2:** 117. 1821.
   Plectocephalus americanus Don in Sweet, Brit. Flower Gard. **2:** *pl. 51.* 1823–29.
   Type locality: "On the banks of streams, and in denudated alluvial situations, throughout the plains or prairies of the upper part of Arkansas territory."
   Range: Arkansas and Louisiana to Arizona and southward.
   New Mexico: Socorro; Mogollon Mountains; Mangas Springs; G O S Ranch; Mesilla; White Mountains; mountains north of Santa Rita. Moist slopes and along streams, in the Upper Sonoran and Transition zones.

132. CIRSIUM Hill. Thistle.

Coarse biennial or perennial herbs with prickly, often pinnatifid, sessile, alternate leaves and large heads, the latter solitary, racemose, or clustered at the ends of the branches; flowers all tubular, perfect; bracts of the involucre imbricated in many ranks, mostly tipped with prickles; receptacle clothed with soft bristles; achenes oblong or obovoid, compressed, smooth; pappus of numerous plumose bristles united into a ring at the base, deciduous.

KEY TO THE SPECIES.

Bracts deep purple, reflexed for half their length, not arachnoid . 1. *C. vinaceum.*
Bracts not purplish or if so only at the very tips, when reflexed more or less arachnoid.
  Plants acaulescent.. 2. *C. acaulescens.*
  Plants not acaulescent.
    At least some of the bracts pectinate-ciliate with weak spines; flowers greenish yellow.

None of the bracts dilated at the apex, all with long-attenuate tips, conspicuously arachnoid, in many series .... 3. *C. pallidum.*

At least some of the bracts dilated at the apex, not conspicuously arachnoid, often nearly glabrous, in few series.

Inner bracts conspicuously elongated, with narrow, mostly coriaceous tips; heads small, 2 cm. broad, mostly clustered at the ends of the branches .... 4. *C. inornatum.*

Inner bracts not conspicuously elongated, with broad green tips; heads large, 3 cm. broad, mostly solitary .... 5. *C. gilense.*

None of the bracts pectinate-ciliate; flowers variously colored.

Bracts not with a dorsal glandular ridge.

Bracts squarrose, at least the lower reflexed; largest heads 6 to 7 cm. in diameter .... 6. *C. neomexicanum.*

Bracts not squarrose, all erect or appressed; heads all less than 5 cm. in diameter.

Style produced at the tip only 4 to 6 times its diameter above the conspicuous node; corollas bright carmine; leaves tomentose on both sides .... 7. *C. arizonicum.*

Style produced many times its diameter beyond the inconspicuous node; corollas purplish or whitish; leaves glabrate, at least above.

Inner bracts much elongated, the tips not dilated; heads small, 2 cm. in diameter or less .... 8. *C. pulchellum.*

Inner bracts but little elongated, somewhat dilated at the tips; heads large or small.

Flowers purplish; heads small, about 2 cm. in diameter, solitary; inner bracts with purple tips ..... 9. *C. wheeleri.*

Flowers whitish; heads larger, mostly 3 to 4 cm. broad, clustered at the ends of the branches; bracts not with purple tips .... 10. *C. coloradense.*

Bracts with a dorsal glandular ridge.

Bracts scabrous-ciliate, purplish .... 11. *C. grahami.*

Bracts not scabrous-ciliate, usually not at all purplish.

Plants glabrate.

Stems winged below; leaves thin, usually merely dentate with weak spines; heads broadly campanulate .... 12. *C. wrightii.*

Stems not winged; leaves thick, pinnatifid, with stout spines; heads narrowly campanulate .... 13. *C. calcareum.*

Plants permanently tomentose.
Involucres less than 3 cm. wide.
Leaves very spiny; bracts scarcely if at all tomentose on the margins, the tips 5 mm. long or more; leaves soon glabrate on the upper surface....................14. *C. perennans.*
Leaves sparingly spiny; bracts strongly tomentose on the margins, the tips 3 mm. long or less; leaves usually permanently tomentose on both surfaces..........15. *C. undulatum.*
Involucres 4 to 6 cm. wide.
Spines of the involucral bracts usually about 5 mm. long; leaves very broad; corollas usually carmine.16. *C. megacephalum.*
Spines of the bracts 1 cm. long; leaves narrow; corollas purplish rose, often white....................17. *C. ochrocentrum.*

1. **Cirsium vinaceum** Woot. & Standl.
Carduus vinaceus Woot. & Standl. Contr. U. S. Nat. Herb. **16:** 196. 1913.
TYPE LOCALITY: In the Sacramento Mountains near Fresnal, New Mexico. Type collected by Wooton, July 12, 1899.
RANGE: Known only from type locality, in the Transition Zone.

2. **Cirsium acaulescens** (A. Gray) Daniels, Univ. Mo. Stud. Sci. **2**[2]: 254. 1911.
Cnicus drummondii acaulescens A. Gray, Proc. Amer. Acad. **10:** 40. 1874.
Carduus acaulescens Rydb. Bull. Torrey Club **28:** 508. 1901.
TYPE LOCALITY: Rocky Mountains.
RANGE: Wyoming to northern New Mexico.
NEW MEXICO: Taos Mountains; Costilla Valley. Meadows.

3. **Cirsium pallidum** Woot. & Standl.
Carduus pallidus Woot. & Standl. Contr. U. S. Nat. Herb. **16:** 195. 1913.
TYPE LOCALITY: In the Pecos River National Forest near Winsors Ranch, New Mexico. Type collected by Standley (no. 4357).
RANGE: New Mexico and southern Colorado.
NEW MEXICO: Winsors Ranch; Tunitcha Mountains; Chama; mountains west of Las Vegas; Baldy; White and Sacramento mountains. Along streams and in wet meadows, in the Transition and Canadian zones.

4. **Cirsium inornatum** Woot. & Standl.
Carduus inornatus Woot. & Standl. Contr. U. S. Nat. Herb. **16:** 195. 1913.
TYPE LOCALITY: Sacramento Mountains near Cloudcroft, New Mexico. Type collected by Wooton, August 24, 1901.
RANGE: Known only from type locality.

5. **Cirsium gilense** Woot. & Standl.
Carduus gilensis Woot. & Standl. Contr. U. S. Nat. Herb. **16:** 195. 1913.
TYPE LOCALITY: In the Mogollon Mountains on the West Fork of the Rio Gila, Socorro County, New Mexico. Type collected by Metcalfe (no. 377).
RANGE: Known only from type locality.

6. **Cirsium neomexicanum** A. Gray, Pl. Wright. **2:** 101. 1853.
Cnicus neomexicanus A. Gray, Proc. Amer. Acad. **10:** 45. 1874.
Carduus neomexicanus Greene, Proc. Acad. Phila. **1892:** 362. 1893.

TYPE LOCALITY: Organ Mountains, New Mexico.

RANGE: Colorado to Arizona and New Mexico.

NEW MEXICO: Santa Fe; Cliff; San Luis Pass; Carrizalillo Mountains; Organ Mountains. Plains and dry hills, in the Upper Sonoran Zone.

7. **Cirsium arizonicum** (A. Gray) Petrak, Bot. Tidsskr. **31**: 68. 1911.

Cnicus arizonicus A. Gray, Proc. Amer. Acad. **10**: 44. 1874.

Carduus arizonicus Greene, Proc. Acad. Phila. **1892**: 362. 1893.

TYPE LOCALITY: "Arizona and S. Utah."

RANGE: Utah to Arizona and New Mexico.

NEW MEXICO: San Luis Mountains. Low hills and plains.

8. **Cirsium pulchellum** (Greene) Woot. & Standl.

Carduus pulchellus Greene; Rydb. Colo. Agr. Exp. Sta. Bull. **100**: 401. 1906.

TYPE LOCALITY: Piedra, Colorado.

RANGE: Southern Colorado to New Mexico.

NEW MEXICO: Tunitcha Mountains; Cloudcroft. Mountains, in the Transition Zone.

9. **Cirsium wheeleri** (A. Gray) Petrak, Bot. Tidsskr. **31**: 67. 1911.

Cnicus wheeleri A. Gray, Proc. Amer. Acad. **19**: 56. 1883.

Carduus wheeleri Heller, Cat. N. Amer. Pl. 7. 1898.

TYPE LOCALITY: Near Camp Apache, Arizona.

RANGE: Mountains of southern Arizona and southwestern New Mexico.

NEW MEXICO: Luna; West Fork of the Gila; north of Ramah. Transition Zone.

10. **Cirsium coloradense** (Rydb.) Cockerell, Univ. Mo. Stud. Sci. **2**^{2}: 254. 1911.

Carduus coloradensis Rydb. Bull. Torrey Club **32**: 132. 1905.

TYPE LOCALITY: Pagosa Springs, Colorado.

RANGE: Mountains of Colorado and northern New Mexico.

NEW MEXICO: North of Ramah; Rio Mora; Tesuque. Transition Zone.

11. **Cirsium grahami** A. Gray, Pl. Wright. **2**: 102. 1853.

Cnicus grahami A. Gray, Proc. Amer. Acad. **19**: 57. 1883.

Carduus grahami Greene, Proc. Acad. Phila. **1892**: 363. 1893.

TYPE LOCALITY: Low ground, in valleys between the Sonoita and the San Pedro, Sonora.

RANGE: Mountains of southern Arizona and southwestern New Mexico.

NEW MEXICO: Mogollon Mountains. Transition Zone.

12. **Cirsium wrightii** A. Gray, Pl. Wright. **2**: 101. 1853.

Cnicus wrightii A. Gray, Proc. Amer. Acad. **10**: 41. 1874.

Carduus wrightii Heller, Cat. N. Amer. Pl. 8. 1898.

TYPE LOCALITY: "Around springs near San Bernandino, on the borders of New Mexico and Sonora."

RANGE: Western Texas to southern Arizona.

NEW MEXICO: White Mountains; Roswell; Fresnal. Mountains, in the Upper Sonoran and Transition zones.

13. **Cirsium calcareum** (Jones) Woot. & Standl.

Cnicus calcareus Jones, Proc. Calif. Acad. II. **5**: 704. 1895.

TYPE LOCALITY: Cainville, Utah.

RANGE: Utah to northern Arizona and New Mexico.

NEW MEXICO: Carrizo Mountains (*Standley* 7401). Dry hills and canyons, in the Upper Sonoran Zone.

14. Cirsium perennans (Greene) Woot. & Standl.

Carduus perennans Greene, Bull. Torrey Club **25:** 125. 1898.

TYPE LOCALITY: White Mountains, New Mexico. Type collected by Wooton (no. 326).

RANGE: Mountains of southern New Mexico.

NEW MEXICO: Eagle Peak; White and Sacramento mountains; Organ Mountains. Transition Zone.

15. Cirsium undulatum (Nutt.) Spreng. Syst. Veg. **3:** 374. 1826.

Carduus undulatus Nutt. Gen. Pl. **2:** 130. 1818.

Cnicus undulatus A. Gray, Proc. Amer. Acad. **10:** 42. 1874.

TYPE LOCALITY: "On the calcareous islands of Lake Huron, and on the plains of Upper Louisiana."

RANGE: British America to Utah and Texas.

NEW MEXICO: Tunitcha Mountains; Chama; Tierra Amarilla; Pecos; Winsors Ranch; Rio Mora; Santa Fe; Anton Chico; White Mountains. Plains and hills, in the Upper Sonoran and Transition zones.

16. Cirsium megacephalum (A. Gray) Cockerell, Univ. Mo. Stud. Sci. **2**[2]**:** 254. 1911.

Cnicus undulatus megacephalus A. Gray, Proc. Amer. Acad. **10:** 42. 1874.

Carduus megacephalus Smyth, Trans. Kans. Acad. **16:** 160. 1899.

TYPE LOCALITY: Texas.

RANGE: Idaho and South Dakota to Arizona and Texas.

NEW MEXICO: Mimbres; Mangas Springs; Mogollon Mountains; Santa Rita; San Luis Mountains; Dog Spring. Dry hills and plains, in the Lower and Upper Sonoran zones.

17. Cirsium ochrocentrum A. Gray, Mem. Amer. Acad. n. ser. **4:** 110. 1849.

Cnicus ochrocentrus A. Gray, Proc. Amer. Acad. **19:** 57. 1883.

Carduus ochrocentrus Greene, Proc. Acad. Phila. **1892:** 363. 1893.

TYPE LOCALITY: Mountain sides around Santa Fe, New Mexico. Type collected by Fendler (no. 486).

RANGE: Nebraska and Colorado to Arizona and New Mexico.

NEW MEXICO: Pecos; Sierra Grande; San Juan; Nara Visa; Kennedy; Chiz; Ramah; Socorro; Mangas Springs; Organ Mountains; Gray; Tularosa; Roswell. Plains and low hills, chiefly in the Upper Sonoran Zone.

133. ARCTIUM L. BURDOCK.

Coarse biennial herb with large, cordate-ovate, petiolate, more or less tomentose basal leaves and rather small, clustered heads of purple flowers; involucre globose, the much imbricated, many-ranked bracts with filiform hooked tips; receptacle bristly; achenes oblong, compressed, transversely rugose; pappus of numerous short bristles.

1. Arctium minus Schkuhr, Bot. Handb. **3:** 49. 1803.

TYPE LOCALITY: Germany.

NEW MEXICO: Fruitland (*Standley*).

A common weed in most parts of the eastern United States, introduced from Europe. So far it has been noted in only this one locality in New Mexico, where it is abundant in cultivated fields and along ditch banks.

SUMMARY OF THE LARGER GROUPS, WITH NUMBERS OF GENERA AND SPECIES.

PTERIDOPHYTA.

	Genera.	Species.
Filicales:		
Polypodiaceae	13	31
Salviniales:		
Marsileaceae	1	1
Salviniaceae	1	1
Equisetales:		
Equisetaceae	1	4
Lycopodiales:		
Selaginellaceae	1	5
Total Pteridophyta	17	42

GYMNOSPERMAE.

	Genera.	Species.
Pinales:		
Pinaceae	4	14
Juniperaceae	2	7
Gnetales:		
Ephedraceae	1	4
Total Gymnospermae.	7	25

MONOCOTYLEDONES.

	Genera.	Species.
Pandanales:		
Typhaceae	1	1
Naiadales:		
Potamogetonaceae	2	5
Naiadaceae	1	1
Alismales:		
Juncaginaceae	1	2
Alismaceae	3	3
Poales:		
Poaceae	74	270
Cyperaceae	9	70
Arales:		
Lemnaceae	2	4
Xyridales:		
Commelinaceae	2	5
Pontederiaceae	1	1
Liliales:		
Calochortaceae	1	3
Melanthaceae	4	7
Juncaceae	2	22
Dracaenaceae	3	12
Convallariaceae	5	6
Alliaceae	3	13
Liliaceae	4	4
Asphodelaceae	1	1
Amaryllidales:		
Amaryllidaceae	2	6
Iridaceae	3	5
Orchidales:		
Orchidaceae	10	18
Total Monocotyledones	134	459

DICOTYLEDONES.

	Genera.	Species.
Piperales:		
Saururaceae	1	1
Salicales:		
Salicaceae	2	24
Juglandales:		
Juglandaceae	1	2
Fagales:		
Betulaceae	3	4
Fagaceae	1	24
Urticales:		
Ulmaceae	1	1
Moraceae	1	1
Urticaceae	3	5
Cannabinaceae	1	1
Santalales:		
Loranthaceae	2	10
Santalaceae	1	1
Aristolochiales:		
Aristolochiaceae	1	1
Polygonales:		
Polygonaceae	9	70
Chenopodiales:		
Chenopodiaceae	12	42
Amaranthaceae	8	24
Corrigiolaceae	1	2
Allioniaceae	13	46
Phytolaccaceae	1	1
Aizoaceae	3	4
Portulacaceae	6	20
Alsinaceae	8	31
Silenaceae	5	11
Ranales:		
Ceratophyllaceae	1	1
Ranunculaceae	16	66
Berberidaceae	2	6
Papaverales:		
Papaveraceae	3	7
Fumariaceae	1	4
Brassicaceae	30	101
Capparidaceae	4	6
Resedaceae	1	1
Rosales:		
Crassulaceae	3	9
Saxifragaceae	5	15
Parnassiaceae	1	2
Hydrangeaceae	4	9
Grossulariaceae	2	10
Platanaceae	1	1
Rosaceae	21	63
Malaceae	4	13
Amygdalaceae	3	10
Mimosaceae	7	17
Cassiaceae	4	15
Krameriaceae	1	2
Fabaceae	32	189
Geraniales:		
Geraniaceae	2	9
Linaceae	2	8
Oxalidaceae	2	8

Geraniales—Continued.	Genera.	Species.
Zygophyllaceae	3	7
Koeberliniaceae	1	1
Malpighiaceae	1	1
Rutaceae	3	4
Simarubaceæ	1	1
Meliaceae	1	1
Polygalales:		
Polygalaceae	2	10
Euphorbiales:		
Euphorbiaceae	15	52
Callitrichaceae	1	1
Sapindales:		
Anacardiaceae	4	12
Celastraceae	3	3
Aceraceae	2	5
Sapindaceae	2	2
Rhamnales:		
Rhamnaceae	4	9
Vitaceae	3	3
Malvales:		
Malvaceae	14	44
Hypericales:		
Elatinaceae	1	1
Tamaricaceae	1	1
Fouquieriaceae	1	1
Frankeniaceae	1	1
Hypericaceae	1	1
Violaceae	2	13
Opuntiales:		
Loasaceae	4	18
Cactaceae	5	67
Thymelaeales:		
Elaeagnaceae	1	2
Myrtales:		
Epilobiaceae	16	62
Gunneraceae	2	2
Umbellales:		
Cornaceae	2	3
Hederaceae	1	1
Apiaceae	26	36
Ericales:		
Monotropaceae	2	2
Pyrolaceae	3	7
Ericaceae	2	4
Vacciniaceae	1	2
Primulales:		
Plumbaginaceae	1	1
Primulaceae	7	20
Ebenales:		
Sapotaceae	1	1
Oleales:		
Oleaceae	4	9

Gentianales:	Genera.	Species.
Gentianaceae	10	21
Asclepiadales:		
Apocynaceae	2	11
Asclepiadaceae	7	27
Polemoniales:		
Dichondraceae	1	2
Cuscutaceae	1	4
Convolvulaceae	5	18
Polemoniaceae	9	44
Hydrophyllaceae	5	26
Ehretiaceae	3	4
Heliotropaceae	2	3
Boraginaceae	13	41
Verbenaceae	3	13
Menthaceae	20	55
Solanaceae	10	39
Scrophulariaceae	22	100
Acanthaceae	6	6
Pinguiculaceae	1	1
Orobanchaceae	3	4
Bignoniaceae	2	2
Martyniaceae	1	3
Plantaginales:		
Plantaginaceae	1	6
Rubiales:		
Rubiaceae	5	21
Caprifoliaceae	6	13
Adoxaceae	1	1
Campanulales:		
Cucurbitaceae	8	12
Campanulaceae	2	3
Lobeliaceae	1	2
Valerianales:		
Valerianaceae	1	3
Asterales:		
Cichoriaceae	16	51
Mutisiaceae	3	5
Ambrosiaceae	6	18
Asteraceae	133	511
Total Dicotyledones	690	2,377

SUMMARY.

	Genera.	Species.
Pteridophyta	17	42
Spermatophyta:		
Gymnospermae	7	25
Angiospermae—		
Monocotyledones	134	459
Dicotyledones	690	2,377
Total for New Mexico	848	2,903

GEOGRAPHIC INDEX.

ABIQUIU. Southeastern Rio Arriba County. Alt. 1,808 meters.

ABIQUIU PEAK. Near town of Abiquiu. Alt. 3,428 meters.

ACOMA. Central Valencia County. Alt. 1,925 meters.

ADEN. Station on the Southern Pacific Railroad in western Dona Ana County. Alt. 1,338 meters.

AGRICULTURAL COLLEGE. Central Dona Ana County. Alt. 1,175 meters. The post-office is now State College.

AGUA AZUL. Southern McKinley County. Alt. 2,037 meters.

AGUA CHIQUITA. A stream in eastern Otero and southwestern Chaves counties, a tributary of the Penasco.

AGUA FRIA. A spring and a small stream in northern Valencia County. Alt. 1,977 meters.

ALAMILLO. Northeastern Socorro County. Alt. 1,413 meters.

ALAMOCITAS CANYON. Northern Valencia County.

ALAMOGORDO. Northern Otero County. Alt. 1,312 meters.

ALAMO VIEJO. Southeastern Grant County.

ALBERT. Southwestern Union County. Alt. 1,433 meters.

ALBUQUERQUE. Bernalillo County. Alt. 1,504 meters.

ALGODONES. Southeastern Sandoval County. Alt. 1,555 meters.

ALIZO. See Alizo Creek. Locally called "Celees."

ALIZO CREEK. Extreme western Socorro County.

ALMA. Southwest corner of Socorro County. Alt. 1,677 meters.

AMPERSAND CREEK. Near Fort Bayard, northeastern Grant County.

ANCHO. Northwestern Lincoln County. Alt. 1,864 meters.

ANGUS. Southwestern Lincoln County. Alt. 2,135 meters.

ANIMAS CREEK. Western Sierra County, heading in the Black Range.

ANIMAS MOUNTAINS. Southern Grant County.

ANIMAS PEAK. Southern Grant County. Alt. 1,862 meters.

ANIMAS VALLEY. Southwestern Grant County.

ANNISTON. Quay County.

ANTHONY. Southeastern Dona Ana County, on the Texas line. Alt. 1,151 meters. The railroad station is called La Tuna.

ANTON CHICO. Northwest corner of Guadalupe County. Alt. 1,638 meters.

APACHE CREEK. Northeastern Grant County.

APACHE MOUNTAINS. Southern Grant County.

APACHE SPRING. Western Socorro County, north of Reserve.

APACHE TEJU. Eastern Grant County. Alt. 1,671 meters.

ARCH. Eastern Roosevelt County.

ARROYO HONDO. Southern Taos County.

ARROYO PECOS. Near Las Vegas, San Miguel County.

ARROYO RANCH. Near Roswell, Chaves County.

ARTESIA. Northwestern Eddy County. Alt. 1,030 meters.

ASCARATE RANCH. In the Mesilla Valley, southern Dona Ana County.

ATARQUE DE GARCIA. Extreme southwestern corner of Valencia County.

AZTEC. Northeastern San Juan County. Alt. 1,705 meters.

BAD LANDS. Near Tiznitzin, central San Juan County.

BALDY. Post office in western Colfax County. Alt. 2,943 meters.

BARKERS LAKE. In the Las Vegas Range, northwestern San Miguel County.

BARRANCA. Southwestern Taos County. Alt. 2,115 meters.

BAYARD STATION. Western Grant County. Alt. 1,763 meters.

BEAR CANYON. Sandia Mountains, Bernalillo County. This is a common name in Mexico, and a Bear Canyon can be found in almost every range.

BEAR CREEK. Lincoln National Forest, Lincoln County.

BEAR MOUNTAIN. Northeastern Grant County, northwest of Silver City. Alt. 2,464 meters.

BEAR RIDGE. Zuni Mountains, northern Valencia County.

BEAR TOOTH RIDGE. Near Fort Bayard, northeastern Grant County.

BEATTYS CABIN. In the Pecos National Forest in either Mora or San Miguel County.

BEENHAM. Central Union County.

BELEN. Southwestern Valencia County. Alt. 1,459 meters.

BELL. Northeastern Colfax County. Alt. 2,464 meters.
BELL RANCH. Eastern San Miguel County.
BEN MOORE. Eastern Grant County near Santa Rita. The name is no longer in use.
BERENDO CREEK. Southern Sierra County, heading in the Black Range.
BERENDO RIVER. Near Roswell, western Chaves County.
BERNAL. Alt. 1,850 meters. See Ojo de Bernal.
BERNALILLO. Southeastern Sandoval County. Alt. 1,555 meters.
BEULAH. Northwestern San Miguel County. Alt. 2,410 meters.
BIG HATCHET MOUNTAINS. Southeastern Grant County.
BISHOPS CAP. Peak at the south end of the Organ Mountains, Dona Ana County.
BLACK LAKE. Southwest corner of Colfax County.
BLACK RANGE. In eastern Grant and western Sierra County. The highest peak has an altitude of 3,106 meters.
BLACK ROCK. On the Zuni Reservation, southwestern McKinley County. Alt. 1,952 meters.
BLAZERS MILL. In the Mescalero Apache Reservation on Tularosa Creek, Otero County.
BLOOMFIELD. Northeastern San Juan County. Alt. 1,662 meters.
BLUE CREEK. Western Grant County.
BLUE WATER. Northern Valencia County. Alt. 1,986 meters.
BLUE WATER RIVER. Northern Valencia County.
BONITO CREEK. Southern Lincoln County, heading in the White Mountains.
BOSQUE SECO. In the Mesilla Valley, southern Dona Ana County. Alt. 1,162 meters.
BOULDER LAKE. Central Rio Arriba County.
BOX CANYON. East of Santa Fe, Santa Fe County. This name is used extensively throughout the region for any narrow canyon with high vertical walls.
BOX S SPRING. Southern McKinley County, south of Fort Wingate.
BRAZOS. Northeastern Rio Arriba County.
BRAZOS CANYON. Northeastern Rio Arriba County.
BREMONDS RANCH. Near Roswell, Chaves County.
BRICE. Southwestern Otero County.
BROCKMANS RANCH. Eastern Grant County.
BUCHANAN. Southern Guadalupe County.
BUEYEROS. Southern Union County.
BULLARDS PEAK. Northern Grant County.
BULL CAMP. In the Gallinas Mountains, north-central Socorro County.
BURRO MOUNTAINS. Central Grant County. The highest peak has an altitude of about 2,460 meters.
CABRA SPRING. Southern San Miguel County.
CACTUS FLAT. Northwestern Grant County.
CAMBRAY. Eastern Luna County. Alt. 1,288 meters.
CAMERON CREEK. Near Fort Bayard, eastern Grant County.
CANADA ALAMOSA. A narrow valley above Monticello in northern Sierra County. Alt. 1,995 meters.
CANADA CREEK. A stream running through the Canada Alamosa.
CANADIAN RIVER. Running through Colfax, Mora, San Miguel, and Quay counties.
CANJILON. Eastern Rio Arriba County.
CANYONCITO. Eastern Santa Fe County. Alt. 2,090 meters.
CANYON LARGO. In eastern San Juan and western Rio Arriba counties.
CANYON DE CHELLY. The head of this canyon is in the mountains of western San Juan County. The main part of the canyon lies in Arizona.
CAPITAN MOUNTAINS. Southern Lincoln County. The highest point, Capitan Peak, has an altitude of 3,057 meters.
CARLISLE. Northwestern Grant County.

CARLSBAD. Eddy County. Alt. 946 meters. Formerly known as Eddy.

CARPENTER CREEK. At the south end of the Black Range.

CARRIZALILLO MOUNTAINS. Southwest corner of Luna County.

CARRIZO CREEK. Northeastern Otero County.

CARRIZO MOUNTAINS. Northwestern corner of San Juan County, extending into Arizona.

CARRIZOZO. Southwestern Lincoln County. Alt. 1,656 meters.

CARSON NATIONAL FOREST. Comprises a large part of eastern Rio Arriba County, extending to Taos and Sandoval counties.

CATSKILL. Northwestern Union County.

CAUSEY. Southeast corner of Roosevelt County.

CAVE CANYON. Near Mangas Springs, Grant County.

CEBOLLA. Western Mora County. Alt. 2,287 meters.

CEBOLLA SPRINGS. Eastern Valencia County.

CEDAR CREEK. Near Mescalero Agency, Otero County.

CEDAR HILL. Northwestern San Juan County. Alt. 1,830 meters.

CEDAR SPRING. Florida Mountains, Luna County.

CENTRAL. Eastern Grant County. Alt. 1,820 meters.

CERRILLOS. Santa Fe County. Alt. 1,788 meters.

CHAMA. Northeastern Rio Arriba County. Alt. 2,393 meters.

CHAMA RIVER. Rio Arriba County; a tributary of the Rio Grande.

CHAMBERINO. Southeastern Dona Ana County. Alt. 1,147 meters.

CHAMITA. Southeastern Rio Arriba County. Alt. 1,715 meters.

CHAVES. Southeastern corner of Valencia County. Alt. 1,545 meters.

CHEROKEE BILL SPRING. On the Mescalero Apache Reservation near Ruidoso post-office.

CHICORICO CANYON. Near Raton, Colfax County.

CHINCHERITAS MOUNTAINS. A low range in western Socorro County between Datil and Quemado.

CHIZ. A small Mexican town in northwestern Sierra County near Fairview.

CHLORIDE. Northwestern Sierra County.

CHOSA SPRING. An alkali spring in northwestern Otero County near Tularosa.

CHUPADERO MESA. Northeastern Socorro County.

CHUSCA CANYON. Southwestern San Juan County.

CHUSCA MOUNTAINS. On the Navajo Reservation in southwestern San Juan and northwestern McKinley counties. The highest peaks have an altitude of 2,625 meters.

CIENAGA RANCH. Southeastern Grant County.

CIMARRON. Western Colfax County. Alt. 1,947 meters.

CIMARRON RIVER. Northern Union County. Another stream of the same name is in Colfax County.

CITY OF ROCKS. An area of peculiarly sculptured rocks northeast of Hudson in southeastern Grant County.

CLARK. Southern Santa Fe County. Alt. 1,852 meters.

CLAYTON. Eastern Union County. Alt. 1,544 meters.

CLEMOW. Socorro County, between Magdalena and Socorro. Alt. 1,649 meters.

CLEVELAND. Western Mora County. Alt. 2,165 meters.

CLIFF. Northwestern Grant County. Alt. 1,379 meters.

CLOUDCROFT. Northeastern Otero County. Alt. 2,745 meters.

CLOVERDALE. Southwestern Grant County.

CLOVIS. Southeastern Curry County.

COBRE. See Santa Rita.

COHNOS. Colfax County.

COLD SPRING CANYON. Sacramento Mountains, northeastern Otero County.

COLFAX. Colfax County.
COLORADO. Northwestern Dona Ana County.
CONDES CAMP. Western Grant County on the Arizona line.
COOK. Northern Luna County.
COOKS PEAK. Northern Luna County. Alt. 2,532 meters.
COOKS SPRING. Northern Luna County. Alt. 1,319 meters.
COOLIDGE. Southern McKinley County east of Gallup. Alt. 2,128 meters. Now known as Guam.
COONEY. Southwestern Socorro County. Alt. 1,784 meters.
COPPER CANYON. Magdalena Mountains, Socorro County.
COPPER FLAT. Near Fort Bayard, Grant County.
COPPER MINES. See Santa Rita del Cobre.
COSTILLA PASS. Northeastern Taos County. Alt. 3,107 meters.
COSTILLA RANGE. Northeastern Taos County. Costilla Peak has an altitude of 3,853 meters.
COWLES. On the Pecos River in northwestern San Miguel County. Alt. 2,440 meters.
COX CANYON. In the Sacramento Mountains, Otero County.
COYOTE CREEK. Western Mora and southwestern Colfax counties.
COYOTE SPRING. Near Carrizozo, southwestern Lincoln County.
CRAINS RANCH. In Mule Creek basin, northwestern Grant County.
CRATERS. Central Valencia County.
CRAWFORDS RANCH. On Deer Creek, southwestern Grant County.
CREWS MESA. Near Beulah, northwestern San Miguel County.
CROSS L RANCH. In Cimarron Canyon, northeastern Union County.
CUBERO. Northeastern Valencia County. Alt. 1,802 meters.
CUCHILLO. Northern Sierra County.
CUCHILLO NEGRO. Small stream in northern Sierra County.
CUEVA. In the west side of the Organ Mountains, Dona Ana County.
DAILEY CANYON. Near Beulah, northwestern San Miguel County.
DARK CANYON. In the Guadalupe Mountains, southwestern Eddy County. The name is often used throughout the State. Another Dark Canyon is in the White Mountains in northeastern Otero County.
DATIL. Central Socorro County. Alt. 2,348 meters.
DATIL MOUNTAINS. Western Socorro County. One of the highest peaks, measured by the Wheeler Survey, has an altitude of 2,879 meters.
DAYTON. Northwestern Eddy County. Alt. 1,006 meters.
DEER CREEK. Extreme southwestern Grant County, near the Mexican Boundary.
DEFIANCE. In Arizona just across the line from northwestern McKinley County. Alt. 2,148 meters.
DELAWARE CREEK. A tributary of the Pecos on the southern edge of Eddy County.
DEL NORTE. See Rio Grande.
DEMING. Central Luna County. Alt. 1,316 meters.
DESERT. Southwestern Otero County. Alt. 1,240 meters.
DES MOINES. Northwestern Union County. Alt. 2,023 meters.
DEVILS PARK. In the Mogollon Mountains north of Cooney, western Socorro County.
DEXTER. Southern Chaves County. Alt. 1,052 meters.
DIAMOND A WELLS. Grant County, south of Silver City.
DOG CANYON. Guadalupe Mountains, southwestern Eddy County.
DOG MOUNTAINS. Southern Grant County.
DOG SPRING. Southeast corner of Grant County. Alt. 1,432 meters.
DONA ANA. Central Dona Ana County. Alt. 1,190 meters.
DONA ANA MOUNTAINS. Central Dona Ana County.
DORA. Southeastern Roosevelt County.

DRIPPING SPRING. On the west side of the Organ Mountains, Dona Ana County, at Van Pattens Camp.

DRY CREEK. Southwestern Socorro County. There is a Little Dry and a Big Dry Creek in the region.

DUCK CREEK. Northwestern Grant County.

DULCE. Northern edge of Rio Arriba County. Alt. 2,062 meters. Agency of the Jicarilla Apache Reservation.

DULCE LAKE. Northern Rio Arriba County.

DURAN. Southeastern Torrance County. Alt. 1,830 meters.

DURFEYS WELL. South edge of San Augustine Plains, west and a little south of Magdalena, Central Socorro County.

DWYER. Eastern Grant County.

EAGLE CREEK. In the White Mountains, southern Lincoln County.

EAGLE PEAK. Northeastern Grant County. Alt. 2,986 meters.

EAST CANYON: Northeastern Grant County.

EAST FORK OF THE GILA. Southwestern Socorro County.

EAST LAS VEGAS. Western San Miguel County. Alt. 1,947 meters.

EAST VIEW. Western Torrance County. Alt. 2,135 meters.

EDDY. Now known as Carlsbad.

EL CEDRO. Manzano Mountains, eastern Bernalillo County.

ELIDA. Southwestern Roosevelt County. Alt. 1,325 meters.

ELIZABETHTOWN. Western Colfax County. Alt. 2,582 meters.

ELK. Southwestern Chaves County.

ELK CANYON. Sacramento Mountains, northeastern Otero County.

ELK MOUNTAIN. Northwest corner of San Miguel County. Alt. 3,507 meters.

ELLIS RANCH. In the Sandia Mountains, southeastern Sandoval County.

EL RITO. There are two El Ritos in New Mexico: One in eastern Rio Arriba County—Wooton's records in 1904; the other in eastern Valencia County—Rusby's records, and Wooton's in 1892 and 1906. Alt. of latter 2,071 meters.

EL RITO CREEK. Southeastern Rio Arriba County.

EL RITO DRAW. Southeastern Rio Arriba County.

EL VADO. Central Rio Arriba County.

EMBUDO. Southeastern Rio Arriba County. Alt. 1,770 meters.

EMORY PEAK. Southern Grant County.

EMORY'S 5TH MONUMENT. Near Columbus, southern edge of Luna County.

EMORY SPRING. Southern Grant County.

ENDEE. Eastern Quay County.

ENGLE. Eastern Sierra County. Alt. 1,458 meters.

ENSENADA. Northern Rio Arriba County.

ENSENADA CREEK. Northern Rio Arriba County.

ESPANOLA. Southeastern Rio Arriba County. Alt. 1,700 meters.

ESTANCIA. Central Torrance County. Alt. 1,885 meters.

FAIRVIEW. Northwestern Sierra County. Alt. 1,891 meters.

FALVES. A ranch in the Sacramento Mountains near Cloudcroft.

FARMINGTON. Northern San Juan County. Alt. 1,525 meters.

FELIX. Southwestern Chaves County.

FIERRO. Northeastern Grant County. Alt. 2,030 meters.

FILMORE CANYON. In the west side of the Organ Mountains, Dona Ana County.

FITZGERALD CIENAGA. A marshy flat in the mountains north of Reserve, western Socorro County.

FLORA VISTA. Northern San Juan County. Alt. 1,676 meters.

FLORIDA MOUNTAINS. Central Luna County. Florida Peak has an altitude of 2,225 meters.

FOLSOM. Northwestern Union County. Alt. 1,955 meters.

Fort Bayard. Eastern Grant County. Alt. 1,850 meters.

Fort Craig. On the Rio Grande, southern Socorro County. Alt. 1,357 meters.

Fort Cummings. Northern Luna County. Alt. 1,457 meters.

Fort Defiance. In Arizona, just across the line from northwestern McKinley County. Alt. 2,148 meters.

Fort Filmore. Southern Dona Ana County.

Fort Selden. On the Rio Grande, central Dona Ana County. Alt. 1,206 meters.

Fort Stanton. Southern Lincoln County. Alt. 1,875 meters.

Fort Tularosa. Western Socorro County. Alt. 2,055 meters.

Fort Union. Southern Mora County. Alt. 2,047 meters.

Fort Wingate. Southern McKinley County. Alt. 2,146 meters.

Franeys Peak. Near Fort Bayard, eastern Grant County.

Franklin Mountains. Southeastern Dona Ana County. Most of the range lies in Texas, only the north end extending into New Mexico.

Fray Cristobal. Range in northeastern Sierra County. Alt. 2,027 meters.

Fresnal. In the Sacramento Mountains, Otero County. Now known as Wooten.

Fresnal Creek. In the Sacramento Mountains, Otero County.

Frio Draw. Northern Guadalupe County.

Frisco. Western Socorro County. Alt. 1,735 meters. The name is applied to a little settlement and to a river, the latter usually written San Francisco on the maps, but never referred to by that name.

Fruitland. Northern San Juan County. Alt. 1,500 meters.

Fulton. Western San Miguel County.

Gage. Western Luna County. Alt. 1,369 meters.

Galisteo. Santa Fe County. Alt. 1,865 meters.

Gallinas Canyon. Western San Miguel County. The canyon through which runs the Gallinas River.

Gallinas Mountains. Torrance and Lincoln counties. The highest peak has an altitude of 2,974 meters.

Gallinas Planting Station. In the Las Vegas Mountains, northwestern San Miguel County.

Gallinas River. San Miguel County.

Gallo Spring. Western Socorro County. Alt. 2,314 meters.

Gallup. Western McKinley County. Alt. 1,976 meters.

Garfield. Northwest corner of Dona Ana County.

Gavilan Creek. In the White Mountains, southern Lincoln County.

Georgetown. Eastern Grant County. Alt. 1,969 meters.

Gila. Northern Grant County. Alt. 1,232 meters.

Gila Hot Springs. On the Gila River, northwestern Grant County.

Gila National Forest. Includes the Mogollon, Black, and Burro ranges in Socorro, Sierra, and Grant counties.

Gila River. In western Grant and Socorro counties.

Gilmores Ranch. On Eagle Creek in the White Mountains, southern Lincoln County. Alt. 2,257 meters.

Glencoe. Southeastern Lincoln County. Alt. 1,708 meters.

Glenwood. On the Rio Frisco, southwestern Socorro County. Alt. 1,525 meters.

Glorieta. Eastern Santa Fe County. Alt. 2,262 meters.

Glorieta Mountains. Mountains near Glorieta, Santa Fe County.

Goat Mountain. At Raton, northern Colfax County.

Gold Gulch. Northern Grant County.

G O S Ranch. Northeastern Grant County.

Graham. Southwestern Socorro County. Alt. 1,677 meters.

Grand Canyon of the Gila. Northwestern Grant County.

Grant. Northern Valencia County. Alt. 1,965 meters.

GRANTS STATION. See Grant.

GRASS MOUNTAIN. East of the Pecos River, in the mountains of extreme northwestern San Miguel County.

GRAY. Southern Lincoln County.

GRAYS PEAK. In the Capitan Mountains, southeastern Lincoln County.

GREENWOOD CANYON. Near Mangas Springs, northern Grant County.

GROSSTEDT PLACE. Near Gallo Spring, in western Socorro County.

GUADALUPE CANYON. Southwest corner of Grant County.

GUADALUPE MOUNTAINS. In southeastern Otero and southwestern Eddy counties.

GUADALUPE PASS. Southwest corner of Grant County.

GUADALUPITA. Northwestern Mora County. Alt. 2,341 meters.

HACHITA. Southwestern Grant County. Alt. 1,376 meters.

HADLEYS SPRINGS. Southeastern Sierra County.

HANOVER MOUNTAIN. Eastern Grant County, near Hanover. Alt. 2,256 meters.

HARRINGTON RANCH. Near Fort Bayard, northeastern Grant County.

HARRISONS RANCH. About half a mile south of Pecos, western San Miguel County.

HARVEYS UPPER RANCH. In the Las Vegas Range, northwestern San Miguel County. Alt. 2,928 meters.

HATCHET RANCH. Southern Grant County, southwest of Hachita.

HEAD AND WILSON RANCH. Northwestern Grant County, north of Carlisle.

HEBRON. Northern Colfax County. Alt. 1,880 meters.

HELL CANYON. Magdalena Mountains, Socorro County.

HERMANAS. Southwestern Luna County. Alt. 1,357 meters.

HERMITS PEAK. In the Las Vegas Range, northwestern San Miguel County. Alt. 3,111 meters.

HERMOSA. Western Sierra County.

HESS RANCH. In the Guadalupe Mountains, southwestern Eddy County.

HIGHROLLS. In the Sacramento Mountains, northern Otero County. Alt. 1,982 meters.

HILLSBORO. Southwestern Sierra County. Alt. 1,593 meters.

HILLSBORO PEAK. In the Black Range, western Sierra County. Alt. 3,068 meters.

HILLSBORO ROAD. The wagon road from Arrey to Hillsboro, southwestern Sierra County.

HOLTS RANCH. See Lone Pine.

HOLY GHOST CREEK. A tributary of the Pecos, in the mountains of the northwest corner of San Miguel County.

HONDO CANYON. On the east side of the White Mountains, southeastern Lincoln County.

HONDO HILL. On the east side of the White Mountains, southeastern Lincoln County.

HOP CANYON. Magdalena Mountains, Socorro County.

HOPE. Northwestern Eddy County.

HOPKINS MILL. Near Arrey, in the Rio Grande Valley, southern Sierra County.

HORACE. Northern Valencia County. Alt. 1,952 meters.

HORSE CAMP. Northeast of Fairview, about on the line between Sierra and Socorro counties.

HORSE SPRING. Western Socorro County. Alt. 2,149 meters.

HORSETHIEF CANYON. In the mountains of Pecos National Forest.

HURRAH CREEK. Northern Guadalupe County.

ICE CANYON. In the west side of the Organ Mountains, southeastern Dona Ana County.

INDIAN BUTTE. San Mateo Mountains, Socorro County.

INDIAN CANYON. In the Animas Mountains, southern Grant County.

INDIAN CREEK. A tributary of the Pecos, in northwestern San Miguel County.

INSCRIPTION ROCK. Northwestern Valencia County. Alt. 2,207 meters.

IRON CREEK. In the south end of the Black Range.
JAMES CANYON. In the Sacramento Mountains, Otero County.
JARILLA. Southern Otero County.
JARILLA JUNCTION. Southern Otero County. Alt. 1,259 meters.
JEMEZ MOUNTAINS. Rio Arriba and Sandoval counties. The highest peak has an altitude of 2,939 meters.
JEWETT. On the San Juan River, northern San Juan County. Sometimes known as Liberty.
JEWETT SPRING and JEWETT GAP. Western Socorro County, north of Reserve.
JICARILLA MOUNTAINS. On the Jicarilla Apache Reservation, Rio Arriba County.
JOHNS CANYON. Near Beulah, northwestern San Miguel County.
JOHNSONS BASIN. Extreme western Socorro County, near the Arizona line.
JOHNSONS MESA. Northeastern Colfax County.
JORNADA DEL MUERTO. A wide, sandy plain east of the Rio Grande in Socorro, Sierra, and Dona Ana counties.
JOSEPH. Western Socorro County. Alt. 2,059 meters.
JUNIPER SPRING. Southwest corner of Grant County.
KELLY. Central Socorro County. Alt. 2,287 meters.
KELLYS RANCH. On the Frisco River between Alma and Reserve, western Socorro County.
KELLYS SPRING. In the Lincoln National Forest.
KENNEDY. Central Santa Fe County. Alt. 1,832 meters.
KIEHNES RANCH. On the Rio Negrito, western Socorro County.
KINGMAN. Western edge of San Miguel County. Alt. 2,076 meters.
KINGSTON. Southwestern Sierra County. Alt. 1,982 meters.
KNOWLES. Northeast corner of Eddy County.
LA CUESTA. A little town on the Pecos River a few miles above Anton Chico, northwestern Guadalupe County.
LA CUEVA. Near Las Vegas, San Miguel County.
LACY. Northern Roosevelt County.
LAGUNA. Eastern Valencia County. Alt. 1,790 meters.
LAGUNA COLORADO. Southeastern San Miguel and northeastern Guadalupe counties.
LA JARA LAKE. On the Jicarilla Apache Reservation, northern Rio Arriba County.
LAKE ARTHUR. Southern Chaves County. Alt. 1,027 meters.
LAKE LA JARA. On the Jicarilla Apache Reservation, Rio Arriba County.
LAKE PEAK. In the Santa Fe Mountains, northeastern Santa Fe County. Alt. 3,784 meters.
LAKE VALLEY. Southern Sierra County. Alt. 1,651 meters.
LAKEWOOD. Eddy County. Alt. 1,006 meters.
LA LANDE. Northwestern Roosevelt County. Alt. 1,220 meters.
LA LUZ. Northern Otero County. Alt. 1,470 meters.
LA LUZ CANYON. Northern Otero County.
LAMY. Santa Fe County. Alt. 1,970 meters.
LAS ANIMAS. Southern Grant County.
LAS CRUCES. Central Dona Ana County. Alt. 1,181 meters.
LAS HUERTAS CANYON. In the Sandia Mountains, southeastern Sandoval County
LAS LAGUNITAS. Near Las Vegas, San Miguel County.
LAS PALOMAS. On the Rio Grande, central Sierra County. Alt. 1,258 meters.
LAS PALOMAS HOT SPRINGS. Central Sierra County. Alt. 1,258 meters.
LAS PLAYAS SPRINGS. Southern Grant County.
LAS VEGAS. Eastern San Miguel County. Alt. 1,947 meters.
LAS VEGAS HOT SPRINGS. Eastern San Miguel County. Alt. 2,047 meters.
LAS VEGAS RANGE. Northwestern San Miguel County.
LAVA. Southeastern Socorro County. Alt. 1,435 meters.

LEACHS POST OFFICE. Southeastern Roosevelt County.
LEMITAR. Eastern Socorro County. Alt. 1,407 meters.
LEMITAR SPRING. Eastern Socorro County.
LEOPOLD. East of the Burro Mountains, Grant County.
LINCOLN. Southeastern Lincoln County.
LINCOLN NATIONAL FOREST. In Lincoln and Torrance counties.
LITTLE BUCK MINE. Organ Mountains, Dona Ana County.
LITTLE BURRO MOUNTAINS. Northern Grant County.
LITTLE CREEK. In the White Mountains, southern Lincoln County.
LITTLE FLORIDA MOUNTAINS. Luna County.
LIVERMORE SPRING. Southern Grant County.
LLANO ESTACADO. Plains covering most of eastern New Mexico east of the Pecos River as far north as Quay County.
LOGAN. Northern Quay County. Alt. 1,166 meters.
LONE MOUNTAIN. Eastern Grant County. Alt. 1,921 meters.
LONE PINE. Southwestern Socorro County. Alt. 1,826 meters.
LONG CANYON. Northwestern Union County near Folsom.
LONGS RANCH. Southeastern Roosevelt County.
LOOKOUT MINES. At the south end of the Black Range, southwestern Sierra County.
LORDSBURG. Central Grant County. Alt. 1,296 meters.
LOS LUNAS. Eastern Valencia County. Alt. 1,474 meters.
LOS MOROS. This is probably meant for El Moro or Inscription Rock. It is a designation used on some of Bigelow's labels.
LOS PILARES. Some beautifully sculptured rocks in western Valencia County near Atarque de Garcia.
LOST RIVER. Near Tularosa, northeastern Otero County.
LOWER PLAZA. Western Socorro County. Alt. 1,720 meters. Another name for Frisco.
LUERA SPRING. Central Socorro County. Alt. 2,314 meters.
LUNA. Western edge of Socorro County. Alt. 2,287 meters.
LUNA VALLEY. Western Socorro County.
LYNN. Northern edge of Colfax County. Alt. 2,291 meters.
MADERA. In the Sandia Mountains.
MAGDALENA. Central Socorro County. Alt. 1,998 meters.
MAGDALENA MOUNTAINS. Central Socorro County. The highest peak has an altitude of 3,293 meters.
MALAGA. Southern Eddy County. Alt. 915 meters.
MALONES CROSSING. Otero County, northeast of the White Sands.
MALONES RANCH. Otero County, northeast of the White Sands.
MANGAS CANYON. Northern Grant County.
MANGAS CREEK. Northern Grant County.
MANGAS SPRINGS. Northern Grant County. Alt. 1,464 meters.
MANGUITAS SPRING. Western Socorro County.
MANZANARES. Western San Miguel County. Alt. 2,004 meters.
MANZANARES VALLEY. Western San Miguel County.
MANZANO MOUNTAINS. In Socorro, Torrance, and Valencia counties. The highest peak has an altitude of 3,076 meters.
MARTIN and SLOAN RANCH. In western Grant County near Red Rock.
MAXWELL CITY. Central Colfax County. Alt. 1,849 meters.
MCCARTHYS STATION. Northeastern Valencia County. Alt. 1,873 meters. Doctor Rusby cites what is probably the same place as McArtys Ranch on some labels.
MCCLURES RANCH. Southwestern Socorro County northeast of Mogollon.

McIntosh. Northwestern Torrance County. Alt. 1,872 meters.

McKinneys Park. A long sloping timbered area constituting the last approach to Mogollon Peak, southwestern Socorro County.

McKnight Canyon. Northeastern Grant County, a tributary of the Rio Mimbres.

Melrose. Western Curry County. Alt. 1,495 meters.

Mesa Redonda. South of Tucumcari in central Quay County.

Mescalero Agency. Northeastern Otero County. Alt. 1,975 meters.

Mescalero Apache Reservation. Northeastern Otero County in the White and Sacramento mountains.

Mesilla. Central Dona Ana County. Alt. 1,174 meters.

Mesilla Park. Central Dona Ana County. Alt. 1,179 meters.

Mesilla Valley. The Valley of the Rio Grande in Dona Ana County.

Mesquite Lake. Southeastern Dona Ana County, in the old bed of the Rio Grande. Alt. 1,164 meters.

Middle Fork of the Gila. Southwestern Socorro County in the Mogollon Mountains.

Miller Hill. Near Mangas Springs, northwestern Grant County.

Mimbres. Eastern Grant County. Alt. 1,323 meters.

Mimbres Mountains. Eastern Grant and southwestern Sierra counties.

Mimbres River. Grant and Luna counties.

Mineral Creek. Western Sierra County.

Mineral Hill. Northwestern San Miguel County.

Modoc. On the west side of the Organ Mountains, southeastern Dona Ana County.

Mogollon. Southwestern Socorro County. Alt. 1,830 meters.

Mogollon Creek. Northern Grant and southwestern Socorro counties, in the Mogollon Mountains.

Mogollon Mountains. Socorro and Grant counties. The highest peak has an altitude of 3,202 meters.

Mogollon Road. A road through the Mogollon Mountains originally designed to connect Mogollon with Magdalena, southwestern Socorro County.

Monica Canyon. In the San Mateo Mountains, central Socorro County.

Monica Spring. At the north end of the San Mateo Mountains, central Socorro County. Alt. 2,319 meters.

Monument 40. See Upper Corner Monument.

Monument Rock. In Santa Fe Canyon, about 9 miles above Santa Fe.

Mora. Western Mora County. Alt. 2,135 meters.

Mora Creek. In Mora County, flowing into the Canadian.

Mora River. A tributary of the Pecos, northwestern San Miguel County.

Moreno Valley. Western Colfax County.

Moriarity. Northern Torrance County. Alt. 1,892 meters.

Mountainair. Western Torrance County. Alt. 1,997 meters.

Mount Joe. Lincoln National Forest.

Mount Sedgwick. In the Zuni Mountains, Valencia County. Alt. 2,851 meters.

Mule Creek. Northwestern Grant County.

Mule Spring. Eastern Grant County. Alt. 1,611 meters.

Nacimiento Mountain. Southern edge of Rio Arriba County. Alt. 3,064 meters.

Nambe. Northern Santa Fe County. Alt. 1,844 meters.

Nambe Valley. Northern Santa Fe County.

Nara Visa. Northeastern Quay County. Alt. 1,276 meters.

Navajo Church. An isolated rock in southern McKinley County.

N Bar Ranch. Western Socorro County, southeast of Reserve.

Negrito Creek. A tributary of the Rio San Francisco in western Socorro County.

N H Ranch. Western Socorro County, north of Reserve.

Nogal. Southern Lincoln County.

NOGAL CANYON. Southern Lincoln County.

NOONDAY CAMP. About 15 miles north of Salt Lake, southwest corner of Valencia County.

NORIA. South edge of San Augustine Plains, south central Socorro County. Alt. 1,254 meters.

NORTH PERCHA CREEK. Southern Sierra County, heading in the Black Range.

NUTRIA. On the Zuni Reservation, southwestern McKinley County. Alt. 2,115 meters. Another place of the same name is in Rio Arriba County, south of Tierra Amarilla.

NUTRIA RIVER. On the Zuni Reservation, southwestern McKinley County.

NUTRITAS CREEK. Central Rio Arriba County.

NUTT. Northeastern Luna County. Alt. 1,430 meters.

NUTT MOUNTAIN. Northeastern Luna County.

OAK CANYON. Near Folsom, Northwestern Union County.

OCATE CREEK. Mora County.

OGLE. Western Quay County.

OJO CALIENTE. North end of Canada Alamosa, north line of Sierra County.

OJO CALIENTE. On the Zuni Reservation, southwestern McKinley County. Alt. 1,921 meters.

OJO DE BERNAL. On the old Santa Fe Trail between Las Vegas and Pecos, a few miles east of the Pecos River.

OJO DE GAVILAN. Western Grant County.

OJO DEL MUERTO. Eastern Luna County.

OJO DE VACA. Northwestern Luna County. Alt. 1,522 meters.

OLD ALBUQUERQUE. Bernalillo County.

OLD TIPTOP. The highest peak of the Organ Mountains, Dona Ana County. Alt. 2,777 meters.

OLLA. At the upper end of the Jornada del Muerto.

ORGAN MOUNTAINS. Eastern Dona Ana County. The highest peak has an altitude of 2,777 meters.

ORGAN PASS. The pass between the Organ and San Andreas mountains, Dona Ana County. Alt. 1,667 meters.

ORGAN PEAK. One of the principal peaks of the Organ Mountains, Dona Ana County.

OROGRANDE. Southern Otero County. Alt. 1,258 meters.

OSCURO MOUNTAINS. Eastern Socorro County. The highest peak has an altitude of 2,663 meters.

OTIS. Central Eddy County. Alt. 953 meters.

OTTO. Southern Santa Fe County. Alt. 1,891 meters.

PAJARITO PARK. Northeastern Sandoval County.

PALOMAS. Central Sierra County. Alt. 1,268 meters. This is also known as Las Palomas, and is the Pigeon Creek of Emory's Reconnoissance.

PAQUATE. Northeastern Valencia County. Alt. 1,891 meters.

PARKERS WELL. East of the Organ Mountains, Dona Ana County.

PARK VIEW. Northern Rio Arriba County.

PATTERSON. Western Socorro County.

PECOS. Western San Miguel County. Alt. 1,942 meters.

PECOS BALDY. Southwest corner of Mora County, in the Pecos National Forest. Alt. 3,784 meters.

PECOS NATIONAL FOREST. Includes parts of Taos, Mora, Santa Fe, and San Miguel counties.

PECOS RIVER. The second largest stream of the State, heading in the Santa Fe and Las Vegas mountains, flowing south through San Miguel, Guadalupe, Chaves, and Eddy counties.

PECOS RUINS. The remains of the old pueblo of Pecos, southwest of the present town of that name, western San Miguel County.

PELMANS WELL or PELMANS RANCH. On the plains near the White Sands, western Otero County.

PENA BLANCA. Low mountain at the south end of the Organ Mountains, Dona Ana County.

PERCHA CREEK. Southern Sierra County, heading in the Black Range.

PERICO. Eastern Union County.

PERICO CREEK. Eastern Union County.

PESCADO SPRING. Southern McKinley County. Alt. 1,996 meters.

PICACHO. Southeastern Lincoln County.

PICACHO MOUNTAIN. An isolated mountain in central Dona Ana County. Alt. 1,472 meters.

PINE CIENAGA. Northwestern Grant County.

PINO CANYON. In the Sandia Mountains.

PINOS ALTOS. Northern Grant County. Alt. 2,088 meters.

PINOS ALTOS MOUNTAINS. Northern Grant County. The highest peak has an altitude of 2,479 meters.

PITT LAKE. Northern Guadalupe County.

PLACITAS. Southeastern Sandoval County. Alt. 1,982 meters.

PLAYAS VALLEY. Southern Grant County.

PLAZA LARGA. Central Quay County.

PLEASANTON. Southwest corner of Socorro County. Alt. 1,372 meters.

POJOAQUE. Northern Santa Fe County. Alt. 1,754 meters.

POLK CANYON. Near Mangas Springs, northern Grant County.

PONCHUELO CREEK. A tributary of the Pecos in the mountains of the northwest corner of San Miguel County.

PONIL CREEK. Central Colfax County.

PORTALES. Eastern Roosevelt County. Alt. 1,220 meters.

PROVIDENCIA LAKE. Central Luna County.

PUERCO RIVER. A tributary of the Rio Grande, running through Socorro, Valencia, and Sandoval counties. There is another stream of the same name in McKinley County.

PUERTECITO. Northwestern Socorro County.

QUAY. Central Quay County.

QUEEN. In the Guadalupe Mountains, southwestern Eddy County. Alt. 1,800 meters.

QUEMADO. Northwestern Socorro County. Also known as Rito Quemado.

RAMAH. Southwestern McKinley County. Alt. 2,135 meters.

RANCHOS DE TAOS. Central Taos County.

RATON. Northwestern Colfax County. Alt. 2,020 meters.

RATON MOUNTAINS. Northwestern Colfax County, lying partly in Colorado. The highest peak has an altitude of 2,544 meters.

RED LAKE. Southern Roosevelt County.

REDLANDS. Southeast corner of Roosevelt County.

REDROCK. Northwestern Grant County.

REDSTONE. Eastern Grant County.

RESERVE. Western Socorro County.

RILEYS RANCH. On the west side of the Organ Mountains, southeastern Dona Ana County; now abandoned.

RINCON. Northwestern Dona Ana County. Alt. 1,232 meters.

RIO ALAMOSA. Southern Socorro and northern Sierra counties.

RIO APACHE. Western Socorro County. The same as Apache Creek.

RIO BLANCO. Eastern San Juan County.

RIO BONITO. In the White Mountains, southern Lincoln County.

RIO DEL NORTE. See Rio Grande.

RIO DE SANTA FE. A small stream running down from the Santa Fe Mountains, through Santa Fe.

RIO FRESNAL. In the Sacramento Mountains, Otero County.

RIO FRISCO. Western Socorro County.

RIO GILA. One of the principal streams of the State, heading in the Mogollon Mountains, flowing through Socorro and Grant counties.

RIO GRANDE. The principal river of the State, flowing across it from the middle of the north boundary to about the middle of the southern boundary.

RIO HONDO. Western San Miguel County. Another of the same name is in Lincoln and Chaves counties.

RIO LAGUNA. Eastern Valencia County.

RIO MIMBRES. In Grant and Luna counties.

RIO NAMBE. Northern Santa Fe County, heading in the Santa Fe Range.

RIO NEGRITO. Western Socorro County.

RIO PECOS. See Pecos River.

RIO PUEBLO. Southern Taos County.

RIO PUERCO. See Puerco River.

RIO RUIDOSO. In the White Mountains, southern Lincoln County.

RIO SAN FRANCISCO. Western Socorro County.

RIO SAN JOSE. Eastern Valencia County.

RIO SECO. South of the Gallinas Mountains, Lincoln County.

RIO TESUQUE. Northern Santa Fe County, heading in the Santa Fe Range.

RIO ZUNI. On the Zuni Reservation, southwestern McKinley County.

RITO QUEMADO. Northwestern Socorro County. Alt. 2,082 meters.

RIVERA. Western San Miguel County.

ROBS CANYON. Western Sierra County.

ROCIADA. Northwestern San Miguel County. Alt. 2,257 meters.

ROCKY CANYON. Northeastern Grant County.

ROPES SPRING. San Andreas Mountains, Dona Ana County.

ROSA. Northeast corner of San Juan County. Alt. 1,830 meters.

ROSWELL. Chaves County. Alt. 1,085 meters.

ROUND MOUNTAIN. On Tularosa Creek above Tularosa, northern Otero County. This is the locality referred to in the citation of specimens, unless otherwise stated. There is another Round Mountain on the upper Pecos River above Winsors Ranch, northwest corner of San Miguel County. Alt. of the latter 3,385 meters.

ROWE. Western edge of San Miguel County. Alt. 2,078 meters.

ROY. Eastern Mora County. Alt. 1,795 meters.

RUIDOSO. In the White Mountains, on the southern edge of Lincoln County. Alt. 2,059 meters.

RUIDOSO CREEK. In the White Mountains, southern Lincoln County.

SABINAL. Northeastern Socorro County. Alt. 1,447 meters.

SACRAMENTO MOUNTAINS. Eastern Otero County

SADDLE ROCK CANYON Near Mangas Springs, western Grant County.

ST. VRAIN. Southern Curry County. Alt. 1,298 meters.

SALADO CANYON. Southern Lincoln County.

SALAZAR. Southwestern Sandoval County.

SALINAS. Northern Otero County.

SALT LAKE. Northwestern Socorro County.

SAN ANDREAS MOUNTAINS. Southeastern Socorro and northwestern Dona Ana counties.

SAN ANTONIO. Eastern Socorro County. Alt. 1,378 meters.

SAN AUGUSTINE PASS. North end of the Organ Mountains, Dona Ana County.
SAN AUGUSTINE PLAINS. Central Socorro County. Alt. 2,060 meters.
SAN AUGUSTINE RANCH. On the east side of the Organ Mountains, Dona Ana County.
SANDIA MOUNTAINS. Bernalillo and Sandoval counties. The highest peak has an altitude of 3,236 meters.
SAN FRANCISCO MOUNTAINS. Western Socorro County.
SAN IGNACIO. Western San Miguel County.
SAN JUAN. Southeastern Rio Arriba County. Alt. 1,708 meters.
SAN JUAN RIVER. Northern San Juan County.
SAN LORENZO. Northeastern Grant County. Alt. 1,862 meters.
SAN LUIS MOUNTAINS. Southern Grant County.
SAN LUIS PASS. Southern Grant County.
SAN MARCIAL. Southeastern Socorro County. Alt. 1,354 meters.
SAN MATEO MOUNTAINS. Southern Socorro County.
SAN MATEO PEAK. At the south end of the San Mateo Range, Socorro County. Alt. 3,114 meters.
SAN MATEO SPRING. San Mateo Mountains, Socorro County.
SAN MIGUEL. Western San Miguel County. Alt. 1,836 meters.
SAN MIGUEL. Southern Dona Ana County.
SAN RAFAEL. Northern Valencia County. Alt. 1,985 meters.
SANTA ANTONITA. Sandia Mountains.
SANTA BARBARA. Northern Dona Ana County.
SANTA CLARA CANYON. Northeast corner of Sandoval County.
SANTA FE. Santa Fe County. Alt. 2,117 meters.
SANTA FE CANYON. Canyon of Santa Fe Creek.
SANTA FE CREEK. A small stream heading in the Santa Fe Range and flowing through Santa Fe.
SANTA FE MOUNTAINS. Santa Fe and San Miguel counties.
SANTA RITA. Eastern Grant County. Alt. 1,879 meters. This was originally known as Santa Rita del Cobre. In the earlier reports it is called Copper Mines or Cobre.
SANTA RITA MOUNTAIN. Northeastern Grant County.
SANTA ROSA. Central Guadalupe County. Alt. 1,407 meters.
SANTO DOMINGO. Eastern Sandoval County. Alt. 1,583 meters.
SAPELLO CANYON. Northwestern San Miguel County in the Las Vegas Range.
SAWYERS PEAK. Black Range, western Sierra County.
SCHOOLHOUSE CANYON. In the White Mountains, southwestern Lincoln County.
SHALAM. Central Dona Ana County.
SHALAM HILLS. Mountains west of the Rio Grande, opposite Shalam, Dona Ana County.
SHEEP MOUNTAIN. In the San Andreas Mountains, northeastern Dona Ana County.
SHINGLE CANYON. Northeastern Grant County.
SHIPROCK. On the San Juan River, northwestern San Juan County. Alt. 1,425 meters.
SIA. Central Sandoval County.
SIERRA BLANCA. White Mountains, Lincoln County.
SIERRA DE LAS ANIMAS. Southern Grant County.
SIERRA GRANDE. An isolated peak in northwestern Union County.
SILVER CITY. Northern Grant County. Alt. 1,768 meters.
SILVER CITY DRAW. Northern Grant County.
SILVER SPRING CANYON. Sacramento Mountains, Otero County.
SIXTEEN SPRING CANYON. Sacramento Mountains, eastern Otero County.
SOCORRO. Eastern Socorro County. Alt. 1,393 meters.

52576°—15——49

SOCORRO MOUNTAINS. Eastern Socorro County. Alt. 2,220 meters.
SOLEDAD CANYON. In the Organ Mountains, Dona Ana County.
SOUTH BERENDO CREEK. Near Roswell, Chaves County.
SOUTH SPRING. Near Roswell, Chaves County.
SOUTH SPRING RIVER. Near Roswell, Chaves County.
SPIRIT LAKE. In the Pecos National Forest, Santa Fe or San Miguel County.
SPRINGER. Southern Colfax County. Alt. 1,759 meters.
STANLEY. Southern Santa Fe County. Alt. 1,927 meters.
STAR PEAK. In the Black Range, western Sierra County.
STEINS PASS. Western Grant County. Alt. 1,327 meters.
STINKING LAKE. On the Jicarilla Apache Reservation, Rio Arriba County. Alt. 2,190 meters.
STRAUSS STATION. Southeastern Dona Ana County. Alt. 1,245 meters.
SULLIVANS HOLE. Southern Socorro County.
SULPHUR SPRINGS. Southwestern San Juan County.
SUWANEE. Eastern Valencia County.
SWAN CANYON. In the Burro Mountains, Grant County
SYCAMORE CREEK. Northern Grant County.
TAOS. Central Taos County. Alt. 2,130 meters.
TAOS MOUNTAINS. Eastern Taos County. Taos Peak has an altitude of 4,011 meters.
TEEL. Eastern Grant County.
TELEGRAPH MOUNTAINS. West of Mangas Springs, northwestern Grant County.
TESUQUE. Northern Santa Fe County. Alt. 1,904 meters.
TEXICO. Southeastern Curry County on the Texas boundary line. Alt. 1,262 meters.
TEXLINE. In Texas, but on the boundary line, east of Clayton. Alt. 1,433 meters.
THOREAU. Southern McKinley County. Alt. 2,135 meters.
THORNTON. Eastern Sandoval County. Alt. 1,603 meters.
THREE RIVERS. North edge of Otero County. Alt. 1,387 meters.
TIERRA AMARILLA. Northern Rio Arriba County. Alt. 2,277 meters.
TIERRA BLANCA. Southwestern Sierra County.
TIJERAS. In the Sandia Mountains, Bernalillo County. Alt. 1,895 meters.
TIJERAS CANYON. In the Sandia Mountains, Bernalillo County.
TIZNITZIN. Southern San Juan County. Alt. 1,769 meters.
TOBOGGAN. Sacramento Mountains, Otero County. Alt. 2,351 meters.
TORRANCE. Southeastern Torrance County. Alt. 1,961 meters.
TORREVIOS. Fifty miles northwest of Magdalena.
TORTUGAS. A low, isolated mountain southeast of Las Cruces, Dona Ana County.
TRES HERMANAS. Three isolated peaks in southern Luna County. Alt. 2,181 meters.
TRINCHERA PASS. Northwestern Union County. Alt. 2,159 meters.
TROUT SPRING. West of Las Vegas, western San Miguel County.
TRUCHAS PEAK. Southeastern corner of Rio Arriba County. Alt. 4,012 meters.
TRUJILLO CREEK. Southern Sierra County.
TUCUMCARI. Central Quay County. Alt. 1,276 meters.
TUCUMCARI MOUNTAIN. Central Quay County.
TUERTO MOUNTAIN. Near Santa Fe, Santa Fe County.
TULAROSA. Northern Otero County. Alt. 1,325 meters.
TULAROSA CREEK. Northern Otero County.
TULAROSA RIVER. Western Socorro County.
TUNITCHA MOUNTAINS. Western San Juan County.
TURNERS RANCH. Northwest of Tularosa, northwestern Otero County.
TURQUOISE. Western Otero County. Alt. 1,247 meters.
UPPER CORNER MONUMENT. Southwest corner of Luna County, one of the monuments marking the Mexican Boundary.

UPPER PECOS RIVER. That part of the Pecos lying in the mountains above the town of Pecos.

UTE PARK. Western Colfax County. Alt. 2,350 meters.

VALLE SANTA ROSA. In the Jemez Mountains, southern Rio Arriba County.

VALVERDE. Southern Socorro County.

VAN PATTENS CAMP. On the west side of the Organ Mountains, Dona Ana County.

VEGA DE SAN JOSE. Northwestern Valencia County.

VERMEJO PEAK. Northwestern Colfax County.

VICTORIA. Southern Dona Ana County. Alt. 1,153 meters.

V PASTURE. On Ruidoso Creek northwest of Ruidoso Post Office, southern Lincoln County.

V RANCH. East of Alto, southern Lincoln County.

WAGON MOUND. Central Mora County. Alt. 1,884 meters.

WARM SPRING. Near Faywood, southeastern Grant County.

WATER CANYON. Magdalena Mountains, central Socorro County.

WATROUS. Southern edge of Mora County. Alt. 1,951 meters.

WEST FORK OF THE GILA. Southwestern Socorro County, in the Mogollon Mountains.

WHEELER PEAK. Western Colfax County. Sometimes confused with Taos Peak.

WHEELERS RANCH. On Apache Creek, western Socorro County.

WHITE MOUNTAIN PEAK. North edge of Otero County. Alt. 3,624 meters.

WHITE MOUNTAINS. In southern Lincoln and northern Otero counties.

WHITE OAKS. Western Lincoln County. Alt. 1,973 meters.

WHITE SANDS. Western Otero County. Alt. 1,185 meters.

WHITE WATER. In Chihuahua, near the southern edge of Grant County, New Mexico.

WILLARD. Central Torrance County. Alt. 1,858 meters.

WILLOW CREEK. A tributary of the Gila River, on the east slopes of the Mogollon Range, southwestern Socorro County.

WILLOW CREEK. A tributary of the Chama River near Chama, Rio Arriba County.

WINGFIELDS RANCH. On Ruidoso Creek in the White Mountains, southern Lincoln County.

WINSOR CREEK. A small tributary of the Pecos in the mountains of the northwest corner of San Miguel County.

WINSORS RANCH. On the Upper Pecos near the northwest corner of San Miguel County. Alt. 2,562 meters.

WINTER FOLLY. Sacramento Mountains, Otero County.

WOLF CREEK. Southern Mora County.

ZUNI. Southwest corner of McKinley County.

ZUNI MOUNTAINS. McKinley and Valencia counties.

ZUNI RESERVATION. Southwestern McKinley County.

LIST OF NEW GENERA, SPECIES, AND HYBRIDS, AND NEW NAMES.

Page.

Oreolirion arizonicum (Rothr.) Bicknell 147

Sisyrinchium arizonicum Rothr.

Quercus emoryi×pungens Woot. & Standl 169

Quercus grisea×emoryi Woot. & Standl 170

Quercus arizonica×grisea Woot. & Standl 171

Razoumofskya microcarpa (Engelm.) Woot. & Standl 179

Arceuthobium douglasii microcarpum Engelm.

Alsine cuspidata (Willd.) Woot. & Standl 236

Stellaria cuspidata Willd.

Radicula terrestris (R. Br.) Woot. & Standl 284

Nasturtium terrestre R. Br.

Chamaecrista wrightii (A. Gray) Woot. & Standl 335

Cassia wrightii A. Gray.

Astragalus proximus (Rydb.) Woot. & Standl 366

Homalobus proximus Rydb.

Astragalus impensus (Sheld.) Woot. & Standl 369

Astragalus viridis impensus Sheld.

Oxytropis richardsoni (Hook.) Woot. & Standl 370

Oxytropis splendens richardsoni Hook.

Oxytropis pinetorum (Heller) Woot. & Standl 371

Aragallus pinetorum Heller.

Oxytropis vegana (Cockerell) Woot. & Standl 371

Aragallus pinetorum veganus Cockerell.

Polygala parvifolia (Wheelock) Woot. & Standl 392

Polygala lindheimeri parvifolia Wheelock.

Tithymalopsis strictior (Holzinger) Woot. & Standl 396

Euphorbia strictior Holzinger.

Stillingia smallii Woot. & Standl 405

Stillingia salicifolia Small, 1903, not Baill. 1865.

Trionum trionum (L.) Woot. & Standl 417

Hibiscus trionum L.

Hibiscus involucellatus (A. Gray) Woot. & Standl 417

Hibiscus denudatus involucellatus A. Gray.

Sidanoda (Robinson) Woot. & Standl 427

Anoda section *Sidanoda* Robinson.

Sidanoda pentaschista (A. Gray) Woot. & Standl 427

Anoda pentaschista A. Gray.

Echinocereus rosei Woot. & Standl 457

Echinocereus polyacanthus Engelm. err. det. Standley, 1908.

Raimannia mexicana (Spach) Woot. & Standl 470

Oenothera mexicana Spach.

Acerates ivesii (Britton) Woot. & Standl 509

Asclepias viridiflora ivesii Britton.

Nyctelea micrantha (Torr.) Woot. & Standl 535

Phacelia micrantha Torr.

Diapedium torreyi (A. Gray) Woot. & Standl 598

Dicliptera torreyi A. Gray.

Page.

Proboscidea louisiana (Mill.) Woot. & Standl. ... 602
Martynia louisiana Mill.

Proboscidea parviflora (Wooton) Woot. & Standl. ... 602
Martynia parviflora Wooton.

Erigeron grayi Woot. & Standl. ... 679
Erigeron stenophyllus A. Gray, 1856, not Nutt. 1847.

Brachyactis woodhousei (Wooton) Woot. & Standl. ... 682
Aster woodhousei Wooton.

Ratibida columnifera (Nutt.) Woot. & Standl. ... 706
Rudbeckia columnifera Nutt.

Ratibida columnifera pulcherrima (DC.) Woot. & Standl. ... 706
Obeliscaria pulcherrima DC.

Villanova biternata (A. Gray) Woot. & Standl. ... 725
Bahia biternata A. Gray.

Mesadenia decomposita (A. Gray) Standley ... 749
Cacalia decomposita A. Gray.

Cirsium vinaceum Woot. & Standl. ... 751
Carduus vinaceus Woot. & Standl.

Cirsium pallidum Woot. & Standl. ... 751
Carduus pallidus Woot. & Standl.

Cirsium inornatum Woot. & Standl. ... 751
Carduus inornatus Woot. & Standl.

Cirsium gilense Woot. & Standl. ... 751
Carduus gilensis Woot. & Standl.

Cirsium pulchellum (Greene) Woot. & Standl. ... 752
Carduus pulchellus Greene.

Cirsium calcareum (Jones) Woot. & Standl. ... 752
Cnicus calcareus Jones.

Cirsium perennans (Greene) Woot. & Standl. ... 753
Carduus perennans Greene.

INDEX.

[Synonyms in italic.]

	Page.
Abies	34–35
Abronia	223–224
Abutilon	419–420
Acacia	328–329, *330*
Acalypha	403
Acamptoclados	84
Acanthaceae	597–598
Acanthochiton	213
Acanthus family	597–598
Acer	410–411
Aceraceae	410–411
Acerates	508–509
Achillea	733
Achroanthes	151
Achyranthes	*214*, *215*
Achyropappus	724–725
Aciphyllaea	716
Acleisanthes	225
Acmispon	348
Acomastylis	317
Aconitum	247
Acrolasia	435–436
Acrostichum	20, 25
Actaea	249
Actinella	727, 729, 730, 731, 732
Actinomeris	714
Acuan	330–331
Adder's mouth	151
Adelia	496
Adenostegia	590–591
Adiantum	21
Adoxa	612
Adoxaceae	612
Agastache	562–564
Agave	145–147
Agoseris	624–626
Agrimonia	309
Agrimony	309
Agropyron	104–106
Agrostemma	242
Agrostis	*70*, *76*, *77*, 78–79
Ailanthus	390
Aira	*82*, *92*, *93*
Aizoaceae	228–229
Alcohol, manufacture from Dasylirion	138
Alder	164
Alfalfa	343
Alfileria	381
Alisma	42
Alismaceae	42
Alismales	41–42
Allenrolfea	201
Alliaceae	140–143
Alligator juniper	36
Allionia	*218*, 219–221
Allioniaceae	216–228
Allioniella	222
Allium	140–143
Allocarya	547
Allosorus	24
Almond family	324–327
Alnus	164
Alopecurus	73–74, *77*
Alpine bistort	197
speedwell	589
Alsike	345
Alsinaceae	234–240
Alsine	235–236
Alsinella	238
Alsinopsis	240
Alternanthera	*213*, 215
Alum root	294
Amaranth	210
family	209–216
globe	214
Amaranthaceae	209–216
Amaranthus	210–213
Amarella	501–502
Amaryllidaceae	145–147
Amaryllidales	145–148
Amaryllis family	145–147
Amauria	*725*
Amblogyne	211
Ambrosia	635, *636*
Ambrosiaceae	631–636
Amelanchier	322–323
Amellus	*659*, *664*
American brooklime	589
Amole	136, 146
Amorpha	349
Ampelopsis	*416*
Amphiachyris	656
Amphilophis	50–51
Amsonia	504
Amygdalaceae	324–327
Anacardiaceae	405–409
Anagallis	492
Anantherix	508
Anaphalis	695
Androcera	574
Andropogon	*49*, *50*, 51, *52*
Andropus	530–531
Androsace	493–494
Anemone	255
Anemopsis	154
Anethum	*485*
Angelica	481
Angiospermae	39–753
Angle-pod	507
Anisacanthus	597–598
Anisolotus	346–347
Anoda	426, *427*
Anogra	467–469
Antelope horns	508

	Page.
Antennaria	695–697
Anthemideae	637, 645
Anthemis	733–734
Anthericum	144–145
Anthopogon	501
Anticlea	129
Antirrhinum	*577*, 578, *579*
Anulocaulis	226
Apache plume	317
Aparejo grass	71
Aphanostephus	691
Aphora	402
Aphyllon	600
Apiaceae	475–485
Apinus	32
Apium	478
Aplopappus	*659, 664, 665, 666, 667*
Apocynaceae	504–506
Apocynum	505–506
Apodanthera	615
Apple family	321–324
Aquilegia	248–249
Arabis	*269*, 279–280
Aragallus	370, 371
Arales	124–125
Aralia	475
Arbutus	488–489
Arceuthobium	*178, 179*
Archemora	484
Arctium	753
Arctostaphylos	489
Arenaria	238–239, *240*
Argemone	261
Argentina	314–315
Argyrothamnia	402
Aristida	62–65
Aristolochia	181
Aristolochiaceae	181
Aristolochiales	181
Arizona cypress	35
fescue	102
oak	170
pine	32
yellow pine	33
Arnica	740
Arrow grass	41
family	41
Arrowhead	42
Arrow-wood	693
Artemisia	734–739
Arundo	*80*, 89–90
Asclepiadaceae	506–513
Asclepiadales	503–513
Asclepias	509–512
Asclepiodora	508
Ash	496
mountain	324
Asparagus	139
Aspen	155
Asphodel family	144–145
Asphodelaceae	144–145
Aspidium	25, 26
Asplenium	24–25
Aster	*680*, 682–686, *687, 688, 689, 690, 692*
family	637–753
spiny	686
Asteraceae	637–753
Asterales	618–753
Astereae	637, 638
Astragalus	356–369, *370*
Astrophyllum	390
Atamasco lily	147
Atamosco	147
Atelophragma	358, 363
Atenia	479
Atheropogon	86, 87
Athyrium	25
Atragene	258
Atriplex	201–205
Aulospermum	481–482
Avena	81
Avens	317
Azolla	27
Baccharis	671–672
Bachelor's button	214
Baeria	723–724
Bahia	724, *725*
Bailey, Vernon	11
Baileya	718
Ball, C. R	11
Balsam fir	35
Barberry	258
family	258–260
Bark of yellow pine, use by Indians	32
Barley, meadow	107
wall	107
Barnhart, J. H	11
Barnyard grass	59
Bartlettia	749
Bartonia	433, 434
Basil, wild	556
Basketry, use of Martynia fiber	601
use of Nolina	138
use of Yucca leaves	137
Bastard toadflax	181
Batanthes	526, 527
Batidaea	320
Batis	*199*
Batrachium	250–251
Bean	377
coral	376
Metcalfe	378
wild	378
Bearberry	489
Beardgrass	77
Beard-tongue	579
Beargrass	137
Bebbia	740
Beckmannia	84
Bedstraw	604
northern	605
sweet-scented	606
Bee plant, Rocky Mountain	290
Beech family	164–174
Beggar-ticks	703
Belvisia	25
Bentgrass, water	78
Berberidaceae	258–260
Berberis	258, *259, 260*
Berlandiera	699–700
Bermuda grass	84
Bernardia	405

	Page.
Berthelotia	693
Berula	479–480
Besseya	590
Betula	163
Betulaceae	163–164
Bidens	*702*, 703–705
Bigelovia	*661, 662, 663, 665*
Bignonia	*601*
family	600–601
Bignoniaceae	600–601
Bilderdykia	197
Bindweed	519
black	197
Biological Survey, botanical collections	10
Birch	163
family	163–164
Bird-of-paradise flower	334
Bird's-foot trefoil	346
Birthwort family	181
Biscutella	*269*
Bistort	197
alpine	197
Bistorta	197
Bitter dock	193
Black bindweed	197
grama	87
locust	356
medic	343
nightshade	573
oak	169
raspberry	320
willow	161
Black-eyed Susan	707
Bladder-pod	274
double	270
Bladderwort	599
family	599
Blanket flower	719
Blazing star	649
Blepharipappus	698
Blepharoneuron	74–75
Blite	206
Blitum	206
Blue curls	554
flag	148
grama	87
toadflax	577
Bluebell	616
family	616–617
Blueberry	489
Blue-eyed grass	147
Bluegrass	98
English	100
Kentucky	100
Bluestem, Colorado	106
Bluets	607
Blueweed	710
Boebera	715–716
Boehmeria	176
Boerhaavia	*225*, 226–228
Bommeria	19
Borage family	538–548
Boraginaceae	538–548
Borreria	606
Bossekia	320, 321
Botany of New Mexico, bibliography	9
of Western Texas	9
Bouncing Bet	240
Bouteloua	85–87
Bouvardia	606
Box elder	411
Brachiaria	*57*
Brachyactis	682
Brachylobus	284
Brachyris	656
Bracken	21
Bracted orchis	152
Brainerd, Ezra	11
Brake, cliff	23
Brassica	281, *288*
Brassicaceae	263–289
Brayulinea	216
Brickellia	647, 651, 652, 653
Brigham Young weed	38
Brittle fern	26
Britton, N. L	11
Brittonastrum	563, 564
Briza	*93*
Brizopyrum	*94*
Brodiaea	*143*
Brome grass	95
Bromus	95–97
Brooklime, American	589
Brookweed	491
Broomrape	599
family	599
Broussonetia	339
Brownweed	658
Buchloe	83
Buckeye, New Mexican	412
Buckthorn	413, 414, 495
family	413–415
Buckwheat	197
family	181–197
Buffalo berry	458
bur	574
grass	82
Bugleweed	555
Bugloss, viper's	547
Bugseed	205
Bulbills	82–83
Bulbostylis	748
Bulrush	114
Bumelia	495
Bunch-flower family	128–130
Bunchgrass	76
sand	73
Bur cucumber, one-seeded	614
Burdock	753
Burnut	386
Burrielia	724
Burro weed	201
Bursa	273
Buttercup	251
Butterfly weed	511
Cabbage, skunk	128
Cacalia	749
Cachanilla	693
Cactaceae	436–458
of New Mexico, bulletin	11
Cactus	*441, 444, 451*
candelabrum	443
cane	443
family	436–458

	Page.
Cactus, rainbow	456
tree	443
Caesalpinia	334
Calamagrostis	79-80
Calamintha	557
Calamovilfa	80
Calandrinia	*233*
Calceolaria	431
California poppy	262
Calliandra	328
Calligonum	204
Calliopsis	703
Callirrhoe	418
Callisteris	526, 527
Callitrichaceae	405
Callitriche	405
Calochortaceae	127-128
Calochortus	127-128
Caltha	249
Caltrop family	385-387
Calycoseris	627
Calymenia	220, 221
Calypso	150
Camass, death	130
Camelina	288, 289
Camote de ratón	333
Campanula	616
Campanulaceae	616-617
Campanulales	612-617
Campion, moss	241
Cañaigre	191
Canary grass	61
Cañatillo	38
Cancer root	600
Candelabrum cactus	443
Candy, manufacture from *Echinocactus wislizeni*	453
Cane cacti	437
cactus	443
Cannabinaceae	177
Canteloupe	616
Cantua	*528*
Caper family	289-291
Capnoides	262-263
Capparidaceae	289-291
Caprifoliaceae	608-612
Capriola	84
Capsella	273
Cardamine	276
Cardaria	271
Carduus	*751, 752, 753*
Carex	116-124
Carlowrightia	598
Carminatia	646
Carpetweed	228, 229
family	228-229
Carphephorus	*740*
Carpochaete	648
Carrizo	89
Carrot	484
Carum	479
Caryopitys	31
Cashew family	405-409
Cassia	334-335
Cassiaceae	332-335
Castilleja	591-594
Catalpa	600
Catchfly	241
night-flowering	242
sleepy	241
Cat-claw	331
Cathartolinum	382-383
Catnip	564
Cat-tail	39
family	39
Caulinia	41
Ceanothus	413-414
Cedar	36
salt	427
Cedronella	563, 564
Celastraceae	409-410
Celery	478
Celtis	174-175
Cenchrus	61
Centaurea	749
Centaurium	498-499
Centaury	498
Century plant	145
Cerastium	236-237
Cerasus	*326*, 327
Ceratophyllaceae	243
Ceratophyllum	243
Cercocarpus	318-319
use in dyeing	164
Cereus	*455, 456, 457, 458*
Cevallia	432
Chaenactis	723
Chaetochloa	60
Chamaecrista	335
Chamaenerion	463
Chamaesaracha	*571*, 574
Chamaesyce	397-401
Chamiso	204
Chaptalia	630
Charlock	281
Cheat	97
Cheilanthes	*20*, 21-23
Cheiranthus	*282, 283*
Cheirinia	281-283
Chelone	*582*
Chenopodiaceae	198-209
Chenopodiales	198-242
Chenopodina	200
Chenopodium	206-209
Cherry	327
Chestnut oak	171
Chickweed	236
family	234-240
mouse-ear	236
Chico bush	199
Chicory	622
family	618-630
Chilopsis	601
Chimaphila	487
China-berry	390
family	390
Chloris	*83, 85, 86*, 87-88
Chokecherry	325
Chollas	437
Chondrophylla	502

	Page.
Chondrosium	86, 87
Chrysactinia	714
Chrysanthemum	734
Chrysocoma	663
Chrysoma	666
Chrysopsis	658–660
Chrysothamnus	660–663
Chthonia	715
Chuchupate	480
Cichoriaceae	618–630
Cichorium	622
Cicuta	479
Ciénaga, definition of	28
Cienaguilla, definition of	28
Cincinalis	20
Cinna	78
Cinquefoil	310
shrubby	316
Circaea	460
Cirsium	749–753
Cissus	416
Citrullus	615–616
Cladium	112
Cladothrix	213–214
Clammy weed	289
Clappia	719
Clavigera	651
Claytonia	233
Clematis	*255*, 256, *257*, *258*
Clementsia	292
Cleome	*289*, *290*
Cleomella	290
Cliff brake	23
Clinopodium	556–557
Cloak fern	19
Closed gentian	502
Clover	344
dodder	515
prairie	355
red	345
sweet	343, 344
white	345
Cnemidophacos	359, 364
Cnicus	*751*, *752*, *753*
Cockerell, T. D. A.	10, 11
Cockle, corn	242
Cocklebur	634
Coelestina	648
Coeloglossum	152
Cogswellia	484
Cohosh	249
Coldenia	*536*
Coleosanthus	650–653
Collinsia	579
Collomia	521, *523*, *528*
Cologania	378–379
Colorado blue spruce	34
bluestem	106
rubber plant	727
Columbine	248
Comandra	181
Commelina	125–126
Commelinaceae	125–126
Common groundsel	743
mallow	419
plantain	603
Common poppy	262
purslane	230
ragweed	635
sunflower	711
Conanthus	*530*, *531*
Condalia	413
Cone flower	705, 707
Conioselinum	480
Conobea	588
Conoclinium	647
Conopholis	599
Conringia	288
Convallaria	*139*
Convallariaceae	138–140
Convolvulaceae	515–519
Convolvulus	*518*, 519
Conyza	*673*
Conyzella	673
Coral bean	376
root	150
Corallorhiza	150
Cordylanthus	591
Coreopsis	*702*, 703
Coriander	484
Coriandrum	484–485
Corispermum	205–206
Cork-bark fir	34
Corn cockle	242
Cornaceae	474–475
Cornel	474
Cornucopiae	*79*
Cornus	474
Corrigiolaceae	216
Corydalis	263
Cosmidium	702
Cosmos	701–702
wild	723
Cota	703
Cotton grass	116
Cottonwood	155
mountain	156
valley	155
Coulter, J. M.	9
Coulter's Rocky Mountain Flora	10
Covillea	386
Cow parsnip	484
Cowania	318
Cowherb	241
Crabgrass	54
Texan	85
Cranesbill	380
family	379–381
Crassina	700–701
Crassulaceae	292–294
Crataegus	323–324
Creosote bush	386
Crepis	*628*, 629–630
Cress, hoary	271
penny	272
rock	279
water	284
yellow	284
Cressa	516
Cristaria	*425*
Crotalaria	339
Croton	402–403

Page.
Crowfoot family 243-258
grama 88
water 250
Crownbeard 713
Crown-of-thorns 388
Crunocallis 233
Crusea 606
Crypsis *90*
Cryptanthe 546-547
Cubiertas 453
Cucumis 616
Cucurbita 615
Cucurbitaceae 612-616
Cudweed 694
Cupressus 35-36
Curled mallow 419
Currant 301
golden 302
Cuscuta 514-515
Cuscutaceae 514-515
Cycloloma 206
Cymopterus *481*, 482
Cynareae 637, 645
Cynodon 84
Cynoglossum 539
Cynomarathrum 485
Cynthia 621-622
Cyperaceae 110-124
Cyperus 110-112
Cyphomeris 225
Cypress 35
Arizona 35
Cypripedium 149-150
Cystium 357, 361
Cystopteris 26
Cytherea 150
Dactylis 97-98
Dactylophyllum 529-530
Dagger 137
Daisy, ox-eye 734
Dalea 351, 352, 353, 354, 355
Dandelion 626
false 624
Danthonia 80-81
Daphnidostylis 489
Dasiphora 316
Dasylirion 138
Dasyochloa 90
Dasystephana 502-503
Dasystoma 590
Datil 137
Datura 567
Daucus *478*, 484
Dayflower 125
Death camass 130
Deer's ears 499
Delphinium 244-247
Deschampsia 81-82
Descriptions of new plants from New Mexico, publication 11
Desert willow 601
Desmanthus 331
Desmodium 372
Devil claws 601
horns 601

Page.
Devil's pincushion 453
Deweya *479*
Deyeuxia 80
Diapedium 598
Dichelostemma 143
Dichondra 514
family 514
Dichondraceae 514
Dichrophyllum 396
Dicliptera 598
Dicoria 632
Dicranocarpus 698
Dieteria 665, 688, 689
Digitaria 54
Diholcos 365
Dinebra 86
Diotis *201*
Dipetalia 291
Diplachne 83, 84
Diplopappus 660, 690
Dipteracanthus 598
Dipterostemon 143
Disaccanthus 269
Disella 424
Disporum 140
Distasis 678, 680
Distegia 612
Distichlis 98
Ditaxis 401-402
Dithyraea 269-270
Dock 191
bitter 193
golden 193
pale 192
yellow 192
Dodder 514
clover 515
family 514-515
Dodecatheon 492
Dogbane 505
family 504-506
Dogwood family 474-475
Dolicholus 376
Dondia 200-201
Donia 655
Double bladder-pod 270
Douglas spruce 35
Draba 276-279
Dracaenaceae 135-138
Dracocephalum *562*
Dracopis 706
Dragon-head 562
Drejera *597*
Dropseed 75
Nealley's 77
sandhill 76
Drosace 494-495
Drymaria 234-235
Drymocallis 316
Dryopetalon 288
Dryopteris 25
Duckweed 124
family 124-125
Dugaldea 726
Dulce de viznaga 453

	Page.
Dyeing, use of Alnus	164
use of Cercocarpus	164
use of Chrysothamnus	660
use of Rumex	192
Dysodia	*716*, *717*
Eatonia	92
Ebenales	495
Echeandia	*144*
Echinocactus	451–454, *456*
Echinocereus	454–457
Echinochloa	59
Echinocystis	613, 614
Echinopepon	613
Echinospermum	540
Echium	547–548
Eclipta	705
Eddya	536–537
Edwinia	298–299
Egletes	*691*
Ehretiaceae	535–537
Elaeagnaceae	458
Elaeagnus	458
Elaterium	*614*
Elatinaceae	427
Elatine	427
Elderberry	609
Eleocharis	112–114, *115*
Elephant, little red	595
Elephantella	595
Eleusine	*83*
Elkslip	249
Ellis, Miss Charlotte C	10
Elm familv	174–175
Elymus	*105*, 108–109
Elyna	124
Elyonurus	49
Elytraria	597
Encelia	712
Enchanter's nightshade	460
Engelmann spruce	34
Engelmannia	700
English bluegrass	100
ribgrass	603
walnut	161
Enomegra	261
Ephedra	38–39
Ephedraceae	38–39
Epicampes	74
Epilobiaceae	459–473
Epilobium	463–464
Epipactis	152
Equisetaceae	28–29
Equisetales	28–29
Equisetum	28–29
Eragrostis	*84*, 93–94, *101*
Eremocarya	544
Eriastrum	520–521
Ericaceae	488–489
Ericales	486–490
Erigéron	*673*, 674–682, *693*
Eriocarpum	664, 665
Eriochloa	54–55
Eriocoma	73
Eriodictyon	530
Eriogonum	182–190
Eriogynia	310
Erioneuron	90
Eriophorum	116
Eritrichium	544, *545*, *546*, *547*
Erodium	381
Eruca	281
Eryngium	477–478
Erysimum	*282*, *283*, *285*
Erythraea	499
Erythremia	629
Erythrina	376
Erythrocoma	317
Eschenbachia	673
Eschscholzia	262
Estafiata	738
Eucephalus	686
Euchroma	594
Euklisia	268–269
Eupatorieae	637
Eupatorium	646–647, *649*, *651*
Euphorbia	*394*, *395*, *396*, *397*, *398*, *399*, *400*, *401*
Euphorbiaceae	393–405
Euphorbiales	392–405
Euphrosyne	633
Euploca	537
Eurotia	201
Eustoma	499
Euthamia	667–668
Eutoca	533
Evax	693
Evening primrose	469
family	459–473
Everlasting, pearly	695
Evolvulus	516
Eysenhardtia	349
Fabaceae	336–379
Fagaceae	164–174
Fagales	163–174
Fagopyrum	197
Fall witch grass	54
Fallugia	317–318
False dandelion	624
flax	288
foxglove	590
hellebore	129
indigo	349
needle grass	89
nettle	176
oats	82
Solomon's seal	139
Fendler oak	167
Fendlera	299–300
Fendlerella	299
Fennel	485
Fern, brittle	26
cloak	19
family	18–27
lady	25
lip	21
maiden-hair	21
male	25
Venus-hair	21
Fernald, M. L	11
Fescue	101
Arizona	102
meadow	103
Festuca	*83*, *96*, 101–103, *105*

	Page.
Fetid marigold	715
Figwort	578
family	575–597
Filaginopsis	693
Filaree	381
Filicales	18–27
Filix	26
Fimbristylis	*114*
Fir	34
balsam	35
cork-bark	34
Fireweed	463
Flag, blue	148
Flaveria	726–727
Flax, false	288
family	381–383
wild	382
yellow	382
Fleabane	674
marsh	693
Flora of Colorado	10
of New Mexico, plan	11
preparation	10
Flourensia	711–712
Flowering ash	496
Flower-of-an-hour	417
Flower-of-Parnassus	298
family	298
Fly honeysuckle	611
Foeniculum	485
Forestiera	495–496
Forget-me-not, mountain	544
Forsellesia	409–410
Fouquieria	428
Fouquieriaceae	428
Four-o'clock	221
family	216–228
Foxglove, false	590
Foxtail, green	60
marsh	73
pine	33
slender	74
Fragaria	315
Frankenia	428
family	428
Frankeniaceae	428
Franseria	636
Frasera	499–500
Fraxinus	496
Fremontia	199
Fringed gentian	501
Fritillaria	144
Froelichia	214
Fumariaceae	262–263
Fumitory family	262–263
Gaertneria	636
Gaillardia	719–720, *732*
Galactia	376–377
Galinsoga	701
Galium	604–606
Galleta grass	53
Galpinsia	465–466
Gambel oak	174
Garrambullo	442
Garrya	474–475
Gaura	461–462

	Page.
Gaurella	473
Gayoides	419
Gayophytum	464
Gentian	501
closed	502
family	497–503
fringed	501
Gentiana	*501, 502, 503*
Gentianaceae	497–503
Gentianales	497–503
Geoprumnon	357, 361
Geraniaceae	379–381
Geraniales	379–390
Geranium	380–381
Gerardia	*590*
Germander	553
Geum	317
Giant hyssop	562
reed	89
Gilia	*521, 523*, 525–529, *530*
Globe amaranth	214
Glossopetalon	410
Glyceria	103
Glycyrrhiza	371
Gnaphalieae	637, 640
Gnaphalium	694–695
Gnetales	38–39
Golden currant	302
dock	193
Goldenrod	668
Goldman, E. A	11
Golondrina	397
Gomphocarpus	513
Gomphrena	214–215
Gonolobus	*507*
Gonopyrum	193
Goodyera	151
Gooseberry	303
family	301–304
Goosefoot	206
family	198–209
maple-leaved	208
oak-leaved	208
Goosegrass	606
Gossypianthus	215
Gourd	615
family	612–616
Grama, black	87
blue	87
crowfoot	88
grass	85
hairy	87
sandhill	86
six-weeks	86
tall	85
Grape	415
family	415–416
Oregon	259
Grass, aparejo	71
arrow	41
barnyard	59
Bermuda	84
blue-eyed	147
brome	95
buffalo	82
canary	61

	Page.
Grass, cotton	116
fall witch	54
false needle	89
family	43-109
galleta	53
grama	85
hair	79
Hungarian brome	97
Indian	52
Johnson	52
joint	55
June	92
manna	103
meadow	104
melic	95
mesquite	69
mutton	101
New Mexican porcupine	66
nut	112
orchard	97
panic	56
pigeon	60
pinyon	66
porcupine	65
purple hair	69
purple needle	64
reed canary	61
ring	70
rye	106
sage	49
sand	80
six-weeks needle	63
sleepy	67
slender wheat	105
slough	84
southern canary	61
squirrel-tail	107
stink	93
switch	58
tall sage	51
Texas curly mesquite	53
tobosa	53
vanilla	61
vine mesquite	57
wheat	104
wild oat	80
wood reed	78
Grasses of New Mexico, bulletin	11
Gratiola	586
Greasewood	199, 386
Great ragweed	635
Green foxtail	60
Greene, E. L	11
Greggia	270
Grindelia	653-655
Grossularia	303-304
Grossulariaceae	301-304
Ground-cherry	570
Groundsel, common	743
Guardiola	698
Guayule	699
Guilleminea	216
Gum plant	653
Gunneraceae	473
Gutierrezia	656-658

	Page.
Gymnogramme	19
Gymnolomia	707-708
Gymnopteris	*19*
Gymnosperma	656
Gymnospermae	30-39
Gyrostachys	154
Habenaria	*152, 153, 154*
Hackberry	174
Hair grass	79
Hairy grama	87
Halenia	499
Halerpestes	253
Halostachys	*201*
Hamosa	357, 361
Haploesthes	748
Hare's-ear mustard	288
Hartmannia	470-471
Hawkweed	627
Hawthorn	323
Heath family	488-489
Hedeoma	*556*, 557-558
Hederaceae	475
Hedge hyssop	586
mustard	286
nettle	560
Hediondilla	386
Hedyotis	*607*
Hedysarum	*371*, 373
Helenieae	637, 643
Helenium	726
Heliantheae	637, 641
Helianthella	712
Helianthus	710-711, *712*
Heliomeris	708
Heliopsis	701
Heliotropaceae	537-538
Heliotrope family	537-538
Heliotropium	537-538
Hellebore, false	129
Helleborine	152
Hemicarpha	116
Hemiptilium	620
Hemlock, water	479
Hemp family	177
Henbit	565
Hendecandra	402
Heracleum	484
Herbarium material studied	10
Herrick, C. L	10
Herrickia	686
Hesperanthes	144
Hesperaster	433, 434, 435
Hesperidanthus	267
Heteranthera	126
Heteropogon	52
Heterospermum	*698*, 705
Heterotheca	658
Heterothrix	268
Heuchera	294-296
Hibiscus	417
Hickory nut	161
Hieracium	627-628
Hierochloe	61
Hilaria	53-54
Hippophae	*458*

Page.

Hippuris 473
Hoary cress 271
Hoffmanseggia 332–333
Hog millet 58
Holcus 51–52, *61*
Holodiscus 310
Homalobus 365, 366
Honeysuckle 611
  family 608–612
  fly 611
Hop 177
  hornbeam 163
Hordeum 106–107
Horehound 556
Horned pondweed 40
Hornwort 243
  family 243
Horsemint 559
Horsetail 28
  family 28–29
Horseweed 673
Hosackia 346, 347, 348
Hound's-tongue 539
Houstonia 607–608
Houttuynia *154*
Huckleberry family 489–490
Humulus 177
Hungarian brome grass 97
Hybanthus *431*
Hydrangea family 298–301
Hydrangeaceae 298–301
Hydrophyllaceae 530–535
Hydrophyllum 535
Hymenatherum 716, 717
Hymenoclea 634
Hymenopappus 720–722, *723*
Hymenothrix 722–723
Hymenoxys 727–730, *731*
Hypericaceae 428
Hypericales 427–431
Hypericum 428
Hypopitys 486
Hyssop, giant 562
  hedge 586
Ibervillea 614–615
Ibidium 154
Ilex *418*
Illecebrum *216*
Indian grass 52
  mallow 419
  mustard 281
  pipe family 486
  tobacco 695
Indigo, false 349
Inula *658, 690*
Ionidium 431
Ionoxalis 384–385
Ipomoea 517–519
Ipomopsis 527, 528
Iridaceae 147–148
Iris 148
  family 147–148
Ironweed 645
Ironwood 495
Isocoma 665–666
Isolepis 114

Page.

Iva 633–634
Ivesia 313
Ivy family 475
Jaboncillo 412
Jack pine 32
Jamesia 299
Janusia 388
Jatropha 401
Jerusalem oak 208
Joe Pye weed 646
Johnson grass 52
Joint-fir family 38–39
Joint grass 55
Juglandaceae 161–163
Juglandales 161–163
Juglans 162–163
Juncaceae 130–135
Juncaginaceae 41
Junco 387
Juncoides 134–135
Juncus 130–134, *135*
June grass 92
Jungle rice 59
Juniper 36
  alligator 36
  family 35–38
  mistletoe 180
  one-seeded 37
  Rocky Mountain 37
  Utah 37
Juniperaceae 35–38
Juniperus 36–38
  use in dyeing 164
Kallstroemia 386–387
Keerlia *691*
Kentrophyta 359, 369
Kentucky bluegrass 100
Kidney diseases, use of Ephedra 38
Knotweed 193
Kobresia 124
Kochia 209
Koeberlinia 387–388
Koeberliniaceae 387–388
Koeleria 92–93
Krameria 336
Krameriaceae 336
Krynitzkia 545, 546, 547
Kuhnia 649–650
Kunzia 318
Laciniaria 649
Lactuca 622–623
Lady fern 25
Lady's-slipper 149
Lamb's quarters 206, 208
Lamium 565
Laphamia 719
Lappago 52
Lappula 539–541
Larkspur 244
Larrea 386
Lathyrus 375–376
Lavauxia 472–473
Layia *698*
Leather flower 256
Lechuguilla 146
Legouzia 616

	Page.
Leiostemon	586
Lemita	406
Lemna	124–125
Lemnaceae	124–125
Leontodon	*626*
Lepachys	706
Lepargyrea	458
Lepidium	271–272
Leptasea	297
Leptilon	673
Leptocaulis	478, 479
Leptochloa	83–84
Leptodactylon	522–523
Leptoloma	54
Lepturus	*85*
Lesquerella	274–276
Lettuce	622
Leucampyx	723
Leucelene	690
Leucocrinum	143
Leucophyllum	577
Leucosyris	686–687
Lewisia	233
Liatris	649
Licorice, wild	371
Life zones of New Mexico, account of	11
Ligusticum	480
Liliaceae	143–144
Liliales	127–145
Lilium	144
Lily	144
Atamasco	147
family	143–144
Mariposa	127
white mountain	143
Lily-of-the-valley family	138–140
Limnorchis	152–154
Limoncillo	714
Limonium	490
Limosella	588
Linaceae	381–383
Linanthus	529, *530*
Linaria	577
Lindenia	225
Linnaea	610
Linosyris	661, 662, 665, 666
Linum	382, *383*
Lip fern	21
Lippia	550, *551*
Lisianthus	*499*
Listera	152
Lithophragma	297
Lithospermum	542–543, *547*
Little red elephant	595
Lizard's-tail family	154
Lloydia	144
Loasa family	431–436
Loasaceae	431–436
Lobelia	617
family	617
Lobeliaceae	617
Loco weed	361, 362, 370, 371
Locust	356
Lolium	106
Lomatium	484
Lonicera	611, *612*

	Page.
Loosestrife	459, 491
family	459
Lophotocarpus	42
Loranthaceae	177–181
Lote bush	413
Lotus	*346, 347, 348*
Lousewort	595
Low speargrass	99
Lowellia	717
Lungwort	541
Lupine	340
Lupinus	340–343
Luzula	134, 135
Lychnis	*242*
Lycium	568–569
Lycopersicum	566
Lycopodiales	29–30
Lycopus	555
Lycurus	62
Lygodesmia	628–629
Lysimachia	*491*
Lythraceae	459
Lythrum	459
Macdougalia	727
Machaeranthera	687–690
Mackenzie, K. K	11, 116
Macrocalyx	535
Macromeria	543–544
Macronema	*667*
Macrorhynchus	625
Madder family	603–608
Madia	697–698
Madronella	556
Maguey	145
Mahogany, mountain	318
Maiden-hair fern	21
Malaceae	321–324
Malacothrix	624
Male fern	25
Mallow	419
family	416–427
rose	417
Malpighiaceae	388
Malva	*418*, 419, *424*
Malvaceae	416–427
Malvales	416–427
Malvastrum	*422*, 425
Mamillaria	447–451, *453*, *457*
Manna grass	103
Manzana de puya larga	324
Manzanita	489
Maple	410
family	410–411
Maple-leaved goosefoot	208
Marah	613
Marestail	473
Margaranthus	569–570
Marigold, fetid	715
Marilaunidium	531–532
Mariola	699
Mariposa lily	127
family	127–128
Marrubium	556
Marsh elder	633
fleabane	693
foxtail	73

	Page.
Marsilea	27
Marsileaceae	27
Martynia	*601, 602*
Martyniaceae	601–602
Maruta	734
Matrimony vine	569
Maurandia	578
Mayweed	734
Meadow barley	107
fescue	103
grass	104
rue	254
Mearns, E. A.	10
Medic, black	343
Medicago	343
Megarrhiza	613
Meibomia	371–372
Melampodium	698–699
Melandryum	241, 242
Melanobatus	320
Melanthaceae	128–130
Melia	390
Meliaceae	390
Melic grass	95
Melica	95
Melilotus	343–344
Melissa	*557*
Melón loco	615
Meloncilla	424
Melosmon	553
Menodora	497
Menodoropsis	497
Mentha	554–555
Menthaceae	551–566
Mentzelia	*433, 434, 435*, 436
Mercury, three-seeded	403
Meriolix	466–467
Mertensia	541–542
Mesadenia	749
Mescal	145, 146
Mescalero Apaches, derivation of name	146
Mesquite	330
grass	53, 57, 69
Metcalfe bean	378
Mexican Boundary Survey	10
rubber plant	699
saltgrass	94
white pine	33
Micrampelis	*613*
Micranthes	296–297
Microsteris	523
Microstylis	151
Mignonette family	291
Milfoil, water	473
Milium	*58*
Milk pea	376
Milkweed	509
family	506–513
swamp	511
Milkwort	391
family	390–392
Milla	143
Millet, hog	58
Mimosa	*329, 330*, 331–332
family	327–332
Mimosaceae	327–332
Mimulus	586–588
Mint	554
family	551–566
round-leaved	555
Mirabilis	*220*, 221, *222*
Mistletoe	179
family	177–181
juniper	180
Mock orange	300
Moehringia	238
Moldavica	562
Mollugo	228–229
Monarda	559–560
Monardella	*556*
Moneses	488
Monkey flower	586
Monkshood	247
Monnina	391
Monocotyledones	39–154
Monolepis	206
Monotropaceae	486
Montia	*233*
Moraceae	175
Mormon tea	38
Morning-glory	517
family	515–519
Morongia	330
Mortonia	410
Morus	175
Moschatel	612
family	612
Moss campion	241
Mountain ash	324
cottonwood	156
forget-me-not	544
mahogany	318
rice	73
sorrel	190
timothy	73
Mouse-ear chickweed	236
Mousetail	249
Mud plantain	126
Mudwort	588
Muérdago	180
Muhlenbergia	68–72
Mulberry	175
family	175
Mulford, Miss A. I.	10
Mullein	577
Munroa	90
Musk-root	612
Mustard	281
family	263–289
hare's-ear	288
hedge	286
tansy	286
Mutisiaceae	630–631
Mutton grass	101
Myosotis	*546*
Myosurus	249–250
Myriophyllum	473
Myriopteris	22
Myrrhis	*478*
Myrtales	459–473
Myzorrhiza	599–600
Naiadaceae	41

Page.
Naiadales 39–41
Naias 41
family 41
Nama *531*
Nash, George V 11
Nasturtium 284, 285
National Herbarium, New Mexican collections 10
Navaho tea 702
Navarretia *521*, *529*
Nazia 52–53
Nealley's dropseed 77
Needle grass 62
Negundo 411
Nelson, Aven 10
Nemoseris 621
Nepeta 564
Nerisyrenia 270
Nettle 176
false 176
family 176–177
New Mexican buckeye 412
locust 356
porcupine grass 66
ranges of species, basis of 10
New Mexico Agricultural College, herbarium 10
New York Botanical Garden, herbarium material studied 10
Nicolletia 715
Nicotiana 567–568
Nigger weed 423
Night-flowering catchfly 242
Nightshade, black 573
enchanter's 460
family 566–574
Ninebark 309
Nogal 163
Nolina 137–138
Nopales 437
Northern bedstraw 605
Notholaena 19–21
Nut grass 112
Nuttallia 432–435
Nyctaginia 224
Nyctelea 535
Oak 164
Arizona 170
black 169
chestnut 171
Fendler 167
Gambel 174
Jerusalem 208
poison 408
shinnery 172
white-leaf 167
Oak-leaved goosefoot 208
Oats 81
false 82
wild 81
Obeliscaria 706
Obione 203, 204, 205
Ocotillo 428
family 428
Odostemon 259–260
Oenothera *465*, *466*, *467*, *468*, 469–470, *471*, *472*, *473*
Oldenlandia *607*, 608
Oleaceae 495–497
Page.
Oleales 495–497
Oleaster 458
family 458
Oligoneuron 666
Olive family 495–497
Onagra 470
One-flowered wintergreen 488
One-seeded bur cucumber 614
juniper 37
Onion 140
family 140–143
Onobrychis 371
Onosmodium 543, *544*
Ophrys 152
Opulaster 309
Opuntia 437–447
Opuntiales 431–458
Orchard grass 97
Orchidaceae 148–154
Orchidales 148–154
Orchis *152*
bog 152
bracted 152
family 148–154
Oregon grape 259
Oreobatus 320–321
Oreobroma 233–234
Oreocarya 544–546
Oreochrysum 667
Oreolirion 147
Oreosciadium 479
Oreoxis 480
Orobanchaceae 599–600
Orobanche *600*
Orpine family 292–294
Orthocarpus 594–595
Oryzopsis *66*, 72–73
Osmorrhiza 478
Ostrya 163
Othake 722
Owl's claws 726
Oxalidaceae 384–385
Oxalis *384*, *385*
Ox-eye 701
daisy 734
Oxybaphus 220, 221, 222
Oxygraphis 253
Oxypolis 483–484
Oxyria 190
Oxytropis 370–371
Pachistima 410
Pachylophus 471
Pachypodium 267
Padus 325–326
Painted cup 591
Palafoxia *722*
Pale dock 192
Palma 137
Palmilla 136
Pandanales 39
Panic grass 56
Panicularia 103, *104*
Panicum *54*, 56–59, *60*, *84*
Papaver 262
Papaveraceae 260–262
Papaverales 260–291
Pappophorum 88–89

	Page.
Parietaria	177
Parnassia	298
Parnassiaceae	298
Paronychia	*215*, 216
Parosela	350-354
Parrasia	270
Parryella	350
Parsley family	475-485
Parsnip	485
cow	484
Parthenium	699
Parthenocissus	415-416
Partridge pea	335
Paspalum	55
Pasque flower	255
Pastinaca	485
Pea family	336-379
milk	376
partridge	335
wild	375
Pearly everlasting	695
Pecan	161
Pectis	714-715
Pedicularis	595-597
Pellaca	23-24
Pellitory	177
Peniocereus	458
Penny cress	272
Pennyroyal	557
Pentstemon	579-585, *586*
Peplis	*427*
Peppergrass	271
Peramium	150-151
Peraphyllum	324
Perezia	630-631
Pericome	719
Peritoma	290
Perityle	719
Persicaria	195-196
Petalostemum	355-356
Peteria	356
Petradoria	671
Petrophyton	310
Petunia	568
Phaca	357, 365, 366, 367, 368
Phacelia	532-535
Phacopsis	358, 364
Phalaris	61, *84*
Phanerophlebia	26
Pharbitis	518, 519
Pharnaceum	*228*
Phaseolus	377-378
Phellopterus	481
Philadelphus	300-301
Phileozera	729
Philibertella	507-508
Philibertia	508
Phleum	73
Phlox	523-525
family	519-530
Phoradendron	179-181
Phragmites	89
Phyla	550-551
Phyllanthus	401
Phymosia	420
Physalis	*569*, 570-572
Physaria	270
Phytogeography of New Mexico	11
Phytolaccaceae	228
Picea	33-34
Pickerel-weed family	126
Picradenia	729
Picradeniopsis	724
Picrothamnus	734
Pigeon grass	60
Pigweed	210
Pimpernel	492
Pinaceae	30-35
Pinales	30-38
Pincushion cactus	447
Pine	31
Arizona	32
Arizona yellow	33
family	30-35
foxtail	33
jack	32
Mexican white	33
Texas	32
white	32
yellow	32
Pinedrops	486
Pinesap	486
Pinguiculaceae	599
Pink family	240-242
Pinus	31-33, *35*
Pinyon	31
grass	66
Piperales	154
Pipsissewa	487
Plagiobothrys	544
Plantaginaceae	602-603
Plantaginales	602-603
Plantago	602-603
Plantain	602
common	603
family	602-603
mud	126
rattlesnake	150
water	42
Platanaceae	304
Platanus	304
Platyschkuhria	725
Plectocephalus	749
Pleuraphis	53
Pleurogyna	503
Pleurophragma	267-268
Pluchea	693
Plum	327
Plumbaginaceae	490
Plumbago family	490
Poa	*93*, *94*, 98-101, *103*, *104*
Poaceae	43-109
Poales	42-124
Podosaemum	70, 72
Podostemma	512
Poinciana	334
Poinsettia	394-395
Poison oak	408
Pokeweed family	228
Polanisia	289
Polemoniaceae	519-530
Polemoniales	513-602

	Page.
Polemonium	521–522, *535*
Poliomintha	556
Polygala	391–392
Polygalaceae	390–392
Polygalales	390–392
Polygonaceae	181–197
Polygonales	181–197
Polygonella	*193*
Polygonum	193–195, *196*, *197*
Polypappus	693
Polypodiaceae	18–27
Polypodium	*25*, *26*, 27
Polypody	27
Polypogon	*70*, 77–78
Polypteris	722
Pomaria	333
Pondweed	40
family	39–41
horned	40
Poñil	318
Pontederia	126
Pontederiaceae	126
Popotillo	38
Poppy	262
California	262
common	262
family	260–262
prickly	261
Populus	155–156
Porcupine grass	65
Porophyllum	714
Portulaca	230, *232*
Portulacaceae	229–234
Potamogeton	40
Potamogetonaceae	39–41
Potato, wild	573
Potentilla	310–314, *315*, *316*
Prairie clover	355
Prenanthella	629
Prenanthes	*621*, *628*
Prickly lettuce	622
pear	437
poppy	261
Primrose	492
family	490–495
Primula	492–493
Primulaceae	490–495
Primulales	490–495
Proboscidea	601–602
Prosartes	140
Prosopis	330
Prunella	565
Prunus	*326*, 327
Psathyrotes	748
Psedera	416
Pseudocymopterus	482–483
Pseudotsuga	35
Psilactis	690
Psilostrophe	718
Psoralea	348–349, *352*, *354*
Ptelea	388–389
Pteridium	21
Pteridophyta	18–30
Pteris	*24*
Pterospora	486
Pteryxia	485
Ptilocalyx	536
Ptiloria	620–621
Puccinellia	104
Puccoon	542
Pulque	145
Pulsatilla	255
Purple hair grass	69
needle grass	64
Purshia	318
Purslane	230
common	230
family	229–234
sea	229
speedwell	589
Pyrola	487–488
Pyrolaceae	486–488
Pyrrhopappus	624
Pyrrocoma	667
Quaking aspen	155
Quamoclidion	222
Quamoclit	517
Quassia family	390
Queen's delight	404
Quelite salado	200
Quercus	164–174
Quincula	569
Rabbit brush	660
Radicula	283–285
Radish	280
Rafinesquia	621
Ragweed	635, 636
family	631–636
Raimannia	470
Rainbow cactus	456
Ranales	243–260
Ranunculaceae	243–258
Ranunculus	251–253
Raphanus	280
Raspberry	319, 320
black	320
Ratibida	705–706
Rattle, yellow	595
Rattlebox	339
Rattlesnake plantain	150
Rayless goldenrod	665
Razoumofskya	178–179
Red clover	345
Red columbine	248
raspberry	320
Redtop	79
Reed bentgrass	79
canary grass	61
giant	89
Relbunium	*605*
Resedaceae	291
Resedella	291
Resin of yellow pine, use by Indians	32
Resurrection plant	29
Reverchonia	401
Rhamnaceae	413–415
Rhamnales	412–416
Rhamnus	414–415
Rhinanthus	595
Rhodiola	293
Rhoeidium	408
Rhombolytrum	91

Page.
Rhus ... *407*, 408–409
Rhynchosia ... 376
Ribes ... 301–303, *304*
Ribgrass, English ... 603
Rice, mountain ... 73
Ridellia ... 718
Ring grass ... 70
Rivina ... 228
Robinia ... 356
Robinson, B. L. ... 11
Rock cress ... 279
Rocky Mountain bee plant ... 290
  columbine ... 248
  juniper ... 37
Roripa ... 284, 285
Rosa ... 306–309
Rosaceae ... 305–321
Rosales ... 291–379
Rose ... 306
  family ... 305–321
  mallow ... 417
Round-leaved mint ... 555
Rubacer ... 320
Rubber plant, Colorado ... 727
  Mexican ... 699
Rubiaceae ... 603–608
Rubiales ... 603–612
Rubus ... 319–320, *321*
Rudbeckia ... *706*, 707
Rue family ... 388–390
  meadow ... 254
Ruellia ... 598
Rulac ... 411
Rumex ... *190*, 191–193
Rush ... 130
  family ... 130–135
  spike ... 112
  wood ... 134
Russian thistle ... 199
Rutaceae ... 388–390
Rutosma ... 388
Rydberg, P. A. ... 10, 11
  work upon Astragalus ... 357
  work upon Quercus ... 164
Rydbergia ... 730
Rye grass ... 106
  wild ... 108
Sabina ... 37
Sage ... 560
  grass ... 49
Sagebrush ... 204, 734
Sagina ... 238
Sagittaria ... 42
St. Johnswort ... 428
  family ... 428
Salicaceae ... 154–161
Salicales ... 154–161
Salix ... 156–161
Salomonia ... 139
Salpingia ... 466
Salsify ... 621
Salsola ... 199–200, *206*
Salt bush ... 201
  cedar ... 427
Saltgrass ... 98
  Mexican ... 94

Page.
Salvia ... 560–562
Salviniaceae ... 27
Salviniales ... 27
Sambucus ... 609–610
Samolus ... 491
Sand bunchgrass ... 73
  grass ... 80
  plum ... 327
  verbena ... 223
Sandalwood family ... 181
Sandbar willow ... 159
Sandbur ... 61
Sandhill dropseed ... 76
  grama ... 86
Sandwort ... 238
Sanfoin ... 371
Sanicula ... 477
Santalaceae ... 181
Santalales ... 177–181
Sanvitalia ... 700
Sapindaceae ... 412
Sapindales ... 405–412
Sapindus ... 412
Sapium ... *404*
Sapodilla family ... 495
Saponaria ... 240, *241*
Sapotaceae ... 495
Sarcobatus ... 199
Sarcostemma ... *508*
Sartwellia ... 727
Saururaceae ... 154
Savastana ... 61–62
Sawgrass ... 112
Saxifraga ... 296, *297*
Saxifragaceae ... 294–297
Saxifrage family ... 294–297
Scarlet hedge nettle ... 565
Schaueria ... *598*
Schedonardus ... 85
Schizachyrium ... 49–50
Schkuhria ... 720, *725*
Schmaltzia ... 406–408, *409*
Schoenocaulon ... 129
Schoenocrambe ... 285
Scirpus ... *113*, 114–116
Scleropogon ... 89
Scouring rush ... 29
  smooth ... 28
Screw bean ... 329
Scrophularia ... 578–579
Scrophulariaceae ... 575–597
Scutellaria ... 555–556
Sea purslane ... 229
Sedge ... 116
  family ... 110–124
Sedum ... *292*, 293–294
Selaginella ... 29–30
  family ... 29–30
Selaginellaceae ... 29–30
Selenia ... 266
Self-heal ... 477, 565
Selinocarpus ... 224–225
Senecio ... 740–748
Senecioneae ... 637, 645
Senna ... 334
  family ... 332–335

Page.
Sensitive brier 330
Sericotheca 310
Service berry 322
Seseli *485*
Sesleria *83*
Sesuvium 229
Setaria 60
Shad scale 204
Sheep sorrel 191
Sheepweed 658
Shepherdia 458
Shepherd's-purse 273
Shinleaf 488
Shinnery oak 172
Shin-oak 172
Shoestrings 349
Shooting star 492
Shrubby cinquefoil 316
 trefoil 388
Sibbaldia 315-316
Sicydium *615*
Sicyos 614
Sida *418, 419, 423, 424*, 425-426
Sidalcea 418
Sidanoda 427
Sideranthus 663-665
Sieglingia 90, 91
Sieversia 316-317, *318*
Silenaceae 240-242
Silene 241-242
Simarubaceae 390
Simsia *712*
Sinapis *281*
Sisymbrium *284*, 285-286, *287*
Sisyrinchium 147-148
Sitanion 107-108
Sitilias 624
Sium *479*
Six-weeks grama 86
 needle grass 63
Skullcap 555
Skunk cabbage 128
Sleepy catchfly 241
 grass 67
Slender foxtail 74
 wheat grass 105
Slough grass 84
Small, J. K 11
Smartweed 195
Smilacina 139
Smooth scouring rush 28
Snakeweed 656, 658
Sneezeweed 726, 733
Snowberry 610
Snow-on-the-mountain 396
Soap, substitute 136
Soapberry 412
 family 412
Soapweed 136
Solanaceae 566-574
Solanum 572-573, *574*
Solidago *667*, 668-671
Solomon's seal 139
 false 139
Sonchus *622*, 623
Sophia 286-288

Page.
Sophora 339
Sorbus 324
Sorghastrum 52
Sorghum 52
Sorilla 390
Sorrel, mountain 190
 sheep 191
Sotol 138
Southern Canary grass 61
Sow thistle 623
Speargrass, low 99
Spearmint 554
Spectacle-pod 269
Specularia 616-617
Speedwell 588
 alpine 589
 purslane 589
 thyme-leaved 590
 water 589
Spergula *238*
Spermacoce *606*
Spermatophyta 30-753
Spermolepis 478-479
Sphaeralcea 420-424
Sphaerostigma 465
Sphenopholis 92
Spiderwort 126
 family 125-126
Spike rush 112
Spiny aster 686
Spiraea *309, 310, 321*
Spirodela 124
Spirostachys *201*
Spleenwort 24
Sporobolus *71, 74*, 75-77
Sprangle 83
Spring beauty 233
Spruce 33
 Colorado blue 34
 Douglas 35
 Engelmann 34
Spurge 397
 family 393-405
Squaw root 599
Squirrel-tail grass 107
Stachys 565-566
Staff-tree family 409-410
Standley, Paul C., botanical work in New Mexico 5
Stanleya 266-267
Stanleyella 267
Star thistle 749
Starwort 235
 water 405
State flower of Colorado 248
Statice 490
Stegnocarpus 536
Steironema 491
Stellaria 235, 236
Stenactis 681
Stenandrium 598
Stenogonum *187*
Stenolobium 601
Stenophyllus 114
Stenotus 666, *667*
Stephanomeria 620

	Page.
Stevia	648, *722*
Stickseed	539
Stillingia	404–405
Stink grass	93
Stipa	65–67, *73*
Stonecrop	293
Strawberry	315
Streptanthus	*267*, *268*, *269*
Streptopus	139–140
Strombocarpa	329–330
Strophostyles	378
Stylocline	693
Suaeda	200, 201
Sumac	408
Sunflower, common	711
Svida	474
Swamp milkweed	511
Sweet cicely	478
clover	343, 344
Sweet-scented bedstraw	606
Swertia	500
Switch grass	58
Sycamore	304
family	304
Symphoricarpos	610–611
Syntherisma	54
Synthyris	*590*
Tagetes	716
Talinopsis	230–231
Talinum	231–233
Tall grama	85
sage grass	51
Tamaricaceae	427–428
Tamarix	427–428
family	427–428
Tanacetum	734
Tanning, use of cañaigre	192
Tansy	734
mustard	286
Taraxacum	626–627
Tar-bush	711
Tarweed	697
Tasajilla	442
Tea, Mormon	38
Navaho	702
Teloxys	*208*
Tequila	145
Tessaria	693
Tetraclea	553–554
Tetradymia	739–740
Tetraneuris	730–733
Teucrium	553
Texan crabgrass	85
Texas curly mesquite grass	53
pine	32
timothy	62
Thalesia	600
Thalictrum	254
Thaspium	*483*
Thelesperma	702–703
Thelypodium	*267*, 268, *285*
Thermopsis	338–339
Thimble-berry	320
Thistle	261, 749
Russian	199
sow	623
star	749
Thlaspi	272–273
Thorn-apple	567
Three-seeded mercury	403
Thymelaeales	458
Thyme-leaved speedwell	590
Thymophylla	716–717
Thysanocarpus	266
Tick trefoil	371
Tiedemannia	484
Tigarea	318
Timothy	73
mountain	73
Texas	62
Tiniaria	197
Tissa	235
Tithymalopsis	396
Tithymalus	395–396
Tium	358, 363, 369
Toadflax, bastard	181
blue	577
Tobacco	567
Indian	695
Tobosa grass	53
Tomatilla	568
Tomato	566
Tonestus	667
Tornillo	329
Touterea	433, 434, 435
Townsendia	691–693
Toxicodendron	408
Toxicoscordion	130
Trachypogon	49
Tradescantia	126
Tragacantha	364
Tragia	404
Tragopogon	621
Tragus	52
Trautvetteria	258
Tree cactus	443
Tree-of-heaven	390
Trefoil, bird's-foot	346
shrubby	388
tick	371
Trianthema	229
Triathera	86
Tribulus	386
Trichachne	54
Trichloris	88
Trichophyllum	724
Trichostema	554
Tricuspis	91
Tridens	91–92
Tridophyllum	312
Trifolium	*343*, 344–346
Triglochin	41
Trigonella	*348*
Triodia	*90*, *91*
Trionum	417
Tripsacum	49
Tripterocalyx	222–223
Trisetum	82
Triticum	*105*
Trixis	631
Trompillo	573
Troximon	625, 626
Tubiflora	597
Tumbleweed	213

Page.

Tuna........ 437
Turritis........ 280
Twayblade........ 152
Twin flower........ 610
Twisted-stalk........ 139
Typha........ 39
Typhaceae........ 39
Tyria........ 405
Ulmaceae........ 174–175
Umbellales........ 474–485
Ungnadia........ 412
Unicorn plant........ 601
family........ 601–602
Uniola........ *98*
Urachne........ 73
Uralepis........ 83, 90, 101
Uropappus........ 622
Urtica........ 176
Urticaceae........ 176–177
Urticales........ 174–177
Utah juniper........ 37
Utricularia........ 599
Uvularia........ *140*
Vaccaria........ 241
Vacciniaceae........ 489–490
Vaccinium........ 489–490
Vagnera........ 139
Valerian........ 617
family........ 617
Valeriana........ 617–618
Valerianaceae........ 617–618
Valerianales........ 617–618
Valley cottonwood........ 155
Valota........ 54
Vanilla grass........ 61
Vaseya........ 70
Vauquelinia........ 321
Velas de coyote........ 443
Venereal diseases, use of Ephedra........ 38
Venus-hair fern........ 21
Venus's looking-glass........ 616
Veratrum........ 128–129
Verbascum........ 577
Verbena........ 548–550, *597*
sand........ 223
Verbenaceae........ 548–551
Verbesina........ *705*, 713–714
Vernonia........ 645–646
Veronica........ 588–590
Vernonieae........ 637
Vervain........ 548
family........ 548–551
Vesicaria........ *275*, *276*
Vetch, wild........ 373
Viborquia........ 349
Vicia........ 373–375
Viguiera........ 709
Vilfa........ 71, 74, 76
Villanova........ 725–726
Vincetoxicum........ 507
Vine mesquite grass........ 57
Viola........ 429–431
Violaceae........ 428–431
Violet........ 429
family........ 428–431
wood-sorrel........ 384
Viorna........ 256–257

Page.

Viper's bugloss........ 547
Virginia creeper........ 415
Virgin's bower........ 256, 258
Viscum........ *181*
Vitaceae........ 415–416
Vitis........ 415, *416*
Viznaga........ 452
Wahlbergella........ 242
Wall barley........ 107
Wallflower, western........ 281
Walnut........ 162
English........ 161
family........ 161–163
Washingtonia........ 478
Water bentgrass........ 78
cress........ 284
crowfoot........ 250
hemlock........ 479
milfoil........ 473
family........ 473
plantain........ 42
family........ 42
speedwell........ 589
starwort........ 405
family........ 405
Waterleaf........ 535
family........ 530–535
Watermelon........ 615
Waterwort........ 427
family........ 427
Wedelia........ 218
Wedeliella........ 218
Western black willow........ 160
ragweed........ 635
wallflower........ 281
Wheat grass........ 104
White clover........ 345
mountain lily........ 143
pine........ 32
White-leaf oak........ 167
Whitlow grass........ 276
Whitlow-wort........ 216
family........ 216
Wild basil........ 556
bean........ 378
cosmos........ 723
flax........ 382
licorice........ 371
oat grass........ 80
oats........ 81
pea........ 375
plum........ 327
potato........ 573
rye........ 108
vetch........ 373
Willow........ 156
black........ 161
desert........ 601
family........ 154–161
sandbar........ 159
western black........ 160
Willow-herb........ 463
Winter fat........ 201
Wintergreen........ 487
family........ 486–488
one-flowered........ 488
Wiregrass........ 64

Page.

Wislizenia 290-291
Withania *574*
Wood reed grass 78
  rush 134
Woodsia 26
Wood-sorrel family 384-385
  violet 384
  yellow 385
Wooton, E. O., botanical work in New Mexico 5
  herbarium of 10
Wootonella 713
Wootonia 698
Wormwood 734
Wyethia 709
Wyomingia 677
Xanthium 634-635
Xanthocephalum 656
Xanthoxalis 385
Ximenesia 713
Xylophacos 359, 364, 365
Xylosteon 611-612
Xyridales 125-126
Yarrow 733
Yellow columbine 248

Page.

Yellow cress 284
  dock 192
  flax 382
  pine 32
  rattle 595
  sweet clover 343
  weed 658
  wood-sorrel 385
Yerba del negro 423
  de víbora 658
  mansa 154
  parda 710
Yucca 135-137
  family 135-138
Zaluzania 708
Zanichellia 40-41
Zapania 551
Zauschneria 464
Zephyranthes 147
Zinnia 700, 701
Zizyphus 413
Zonal distribution of New Mexican plants 11
Zygadenus *129, 130*
Zygophyllaceae 385-387
Zygophyllidium 397

○

Randall Library – UNCW
QK176 .W73 1972 NXWW
Wooton / Flora of New Mexico,
304900155172.